D1029181

THE CAMBRIDGE HISTORY OF SCIENCE

VOLUME 4

Eighteenth-Century Science

This volume offers to general and specialist readers alike the fullest and most complete survey of the development of science in the eighteenth century, exploring the implications of the "Scientific Revolution" of the previous century and the major new growth points, particularly in the experimental sciences. It is designed to be read as both a narrative and an interpretation, and also used as a work of reference. Although prime attention is paid to Western science, space is also given to science in traditional cultures and to colonial science. The coverage strikes a balance between analysis of the cognitive dimension of science itself and interpretation of its wider social, economic, and cultural significance. The contributors, world leaders in their respective specialties, engage with current historiographical and methodological controversies and strike out positions of their own.

Roy Porter (1946–2002), Professor Emeritus of the Social History of Medicine at the Wellcome Trust Centre for the History of Medicine at University College London, was educated at Cambridge University. He was the author of more than 200 books and articles, including *Doctor of Society: Thomas Beddoes and the Sick Trade in Late Enlightenment England* (1991), *London: A Social History* (1994), *"The Greatest Benefit to Mankind": A Medical History of Humanity* (1997), and *Bodies Politic: Disease, Death and Doctors in Britain, 1650–1900* (2001). He was a coauthor of *The History of Bethlem* (1997).

THE CAMBRIDGE HISTORY OF SCIENCE

General editors
David C. Lindberg and Ronald L. Numbers

VOLUME 1: *Ancient Science*
Edited by Alexander Jones

VOLUME 2: *Medieval Science*
Edited by David C. Lindberg and Michael H. Shank

VOLUME 3: *Early Modern Science*
Edited by Lorraine J. Daston and Katharine Park

VOLUME 4: *Eighteenth-Century Science*
Edited by Roy Porter

VOLUME 5: *The Modern Physical and Mathematical Sciences*
Edited by Mary Jo Nye

VOLUME 6: *The Modern Biological and Earth Sciences*
Edited by Peter Bowler and John Pickstone

VOLUME 7: *The Modern Social Sciences*
Edited by Theodore M. Porter and Dorothy Ross

VOLUME 8: *Modern Science in National and International Context*
Edited by David N. Livingstone and Ronald L. Numbers

David C. Lindberg is Hilldale Professor Emeritus of the History of Science at the University of Wisconsin–Madison. He has written or edited a dozen books on topics in the history of medieval and early modern science, including *The Beginnings of Western Science* (1992). He and Ronald L. Numbers have previously coedited *God and Nature: Historical Essays on the Encounter between Christianity and Science* (1986) and *Science and the Christian Tradition: Twelve Case Histories* (2003). A Fellow of the American Academy of Arts and Sciences, he has been a recipient of the Sarton Medal of the History of Science Society, of which he is also past-president (1994–5).

Ronald L. Numbers is Hilldale and William Coleman Professor of the History of Science and Medicine at the University of Wisconsin–Madison, where he has taught since 1974. A specialist in the history of science and medicine in America, he has written or edited more than two dozen books, including *The Creationists* (1992) and *Darwinism Comes to America* (1998). A Fellow of the American Academy of Arts and Sciences and a former editor of *Isis,* the flagship journal of the history of science, he has served as the president of both the American Society of Church History (1999–2000) and the History of Science Society (2000–1).

5231871

Q
125
C36
.2003

THE CAMBRIDGE HISTORY OF SCIENCE

VOLUME 4

Eighteenth-Century Science

Edited by
ROY PORTER

DISCARDED

NORMANDALE COMMUNITY COLLEGE
LIBRARY
9700 FRANCE AVENUE SOUTH
BLOOMINGTON, MN 55431-4399

MAR 1 1 2004

CAMBRIDGE
UNIVERSITY PRESS

PUBLISHED BY THE PRESS SYNDICATE OF THE UNIVERSITY OF CAMBRIDGE
The Pitt Building, Trumpington Street, Cambridge, United Kingdom

CAMBRIDGE UNIVERSITY PRESS
The Edinburgh Building, Cambridge CB2 2RU, UK
40 West 20th Street, New York, NY 10011-4211, USA
477 Williamstown Road, Port Melbourne, VIC 3207, Australia
Ruiz de Alarcón 13, 28014 Madrid, Spain
Dock House, The Waterfront, Cape Town 8001, South Africa

http://www.cambridge.org

© Cambridge University Press 2003

This book is in copyright. Subject to statutory exception
and to the provisions of relevant collective licensing agreements,
no reproduction of any part may take place without
the written permission of Cambridge University Press.

First published 2003

Printed in the United States of America

Typeface Adobe Garamond 10.75/12.5 pt.   *System* QuarkXPress 4.04 [AG]

A catalog record for this book is available from the British Library.

Library of Congress Cataloging in Publication Data
(Revised for volume 4)
The Cambridge history of science
p. cm.
Includes bibliographical references and indexes.
Contents: – v. 4. Eighteenth-century science / edited by Roy Porter
v. 5. The modern physical and mathematical sciences / edited by Mary Jo Nye
ISBN 0-521-57243-6 (v. 4)
ISBN 0-521-57199-5 (v. 5)
1. Science – History.   I. Lindberg, David C.   II. Numbers, Ronald L.
Q125 .C32   2001
509 – dc21
2001025311

ISBN 0 521 57243 6 hardback

Roy Porter, Professor Emeritus of the Social History of Medicine at the Wellcome Trust Centre for the History of Medicine at University College London, died unexpectedly on March 3, 2002, and was, sadly, unable to see the publication of this volume. His contributions to the fields of the history of medicine, science, and the Enlightenment were numerous, important, and far-reaching. His loss is mourned by historians of science and others who had the chance to encounter his sharp intellect and robust character.

CONTENTS

List of Illustrations	*page* xvii
Notes on Contributors	xxi
General Editors' Preface	xxix

1   **Introduction**   1
     ROY PORTER

PART I. SCIENCE AND SOCIETY

2   **The Legacy of the "Scientific Revolution":**
    Science and the Enlightenment   23
     PETER HANNS REILL
    The Scientific Revolution, Mechanical Natural Philosophy,
      and the Enlightenment   25
    The Mid-Century Skeptical Critique of Mechanical
      Natural Philosophy   28
    Vitalizing Nature: A Late Enlightenment Response
      to Skepticism   32
    Conclusion: Between Enlightenment Vitalism and
      Romantic *Naturphilosophie*   41

3   **Science, the Universities, and Other Public Spaces:**
    Teaching Science in Europe and the Americas   44
     LAURENCE BROCKLISS
    Around 1700   44
    Science in the University in the Eighteenth Century:
      Creating Space   52
    Science in the University in the Eighteenth Century:
      The Curriculum   59
    The Expansion in Provision   73
    Conclusion   79

4   **Scientific Institutions and the Organization of Science**   87
        JAMES MCCLELLAN III
        The "Organizational Revolution" of the
            Seventeenth Century   87
        The Age of Academies   90
        The Periodical Journal   95
        Universities and Colleges   96
        Observatories   98
        Scientific Institutions and European Expansion   100
        Botanical Gardens   101
        Organized Science in Society   103
        A Nineteenth-Century Postscript   105

5   **Science and Government**   107
        ROBERT FOX

6   **Exploring Natural Knowledge: Science and the Popular**   129
        MARY FISSELL AND ROGER COOTER
        Newtonianism   134
        Agriculture   139
        Medicine   146
        Botany   151
        Conclusion   156

7   **The Image of the Man of Science**   159
        STEVEN SHAPIN
        The Godly Naturalist   162
        The Moral Philosopher   164
        The Polite Philosopher of Nature   167
        Conclusion: The Civic Expert and the Future   178

8   **The Philosopher's Beard: Women and Gender in Science**   184
        LONDA SCHIEBINGER
        Institutional Landscapes   184
        "Learned Venuses," "Austere Minervas," and
            "Homosocial Brotherhoods"   192
        The Science of Woman   197
        Gendered Knowledge   201
        Beyond Europe   207
        Past and Future   210

9   **The Pursuit of the Prosopography of Science**   211
        WILLIAM CLARK
        What Is Prosopography?   212
        Prosopography in the History of Science   213
        Students   214
        Jesuits   218
        European National and Provincial Communities of Science   220
        France   222

Great Britain 225
The Austro-German Lands 227
Women 232
The Scientific Community of the Eighteenth Century 233
Enlightened Prosopography 235

PART II. DISCIPLINES

10   **Classifying the Sciences** 241
    RICHARD YEO
    Classification in Practice 245
    Maps of Sciences in Encyclopedias 249
    Baconian Division of the Sciences 253
    Harris's *Lexicon Technicum* 254
    Chambers's *Cyclopaedia* 256
    The Encyclopédie 260
    The Demise of Maps of Knowledge in Encyclopedias 263
    Conclusion 266

11   **Philosophy of Science** 267
    ROB ILIFFE
    Approaches to Natural Philosophy in the
      Seventeenth Century 268
    The Heritage of Newton 272
    Metaphysics, Theology, and Matter Theory 275
    Methodology 280
    Conclusion 283

12   **Ideas of Nature: Natural Philosophy** 285
    JOHN GASCOIGNE
    The Establishment of Newtonianism within Britain 289
    The Diffusion of Newton's Work on the Continent 295
    Conclusion 303

13   **Mathematics** 305
    CRAIG FRASER
    The Century of Analysis 307
    Leonhard Euler 307
    Joseph Louis Lagrange 320
    Robert Woodhouse and George Peacock 325
    Conclusion 327

14   **Astronomy and Cosmology** 328
    CURTIS WILSON
    The Astronomy of the Solar System in 1700:
      Newton's First Efforts to Derive Precise
      Astronomical Predictions 329
    The Figure of the Earth 332

The First Analytical Formulation of the Perturbational
Problem: Euler — 334
Star Positions and Physical Theory: Bradley, d'Alembert,
and Euler — 338
The Lunar Problem: Clairaut, Euler, d'Alembert,
and Mayer — 339
The Return of Halley's Comet in 1759 — 342
The Transits of Venus of 1761 and 1769 — 343
Secular and Long-Term Inequalities — 344
Cosmology and the Nebular Hypothesis — 348
Conclusion: The Laplacian Synthesis in the 1790s and Later — 351

15 **Mechanics and Experimental Physics** — 354
    R. W. HOME
    Mechanics — 360
    Experimental Physics — 363
    Toward a Quantified Physics — 371

16 **Chemistry** — 375
    JAN GOLINSKI
    Discipline and Enlightenment — 377
    The Philosophy of Matter — 381
    Affinities and Composition — 384
    Gases and Imponderables — 388
    The Making of a Revolution — 392

17 **The Life Sciences** — 397
    SHIRLEY A. ROE
    The Rise of Newtonian Physiology — 400
    Animism, Vitalism, and the Rejection of Mechanism — 404
    Mechanistic Preformation — 406
    Organisms at the Borders — 409
    Generation through Newtonian Forces — 411
    The Resurgence of Preexistence Theories — 413
    The Rise of Materialism — 414
    Conclusion — 416

18 **The Earth Sciences** — 417
    RHODA RAPPAPORT
    Fossils and the Flood — 419
    Buffon's Synthesis at Mid-Century — 421
    New Approaches at Mid-Century — 423
    The Roles of Fire and Water in Earth Science — 426
    Fossils, Time, and Change — 431

19 **The Human Sciences** — 436
    RICHARD OLSON
    Notions of "Science" in the Human Sciences — 437
    Notions of "Human" in the Human Sciences — 440

The Reservoir of Human "Experiments": History and
Travel Accounts 442
Legal Localism, Moral Philosophy, and Philosophical
History: The Triumph of Environmentalism and the
Stadial Theory of Social Change 444
Race and the Place of Humans in the Natural Order:
The Background to Physical Anthropology 450
Enriching the State and Its Citizens: Cameralism and
Political Economy 451
Quantification in the Human Sciences 456
Sensationalist/Associationist Psychology, Utility, and
Political Science 457
General Evaluation of Eighteenth-Century
Human Sciences 461

20   **The Medical Sciences** 463
     THOMAS H. BROMAN
     The Shape of Medical Education 465
     Physiology 468
     Pathology 476
     Conclusion: The Medical Sciences in the 1790s 481

21   **Marginalized Practices** 485
     PATRICIA FARA
     Rhetorics of Enlightenment 486
     Animal Magnetism 492
     Physiognomy 495
     Astrology 497
     Alchemy 499
     Hutchinsonianism 503
     Conclusion 506

PART III. SPECIAL THEMES

22   **Eighteenth-Century Scientific Instruments
     and Their Makers** 511
     G. L'E. TURNER
     The Role of Apparatus in Lectures 521
     Instruments in Scientific Research 522
     Methods, Materials, and Makers 525
     The Instrument Trade in Europe and North America 531
     A Scientific Collaboration 534

23   **Print and Public Science** 536
     ADRIAN JOHNS
     Cultures of Print at the Onset of Enlightenment 536
     Property and Piracy in the Production of Enlightenment 540

Reading and the Redefinition of Reason   550
Authorship, Genius, and the End of Enlightenment   555

24   **Scientific Illustration in the Eighteenth Century**   561
     BRIAN J. FORD
     Illustration before the Eighteenth Century:
      A Tradition of Obscurantism   563
     A Respite of Realism   564
     From Wood to Metal Engraving   568
     Early Technical Problems   572
     Acknowledged and Unacknowledged Reuse   572
     Zoology: A New Realism   574
     New Studies in Human Anatomy   577
     A New View: Microscopy   579
     New Technology for a New Century   582

25   **Science, Art, and the Representation of the**
     Natural World   584
     CHARLOTTE KLONK
     The Archive of Nature   585
     History Painting and Cosmogonies   587
     Nature's Long History and the Emergence of the Sublime   592
     Beyond the Immediately Observable: Geological Sections
      and Diagrams   609

26   **Science and Voyages of Discovery**   618
     ROB ILIFFE
     The Background to Scientific Voyages   621
     The Importance of Venus   622
     Imperial Voyaging   624
     Terra Australis: Cook's First Two Voyages   626
     The Northwest Passage: Cook's Final Voyage   631
     Implications of Cook's Voyages: Longitude and Scurvy   634
     After Cook   638
     Spanish Voyages   641
     Conclusion   644

PART IV. NON-WESTERN TRADITIONS

27   **Islam**   649
     EMILIE SAVAGE-SMITH
     Military Technology and Cartography   653
     Mechanical Clocks and Watches   655
     The Printing Press   656
     Astronomy   659
     Medicine   661
     European Interest in the Middle East   665
     The Intermingling of Traditions   666

28 **India**   669
    DEEPAK KUMAR
    The Three Shades of Opinion   671
    Astronomy   674
    A Lone Light   675
    Maqul in Education   678
    Medicine: Its Texts and Practices   680
    Tools and Technologies   683
    Reflections   685

29 **China**   688
    FRANK DIKÖTTER
    Jesuit Science   688
    Evidential Scholarship   691
    Medicine   695

30 **Japan**   698
    SHIGERU NAKAYAMA
    Science as an Occupation   699
    The Ban on Western Scientific Knowledge   700
    Translations of Western Works   702
    The Independent Tradition of Mathematics   703
    Mathematics as an Occupation   705
    Publication in Mathematics   705
    Astronomy within the Traditional Framework   706
    Astronomy as an Occupation   708
    Publication in Astronomy   709
    Introduction of Copernicanism and Newtonianism   709
    Physicians as Intellectual Connoisseurs   710
    From the Energetic to the Solidist View of the Human Body   711
    The Medical Profession as an Occupation   713
    Materia Medica   715
    Conclusion   716

31 **Spanish America: From Baroque to Modern
    Colonial Science**   718
    JORGE CAÑIZARES ESGUERRA
    Early Institutions   719
    Patriotic, Neoplatonic, and Emblematic Dimensions   722
    In Service to Crown and Commerce   730
    Travelers and Cultural Change   735
    A Unifying Theme   737

PART V. RAMIFICATIONS AND IMPACTS

32 **Science and Religion**   741
    JOHN HEDLEY BROOKE
    The Diversity of Natural Religion   743

Relating the Sciences to Religion   744
Science and Secularization   749
Providence and the Utility of Science   753
Religion and the Limitations of Reason   755
The Legacy of Enlightenment Critiques   758

33 **Science, Culture, and the Imagination:**
    Enlightenment Configurations   762
    GEORGE S. ROUSSEAU
    A Century of Change   762
    Doctrines of Optimism   764
    Parallel Mental Universes   771
    Optimism and Doubt   774
    Forms of Representation   777
    Science and Reverie   782
    Progresses to Perfection   785
    The Imaginations of Consumers   790

34 **Science, Philosophy, and the Mind**   800
    PAUL WOOD
    Seventeenth-Century Exemplars   802
    Newtonian Legacies   809
    Quantification   814
    Anatomizing the Mind   817
    The Natural History of Human Nature   819
    Conclusion   824

35 **Global Pillage: Science, Commerce, and Empire**   825
    LARRY STEWART
    The Progress of Trade and Learning   825
    Merchants and Imperial Science   828
    The Botanic Empire   833
    The Transport of Nature   838
    Instruments of Empire   841
    Conclusion   843

36 **Technological and Industrial Change:**
    A Comparative Essay   845
    IAN INKSTER
    Europe: The Strength of Weak Ties   846
    The Case of Britain   853
    European Limit: Russia and Technological Progress   858
    Beyond Europe, I: Japan   866
    Beyond Europe, II: India and China   871
    Conclusions   878

Index   883

ILLUSTRATIONS

8.1 Laura Bassi, professor of Newtonian physics and
mathematics at the University of Bologna   *page* 186

8.2 Astronomers Elisabetha and Johannes Hevelius   191

8.3 "Academy of Sciences, Arts, and Trades," the
frontispiece to Diderot and d'Alembert's *Encyclopédie*   194

8.4 The French anatomist Marie-Geneviève-Charlotte
Thiroux d'Arconville's female skeleton compared to
that of an ostrich   199

8.5 "Carl Linnaeus's Classes or Letters" illustrating
Linnaeus's sexual system   204

8.6 "Nature," from Charles Cochin and Hubert François
Gravelot, *Iconologie par figures: ou Traité complet de
allégories, emblêmes & etc.*   208

8.7 Merian's *flos pavonis*   209

10.1 The "View of Knowledge" in the Preface of
Chambers's *Cyclopaedia*   257

10.2 The classification of knowledge given in d'Alembert's
Preliminary Discourse (1751)   261

13.1 Varignon and the "Courbe généatrice"   310

13.2 L'Hôpital and the center of curvature   313

13.3 L'Hôpital and second-order differentials   314

16.1 The "Table of different relationships observed between
different substances," submitted to the Paris Academy
of Sciences by E. F. Geoffroy in 1718   385

22.1 An air pump made for Jean-Antoine Nollet
(1700–1770)   515

22.2 The Oval Room in the Teyler Museum, Haarlem,
the Netherlands   518

22.3 A range of instruments produced by the German
instrument-maker Georg Friedrich Brander (1713–1783)   520

22.4 The observer's room of the Radcliffe Observatory,
      Oxford   524
22.5 A trade card of Dudley Adams (1762–1830)   530
24.1 Plumier's study of American ferns (1705)   567
24.2 *Crinum,* wood-cut published by Olof Rundbeck in the
      Campi Elyssi of 1701   569
24.3 The horse skeleton by Stubbs (1776)   577
24.4 Trembley's study of *Hydra* (1744)   581
24.5 Joseph Priestley and oxygen (1774)   582
25.1 Frontispiece from Thomas Burnet, *Telluris Theoria
      Sacra* (1689)   589
25.2 Nicholas Blakey, illustration for "De la formation des
      Planètes" (1749)   591
25.3 Jacques de Sève, vignette for "Histoire naturelle
      de l'homme" (1749)   592
25.4 Melchior Füßli and Johann Daniel Preißler, *Genesis
      Cap. I. v. 9. 10. Opus Tertiae Diei,* engraving (1731)   595
25.5 Melchior Füßli, *Genesis Cap. VII, v. 21. 22. 23.
      Cataclysmi Reliquia,* engraving (1731)   596
25.6 Melchior Füßli, *Planten Bruck,* engraving from Johann
      Jakob Scheuchzer (1708)   598
25.7 Théodore de Saussure, *Vue de l'aiguille du Géant, prise
      du côté de l'Ouest,* engraving (1796)   600
25.8 Marc Théodore Bourrit, *Vue Circulaire des Montagnes
      qu'on découvre du sommet du Glacier de Buet,*
      engraving (1779)   602
25.9 Pietro Fabris, *View of the great eruption of Vesuvius from
      the mole of Naples in the night of the 20th Oct.
      1767* (1776)   605
25.10 Edwin Sandys, *A true Prospect of the Giants Cawsway
      near Pengore-Head in the County of Antrim,*
      engraving (1697)   608
25.11 Susanna Drury, *The West Prospect of the Giant's Causway
      in the County of Antrim in the Kingdom of
      Ireland* (1743/4)   610
25.12 Johann Friedrich Wilhelm Charpentier, *Petrographische
      Karte des Churfürstentums Sachsen und der Incorporierten
      Lande,* hand-colored engraving (1778)   611
25.13 Section of stratification on the southern edge of the
      Harz mountains, engraving, from Johann Gottlob
      Lehmann (1759)   614
25.14 *Volcan de la Première Époque, Volcan de la Seconde &
      Troisième Époque,* engraving (1718)   615

31.1 Hydrocamel represented in the lake of Mexico,
      map (1700)   724
31.2 Frontispiece to Cabrera y Quintero's *Escudo de Armas
      de México* (1746)   726
31.3 Frontispiece of a thesis defense dedicated to the French
      academicians Bouguer, La Condamine, and
      Godin (1742)   736
33.1 Robert Pine, portrait of a deranged or possessed woman,
      late eighteenth-century engraving by William Dickenson   779

NOTES ON CONTRIBUTORS

LAURENCE BROCKLISS is Fellow and Tutor in Modern History at Magdalen College, Oxford, and Reader in Modern History at the University of Oxford. He has published widely on education and science and medicine in early modern France and was the second editor of the journal *History of Universities*. His books include *French Higher Education in the Seventeenth and Eighteenth Centuries: A Cultural History* (1987) and the coauthored *The Medical World of Early Modern France* (1997). He is currently writing a book on the Enlightenment in Provence.

THOMAS H. BROMAN is Associate Professor of the History of Science and the History of Medicine at the University of Wisconsin–Madison. He is the author of *The Transformation of German Academic Medicine, 1750–1820* (1996). He is currently researching the history of the periodical press in eighteenth-century Germany.

JOHN HEDLEY BROOKE is Andreas Idreos Professor of Science and Religion and Director of the Ian Ramsey Centre at Oxford University. A former editor of the *British Journal for the History of Science,* he was president of the British Society for the History of Science from 1996 to 1998. His book *Science and Religion: Some Historical Perspectives* (1991) was awarded the Watson Davis Prize of the History of Science Society and a Templeton Prize for outstanding book in the field of science and religion. His research interests also include the history of chemistry, the theme of his more recent book, *Thinking About Matter* (1995). In 1995, jointly with Professor Geoffrey Cantor, he gave the Gifford Lectures at Glasgow University, which were published as *Reconstructing Nature: The Engagement of Science and Religion* (1998).

JORGE CAÑIZARES ESGUERRA is Associate Professor of History at the State University of New York at Buffalo. He is the author of the multiple-prize-winning *How to Write the History of the New World: Histories, Epistemologies, and Identities in the Eighteenth-Century Atlantic World* (2001).

WILLIAM CLARK currently teaches at the Department of the History and Philosophy of Science, Cambridge University. He works mostly on early modern German science and scholarship and is coeditor with Jan Golinski and Simon Schaffer of *The Sciences in Enlightened Europe* (1999).

ROGER COOTER is Director of the Wellcome Unit for the History of Medicine, University of East Anglia, Norwich. The author of *The Cultural Meaning of Popular Science* (1984) and *Surgery and Society in Peace and War* (1993), he has also edited volumes on the history of child health, alternative medicine, accidents in history, war and medicine, and, most recently, medicine in the twentieth century. He has written widely on science in popular culture.

FRANK DIKÖTTER is Senior Lecturer in the History of Medicine and Director of the Contemporary China Institute at the School of Oriental and African Studies, University of London. He has published a number of books and articles on cultural history that are directly related to science in modern China, including *The Discourse of Race in Modern China* (1992) and *Sex, Culture and Modernity in China* (1995). His latest monograph is titled *Imperfect Conceptions: Medical Science, Birth Defects and Eugenics in China* (1998). He is currently working on science, crime, and punishment in the republican period.

PATRICIA FARA is an Affiliated Lecturer in the History and Philosophy of Science Department at Cambridge University and a Fellow of Clare College, Cambridge. Her most recent book is *Sympathetic Attractions: Magnetic Practices, Beliefs, and Symbolism in Eighteenth-Century England* (1996). She has also published a book on computers (1982) and coedited the essay collections *The Changing World* and *Memory* (1996 and 1998). Her new book, *Newton: The Making of Genius,* discusses changing concepts of genius and how Isaac Newton has been constructed as a scientific and national hero since the end of the seventeenth century.

MARY FISSELL is Associate Professor of the History of Medicine at the Johns Hopkins University, Baltimore. She is the author of *Patients, Power, and the Poor in Eighteenth-Century Bristol* (1991) and a wide range of articles on the cultural history of science, medicine, and the vernacular. She is completing a study of women and popular medical books in the eighteenth century and is involved in a new project on the cultural construction of vermin.

BRIAN J. FORD is Royal Literary Fellow at the Open University. He is also a Fellow of Cardiff University and a member of the University Court. Among his many books is *Images of Science: A History of Scientific Illustration* (1993).

Ford is a Fellow of the Institute of Biology, where he chairs the History Network, and a Fellow and Honorary Surveyor of Scientific Instruments of the Linnean Society. He is a popular lecturer and broadcaster on radio and television, and he lives in Cambridgeshire.

ROBERT FOX is Professor of the History of Science at the University of Oxford. His research interests include the history of the physical sciences in the eighteenth and nineteenth centuries, the relations among technology, science, and industry in modern Europe, and the social and cultural history of science in nineteenth-century France.

CRAIG FRASER teaches all aspects of the history of mathematics at the University of Toronto's Institute for the History and Philosophy of Science and Technology. His research has centered on the history of analysis and mechanics in the eighteenth and nineteenth centuries, with particular emphasis on the calculus of variations and the conceptual foundations of exact science. He is also interested in the history of modern cosmology, particularly the relations between theoretical and observational cosmology during the period 1915–50.

JOHN GASCOIGNE was educated at Sydney, Princeton, and Cambridge Universities and has taught at the University of New South Wales since 1980. He is the author of *Cambridge in the Age of the Enlightenment: Science, Religion and Politics from the Restoration to the French Revolution* (1989) and of a two-volume work on Joseph Banks and his intellectual and political milieu (1994 and 1998).

JAN GOLINSKI is Professor in the History Department and the Humanities Program at the University of New Hampshire. He is the author of *Science as Public Culture: Chemistry and Enlightenment in Britain, 1760–1820* (1992) and *Making Natural Knowledge: Constructivism and the History of Science* (1998). He coedited *The Sciences in Enlightened Europe,* with William Clark and Simon Schaffer, and is currently writing a cultural history of the weather in the eighteenth century.

R. W. HOME studied physics and then the history and philosophy of science at the University of Melbourne before completing a Ph.D. in the history and philosophy of science at Indiana University. He has been Professor of History and Philosophy of Science at the University of Melbourne since 1975. He has published extensively on the history of eighteenth-century physics and, more recently, on the history of science in Australia.

ROB ILIFFE completed a Ph.D. in the history of science at Cambridge University and is currently Senior Lecturer in the Centre for History of Science, Technology and Medicine at Imperial College, London. He has published a number of articles on the history of science between 1500 and 1800, and he is currently Editorial Director of the Newton Project and editor of the journal *History of Science.*

IAN INKSTER is Research Professor of International History at Nottingham Trent University and Permanent Visiting Professor of European History at the Institute of European Studies, Fo Kuang University, Taiwan. Recent books include *Science and Technology in History* (1991), *Clever City* (1991), *Scientific Culture and Urbanisation in Industrialising Britain* (1997), *Technology and Industrialisation* (1998), and *Japanese Industrialisation 1603–2000* (2000).

ADRIAN JOHNS is Associate Professor of History at the University of Chicago. He is the author of *The Nature of the Book: Print and Knowledge in the Making* (1998).

CHARLOTTE KLONK is Research Fellow in the Department of History of Art at the University of Warwick. She is the author of *Science and the Perception of Nature: British Landscape Art in the Late Eighteenth and Early Nineteenth Centuries* (1996) and is presently writing a book on European art museums and their spectators.

DEEPAK KUMAR teaches the history of education at the Zakir Husain Centre of Educational Studies, Jawaharlal Nehru University, New Delhi. He is the author of *Science and the Raj 1857–1905* (1995) and coeditor of *Technology and the Raj* (1995).

JAMES MCCLELLAN III is Professor of History of Science at Stevens Institute of Technology in Hoboken, New Jersey. His research interests center on European, notably French, scientific institutions, the scientific press, and science and the French colonial enterprise in the eighteenth century.

SHIGERU NAKAYAMA is Professor, STS Centre, Kanagawa University. His Ph.D. was in the history of science and learning at Harvard (1959). His publications include *Science, Technology and Society in Postwar Japan* (1991), *A History of Japanese Astronomy* (1969), and *Academic and Scientific Traditions in China, Japan, and the West* (1984).

RICHARD OLSON received his Ph.D. in the history of science from Harvard University in 1967. He is currently Professor of History and Willard W. Keith Fellow in the Humanities at Harvey Mudd College. His recent works include *The Emergence of the Social Sciences: 1642–1792* (1993) and volumes one and two of *Science Deified and Science Defied* (1982 and 1990). He is currently working on volume three, which focuses on nineteenth-century scientism.

ROY PORTER (1946–2002) was Professor Emeritus of the Social History of Medicine at the Wellcome Trust Centre for the History of Medicine at University College London. Recent books included *Doctor of Society: Thomas Beddoes and the Sick Trade in Late Enlightenment England* (1991), *London: A Social History* (1994), *"The Greatest Benefit to Mankind": A Medical History of Humanity* (1997), *Enlightenment: Britain and the Creation of the Modern World*

(2000), and *Bodies Politic: Disease, Death and Doctors in Britain, 1650–1900* (2001). He was a coauthor of *The History of Bethlem* (1997). His interests included eighteenth-century medicine, the history of psychiatry, and the history of quackery.

RHODA RAPPAPORT, Professor Emeritus of History at Vassar College, is the author of articles on various topics in eighteenth-century geology and of the book *When Geologists Were Historians, 1665–1750* (1997).

PETER HANNS REILL is Professor of History at the University of California, Los Angeles, and Director of UCLA's Center for Seventeenth and Eighteenth Century Studies and of the William Andrews Clark Memorial Library. He has published in the areas of eighteenth-century German intellectual history, the history of historical writing, and the history of science and the humanities. In addition to numerous articles in these areas, he has written *The German Enlightenment and the Rise of Historicism* (1975) and edited or coedited *Aufklärung und Geschichte: Studien zur deutschen Geschichtswissenschaft im 18. Jahrhundert, The Encyclopedia of the Enlightenment* (1996), and *Visions of Empire: Voyages, Botany, and Representations of Nature* (1996). He has held fellowships from the Fulbright Commission, the Max-Planck-Institute for History, the Institute for Advanced Studies in Berlin, and the Guggenheim Foundation.

SHIRLEY A. ROE is Professor of History at the University of Connecticut. She is author of *Matter, Life, and Generation: Eighteenth-Century Embryology and the Haller-Wolff Debate* (1981), editor of *The Natural Philosophy of Albrecht von Haller* (1981), and coeditor (with Renato G. Mazzolini) of *Science Against the Unbelievers: The Correspondence of Bonnet and Needham, 1760–1780* (1986). She is currently completing a book on eighteenth-century biological materialism and its social/political context.

GEORGE S. ROUSSEAU spent most of his career at UCLA and is currently Research Professor of English, De Montfort University, Leicester. He is author (with Marjorie Hope Nicolson) of *This Long Disease My Life: Alexander Pope and the Sciences* (1968) and editor of *Organic Form: The Life of an Idea* (1972), *Goldsmith: The Critical Heritage* (1974), *The Letters and Private Papers of Sir John Hill* (1981), and a trilogy of books about knowledge in the Enlightenment titled *Perilous Enlightenment: Pre- and Post-Modern Discourses – Sexual, Historical* (1991), *Enlightenment Crossings: Pre- and Post-Modern Discourses – Anthropological* (1991), and *Enlightenment Borders: Pre- and Post-Modern Discourses – Medical, Scientific* (1991).

EMILIE SAVAGE-SMITH is Senior Research Associate at the Oriental Institute, University of Oxford. She has published studies on a variety of medical and divinatory practices in the Islamic world, as well as on celestial globes and mapping. Her most recent book (with coauthor Francis Maddison) is *Science,*

Tools & Magic, 2 vols. (The Nasser D. Khalili Collection of Islamic Art, XII) (1997).

LONDA SCHIEBINGER is Edwin E. Sparks Professor of the History of Science at Pennsylvania State University. She is author of *The Mind Has No Sex? Women in the Origins of Modern Science* (1989), the prize-winning *Nature's Body: Gender in the Making of Modern Science* (1993), and *Has Feminism Changed Science?* (1999), and she is editor of *Feminism and the Body* (2000). Her current research explores gender in the European voyages of scientific discovery.

STEVEN SHAPIN teaches in the Department of Sociology and the Science Studies Program at the University of California, San Diego. Among his books are *A Social History of Truth: Civility and Science in Seventeenth-Century England* (1994) and *The Scientific Revolution* (1996). He is coeditor of *Science Incarnate: Historical Embodiments of Natural Knowledge* (1998).

LARRY STEWART teaches the history of science at the University of Saskatchewan, Canada. He is the author of *The Rise of Public Science: Rhetoric, Technology and Natural Philosophy in Newtonian Britain, 1660–1750* (1992).

G. L'E. TURNER, D.Sc., D.Litt., is Visiting Professor of the History of Scientific Instruments at Imperial College, London. He holds a Leverhulme Emeritus Fellowship. His books include *Scientific Instruments and Experimental Philosophy 1550–1850* (1991) and *The Practice of Science in the Nineteenth Century: Teaching and Research Apparatus in the Teyler Museum* (1996).

CURTIS WILSON held a tutorship at St. John's College, Annapolis, Maryland, during the years 1948–66 and 1973–88; he was Dean of the College in 1958–62 and 1973–7. During the years 1966–73 he was a Visiting Associate Professor and then Professor in the Department of History at the University of California, San Diego. Among his works on the history of astronomy are *Astronomy from Kepler to Newton: Historical Studies* (1989) and chapters in *Planetary Astronomy from the Renaissance to the Rise of Astrophysics,* eds. René Taton and Curtis Wilson (Part A, 1989; Part B, 1995).

PAUL WOOD is a member of the Department of History and Director of the Humanities Center, University of Victoria, Canada. His research has focused primarily on the intellectual milieu of the Scottish Enlightenment. His most recent book, *Thomas Reid on the Animate Creation: Papers Relating to the Life Sciences* (1995), is an edition of the hitherto unpublished manuscripts on natural history, physiology, and materialism of the Scottish common sense philosopher Thomas Reid. He is currently editing Reid's collected correspondence for the Edinburgh Edition of Thomas Reid.

RICHARD YEO is Reader in the History and Philosophy of Science at Griffith University, Brisbane, Australia. He studied history and psychology at Sydney University, where he earned a Ph.D. on natural theology and the philosophy of knowledge in nineteenth-century Britain. His recent publications include *Defining Science: William Whewell, Natural Knowledge and Public Debate in Early Victorian Britain* (1993) and *Encyclopaedic Visions: Scientific Dictionaries and Enlightenment Culture* (2001).

GENERAL EDITORS' PREFACE

In 1993, Alex Holzman, former editor for the history of science at Cambridge University Press, invited us to submit a proposal for a history of science that would join the distinguished series of Cambridge histories launched nearly a century ago with the publication of Lord Acton's fourteen-volume *Cambridge Modern History* (1902–12). Convinced of the need for a comprehensive history of science and believing that the time was auspicious, we accepted the invitation.

Although reflections on the development of what we call "science" date back to antiquity, the history of science did not emerge as a distinctive field of scholarship until well into the twentieth century. In 1912 the Belgian scientist-historian George Sarton (1884–1956), who contributed more than any other single person to the institutionalization of the history of science, began publishing *Isis*, an international review devoted to the history of science and its cultural influences. Twelve years later he helped to create the History of Science Society, which by the end of the century had attracted some 4,000 individual and institutional members. In 1941 the University of Wisconsin established a department of the history of science, the first of dozens of such programs to appear worldwide.

Since the days of Sarton historians of science have produced a small library of monographs and essays, but they have generally shied away from writing and editing broad surveys. Sarton himself, inspired in part by the Cambridge histories, planned to produce an eight-volume *History of Science*, but he completed only the first two installments (1952, 1959), which ended with the birth of Christianity. His mammoth three-volume *Introduction to the History of Science* (1927–48), a reference work more than a narrative history, never got beyond the Middle Ages. The closest predecessor to *The Cambridge History of Science* is the three-volume (four-book) *Histoire Générale des Sciences* (1957–64), edited by René Taton, which appeared in an English translation under the title *General History of the Sciences* (1963–4). Edited just before the late-twentieth-century boom in the history of science, the Taton set quickly became dated.

During the 1990s Roy Porter began editing the very useful Fontana History of Science (published in the United States as the Norton History of Science), with volumes devoted to a single discipline and written by a single author.

The Cambridge History of Science comprises eight volumes, the first four arranged chronologically from antiquity through the eighteenth century, the latter four organized thematically and covering the nineteenth and twentieth centuries. Eminent scholars from Europe and North America, who together form the editorial board for the series, edit the respective volumes:

Volume 1: *Ancient Science,* edited by Alexander Jones, University of Toronto

Volume 2: *Medieval Science,* edited by David C. Lindberg and Michael H. Shank, University of Wisconsin–Madison

Volume 3: *Early Modern Science,* edited by Lorraine J. Daston, Max Planck Institute for the History of Science, Berlin, and Katharine Park, Harvard University

Volume 4: *Eighteenth-Century Science,* edited by Roy Porter, Wellcome Trust Centre for the History of Medicine at University College London

Volume 5: *The Modern Physical and Mathematical Sciences,* edited by Mary Jo Nye, Oregon State University

Volume 6: *The Modern Biological and Earth Sciences,* edited by Peter Bowler, Queen's University of Belfast, and John Pickstone, University of Manchester

Volume 7: *The Modern Social Sciences,* edited by Theodore M. Porter, University of California, Los Angeles, and Dorothy Ross, Johns Hopkins University

Volume 8: *Modern Science in National and International Context,* edited by David N. Livingstone, Queen's University of Belfast, and Ronald L. Numbers, University of Wisconsin–Madison

Our collective goal is to provide an authoritative, up-to-date account of science – from the earliest literate societies in Mesopotamia and Egypt to the beginning of the twenty-first century – that even nonspecialist readers will find engaging. Written by leading experts from every inhabited continent, the essays in *The Cambridge History of Science* explore the systematic investigation of nature, whatever it was called. (The term "science" did not acquire its present meaning until early in the nineteenth century.) Reflecting the ever-expanding range of approaches and topics in the history of science, the contributing authors explore non-Western as well as Western science, applied as well as pure science, popular as well as elite science, scientific practice as well as scientific theory, cultural context as well as intellectual content, and the dissemination and reception as well as the production of scientific knowledge. George Sarton would scarcely recognize this collaborative effort as the history of science, but we hope we have realized his vision.

David C. Lindberg
Ronald L. Numbers

I

INTRODUCTION

Roy Porter

"*Was ist Aufklärung?*" asked Immanuel Kant in 1784, and the issue has remained hotly debated ever since.[1] Not surprisingly, therefore, if we now pose the further question "What was Enlightenment *science?*" the uncertainties are just as great – but here the controversies assume a different air.

Studies of the Enlightenment proper paint the Age of Reason in dramatic hues and reflect partisan viewpoints: some praise it as the seedbed of modern liberty, others condemn it as the poisoned spring of authoritarianism and alienation.[2] Eighteenth-century science, by contrast, has typically been portrayed in more subdued tones. To most historians it lacks the heroic quality of what came before – the martyrdom of Bruno, Galileo's titanic clash with the Vatican, the "new astronomy" and "new philosophy" of the "scientific revolution," the sublime genius of a Descartes, Newton, or Leibniz.[3] After that

---

[1] James Schmidt (ed.), *What Is Enlightenment? Eighteenth Century Answers and Twentieth Century Questions* (Berkeley: University of California Press, 1996).

[2] Partisans of the Enlightenment include Peter Gay, who in his *The Enlightenment: An Interpretation*, vol. II: *The Science of Freedom* (London: Weidenfeld and Nicolson, 1970), applauds the *philosophes'* championing of science: "The philosophes seized upon the new science as an irresistible force and enlisted it in their polemics, identifying themselves with sound method, progress, success, the future" (p. 128). Eric Hobsbawm has recently written, "I believe that one of the few things that stands between us and an accelerated descent into darkness is the set of values inherited from the eighteenth-century Enlightenment," *On History* (London: Weidenfeld and Nicolson, 1997), p. 265.

Suspicious of the Enlightenment have been Lester G. Crocker, *An Age of Crisis: Man and World in Eighteenth Century French Thought* (Baltimore, MD: Johns Hopkins University Press, 1959); J. L. Talmon, *The Rise of Totalitarian Democracy* (Boston: Beacon Press, 1952); and Max Horkheimer and Theodor T. Adorno, *Dialectic of Enlightenment* (New York: Herder and Herder, 1972), all of whom read fascism and totalitarianism back into Enlightenment rationality.

[3] The drama of seventeenth-century developments is, of course, registered in the term "scientific revolution." Their "revolutionary" nature was expressed by many Enlightenment commentators, notably Fontenelle, although "the scientific revolution" is a modern coinage often challenged today. See I. Bernard Cohen, *Revolution in Science* (Cambridge, MA: Harvard University Press, 1985); Cohen, "The Eighteenth-Century Origins of the Concept of Scientific Revolution," *Journal of the History of Ideas*, 37 (1976), 257–88; H. Floris Cohen, *The Scientific Revolution: A Historiographical Inquiry* (Chicago: University of Chicago Press, 1994); John Henry, *The Scientific Revolution and the Origins of Modern Science* (London: Macmillan, 1997); David C. Lindberg, "Conceptions of the Scientific

age of heroes, the eighteenth century has been chid for being dull, a trough between the peaks of the "first" and the "second" scientific revolution, a lull before the storm of the Darwin debate and the astounding breakthroughs of nineteenth-century physics. At best, dwarves were perched on giants' shoulders. "The first half of the eighteenth century was a singularly bleak period in the history of scientific thought," judged Stephen Mason; the age was marked, thought H. T. Pledge, by "an element of dullness," due in part to its "too ambitious schemes" and its "obstructive crust of elaboration and formality."[4] "The lost half century in English medicine" was William Lefanu's corresponding label for the post-1700 era, whereas another medical historian, Fielding H. Garrison, characterized the entire century as an "age of theories and systems," bedeviled by a "mania for sterile, dry-as-dust classifications of everything in nature" – one fortunately succeeded by an era that brought "The Beginnings of Organized Advancement of Science."[5]

Given such judgments, it is not surprising that muted terms such as "consolidation" have come to mind for characterizing the natural sciences in the eighteenth century. Conceding that "when Newton died [1727] the great creative phase of the scientific revolution was already finished," Rupert Hall nevertheless stressed that "its acceptance and assimilation were still incomplete": such were the bread-and-butter tasks remaining for the eighteenth century to accomplish.[6]

Casting the job of "completion" in an altogether more positive light, however, Laurence Brockliss contends in his contribution to this volume (Chapter 3) that "if the Scientific Revolution is seen as a broader cultural moment whereby the Galilean/Newtonian mathematical and phenomenological approach to the natural world became part of the mind set of the European and American elite, then that Revolution occurred *in the eighteenth century* (pre-

---

Revolution from Bacon to Butterfield: A Preliminary Sketch," in David C. Lindberg and Robert S. Westman (eds.), *Reappraisals of the Scientific Revolution* (Cambridge University Press, 1990), pp. 1–26; Roy Porter and Mikuláš Teich (eds.), *The Scientific Revolution in National Context* (Cambridge University Press, 1992); John A. Schuster, "The Scientific Revolution," in R. C. Olby, G. N. Cantor, J. R. R. Christie, and M. J. S. Hodge (eds.), *Companion to the History of Modern Science* (London: Routledge, 1990), pp. 217–43. From a more philosophical standpoint, the writings of T. S. Kuhn remain stimulating: *The Structure of Scientific Revolutions* (Chicago: University of Chicago Press, 1962). Michael Fores's "Science and the 'Neolithic Paradox,'" *History of Science*, 21 (1983), 141–63, and his "Newton on a Horse: A Critique of the Historiographies of 'Technology' and 'Modernity,'" *History of Science*, 23 (1985), 351–78, attack the "myth" of the scientific revolution; Steven Shapin's *The Scientific Revolution* (Chicago: University of Chicago Press, 1996) – a work with an exceptionally fine bibliographical essay – opens provocatively: "There was no such thing as the Scientific Revolution, and this is a book about it" (p. 1).

[4] S. F. Mason, *Main Currents in Scientific Thought* (London: Routledge, 1956), p. 223; H. T. Pledge, *Science since 1500* (London: HMSO, 1939), p. 101.

[5] Fielding H. Garrison, *An Introduction to the History of Medicine* (Philadelphia: Saunders, 1917), p. 303; W. R. LeFanu, "The Lost Half Century in English Medicine, 1700–1750," *Bulletin of the History of Medicine*, 66 (1972), 319–48. For correctives, see the essay in this volume (Chapter 20) by Thomas Broman and also W. F. Bynum, "Health, Disease and Medical Care," in G. S. Rousseau and Roy Porter (eds.), *The Ferment of Knowledge* (Cambridge University Press, 1980), pp. 211–54.

[6] A. R. Hall, *The Scientific Revolution, 1500–1800* (London: Longman, 1954), p. iii.

dominantly outside the English-speaking world after 1750)."[7] And in a similar way, Margaret Jacob has pictured the century as the era when "scientific knowledge became an integral part of Western culture" or in other words became "public knowledge."[8] "Acceptance" and "assimilation" thus may be highly apposite epithets for eighteenth-century science, especially if they are intended not to excuse drabness but to highlight transformative processes. The incorporation of science into modernity was at least as momentous as the dazzling innovative leaps of a Kepler or Harvey; it certainly presents the historian with taxing problems to explain.[9]

It is important, in any case, that talk of "assimilation" and "consolidation" should not convey the false impression that all the great breakthroughs of early modern natural science had already been achieved by 1700 and that what remained was no more than a matter of dotting i's and crossing t's – or, in Kuhnian parlance, the pursuit of normal science within well-established paradigms.[10] We should not minimize the still inchoate condition in 1700 even of those sciences intimately associated with Newton, Huygens, Leibniz, and the other pioneers of a new mathematical physics; nor, indeed, should we forget that, at the turn of the century, Leibniz still had sixteen years to live and Newton twenty-seven, or that Newton's *Opticks* had not even been published. God may have said, "*Let Newton be, and all was light,*" but the light Newton had shed by 1700 was more like the first rays of dawn than the dazzle of the noonday sun. Although Simon Schaffer has well observed that, by the nineteenth century, "it became possible to see Newtonianism as the common sense of the physical sciences,"[11] that would be an anachronistic judgment if applied to its predecessor, for although Newton has often been "celebrated as bringing the mechanical philosophy to perfection," that was

---

[7] Italics added. Cf. Henry, *The Scientific Revolution*, p. 96, writing in the same mode about the eighteenth century: "It is possible to conclude that the very fact that they now saw natural philosophy in this way, and dared to hope that it might be used to establish laws for the correct ordering and running of society, is in itself indicative that a revolution in the ordering of knowledge had indeed taken place. The scientific revolution was complete."

[8] Margaret C. Jacob, *The Cultural Meaning of the Scientific Revolution* (New York: Knopf, 1988), p. 3. For "public knowledge," see Larry Stewart, *The Rise of Public Science: Rhetoric, Technology, and Natural Philosophy in Newtonian Britain, 1660–1750* (Cambridge University Press, 1992).

[9] Highly influential has been Steven Shapin's and Simon Schaffer's *Leviathan and the Air Pump* (Princeton, NJ: Princeton University Press, 1985). This book raised, and attempted via a concrete case study to resolve, the crucial question of how the new science established its truth status, a problem to which Shapin returned in *A Social History of Truth: Civility and Science in Seventeenth-Century England* (Chicago: University of Chicago Press, 1994).

[10] T. S. Kuhn, *The Structure of Scientific Revolutions*. On the geographico-cultural diffusion of science, see, for instance, Henry Guerlac, *Newton on the Continent* (Ithaca, NY: Cornell University Press, 1981); Guerlac, "Where the Statue Stood, Divergent Loyalties to Newton in the Eighteenth Century," in Earl R. Wasserman (ed.), *Aspects of the Eighteenth Century* (Baltimore, MD: Johns Hopkins University Press, 1965), pp. 317–34; A. Rupert Hall, "Newton in France: A New View," *History of Science*, 13 (1975), 233–50.

[11] Simon Schaffer, "Newtonianism," in Olby et al. (eds.), *Companion to the History of Modern Science*, pp. 610–26, especially p. 611. For comment on the Pope quotation, see Steven Shapin, "Social Uses of Science," in Rousseau and Porter (eds.), *The Ferment of Knowledge*, pp. 93–142, especially p. 95.

hardly so, insists Steven Shapin.[12] The Lucasian Professor bequeathed as many problems as solutions, and, as Curtis Wilson's discussion of astronomy and cosmology (Chapter 14 in this volume) demonstrates, eighteenth-century astro-physicists were still making striking innovations – observational, computational, and theoretical.[13]

Even more remarkable, perhaps, and often interlinked, were contemporary developments in mathematics. To many European practitioners, Newton's methods appeared radically wanting. While British mathematicians were treading water, hampered by the clumsy Newtonian "fluxion" procedures, the Bernoullis, Maupertuis, Euler, Clairaut, d'Alembert, Lagrange, Laplace, and other Continental mathematicians, many of whom were closely linked with the Berlin, St. Petersburg, and Paris academies, made brilliant advances. Innovative techniques in analysis spurred the application of mathematics to many problems, including the motion of rigid bodies, vibration, hydromechanics, and tension; and conservation laws were developed that theorized the cosmos in terms alien to the cosmology of divine intervention championed by Newton, pointing toward Laplace's nebular hypothesis. Surveying the vis viva controversy and the strides made by rational mechanics, John Henry has recently confirmed that "eighteenth century developments in mathematics perhaps owe more to the achievements of Leibniz and the Bernoulli brothers, than to Newton, whose dominion over British mathematicians seems to have led to a noticeable decline."[14]

Moreover, the headway made by eighteenth-century mathematics was far from confined to the internal and technical achievements that form the core of Craig Fraser's contribution to this volume (Chapter 13). In the *Preliminary Discourse* to the *Encyclopédie*, Jean d'Alembert proclaimed mathematics to be the basis of all physical science:

> The use of mathematical knowledge is no less considerable in the examination of the terrestrial bodies that surround us [than it is in astronomy]. All the properties we observe in these bodies have relationships among themselves that are more or less accessible to us. The knowledge or the discovery of these relationships is almost always the only object that we are permitted to attain, and consequently the only one that we ought to propose for ourselves.[15]

Corroborating Margaret Jacob's claim that in the eighteenth century "scientific knowledge became an integral part of Western culture," historians have stressed the permeation of the *"esprit géometrique"* (or "calculating spirit")

---

[12] Shapin, *The Scientific Revolution*, p. 157.
[13] See also J. D. North, *The Fontana History of Astronomy and Cosmology* (London: Fontana, 1994).
[14] Henry, *The Scientific Revolution*, p. 94.
[15] Quoted in Thomas L. Hankins, *Science and the Enlightenment* (Cambridge University Press, 1985), pp. 46–7.

into everyday life, from life insurance to gambling and other situations in which the determination of probabilities became pressing.[16]

Nor was that all. As signaled many years ago by Herbert Butterfield's notoriously question-begging chapter heading, "The Postponed Scientific Revolution in Chemistry," one field that proved exceptionally innovative – in new experimental practices, practical discoveries, and theoretical reconceptualization – was chemistry. In his article in this volume (Chapter 16), Jan Golinski underscores the significance of the dramatic recognition that the atmosphere was not a uniform physical state but a mix of separate gases with distinct chemical properties. In that light he reassesses Butterfield's claim that Lavoisierian chemistry constituted the concluding chapter of the seventeenth-century "scientific revolution."[17]

Meanwhile, new specialties were taking shape, so that by the turn of the nineteenth century, as is shown here by Rhoda Rappaport (Chapter 18) and Shirley Roe (Chapter 17), terms such as "geology" and "biology" had been minted and were soon to become standard labels for emergent disciplinary domains. Aspects of the physical sciences amenable to experimental inquiry – notably magnetism, electricity, optics, fluid mechanics, pneumatics, the study of fire, heat, and other subtle or imponderable fluids, meteorology, strength of materials, hydrostatics and hygrometry, to list only the most prominent – took striking steps forward: as Rod Home emphasizes (Chapter 15), understanding of magnetism and electricity changed radically between 1700 and 1800. It ceased to be plausible to view physics, in the traditional, Aristotelian manner, primarily as a branch of philosophy: by 1800 true physics meant experimental physics.[18]

Even in well-plowed fields of inquiry such as natural history, remarkable changes can be seen. It was at this time, for instance, that plant sexuality was first fully established as the foundation for botanical thinking within the new and enduring taxonomic system developed by Linnaeus. The first evolutionary theories were advanced, associated (obliquely) with Buffon and (explicitly)

---

[16] For the calculating spirit and its applications in the realms of probability, see Gerd Gigerenzer, Zeno Swijtink, Theodore Porter, Lorraine Daston, John Beatty, and Lorenz Krüger, *The Empire of Chance* (Cambridge University Press, 1989); Ian Hacking, *The Emergence of Probability* (Cambridge University Press, 1975); Hacking, *The Taming of Chance* (Cambridge University Press, 1990); Lorraine Daston, *Classical Probability in the Enlightenment* (Princeton, NJ: Princeton University Press, 1988); Daston, "The Domestication of Risk: Mathematical Probability and Insurance 1650–1830," in Lorenz Krüger, Lorraine Daston, and M. Heidelberger (eds.), *The Probabilistic Revolution* (Ann Arbor: University of Michigan Press, 1987), pp. 237–60; Tore Frängsmyr, J. L. Heilbron, and Robin E. Rider (eds.), *The Quantifying Spirit in the 18th Century* (Berkeley: University of California Press, 1990).

[17] Herbert Butterfield, *The Origins of Modern Science, 1300–1800* (London: Bell, 1949). On that "postponed revolution," see William H. Brock, *The Fontana History of Chemistry* (London: Fontana, 1992), p. 86: "Lavoisier's synthesis of constitutional ideas and experiment appears as impressive as the work of Newton in physics the century before." For a different notion of "revolutionary" chemistry, see Archibald Clow and Nan L. Clow, *The Chemical Revolution: A Contribution to Social Technology* (London: Batchworth Press, 1952).

[18] For a recent study see, for instance, Patricia Fara, *Sympathetic Attractions: Magnetic Practices, Beliefs and Symbolism in Eighteenth-Century England* (Princeton, NJ: Princeton University Press, 1996).

with Erasmus Darwin and Lamarck.[19] It is not crudely Whiggish or merely celebratory of the so-called "forerunners of [Charles] Darwin"[20] to insist that theorists of life were finding that the static, hierarchical, and Christian Chain of Being no longer possessed explanatory power and that the living needed to be conceptualized within a more dynamic framework and an extended timescale. In short, wherever one looks, there was, during the eighteenth century, no stalling in scientific theory or practice, no shortage of what (depending on which philosophies or sociologies of science we adopt) we can call the "discovery," "invention," or "construction" of new knowledge.[21]

It would be wrong, however, to imply that eighteenth-century science deserves study solely for, and in respect of, its conceptual innovativeness. And this point leads us back to the notion of "consolidation." Gradually, unevenly, but, perhaps, inexorably, the production of knowledge about Nature and the casting of discourse in natural terms were playing increasingly prominent roles in culture, ideology, and society at large. Natural philosophers and historians were claiming their place in the sun alongside churchmen and humanists. Gentlemen of science – and, as Londa Schiebinger documents in this volume (Chapter 8), a handful of ladies, too – were winning admittance into the Republic of Letters and were changing its complexion in the process.[22] Furthermore, as Robert Fox (Chapter 5) and Rob Iliffe (Chapter 26) substantiate, governments were increasingly employing experts as administrators, explorers, civil and military engineers, propagandists, and managers of natural resources. Science was held to provide the knowledge base necessary for "enlightened absolutism," above all through statistics (*Statistik:* state information) and political arithmetic; scientific experts would be brokers in the Baconian marriage of knowledge and power. Looking back, historians might variously interpret such developments as progressive or, on the other hand, as acts of social policing; but, either way, natural knowledge acquired an enhanced public prominence during the last years of the *ancien régime,* mediating values and visions. Despite their radically disparate philosophical allegiances, the deeply pious Joseph Priestley and the *philosophe* Condorcet were both looking, during the French Revolutionary era, to a future society transformed by scientific discoveries and scientific rationality – one marked not merely by material improvements but by the perfectibility of humankind in a new heaven on Earth.[23]

---

[19] Jacques Roger, "The Living World," in Rousseau and Porter (eds.), *The Ferment of Knowledge,* pp. 255–84; Maureen McNeil, *Under the Banner of Science: Erasmus Darwin and His Age* (Manchester: Manchester University Press, 1987).

[20] Bentley Glass, Owsei Temkin, and William L. Straus (eds.), *Forerunners of Darwin, 1745–1859* (Baltimore, MD: Johns Hopkins University Press, 1968).

[21] For debates about scientific "knowledge," see Helge Kragh, *An Introduction to the Historiography of Science* (Cambridge University Press, 1987).

[22] Anne Goldgar, *Impolite Learning: Conduct and Community in the Republic of Letters 1680–1750* (New Haven, CT: Yale University Press, 1995).

[23] For state-employed experts, see Ken Alder, *Engineering the Revolution: Arms and Enlightenment in*

Some measure of science's growing authority is evident, as G. S. Rousseau observes in his discussion of literary responses (Chapter 33), in the vehemence of the Romantic revolt against it. The antiscience satires of the Augustan era – poking fun at virtuosi who peered down telescopes and mistook flies for elephants on the moon – give the impression that, around 1700, humanists still hardly discerned a scientific "threat." Indeed, many men of letters – not least, as we have already seen, Alexander Pope – were notably fulsome about scientific advances:

> Newton, pure Intelligence, whom God
> To Mortals lent, to trace his boundless Works
> From laws sublimely simple

sang James Thomson. Humanists were prominent in the dissemination of the sublime truths of the new science. In 1686, for instance, Bernard de Fontenelle produced his famous dialog *On The Plurality of Worlds* – the first work in France that made science both intelligible and entertaining to the general reading public. The man of letters thus conferred his blessing upon natural science, preparing the way, so to speak, for the cultural displacement of Christianity.

In stark contrast, there was something quite new in the venom of William Blake, directed in the late eighteenth century at the infernal trinity of Bacon, Locke, and Newton, as also in Charles Lamb's notorious toast to Newton's health "and confusion to mathematics," or, in its subtler manner, Goethe's formulation of an alternative to the mechanistic reductionism he deplored in Newtonianism. Mechanical science, judged Romantic critics, was turning into a veritable Frankenstein's monster.[24]

Perhaps the most telling index of this eighteenth-century "consolidation" of science is its embodiment in permanent institutional form. In earlier generations, natural knowledge had possessed few stable specialist platforms, and none unique unto itself. Most adepts had had to carve out a personal niche, be it at court, in the Church, or in academe; a few, such as Tycho Brahe, had been able to draw on private wealth, while others, such as Paracelsus, had lived

---

France, 1763–1815 (Princeton, NJ: Princeton University Press, 1997). For utilitarianism and progress see David Spadafora, *The Idea of Progress in Eighteenth Century Britain* (New Haven, CT: Yale University Press, 1990); Sidney Pollard, *The Idea of Progress: History and Society* (London: Watts, 1968); Keith Michael Baker, *Condorcet: From Natural Philosophy to Social Mathematics* (Chicago: University of Chicago Press, 1975).

[24] See the essay by Rousseau in this volume (Chapter 33) and also Gillian Beer, "Science and Literature," in Olby et al. (eds.), *Companion to the History of Modern Science*, pp. 783–98; Andrew Cunningham and Nicholas Jardine (eds.), *Romanticism and the Sciences* (Cambridge University Press, 1990); Marjorie Hope Nicolson, *Newton Demands the Muse: Newton's "Opticks" and the Eighteenth-Century Poets* (Princeton, NJ: Princeton University Press, 1946); Nicolson, *Science and Imagination* (Ithaca, NY: Cornell University Press, 1956); Joseph M. Levine, *The Battle of the Books: History and Literature in the Augustan Age* (Ithaca, NY: Cornell University Press, 1992). Thomson is quoted in Colin Russell, *Science and Social Change in Britain and Europe, 1700–1900* (New York: St. Martin's Press, 1983), p. 16.

hand-to-mouth. Although some educational foundations, as Brockliss here demonstrates, had given a modicum of encouragement to scientific and medical studies, the natural sciences could never become dominant in the traditional university system, whose rationale lay in training the clergy, a goal later supplemented by the aim of educating gentlemen or civil servants. In any case, by the 1700s universities were generally stagnating, although, of course, thanks to the Humboldtian reforms, they were to enjoy a surprising nineteenth-century resurrection.[25]

The precariousness of traditional institutional backing for science was alleviated during the eighteenth century. Many European rulers, with an eye, as Fox shows, to both practicality and prestige, made it their business to create state support programs for savants through such official bodies as the French Académie Royale des Sciences. Scientific academies, notably those in Paris, St. Petersburg, and Berlin, established clutches of permanent, state-funded posts for men of science; they might be seen as early engines of collective scientific research. In addition, scientific societies sprang up, national and local, formal and unofficial, practical and ornamental, closed and open. In his discussion in Chapter 4, James McClellan speaks of the sprouting of around a hundred of them by the close of the century, from Boston to Brussels, from Trondheim to Mannheim. Through such developments, the eighteenth century constituted, he contends, a "distinct era in the organizational and institutional history of European science," corroborating the view earlier canvassed that "the scientific enterprise became newly solidified in the eighteenth century."[26]

Leading lights in such academies also played other parts in spreading and seeding the natural sciences, for example among the wider circles of the salons. In France this was initially thanks to the efforts of Fontenelle, the perpetual secretary of the French Academy from 1699 to 1741, and also Voltaire, who popularized Newtonianism for French readers. "It was said of Socrates," wrote Joseph Addison, cofounder of the *Spectator,*

> that he brought Philosophy down from Heaven to inhabit among Men; and I shall be ambitious to have it said of me, that I have brought Philosophy out of Closets and Libraries, Schools and Colleges, to dwell in Clubs and Assemblies, at Tea-Tables and in Coffee Houses.[27]

[25] Laurence Brockliss, *French Higher Education in the Seventeenth and Eighteenth Centuries: A Cultural History* (Oxford: Clarendon Press, 1987); H. de Ridder-Symoens (ed.), *A History of the University in Europe,* vol. 2: *Universities in Early Modern Europe (1500–1800)* (Cambridge University Press, 1996); John Gascoigne, *Cambridge in the Age of the Enlightenment: Science, Religion and Politics from the Restoration to the French Revolution* (Cambridge University Press, 1989); Roger Emerson, "The Organization of Science and Its Pursuit in Early Modern Europe," in Olby et al. (eds.), *Companion to the History of Modern Science,* pp. 960–79.

[26] See also James E. McClellan III, *Science Reorganized: Scientific Societies in the Eighteenth Century* (New York: Columbia University Press, 1985).

[27] In D. F. Bond (ed.), *The Spectator,* 5 vols. (Oxford: Clarendon Press, 1965), vol. 1, p. 44. For Fontenelle, see Herbert Butterfield, *The Origins of Modern Science, 1300–1800* (London: Bell, 1949, 1950),

Alongside Addisonian moral and social philosophy, science too was infiltrating elite centers of social intercourse.

And if science was a growing presence within what Jürgen Habermas has styled the "public sphere" – in societies and salons, in lecture courses and museums – it was equally becoming established in the mind, as an ideological force and a prized ingredient in the approved cultural diet.[28] Controversies rage among historians – they are assessed in Chapter 6 by Mary Fissell and Roger Cooter – as to how best to interpret the outreach of science: "diffusion," "trickle down," and "social control" explanatory models have all been proposed, and in their turns severely criticized (here the "fried-egg" paradigm is the prime target for attack).[29]

"Supply and demand" models clearly beg many questions, but they at least have the virtue of recognizing that, in advanced regions of Europe, something like a marketplace in ideas had emerged. Consumers might buy into whichever aspects of science they chose, be they demonstrations in chemistry, or microscopes, or popular books such as Algarotti's *Newtonianism for the Ladies.* And the promoters of science were obliged to adjust their goods to what the market would bear: failure to do so could be disastrous, as is evident from the bankruptcies reported in Gerard Turner's account of the boom-and-bust trade in scientific apparatus (Chapter 22).

In complementary ways, Larry Stewart (Chapter 35) and Rob Iliffe trace the rise of the tangible empire of science, through exploration and colonization, and thereby provide further insights into its growing ideological hegemony.[30] "A comprehension of the power of Newton's natural philosophy, therefore, expanded beyond the colleges, or Crane Court [i.e., the Royal Society], or even beyond the subculture of instrument makers in Fleet Street," Stewart has elsewhere contended, discussing science's broadening appeal:

p. 144. For science within public culture, see Thomas Broman, "The Habermasian Public Sphere and 'Science *in* the Enlightenment'," *History of Science,* 36 (1998), 123–49, and J. Habermas, *The Structural Transformation of the Public Sphere: An Inquiry into a Category of Bourgeois Society* (Cambridge, MA: MIT Press, 1989). Valuable work on public science has been published by Simon Schaffer: "Natural Philosophy and Public Spectacle in the Eighteenth Century," *History of Science,* 21 (1983), 1–43; Schaffer, "The Consuming Flame: Electrical Showmen and Tory Mystics in the World of Goods," in John Brewer and Roy Porter (eds.), *Consumption and the World of Goods in the 17th and 18th Centuries* (London: Routledge, 1993), pp. 489–526. For Voltaire, see F. M. Voltaire, *Letters Concerning the English Nation* (London: printed for C. Davis and A. Lyon, 1733).

[28] Jan Golinski, *Science as Public Culture: Chemistry and Enlightenment in Britain, 1760–1820* (Cambridge University Press, 1992); Stewart, *The Rise of Public Science;* Larry Stewart, *Industry and Enterprise in Britain: From the Scientific to the Industrial Revolution 1640–1790* (London: Athlone Press, 1995).

[29] See in particular Roger Cooter and Stephen Pumfrey, "Separate Spheres and Public Places: Reflections on the History of Science Popularization and Science in Popular Culture," *History of Science,* 32 (1994), 237–67.

[30] Larry Stewart, "Public Lectures and Private Patronage in Newtonian England," *Isis,* 77 (1986), 47–58; Stewart, "The Selling of Newton: Science and Technology in Early Eighteenth-Century England," *Journal of British Studies,* 25 (1986), 179–92. The ideological uses of science have especially been emphasized by Margaret C. Jacob: *The Newtonians and the English Revolution 1689–1720* (Hassocks, Sussex: Harvester Press, 1976), p. 18:

... the social fluidity and the commerce of rationality in enlightenment England rested on the presumptions of the concrete, the practical and the entertaining. The efforts of those like Joseph Addison and Richard Steele in the coffeehouses, the *Spectator* could easily reflect, forced 'Philosophy out of Closets and Libraries, Schools and Colleges'. The sanction of natural philosophy came to rest in a far wider community than literacy alone might lead us to expect; the liberty of the coffeehouses may have been one reason, just as certainly as the rising cult of money was another.[31]

Within the Enlightenment project, the discourses of philosophy, poetry, religion, and politics appropriated the scientific methods and models associated with Bacon and Descartes, Galileo and Gassendi, and, above all, Newton. There were Newtonian poems galore, Newtonian theories of government, corpuscularian models of society, of political economy, of the mind and the passions, all disseminated by magazines and spread through provincial assemblies from Newcastle to Naples. Although such revisionist historians as J. C. D. Clark have recently questioned the importance of natural science to the consciousness of the age, E. P. Thompson was surely nearer the mark in maintaining that "the bourgeois and the scientific revolutions in England . . . were clearly a good deal more than just good friends"; the same holds for the relations between science and polite society in the Dutch Republic, the German principalities, the Italian duchies, and the Swiss cantons.[32]

Although the Enlightenment assuredly involved far more than the uptake of natural science, it would have been unthinkable without the surge of confidence in human powers over Nature conferred by the new philosophy. For the *philosophes,* scientific inquiry was the new broom *par excellence* that would sweep mystifications and obscurantism aside, removing the mumbo-jumbo of the Church and the "feudal" ways that kept the masses poor, hungry, and oppressed – that much is evident from a glimpse at any of the twenty-eight

This social explanation for the triumph of Newtonianism in the late seventeenth century stresses what previous commentators have ignored – its usefulness to the intellectual leaders of the Anglican church as an underpinning for their vision of what they liked to call the "world politick." The ordered, providentially guided, mathematically regulated universe of Newton gave a model for a stable and prosperous polity, ruled by the self-interest of men.

The idea of the empire of humankind over nature, central to the understanding of eighteenth-century science, is well expressed by Paula Findlen's title: *Possessing Nature: Museums, Collecting, and Scientific Culture in Early Modern Italy* (Berkeley: University of California Press, 1994). See also James A. Secord, "Newton in the Nursery: Tom Telescope and the Philosophy of Tops and Balls, 1761–1838," *History of Science,* 23 (1985), 127–51.

[31] Stewart, *The Rise of Public Science,* pp. xxxi–xxxii.

[32] E. P. Thompson, "The Peculiarities of the English," in his *The Poverty of Theory and Other Essays* (London: Merlin Press, 1978), pp. 35–91, especially p. 60. Contrast J. C. D. Clark, *Revolution and Rebellion: State and Society in England in the Seventeenth and Eighteenth Centuries* (Cambridge University Press, 1986).

For science and the *philosophes,* see Hankins, *Science and the Enlightenment,* and Dorinda Outram, *The Enlightenment* (Cambridge University Press, 1995). Colm Kiernan's *Enlightenment and Science in Eighteenth-Century France,* 2nd ed. (*Studies on Voltaire and the Eighteenth Century,* 59, Genève, 1968), although stimulating, employs a Lovejoyan "history of ideas" approach that now looks very dated.

volumes of the *Encyclopédie*. Indeed, as Richard Yeo stresses in Chapter 10, perhaps the prime impulse behind the encyclopedic project was the dissemination of scientific knowledge. The increasingly rapid accretion of such information created the need – or at least provided the rationale – for new encyclopedias and for updated editions of the old ones. Charles Lamb's droll confession that he was "a whole encyclopaedia behind the rest of the world" does not simply measure the *depths* of his ignorance but shows that what he knew was hopelessly *out of date.*[33]

Promoters of science and the Enlightenment should not, to be sure, be taken at their own estimations. The natural sciences always came gift-wrapped in ideology, a point well made by Reill in his historiographical discussion of science and the Enlightenment in Chapter 2. The voice of "science" might bolster elite culture, while discrediting the beliefs and behaviors of the pious, the poor, and the plebs, of women and the marginalized.[34] In certain situations, science – indeed the "social sciences" (a phrase popularized by Turgot, Condorcet, and their circles) – declared that belief in witchcraft was mere superstition; in others it pronounced the superiority of the white man or pronounced upon the hysterical tendencies of the female nervous system. The new techniques – statistical enumeration, biopolitical surveys – applied to specific "social uses,"[35] staked claims to authority on the basis of the physical sciences, as Richard Olson stresses in discussing the underpinnings of the human sciences in Chapter 19. Nor were the weapons of science available only to "progressives." In his *Essay on the Principle of Population* (1798), "Parson" Malthus was confident that he could demolish the foolish perfectibilism of the French revolutionaries with some tabulations of data and a simple equation.[36]

It had been Newton himself who had ventured in the *Opticks* (1704) that

---

[33] Lamb is quoted in Richard Yeo, *Encyclopaedic Visions: Scientific Dictionaries and Enlightenment Culture* (Cambridge University Press, 2001).

[34] The ideological appropriations of Newtonianism, in support of the social and political order dominant after the "glorious revolution" of 1788–9, have been emphasized by Margaret Jacob, *The Cultural Meaning of the Scientific Revolution;* Jacob, *The Politics of Western Science, 1640–1990* (Atlantic Highlands, NJ: Humanities Press International, 1994); Jacob, "Reflections on the Ideological Meaning of Western Science from Boyle and Newton to the Postmodernists," *History of Science,* 101 (1995), 333–57; Jacob, *The Radical Enlightenment: Pantheists, Freemasons and Republicans* (London: Allen & Unwin, 1981).

[35] Steven Shapin, "Social Uses of Science," in Rousseau and Porter (eds.), *The Ferment of Knowledge,* pp. 93–142.

[36] For probability, see note 16. For science in political theory, see I. Bernard Cohen, *Science and the Founding Fathers: Science in the Political Thought of Thomas Jefferson, Benjamin Franklin, John Adams, and James Madison* (New York: W. W. Norton, 1995); Dorinda Outram, "Science and Political Ideology, 1790–1848," in Olby et al. (eds.), *Companion to the History of Modern Science.* For political arithmetic, see Andrea Rusnock, "Biopolitics: Political Arithmetic in the Enlightenment," in William Clark, Jan Golinski, and Simon Schaffer (eds.), *The Sciences in Enlightened Europe* (Chicago: University of Chicago Press, 1999). For the social sciences, see Richard Olson, *The Emergence of the Social Sciences, 1642–1792* (New York: Twayne, 1993). On Malthus, see Roy Porter, "The 1790s: 'Visions of Unsullied Bliss,'" in Asa Briggs and Daniel Snowman (eds.), *Fins de Siècles: The Changing Sense of an Ending* (New Haven, CT: Yale University Press, 1996), pp. 125–56.

by the perfection of natural philosophy, "the bounds of Moral Philosophy will also be enlarged." Projects taking their cue from this declaration enjoyed great prestige; it was not only David Hume but also many other savants who aspired to be the "Newton of the moral sciences." Hence it comes as no surprise that, by the 1790s, Edmund Burke, recoiling from the accursed atrocities of the French Revolution, could lament that "the age of chivalry is gone – that of sophisters, economists, and calculators, has succeeded; and the glory of Europe is extinguished for ever."[37]

Addressing such developments – the rise of those very economists and calculators – some of the most important recent work on Enlightenment science has explored the recruitment of science as a disciplinary and regulative authority. In a series of works, Michel Foucault analyzed the role played by scientific rationality in creating new regimes and technologies of power, often for the management of populations and environments.[38] For their part, feminists have maintained that the models and metaphors of mechanical science lent themselves to doctrines of male domination.[39] E. P. Thompson showed how the new science of political economy was used to discredit the traditional "moral economy,"[40] and many studies have explored how natural knowledge was conscripted to nullify popular and folk knowledges.[41]

The potency of science and its ideological uses – or "abuses" – must not, however, be exaggerated, and we must be careful not to predate its hegemony; after all, the English language had no need for the very word "scientist" until well into the next century.[42] As John Brooke (Chapter 32) and many other authors in this volume emphasize, it would be utterly anachronistic to imply a Grand Canyon or a polarity between the investigation of Nature and the contemplation of God, just as it would be simplistic to assume a preordained transition from a religious cosmos, in whose workings a personal God intervened, to a later naturalistic one, governed exclusively by natural laws.[43] Instead of any such teleological or evolutionary readings, the challenge the scientific

---

[37] Edmund Burke, *Reflections on the Revolution in France* (1790).

[38] Michel Foucault, *Power/Knowledge: Selected Interviews and Other Writings 1972–1977* (Brighton: Harvester Press, 1977); Foucault, *Discipline and Punish: The Birth of the Prison* (Harmondsworth: Penguin, 1979).

[39] Carolyn Merchant, *The Death of Nature: Women, Ecology and the Scientific Revolution* (San Francisco: Harper and Row, 1980); Brian Easlea, *Science and Sexual Oppression: Patriarchy's Confrontation with Woman and Nature* (London: Weidenfeld and Nicolson, 1981); J. R. R. Christie, "Feminism and the History of Science," in Olby et al. (eds.), *Companion to the History of Modern Science*, pp. 100–9; L. J. Jordanova, *Sexual Visions: Images of Gender in Science and Medicine Between the 18th and the 20th Centuries* (Madison: University of Wisconsin Press, 1989); Sandra Harding and Jean F. O'Barr (eds.), *Sex and Scientific Inquiry* (Chicago: University of Chicago Press, 1987).

[40] E. P. Thompson, "The Moral Economy of the English Crowd," *Past and Present*, 1 (1971), 76–136.

[41] See, for instance, Patrick Curry, *Prophecy and Power: Astrology in Early Modern England* (Cambridge: Polity Press, 1989).

[42] Sydney Ross, "*Scientist:* The Story of a Word," *Annals of Science*, 18 (1962), 65–85.

[43] John Hedley Brooke, "Science and Religion," in Olby et al. (eds.), *Companion to the History of Modern Science*, pp. 763–82; Brooke, "Science and the Secularisation of Knowledge: Perspectives on Some Eighteenth-Century Transformations," *Nuncius*, 4 (1989), 43–65; Brooke, *Science and Religion: Some Historical Perspectives* (Cambridge University Press, 1991).

enterprise presents to the historian lies in explaining uneven development and resistance. After all, the French Revolution closed down the Academy of Science ("The Republic has no need of *savants*," gloated the Jacobins), guillotined two of the nation's premier men of science – Lavoisier the chemist and Baily the astronomer – and hounded to death science's leading spokesman, Condorcet. Science, in other words, was not the spirit of the future mounted majestically on an iron horse; rather, it was a resource with multiple uses, and foes no less than friends. Hence, if one thing characterizes the mood of contemporary scholarship and so the tone of this book, it is a sensitivity to the need to set science in context. Science, maintained Steven Shapin twenty years ago, will be misunderstood unless strictly interpreted in contexts of use: it is a lesson that fortunately has been heeded.[44]

It must be stressed that "science" never presented a united front. However much *savants* liked to pretend in their propaganda that science was the candid, cosmopolitan, and liberal pursuit of natural truth, the actuality was otherwise. Protagonists might claim that the "sciences are never at war,"[45] but practitioners of science formed cliques like those of any other profession or pursuit; rival camps slugged it out in every field of inquiry – Newtonians versus Leibnizians, Neptunist versus Plutonist geologists – and splits were often exacerbated by religious, linguistic, and patriotic allegiances. Secretiveness, jealousy, and rivalry were inflamed by priority disputes, ferocious battles over the ownership of discoveries and inventions, and other claims to scientific property.[46] The battle lines in chemistry and the science of life, here documented by Jan Golinski and Shirley Roe, largely followed national loyalties.[47]

As Patricia Fara shows in Chapter 21, there were equally fierce boundary disputes respecting the legitimation and policing of particular sciences and concerning the marginalization and anathematization of practices as pseudosciences, showmanship, swindles, and spectacle.[48] As Charles Gillispie long ago

---

[44] On science in the French Revolution, see Charles C. Gillispie, *Science and Polity in France at the End of the Old Regime* (Princeton, NJ: Princeton University Press, 1980); Dorinda Outram, "The Ordeal of Vocation: The Paris Academy of Sciences and the Terror, 1793–95," *History of Science,* 21 (1983), 251–73; Shapin, "Social Uses of Science," in Rousseau and Porter (eds.), *The Ferment of Knowledge,* pp. 93–142, especially p. 95.

[45] Sir Gavin de Beer, *The Sciences Were Never at War* (London: Thomas Nelson and Sons, 1960); A. R. Hall, *Philosophers at War: The Quarrel Between Newton and Leibniz* (Cambridge University Press, 1980).

[46] For "scientific property," see Rob Iliffe, "'In the Warehouse': Privacy, Property and Priority in the Early Royal Society," *History of Science,* 30 (1992), 29–62.

[47] See also, for chemistry, C. E. Perrin, "Revolution of Reform: The Chemical Revolution and Eighteenth Century Concepts of Scientific Change," *History of Science,* 25 (1987), 395–423; for the life sciences, Jacques Roger, *Les sciences de la vie dans la pensée française du XVIIIè siècle* (Paris: A. Colin, 1971); Roger, "The Living World," in Rousseau and Porter (eds.), *The Ferment of Knowledge,* pp. 255–84.

[48] See also Gloria Flaherty, "The Non-Normal Sciences: Survivals of Renaissance Thought in the Eighteenth Century," in Christopher Fox, Roy Porter, and Robert Wokler (eds.), *Inventing Human Science: Eighteenth-Century Domains* (Berkeley: University of California Press, 1995), pp. 271–91; Robert Darnton, *Mesmerism and the End of the Enlightenment in France* (Cambridge, MA: Harvard University Press, 1968).

observed, the eighteenth century brought profound struggles between those
who were convinced that mathematical physics must provide the template for
true science and scientific romantics such as Diderot – or, later, the *Naturphilosophen*, – touting more holistic, vitalistic, and subjective versions of nature.[49]

In particular, the notion of some unifying metaphysical "Newtonian" umbrella under which all the sciences could shelter – Schaffer has dubbed it the
myth of the "alleged coherence of a single 'Newtonianism'"[50] – comes under
repeated fire in the chapters that follow.[51] There were numerous distinctive
philosophies of nature, insists John Gascoigne (Chapter 12). Although many
people aspired to be recognized as the Newton of the moral sciences, and the
dream of establishing the definitive scientific method carried great appeal,
there remained, insists Paul Wood (Chapter 34), competing models for a science of mind or human nature, and far from all of them were Newtonian:
"Newton's impact has been exaggerated," he concludes, "and his writings
were read in such radically different ways that it is difficult to identify a unified Newtonian tradition in the moral sciences."

Central to the problems of comprehending eighteenth-century natural
science is the question as to the species of knowledge it was supposed to constitute. The term typically deployed in the early modern era for such inquiries
was "natural philosophy" – as in Newton's *Principia mathematica philosophiae
naturalis* (1687) – this being regarded as a system of concepts mediating between matters of fact and philosophy and leading, by implication, "through
Nature up to Nature's God." The term "natural philosophy" and the ideal it
embodied remained widespread. But there are grounds for questioning the
challenging view recently advanced by Andrew Cunningham that the framework set by natural philosophy remained dominant into the nineteenth century – and hence also his inference that it is anachronistic to speak at all of
eighteenth-century "science" in the modern sense.[52] Much evidence adduced
in this volume suggests that the balkanization of specialist disciplines was

[49] Charles C. Gillispie, *The Edge of Objectivity: An Essay in the History of Scientific Ideas* (Princeton, NJ: Princeton University Press, 1960), pp. 187f., 192f.

[50] Simon Schaffer, "Natural Philosophy," in Rousseau and Porter (eds.), *The Ferment of Knowledge*, pp. 55–91, especially p. 58.

[51] For affirmations of coherent Newtonian traditions, see R. E. Schofield, *Mechanism and Materialism: British Natural Philosophy in an Age of Reason* (Princeton, NJ: Princeton University Press, 1970); Arnold Thackray, *Atoms and Powers: An Essay on Newtonian Matter Theory and the Development of Chemistry* (Cambridge, MA: Harvard University Press, 1977); P. M. Heimann and J. E. McGuire, "Newtonian Forces and Lockean Powers: Concepts of Matter in Eighteenth-Century Thought," *Historical Studies in the Physical Sciences*, 3 (1971), 233–306; Heimann, "Newtonian Natural Philosophy and the Scientific Revolution," *History of Science*, 11 (1973), 1–7; Heimann, "'Nature Is a Perpetual Worker': Newton's Aether and Eighteenth-Century Natural Philosophy," *Ambix*, 20 (1973), 1–25; Heimann, "Voluntarism and Immanence: Conceptions of Nature in Eighteenth-Century Thought," *Journal of the History of Ideas*, 39 (1978), 271–83; Peter Harman, *Metaphysics and Natural Philosophy* (Brighton: Harvester Press, 1982).

[52] Andrew Cunningham, "Getting the Game Right: Some Plain Words on the Identity and Invention of Science," *Studies in the History and Philosophy of Science*, 19 (1988), 365–89; Brooke, *Science and Religion*.

already undermining any authentic notion of a unifying natural philosophy. "Overall, then," maintains Gascoigne, stressing not the resilience of natural philosophy but its breakup,

> the eighteenth century sees the transition from natural philosophy as a branch of philosophy to the beginnings of an array of scientific disciplines that largely undermined the assumption of a unified view of nature on which the enterprise of natural philosophy had traditionally been based.

Finally, some words are needed about the aims, intentions, and scope of this book. There is no single "natural" way to cut up the knowledge cake in a volume such as this – a dilemma that amusingly reflects that faced by the compilers of eighteenth-century encyclopedias. After consulting with colleagues, I have exercised the editor's prerogative and have chosen to follow a fairly traditional division of topics, giving some priority to separate disciplines, with chapters on chemistry, astronomy, medicine, mathematics, and so forth. That such a partition is not anachronistic is confirmed by Richard Yeo's account of Enlightenment "maps of knowledge."[53] I have chosen to employ such divisions largely because I believe (and here I endorse the views of the old encyclopedia editors) that these will prove more lastingly convenient to readers and students than alternative thematizations in tune with the academic fashions of the late 1990s.

It is the aim of this volume – and of this Cambridge series as a whole – to provide critical syntheses of the best modern thinking. As one would expect from a team of leading scholars, there is a great deal that is original in the essays that follow; but the prime aim has not been to fly speculative kites or proselytize for a party line. Rather, the emphasis has been upon providing balanced interpretations backed by basic information in a book that can double as a reference text.

No apology is needed for telling the stories of eighteenth-century science. Some twenty years ago Susan Cannon griped at our ignorance about even the basics:

> For the history of science and the history of ideas in the 18th century you can trust almost no one. The amount of 'hard' history of science for that period is so lacking that one simply leaps . . . from Newton in optics to Young in optics . . . It is no reproach to my friends who are trying to do something with the 18th century to tell them that their labors have not yet reached the point at which a 19th century historian can confidently go ahead from the stable platform they had erected.[54]

[53] For contemporary knowledge maps, see Richard Yeo, "Reading Encyclopedias: Science and the Organization of Knowledge in British Dictionaries of Arts and Sciences, 1730–1850," *Isis*, 82 (1991), 24–49; Yeo, "Genius, Method and Mortality: Images of Newton in Britain, 1760–1860," *Science in Context*, 2 (1988), 257–84.

[54] Susan Faye Cannon, *Science in Culture: The Early Victorian Period* (New York: Science History Publications, 1978), pp. 133–4.

Since then, researchers have beavered away, but most of their findings have appeared in scholarly journals and specialized monographs, often expensive and of limited circulation. Hence, one prime service that this volume can provide is that of synthesis, assimilating the research of recent decades, interpreting it and distilling it into a form that is accessible to students and nonspecialists as well as to researchers in the field. Hopefully, the "stable platform" will at last have been built.

Surprising though this may seem, this is a novel endeavor. It has been a long time since a weighty general account of eighteenth-century science has appeared. *The Beginnings of Modern Science: From 1450 to 1800,* a work edited by René Taton, which appeared in French in 1958 and in English translation seven years later, was the last big general text that included a substantial section on the eighteenth century. But that work is essentially a compilation, and its interpretations now appear horribly dated, mainly on account of their pervasive positivistic bias: "18th-century science," we there read, "was largely responsible for the rise of rationalism and for the shedding of much theological lumber."[55] One welcome exception to the dearth of treatments has been Thomas Hankins's *Science and the Enlightenment* (1985), but that work is tantalizingly brief.[56] This lack of modern texts on the eighteenth century may be regarded as, in part, an accidental by-product of the vagaries of historical periodization. The original edition of Rupert Hall's pioneering *The Scientific Revolution, 1500–1800* (1954) thus nominally envisaged that "revolution" as going right up to 1800, although disproportionately little space was actually devoted to the eighteenth century. In his 1983 rewriting of the book, Hall chose to truncate his terminal date to 1750 and predictably devoted even less space to the eighteenth century – just 10 pages out of a 350-page book.[57]

It is worth drawing attention to the originality of some of the topics and interpretative thrusts contained within the present work, reflecting as they do the revitalization of the field in recent years. Some twenty years ago, *The Ferment of Knowledge* attempted a historiographical survey of eighteenth-century science and pointed to research opportunities.[58] Comparison with that volume is instructive. The present book gives prominence to many research areas energetically developed since then. It also – through its silences – negatively indicates other concerns that have dissolved away. The "internalist

---

[55] René Taton (ed.), *The Beginnings of Modern Science: From 1450 to 1800,* trans. A. J. Pomerans (London: Readers Union/Thames and Hudson, 1965), p. 578.

[56] Hankins, *Science and the Enlightenment.*

[57] A. R. Hall, *The Scientific Revolution, 1500–1800.* Hall now explains: "I now omit the successor phases of the eighteenth century in which the sciences of chemistry and electricity received their first coherent forms" (p. vii).

[58] Rousseau and Porter (eds.), *The Ferment of Knowledge.* Where the historiography once stood can be gathered from Paul T. Durbin (ed.), *A Guide to the Culture of Science, Technology and Medicine* (New York: Free Press, 1980); Pietro Corsi and Paul Weindling (eds.), *Information Sources in the History of Science and Medicine* (London: Butterworth Scientific, 1983).

versus externalist" controversy, for instance, once so noisy and bitter, is now, with universal acceptance among professional historians of the social production of knowledge, a dead letter.[59]

This volume offers a generous representation of non-Western science – something absent from *The Ferment of Knowledge* – with chapters by Emilie Savage-Smith on Islam (Chapter 27), Frank Dikötter on China (Chapter 29), Shigeru Nakayama on Japan (Chapter 30), and Deepak Kumar on India (Chapter 28), as well as Jorge Cañizares Esguerra's reading of the ambiguous quasi-colonial context of Latin America (Chapter 31). The lion's share of the book is, however, given over to "Western" science, which essentially – despite Benjamin Franklin et al. – means "Old World" science. Should Europe be so privileged? The pros and cons may be debated endlessly. It must be said, however, that European science was undergoing far more dynamic developments than the other non-Western traditions here surveyed and that it was Western science that strode imperialistically over the rest of the world, as Ian Inkster emphasizes in a bold essay (Chapter 36) that examines technological, scientific, and economic nodal points comparatively in East and West.[60]

As already noted, the "social uses" of science and the strategies underpinning them here receive far more attention than was common twenty years ago, when Shapin could claim, albeit tendentiously, that "social uses have not . . . greatly interested historians of science."[61] This situation has dramatically changed; indeed, Fissell and Cooter claim, perhaps equally tendentiously, in their exploration of the "sites and forms" of natural knowledge that "while thirty years ago much attention was paid to the intricacies of Isaac Newton's thought, now historians explore the social uses of such thought."

Certain areas of inquiry that were long neglected or treated perfunctorily have now been revitalized. As is reflected in Gerard Turner's contribution, study of scientific collections and instruments has moved out of the museum and away from its artifacts per se, into a wider probing of the social functioning of science within material culture.[62] It is no accident that Turner launches his essay with the clarion call by the chemist James Keir: "The diffusion of a general knowledge and of a taste for science, over all classes of men, in every

59 Steven Shapin calls the distinction "rather silly": *The Scientific Revolution*, p. 9. See also Shapin, "Discipline and Bounding: The History and Sociology of Science as Seen Through the Externalism and Internalism Debate," *History of Science*, 30 (1992), 333–69.

60 Cognitive imperialism has been widely discussed. See Edward W. Said, *Culture and Imperialism* (London: Chatto and Windus, 1993), and the chapter on the nonrevolution in science outside Europe in H. Floris Cohen, *The Scientific Revolution*. For modern notions of culture, see Clifford Geertz, *The Interpretation of Cultures* (New York: Basic Books, 1973).

61 Shapin, "Social Uses of Science," 93. On p. 118 of that essay, Shapin refers readers to a "forthcoming" book of his on *The Social Use of Nature*. It is a great shame that this has not appeared.

62 Ann Bermingham and John Brewer (eds.), *The Consumption of Culture, 1600–1800: Image, Object, Text in the 17th and 18th Centuries* (London: Routledge, 1995); John Brewer and Roy Porter (eds.), *Consumption and the World of Goods in the 17th and 18th Centuries* (London: Routledge, 1993); John Brewer and Susan Staves (eds.), *Early Modern Conceptions of Property* (London: Routledge, 1995).

nation of Europe, or of European origin, seem to be the characteristic features of the present age."[63] The same can be said for assessments of the visual expressions of science. As Brian Ford (Chapter 24) and Charlotte Klonk (Chapter 25) show in their complementary pieces, botanical illustrations and landscape art were enjoying something of a Golden Age by appealing to a rising appreciation of Nature and the natural – categories that, as Brooke reminds us, straddled the religious, the esthetic, and the rational.[64]

Perhaps most noticeably of all, prompted first by structuralist interest in discourse analysis and then by postmodernist preoccupations with textuality, attention has recently been directed to the media of science communication and the rhetoric of scientific truth. Condillac famously asserted that "the art of reasoning is nothing more than a language well arranged," and the question of scientific discourse was likewise central to Lavoisier, whose *Méthode de nomenclature chiminque* (1787) asserted that

> A well-composed language, adapted to the natural and successive order of ideas will bring in its train a necessary and immediate revolution in the method of teaching. The logic of the sciences is thus essentially dependent on their language.[65]

Such Enlightenment concerns speak directly to our current fascination with the power of words to make and remake worlds, as is registered in Adrian Johns's evaluation of the impact of print culture upon the authority and accreditation of science (Chapter 23).[66]

---

[63] James Keir, Preface to *The First Part of a Dictionary of Chemistry* (Birmingham: printed by Pearson and Rollason for Elliot and Kay, 1789).

[64] For art and science, see also Barbara Stafford, *Imaging the Unseen in Enlightenment Art and Medicine* (Cambridge, MA: MIT Press, 1992); Stafford, *Artful Science: Enlightenment, Entertainment, and the Eclipse of Visual Education* (Cambridge, MA: MIT Press, 1994).

[65] Quoted in Hankins, *Science and the Enlightenment,* p. 109.

[66] On the problem of writing science, see the excellent discussion by Shapin, *The Scientific Revolution,* p. 108:

> Such means were found in the forms of scientific communication itself. Experience might be extended and made public by *writing* scientific narratives in a way that offered distant readers who had not directly witnessed the phenomena – and probably never would – such a vivid account of experimental performances that they might be made into *virtual witnesses.* Most practitioners who took Boyle's factual particulars into their stock of knowledge did so not through direct witnessing or through replication but through reading his reports and finding adequate grounds to trust their accuracy and veracity. As Boyle said, his narratives (and those that competently followed the style he recommended) were to be "standing records" of the new practice, and readers "need not reiterate themselves an experiment to have as distinct an idea of it, as may suffice them to ground their reflexions and speculations upon." Virtual witnessing involved producing in a reader's mind such an image of an experimental scene as obviated the necessity for either its direct witness or its replication.

> On the language of science, see Maurice P. Crosland, *Historical Studies in the Language of Chemistry* (London: Heinemann, 1962); J. V. Golinski, "Language, Discourse and Science," in Olby et al. (eds.), *Companion to the History of Modern Science,* pp. 110–26; L. J. Jordanova (ed.), *Languages of Nature: Critical Essays on Science and Literature* (London: Free Association Books, 1986). For language theories, see Hans Aarsleff, *From Locke to Saussure: Essays on the Study of Language and Intellectual History* (Minneapolis: University of Minnesota Press, 1982); Brian Vickers and Nancy S. Struer,

The insertion of science within the social fabric has also been undergoing considerable rethinking. William Clark here appraises research into prosopography (Chapter 9), and the connected questions of the identity and representation of the savant are taken up by Steven Shapin (Chapter 7) and Londa Schiebinger (Chapter 8).[67] Between them, this trio of contributions transcends what was in danger of becoming the hackneyed topic of "professionalization," so often warped by the presentist fixations of the sociology of professions.

Although comprehensiveness cannot be a sane aim in a volume of this size,[68] an attempt has been made to strike a balance between knowledge and society, between topics primarily cognitive and others more culturally oriented. Attention is given to the material culture of science (books, illustrations, communication and societies), to science's interplay with other discourses

Rhetoric and the Pursuit of Truth: Language Change in the Seventeenth and Eighteenth Centuries: Papers Read at a Clark Library Seminar 8 March 1980 (Los Angeles: William Andrews Clark Memorial Library, 1985).

On book culture, see Roger Chartier, *L'Ordre de livres: lecteurs, auteurs, bibliothèques en Europe entre XIVe et XVIIIe siècle* (Aix-en-Provence: Alinea, 1992), trans. Lydia Cochrane as *The Order of Books: Readers, Authors and Libraries in Europe between the 14th and 18th Centuries* (Cambridge: Polity Press, 1994); Elizabeth Eisenstein, *The Printing Press as an Agent of Change,* 2 vols. (Cambridge University Press, 1979); William Eamon, "From the Secrets of Nature to Public Knowledge," in Lindberg and Westman (eds.), *Reappraisals of the Scientific Revolution,* pp. 333–65; Paolo L. Rossi, "Society, Culture and the Dissemination of Learning," in Stephen Pumfrey, Paolo L. Rossi, and Maurice Slavinski (eds.), *Science, Culture and Popular Belief in Renaissance Europe* (Manchester: Manchester University Press, 1991), pp. 143–75; Robert Darnton, "The High Enlightenment and the Low Life of Literature in Pre-Revolutionary France," *Past and Present,* 51 (1971), 81–115; Darnton, *The Business of Enlightenment: A Publishing History of the Encyclopedie, 1775–1800* (Cambridge, MA: Harvard University Press, 1979); Darnton, *The Literary Underground of the Old Regime* (Cambridge, MA: Harvard University Press, 1982). See also Michel Foucault, "What Is an Author?" in *Language, Counter-Memory, Practice: Selected Essays and Interviews,* ed. Donald F. Bouchard, trans. Donald F. Bouchard and Sherry Simon (Ithaca, NY: Cornell University Press, 1977), pp. 113–38.

[67] For prosopography, see Steven Shapin and Arnold Thackray, "Prosopography as a Research Tool in the History of Science: The British Scientific Community, 1700–1900," *History of Science,* 12 (1974), 1–28; Lewis Pyenson, "Who the Guys Were," *History of Science,* 15 (1977), 155–88. On the man and woman of science, see Londa Schiebinger, *The Mind Has No Sex? Women in the Origins of Modern Science* (Cambridge, MA: Harvard University Press, 1989); Paula Findlen, "A Forgotten Newtonian: Women and Sciences in the Italian Provinces," in Clark, Golinski, and Schaffer (eds.), *The Sciences in Enlightened Europe;* Steven Shapin, "History of Science and Its Sociological Reconstructions," *History of Science,* 20 (1982), 157–211; Shapin, "'A Scholar and a Gentleman': The Problematic Identity of the Scientific Practitioner in Early Modern England," *History of Science,* 29 (1991), 279–327.

For understanding the "man of science," eloges have proved highly fruitful. See Dorinda Outram, "The Language of Natural Power: The *Éloges* of Georges Cuvier and the Public Language of Nineteenth Century Science," *History of Science,* 16 (1978), 153–78; Charles B. Paul, *Science and Immortality: The Eloges of the Paris Academy of Science (1699–1791)* (Berkeley: University of California Press, 1980); George Weisz, "The Self-Made Mandarin: The *Éloges* of the French Academy of Medicine, 1824–47," *History of Science,* 26 (1988), 13–40.

[68] Different readers will deplore different gaps; it would, for instance, have been desirable to have had chapters on such topics as geography, meteorology, botany, engineering, and so forth, although space limitations have precluded such comprehensiveness. The editor is well aware that such gaps exist, some of them due to commissioned contributors having to drop out unavoidably at the last moment. The editor is also aware of the heavy Anglo-American bias among the contributors. This was not his intention, and invitations to contribute were extended to many scholars from beyond the Anglo-Saxon heartland.

(religion, literature, art), and to the symbiosis of science with economy, society, and the state. In the end, the value of this volume will rest not so much upon the inclusion or exclusion of a particular heading in the Contents list but rather in the success of the authorial team in engaging with key issues and forging wider connections.

Part I

SCIENCE AND SOCIETY

2

THE LEGACY OF THE "SCIENTIFIC REVOLUTION"

Science and the Enlightenment

Peter Hanns Reill

If there is one characterization of the Enlightenment that appears as a truism, it is the assertion that the Enlightenment adopted, extended, and completed the intellectual and social project usually characterized as the "Scientific Revolution," a movement forged by Johannes Kepler (1571–1630) and Galilei Galileo (1564–1642), developed by René Descartes (1596–1650) and Gottfried Wilhelm Leibniz (1646–1716), and completed by Isaac Newton (1642–1727). In this view, the Enlightenment becomes both the inheritor of this legacy and its most persistent and dogmatic trustee. Because the Enlightenment is often seen as an age in which a "scientific paradigm" is accepted and transformed into "normal science,"[1] the history of Enlightenment science has often been considered "a tiresome trough to be negotiated between the peaks of the seventeenth and those of the nineteenth century; or as a mystery, a twilight zone in which all is on the verge of yielding."[2] For many recent commentators even the twilight zone has been dispelled, revealing clear and close links between Enlightenment science and the "rational" imperatives of the Scientific Revolution, establishing the Enlightenment as the prototypical era in which scientific and instrumental reason became a defining characteristic of modern culture. These linkages between the Scientific Revolution, Enlightenment science, and a negative evaluation of modernity were first drawn by some intellectuals horrified by the destructiveness of modern civilization at the end of the Second World War. Max Horkheimer, for example, claimed in 1946 that "the collapse of a large part of the intellectual foundation of our civilization is . . . the result of technical and scientific progress."[3] He located the origins of this demise – whose process he characterized as "the self-destructive tendency

---

[1] The two terms "paradigm" and "normal science" are central to Kuhn's interpretation of the dynamics of scientific revolution. Thomas S. Kuhn, *The Structure of Scientific Revolutions,* 3rd ed. (Chicago: University of Chicago Press, 1996).

[2] G. S. Rousseau and Roy Porter (eds.), *The Ferment of Knowledge: Studies in the Historiography of Eighteenth-Century Science* (Cambridge University Press, 1980), p. 2.

[3] Max Horkheimer, "Reason Against Itself: Some Remarks on Enlightenment," in James Schmidt (ed.),

of Reason" – in the Enlightenment. This line of analysis, further elaborated by Horkheimer and Theodore Adorno in the *Dialectic of the Enlightenment,* was later expanded and amplified by many commentators: postmodernists who rebel against the so-called hegemony of enlightenment rationality and analyze what Michel Foucault called the knowledge/power dyad that gave rise to the intrusive, all-controlling panopticom of modern social control;[4] some feminists who decry the Enlightenment's supposed elevation of universality over distinctness;[5] and "converted" philosophers of science, such as Stephen Toulmin, who seek to uncover modernity's dangerous and outmoded hidden agenda by searching out the political and social forces that led to its inception.[6] Despite the vast differences separating these critics and the multiple tones of major and minor that they sound, the indictment is clear. The Enlightenment in its fascination with science and universalizing reason sired such movements as gender and racial discrimination, colonialism, and totalitarianism.

These are strong words. For historians of the Enlightenment there seems to be a radical breech between what is meant by the central signifiers in this critique and what the historians perceive. Clearly, the major focus in such attacks is the Enlightenment's supposed worship of science, reason, and universality, of a form of knowledge/power that is invariably characterized in the singular. It is obvious what that singular suggests: the triumph in and by the Enlightenment of a mathematically based science, founded on certain essential presuppositions concerning matter, method, and explanation whose reign has lasted until today. Toulmin described this macro-historical movement as follows:

> In choosing the goals of modernity, an intellectual and practical agenda that
> . . . focused on the seventeenth-century pursuit of mathematical exactitude
> and logical rigor, intellectual certainty and moral purity, Europe set itself on
> a cultural and political road that has led both to its most striking technical
> successes and to its deepest human failures.[7]

Yet when one begins to query what was really implied beneath this all-powerful engine of cultural and social change, the picture becomes much more hazy, complicating and confusing the new anti-Enlightenment master nar-

---

What Is Enlightenment? Eighteenth-Century Answers and Twentieth-Century Questions (Berkeley: University of California Press, 1996), p. 359.

[4] This is most clearly argued in Foucault's later works; see especially Michel Foucault, *Discipline and Punish: The Birth of the Prison,* trans. Alan Sheridan (New York: Random House, 1979).

[5] The classic critique of modern science from a radical feminist position was provided by Carolyn Merchant, *The Death of Nature: Women, Ecology and the Scientific Revolution* (New York: Harper and Row, 1980); for more recent critiques, see Noami Schor, "French Feminism Is a Universalism," *Differences: A Journal of Feminist Cultural Studies* (1995); Robin May Schott, "The Gender of the Enlightenment," in Schmidt, *What Is Enlightenment?* pp. 471–87.

[6] Stephen Toulmin, *Cosmopolis: The Hidden Agenda of Modernity* (New York: Free Press, 1990).

[7] Ibid., p. x.

ratives that are being forged and opening fascinating alternatives to evaluate what is often called the Enlightenment project. As studies increasingly question the uniformity of the Scientific Revolution, it is becoming apparent that if there is a legacy, it is extremely complex, contradictory, and rich in various interpretations.

THE SCIENTIFIC REVOLUTION, MECHANICAL NATURAL PHILOSOPHY, AND THE ENLIGHTENMENT

This is certainly true for the manner in which nature was interpreted in the Enlightenment and the way in which those interpretations were deployed in discourses dealing with human activities. Recently, historians of eighteenth-century science have begun to question the assumption that the natural philosophy of the period can be reduced to what is often called mathematical mechanism.[8] It is usually conceded that during the first half of the Enlightenment, roughly from the late 1680s to the 1740s, this form of natural philosophy, expressed in a myriad of sometimes conflicting forms, displaced traditional Aristotelian natural philosophy. During that period, the central project of mechanical natural philosophy was to incorporate the methods and assumptions of formal mathematical reasoning into explanations for natural phenomena. Its overriding impulse was to transform contingent knowledge into certain truth, to reduce the manifold appearances of nature to simple principles. In this process, leading proponents of the mechanical philosophy of nature proposed a new definition of matter, established methodological and explanatory procedures to incorporate this definition into a viable vision of science, and evolved an epistemology that authorized these procedures. Matter's essence was streamlined and simplified: it was defined as homogeneous, extended, hard, impenetrable, movable, and inert. The result, in Horkheimer's words, was that "Nature lost every vestige of vital independent existence, all value of its own. It became dead matter – a heap of things."[9]

In many ways, this description does indeed characterize some of the dominant movements of the late seventeenth and early eighteenth centuries. Driven by searing social and political rifts, aware of the terrifying results of sectarian dispute, and desiring safety and peace, many leading natural philosophers sought to construct a new world view that elevated uniformity and regularity into a scientific synthesis powerful enough to overthrow both the reigning academic system derived from Aristotelianism and scholasticism and the socially and politically dangerous hermetic, alchemic, and natural

---

[8] For an excellent analysis of this tendency, see Simon Schaffer, "Natural Philosophy," in Rousseau and Porter (eds.), *The Ferment of Knowledge*, pp. 53–91.

[9] Horkheimer, "Reason Against Itself," p. 361.

magical traditions, themselves forged as alternatives to the prevailing system.[10] The central issue revolved around the definition of matter. The basic question – was matter living or dead, inert or active, imbued with appetites and desires or passive – touched upon essential elements of late seventeenth- and early eighteenth-century religious, cultural, and political life.

Although early modern Aristotelian natural philosophy differed radically from the hermetic/alchemical/natural magical traditions, both had proposed a definition of matter, which assumed it to be animate and endowed with qualities, appetites, sympathies, and desires. Mechanical natural philosophy banned these qualities from the essential realm of matter: at best they were deemed accidental, at worst "occult qualities," dismissed by Newton as being the misplaced attempt to explain the inexplicable.

> The Aristotelians gave the name 'occult qualities,' not to manifest qualities, but to such qualities only as they supposed to lie hid in bodies and to be the unknown causes of gravity, and of magnetic and electric attractions, and of fermentations. . . . Such occult qualities put a stop to the improvement of natural philosophy, and therefore of late years have been rejected. To tell us that every species of things is endowed with an occult specific quality by which it acts and produces manifest effects is to tell us nothing.[11]

Matter's essence was streamlined and simplified. It was characterized by the "two catholic principles" of extension and motion.[12] Observable difference in matter could now be explained by differences in shape and size and by the motions of its particles or constituent parts. Motion was defined as the result of a force or action imposed on matter by an outside agent. Either at rest or in motion, matter tended to remain in that state until something else intervened. In short, the idea of inertia became one of the pillars supporting the mechanical philosophy of nature. Leibniz made this clear: "Whatever takes place in matter arises in accordance with laws of change from the preceding condition of matter. And this is what those who say that everything corporeal can be explained mechanically hold, or ought to hold."[13] Hence, in all analyses of motion, the relations of cause and effect were considered to be directly proportional. A fixed and knowable relation between them could be established.

Given this definition of matter, mechanical natural philosophers were able to evolve a new research program and explanatory strategy that was both convincing and capable of further extension. Science was directed toward estab-

[10] For the radical implications of the hermetic, natural magic tradition, see Frances A. Yates, *The Rosicrucian Enlightenment* (London: Routledge, 1972).

[11] Isaac Newton, Query 31 of the Optics, in H. S. Thayer (ed.), *Newton's Philosophy of Nature: Selections from His Writings* (New York: Hafner Press, 1953), p. 168.

[12] Robert Boyle, quoted by Richard S. Westfall, *The Construction of Modern Science: Mechanism and Mechanists* (Cambridge University Press, 1977), p. 66.

[13] Gottfried Wilhelm Leibniz, quoted in L. J. Rather and J. B. Frerichs, "The Leibniz-Stahl Controversy – I. Leibniz' Opening Objections to the *Theorie medica vera*," *Clio Medica*, 3 (1968), 24.

lishing a comprehensive system of measure and order, a universal *mathesis*. Mathematics became the privileged language of natural philosophy; more than that, it was assumed to be its ideal form of exposition. In the hierarchy of knowledge, the place occupied by any specific form of knowledge was established by the degree to which its subject matter was capable of being treated in a manner guided by mathematical principles. Despite the considerable differences between even the better-known proponents of mechanical natural philosophy – Descartes, Leibniz, Pierre Gassendi (1592–1655), Marin Mersenne (1588–1648), Robert Boyle (1627–1691) and Newton – most aspired to achieve a mathematical explanation of the universe. Those attempting to use mathematics as a model on which to construct reality held that only through such a procedure could self-evident, certain knowledge be established.[14] A mathematical description of reality was seen as the way to escape the perceived horrors of contingent – and hence, unsure – knowledge.

This project was authorized by an epistemology proclaimed most clearly by Descartes. It was grounded on the radical distinction between mind and matter and, by extension, between observer and observed. Despite the considerable differences separating the great proponents of the mechanical philosophy of nature, none was willing to deny the Cartesian duality,[15] for, without it, the certainty to which mechanism aspired could not be ensured. Only when Nature – in both form and motion – could be considered as the "radically other" could it be treated as pure object.

Within this general epistemological frame, Newton offered a variant that displaced Cartesian and to a lesser extent Leibnizian methodological procedures. In his critique of "hypothetical reasoning," Newton proposed what later was called the "experimental method," arguing for a close correlation between experiment and explanatory procedures. But even though he "feigned no hypotheses," Newton's method depended on the organizing power of mathematical logic. He proceeded by a process of radical reduction that in its extreme denied commonly observed reality and seemed, at times, to dissolve materiality itself. This was especially true for the *Principia,* where, according to Arnold Thackray, Newton's view of the universe was "an almost matterless entity, sustained by God's will, regulated by his divine intervention and operating through anti-material forces."[16] Thus, despite the great differences separating Newton from Descartes and Leibniz, his general approach affirmed

---

[14] Some mechanical philosophers such as Boyle did not subscribe to what I would call the strong program of mechanical natural philosophy – namely, to mathematize nature. Instead, they used mathematics as quantification. It served as a tool of discovery and proof. Its *metier* was the investigation of individual facts and not the construction of a coherent world picture.

[15] Even Leibniz leaves the split intact, although he postulates the idea of the preestablished harmony in which mind and matter proceed along parallel paths; however, they never interact. This was made especially clear in his dispute with Georg Stahl, who saw a direct connection between both. Rather and Frerichs, "The Leibniz-Stahl Controversy," p. 26.

[16] Arnold Thackray, "Matter in a Nut-Shell: Newton's Optics and Eighteenth-Century Chemistry," *Ambix,* 15 (1968), 44.

the essential principles of the mechanical philosophy of nature so intimately associated with the "Scientific Revolution."

In this language of nature, things were either identical to each other or they were different. All intervening, mediating connections were negated. A direct relationship between the name and the named, the sign and the signified, was established. Signs – once hieroglyphs of active matter – became transformed into arbitrary, yet specific, symbols that could be ordered, arranged, and manipulated by sovereign human reason, freed, by definition, from the contingencies of matter. Within the years between the deaths of Descartes and Newton, the new mechanical philosophy of nature had not only demonstrated its ability to account for many of nature's puzzles in a surer and simpler way but had also proved itself capable of being employed to undergird the existing religious, social, and political system of the time.

In most histories of Enlightenment science, the master narrative recounts the triumph and spread of the Newtonian form of this language of nature. An excellent example of this approach is provided in this volume by John Gascoigne in his essay, "Ideas of Nature: Natural Philosophy." However, mechanical natural philosophy, including its various Newtonian varieties, never totally vanquished the contending traditions it had sought to exterminate – namely, animism, alchemy and derivatives, and varieties of Paracelsian thought. These were carried on and developed by thinkers in all parts of Europe and sometimes remained embedded within popular traditions and practices. During the last half of the century, variations of these traditions would be resurrected and reformulated to criticize some of the essential principles of mechanical natural philosophy. This occurred when the universality of mechanical principles was either questioned or openly attacked.

THE MID-CENTURY SKEPTICAL CRITIQUE OF MECHANICAL NATURAL PHILOSOPHY

By the mid-eighteenth century, some of the core assumptions of the mechanical philosophy of nature were no longer considered satisfying or self-evident to a small but increasing number of scholars and writers. For many younger intellectuals, mechanism's very success had made it suspect, for, as Margaret Jacob and Aram Vartanian have shown, the brave new world of seventeenth-century mechanism was very easily adapted to serve as support for political absolutism, religious orthodoxy, and established social hierarchies.[17] For many mid-century French thinkers, for example, mechanism was associated with the system created by Louis XIV, and by then "Louis Quartozean culture

---

[17] Margaret C. Jacob, *The Radical Enlightenment: Pantheists, Freemasons, and Republicans* (London: Allen & Unwin, 1981); Aram Vartanian, *La Méttrie's L'Homme Machine: A Case Study in the Origins of an Idea* (Princeton, NJ: Princeton University Press, 1960).

appeared antiquated and oppressive."[18] Increasingly, the terms "machine" and "mechanism" became associated with despotism and dead, confining uniformity. Immanuel Kant (1724–1804) provides an example. In his essay *What Is Enlightenment?* and in the *Critique of Judgment*, the machine metaphor is employed to criticize the absolute state.

> The *Aufklärung* called for a government in which the person is "*more than a machine.*" In the *Critique of Judgment* the "monarchical state" is referred to as a "mere machine" if it is ruled "by a single absolute will." In contrast, a monarchical state ruled "according to internal *Volksgesetzen*" is designated an "animated body" [*beseelten Körper*].[19]

Dissatisfaction with the social and political world in which the mechanical philosophy thrived easily spilled over into a critique of the order of things propounded by philosophy. This dissatisfaction was signified by an emerging crisis of assent, expressed in a wave of mid-century skepticism directed against the spirit of systems, against a one-sided reliance on abstract and hypothetical reasoning in constructing a coherent picture of reality. In one sense this can be seen as the logical extension of Newton's "experimental method," although it differed from it by including mathematical explanations of nature under the heading of abstract and hypothetical reasoning. For leading thinkers of the late Enlightenment, abstract philosophy was deemed incapable of accounting for nature's vast variety. David Hume (1711–1776) announced this theme in the opening paragraph of his essay "The Skeptic."

> There is one mistake, to which philosophers seem liable, almost without exception; they confine too much their principles, and make no account of that vast variety, which nature has so much affected in all her operations. When a philosopher has once laid hold of a favorite principle, which perhaps accounts for many natural effects, he extends the same principle over the whole of creation, and reduces to it every phenomenon, though by the most violent and absurd reasoning. Our own mind being narrow and contracted, we cannot extend our conception to the variety and extent of nature; but imagine, that she is as much bounded in her operations, as we are in our speculation.[20]

Hume's skeptical analysis of causation was only one instance, although probably the most radical and least widely spread, of the reevaluation of mechanical natural philosophy. Georges-Louis Le Clerc, Comte de Buffon (1707–1788) offered a more acceptable critique of the introduction of mathematical

---

[18] Robert Darnton, *The Forbidden Best-Sellers of Pre-Revolutionary France* (Princeton, NJ: Princeton University Press, 1995), p. 196.

[19] Peter Burg, *Kant und die Französische Revolution* (Berlin: Duncker & Humblot, 1974), pp. 176–7.

[20] David Hume, *The Philosophical Works*, eds. Thomas Hill Green and Thomas Hodge Grose, 4 vols. (London, 1883), 3:213–14. For a recent discussion of skepticism in the eighteenth century, see Richard Popkin and Johan Van der Zande, *Skepticism in the Late Eighteenth and Early Nineteenth Century* (Dordrecht: Kluwer, 1998).

principles into the core of natural philosophical reasoning. In the introductory essay to his magisterial *Histoire naturelle* (1749–89), the most widely read work on natural philosophy in the latter half of the century, Buffon drew a distinction between abstract and physical truths. The first were products of human invention: they were imaginary, creations of the ratio. The second were real: they existed in nature and were the object of human inquiry. Mathematical proofs belonged to the first category. In fact, they were its prototype. They were founded on arbitrarily accepted logical principles. These, in turn, were used to generate equally arbitrary, although more complex, principles. All were joined by a method of definition whereby consistency was maintained by rigorously excluding anything that did not agree with the first abstract principle. Buffon considered a mathematical proof sterile, incapable of affirming anything other than its initial starting point. Mathematical systems were hermetically sealed, closed forever to the realities of observable nature.

> It is enough to have proven that mathematical truths are merely truths of definition or, if you will, different expressions of the same thing, and they are only truths relative to these same definitions that we have discussed. For this reason, they have the advantage of always being exact and demonstrative, but also abstract, intellectual and arbitrary.[21]

Physical truths, in contradistinction, were based on things that have actually occurred. "They do not depend at all on us."[22] To understand physical truths, the researcher must compare and observe similar sets of past occurrences. Science, according to Buffon, was the description and understanding of real things that have taken place in the world. Buffon characterized the different forms of knowledge as follows:

> In Mathematics, one supposes; in the natural sciences one poses a question and establishes truth. The former deals with definitions, the latter with facts. One moves from definition to definition in the abstract sciences, and from observation to observation in the real sciences; in the first, one finds self-evident knowledge, in the second, certainty.[23]

For both Buffon and Hume, understanding connections in nature was based on repeated historical observations of succession. In Hume's definition, cause "*is an object, followed by another, and where all the objects, similar to the first, are followed by objects of the second.*"[24] In late eighteenth-century terms, the

---

[21] George-Louis Leclerc, Comte de Buffon, *Histoire naturelle, générale et particulière*, 36 vols. (Paris, 1749–89), 1:54.

[22] Ibid., 1:54.

[23] Ibid., 1:55. In French the last sentence reads, "dans les premières on arrive à l'evidence, dans les dernières à la certitude." Instead of translating *l'evidence* simply as "evidence," I have called it "self-evident knowledge," which sums up the eighteenth-century understanding of this word.

[24] Hume, *The Philosophical Works*, 4:63.

new science was to be a science of facts, observation, and controlled inference. Its ideal expository form was a historical narrative.

Here, we encounter a thoroughgoing reversal of intellectual priorities. Hume and Buffon stand late seventeenth-century mechanical-mathematical natural philosophy on its head. According to the leading late seventeenth- and early eighteenth-century proponents of mechanism, history was the lowest form of knowledge. It was the knowledge of individual facts; whatever order one imposed on them was, at best, pragmatic, lacking the definitional clarity of a mathematical demonstration. Since history could not banish contradiction from its realm, it was deemed incapable of ever aspiring to certain truth: it was condemned to wallow in the morass of contingent knowledge. Although acknowledged as a form of understanding, it was considered a lesser being in the hierarchy of knowledge.[25] Knowledge of facts was sometimes deemed the starting point for sound natural philosophy, but history could at best provide the material that was later to be reshaped by the ordering power of universal *mathesis,* under whose aegis contradiction vanished before the piercing rays of human reason. History served as handmaiden to discursive logic and mathematical analyses, the appointed sovereigns of human understanding. For Buffon and Hume, the opposite was true. What was real was contingent. The rest was delusion, human hubris elevated to a scientific ideal.

By the elevation of the contingent over the coherent, it soon became a commonplace that all human knowledge was extremely constricted, because of both its reliance on sense impressions and its limited scope. If humans were endowed with reason, its power to pierce the veil of the unknown was greatly circumscribed. At the same time, many late Enlightenment thinkers surrendered the idea that nature's operations could be comprehended under the rubric of a few simple, all-encompassing laws. "Variety" and "similarity" replaced "uniformity" and "identity" as the terms most associated with nature's products. Hume made this clear in his *Enquiry Concerning Principles of Morals* (1751), where he denied all concepts of inherent identities. What is identical appears so only because we have been accustomed by habit to consider it so. "But there is nothing in a number of instances, different from every single instance, which is supposed to be exactly similar; except only, that after a repetition of similar instances, the mind is carried by habit, upon the appearance of one event, to expect its usual attendant, and to believe, that it will exist."[26] Nature not only was seen as complex but also was considered to be in continuous movement. As one anonymous French author stated, "The world is a theater of continual revolutions,"[27] in which new ones replace old forms of existence. Qualitative, directional change over time was deemed natural to

---

[25] This is the definition Christian Wolff gives of history. *Gesammelte Werke* 1 ABt., *Deutsche Schriften* (Hildesheim: Georg Olms, 1965), 1:115.

[26] Hume, *Philosophical Works*, 4:62.

[27] Anonymous, *Traité des Extremes ou élements de la science de la réalité* (Amsterdam, 1768), p. 232.

organized bodies. But this "progressive" development was not continuous. It proceeded through a series of drastic changes, "revolutions" in the "economy of nature," whereby outward form was changed drastically, followed by a gradual development in the newly formed shape. There was a continuous interplay between free creation and regular development. These three assumptions – the limiting of reason's competence, producing a wide-ranging epistemological modesty; the expansion of nature's complexity; and the historization of nature – set a new agenda for late Enlightenment natural philosophers. To paraphrase Hume, they were required to rethink the meaning of the terms *"power, force, energy* and *connexion."*[28]

VITALIZING NATURE: A LATE ENLIGHTENMENT RESPONSE TO SKEPTICISM

Generally, one can discern two broad, late eighteenth-century strategies designed to satisfy the objections raised by the skeptical critique of reductive rationalism and uniformity. The first, and best known, was formulated by neo-mechanists such as Jean Le Rond d'Alembert (1717–1783), Joseph-Louis Lagrange (1736–1813), Pierre-Simon, Marquis de Laplace (1749–1827) and Marie-Jean-Antoine-Nicolas Caritat, Marquis de Condorcet (1743–1794). These thinkers usually focused on the physical sciences, although they often extended their scrutiny to the nascent fields of inquiry that acquired the name "social sciences" in the late eighteenth century. Although retaining the mechanists' definition of matter as inert, they limited the role of mathematics in describing nature to that of an instrument of discovery instead of considering it a model of reality. In so doing, they put aside those debates concerning the ultimate composition of matter (was it made up of atoms, monads, or immaterial points)[29] or the definition of force (the *vis viva* controversy)[30] that had animated early eighteenth-century thinkers. Rather, they developed the

---

[28] Hume, *Philosophical Works,* 4:51.

[29] The disinclination to engage in the regnant questions of the early eighteenth century was made evident by the German mathematician W. J. C. Karstens in his discussion of the earlier disputes concerning matter, where he dismissed the whole controversy concerning these issues as useless. W. J. G. Karstens, *Physische-chemische Abhandlung, durch neuere Schriften von hermetischen Arbeiten und andere neue Untersuchungen veranlasset,* 2 vols. (Halle, 1786, 1787), 2:69.

[30] On the vis viva controversy, see Thomas L. Hankins, "Eighteenth-Century Attempts to Resolve the *Vis Viva* Controversy," *Isis,* 56 (1965); Carolyn Iltis (Merchant), "D'Alembert and the *Vis Viva* Controversy," *Studies in History and Philosophy of Science,* 1 (1970); Iltis, "The Decline of Cartesianism in Mechanics: The Leibnizian-Cartesian Debates," *Isis,* 64 (1973); Iltis, "The Leibnizian-Newtonian Debates: Natural Philosophy and Social Psychology," *The British Journal for the History of Science,* 6 (1973); Iltis, "Madam du Chatelet's Metaphysics and Mechanics," *Studies in History and Philosophy of Science,* 8 (1977); David Papineau, "The Vis Viva Controversy: Do Meanings Matter?" *Studies in History and Philosophy of Science,* 8 (1977); Giorgio Tonelli, "Analysis and Syntheses in XVIIIth Century Philosophy Prior to Kant," *Archiv für Begriffsgeschichte,* 20 (1976); Tonelli, "Critiques of the Notion of Substance Prior to Kant," *Tijdschrift voor Philosophie,* 23 (1961); Tonelli, "The Philosophy of d'Alembert: A Sceptic beyond Sceptism," *Kantstudien,* 67 (1976).

mathematics of probability as the surest guide to direct observational reason, while maintaining a strong epistemological modesty concerning the truth claims of these activities. They sought to evolve a science of "facts" that was linked and guided by the operations of probabilistic reasoning. Since this is a well-known episode in the history of eighteenth-century science, I here concentrate on the second response to the skeptical critique.[31]

This latter approach was proposed by a loose group of thinkers, less frequently studied though extremely numerous, whom I call, for want of a better term, Enlightenment vitalists. Their inquiries usually centered on the fields of chemistry, geology, the life sciences, medicine, and natural history, disciplines that became the premier areas of study for late Enlightenment naturalists. Like the neo-mechanists, they too were committed to evolving a science of facts guided by a form of observational and combinatorial reason, but unlike the neo-mechanists, the Enlightenment vitalists also sought to reformulate the concept of matter in their construction of a science that respected natural variety, dynamic change, and the epistemological consequences of skepticism.

For the vitalists, the basic failure of mechanism was its inability to account for the existence of living matter. This had led mechanists to posit a radical separation between mind and matter that only the intervention of God could heal, either as the universal occasion for all phenomena or as the creator of a preestablished harmony between mind and matter. This mind/body dichotomy was, according to Stephen Toulmin, the "chief girder in the framework of Modernity, to which all the other parts were connected."[32] Enlightenment vitalists sought to dissolve this dichotomy, to dismantle modernity's girder, by positing the existence in living matter of active or self-activating forces, which had a teleological character. Living matter was seen as containing an immanent principle of self-movement whose sources lay in these active powers, which resided in matter itself. Thus, we encounter natural philosophers populating the world of matter with a host of forces – such as elective affinities, vital principles, sympathies and formative drives – reminiscent of the living world of Renaissance natural philosophy. Rather than considering Nature to be Horkheimer's "heap of things," Enlightenment vitalists envisioned it as a teeming interaction of active forces revolving around each other in a developmental dance. The German physiologist, comparative anatomist, and anthropologist Johann Friedrich Blumenbach (1752–1840) provided a typical example. In the complex composition of organized matter (the term usually assigned to living matter), he discerned a number of "common or general vital energies that exist more or less, in almost all, or at least

---

[31] For this development see Eric Brian, *La mesure de l'Etat: adminstrateurs et geometres au XVIIIe siecle* (Paris: Albin Michel, 1994); Keith Baker, *Condorcet: From Natural Philosophy to Social Mathematics* (Chicago: University of Chicago Press, 1975); Lorraine Daston, *Classical Probability in the Enlightenment* (Princeton, NJ: Princeton University Press, 1988).

[32] Toulmin, *Cosmopolis*, p. 108.

in a great many parts of the body."[33] The foremost of these was the formative drive *(Bildungstrieb)*, which Blumenbach defined as a power that directs the formation of bodies, prevents them from destruction, and compensates them through reproduction from any mutilations the body may incur.[34] According to Blumenbach, the *Bildungstrieb* was an "occult power," similar in one sense to gravity: it could not be seen directly. But, unlike gravity, it also could not be measured. It could be recognized only by its effects.[35] In addition to these general vital powers, Blumenbach posited another vital energy, "namely the *vita propria,* or specific life: under which denomination I mean to arrange such powers as belong to certain parts of the body, destined for the performance of peculiar functions."[36] According to him, "virtually every fibral in the living body possessed a vital energy inherent in itself."[37] In short, in Blumenbach's vision, one typical of Enlightenment vitalism, all the strictures condemned by Newton concerning occult qualities were reintroduced into the life sciences. An organized body consisted of a complex conjuction of energies and forces of varying intensities and functions that could not be reduced to a single dominating principle. It was a constituent assembly of forces, operating through cooperation rather than by direction from a single sovereign authority.

Blumenbach's deployment of the concept of occult powers is indicative of the strategy adopted by many Enlightenment vitalists in designing a theory of science that contested some of mechanism's essential principles. As Steven Shapin remarked, the turn away from "Newtonian theories which required external animating spiritual agencies to those which placed the principle of animation and pattern within the natural entities" was widespread in the latter half of the century.[38] To authorize this move, the founders of Enlightenment vitalism pursued a two-part program. The first entailed the rehabilitation of an "ancient" tradition to counter the claims of mechanism; it was combined with those ideas of mechanism still consistent with or not contradictory to the new language of nature. The second part was to evolve a unique explanatory and methodological field that differentiated the "vital" sciences of chemistry, geology, the life sciences, medicine, and natural history from physics without directly challenging the principles on which the latter was constructed. A new grammar, vocabulary, and epistemology for these sciences was thereby developed, establishing, in the process, independent disciplinary matrixes for the fields being pursued.

In the first instance, a new pantheon of scientific precursors was created

[33] Johann Friedrich Blumenbach, *Elements of Physiology,* Charles Caldwell (trans.), 2 vols. (Philadelphia, 1795), 1:33.

[34] Ibid, 1:22.

[35] Blumenbach, *Ueber den Bildungstrieb,* 2nd ed. (Göttingen, 1791), pp. 33–4. The leading French life scientist of the period, Paul Barthez, followed the same strategy, calling his concept of the *principe vital* an occult force, identical to Blumenbach's. Paul Barthez, *Nouvelle Méchanique des mouvements de l'homme et des animaux* (Carcassone, 1798), p. v.

[36] Blumenbach, *Elements of Physiology,* 1:34.   [37] Ibid, 1:22.

[38] Shapin in *Ferment of Knowledge,* pp. 117–18.

and placed beside Newton as exemplars for the development of correct science. They included Francis Bacon (1561–1626) for science in general, Hippocrates and Pliny for natural history and medicine, and Philipus Theophrastus, Bombast von Hohenheim, Paracelsus (1493–1541), Jean Baptist van Helmont (1579–1644), and Franciscus Mercurius van Helmont (1614–1699) for chemistry. To these was added Georg Ernst Stahl (1659?–1734), a more modern naturalist, originally demeaned by the mechanists but elevated in the latter half of the century to a status almost equivalent to Newton's. Stahl's importance in the history of late eighteenth-century science has virtually been forgotten. In most standard accounts he appears briefly as the formulator of the phlogiston theory, which was finally overthrown by Antoine Laurent Lavoisier (1743–1794), ushering in what is usually called the Chemical Revolution and the beginning of modern chemistry. But for many thinkers of the latter half of the eighteenth century, Stahl's theories seemed to offer a compelling alternative to mechanism, a starting point from which they could develop their approaches in many different and fruitful directions.

Stahl drew a sharp distinction between a "mechanical Body" (*corpus mechanicum*) and a living system. A living system, which constituted an "*oeconomia vitae,* had its own laws, its own goals, its uses and effects."[39] Although a living body acted according to mechanical means, its mode of action surpassed physical-mechanistic necessity. The major error of the mechanists was to conflate the two: "necessity was," for them, "too closely connected to passive contingency [*Contingentia passiva*]." Rather than being merely passive, vital matter was controlled by "a higher principle," with its own self-prescribed goal. Goal or telos was, Stahl argued, an integral part of the "living economy" of nature. It assumed the existence of an active moral principle in nature, a "*Principium moraliter activum.*"[40]

Mechanism also erred, Stahl claimed, in its definition of elementary matter. It was useless to consider bodies as aggregates of minute homogeneous elementary particles.[41] In the phenomenal world, matter is always conjoined. "Nowhere in nature do there exist elementary bodies which our senses [*sens*] are able to perceive. Everything which we see, taste, feel, or touch is mixed [*mixte*], compounded [*composé*]."[42] Perceptible matter therefore was heterogeneous. Instead of homogeneous particles, Stahl argued that there existed in the perceptible world basic elements, each with its own qualities or essences. These elements joined with each other and with other combinations in a variety of ways, forming a complex gradation of species that could be classified according to degrees of resemblance or similarity. Because there were no such things as isolated, uniform building blocks of nature, all of nature was

[39] Georg Ernst Stahl, *Sudhoffs Klassiker der Medizin* (Leipzig: Johann Ambrosius Barth, 1961), 36:52.
[40] Ibid, p. 50.
[41] Hélène Metzger, *Newton, Stahl, Boerhaave et la Doctrine Chimique* (Paris: Félix Alcan, 1930), pp. 102–6.
[42] Stahl, quoted by Metzger, *Newton, Stahl, Boerhaave,* p. 118.

connected through sympathies, rapports, or affinities. In this world of interacting forces, each appearance was unique, possessing an individual character
created by a harmonic resolution of related sympathies.

In this exposition Stahl employed two explanatory figures that later eighteenth-century vitalists would adopt and expand. The first was the definition
of harmony as the product of active forces. The second was the inner/outer
topos in which the hidden unseeable was considered the real; the immediately
observable was only a representation of the real. The external was a sign of
something else. Hence, the path to understanding reality required a mode of
perception that transcended both abstract rationalism and simple empiricism.
One had to delve below the world of transparency to approach the inner core
of reality, itself animated by an active principle. This was possible because of
the similarity between the observed and the observer. Since humans were
endowed with souls, it was within their scope sympathetically to understand
the operation of the soul in other bodies and to intuit the operation of active
forces in nature. This process required humans first to understand themselves. Knowledge of the living economy of nature began with the act of self-
investigation.[43]

Stahl's doctrine became a rallying point for Enlightenment vitalists because
of its epistemology, its theory of matter, its concentration on active forces or
principles, and its linkage of spirit and body. These positions formed the basis for what Robert Siegfried, Betty Jo Dobbs, and J. B. Gough consider the
essence of the "Stahlian Revolution" in chemistry, namely, the emphasis on
composition and the attendant procedures of analysis and synthesis as
guiding the chemical endeavor.[44] It is a judgment that the young eighteenth-
century German Franz Xaver Baader (1765–1841) shared. In his analysis of the
question of "matter of heat," Baader argued that the adoption of Stahlian
chemistry in the 1740s had broken the spell of the "Methodo scientifico-
mathematica" and had brought down the "little houses of cards of mechanical
vibration, collision and pressure."[45] According to Baader, Lavoisier's work
affirmed and refined the basic lines of chemical argument that was initiated
by Stahl and further refined and developed by such critics of mechanism as
Torbern Olof Bergman (1735–1784), Joseph Black (1728–1799), Joseph Priestley
(1735–1804), and Carl Wilhelm Scheele (1742–1786). Stahl's contemporary
importance for the life sciences was equally significant. His methods, modeled after Hippocrates, were adopted in the 1740s in Montpellier and later in

---

[43] Stahl, *Sudhoffs Klassiker der Medizin*, p. 37.

[44] The position was first put forward by Robert Siegfried and Betty Jo Dobbs in "Composition: A
Neglected Aspect of the Chemical Revolution," *Annals of Science*, 24 (1968), 275–93. J. B. Gough
develops this argument in his article "Lavoisier and the Fulfillment of the Stahlist Revolution," in
Arthur Donovan (ed.), *The Chemical Revolution: Essays in Chemical Revolution. Reinterpretation.
Osiris*, 2nd ser., vol. 4 (1988).

[45] Franz Xaver Baader, *Vom Wärmestoff, seine Verteilung, Bindung und Entbindung vorzüglich beim Bren-
nen der Körper* (Vienna, 1786), p. 26.

Edinburgh, Bologna, and Göttingen, Europe's leading centers of medical education. Although his principles were drastically revised during the late eighteenth century, his contributions were widely acknowledged. Thus, the physician Pierre Roussel (1742–1802) claimed that the life sciences had been revolutionized around the mid century mark by the medical men of Montpellier and Paris, who, "rejecting the power of established authority," transformed the study of physiology, natural history, and anatomy.[46] Pierre-Jean-Georges Cabanis (1757–1808) concurred. In his *Coup d'Oeil sur les Revolutions et sur la Réforme de la Médecine* (1804), Cabanis characterized Stahl as "one of those extraordinary geniuses which nature brings forward from time to time to renew the sciences" and considered him the equal of Hippocrates, Bacon, and Newton.[47]

The reintroduction into nature of active, goal-directed living forces suggested by Stahl and then implemented by people such as Blumenbach in Germany, John Hunter (1728–1793) in Great Britain, and Paul-Joseph Barthez (1734–1806) in France led Enlightenment vitalists to reassess the basic methodological and analytic categories of scientific investigation and explanation. The new conception of matter dissolved the strict mechanistic distinction between observer and observed; as a result, relation, *rapport*, or *Verwandschaft* replaced aggregation as one of the defining principles of matter. Identity and noncontradiction were replaced by degrees of relation and similarity. The world of living matter consisted of a circle of relations, which, looking at it from the human vantage point, radiated out to touch all forms of matter. Thus, the constituent parts of living matter formed a "synergy" in which each conjoined particle was influenced by each other particle and the *habitus* in which it existed.[48] By emphasizing the centrality of interconnection, Enlightenment vitalists modified the concept of cause and effect. In the world of living nature, each constituent part of an organized body was both cause and effect of the other parts. Reciprocal interaction became the primary relationship in living systems. Furthermore, with the reintroduction of the centrality of goal into living nature, Enlightenment vitalists made goal the efficient cause of development. An explanation for something's existence took the form of a narrative modeled on the concept of stage-like development or epigenesis, in which a body evolves through stages from a point of creation effected by the merging of male and female seminal fluids. Unique creation and true qualitative transformation were central to the vitalists' vision of living nature.[49]

---

[46] I have had access only to the German translation. Pierre-Jean-Georges Cabanis, *Physiologie des weiblichen Geschlechts,* trans. Christian Michaelis (Berlin, 1786), pp. xiii–xv.

[47] Pierre-Jean-Georges Cabanis, *Coup d'Oeil sur les Revolutions et sur la Réforme de la Médecine* (Paris, 1804), p. 146.

[48] The term "synergy" was coined by Georg Stahl and then used extensively by Paul Barthez in his theory of vital physiology, especially in his *Nouveaux élements de la science de l'homme,* 2 vols. (Montpellier, 1778).

[49] For Kant's debt to this explanatory model, see Wolfgang Krohn and Günther Küppers, "Die natürlichen

These shifts in natural philosophic assumptions challenged Enlightenment vitalists to construct an epistemology capable of justifying and validating them. True to the skeptical critique of causation and forces, the vitalists agreed that active life forces could not be seen directly, nor could they be measured. They were, as Blumenbach called them, "occult powers" in the traditional sense of the term, and not as modified by Newton, who insisted on their quantification. At best they were announced by outward signs, whose meaning could be grasped only indirectly. This language of nature underscored the topos, championed by Stahl, of locating real reality as something that lurked within a body. That which was immediately observable was considered superficial. Understanding entailed a progressive descent into the depths of observed reality, using signs as the markers to chart the way. Thus, Enlightenment vitalists reintroduced the idea of semiotics as one of the methods to decipher the secrets of nature.

The basic epistemological problem was to understand the meaning of these signs and to understand how to perceive the interaction of the postulated individual yet linked active forces, powers, and energies without collapsing one into the other. To resolve this problem, Enlightenment vitalists called for a form of understanding that combined the individualized elements of nature's variety into a harmonic conjunction that recognized both nature's unity and nature's diversity. The methods adopted to implement this program were analogical reasoning and comparative analysis.

Analogical reasoning became the functional replacement for mathematical analysis. With it, one could discover similar properties or tendencies between dissimilar things that approximated natural laws without dissolving the particular in the general. The fascination with analogies was strengthened by a general preference for functional analysis, in which actual outward form was subordinated to activity. Comparative analysis reinforced the concentration on analogical reasoning. It allowed one to consider nature as composed of systems having their own character and dynamics, yet demonstrating similarities not revealed by the consideration of outward form. The major task of comparative analysis was to see similarities and differences and mediate between them, finding analogies that were not immediately apparent. In this, Enlightenment vitalists thought they were returning to methods pioneered by Bacon, which, they believed, correctly mirrored nature's path. The German physiologist Carl Friedrich Kielmeyer (1765–1844) defined this approach as follows: "Manifoldness within unity was nature's plan in its formation [*Bildungen*]; the undivided capacity in humans to see similarities and differences is therefore also the interpretative Organon" of correct scientific method.[50]

Ursachen der Zwecke: Kants Ansätze der Selbstorganisation," *Selbstorganisation: Jahrbuch für Komplexität in den Natur-, Sozial- und Geisteswissenschaften,* 3 (1992), 7–15.

[50] Carl Friedrich Kielmeyer, *Gesammelte Schriften: In Natur und Kraft. Die Lehre von der Entwicklung organische Naturlehre,* F. H. Holler (ed.) (Berlin: Keiper, 1938), p. 125.

However, in pursuing a program based on analogical reasoning and comparative analysis, a further epistemological problem arose. If nature was unity in diversity, how could one choose which element to emphasize? When should one concentrate on the concrete singularity, and when should one cultivate generalizing approaches? The proposed answer was to do both at once, allowing the interaction between them to produce a higher form of understanding than provided either by simple observation or by discursive, formal logic. This type of understanding was called divination, intuition, or *Anschauung*. Its operation was based on the image of mediation, of continually moving back and forth from one to the other, letting each nourish and modify the other. Buffon described this practice in the introduction to his *Histoire naturelle*.

> The love of the study of Nature supposes two seemingly opposite qualities of the mind: the wide-ranging view [*coup-d'oeil*] of an ardent mind that embraces everything with one glance, and the detail-oriented laboring instinct that concentrates only on one element.[51]

Cabanis's 1804 eloge to Stahl, quoted earlier, deployed the Buffonian ideal to describe Stahl's genius. Stahl "possessed a rapid and vast *coup-d'oeil* capable of overseeing the whole" combined with "that patient observation which scrupulously pursues minute details."[52] Still later, in 1822, Wilhelm von Humboldt (1767–1835) attested to the appeal of this logic of mediation in his description of how one obtained historical knowledge.

> Thus two methods have to be followed simultaneously in the approach to historical truth; the first is the exact, impartial, critical investigation of events; the second is the connecting of the events explored and the intuitive understanding of them which could not be reached by the first.[53]

He summed up this approach by concluding that "observational understanding [*beobachtende Verstand*] and the poetic power of imagination must stand together in a harmonic conjunction."[54] A further proof of the extent to which this epistemological model captured the imagination of late Enlightenment thinkers can be seen in the review of Moses Mendelssohn's (1729–1786) *Morgenstunden* by the German philosopher Johann Georg Heinrich Feder (1740–1821), the spiritual leader of the Bavarian Illuminati. Feder considered the work excellent because Mendelssohn had followed "the middle way, upon which alone thorough understanding can arise, the way of the painstaking observation of inner and outer nature and of careful analogical suppositions."[55]

---

[51] Buffon, *Histoire naturelle*, 1:4.   [52] Cabanis, *Coup d'Oeil*, p. 146.
[53] Wilhelm von Humboldt, "On the Historian's Task," *History and Theory,* 6 (1967), 59.
[54] Von Humboldt, in *Wilhelm von Humboldt Werke*, eds. Andreas Flitner and Klaus Giel, 5 vols. (Stuttgart: J.G. Cotta'sche Buchhandlung, 1980–1), 1:377.
[55] Johann Georg Heinrich Feder, *Göttingische Anzeigen von gelehrten Sachen*, (1786), p. 66.

By analogy, this act of mediation was supposedly mirrored in the physical world through the action of the life forces. Thus, for example, Blumenbach argued that the *Bildungstrieb* successfully mediated between the "two principles . . . that one had assumed could not be joined, the teleological and the mechanical."[56] Friedrich Schiller (1759–1805), who was trained as a physician, made a similar claim in his first medical dissertation, written in 1779, for a force that mediated between mind and matter. He described it as

> a force [that] in fact exists between matter and mind. This force is quite distinct from the world and the mind. If I remove it, the world can have no effect on the mind. And yet the mind still exists, and the objects still exist. Its disappearance has created a rift between world and mind. Its presence illuminates, awakens, animates everything about it.

It was, he claimed, "a force, which is spiritual on the one hand, and material on the other, an entity that is penetrable on the one hand and impenetrable on the other."[57] Correct understanding formed an analogue to this force as it moved from the concrete to the intellectual and back.

In this movement, however, understanding passed through a third, hidden, and informing agent that was, in effect, the ground on which all reality rested. In eighteenth-century language, this hidden middle element, opaque, unseeable, and yet essential, was called by such terms as the "internal mould" (Buffon), "prototype" (Jean Baptiste Robinet [1735–1820]), *Urtyp* (Johann Wolfgang von Goethe [1749–1832]), *Haupttypus* (Johann Gottfried Herder [1744–1803]), or schema (Kant). Some writers used the image of a magnetic field to give it visual representation. It was constituted by the magnetic poles and yet united them without submerging them in a reductive unity. The area of its greatest effect was the middle, where the field encompassed the largest area.

For us, this model of apprehension is difficult to understand, for it flies in the face of what we consider rational, logical, or scientific. I believe that it points to an attempt to incorporate the skeptical critique of rationalism by seeking to go beyond binary systems of logic and explanation. Binary systems assume that the distance between signifier and signified can be collapsed, that reason can look at the world and it would look back reasonably. What these late Enlightenment thinkers seemed to prefer was a ternary system. This system which introduced something between sign and signified, through which, in Kant's definition of the "schemata," everything was refracted; but this system could never be seen, grasped, or directly identified.[58] In short, these

---

[56] Blumenbach, *Ueber den Bildungstrieb* (Göttingen: Vandenhoek & Ruprecht, 1791), pp. 65–6.

[57] Friedrich Schiller, quoted in Kenneth Dewhurst and Nigel Reeves (eds.), *Frederick Schiller: Medicine, Psychology and Literature with the First English Edition of the Complete Medical and Physiological Writings* (Berkeley: University of California Press, 1978), p. 152.

[58] Kant defines the *schemata* as that which "underlies our pure sensible concepts." Although it made sensible understanding possible, its operations remained a mystery. "This schematism of the understanding, in its application to appearances and their mere form, is an art concealed in the depths of the human soul, whose real modes of activity nature is hardly likely ever to allow us to discover,

thinkers were arguing for a harmonic view of nature that organized reality around the figures of ambiguity and paradox central to the skeptical stance, a position that was reluctant to reduce one thing to another but allowed them to be allied to each other. This harmonic ideal often was expressed through the use of creative oxymorons such as Buffon's "internal mould" or Schiller's concept of "material ideas," which verbally reconstructed the paradoxical *rapports.*

But how did the Enlightenment vitalists validate this theory of understanding? What allowed them to proclaim that the tools of analogical reasoning; comparison; and internal, intuitive understanding were scientifically objective? The problem was especially acute because of the blurring between object and observer. But it was precisely this mingling that served as the justification for this approach to science. It was argued that because humans were part of living nature, they could, through the act of sympathetic understanding, acquire a living knowledge of nature's processes. Similarity and relationship were the vehicles of understanding, which by passing through the extended middle ensured the truth-values of these endeavors.

This harmonic view of reality formed the core and essence of the late Enlightenment vitalistic vision of nature and humanity, differentiating it from early eighteenth-century mechanism and later Romantic *Naturphilosophie;* this view accounted for its fascination with extremes – boundaries and limits – and its hoped-for mediations. It was not a dualistic vision of nature and humanity, for real reality always lay between both. Harmony, the joining of opposites within an expanded middle generated by reciprocal interaction, served as the norm and desired end of each natural process, although that dynamic was continually in motion, leading to ever-changing harmonic combinations. Living nature, then, was the place where freedom and determinism merged. Its description invoked images and metaphors either drawn from the moral sphere or directly applicable to it. Horkheimer claimed that "the inner logic of science itself tends towards the idea of one truth which is completely opposed to the recognition of such entities as the soul and the individual."[59] The science envisioned by Enlightenment vitalists sought to reintroduce entities such as soul and individuality into the inner core of scientific thinking.

CONCLUSION: BETWEEN ENLIGHTENMENT VITALISM AND ROMANTIC *NATURPHILOSOPHIE*

Enlightenment vitalism was nourished by and within the late eighteenth century skeptical critique of absolute solutions and reductive rationalism. Like its

and to have open to our gaze." Immanuel Kant, *Critique of Pure Reason,* trans. Norman Kemp Smith (New York, 1965), B 180–1 /A 141.
[59] Horkheimer, "Reason Against Itself," p. 364.

neo-mechanist counterpart, it was founded on a deeply held epistemological modesty that was willing to suspend absolute judgments in favor of conditional ones. It could thrive as long as ambiguity and paradox were seen as productive and not considered either dangerous or ineffective. With the tensions generated during the era of the French Revolution and Napoleonic wars, that epistemological modesty was shattered by the desire for absolute answers. Disdainful of science "stuck in the rubbish dump of sensory reflection,"[60] Romantic *Naturphilosophie* aimed at a new universal *mathesis,* a totalizing vision that – given the stress of decades of warfare, of social and emotional uncertainty, of a loss of faith in the complex aspirations of the late Enlightenment – led many young men and women to yearn for absolute answers that relegated the mundane world to an epiphenomenon and asserted spirit as the true essence of reality. If Enlightenment vitalism sought to limit mechanism's rule, *Naturphilosophie* desired to destroy it. As Friedrich Wilhelm Joseph Schelling (1755–1854) proclaimed, to philosophize about nature meant to "lift it out of the dead mechanisms where it bashfully appeared and to animate it, so to speak, by freedom, to elevate it to its own, free development."[61] Spirit, freedom, and active force were seen as one. "All original [*ursprünglichen*], that is, all dynamic, natural manifestations [*Erscheinungen*] must be explained by forces, which exist in matter even when it is at rest (for there is also movement in rest, this is the basic postulate of dynamic philosophy)."[62]

This new adventure of reason sought to unite what Enlightenment vitalism had sundered: to recapture on a different level the universal vision that had driven the philosophies of Plato, Pythagoras, Plotinus, Descartes, and Leibniz; to unite spirit and matter into a uniform, consistent whole, devoid of leaps in nature, empty space, and the distinctions between living and dead matter; to launch a full-scale attack on contingency; and to chart the history of the universe from the beginning of time to its end, considered as a living essence developing from absolute, inherent spiritual principles. It offered a new "creation myth" formulated in the language of the most "advanced" contemporary "sciences" – the disciplines central to Enlightenment vitalism – yet aspiring, at the same time, to transcend the explanatory limits that had been imposed on them. It aimed "at a total history, one that would encompass the entire differentiation of the cosmos from the original oneness, through the formation of the solar system and the earth, the proliferation of the three kingdoms of nature . . . to the culmination of the universe in humankind."[63] This all-encompassing view, symbolized by Lorenz Oken's (1779–1851) invocation of

---

[60] Nicholas Jardine, "Naturphilosophie and the Kingdoms of Nature," in Nicholas Jardine, J. A. Secord, and Emma C. Spary (eds.), *Cultures of Natural History* (Cambridge University Press, 1996), p. 233.

[61] Friedrich Wilhelm Joseph Schelling, *Erster Entwurf eines Systems der Naturphilosophie für Vorlesungen* (Jena and Leipzig, 1799), p. 6; *Friedrich Wilhelm Joseph von Schellings Sämmtliche Werke,* 14 vols. (Stuttgart, 1858), 2:13.

[62] The original 1799 edition does not have this aside. It is contained in the *Werke,* 2:25.

[63] Jardine, "Naturphilosophie," p. 232.

the Pythagorean injunction "Geometria est Historia,"[64] elevated the results of reflective introspection – authorized by the philosophic concept of identity (*Identitätsphilosophie*) – to the status of universal truths of nature. The late Enlightenment's epistemological modesty was sacrificed on the altar of certainty. Rather than juxtapose and harmonize the contending activities of precise observation and imaginative reconstruction, the *Naturphilosophen* desired to merge science and philosophy into a new cultural, scientific, and esthetic synthesis, a type of synthesis seen by postmodernists as characterizing the Enlightenment project.

This is the ultimate irony, for a careful look at the late Enlightenment might reveal a way of thinking and doing that is much more sympathetic to postmodernism than to Romanticism. In its endeavor to view nature not just as a "heap of things," to create a place for the soul and the individual, to avoid the rush to reductionism, and to recognize the epistemological value of ambiguity and paradox, the late Enlightenment, at least in part, envisioned an order of things that stood in stark contrast to the instrumental reason often associated with it. If there is such a thing as the Enlightenment project, it included a healthy respect for differentness, free movement, and creation. Adam Ferguson (1723–1816), an avid reader of natural philosophy, made this explicit in 1767.

> Our notion of order in civil society is frequently false: it is taken from the analogy of subjects inanimate and dead; we consider commotion and action as contrary to its nature; we think it consistent only with obedience, secrecy, and the silent passing of affairs through the hands of a few: The good order of stones in a wall, is their being properly fixed in places for which they are hewn; were they to stir the building must fall: but the order of men in society, is their being placed where they are properly qualified to act. The first is a fabric made of dead and inanimate parts, the second is made of living and active members. When we seek in society for the order of mere inaction and tranquillity, we forget the nature of our subject, and find the order of slaves, not of free men.[65]

---

[64] Lorenz Oken, *Abriss der Naturphilosophie: Bestimmt zur Grundlage seiner Vorlesungen über Biologie* (Göttingen, 1805), p. 1.

[65] Adam Ferguson, *An Essay on the History of Civil Society,* ed. Duncan Forbes (Edinburgh: Edinburgh University Press, 1966), pp. 268–9.

3

SCIENCE, THE UNIVERSITIES, AND OTHER PUBLIC SPACES
Teaching Science in Europe and the Americas

Laurence Brockliss

To date there has been little detailed research into the history of institutionalized science teaching in the eighteenth century, apart from work done on the
British Isles, France, and the Netherlands. The paucity of data reflects the
fact that until recently historians of eighteenth-century natural philosophy
have taken little interest in the history of science in the classroom, assuming
the subject of small importance. This chapter aims to demonstrate that such
a judgment is misguided even if the conclusions of such a study must necessarily be provisional. The history of science teaching in the Age of Reason
throws light on the speed and manner with which new theories and discoveries became part of the European cultural inheritance. More important, it also
advances our understanding of the way in which distinctive natural sciences
came to be defined and stabilized and distinctive national scientific traditions
began to emerge at the end of the period.

AROUND 1700

Traditionally, public teaching in the natural sciences was the preserve of the
universities, where the resposibility for teaching the gamut of human knowledge was divided among the faculties of arts, theology, law (sometimes divided
into separate canon and civil law faculties), and medicine. By 1700, after three
centuries of expansion, the number of Europe's universities had grown from
40 to some 150, and they were to be found in all parts of the continent except Russia. A further fifteen or so universities or university colleges had also
been founded in the New World, including three in the then English North
American colonies: Harvard, Yale, and the College of William and Mary at
Williamsburg. By the turn of the eighteenth century, however, the universities
no longer had a monopoly on science teaching, for in a number of countries
instruction had been relocated in municipal colleges. These had been initially
founded as feeder schools for the local university, providing instruction in

Latin and Greek grammar and rhetoric, the languages of university learning, but in the sixteenth and seventeenth centuries they had frequently usurped the province of the university and had begun to teach philosophy and mathematics, too. What distinguished these institutions from universities was that they were not empowered to grant degrees.

As a result of this development, the provision of institutionalized science teaching across the European world was very uneven. In the British Isles, where the grammar and Scottish burgh schools stuck to their last, or in the English-speaking colonies, where schools of any kind were few and far between, science was publicly taught only in the universities and university colleges. A similar situation pertained in the Spanish/Austrian Netherlands and the other parts of Protestant northern Europe. In Catholic Latin Europe and the Spanish and Portuguese colonies, in contrast, instruction in the natural sciences was widely available in the university feeder schools, so provision was much more plentiful. In France, for instance, philosophy was no longer taught in the 20-odd universities at all but in some 100 *collèges de plein exercice.* On the other hand, the density of the provision had little effect on the social character of the student populaton that attended these public courses. Broadly speaking, access to public science teaching was limited everywhere to relatively affluent males in their late teens who were destined for one of the three professional careers that university education primarily existed to serve: the Church, law, and medicine.[1]

Within the university and college system the study of the natural world in 1700 was divided into three separate subject areas or distinctive *scientiae.* Principally, the natural sciences fell under the head of philosophy, which comprised the four subsciences of logic, ethics, physics, and metaphysics. The order in which the last three were taught changed over the centuries, but logic was always studied at the beginning of the course because it provided the analytical tools for an understanding of the other philosophical sciences. Physics, or the science of natural bodies, *corpora naturalia,* was thus as much a logical science as were ethics and metaphysics. There was no epistemological distinction between them. Physics and metaphysics in particular were customarily seen as intimately connected to the extent that the former provided evidence of divine goodness whereas the latter demonstrated God's existence and attributes.

The classroom science of physics at the beginning of the eighteenth century was a causal and deductive science: its purpose was to explain observed natural phenomena in terms of unimpeachable fundamental principles about the nature of matter through constructing water-tight causal chains. In this sense it was still an Aristotelian science whose epistemology was drawn

---

[1] Willem Frijhoff, "Patterns," in *A History of the University in Europe* (Cambridge University Press, 1992–), vol. 2: *Universities in Early Modern Europe (1500–1800),* ed. Hilde de Ridder-Symoens (1995), especially pp. 90–105 (tables and patterns); Roger Chartier, Marie-Madeleine Compère, and Dominique Julia, *L'Education en France du XVIe au XVIIIe siècle* (Paris: SEDES, 1976), chaps. 5–6.

primarily from the *Posterior Analytics*. It was an Aristotelian science, too, in that its subject matter was largely determined by Aristotle's surviving works on natural philosophy. The course would proceed from the general to the particular, beginning by introducing students to the chief themes in Aristotle's *Physics* and then moving on to investigate topics in the *De caelo*, the *De generatione et corruptione*, the *De meteorologia*, the *De anima*, and the *De parvis naturalibus*. Consequently, by the end of the course, which was usually a year in length, the student would have been instructed in the principles of matter and motion, the structure of the superlunary world, the process of change and decay on Earth, the characteristics of inanimate terrestial phenomena, and the mysteries of life – human, animal and vegetable.

In many parts of Catholic Europe and throughout Spanish and Portuguese America, the content as well as the structure of the physics course was equally Aristotelian. This was especially the case in the large number of colleges and universities controlled by the Society of Jesus. This did not mean that Jesuit and other Aristotelian professors taught a physics completely oblivious of contemporary developments in the natural sciences: sixteenth- and seventeenth-century Aristotelianism was a vibrant and eclectic physical philosophy that successfully incorporated most of the new observational discoveries.[2] It meant, rather, that Jesuit physics remained wedded to the Thomist Aristotelian position that natural bodies were the amalgamation of matter and form, that forms were immaterial, and that only formalistic and qualitative explanations of natural phenomena were legitimate.

On the other hand, in the Protestant world and in Catholic colleges and universities where philosophy teaching was in the hands of lay or secular professors, the traditional Aristotelian kernel of the physics course had been already or was in the process of being jettisoned. Instead, the professors had largely embraced some form of the new mechanical philosophy. The large majority were, broadly speaking, Cartesians and taught their pupils that the universe was a plenum in which natural phenomena, both sub- and superlunary, could be explained almost entirely in terms of indefinitely divisible particulate matter in motion. Only human beings (who could themselves move as well as be moved) and perhaps animals had superadded immaterial forms or souls, but even they, physiologically-speaking, were machines. In France, Catholic secular professors at the University of Paris followed closely Descartes's formulation of his mechanical philosophy in his 1644 *Principia*. In the Protestant universities of northern Germany, on the other hand, the first two decades of the eighteenth century saw the rapid dissemination of an eclectic form of

[2] Charles B. Schmitt, "Towards a Reassessment of Renaissance Aristotelianism," *History of Science*, 11 (1973), 159–93; Christia Mercer, "The Vitality and Importance of Early Modern Aristotelianism," in Tom Sorrell (ed.), *The Rise of Modern Philosophy: The Tension Between the New and Traditional Philosophy from Machiavelli to Leibniz* (Oxford: Clarendon Press, 1993); Laurence W. B. Brockliss, *French Higher Education in the Seventeenth and Eighteenth Centuries: A Cultural History* (Oxford: Clarendon Press, 1987), pp. 337–50.

Cartesian physics that drew on Leibniz's theory of monads, at least for its account of organic matter. This German variant was the creation of Christian Wolff (1679–1754) who took over the chair of natural philosophy and mathematics at the new Prussian Pietist university of Halle (founded in 1693) in 1706.

Very few mechanist professors, Catholic or Protestant, accepted the Gassendist variant of the mechanical philosophy, which argued that the universe was formed from indivisible atoms whirling around in a vacuum. This was partly because Gassendist atomism was too closely associated with Epicurian materialism but also because Gassendi seemed inconsistent and endowed his atoms with nonmechanical attributes.[3] Not surprisingly, then, outside the English-speaking world, no professor of physics accepted Newton's development of Gassendist mechanism either. Mechanist professors on the European continent, if they discussed Newton's work at all, found the concept of a two- or multiple-force universe impossible to comprehend: all motion (visible or invisible) had to be by physical contact. Even physics teachers in the British Isles found Newton's work difficult to understand. By the 1690s his theory of universal gravitation, as well as his work on light and color, was being discussed by professors of philosophy in the Scottish universities, in particular at Edinburgh, but it took another decade for Newton's critique of Descartes's vortexes to be sympathetically received. Scottish professors at the turn of the eighteenth century preferred to attempt to accommodate the Englishman's discoveries to Cartesian plenism and were reluctant to abandon an impulsionist physics. In 1704 the Edinburgh professor Charles Erskine (1680–1763) produced a set of physical theses that were enthusiastically Newtonian. Nonetheless, he could still declare, "Leibniz has shown beyond doubt that gravity derives from the impulse of the surrounding fluid, as do magnetic actions; this is quite clear from his investigations into the causes of celestial motions."[4]

The emergence of a strong Cartesian presence in college and university classrooms around 1700 did not really signify that Europe's professors of physics were dividing into ancients and moderns. In fact, the Cartesian course in many respects was traditionalist. Cartesian, as much as Aristotelian, physics

---

[3] Edward G. Ruestow, *Physics at Seventeenth- and Eighteenth-Century Leiden: Philosophy and the New Science in the University* (The Hague: M. Nijhoff, 1973), chap. 4; Michael Heyd, *Between Orthodoxy and Enlightenment: Jean-Robert Chouet and the Introduction of Cartesian Science in the Academy of Geneva* (The Hague: M. Nijhoff, 1982), especially chap. 4; Brockliss, *French Higher Education*, pp. 350–9; Laurence W. B. Brockliss, "Descartes, Gassendi, and the Reception of the Mechanical Philosophy in the French Collèges de Plein Exercice, 1640–1730," *Perspective on Science*, 3 (1995), 450–79; Geert Vanpaemel, *Echo's van een wetenschappelijke revolutie. De mechanistische natuurwetenschap aan de Leuvense Artesfaculteit, 1650–1797* (Verhandelingen van de koninklijke Academie voor Wetenschappen, Letteren, en Schone Kunsten van België, Klasse der Wetenschappen, 173; Brussels: Paleis der Academiën, 1986), chaps. 3–6; Brendan Dooley, "Science Teaching as a Career at Padua in the Early Eighteenth Century: The Case of Giovanni Poleni," *History of Universities*, 4 (1984), especially pp. 131–5.

[4] C. M. Shepherd, "Newtonianism in the Scottish Universities in the Eighteenth Century," in R. H. Campbell and Andrew S. Skinner (eds.), *The Origins of the Scottish Enlightenment* (Edinburgh: John Donald, 1982), chap. 4, especially p. 74.

was a causal and verbal science based squarely on the study of Aristotelian logic. Moreover, in its classroom version, even the vocabulary of Cartesian physics retained many Aristotelian vestiges. Thus, the Paris professor Jerome Besoigne (1686–1763), in a course delivered at the Collège du Sorbonne-Plessis in 1713–14, could still use the term "substantial form," merely giving it a Cartesian gloss: "Substantial or essential forms of bodies should be understood as nothing other than a certain disposition of the whole body and its parts, or a congeries of accidents and qualities." Besoigne could also declare that this was Aristotle's own understanding of the concept: it was the Stagyrite's Peripatetic followers who had invented the idea of nonmaterial substantial forms added to matter.[5]

The Cartesian courses, furthermore, had no mathematical content, and no attempt was made to enliven the traditional professorial dictation with experiments. Admittedly, some Protestant Cartesian professors, such as M. G. Loescher (d. 1735) at Wittenberg, described their course as one in experimental physics, but the reality was different. Like other contemporary professors (both Cartesian and Aristotelian), such professors illustrated their course by *describing* experiments that confirmed their position: they did not themselves *perform* them. Cartesian physics was a completely new type of physics in only one respect: it emphasized that physics was a practical science. Aristotelians always argued that natural philosophy was a theoretical subject. In constrast, professors like Loescher took up the utilitarian rhetoric of the experimental philosophers. In his 1714 inaugural lecture Loescher argued that a knowledge of physics would eventually aid the progress of all the arts necessary for human existence.[6]

Both the Aristotelian and the Cartesian classroom course of physics at the turn of the eighteenth century, then, only partially reflected the concerns of the new science. Most adepts of the experimental philosophy, whatever their natural philosophical allegiance, were primarily interested in the production of natural effects or "matters of fact." The growing concern of its leading practitioners was not the creation of a traditional causal physics but rather the careful measurement and observation of natural phenomena in the hope of discovering mathematically describable laws underpinning their regular behavior. Nonetheless, the work of the contemporary experimental philosopher did find its way more directly into the classroom to the extent that it was taught as part of a course in mathematics. Although mathematics as a subject was deemed distinct from philosophy and subordinate to it – in that

---

[5] Bibliothèque de Sainte-Geneviève, Paris, MS 2081, "Physica," fos. 72, 75.

[6] M. G. Loescher, "Oratio inauguralis habita die 2. ian. A. MDCCXIV de physica ad rem publicam accommodanda," in Loescher, *Physica experimentalis compendiosa in usum juventutis academicae adornata . . .* (Wittenberg: G. Zimmermann, 1715), especially pp. 206–8. Descartes had particularly emphasized the utility of his philosophy for medicine in the sixth part of his *Discours de la méthode* (1637): see Descartes, *Oeuvres philosophiques*, ed. F. Alquié, 3 vols. (Paris: Garnier, 1963–73), vol. 1, p. 649.

it dealt with the natural body in the abstract – it had been an important part of the arts curriculum in the medieval university, embracing astronomy, optics, and music.[7] This tradition of teaching applied as well as theoretical mathematics was continued in the sixteenth and seventeenth centuries. Although many of the courses given in the colleges and universities were completely elementary and embraced only the first books of Euclid, some institutions developed the medieval inheritance further and in the seventeenth century began to offer lectures on the latest work in astronomy, optics, harmonics, and dynamics. The Jesuits were particularly important here in that many of their courses in mathematics were devised with prospective army and naval officers in mind, members of the nobility who (on the European continent at least) did not traditionally attend college and university and whose scientific knowledge of the natural world was gained (if at all) from books rather than lectures. Although the Order was Aristotelian and anti-Copernican, their professors of mathematics, especially in France, were free to develop a noncausal science of practical mathematics that gave their limited audience a solid grounding in the sciences of ballistics, fortification, and navigation and even introduced them to new subjects such as electricity and magnetism that they themselves helped to develop. Typical was the textbook published by the Paris-based Jesuit Louis Bertrand Castel (1688–1757) in 1728 under the title *Mathématique abrégée universelle.*[8]

The teaching of mathematics played a particularly important role in the dissemination of the new science in Great Britain. By the end of the seventeenth century there were endowed chairs at Cambridge, St. Andrews, Edinburgh, Glasgow, and Oxford, where the two Savilian chairs in geometry and astronomy had been founded in 1619. At Oxford and Cambridge, too, many colleges provided lectures in mathematics from the time of Elizabeth. By and large the teaching was in the hands of dedicated and proficient mathematicians who provided effective tuition in both theoretical and practical mathematics. Newton, holder of the Cambridge Lucasian chair (founded in 1663), was only the most exceptional of a bevy of talented mathematicians occupying university posts in the British Isles in the late seventeenth and eighteenth centuries. It was these professors – especially the members of the Gregory dynasty who taught mathematics at St. Andrews, Edinburgh, and Oxford – who first unequivocally championed Newtonian physics in the universities. Mathematically adept, they were able to follow the argument in Newton's *Principia* and grasp his critique of Descartes's impulsionist explanation of planetary

[7] John North, "The *Quadrivium*," in Hilde de Ridder-Symoens (ed.), *A History of the University in Europe*, vol. 1: *Universities in the Middle Ages* (1992), pp. 337–59.
[8] François de Dainville, "L'Enseignement des mathématiques dans les collèges jésuites de France du seizième au dix-huitième siècle," *Revue d'histoire des sciences*, 7 (1954), 6–21, 109–23; Brockliss, *French Higher Education*, pp. 381–3, 386–8. On Jesuit science *tout court*, see especially Steven J. Harris, "Transposing the Merton Thesis: Apostolic Spirituality and the Establishment of the Jesuit Scientific Tradition," in Rivka Feldkay and Yehuda Elkana (eds.), *"After Merton": Protestant and Catholic Science in Seventeenth-Century Europe*, special issue of *Science in Context*, 3:1 (1989), 29–66.

motion. Because theirs was an analytical and not a causal science, they were free to embrace Newton's concept of a universe of different forces without having to trouble themselves about its epistemological status. Unlike their colleagues in philosophy, they were not constrained by the need to accommodate Newtonian physics with a priori mechanist principles and could teach his mathematical philosophy technically and coherently.[9]

Yet if some college and university mathematics courses by the turn of the eighteenth century were sufficiently sophisticated, especially in the British Isles, to ensure that students with a mathematical bent could obtain a good idea of contemporary developments in mathematical physics, they were no more likely than courses in physics to introduce their auditors to the experimental philosophy *tout court.* They were principally courses in geometry and its many practical applications and were taught in Latin with little recourse to visual aids beyond the occasional printed diagram. To experience nature being put to the question in an official course given in the university world in 1700, a student would have had to transfer to the faculty of medicine and attend the lectures in anatomy, botany, and chemistry. Whereas physics was a causal and mathematics an analytical science, anatomy, botany, and chemistry were simple descriptive sciences taught by dissection and demonstration. In complete contrast to lectures in physics and mathematics, the emphasis was on visual learning. Indeed, the atmosphere bordered on the theatrical, and demonstrations, especially dissections, were commonly attended by interested laymen as well as medical students.

However, in many faculties of medicine the quality of the teaching was poor and the value of the experience, even as entertainment, limited. Anatomy and botany were new subjects that had become firmly established in the medical curriculum only in the seventeenth century as part of a novel interest in giving medical students a visual acquaintanceship with the internal and external parts of the human body and the structure of the plants traditionally deployed in healing its ills. Chemistry was even newer, a branch of the practical medical science of pharmacy and initially little more than the art of distilling and manufacturing the chemical remedies introduced into the pharmacopeia by the Paracelsians. Taught first as a separate subject in the early seventeenth century at Marburg in Hesse–Cassel (where the duke was an adept), it was established as a distinctive area of study only in a handful of universities at the turn of the eighteenth century, including Oxford, Cambridge, and Montpellier.

[9] Mordechai Feingold, *The Mathematicians' Apprenticeship: Science, Universities and Society in England, 1560–1640* (Cambridge University Press, 1984), especially chap. 1; Feingold, "The Mathematics Sciences and New Philosophies," in *The History of the University of Oxford* (Oxford: Clarendon Press, 1984–), vol. 4: *The Seventeenth Century* (1997), Nicholas Tyacke (ed.), chap. 6; Ronald G. Kant, "Origins of the Enlightenment in Scotland: The Universities," in Campbell and Skinner (eds.), *The Scottish Enlightenment,* pp. 45–7; Shepherd, "Newtonianism," especially p. 82.

Because the core subjects of the medical curriculum – physiology, pathology, and therapeutics – were, like physics, deemed to be causal sciences, they were taught without visual aids; anatomy, botany, and chemistry-cum-pharmacy had little connection with traditional medical learning and were given a low status. As a result, the three sciences were usually taught by junior professors with little experience. It was typical that Herman Boerhaave (1668–1738), the most influential figure in the theory and practice of medicine in the first half of the eighteenth century, should have begun his teaching career at the University of Leiden in 1709 as professor of botany, a discipline about which he knew nothing. Furthermore, the three sciences were frequently taught in run-down premises. Some universities, notably Padua, Montpellier, Uppsala, and Leiden, had purpose-built dissecting theaters, but many did not. In the early seventeenth century Jean Riolan II (1580–1657), purportedly the best anatomist in Europe, performed dissections at the University of Paris in the open air. Even universities with good facilities had difficulty obtaining bodies and botanical specimens. By 1700 there was only a handful of functioning botanical gardens outside the Italian peninsula, and many of them, such as Oxford's, grew vegetables for consumption as much as plants for study.[10]

At the beginning of the eighteenth century, therefore, the experimental philosophy in its Baconian guise had only the slimmest of footholds in the official curriculum of institutions of higher education. Students had little chance to witness experiments and demonstrations and usually no opportunity to perform them. Students at the University of Paris were luckier than most. Although the facilities of the Paris Faculty of Medicine were poor, interested students had the chance to follow practical courses in anatomy, botany, and chemistry at the city's *Jardin du roi,* an independent institution founded through the efforts of Gui de la Brosse in the 1630s. Here, it was even possible to gain hands-on experience.[11]

[10] Charles B. Schmitt, "Science in the Italian Universities in the Sixteenth and Early Seventeenth Centuries," in Maurice P. Crosland (ed.), *The Emergence of Science in Western Europe* (London: Macmillan, 1975); Edgar A. Underwood, "The Early Teaching of Anatomy at Padua with Special Reference to a Model of the Paduan Anatomy Theatre," *Annals of Science,* 19 (1963), 1–26; Brockliss, *French Higher Education,* pp. 394–5, 397–9; Lucy S. Sutherland and Leslie G. Mitchell (eds.), *The History of the University of Oxford,* vol. 5: *The Eighteenth Century* (1986), pp. 711–23; Robert T. Gunther, *Early Science at Cambridge* (London: facsimile ed., 1969), pp. 221–4; Theodor Lunsingh-Scheuleer, "Un ampithéâtre d'anatomie moralisée," in Lunsingh-Scheuleer and Guillame H. M. Posthumus Meyjes (eds.), *Leiden University in the Seventeenth Century: An Exchange of Learning* (Leiden: Brill, 1975); Christoph Meinel, "'Artibus Academicis Inseranda': Chemistry's Place in Eighteenth and Early Nineteenth Century Universities," *History of Universities,* 7 (1988), 92–3; Maarten Ultee, "The Politics of Professorial Appointment at Leiden, 1709," *History of Universities,* 9 (1990), 167–94 (for Boerhaave).

[11] Yves Laissus, "Le Jardin du roi," in René Taton (ed.), *L'Enseignement et diffusion des sciences au dix-huitième siècle* (Paris: Hermann, 1963); Rio C. Howard, "Gui de la Brosse: The Founder of the Jardin des Plantes," Ph.D. dissertation, Cornell University, Ithaca, New York, 1974.

SCIENCE IN THE UNIVERSITY IN THE
EIGHTEENTH CENTURY: CREATING SPACE

In most important respects, the structure of science teaching in the colleges and universities of Europe and the Americas changed little in the eighteenth century. To begin with, the numbers studying the natural sciences within the system probably stagnated. Despite the rapid rise in the European population on both sides of the Atlantic, the overall number of colleges and universities grew only slightly and attendance rolls generally fell.[12] Only in the English-speaking world did the system visibly expand. In the British Isles the establishment of the new dissenting academies, notably the foundations at Warrington and Hackney, challenged the monopoly of the universities by creating a nonconformist equivalent of the French *collège de plein exercice,* albeit temporarily, inasmuch as most of them closed in the early nineteenth century. In North America, similarly, denominational rivalry as much as facts of geography led to a rapid expansion in the number of colleges-cum-universities in the second half of the century, starting with the foundation of Princeton in 1746. Nineteen such institutions existed in the new United States by 1790, although, apart from Yale, they attracted few students. On the Continent, by contrast, the one significant foundation during the century was the Hanoverian University of Göttingen, opened in 1733. As the system continued to be the preserve of the sons of the elite, the window of opportunity for the study of science was seldom opened more widely to humbler students or women. Even in the British American colonies, attending college was expensive at £10 to £20 per annum.[13]

The organization of the college and university curricula equally remained much the same. Students continued to prepare for their university studies by immersing themselves in the lengthy study of classical languages. For the most part, the university presented the same four- or five-faculty facade to the world that it had always done. The study of philosophy in the faculty of arts continued to cover the four traditional philosophical sciences. Above all, despite the far-reaching developments in contemporary natural science, few serious attempts were made to separate physics formally from the other parts of philosophy. A student's primary introduction to the natural world usually continued to come through the study of physics as part of philosophical studies. Indeed, all the philosophical sciences often continued to be taught by the same professor. If, in Scotland, specialist professorships in the different philosophi-

---

[12] The most careful study of the decline is Willem Frijhoff, "Surplus ou déficit: Hypothèses sur le nombre réel des étudiants en Allemagne à l'époque moderne (1576–1815)," *Francia,* 7 (1979), 173–218.

[13] Herbert McLachlan, *English Education under the Test Acts* (Manchester: Manchester University Press, 1931); Daniel Boorstin, *The Americans,* 1: *The Colonial Experience* (Harmondsworth: Penguin, 1958), pp. 209–10, 436–9 (bibliographical information about the early history of individual American colleges); Henry May, *The Enlightenment in America* (Oxford: Oxford University Press, 1976), p. 381, n. 11 (student numbers in 1783).

cal sciences were slowly established in the first half of the eighteenth century, except at King's College Aberdeen, this had not yet occurred in the French *collèges de plein exercice* on the eve of the French Revolution. Nor had specialization advanced very far in the North American colleges-cum-universities by the end of the century. On the eve of Independence, only Harvard had a separate, endowed professorship in mathematics and natural philosophy (the Hollis professorship, established in 1727). In the other colleges, it was commonplace for one teacher or tutor to give instruction in the gamut of the arts and sciences. For instance, Jefferson's tutor at William and Mary in the early 1760s taught him ethics, rhetoric, and *belles-lettres* as well as natural philosophy.[14]

The general absence of structural change, in particular the failure to give natural philosophy a larger or more distinctive role in the arts curriculum, reflected the fact that in the eyes of the Church and state the purpose of the college and university system remained unchanged. As in previous centuries, the system was intended to produce effective members of the three traditional professions of the Church, the bar, and medicine, especially the first two. To the extent that most figures of authority (and most members of the elite) felt that this end was best achieved by giving prospective entrants a solid classical education, a general knowledge of the different branches of philosophy, and a period of professional training, the place of natural philosophy in the curriculum was unlikely to be greatly extended. A year's study in the science of physics was quite sufficient.

In contrast to previous centuries, however, this establishment view did not go unchallenged. As the eighteenth century progressed, a bevy of radical educational commentators, impressed by the contemporary achievements of the new science, increasingly voiced the need for a more science-oriented arts curriculum. While continuing to uphold the narrow professional purpose of university and college studies, they began to promote the particular educational value of the study of mathematics and natural philosophy. In the middle of the century the lead was taken by the French *philosophes,* but by the 1790s voices were being raised all over Europe in favor of curricular reform. One of the first and most trenchant attacks on the existing system was penned by the mathematician-*philosophe* Jean Le Rond d'Alembert (1717–1783) and appeared under the heading "collège" in the third volume of the *Encyclopédie.* D'Alembert tested the current arts curriculum on the anvil of utility and found it completely wanting:

---

[14] Sir Alexander Grant, *The Story of the University of Edinburgh During its First Three Hundred Years,* 2 vols. (London: Longman, 1884), vol. I, pp. 262–4; Paul B. Wood, *The Aberdeen Enlightenment: The Arts Curriculum in the Eighteenth Century* (Aberdeen: Aberdeen University Press, 1993), p. 89; Brockliss, *French Higher Education,* pp. 193–4; I. Bernard Cohen, *Science and the Founding Fathers: Science in the Political Thought of Thomas Jefferson, Benjamin Franklin, John Adams and James Madison* (New York: W. W. Norton, 1995), pp. 68–72.

Why pass six years in learning, be it well or badly, a dead language? . . . This time would be better spent in learning the rules of one's own tongue, of which one is totally ignorant on leaving college. . . . In philosophy, logic should be limited to a couple of lines, metaphysics to an abridgement of Locke, philosophical ethics to the works of Seneca and Epictetus, Christian ethics to the Sermon on the Mount, and physics to experiments and geometry, which is the best of all logics and physics.[15]

In most parts of Europe and the Americas these demands fell on deaf ears, no more so than in pre-Revolutionary France. In general, little structural curricular change that gave a greater role or distinction to mathematics and natural philosophy was effected anywhere before the end of the century. The most that was achieved to undermine the dominance of traditional classical education was the acceptance that instruction in the natural sciences and medicine was better undertaken in the vernacular rather than in Latin.[16] Nevertheless, in a few countries, primarily those in which reform was placed squarely on the government agenda as a result of real or perceived national weakness, the *philosophe* critique was viewed more sympathetically.

As early as 1745 in Sweden, the dominant Hat party, anxious to secure the state against further encroachments from its aggressive neighbors, set up an educational commission that suggested a complete restucturing of the university system and the emancipation of natural philosophy. After prophylactic studies in a faculty of arts, reduced to the study of logic, metaphysics, ethics, and Latin rhetoric, students were to enter one of four completely renamed faculties according to their chosen careers. Henceforth there were to be two entirely separate faculties of mathematics and physics: the one training military officers and land-surveyors, and the other medical practitioners.

Although the Hats' plans were never realised, their initiative eventually bore fruit on the other side of the Baltic. Thirty years later, even more radical reform plans were drawn up by the Polish Diet's Committee for National Education. Spurred into action by the First Partition of the country in 1772, the Polish elite responded by modernizing the educational system. Under a plan of 1777 the subjects of the traditional arts curriculum of the two universities of Cracow and Vilna were to be redistributed between the new faculties of physics and moral philosophy, and the new physics faculty comprised mathematics as well as medicine. In addition, the curriculum of the univer-

---

[15] Jean Le Rond d'Alembert, "Collège," in Denis Diderot (ed.), *Encyclopédie, ou dictionnaire raisonné des sciences, des arts et des métiers,* 17 vols. (Paris and Neutchâtel: Briasson et al and Samuel Faulche, 1751–65), vol. 3, pp. 636–7. Many examples of the critical literature outside France are detailed in James A. Leith (ed.), *Facets of Education in the Eighteenth Century* (Studies on Voltaire and the Eighteenth Century, 167: Oxford: The Voltaire Foundation, 1977), passim; see also Charles E. McClelland, *State, Society and University in Germany, 1700–1914* (Cambridge University Press, 1980), chaps. 2 and 3.

[16] Dominique Julia, "Une réforme impossible: le changement du cursus dans la France du 18e siècle," *Actes de la Recherche en Sciences Sociales,* 47–8 (1983), 53–76; Brockliss, *French Higher Education,* pp. 191–2.

sity feeder schools was to be reorganized so that a much greater time was given to natural science at the expense of classical languages. Unlike the Swedish plan, the Polish proposals were actually put into practice. There again, they did not last beyond the final partition of the country and its disappearance from the map in 1796.

Even the more modest contemporaneous attempt by one of the partitioning powers, the Austrian Habsburgs, to increase the importance of natural science merely at the college level also ran quickly into the sand. An initial attempt to enliven the college curriculum was undertaken by Maria Theresa as part of a general reform of elementary and collegiate education in 1774–7, but little seems to have been achieved. Ten years later, after Joseph von Sonnenfels (1733–1817), a professor of law at the University of Vienna, delivered to Joseph II a damning indictment of the antediluvian state of Austrian public education, the reform of the curriculum had to be launched all over again, only to be stymied again by the general reaction against enlightened reform that set in following the Emperor's death in 1790.[17]

In fact, the only radical and permanent structural shake-up of the system before the Revolutions of 1848 occurred in France and the Netherlands in the maelstrom of the Napoleonic era and its immediate aftermath. The French Revolutionaries, influenced by the Enlightenment critique, abolished all the country's colleges and universities on the grounds that they were corporative and elitist institutions offering an outdated curriculum. The colleges were temporarily replaced in 1795 by a new type of feeder school, the *école centrale*, which placed a novel emphasis on mathematical and scientific education. However, when these schools proved unpopular with parents, they were quickly replaced in turn by the *lycée*, which offered much the same curriculum as before. The universities, on the other hand, were permanently replaced in 1808 by a single institution, the *Université impériale*. This was a national umbrella organization administering multiple faculty sites, including for the first time in Europe *separate* faculties of arts and sciences. The same system was adopted by the new state of Belgium-Holland in 1815, but not until the mid-nineteenth century were the arts and sciences divided in the universities of other countries.[18]

On the other hand, although there was no general enhancement of the

[17] Sten Lindroth, *A History of the University of Uppsala, 1477–1977* (Uppsala: Uppsala University, 1976), pp. 98–100; Grzegorz L. Seidler, "Reform of the Polish School System in the Era of the Enlightenment," in Leith (ed.), *Facets*, pp. 348–54; B. Becker-Cantarino, "Joseph von Sonnenfels and the Development of Secular Education in Eighteenth-Century Austria," ibid., pp. 41–6; Matyas Bajko, "The Development of Hungarian Formal Education in the Eighteenth Century," ibid., pp. 212–16; Robert J. Kerner, *Bohemia in the Eighteenth Century* (New York: AMS Press, 1932), chap. 11.

[18] Laurence W. B. Brockliss, "The European University in the Age of Revolution, 1789–1850," in Michael G. Brock and Mark C. Curthoys (eds.), *The History of the University of Oxford*, vol. 6: *Nineteenth-Century Oxford, Part 1* (1997), chap. 2. The best study of the curriculum of the *écoles centrales* is Sergio Moravia, *Il tramonto dell' illuminismo: filosofia e politica nella società francese, 1771–1810* (Bari: Laterzo, 1968), pp. 347–69.

curricular importance and status of the natural sciences before the mid-nineteenth century, the conditions for its emancipation were being laid in a different way as the eighteenth century progressed. Broadly speaking, before 1700 theology was the queen of the sciences and the faculty of theology the most important university faculty, even if in terms of student numbers it played second fiddle to the faculty of law. At this date there were almost no students of medicine. The philosophy course as a whole, then, was primarily propaedeutic to the study of theology, all the more in that it was often possible to enter the law faculty without a degree in arts. Consequently, the philosophy curriculum, including the course in physics, was supposed to contain nothing inconsistent with religious orthodoxy but provide the concepts and logical tools by which many of the truths of Revelation might be rationally underpinned and a deeper, non-Biblical understanding of God and His creation developed. Obviously, were philosophy ever to lose this subordinate and dependent role, it would only hasten its bifurcation into separate, distinctive disciplines.

It is no surprise, then, that it was in Revolutionary and Napoleonic France that the emancipation of the natural sciences was first permanently achieved because it was in post-Revolutionary France that educationalists were first confronted with creating *ex novo* a college and university system that was not primarily directed toward the study of theology.[19] However, long before the French Revolution, there were signs in the universities of Protestant northern Germany that the relationship between philosophy and theology was no longer so close. A key moment was the foundation of the University of Halle in 1693, for the Prussian university was the brainchild of the Pietist August Hermann Francke (1663–1727) who deplored theological rationalization and wanted to rebuild Lutheranism as the religion of the Bible and the spirit. As Pietism in the first half of the eighteenth century became the dominant force in Germany's Lutheran universities, so the connection between philosophy and theology was steadily eroded. Admittedly, Pietist theologians did not approve of philosophical freedom, even if they had a limited use for the philosopher's tools. As a result of Pietist pressure, Wolff, for instance, was forced to leave Halle for Marburg in 1731 after a ten-year battle over his admiration for Confucian ethics. Nonetheless, the wheel of emancipation kept turning in the Protestant north with the foundation of the University of Göttingen a few years later. The Hanoverian university was the first to be founded as a multiconfessional institution. Although it boasted a faculty of theology, its members were forbidden to establish and enforce a party line.[20]

---

[19] The reconstruction of higher education mooted by the Revolutionaries and eventually carried out by Napoleon did find room for theology schools and faculties, but they were accorded little significance.

[20] Wilhelm Schrader, *Geschichte der Friedrichs-Universität zu Halle*, 2 vols. (Berlin, 1894), vol. 1; Emil F. Rössler, *Die Gründung der Universität Göttingen* (Göttingen: Vandenhoeck and Ruprecht, 1855), documents and commentary.

Consequently, from its foundation, Göttingen's faculty of philosophy had an enviable freedom. The 1737 statutes specifically allowed professors to choose their own textbooks and organize their courses as they would, provided that they taught nothing contrary to religion, morals, and the state. Not surprisingly, the university attracted a constant stream of dynamic and talented teachers, especially in medicine, natural philosophy, and mathematics. Its medical professors included the botanist and physiologist Albrecht von Haller (1708–1777) and the naturalist Johann-Friedrich Blumenbach (1752–1841). Among its philosophy professors was the electrical experimenter and geologist Georg Christoph Lichtenberg (1742–1799). Because many of the science and medicine professors were active researchers, Göttingen soon had its own scientific society with its own published transactions, distinct from the Elector's academy to which the professors also belonged. In an important respect, Göttingen was the first research university whose professors were as dedicated to publication as teaching. Many of the professors, though, such as the mathematician Abraham-Gotthelf Kaestner (1719–1800), specialized in writing textbooks and not in creative research. The accolade, then, should be bestowed with reservation. Moreover, toward the end of the century other north German universities were beginning to become centers of scientific research. Göttingen had its imitators. Thus at Helmstedt and Leipzig, professors Johann Friedrich Pfaff (1765–1825) and Karl Friedrich Hindenberg (1741–1808) established a combinatorial school of mathematical analyis and launched the first-ever mathematical journal, *Archiv der reinen und angewandten Mathematik.* Similarly, the chemist Lorenz Crell (1744–1816) also used Helmstedt to organize the nascent German chemical community and publish his chemistry periodical.[21]

Clearly, professors of science and mathematics in the north German universities did not have to await the institutional separation of the moral and natural sciences before they began developing intra-, cross-, and extrafaculty ties. By the end of the eighteenth century natural philosophy and specific natural sciences were beginning to acquire an identity and status of their own *within* the existing university structure. The professors' activities were only further encouraged by the epistemological writings of one of their philosophical colleagues at the Prussian university of Königsberg, Immanuel Kant (1724–1804), who himself had spent his first years in academic life teaching physics. In the course of three classic treatises, beginning with the *Critique of Pure Reason* in 1781, Kant demonstrated to his compatriots the epistemological distinctiveness of the moral and physical sciences. With his philosophical ideas

[21] Rainer A. Müller, *Geschichte der Universität von der mittelalterlichen Universitas zur deutschen Hochschule* (Munich: Callway, 1990), p. 63; McClelland, *State, Society and University,* pp. 37–45; R. Steven Turner, "University Reformers and Professorial Scholarship in Germany, 1760–1806," in Lawrence Stone (ed.), *The University in Society,* 2 vols. (Princeton, NJ: Princeton University Press, 1974). There is no detailed recent history of Göttingen, only the brief G. Meinhardt, *Die Universität Göttingen: ihre Entwicklung und Geschichte von 1734–1974* (Göttingen: Musterschmidt, 1977).

gradually gaining ground in Protestant and even in parts of Catholic Germany, the sage of Königsberg went one step further in 1798 and, in his *Streit der Facultäten,* cut the faculty of philosophy completely adrift from the others. The curriculum of the professional faculties, he declared, could be supervised by the state since the state, through its duty to maintain the security of its citizens, had an interest in the ideology of their products. The state, however, had an equal duty to give professors in the philosophy faculty total freedom.

> It is absolutely essential that within the university there is a faculty involved in public scientific instruction [using the term in its widest sense] which, being independent of the orders of the government, has the liberty, if not to give orders, at least to give judgements on everything of scientific interest, that is to say on truth. In this faculty reason must have the authority to speak openly, for without this liberty, truth cannot be made manifest (and this will be prejudicial even to the government). Reason, moreover, is free by its very nature and can welcome no order directing something to be received as the truth (no *crede,* simply a free *credo*).[22]

With the triumph of Kantianism in the north German universities at the turn of the nineteenth century, the umbilical cord connecting the faculties of philosophy and theology was permanently and significantly severed in a number of Christian states. Ten years later, Kant's Idealist disciple, the Prussian minister of education, Wilhelm von Humboldt (1767–1835) completed the emancipatory revolution by arguing in favor of professorial freedom in each of the faculties, even theology. No more than Kant did Humboldt think of altering the structure of the traditional university. When he established the University of Berlin in 1810 as the first university whose professors were statutorily expected to prosecute research as well as teach, he maintained the traditional fourfold faculty division. The moral and natural sciences continued to be lumped together under a single institutional umbrella, but in a new environment where subject professors were free to teach what they liked and philosophy students free to attend any classes they pleased.[23]

This traditional structure was to last in most German universities until the end of the First World War. Although few professors in nineteenth-century German philosophy faculties ever embraced Humboldt's belief that the discipline remained a unity despite the epistemological divide, scientists and

---

[22] Immanuel Kant, *Streit der Facultäten,* ed. Klaus Reich (Hamburg, 1959), p. 12. A good introduction to Kant's epistemology is Stephen Körner, *Kant* (Harmondsworth: Penguin, 1959).

[23] Otto Vossler, "Humboldts Idee der Universität," *Historisches Zeitung,* 178 (1954); Clemens Menze, *Die Bildungsreform Wilhelm von Humboldts* (Hanover: Schroedel, 1975); Eduard Spranger, *Wilhelm von Humboldt und die Reform des Bildungswesens* (Tübingen: Niemayer, 1960). Humboldt's role in the foundation of the University of Berlin was not quite so prominent as is usually thought: see U. Muhlack, "Die Universitäten in Zeichen von Neuhumanismus und Idealismus," in Peter Baumgart and Notker Hammerstein (eds.), *Beiträge zu Problemen deutscher Universitätsgründungen der frühen Neuzeit* (Nendeln: KTO Press, 1978).

artists quickly learned to live with the structure, building an independent identity for their subject through the creation of the professorial research seminar. This was another feature of the modern university whose origins can be traced to Göttingen in the second half of the eighteenth century, but before the Humboldtian reforms it had been used to train classics teachers and not natural scientists. In the 1820s and 1830s, mathematics and natural philosophy seminars began to mushroom all over northern Germany with the result that science and scientific research were firmly institutionalized within the university system.[24] If in most parts of Europe and the Americas, the relatively inchoate and limited role of natural philosophy in the arts curriculum of the eighteenth century ineluctably encouraged the development in the long term of separate faculties of science, so too the higher and more independent profile accorded the subject in Lutheran Germany played a crucial role in the survival of a more traditional organizational model in the institutional heartland of nineteenth-century European science.

SCIENCE IN THE UNIVERSITY IN THE EIGHTEENTH CENTURY: THE CURRICULUM

The fact that outside northern Germany little more space or dignity was afforded the natural sciences in the college and university curriculum before the French Revolution did not mean that science teaching itself was moribund. In fact, the institutional inertia masked profound changes in its content and articulation, something critics of the university system either willfully or unwittingly ignored. In the eighteenth century, the colleges and universities everywhere successfully accommodated their courses in scientific subjects to current scientific fashion. Frequently change was wrought unsolicited from within. The colleges and universities could read the runes. In an age that saw the new experimental philosophy increasingly lionized by the state and given its own institutional identity in the form of the scientific academy, it was only prudent to pay greater heed to contemporary scientific culture. Professors charged with teaching the natural sciences did not need to be practicing experimental philosophers themselves (although some were) to see the wisdom of keeping abreast of contemporary scientific developments if they wanted to attract an audience and keep their pupils' respect. Sometimes, however,

---

[24] W. Erben, "Enstehung der Universitätsseminare," *Internationale Monatsschrift für Wissenschaft, Kunst und Technik* (1913), 1247–64, 1335–47 (still the essential point of departure); R. Steven Turner, "The Growth of Professorial Research in Prussia, 1818 to 1848," *Historical Studies in the Sciences*, 3 (1971), 143–50; C. Jungnickel, "Teaching and Research in the Physical Sciences and Mathematics in Saxony, 1820–1850," ibid., 15 (1985), 3–47; relevant essays in Kathryn M. Olesko (ed.), *Science in Germany: The Intersection of Institutional and Intellectual Issues, Osiris*, 2nd ser., vol. 5 (Philadelphia: Sheridan Press, 1989); relevant essays in Gert Schubring (ed.), *"Einsamkeit und Freiheit" neu besichtigt. Universitätsreformen und Disziplinenbildung in Preussen als Modell für Wissenschaftspolitik im Europa des 19. Jahrhunderts* (Stuttgart: Franz Steiner, 1991).

especially in parts of the Catholic world where the regular orders controlled science teaching and concerns about the maintenance of religious orthodoxy were particularly strong, change was ultimately foisted on the university system from without. The state might have been little interested in seriously restructuring the universities and colleges, but by the final third of the century rulers were no longer happy to think that the professional elite, especially medical practitioners, were being reared on an antediluvian science.

Change was especially profound with regard to the teaching of physics. However little attention was paid to Newton at the beginning of the century, this was not the case by the end. By the last decades of the century, everywhere across the continent courses were taught that accepted Newton's theory of universal gravitation, his championship of a multiple-force universe, and his radical phenomenological approach to the study of the natural world. As a result, in a complete breach with five centuries of tradition, college and university physics had ceased for the most part to be a science of causes, devoted to relating natural effects to fundamental principles, and had instead become an analytical science concerned with explicating the mathematical laws governing the behavior of natural bodies. It was thus a completely new type of physics, now much more closely in tune with the scientific epistemology of the continent's contemporary, post-Newtonian experimental philosophers. It was also clearly a physics that had no relation to the other three parts of the philosophy course. The introduction of a phenomenological physics emphasized – a century before the formal creation of separate faculties of arts and sciences – that the traditional conception of philosophy as a united, coherent, quadrapartite discipline no longer obtained.

The revolution can be dated generally to the third quarter of the eighteenth century. Until then, most physics courses, outside Britain, the North American colonies, and the Netherlands (where a Newtonian physics was being taught by the second decade of the century), continued to be dominated by Cartesian principles or even, where the Jesuits ruled the roost, quasi-Aristotelian ones.[25] As late as the 1730s, for instance, Charles-Ariège Barloeuf (1715–after 1762), a professor at the Jesuit college at Caen, maintained the Aristotelian distinction between the sub- and superlunary world and would accept mechanist explanations only for the behavior of terrestrial natural phenomena. Even in Britain and the Netherlands, a much more traditionalist physics was still expounded in the first half of the century in conjunction with Newtonian courses. At Oxford and Cambridge, where the colleges often laid on their own courses in natural philosophy, a number of tutors continued to support a causal, nonmathematical physics. The set physics textbook that Christ Church students at Oxford were examined on in college from 1717 to 1755 and again

---

[25] There are only three detailed accounts of the establishment of Newtonian physics in the classroom on the continent: Brockliss, *French Higher Education*, chap. 8, sections 3–4; Vanpaemel, *Echo's van een wetenschappelijke revolutie*, especially chap. 7; and Ruestow, *Physics at Leiden*, chap. 7.

from 1760 to 1762 was the early seventeenth-century *Enchiridium Physicum* of the Dane Caspar Bartholinus (1585–1630)![26]

Given the tenacity with which members of the French *Académie des Sciences* in the first three decades of the eighteenth century attempted to find a mathematical defense of Cartesian vortex theory and the Gallocentric nature of contemporary European culture, it is unsurprising that Newton's phenomenological physics was slow to take root in the Continent's colleges and universities. Nor is it surprising, given the *Académie's* authority, that once its young Turks, such as Clairault, had pledged their support to the Newtonian approach, the revolution was effected rapidly after 1750.[27] In Catholic Europe the introduction of a Newtonian physics was made easier by the expulsion of the Jesuits from individual states, beginning with Portugal in 1758, and the Order's complete abolition in 1774. Although the Jesuits, too, had apparently begun to embrace Newtonianism in the third quarter of the century – doubtless under the influence of their own Rome-based Newtonian mathematician, Roger Boscovich (c. 1711–1787) – their courses were continually accused of being out-of-date. The removal of the Society from its dominant role in philosophy teaching in Catholic Europe at the very least forced the secular authorities to consider the question of the content and articulation of the physics course, even if they remained unmoved to calls for a complete overhaul of the undergraduate curriculum.

As a result, under state initiative Newtonianism eventually came to be taught in the final quarter of the century even in the Portuguese, Spanish, and Austrian empires, areas that had hitherto shown minimal interest in attractionist physics. In Eastern Europe the Habsburgs found a useful ally in another regular order, the Piarists, who had gained a reputation for being interested in modern science and already controlled a large number of municipal colleges before the Jesuits' expulsion, especially in Hungary. In Portugal, the royal minister, Pombal, promulgated new statutes for the University of Coimbra in 1772 that outlawed Aristotelianism and instituted a course of physics around Newtonian principles. In Spain from the early 1770s, the Bourbon king Charles III oversaw a painstaking curricular revision of each university in turn.[28] By the late 1780s reform had reached as far as Peru. In 1787 Toribio Rodríguez de Mendoza (1750–1825), the rector of the college of San Carlos at Lima, introduced a new plan of studies that ordered the teaching of Newton

[26] Bibliothèque Municipale Vire, MS A48, Bessore, "Physica particularis"; Gunther, *Science at Cambridge*, p. 50; Edward G. W. Bill, *Education at Christ Church Oxford, 1660–1800* (Oxford: Clarendon Press, 1988), pp. 308–10.

[27] The best introduction to the *Académie's* attitude toward Newton remains Pierre Brunet, *L'Introduction des théories de Newton en France au XVIIIe siècle*, vol. 1: *Avant 1738* (Paris: Albert Blanchard, 1931).

[28] Bajko, "Hungarian Formal Education," pp. 197–8; José Ferreira Carrato, "The Enlightenment in Portugal and the Educational Reforms of the Marquis of Pombal," in Leith (ed.), *Facets*, pp. 385–93; *Estatutos da Universidade de Coimbra (1772)*, facsimile ed., 3 vols. (Coimbra: Universidade de Coimbra, 1972); Mariano Peset, *Carlos III y la legislación sobre universidades* (Madrid: Ministerio de Justicia, 1988).

as the only acceptable modern natural philosopher. Descartes, Gassendi, and their followers were henceforth to be excluded from the classroom because of their doctrinaire spirit. But Newton was different:

> The system of this wise Englishman is not founded on arbitrary hypotheses but incontestable principles. Daily they are confirmed by experience and they are totally consistent with observations made before and after his life. For this reason Diderot and D'Alembert claim the system to be true and demonstrated, and all the wise men of Europe have declared their allegiance to it.[29]

However, even where professors declared themselves to be Newtonians, they were frequently reluctant to disgard completely the traditional commitment to a science of causes. The first French professor of philosophy known to have embraced Newtonian physics, Pierre Sigorgne (1719–1809) at the Paris Collège du Sorbonne-Plessis, was a thoroughgoing phenomenologist. But many of his successors in other French *collèges de plein exercice*, such as the Oratorian Joseph Valla, who taught in his Order's colleges at Soissons and Lyons, retained the hope that it would be possible one day to play God and understand the fundamental principles of Nature. Valla (d. 1790) remained committed to the logical necessity of a single-force impulsionist universe:

> Although the Newtonian hypothesis more accurately explains the motion of the celestial bodies than others which have preceded it, still the fundamental principle by which everything is moved remains doubtful and uncertain. For what is attributed to mutual attraction, could be the primitive effect of some impulsion. Even if the motion cannot be successfully explained by a law of impulsion, does this mean it is inexplicable thereby? By accepting a demonstration of this kind [i.e. mutual attraction], we must admit a new and scarcely intelligible principle, especially when nature only operates by simple causes, however fecund its effects.[30]

More important, it remained possible in some universities, notably in Castile, to receive tuition in the traditional physics, even at the turn of the nineteenth century. When the Spanish crown initiated the reform of the Castilian universities in the late 1760s, the theologians seem to have campaigned strongly against the bifurcation of philosophy into two separate subject areas. As a result, a compromise was reached. When the country's principal university, Salamanca, received new statutes in 1771, two courses in physics were established: one in modern physics for prospective medical students, and one in traditional Thomist physics for theologians built around the textbook

---

[29] Antonio E. Ten, "El convictorio carolino de Lima y la introducción de la ciencia moderna en el Perú virreinal," in *Universidades espanolas y americanas. Epoca colonial,* foreword by Mariano Peset (Valencia: CSIC, 1987), p. 527.

[30] Joseph Valla, *Institutiones philosophicae . . . ad usum scholarum,* 6 vols. (Lyons: Perisse, 1780), vol. 4, p. 448.

of the mid-seventeenth-century French Dominican Antoine Goudin (dates unknown).[31]

Moreover, the new course in physics that was instituted throughout Europe was in many, probably most, cases a course in mathematical physics *manqué*. The professors might have embraced Newton's physics, but they did not generally expound his theories using the mathematical approach of the *Principia*. Rather, they eschewed mathematical analysis in their demonstrations in favor of illustrating the solidity of Newtonian physics by experiments. The use of the illustrative experiment, it will be recalled, had no place in the traditional physics course, no more than had mathematics, although experiments were often referred to in the professorial exposition. However, from the moment that a Cartesian physics began to be taught in the classroom in the second half of the seventeenth century, the occasional professor began to offer an extracurricular course in experimental physics in addition to his official lectures. The first such course seems to have been offered at Würzburg as early as the 1660s, and by the turn of the eighteenth century they were commonplace. However, at this date, the instruction was not connected in any logical way with the official course: the teachers concentrated on showing off the versatility of particular pieces of apparatus, such as the ubiquitous air pump, and the courses were frequently given by Peripatetic outsiders, such as Pierre Polinière (1671–1734), who made the rounds of the Paris colleges.[32]

The first professor to think of using experiments as a way of illustrating a complete and coherently organized course of physics seems to have been the Frenchman Jacques Rohault (d. 1672), who gave private lectures on Cartesian natural philosophy in a number of towns in France around 1670, thereby making one of the first breaches in the university monopoly of physics teaching. Within the university world, Rohault's earliest imitator was probably John Keill (1671–1721), a Scotsman who devised an experimental course in Newtonian physics, which he taught at Oxford perhaps from as early as 1694 until 1709. The most influential figure in the creation of the new pedagogical genre, however, was indisputably another passionate Newtonian, Willem Jacob van 's Gravesande (1688–1742), who held the chair of mathematics and astronomy at Leiden from 1717. Although Rohault and numerous later university professors of experimental physics published their courses, it was 's Gravesande's physics textbook, published in Latin in 1720–1, that really formalized the structure of the new course. The English translation alone, which was the work of Keill's Oxford successor, John Theophilus Desaguliers (1683–1744), had

---

[31] George M. Addy, *The Enlightenment in the University of Salamanca* (Durham, NC: Duke University Press, 1966), pp. 104–5, 111.

[32] Ruestow, *Physics at Leiden*, pp. 96–8, 103, 114 (general comments on the development); Brockliss, *French Higher Education*, p. 189; Marta Cavazza, "Orti botanici, teatri anatomici, osservatori astronomici, musei e gabinetti scientifici," in G. P. Brizzi and J. Verger (eds.), *Le università dell'Europa. Le scuole e i maestri. L'Età moderna* (Milan: Amilcare Pizzi, 1995), p. 86; Geoffrey Smith, *Science for a Polite Society: Gender, Culture and the Demonstration of Enlightenment* (Oxford: Westview Press, 1995), pp. 194–208 (on Polinière).

gone through six editions by mid-century. It was replaced only in the second half of the century by the equally successful posthumous publication in 1762 of the *Introductio ad philosophiam naturalem* of 's Gravesande's Leiden successor, Pieter Musschenbroek (1692–1761), discoverer of the Leiden jar.[33]

As a result of this development, chairs in natural philosophy (where they existed) were frequently rechristened chairs in experimental physics, and colleges were forced, often for the first time, to purchase a collection of physical apparatus and set aside laboratory space. As was only to be expected, this institutionalization of the new course seldom occurred before the middle decades of the century. In northern Italy, for instance, the first university to establish a chair in experimental physics and endow a *cabinet de physique* seems to have been Pavia in 1730. The next was Padua in 1738, where the new course was entrusted to Giovanni Poleni (1683–1761), a professor who had long shown an interest in providing visual tuition in the natural sciences. At that date the Paduan laboratory was purportedly the best equipped in Europe. Other neighboring universities gradually followed suit – Pisa (1746), Turin (1748), Modena (1760) and Parma (1770) – wheras in 1787 the facilities at Pavia, where the chair was held from 1778 by the young experimenter, Alessandro Volta (1745–1827), were improved with the opening of a purpose-built physics theater.[34]

The effectiveness of the new course seems to have depended primarily on the quality of the physical apparatus. Because instruments were expensive and good instrument-makers rare (a favorite source was Musschenbroek's brother), many smaller institutions could afford only modest collections. This was particularly the case for the university colleges in North America, where only Harvard had a good instrument collection from an early date. Princeton's cabinet may have boasted the famous Rittenhouse orrery, pride of American scientific technology, but otherwise the cupboard was virtually bare, as the trustees continually bemoaned. As late as 1796 the college had to launch a modest appeal (subscribed to by two famous alumni: Madison and Burr) to provide the college with much-needed chemical equipment. Even good equipment, however, did not always guarantee a successful lecture course. Sometimes the professor was simply an incompetent experimenter, however knowledgeable in other respects, as revealed by Jeremy Bentham's verdict on the course given by the astronomer Nathaniel Bliss (1700–1764) at Oxford in 1763.

> Mr Bliss seems to me to be a very good sort of Man, but I doubt he is not very well qualified for his Office, in the practical Way I mean, for he is oblig'd

---

[33] Trevor McClaughlin, "Le concept de science chez Jacques Rohault," *Revue d'histoire des sciences,* 30 (1977), 225–40; Geert Vanpaemel, "Rohault's *Traité de physique* and the Teaching of Cartesian Physics," *Janus,* 71 (1984), 31–40; *History of Oxford,* vol. 5, p. 671; Ruestow, *Physics in Leiden,* chap. 7 passim.

[34] Cavazza, "Orti," pp. 86–7; P. Vaccari, *Storia dell' università di Pavia* (Pavia: Università di Pavia, 1957), chaps. 7 and 8 passim; M. Cecilia Ghetti, "Struttura e organizzazione dell'università di Padova della metà del 1700 al 1797," *Quaderni per la storia dell'Università di Padova,* 16 (1983), 87–8.

to make excuses for almost every experiment they not succeeding according to expectation; in the speculative part, I believe he is by no mean deficient.[35]

It was understandable that as universities all over Europe adopted Newton's phenomenological physics the new course would have been given an experimental rather than a mathematical bias. Mathematics still remained the Cinderella in the undergraduate curriculum. Despite the existence of permanent courses in mathematics everywhere, there was seldom an insistence that the average student should gain any more than a rudimentary knowledge of the subject. 's Gravesande himself thought that students should learn only bits of geometry and algebra in the faculty of arts. Evidently, then, most philosophy students could never have followed a mathematical presentation of the Newtonian natural science. This was presumably why Keill developed the new genre at Oxford in the first place. David Gregory (1661–1708) and later Savilian professors of astronomy may have provided a sophisticated mathematical rendering of Newton's *Principia,* but they can have had few auditors. It is indicative that James Bradley (1693–1762), who gained the Savilian chair in 1721, felt the need to offer a separate and very popular course in experimental physics, which he taught from 1729 to 1760 in the Ashmolean Museum.[36]

All the same, there were places in Europe where arts students were introduced to the new physics course primarily mathematically. This may have been the case at eighteenth-century Cambridge. As at Oxford, a mathematical Newtonian natural philosophy seems to have been taught at the Fenland University throughout the century by the titular *mathematics* professor – namely, Newton's successors in the Lucasian chair, such as the blind Nicholas Saunderson (1682–1739). In contrast to Oxford, however, it also is possible that Cambridge students were taught physics in a similar fashion as part of their college-based studies. Eighteenth-century Cambridge was a peculiar university in that students in the arts received little instruction in logic, physics, and metaphysics as these sciences of philosophy were usually understood; instead, they were predominantly fed an unprecedented and relentless diet of ethics and mathematics. This development culminated in 1753 in the establishment of the classified mathematical tripos. Thereafter, candidates for the B.A., after being perfunctorily quizzed on ethics and mathematics, could also take a longer, written examination for an honors degree in the second science and compete for the coveted title of Senior Wrangler. Because the tripos examination included mathematical physics as much as mathematics, presumably the college course in mathematics, too, offered the oportunity to study the

[35] Timothy L. S. Sprigge et al. (eds.), *The Correspondence of Jeremy Bentham,* vol. 1 (London: Athlone Press, 1968), p. 67; Cohen, *Science and the Founding Fathers,* pp. 262–9; Boorstin, *Colonial Experience,* pp. 279–80; I. Bernard Cohen, *Some Early Tools of American Science: An Account of the Early Scientific Instruments and Mineralogical and Biological Collections in Harvard University* (Cambridge, MA: Harvard University Press, 1950).

[36] *History of Oxford,* vol. 5, pp. 650, 674; Ruestow, *Physics at Leiden,* p. 135.

science's application to natural philosophy. Only the brightest college students might have studied Newton's *Principia,* but many others would have conned the textbooks of mathematical physics, such as the *Ordo institutionum physicarum,* published in 1743 by Thomas Rutherford (1712–71) of St. John's.[37]

Admittedly, only a handful of Cambridge students ever took honors in the eighteenth century. In 1750 apparently only twenty candidates for the B.A. were proficient enough mathematicians to demonstrate the principal propositions in the *Principia.* The role of the new mathematical physics in the education of eighteenth-century Cambridge students should therefore not be exaggerated. Indeed, perhaps, given the consumer-oriented nature of the college tutorial system, the majority of students received little tuition in mathematical physics and, like their Oxford counterparts, picked up the essentials of Newtonian natural philosophy by attending public or private experimentally based courses in natural philosophy.[38]

The position was very different in France. There in the second half of the eighteenth century the traditional course of natural philosophy was replaced by a mathematical physics – not just in one or a handful of the country's 100 *collèges de plein exercice* but in them all – and the new course was taken by 2,500 students per annum. So that philosophy students could cope with the demands of this new course, the first three months of the physics year were devoted to a crash course in mathematics: in a matter of weeks, students were taken from the principles of arithmetic to the principles of calculus. It was thus that the young Pierre Simon de Laplace (1749–1827) was introduced to Newtonian physics by Christophe Gabled (1734–82) at the Caen Collège des Arts in the 1760s.[39]

In France, as a result, courses in experimental physics were normally taught only outside the main curriculum, customarily in the vacation *after* students had finished their physics year. The course, too, was clearly meant to entertain as much as edify, for it was usually open to members of the public, including women. No more than a handful of institutions, therefore, ever appointed a specific professor of experimental physics, the two most important chairs being founded at the Paris Collège de Navarre in 1753 and the independent Collège Royal (today the Collège de France) in 1769. Indeed, as at the turn of the eighteenth century, a French course in experimental physics might still be given by an outsider, although by the 1780s most colleges had established

[37] Denys A. Winstanley, *Unreformed Cambridge: A Study of Certain Aspects of Eighteenth-Century Cambridge* (Cambridge University Press, 1953); Gunther, *Cambridge Science,* pp. 34–63 passim; Elisabeth Leedham-Green, *A Concise History of the University of Cambridge* (Cambridge University Press, 1996), pp. 123–7. The Plumian chair in astronomy, established in 1707, seems to have become dedicated to experimental physics over the century.

[38] John Green, *The Academic: Or a Disputation on the State of the University of Cambridge* (London: C. Say, 1750), pp. 23–5.

[39] Brockliss, *French Higher Education,* pp. 189–90, 379–85.

their own *cabinets de physique,* and the professors of philosophy performed the experiments.[40]

The development of a new conception of university physics in the course of the eighteenth century, regardless of the manner in which this Newtonian phenomenological natural philosophy actually came to be taught, also brought with it a redefinition of the content of the science. At the turn of the eighteenth century university physics in the Aristotelian tradition embraced the study of the natural world *tout court.* The new course, in contrast, was much more narrowly defined. In essence, it became a course in the study of the visible motions of inert natural phenomena and embraced the subsciences of mechanics, statics, dynamics, optics, acoustics, and astronomy. To this list professors of an experimental physics added the new, highly theatrical sciences of magnetism and electricity, which began to be mathematicized only at the turn of the nineteenth century.[41] The physics course, then, no longer contained the study of the stucture of terrestrial matter, meteorology (in the widest sense of term), and living organisms. A new type of physics had appeared that corresponded (albeit loosely) to the topics covered by Aristotle's own book of *Physics* and his *De caelo* but ignored the content of his other works. It was a course of physics, too, that bore a striking resemblance, however the science was actually taught, to large parts of the traditional course in practical mathematics. Consequently, by the end of the century in many universities and colleges, if not everywhere, professors of mathematics confined their attention to pure mathematics: they taught techniques (including conic sections and calculus) that could be applied to the study of the natural world, but they themselves no longer taught a mathematical physics under the guise of practical mathematics. Now that physics was an analytical, and not a causal, science, their previous role was redundant. When the University of Coimbra was reorganized in 1772 this new dispensation, uniquely, was even institutionalized. Pombal established a separate faculty of mathematics distinct from the faculty of philosophy. In this case, the professorial remit did include the teaching of practical mathematics, but only so far as it related to subjects such as architecture and design.[42]

The physics taught in the classroom was thus a truncated form of the traditional course, even if it were up-to-date. By being redefined as a

---

[40] Jean Torlais, "La physique expérimentale," in René Taton (ed.), *Enseignement et diffusion des sciences au dix-huitième siècle* (Paris: Hermann, 1963), pp. 619–45; Chartier, Compère, and Julia, *L'Education en France,* pp. 274–5; Roderick W. Home, "The Notion of Experimental Physics in Eighteenth-Century France," in J. C. Pitt (ed.), *Change and Progress in Modern Science* (University of Western Ontario Series in Philosophy of Science, 27; London, Ontario: D. Reidel, 1985). Good documentation survives about setting up a new *cabinet de physique* at Bourges in 1779: see Archives Départmentales du Cher, D 358 (college accounts).

[41] A physics course in France did contain brief remarks on electricity and magnetism; at this juncture the course became descriptive rather than analytical.

[42] Carrato, "Enlightenment in Portugal," pp. 380–90.

phenomenological and mathematical science that studied forces, it could no longer contain those aspects of the natural world that had not yet been successfully brought under the Newtonian umbrella because they dealt with internal and thus invisible and unmeasurable changes rather than superficial and visible ones. Despite Newton's own optimism about the future extension of his research program, large areas of the natural world remained closed to Newtonian analysis, especially when their study involved change across long intervals of time. These other parts of the traditional course usually ceased to be covered by the arts curriculum. Only very occasionally were they taught as separate subjects under a new or nearly new disciplinary heading such as chemistry, geology, or natural history.

Courses in such subjects were generally established under the arts umbrella only in Sweden and northern Italy and in the universities and university colleges of England and North America where the faculty structure had broken down or had yet to come into existence. Cambridge, uniquely, had a chair in geology (the Woodwardian) as early as 1731. In Sweden, chairs in chemistry were founded in all the country's arts faculties in the second half of the century, starting with the appointment of the mineralogist Johann Gottschalk Wallerius (1709–1785) to one at Uppsala in 1750. The arts faculty at Padua had a chair in chemistry from 1759 and another in experimental agriculture from 1761, whereas a chair in natural history was founded at Pavia in 1771 and entrusted to the zoologist Lazaro Spallanzani (1729–1799).[43]

In general, students who wanted to find out about the forgotten parts of the traditional physics course had to migrate to a medical faculty, where, it will be recalled, one of these subjects, chemistry, was already being taught descriptively in 1700. Over the eighteenth century there was a veritable explosion in the number of chemistry courses given in medical faculties. At the turn of the century hardly any faculty offered permanent tuition in the subject; on the eve of the French Revolution faculty chairs were commonplace. In Germany, for instance, there were only six medical professors entrusted with teaching the subject in 1720; by 1780 there were twenty-eight. In western Europe, only Spain remained poorly provided. The courses, too, became lectures in the general theory and practice of chemistry, in which as much attention might be paid to the subject's agricultural as to its pharmaceutical use. In the hands of professors such as William Cullen (1710–1790) and Joseph Black (1728–1799) at Edinburgh, chemistry in the medical faculties became a novel analytical university science devoted to the structure of matter; its principles, like those of physics, were explained through a series of illustrative experiments.[44]

[43] Gunther, *Cambridge Science*, pp. 424–35; Lindroth, *Uppsala*, p. 102; Vaccari, *Pavia*, pp. 174, 177–81; Ghetti, "Struttura di Padova," pp. 87, 89; Meinel, "Chemistry's Place in Universities," pp. 98, 104.

[44] Karl Hufbauer, *The Formation of the German Chemical Community* (Berkeley: University of California Press, 1982), pp. 33–4; Alberto Elena, "Science in Spanish Universities," in Giuliano Pancaldi

The teaching of natural history, on the other hand, was a novel departure for the faculties of medicine whose traditional interest in nonhuman life forms had been limited to the plant realm. From the mid-eighteenth century, however, a number of chairs in botany were given an extended brief. In a handful of faculties separate chairs of natural history were created, whose professors gave courses in a variety of subjects that would eventually become the distinctive disciplines of meteorology, hydrography, geology, mineralogy, and zoology. Because all these new courses were laboratory-based and instruction was heavily dependent on visual aids, their establishment usually required a heavy financial commitment on the faculty's part. Botanical gardens had to be relaid or founded for the first time, and a natural-history cabinet, analogous to the physics professors' collection of instruments, created and housed. The fullest collections, such as the one built by John Walker (1731–1803), professor of natural history at Edinburgh from 1779, were gradually rechristened museums and became centers of private lay contemplation as much as professorial repositories. Nonetheless, the heavy investment frequently bore a rich fruit, for many eighteenth-century professors of botany and natural history were highly creative scientists who added more than most university post-holders to the sum of knowledge. This was particularly true of the Uppsala professor Carl Linnaeus (1707–1778), who relatively early in his life devised a revolutionary system of botanical classification, based on the sexual characteristics of plants, which was adopted by virtually every university naturalist in the second half of the century. Arguably, the Uppsala professor's achievement was possible only because of his access to the University's botanical garden, where from the late 1720s he taught, researched, organized a team of assistants, and even lived for many years surrounded by his herbarium and his collection of shells.[45]

The growing number of courses – in what were to be later known as the earth and life sciences, especially courses in chemistry – that became available in faculties of medicine over the century did not merely reflect their establishment in long-existent centers of medical teaching. Rather, it reflected the dramatic expansion in the number of active faculties of medicine. At the beginning of the eighteenth century there was only a handful of dynamic faculties, notably Padua, Paris, Montpellier, and Leiden. Most faculties of medicine were dormant, with a handful of students at best and one or two professors. As the century progressed, however, medicine became an entrenched subject in many more universities – even the University of Buda (formerly at Nagyzombat) had a faculty from 1769 – and a number of these new or

---

(ed.), *Le università e le scienze. Prospettive storiche e attuali* (Bologna: Aldo Martello, 1993), p. 109; John R. Christie, "The Rise and Fall of Scottish Science," in Crosland (ed.), *Emergence*, pp. 111–26; Arthur L. Donovan, *Philosophical Chemistry and the Scottish Enlightenment: The Doctrines and Discoveries of William Cullen and Joseph Black* (Edinburgh: Edinburgh University Press, 1975).

[45] Grant, *Edinburgh*, 1, appendix k; Lindroth, *Uppsala*, pp. 94, 130; Tore Frängsmyr (ed.), *Linnaeus: The Man and His Work* (Berkeley: University of California Press, 1983).

rejuvenated faculties became important centers of medical learning, in particular Halle, Göttingen, Edinburgh, Vienna, and Valencia.[46]

The expansion in the number of operational medical faculties reflected in
turn the growing demand for trained medical practitioners in an age increasingly obsessed with health and longevity. It must be stressed, though, that this
was a European phenomenon not reflected on the other side of the Atlantic.
In Spanish America the medical faculties remained moribund. The University
of Mexico had three medical chairs from 1620, but no dissections seem to
have been performed consistently in the eighteenth century before the 1770s;
at the turn of the nineteenth century, the medical chairs of the University of
Lima were empty. In North America the situation was little better. If medicine was being taught in a number of university colleges by 1800, no instruction at all in the subject was being given before the foundation of a chair at
the College of Philadelphia in 1765.[47]

The rapid expansion in the number of faculties of medicine, however,
scarcely explains why they became the centers for teaching the earth and life
sciences in the eighteenth century. In part, this must reflect the fact that all
except chemistry were still primarily descriptive sciences and their practitioners' primary object was to perfect systems of classification in imitation of
those already developed in botany. The development made sense, too, given
the fact that most of the subjects forming the science of natural history were
taught visually with the help of specially selected and prepared specimens.
Consequently, they were pedagogically closely allied to anatomy, another "ancillary" medical science of growing importance in the eighteenth century, in
which lectures were illustrated not only through the dissection of cadavers
(still, in the eighteenth century, often difficult to acquire) but also through
pickled bodily parts that were stored cheek by jowl with objects of natural
history in the nascent university museum.[48]

The association between medicine and the burgeoning sciences of chemistry and natural history, moreover, was strengthened further toward the
end of the eighteenth century as the leading faculties restructured the medical

[46] Bajko, "Hungarian Formal Education," p. 190; Frank T. Brechka, *Gerard Van Swieten and His World,*
1700–1772 (The Hague: M. Nijhoff, 1970), pp. 134–70 (on the reform of the Viennese faculty); Lisa
Rosner, *Medical Education in the Age of Improvement: Edinburgh Students and Apprentices, 1760–1820*
(Edinburgh: Edinburgh University Press, 1991); Salvador Albinana, *Universidad e Ilustración: Valencia*
en la época de Carlos III (Valencia: University of Valencia Press, 1988). There are no modern studies
of the Halle and Göttingen faculties.

[47] John T. Lanning, *Academic Culture in the Spanish Colonies* (Port Washington, NY: Kennikat Press,
1940), pp. 109, 131–2; Lanning, *The Royal Protomedicato: The Regulation of the Medical Professions in*
the Spanish Empire (Durham, NC: Duke University Press, 1985), p. 327; Gregorio Weinberg, "The
Enlightenment and Some Aspects of Culture and Higher Education in Spanish America," in Leith
(ed.), *Facets,* p. 512; Boorstin, *Colonial Experience,* chaps. 34–7 (on the nonacademic nature of pre-
Revolutionary American medicine).

[48] On the difficulty of finding bodies, even at a leading medical center such as Montpellier, see Jean-
Antoine Chaptal, *Mes Souvenirs,* ed. A. Chaptal (Paris: E. Plon, 1893), pp. 16–17.

curriculum in the light of Enlightenment demands that medicine become a much more evidence-based science. In the first part of the century anatomy and botany were still seen as the handmaidens of surgery and pharmacy, and they had little pedagogical connection with the study of the component parts of theoretical medicine – physiology, hygiene, semiotics, pathology, and therapeutics. Although many faculty professors, such as Alexander Munro Primus (1697–1767) at Edinburgh, Bernard Siegfried Albinus (1697–1770) at Leiden, and the Dane Jacques-Bénigne Winslow (1669-1760) at the Paris Jardin du Roi, were now skillful and innovative hands-on anatomist-physicians, their lessons in the main continued to be primarily given for the benefit of trainee surgeons. Inevitably, the author of the most successful anatomy textbook in the eighteenth century, the Altdorf and Helmstedt professor Lorentz Heister (1683–1758), was also the author of an equally renowned surgical manual, which would eventually be translated into Japanese![49]

From 1750, however, Enlightened medical reformers everywhere in Europe began to agitate for the proper integration of the ancillary medical sciences into the faculty curriculum. While also insisting that the facilities for teaching these sciences continue to be improved, the reformers principally sought to make their study *de rigueur* for the tyro medical student. These developments culminated in a report presented to the French National Assembly in 1790 by one of France's leading Enlightened physicians, Félix Vicq d'Azyr (1748–1794), in which it was suggested that medical training should be lengthened to six to seven years from the traditional three to four, so that room could be made for a compulsory two-year introductory course in anatomy, botany, chemistry, mineralogy, and zoology. Vicq's reform plans, with certain modifications, informed the structure of the curriculum of the new Paris medical school, eventually established in 1795, which dominated medical science in the first three decades of the nineteenth century. In the new school not only were students obliged to study a wide range of ancillary sciences, including natural history and chemistry, but also the central importance of anatomy in particular to medical training was emphasized by its curricular association with physiology. Traditionally taught in France and elsewhere as two separate subjects,

---

[49] Heister had studied under the Dutch anatomist Frederick Ruysch (1638–1731), who was the anatomy lecturer of the Amsterdam guild of surgeons. There is no overview of eighteenth-century faculty anatomy teaching, but many details can be garnered from studies of individual faculties or universities – for example, Charles Webster, "The Medical Faculty and the Physic Garden," in *History of the University of Oxford*, vol. 5, pp. 703–7; Bill, *Education at Christ Church*, pp. 313–26; Christopher Lawrence, "Ornate Physicians and Learned Artisans: Edinburgh Medical Men, 1726–1776," in W. F. Bynum and Roy S. Porter (eds.), *William Hunter and the Eighteenth-Century Medical World* (Cambridge University Press, 1985), chap. 6 passim; Brockliss, *French Higher Education*, pp. 74–5, 396–9; Louis Dulieu, *La Medecine à Montpellier*, 3. *L'Epoque Classique*, 2 pts., 1 vol. (Avignon: Presse universelle, 1983–6), pt. i, pp. 332–62; Addy, *Salamanca*, pp. 103–8, 123, 176; Michael E. Burke, *The Royal College of San Carlos: Surgery and Spanish Medical Reform in the Late Eighteenth Century* (Durham, NC: Duke University Press, 1977), pp. 47–54. The Paris faculty was precocious in demanding from the 1730s that its graduands take a practical anatomy and surgery examination.

in early nineteenth century Paris anatomy and physiology were treated as a single discipline and entrusted to a single professor.[50]

The nonmathematized sciences, then – to the extent that they dealt with the structure of matter, the history of the Earth, and the classification of its life forms – were deemed as the eighteenth century progressed to be best studied as part of medicine rather than natural philosophy. This belief must have carried even more conviction among the many medical scientists of the second half of the eighteenth century who embraced vitalist medical philosophies. The mechanical philosophers of the preceding century had looked forward to the time when medicine, too, would become a mechanical science. Descartes's mechanical account of human physiology, the posthumously published *De l'homme,* had actually been written before he composed his major statement of mechanical physics, the 1644 *Principia.* Satisfactorily reducing life to a mechanism, however, proved impossible, and from the mid-eighteenth century professors at most leading faculties, such as Paul-Joseph Barthez (1734–1806) at Montpellier, were arguing that organic matter operated according to its own, mathematically irreducible laws. The new medical philosophy was succinctly summarized by one of Barthez's pupils, Jean-Antoine-Claude Chaptal (1756–1832), the chemist and Napoleon's interior minister.

> The laws of mechanics, hydraulics and chemical affinities act on all matter; but in the case of the animal economy, they are so completely subordinate to the laws of vitality that their effect is almost nil; and dependent on the intensity of that vitality, so living phenomena distance themselves further and further from the results calculated according to those [physical] laws.[51]

Although this was never a universal belief among medical scientists – Vicq d'Azyr, for one, was a strenuous opponent – it probably was the majority perception at the turn of the nineteenth century, promoted by the leading lights of medical science, such as the iconic hero of the new Paris school, the short-lived anatomist Marie-François-Xavier Bichat (1771–1802).[52]

This being so, it was inevitable that the medical faculties would draw into their orbit the unmathematized and perhaps unmathematicizable parts of

---

[50] Félix Vicq d'Azyr, "Nouveau Plan de constitution pour la médecine en France," in *Histoire et mémoires de la Société Royale de Médecine: Années 1787–1788,* 9 (Paris, 1790), pp. 17, 19, 41–5 (contains a description of restructured medical courses already established in other countries); Erwin Ackerknecht, *Medicine at the Paris Hospital, 1794–1848* (Baltimore, MD: Johns Hopkins University Press, 1967), p. 35.

[51] Chaptal, *Souvenirs,* pp. 19–20.

[52] Elizabeth Haigh, *Xavier Bichat and the Medical Theory of the Eighteenth Century* (*Medical History,* suppl. 4; London: Wellcome Institute for the History of Medicine, 1984); François Duchesneau, *La Physiologie des lumières: empirisme, modèles et théories* (The Hague: M. Nijhoff, 1982); Laurence W. B. Brockliss and Colin Jones, *The Medical World of Early Modern France* (Oxford: Clarendon Press, 1997), chap. 7, sect. B and C; Michel Foucault, *Naissance de la clinique* (Paris: PUF, 1963); John E. Lesch, *Science and Medicine in France: The Emergence of Experimental Physiology, 1790–1855* (Cambridge, MA: Harvard University Press, 1984); Elizabeth A. Williams, *The Physical and the Moral: Anthropology, Physiology and Philosophical Medicine in France, 1750–1850* (Cambridge University Press, 1994).

natural philosophy, regardless of these subjects' utility for medical students. Abandoning the old Aristotelo-Galenic and mechanist attempt to explain the internal and external cause of disease, vitalists were particularly important in developing the new classificatory studies of morbid anatomy and nosology which had close affinities with contemporary botany and zoology. At the same time, the vitalists' view that the forces governing physiological changes were *sui generis* gelled neatly with the antimechanist prejudices of many chemists and proto-evolutionist geologists, such as George Louis Leclerc, Comte de Buffon (1707–1788), intendant of the Paris *Jardin du roi,* who equally felt that the reactions and terrestrial changes they studied could not be reduced to mathematical laws.[53]

THE EXPANSION IN PROVISION

The eighteenth century not only saw important changes in the content and articulation of science teaching in the traditional system of higher education, but also witnessed a dramatic expansion in the overall provision of instruction.

In the first place, this was the consequence of the establishment of a plethora of publicly and privately funded alternative centers of institutionalized learning, where scientific subjects had a much higher and often dominant profile. Most of these new schools had a mathematical bias and reflected the growing demand for some form of training in practical and theoretical mathematics among entrants to careers whose adepts had seldom or never graced the portals of college or university. In the eighteenth century, military and naval officers, officers in the merchant marine, accountants, surveyors, engineers, merchants, and even artists (a rapidly expanding group in a consumerist age) began to seek formal education in mathematical science, in part as a means more to clearly define and advance their incipient professional identity. Because the practitioners of these arts were unwilling to gain this knowledge through attending traditional institutions of learning (with their emphasis on a lengthy formation in classical culture), a completely new network of largely private schools and academies sprang up to serve their needs. Little is known in detail about the curriculum of these schools, but it is clear that they offered a "modern" alternative to the classical diet of the university's feeder schools. Most of them provided tuition in modern languages as well as scientific subjects, but a handful – such as Berthaud's academy in Paris, which was attended by the Jacobin Lazare Carnot – were mathematical crammers. Some, too, offered highly specialized training, such as the muncipally-funded French *écoles de dessin,* which aimed to boost the quality of the country's

---

[53] Jacques Roger, *Buffon: Un philosophe au jardin du roi* (Paris: Fayard, 1989).

decorative arts by giving young artisans lessons *inter alia* in perspective and anatomy.[54]

In the second half of the eighteenth century, this new alternative network of much more scientifically oriented educational institutions received the imprimatur of the state with the establishment of a number of state-funded *collegia nobilium* and specialist mathematical schools. As we have seen, the Jesuits in particular had begun to offer instruction in practical mathematics to army and naval officers in the preceding century. Less rigorous education along these lines was also available in the many, usually ephemeral, privately run noble academies that sprang up across the continent in the seventeenth century to teach the wealthiest members of the nobility the courtly and military arts.[55] In the eighteenth century, the state came to see institutionalized military training as of vital importance and began to found its own military academies to provide subsidized education especially for the poorer nobility, who were the mainstay of the officer corps. Classic examples were the Polish Knights' School (where the patriot Kosciusko was trained), the Austrian *Theresianum* and other Viennese state-sponsored noble schools, and Pombal's short-lived Portuguese noble academy.[56] Furthermore, from 1750, some states began to set up more specialized and technical schools to train artillery officers as well as military, civil, and mining engineers. The most famous mining academy was established in 1765 at Freiberg in Saxony and for forty years included among its professors the geologist Abraham Gottlob Werner (1750–1817). By the turn of the nineteenth century there was even a flourishing mining school in Mexico City, much praised by the naturalist Alexander von Humboldt (1769–1859).[57] In the late eighteenth century, the system of government-sponsored technical schools was most advanced in France, where the state had established a prestigious officer-training school, the *Ecole militaire* (1751), an artillery academy at La Fère (1758), specialist military, civil, and mining engineering schools – the *Ecole des Ponts et Chaussées* (1747), the *Ecole de Mézières* (1748), and the *Ecole de Mines* (1783) – and thirteen noble feeder academies (1776). Unlike most of France's higher-educational institutions, these state-run academies survived the Revolution, albeit in modified form, and the system was perfected in 1795 with the foundation of the Paris *Ecole Polytechnique,* an

---

[54] Nicholas Hans, *New Trends in Education in the Eighteenth Century* (London: Routledge, 1951); Philippe Marchand, "Un modèle éducatif original à la veille de la Révolution," *Revue d'histoire moderne et contemporaine*, 22 (1975), 549–67; V. Advielle (ed.), *Journal professionel d'un maître de pension de Paris au XVIIIe siècle* (Pont-l'Evêque, 1888) on Berthaud; Harvey Chisick, "Institutional Innovation in Popular Education in Eighteenth-Century France: Two Examples," *French Historical Studies*, 10 (1977), especially 47–58, on the *école de dessin* at Amiens.

[55] Norbert Conrads, *Ritterakademien der Frühen Neuzeit. Bildung als Standesprivileg im 16. und 17. Jahrhundert* (Göttingen: Vandenhoek and Ruprecht, 1982); none was founded in the British Isles.

[56] Seidler, "Polish School System," pp. 340–1; Helmut Engelbrecht, *Geschichte der österreichischen Bildungswesen: Erziehung und Unterricht auf dem Boden Österreichs*, 3 vols. (Vienna: Österreich Bundesverlag, 1982–4), vol. 3, pp. 181–5; Carrato, "Enlightenment in Portugal," pp. 371–5.

[57] José-Luis Peset, "Los originés de la ensenenza téchnica en América: el collégio de minéria de Mexico," in *Universidades espanolas.*

advanced military technical college that prepared scholarship boys for even more specialist training.[58] Among the great powers, only Britain did not invest in these new professional training schools.

These new state academies gave instruction primarily in theoretical and practical mathematics. To this extent they complemented the courses in mathematics given in the colleges and universities, but their teaching was at a much higher level and the range of subjects more extended. Understandably, in the mining academies, the new, if still unstable, science of chemistry, unshackled from its traditional association with faculties of medicine, had as important a place in the curriculum as practical mathematics. Chemistry teaching, too (along with natural history), was often institutionalized in another set of state or municipally financed institutions established in the course of the century: the schools and colleges of surgery and pharmacy. Set up in most European states (but again not in Britain) to provide high-quality formal education for tyro (barber) surgeons and apothecaries, who had been traditionally trained through a system of apprenticeship, a number of these schools, especially in France and Spain, were important centers of scientific education by the end of the century. In 1811 the model crossed the Atlantic when the college of San Fernando was founded in Lima thanks to successful lobbying by José Hipólito Unánué, the leading medical reformer in Spanish America.[59]

In the second place, and of even greater significance, the expansion in the provision of science teaching in the eighteenth century was the result of the efflorescence of an ever-growing, diffuse, and complex network of impermanent and *ad hominem* lecture courses. Here again the state played an important sponsorship role. The scientific academies – some eighty in number – that were gradually established under the state's aegis in the course of the eighteenth century were intended to be embryonic research institutes where experimental philosophers could meet to exchange opinions and vet new scientific ideas. They were not intended to be centers of teaching. Indeed, their creation was meant to institutionalize the dichotomy between creative research and the dissemination of scientific knowledge that had implicitly emerged in the seventeenth century in light of the university's guarded reception of the

---

[58] G. Serbos, "Ecole Royale des Ponts et Chaussées"; A. Birembaut, "L'Enseignement de la minéralogie et des techniques minières"; René Taton, "Ecole Royale de Génie de Mézières"; Roger Hahn, "L'Enseignement scientifique aux écoles militaires et d'artillérie": all in *Enseignement et diffusion des sciences au XVIIIe siècle*, pp. 345–63, 365–418, 513–45, 559–66; R. Laulan, "L'Enseignement à l'Ecole Royale Militaire de Paris, de l'origine à la réforme du comte de Saint-Germain," *Information historique*, 19 (1957), 152–8; Terry Shinn, *L'Ecole Polytechnique 1794–1914* (Paris: Presses de la Fondation nationale des Sciences Politiques, 1980); Bruno Belhoste, "Les Origines de l'Ecole Polytechnique. Des anciennes écoles d'ingénieurs à l'Ecole Centrale des Travaux publics," in Dominique Julia (ed.), *Les Enfants de la Patrie*, special no. of *Histoire de l'Education*, 42 (1989), 13–54; Ivor Grattan-Guinness, "*Grandes Ecoles, Petite Université*: Some Puzzled Remarks on Higher Education in Mathematics in France, 1795–1840," *History of Universities*, 7 (1988), 197–225.

[59] Brockliss and Jones, *Medical World of Early Modern France*, pp. 506–9; Toby Gelfand, *Professionalizing Modern Medicine: Paris Surgeons and Medical Science and Institutions in the Eighteenth Century* (London: Greenwood, 1980), especially chaps. 4–5; Burke, *The Royal College of San Carlos: Surgery and Spanish Medical Reform*, pp. 59–63 and chap. 4; Lanning, *Royal Protomedicato*, pp. 332–49.

Scientific Revolution. Nonetheless, many academies did begin to sponsor instruction in the sciences. Only one, the St. Petersburg Academy (established in 1724), actually became a quasi-university (a reflection of the absence of the institution in Russia before the creation of the University of Moscow in 1755), but it was commonplace for academies to patronize courses in the descriptive sciences of anatomy, botany, and chemistry, again encroaching on the traditional monopoly of the medical faculty. The academy at Bologna, founded in 1714 with five professorial chairs, was particularly important in this respect. Thus the academies were able to establish their credentials in the nascent public sphere and shed an otherwise elitist and self-referential image. By offering courses in the more accessible sciences, which pandered to the growing amateur Rousseauian interest in botany and provided artisan entrepreneurs with information about the industrial uses of chemistry, the academies proved themselves to be ardent champions of public utility.[60] Public courses in chemistry particularly abounded. By the end of the eighteenth century, they were not only organized by scientific academies but also by other state institutions. In France on the eve of the Revolution, the chemist and dye manufacturer Chaptal was employed to teach the subject by the Estates of Languedoc. In Spain, chemistry courses were sponsored by a number of the new economic societies set up in the reign of Charles III (d. 1788).[61]

However, the public courses patronized by the state were only the tip of the iceberg. The key to the development was the explosion in the number of private courses offered. Some private tuition in the sciences had always existed – medieval alchemists as much as Renaissance experimental philosophers had trained assistants – but there is no evidence before the final quarter of the seventeenth century that the relationship was any other than one of master and apprentice. It was only from about 1670 that the first signs of a much less intimate form of fee-based private instruction made its appearance, as was noted earlier in the case of Rohault. In the eighteenth century, this trickle became a flood. At least in the wealthier parts of Europe, the provision of private science courses was one of the most dynamic sectors of the service economy, as an ever-growing band of entrepreneurial teachers attempted to benefit from (but also stimulate) the new Enlightenment-generated interest in scientific knowledge.

Private courses principally took two forms. On the one hand, private tuition was provided in a range of ancillary medical subjects for tyro physicians,

---

[60] James E. McClellan III, *Science Reorganized: Scientific Societies in the Eighteenth Century* (New York: Columbia University Press, 1985); Daniel Roche, *Le Siècle des lumières en province: Académies et académiciens provinciaux, 1680–1789*, 2 vols. (Paris: Mouton, 1978), vol. 1, pp. 124–31; G. A. Tishkin, "A Female Educationalist in the Age of the Enlightenment: Princess Dashkova and the University of St. Petersburg," *History of Universities*, 13 (194), 137–50; Marta Cavazza, "L'insegnamento delle scienze sperimentali nell'istituto delle scienze di Bologna," in Pancaldi (ed.), *Le università e le scienze*, pp. 155–79.

[61] Chaptal, *Souvenirs*, pp. 25–7, 32–3; Robert J. Shafer, *The Economic Societies in the Spanish World, 1753–1821* (Syracuse, NY: Syracuse University Press, 1958).

surgeons, apothecaries, and even midwives as a supplement to the existing institutionalized provision (both new and old). Although the medical faculties and the new surgical and pharmaceutical colleges paid much more attention to these subjects in the course of the century, they nonetheless seldom offered students the chance to gain any "hands-on" experience. Private teachers, who were usually leading local practitioners, supplied a deficiency that the faculties would not or could not (from lack of facilities) supply. In the first half of the eighteenth century the Mecca for private instruction in anatomy and surgery especially was Paris, where a number of hospital surgeons with easy access to cadavers offered students from all over Europe the chance to learn the art of dissection. The most famous teacher was the surgeon-general at the Charité hospital, Henri-François Ledran (1685–1770), whose pupils included the future Göttingen professor, naturalist, and physiologist Albrecht von Haller. In the second half of the eighteenth century, even though Paris remained an important center of extracurricular medical education, the torch was passed to London. In the reign of George III the British capital was awash with private anatomy schools, none more famous than the Great Windmill Street Academy of the Paris-trained William Hunter (1718–1783).[62]

On the other hand, and more important in the context of the dissemination of scientific knowledge *tout court,* private courses were also offered in physics in ever-growing numbers. These courses were aimed specifically at those who had not had the opportunity of studying the subject in a public institution and were intended for women as well as men. Short – they seldom lasted more than a couple of weeks – and nonrecurrent, private courses were offered in any town where an entrepreneur felt there was a large enough pool of affluent subscribers, with the result that in England they were more likely to be proposed in Bath than Birmingham. From the beginning, private lecturers were peripatetics who traveled from town to town to ply their trade, even if the most successful were able to provide themselves eventually with permanent premises in large cities. A few were international celebrities, notably the Englishman Stephen Demainbray (1710–1782), who gave private courses in physics on both sides of the Channel and boasted his membership in a clutch of foreign academies. At least one lecturer was a woman – the Bolognese Laura Bassi (d. 1778), who taught physics in her own home from 1750 until her death.[63]

---

[62] Susan C. Lawrence, *Charitable Knowledge: Hospital Pupils and Practitioners in Eighteenth-Century London* (Cambridge University Press, 1996); Gelfand, *Professionalizing Modern Medicine,* especially chaps. 6–7; Urs Boschung (ed.), *Johannes Gessners Pariser Tagebuch 1727* (Bern: Hans Huber, 1985); Roy Porter, "William Hunter: A Surgeon and Gentleman"; Toby Gelfand, "'Invite the Philosopher as Well as the Charitable': Hospital Teaching as Private Enterprise in Hunterian London," both in Bynum and Porter (eds.), *William Hunter and the Eighteenth-Century Medical World,* pp. 7–34, 129–51.

[63] Roy S. Porter, "Science, Provincial Culture and Public Opinion in Enlightenment England," *British Journal for Eighteenth-Century Studies,* 3 (1980), 20–46; Larry Stewart, *The Rise of Public Science: Rhetoric, Technology, and Natural Philosophy in Newtonian Britain, 1660–1750* (Cambridge University Press, 1992); J. Golinski, *Science as Public Culture: Chemistry and Enlightenment in Britain, 1760–1820*

As with the course of physics that evolved in the universities as the eighteenth century progressed, the content of these private courses was generally restricted to the subject areas of modern physics. A typical series of lectures would begin with mechanics and hydraulics and proceed through hydrostatics, pneumatics, magnetism, and electricity to astronomy. Attention was seldom paid, except by specialist lecturers such as the English itinerant Peter Shaw (1694–1763), to chemistry or the life sciences. The limited powers of concentration and the mathematical illiteracy of their intended audience, however, ensured that these private courses in physics would be extremely down-market events. Like the most common form of the university course, the private courses were inevitably lectures on experimental, and not mathematical, physics. But they were much more simplistic, theatrical, and colorful than their university counterpart. Although teachers, such as the London-based FRS James Ferguson, stressed in their prospectuses that their intention was to instruct their audience, they also emphasized that they aimed to please. Ferguson, for instance, in 1767 promised his audience that the part of his course dealing with electricity would contain a "variety of curious and entertaining experiments," which would include "giving gentle shocks, turning little Mills with Paper Vanes, striking Holes through Cards, electrifying *plus* and *minus,* Ringing of Bells, causing Cork Balls with linen threads for legs to move like spiders."[64]

It was only at the turn of the nineteenth century that the private course metamorphosed into a more serious form of instruction with the foundation in London of the Harvard-educated Count Rumford's Royal Institution. Staffed by scientists of the eminence of Humphry Davy (1778–1829) and Thomas Young (1773–1829), it offered courses to the general public that were far more demanding than the fare peddled by private lecturers in physics.[65]

As a result of the burgeoning number of private and public courses in science, a much larger section of the population, both male and female, must have received some form of tuition in the subject than would otherwise have been possible. Nonetheless, the primary beneficiaries of this expansion in provision would still have belonged to the elite. Private lectures in physics in particular were only for the socially select: in England they never cost less

---

(Cambridge University Press, 1992); Sutton, *Science for a Polite Society,* chaps. 6–8; Alan Q. Morton and James A. Wess, *Public and Private Science: The King George III Collection* (Oxford: Oxford University Press in association with the Science Museum, 1996) on Demainbray; John R. Millburn, *Benjamin Martin: Supplement* (London: Vade-Mecum Press, 1993) on another English itinerant professor; Jean Torlais, *L'Abbé Nollet: un physicien au siècle des lumières* (Paris: Societé des journeaux et publications du Centre, 1954) for a French example; Cavazza, "L'insegnamento delle scienze," p. 156 for Bassi.

[64] Museum of the History of Science, Oxford, collection of course prospectuses (chiefly English), unclassified: *Syllabus of Lectures* by James Ferguson FRS. A good idea of the content of these private courses can be gleaned from the textbooks their authors often produced: e.g., Stephen Demainbray, *A Short Account of a Course of Natural and Experimental Philosophy* (London, 1754).

[65] Morris Berman, *Social Change and Scientific Organization: The Royal Institution, 1799–1844* (London: Heinemann, 1978).

than a guinea (21 shillings or £1 1s 0d) per series. But even private lectures in medicine were beyond the pocket of all but the most affluent students. When Guillaume-François Laënnec, the uncle of René-Théophile-Hyacinthe (1781–1826), inventor of the stethoscope, studied in Paris in the 1760s, the average cost of attending an extracurricular course was £3 to £4 sterling.[66] Not surprisingly, the first generation of British engineers, such as Thomas Telford and George Stephenson, were autodidacts.

It was only at the turn of the nineteenth century, at the moment "the people" appeared on the political stage for the first time in the French Revolution, that serious attempts were made to provide scientific education for the working class. In Glasgow in 1799, as a result of a bequest left by the University's professor of natural philosophy John Anderson (1726–1796), the Andersonian Institution was opened. It offered courses similar to those provided by the University but with a more practical orientation. More important, in 1817 the London Mechanics' Institute was set up, with the deliberate aim to disseminate scientific knowledge among the working class. Although short-lived itself, it was the prototype for the scores of Mechanics' Institutes that were created throughout Great Britain in the 1820s. Other countries slowly followed Britain's lead. In the 1790s, attempts by French educational reformers to devise a new college and university curriculum that would be both science-oriented and open to all those able to benefit from higher study failed miserably. With the Restoration, the state adopted the British solution and fell back on creating alternative institutions for working-class technical education, the most notable being the Paris *Conservatoire des Arts et Métiers,* where two thousand were attending evening classes by 1824.[67]

CONCLUSION

Enough has been said to demonstrate that science teaching underwent a dramatic upheaval in the eighteenth century. To begin with, both the articulation and the content of the traditional university and college course in the science of physics were refashioned. Natural philosophy ceased to be conceived as a causal science, and physics was divided into two sets of scientific disciplines whose subject matter much more accurately reflected the state of contemporary experimental philosophy. One set was still called physics and

---

[66] Alfred Rouxeau, *Un étudiant en médecine quimpérois: Guillaume-François Laënnec aux derniers jours de l'ancien régime* (Nantes: L'Imprimerie du "Nouvelliste," 1926), pp. 35–78 passim.

[67] John Fletcher Clews Harrison, *Learning and Living, 1790–1860: A History of the Adult Education Movement* (London: Routledge, 1961), pp. 57–74; Robert Roswell Palmer, *The Improvement of Humanity: Education and the French Revolution* (Princeton, NJ: Princeton University Press, 1985), especially chap. 4; Frederick B. Artz, *The Development of Technical Education in France, 1500–1850* (Cambridge, MA: Society for the History of Technology, 1966), especially pp. 216–7; André Prévot, *L'Enseignement technique chez les Frères des écoles chrétiennes aux XVIIIe et XIXe siècles* (Paris: Ligel, 1964).

was taught in the arts or philosophy faculty, but the subject was much reduced in its scope and essentially covered the material contained in Aristotle's
Physics and *De caelo*, now studied mathematically and phenomenologically
under the influence of Newton. The other set consisted of a group of ill-
defined and as yet unstable sciences corresponding to the material in Aristotle's
other books: chemistry, anatomy, geology, mineralogy, botany, and zoology
(the last four often subsumed under the heading of natural history), which were
not mathematicized and tended to be taught for epistemological and utilitarian reasons in the medical faculty.

At the same time, the proportion of the population who had the opportunity to receive tuition in the natural sciences grew exponentially over the
century. At the turn of the eighteenth century the teaching of natural philosophy was almost totally confined to the university world, and few people
beyond male members of the professional elite (clergymen, lawyers, and physicians) could have formally studied the subject. As the century progressed, an
increasing number of alternative educational institutions (both public and
private) were established that offered tuition in the new physics and the
embryonic earth and life sciences, and an even wider (and female as well as
male) constituency was reached through the explosion in the number of private, ephemeral, and peripatetic courses. The turn of the century, too, saw
the first attempts to provide a scientific education for the working class.

Several points can be made about the deeper significance of this twin development. First, the changes ensured that the new science moved from the
periphery to the center of European culture. Although historians of science
have always privileged the seventeenth century as the era of the Scientific
Revolution, it is quite clear that this assumption can hold water only as long
as attention is concentrated on the activities of the small group of experimental philosophers, astronomers, and mathematicians whose investigations
laid the foundations of modern science. If the Scientific Revolution is seen
as a broader cultural period in which the Galilean/Newtonian mathematical
and phenomenological approach to the natural world became part of the
mindset of the European and American elite, then that Revolution occurred
in the eighteenth century (predominantly outside the English-speaking world
after 1750). In the seventeenth century, experimental philosophers were marginalized men often subject to ridicule, striving to gain the approbation of
the prince and court and to legitimize their activities by aping courtly modes
of behavior in their investigative practices.[68] In the eighteenth century their
successors moved out of the shadows and, basking in the iconic status posthumously achieved all over Europe by Newton, obtained an enviable social cachet.

---

[68] Mario Biagioli, "Scientific Revolution, Social Bricolage and Etiquette," in Roy Porter and Mikuláš
Teich (eds.), *The Scientific Revolution in National Context* (Cambridge University Press, 1992), chap. 1;
Steven Shapin, *A Social History of Truth: Civility and Science in Seventeenth-Century England* (Chi-
cago: University of Chicago Press, 1994); Laurence W. B. Brockliss, "Civility and Science: From Self-
Control to Control of Nature," *Sartoniana*, 10 (1997), 43–73.

That this change took place and that the experimental philosophy became the self-conscious hallmark of European civilization can be attributed to an important degree to its permeation of the classroom. Although the seventeenth-century university world may not have been as hostile to the new science as was once thought, it accepted the work of experimental philosophers only on its own terms.[69] The revolution in science teaching that occurred in the eighteenth century, although obviously itself made possible only by a more positive attitude toward the new science on the part of Church and state, helped to create a social elite who worshipped at the altar of the experimental philosophy. Except in France for a few brief years during the Revolution, the acceptance of Newtonian science in the classroom did not displace the primacy of classical culture. However, it did make the cultivation and pursuit of the new science, albeit its nonmathematicized aspects and often at second hand, an acceptable activity for an educated person. The natural-history cabinet, as well as the cabinet of antiquities, could henceforth be safely placed in the library without offending one's friends.[70]

Moreover – to emphasize further the creative role of the university world – the dissemination of the new science, especially its popularization beyond a university and college audience, was greatly facilitated by the way in which the new physics was presented. In the first instance, the construction of a specifically experimental course in Newtonian physics was necessitated by the limited mathematical knowledge of traditional students. In the longer term, however, 's Gravesande and his colleagues created a teaching format that could easily be adapted outside the university world to bring the new physics (and the other sciences) to a wider audience. The new format was essentially theatrical: it made science education an entertainment and was consequently a suitable vehicle for introducing the new science to ladies and gentlemen reared to find happiness in diversion rather than study and taught that excessive concentration was the mark of pedantry. The development of a classroom experimental physics, therefore, was a key moment in making the new science socially respectable. A recondite, cerebral, and highly unconventional way of approaching the natural world (whose adepts could not even demonstrate its much-vaunted utility) was cleverly packaged in an age of consumption as just another consumer product to delight and titillate (irrespective of the pious protestations of professorial entrepreneurs to the contrary).

Thus, the institutionalization of the new science in the classroom formed an essential bridge between the world of the experimental scientist and contemporary elite culture. At the same time, and not just through the posts

---

[69] The most recent discussion is Roy Porter, "The Scientific Revolution and Universities," in Ridder-Symoens (ed.), *History of the University in Europe,* vol. 2, chap. 13.

[70] On the development of collecting across the seventeenth and eighteenth centuries, see Krzystof Pomian, *Collectors and Curiosities: Paris and Venice 1500–1800,* English trans. (Cambridge University Press, 1990); Patricia E. Kell, "British Collecting, 1656–1800: Scientific Enquiry and Social Practice," D. Phil. dissertation, Oxford University, 1996.

established in the new state academies, this institutionalization created a space in which scientists could work. As the new courses in the different branches of natural philosophy came to be taught increasingly by specialist professors, so university and college posts began to be filled for the first time by significant numbers with working scientists. Although the extent to which the seventeenth-century experimental philosopher worked outside the university world can be exaggerated, figures such as Newton were always exceptional. In the eighteenth century, in contrast, scientists were found in growing numbers within the university system, especially holding chairs in mathematics and medicine, but even as professors of experimental philosophy. As we saw, the electrical experimenter Alessandro Volta taught experimental philosophy for many years at the University of Pavia and made good use of its well-equipped physics laboratory to pursue his own work as well as to illustrate his lectures. He saw no contradiction between his roles as researcher and teacher, begging the state to meet his requests for more space and money so that "[I] may devote all my talents to promoting the science I profess and the instruction of young students in it."[71]

Astronomers in particular began to covet posts in the most famous universities as the eighteenth century wore on, and as a number of institutions invested in purpose-built observatories. Before 1700, college and university teachers with astronomical interests had made their observations from the top of the highest available building, although crude observatories were established at Leiden (1632), Copenhagen (1641), and Utrecht (1642). By the end of the eighteenth century, observatories had been erected in at least another twelve universities, including Uppsala, built for Anders Celsius (1701–1744) in 1741; Göttingen and Prague (1751), Oxford (1773), Dublin (1783, the Dunsink Observatory), and Coimbra (1792). In 1800 the best equipped of them all was the Oxford Radcliffe Observatory, installed at the cost of £28,000. Significantly, the university building program outpaced those of other foundations: only two or three observatories of importance were attached to the new scientific academies, and only seven or so royal observatories were built in the course of the century to add to the two extant in 1700 (Paris and Greenwich). Although few university astronomers were fortunate enough to hold specialist chairs in astronomy – most of them were professors of mathematics – they could pursue their observations at the new observatories unhindered. Unlike the physics, chemistry, and natural-history laboratories-cum-museums, the observatories were not primarily teaching spaces. Indeed, they seem to have had no teaching role at all but rather were erected purely to enhance a university's international prestige and emphasize its "modernity."[72]

---

71 Cavazza, "Orti botanici," p. 87.
72 Lindroth, *Uppsala*, pp. 102, 132; Gerard L'E. Turner, "The Physical Sciences," in Brock and Curthoys (eds.), *The History of the University of Oxford*, vol. 5, pp. 679–81; Robert B. McDowell and David Webb, *Trinity College Dublin 1592–1952: An Academic History* (Cambridge University Press, 1982), pp. 64–5. Other information supplied by courtesy of Roger Hutchins.

Admittedly, the connection between science teaching and scientific research still remained casual. Most scientists continued to have nonacademic jobs or no job at all, famous cases in point being the French tax farmer Antoine-Louis Lavoisier (1743–1827) and the anatomist *Wunderkind* Bichat. Many professors of experimental philosophy, too, could be of poor quality, such as the uncharismatic pedant Dr. Forrest, who bored the future Lord Chancellor Campbell at St. Andrews in the 1790s. Indeed, even when a leading scientist did hold an academic post, his talents might not always be effectively used. When the future chemist Chaptal attended the medical faculty at Montpellier, he found to his chagrin that the country's leading chemist, Gabriel-François Venel (1723–1775), was giving a public course in hygiene and not his specialty.[73] Nonetheless, the opportunity for combining teaching and research, now that most universities and many colleges had relatively well-equipped observatories, laboratories, and anatomy theaters, was certainly there.

Furthermore, the institutionalization of the new science in the classroom also played a significant part in the evolution of the experimental philosophy into a series of discrete sciences. If eighteenth-century institutions of higher learning did not originate the designation of the subject areas into which the natural sciences have come to be divided, they certainly helped to define and confirm their epistemology, content, and direction. For two thousand years the categorization of natural philosophy and mathematics had been determined by the surviving oeuvre of the Greeks: Aristotle, Euclid, and, belatedly in the Renaissance, Archimedes, who defined how these sciences were divided. By the eighteenth century the traditional boundaries had dissolved, but new ones commanding general consent had yet to be erected: the term "experimental philosophy" covered a multitude of methodologies and research areas. It was primarily through the unambiguous and prescriptive reconstruction of scientific space in the professorial course and textbook that the relationship among natural philosophy, mathematics, medicine, and their different branches was eventually renegotiated. The establishment of the new science in the classroom was thus an essential step in the emergence of the natural sciences as we know them today.

Above all, the new science of physics was a creation of the university world. Had it been left to the new scientific academies, the term might have remained forever associated with old-style natural philosophy and gradually disappeared from the lexicon. When the French *Académie des Sciences* was divided into sections in 1699, the new science was divided into six categories: three mathematical (geometry, astronomy, and mechanics), and three medical (anatomy, chemistry, and botany). Pointedly, physics was not among them. In 1795, on the other hand, when the Academy was reconstituted as the First Section of the Institut, the new sectionalization of scientific knowledge closely mirrored

---

[73] Hon. Mrs. Hardcastle (ed.), *Life of John, Lord Campbell,* 2 vols. (London: John Murray, 1881), vol. 1, p. 21; Chaptal, *Souvenirs,* pp. 15–16.

developments in the university world. Not only was physics constituted as a separate section, but also a clear distinction was made between the mathematical sciences to which physics belonged and the experimental or classificatory sciences that comprised the subjects taught in the medical faculty.[74]

Of course, the rearranging and repositioning of the branches of natural philosophy effected in the world of learning by 1800 did not completely anticipate the institutionalized division of the natural sciences today. Biology, as a separate science, had yet to be constituted, and the nonmathematicized sciences had to be liberated from the grip of the medical faculty. Their moment of release, however, was not to be long in coming. In the first half of the nineteenth century, as vitalist ideas began to go out of fashion in medical circles, so too did the associated belief that the medical and physical sciences were epistemologically distinct. There was no longer any generic reason why chemistry and the other "ancillary medical sciences" should find their exclusive home in the medical faculty. At the same time, there were good practical reasons why chemistry in particular should be moved to another faculty. Long before the end of the eighteenth century it was recognized that the subject had an industrial and agricultural, and not just a pharmaceutical, application. Inevitably, in many early nineteenth-century medical faculties, such as that of Glasgow, the course in chemistry had to be tailored to meet the needs of a disparate and largely nonmedical clientele. In addition, professors in these ancillary sciences began to dislike their dependent status. In Germany under the influence of Humboldtian ideas about the value of pure research, chemistry professors in particular began to resent the way that their subject was conceived simply as the handmaiden of medicine.[75]

The way was open therefore for all the different branches of the natural sciences again to come together under the same institutional category, a marriage made all the easier by the creation of separate faculties of science. Chemistry would move out of the medical faculty in Belgium-Holland, for instance, as soon as the faculties of philosophy were split in two in 1815. By 1850, then, if not 1800, the modern positioning of the natural sciences was firmly in place. By that date, too, the first university courses in engineering science were being established, such as those created in England in 1838 in the newly founded university colleges of Durham and King's London.[76] Of the ancillary medical sciences, only anatomy remained permanently in the medical curriculum. This reflected the fact that from the end of the eighteenth century, especially with the creation of the new medical schools in Revolutionary

---

[74] Maurice Crosland, *Science under Control: The French Academy of Sciences 1795–1914* (Cambridge University Press, 1992), especially pp. 124–33.

[75] Derek Dow and Michael Moss, "The Medical Curriculum at Glasgow in the Early Nineteenth Century," *History of Universities*, 7 (1988), 227–57; Meinel, "Chemistry in Universities," pp. 116–18.

[76] H. A. M. Snelders, "Chemistry at the Dutch Universities, 1669–1900," *Academia Analecta: Mededelingen van de koninklijke Academie voor Wetenschappen, Letteren en Schone Kunsten van België, Klasse der Wetenschappen*, 48:4 (1986), 59–75; Robert A. Buchanan, *The Engineers: A History of the Engineering Profession in Britain, 1750–1914* (London: Jessica Kingley, 1989), pp. 165–70.

France, anatomy was no longer taught as a distinct, descriptive science but as an integral part of the study of physiology and pathology.

Most important, the developments in university science teaching outlined in this chapter also helped to cement specifically national traditions in science, something that profoundly affected the way science, particularly physical science, was done in the first part of the nineteenth century, especially in France. The fact that the new physics was taught as an experimental, and not a mathematical, subject may have ensured that the new science became part of the mindset of Europe's elite. It also ensured, however, that the pool of mathematically trained scientists able to continue the work of Newton and his followers in the second half of the eighteenth century would still be small. Very few students who emerged from the colleges and universities of Europe would have been much more mathematically literate than their seventeenth-century predecessors. Even in the one eighteenth-century British university – Cambridge – where there was a genuine enthusiasm for a mathematical physics, the number of students gaining honors in the mathematical tripos each year was only a small proportion of the total. The only country where the whole student body was subjected to an intense diet of mathematics and mathematical physics was, as we have seen, France, where the college course in natural philosophy had been restructured on real Newtonian lines and the new state military academies taught a sophisticated mathematical physics.

The reason for this singular development can only be guessed at. It is almost certainly connected in some way with the French *Académie des Science*'s commitment in the first three decades of the eighteenth century to finding a mathematical defense of Cartesian vortexes. Although the attempt to defend a French national icon gloriously failed, it seems to have created a French tradition of mathematical physics, which was adopted at college level after the Academy threw in the towel and accepted that vortex planetary motion could not be made to fit Kepler's and Newton's laws. Yet if the reason for the French classroom's love affair with mathematical physics cannot be fully explained, its consequence is clear. From 1790 to 1830, the French contribution to the advance of mathematical physics (and the mathematization of other sciences, especially chemistry) far surpassed the efforts of other nations. Historically, this has been attributed to many factors: Revolutionary support for science, the creation of new institutions of learning (the *grandes écoles*) where scientists could find employment, even the simple patronage power of Laplace and Claude-Louis Berthollet (1748–1822) who had the influence thereby to impose their own Newtonian agenda on the French scientific community.[77] All these factors undoubtedly played a part, but the one neglected aspect in the

---

[77] Joseph Ben-David, "The Rise and Decline of France as a Scientific Centre," *Minerva*, 8 (1970), 160–79; Robert Fox, "Scientific Enterprise and the Patronage of Research in France," *Minerva*, 11 (1973), 442–73; Dorinda Outram, "Politics and Vocation: French Science, 1793–1830," *British Journal for the History of Science*, 13 (1980), 27–44.

explanation of French scientific hegemony is the role of the *ancien régime* college physics course. It is entirely forgotten that the first generation of this great era of French physics, above all Laplace, its *metteur en scène,* had been educated in the pre-Revolutionary *collèges de plein exercice.* It was there that these students gained their expertise in and enthusiasm for a mathematical physics. There must have been many college students who were completely at sea in their physics year, as they struggled to cover mathematics in three months and the whole of the new physics in six. But the sheer number of mathematically literate students that emerged from French colleges in the last thirty years of the *ancien régime* guaranteed that the ground was prepared for the efflorescence of French mathematical physics in the benign climate of the Revolution (and Laplace, of course, had already made his mark on the discipline before 1789).

No other country could boast such a mathematically literate professional elite at the end of the eighteenth century, so no other country could have produced such a galaxy of mathematical physicists. Indeed, in Britain's case the concomitant commitment of the large majority of its teachers of the new physics (both public and private) to an experimentalist presentation of Newton possibly helped to cement a rival experimental tradition. Britain in the late eighteenth and the first part of the nineteenth centuries did not produce innovative mathematical physicists but in figures such as Young, Davy, and Michael Faraday (1791–1867), it did possess a coterie of eminent experimentalists who powerfully contributed in a different way to the advance of their discipline. It is difficult to believe that their appearance was coincidental. An experimental tradition in Great Britain, of course, can be traced back to Francis Bacon (1561–1626) and Robert Boyle (1627–91), but Newton himself and his immediate followers such as the Gregories, Roger Cotes (1682–1716), and Colin Maclaurin (1698–1746, professor of mathematics at Edinburgh), had met that tradition head on and had aimed to establish the supremacy of a mathematical physics. Arguably, that they failed to do so was closely related to the way in which Newtonianism was packaged for the consumer and the manner in which physics as a science consequently came to be conceived.

An understanding of the way in which the new science was established in the universities and colleges of Europe in the eighteenth century is crucial if we are to comprehend the distinctive contribution of different nations to modern science. At the very least, the manner of its establishment in France and Britain helped to promote two rival traditions of physics, one mathematical and one experimental, which have affected the two countries' approaches to natural science ever since.[78]

---

[78] This argument is developed further in Laurence W. B. Brockliss, "L'Enseignement des mathématiques sous l'ancien régime et la fecondité de la science française à l'époque révolutionnaire et napoléonienne" (unpublished paper, 1994).

4

SCIENTIFIC INSTITUTIONS AND THE ORGANIZATION OF SCIENCE

James McClellan III

The eighteenth century represents a distinct era in the organizational and institutional history of European science. Growing out of an "organizational revolution" that accompanied the intellectual transformations of science in the sixteenth and seventeenth centuries, the scientific enterprise became newly solidified in the eighteenth century. Indicative of this solidification, European governments increasingly supported and structured novel social and institutional forms for eighteenth-century science. Governments moved to support science for the perceived usefulness of expert knowledge of nature.

Science reorganized in the eighteenth century centered on national academies of science modeled after the Royal Society of London (1662) and the French Académie Royale des Sciences (1666). It also involved observatories, botanical gardens, and new forms of publication and scientific communication. This characteristic Old-Regime style of organized and institutionalized science matured over the course of the eighteenth century and was replaced in the nineteenth century by an equally distinct form for organized science that came to involve specialized societies, disciplinary journals, and a revived university system.

THE "ORGANIZATIONAL REVOLUTION" OF THE SEVENTEENTH CENTURY

The organizational and institutional character of science in the eighteenth century developed from seventeenth-century antecedents and an "organizational revolution" that formed part of the Scientific Revolution.[1]

---

[1] On the "Organizational Revolution," see James E. McClellan III, *Science Reorganized: Scientific Societies in the Eighteenth Century* (New York: Columbia University Press, 1985), chap. 2; a still-serviceable older literature includes Martha Ornstein, *The Role of Scientific Societies in the Seventeenth Century* (Reprint: New York: Arno Press, 1975; original ed., 1928); and Harcourt Brown, *Scientific Organizations in Seventeenth Century France* (Reprint: New York: Russell & Russell, 1967; original ed., 1934).

Although the medieval university continued to provide an institutional basis for science and natural philosophy in the seventeenth and eighteenth centuries, by and large contemporary universities – bastions of Aristotelianism – declined as the institutional loci of scientific novelty in the seventeenth century, and new, complementary venues arose to channel the new science. In particular, the courts of late Renaissance princes provided centers and sources of patronage for many men of science and their works.[2] Galileo's leaving the University of Padua, where he taught for eighteen years, for the Medici court in Florence in 1610 is emblematic of this institutional shift. The careers of Copernicus, Tycho, Kepler, Descartes, and many other scientific luminaries likewise illustrate the turning away from universities that was characteristic of the institutional history of science in the seventeenth century.

The creation of new "Renaissance"-style science academies formed a significant part of these trends.[3] Growing out of earlier learned associations devoted to language and literary studies, new organizations such as the *Accademia dei Lincei* (Rome, 1603–1630) and the *Accademia del Cimento* (Florence, 1657–1667) took up scientific investigations and became rallying points for their members. Renaissance-type academies exemplify changed conditions for the organization of science in the seventeenth century; but they were neither formally chartered nor state-supported institutions, and because they depended on the patronage of their noble sponsors, "Renaissance" academies for the most part were transitory organizations. By the same token, the number of Renaissance-style academies devoted to science actually increased in the second half of the seventeenth century, and one, the *Academia Naturae Curiosorum* (1677), remained a high-status institution of science and medicine throughout the eighteenth century.

The movement to recast the organizational and institutional bases of science in the seventeenth century culminated in the creation of the Royal Society of London and the French Académie Royale des Sciences in the 1660s.[4]

---

[2] See the collection edited by Bruce T. Moran, *Patronage and Institutions: Science, Technology, and Medicine at the European Court, 1500–1750* (Rochester, NY: Boydell Press, 1991). See also Roger L. Emerson, "The Organisation of Science and Its Pursuit in Early Modern Europe," in R. Olby, G. Cantor, J. Christie, and A. Hodge (eds.), *Companion to the History of Modern Science* (London: Routledge, 1990), pp. 960–79.

[3] On Renaissance-style academies, see note 1 and Michele Maylender, *Storia delle Accademie d'Italia*, 5 vols. (Bologna: L. Cappelli, 1926–30; reprint Rome: Arnaldo Forni, s.d.). See also W. E. Knowles Middleton, *The Experimenters: A Study of the Accademia del Cimento* (Baltimore, MD: Johns Hopkins University Press, 1971); and Eric W. Cochrane, *Tradition and Enlightenment in the Tuscan Academies, 1690–1800* (Chicago: University of Chicago Press, 1961).

[4] The literature on the Paris Academy and Royal Society is well developed. On the Paris Academy, see Roger Hahn, *The Anatomy of a Scientific Institution: The Paris Academy of Sciences, 1666–1803* (Berkeley: University of California Press, 1971; paperback ed., 1986); Éric Brian and Christiane Demeulenaere-Douyère (eds.), *Histoire et mémoire de l'Académie des sciences: Guide de recherches* (London: Tec & Doc, 1996); and David J. Sturdy, *Science and Social Status: The Members of the Académie des Sciences, 1666–1750* (Woodbridge, Suffolk, UK: Boydell Press, 1995). On the early Academy, see Alice Stroup, *A Company of Scientists: Botany, Patronage, and Community at the Seventeenth-Century Parisian Royal Academy of Sciences* (Berkeley: University of California Press, 1990); David Lux, *Patronage and Royal*

Several features distinguish the Royal Society, the Paris Academy, and successor societies from previous institutional settings for science. Eighteenth-century-style scientific societies were official corporate bodies with charters issued by the nation-state or other governing authority. To varying degrees societies received financial support from the state, and they reciprocally performed official functions as part of formal or informal government bureaucracies. They had patrons, but the role of the patron declined to insignificance. They devoted themselves explicitly to research and the advancement of the natural sciences. Unlike universities, their scientific commitments were not subservient to other institutional goals, and they essentially did no teaching. Founded in Berlin in 1700 with G. W. Leibniz as its President, the Prussian *Societas Scientarium* was the next major science academy to appear after the Royal Society and the Paris Academy, and the number and importance of scientific societies grew thereafter.

Coincident with these developments, new mechanisms arose in the seventeenth century for communicating science. Previously, the printed book, private correspondence, and personal travel represented the chief means by which scientific communities exchanged news and information. Tellingly, formal correspondence networks came to augment these traditional modes; the circle that arose around Théophraste Renaudot in the 1630s is a noted example. But the creation of institutionalized networks of correspondence associated with the emerging scientific societies represents an even more potent innovation. In the 1660s and 1670s, for example, Henry Oldenburg single-handedly invigorated scientific exchange across Europe through his wide-ranging network of correspondents, a role strengthened by his institutional position as secretary of the Royal Society of London.[5]

Science in Seventeenth-Century France: The Académie de Physique in Caen (Ithaca, NY: Cornell University Press, 1989); and Marie-Jeanne Tits-Dieuaide, "Les savants, la société et l'état: À propos du «revouvellement» de l'Académie royale des sciences (1699)," *Journal des Savants*, Janvier-Juin 1998, 79–114. An important, if overlooked, article is Rhoda Rappaport, "The Liberties of the Paris Academy of Sciences, 1716–1785," in Harry Woolf (ed.), *The Analytic Spirit: Essays in the History of Science in Honor of Henry Guerlac* (Ithaca, NY: Cornell University Press, 1981).

On the Royal Society, see Michael Hunter, *Establishing the New Science: The Experience of the Early Royal Society* (Woodbridge, Suffolk, UK: Boydell Press, 1989) and his *The Royal Society and Its Fellows, 1660–1700: The Morphology of an Early Scientific Institution* (Chalfont St. Giles, Bucks, UK: British Society for the History of Science, 1982); Richard Sorrenson, "Towards a History of the Royal Society in the Eighteenth Century," *Notes and Records of the Royal Society of London*, 50 (1996), 29–46; David P. Miller, "'Into the Valley of Darkness': Reflections on the Royal Society in the Eighteenth Century," *History of Science*, 27 (1989), 155–66; Marie Boas Hall, *Promoting Experimental Learning: Experiment and the Royal Society, 1660–1727* (Cambridge University Press, 1991). Still valuable among an older literature is Henry Lyons, *The Royal Society, 1660–1940: A History of Its Administration under Its Charters* (Reprint: New York: Greenwood Press, 1968; original ed., 1944).

[5] On these points see John L. Thornton and R. I. J. Tully, *Scientific Books, Libraries & Collectors: A Study of Bibliography and the Book Trade in Relation to Science* (London: The Library Association, 1971), chaps. 1–4; David A. Kronick, *A History of Scientific & Technical Periodicals: The Origins and Development of the Scientific and Technical Press, 1665–1790*, 2nd ed. (Metuchen, NJ: Scarecrow Press, 1976), chaps. 1–4; *The Correspondence of Henry Oldenburg*, 13 vols., ed. and trans. A Rupert Hall and Marie Boas Hall (vols. 1–9: Madison: University of Wisconsin Press, 1965–73; vols. 10–11: London: Mansell; vols. 12–13: London, Taylor & Francis, 1983–6); Howard M. Solomon, *Public Welfare, Sci-*

The appearance of the scientific journal in the 1660s marked a final novelty in the "organizational revolution" of the seventeenth century. The *Journal des Sçavans* was issued from Paris in 1665, followed in the same year by the *Philosophical Transactions* of the Royal Society of London. Scientific journals offered more timely publication than books and wider access than correspondence. In effect, journals created the scientific paper as the standard unit for publishing the results of scientific research. The institutional sponsorship of the *Philosophical Transactions* by the Royal Society set a precedent that linked the new scientific periodicals with the scientific societies, a precedent that was almost universally taken up by scientific societies in the eighteenth century.

The "organizational revolution" of the seventeenth century effected fundamental changes in the organizational and institutional character of contemporary science. Already the more direct intervention of the state can be seen. But the full effect of these changes was not felt until the next century and the flowering of a distinctly eighteenth-century style for the organization and pursuit of science and natural knowledge.

THE AGE OF ACADEMIES

Learned societies modeled after the Royal Society of London and the Académie Royale des Sciences in Paris formed the backbone of organized and institutionalized science in the eighteenth century, and indeed, the century has been labeled "the Age of Academies."[6]

The number of official scientific societies grew exponentially after 1700 as part of a European-wide institutional movement. The period through roughly 1750 witnessed the creation of the leading national scientific societies: London (1662), Paris (1666), Berlin (1700), St. Petersburg (1724), and Stockholm (1739). Major provincial and regional societies arose at this time in Montpellier (1706), Bordeaux (1712), Bologna (1714), Lyons (1724), Dijon (1725/1740), Uppsala (1728), and Copenhagen (1742). The second half of the century saw the appearance of societies in lesser European states and provinces: in Göttingen (1752), Turin (1757), Munich (1759), Mannheim (1763), Barcelona (1764),

---

ence, and Propaganda in Seventeenth Century France: The Innovations of Théophraste Renaudot* (Princeton, NJ: Princeton University Press, 1972).

[6] Bernard de Fontenelle coined the expression in the eighteenth century. See Roger Hahn, "The Age of Academies," in Tore Frängsmyr (ed.), *Solomon's House Revisited: The Organization and Institutionalization of Science* (Canton, MA: Science History Publications, 1990), pp. 3–12; McClellan, *Science Reorganized*, chap. 1. See also Robin E. Rider, "Bibliographical Afterword," in Tore Frängsmyr, J. L. Heilbron, and Robin E. Rider, (eds.), *The Quantifying Spirit in the 18th Century* (Berkeley: University of California Press, 1990), pp. 381–96, especially 387–8; and Mary Terrall, "The Culture of Science in Frederick the Great's Berlin," *History of Science* 28 (1990), 333–64. Harry Redner provides an odd but interesting perspective in "The Institutionalization of Science: A Critical Synthesis," *Social Epistemology,* 1 (1987), 37–59. The unsurpassed source documenting French academies in the eighteenth century remains Daniel Roche, *Le siècle des lumières en province: Académies et académiciens provinciaux, 1680–1789,* 2 vols. (Paris: Mouton, 1978).

Brussels (1769), Padua (1779), Edinburgh (1783), and Dublin (1785), among other locales. The learned society movement became such an institutional trend that, in the case of the *Hollandsche Maatschappij der Wetenschappen* (1752), for example, the lack of a comparable local institution provided an incentive to create one! By 1789, some seventy formally chartered scientific societies spread across Europe, from the *Kongelige Norske Videnskabers Selskab* (1760) in Trondheim in the north to the *Reale Accademia delle Science e Belle-Lettere* (1778) in Naples in the south, from the *Academia Scientiarum Imperialis* (1724) in St. Petersburg in the east to the *Academia real des ciências de Lisboa* (1779) in the west.

Among at least a certain class of urban dwellers, the formation of learned societies represented an expression of contemporary sociability, and, complementing the elite organizations, dozens of unofficial organizations augmented the set of formally chartered institutions. Some, such as the *Naturforschende Gesellschaft* of Danzig (1743), remained private but of a high status. Many others were on their way to formal recognition by the end of the century. And many, more ephemeral societies, as recent research has shown, spread across England, the Low Countries, Germany, and Italy, thereby bringing the world of science and polite learning to urban centers and literate communities of all sizes.[7] In Britain, a distinctive form of provincial society, the Literary and Philosophical Society, appeared toward the end of the eighteenth century; Literary and Philosophical societies in Manchester (1781), Derby (1783), and Newcastle-Upon-Tyne (1793) are early instances, and their numbers grew in the nineteenth century. The private Lunar Society of Birmingham (1766–1791) possessed a remarkable membership that included Joseph Priestley, James Watt, Matthew Boulton, and Erasmus Darwin. In France, especially as the Revolution approached, a series of popular *musées* and *lycées* emerged to communicate discoveries from the learned world.[8]

Why would virtually every Western polity – from the Holy Roman Empire

[7] In addition to Roche, *Le siècle des lumières*, see Gwendoline Averley, "English Scientific Societies of the Eighteenth and Early Nineteenth Centuries" (Ph.D., Council for National Academic Awards, UK, 1989) (summary in *Dissertation Abstracts International* (1991), vol. 51, p. 2854A). Henry Lowood, *Patriotism, Profit, and the Promotion of Science in the German Enlightenment: The Economic and Scientific Societies, 1760–1815* (New York: Garland Publishing, 1991); Karl Hufbauer, *The Formation of the German Chemical Community (1720–1795)* (Berkeley: University of California Press, 1982), especially Appendix II; W. W. Mijnhardt, *Tot Heil van't Menschdom: Culturele genootschappen in Nederland, 1750–1815* (Amsterdam: Rodopi, 1988); Amedo Quodam, "La sienze e l'Accademie," in Laetitia Boehm and Enzio Raimondi (eds.), *Università, Accademie e Società scientifiche in Italia e in Germania dal Cinquecento al Settecento* (Bologna: Il Molino, 1981), pp. 21–69; Ugo Baldini and Luigi Besana, "Organizzazione e funzione delle accademie," in Gianni Micheli (ed.), *Storia d'Italia. Annali 3: Scienza e tecnica nella cultura e nella società dal Rinascimento a oggi* (Turin: G. Einaudi, 1980), pp. 1323–30; Brendan Dooley, *Science, Politics, and Society in Eighteenth-century Italy: The Giornale de' letterati d'Italia and Its World* (New York: Garland Publishing, 1991); see also the essay review by Paula Findlen, "From Aldrovandi to Algarotti: The Contours of Science in Early Modern Italy," *British Journal for the History of Science*, 24 (1991), 353–60.

[8] McClellan, *Science Reorganized*, pp. 148–9, Appendix 2; Robert E. Schofield, *The Lunar Society of Birmingham: A Social History of Provincial Science and Industry in Eighteenth Century England* (Oxford: Clarendon Press, 1963); Hahn, *Anatomy*, p. 107.

to the Commonwealth of Pennsylvania – charter a scientific society? The primary answer concerns the perceived usefulness of these institutions. In a quid pro quo exchange between state and institution, scientific societies delivered technical expertise in support of governance. In return, scientific societies received recognition, aid, and a modicum of independence to govern their own affairs.[9] The Paris Academy, for example, judged patent claims. The Royal Society of London provided occasional expert opinion to the British government on matters such as protecting buildings against lightning strikes. A lesser society might aid local authorities in regional development. To this end, the Bordeaux Academy, for example, published a six-volume natural-history survey of the surrounding province of Guyenne (1715–1739). In return, societies received formal recognition, legal existence, and often financial support. By and large they were also free to elect (and police) their own members, to publish freely, and to initiate scientific projects.

One needs to situate eighteenth-century scientific societies within the larger context of contemporary learned organizations. The most prominent science societies usually devoted themselves exclusively to the natural sciences, but many, particularly provincial organizations, also incorporated other disciplinary interests, such as *belles-lettres*. Reorganized by Frederick II in 1744, the Prussian *Académie royale des sciences et belles-lettres,* for example, came to include a section devoted to speculative philosophy! Eighteenth-century scientific societies are thus to be ranged alongside language academies (such as the *Académie française,* 1635), *belles-lettres* and literary associations, societies devoted to technology and the mechanical arts (e.g., the Royal Society of Arts, London, 1754), fine-arts and architecture societies, medical and surgical societies, agricultural societies, economic-development societies, and a variety of other specialized organizations. In the eighteenth century the norm for the organization of cultural pursuits, not least science, was the form of the learned society.

A useful distinction can be drawn between academies and societies per se, as exemplified by the prototypes of the Royal Society of London and the Paris Academy of Sciences. Generally speaking, societies had a larger, less structured membership, received less government support, and thought of themselves as more "independent" than their sister academies. The Royal Society, for example, did not receive regular government funding, and it depended on the dues of its members for its ordinary operations. The Royal Society averaged approximately 325 Fellows, the vast majority of whom were amateurs and purely social members. Sustained scientific work was not possible at the weekly meetings of the Royal Society, and the 21-member governing Council conducted

---

[9] On this point see Charles C. Gillispie, *Science and Polity in France at the End of the Old Regime* (Princeton, NJ: Princeton University Press, 1980), passim and "Conclusion"; McClellan, *Science Reorganized,* pp. 25–34; see also Robin Briggs, "The Académie Royale des Sciences and the Pursuit of Utility," *Past and Present,* 131 (1991), 38–88.

the real business of the institution at its monthly meetings. In contrast, academies were more clearly institutions of state, with a smaller, more restricted, and often paid membership and with more plainly defined official duties. The French government, for example, provided the Paris science academy with quarters, funds for its operations, and pensions for its top grades of members. The Paris Academy met twice weekly, and with a resident membership of about forty-five committed men of science, its meetings were comparatively substantive and effective. (An institution's name, incidentally, is not a reliable guide to its type; the *Société royale des sciences* of Montpellier, for example, was an academy!)

Although the differences between academies and societies are real, it goes too far to distinguish institutions categorically. A more accurate view sees academies and societies as functionally similar but characteristic of two different cultural spheres: the society form is typical of maritime, Protestant, relatively more democratic Europe; the academy form is typical of Continental, Catholic, and relatively more authoritarian regimes. In the final analysis, rather than distinguish academies and societies, it proves more useful to rank institutions, regardless of type, into hierarchical categories downward from national organizations, through regional, provincial, and local associations, to the most ephemeral groupings of amateurs.

Academies and societies fostered the natural sciences in the eighteenth century in a variety of ways. Members presented the results of their research at society meetings. Learned society proceedings, typified by the annual *Histoire et mémoires* of the Paris Academy, quickly became the primary vehicles for the publication of research. Academies actively directed research by funding thousands of prize contests that offered financial rewards and publication outlets for work on topics set by sponsoring institutions. The question posed by the Paris Academy for 1737 on the nature of fire, with Voltaire and Mme de Châtelet as laureates, is a famous example. Institutions also undertook research projects directly; the expeditions sent to Lapland and Peru by the Paris Academy in the 1730s to measure the shape of the Earth and to adjudicate disputes between Newtonians and Cartesians are celebrated examples. Eighteenth-century scientific societies also undertook common projects. Led by the scientific societies, coordinated efforts to observe the transits of Venus in 1761 and 1769 were the largest scientific ventures of the eighteenth century. The initiative sponsored by the Meteorological Society of Mannheim (1780–1795) to collect weather data from around the world is a lesser-known but equally ambitious institutional undertaking.[10]

[10] McClellan, *Science Reorganized*, chap. 6; John L. Greenberg, *The Problem of the Earth's Shape from Newton to Clairaut: The Rise of Mathematical Science in Eighteenth-Century Paris and the Fall of "Normal" Science* (Cambridge University Press, 1995); David C. Cassidy, "Meteorology in Mannheim: The Palatine Meteorological Society, 1780–1795," *Sudhoffs Archiv* 69 (1985), 8–25. Regarding the Venus observations, still unsurpassed is Harry Woolf, *The Transits of Venus: A Study of Eighteenth-Century Science* (Princeton, NJ: Princeton University Press, 1959).

At the mid-eighteenth-century mark, European scientific societies began to formalize interinstitutional contacts (notably through the regular exchange of publications) and to coalesce into a European-wide system of institutions. Reciprocal elections of honorary and corresponding members reinforced these ties, and collaborative projects in the second half of the century strengthened the reality of the international network of academies and societies that spanned eighteenth-century Europe. In this spirit one needs mention several initiatives to link groups of societies formally. Condorcet's plan of the mid-1770s to unite French provincial academies failed, but a provincial effort led by the Arras Academy beginning in 1785 succeeded. A successful association of German academies dates from 1794.[11]

Sociologically, eighteenth-century scientific societies defined local and international scientific communities, and the number and quality of learned society memberships bespoke a person's status in the contemporary world of science. In a handful of instances the scientific societies provided the institutional and economic wherewithal for the pursuit of full-time careers in the sciences. The case of the mathematical physicist Leonard Euler is revealing. Euler spent his entire professional life within the confines of the scientific academies. His career trajectory began at the St. Petersburg Academy (1727–41), continued at the Berlin Academy (1741–66), and ended back at St. Petersburg (1766–83). The case of J.-L. Lagrange, who moved upward from the science academy in Turin to positions in Berlin and then in Paris, also illustrates that contemporary academies formed an institutional basis for professional careers in science in the eighteenth century.[12]

The ideology of the day made academies and societies "the diverse colonies of the Republic of Letters," and, indeed, much of their collective activity in the common culture of the times consolidated men and institutions of science into a transnational unity. By the same token, other forces operated against Enlightenment cosmopolitanism: national economic interests of governments, regionalism, and particularism that (especially in Italy) set learned associations against one another, language barriers, and differences of religious confession – all acted centrifugally to weaken the Republic of Letters and the contemporary international system of scientific societies.[13]

---

[11] Roche, *Le siècle des lumières,* vol. 1, pp. 68–74; McClellan, *Science Reorganized,* pp. 182–7.

[12] McClellan, *Science Reorganized,* chap.7; John Gascoigne, "The Eighteenth-Century Scientific Community: A Prosopographical Study," *Social Studies of Science,* 25 (1995), pp. 575–81. Contrast Roger Hahn, "Scientific Careers in Eighteenth-Century France," in M. Crosland (ed.), *The Emergence of Science in Western Europe* (New York: Science History Publications, 1976), pp. 127–38, and Hahn, "Scientific Research as an Occupation in Eighteenth-Century Paris," *Minerva,* 13 (1975), 501–13; see also Charles B. Paul, *Science and Immortality: The Éloges of the Paris Academy of Sciences, 1699–1791* (Berkeley: University of California Press, 1980), chap. 5.

[13] On these points see James E. McClellan III, "L'Europe des Académies: Forces centripètes, forces centrifuges," *Dix-Huitième Siècle,* 25 (1993), 155–65; and Lorraine Daston, "The Ideal and Reality of the Republic of Letters in the Enlightenment," *Science in Context,* 4 (1991), 367–86.

THE PERIODICAL JOURNAL

The periodical journal represents a significant element in the structure of organized science in the eighteenth century, and, as mentioned, the proceedings of scientific societies provided the main medium for the publication of eighteenth-century science.[14] The Paris Academy and its sister academies typically issued whole volumes of memoirs on an annual or more extended basis. The *Philosophical Transactions* of the Royal Society appeared approximately quarterly, and a handful of other organizations occasionally produced quarterly or trimestral numbers. The periodical publications of the scientific societies were specialized in the sense that they concerned themselves with the sciences, but they were not limited to any one specialty or discipline. Scientific society memoirs do not exhaust the scope of the contemporary scientific press, and some independent periodicals, such as the *Journal des Sçavans,* the Jesuit *Mémoires de Trévoux,* Pierre Bayle's *Mémoires de la République des Lettres,* or Leibniz's *Acta Eruditorum* provided vital means by which readers across Europe learned of developments in the world of science and natural philosophy. The publications of the scientific societies remained paramount, however, as the loci for initial publication of scientific research by the most renowned practitioners. The rest of the contemporary scientific press primarily published derivative material. In other words, original scientific papers most often appeared in the publications of the scientific societies. Societies also systematically distributed their volumes among themselves, and that made their publications more readily available to other societies' members, precisely the audience with the greatest interest in the output of the scientific societies.[15]

Although the publications of the scientific societies dominated the world of scientific publishing in the eighteenth century, problems beset the contemporary modus operandi, and those problems mounted as the eighteenth century wore on and as the pace of scientific activity increased.[16] Language barriers posed one difficulty, as Latin publication declined and as mainstream scientists found the work of their colleagues writing in vernacular languages such as English or Swedish less accessible. In response, several academies and societies undertook translation activities within their own confines. Similarly, the foreign series of the *Collection Académique* appeared in thirteen volumes in Paris from 1755 to 1779. As its name suggests, its purpose was to make

---

[14] See Thornton and Tully, *Scientific Books;* Kronick, *Scientific & Technical Periodicals.*

[15] The libraries of Old-Regime academies and societies have yet to be systematically studied. The Academy of Sciences met in the Bibliothèque du Roi, giving Parisian academicians access to significant bibliographic and scientific resources. Beyond that, contemporary university libraries are not known for having strong collections, and access to institutional and private libraries was generally limited.

[16] On these points, see James E. McClellan III, "The Scientific Press in Transition: Rozier's Journal and the Scientific Societies in the 1770s," *Annals of Science,* 36 (1979), 425–49.

available the transactions of the chief foreign learned societies in the universal language of French.

Delays in the publication of society memoirs, however, proved the most aggravating problem. At one point, for example, the time lag between the reading of a paper in the Paris Academy and its appearance in the Academy's *Mémoires* mounted to seven years; the average was three years, an increasingly unacceptable delay. Against this background, one periodical stands out: the *Observations sur la physique, sur l'histoire naturelle et sur les arts,* or *Rozier's Journal* as it is known. The Abbé François Rozier published this journal in Paris from 1772. Notably, Rozier's *Journal* appeared monthly and brought to its readers unprecedentedly current news of the world of science. Rozier was especially intent on providing information to active researchers, a feature that distinguished his journal from derivative publications and, arguably, from institutional proceedings that may have had more archival functions. Indicative of the centrality of the existing scientific societies, however, Rozier did not launch his enterprise in opposition to their publications or procedures. Rather, blessed by the Paris Academy, he enlisted the support of learned societies across Europe and America, and he exploited their distribution system for the international dissemination of his journal.

The distinctive eighteenth-century form of periodical publication of scientific research remained in place until the last years of the century. Only at that time did the disciplinary journal per se begin to make its appearance. The publication of Crell's *Chemische Journal* (1778), Curtis' *Botanical Magazine* (1787), the *Annales de Chimie* (1789), and the *Annalen der Physik* (1799) signaled a new mode for scientific publication and, indeed, for organized science as a whole as it moved into the nineteenth century.

UNIVERSITIES AND COLLEGES

Although academies and societies became the foremost active centers of science, from a global point of view traditional universities and colleges continued to provide important, perhaps the most important, institutional bases for the organization of science in the eighteenth century. As Laurence Brockliss details in Chapter 3 of this volume, eighteenth-century universities and colleges by and large retained their medieval intellectual and institutional character.[17] They were first and foremost pedagogical institutions, and they

---

[17] See Chapter 3 in this work, "Science, the Universities and other Public Spaces: Teaching Science in Europe and the Americas." See also Laurence Brockliss, *French Higher Education in the Seventeenth and Eighteenth Centuries* (Oxford: Oxford University Press, 1987), especially chap. 7; and Hilde de Ridder-Symoens (ed.), *A History of the University in Europe, Volume Two: Universities in Early Modern Europe (1500–1800)* (Cambridge University Press, 1996). In this context Michael Heyd, *Between Orthodoxy- and the Enlightenment: Jean-Robert Chouet and the Introduction of Cartesian Science in the Academy of Geneva* (The Hague: M. Nijhoff, 1982) and Edward Grant Ruestow, *Physics at Seventeenth- and Eighteenth-Century Leiden: Philosophy and the New Science in the University* (The

were not, generally speaking, progressive centers of research or innovation for science in the eighteenth century. By the same token, universities served as essential "gatekeepers" to the world of contemporary science in that virtually everyone who published a scientific paper in the eighteenth century at some point had matriculated at a university. Universities taught and exposed students to the natural sciences, and, overwhelmingly, future scientific society members first encountered the world of learning through the universities. The rarity of exceptions, such as the Dutch draper and microscopist Anton van Leuuwenhoek (1632–1723), truly "proves" this rule. Similarly, even though vernacular languages gained greater currency in the eighteenth century, as noted, Latin remained the mother tongue of the university and an indispensable entrée into the world of contemporary science and its sources.

A handful of progressive institutions came to incorporate attitudes and individuals sympathetic to the spirit and content of cutting-edge science. Experimental and mathematical Newtonianism gradually penetrated university culture, and the Dutch universities, in particular, gained a reputation as advanced centers of scientific pedagogy, notably on account of the work and writings of Willem Jacob van 's Gravesande (1688–1742) at Leiden, who advocated the teaching of natural philosophy through the use of experiments. In France the Collège Royal (1530) underwent a series of reforms (particularly in 1774) that created scientific chairs in anatomy, astronomy, botany, chemistry, mathematics, mechanics, and experimental and "universal" physics. These reforms made the Collège Royal the foremost seat of advanced learning and instruction in the sciences in France.[18] Scottish universities also became known as similarly liberal institutions, and, with professors such as Joseph Black at Edinburgh, students flocked there from all over Europe, particularly to study medicine. As the cases of 's Gravesande, Black, and, earlier, Newton likewise make clear, universities, as they had for centuries, provided positions for the scientific professoriate. Although eighteenth-century professors as a group were not especially distinguished, many, such as Albrecht von Haller at Göttingen and Linneaus at Uppsala, used university positions – often in medical faculties – to pursue distinguished careers in the sciences.

In several cases science academies arose in university settings and were grafted to traditional university structures. The Institute of Bologna, which

Hague: M. Nijhoff, 1973), are still worth consulting. Although primarily concerned with the seventeenth century, John Gascoigne, "A Reappraisal of the Role of the Universities in the Scientific Revolution," in David C. Lindberg and Robert S. Westman (eds.), *Reappraisals of the Scientific Revolution* (Cambridge University Press, 1990), pp. 207–60, illuminates the issues under consideration here.

18 On these points, see J. L. Heilbron, *Electricity in the 17th & 18th Centuries* (Berkeley: University of California Press, 1979), pp. 14, 142; Thomas L. Hankins, *Science and the Enlightenment* (Cambridge University Press, 1985), pp. 46–50; A. Rupert Hall, "'s Gravesande, Willem Jacob," *Dictionary of Scientific Biography*, V, 509–11. On the Collège Royal, see Gillispie, *Science and Polity*, pp. 130–43; Jean Torlais in René Taton (ed.), *Enseignement et diffusion des sciences en France au XVIIIe siècle* (Paris: Hermann, 1964 [reprint, 1986]), pp. 261–86.

functioned as the "research" arm of the University of Bologna, incorporated the renowned Bolognese Academy of Sciences (1714); science professors at the University served as academicians within the Academy. Similarly, at St. Petersburg, academicians held dual positions at the Imperial Academy and at the associated university and gymnasium. In several other university towns, academies and universities became closely intertwined. The University of Göttingen, for example, and the *Königliche Societät der Wissenschaften* (1752) enjoyed especially close relations, as did the University in Montpellier and the *Société royale des sciences*. In Paris, the Collège Royal and the Academy of Sciences shared an overlapping membership and numerous links.[19] Thus, although academies in many ways supplanted universities as the vital core of scientific activity in the eighteenth century, as far as the overall organization of science is concerned, academies complemented universities more than competed with them.

OBSERVATORIES

Astronomical observatories formed another pillar on which institutionalized science rested in the eighteenth century.[20] The institution of the observatory had arisen earlier in the Islamic world, and because of high costs entailed in buildings, equipment, and staff, observatories required substantial patronage. Tycho Brahe's Uraniborg and Stjerneborg, erected on the Danish island of Hveen in the late sixteenth century, make the point with regard to Renaissance Europe. Tycho's great installations derived entirely from the patronage of the Danish King, Frederick II, and Tycho boasted that one of his instruments cost more than the annual salary of a university professor! Given the prevailing "Renaissance" model for organized science, when royal patronage was withdrawn, Tycho had to move on, in this case to Prague and the Imperial Court there.[21]

Characteristically, observatories in the eighteenth century did not depend

---

[19] Sturdy, *Science and Social Status*, pp. 9–10; Hahn, *Anatomy*, pp. 72ff.; McClellan, *Science Reorganized*, pp. 12–13.

[20] A comprehensive study of eighteenth-century observatories remains to be written. Starting points for such a study include Claire Inch Moyer, *Silver Domes: A Directory of Observatories of the World* (Denver, CO: Big Mountain Press, 1955), and C. André, G. Rayet, and A. Angot, *L'Astronomie pratique et les observatoires en Europe et en Amérique, depuis le milieu du XVIIe siècle jusqu'à nos jours* (Paris: Gauthier-Villars, 1874–8). On the better-known French case, see Roger Hahn, "Les observatoires en France au XVIIIe siècle," in Taton (ed.), *Enseignement et diffusion*, pp. 653–65; on the Parisian *Observatoire*, see Gillispie, *Science and Polity*, pp. 99–130.

[21] On Tycho and his career, see Victor E. Thoren, *The Lord of Uraniborg: A Biography of Tycho Brahe* (Cambridge University Press, 1990), and John North, *The Norton History of Astronomy and Cosmology* (New York: W. W. Norton, 1994). On the costs of Tycho's instruments, see Ann Blair, "Tycho Brahe's Critique of Copernicus and the Copernican System," *Journal of the History of Ideas*, 51 (1990), 355–77, here 369.

on Renaissance-style court patronage but became institutions incorporated directly into the apparatus of state. Again, monarchical authority in France and England led the way by founding the *Observatoire royal* in Paris in 1668 and the Royal Observatory at Greenwich in 1675. Other national observatories followed in Berlin (1708), St. Petersburg (1725), and Stockholm (1753), and major regional observatories were established in Bologna (1723), Uppsala (1742), Marseilles (1749), Cádiz (1753), Milan (1760), Padua (1767), and Mannheim (1774). A larger number of private facilities complemented these official ones, including a series of stations staffed by the Jesuits. One hundred thirty observatories dotted the globe at the end of the eighteenth century.[22]

National observatories brought institutionalized astronomy into state service. Not surprisingly, navigational matters and the problem of longitude in particular provided the explicit rationale for creating the Greenwich and Paris observatories.[23] The Paris observatory, home to four generations of the Cassini family dynasty, became the institutional seat for the related, century-long project to map the kingdom of France.[24] The Paris observatory and other national observatories similarly produced ephemerides, calendars, almanacs, and related astronomical and nautical works of obvious utility. In contrast to the almost universal practice for patronized astronomy prior to the eighteenth century, the observatories did not, as far as one can tell, produce horoscopes.

The leading observatories typically effected close, and many times formal, connections to scientific societies and vice versa. The Royal Society of London came to exercise supervisory control as "Visitors" to the Royal Observatory at Greenwich. In Paris royal astronomers monopolized the astronomy section of the Academy of Sciences, and the Academy's *Mémoires* became, in essence, the publication arm of the royal *Observatoire*. In Berlin and St. Petersburg, the state observatories were formally affiliated with their companion scientific societies, the royal or imperial astronomer doing double duty at the observatory and the academy. In France, provincial academies in Dijon, Marseilles, Montpellier, Toulouse, and elsewhere came to administer local observatories attached to them. In 1784 the English and French national observatories and learned societies began a cooperative project of coordinating the meridians at Greenwich and Paris. The mutual benefit to the astronomers and their respective governments should be obvious.

---

[22] André, *L'Astronomie pratique*, vol. 1, p. v.

[23] On this point see Dava Sobel, *Longitude: The True Story of a Lone Genius Who Solves the Greatest Scientific Problem of His Time* (New York: Penguin Books, 1996), pp. 28–31, and William J. H. Andrewes (ed.), *The Quest for Longitude: The Proceedings of the Longitude Symposium, Harvard University, November 4–6, 1993* (Cambridge, MA: Collection of Historical Scientific Instruments, Harvard University, 1996), passim.

[24] See especially Josef W. Konvitz, *Cartography in France, 1660–1848: Science, Engineering and Statecraft* (Chicago: University of Chicago Press, 1987).

SCIENTIFIC INSTITUTIONS AND EUROPEAN EXPANSION

Geographically, European scientific institutions in the eighteenth century were not limited to Europe, and as European powers increasingly made their presence felt on a world scale, they transplanted institutional models – scientific and otherwise – to their overseas possessions. Science became an instrument of eighteenth-century European colonial expansion.

Western-style colleges and universities were established outside Europe, including the University of San Marcos in Lima, Peru (1551), and the *Real y Pontificia Universitad de Mexico* (1551/1553). In addition to an anatomical theater and science professorships in Lima, an observatory in Santa Fe de Bogotà, and a scientific and technical press, in 1792 Spanish colonial authorities established the famous School of Mines in Mexico, which taught advanced science and helped train cadres of technical specialists.[25] French and Portuguese mercantilist policies outlawed the creation of secondary schools outside their respective home countries, but, paralleling other differences in national style, several such pedagogical institutions arose in the British colonies of North America; at Harvard College (1663), the College of William and Mary (1693), Yale (1701), Princeton (1746), and King's College/Columbia (1754).

Colonial scientific societies also emerged in extra-European contexts. The most famous was the American Philosophical Society (APS) for Promoting Useful Knowledge (Philadelphia, 1768). This was Benjamin Franklin's society. The APS published three substantial volumes of *Transactions* in the eighteenth century and participated in the life of contemporary science, but it enjoyed a greater international reputation than perhaps it deserved on account of its association with the great man. Following the American Revolution, the American Academy of Arts and Sciences was founded in Boston in 1780. Another society-type institution, it, too, published *Memoirs* and functioned on the level of a typical European provincial society. Elsewhere in North America, short-lived private societies appeared in Virginia (1772, 1786), New York (1784), Connecticut (1786), and Kentucky (1787).[26]

In South America a private *Academia Scientifica* existed in Rio de Janeiro in the 1770s; although of passing importance, it did maintain relations with the state science academy in Sweden. More impressive, the French government granted letters patent founding the *Société royale des sciences et des arts* in colonial Haiti (1784), then Europe's richest and single most important colony.

---

[25] David Wade Chambers, "Period and Process in Colonial and National Science," in Nathan Reingold and Marc Rothenberg (eds.), *Scientific Colonialism: A Cross-Cultural Comparison* (Washington, DC: Smithsonian Institution Press, 1987), pp. 297–321, especially pp. 300–5; see also Antonio E. Ten, "Ciencia Y Universidad en la America hispanana: La Universidad de Lima," in Antonio Lafeunte and José Sala Català (eds.), *Ciencia colonial en América* (Madrid: Alianza Editorial, 1992), pp. 162–91; Anthony Pagden, "Identity Formation in Spanish America," in Nicolas Canny and Anthony Pagden (eds.), *Colonial Identity in the Atlantic World, 1500–1800* (Princeton, NJ: Princeton University Press, 1987), pp. 51–93, especially pp. 85–9.

[26] McClellan, *Science Reorganized,* pp. 140–5 and appendixes.

This little-known institution worked diligently with the Paris Academy of Sciences and other agencies of French government to promote the success of French colonial development. On the island of Java in the East Indies, Dutch colonial authorities officially incorporated the *Bataviaasch Gnootschap van Kunsten en Wetenschappen* in 1778; it became closely connected with senior Dutch societies in Europe.[27]

As detailed in the next section, France, Spain, Britain, and the Netherlands also established colonial botanical gardens and linked them to scientific, economic, and governmental centers in Europe. The examples of colonial universities, technical colleges, learned societies, and botanical gardens make clear that the institutional expansion of science outside Europe in the eighteenth century took place in the general context of European colonial expansion and with the goal of facilitating that expansion.

BOTANICAL GARDENS

Botanical gardens provide a final formal setting to be considered in this survey of scientific institutions and the organization of science in the eighteenth century. And many of the themes sounded to this point are heard again in connection with botanical gardens: increased importance of state support, increased emphasis on the social utility of science, and increased professionalization of scientific cadres.

Europe possessed sixteen hundred botanical gardens of several different types at the end of the eighteenth century.[28] The oldest and least important type was the medical or pharmacy garden. These were associated with universities and, more in particular, with medical faculties, and their roots extended back to the late Middle Ages. Medical professors controlled herb and apothecary gardens, and the scientific study of plants in these gardens was subordinated to pharmaceutical applications in materia medica. The number and importance of pharmacy gardens decreased in the eighteenth century, especially in comparison to other types.

The most numerous and renowned were the scientific gardens, of which the Jardin du Roi in Paris (1635) and the Royal Gardens at Kew (1759) are the most prominent examples. These installations were creations of the state and not the university, and they were headed by botanists and taxonomists. As

---

[27] James E. McClellan III, *Colonialism and Science: Saint Domingue in the Old Regime* (Baltimore, MD: Johns Hopkins University Press, 1992), p. 5 and Part III; McClellan, *Science Reorganized,* pp. 125, 145, and appendixes; Jean Gelman Taylor, *The Social World of Batavia: European and Eurasian in Dutch Asia* (Madison: University of Wisconsin Press, 1983). The Asiatic Society of Calcutta (1784) needs to be considered in this context.

[28] On botanical gardens, see Lucile H. Brockway, *Science and Colonial Expansion: The Role of the British Royal Botanic Gardens* (New York: Academic Press, 1979); Yves Laissus, "Le Jardin du Roi," in Taton (ed.), *Enseignement et diffusion,* pp. 287–342; McClellan, *Colonialism and Science,* chap. 9; Gillispie, *Science and Polity,* pp. 143–84; Sturdy, *Science and Social Status,* pp. 6–9.

national gardens, the larger ones, such as Paris and Kew, became international centers for botanical research, to which specimens were sent from the frontiers of exploration and colonial settlement. Although the notion existed that economically useful results might be forthcoming from acclimatization and other botanical experiments, the primary rationale for these scientific gardens (at least in the minds of the scientific staff) was the disinterested study and classification of the vegetable kingdom. Scientific gardens also did considerable teaching in the scientific aspects of botany and related areas of knowledge, including chemistry, anatomy, and geology.

As additional elements of state bureaucracies, the leading scientific botanical gardens developed close connections with their associated scientific societies. In the paradigmatic Parisian case again, the senior staff at the Jardin du Roi held ranking positions in the botany section of the Academy of Sciences, just as astronomers at the Observatoire dominated the astronomy section.

Comparatively unheralded, but even more indicative of the tenor of the times, a third type of garden arose toward the end of the eighteenth century: the applied botanical or economic garden. Economic gardens were devoted, not to the formal scientific study of the plant world, but rather to the active exploitation of potentially useful and economically beneficial commodity products. Typically, these gardens arose on the colonial periphery of far-flung French, British, and Dutch empires, and they enjoyed less direct supervision from the main centers in Paris, Kew, or Amsterdam. For example, the Dutch founded botanical gardens at Capetown in 1694, and other eighteenth-century Dutch stations existed in Ceylon (Sri Lanka) and in Batavia. The British created a substantial network of colonial gardens satellite to Kew: St.-Vincent in the West Indies (1764), Jamaica (1775), Calcutta (1786), Sydney (1788), and Penang Malaya (1800). The French established a similarly extensive set of colonial gardens linking stations in Guadeloupe (1716), Martinique, and St.-Domingue (Haiti) (1777) in the Caribbean region, Cayenne in South America, and Île de France (later Mauritius – three gardens: 1735, 1748, and 1775) and Île Bourbon (later Réunion; 1767) in the Indian Ocean. These colonial gardens transshipped products, including sugar cane, vanilla, and the breadfruit plant – the latter thought especially useful for provisioning slaves working West Indian plantations. In France itself, the royal botanical garden at Nantes (a coopted university pharmacy garden) and the Jardin du Roi in Paris provided the main metropolitan hubs for these activities. Characteristically, the French economic gardens received directions primarily from the Ministry of the Navy and less from the Academy of Sciences and the Jardin du Roi, and professional gardeners commissioned as part of a botanical service assumed a greater importance in the applied botanical gardens of the later eighteenth century than did scientific academicians or academicians in training. A similar pattern is apparent in the Spanish world with the creation of the Royal Botanical Garden in Mexico in 1786.

ORGANIZED SCIENCE IN SOCIETY

European science in the eighteenth century evoked multifaceted social responses that affected high and low culture and social centers and peripheries. It would be misleading to limit discussion of the organization of science in the eighteenth century solely to its institutional aspects. Other chapters in this volume investigate the multiple social impacts of eighteenth-century science. A few comments vis-à-vis the place of organized science in society may not be out of place here.

The world of organized science in the eighteenth century was almost exclusively male. To be sure, Mme de Châtelet was hardly a token figure. She was as knowledgeable of contemporary science as anyone, and her translation of Newton's *Principia* (partial edition, 1756; posthumous full edition, 1759) to this day remains the vehicle by which French readers scale those empyrean heights.[29] A few Italian women, such as the mathematician Maria Gaetana Agnesi (1718–1799) and Laura Bassi (1711–1778), professor of natural philosophy at the University of Bologna, managed to carve out active research careers in the male world of contemporary science, often with the backing of local towns and universities. A slightly larger number of privileged women had contact with contemporary science through salon culture of the time. Their talents and accomplishments notwithstanding, women scientists and amateurs were exceptions and cultural ornaments. A curious corollary of this gender division holds that because official science was male it also excluded those *men* who were not the interpreters of Nature and ennobled the manly heroes who were.[30]

In the eighteenth century, science ignited the popular imagination as never before. Ballooning, Mesmerism, the lightning rod, and the heroics of scientific travelers provoked responses from all levels of society, from the peasants who pitchforked alien balloons landing in their pastures to the highbrows who took courses on experimental physics or sought cures around Mesmer's tub.[31]

---

[29] See Esther Ehrman, *Mme de Châtelet: Scientist, Philosopher and Feminist of the Enlightenment* (Leamington Spa: Berg, 1986); Mary Terrall, "Emilie du Châtelet and the Gendering of Science," *History of Science,* 33 (1995), 283–310; René Taton, "Châtelet, Marquise du," *Dictionary of Scientific Biography,* III, 215–17.

[30] Mary Terrall, "Gendered Spaces, Gendered Audiences: Inside and Outside the Paris Academy of Sciences," *Configurations,* 2 (1995), 207–32, makes this last point. See also Edna E. Kramer, "Agnesi, Maria Gaetana," *Dictionary of Scientific Biography,* I, 75–7. Paula Findlen, "Science as a Career in Enlightenment Italy: The Strategies of Laura Bassi," *Isis,* 84 (1993), 441–69. Geoffrey V. Sutton, *Science for a Polite Society: Gender, Culture, and the Demonstration of Enlightenment* (Boulder, CO: Westview Press, 1995), presents the best information about contemporary salon and related scientific cultures, and he argues a strong case for the active participation of women in contemporary scientific practice in and around formal institutions.

[31] On popular science movements in the eighteenth century, see Charles Coulston Gillispie, *The Montgolfier Brothers and the Invention of Aviation* (Princeton, NJ: Princeton University Press, 1983); Robert Darnton, *Mesmerism and the End of the Enlightenment in France* (Cambridge, MA: Harvard University Press, 1968); Lindsay Wilson, *Women and Medicine in the French Enlightenment: The Debate over Maladies des Femmes* (Baltimore, MD: Johns Hopkins University Press, 1993).

In the present context, the role of institutions as mediators of these popular fads needs emphasis. In France, committees of the royal academies of science and medicine proscribed Mesmer. The Academy of Sciences quickly took control of ballooning trials in the French capital, and in the provinces academies in Lyon, Dijon, Marseilles, Bordeaux, and Besançon similarly assumed institutional authority over the lighter-than-air phenomenon.

The connection between science and technology in the eighteenth century is relevant to this discussion. In the realm of scientific instruments (creating, for example, the chronometer or achromatic lenses), technology and the crafts impinged crucially on the world of eighteenth-century science. Taking the lead from Bacon and Descartes, the *ideology* of the times emphasized the utility of the sciences applied to practical ends. The *Encyclopédie* of Diderot and d'Alembert sought to break down guild secrecy and to spread rational manufacture by illustrating the details of craft procedures. Acting on similar ideological commitments, the Paris Academy of Sciences published its famous technological series, the *Description des arts et métiers.* This set of seventy-four technical treatises appeared between 1761 and 1782 and has been characterized as "the largest body of technological literature that had ever been produced."[32] And, to some small extent – as in the case of lightning rods, for example, or inoculation against smallpox – discoveries in science and medicine found applications in the everyday world.

Regarding the key case of England in the eighteenth century, recent research has identified the ways science diffused socially to British artisans and entrepreneurs whose activities fomented the Industrial Revolution.[33] Those ways included the mechanism of the public lecture, the ideology of useful knowledge, the paradigm of mechanics, and scientific rationalism and systematic experimentation as potent examples for effecting change. Those influences notwithstanding, the lack of direct involvement of organized scientific institutions – not to say scientific ideas – in the early Industrial Revolution is striking. Almost all the engineers and technologists who got their hands dirty in the early Industrial Revolution worked at great social and intellectual removes from the refined scientific world in London. The exceptions, such as James Watt or Josiah Wedgwood, bridged the gap between contemporary science and technology only socially and not by dint of any push to further industrialization through applied science. The Royal Society admitted Watt in 1785, for example, not as an industrial pioneer for his having invented the

---

[32] See Gillispie, *Science and Polity,* pp. 337–56, quotation at, p. 344; see also Charles Coulston Gillispie (ed.), *A Diderot Pictorial Encyclopedia of Trades and Industry: Manufacturing and Technical Arts in Plates from l'Encyclopédie,* 2 vols. (New York: Dover Books, 1959).

[33] Margaret Jacob, *Scientific Culture and the Making of the Industrial West* (Oxford: Oxford University Press, 1997); Jan Golinski, *Science as Public Culture: Chemistry and Enlightenment in Britain, 1760–1820* (Cambridge University Press, 1992); Larry Stewart, *The Rise of Public Science: Rhetoric, Technology, and Natural Philosophy in Newtonian Britain, 1660–1750* (Cambridge University Press, 1992); see also Briggs, "The Académie Royale des Sciences."

separate condenser for the steam engine but rather as someone who had made natural philosophical contributions concerning the nature of water. The unschooled Wedgwood became FRS in 1783, not for industrializing pottery manufacture in England, but for his invention of a device for measuring high temperatures.[34] Plainly and tellingly, the early Industrial Revolution developed without significant input from eighteenth-century academies or universities.

A NINETEENTH-CENTURY POSTSCRIPT

As in so much else, the French Revolution marked the end of an era in the organizational and institutional history of science. A second "organizational" revolution unfolded on the other side of 1815 that recast the scientific enterprise into new, more recognizably modern forms.[35] Several features characterize the revised state of affairs that developed in the nineteenth century. Most notably, learned scientific societies – the heart and soul of the Old-Regime system – declined in relative importance as the leading institutions for promoting science. Specialized and discipline-oriented organizations, of types such as the Geological Society of London (1807) and the Royal Astronomical Society (1831), increasingly came to supplant the umbrella scientific society as foci for practicing communities of scientists. The major national academies continued to exist, but, certain exceptions aside (notably in Russia), their function became more that of honorary organizations recognizing scientific accomplishment and reputation achieved earlier and elsewhere. By dint of sheer numbers, publication in the disciplinary journals of specialist societies likewise came to overshadow publication in the proceedings of the scientific societies as the main sources to which scientists went for information and to present the results of their research. The founding of distinctively professional organizations for science, modeled on the *Deutsche Naturforscher Versammlung* (1822) and British Association for the Advancement of Science (1831), also indicates the appearance of a new mode for organized science. Similarly, the

---

[34] See articles on Watt and Wedgwood by Harold Dorn, *Dictionary of Scientific Biography*, XIV, 196–9 and 213–14; see also Charles Weld, *A History of the Royal Society*, vol. 2 (New York: Arno Press, 1975; original ed., 1848), pp. 176–85.

[35] On the second "organizational revolution," consult McClellan, *Science Reorganized*, Epilogue; Maurice Crosland, *Science Under Control: The French Academy of Sciences, 1795–1914* (Cambridge University Press, 1993), and Robert Fox and George Weisz (eds.), *The Organization of Science and Technology in France, 1808–1914* (Cambridge University Press, 1980). See also Morris Berman, *Social Change and Scientific Organization: The Royal Institution, 1799–1844* (Ithaca, NY: Cornell University Press, 1978); Jack Morrell and Arnold Thackray, *Gentlemen of Science: The Early Years of the British Association for the Advancement of Science* (Oxford: Clarendon Press, 1981); Roy MacLeod and Peter Collins (eds.), *The Parliament of Science: The British Association for the Advancement of Science, 1831–1981* (Northwood, Middlesex: Science Reviews, 1981); Marie Boas Hall, *All Scientists Now: The Royal Society in the Nineteenth Century* (Cambridge University Press, 1984); and the essay review by David Philip Miller, "The Social History of British Science: After the Harvest?" *Social Studies of Science*, 14 (1984), 115–35.

revitalizing of German universities as centers for research as well as teaching charted an influential new course for science in society. The creation of the university teaching laboratory, beginning with Leibig's chemistry laboratory at the University of Geissen in 1826, was likewise a significant feature of universities as revived centers for organized and institutionalized science. Along these lines, the coining of the English word "scientist" in 1834 is rightly taken to indicate how much circumstances had changed for science as the nineteenth century wore on. All these changes underscore the distinctive character of organized science in the eighteenth century and indicate how antiquated the previous "Age of Academies" had become.

5

SCIENCE AND GOVERNMENT

Robert Fox

The engagement of eighteenth-century governments and monarchs in the patronage of science had predominantly utilitarian motives. It was inspired, to very different degrees in different countries, both by a belief in the value of scientific knowledge for manufacturing, agriculture, medical improvement, public works, and warfare and by a perception of science as a form of culture whose promotion would lend luster to any regime seeking to parade its adjustment, however cautious, to the beneficent forces of enlightenment and modernity. Some of these motives had already borne scientific fruit in the seventeenth century, tentatively in England, where Charles II's patronage of the Royal Society had been no more than nominal, and in a far more concrete fashion in France in the new and existing institutions that were supported under the influence of Louis XIV's minister Colbert.[1] From its foundation in 1666, the Académie Royale des Sciences was an instrument of the state: its members received material support, in the form of salaries and facilities, and in return the monarchy looked for a source of glory that would outshine the Royal Society in London and for services and expert advice of the kind it requested and received on the water supply to Versailles and on the inventions and machines that were routinely submitted to the Académie for judgment.[2] It was with a similar aspiration to bind the interests of science to those of the state that Colbert commissioned Claude Perrault to design the

---

[1] For surveys of the early scientific societies of the seventeenth century, see Martha Ornstein, *The Role of Scientific Societies in the Seventeenth Century* (Chicago: University of Chicago Press, 1928); James E. McClellan III, *Science Reorganized: Scientific Societies in the Eighteenth Century* (New York: Columbia University Press, 1985), pp. 41–66; and several of the early chapters of David C. Goodman and Colin A. Russell (eds.), *The Rise of Scientific Europe 1500–1800* (Sevenoaks: Hodder & Stoughton and The Open University, 1991).

[2] Roger Hahn, *The Anatomy of a Scientific Institution: The Paris Academy of Sciences, 1666–1803* (Berkeley: University of California Press, 1971), especially pp. 19–26, and Alice Stroup, "The Political Theory and Practice of Technology under Louis XIV," in Bruce T. Moran (ed.), *Patronage and Institutions: Science, Technology, and Medicine at the European Court 1500–1750* (Rochester, NY: Boydell Press, 1991), pp. 211–34.

Observatory of Paris in the late 1660s. In its utility, as in its physical grandeur and the facilities it offered, the new building would reflect the glory of the Roi-Soleil, outstripping the observatories of England, Denmark, and China and providing both a setting for all the activities of the Académie (a function that, in the event, it never fulfilled) and a focus for the astronomical, geodesic, and meteorological work for which a century later the institution would become famous.[3] Older royal institutions, too, came under Colbert's wing. In this process, the Collège Royal, a sixteenth-century foundation that offered public lectures in a range of scientific and scholarly disciplines, and the Jardin du Roi, a botanical garden created in 1635 to cater for a facet of scientific training that the Faculty of Medicine was manifestly failing to provide, both assumed a new importance as contexts in which the conception of science as a proper responsibility of the state was reinforced.

The proliferation of academies across Europe during the eighteenth century diffused the ideal of governmental involvement in the patronage of science. In practice, however, few academies enjoyed the degree of support and proximity to the seats of political power that distinguished the Parisian Académie des Sciences in its early years, even though certain of their champions saw a close integration with the state as essential. When Gottfried Wilhelm von Leibniz elaborated his plan for a royal academy in Berlin, for example, he certainly envisaged an institution that would be, if anything, even closer to the Prussian court than the Académie was to Versailles.[4] But the reality that followed the creation of the Societas regia scientiarum in 1700, under the auspices of the ambitious Elector who was soon to become King Frederick I of Prussia, fell far short of Leibniz's vision. It was eight years before an adequate observatory was provided, and while the Societas performed its main public duty of publishing an official almanac at the time of the delicate move from the Julian to the Gregorian calendar, it suffered from a court that sought to exercise control (by the introduction of officials of its own choosing) without providing the level of financial support that had been anticipated. It was only from 1746, with Frederick II (the Great) on the throne, that the material well-being and intellectual autonomy of the Académie Royale des Sciences et Belles-Lettres de Prusse, as the Societas now became, were assured.

In Berlin, Frederick created what contemporaries saw for more than a decade as an ideal structure that fostered governmental involvement without undue intrusiveness. Practical services, in the form of advice on the calendar and inventions, were expected, but when Pierre-Louis Maupertuis and Leonard Euler were brought from Paris and St. Petersburg, respectively, they came as men of science whose distinction alone justified their presence. From 1746 until

---

[3] Charles Wolf, *Histoire de l'Observatoire de Paris de sa fondation à 1793* (Paris: Gauthier-Villars, 1902), pp. 2–4.
[4] McClellan, *Science Reorganized,* pp. 68–74, and Adolf Harnack, *Geschichte der Königlich Preussischen Akademie der Wissenschaften zu Berlin,* 3 vols. (Berlin: Reichsdruckerei, 1900), vol. 1, pp. 27–492.

his death in 1759 Maupertuis, as president, guided the Berlin Academy in directions that allowed it to become at once a national symbol of the benefits of enlightened monarchy and a constituent of the international Republic of Letters. In pursuit of its latter, international role, the Academy adopted French as its official language and inaugurated prize competitions open to all comers (including d'Alembert, who won the first competition, on the cause of winds). It also launched an annual volume of proceedings and memoirs, the *Histoire,* that allowed it to engage in the exchange of publications with other academies and so to blur the boundaries between national interest and the universalism not only of science but also of the areas of nonscientific scholarship that were represented in the Academy.

Despite the promising rehabilitation of the Berlin Academy during Maupertuis's presidency, the institution was soon to experience the darker as well as the benign side of patronage by the state. Following Maupertuis's death, Frederick II assumed a degree of personal control, in the appointment of new members and the interactions with other academies, that has been held at least partially responsible for the Academy's diminished international prominence between the 1760s and Frederick's death in 1786.[5] Other academies too were affected by the irregularity and changing priorities of the patronage they received from their various governments. For Czar Peter the Great, the Imperial Russian Academy of Sciences, founded in St. Petersburg in 1724, shortly before his death, was only one element in a much broader movement to break his country's isolation from the West and to achieve modernization through the advancement of modern knowledge, in particular of science and technology.[6] Governmental control was strict, and the introduction of sixteen distinguished members from abroad reflected the calculated priorities of national policy as well as the necessity of importing men of ability in a previously backward country dominated by a conservative Byzantine church. The imposed internationalism of the St. Petersburg Academy created difficulties: the national groups within it – mainly French, German, and Russian – did not always work well together, and the failure of most of the foreign members to master Russian (Euler being a notable exception) meant that their critics could easily charge them with a preference for addressing one another and the learned world at large (usually in Latin) rather than addressing a nation in need of the kind of cultural and technical improvements that Peter expected of them.

After Peter was gone, continued closeness to the government engendered an instability comparable to that of the Berlin Academy some years later. Court influence often took the form of ignorant administrative busybodying, and the struggle for intellectual autonomy, reinforced by the very real scientific achievements of such men as Euler and Daniel Bernoulli but repeatedly

---

[5] McClellan, *Science Reorganized,* pp. 176–8.
[6] Ibid, pp. 74–83, and Alexander Vucinich, *Science in Russian Culture: A History to 1860* (London: Peter Owen, 1965), pp. 38–122.

undermined by phases of intrusive political conservatism, had only partial success in the form of the charter that Czarina Elizabeth granted the Academy in 1747. Faced with new regulations that sought to direct attention from the theoretical work pursued by the most eminent foreign academicians toward activities more relevant to the material needs of Russia, Euler saw no alternative to withdrawal.

The tribulations of the Berlin and St. Petersburg academies illustrate with brutal clarity that wherever governmental involvement was strong, there lurked the threat of an inhibiting subservience to a politically motivated conception of the national interest. Despite the dangers of unwelcome interference, however, some measure of recognition by the state, extending to the allocation of a formal public role if not to lavish material support, was virtually essential if an academy was to prosper. Too often, though, recognition went little further than the granting of a name. In Sweden, the fine-sounding title of Societas regia literaria et scientiarum Sueciae in Uppsala lent dignity. But it could not conceal the fact that the society remained, like the informal group from which it sprang, little more than a coterie based in the University of Uppsala.[7] Nor could it prevent the decline that set in a decade after the society received its title and royal recognition in 1728. Similarly, the mere granting of the title "Royal" to the Vetenskapsakademien of Stockholm in 1741 did little for an institution that had begun its existence two years earlier as an independent body without any bonds to government.[8] What did transform the Academy, on the other hand, was a parliamentary decision of 1747 to grant it the exclusive right to publish the national almanac.[9] This helped to bring the institution to the center of Swedish life and to foster its initial aspiration to advance the nation's economy and well-being in a period – the so-called Era of Liberty that began with the death of the last absolute monarch, Charles XII, in 1718 – in which mercantilism and utilitarianism converged to advance the interests of science. Since the almanac sold 135,000 copies in its first year (1749) and well over twice that number annually by 1785, the decision also presented the Academy with a bestseller that ensured a substantial income and allowed it to embark, independently, on the construction and fitting out of its own fine observatory. Opened in the presence of the king and queen in 1753, the observatory was the focus for regular expenditure over the years and for a particularly handsome donation of instruments from the royal collection by

[7] McClellan, *Science Reorganized*, pp. 83–6. On the Society and more generally the background to science in eighteenth-century Sweden, see also Colin A. Russell, "Science on the Fringe of Europe: Eighteenth-Century Sweden," in Goodman and Russell (eds.), *The Rise of Scientific Europe*, pp. 305–32.

[8] McClellan, *Science Reorganized*, pp. 86–9. The standard history of the society covering this period is Sten Lindroth, *Kungl. Svenska Vetenskapsakademiens Historia 1739–1818*, 2 vols. (Stockholm: Kungl. Vetenskapsakademien, 1967); see especially the two parts (continuously paginated) of vol. 1.

[9] Lindroth, *Kungl. Svenska Vetenskapsakademiens Historia*, vol. 1, pp. 823–67; McClellan, *Science Reorganized*, pp. 88–9, and Ulf Sinnerstad, "Astronomy and the First Observatory," in Tore Frängsmyr (ed.), *Science in Sweden: The Royal Swedish Academy of Sciences 1739–1989* (Canton, MA: Science History Publications, 1989), pp. 45–71 (48–50).

the enlightened, well-traveled King Gustav III in 1772.[10] So long as Gustav, the "crowned democrat" and admirer of the French *philosophes,* was on the throne, science, like other cultural pursuits, was well served, in the context of a policy that yoked, on the one hand, the strengthening of the monarchy and the defense of Sweden against the expansionist menace of the Russian Empire to, on the other hand, the promotion of Enlightenment thought, in particular in the forms in which it had emanated from France. But Gustav's death in 1792 and the subsequent weakening of royal favor provided yet another illustration of the vulnerability of state-sponsored science. Although it is true that Sweden's diminished position as a scientific nation by the end of the century had other causes as well, the indifference of Gustav III's successor, his son Gustav IV, clearly played a part.[11]

The accelerating pace with which academies were founded from the 1740s and the resulting diversity make it difficult to move from the specific instances already mentioned to a generalization about the role of government in a movement that now swept from the major European capitals through provincial France, the German-speaking parts of central Europe, and the Italian peninsula, as well as (more unevenly) Scandinavia, Britain, North America, and Iberia.[12] But monarchs virtually everywhere were readier than ever to pay at least lip service to the convergence of *potentia* and *scientia* by acting as patrons or protectors, granting royal letters patent (given particularly freely to the provincial French academies),[13] and looking to the institutions under their sway for evidence of the kind of usefulness that was appropriate to social and economic circumstances very different from those of the first half of the century. The quickening pace of industry presented the most enticing challenge, although it proved to be one to which the system of state-supported academies offered a disappointing response. They could cope well enough with the proffering of advice on mechanical inventions and improvements in traditional machinery and on agricultural implements and practices; all these called for a relatively modest level of scientific input and rested on a large existing stock of craft knowledge that changed slowly. But it proved far more difficult to harness the science of the academicians to the understanding and improvement of the new areas of manufacturing in textiles, chemicals, and metallurgy.

In this respect, the Royal Academy of Turin was the setting for a revealing disappointment. Founded in 1759 as a private society (*società privata*) but

---

[10] Sinnerstad, "Astronomy and the First Observatory," pp. 55.

[11] For a comment on the possible causes of the decline of Swedish science after the 1770s, see Russell, "Science on the Fringe of Europe," pp. 330–1.

[12] The movement and the circumstances that distinguish it from the earlier development of academies and societies during the first half of the century are discussed in McClellan, *Science Reorganized,* pp. 109–51.

[13] On the academies of provincial France, see Daniel Roche, *Le siècle des lumières en province: Académies et académiciens provinciaux, 1680–1789,* 2 vols. (Paris: Mouton, 1978).

bound ever more closely to the Piedmontese throne by its transformation into the Società Reale della Scienze (by Vittorio Amedeo II in 1759) and then into the Accademia Reale delle Scienze (by Vittorio Amedeo III in 1783), the Academy was asked, in 1789, to undertake an investigation of the processes of dyeing, in particular on wool.[14] The request, transmitted by Count Graneri, the king's newly appointed chief minister and leading advocate of free trade, stressed the importance of reducing Piedmont's dependence on foreign markets and rehearsed the benefits that were to be anticipated if only savants would "deign to enter the workshop in order to combine practice with theory, instead of leaving it to artisans."[15] The union of patriotic sentiment with a vision of an economy invigorated by science evoked a ready response from members who, for thirty years since the founding of the *società privata,* had repeatedly sought an involvement in the technological improvement of their country. For eighteen months, a committee of nine academicians with appropriate interests (representing almost half of the Academy's total resident membership) addressed the task of publishing a comprehensive digest of the art of dyeing and of the legislative and economic context as it affected Piedmont. The plans were grandiose: a library of books and journals on dyeing, mainly in Italian and French, was assembled, a questionnaire was distributed to manufacturers and artisans involved in all the stages of the production and finishing of woolen goods, and a laboratory was fitted out. The reality, though, fell far short of the high initial expectations. The laboratory was never used, the gathering of information about practices proved far more difficult than Graneri's initial request had anticipated, and the chemical knowledge that the academicians had at their disposal proved impotent before the complexities of the technology they were seeking to understand and advance.

The large quantity of accumulated notes and draft reports indicates the seriousness with which the inquiry was pursued. But the abrupt cessation of the work in 1791 inexorably signaled its failure. At least in the local Piedmontese context, the cost of the failure was high. The government's overriding aim of engaging science in the promotion of the use of locally grown woad as a substitute for imported indigo (the coloring material for the blue military uniform of Piedmont) had been poorly served, and the academicians' hopes of demonstrating the importance, for the Piedmontese economy, of their engagement in the international world of learning had come to nothing.

What occurred in Turin is telling as an example of the late flowering of governmental confidence in an academy as a potential servant of the national

---

[14] Luisa Dolza, "Dyeing in Piedmont in the Late Eighteenth Century," *Archives internationales d'histoire des sciences,* 46 (1996), 75–83, and Dolza, "The Struggle for Technological Independence: Textiles and Dyeing in Eighteenth-century Piedmont" (University of Oxford M.Litt. thesis, 1995), especially chap. 3, which deals with the academy's investigation into dyeing. On the history of the Turin Academy, see also *Tra Società e Scienza. 200 Anni di Storia dell'Accademia delle Scienze di Torino. Saggi, Documenti, Immagini* (Turin: Umberto Allemandi, 1988).

[15] Dolza, "The Struggle for Technological Independence," pp. 89.

interest (manifested most enduringly in Vittorio Amedeo III's installation of the Academy in a fine seventeenth-century palace in 1786). However, it also points to the difficulty, in practice, of achieving the union of understanding and utility that was expressed in the Academy's motto, "veritas et utilitas." Examples of a similar disparity between aspiration and realization can be found in many other parts of Europe. Nevertheless, what James McClellan III has called the "scientific society movement" of the later eighteenth century continued and in certain cases even prospered.[16] Several academies with national status, for example, maintained a significant public function in the editing of almanacs and in the approval of inventions and the assessment of requests for patents and other forms of privilege, and most of them laid implicit claim to an economic and patriotic role by redoubling their efforts in the mounting of prize competitions on applied subjects. But the accumulating record of disappointment in the attempts to apply science, reinforced by the growing indifference of manufacturers, agriculturalists, and men of science toward the prizes and other incentives that the academies offered, inexorably exposed the fragility of the academicians' utilitarian rhetoric. The greatest challenge to the status of academies, however, arose from the changing nature of the technological innovations that characterized the incipient Industrial Revolution, especially in large machinery and chemical and metallurgical processes.

In Britain, where the impact of the Industrial Revolution was greatest, the governmental structures that might have responded to the new challenge were few and weak. John Theophilus Desaguliers was just one of a number of individual Fellows of the Royal Society who displayed an interest in manufacturing and the education of artisans almost from the time of his election to the society in 1714.[17] But the tone of the society remained metropolitan and aristocratic, and even in the years of its intellectual reinvigoration under the long presidency of Sir Joseph Banks (1778–1820), the concern for industrial technology remained muted. In this respect, the Royal Society of Edinburgh, founded in 1783, was no different,[18] and only the Society of Arts, from its foundation in London in 1754, offered a national setting in which the interests of manufacturing and commerce could be aired. However, like the local literary and philosophical societies that were established in the industrial North and Midlands in the late eighteenth and early nineteenth centuries, the Society of Arts fulfilled this role in the absence of any royal or other state recognition, the prefix "Royal" only being added in 1908.[19]

In continental Europe, the national academies, with their strong traditions

[16] McClellan, *Science Reorganized*, pp. 109–51.

[17] Larry Stewart, *The Rise of Public Science: Rhetoric, Technology, and Natural Philosophy in Newtonian Britain, 1660–1750* (Cambridge University Press, 1992), especially pp. 213–54.

[18] Steven Shapin, "Property, Patronage, and the Politics of Science: The Founding of the Royal Society of Edinburgh," *The British Journal for the History of Science*, 7 (1974), 1–41.

[19] Henry Trueman Wood, *A History of the Royal Society of Arts* (London: John Murray, 1913), pp. 20–2, and Derek Hudson and Kenneth H. Luckhurst, *The Royal Society of Arts 1754–1954* (London: John Murray, 1954), pp. 3–18.

of involvement in technology and the applied aspects of science, could not stand aloof from the quickening pace and growing diversity of the utilitarian demands that were made on them. The Académie Royale des Sciences in Paris was the most notable academy that felt the pressure of these new circumstances. Especially in the 1770s and 1780s, its space, time, and facilities were all placed under great strain as academicians struggled to cope with a rising tide of piecemeal requests for advice from local administrations and courts.[20] Ministries, too, contributed to the strain, in part through their own piecemeal requests for advice but also through the expectation of enlightened ministers and officials of the mid and later eighteenth century, such as Daniel Trudaine, Anne-Robert-Jacques Turgot, and Jean-François Tolozan, that academicians would participate in the escalating and inevitably more bureaucratic rationalization of the areas of technology that lay under immediate state control. The expectation had important consequences for the place of the Académie in late *ancien régime* society; although it perpetuated the Colbertian ideal of closeness between the worlds of government and of learning in France, it also eroded the academicians' rather retired, formal position in the state's provision for technological efficiency. Now, savants whose services were sought were likely to find themselves responding to the calls upon their time and sense of duty in settings closer to the scene of production than to the quiet rooms in the Louvre in which they had traditionally formulated their judgments. This adjustment in the location of government-sponsored science was part and parcel of the state's steadily growing involvement in the diverse group of designated *manufactures royales* that had begun in the seventeenth century. As early as the 1660s, the Manufacture Royale des Glaces (for mirrors) at St.-Gobain and the tapestry and carpet-making enterprise of the Gobelins had come under the complete control of the Crown. But during the eighteenth century, the network of *manufactures royales* had been significantly extended, most notably in 1745, when the manufacture of porcelain too became an activity of the state, first in the château of Vincennes, at the eastern extremity of Paris, and then at Sèvres.[21]

The scientific importance of these factories, in particular the Gobelins and the porcelain factories at Vincennes and Sèvres, lay in their practice of engaging chemists as expert advisers. The appointment of the academician Jean Hellot – who worked first at Vincennes (from 1751) and then at Sèvres (from the factory's reestablishment there in 1756 until his death in 1766) – and of his assistant and later successor, Pierre-Joseph Macquer, who was employed at Sèvres until his death in 1784, were landmarks in this respect, albeit landmarks of significantly diverse character.[22] Hellot's approach was characteris-

---

[20] Hahn, *The Anatomy of a Scientific Institution*, pp. 116–24.
[21] Charles Coulston Gillispie, *Science and Polity in France at the End of the Old Regime* (Princeton, NJ: Princeton University Press, 1980), pp. 390–413.
[22] Ibid., pp. 401–6.

tic of an earlier conception of a proper scientific engagement, being resolutely empirical and tied to the practices of the shop floor. Macquer's, by contrast, was experimental. Its objectives embraced the precise chemical analysis of clays and other ingredients and the establishment of at least a rough and ready foundation in theory that would account for the properties of the various kinds of porcelain and prove equal to the teasing but eventually successful quest for a high-quality hard porcelain made from a French alternative to the imported kaolin that came in uncertain quantities from Saxony.

Both Hellot and Macquer also held advisory positions as dye chemists at the Gobelins, where (in the absence of adequate records) it must be assumed that a similar contrast existed between their interests, respectively, in the observation of day-to-day practice and in the development of a more "scientific" approach.[23] At the Gobelins, however, the nature of the technologies that were used meant that recourse to theory and experiment was more difficult than it was at Sèvres and that the craft tradition was correspondingly more resilient. Nevertheless, Hellot published an important book in 1750 in which he advanced a physical theory of the mechanisms that bound the particles of color to the fibers being dyed.[24] Later, following Macquer's death in 1784, the intellectual challenge and (it must be said) an income of six thousand livres a year were sufficient to induce Claude-Louis Berthollet to accept the position of *directeur des teintures* at the Gobelins and so to divert him from a career in medicine to one that was to take him, during the Empire of Napoleon I, to a position of preeminence in the community of French chemists.[25] In this rise, Berthollet's two-volume *Eléments de l'art de la teinture* (1791), a work required by his letter of appointment from Louis XVI's minister of state Charles-Alexandre Calonne,[26] was an important stepping-stone. The book was very much the work of a savant. It did nothing to conceal the distance that separated its contents from the rule-of-thumb realities of the dyeshop and so to bridge a gap that Berthollet blamed on the mystery that dyers themselves maintained – a mystery that his fellow chemist Jean-Antoine Chaptal analyzed at about the same time in terms of the prejudices engrained in the minds of artisans who saw the chemist as a "dangerous innovator."[27] Instead of practical advice and recipes, Berthollet offered a systematic description of the properties of the fibers and reagents involved in dyeing and

---

[23] Ibid., pp. 407–10.

[24] Jean Hellot, *L'art de la teinture des laines et des étoffes de laine en grand et petit teint* (Paris, 1750).

[25] Michelle Sadoun-Goupil, *Le chimiste Claude-Louis Berthollet (1748–1822). Sa vie, son oeuvre* (Paris: Librairie Philosophique J. Vrin, 1977), pp. 21–4. See also Sadoun-Goupil, "Science pure et science appliquée dans l'oeuvre de Claude-Louis Berthollet," *Revue d'histoire des sciences,* 27 (1974), 127–45, especially 127–33 and 140–5.

[26] Calonne to Berthollet, 24 February 1784, quoted in Sadoun-Goupil, *Le chimiste Claude-Louis Berthollet,* p. 22.

[27] J. A. Chaptal, *Elemens de chimie,* 3 vols. (Montpellier, 1790), vol. I, p. lii. On Berthollet's position with respect to the scientific and artisanal traditions, see Barbara Whitney Keyser, "Between Science and Craft: The Case of Berthollet and Dyeing," *Annals of Science,* 47 (1990), 231–60.

bleaching; using the doctrine of affinities, he went as far as contemporary theory would allow in the explanation of the chemical processes at work. Despite its origins in a commission by the government and Berthollet's statement that he had sought to place himself "entre les physiciens & les artistes,"[28] the *Eléments* was addressed less to the shop floor than to the appropriate sectors of the international community of chemists, who saw to its translation into English and Spanish.

Although France stood out among European countries for the extent and intimacy of governmental involvement in manufacturing,[29] states everywhere maintained some presence in the production of either finished goods or raw materials for industrial use. The nature of the presence – and of the scientific and technological expertise that was provided in support of it – varied greatly. At one extreme lay French interventionism, manifested not only in the state-owned enterprises but also in the structures for more remote forms of encouragement from which private factories benefited: in these suppler structures, the inventor and builder of automata, Jacques Vaucanson, exercised a powerful influence both through the advice he gave and through the looms and other mechanical devices (mainly for the production of silk) that he himself perfected in his capacity as the state's senior inspector of manufactures for more than forty years from 1741.[30] At the other extreme was Britain, where the responsibility of the state was seen to lie in little more than the provision of a patent system that, however imperfectly, would protect the innovations of inventors and private industrialists.[31] Between those extremes were cases, such as those of Spain, Sweden, and the more industrially active German states, in which the mercantilist tendencies of governments had significant consequences for the relations between science and industry.

Of all European countries, Bourbon Spain came closest to France in the degree of governmental intervention in manufacturing and the scientific and technical support that it required. After the accession of the first Bourbon king, the French-born Felipe V, in 1701, royal manufactures multiplied under the aegis of a coordinating committee, the Junta General de Comercio y Moneda, in pursuit of a mercantilist economic policy on Colbertian lines.[32]

---

[28] Claude-Louis Berthollet, *Eléments de l'art de la teinture*, 2 vols. (Paris, 1791), vol. 1, p. xlii.

[29] For another good example of the French state's attempts to engage scientific expertise in the advancement of technology, see the discussion of Lavoisier's activity in the production of gunpowder in the later years of the *ancien régime*, in Patrice Bret, "Lavoisier et l'apport de la chimie académique à l'industrie des poudres et salpêtres," *Archives internationales d'histoire des sciences*, 46 (1996), 57–74.

[30] André Doyon and Lucien Liaigre, *Jacques Vaucanson: Mécanicien de génie* (Paris: Presses Universitaires de France, 1966), pp. 175–253.

[31] Christine MacLeod, *Inventing the Industrial Revolution: The English Patent System, 1660–1800* (Cambridge University Press, 1988), pp. 115–200.

[32] On the relations among science, manufacturing, and the enlightened Spanish monarchy, see several contributions, especially those in section I ("La política científica ilustrada"), Joaquín Fernández and Ignacio González Tascón (eds.), *Ciencia, técnica y estado en la España ilustrada* (Madrid: Ministerio de Educación y Ciencia, 1990). On the particular case of dyeing, bleaching, and calico printing, I

From the start, but especially during the reign of Carlos III (1759–1788), dyeing and the printing of fabrics were a main (although by no means exclusive) focus, and dyers and colorists were routinely enticed to senior, well-paid positions in the relevant factories. One of the earliest of these new arrivals was an Irish dyer, Michael Stapleton, who was appointed Tintorero Mayor at the Real Fábrica de Paños (wool) in Guadalajara near Madrid in 1725; thereafter, until the late eighteenth century, Guadalajara continued to attract technical experts from abroad. Although similar appointments at royal manufactures in Talavara, Avila, and Madrid show that the case of Guadalajara was by no means unique, what happened there served as a model of a coordinated industrial enterprise committed to modernization. It pursued its aims not only through the foreigners it engaged but also by receiving visitors, encouraging its own employees and apprentices (*pensionados*) to travel abroad, and creating its own school of dyeing and chemistry.

The aspiration of the Guadalajara factory for technological self-sufficiency was an elaborate expression of a broader governmental policy, which consistently gave a high priority to the fashioning, in every trade, of a work force that would match those of the most advanced nations in its command of both the practices and the science of its craft. Agriculture, as much as manufacturing industry, was perceived as the likely beneficiary of the policy. This, allied to the Bourbon monarchy's special concern for the textile industry and in particular for dyeing and printing, served to reinforce the privileged place of chemistry among the auxiliary sciences that were fostered, both through "in house" instruction of the kind that was provided at Guadalajara and through other institutions, such as the chemical laboratory that was opened at the royal artillery school of Segovia in the late 1780s. It was here and, from 1799, in a post at the well-equipped new laboratory in Madrid, both financed by the government, that Joseph-Louis Proust, the outstanding chemist of late eighteenth-century Spain, performed his important work on definite proportions. The period has obvious significance as the one in which the new French chemistry of Lavoisier entered Spain, mainly through the schools in Madrid and an important link between the chemists of Montpellier and Barcelona.[33]

rely heavily on Agustí Nieto-Galan, "Dyeing, Calico Printing and Technical Exchanges in Spain: The Royal Manufactures and the Catalan Textile Industry, 1750–1820," in Robert Fox and Agustí Nieto-Galan (eds.), *Natural Dyestuffs and Industrial Culture in Europe, 1750–1880* (Canton, MA: Science History Publications, 1999), pp. 101–28. On the role of the state in calico printing, see James K. J. Thomson, "State Intervention in the Catalan Calico-printing Industry in the Eighteenth Century," in Maxine Berg (ed.), *Markets and Manufacture in Early Industrial Europe* (London: Routledge, 1991), 57–89, and Thomson, *A Distinctive Industrialization: Cotton in Barcelona, 1728–1832* (Cambridge University Press, 1992), especially pp. 40–6, 67–72, 132–8, 157–9, and 202–5.

[33] Ramón Gago, "The New Chemistry in Spain," *Osiris*, 2nd ser. 4 (1988), 169–92; Agustí Nieto-Galan, "Un projet régional de chimie appliquée à la fin du XVIIIe siècle. Montpellier et son influence à l'école de Barcelone: Chaptal et Francesc Carbonell," *Archives internationales d'histoire des sciences*, 44 (1994), 38–62; and Nieto-Galan, "The French Chemical Nomenclature in Spain: Critical Points, Rhetorical Arguments, and Practical Uses," in Bernadette Bensaude-Vincent and Ferdinando Abbri

But with the new chemistry, there also came a knowledge of Berthollet's theoretical treatment of bleaching and dyeing. This became known from 1795
through the Spanish edition of the *Eléments de l'art de la teinture*.[34] The translation was the work of Domingo García Fernández, a pupil of Chaptal in Montpellier who went on to occupy a number of important positions in the extensive technical administration of the state, notably (at different times in his long
career) as Director of the Ministry of Finance's glass factory at San Ildefonso,
General Director of the Royal Manufactures of Gunpowder and Saltpeter,
and Director of Mining in Almadén.[35] These and other contacts between
France and Spain certainly make it hard to sustain the traditional view of the
Spanish scientific community as isolated and inactive; equally they point to
the role of the monarchy in reconciling the national economic interest with
an openness to the most progressive currents in science internationally.

In Sweden, too, chemistry was the main beneficiary of a governmental
concern for economic improvement, in this case a concern that went back
further than it did in Spain. The concern was first manifested in the 1630s,
when the government of the day created a powerful Board of Mines primarily
to control the expansion of the copper mine at Falun but also to regulate all
aspects of the mining industry.[36] From the start, the Board could call on a
Chamber of Assaying and an associated Laboratorium chymikum, both of
them in Stockholm, where the core work of assaying was gradually extended
to include a wide range of work in analysis, metallurgy, and other branches of
chemical technology. Urban Hiärne (who ran the laboratory from 1683 to 1719)
and Georg Brandt (who did so from 1730 until his death in 1768) had an especially important role in this broadening of activity and in preparing the ground
for a flowering of Swedish chemistry that profitably obscured the boundaries
between academic chemistry and the chemistry of the mining industry and
between theory and description. Torbern Bergman and Carl Wilhelm Scheele,
the two leading Swedish chemists of the 1770s and early 1780s, were the most
distinguished representatives of this chemical Golden Age. The role of the
Board of Mines in what occurred cannot be overstated, not least because of
the benefits that it brought to disciplines other than chemistry. From 1700,
for example, the range of the Board's activities was extended by the addition

---

(eds.), *Lavoisier in European Context: Negotiating a New Language for Chemistry* (Canton, MA: Science
History Publications, 1995), pp. 173–91.

[34] Claude-Louis Berthollet, *Elementos del arte de teñir*, trans. Domingo García Fernández, 2 vols. (Madrid,
1795–6).

[35] Gago, "The New Chemistry in Spain," pp. 177–8.

[36] Anders Lundgren, "The New Chemistry in Sweden: The Debate That Wasn't," *Osiris*, 2nd ser. 4
(1988), 146–68. See also, for helpful insights into early industrialization in Sweden, Svante Lindqvist,
Technology on Trial: The Introduction of Steam Power Technology into Sweden, 1715–1736 [Uppsala
Studies in History of Science 1] (Stockholm: Almqvist & Wiksell, 1984). A general survey of the relations among science, industry, and the state is Russell, "Science on the Fringe of Europe." For a
popular but useful account of the Falun mine and its place in Swedish life in this period, see Sven
Rydberg, *The Great Copper Mine: The Stora Story* (Hedemora: Stora Kopparbergs Bergslags AB in
collaboration with Gidlunds Publishers, 1988), pp. 85–160.

of a Mechanical Laboratory at Falun based on a fine collection of engineering models, and the science of machines began to flourish as a Swedish speciality. Here, Christopher Polhem, the technical director of the copper mine and "Archimedes of the North," performed some of his most important experiments in hydrodynamics while also advising on practical aspects of mining technology throughout Sweden.[37]

It was not only in Sweden that mining served as a powerful incentive for governmental investment in science and technology. In countries with large mineral deposits, national administrations for the control of mining were already common in the seventeenth century. But it was in the eighteenth century that the importance of training scientifically informed administrators and technical officials was formally recognized by the creation of mining academies in which science was often able to flourish alongside the main task of training in the more applied aspects of the curriculum. Although important academies were established in Ekaterinburg in Siberia, St. Petersburg, and Paris (the prestigious Ecole Royale des Mines) between 1763 and 1783, the new academies tended to cluster in central Europe. Among the most notable of them was the Freiberg Mining Academy in Saxony.[38] Opened in 1765 as part of a programme of economic rehabilitation after the Seven Years' War, the Academy was able to build on a tradition of technological improvement, encouraged and supported by the Elector of Saxony, that went back to the sixteenth century. Because of this tradition and the courtly patronage it enjoyed, it had no difficulty in acquiring fine teaching staff, such as Christlieb Ehregott Gellert, an experienced analyst and authority on machinery and smelting (and the first Professor of Metallurgical Chemistry). It also gained a reputation that attracted students from across the continent as well as those from Saxony (whose expenses were met by the state).

The reputation of the Freiberg Academy rested not only on the prestigious careers to which it gave access but also on its prominence in the international world of science. No one contributed more to that prominence than Abraham Gottlob Werner. This early student at the Academy completed his preparation with legal study at the University of Leipzig before returning to Freiberg in 1775 as Professor of Mining and Mineralogy and Curator of the Academy's collection of minerals. Werner's main qualification for the post was an important work on the classification of fossils that he had published at Leipzig.[39] But once in Freiberg, he built his reputation and that of the Academy less on his publications than on his skill as a teacher and on the vast correspondence that he maintained with mineralogists and geologists throughout Europe. At the end of more than forty years at the Academy and almost as long as

---

[37] Lindqvist, *Technology on Trial*, pp. 67–79.

[38] On the Freiberg Academy and more generally on the relations between chemistry and mining in central Europe, see Gerrylynn K. Roberts, "Establishing Science in Eighteenth-Century Central Europe," in Goodman and Russell (eds.), *The Rise of Scientific Europe*, pp. 361–86 (375–85).

[39] Abraham Gottlob Werner, *Von den äusserlichen Kennzeichen der Fossilien* (Leipzig, 1774).

inspector of mines in the Saxon Mining Service, Werner had made a signal mark on both the Saxon economy (reflected in the quickening pace of the production and exportation of silver, lead, and other metals) and the discipline of geology, in which he developed a "Neptunist" account of the earth's history that continued to be widely discussed well into the nineteenth century.[40]

The mining academies of central Europe provide compelling evidence of the confidence of governments in the capacity of scientific intervention to enhance the returns on the mineral resources on which their economies largely depended. What occurred in Saxony had its equally beneficent counterpart in the Austro-Hungarian empire, where the decision to improve the faltering performance of the mines of Schemnitz (now Banská Stiavnica in Slovakia) by a resolute educational initiative was taken by the government, working through an imperial commission. As in Freiberg, the immediate incentive was the quest for recovery after the Seven Years' War. Accordingly, it was money from the central government, administered in this case by the Imperial Mining Chamber in Vienna, that made it possible first to establish the curriculum of Schemnitz's modest mining school as a practical complement to the theoretical syllabuses of the University of Prague. Then, in 1770, the school was reconstituted as an independent Mining Academy with a three-year syllabus and specialist divisions devoted to mathematics, chemistry and metallurgy, and the sciences of mining.[41] Especially in the division for chemistry and metallurgy, notable work, both scientific and technological, was done under Nicholas Jacquin, J. A. Scopoli, and then Anton Reprecht von Eggesberg. After a promising start, however, the reputation of Schemnitz failed to keep pace with that of Freiberg, largely because of the physical distance (200 kilometers, representing a journey of three days) and the degree of incomprehension of the day-to-day realities of mining that separated the Academy from the administration in Vienna. Also, the shifting political priorities of the Empire and the ambitions of some of the ablest professors to leave the remote, mountainous region in which Schemnitz was situated engendered damaging instability. By the turn of the century, the brief Golden Age that the Academy enjoyed in the 1770s and 1780s was a thing of the past.

The later history of the Schemnitz Academy points again to the element of vulnerability that was always present in institutions that depended directly on the support of government. Favor could quickly turn to indifference, and on occasion politically or ideologically motivated interference could rob an institution of a teacher whose opinions earned disapproval: the power of the King of Prussia, Frederick William I, to dismiss the mathematician and philosopher Christian Wolff from his chair at the University of Halle in 1723, fol-

---

[40] Rachel Laudan, *From Mineralogy to Geology: The Foundations of a Science, 1650–1830* (Chicago: University of Chicago Press, 1987), pp. 87–112.

[41] Roberts, "Establishing Science in Eighteenth-Century Central Europe," pp. 380–3, and D. M. Farrer, "The Royal Hungarian Mining Academy, Schemnitz: Some Aspects of Technical Education in the Eighteenth Century" (University of Manchester M.Sc. thesis, 1971).

lowing a charge of atheism, and to banish him from the country delivered a very public reminder of the fate that could follow serious royal censure. Such cases were rare, however, and the diffuseness of the boundaries between science and the economic or strategic interests of the state was far more often a source of good.

Academies for civil and military engineering provide further abundant evidence of the scientific as well as the technological benefits that could flow from a well-founded system of state-controlled instruction. The consistent support that governments of the *ancien régime* bestowed on the Ecole des Ponts et Chaussées from its foundation in 1747 allowed the school, which was the first in a long line of advanced engineering schools in France, to develop as a source not only of highly trained men for the state administration responsible for roads, bridges, harbors, and (from the late eighteenth century) canals but also as an institution in which mathematics and mechanics could be pursued in their theoretical as well as their practical aspects.[42] The impact of the school and the state corps des Ponts et Chaussées that it fed was out of all proportion to their size. Of 387 students admitted between 1769 and 1788, only 141 were commissioned into the corps.[43] Moreover, the school had no permanent teaching staff. Instead, for almost half a century Jean-Rodolphe Perronet, the first head of the school (and the corps) administered a system of instruction in which the ablest senior pupils taught their juniors. In addition, all pupils attended courses at other Parisian institutions, such as the Jardin du Roi, the Collège de France, the recently founded Ecole des Arts (a private architectural school), and the school attached to the Académie d'Architecture, as a complement to the study of such textbooks as Alexis-Claude Clairaut's *Elémens d'algèbre* (1746) and Charles Bossut's treatises on mechanics. Competition, fostered by a system of prizes that gradually gave way to a greater emphasis on examinations, was another essential pedagogical tool, one that constantly stretched the pupils to the limit of their capacity and maintained relentlessly high standards in the passage from the school to an appointment in the corps.[44] The quality of the roads near Narbonne, which the English traveler Arthur Young described as "stupendous works" in 1787, clearly cannot be explained entirely by the influence of the relatively few pupils of the school who went on to become fully fledged *ingénieurs des Ponts et Chaussées*:[45] a lesser hierarchy of *inspecteurs* and *sous-ingénieurs* awaiting

---

[42] Antoine Picon, *L'invention de l'ingénieur moderne: L'Ecole des Ponts et Chaussées 1747–1851* (Paris: Presses de l'Ecole Nationale des Ponts et Chaussées, 1992), Livre 1 ("Les ingénieurs des lumières") on the eighteenth century. For a good, briefer account, see Gillispie, *Science and Polity in France,* pp. 479–98. The designation "Royale" was granted in 1755.

[43] Statistical information on the admissions to the Ecole des Ponts et Chaussées and to the corps is conveniently gathered in Picon, *L'invention de l'ingénieur moderne,* pp. 731–3.

[44] For an excellent study of the *concours* in architecture, mapping, and design, see Picon, *L'invention de l'ingénieur moderne,* pp. 149–207 and 734–5.

[45] Young's comment is quoted in Gillispie, *Science and Polity in France,* p. 495. For the original, see Arthur Young, *Travels, during the Years 1787, 1788 and 1789. Undertaken more particularly with a View of*

promotion and a much larger body of *conducteurs* and other assistants (many of them former pupils of the school who had not completed the course) were also essential to the technical excellence that French civil engineering achieved. But the system of which the school and corps were an essential part was unanimously recognized as a success, marked by such monuments as the daring arches of Perronet's Louis XVI bridge in Paris and Louis-Alexandre Cessart's retrospective *Description* of his work in hydraulic engineering, of which his scheme for the creation of an artificial port and an offshore seawall at Cherbourg was the most daring illustration.[46]

It is characteristic of the Ecole des Ponts et Chaussées that its legacy took a predominantly material form. Among its graduates during the eighteenth century, only Gaspard Riche de Prony, who studied at the school from 1776 to 1780, achieved a scientific reputation that transcended the realm of civil engineering and the areas of architecture into which members of the corps des Ponts et Chaussées extended their brief. In this respect, schools of military engineering were scientifically more fertile, at least in France. One reason for this, as Charles Gillispie has suggested, may be that, in military engineering, the separation between the schools and the world of practice was greater than it was in civil engineering.[47] Indeed, those who taught at the royal engineering school, the Ecole royale du Génie, that was founded at Mézières in 1748 often saw a teaching appointment as a welcome means of escaping from the rigors of normal duty. There, as to a lesser extent at the artillery school at La Fère (transferred to Bapaume in 1766), a staff well versed in both the techniques and the underlying theory of military engineering and gunnery offered training that opened the way to a fine military career and even in some cases – such as those of Charles-Augustin Coulomb, a pupil at Mézières in 1760–1, and Lazare Carnot, a decade later – scientific eminence.[48] The fact that a high proportion of the entrants who embarked on the two-year course, amounting to more than two-thirds in the last decade of the ancien régime, were the sons of noble families was a constant threat to the seriousness of the school. But it does not seem to have detracted significantly from the institution's reputation. Moreover, the aristocratic tone was perfectly compatible with the admission of candidates from socially less elevated backgrounds; one, Gaspard Monge, was to become the outstanding exemplar of the scientific tradition of Mézières from the time he was appointed professor of mathematics there

ascertaining the Cultivation, Wealth, Resources, and National Prosperity, of the Kingdom of France (Bury St. Edmunds, 1792), vol. 1, p. 20.

[46] Louis-Alexandre de Cessart, *Description des travaux hydrauliques de Louis-Alexandre de Cessart*, 2 vols. plus 1 vol. of plates (Paris: A. A. Renouard, 1806–8).

[47] Gillispie, *Science and Polity in France*, p. 506. On Mézières and the other military schools, see pp. 506–48.

[48] On the schools at Mézières and La Fère and the training offered there and elsewhere to military engineers and artillery officers, see Roger Hahn, "L'enseignement scientifique aux écoles militaires et d'artillerie," and René Taton, "L'Ecole royale du Génie de Mézières," in *Enseignement et diffusion des sciences en France au XVIIIe siècle* (Paris: Hermann, 1964), pp. 513–45 and 559–615.

in 1769, when he was still in his early twenties. Monge's election to the Académie des Sciences with the rank of *adjoint géomètre* in 1780 and the fame that came to him from the 1790s as the creator of the discipline of descriptive geometry recognized his contribution as a mathematician rather than as a military engineer, but his work bore the indelible mark of the teaching he had undertaken as professor (although with increasing reluctance, it must be said) in such practical subjects as drawing, cartography, and surveying.[49]

Despite the visibility of Mézières and the Ecole des Ponts et Chaussées and the admiration they attracted abroad, the immediate impact outside France of the French model of specialized professional schools under close state control was limited. Britain, for example, remained loyal to apprenticeship and learning on the job until the mid-nineteenth century. Sweden, on the other hand, was typical of a number of countries in which indigenous traditions in technical education and research, such as those administered by the government's Board of Mines since the seventeenth century, made the borrowing of foreign models unnecessary. And even in Spain, where France was always a natural object of attention, emulation was slow and imperfect. It was not until the very end of the eighteenth century that the French model began to take root there under the influence of Agustín de Betancourt, who had studied at the Ecole des Ponts et Chaussées between 1784 and 1791 as one of a number of scholarship holders supported by the Spanish monarchy.[50] In the event, the first Escuela de Caminos y Canales, founded in Madrid in 1802 following a reorganization of the Real Gabinete de Máquinas (in which scale models and other materials that the scholarship holders had brought back from France had been used for the instruction of engineers since 1792), was short-lived: it succumbed to the disruption of the War of Independence (1808–14) and, after another failed attempt in the 1820s, was properly constituted only in 1834. Even before 1834, however, the long years of halting preparation and false starts brought considerable benefits, above all in the Spanish translations of such works as Monge's *Géométrie descriptive* and Louis B. F. Francoeur's *Traité de mécanique élémentaire,* which helped to strengthen a bridge between the French and Spanish communities in mathematics and engineering comparable to the one that chemists had recently begun to build between Montpellier and Barcelona.

The ease with which knowledge and practices in science, technology, and education passed between nations is a leading characteristic of the eighteenth century. Books and instruments were traded freely across national boundaries; translations of major works were frequent, especially in the later years of the

---

[49] René Taton, *L'oeuvre scientifique de Monge* (Paris: Presses Universitaires de France, 1951), especially pp. 10–19, 50–100, and 352–75.

[50] Santiago Riera i Tuèbols, "Industrialization and Technical Education in Spain, 1850–1914," in Robert Fox and Anna Guagnini (eds.), *Education, Technology and Industrial Performance in Europe, 1850–1939* (Cambridge University Press, and Paris: Editions de la Maison des Sciences de l'Homme, 1993), pp. 146–70.

century; and monarchs and governments made their contribution by the support that most of them gave to the publications of the academies and national societies they helped to sustain. Personal mobility, too, was greater than ever before. With increasing frequency, scientists and engineers traveled abroad, often with the aid of their governments, to gather information on industrial and military matters: the journeys of Gabriel Jars to the mines of central Europe and Britain in the 1750s and 1760s and of several French military engineers who went to Britain in the 1780s, for example, were undertaken as systematic fact-finding missions amounting to technological espionage.[51] But the freedom with which knowledge could be gathered, even by the most determined inquirers, had its limits. The caution of James Watt and Matthew Boulton in the information they were willing to divulge to visitors who saw their steam engines under construction or at work was a typical response where economic advantage was involved,[52] and military and naval installations always remained sensitive areas.

A well-honed rhetoric stressed the universal character of the Republic of Letters and the principle that knowledge was, or should be, open to all, but it could not conceal the element of national interest that, to varying degrees, fired the majority of the initiatives to which governments gave their material backing. An administration's association with an academy, observatory, or botanical garden would lend the aura of enlightenment at a time when absolute rule was falling into disrepute; the promotion of schools of engineering would serve obvious strategic and economic ends; and state-sponsored visits abroad would help to prevent a rival nation from gaining an unobserved advantage. Such considerations were bred of the competitiveness that, in an age dominated by mercantilist thinking, drove national policy making.

As the century passed, the effects of this rivalry had increasingly grandiose consequences. Even in Britain, where governmental support for science tended to be modest, King George III began the construction of his own observatory at Richmond in 1768 and sustained the work of what subsequently became known as Kew Observatory, notably by the appointment of Stephen Demainbray, who held the post of superintendent (albeit largely as a sinecure) until

---

[51] Margaret Bradley, "Engineers as Military Spies? French Engineers Come to Britain, 1780–1790," *Annals of Science*, 49 (1992), 137–61. For an excellent brief account of industrial espionage in the eighteenth century, see John R. Harris's Rolt Memorial Lecture of 1984, reprinted in Harris, *Essays in Industry and Technology in the Eighteenth Century: England and France* (Aldershot: Variorum, 1992), 164–75. Jars's travels are described in Gabriel Jars, *Voyages métallurgiques ou recherches et observations sur les mines et forges de fer . . . faites en 1758, jusques & compris 1759, en Allemagne, en Suede* [sic], *Angleterre . . . & en Hollande*, 3 vols. (Lyon, 1774–81). For a biographical sketch of Jars, see the eloge of him by Granjean de Fouchy, ibid., vol. 1, pp. xxi–xxviii.

[52] For a vivid illustration of the difficulty that visitors had in securing information, see the account of Agustín de Betancourt's visits to the Soho works in Birmingham and Albion Mills in London in 1788 in Jacques Payen, *Capital et machine à vapeur: Les frères Périer et l'introduction en France de la machine à vapeur de Watt* (Paris: Mouton, 1969), pp. 157–9. Betancourt had been sent by the court of Spain to assemble a collection of models for the instruction of artisans in the principles of hydraulics.

his death in 1782.[53] Still more ambitious was the series of three costly voyages of exploration in the Pacific Ocean that Captain James Cook commanded from 1768 to 1771, from 1772 to 1775, and from 1776 to 1780, during the last of which he met his death.[54] Financed by the Admiralty at the request of King George III (himself the possessor of a fine collection of instruments),[55] the voyages were seen as having significant scientific objectives, including (at the Royal Society's request and with the aid of a personal donation of £4000 from the king) the plan of observing the transit of Venus in 1769.[56] The discovery of unknown islands, plants, and peoples also helped to justify the large investment that the expeditions demanded. The motives here were mixed: territorial aspirations, curiosity about exotic lands and cultures, and the quest for an understanding of nature all vied with one another. But Cook's voyages, like that of the French explorer Louis Antoine de Bougainville to Tahiti between 1766 and 1769,[57] provide ample evidence of the extent to which science could benefit from such an amalgam of incentives.

Sailing in distant, little-known seas put navigational techniques and maps to the rudest test and stimulated further improvement.[58] Cook's second voyage, for instance, is notable for the trial and vindication of John Harrison's marine chronometer as an aid in the determination of longitude: four "watch machines," including a very successful copy of Harrison's "H.4" timekeeper made by Larcum Kendall, were taken on the voyage, along with other apparatus for astronomical observation.[59] Travel and the concern for national interests that helped to give it purpose also stimulated the need for improved charts, hastening the transformation in the accuracy of maps, of land and sea, that gathered pace in most European countries about mid-century. In this transformation, strong official backing, stimulated as much by a desire for

[53] On Demainbray and his career at the observatory, see Alan Q. Morton and Jane A. Wess, *Public & Private Science: The King George III Collection* (Oxford: Oxford University Press in association with the Science Museum, 1993), pp. 89–127.

[54] On the voyages and the background to them, see John Cawte Beaglehole, *The Life of Captain James Cook* (London: Adam & Charles Black, 1974).

[55] The collection is now handsomely displayed in the Science Museum, London. See Morton and Wess, *Public & Private Science.*

[56] Harry Woolf, *The Transits of Venus: A Study of Eighteenth-Century Science* (Princeton, NJ: Princeton University Press, 1959), pp. 161–70.

[57] The voyage is described in Louis Antoine, Count Bougainville, *Voyage autour du monde par la frégate du roi la Boudeuse, et la flûte l'Etoile, en 1766, 1767, 1768, & 1769* (Paris, 1771).

[58] For a study of the effect of voyages on conceptions of the world and of the nature of geography in the eighteenth century, written with special although not exclusive reference to France, see Numa Broc, *La géographie des philosophes: Géographes et voyageurs français au XVIIIe siècle* (Paris: Editions Ophrys, 1975).

[59] On H.4 and Harrison's long and only partially successful struggle to secure its recognition as deserving of the prize of £20,000 that the British government, through the Board of Longitude, had offered in 1714 for an accurate method of determining longitude at sea, see Anthony G. Randall, "The Timekeeper That Won the Longitude Prize," in William J. H. Andrewes (ed.), *The Quest for Longitude* (Cambridge, MA: Collection of Scientific Instruments, Harvard University, 1996), pp. 236–54.

administrative orderliness at home as by territorial aspirations abroad, made France the unrivaled pacemaker. Work on the 181 sheets of the *carte topographique de la France* – organized by César-François Cassini de Thury, the director of the Paris Observatory, under the aegis of the Académie des Sciences and financed by the government (in response to the wishes of Louis XV) – began in 1750 and proceeded rapidly until Cassini's death in 1784 (by which time almost 90 percent of the country had been covered on a scale of 1/86,400).[60] In the detail it displayed, it was not comparable with the map, on the scale of 1/80,000, that replaced it between 1818 and 1866, but it stood as an impressive monument to the work of Cassini and the *ingénieurs géographes* who were employed under him.[61] The French also excelled in maps with a more scientific purpose: here the *Atlas et description minéralogiques de la France,* on which the mineralogist Jean-Etienne Guettard and his later collaborators Antoine Lavoisier and Antoine Monnet worked, with substantial ministerial support, from 1746 until 1780, was the century's outstanding (although unfinished) achievement in geological cartography.[62] The British contribution, by contrast, tended to be directed more strongly to marine cartography, as befitted a leading maritime power, whereas the mapping of Britain itself did not advance significantly until the establishment of the Ordnance Survey in 1791.

This growth of investment in exploration and mapping is another facet of a far broader extension of what were perceived as the interests and responsibilities of governments and monarchs during the eighteenth century. National rivalries of the kind that intensified British and French concern with the South Pacific, the reform movements that fed on and fostered an increasing openness to Enlightenment thought, the recognition of the value of a better mathematical and technical training for military and naval officers, and the stirrings of what gradually took shape as the first industrial revolution all stimulated changes that had consequences for the relations between science and government. Amid a cluster of such diverse causes and motives, the advancement of science was seldom conceived as an end in itself, but science and the communities that pursued it were consistently, if unsystematically, the beneficiaries. In some cases, such as that of Prussia during the reign of Frederick the Great,

---

[60] Henry Marie Auguste Berthaut, *La carte de France 1750–1898,* 2 vols. (Paris: Imprimerie du Service Géographique [de l'Armée], 1898–9), vol. 1, pp. 48–70.

[61] On the later map, see ibid., vol. 1, pp. 171–337. For a history of the *ingénieurs géographes,* a group that underwent many vicissitudes as the needs of the state changed, see ibid., vol. 1, pp. 71–170, and Berthaut, *Les ingénieurs géographes militaires 1624–1831,* 2 vols. (Paris: Imprimerie du Service Géographique [de l'Armée], 1902), especially (for the eighteenth century) vol. 1, pp. 1–120.

[62] Rhoda Rappaport, "The Geological Atlas of Guettard, Lavoisier, and Monnet: Conflicting Views on the Nature of Geology," in Cecil J. Schneer (ed.), *Toward a History of Geology* (Cambridge, MA: MIT Press, 1969), pp. 272–87. The main source of support was put in place in 1766 by Henri L.-J.-B. Bertin, then minister of state under Louis XV with special responsibility for mining, manufacturing, agriculture, and trade.

what occurred could be regarded, at least in its general form, as a continuation of an old tradition of courtly patronage going back to the Renaissance. But the level of support given by the new, more formally constituted structures of the eighteenth century was at once more substantial and more secure than a court alone could provide.

The unprecedented prominence that science, mathematics, and technology had come to occupy in the administrative mechanisms of most European states by the later eighteenth century laid important foundations for the age of professional science that emerged a century or so later. The increase in the possibilities of employment created opportunities for scientific career-making that in 1780 or 1800 were far richer than they had been in the first quarter of the eighteenth century. Even in the universities, which – in many countries and most conspicuously in France – had viewed any idea of governmental intervention with suspicion, enlightened ideals gave science a new prominence. The case of Portugal, where the far-reaching reform of the University of Coimbra in 1772 resulted in the establishment of an observatory in the Faculty of Mathematics and of a chemical laboratory, a cabinet of physics, a museum of natural history, and a botanical garden in the Faculty of Philosophy, illustrates the trend very well.[63] It also demonstrates the capacity of a resolute administration, led in this case by the Marquess of Pombal as prime minister and backed by King José I, to effect change in an essentially conservative institution. No less significantly, it is characteristic of the later eighteenth century that the change and the secularizing tendency it represented were sufficiently resilient to withstand, albeit not wholly unscathed, the death of José I in 1777, the fall of Pombal in the following year, and the renewed influence of sections of the aristocracy and the clergy that opposed the commitment to modernization inherent both in what had occurred and more broadly in the aspirations of the network of internationally minded *estrangeirados* to which Pombal belonged.

It was their general stability that gave the proliferating structures for education, research, and the practice of science and technology their importance for the developments that were to follow as institutionalization accelerated during the nineteenth century. In France, even the turmoil of the Revolution and the Reign of Terror between 1789 and 1794 caused only a temporary interruption in the work of the schools, academies, and other bodies inherited from the *ancien régime* (the French universities, which did not reopen after their closure in 1793, being the only conspicuous exception). There is abundant evidence that during the eighteenth century dependence on governmental or

---

[63] The comments that follow are based on Ana Simões, Ana Carneiro, and Maria Paula Diogo, "Constructing Knowledge: Eighteenth-century Portugal and the New Sciences," in Kostas Gavroglu (ed.), *The Sciences in the European Periphery during the Enlightenment* [*Archimedes*, vol. 2] (Dordrecht: Kluwer, 1999), pp. 1–40.

royal patronage never entirely lost its element of insecurity, and individual astuteness remained as important as ever if the opportunities for personal advancement in science that such patronage made possible were to be exploited to the full. Nevertheless, by the end of the century those opportunities were firmly rooted in the normal structures of a modern state, and careers and intellectual strategies could be planned accordingly.

6

EXPLORING NATURAL KNOWLEDGE
Science and the Popular

Mary Fissell and Roger Cooter

Today, science is something we think we recognize when we see it; it is a part of our cultural landscape. Regarded as easily distinguished from religion, it involves the production of new knowledge rather than the reproduction of faith. Science's stated mission is to tell truths about the natural world – truths produced by trained scientists working in specific fields. There is much argument about details, but a single method is held to lie at the heart of its production.

The processes by which new scientific knowledge is diffused or reformulated for different audiences are also generally regarded as unproblematic. First elaborated and validated in specialist journals, scientific ideas are usually thought to make their way into undergraduate textbooks and subsequently, or simultaneously, undergo popularization or reframing for a wide audience. Newspapers, magazines, television, and radio help perform the task. Ultimately, a few scientific ideas become so widespread that they can be referred to in the shorthand of jokes or cartoons.

This commonsense model of the production and diffusion of scientific knowledge is something like a fried egg, sunny-side up. At the center, the self-contained yolk represents new knowledge generated by scientists. Surrounding this is a penumbra of ever-thinning white, representing diffusion. Finally, the crackly bits at the outer edge of the white – those jokes and catchphrases – barely resemble the self-contained yolk. As another historian has described it, the transfer of scientific knowledge is often seen simplistically as moving from areas of high truth concentration to those of low truth concentration.[1] It is

---

[1] Steven Shapin, "'Nibbling at the Teats of Science': Edinburgh and the Diffusion of Science in the 1830s," in Ian Inkster and Jill Morrell (eds.), *Metropolis and Province: Science in British Culture, 1780–1850* (London: Hutchinson, 1983), p. 151. See also Roger Cooter and S. Pumfrey, "Separate Spheres and Public Places: Reflections on the History of Science Popularization and Science in Popular Culture," *History of Science*, 32 (1994), 237–67.

For their valuable comments on an earlier version of this chapter, we are grateful to Rob Iliffe, Thomas Kaiserfeld, Bill Luckin, Jack Morrell, Simon Nightingale, and Simon Schaffer.

as though natural knowledge effortlessly flows from center to periphery, as if there were no energy costs, no resistances.

But for the eighteenth century, as for other periods, neither "science" nor this common-sense model of scientific diffusion is very helpful. Then, no one made a living doing scientific research. Indeed, the word "scientist" had not been coined. Few people made an absolute divide between religion, the stuff of belief, and what was called natural philosophy, the stuff of experiment and analysis. Nor was natural history fully divisible from religion. Natural theology, or the study of the relationships between God and the natural world, continued to be pursued well into the nineteenth century.

It is no easy matter, therefore, to address "science" and the processes of its "popularization" for the eighteenth century. In almost every respect the terms are anachronistic and misleading. A part of the purpose of this chapter is to indicate how such analytical categories fail to provide sufficiently complex and inclusive historical accounts. Put otherwise, we seek to illustrate the faultiness of the fried-egg model. At the same time we submit alternative means of comprehending and analyzing popular natural knowledges in the eighteenth century. Thus, instead of retrospectively defining "science," we borrow the seventeenth-century view of the subject held by the English natural philosopher Robert Hooke (1635–1703). In 1663, Hooke declared the new Royal Society's mission to include "the knowledge of naturall things, and all useful: Arts, Manufactures, Mechanick practices, Engines and inventions by Experiment." This broad remit enables us to avoid all-too-easy twentieth century assumptions about the differences between "popular" and "professional," "non-science" and "science." It allows us to cast our net wide in search of how and where knowledges of the natural world were created, discussed, and deployed. For purposes of analysis, we can consider as comparable the historically nameless women who sold their herbs and expertise to apothecaries, and the clergyman Gilbert White (1720–1793) of Selborne patiently recording the changing details of field and forest near his home.[2] We can explore a rowdy coffeehouse gathering of London artisans watching an itinerant lecturer stage a miniature earthquake to demonstrate God's providential design on the same terms as we can investigate a meeting of pious Swedish businessmen discussing a problem with a Newcomen engine in one of their mines or manufactories.[3] Or we can pose questions about the nature of humanity by reading a cheap pamphlet trumpeting a monster birth, or Lord Monboddo's (1714–1799) philosophical examination of the links between man and ape.[4] Although we cannot suppose, for example, that savants in Saint Domingue experimenting with ballooning, electricity, and Mesmerism did so in the same ways and with

[2] David E. Allen, *The Naturalist in Britain* (Harmondsworth: Penguin, 1978); Allen, "Natural History in Britain in the Eighteenth Century," *Archives of Natural History,* 20 (1993), 333–47.
[3] Svante Lindqvist, *Technology on Trial: The Introduction of Steam Power Technology into Sweden, 1715–1736* (Uppsala: Almqvist & Wiksell, 1984).
[4] E. L. Cloyd, *James Burnett, Lord Monboddo* (Oxford: Clarendon Press, 1972).

the same understandings as their contemporary sansculottes in Paris, we need not claim the superiority of the one over the other.[5]

Of course, when Robert Hooke listed the topics of interest to the Royal Society, he was also staking a claim to the Society's role as maker and validator of natural knowledge. He was, if you will, placing the Society in the middle of the egg yolk. However, while borrowing his description, we seek to avoid any easy assumptions about the relationships between center and periphery, yolk and white. Just as our current definition of science does not work well for the eighteenth century, neither do the social relationships implied in the fried-egg model of production and diffusion tell us much about the past. We therefore focus, instead, on the *sites* and *forms* of natural knowledge – that is, where the knowledge was produced and in what modes it was performed or enacted. Today, an inference is often made between form/site and social location. Laboratories are sites for scientists, whereas cat shows are sites for cat fanciers. The form of the scientific journal belongs to a research scientist, whereas a TV nature program belongs to the viewing public. Such assignments may not be very helpful in understanding science today; certainly, they are inappropriate to the past. As various cultural historians have shown, the identification of particular sites and forms as "popular" misreads historical relations among forms, sites, and social locations. We cannot, for example, assume that a small, cheap pamphlet belonged only to the "lower" sort of the reading public; it might have been read by an apprentice or declaimed aloud in an alehouse, but it might as well have been perused by an aristocrat.

In discussing the sites and forms of natural knowledge, we also seek to avoid privileging the cognitive content of knowledge over its social and cultural locations. The setting within which a piece of natural knowledge was produced or discussed is as important as its content – indeed, form and content are not easily divisible. An idea published in a cheap pamphlet is not the same as an idea propounded in a gentleman's drawing room, no matter how similar their cognitive content might seem. Thus, just as we cannot simply ascribe readership from social location, so we cannot assume that the cheap pamphlet was merely a popularized or watered-down version of the drawing-room discussion.

It is therefore to the relationships among the forms, sites, and social meanings of natural knowledge in the eighteenth century that we seek to draw attention. But it is necessary to be selective. The sites of natural knowledge in the eighteenth century were as diverse as the forms were varied. In addition to gentlemen's drawing rooms, the sites included coffeehouses, farms, taverns, churches, reading rooms, and cottages, among others. The forms include printed works, such as encyclopedias, magazines, children's books, and letters,

[5] James E. McClellan III, *Colonialism and Science: Saint Domingue in the Old Regime* (Baltimore, MD: Johns Hopkins University Press, 1992). On the Parisian sansculottes, see Robert Darnton, *Mesmerism and the End of the Enlightenment in France* (New York: Schocken Books, 1970).

as well as oral forms, such as sermons, lectures, and dialogs. They also include material forms, such as cows, flowers, and mechanical hoes.

In this chapter we concentrate on four of the territories of eighteenth-century natural knowledge. In each we take up a different analytical theme. Newtonianism, our first topic, serves primarily to illustrate some of the pitfalls of a hierarchical model of natural knowledge, even one broadened by consideration of popularization. A discussion of agricultural technologies then enables us to consider some of the ways in which economic tensions shaped natural knowledge. An analysis of medical books written for lay people, coupled with a spectacular medical incident, permits us to examine the ways in which natural knowledges circulated within the eighteenth century. Here, in particular, we focus on the concept of appropriation as a way to interpret such circulation. Finally, botany, or the natural knowledge of plants, provides us with the basis for discussing two current historical models of cultural change: commodification and the reform of popular culture.

Many of our examples are drawn from Britain. In part, this emphasis reflects historiographic trends. The extensive range of natural knowledges and practices is simply better documented for Britain than for anywhere else. We know about women and natural philosophy in Italy, France, and England,[6] but there is little available historical scholarship on, say, seedsmen or on professional gardeners for Italy or German-speaking countries. Histories of science that are focused on Italy, France, Germany, Spain, or any other Continental country, are often written by historians living in those countries who, until recently, have concentrated on the kinds of scientific activities that are still validated today. In part this may be because in France and elsewhere on the Continent the *Annales* historiography was far less successful than in Britain or America in institutionalizing a cultural history of natural philosophy. Thus, we know a great deal about certain significant male thinkers, from Goethe to Linnaeus, and about their various influences, both national and supranational. We know far less, however, about more humble practitioners of natural knowledge, who are rarely cast as the symbolic forebears of today's scientists.

Historical as well as historiographical grounds justify our British weighting, for there were important differences between Britain and many Continental countries (including their colonies) in the eighteenth century. Protestantism, or rather a set of assumptions about the extent of God's role in day-to-day human lives, was one such difference. Queen Anne (1664–1714) was the last British monarch to touch for the "King's Evil" (scrofula), or to invoke healing powers derived from the sacred nature of the throne. But on the Continent, healing shrines were still sanctioned by the state. In France, until the Revolution, the Royal touch was practiced. Britain's lack of a fully dominant

---

[6] See Londa Schiebinger, *The Mind Has No Sex? Women in the Origins of Modern Science* (Cambridge MA: Harvard University Press, 1989), and Lorraine Daston, "The Naturalized Female Intellect," *Science in Context*, 5 (1992), 209–35.

court culture is a related difference. To be sure, English monarchs surrounded themselves with followers and elaborate court procedures and rituals, as on the Continent. But many other sites vied for cultural authority. In France some natural knowledges could be wholly embedded in court culture; as we discuss later, a published dialog about cosmology was entirely cast within the highly articulated and polished tropes of courtly speech. In contrast, the English text that most resembles its French counterpart in topic and intended audience was full of examples drawn from everyday experience. In the English case, natural knowledge did not function only or predominantly as cultural ornament; it could also provide a basis for transforming the material world.

In Paris, court culture produced a steady market for luxury goods, and many Parisian craftsmen directed their efforts toward ever-finer brocades or highly elaborate umbrellas or other high-end consumer goods. In contrast, England in the eighteenth century was becoming a culture of consumption on a wide scale. Historians have analyzed the ways in which ever-more-differentiated consumer goods, from tea to the china from which it was drunk, became standard in middle-class homes and fostered certain kinds of economic development. If, for a moment, we consider natural knowledge to be a commodity, the differences between England and France are striking. In Paris, natural knowledge was performed and consumed in salons – decorative polite meetings hosted by women but frequented by men. By and large, such salons were the purview of the upper classes.[7] In England, however, natural knowledge was a commodity consumed in a wide variety of polite locations, from coffeehouses and provincial societies to children's nurseries. This difference also shaped the ways in which science was gendered. In Paris, women functioned as the arbiters of taste and refinement in their salons; the natural knowledge performed in these social settings was thus somewhat feminized. In England, coffeehouses were often masculine places, as were some provincial societies, whereas other sites, such as gardens, were not necessarily gender-specific. Women who translated scientific works into English, such as the bluestocking Elizabeth Carter (1717–1806), did not feminize natural knowledge, nor were they arbitrating the polite social relations that characterized the culture of the salon.

In what follows, then, British exemplification of natural knowledge should not be read as merely reflecting personal bias. Rather, it should serve as a reminder of crucial historical and historiographical differences between contexts – then as now. Contexts clearly matter for any historical discussion, and

---

[7] Geoffrey V. Sutton, *Science for a Polite Society: Gender, Culture, and the Demonstration of Enlightenment* (Boulder, CO: Westview Press, 1995). This portrayal of France versus England should not be thought complete. For instance, an understanding of military technology as a commodity, and its relationships to the state, might provide a rather different comparison of the two countries' forms and sites of natural knowledge. See, for example, Charles C. Gillispie, *Science and Polity in France at the End of the Old Regime* (Princeton, NJ: Princeton University Press, 1980), and Ken Alder, *Engineering the Revolution: Arms and Enlightenment in France, 1763–1815* (Princeton, NJ: Princeton University Press, 1997).

it would be hard to deny that our own Anglo-American historiographical context has not conditioned the analytical frames that we seek to elaborate in this chapter. That said, we trust that the primary analytical purpose of the chapter can be nurtured through the examples and that examples (and counter-examples) from other places may be fostered through the analysis.

NEWTONIANISM

Over the past few decades, the history of science has moved from the study of great men to the analyses of the social contexts and constructions of science. Emblematic of this shift is the growth of interest in Newtonianism.[8] Whereas thirty years ago much attention was paid to the intricacies of Isaac Newton's thought, now historians explore the social uses of such thought. Here, we examine the career of Newtonianism – or the careers of Newtonianisms – and suggest that the fried-egg model of knowledge production and diffusion may serve to foreshorten our understandings of the social meanings attached to the name of Newton.

One of the first studies of Newtonianism in its social context was that by Margaret Jacob, which focused on a group of Anglican clergymen who preached a series of sermons endowed through the will of Robert Boyle (1627–1691).[9] The Boyle lecturers, Jacob showed, did not see their purpose as popularizing Newton nor as creating a distinct Newtonianism. Rather, in the course of their battles within the Anglican church, as well as those waged against atheists and deists, they found in Newton's view of the universe the ingredients for a powerful natural theology. They argued that the universe was governed by divine providence – a providence that coexisted with natural laws such as gravity and motion – and that this governance made for an orderly and predictable world. As one Boyle lecturer noted, "What a noble Contrivance this [gravity] is of keeping the several Globes of the Universe from shattering to Pieces."[10] These sermons showed how Newton's account of the mechanics of a universe governed by laws that did not vary could be made into the natural correlate of a stable, prosperous, well-governed, and hierarchical social structure that the Boyle lecturers sought to reproduce.

In moving from Newton's study to the Newtonian pulpit, Jacob and other historians have worked mainly from printed sermons, largely overlooking the fact that sermons are usually presented first as oral performances. The Boyle

---

[8] On the various versions of Newtonianism and their history, see Simon Schaffer, "Newtonianism," in R. C. Olby et al. (eds.), *Companion to the History of Modern Science* (London: Routledge, 1990), pp. 610–26.

[9] Margaret C. Jacob, *The Newtonians and the English Revolution, 1689–1720* (Hassocks, Sussex: Harvester Press, 1976).

[10] William Derham, *Physico-Theology: or, A Demonstration Of The Being And Attributes Of God, From His Works Of Creation* [1714], quoted in Jacob, *The Newtonians and the English Revolution*, p. 180.

lectures were deliberately rotated among different London churches in order to reach wide audiences. Indeed, the first lecturer, Richard Bentley (1662–1742), sought to change the date of a sermon to "December when ye Town would be very full, [instead of] in September when it is always thinner."[11] Although we cannot recover these oral performances, doubtless they were differently nuanced from the printed works available to us today. Even a simple tone of voice could carry much meaning. When Voltaire met Boyle lecturer Samuel Clarke (1675–1729) in 1726, he was struck by Clarke's reverent mode of uttering the name of God, a habit that Clarke professed to have learned from Newton himself.

Sermons were not the only public oral presentations of Newtonian natural philosophy. Increasingly, the inhabitants of London and of provincial towns were able to attend science lectures, open to anyone who could pay the admission fee. Coffeehouses, schools of writing, and provincial societies all hosted such lectures. In 1725, for example, the Spalding Gentlemen's Society, located in the small market town of Spalding, Lincolnshire, enjoyed a series of natural-philosophical lectures by Jean Theophilus Desaguliers (1683–1744). Desaguliers, a Huguenot refugee and Freemason, had been employed by the Royal Society to do the skilled manual work of experiments and demonstrations. He molded these skills into a very successful career as a lecturer, marrying the elegance of Newtonian principles with the mechanical practicalities of steam engines and water pumps. Although his lectures suggested an easy progression from abstract principles to practical machines, the relationship between the two may have been more complex. As Larry Stewart has argued, the success of such lectures depended more on their practical mechanical content than on any Newtonianism; indeed, such lectures may have helped to create a broad acceptance of Newtonianism and natural philosophy by means of the practical projects with which they were associated in lectures.[12] Combining a range of opportunities and interests, Desaguliers helped to forge the new occupation of natural philosophy lecturer. In 1734, he could not "help boasting of the 11 or 12 Persons who performed Experimental courses at this time in England and other parts of the world [because] I have had the honour of having eight of them as my scholars."[13] By 1740, science lecturing had become a recognized occupation, and the public had a wide range of lecturers and lectures from which to choose.

Initially, science lecturers often used explicitly Newtonian principles to structure their presentations. Crucial to their lectures was the performance of experiments and dramatic demonstrations of scientific principles that governed

[11] Quoted in Jacob, *The Newtonians and the English Revolution*, pp. 148–9.
[12] Larry Stewart, *The Rise of Public Science: Rhetoric, Technology, and Natural Philosophy in Newtonian Britain, 1660–1750* (Cambridge University Press, 1992).
[13] Jean Theophilus Desaguliers, *A Course of Experimental Philosophy*, vol. 1 (London, 1734), fol. C verso, quoted in Stephen Pumfrey, "Who Did the Work? Experimental Philosophers and Public Demonstrators in Augustan England," *British Journal for the History of Science*, 28 (1995), p. 135.

the natural world. However, as more and more lecturers competed for custom, their lectures offered an increasingly broad array of interpretations of the natural world as well as dramatic entertainment. Desaguliers and his contemporaries created a Newtonianism that was fully consonant with the social elite's ideas about natural theology's relationship to political stability – in the words of Alexander Pope, "what ever is, is right." But later lecturers explored other visions. As Simon Schaffer has suggested, representations of natural philosophy could not be guaranteed to underwrite social stability.[14] The wonders of nature manifested through earthquakes, lightning, electricity, and magnetism were readily adapted to radical causes and commercial spectacle.

The career of Philippe Jacques De Loutherbourg (1740–1812) illustrates the heterodox nature of popular natural philosophy in the later eighteenth century. De Loutherbourg gave demonstrations of electrical and other natural-philosophical wonders, briefly ran a clinic for electrical healing, and was connected with a variety of London radicals. While Desaguliers had packaged his natural philosophy as useful commercial knowledge, De Loutherbourg drew upon his experience in the production of commercial spectacle. He was well known for his innovative stage spectaculars, which illustrated the wonders of new technologies as well as those of natural history. For example, in a 1785 pantomime about Captain Cook's voyage to Tahiti, De Loutherbourg incorporated the flying balloon, invented only two years earlier. At that time, pantomimes were forbidden by law to make use of spoken dialog; thus, in this particular performance, De Loutherbourg relied on "Tahitian" music and accompanying songs (the meaning of the Tahitian words being footnoted in the program). To us, the balloon, the scantily clad actresses, and the huge painting of the apotheosis of Captain Cook that descended onto the stage are part of the world of entertainment. But the critic from *The Times* described the performance as "a spectacle worthy of the contemplation of every rational being, from infant to the aged philosopher. A spectacle that holds forth the wisdom and dispositions of Providence in the strongest view."[15] Clearly, the form of the popular science "lecture" could be improvement and amusement at the same time.

The theater as a site of, and form for, natural knowledge also serves to remind us that such knowledge reached beyond exclusively male preserves. Indeed, some forms of natural knowledge, notably books and magazines, were specifically intended for women. The books were often in a form no longer associated with natural knowledge: the dialog, an implicitly condescending

[14] Simon Schaffer, "Natural Philosophy and Public Spectacle in the Eighteenth Century," *History of Science,* 21 (1983), 1–43.

[15] Quoted in Greg Dening, *Mr Bligh's Bad Language* (Cambridge University Press, 1992), p. 270. De Loutherbourg represents sites and forms of knowledge very different from those of Desaguliers. His interests in alchemy and Freemasonry, for example, construed knowledge as occult, attained through semimystical means rather than public demonstration. See Stephen Daniels, "Loutherbourg's Chemical Theatre: Coalbrookdale by Night," in John Barrell (ed.), *Painting and the Politics of Culture: New Essays on British Art 1700–1850* (Oxford: Oxford University Press, 1992), pp. 195–230.

idiom that became the favorite way to explain natural knowledge to women in the eighteenth century.

Two such books can also serve to illustrate the ways in which sites and forms of knowledge varied from country to country. The first is Francesco Algarotti's (1712–1764) *Sir Isaac Newton's Philosophy Explain'd for the Use of the Ladies* (1737). It was published in Italian but was quickly translated into French, English, German, Dutch, and other European languages.[16] Algarotti's book was explicitly intended to be an argument against the other book we wish to consider here, Bernard le Bovier de Fontenelle's *Conversations on the Plurality of Worlds,* which appeared in French in 1686 and was also widely translated.[17] Fontenelle (1657–1757) continued to produce revised editions until 1742; although his *Conversations* were resolutely Cartesian, in later life he increasingly embraced certain Newtonian ideas.[18] Fontenelle conducts his dialog with a Marquise; both participants flirt with each other, play with literary conventions such as the pastoral, and conduct themselves within the highly artificial modes of discourse of the Court. At the beginning of the book, Fontenelle describes the natural world as an opera, so cunningly contrived that it is almost impossible to see the theatrical devices, such as sets, lighting, and special effects, that dazzle the spectator. In other words, in this very Parisian book, natural knowledge is a cultural ornament, a mode of interaction between aristocratic men and women. Not surprisingly (since Fontenelle had already written plays and poetry), his dialog functions as another polite diversion.

Algarotti, too, would go on to publish books on painting, opera, and other polite subjects. But his mode of presentation, *Newton's Philosophy Explain'd,* was very different from that of Fontenelle. While Fontenelle's Marquise is presumed to know nothing about natural philosophy, Algarotti's partner is already familiar with Cartesianism; Algarotti wishes to persuade her that it is nothing more than a "philosophical romance." Rather than play with elaborate court rhetoric, Algarotti uses familiar objects in his character's upper-class home, such as pink face powder and paintings hung on the wall, to explain the basics of Newtonianism. Here, natural knowledge can speak to everyday

[16] Marta Feher, "The Triumphal March of a Paradigm: A Case Study of the Popularization of Newtonian Science," *Tractix,* 2 (1990), 93–110.

[17] This discussion is drawn from Mary Terrall, "Émilie du Châtelet and the Gendering of Science," *History of Science,* 33 (1995), 283–310; Terrall, "Gendered Spaces, Gendered Audiences: Inside and Outside the Paris Academy of Sciences," *Configurations,* 2 (1995), 207–32; Paula Findlen, "Translating the New Science: Women and the Circulation of Knowledge in Enlightenment Italy," *Configurations,* 3 (1995), 167–206; Aileen Douglas, "Popular Science and the Representation of Women: Fontenelle and After," *Eighteenth-Century Life,* 18 (1994), 1–14; Erica Harth, *Cartesian Women: Versions and Subversions of Rational Discourse in the Old Regime* (Ithaca, NY: Cornell University Press, 1992). See also Alice N. Walters, "Conversation Pieces: Science and Politeness in Eighteenth-Century England," *History of Science,* 35 (1997), 121–54.

[18] See Henry Guerlac, *Newton on the Continent* (Ithaca, NY: Cornell University Press, 1981), and Fielding H. Garrison, "Fontenelle as a Popularizer of Science," in Garrison (ed.), *Contributions to the History of Medicine* (New York: Hafner, 1966), pp. 855–72.

objects, in contrast to Fontenelle's teasing of his Marquise by offering her imaginary elephants to hold up the earth in that empty Cartesian outer space. The less-artificial and less-condescending tone of Algarotti's dialog is emblematic of the spaces in learned Italian culture for women; unlike in other European countries, in Italy some exceptional women were admitted to scientific academies and even universities.[19] Thus, different countries created natural knowledge for women in ways congruent with those cultures' characteristic sites and forms of natural knowledge. Whereas France was courtly and mannered, England was characterized by the sociability of "middling sorts."

As English mothers consumed varieties of Newtonianism over tea and fathers considered others in coffeehouses, their children might encounter Newton in the nursery. James Secord has analyzed a small book titled *The Newtonian system of philosophy, adapted to the capacities of young gentlemen and ladies* (1761), which in many ways translated the world of science lecturing and provincial learned societies into the newly commercialized world of childhood.[20] *The Newtonian System* is presented as lectures given to the Lilliputian Society by young "Tom Telescope" in the manner of Desaguliers's lectures to the gentlemen of Spalding. Like Algarotti, Tom Telescope uses everyday objects to illustrate natural laws. And as with De Loutherbourg, Tom Telescope turns natural philosophy into spectacle that is comprehended by his viewers visually, sensually, and aurally. A candle, a cricket ball, and a fives ball illustrate the workings of an eclipse. The text is more dialog than lecture; Tom Telescope is always interrupted by his child listeners and then engages with them in didactic conversations. In fact, the book may be doubly oral: not only is it in lecture and dialog form, but it was probably also intended to be read aloud like other children's books.

Tom Telescope thus adopts many of the forms in which Newtonianisms were being presented in eighteenth-century Britain. And as with others, Tom Telescope's Newtonianism serves social functions. As with the Boyle lectures, it situates the listeners in an orderly universe governed by natural laws. As Tom Telescope puts it, "A man may even at home and within himself see the wonders of God in the Works of the Creation." Tom Telescope's version also points to deference and hierarchy as the modes of conduct appropriate to the social correlate of that orderly natural world. Unlike other expositions of Newtonianisms, however, this one was a best seller. As many as thirty-five thousand copies were issued in eighteenth-century England alone.

All these Newtonianisms contained elements of amusement, and some may even have been written half tongue-in-cheek. Among the latter may have been the pamphlet *A Philosophical Essay Upon Actions on Distant Subjects. Wherein*

---

[19] Paula Findlen, "Science as a Career in Enlightenment Italy: The Strategies of Laura Bassi," *Isis,* 84 (1993), 440–69.

[20] James A. Secord, "Newton in the Nursery: Tom Telescope and the Philosophy of Tops and Balls, 1761–1838," *History of Science,* 23 (1985), 127–51.

are clearly Explicated According to the New Philosophy and Sir Isaac Newton's Laws of Motion, All Those Actions Usually Attributed to Sympathy and Antipathy . . . (3rd edition 1715), which was distributed gratis by a medical entrepreneur interested in hawking the Anodyne Necklace, a time-honored remedy for teething infants. Dedicated to the Royal Society (possibly both as a puff to the gullible and as wit to the wise), the pamphlet posed the question of how a coral necklace around a baby's neck could affect its teeth, analyzing the question in terms of action at a distance – an issue central to Newton's theory of the universe. Atoms of rose coral could operate "sympathetically" on red gums, the author postulated, relating this to an exploration of questions such as why dogs barked at strangers; why one person's yawn sets off a chain of yawns among companions; and how jarring sounds set the teeth on edge. Where popular science lecturers pointed to the power of Newtonianism in order to sell their expertise, this medical entrepreneur poked fun at the power of Newtonianism to explain the natural world. The reader of, or listener to, the pamphlet needed to know only a little Newton in order to laugh at (or embrace) these absurd "actions at a distance."

All five of the Newtonianisms we have referred to – Boyle lectures, popular science lectures and performances, Newton for ladies, Newton for children, and would-be Newtonian explanations of (or appropriations for) age-old remedies – can be understood as "popularizations" of the thought of Isaac Newton. But a consideration of Algarotti's text, or Secord's analysis of Tom Telescope, or the Anodyne Necklace pamphlet, suggests that the name Newton and the ideas credited to him have more complex relationships to the historical actor Isaac Newton than any current model of popularization would permit. Recent work, such as Larry Stewart's *The Rise of Public Science* (1992), has done much to broaden our understanding of the social roles of Newtonianisms in Augustan England; indeed, Stewart might be said to have escaped the fried-egg model by emphasizing the ways in which Newton's thought was merely part of the package sold by science lecturers – and not necessarily the most important part. But most historical analyses of eighteenth-century science continue to be governed by hierarchical models of center and periphery.

In what follows, we explore the sites and forms of three other kinds of natural knowledge in order to broaden discussion beyond the popularization model. All three of these natural knowledges flourished in the eighteenth century, but they have seldom been studied by historians of science. Indeed, historians of science have often failed to notice them because they do not conform to conventional ideas of what "science" is.

AGRICULTURE

Much of the eighteenth-century activity retrospectively labeled "science" took place in towns and cities. London's role as the center of print culture and the hub of a commercial empire has much to do with this urban focus. Those historians who have amplified the connections between Newtonianism and the commercial (even industrializing) culture of eighteenth-century Britain have rightly pointed out that provincial cities and even small towns often replicated the culture of the metropolis, as in the case of the Spalding Gentlemen's Society hearing popular science lecturers. But eighteenth-century Britain was, like the rest of Europe, predominantly rural. Social relations were dominated by aristocratic or gentry landowners and their tenant farmers and agricultural laborers. Here, no less than in the city, natural knowledges abounded, and their sites and forms were equally as varied as those of Newtonianism. In this section we focus on agriculture as a way of exploring our second analytical theme: the relationships between natural knowledge and economic interests. Attention must be paid both to the straightforward understanding of knowledge as property (which could be bought and sold) and to those instances in which references to economic interests were studiously avoided, such as in appeals to "the public good," or to gentlemanly reticence about appearing in print. The latter were as important among the sites and forms of agricultural knowledge as the explicit economies of knowledge as property.

Agricultural knowledge was produced in technical illustrations, periodicals, books, letters, conversations, and material objects, such as machines and even farm animals. As with other kinds of natural knowledge, tensions existed between those forms that were the property of an individual and those that claimed to be open to all for the public good. That individuals or projects can rarely be categorized as wholly private property or wholly public is illustrated in the disputes between the agriculturists Jethro Tull (1674–1741) and Stephen Switzer (1682?–1745).

Tull was a gentleman farmer who developed a new agricultural machine, the seed drill, in response to frustration at failing to persuade his laborers to plant seeds in his preferred way. Not surprisingly, the agricultural laborers did not respond favorably to Tull's invention. But Tull went on to develop increasingly intensive methods of farming that centered on improved hoeing practices. By the first decade of the eighteenth century, aristocrats and gentlemen farmers began to visit Tull to talk with him about his methods. Tull saw no reason to circulate his knowledge in any form more public than that of word of mouth. It was not until 1731 that he wrote *Horse-Hoeing Husbandry*, claiming, like many other gentry authors, that he published only because of the solicitations of his noble visitors.[21]

Such a public form, however, proved troublesome to Tull. He was attacked by the agriculturist Switzer, who accused him of plagiarizing agricultural innovations from earlier writers. Switzer's career as an agricultural improver was

---

[21] N. Hidden, "Jethro Tull I, II, and III," *Agricultural History Review*, 37 (1989), 26–35; and G. E. Fussell, *More Old English Farming Books from Tull to the Board of Agriculture, 1731–1793* (London: Crosby, Lockwood & Son, 1950).

very different from that of Tull. Tull had graduated from Oxford and had been called to the bar before becoming a gentleman farmer. Switzer, on the other hand, had earned his living as a gardener to aristocrats. Later, he went into business as a seedsman selling his wares "at the sign of the Flower Pot" in Westminster Hall. He also founded and edited the monthly *Practical Husbandman and Planter,* where he attacked Tull not only for pinching earlier ideas but also for his denigration of the farming techniques contained in Virgil's *Georgics.* Switzer leapt to the defense of the Roman writer's reputation as an agricultural expert.

At first glance, it might seem that Tull was an old-fashioned gentleman farmer and Switzer a forward-looking entrepreneur. However, Switzer's advocacy of a form of intellectual property must be balanced by his defense of classical authority. One must be aware also that Switzer worked within a time-honored system of aristocratic patronage; he was employed as a gardener by noblemen and dedicated his books to those patrons. Tull, on the other hand, sought to remedy disciplinary problems with laborers by technological innovation. The sites and forms of these men's natural knowledges varied: for Tull, his own farm was the site for knowledge production and deployment, whereas Switzer adopted aristocratic patronage as well as the running of a business. Both men used the same form, that of a book on agricultural technique, but they arrived at that mode from different experiences. For decades, the forms of Tull's knowledge were the mechanical device of the seed drill itself and his conversations with visitors. Ultimately that knowledge was reformulated as a printed book. Switzer also translated one type of natural knowledge into others. Initially, the gardens he designed for his patrons were themselves forms of knowledge, which he then transmuted into print in books and magazines aimed at fairly broad audiences.

If the lives of Jethro Tull and Stephen Switzer reveal two kinds of relationships between an individual and natural knowledge, the wave of agricultural improvement associated with the Scottish Enlightenment illuminates others. Neither Tull nor Switzer saw his work as directed toward the general public good, but that was the stated intention of the Scots improvers. They had need of such rhetoric in the wake of the Jacobite Rebellion of 1745 and the demise of the old highland culture in which small farmers were tied to chieftans through a semifeudal system of mutual obligations and landownership. In this context, the language of "public good" concealed the recent historical upheaval in which agriculture came to be embedded in, and expressive of, the social relations of a new political economy. As Lord Kames (1696–1782) proclaimed in *The Gentleman Farmer, Being An Attempt To Improve Agriculture By Subjecting It To The Test Of Rational Principles* (1776), "No other occupation rivals agriculture, in connecting private interests with that of the public."[22]

---

[22] Lord Kames, pp. xvii-xviii, as quoted in Jan Golinski, *Science as Public Culture: Chemistry and Enlightenment in Britain, 1760–1820* (Cambridge University Press, 1992), p. 35. On Kames, see also William

The noted Scottish doctor and chemist William Cullen (1710–1790), for whom Kames served as patron, understood "public" and "private" agricultural knowledge in a number of ways. From his early days as a practitioner through his highly successful professorial career in Edinburgh, Cullen gave lectures on agriculture, focusing on its chemical aspects.[23] Cullen told Lord Kames that he was introducing a discussion of agriculture into his medico-chemical lectures in 1749, acknowledging that the subject was rarely discussed in an academic setting. In his lectures, Cullen cautiously employed a rhetoric of theory and practice, suggesting to his audience that an understanding of first principles would enable an adoption of practical measures. However, as he knew from his personal experience of farming, such a transition was not always easy. Indeed, Cullen was so diffident about the ability to improve agriculture through the explication of first principles that in 1768 he gave a series of non-medical agricultural lectures to a trusted audience composed only of invited friends. Like Tull, Cullen felt no need to translate the oral form of lectures into print. It was not until 1796, after his death, that the 1768 lectures were published.

The forms and sites of agricultural knowledge production and performance clearly bear some resemblance to those discussed earlier in connection with Newtonianisms. Thus, the knowledges were often oral performances in lecture form, although unlike the Newtonians, Cullen did not need to persuade his audience that his agricultural knowledge was a valuable practical commodity for hire. Cullen thus stands at a midpoint between those entrepreneurial lecturers who translated natural knowledge into direct personal gain, on the one hand, and those agricultural improvers who sought to spread their knowledges wide in order to convert farmers' methods for the greater public good. Of course, not all farmers (nor farmhands) wished to be persuaded; some landlords wrote clauses into their leases specifying that tenants must now follow the directions of the land steward.

Different again was Arthur Young (1741–1820), who argued for agricultural improvement as a public good in terms of political economy. After farming for four years, Young published *The Farmer's Letters to the People of England: Containing the Sentiments of a Practical Husbandman on Various Subjects of the Utmost Importance* (1767), a book he later considered almost totally inaccurate. Young continued to run a variety of farms, often unsuccessfully, and briefly served as the land agent for an aristocratic landowner in Ireland. But increasingly, he made his career as a writer and agricultural expert. Rather than rely on the patronage of aristocratic landowners, he constructed his expertise within the framework of political patronage, especially that of Pitt the Younger.

C. Lehmann, *Henry Home, Lord Kames, and the Scottish Enlightenment* (The Hague: M. Nijhoff, 1971), and Ian S. Ross, *Lord Kames and the Scotland of His Day* (Oxford: Clarendon Press, 1972).

[23] Charles Withers, "William Cullen's Agricultural Lectures and Writings and the Development of Agricultural Science in Eighteenth-Century Scotland," *Agricultural History Review*, 37 (1989), 144–56; and Golinski, *Science as Public Culture*.

Young published his accounts of tours of various regions of England, Ireland, and parts of Europe to great success; indeed, well-to-do German and Russian farmers came to visit the English improvers he referred to in his books.[24] Young's form of natural knowledge blended travel anecdote with careful observation of grain prices, population, costs of produce, and other details of political economy. It emphasized the importance of direct observation and careful compilation of facts. To this same end, in 1784, Young founded *The Annals of Agriculture,* which he published continuously for the next twenty-five years. Young himself wrote between a quarter and a third of the journal's contents; contributors included a wide array of agriculturists, political economists, and natural philosophers, among them Jeremy Bentham (1748–1832), Frederick Morton Eden (1766–1809), Joseph Priestley (1733–1804), John Symonds (1730–1807), and Thomas William Coke of Holkham (1752–1842) as well as a sprinkling of noble lords and even royalty, including George III (1738–1820), who wrote under the pseudonym of his Windsor shepherd Ralph Robinson. Young regretted that the circulation of the journal hovered at no more than 350 copies. But from another perspective we can understand the *Annals* as a new and successful location for the practice of natural knowledge. If urban and provincial centers produced sites such as coffeehouses and gentlemen's societies, Young created a community of agricultural improvers who were linked, not in a specific geographical location but rather by means of the journal. He forged a community of interest from an array of landowners, economists, and agriculturists who shared concerns about specifics of varieties of grasses, livestock, drainage, and so on with a strong view of agriculture's social role. For example, Young discussed rural poverty and enclosures in both moral and practical terms. It has been argued that in so doing, Young "reported on, related to, and generated the ideology for, a small, progressive, agrarian elite."[25] This cultural production was quite different from that of Tull, Switzer, or Cullen.

Young's emphasis on political economy and the ease with which he moved in social and political circles in London enabled him to become a new type of expert. For example, in 1788 he was deputed by the wool-growers of Suffolk to support a petition against the wool bill. He testified to both houses of Parliament, lobbied politicians, and wrote two pamphlets on the subject. Although the wool bill was passed (and Young burned in effigy in Norwich) his career as an agricultural expert flourished.

Another site where Young practiced natural knowledge was at the Board of Agriculture, founded in 1793, of which he was the first secretary. Young had obtained the post by means of Pitt's patronage, and he created a distinct form

[24] Joan Thirsk, "Agricultural Innovations and Their Diffusion," in Thirsk (ed.), *The Agrarian History of England and Wales: Volume V, 1640–1750* (Cambridge University Press, 1985), p. 574.
[25] Maureen McNeil, *Under the Banner of Science: Erasmus Darwin and His Age* (Manchester: Manchester University Press, 1987), p. 183.

for natural knowledge: the agricultural survey.[26] He organized and wrote a number of county surveys of agriculture, such as his *General View of the Agriculture of the County of Suffolk* (1794). Whereas Tull wrote on agriculture in terms of specific practices, Young blended that type of writing with political economy. It was his aim to make land yield maximum profit – to "turn sand into gold."

In 1770, Young "discovered" Robert Bakewell (1725–1795), whose modes of natural knowledge were radically different from those of Young. At this point Bakewell had been breeding livestock for twenty-five years, and word of his successes had begun to circulate. But Bakewell kept his knowledge secret; he revealed neither his initial breeds nor the more recent ancestry of his animals. For Bakewell, natural knowledge was private property; rumor had it that he kept no breeding records and confided only in his elderly shepherd. He profited from his ingenuity by hiring out his animals for breeding purposes; one of his rams was rented out for one thousand guineas per season! It was also rumored that before selling his old sheep to a butcher he infected them with sheep rot to make sure that no one else could breed from them.

As with Tull, for Bakewell the site for the production of new natural knowledge was the farm. But while Tull met with a succession of aristocratic visitors and eventually published his methods, Bakewell remained silent. For him the form of natural knowledge was the animal itself; he understood himself as making "the best machine for converting herbage into money."[27]

Agricultural natural knowledges were thus produced and reproduced at a range of sites and in a variety of forms. The sites include Bakewell's and Tull's farms, Cullen's lecture theater, Switzer's seed shop, and Young's *Annals*. The forms include Tull's conversations with his aristocratic visitors, Bakewell's guarded discussions of his methods, and Cullen's lectures to selected friends – all oral forms of the practice of natural knowledge. Switzer's and Tull's "how-to" books on agriculture and gardening stand in contrast to Young's range of forms, including books structured as letters, descriptions of travel combined with political economics, and surveys. Finally, there are the forms of natural knowledge that were not expressed in words, spoken or written: the material objects of a sheep, a new plant, or a seed drill.

This panoply of forms and sites does not fit into any tidy social classification. Arthur Young relied on political patronage; Switzer made use of aristocratic master/servant patronage; and Cullen enjoyed the support of Lord Kames. Young published contributions by George III, but Bakewell took tea with "Farmer George." Both Switzer and Young wrote and published journals,

[26] This form of knowledge was originally the idea of John Sinclair (1754–1835), the first President of the Board of Agriculture, and it was largely based on Sinclair's multivolume *Statistical Account of Scotland* (1791–9). See Rosalind Mitchison, *Agricultural Sir John: The Life of Sir John Sinclair of Ulbster 1754–1835* (London: G. Bles, 1962). See also John Gascoigne, *Joseph Banks and the English Enlightenment: Useful Knowledge and Polite Culture* (Cambridge University Press, 1994), chap. 5.

[27] Harriet Ritvo, *The Animal Estate* (Harmondsworth: Penguin, 1990), p. 66.

each one a mode of personal self-promotion although quite different in social function. An understanding of these varied agricultural knowledges points toward a tension between ideas as private property and ideas as public good. Again, however, no easy progression can be traced. Switzer and Bakewell understood their natural knowledge as an individual commodity that could be sold in the forms of plants, garden designs, and animals. Tull did not seek to profit from his farming innovations and was appalled when confronted by charges of plagiarism grounded in a notion of ideas as property. But neither did Tull adopt the rhetoric of public good, as deployed skillfully by Cullen and Young. Yet Cullen and Young did not equate the "public good" with any kind of vision of "knowledge for all." Instead, they placed themselves as experts who would produce and manipulate certain natural knowledges and then offer them to a select audience. Although they did not profit directly from such expertise, they accrued a sort of intellectual capital that could be translated into personal profit, as in Young's salary as secretary to the Board of Agriculture. The "public good" served as a kind of veiling of economic interests, even if tenant farmers who used the natural knowledges produced by the deployers of "public good" rhetorics were equally eager to see their activities definable in terms of units of output per units of input. In other words, agricultural knowledge cannot be understood simply (or linearly) in terms of "secret knowledge" that might produce financial gain, nor (anachronistically) in terms of "property rights" residing in patents and authorship. More-entangled sets of social and economic relations were involved across a diversity of forms and sites of knowledge production and reproduction.

Thus, for agricultural natural knowledges the fried-egg model serves as poorly as it does for Newtonianisms, but for additional reasons. While our discussion of the social deployments and meanings of Newtonianism pointed to the limits of thinking in terms of hierarchical (high-to-low and center-to-periphery) diffusions of natural knowledge, the discussion here on the plurality of sites and forms of agricultural knowledge underlines the poverty of thinking in terms of theory-to-practice diffusion (a frequent corollary to the fried-egg model) and the poverty of contemplating such natural knowledge in terms of straightforward or unmediated economic coordinants and interests. At its most basic, the center-to-periphery aspect of the model, in which diffusion seems to happen without much difficulty or resistance, cannot encompass the complex and sometimes contradictory social relations of eighteenth-century agriculture. Jethro Tull's recalcitrant laborers, or the tenants forced to sign leases that guaranteed their use of new methods of farming, were not the passive recipients of "better" knowledge deriving from the center. (On the contrary, it was often their knowledge that the elite appropriated.) Likewise, both William Cullen's and Arthur Young's dismal personal experiences of farming suggest that any model that portrays the flow of theory to practice as easy or unproblematic cannot account for the social relations of eighteenth-century agriculture. Indeed, in his own way, each of these advocates of theory-driven knowledge had to admit that he could not always connect abstraction to practices. In short, the fried-egg model of popularization does not facilitate a sufficiently complex analysis of the many kinds of agricultural natural knowledge discussed here.

MEDICINE

Like agriculture, medicine was a natural knowledge that might be described as "for the public good," and, like agriculture, medicine was conducted within a realm of explicit economic relationships. In what follows, we analyze "popular" medicine – that is, medical writings intended for nonmedical readers. Looking to the popular medical books of John Wesley (1703–1791) and William Buchan (1729–1805) and to the strange story of Mary Toft, the "rabbit breeder," we concentrate here on the circulation and appropriations of this knowledge – its fluidity – as a further means of avoiding some of the wider assumptions of the production-and-diffusion model of popularization. We use "circulation" to emphasize that natural knowledges were not made by one group and then handed down to another. Rather, all natural knowledges, whether about matter theory or manure, were constructed and enacted socially; although knowledges circulated, they did not do so in an abstract fashion. We use the term "appropriation" to refer to the cultural acquisition of knowledges, or the ways in which they are borrowed from one social setting and reformulated in another.[28]

Knowledge of health and healing was widespread in the eighteenth century, extending far beyond the purviews of physicians, surgeons, apothecaries, and midwives. One of the century's most frequently reprinted medical manuals was John Wesley's *Primitive Physick*, first published in 1747. It is a straightforward text: an introduction and an ailment-by-ailment list of remedies. The introduction situated healing within a religious context; Wesley, the founder of Methodism, advocated a pure life and would go on to coin the phrase "cleanliness is next to godliness." His remedies were mostly uncomplicated preparations of readily available herbs.

An analysis of Wesley's medical advice suggests the ways in which natural knowledges were appropriated among social groups and locations. For instance, Wesley's insistence that simple remedies were better than the polypharmacy of learned physicians, and his emphasis on the importance of locally available herbs, looked back to the mid-seventeenth century when Nicholas Culpeper (1616–1654) had written his *English Physician*. Wesley's bald list of remedies

---

[28] As elaborated by Roger Chartier in his essay "Culture as Appropriation: Popular Cultural Uses in Early Modern France," in S. L. Kaplan (ed.), *Understanding Popular Culture: Europe from the Middle Ages to the Nineteenth Century* (Berlin: Mouton, 1984).

also drew on traditional remedies. Although he never said as much, he often employed the "doctrine of signatures," the belief that a plant's shape or color indicated its healing powers. The doctrine of signatures was predicated on the belief that the world was designed by God for human use; He had made plants that revealed their healing properties as a help to man.

However, we cannot understand Wesley's medicine simply as the unproblematic leftovers of a worldview that had been commonplace a century earlier – any more than we could describe his religion in this way. The other key to Wesley's text lies in the writings of George Cheyne (1671–1743), who had advocated the sparse diet and cold bathing regimen to which Wesley referred in his introduction. Cheyne was a complex figure, a Newtonian physician criticized by Newtonians, and a grossly obese man who bounced between overeating and the strict dietary regimes he advocated in print.[29] He posited a highly mechanical theory of the body wherein quantifiable fluids and canals operated according to the same kinds of natural laws as governed the universe. But he also subscribed to an increasingly mystical and even millenarian set of religious beliefs that did not sit well with many Newtonians. Wesley ignored Cheyne's valorization of nervous illness among the well-to-do, borrowing Cheyne's emphasis on diet and regimen and combining it with his own desires for a purified and simple religion and medicine.

In later editions of *Primitive Physick*, Wesley added two sections. The first was on the practice of cold bathing, something advocated by a range of medical men in the middle of the eighteenth century. The second was a section on the use of electricity in medicine, a subject also then much in vogue. Learned disquisitions drew analogies between electricity and the invisible workings of the nervous system, but at the same time electricity was displayed theatrically (like, and often along with, Mesmerism) as an entertaining and strange testament to the active powers of the universe. Wesley's treatment of the subject was noncontroversial; as with herbal remedies, he wished to convey what he understood as useful healing practices without explicit theorizing.

Thus, Wesley's text illustrates the ways in which various natural knowledges circulated among a range of social locations. The form of his book similarly reveals the fluidity of natural knowledges and the ease with which they might be appropriated. Wesley was not a physician, nor had he received any formal medical instruction. But like many of his contemporaries, he collected and shared recipes for remedies. As with manuscript notebooks in which such remedies were recorded, Wesley assumed a working knowledge of various types of illnesses and a familiarity with herbs and their preparations. An analogy

---

[29] Anita Guerrini, "Isaac Newton, George Cheyne and the *Principia Medicinae*," in Roger French and Andrew Wear (eds.), *The Medical Revolution of the Seventeenth Century* (Cambridge University Press, 1989), pp. 222–45; Guerrini, "A Case History as Spiritual Autobiography: George Cheyne and 'Case of the Author,'" *Eighteenth-Century Life*, 19 (1995), 18–27; and G. S. Rousseau, "Mysticism and Millenarianism: 'Immortal Dr. Cheyne,'" in Richard Popkin (ed.), *Millenarianism and Messianism in the Enlightenment* (Berkeley, CA: Clarke Library, 1990).

between forms of knowledge can also be drawn between Wesley's preaching and his *Primitive Physick*. His open-air preaching reached huge crowds, including many working-class people who rarely attended their own parish churches. Similarly, his book was inexpensive and simple in format and therefore easily adopted by people who had neither time nor money for physicians' elaborate therapeutics. Indeed, the most important component of Wesley's natural knowledge was Methodism itself, which was not then formally separated from Anglicanism. The "primitive" in the title of his book expressed Wesley's desire to return to a more purified or less mediated set of religious practices. The book itself was published at the Foundry in Moorfields, London, the site of Wesley's preaching in London. Editions were also published in cities such as Bristol, which had strong links to Wesley. Undoubtedly, many people bought the book as a marker of their commitment to Wesley's religious views. Thus, among the sites of Wesley's natural knowledge can be included the thousands and thousands of households where *Primitive Physick* was owned and read.

In 1769, William Buchan published his six-shilling *Domestic Medicine*, which was to become the most popular book of its type in the later eighteenth century. As with *Primitive Physick*, Buchan's text reveals the ways in which natural knowledges of the body circulated among a variety of social locations. As Charles Rosenberg has noted, Buchan's text appropriates traditional understandings of the body to ideas and practices that were particularly appealing to the "middling sorts" of the later eighteenth century.[30] Although Buchan explicitly criticizes folk medicine and lay practice, he nonetheless includes a number of time-honored remedies in his text, placing them within his own Enlightenment framework. He assumes, for example, that people will have their blood let and order purges according to their own assessments of their health. He does not provide a wealth of diagnostic material; rather, like Wesley, he assumes that his readers possess certain levels of medical knowledge. However, Buchan differs from Wesley in his creation of boundaries between different knowledges; his title, *Domestic Medicine*, implies that there is in fact a nondomestic medicine. On the other hand, Wesley's title asserts the importance of the religious philosophy underlying his work.

Both Wesley and Buchan situate their medicine in a moral framework, but Buchan insists on a profoundly secular version of morality. Not for him the easy equation of primitive religion and primitive physick. Indeed, between the second and third editions of *Domestic Medicine*, Buchan removed almost all references to religion, even transforming "Jesuit's bark" to "Peruvian bark." Instead, Buchan emphasizes the responsibilities of the middling sorts for the maintenance of their own health; they are neither the indolent aristocracy,

---

[30] Charles Rosenberg, "Medical Text and Social Context: Explaining William Buchan's *Domestic Medicine*," *Bulletin of the History of Medicine*, 57 (1983), 22–42. See also Christopher Lawrence, "William Buchan: Medicine Laid Open," *Medical History*, 19 (1975), 20–35.

prone to ailments caused by luxury and indulgence, nor the peasant or industrial worker whose income would not permit employment of the proper means to maintain health.

Another way in which Buchan sets boundaries is by his insistence that certain ailments can be treated only by a doctor. He chides those who hesitate to call in a physician, charging that many people wait too long, until even a physician is hard-pressed to save the patient. One of the primary agendas of the text is to distinguish between those ailments that are appropriately treated domestically and those that require a doctor's attendance. Part of this boundary is gendered. Buchan repeatedly denigrates the medical knowledge of nurses, midwives, and old women. He wrests the medical care of infants and children from mothers to physicians, warning of the dire consequences of failing to call in the doctor promptly. Nor will just any doctor do. Buchan creates divisions between appropriately trained doctors, such as himself, and quacks and charlatans of whom patients are all too fond.

Thus, Buchan's text points both to the appropriation of traditional understandings of the body and to the medical profession's attempts to assign certain types of knowledge to itself. Like the agriculturist Arthur Young, Buchan appeals to the public good in his construction of boundaries. He inveighs against infant deaths, claiming (as the French Physiocrats did) that more attention should be paid to the health of children than to the treatment of the elderly, since the nation's future lies with its children. As with Young, these appeals to the public good placed natural knowledge in the service of morality at the same time that they concealed some of the economic relationships that structured the construction and deployment of such knowledges. Both men can be understood as contributing to an Enlightenment critique of vernacular natural knowledge that assigned many beliefs and practices to the realm of superstition and ignorance. However, such a move was not easily accomplished, and, especially in the case of medicine, various natural knowledges continued to circulate among various social locations. Nowhere is this better illustrated than in the strange case of Mary Toft, the "rabbit breeder" of Godalming, Surrey.

The rudiments are as follows. In October 1726, Mary Toft, the illiterate wife of a poor cloth worker, gave birth to a rabbit and then a number of subsequent rabbits or parts of rabbits. News of her amazing production spread quickly; she was visited in Godalming by leading doctors and then brought to London in early December. After a porter was discovered red-handed bringing her a rabbit, Toft was subjected to the interrogation and threats of doctors and a magistrate before finally admitting that the rabbit births had been a hoax.

Among the doctors who visited Mary Toft in Godalming was Nathanael St. André, surgeon to the King's household. When Toft went into labor again, he delivered the head of a rabbit, as well as other parts, and was convinced that she had truly produced a monstrous birth. In fact, Toft later confessed that she and her family had dreamed up the hoax solely to make money by exhibiting her and the monstrous births. What was initially "delivered" had actually been a cat, whose guts had been replaced by the spinal cord of an eel upon which the Tofts had dined. It seems that it was only when a further cat was unavailable for the hoax that the family turned to rabbits.[31]

Why would St. André, or any other trained medical man, believe that Toft had genuinely conceived and given birth to rabbits? In part, the answer lies in the natural knowledges shared by an illiterate peasant and a "learned" medical man. Mary Toft crafted her story in accordance with those knowledges; she claimed that while newly pregnant, she had been scared by a rabbit while working in the fields. She then dreamed of rabbits and longed for rabbit meat. The extraordinary power of the maternal imagination on the shape of a fetus had long been part of academic medical discussions of pregnancy and birth. Toft's knowledge of reproduction had enough in common with that of learned medical men that she was able to fool at least some of them. The form of Toft's knowledge was both narrative (the story about the rabbit) and material (the bodies of rabbits to which she allegedly gave birth). Doctors interrogated this material form by means of their own knowledge and practices. They dissected the rabbit bodies, looking both at the lungs (to determine if the animals had ever breathed) and at the gut to look for fecal pellets, which fetal rabbits do not produce. However, their findings and interpretations were contentious and conflict-ridden.

The Toft story illustrates a double appropriation. On the one hand, Toft herself constructed a natural knowledge of reproduction from the various constituent parts available to her. Then the tale of Toft, after rapidly circulating in pamphlets and satires, was itself appropriated. It is impossible to separate fully these two appropriations, since we know about the first only as a result of the second. Nevertheless, it is clear that the second – the appropriation of the story of Mary Toft – was made to serve many purposes. For example, William Whiston (1667–1752), the renegade Newtonian, argued that the births fulfilled the Biblical prophecy of Esdras's foretelling of the Final Judgment. Even after the hoax was revealed, Whiston continued to argue that Toft had indeed given birth to rabbits. On the other hand, the physician James Blondel (d. 1734) was sufficiently provoked by the Toft incident to write a scholarly attack on the idea of "maternal imagination." Meanwhile, pamphleteers transformed the Toft story into highly sexualized satires of boundless female desire. William Hogarth (1697–1764) represented the Toft story visually at least twice. At the height of the controversy, he published an engraving titled "Cunicularii, or the Wise Men of Godliman in Consultation," which depicted a dozen rabbits frolicking on the floor while the "wise" doctors confer, oblivious to the fact that Toft's husband and mother-in-law appear

[31] Dennis Todd, *Imagining Monsters: Miscreations of the Self in Eighteenth Century England* (Chicago: University of Chicago Press, 1995).

to be passing a concealed object (a rabbit?) to Toft. Thirty-five years later, Hogarth reworked this image in a print called "Credulity, Superstition and Fanaticism." Here Hogarth anachronistically places Toft in a Methodist meeting house, along with other examples of hoaxes, such as ghosts and a boy who vomited rags, pins, and nails. In this representation, Hogarth transforms Toft into an exemplar of "enthusiasm," a mode of religious practice that exults imagination and physical manifestations of the divine.

Thus, the case of Mary Toft suggests that no easy equations exist between forms, sites and contents of natural knowledge. The natural knowledge deployed by Toft and her family for financial gain was equally at home among learned medical men. The material form of natural knowledge – the bodies of the rabbits themselves – was easily translated into anatomical interrogation. Indeed, the same test performed on the rabbits' lungs was later the subject of a learned inquiry by William Hunter (1718–1783), in an investigation of its use for humans in suspected cases of infanticide. The ease with which various medical knowledges mixed, overlapped, and interacted with each other points to the difficulty experienced by Buchan when he attempted to divide some medical knowledge from others. For most of the eighteenth century, however, for most people the problem simply didn't exist; modern fried-egg conceptions of how natural knowledge *ought* to behave were irrelevant.

BOTANY

Natural knowledges that focused on plants were much less well integrated than those of medicine, and practitioners may have had less in common with one other – or less shared knowledge – than was the case in medicine. During the eighteenth century, knowledge and practice concerning plants (which were increasingly collected under the rubric of "botany") changed in a variety of ways. In this section we concentrate on two further analytic categories to pursue some of the similarities as well as the differences among knowledges about plants. The first is the reform of popular culture, the slow process by which elites and middling sorts sought to define themselves as culturally different from the lower orders.[32] Second, we focus on "commodification" to explore the economic relationships that helped to structure ideas and practices centered on plants.

One kind of botany was the common property of many social groups: the knowledge of plants useful to humans. Country dwellers were familiar with cutting reeds for thatching, collecting thistledown for stuffing pillows, and using horsetail to scour pots and pans. As indicated in the preceding section, knowledge of healing plants was extensive among laborers, artisans, and rural

[32] Jonathan Barry and Chris Brooks (eds.), *The Middling Sort of People: Culture, Society and Politics in England, 1500–1800* (London: Macmillan, 1994).

folk. Indeed, it was sometimes acknowledged that country people knew more about plants than their betters. As a boy, Joseph Banks (1743–1820), the future president of the Royal Society, paid herbwomen to teach him the names of flowers. William Curtis (1746–1799), later to found the *Botanical Magazine,* became interested in flowers during conversations with an ostler who studied herbals. As Curtis's example suggests, we cannot assume that the botanical knowledges possessed by workers were only transmitted orally or that they represent some sort of traditional oral wisdom handed down through generations. Rather, there was a kind of continual circulation among local customs and books such as Culpeper's *English Physician,* which itself drew on oral knowledge.

One of the tidiest examples of the "reform of popular culture" is the changes that occurred in botanical nomenclature, illuminated by Keith Thomas.[33] Gradually over the seventeenth and eighteenth centuries, many of the names used by herbwomen and country dwellers were abandoned or changed by their betters. Names that were too crudely anatomical or magical were replaced by more genteel ones. Plants such as black maidenhair, naked ladies, priests bollocks, and horse pistle were rechristened. This gentrification of plant nomenclature was followed by a different wave of concern about the indelicacy of flower names. Late eighteenth-century botanists took fright at the thought of ladies studying Linnaean nomenclature based on the sexual parts of plants. Thus, gradually, people who considered themselves "polite" did not have available the utilitarian and magical knowledges of plants deployed by working-class men and women. Of course, workers were themselves in a dialectical relationship with their betters. As ladies and gentlemen came to speak of plants in the language of Linnaeus, working men's botanical societies gradually adopted Latin plant names in addition to the centuries-old vernacular names.

Linnaean nomenclature was first introduced in Britain in the Latin text of Linnaeus himself, his *Species Plantarum* (1753). Non-Latin readers acquired the nomenclature through books such as James Lee's (1715–1795) *Introduction to Botany* (1760), a translation of a text by Linnaeus, or William Withering's (1741–1799) *A Botanical Arrangement* (1776). Withering's learned Linnaean text was in turn superseded by different forms, such as the handbook and illustrated works that depicted the sexual parts of flowers so that readers could learn to apply Linnaean concepts. Such concepts were even featured on playing cards, such as those produced by James Sowerby (1757–1822), in which cards contained engravings of parts of plants with botanical questions and answers.

Over the course of the eighteenth century, women became increasingly important producers and consumers of botanical knowledge. Unlike herbwomen, genteel ladies studied plants as a form of polite recreation. Collecting plants, reading about them, and drawing them were types of social distinction

[33] Keith Thomas, *Man and the Natural World: Changing Attitudes in England 1500–1800* (Harmondsworth: Penguin, 1984).

as well as natural knowledge. In addition this knowledge of plants might take on a moral purpose. Priscilla Wakefield's (1751–1832) *An Introduction to Botany* (1796) urged that botany "become a substitute for some of the trifling, not to say pernicious objects, that too frequently occupy the leisure of young ladies of fashionable manners."[34] Just as Tom Telescope's lectures were given to entice his playmates from the evils of cardplaying, so Wakefield's text (written in an epistolary form) provided young ladies with an improving pastime.

Ladies may have studied botany only in the parlor, over the breakfast table, or on decorous country walks, but the forms of their study were highly varied. Playing cards, large folio volumes full of engravings, smaller handbooks, dialogs and amateur watercolors all purveyed this natural knowledge. Even poetry was a form for botanical knowledge; Erasmus Darwin's (1731–1802) *Love Among the Plants* (1791) is well known for its highly sexualized version of Linnaean classification.[35] Less well known is Frances Arabella Rowden's (fl. 1801–1821) *A Poetical Introduction to the Study of Botany* (1801). Based on Darwin's poem, it transformed his luxuriant sexual metaphors into female-centered images of delicacy and maternity.

The reform of popular culture manifested in the creation of genteel botany was only one factor affecting the natural knowledge of plants in the eighteenth century. The other lies in the increasing commodification of gardening and horticulture from the late seventeenth century. Although grains had long been commodified and sold for the baking of bread and the making of gruels, other plants became much more closely linked to the market in this period. Many English towns maintained the "assize of bread" well into the eighteenth century, in which a town's governors set prices for various types of bread sold in their jurisdiction. Starting with the Dutch tulip mania of the seventeenth century, however, decorative plants became commodities traded in the open market, at prices set by supply and demand, rather than by local elites. Here, too, the relationships between social classes were structured in part by differing access to material resources. Aristocrats could compete with one another by hiring knowledgeable gardeners and building high-tech greenhouses, whereas working men grew single flowers and put them into competition in flower shows. Aristocrats did not directly commodify individual plants or species, but workers could directly sell rare plants and seeds for tidy profits. Aristocrats, of course, were players in this market, too, but they usually bought rather than sold.

---

[34] As quoted in Ann B. Shteir, *Cultivating Women, Cultivating Science: Flora's Daughters and Botany in England, 1760–1860* (Baltimore, MD: Johns Hopkins University Press, 1996), p. 86. See also Shteir, "Linnaeus's Daughters: Women and British Botany," and Shteir, "Botany in the Breakfast Room," both in Pnina Abir-Am and Dorinda Outram (eds.), *Uneasy Careers and Intimate Lives: Women in Science, 1789–1979* (New Brunswick, NJ: Rutgers University Press, 1987).

[35] Janet Browne, "Botany for Gentlemen: Erasmus Darwin and the Loves of Plants," *Isis*, 80 (1989), 593–620, and Londa Schiebinger, "The Private Life of Plants: Sexual Politics in Carl Linnaeus and Erasmus Darwin," in Marina Benjamin (ed.), *Science and Sensibility: Gender and Scientific Enquiry, 1780–1945* (Oxford: Blackwell, 1991), pp. 121–43.

The new economic relations also helped to structure botanical knowledges. As at Stephen Switzer's stand at the sign of the flowerpot, metropolitan seedsmen sold an ever-wider array of seeds and plants. The variety of local plant names made the work of seedsmen difficult, since they could be accused of fraud when the name for a plant was not the same as that used by a customer. In response to this problem, a London Society of Gardeners published *Catalogus Planatarum* in 1730 in an attempt to standardize names. As the number of available plants had increased massively, the problem had become acute. From the sixteenth century the introduction of new species from other parts of Europe, the Middle East, and the New World created an ever-larger array of gardening possibilities. It is estimated that in 1500 England had perhaps two hundred kinds of cultivated plants; by 1839 the number had grown to about eighteen thousand. By 1705, the Brompton Nursery in London alone had some ten million different plants for sale. Gooseberry plants, in particular, flourished; more than three hundred varieties were on sale in England by 1780. Distinctions of taste grew accordingly.

Plants went in and out of fashion. A new flower might be introduced at an exorbitant price, only to fall in cost and cachet as it became more widely known. The most extravagant example is the Dutch tulip mania of the seventeenth century, but the eighteenth century saw many plants move from rarity to commonplace to obsolescence. The multiplicity of plants was created by consumer demand as gardens became sites of emulation. And with emulation came the professional gardener. Although men and women had been hired to work in gardens since at least the Middle Ages, it was only in the late seventeenth and early eighteenth centuries that the professional gardener fully came into being. Like the land steward on improvement-oriented farms, the new gardener supervised the work of others and sold his expertise in the form of garden designs and specialized knowledge in the care of exotic or unusual plants. Such men (and all of them were, it seems, men) could command salaries undreamed of by the manual laborers under their supervision. At the very top of the heap were men such as Henry Wise (1653–1738), gardener to Queen Anne and George I (1660–1727), who enjoyed £1600 per annum.[36] Such sums were reserved for royalty, but already, by the 1680s, the gardener at Lyme Hall in Cheshire was earning £60 a year, an income on a par with a well-paid clergyman.[37]

Emulation and interest in unusual plants were by no means confined to the upper and middle classes. One of the most important sites for the production of horticultural knowledge was florist societies, the members of whom were usually from the lower middle and artisan classes. Within these societies natural knowledge was structured by the pursuit of both cultural distinction

[36] David Green, *Gardener to Queen Anne: Henry Wise and the Formal Garden* (Oxford: Oxford University Press, 1956).
[37] Thomas, *Man and the Natural World*, p. 225.

and financial gain. "Florists" were those who bred new and startling varieties of the so-called florists flowers: hyacinths, pinks, tulips, ranunculi, anemones, auriculas, narcissi, and carnations. In many cities and towns, florist societies sponsored feasts and flower shows – usually one in the spring for auriculas and another at midsummer for carnations. As garden writer William Hanbury (1725–1778) observed in 1770, many flower shows were won by "small tradesmen, weavers, or the like" because such men had the requisite habits of industry for the laborious production of a perfect bloom.[38] A pair of millers, for example, bred an auricula with 123 blossoms on one stem. Indeed, it was a commonplace in the later eighteenth century that the best pinks and auriculas were grown by the weavers of Spitalfields, Manchester, and Paisley. Anne Secord has analyzed how nineteenth-century artisan botanists produced a distinctive natural knowledge framed within emergent class and gender identities, and she has specified pubs as the crucial site for such activity.[39] Some of these features apply equally well to eighteenth-century florists. Then, too, florists' societies often met in taverns or inns. Sometimes innkeepers sponsored flower shows, charging an entry fee that entitled the florist to see flowers and drink beer. And their knowledge of plants, although different from that of the later artisan botanists, was similarly framed by social identity. For instance, the masculinity of such groups was expressed in the frequently used name "Sons of Flora" for a local society.

Insofar as the material forms of rare flowers were produced by florists in a world structured by fashion and emulation, their producers differed little from their social betters who competed among themselves to acquire rarities. For instance, aristocrats in the early eighteenth century battled to produce pineapples, a tricky fruit that required careful hothouse management, but the results of which could be presented at banquets and dinners. However, another view of florists might emphasize the commodification of flowers. Although prizes at flower shows were relatively small, measured in pence or shillings, a spectacular new auricula or hyacinth could command considerable sums in the marketplace. In the 1770s, many hyacinths could be had for a few pence each, but the one called Black Flora cost twenty guineas. Likewise, hundreds of varieties of auriculas were available commercially, and it seems that there was always room for more.

Thus, botanical knowledges, too, varied considerably in forms and sites. The material form of a plant itself was employed by aristocrat and miller alike. Latin texts, playing cards, and gardening catalogs each functioned as printed forms of botanical knowledge, and taverns and drawing rooms, stableyards and seed stands were all sites for the expression and reproduction of that knowledge. As with natural knowledges of agriculture and medicine, the

---

[38] Thomas, *Man and the Natural World,* p. 229.
[39] Anne Secord, "Science in the Pub: Artisan Botanists in Early Nineteenth-Century Lancashire," *History of Science,* 32 (1994), 269–315.

continuities and changes in the sites and forms of botanical knowledge cannot be fully accounted for either in terms of the reform of popular culture or in terms of the rise in consumerism. Nevertheless, both models offer more dynamic accounts than those grounded in passive notions of popularization.

CONCLUSION

One of the stated aims of this chapter has been to explore natural knowledge in its broadest sense. We have sought to destabilize two related hierarchies: that of abstract ideas over practices, and that of popularization – what we have irreverently dubbed the fried-egg model. Although in recent years a focus on science in context has nibbled away at the older historiographical primacy of the ideas of great men, almost unconsciously historians have continued to reproduce the kind of structure that places a Newton ahead of Newtonianism. The second problem, that of popularization, is similar to the first: historians have assumed a kind of downward or outward flow of knowledge from scientist to public.

In questioning this model, we have drawn on the insights of those historians and sociologists of science who have insisted on a symmetrical approach to all forms of natural knowledge, be they abstract theory or everyday practices. We have not therefore privileged one type over another and have sought to incorporate into this analytical frame much that historians of science have explicitly or implicitly deprivileged through exclusion. And since historians of popular culture have reminded us of the difficulties inherent in conceptualizing a "popular culture,"[40] we have endeavored to avoid any easy assignments of "popular" or nonpopular.

Instead, we have stressed an understanding of the sites and forms of natural knowledge. Ways of knowing are socially and geographically specific, but the one cannot be read from the other – or vice versa. Always contingent, those relationships must not be assumed but rather must be seen as open to historical investigation. To talk of sites is to employ both literal geography and that of metaphor. As we have indicated, coffeehouses, barnyards, fields, and ladies' drawing rooms were all important sites for the construction and display of natural knowledge. But sites might also be constituted through social identity or common interest, as in the case of the Spalding Gentlemen's Society or Arthur Young's *Annals of Agriculture*. Self-identification as a gentleman and the disposable income required for subscription served to bound these particular sites. In Spalding, that identification resembles the social world created by those who penned such journals as the *Spectator* and the *Gentleman's*

---

[40] See here, especially, Tim Harris's introduction to his edited collection, *Popular Culture in England, c. 1500–1850* (London: Macmillan, 1995).

Magazine – of genteel, polite men interested in various forms of natural and literary knowledge. As for Young's magazine, perhaps the crucial characteristics were a perception of oneself as "improving" and attached to the land.

Forms, we have argued, can be categorized as oral, printed, or material. Thus, Robert Bakewell's sheep or a miller's 123-bloom auricula are material forms of knowledge that performed acts of communication in that culture. Like material forms, oral ones were significant throughout the eighteenth century in the wide range of lectures, personal conversations, and sermons. It is only by means of print or manuscript, text or illustration, that we can know about oral or material forms. But the boundaries among all these were rarely absolute: William Hanbury learned about flowers in conversation with an ostler, but the ostler had read his Culpeper. Pictures of Toft and her rabbits, or paintings of new improved livestock, were as important as the material objects they purported to represent. So, too, print often employed oral forms, as with Tom Telescope's lectures or Algarotti's dialogs.

The emphasis on sites and forms challenges any easy model of popularization. It enables us to see natural knowledges as cultural products both shaping and shaped by longer currents of British history – as no different in this respect from choral music, or bearbaiting, or the brewing of beer. Three themes in particular can help us integrate the sites and forms of natural knowledge into larger patterns of historical change. All three are interrelated, and they all derive from some of the historiographical emphases of the past few decades.

The first is commodification, the process by which things become identified as items that can be bought and sold. Analysis of this process is closely related to an interpretation of eighteenth-century Britain as a consumer culture. Here, historians have shifted from a focus on production, be it of medicines or portraits, to one of consumption that poses questions about demand and fashion. Livestock and flowers represent the most obvious cases of commodification, where ever-finer distinctions of taste were in a reciprocal relation to the multiplication of specific products.

However, natural knowledge was not commodified only by means of material goods. One of the distinctive features of the eighteenth century is the increased possibilities that existed for people to market themselves based on their command of natural knowledge. Desaguliers, for instance, put his natural knowledge to commercial use, advising on waterworks, steam engines, and the like. In addition, land stewards, hired by improvement-minded landlords, were able to transform their mastery of details of agricultural theory and practice into paid employment. The commodification of natural knowledge was not, of course, a process peculiar to the eighteenth century. Herbwomen had been marketing their knowledge for decades if not centuries, and engineers from the Netherlands and Germany had plied their trade in England for a century or more. What is unique and peculiar to the eighteenth century is the growth in the numbers and kinds of respectable, middling sorts of purveyors of natural knowledge: the gardener at Lyme Hall, for instance, earning a clergyman's salary.

These entrepreneurs functioned in a metropolitan and provincial world in which natural knowledge was both entertainment and improvement. Historians have characterized this world as one of sociability, the second theme that highlights chronological change. The multitude of gatherings – coffeehouses, provincial societies such as that in Spalding, or the many Sons of Flora – were part and parcel of a larger wave of sociability. Few such sites existed in the previous century, and natural knowledge as a polite means of interaction was both a stimulus to and an enactment of sociability.

If sociability drew people together in new configurations, our third theme refers to modes of social difference. For along with the proliferation of sites and forms came increasingly well articulated distinctions among them. Whether we focus on economic distinctions and the concern with rank or emphasize the cultural correlatives of such differences, the eighteenth century can be understood as a period of increasing distance between social groups – distances that would harden into strong class distinctions in the following century. However, in the eighteenth century, social groupings and natural knowledges were polymorphous. There were very few "alternative" sciences of the sort that rose to a degree of prominence in the nineteenth century. The carefully constructed epistemological and social challenges to scientific and social orthodoxy that characterized phrenology or artisan pub botany, for example, have few analogues a century earlier. Rather, we see the gradual segregation of certain forms and sites. Culpeper's herbal, for example, was used by ladies and shoemakers alike in 1700. By 1800 it was the bible of working-class herbalism, but it had been supplanted in middle- and upper-class homes by forms such as illustrated botanical texts addressed to the ladies or Buchan's class-specific advice about when to call the doctor.

Thus, an emphasis on the sites and forms of natural knowledges in eighteenth-century Britain not only provokes questions about the nature and breadth of such knowledges but also provides a means for apprehending their histories. This model is richer and more sensitive to social, economic, and cultural change than that afforded by the retrospectively constructed, hierarchically simplistic fried-egg model. The privileged epistemological status often attached to twentieth-century science is, in many ways, a product of early-modern struggles and negotiations. To read those epistemological claims back into the very period that produced them is to efface those struggles and negotiations; it is to decide in advance that contingency and the play of complex historical forces had no role in moving from the natural knowledges of the eighteenth century to the science of today.

7

THE IMAGE OF THE MAN OF SCIENCE

Steven Shapin

The relations between the images of the man of science and the social and cultural realities of scientific roles are both consequential and contingent. Finding out "who the guys were" (to use Sir Lewis Namier's phrase) does indeed help to illuminate what kinds of guys they were thought to be, and, for that reason alone, any survey of images is bound to deal – to some extent at least – with what are usually called the realities of social roles.[1] At the same time, it must be noted that such social roles are always very substantially constituted, sustained, and modified by what members of the culture *think* is, or should be, characteristic of those who occupy the roles, by precisely whom this is thought, and by what is done on the basis of such thoughts. In sociological terms of art, the very notion of a social role implicates a set of norms and typifications – ideals, prescriptions, expectations, and conventions thought properly, or actually, to belong to someone performing an activity of a certain kind. That is to say, images are part of social realities, and the two notions can be distinguished only as a matter of convention.

Such conventional distinctions may be useful in certain circumstances. So-cial action – historical and contemporary – very often trades in juxtapositions between image and reality. One might hear it said, for example, that modern American lawyers do not really behave like the high-minded professionals

[1] For introductions to pertinent prosopography, see, e.g., Robert M. Gascoigne, "The Historical De-mography of the Scientific Community, 1450–1900," *Social Studies of Science*, 22 (1992), 545–73; John Gascoigne, "The Eighteenth-Century Scientific Community: A Prosopographical Study," *Social Studies of Science*, 25 (1995), 575–81; Steven Shapin and Arnold Thackray, "Prosopography as a Re-search Tool in History of Science: The British Scientific Community, 1700–1900," *History of Science*, 12 (1974), 1–28; and especially William Clark, "The Prosopography of Science," in this volume. The use of anything but this gendered language to designate the eighteenth-century "man of science" would be historically jarring. The system of exclusions that kept out the vast numbers of the unlettered also kept out all but a very few women. Women's role in eighteenth-century science is surveyed in this volume by Londa Schiebinger.

This chapter was substantially written while the author was a Fellow of the Center for Advanced Study in the Behavioral Sciences, Stanford, California. He thanks the Center, and the Andrew W. Mellon Foundation, for their support.

portrayed in official propaganda, and statements distinguishing image and reality in this way thus present themselves as real to those who wish to understand contemporary American society. But such a disjunction constitutes a new image to be contrasted with the old, perhaps one portraying lawyers to be as venal as car salespeople, and those who deal with lawyers on that basis help to constitute new social realities. Nor – for historians or sociologists – is there some methodological sin that is inherently attached to asking, for example, whether eighteenth-century men of science – individually or collectively – "really" possessed the range of virtues, vices, or capacities widely attributed to them, just so long as we appreciate that any social role is constituted through *some* set of beliefs about what its members are like and should be like.

So it is not proper – either in historical or in sociological practice – to speak of social roles without what might be termed their "characterology." What typifications attached to the person of the eighteenth-century man of science? What virtues, vices, dispositions, and capacities was such a person thought to possess, and in what combinations? What relationships were there between the socially recognized characters of the man of science and those attached to other social roles? What variation was there in the characterology of the man of science? Was there one settled image of the man of science, or were there several, possibly conflicting characterizations, attached to different versions of his identity – the mathematician, the philosopher, the ornithologist, and so on – or expressing different sensibilities toward what these people were like *tout court*?[2] Qualified in these ways, characterology can be used as a pertinent organizing principle for a survey of images of the man of science in the eighteenth century.

Yet before that characterology can even be presented, a possible preconception about eighteenth-century social roles should be confronted and dismissed. At neither end of the eighteenth century did the role of the "man of science" exist as a coherent and distinctive social kind. In the late seventeenth century the pursuit of natural knowledge took place within a wide variety of existing social roles. The typifications and expectations bearing on those who happened to pursue different sorts of natural knowledge within those roles were not those of the professional scientist – that social kind did not, of course,

---

[2] Although characterology dates back to antiquity – in the work of Theophrastus – the delineation of "characters" has the advantage of being a revived early modern usage as well. A number of sixteenth-, seventeenth-, and early eighteenth-century men of letters compiled the "characters" of, for example, "the philosopher," "the mathematician," "the School-man," "the scholar," "the courtier," etc., as well as the more traditional allegorical embodiment of the virtues and vices; see, e.g., Joseph Hall, *Characters of Virtues and Vices* (London, 1608); Samuel Butler, *Characters and Passages from Note-Books*, ed. A. R. Waller (Cambridge University Press, 1908); John Earle, *Micro-cosmographie, or, A Piece of the World Characteriz'd in Essays and Characters* (London, 1650); Jean de la Bruyère, *Characters*, trans. Henri Van Laun (London: Oxford University Press, 1963; orig. publ. 1690). And for practical uses of characterology in eighteenth-century social relations, see, e.g., Philip Dormer Stanhope, Earl of Chesterfield, *Letters to His Son*, ed. Oliver H. Leigh, 2 vols. (New York: Tudor Publishing, n.d.; orig. publ. 1774), vol. 1, pp. 105–6, 387; vol. 2, p. 16.

exist – but rather were predominantly those of what might be called the host social role. The roles of the university professor, the physician or surgeon, the gentleman, the courtier, the crown or civil servant, the cleric, and many others were each accompanied by a set of widely understood, and relatively coherent, characters, conventions, and expectations, and it was these that colored whatever pursuit of natural knowledge might happen to occur within such roles. That is to say, the images of eighteenth-century men of science – in all their variety – were very significantly shaped by appreciations of what was involved in the host roles: what sorts of people occupied such roles, with what characteristics and capacities, doing what sorts of things, and acquitting what sorts of recognized social functions, with what sorts of value attached to such functions?[3]

Moreover, in the eighteenth century it did not necessarily follow from an individual's being recognized as having produced natural knowledge of great scope or acknowledged quality that such an individual was clearly *identified* – by his contemporaries or even by himself – with a distinct intellectual role, still less with the role of the man of science. In the middle of the seventeenth century Blaise Pascal gave up natural philosophy and mathematics for a higher religious calling, as did the Dutch entomologist Jan Swammerdam some years later. At the turn of the century Isaac Newton, having exchanged his mathematical professorship for the administration of the Royal Mint, insisted that correspondents recognize he was not "*trifling* away" his time "about Mathematical things" when he "should be about ye Kings business."[4] Nor, despite Gottfried Wilhelm Leibniz's European reputation as mathematician and philosopher, did this count for much with his Hanoverian employers, who demanded, at the end of his life, that he devote himself wholly to completing a politically useful dynastic history. In America, Benjamin Franklin achieved international celebrity as inventor of the lightning rod and, to a lesser extent, as a theorist of electricity, but, so far as his local culture was concerned, he was identified primarily as a printer, a businessman, a diplomat, and a statesman. To *do* science – as current sensibilities recognize it – was not necessarily the same thing as to *be* a man of science, to occupy that social role. What historians recognize as crucially important scientific research might be, in contemporary terms, only a moment or an element – among others – in a life fundamentally shaped by other concerns and lived out within other identities. This is just another way of noting the disjunction between activities,

---

[3] The methodological framework for this way of conceptualizing the social role of the man of science is sketched in Steven Shapin, "The Man of Science in the Early Modern Period," in Lorraine Daston and Katharine Park (eds.), *The Cambridge History of Science*, vol. 3: *Early Modern Science* (Cambridge University Press, forthcoming), which can be treated as preface to the present chapter. See also Roy Porter, "Gentlemen and Geology: The Emergence of a Scientific Career, 1660–1920," *The Historical Journal*, 21 (1978), 809–36, at 809–15.

[4] Isaac Newton to John Flamsteed, 6 January 1698/99, in Newton, *The Correspondence of Isaac Newton*, eds. H. W. Turnbull, J. D. Scott, A. R. Hall, and Laura Tilling, 7 vols. (Cambridge University Press, 1959–77), vol. 4, p. 296.

identities, and roles that was characteristic of virtually all scientific activity until the professionalized arrangements of the twentieth century.[5] The cultural character of scientific work, and of scientific workers, was taken substantially from practitioners' identification with established host roles.

This state of affairs obtained at both ends of the eighteenth century, and it persisted into much of the nineteenth. Yet a series of subtle and consequential changes was being effected from about the 1680s to about the 1820s – changes that were partly shifts in concrete social realities and partly shifts in social aspirations and in cultural images of what it was to do science. It was these changes that by the 1830s inspired systematic and public agitation for the professionalization of science and that allowed such agitation to be regarded as meaningful, if not as wholly and effectively persuasive. Those changes were more closely associated with some characters of the man of science than with others, and I shall be using the notion of characters to indicate both those structures and images that were conserved across our period and those that experienced the sorts of changes whose social and cultural significance became most clear in nineteenth-century professionalizing movements. The characters I shall treat in this survey are the Godly Naturalist, the Moral Philosopher, and the Polite Philosopher. By way of conclusion, I shall make some briefer remarks about the developing eighteenth-century character of the Civic Expert.[6]

THE GODLY NATURALIST

The roles of the pious naturalist and, more specifically, of the parson-naturalist, were thickly populated and culturally understood throughout the period, especially, but not exclusively, in Protestant culture. The Renaissance argument that God had written two books by which His existence, attributes, and intentions might be known – Scripture and the Book of Nature – continued in currency in the developing culture of "natural theology." Sentiments that inspired parson-naturalists, and even parson-experimentalists, at the time of the Glorious Revolution were still vigorous at the time of the French Revolution. The "argument from design" (inferring God's existence,

---

[5] The point being made here is not at all the same as the traditional distinction between "professionalism" and "amateurism," if the latter is taken to indicate a less than wholehearted or serious commitment to science. At issue here are not distinctions in seriousness or quality but rather differences in personal and cultural identities and their contemporary cultural consequences. See Porter, "Gentlemen and Geology," pp. 814–15.

[6] Although this set of characters emerges "naturally" from recognized eighteenth-century cultural discourse, it necessarily reflects both a late twentieth-century historian's selective criteria of significance and the pragmatic constraints of such a survey. Were there space enough, other characters would merit extended discussion, e.g., those associated with medical practice and with the popularization of science. And, almost needless to say, a given individual might be described through the repertoires of more than one recognized character. Indeed, for such celebrated individuals as Newton, there was typically a cultural contest over which character best described him, what kind of person he *was*.

wisdom, benevolence, and power from the evidence of contrivance in organic and material nature) seemed overwhelmingly persuasive to such English clerics as the Reverend John Ray in the 1690s, the Reverend Stephen Hales in the 1720s, the Reverend Gilbert White of Selborne in the 1780s, and the Reverend William Paley in the 1800s, and to such French divines as the Abbé Noel Antoine Pluche, whose *Spectacle de la nature* was an international bestseller from the 1730s. And even though many parishioners undoubtedly found the spectacle of a bird-watching and bug-hunting vicar mildly amusing – the Scottish parishioners of the natural historian John Walker called him "the mad minister from Moffat" – valued natural-theological sensibilities were available to offset any appearance of culpable oddness in such pursuits. The naturalist-parson belonged to the century's inventory of recognized characters, and the scientific portion of his activities was understood to flow from some version of what it was to *be* a minister. And, in the parson's self-understanding, doing science might not be a mere avocation; it might be counted as a legitimate and important part of his priestly vocation. The parson-naturalist's scientific inquiries were surrounded by the aura shed by his priestly role.[7]

Nor were natural theological justifications and motives confined to men of the cloth. They were widely available to explain what sort of thing one was doing when one was doing science, what kind of person one was, what place and value science had in the overall culture, and what role in the social system was supposed to be occupied by those engaged in the pursuit of natural knowledge. So whatever was understood about the virtues and capacities of the priest was available to understand those godly investigators who were called "priests of nature."[8] These justifications and appreciations were a ubiquitous feature of eighteenth-century culture, again especially in Britain, and they might be importantly expressed by the occupants of a great range of roles: the university professor, the academician, the medical man, the gentleman, the instrument-maker, and the popular lecturer, writer, and showman, as well as by those whose roles were contained within formal religious institutions.

The aura of holiness that "naturally" surrounded the priest might also be discerned around a range of ostensibly secular practitioners. In the Netherlands the draper-microscopist Antoni van Leeuwenhoek saw the wisdom of God in the architecture of even the tiniest of his creatures. In America, the Quaker botanist John Bartram announced that it was through the "telescope" of nature that "God in his glory" could be seen. In Sweden Carl Linnaeus was described as "a second Adam," giving species their proper names and conceiving

---

[7] For a survey of forms of natural theology through this period and beyond, see John Hedley Brooke, *Science and Religion: Some Historical Perspectives* (Cambridge University Press, 1991), especially chaps. 4–6; and, for aspects of the British case, see John Gascoigne, "From Bentley to the Victorians: The Rise and Fall of British Newtonian Natural Theology," *Science in Context*, 2 (1988), 219–56.

[8] For this usage in the later seventeenth century, see Harold Fisch, "The Scientist as Priest: A Note on Robert Boyle's Natural Theology," *Isis*, 44 (1953), 252–65, and Simon Schaffer, "Godly Men and Mechanical Philosophers: Souls and Spirits in Restoration Natural Philosophy," *Science in Context*, 1 (1987), 55–85.

of his binomial nomenclature as a "psalter for divine worship": "Man is made for the purpose of studying the Creator's works that he may observe in them the evident marks of divine wisdom." In Germany Leibniz reckoned that there was great religious utility in science, on the condition that natural inquiry was informed by a proper "intellectualist" theology, showing that God's wisdom had created the "best of all possible worlds."[9] In England, the Unitarian chemist Joseph Priestley wrote, "A Philosopher ought to be something greater, and better than another man." If the man of science was not already virtuous, then the "contemplation of the works of God should give a sublimity to his virtue, should expand his benevolence, extinguish every thing mean, base, and selfish in [his] nature."[10]

The culture of natural theology was not uniformly institutionalized and honored. It was never as influential in Catholic as in Protestant cultures. And during the course of the eighteenth century it took notable knocks: in Scotland from David Hume; in Germany from Immanuel Kant; and in France from the *philosophes* and *Encyclopédistes*. Yet wherever the writ of natural theology ran, its sensibilities supported a character of the man of science as godly and the doing of science as the acquittal of religious goals.

THE MORAL PHILOSOPHER

A natural order bearing the sure evidence of divine creation and superintendence was understood to uplift those who dedicated themselves to its study. Godly subject matter made for godly scholars. This was the major way in which the culture of natural theology sustained an image of the man of science as virtuous beyond the normal run of scholars. But eighteenth-century cultures that were not powerfully marked by natural theology also produced pictures of the man of science as specially or uniquely virtuous. The cultural resources for constructing those images and rendering them credible linked the eighteenth century to antiquity as well as to the immediate past.

The eloges presented in commemoration of recently deceased members of the Paris Academy of Sciences offer the eighteenth century's most highly developed and influential portraits of the virtuous man of science. Although a

---

[9] For the religious sensibilities of Swammerdam, Leeuwenhoek, and other Dutch microscopists of the late seventeenth and early eighteenth centuries, see Edward G. Ruestow, *The Microscope in the Dutch Republic: The Shaping of Discovery* (Cambridge University Press, 1996), especially pp. 116–20, 137–45, 166–7, 219–20; for Bartram and Quaker strands of natural theology, see Thomas P. Slaughter, *The Natures of John and William Bartram* (New York: Alfred A. Knopf, 1996), especially chap. 3 (quoting pp. 62–3); for Linnaeus and Leibniz, see remarks in Brooke, *Science and Religion*, pp. 160–3, 197, 231–4 (quoting pp. 162, 197, 232); Lisbet Koerner, *Linnaeus: Nature and Nation* (Cambridge, MA: Harvard University Press, 1999); and Sten Lindroth, "The Two Faces of Linnaeus," in Tore Frängsmyr (ed.), *Linnaeus: The Man and His Work* (Canton, MA: Science History Publications, 1994; orig. publ. 1983), pp. 1–62, especially pp. 11–16.

[10] Joseph Priestley, *The History and Present State of Electricity*, 2 vols., 3rd ed. (London, 1775), vol. 1, p. xxiii.

natural-theological idiom was not strong in that setting, other resources were available to display the superior virtue of the man of science. Many of the more than two hundred eloges composed by Bernard le Bovier de Fontenelle (and his successors Jean-Jacques Dortous de Mairan, Jean-Paul Grandjean de Fouchy, and the Marquis de Condorcet) from 1699 to 1791 drew upon Stoic and Plutarchan tropes to establish both the special moral qualities possessed by those drawn to science and the additional virtues that a life dedicated to scientific truth encouraged in its devotees.[11]

Like many of Plutarch's Greek and Roman heroes, Fontenelle's eighteenth-century men of science were described as embodiments of Stoic fortitude and self-denial. The life of science held out few prospects of material reward and little hope for fame, honor, and the applause of the polite and political worlds. The dedication to truth that drew men to such a life was made manifest by neglect of self and of material self-interest, and by a disregard for public favor and approval. Such power as men of science came to possess was not vaingloriously sought but rather was thrust upon them by patrons who often wanted the material goods understood to flow from scientific knowledge. Moreover, even in the absence of a pronounced natural-theological idiom, it was repeatedly said that the life spent in pursuit of natural knowledge tended to make men humble, serious, simple, and sincere. The immensity, grandeur, and sublimity of nature made modest those who studied it, as did the awareness of the little that was securely known about nature as compared with the vastness of what remained to be known. Sincerity, candor, tranquility, and contentment were naturally instilled in men who lived for the love of nature's truth.[12]

By the 1770s these sentiments were supplemented by Condorcet's Renaissance-humanist preferences for a life of action and civic benevolence. The man of science, in Condorcet's picture, had the capacity to benefit the public realm both materially and spiritually. Condorcet's eloge of Benjamin Franklin accordingly celebrated both Franklin's technological ingenuity and the political reformism that was reckoned to flow from the very nature of modern scientific inquiry. Science would at once produce technological change

---

[11] For these éloges, see especially Charles B. Paul, *Science and Immortality: The Éloges of the Paris Academy of Sciences (1699–1791)* (Berkeley: University of California Press, 1980), on which the following paragraphs largely rely, and, for Georges Cuvier's éloges of the late eighteenth and early nineteenth centuries, see Dorinda Outram, "The Language of Natural Power: The Funeral *Éloges* of Georges Cuvier," *History of Science*, 18 (1978), 153–78. For important treatment of eighteenth- and early nineteenth-century debates over the virtue and mental capacities of Isaac Newton, see Richard Yeo, "Genius, Method, and Morality: Images of Newton in Britain, 1760–1860," *Science in Context*, 2 (1988), 257–84.

[12] In Scotland, Adam Smith was greatly impressed with Fontenelle's eloges. *The Theory of Moral Sentiments* (1759) endorsed the Parisian celebration of mathematicians' and natural philosophers' "amiable simplicity of . . . manners." Their "tranquillity" and their indifference to public opinion flowed from an inner assurance that their claims were both true and important. The same could not be said of "poets and fine writers": Adam Smith, *The Theory of Moral Sentiments*, eds. D. D. Raphael and A. L. Macfie (Oxford: Clarendon Press, 1976), pp. 124–6.

and encourage those mental and moral attributes that would naturalize rational industrial society. "Forever free amidst all manners of servitude, the sciences," Condorcet wrote in the year after the storming of the Bastille, "transmit to their practitioners some of their essence of independence or either fly from countries ruled by arbitrary power or gently prepare the revolution that will eventually destroy it."[13]

The image of the selfless man of science, offering much to the nation and neither receiving nor expecting to receive much in return, was lent credibility by some recognized social circumstances affecting scientific work. In the eighteenth century, as in the seventeenth, a decision to pursue many forms of scientific learning might well be taken against plausible calculations of material self-interest, and often against strong parental wishes or directions. For those lacking independent means, the professions of law, religion, and medicine were understood to ensure an honest and legitimate living. Very many eighteenth-century men of science chose their calling against their fathers' encouragement toward a career at the bar or in the church; in maturity, others managed to combine scientific research with at least nominal legal, administrative, or clerical careers; and many others managed the much easier combination of science and medicine. But social respectability was only dubiously associated with the calling of the practical mathematician or engineer, and it was difficult to envisage clear remunerative and polite career prospects for the physicist, the geographer, the naturalist, or, to a lesser extent, for the astronomer.

If one were battling to rise from the lower orders – as, for example, were the electrician Stephen Gray, the chemist John Dalton, and the geologist William Smith – a career as scientific lecturer, author, or technical consultant might have both its material and its social attractions. If one possessed independent means freeing him from material concerns – as did, for example, the naturalists the Comte de Buffon, the Earl of Bute, and Sir Joseph Banks, the physicist Henry Cavendish, and the geological chemist Sir James Hall – one could afford to adopt an insouciant attitude toward remuneration, toward orthodox notions of cultural respectability, and even toward scientific authorship and the public assertion of property in intellectual goods.[14] But for many in middling social circumstances – from younger sons of the aristocracy to the offspring of the professional and mercantile classes – scientific

---

[13] Condorcet's éloge of Franklin (read 13 November 1790), quoted in Paul, *Science and Immortality*, p. 67; see also Roger Hahn, *The Anatomy of a Scientific Institution: The Paris Academy of Sciences, 1666–1803* (Berkeley: University of California Press, 1971), p. 165; Keith M. Baker, *Condorcet: From Natural Philosophy to Social Mathematics* (Chicago: University of Chicago Press, 1975), especially pp. 293–9.

[14] Porter, "Gentlemen and Geology," p. 815, incisively notes "the lack of pressure to publish" bearing on gentlemen-geologists in the eighteenth century. Indeed, gentlemen-amateurs often worried about the gentility of "appearing in the character of an author." See also David P. Miller, "'My Favourite Studdys': Lord Bute as Naturalist," in Karl W. Schweizer (ed.), *Lord Bute: Essays in Reinterpretation* (Leicester: Leicester University Press, 1988), pp. 213–39, at pp. 215, 218.

inquiry would have to be combined with an adequately remunerated professional or public life. There were many such possible hybrid forms of life in the eighteenth century beyond those attached to the universities and the learned professions: Antoine Laurent Lavoisier famously served as a "tax-farmer"; Leibniz and Johann Wolfgang von Goethe were government officials; Charles Augustin Coulomb worked as a military and civil engineer; and the young Alexander von Humboldt was both a diplomat and a supervisor of mining. For those of intermediate social standing a decision to devote oneself solely or mainly to scientific scholarship might be understood – against this background – as testimony to a particularly selfless and wholehearted kind of dedication. Fontenelle's eloge of the mathematician Michel Rolle notably asserted that "there is between science and wealth an old and irreconcilable distinction," and Condorcet's eloge of another mathematician, Etienne Bézout, explained why his family opposed the young man's scientific vocation: "A father . . . knows that education and enlightenment lead neither to honour nor to fortune." What could account for a commitment to science other than a genuine vocation?[15]

THE POLITE PHILOSOPHER OF NATURE

The same images of vocation, dedication, and detachment that testified to the virtue of the eighteenth-century man of science also constituted a potential handicap to his unconditional membership in polite society and to that society's approval of his activities. Since antiquity, the line between virtuous, holy, or learned disengagement from the conventions of everyday society, on the one hand, and culpable incivility, on the other, had always been subject to contest and conflict. Did the philosopher or learned man fall under the compass of civil and polite society, or did he play by different rules – rules that excused him from obeying society's obligations and expectations? Should the philosophical "citizen of the world" be exempted – wholly or partly – from the responsibilities of mundane citizenship?[16]

Such an exemption created a special cultural space in which the learned

---

[15] Quoted in Roger Hahn, "Scientific Careers in Eighteenth-Century France," in Maurice Crosland (ed.), *The Emergence of Science in Western Europe* (London: Macmillan, 1975), pp. 127–38, on pp. 131–2. Hahn importantly points out (ibid., p. 131) that even in the highly "professionalized" eighteenth-century French setting – where state support of science was at a far higher level than it was in Britain or even Germany – *very* few members of the Paris Academy of Sciences could expect to make a living solely from their state stipends or pensions: "A serious gap existed between what historians refer to proudly as funded government sponsorship of French science, and the life of the individual scientist." For treatment of these issues in the French context, see also Paul, *Science and Immortality,* pp. 69–85, and Maurice Crosland, "The Development of a Professional Career in Science in France," in Crosland (ed.), *The Emergence of Science in Western Europe,* pp. 139–59 (for changes and continuities on either side of the Revolution).

[16] These questions are discussed in Steven Shapin, "'The Mind Is Its Own Place': Science and Solitude in Seventeenth-Century England," *Science in Context,* 4 (1991), 191–218.

man could be recognized and valued, but at the same time it posed a problem for the relations between scholarly society and its civil, gentle, court, or mercantile counterparts. These possibilities and problems were not peculiar to the man of science – in general form they also applied to the logician, rhetorician, and theologian, and to philosophers not primarily concerned with the natural order – although the predicament of the man of science was subject to some special tensions during the course of the Scientific Revolution and Enlightenment.

Seventeenth-century "modern" critics of Scholastic knowledge insisted on its barrenness just as they condemned Scholastic society for its incivility. Criticisms of knowledge and of social forms were strongly linked: the schoolmen's wrangling was said to be so ferocious because – as the current quip has it – so little was at stake. If their inquiries had solid intellectual substance on which to feed, and if the veracity of their claims could be made manifest, then wrangling would truly come to an end. Such moderns as Bacon, Descartes, Hobbes, and Boyle proposed to remedy wrangling through both conceptual and methodological reform. Mechanical metaphors and micro-mechanical explanations might link the natural to the artificial and the natural philosopher to the world of mechanical artifice, thus subjecting intellectual abstraction to the discipline of the concrete and the intelligibly contrived. Correct method would discipline philosophical process and judgment by eliminating or mitigating the role of subjectivities, passions, interests, and cultural conventions. The result would be a new natural philosophy whose products were socially useful and whose practitioners were suitable for membership in civil society. Empirical and experimental methods – favored by the English – would replace Aristotelian "learned gibberish" and dogmatic arrogance with work, fact, and lowered norms of natural-philosophical certainty. Rational methods – preferred by the French – would bind dissension in iron chains of logic, and they were advertised as no less capable of producing useful outcomes. A new utility would rightly attract the esteem of the state and of civil society; a new civility would make the practitioners of natural knowledge fit for the drawing room and the salon.

Such were the claims made by and on behalf of the practitioners of reformed science in the late seventeenth and early eighteenth centuries. Tracing the credibility and consequences of these claims through the eighteenth century is, however, no simple matter. To some extent, natural knowledge had always had a place in courtly and commercial society, and it continued to enjoy that place through the eighteenth century. Wonder, weapons, gadgets, glory, and natural legitimations had long been social desiderata, and these goods might be supplied at least as visibly and efficiently by eighteenth-century scientific practitioners as by their predecessors. Seventeenth-century wondermongers such as Athanasius Kircher and keepers of curious cabinets such as Ulisse Aldrovandi had their eighteenth-century counterparts in such itinerant scientific demonstrators and electrical showmen as Benjamin Martin, Pierre

Polinière, and the Abbé Jean Antoine Nollet, just as Franz Anton Mesmer's spectacular late-eighteenth-century presentation of self was similar to that of the Renaissance theatrical therapist Paracelsus.[17]

Galileo established his practical value to the early seventeenth-century Florentine court with the military compass and the telescope, and his symbolic value through the discovery and naming of the "Medicean stars."[18] To centralizing and imperialist nation-states, his eighteenth-century successors promised – and in many cases delivered – an expanded range of aids to power and glory: cosmological legitimation (as before) but also solutions to the problem of longitude; reliable maps of new colonies; primary surveys of domestic and colonial flora and fauna and techniques for transplanting them around the world; improved agricultural, chemical, ceramic, mining, and metallurgical techniques; better ships, better guns, healthier seamen; and even perpetual motion machines.[19]

---

[17] For Kircher and Aldrovandi, see Paula Findlen, *Possessing Nature: Museums, Collecting, and Scientific Culture in Early Modern Italy* (Berkeley: University of California Press, 1994); see also Krzystof Pomian, *Collectors and Curiosities: Paris and Venice, 1500–1800* (Cambridge: Polity Press, 1990), for contests over the legitimacy of curiosity. For eighteenth-century electricians and natural philosophical showmen, see, e.g., J. L. Heilbron, *Electricity in the 17th and 18th Centuries: A Study of Early Modern Physics* (Berkeley: University of California Press, 1979); Simon Schaffer, "Natural Philosophy and Public Spectacle in the Eighteenth Century," *History of Science*, 20 (1983), 1–43; Schaffer, "The Consuming Flame: Electrical Showmen and Tory Mystics in the World of Goods," in John Brewer and Roy Porter (eds.), *Consumption and the World of Goods* (London: Routledge, 1993), pp. 489–526; Schaffer, "Augustan Realities: Nature's Representatives and Their Cultural Resources in the Early Eighteenth Century," in George Levine (ed.), *Realism and Representation: Essays on the Problem of Realism in Relation to Science, Literature, and Culture* (Madison: University of Wisconsin Press, 1993), pp. 279–318; Roy Porter, "Science, Provincial Culture and Public Opinion in Enlightenment England," *British Journal for Eighteenth-Century Studies*, 3 (1980), 20–46; Geoffrey V. Sutton, *Science for a Polite Society: Gender, Culture, and the Demonstration of Enlightenment* (Boulder, CO: Westview Press, 1995), chaps. 6, 8 (for Polinière and Nollet); Alan Q. Morton (ed.), *Science Lecturing in the Eighteenth Century*, Special Issue of *British Journal for the History of Science*, 28 (March 1995); Alan Q. Morton and Jane Wess, *Public & Private Science: The King George III Collection* (Oxford: Oxford University Press, 1993), chap. 2; Stephen Pumfrey, "Who Did the Work? Experimental Philosophers and Public Demonstrations in Augustan England," *British Journal for the History of Science*, 28 (1995), 131–56; and Larry Stewart, "Public Lectures and Private Patronage in Newtonian England," *Isis*, 77 (1986), 47–58. For Mesmer, see Robert Darnton, *Mesmerism and the End of the Enlightenment in France* (Cambridge, MA: Harvard University Press, 1968).

[18] Mario Biagioli, *Galileo, Courtier: The Practice of Science in the Culture of Absolutism* (Chicago: University of Chicago Press, 1993).

[19] For case studies of eighteenth-century cosmological legitimation, see Steven Shapin, "Of Gods and Kings: Natural Philosophy and Politics in the Leibniz-Clarke Disputes," *Isis*, 72 (1981), 187–215; Simon Schaffer, "Authorized Prophets: Comets and Astronomers after 1759," *Studies in Eighteenth-Century Culture*, 17 (1987), 45–74; Schaffer, "Newton's Comets and the Transformation of Astrology," in Patrick Curry (ed.), *Astrology, Science and Society: Historical Essays* (Woodbridge: Boydell and Brewer, 1987), pp. 219–43; and, for speculations about national variation in requirements for cosmological legitimation, see Mario Biagioli, "Scientific Revolution, Bricolage, and Etiquette," in Roy Porter and Mikuláš Teich (eds.), *The Scientific Revolution in National Context* (Cambridge University Press, 1992), pp. 11–54, at pp. 24–25. For science and naval medicine, see Christopher J. Lawrence, "Disciplining Disease: Scurvy, the Navy, and Imperial Expansion, 1750–1825," in David Philip Miller and Peter Hanns Reill (eds.), *Visions of Empire: Voyages, Botany, and Representations of Nature* (Cambridge University Press, 1996), pp. 80–106. For astronomy and the problem of longitude, see, e.g., David W. Waters, "Nautical Astronomy and the Problem of Longitude," in John G. Burke (ed.), *The Uses of Science in the Age of Newton* (Berkeley: University of California Press, 1983),

So utilitarian images of the man of science were nothing new in the eighteenth century, although the last section of this chapter picks out some subtle and incremental changes affecting these images through this period. Nor was it novel for eighteenth-century advocates to insist that natural knowledge had a proper place in polite culture as supplier of wonders and conversation pieces. Practitioners of natural knowledge in the eighteenth century could still supply marvels, delight, and edifying instruction to polite society. What was new in the early eighteenth century was the insistence that a particular reformed version of natural philosophy had eliminated the disputatious, along with the pedantic, tendencies that had for so long disqualified the scientific practitioner from membership in polite society and his knowledge from a central place in its culture.

From the culture of the mid-seventeenth-century *précieuses* to that of the eighteenth-century *salonnières,* French scientific savants enjoyed some success in making the case for the contribution of science to *politesse* and for the man of (reformed) science as a valued member of polite society. It was, as Geoffrey Sutton nicely puts it, "the philosopher's *honnêteté,* the naturalist's *politesse,* that brought science into elite society" during the last quarter of the seventeenth century and that – together with the developing institutions of natural-philosophical entertainment – sustained the place it had achieved there into the eighteenth century. The presence of significant numbers of women in French places of scientific conversation, entertainment, and instruction was taken as testimony to the innocuousness, and even the politeness, of scientific culture. The Abbé Nollet told potential auditors of his demonstration-lectures that "the path had been cleared by people of condition and merit so respectable" that, as Sutton notes, "no woman needed [to] fear for her reputation by enrolling in the course." By the mid-1730s Madame du Châtelet wrote to a friend that Nollet's lectures were attracting "the carriages of duchesses, peers, and lovely women."[20]

pp. 143–69. For natural history in connection with both utility and politeness, see John Gascoigne, *Joseph Banks and the Enlightenment: Useful Knowledge and Polite Culture* (Cambridge University Press, 1994), and especially Gascoigne, *Science in the Service of Empire: Joseph Banks, the British State and the Uses of Science in the Age of Revolution* (Cambridge University Press, 1998). For perpetual motion devices, see Simon Schaffer, "The Show that Never Ends: Perpetual Motion in the Early Eighteenth Century," *British Journal for the History of Science,* 28 (1995), 157–89. For a range of studies of eighteenth-century science, technology, and the culture of utility, see, e.g., Larry Stewart, *The Rise of Public Science: Rhetoric, Technology and Natural Philosophy in Newtonian Britain, 1660–1750* (Cambridge University Press, 1992), especially pts. 2–3; Jan Golinski, *Science as Public Culture: Chemistry and Enlightenment in Britain, 1760–1820* (Cambridge University Press, 1992); Karl Hufbauer, *The Formation of the German Chemical Community (1720–1795)* (Berkeley: University of California Press, 1982); Myles W. Jackson, "Natural and Artificial Budgets: Accounting for Goethe's Economy of Nature," *Science in Context,* 7 (1994), 409–31; Steven Shapin, "The Audience for Science in Eighteenth Century Edinburgh," *History of Science,* 12 (1974), 95–121 (for agriculture); and Ken Alder, *Engineering the Revolution: Arms and Enlightenment in France, 1763–1815* (Princeton, NJ: Princeton University Press, 1997).

[20] Sutton, *Science for a Polite Society,* pp. 141, 225 (for Nollet and Châtelet); see also Anne Goldgar, *Impolite Learning: Conduct and Community in the Republic of Letters, 1680–1750* (New Haven, CT: Yale University Press, 1995), for the Huguenot scholarly diaspora after the Revocation of the Edict of

But this central role for the man of science in polite society remained rather more an aspiration than a substantial reality in the eighteenth century. Even in France, where the case was more effectively put than elsewhere, the claims of reformed science to an important place in polite culture were not overwhelmingly successful: civil history, *belles-lettres,* rhetoric, ancient and modern languages, genealogy, antiquarianism, geography, and chorography remained far more significant than natural science as polite studies. And in Britain the notion of polite science attracted much skepticism and even ridicule.[21] For one thing, members of polite society could rarely be relied on to observe and appreciate the distinction between reformed science and the Scholastic practice it was supposed to have supplanted. Even if the superior civic virtues of the modern man of science were evident, it was necessary for polite society to encounter such men and to mark the difference in character. The buildup of such patterns of familiarity took time. Eighteenth-century British courtesy texts frequently, and tellingly, missed the distinction between Scholastic and mechanically reformed natural knowledge: it was all metaphysical, all obscure, and all irrelevant to mundane affairs. So far as such handbooks were concerned, the seventeenth-century Scientific Revolution had never happened and the changes that this Revolution was supposed to have effected in the fitness of the man of science for polite society were not worth noticing.[22]

To be sure, those who fashioned polite British opinion occasionally went

---

Nantes; Dena Goodman, "Enlightenment Salons: The Convergence of Female and Philosophic Ambitions," *Eighteenth-Century Studies,* 22 (1989), 329–50; James A. Secord, "Newton in the Nursery: Tom Telescope and the Philosophy of Tops and Balls, 1761–1838," *History of Science,* 23 (1985), 127–51 (for science texts written for the children of the British polite classes); and Alice N. Walters, "Conversation Pieces: Science and Politeness in Eighteenth-Century England," *History of Science,* 35 (1997), 121–54 (for polite science in English domestic settings), especially pp. 130–6 (for women's participation).

[21] For the importance of literary and antiquarian studies even within the eighteenth-century Royal Society of London, see David P. Miller, "'Into the Valley of Darkness': Reflections on the Royal Society in the Eighteenth Century," *History of Science,* 27 (1989), 155–66. Even a late seventeenth-century partisan of reformed science like John Locke was only lukewarm about the place of any form of natural philosophy in the education of a gentleman: see John Locke, *Some Thoughts Concerning Education* (Cambridge University Press, 1899; orig. publ. 1690), pp. 74, 129, 153. The Earl of Chesterfield's detailed mid-eighteenth-century directions for his son's studies mention scientific subjects only very rarely and fleetingly, recommending a few hours turning the pages of a popular astronomy text and the acquisition of "a general knowledge" of practical mathematics relevant to fortification: Chesterfield to his son, 6 December 1748 and 27 April 1749, in Chesterfield, *Letters,* vol. 1, pp. 143–4, 173. Nor did even Joseph Priestley, advocating "a new and better furniture of mind" for those actively engaged in the emerging industrial order, recommend education in the natural sciences for any gentlemen save those whose business might come specially to require the pertinent specialized knowledge: Priestley, "An Essay on a Course of Liberal Education for Civic and Active Life," in John A. Passmore (ed.), *Priestley's Writings on Philosophy, Science, and Politics* (London: Collier-Macmillan, 1965; essay orig. publ. 1765), pp. 285–304, at pp. 286, 294–5.

[22] See, for example, William Darrell, *The Gentleman Instructed, in the Conduct of a Virtuous and Happy Life . . . Written for the Instruction of a Young Nobleman . . . ,* 8th ed. (London, 1723; orig. publ. 1704–12), p. 15; see also Adam Petrie, *Rules of Good Deportment, or of Good Breeding* (Edinburgh, 1720), pp. 46, 58. For early eighteenth-century polite skepticism about the virtues of the reformed man of science, see Steven Shapin, "'A Scholar and a Gentleman,'" *History of Science,* 29 (1991), 279–327.

on record approving the study of nature. Polite people might have drawn polite lessons from nature, although whether these lessons were widely available
is doubtful. Joseph Addison and Richard Steele's *The Spectator,* for example,
insisted that

> a man of polite imagination is let into a great many pleasures that the Vulgar
> are not capable of receiving. . . . It gives him, indeed, a kind of property in
> every thing he sees, and makes the most rude uncultivated parts of nature
> administer to his pleasures: So that he looks upon the world, as it were, in
> another light, and discovers in it a multitude of charms, that conceal them-
> selves from the generality of mankind.[23]

Yet almost all influential eighteenth-century British commentators on genteel society and manners worried about the effect on polite conversation of
too great a commitment to formal, systematic, and "speculative" learning.
Such learning – of whatever sort – was liable to stimulate pedantry, dogma,
obscurity, and the spirit of contention; the Earl of Chesterfield warned that
"deep learning is generally tainted with pedantry, or at least unadorned by
manners."[24] Some writers picked out the special troubles introduced into
polite society by those who made either the minute or the systematically
speculative investigation of nature their particular study. The proper study
of mankind was not stars or starfish, but man.

Early eighteenth-century wits ridiculed the Royal Society's virtuosi and
philosophers for mucking around with "the very dregs of Nature." Boyle's
swilling about in human urine and feces to extract phosphorus and Leeuwenhoek's investigations into the globular structure of mouth-slime elicited a polite retch reflex as well as a smirk.[25] *The Tatler* worried that those who made
minute, trivial, and "despicable" phenomena their objects of study would
themselves become debased and coarsened. Nature offered for study both the
immense and the minute, and polite society was concerned that scientific

[23] Joseph Addison, *The Spectator,* 21 June 1712, in Addison, *Essays of Joseph Addison,* ed. Sir James George
Frazer, 2 vols. (London: Macmillan, 1915), vol. 2, p. 180. For Dr. Johnson's (very general) endorsement
of philosophically modest Baconian practices, see Richard B. Schwartz, *Samuel Johnson and the New
Science* (Madison: University of Wisconsin Press, 1971), especially pp. 68–73 (for his biographical essays on Sydenham and Boerhaave) and pp. 125–45 (for his approval of physico-theology).
[24] Chesterfield to his son, 12 September 1749, in Chesterfield, *Letters,* vol. 1, p. 206. Chesterfield singled
out Maupertuis as a type "one rarely meets with, deep in philosophy and mathematics, and yet *hon-
nête et amiable*" (4 October 1752, in ibid., vol. 2, p. 133), but Chesterfield also offered his son the instructive example of the Earl of Macclesfield, whose astronomical and mathematical expertise was
bested in public argument over calendar reform by the superior rhetorical skills of Chesterfield
himself, no mathematician at all (18 March 1751, in ibid., vol. 1, p. 394). For Continental criticism
of English culture for its lack of "lofty speculation," and for English defense of "coffee table philosophy," see Roy Porter, "The Enlightenment in England," in Porter and Mikuláš Teich (eds.), *The
Enlightenment in National Context* (Cambridge University Press, 1981), pp. 1–18, especially pp. 5–6.
[25] William King, "Useful Transactions in Philosophy, and Other Sorts of Learning . . . ," in King, *Original Works,* vol. 2, pp. 57–178 (orig. publ. 1709), on pp. 98–99 (for Boyle's phosphorus); 103–14, 121–5
(for Leeuwenhoek); 135 (for dregs); King, "A Journey to London, In the Year 1698. After the Ingenious
Method of that made by Dr. Martin Lister to Paris . . . ," ibid. (orig. publ. 1699), vol. 1, p. 198 (for
cats in air-pumps).

scholars had too long made too much of too little.[26] Chesterfield agreed: Fontenelle's popular astronomy was to be preferred to the works of the "insect-mongers, shell-mongers, and pursuers and driers of butterflies."[27] So did the Earl of Shaftesbury: there was nothing about learning in itself that disqualified it from a proper place in the furnishing of a gentleman, but when

> our speculative genius and minute examiner of Nature's works proceeds with equal or perhaps superior zeal in the contemplation of the insect life, the conveniencies, habitations, and economy of a race of shell-fish; . . . he then indeed becomes the subject of sufficient raillery, and is made the jest of common conversations.[28]

The worry here fastened on the effects of scientific inquiry in its minute mode, but many late-eighteenth-century English critics rejected both the pertinence and the propriety of investigating nature, however it was performed and on whatever aspects it happened to focus.[29] The "abstruseness" complained of in both minute and speculative philosophy was added to in the course of the eighteenth century by increasing specialization in almost all the sciences, thereby putting additional pressure on the very idea of the polite man of science taking his conversational part in the general culture. Some societies – notably the Scots – worried about this specialization and its fragmenting effects on social solidarity; others – for example, the French and the Germans – seemed more relaxed about it.[30] The divorce between "the two cultures" was by no means irrevocable by the end of the eighteenth century – a common context significantly endured in many domains – but the withdrawal of the man of science from the general conversation was then well under way. One could not have a polite conversation with an author one could not understand; one could only be lectured at.[31] During the eighteenth century no version of the character of the man of science was immune to polite imputations of abstruseness, pedantry, and incivility. "Nothing," wrote the mental philosopher

---

[26] *The Tatler*, 10–12 January 1709/10, 24–26 August 1710, in Joseph Addison and Richard Steele, *The Tatler*, ed. George A. Aitken, 4 vols. (London: Duckworth, 1898), vol. 3, p. 31; vol. 4, p. 110.

[27] Chesterfield to his son, 6 December 1748, in Chesterfield, *Letters*, vol. 1, pp. 143–4, and, for the particular polite qualifications of astronomy, see Walters, "Conversation Pieces," pp. 124–7.

[28] Anthony Ashley Cooper, 3rd Earl of Shaftesbury, *Characteristics of Men, Manners, Opinions, Times*, ed. John M. Robertson, two vols. in one (Indianapolis: Bobbs-Merrill, 1964; orig. publ. 1711), vol. 2, p. 253. See also Lawrence E. Klein, *Shaftesbury and the Culture of Politeness: Moral Discourse and Cultural Politics in Early Eighteenth-Century England* (Cambridge University Press, 1994), especially chap. 1, and Walters, "Conversation Pieces," pp. 122–3, 126.

[29] See Joseph M. Levine, *Dr. Woodward's Shield: History, Science, and Satire in Augustan England* (Berkeley: University of California Press, 1977), p. 125.

[30] For Scottish anxiety about scientific specialization – continuing into the nineteenth century – see, e.g., George Elder Davie, *The Democratic Intellect: Scotland and Her Universities in the Nineteenth Century* (Edinburgh: Edinburgh University Press, 1961), pt. 2; Shapin, "Audience for Science in Eighteenth Century Edinburgh," especially pp. 99–101, 115–16; and Shapin, "Brewster and the Edinburgh Career in Science," in A. D. Morrison-Low and J. R. R. Christie (eds.), *'Martyr of Science': Sir David Brewster 1781–1868* (Edinburgh: Royal Scottish Museum, 1984), pp. 17–23.

[31] Secord, "Newton in the Nursery," pp. 143–6.

David Hartley, "can easily exceed the Vain-glory, Self-conceit, Arrogance, Emulation, and Envy, that are found in the eminent Professors of the Sciences, Mathematics, [and] Natural Philosophy."[32]

So the image of the polite man of science was indeed systematically presented to gentlemanly society for its acceptance during the eighteenth century. These presentations were part of concerted attempts to justify aspects of scientific inquiry and to show its congruence with the norms and conventions of genteel society. Looked at from the point of view of polite society, however (and especially from its English forms), the credibility of such presentations during the course of the eighteenth century was limited. Yet both the definition and the legitimacy of polite culture were being contested throughout the century. There were major attempts to redefine what it was to be authentically polite, and there were also attempts to reject polite values as a whole. Notions of what science was, what science was for, and who the man of science was all figured in these efforts.

From the late seventeenth century onward, radical "deists" and "freethinkers" appropriated mechanical conceptions of nature to subvert the civic and ecclesiastical hierarchies whose support was one of the explicit purposes of such earlier natural philosophers as Mersenne, Gassendi, Boyle, Ray, and Newton. Natural knowledge was a resource sufficiently plastic in its interpretation to find uses in undermining as well as buttressing existing social inequalities.[33] Hence, the character of the man of science as champion of orthodoxy was joined during the course of the century by the impolite man of science as hero of social reform or revolution, or as antihero of social subversion. In France, Jacobin radicals harvested the crop earlier sown by the *philosophes* and encyclopedists. When science became a tool of ancien régime power, hitherto innocuous images of an open and egalitarian Republic of Science could come to have real political bite.[34] Accordingly, Marat's friend, the radical journalist Jacques-Pierre Brissot, turned the tables on the exclusivity of the Paris Academy in his 1782 book *De la Vérité:*

---

[32] David Hartley, *Observations on Man*, 2 vols. (London, 1749), vol. 2, p. 255; see also Porter, "The Enlightenment in England," 14–15.

[33] For the implication of cosmological ideas in these contests, see, e.g., Margaret C. Jacob, *The Newtonians and the English Revolution 1689–1720* (Ithaca, NY: Cornell University Press, 1976); Jacob, *The Radical Enlightenment: Pantheists, Freemasons, and Republicans* (London: Allen & Unwin, 1981); Jacob, *Living the Enlightenment: Freemasonry and Politics in Eighteenth-Century Europe* (New York: Oxford University Press, 1991); see also Steven Shapin, "Social Uses of Science," in G. S. Rousseau and Roy Porter (eds.), *The Ferment of Knowledge: Studies in the Historiography of Eighteenth-Century Science* (Cambridge University Press, 1980), pp. 93–139; Shapin, "Of Gods and Kings"; and C. B. Wilde, "Hutchinsonianism, Natural Philosophy and Religious Controversy in Eighteenth Century Britain," *History of Science*, 18 (1980), 1–24.

[34] Charles Coulston Gillispie, "The *Encyclopédie* and the Jacobin Philosophy of Science: A Study in Ideas and Consequences," in Marshall Clagett (ed.), *Critical Problems in the History of Science* (Madison: University of Wisconsin Press, 1959), pp. 255–89; see also Hahn, *The Anatomy of a Scientific Institution*, chap. 5.

The empire of science can know neither despots, nor aristocrats, nor electors. . . . To admit a despot, aristocrats, or electors who by edicts set a seal upon the products of geniuses is to violate the nature of things and the liberty of the human mind. It is an affront to public opinion which alone has the right to crown genius.[35]

Who needed the sober method and the arduously acquired expertise of the academies and schools when conceptions of innate and intuitive genius could be touted as guarantors of philosophic truth? In 1793 the Republic's Committee of Public Instruction followed Brissot in announcing that "true genius is almost always *sans culotte*," implying, as Simon Schaffer has written, that genius had been both collectivized and democratized.[36]

English apologists for social stability or for gradual and organic change reckoned that they had learned the lesson: mechanical and experimental philosophy was both protean and powerful; it was likely to do as much social harm as good. Proper science could indeed support proper social order, but the Revolution was, in Edmund Burke's view, rationalism and speculative philosophy gone mad, bad, and dangerous: those who concocted the new French constitution had "much, but bad, metaphysics; much, but bad, geometry; much, but false, proportionate arithmetic."[37] The "wild gas" and the "fixed air" that Burke said were now let loose in France had, in his view, been manufactured domestically as well and, unless vigilance was exercised, were likely to wreak similar effects in Britain.[38]

Burke and his allies marked the radical intellectual egalitarianism and the radical antiauthoritarianism in, for example, these pronouncements of Joseph Priestley: "Any man has as good a power of distinguishing truth from falsity as his neighbours"; "This rapid progress of knowledge will, I doubt not, be the means under God of extirpating all error and prejudice, and of putting an end to all undue and usurped authority in the business of religion as well as of science"; and "The English hierarchy (if there be anything unsound in its constitution) has . . . reason to tremble even at an air pump or an electrical machine."[39] The pneumaticist Thomas Beddoes joined Priestley on a Home

[35] Jacques-Pierre Brissot de Warville, *De la Vérité . . .* (Neufchâtel, 1782), pp. 165–6, quoted in Hahn, *The Anatomy of a Scientific Institution,* p. 153.

[36] Simon Schaffer, "Genius in Romantic Natural Philosophy," in Andrew Cunningham and Nicholas Jardine (eds.), *Romanticism and the Sciences* (Cambridge University Press, 1990), pp. 82–98, on p. 85.

[37] Edmund Burke, *Reflections on the Revolution in France,* ed. Conor Cruise O'Brien (Harmondsworth: Penguin, 1986; orig. publ. 1790), p. 296.

[38] Burke, *Reflections,* p. 90. For Burke against Priestley, see Golinski, *Science as Public Culture,* pp. 176–87, and Maurice Crosland, "The Image of Science as a Threat: Burke versus Priestley and the 'Philosophical Revolution,'" *British Journal for the History of Science,* 20 (1987), 277–307.

[39] Joseph Priestley, *An Examination of Dr Reid's Enquiry into the Human Mind on the Principles of Common Sense* (London, 1774), p. 74, and Priestley, *Experiments and Observations on Different Kinds of Air,* 3 vols. (London, 1774–77), vol. 1, p. xiv (both quoted in Dorinda Outram, "Science and Political Ideology, 1790–1848," in R. C. Olby et al. (eds.), *Companion to the History of Modern Science* (London: Routledge, 1990), pp. 1008–23, at p. 1017); see also Schaffer, "Genius in Romantic Natural

Office list of "Disaffected & seditious persons," likely by his radical teaching to seduce the youth of Oxford.[40] Yet Burke had less to fear from radically impolite British men of science than he thought. Although there were some republican appropriations of science by working-class English Jacobins from the onset of the Revolution to the end of the Napoleonic Wars, effective Home Office policing kept subversion at bay while those members of the British middle classes concerned at all with scientific culture mobilized it less as a subversive resource than as an element in a new conception of what politeness should be.[41]

During the eighteenth century, developments largely outside the two English universities and such metropolitan centers of power as the Royal Society had been gradually drawing science into the heart of an emerging new culture that offered a reformed understanding of what genuine politeness was. Excluded from Oxford, Cambridge, and many traditional venues of professional and political power, English Dissenters – Unitarians, Quakers, Methodists, other nonconforming Protestants and Catholics – developed their own educational institutions and cultural forums. The "dissenting academies" taught scientific subjects and employed notable men of science such as Joseph Priestley and John Dalton to teach them.[42] Informally constituted provincial conversation groups, bringing together progressive Dissenting industrialists and men of science, sprang up from mid-century, their distribution roughly following the contours of industrialization. The Lunar Society of Birmingham (founded in the 1760s) included, among others, Joseph Priestley, the steam engine manufacturers Matthew Boulton and James Watt, the potter Josiah Wedgwood, the physician and chemical manufacturer James Keir, and the

---

Philosophy," pp. 89–90 (for Priestley and Kant on the distribution of philosophic genius); Robert Schofield, *The Enlightenment of Joseph Priestley: A Study of His Life and Work from 1733 to 1773* (University Park: Pennsylvania State University Press, 1998).

[40] Quoted in Trevor H. Levere, "Dr. Thomas Beddoes at Oxford: Radical Politics in 1788–1793 and the Fate of the Regius Chair in Chemistry," *Ambix*, 28 (1981), 61–9, at p. 65. For Beddoes, science, and radical politics, see also Levere, "Dr. Thomas Beddoes (1750–1808): Science and Medicine in Politics and Society," *British Journal for the History of Science*, 17 (1987), 187–204; Dorothy A. Stansfield, *Thomas Beddoes M.D. 1760–1808: Chemist, Physician, Democrat* (Dordrecht: D. Reidel, 1984); Roy Porter, *Doctor of Society: Thomas Beddoes and the Sick-Trade in Late-Enlightenment England* (London: Routledge, 1992); and Golinski, *Science as Public Culture*, chap. 6.

[41] For a relevant survey, see Ian Inkster, "Introduction: Aspects of the History of Science and Science Culture in Britain, 1780–1850," in Inkster and Jack Morrell (eds.), *Metropolis and Culture: Science in British Culture, 1780–1850* (London: Hutchinson, 1983), pp. 11–54; see also Inkster, "London Science and the Seditious Meetings Act of 1817," *British Journal for the History of Science*, 12 (1979), 192–6; J. B. Morrell, "Professors Robison and Playfair, and the *Theophobia Gallica*: Natural Philosophy, Religion and Politics in Edinburgh, 1789–1815," *Notes and Records of the Royal Society*, 26 (1971), 43–63 (for Scottish philosophical anti-Jacobinism and Home Office surveillance of radicals); and Schaffer, "Genius in Romantic Natural Philosophy," pp. 86–7 (for reaction by Burke and Robison to radically impolite natural philosophy).

[42] Nicholas Hans, *New Trends in Education in the Eighteenth Century* (London: Routledge, 1951). These academies were especially numerous in the Midlands and North of England, although there were important ones in the metropolis as well, e.g., the Hackney College.

physician, poet, and natural philosopher Erasmus Darwin.[43] By the 1780s and 1790s provincial scientifically oriented societies (often called "literary and philosophical") had become a common feature of the cultural landscape in the Midlands and North: the Manchester Literary and Philosophical Society was founded in 1781, followed shortly by the Derby Philosophical Society and by similar organizations in Newcastle-upon-Tyne, Liverpool, Leeds, Glasgow, and in many other industrial and mercantile centers.[44]

Early historical interpretations saw such organizations as sites at which useful concrete links were being forged between industry and scientific knowledge and in which seriously impolite conceptions of the man of science were being elaborated. The man of science was here being thrust to the center of the provincial cultural stage, where he could symbolically challenge polite aristocratic and gentlemanly values. The character of the Dissenting provincial man of science would juxtapose hard-nosed utilitarianism to belles-lettristic conversation, radical progressivism to interests in social stability, subversive materialism to orthodox spiritualism, and cultural and political egalitarianism to social hierarchy and deference.

More recent scholarship significantly modifies that picture. There was only a partial discontinuity, it is now considered, between the images and uses of scientific culture in these new cultural forums and those surrounding earlier conceptions of the polite and moral man of science. Indeed, as Arnold Thackray has argued, the centrality of science to these spontaneously produced expressions of provincial and industrial culture did depend "on a particular affinity between progressivist, rationalist images of scientific knowledge and the alternative value system espoused by a group peripheral to English society."[45] If the periphery was here challenging the political and cultural center, and if gentlemanly politeness substantially defined the central value system, then science was a mode of cultural self-expression that could be used

---

[43] Robert S. Schofield, *The Lunar Society of Birmingham: A Social History of Provincial Science and Industry in Eighteenth-Century England* (Oxford: Clarendon Press, 1963); Neil McKendrick, "The Role of Science in the Industrial Revolution: A Study of Josiah Wedgwood as a Scientist and Industrial Chemist," in Mikuláš Teich and Robert M. Young (eds.), in *Changing Perspectives in the History of Science: Essays in Honour of Joseph Needham* (London: Heinemann, 1973), pp. 274–319.

[44] Traditional presumptions about the "decline" and inutility of the eighteenth-century Royal Society of London are both refined and qualified in David Philip Miller, "'Into the Valley of Darkness'"; Miller, "The Usefulness of Natural Philosophy: The Royal Society of London and the Culture of Practical Utility in the Later Eighteenth Century," *British Journal for the History of Science*, 32 (1999), 185–201; Richard Sorrenson, "Towards a History of the Royal Society in the Eighteenth Century," *Notes and Records of the Royal Society of London*, 50 (1996), 29–46; Larry Stewart, "Other Centres of Calculation, or, Where the Royal Society Didn't Count: Commerce, Coffee-Houses and Natural Philosophy in Early Modern London," *British Journal for the History of Science*, 32 (1999), 133–53.

[45] Arnold W. Thackray, "Natural Knowledge in Cultural Context: The Manchester Model," *American Historical Review*, 79 (1974), 672–709, at 678; see also Thackray, "The Industrial Revolution and the Image of Science," in Thackray and Everett Mendelsohn (eds.), *Science and Values: Patterns of Tradition and Change* (New York: Humanities Press, 1974), pp. 3–18; Thackray, *John Dalton: Critical Assessments of His Life and Science* (Cambridge, MA: Harvard University Press, 1972).

symbolically to challenge traditional canons of politeness. Yet, as Thackray has shown, scientific culture in such venues as the Manchester Literary and Philosophical Society was primarily a resource used to redefine rather than to reject the values of politeness. For provincial medical men, organized culture of any sort lent social cachet, and no cultural form was more natural for such men than science. And for those few manufacturers and tradesmen who felt the need for cachet – most did not – science was also an attractive vehicle. The rhetoric of scientific utility linked it to progressive industrial values whereas the rhetoric of scientific politeness offered an access point to English gentility. "A taste for polite literature, and the works of nature and of art," said a self-made Manchester man, "is essentially necessary to form the gentleman." Participation in scientific culture was commended as an alternative to "the tavern, the gaming table or the brothel," and "a relish for manly science" was advertised as "next to religion, the noblest antidote" to "dissipation" and habits "unfavourable to success in business." Natural philosophy, in the Manchester mode, was much more about refinement than revolution, much less about industrial practice than about redefined politeness.[46]

CONCLUSION: THE CIVIC EXPERT AND THE FUTURE

To varying extents each of the characters of the eighteenth-century man of science treated here survived, even flourished, well into the following century. The character of the Godly Naturalist was bruised by Darwinism but did not immediately disappear from the cultural landscape. The character of the Moral Philosopher likewise suffered from the secularization of nature encouraged by scientific naturalism. When nature was no longer conceived as a divinely written book, the study of Nature had diminished power to uplift, and the credibility of ancient conceptions of philosophic disengagement and heroic selflessness was undermined by the professionalization and bureaucratization of scientific research and teaching. Both the receipt of government subvention and the institutionalization of scientific research in the professorial role made it harder to portray the man of science as fulfilling his calling through ascetic self-denial.[47] Similarly, although notions of polite culture continued

---

[46] Thomas Henry, "On the Advantages of Literature and Philosophy," *Manchester Memoirs*, 1 (1785), 7–28, on pp. 9, 11; and Thomas Barnes, "A Plan of Liberal Education," ibid., 2 (1785), 35 (both quoted in Thackray, "Natural Knowledge in Cultural Context," pp. 688–90).

[47] For important treatments of the emergence in eighteenth- and early nineteenth-century German universities of the dual role of the professor – as original researcher as well as teacher – see R. Steven Turner, "The Growth of Professorial Research in Prussia, 1818 to 1848 – Causes and Context," *Historical Studies in the Physical Sciences*, 3 (1971), 137–82; Turner, "University Reformers and Professorial Scholarship in Germany, 1760–1806," in Lawrence Stone (ed.), *The University in Society, Vol. 2* (Princeton, NJ: Princeton University Press, 1974), pp. 495–531; Turner, "The Prussian Universities and the Concept of Research," *Internationales Archiv für Sozialgeschichte der Deutschen Literatur*, 5 (1980), 68–93; also Joseph Ben-David, *The Scientist's Role in Society: A Comparative Study*, new ed. (Chicago: University of Chicago Press, 1984; orig. publ. 1971), chap. 7; see also J. B. Morrell, "The University

in some vigor through the nineteenth century – the heyday of the character of the "scholar and gentleman" – science was no more central to the identity of polite learning among *The Spectator*'s readers of the early twentieth than of the early eighteenth century. Moreover, the power tapped by plugging notions of polite science into gentle and aristocratic culture was gradually diminished by the declining authority of those classes over the past two centuries.

One must, therefore, look elsewhere for a character of the man of science that had its roots in the eighteenth century and reached its fruition in more modern conceptions of the scientist's role. There are many places one might look for such roots, but there is one to which special attention should be drawn, if only because of its apparent mundanity. Long before the eighteenth century, men of science – of various descriptions – had a valued place in both government and commercial enterprises owing to their recognized possession of relevant expertise about the natural world and practical interventions in it. The ancients knew all about the roles of, for example, the mathematically competent military engineer who could design fortifications, the astronomer who made calendars, and the physicians and surgeons who could advise on diet or cut for the stone.

So too did their eighteenth-century counterparts: there is nothing qualitatively new in this period about the character of the man of science as "civic expert." Nor was this character particularly linked to the rhetoric of utility that, from the seventeenth century, picked out the special capacity of some methodologically modernized versions of natural science to contribute to useful outcomes – a rhetoric that might be viewed, as we have seen, with considerable skepticism by other sectors of society. The point here does not hinge on the hoary debate over the relations between scientific theory and technical utility; rather, it concerns the roles and the historical appreciations of scientifically knowing people. And what the eighteenth century witnessed was a vast expansion in the numbers of scientifically trained people employed as civic experts in commerce, the military, and the government settings. The character of the man of science as otherworldly scholar or irrelevant pedant co-existed through the century with his emerging identity as valued civic expert. Sometimes, indeed, these opposing characters were attached to the same person. Who was Ben Franklin – a speculative electrical theorist or the inventor of the lightning rod? Who was Sir Joseph Banks – another collector of curiosities or Britain's national expert adviser on colonial horticulture?

The character of the medical expert needs no special introduction, but eighteenth-century settings in which his expertise was called upon proliferated. The dark satanic mills of the Industrial Revolution generated vast numbers of proletarian casualties that in turn created a demand for infirmaries and for

of Edinburgh in the Late Eighteenth Century: Its Scientific Eminence and Academic Structure," *Isis*, 62 (1971), 158–71.

the physicians and surgeons to staff them. Warfare was, of course, a constant in European history, but an increase in its scale, as well as the expansion of long-distance trade and colonization, likewise produced government demand for naval and military surgeons: the experts who might be able to offer effective prophylactics for scurvy were as valuable to imperial powers as those who offered solutions to the problem of longitude.

The placement of scientifically skilled people in mercantile and industrial enterprises was a matter of state policy in France, whereas in laissez-faire Britain matters took a more circuitous course to a similar recognition of the value of such expertise. Here are a few of many pertinent examples: the geologist James Hutton was also an improving farmer, an innovator in the manufacture of sal ammoniac, and an adviser on the building of the Forth-Clyde canal. The autodidact stratigrapher William Smith established the importance of the fossil record to mining. He was a canal company employee for a number of years at the end of the century and he was encouraged by Sir Joseph Banks to produce a geological map of England and Wales. The chemical expertise of Joseph Black was deployed in furnace construction and glass manufacture and was called upon in connection with bleaching techniques by the Scottish Board of Trustees for Manufactures. The Edinburgh- and Leiden-trained chemist John Roebuck managed an industrial complex that manufactured sulphuric acid, ceramics, and iron. And the story of the relations among Joseph Black, James Watt, and Matthew Boulton in steam-engine manufacture has passed into industrial legend. In France, Coulomb's governmental role as military and civil engineer has already been mentioned. Lavoisier's chemical training was brought to bear on his early official work in factory inspection, in the management of municipal water supplies, and as commissioner in the Royal Gunpowder Administration.

Throughout eighteenth-century Europe and North America, governments increasingly drew on the services of scientifically skilled people and thus helped to constitute the character of the man of science as civic expert. The Swiss anatomist Albrecht von Haller resigned his Göttingen chair to pursue a political career, and for six years he served as director of the Bern saltworks. The Italian natural historian Lazzaro Spallanzani was sent by the Austrian government to visit mines and collect fossils in the Alps. The Croatian natural philosopher Rudjer Boscovich worked as a hydraulic engineer for the Vatican. The mineralogist Abraham Gottlob Werner taught for most of his life in the Saxon mining academy. The young Leibniz was an engineering consultant for the duke of Brunswick-Lüneberg; the young Goethe was a superintendent of mines for the Weimar court; and the young Alexander von Humboldt worked in the Prussian mining service. Everywhere men of science were employed by governments to standardize weights and measures. The vitally important problem of determining longitude at sea was perhaps the most visible instance in which governments acknowledged that their national interests crucially depended on the work of highly skilled men of science, the embodied repositories of esoteric natural knowledge.[48]

However, there was one enterprise of special significance to eighteenth-century patterns constituting the man of science as civic expert, if only because its scale and scope expanded so much during the century. This was the primary survey of the globe, especially in the context of long-distance trade and in imperialist ventures. Here the term "primary survey" includes (i) the compilation and central accumulation of inventories of what natural kinds and phenomena existed in distant parts of the world; (ii) the development of techniques effectively to standardize the representation and retrieval of such information and to ensure its robustness in circulation among those who recorded it, those who wished to gain access to it, and those who wished to use it in practical enterprises; and (iii) the explication of the virtues and values of distant natural kinds and phenomena, possibly, though not necessarily, with respect to the material interests of individual nations. Alexander von Humboldt's isoline mapping program in geophysics is one example of a primary survey, and techniques for representing, orienting, and moving about in a digitized natural world were among its major products.[49] In America, Benjamin Franklin helped raise public subscriptions to support John Bartram's surveying and collecting travels from New York to Florida, and President Thomas Jefferson later commissioned Lewis and Clark to find out what there was in the unknown lands between the settled parts of America and the Pacific.[50]

Consider the questions that might be asked of a botanical expert in this context: What kinds of plants were there in and around Botany Bay in New South Wales? How could one be sure that a species from there was the same as one from Tahiti? What was this particular species good for? And, if it had a commercial value, could it be made to grow in the south of England or in British colonies in the West Indies? Such questions were precisely those that occupied Sir Joseph Banks in the late eighteenth and early nineteenth centuries as he developed both Kew Gardens and his London house in Soho Square into crucially important centers of calculation and accumulation.[51] Could tea be grown economically in the British East Indies, and, if so, where?

[48] See also the involvement of British and French governments in expeditions to observe the transit of Venus in 1761 and 1769: Harry Woolf, *The Transits of Venus: A Study of Eighteenth-Century Science* (Princeton, NJ: Princeton University Press, 1959).

[49] See Michael Dettelbach, "Global Physics and Aesthetic Empire: Humboldt's Physical Portrait of the Tropics," in Miller and Reill (eds.), *Visions of Empire*, pp. 258–92; for plant geography, see Malcolm Nicolson, "Alexander von Humboldt, Humboldtian Science, and the Origins of the Study of Vegetation," *History of Science*, 25 (1987), 167–94.

[50] For materials on men of science and the primary survey of America, see, e.g., Raymond Phineas Stearns, *Science in the British Colonies of America* (Urbana: University of Illinois Press, 1970).

[51] See David Philip Miller's appropriation of a notion from Bruno Latour, in Miller, "Joseph Banks, Empire, and 'Centres of Calculation' in Late Hanoverian London," in Miller and Reill (eds.), *Visions of Empire*, pp. 21–37. For Latour's usage, see Bruno Latour, *Science in Action: How to Follow Scientists and Engineers through Society* (Milton Keynes: Open University Press, 1987), especially pp. 215–57.

The Board of Trade and the East India Company wanted to know, so they drew on the expertise of Joseph Banks. Banks was able to advise them, since he had accumulated and maintained records of trials growing *Camellia sinensis* in English gardens.[52] Banks was, in Daniel Baugh's phrase, a "natural resource imperialist," his expertise available to the British government, military, and trading companies and valued by them for its reliability.[53]

Examples of civic expertise for hire in the context of trade, war, and imperialism could be multiplied indefinitely in a wide range of scientific disciplines: mathematics, astronomy, geography and cartography, geology and mineralogy, meteorology, medicine, chemistry, and physics. Although the role of the man of science as civic expert was not new in the eighteenth century, the numbers occupying that role were increasing along with the expansion of trade, war, and imperialism. The recognized importance of scientific experts followed from their success in constituting themselves and their workplaces as centers of calculation vital to the exercise of long-distance control.

Men of science as civic experts became more numerous during the eighteenth century, and it became increasingly common to hear references to such people. And, as their presence became more usual, the character of the man of science as "useful chap" circulated more widely. The ground was gradually being prepared for the professionalizing movements of the nineteenth century. Governments could plausibly be called on to become the paymasters for scientific inquiry, not because widely persuasive systematic arguments had been made about the ultimate utility of scientific theory but rather because governments could now be reminded of their indebtedness to a corps of skilled experts, many of whom attributed their know-how to their possession of scientific knowledge. Nor was this either a simple or a wholly demand-driven process. The character of the man of science as useless pedant was still available in the early nineteenth century to those resisting the professionalizers' utilitarian rhetoric. Eighteenth-century men of science did respond to demands for their expertise, but they also labored hard to *tell* governments that such expertise was available, authentic, and potent; that they were the people who possessed expertise; and that governments' material interests depended on the nurturing and effective deployment of said expertise. The expertise of men of science and the interests of governments had to be, in many cases, artfully aligned. Such alignment could fail, and superficial appearances of artful alignment may be deceptive. Humboldt, for example, was not in the pay of a naval

---

[52] Miller, "Joseph Banks, Empire," pp. 31–2; also David Mackay, "Agents of Empire: The Banksian Collectors and Evaluation of New Lands," in Miller and Reill (eds.), *Visions of Empire*, pp. 38–57.

[53] Daniel Baugh, "Seapower and Science: The Motives for Pacific Exploration," in Derek Howse (ed.), *Background to Discovery: Pacific Exploration from Dampier to Cook* (Berkeley: University of California Press, 1990), pp. 1–55, at p. 40; see also Simon Schaffer, "Visions of Empire: Afterword," in Miller and Reill (eds.), *Visions of Empire*, pp. 335–52, especially pp. 338–9.

power or its institutions when he developed his techniques for isoline mapping. Function and motive may differ.[54]

Nevertheless, by the end of the eighteenth century a new possibility for the character of the man of science had begun to open up, although the full development of that character was not to occur for many years. The man of science might be conceived of as someone who was neither particularly godly, nor particularly virtuous, nor particularly polite.[55] It could be considered that there was nothing very special about the sorts of people drawn to the study of the natural world, nor anything very special about the effects on character wrought by the study of the natural world. The man of science was not thought to be constitutionally better or worse than other men, nor did his manner of inquiry or object of study make him better or worse than other men. Within his domain of legitimate expertise he knew more, and knew it more reliably. Such men were useful.

---

[54] Dettelbach, "Global Physics," p. 264; but compare British Royal Navy support for Edmond Halley's Atlantic voyages (1698–1701) to produce isoline maps of magnetic variation: Alan Cook, *Edmond Halley: Charting the Heavens and the Seas* (Oxford: Clarendon Press, 1998), chap. 10.

[55] On the decline of virtue in the image of the man of science, see Steven Shapin, "The Philosopher and the Chicken: On the Dietetics of Disembodied Knowledge," in Christopher Lawrence and Shapin (eds.), *Science Incarnate: Historical Embodiments of Natural Knowledge* (Chicago: University of Chicago Press, 1998), pp. 21–50, especially pp. 42–6; and Shapin and Lawrence, "Introduction: the Body of Knowledge," *Science Incarnate,* pp. 1–20, especially pp. 14–15.

8

THE PHILOSOPHER'S BEARD

Women and Gender in Science

Londa Schiebinger

> A woman who . . . engages in debates about the intricacies of mechanics, like the Marquise du Châtelet, might just as well have a beard; for that expresses in a more recognizable form the profundity for which she strives.
>
> Immanuel Kant, 1764

Kant's sentiments reiterated those of the great Carl Linnaeus, who taught in his lectures given at the University of Uppsala in the 1740s that "God gave men beards for ornaments and to distinguish them from women."[1] In the eighteenth century the presence or absence of a beard not only drew a sharp line between men and women but also served to differentiate the varieties of men. Women, black men (to a certain extent), and especially men of the Americas simply lacked that masculine "badge of honor" – the philosopher's beard. As Europe shifted from an estates society to a presumed democratic order, sexual characteristics took on new meaning in determining who would and who would not do science.

INSTITUTIONAL LANDSCAPES

The new sciences of the seventeenth and eighteenth centuries were fostered in a landscape – including universities, academies, princely courts, noble networks, and artisanal workshops – that was expansive enough to include a number of women. In the sustained negotiations over gender boundaries in early modern Europe, it was not at all obvious that women would be excluded from science.[2]

Universities have not been good institutions for women. From the found-

---

[1] Wilfred Blunt, *The Compleat Naturalist: A Life of Linnaeus* (London: William Collins, 1971), p. 157.
[2] Many of the materials in this essay are drawn from Londa Schiebinger, *The Mind Has No Sex? Women in the Origins of Modern Science* (Cambridge, MA: Harvard University Press, 1989).

ing of universities in the twelfth century until late into the nineteenth century, women were proscribed from study. A few exceptional women, however, did study and teach at universities beginning in the thirteenth century, primarily in Italy. These women often flourished in fields, such as physics and mathematics, that today are thought especially resistant to them. The most exceptional woman in this regard was physicist Laura Bassi, who became the second woman in Europe to receive a university degree in 1732 (the first was the Venetian Elena Cornaro Piscopia in 1678) and the first woman to be awarded a university professorship. Celebrated for her work in mechanics, Bassi also became a member of the Istituto delle Scienze in Bologna (Figure 8.1). Like other members she presented annual papers ("On the compression of air," 1746; "On the bubbles observed in freely flowing fluid," 1747; "On bubbles of air that escape from fluids," 1748; and so forth) and received a small stipend. She also invented various devices for her experiments with electricity. The Englishman Charles Burney, who met Bassi during his tour of Italy, found her "though learned, and a genius, not at all masculine or assuming."[3]

The Milanese Maria Gaetana Agnesi, celebrated for her 1748 textbook on differential and integral calculus *Instituzioni analitiche,* was also offered a chair at the University of Bologna. She is often credited with formulating the *versiera,* the cubic curve that (through a mistranslation) has come to be known in English as the "witch of Agnesi."[4] In trying to persuade her to take up a chair of mathematics and natural philosophy, Pope Benedict XIV proclaimed, "From ancient times, Bologna has extended public positions to persons of your sex. It would seem appropriate to continue this honorable tradition."[5] Agnesi accepted this appointment only as an honorary one and, after her father's death in 1752, withdrew from the scientific world to devote herself to religious studies and to serving the poor and aged. By the 1750s, the University of Bologna had offered a position to a third woman, the wax modeler Anna Morandi Manzolini, famous for her anatomical figures showing the development of the fetus in the womb.[6]

---

[3] Charles Burney, *The Present State of Music in France and Italy* (1773), ed. Percy Scholes (London: Oxford University Press, 1959) pp. 159–60.

[4] The curve that bears Agnesi's name had already been described by Pierre de Fermat. Hubert Kennedy, "Maria Gaetana Agnesi," in Louise Grinstein and Paul Campbell (eds.), *Women of Mathematics: A Biobibliographic Sourcebook* (New York: Greenwood Press, 1987), pp. 1–5; Lynn Osen, *Women in Mathematics* (Cambridge, MA: MIT Press, 1974), pp. 33–48, especially 44–5; Edna Kramer, "Maria Gaetana Agnesi," *Dictionary of Scientific Biography,* I, 75–7.

[5] Benedict to Agnesi, September 1750, cited in Alphonse Rebiére, *Les Femmes dans la science,* 2nd ed. (Paris, 1897), p. 11.

[6] Morandi was employed by the university to dissect and prepare bodies in order to teach anatomy to students and curious amateurs. Marta Cavazza, "'Dottrici' e Lettrici dell'Università de Bologna nel settecento," *Annali di Storia delle Università Italiane,* 1 (1997), 120. Maria Dalle Donne held the post of director of the Scuola per levatrici (School of Midwives) from 1804 to 1842 and was, for many years, a member of the Istituto delle Scienze. I thank Dr. Marta Cavazza at the University of Bologna for this information.

Figure 8.1. Laura Bassi, professor of Newtonian physics and mathematics at the University of Bologna from 1732 to 1778. From Alphonse Rebière, *Les Femmes dans la science* (Paris, 1897), facing p. 28. By permission of the Schlesinger Library, Radcliffe College.

The Italian model was not embraced across Europe. Germany experimented with higher education for women, conferring two degrees (at Halle and Göttingen) in the eighteenth century; no degrees were awarded in France or Great Britain. Outside Italy, no women were appointed professors; within Italy, the tradition of women professors did not continue. After about 1800, women were generally proscribed from European institutions of higher learning until the end of the nineteenth and in some cases until the twentieth century. Sofia Kovalevskaia was the next woman to become a professor (of mathematics) within Europe; she was appointed to the University of Stockholm in 1889.

Why did Italy accommodate learned women in ways that other European countries did not? Paula Findlen has suggested that Bassi served to bolster Bologna's flagging patriciate, becoming a "symbol of scientific and cultural regeneration." With Bassi, the city could boast a woman learned beyond any other in Europe. Beate Ceranski concurs that the traditions of Renaissance humanism, in which a woman could be admired for her learning, remained alive in the relatively small Italian city-states; no woman, however – no matter how great her learning – could hold such a position in the larger and more strongly centralized states of France or England, as the example of Gabrielle-Émilie le Tonnelier de Breteuil, Marquise du Châtelet, bears out.[7]

Historians have traditionally focused on the decline of universities and the founding of scientific academies as a key step in the emergence of modern science. Except for a few Italian academies (the Institute of Bologna mentioned earlier and the Accademia de' Ricovrati), the new scientific societies, like the universities, were closed to women. The Royal Society of London, founded in the 1660s and the oldest permanent scientific academy, did not admit the eccentric but erudite Margaret Cavendish, Duchess of Newcastle, although she was well qualified for that position (men above the rank of baron could become members without scientific qualifications). From its founding in 1660 until 1945, the only female member of the Royal Society was a skeleton in its anatomical collection.[8] The Académie Royale des Sciences in Paris, founded in 1666, refused to admit women; even the illustrious Marie Curie (1867–1934) was turned away. The first woman was elected to this academy in 1979. Nor did the Societas Regia Scientiarum in Berlin admit the well-known astronomer Maria Margaretha Winkelmann (1670–1720), who worked at the academy observatory first with her husband and later her son.

The prominence of universities and scientific academies today should not lead us to overemphasize their importance in the past. Several avenues into scientific work existed for women before the stringent formalization of science in the nineteenth century. In the early years of the scientific revolution, women of high rank were encouraged to know something about science. Along with gentlemen virtuosi, gentlewomen peered at the heavens through telescopes, inspecting the moon and stars; they looked through microscopes, analyzing insects and tapeworms. If we are to believe Bernard de Fontenelle,

---

[7] Paula Findlen, "Science as a Career in Enlightenment Italy: The Strategies of Laura Bassi," *Isis*, 84 (1993), 441–69, especially 449; Beate Ceranski, *"Und Sie Fürchtet sich vor Niemandem": Die Physikerin Laura Bassi, 1711–1778* (Frankfurt: Campus Verlag, 1996). See also Paula Findlen, "A Forgotten Newtonian: Women and Science in the Italian Provinces," in William Clark, Jan Golinski, and Simon Schaffer (eds.), *The Sciences in Enlightened Europe* (Chicago: University of Chicago Press, 1999), pp. 313–49.

[8] "A Catalogue of the Natural and Artificial Rarities belonging to the Royal Society, and preserved at Gresham College," in H. Curzon, *The Universal Library: Or, Compleat Summary of Science* (London, 1712), vol. 1, p. 439. Kathleen Lonsdale and Marjory Stephenson were elected to the Royal Society in 1945 (*Notes and Records of the Royal Society of London*, 4 (1946), 39–40). See also Joan Mason, "The Admission of the First Women to the Royal Society of London," *Notes and Records of the Royal Society of London*, 46 (1992), 279–300.

secretary of the Académie Royale des Sciences and *président* of Madame Lambert's salon, it was not unusual to see people in the street carrying around dried anatomical preparations. Especially in Paris, wealthy women were ready consumers of scientific curiosities, collecting everything from conches, stalactites, and petrified wood to insects, fossils, and agates to make their natural history cabinets "the epitome of the universe."[9] In what I have called "noble networks" – of natural philosophers, patrons, and illustrious consumers – well-born women were often able to exchange social prestige for access to scientific knowledge. The physicist Emilie du Châtelet, for example, was able to insinuate herself into networks of scientific men by exchanging patronage for the attention of men of lesser rank but of significant intellectual stature.[10]

Royal women also formed crucial links across Europe as patrons of science. In 1650 Descartes was commissioned by the audacious queen Christina of Sweden to draw up regulations for her scientific academy. Even the highest rank did not, however, insulate women from reproach and ridicule. Many people blamed Christina and the rigors of her philosophical schedule for Descartes's death. For her philosophical prowess, the queen was denounced as a hermaphrodite.[11]

Noble networks also flourished within salons, intellectual institutions or-

[9] P. Remy, *Catalogue d'une Collection de très belles Coquilles, Madrepores, Stalactiques . . . de Madame Bure* (Paris, 1763). On Fontenelle, see Jacques Roger, *Les Sciences de la vie dans la pensée Française du XVIIIe siècle* (Paris, 1963), pp. 165, 181–2; Nina Rattner Gelbart, "Introduction," in Bernard le Bovier de Fontenelle, *Conversations on the Plurality of Worlds*, trans. H. A. Hargreaves (Berkeley: University of California Press, 1990); Aileen Douglas, "Popular Science and the Representation of Women: Fontenelle and After," *Eighteenth-Century Life*, 18 (1994), 1–14; and Geoffrey Sutton, *Science for a Polite Society: Gender, Culture, and the Demonstration of Enlightenment* (Boulder, CO: Westview Press, 1995), chap. 5. Science for ladies remained popular throughout Europe in the eighteenth century. In Italy, the poet Francesco Algarotti published an introduction to Newtonian physics in 1737. In Germany, Johanna Charlotte Unzer published her *Outline of Philosophy for Ladies* (Grundriss einer Weltweisheit für Frauenzimmer) in 1761; in Russia, and from his post at the Academy of Science in St. Petersburg, Leonhard Euler wrote his *Letters to a German Princess on Diverse Points of Physics and Philosophy* in 1768. See also John Harris, *Astronomical Dialogues Between a Gentleman and a Lady* (London, 1719); James Ferguson, *Easy Introduction to Astronomy for Gentlemen and Ladies* (London, 1768); [Lorenz Suckow], *Briefe an das schöne Geschlecht über verschiedene Gegenstände aus dem Reiche der Natur* (Jena, 1770); Pierre Fromageot, *Cours d'études des jeunes demoiselles* (Paris, 1772–5); Jakob Weber, *Fragmente von der Physik für Frauenzimmer und Kinder* (Tübingen, 1779); Christoph Leppentin, *Naturlehre für Frauenzimmer* (Hamburg, 1781); August Batsch, *Botanik für Frauenzimmer* (Weimar, 1795); and Christian Steinberg, *Naturlehre für Frauenzimmer* (Breslau, 1796). See also Gerald Meyer's excellent *The Scientific Lady in England: 1650–1760* (Berkeley: University of California Press, 1955).

[10] For "noble networks," see Schiebinger, *The Mind Has No Sex?* chap. 2. For Châtelet, see René Taton's "Gabrielle-Émilie le Tonnelier de Breteuil, Marquise du Châtelet," *Dictionary of Scientific Biography*, III, 215–17, who provides primary and secondary bibliography; see also Carolyn Iltis, "Madame du Châtelet's Metaphysics and Mechanics," *Studies in History and Philosophy of Science*, 8 (1977), 29–48; Ira O. Wade, *Voltaire and Madame du Châtelet: An Essay on the Intellectual Activity at Cirey* (Princeton, NJ: Princeton University Press, 1941); Elizabeth Badinter, *Emilie, Emilie: L'Ambition féminine au XVIIIe siècle* (Paris, 1983); Linda Gardiner, "Women in Science," in Spencer (ed.), *French Women*, pp. 181–96; and Mary Terrall, "Emilie du Châtelet and the Gendering of Science," *History of Science*, 33 (1995), 283–310.

[11] *Carpenrariana or remarques . . . de M. Charpentier* (Paris, 1724), p. 316; Claude Clerselier, *Lettres de Mr. Descartes* (Paris, 1724), vol. 1, preface.

ganized and run by women. Like the French academies, salons created cohesion among elites, assimilating the rich and talented into the French aristocracy. Although these gatherings were primarily literary in character, science was fashionable at the salons of Madame Geoffrin, Madame Helvétius, and Madame Rochefoucauld; Madame Lavoisier received academicians at her home. There were, however, limits to this type of exchange. In the same way that privilege gave women only limited access to political power and the throne, high social standing gave them only limited access to the world of learning. Because women were barred from the centers of scientific culture – the Royal Society of London or the Académie Royale des Sciences of Paris – their relationship to knowledge was inevitably mediated through a man, whether that man was their husband, companion, or tutor.[12]

It should be noted that ridicule of "learned ladies" appeared in the late seventeenth century along with *virtuosae* themselves. Jean-Baptiste Molière's *Les Femmes Savantes* (1672) was much acclaimed for portraying Cartesian women running mad after philosophy and disrupting established social hierarchies by having no time for marriage or household duties. A husband whose dinner has been neglected rails against his science-minded wife for wanting "to know the motions of the moon, the pole star, Venus, Saturn, and Mars . . . while my food, which I need, is neglected."[13] Fears that learned ladies threatened to disrupt the status quo were justified: it was part of the political program of *salonières* of the seventeenth and eighteenth centuries to eschew traditional forms of marriage and motherhood. With books to read and lectures to attend, upper-class and even middle-class women had shifted the responsibilities of motherhood to wet nurses and governesses. These women's desires to engage, like men, in productive lives free of the cares of parenting came increasingly into conflict with the belief that public employ should be the preserve of men and that women could best serve the nation (and later the race) by producing healthy, and abundant, offspring.

Artisanal workshops served as another avenue into science for eighteenth-century women. Edgar Zilsel was among the first to point to the importance of craft skills for the development of modern science in the West. What Zilsel did not point out, however, is that the new value attached to the traditional skills of the artisan also allowed for the participation of women in the sciences. Women were not newcomers to the workshop; it was in craft traditions that the fifteenth-century writer, Christine de Pizan, had located women's

[12] Carolyn Lougee, *Le Paradis des Femmes: Women, Salons, and Social Stratification in Seventeenth Century France* (Princeton, NJ: Princeton University Press, 1976), pp. 41–53; Alan Kors, *D'Holboch's Coterie* (Princeton, NJ: Princeton University Press, 1976); Charles C. Gillispie, *Science and Polity in France at the End of the Old Regime* (Princeton, NJ: Princeton University Press, 1980), pp. 7, 94; Dena Goodman, "Enlightenment Salons: The Convergence of Feminine and Philosophical Ambitions," *Eighteenth-Century Studies*, 22 (1989), 329–50; Schiebinger, *The Mind Has No Sex?* pp. 30–2; Paula Findlen, "Translating the New Science: Women and the Circulation of Knowledge in Enlightenment Italy," *Configurations*, 2 (1995), 167–206.
[13] Jean-Baptiste Molière, *Les Femmes savantes* (1672), Jean Cordier (ed.) (Paris, 1959), pp. 36–7.

greatest innovations in the arts and sciences: the spinning of wool, silk, and linen and "creating the general means of civilized existence."[14] In the workshop, women's (like men's) contributions depended less on book learning and more on practical innovations in illustrating, calculating, or observing.

Whereas in France women's contributions to the sciences came consistently from women of the aristocracy, in Germany some of the most interesting innovations came from craftswomen. The prominence of artisans in Germany accounts for the remarkable fact that between 1650 and 1710 some 14 percent of all German astronomers were women – a higher percentage even than is true in Germany today (Figure 8.2). Astronomy was not a guild; as I have argued elsewhere, however, the German astronomer of the early eighteenth century bore a close resemblance to the guild master or apprentice, and the craft organization of astronomy gave women a prominence in the field. Trained by their fathers and often observing alongside their husbands, women astronomers in this period worked primarily in family observatories – some built in the attic of the family house, others across the roofs of adjoining houses, still others on city walls. In these astronomical families, the labor of husband and wife did not divide along modern lines: he was not fully professional, working in an observatory outside the home; she was not fully a housewife, confined to hearth and home. Nor were they independent professionals, each holding a chair of astronomy. Instead, they worked as a team and on common problems. They took turns observing so that their observations followed night after night without interruption. At other times they observed together, dividing the work so that they could make observations that a single person could not make accurately. Guild traditions within science allowed women such as the astronomer Maria Margaretha Winkelmann and the celebrated entomologist Maria Sibylla Merian to strengthen the empirical base of science.[15]

A number of other women of lower estates also contributed to science. Midwives, long before the recent enthusiasm for women's health initiatives, took full charge of women's medicine. Wise women developed balms and cordials to prevent disease and cure ills. The eighteenth century was also the time when these aspects of women's traditional knowledges were under attack. In the best-known example, midwives were run out of business, first by those ungainly creatures called "man midwives" and eventually by gynecologists and obstetricians.[16]

---

[14] Christine de Pizan, *The Book of the City of Ladies* (1405), trans. Earl Jeffrey Richards (New York: Persea Books, 1982), pp. 70–80; Edgar Zilsel, "The Sociological Roots of Science," *American Journal of Sociology,* 47 (1942), 545–6; and Arthur Clegg, "Craftsmen and the Origin of Science," *Science and Society,* 43 (1979), 186–201.

[15] Schiebinger, *The Mind Has No Sex?* chap. 3. On Merian, see also Natalie Zemon Davis, *Women on the Margins: Three Seventeenth-Century Lives* (Cambridge, MA: Harvard University Press, 1995).

[16] Jean Donnison, *Midwives and Medical Men: A History of Inter-Professional Rivals and Women's Rights* (London: Heinemann, 1977); Schiebinger, *The Mind Has No Sex?* chap. 4; Ornella Moscucci, *The Science of Woman: Gynaecology and Gender in England, 1800–1929* (Cambridge University Press, 1990);

Figure 8.2. Astronomers Elisabetha and Johannes Hevelius working together with the sextant. From Hervelius's *Machinae coelestis* (Danzig, 1673), facing p. 222. By permission of Houghton Library, Harvard University.

Outside Europe, a number of women aided Europeans' forays into nature, preserving the health and well-being of foreign naturalists by preparing local foods and medicines. Women sometimes also served as local guides for European expeditions; much of the collecting and cataloging for Garcia da Orta's well-known 1563 *Coloquios dos simples e drogas . . . da India,* for example, was

Hilary Marland (ed.), *The Art of Midwifery: Early Modern Midwives in Europe* (London: Routledge, 1993); Adrian Wilson, *The Making of Man Midwifery* (Cambridge, MA: Harvard University Press, 1995); and Nina Rattner Gelbart, *The King's Midwife: A History and Mystery of Madame du Coudray* (Berkeley: University of California Press, 1998).

done by a Konkani slave girl known only as Antonia.[17] In a much celebrated instance, Lady Mary Wortley Montagu served as an international broker for women's knowledges. During her stay in Turkey as the wife of the British Ambassador at Constantinople, Lady Mary learned of an old Greek woman who – with her nutshell and needle – inoculated children against smallpox; Montagu along with her surgeon, Charles Maitland, introduced this practice into England. Montagu's role here may be more that of a mother than a scientist; her willingness to have her own children inoculated convinced many people of the safety of the procedure. Maitland tested the inoculation against smallpox on six prisoners and, by 1723, fifty-one other people, and he wrote several treatises concerning its safety.[18]

In the nineteenth century, the breakdown of the old order (the guild system of artisanal production and aristocratic privilege) closed to women what informal access to science they might have enjoyed. With the privatization of the family and the professionalization of science, women wanting to pursue a career in science had two options. They could attempt to follow the course of public instruction and certification through the universities, as did their male counterparts. Or they could continue to participate within the (now private) family sphere as increasingly invisible assistants to scientific husbands or brothers; this became the normal pattern for women in science in the nineteenth century.[19]

"LEARNED VENUSES," "AUSTERE MINERVAS," AND "HOMOSOCIAL BROTHERHOODS"

In 1985 Evelyn Fox Keller, rephrasing Georg Simmel, declared that science is "masculine," not only in the person of its practitioners but also in its ethos and substance.[20] The elusive and explosive question of the gendering of science, nature, men, and women has been tied for some people to the question of women's access to science, for others to the style of science, and for still others to the content and priorities of science and human knowledge more generally. In the study of conceptions of gender in science, three elements must be distinguished: how gender is defined; how the sex is understood; and how

---

[17] Richard Grove, *Green Imperialism: Colonial Expansion, Tropical Island Edens, and the Origins of Environmentalism: 1600–1860* (Cambridge University Press, 1995), p. 81.

[18] Charles Maitland, *Mr. Maitland's Account of Inoculating the Small Pox* (London, 1722). There was much discussion about who first introduced the smallpox vaccination into Western Europe. In his account, John Andrew claimed that Lady Montagu sent the first report in 1716. James Jurin reported that this type of inoculation had been practiced in Wales from "time out of mind." See Isobel Grundy, *Lady Mary Wortley Montagu: Comet of the Enlightenment* (Oxford: Oxford University Press, 1999).

[19] Pnina Abir-Am and Dorinda Outram (eds.), *Uneasy Careers and Intimate Lives: Women in Science 1789–1979* (New Brunswick, NJ: Rutgers University Press, 1987); Helena Pycior, Nancy Slack, and Pnina Abir-Am (eds.), *Creative Couples in Science* (New Brunswick, NJ: Rutgers University Press, 1996).

[20] The terms featured in the subtitle are Paula Findlen's, "Translating the New Science," p. 171. Evelyn Fox Keller, *Reflections on Gender and Science* (New Haven, CT: Yale University Press, 1985).

actual men and women participated in science. Masculinity and femininity are not characteristics inherent to men or women that have universal meanings above and beyond historical contexts. These terms mean very different things at different times and in different places, and they often refer as much to the manners of a particular class or a particular people as to the characteristics of a particular sex. For the founders of the Royal Society, for example, the much-trumpeted "masculine philosophy" was to be distinctively English (not French), empirical (not speculative), and practical (not rhetorical).[21] "Masculinity" served in this case as a term of approbation and attached only tangentially to men (Figure 8.3).

Scholars have explained the gendering of science in different ways. In her classic 1980 *Death of Nature,* Carolyn Merchant revealed how the rise of a mechanistic worldview entailed the "death of nature." Notions of nature as matter in motion served to weaken moral restrictions embedded in older cosmologies that had forbidden untoward incursions into the belly of "Mother Nature." Merchant focused attention on the rhetorical violence of Francis Bacon's new mechanical (and "masculine") philosophy, which purported to unlock the "secrets . . . in Nature's bosom," to bind "Nature with all her children to [its] service and make her [its] slave."[22] Merchant and much subsequent ecofeminism have emphasized that the newly virile science held devastating consequences for women and for nature, both seen as subordinate females. Although roundly criticized for reinforcing the traditional notion that women belong to nature in ways that men do not, Merchant rightly called attention to the adamant gendering of nature as female in both ancient and modern science traditions.

Others have explained the gendering of science in terms of sexual divisions in physical and intellectual labor. According to this view, science was part of the territory that fell to the masculine party in the broader cultural restructurings of the early modern period. Because science, like any other profession, came to inhabit the public realm, where women (or femininity) dared not tread, science came to be seen as decidedly masculine. As science increasingly lost its amateur status and became a paid vocation, its ties to the public sphere strengthened. Ideologues of the day taught that the public sphere of government and commerce, science and scholarship was founded on the principles of reasoned impartiality – qualities increasingly associated with masculinity. At the same time, the rise of the sentimental family increasingly put the ideal mother in charge of child rearing and moral rectitude. The norms of femininity developed in the late eighteenth century portrayed womanliness as a virtue in the spheres of motherhood and the home but as a handicap in

---

[21] Schiebinger, *The Mind Has No Sex?* chap. 5.
[22] Carolyn Merchant, *The Death of Nature: Women, Ecology, and the Scientific Revolution* (San Francisco: Harper and Row, 1980), pp. 168–72; and Brian Easlea, *Witch-Hunting, Magic, and the New Philosophy* (Hassocks, Sussex: Harvester Press, 1980), pp. 126–9.

FRONTISPICE de L'ENCYCLOPEDIE.

Figure 8.3. "Academy of Sciences, Arts, and Trades," the frontispiece to Diderot and d'Alembert's *Encyclopédie*. In early modern Europe, two allegories vied for power of representation: the feminine "scientia," female muses and otherwordly consorts to the predominantly male practitioners of the sciences; and the new ideal of a "masculine" philosophy, explicitly championed by the Royal Society of London. In this well-known frontispiece, Truth, Reason, Philosophy, Physics, Optics, Botany, and Chemistry are all represented in female form. By permission of the Department of Special Collections and University Archives, Stanford University Libraries.

the world of science.[23] Early modern science thus built the exclusion of actual women, as well as cultural practices and ideals deemed feminine, into what could count as truth.

Yet another well-established tradition fostered the gendering of early modern science: homosociability. David Noble has shown how, following well-established traditions, the presence of learned Venuses or even austere Minervas threatened to disrupt the homosocial bonding that fired many a male intellect. Ancient Hebrew traditions (at least in the interpretation given them by the *Encyclopédie*) held that by virtue of contact with women, men lost the power of prophecy. In Christian traditions of medieval Europe, monastic life – important to the life of the mind – was a celibate one. These traditions continued in universities. Professors at the universities of Oxford and Cambridge were not allowed to marry; until late into the nineteenth century, celibacy was required of all faculty. The perceived dangers of women to the life of the mind – both the threat of carnal desires and the banality of daily bodily maintenance – was so great that a number of philosophers (among them Bacon, Locke, Boyle, Newton, Hobbes, Hume, and Kant) never married. Francis Bacon clearly considered wife and children impediments to great enterprises; Pierre Bayle declared the marriage of a learned woman a waste of national resources. Even Mary Wollstonecraft agreed that unmarried men and women proved the most creative thinkers.[24]

Other scholars have located the gendering of science in the new scientific societies. Steven Shapin has argued that in seventeenth-century England, women, under *covert* first of their fathers and then of their husbands, lacked standing within the economy of civility, the crucial social element that guaranteed truth in the new experimental science. Robert Boyle, an independent gentleman of honor, became the ideal "modest witness" – a faithful and unobtrusive scribe – to natural facts. Women's all-essential modesty, by contrast, was modesty of the opaque and epistemologically polluting body; as Elizabeth Potter has pointed out, women's names never appeared among those attesting to the veracity of experiments, whether or not they were present in cabinets of natural philosophy.[25]

Mary Terrall has similarly focused on the academies, where scientists forged

---

[23] See Maurice Bloch and Jean Bloch, "Women and the Dialectics of Nature in Eighteenth-Century French Thought," in Carol P. MacCormack and Marilyn Strathern (eds.), *Nature, Culture, and Gender* (Cambridge University Press, 1980), pp. 25–41; Joan Landes, *Women and the Public Sphere in the Age of the French Revolution* (Ithaca, NY: Cornell University Press, 1988); Schiebinger, *The Mind Has No Sex?*; and Geneviève Fraisse, *Reason's Muse: Sexual Difference and the Birth of Democracy*, trans. Jane Marie Todd (Chicago: University of Chicago Press, 1994).

[24] David Noble, *A World Without Women: The Christian Clerical Culture of Western Science* (New York: Knopf, 1992); Mario Biagioli, "Knowledge, Freedom, and Brotherly Love: Homosociability and the Accademia dei Lincei," *Configurations*, 2 (1995), 139–66; and Schiebinger, *The Mind Has No Sex?* pp. 151–2.

[25] Steven Shapin, *A Social History of Truth, Civility and Science in Seventeenth-Century England* (Chicago: University of Chicago Press, 1994); Potter in Donna Haraway, *Modest_Witness@Second_Millennium:FemaleMan©_Meets_OncoMouse™* (New York: Routledge, 1997), pp. 26–33.

a masculine identity (as much in France as in England) not only in the absence of women but also as a foil to prominent feminine forms of intellectual
activity, and especially to the world of salons. Members of the prestigious
Parisian Académie Royale des Sciences, as Terrall has argued, portrayed their
labors as a heroic quest for truth requiring strength of mind and also often
of body. Although this image was designed to play to influential female audiences, it also reinforced the exclusion of women; the "doing" of science
became increasingly distinct from the "consuming" of science.[26] Outside the
Académie, Jean-Jacques Rousseau contrasted what he identified as the wom-
anly style of the powerful salons, where "reason is clothed in gallantry," to a
properly vigorous style that was inappropriate for women. Men among themselves would not "humor" one another in dispute; rather, each, feeling himself
attacked by all the forces of his adversary, would feel obliged to use "all his own
force to defend himself."[27] Only through this combative process did Rousseau
believe that the mind gains precision and vigor.

  Did the ardent gendering of scientific culture channel eighteenth-century
women into what we today call the "soft" sciences (the life sciences and natural history) or the "hard" sciences (the physical sciences)? Surprising to modern
eyes, women were as prominent among physicists and mathematicians in the
eighteenth century as among other scientists, except perhaps for botanists.
Of all the sciences recommended for women, botany became the feminine
science *par excellence.* By the nineteenth century, botany's reputation as "un-
manly" – an ornamental branch suitable only for "ladies and effeminate
youths" – was such that it was questioned whether able-bodied young men
should pursue it at all. Hegel even compared the mind of woman to a plant
because, in his view, both were essentially placid. It is not surprising that
botany was thought appropriate for women. Plants had long belonged to
women's domains: peasants and aristocrats alike had worked as healers and
wise women, gathering and cultivating the plants required for domestic medicines. Furthermore, an appreciation of botany posed no threat to orthodoxies
concerning women's nature: a rose was said to mirror the beauty of its devo-
tee, exotic plants were said to flourish under a nurturing female hand, and
the female herself was thought to prosper from the rational pleasures botany
afforded. Although after Linnaeus the study of plants seemed to require more
of a focus on sexuality than might seem suitable to ladies, botany continued
to be advocated (especially in England) as the science leading to the greatest
appreciation of God and his universe.[28]

---

[26] Mary Terrall, "Gendered Spaces, Gendered Audiences: Inside and Outside the Paris Academy of
   Sciences," *Configurations,* 2 (1995), 207–32.
[27] Jean-Jacques Rousseau, *Lettre à M d'Alembert sur les spectacles* (1758), L. Brunel (ed.) (Paris, 1896), p. 157.
[28] Hegel compared the male mind to an animal that acquires knowledge only through much struggle
   and technical exertion. The female mind, by contrast, does not (cannot) rise above its plant-like existence and remains rooted in its *an sich* existence (*Grundlinien der Philosophie des Rechts,* 1821) in
   Werke, ed. Eva Moldenhauer and Karl Michel, 20 vols. (Frankfurt am Main: Suhrkamp, 1969–71),

THE SCIENCE OF WOMAN

At the birth of modern science, the noble networks and artisanal workshops gave women (limited) access to science. Their incursion into serious intellectual endeavor was supported ideologically by the Cartesian wedge driven between mind and body, giving voice to the notion that "the mind has no sex."[29] The expansive mood of the Enlightenment – the feeling that all men are by nature equal – gave women renewed hope that they, too, might begin to share the privileges heretofore reserved for men.

As it emerged toward the end of the eighteenth century, however, the participation of women in normal science was not to be. The exclusion of women from public life required new justifications, based on scientific, and not Biblical, authority. Within the framework of Enlightenment thought, an appeal to natural rights could be countered only by proof of natural inequalities. An individual's place in the *polis* increasingly depended on his or her property holdings and also on sexual and racial characteristics. Science, with its promise of a "neutral" and privileged viewpoint above and beyond the rough-and-tumble of political life, came to mediate between the laws of "nature" and the laws of legislatures. For many, scientists did not have to take a stand in questions of social equalities because "the body spoke for itself."[30]

In this political climate, the eighteenth century witnessed a revolution in "sexual science," the exact study of sexual difference.[31] The revolution was first and foremost a rupture in methodology: Aristotelian and Galenic science had understood divergent sexual temperaments as driven by cosmic principles reduplicating the macrocosm within the microcosm of the individual body.[32] Eighteenth-century science deployed empirical methods to weigh and measure sexual differences in the body. The revolution in sexual science was also marked by what Thomas Laqueur has described as a shift from a one-sex to a two-sex model of difference. The older, one-sex model, favored by Galen

---

vol. 7, pp. 319–20. See also J. F. A. Adams, "Is Botany a Suitable Study for Young Men?" *Science*, 9 (1887), 117–18; Emmanuel Rudolph, "How It Developed That Botany Was the Science Thought Most Suitable for Victorian Young Ladies," *Children's Literature*, 2 (1973), 92–7; and Ann Shteir, *Cultivating Women, Cultivating Science: Flora's Daughters and Botany in England 1760–1860* (Baltimore, MD: Johns Hopkins University, 1996). Lisbet Koerner argues that Linnaeus's new system of botany accommodated women and other lesser-educated folks because, even though it was in Latin, it was useful and simple ("Women and Utility in Enlightenment Science," *Configurations*, 2 (1995), 233–55).

29 François Poullain de la Barre, *De l'Égalité des deux sexes: discours physique et moral* (Paris, 1673). See also Erica Harth, *Cartesian Women: Versions and Subversions of Rational Discourse in the Old Regime* (Ithaca, NY: Cornell University Press, 1992).

30 Samuel Thomas von Soemmerring, *Über die körperliche Verschiedenheit des Negers vom Europäer* (Frankfurt and Mainz, 1785), preface.

31 Cynthia Russett, *Sexual Science: The Victorian Construction of Womanhood* (Cambridge, MA: Harvard University Press, 1989). See also Ludmilla Jordanova, *Sexual Visions: Images of Gender in Science and Medicine between the Eighteenth and Twentieth Centuries* (Madison: University of Wisconsin Press, 1989).

32 Joan Cadden, *Meanings of Sex Difference in the Middle Ages: Medicine, Science, and Culture* (Cambridge University Press, 1993); and Lesley Dean-Jones, *Women's Bodies in Classical Greek Science* (Oxford: Oxford University Press, 1994).

and others, saw male and female genitalia as the same in kind: "All parts that men have, women have too" (including a "spermatical vessel") with the exception that women's are inverted and contained inside the body.[33] Sexual difference was one of degree: woman simply lacked the heat to perfect her organs and thrust them outward from her body. The new "two-sex" model sharply distinguished male and female genitalia; the uterus was no longer configured an inadequate penis but instead was celebrated as a perfect instrument for producing future citizens of the state.[34]

The reevaluation of women's reproductive organs was only one element in a much broader revolution. Sexuality was no longer to be seen as residing exclusively in a "single organ" but, the French physician Pierre Roussel explained in 1775, as extending "through more or less perceptible nuances" into every part of the human body.[35] The first representations of distinctively female skeletons in Western anatomy epitomized this broader revolution (Figure 8.4). The materialism of the age led anatomists to look to the skeleton; as the hardest part of the body it was said to provide a "ground plan" for the body and to give a "certain and natural" direction to the muscles and other parts of the body attached to it.[36] If sex differences could be found in the skeleton, then sexual identity would no longer depend on differing degrees of heat (as the ancients had taught), nor would it be a matter of sex organs appended to a neutral human body (as Vesalius had thought). Instead, sexuality would be seen as penetrating every muscle, vein, nerve, and organ attached to and molded by the skeleton. Although the female skeleton was drawn from nature with painstaking exactitude, great debate erupted over its distinctive features. Political circumstances drew immediate attention to depictions of the skull as a measure of intelligence and the pelvis as a measure of womanliness. The woman's narrow cranium seemed to explain nicely her lesser achievement in science.[37]

By the 1790s, European anatomists presented male and female bodies as

[33] Galen, *On the Usefulness of the Parts of the Body,* trans. Margaret May (Ithaca, NY: Cornell University Press, 1968), vol. 2, pp. 628–9.

[34] Thomas Laqueur, *Making Sex: Body and Gender from the Greeks to Freud* (Cambridge, MA: Harvard University Press, 1990). See also the critical evaluation of Laqueur's work by Katharine Park and Robert Nye, "Destiny Is Anatomy," *The New Republic,* February 18, 1991, 53–7; and Cadden, *Meanings of Sex Differences in the Middle Ages.*

[35] Pierre Roussel, *Système physique et moral de la femme, ou Tableau philosophique de la constitution, de l'état organique, du tempérament, des moeurs, & des fonctions propres au sexe* (Paris, 1775), p. 2. Carl Klose also argued that it is not the uterus that makes woman what she is. Even women from whom the uterus has been removed, he stressed, retain feminine characteristics. See his *Über den Einfluß des Geschlects-Unterschiedes auf Ausbildung und Heilung von Krankheiten* (Stendal, 1829), pp. 28–30. See also Edmond Thomas Moreau, *Quaestio medica: An praeter genitalia sexus inter se discrepent?* (Paris, 1750).

[36] Bernard Albinus, *Table of the skeleton and muscles of the human body* (London, 1749), "Account of the Work."

[37] Schiebinger, *The Mind Has No Sex?* chap. 7; Elizabeth Fee, "Nineteenth-Century Craniology: The Study of the Female Skull," *Bulletin of the History of Medicine,* 53 (1979), 415–33; Stephen Jay Gould, *The Panda's Thumb: More Reflections in Natural History* (New York: W. W. Norton, 1980), chap. 14.

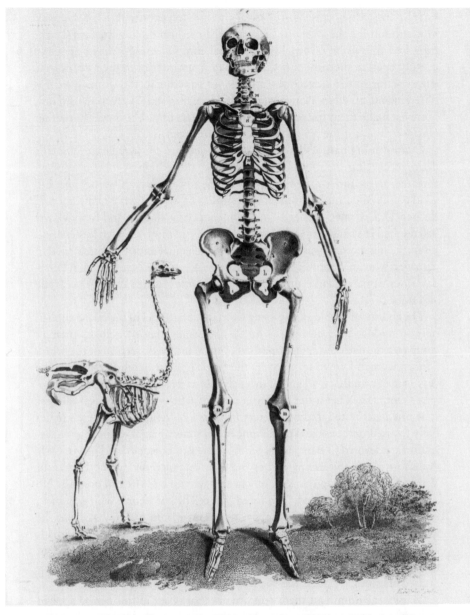

Figure 8.4. The French anatomist Marie-Geneviève-Charlotte Thiroux d'Arconville's female skeleton compared to that of an ostrich; each is remarkable for its large pelvis. From John Barclay, *The Anatomy of the Bone of the Human Body* (Edinburgh, 1829), plate 4. By permission of the Boston Medical Library.

each having a distinct telos – physical and intellectual strength for the man, motherhood for the woman. The Harvard medical doctor Edward Clarke expressed this vision of physical and social complementarity at its apogee a century later: in the same way that "the lily is not inferior to the rose, nor the oak superior to the clover," the man is not superior to the woman; each is different and suited to its own ends.[38] Women's separate perfections did not, however, make them the equals of men in matters of public power but rather destined them for the private sphere and domesticity.

Contravening nature's laws was said to hold dire consequences. Women's desire to develop their intellect was considered the highest form of egoism, threatening to undermine their own health and the health of the race. Dr. Clarke offered examples from clinical studies of women whose education at the new U.S. women's colleges (including Smith, Wellesley, and Bryn Mawr) had resulted in sterility, anemia, menorrhagia, dysmenorrhea, even hysteria and insanity. The message was clear: intensive intellectual endeavor threatened to damage a woman's reproductive organs, causing her ovaries to shrivel. A latter-day Rousseauian, Clarke urged women to revere nature's calling "to cradle and nurse a race."[39]

The abundant ideology idealizing woman as the angel of the home applied only to middle-class Europeans. In 1815, Georges Cuvier, France's premier comparative anatomist, performed his now infamous dissection of the South African woman known to many by the English name Sarah Bartmann. The very name given this woman – Cuvier always referred to her as *Vénus Hotten-totte* – emphasized her sexuality. (Passionate tendencies found in warm climates were often attributed to the planetary influence of Venus.) His interest in her body focused on her sexual parts; nine of his sixteen pages recording the dis-section are devoted to Bartmann's genitalia, breasts, buttocks, and pelvis. Only one short paragraph evaluated her brain. In his memoir on the Hottentot Venus, Cuvier took up the issue of whether science had African origins: "No race of Negro," he declared, "produced that celebrated people who gave birth to the civilization of ancient Egypt, and from whom we may say that the whole world has inherited the principles of its laws, sciences, and perhaps also religion." Without exception, the "cruel law" of nature, he concluded, had "condemned to eternal inferiority those races with a depressed and com-pressed cranium."[40] Such was the fate of Sarah Bartmann.

Neither the dominant theory of race nor that of sex in this period applied to non-European women, particularly those of African descent. Like other

---

[38] Edward Clarke, *Sex in Education: A Fair Chance for Girls* (Boston, 1874), p. 15.

[39] Clarke, *Sex in Education*, pp. 33, 39, 62, 101–2, 136.

[40] Georges Cuvier, "Extrait d'observations faits sur le cadavre d'une femme connue à Paris et à Lon-dres sous le nom de Vénus Hottentotte," *Mémoires du muséum d'histoire naturelle*, 3 (1817), 259–74, especially 272–3. See also Sander Gilman, *Sexuality: An Illustrated History* (New York: Wiley, 1989), and Anne Fausto-Sterling, "Gender, Race, and Nation," in Jennifer Terry and Jacqueline Urla (eds.), *Deviant Bodies: Critical Perspectives on Difference in Science and Popular Culture* (Bloomington: Indi-ana University Press, 1995), pp. 19–48.

females, they did not fit comfortably in the great chain of being, in which primarily males were studied for their comparative superiority. Like other Africans, they did not fit European gender ideals. As a recent book on contemporary black women's studies has put it, all the blacks were men and all the women were white.[41] On both counts – of her sex and her race – Bartmann was relegated to the world of brute flesh. Elite European naturalists who set such store by sexual complementarity when describing their own mothers, wives, and sisters rarely included African women in their new definitions of femininity.

GENDERED KNOWLEDGE

Historians have detailed the accomplishments of women scientists, the exclusion of women from scientific production, the various ideological props and cultural supports justifying that exclusion, the gendering of the persons and cultures of science, and the scientific perusal of female anatomy. Fewer have shown how gender has molded the very content of the sciences. Gender became one potent principle organizing eighteenth-century understandings of the natural world, a matter of consequence in an age that looked to nature as the guiding light for social reform. Let me sketch two examples of how gender molded the results of science. The first is the gendering of Linnean botanical taxonomy, where Europe's tenacious gender roles were overlaid onto unsuspecting plants and their sexual relations.

As extraordinary as it seems today, it was not until the late seventeenth century that European naturalists began recognizing that plants reproduce sexually. The ancient Greeks, it is true, had some knowledge of sexual distinctions in plants: Theophrastus knew the age-old practice of fertilizing date palms by bringing male flowers to the female tree; and Pliny tells us that peasants' agricultural practices recognized sexual distinctions in trees such as the pistachio.[42] Plant sexuality, however, was not a major focus of interest in the ancient world. In this era and throughout the medieval period, plant classification generally emphasized the usefulness of plants to human beings as foods and medicines.

Plant sexuality exploded onto the European stage in the seventeenth and

---

[41] See Gloria Hull, Patricia Bell Scott, and Barbara Smith (eds.), *All the Women Are White, All the Blacks Are Men, But Some of Us Are Brave: Black Women's Studies* (Old Westbury, NY: The Feminist Press, 1982). For the relationship between the science of sex and race, see Nancy Leys Stepan, "Race and Gender: The Role of Analogy in Science," in David Theo Goldberg (ed.), *Anatomy of Racism* (Minneapolis: University of Minnesota Press, 1990), pp. 38–57; and Londa Schiebinger, *Nature's Body: Gender in the Making of Modern Science* (Boston: Beacon Press, 1993), chaps. 4 and 5.

[42] Pliny, the Elder, *Natural History*, trans. H. Rackham (Cambridge, MA: Harvard University Press, 1942), 12, pp. xxxii, 45; A. G. Morton, *History of Botanical Science: An Account of the Development of Botany from Ancient Times to the Present Day* (New York: Academic Press, 1981), pp. 28, 38. For a more detailed discussion of gender in early modern botany, see Schiebinger, *Nature's Body*, chap. 1.

eighteenth centuries for a variety of reasons, including the general interest in sexual differentiation among humans. When sexuality in plants was recognized, everyone wanted to claim the honor of having discovered it. In France, Sébastien Vaillant and Claude-Joseph Geoffroy tussled over priority; in England, Robert Thornton complained that the honor was always given to the French, although properly it belonged to the English. Carl Linnaeus, always keen to reap his due reward for scientific innovation (and not, in fact, the first to describe sexual reproduction in plants), claimed that it would be difficult and of no utility to decide who first discovered the sexes of plants.[43]

Even in this era, interest in assigning sex to plants ran ahead of any real understanding of fertilization, or the "coitus of vegetables," as it was sometimes called. Botanists distinguished certain parts of plants as male and female (as Claude Geoffroy reported) "without knowing well the reason." The English naturalist Nehemiah Grew, the first to identify the stamen as the male part in flowers, developed his notion of plant sexuality from his knowledge of animals. In his 1682 *Anatomy of Plants*, Grew reported that "the attire" (his term for the stamen) resembles "a small penis," the various coverings upon it appear to be "so many little testicles," and the globulets (or pollen) act as "the vegetable sperme." As soon as the plant penis is erected, Grew continues, "this vegetable sperm falls down upon the seed-case or womb, and so touches it with a prolific virtue."[44]

By the early part of the eighteenth century, the analogy between animal and plant sexuality was fully developed. Linnaeus, in his *Praeludia sponsaliorum plantarum*, related the terms of comparison: in the male, the filaments of the stamens are the *vas deferens*, the anthers are the testes, and the pollen that falls from them is the seminal fluid; in the female, the stigma is the vulva, the style becomes the vagina, the tube running the length of the pistil is the Fallopian tube, the pericarp is the impregnated ovary, and the seeds are the eggs. Julien Offray de La Mettrie, along with other naturalists, even claimed that the honey reservoir found in the nectary is equivalent to mother's milk in humans.[45]

Sexual differentials, built on the imperfect analogy between plant and animal life, led to the privileging of certain sexual types over others. Most flowers are hermaphroditic, with both male and female organs in the same individual. As one eighteenth-century botanist put it, there are two sexes (male and female) but three kinds of flowers: males, females, and hermaphrodites or, as they were sometimes called, androgynes. Although most eighteenth-century botanists enthusiastically embraced sexual dimorphism, the conception of

[43] Jacques Rousseau, "Sébastien Vaillant: An Outstanding Eighteenth-Century Botanist," *Regnum Vegetabile*, 71 (1970), 195–228. Giulio Pontedera powerfully rejected the entire notion of plant sexuality in 1720 (*Anthologia, sive de floris natura*). "The Prize Dissertation of the Sexes of Plants by Carolus von Linnaeus," in Robert Thornton, *A New Illustration of the Sexual System of Carolus von Linnaeus* (London, 1799–1807).

[44] Nehemiah Grew, *The Anatomy of Plants* (London, 1682), pp. 170–2.

[45] Linnaeus, *Praeludia sponsaliorum plantarum* (1729; reprinted Uppsala: Almqvist & Wiksell, 1908), section 15; Julien Offray de la Mettrie, *L'Homme plante* (Potsdam, 1748).

plants as hermaphroditic ran into resistance. William Smellie, chief compiler of the first edition of the *Encyclopaedia Britannica* (1771), rejected the whole notion of sexuality in plants as prurient and disapproved of the term "hermaphrodite," noting when using the word that he merely spoke "the language of the system." Smellie denounced Linnaeus for taking his analogy "far beyond all decent limits," claiming that Linnaeus's metaphors were so indelicate as to exceed those of the most "obscene romance-writer."[46]

The ardent sexing of plants coincided with what is commonly celebrated as the rise of modern botanical taxonomy. In the sixteenth and seventeenth centuries, plant materials from the voyages of discovery and newly established colonies flooded Europe (increasing the number of known plants by a factor of 4 between 1550 and 1700), and new methods were developed for organizing these new riches: by 1799, when Robert Thornton published his popular version of the Linnean system, he counted fifty-two different systems of botany. Classification systems were based on different parts of plants. John Ray based his on the flower, calyx, and seed coat; Tournefort, in Paris, grounded his in the corolla and fruit; Albrecht von Haller, taking a very different approach, argued that geography was crucial to an understanding of plant life and that embryogenesis should also be represented in a system of classification. Despite the number and variety of systems, Linnaeus's taxonomy swept away these other systems and, from the 1740s (at least outside France) until the first decades of the nineteenth century, was generally considered the most convenient system of classification.

Linnaeus founded his renowned "Key to the Sexual System" on the *nuptiae plantarum* (the marriages of plants), that is, on the number of husbands (stamen) or wives (pistils) in a particular union. His famous *Systema naturae* divided the vegetable world (as he called it) into *classes* based on the number, relative proportions, and position of the male parts or stamens (Figure 8.5). These classes were then subdivided into some sixty-five *orders* based on the number, relative proportions, and positions of the female parts or pistils. These were further divided into *genera* (based on the calyx, flower, and other parts of the fruit), *species* (based on the leaves or some other characteristic of the plant), and *varieties*.[47]

One might argue that Linnaeus based his system on sexual difference because he was one of the first to recognize the biological importance of sexual reproduction in plants. But the success of Linnaeus's system did not rest on the fact that it was "natural"; indeed Linnaeus readily acknowledged that it was highly artificial. Although focused on reproductive organs, his system did not capture fundamental sexual functions. Rather it focused on purely morphological features (that is, the number and mode of union) – exactly those

---

[46] William Smellie, "Botany," *Encyclopaedia Britannica* (Edinburgh, 1771), vol. 1, p. 653.
[47] Carl Linnaeus, *Systema naturae* (1735), ed. M. S. J. Engel-Ledeboer and H. Engel (Nieuwkoop: B. de Graaf, 1964).

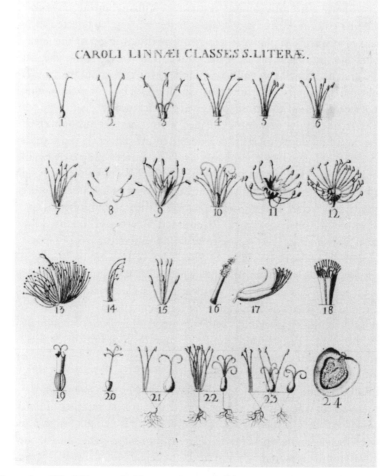

CAROLI LINNÆI CLASSES S.LITERÆ.

Figure 8.5. "Carl Linnaeus's Classes or Letters" illustrating Linnaeus's sexual system. Printed with Linnaeus's *Systema naturae* beginning with the second edition (1737).

characteristics of the male and female organs *least* important for their sexual function.

In view of this fact, it is striking that Linnaeus chose to highlight the sexual parts of plants at all. Furthermore, Linnaeus devised his system in such a way that the number of a plant's stamens (or male parts) determined the *class* to which it was assigned, whereas the number of its pistils (the female parts) determined its *order.* In the taxonomic tree, class stands above order. In other words, Linnaeus gave male parts priority in determining the status of the organism in the plant kingdom. There is no empirical justification for this outcome; rather Linnaeus brought traditional notions of gender hierarchy whole cloth into science. He read nature through the lens of social relations in such a way that the new language of botany incorporated fundamental aspects of the social world as much as those of the natural world. Although today Linnaeus's classification of groups above the rank of genus has been abandoned, his binomial system of nomenclature remains, together with many of his genera and species.

My second example of gender in the content of science comes from zoological nomenclature. In 1758, in the tenth edition of his *Systema naturae,* Carl Linnaeus coined the term *Mammalia* (meaning literally "of the breast") to distinguish the class of animals embracing humans, apes, ungulates, sloths, sea cows, elephants, bats, and all other organisms having hair, three ear bones, and a four-chambered heart. In so doing, he idolized the female mammae as the icon of that class.

Historians of science have taken Linnaeus's nomenclature more or less for granted as part of his foundational work in zoological taxonomy. There was, however, a complex gender politics informing Linnean taxonomy and nomenclature. Why Linnaeus called mammals mammals, I argue, had as much to do with the fact that there is something special about the female breast as with eighteenth-century politics of wet-nursing and maternal breast-feeding and with the contested role of women in both science and the broader culture.

For more than two thousand years most of the animals we now designate as mammals (along with most reptiles and several amphibians) had been called *quadrupeds.*[48] In coining his new term *Mammalia* Linnaeus did not draw from tradition, as was common in this period, but instead devised a wholly new term.

Were there good reasons for Linnaeus to call mammals mammals? Does the longevity of Linnaeus's term reflect the fact that he was simply right, that the mammae, indeed, represent a primary, universal, and unique character of mammals (as would have been the parlance of the eighteenth century)? Yes and no. Linnaeus chose this term even though naturalists in this period did not consider the mammae a universal characteristic of the class of animals he sought to identify (in the eighteenth century, it was commonly accepted that stallions lacked teats). More important, the presence of milk-producing mammae is only one characteristic of mammals, as was commonly known to eighteenth-century European naturalists. Linnaeus could indeed have chosen a more gender-neutral name, such as *Pilosa* (the hairy ones – although hair, and especially beards, was also saturated with gender), for example, or *Aurecaviga* (the hollow-eared ones). Or he could have chosen, perhaps, *Lactentia,* the "sucking ones," which, like the German term *Säugetiere* (suckling animals), nicely universalizes the term inasmuch as male as well as female young suckle at their mothers' breasts.

---

[48] Aristotle, *Generation of Animals,* trans. A. L. Peck (Cambridge, MA: Harvard University Press, 1953), p. lxix; and Pierre Pellegrin, *Aristotle's Classification of Animals: Biology and the Conceptual Unity of the Aristotelian Corpus,* trans. Anthony Preus (Berkeley: University of California Press, 1986). For a more thorough treatment of why mammals are called mammals, see Schiebinger, *Nature's Body,* chap. 2.

If Linnaeus had alternatives, if he could have chosen from a number of equally valid terms, what led him to the term *Mammalia?* Zoological nomenclature – like all language – is to some degree arbitrary; naturalists devise convenient terms to identify groups of animals. But nomenclature is also historic, growing out of specific contexts, conflicts, and circumstances.

Linnaeus created his term *Mammalia* in response to the question of humans' place in Nature. In his quest to find an appropriate term for a taxon uniting humans and beasts, Linnaeus made the breast – and specifically the fully developed female breast – the icon of the highest class of animals. In privileging a uniquely female characteristic in this way, it might be argued, Linnaeus broke with long-standing traditions that saw the male as the measure of all things.[49] It is important to note, however, that in the same volume in which Linnaeus introduced the term *Mammalia,* he also introduced the term *Homo sapiens.*[50] This term was used (as *homo* had been traditionally) to *distinguish* humans from other primates (apes, lemurs, and bats, for example). In the language of taxonomy, *sapiens* is what is known as a "trivial" name. From a historical point of view, however, the choice of the term *sapiens* is highly significant. Reason had traditionally distinguished humans from animals and, among humans, males from females. Thus, within Linnean terminology, a female character (the lactating mammae) ties humans to brutes; a traditionally male character (reason) marks our separateness from brutes.[51]

Linnaeus's fascination with female mammae arose alongside and in step with key political trends in the eighteenth century: the restructuring of child care (the campaigns against wet nurses and midwives) and the restructuring of women's lives as mothers, wives, and citizens. The portrait Linnaeus painted of the naturalness of a mother giving suck to her young fed into movements to undermine the public power of women and to attach a new value to mothering.[52]

Most directly, Linnaeus joined the ongoing campaign to abolish the ancient custom of wet-nursing. Linnaeus – himself a practicing physician – prepared a dissertation against the evils of wet-nursing in 1752. In this treatise, titled "Step Nurse, or a Dissertation on the Fatal Results of Mercenary Nursing," he alluded to his own taxonomy by contrasting the barbarity of women who deprive their children of mother's milk with the gentle care of great beasts –

---

[49] According to Plato, unrighteous and cowardly men returned to earth as women (*Timaeus,* 90e).

[50] Gunnar Broberg (ed.), *Linnaeus: Progress and Prospects in Linnaean Research* (Stockholm: Almqvist & Wiksell, 1980); and Broberg, *Homo Sapiens L.: Studier i Carl von Linnés naturuppfattning och människolära* (The Swedish History of Science Society, 1975).

[51] Genevieve Lloyd, *The Man of Reason: "Male" and "Female" in Western Philosophy* (Minneapolis: University of Minnesota Press, 1984). On the boundary between human and beast, see Julia Douthwaite's study of the wild children: *Exotic Women, Literary Heroines and Cultural Strategies in Ancien Regime France* (Philadelphia: University of Pennsylvania Press, 1992).

[52] Valerie Fildes, *Wet Nursing: A History from Antiquity to the Present* (Oxford: Basil Blackwell, 1988); Hilary Marland (ed.), *The Art of Midwifery: Early Modern Midwives in Europe* (London: Routledge, 1993).

the whale, the fearsome lioness, and fierce tigress – who willingly offer their young the breast.[53]

To champions of enlightenment, the breast became Nature's sign that women belonged in the home (Figure 8.6). It is remarkable that in the heady days of the French Revolution, when revolutionaries marched behind the fierce and bare-breasted Liberty, the maternal breast figured in arguments against women's exercise of civic rights. Delegates to the French National Convention, where many of these decisions were made, declared that Nature had removed women from the political arena. In this case, "the breasted ones" were to be confined to the home.[54]

Linnaeus's term *Mammalia* helped legitimize the restructuring of European society by emphasizing how natural it was for females – both human and nonhuman – to suckle and rear their own children. Linnean systematics, in both his botany and his zoology, had sought to render nature universally comprehensible, yet the categories he devised infused nature with European notions of gender. Linnaeus saw females of all species as tender mothers, a parochial vision he (wittingly or unwittingly) imprinted on Europeans' understandings of nature.

BEYOND EUROPE

Scholars have newly turned their attention away from Europe toward the gendering of knowledge crafted during the expansive voyages of scientific discovery. Moral imperative and scientific warnings kept the vast majority of Europe's women close to home; the German anthropologist Johann Blumenbach was typical in warning that white women taken to very warm climates succumbed to "copious menstruation, which almost always ends, in a short space of time, in fatal hemorrhages of the uterus."[55] There was also the often-expressed fear that women giving birth in the tropics would deliver children resembling the native peoples of those areas. The intense African sun, it was thought, produced black babies regardless of the mother's complexion.

What are the implications of Europe's gendered regimes during the period of initial contact between the world's scientific traditions (many with gendered regimes of their own)? As European naturalists fanned out around the globe collecting strange animals and exotic plants for trading companies and scientific societies, what was overlooked and discarded or picked up and emphasized

[53] Carl Linnaeus, "Nutrix Noverca," respondent F. Lindberg (1752), *Amoenitates academicae* (Erlangen, 1787), in vol. 3. Translated by Gilibert as "La Nourrice marâtre, ou Dissertation sur les suites funestes du nourrisage mercénaire," *Les Chef-d'oeuvres de Monsieur de Sauvages* (Lyon, 1770), vol. 2, pp. 215–44.

[54] Lynn Hunt, *Politics, Culture, and Class in the French Revolution* (Berkeley: University of California Press, 1984), especially part 1.

[55] Johann Blumenbach, *The Natural Varieties of Mankind* (1795), trans. Thomas Bendyshe (1865; New York: Bergman, 1969), p. 212n2. Blumenbach codified notions long current in the culture.

Figure 8.6. Nature portrayed as a young virgin, her breasts dripping with mother's milk. From Charles Cochin and Hubert François Gravelot, *Iconologie par figures: ou Traité complet des allégories, emblêmes &c.* (Paris, 1791), "Nature." By permission of the Pennsylvania State University Libraries.

because gender politics sent into the field mostly unmarried males largely estranged from domestic economies and reproductive regimes? These are questions that remain to be answered. One element that can be identified is a marked disinterest in collecting for the certain aspects of the female side of life; in particular, collecting agencies showed little interest in expanding Europe's pharmacopeia of abortifacients (although they did collect innumerable

Figure 8.7. Merian's *flos pavonis*. The indigenous and slave women in Surinam used the seeds as an abortifacient. Maria Sibylla Merian, *Dissertation sur la generation et les transformations des insectes de Surinam* (The Hague, 1726), plate 45. By permission of the Wellcome Institute Library, London.

menstrual regulators). In a moving passage in her magnificent 1705 *Metamorphosis insectorum Surinamensium*, the German-born naturalist Maria Sibylla Merian, one of the few women to travel on her own to record the bounty of nature, describes how the African slave and Indian populations in Surinam, then a Dutch colony, used the seeds of a plant she identified as the *flos pavonis*, literally "peacock flower (Figure 8.7)," as an abortifacient: "The Indians, who are not treated well by their Dutch masters, use the seeds [of this plant] to abort their children, so that their children will not become slaves like they are. . . . They told me this themselves."[56]

[56] Maria Sibylla Merian, *Metamorphosis insectorum Surinamensium* (1705), ed. Helmut Deckert (Leipzig:

In the explosion of knowledge generally associated with the Scientific Revolution and global expansion, European awareness of herbal antifertility agents, such Merian's *flos pavonis,* declined dramatically. Contrary to other trends, where naturalists assiduously collected local knowledges of plants for medicines and potential profit, there was no systematic attempt to introduce into Europe new and exotic contraceptives and abortifacients gathered from cultures around the globe. Mercantilist policies guiding global expansion did not define trade in such plants as a lucrative or desirable business, nor did the pro-natalist policies of governments encourage the collection of such knowledge.[57] Gender in the emergence of eighteenth-century global science is a topic requiring further research.

PAST AND FUTURE

In the seventeenth and eighteenth centuries, science was a young enterprise that was forging new ideas and institutions. Men of science at this time can be seen as standing at a fork in the road. They could either sweep away traditions of the medieval past and welcome women as full participants in science, or they could reaffirm the traditions of the past and continue to exclude women from rarefied intellectual pursuit. The social and intellectual circumstances directed science down the latter path; paradoxically, the Scientific Revolution participated in the rise of scientific sexism, scientific racism, and, in some cases, the collapse of knowledge systems central to women's health and well-being. The nature of science, however, is no more fixed than is the moral nature of men or women. Understanding the historical circumstances that have distanced women from science and have led to the gendering of aspects of its content can help in the complex task of reworking gender relations in modern science.

Insel Verlag, 1975), commentary to plate no. 45. On Merian, see Margarete Pfister-Burkhalter, *Maria Sibylla Merian: Leben und Werk 1647–1717* (Basel: GS-Verlag, 1980), and Elisabeth Rücker, "Maria Sibylla Merian," *Fränkische Lebensbilder,* 1 (1967), 221–47; Rücker, *Maria Sibylla Merian* (Nuremberg: Germanisches Nationalmuseum, 1967); Schiebinger, *The Mind Has No Sex?* chap. 3; Davis, *Women on the Margins;* Helmut Kaiser, *Maria Sibylla Merian: Eine Biographie* (Dusseldorf: Artemis & Winkler, 1997); and Kurt Wettengl (ed.), *Maria Sibylla Merian, 1647–1717: Artist and Naturalist* (Ostfildern: Hatje, 1998).

[57] Londa Schiebinger, "Lost Knowledge, Bodies of Ignorance, and the Poverty of Taxonomy as Illustrated by the Curious Fate of *Flos Pavonis,* an Abortifacient," in Caroline Jones and Peter Galison (eds.), *Picturing Science, Producing Art* (New York: Routledge, 1998), pp. 125–44.

9

THE PURSUIT OF THE PROSOPOGRAPHY OF SCIENCE

William Clark

The difference of natural talents in different men is, in reality, much less than we are aware of; and the very different genius which appears to distinguish men of different professions . . . is not upon many occasions so much the cause, as the effect of the division of labour. The difference between the most dissimilar characters, between a philosopher and a common street porter, for example, seems to arise not so much from nature, as from habit, custom, and education. . . . By nature a philosopher is not in genius and disposition half so different from a street porter, as a mastiff is from a greyhound . . .[1]

David Sabean remarked a few years ago that Anglo-American sociology faced a crisis, as it had based itself fundamentally on the structures of "social class" – a concept that has now given way nearly completely to the concept of "identity."[2] So many ask now about the historical identity or persona of the scientist but do not seem to want the prosopographer's answer, for that answer has tended to be given in terms of social class and its related sociological notions, such as the division of labor in the scientific community: a Smithian political economy of knowledge. It is interesting, moreover, that although a prosopography of the subjects or "heroes" of knowledge may be at once a rather ancient and a very modern pursuit, its true age, from which it traces its provenance, is the eighteenth century. Our prosopography is kith and kin with the liberal, materialistic, and positivistic social and political philosophy of the eighteenth century.

---

[1] Adam Smith, *An Inquiry into the Nature and Causes of the Wealth of Nations* (1776), Edwin Cannan (ed.), 2 vols. (Chicago: University of Chicago Press, 1976), 1:19–20.

[2] The remark was in conversation; but see David Sabean, *Kinship in Neckarhausen, 1700–1870* (Cambridge University Press, 1998), p. 2.

A fragmentary first draft of this essay was presented on 15 October 1997 at the colloquium of *Abteilung II, Max-Planck-Institut für Wissenschaftsgeschichte,* the discussion at which led to a complete recasting of this article into its current form.

WHAT IS PROSOPOGRAPHY?

"Prosopography is the investigation of the common background characteristics of a group of actors in history by means of a collective study of their
lives."[3] Not only likely to send professors of English and crossword-puzzle
virtuosi scrambling for the OED, mention of "prosopography" can also make
cultural and intellectual historians of science grimace (if not groan). In the
1970s Lawrence Stone, Steven Shapin, Arnold Thackray, and Lewis Pyenson
published articles on prosopography that deserve our attention still. My comments there will be general and are made mindful of our overriding interest
here in the history of science.

Our authors use "prosopography," as I shall do, as a general term for two
sorts of studies that others might separate: collective biographies of groups
(prosopography in the broader sense) and statistical studies of populations
(prosopography in the narrower sense). Collective biographies tend to look
at relatively small, manageable groups, for example, all salaried full members
of the Academy of Sciences in Paris from 1700 to 1789 – fewer than 100 individuals. Statistical studies investigate relatively larger groups, for example,
all publishing Jesuit scientists from 1600 to 1773 – about 1600 individuals. It
seems pedantic to worry about when collective biography becomes statistical
study, especially since academics now use many such techniques – tables, charts,
and so on – for rather small groups. Thus I shall use "prosopography" to cover
the entire spectrum of such studies.

A number of things characterize prosopographical studies. First, they are
centered on individuals in relation to a relevant social group. Relations to
ideas, institutions, and so on are irrelevant or secondary or are derived from
the study of the group. Second, prosopographical studies require delimitation
of the group so that decisions can be resolved regarding whom to count. Such
criteria of delimitation may seem at times arbitrary, but they are essential.
Third, an explicit or implicit prosopographical profile or biographical schema
for the relevant individuals is needed to render collection of data systematic.
So one usually collects names, birth and death dates and places, educational
institutions attended, occupations, and so on. Collective biographies may
collect thick profiles of the relatively small number of individuals, whereas
statistical studies are usually driven to make do with thin profiles of a large
population. Fourth, one is often interested in gaining a better sense of the
relevant group from the prosopography of its members. And fifth, one often
looks to uncover patterns or relations not apparent at the level of intellectual,
institutional, or other such histories.

---

[3] Lawrence Stone, "Prosopography," *Daedalus*, 100 (1971), 46–79, at 46. See also Steven Shapin and
Arnold Thackray, "Prosopography as a Research Tool in the History of Science: The British Scien-
tific Community 1700–1900," *History of Science*, 12 (1974), 1–28; and Lewis Pyenson, "'Who the Guys
Were': Prosopography in the History of Science," *History of Science*, 25 (1977), 155–88.

As collective biography, prosopography is old. It reaches from the ancient "Lives of the Philosophers" to the medieval "Lives of Saints" to early modern "Lexica of the Learned." The eighteenth century saw the rise of lexica or collective biographies restricted to specific national or ethnic groups, although the blossoming of this genre awaited the nineteenth century. Important for us, J. C. Poggendorff's *Biographisch-literarisches Handwörterbuch zur Geschichte der exacten Wissenschaften* began appearing after 1863. Modern prosopographers tend to treat all these sorts of works more as sources than as studies in their own right. So one might use, for example, Jöcher's *Gelehrten-Lexicon* as a prosopographical database for collecting, say, all scholars who published on natural philosophy from 1600 to 1700.

Lexica such as those by Jöcher and Poggendorff moved in the direction of statistical studies, since they reduced scholars or scientists to a "Statistik" or list of dates of birth, professional works, achievements, publications, and so on. Whereas earlier genres, such as the ancient "Lives of Philosophers," made no distinction between private versus public or professional life, eighteenth- and nineteenth-century lexica gave relatively thin profiles of an elite population, omitting most aspects of private life. And it is from such lexica and other similar sources that our modern, statistically oriented prosopographical studies have been allowed to emerge.

PROSOPOGRAPHY IN THE HISTORY OF SCIENCE

Both Stone and Pyenson give special mention to a work by Robert Merton, first published in 1938 and reprinted in 1970.[4] For Stone, Merton's work takes up a position mediate between the two major prosopographical orientations: elites or small groups versus masses or populations. Though studying an elite, Merton produced a statistically based collective biography of the British scientific community in the making. Other prosopographical studies had emerged as a sort of political history of dynasties or small groups of elites and thus focused (and still focus) on the interests and calculations of single actors, families, patronage networks, and so on. Merton's work moved in the direction of a Smithian political economy, since he studied the emergence of a certain personality type and community – the natural philosopher of the new science – as the result of large-scale socioeconomic processes.

Merton's work suffered from a generation of neglect. Apart from noteworthy exceptions, such as Nicholas Hans's *New Trends in Education in the Eighteenth-Century* (1951), contemporary interest in prosopography in the history of science dates from the early 1970s, coinciding with the republication of Merton's work. The renewal of interest occurred against the background

---

[4] Robert K. Merton, *Science, Technology and Society in Seventeenth Century England* (*Osiris*, 4/2, 1938; reprint, New York: Harper, 1970).

of the rise of social history and the sociology of scientific knowledge. Giving early expression to this prosopographical interest in the history of science were Jack Morrell, Steven Shapin, and Arnold Thackray. The journal *History of Science* provided a central site for the propagation of prosopography. Common to such studies was attention to the notion of the scientific community. Prosopography was also touted by Shapin and Thackray as a means to obviate ahistorical or "Whig" approaches to the history of science.

The pursuit of prosopography has continued since the 1970s. But, fragments aside, the prosopography of eighteenth-century science does not exist, and I do not aspire to write it here. My task is to review its current state and enter a plea for its further pursuit. The fortunes of prosopography seem tied to those of social history and sociology. The latter, and especially the sociology of knowledge, may have faced a crisis in the past ten years or so. The same period witnessed a decline in interest in social history and a coeval rise of cultural history.

In the body of this essay, I survey the state of what we know of the prosopography of eighteenth century science. The fragmentary nature of the studies that have been made is reflected within the structure of the essay itself. I look first at two groups: students and Jesuits. Then I turn to three exemplary national settings – France, Great Britain, and the Austro-German lands – and what can be said about them. I thereafter return to a third group: women. This leads to a section on the notion of the eighteenth-century scientific community: in what sense it existed and whether a prosopography of it might be possible. I close with remarks on the eighteenth-century lineaments of our prosopography and its lack of recent popularity.

STUDENTS

The prosopography of early modern and especially of eighteenth-century students has emerged as a battleground. At the risk of becoming a casualty, I shall try to say something about this population, for attendance at institutions of higher learning remained the royal road into a life in science. I restrict attention to enrollments, social class, mobility, and specialization.

A work by Franz Eulenburg long set a framework for studies of enrollments.[5] Eulenburg surveyed German universities. For the period 1700 to 1790, of the thirty-one universities surveyed, he found a curve fluctuating between 4,300 and 4,000 new enrollments per year up to 1750, followed by a long downward trend, with some fluctuations, to just over 2,900 in 1800. This suggested a dismal eighteenth century for student populations. An equally dismal century, with some temporal variation in recovery, emerged in later studies of

---

[5] Franz Eulenburg, *Die Frequenz der deutschen Universitäten von ihrer Gründung bis zur Gegenwart* (1904; reprint Berlin: Akademie, 1994).

English populations. At Oxford, for example, Lawrence Stone found new enrollments hit bottom at 182 in 1750–9, declining from 316 in 1700–9; in 1790–9, at 245, they were moving to a fluctuating increase. Cambridge too showed declining enrollments until the 1760s and then a slow climb back upward.[6]

A revision of the Eulenburg-Stone view appeared in a volume of essays in 1989, in which Dominique Julia and Jacques Revel disputed the relevance of Stone's results outside Oxbridge.[7] Julia and Revel called for a finer-grained analysis, decade by decade, region by region, faculty by faculty. They found in fact a decrease in France only for the theology faculties after 1740. A study of twelve Spanish universities in the same volume found a more or less steady increase in enrollments over the century, rising from 4,263 in 1700 to 7,585 in 1800. Scottish universities also had increasing enrollments, peaking after 1787. For the Germanies, Willem Frijhoff assailed Eulenburg's work head on. Frijhoff deflated Eulenburg's figures pre-1700, rendering the early modern trend slightly positive; however, new enrollments did drop consistently, from around 4,260 in 1706–15, to about 3,040 in 1796–1805.[8]

It seems that we can say that eighteenth-century university populations declined in some places (England and the Germanies) while increasing in others (Scotland and Spain); attention to faculties (in France), moreover, shows opposing trends – for example, graduations in theology declined while those in medicine increased. The numbers at Jesuit institutions and professional schools remain unclear, so there may or may not have been an absolute decline in students.

Let us turn now to social class. Since the Middle Ages, the student body had fallen into three chief parts: nobles, commoners, and paupers. Students were presumed to be male, legitimately born, and Christian. The Reformation and the Counter-Reformation further defined whole colleges and universities in terms of Christian confessions. From 1500 to 1800, the social composition of the university altered: the numbers of plebeians and paupers seems to have consistently declined, and a new category arose in some places: the gentleman.

[6] See John Gascoigne, *Cambridge in the Age of Enlightenment: Science, Religion and Politics from the Restoration to the French Revolution* (Cambridge University Press, 1989), p. 19; Lawrence Stone, "The Size and Composition of the Oxford Student Body 1580–1909," in Lawrence Stone (ed.), *The University in Society* (Princeton, NJ: Princeton University Press, 1974), 2 vols., 1:3–110, at p. 92; on a similar pattern for Castile, see Richard Kagan, "Universities in Castile 1500–1810," in ibid., 2:355–405.

[7] Dominique Julia and Jacques Revel, "Les étudiants et leurs études dans la France moderne," in Dominique Julia, Jacques Revel, and Roger Chartier (eds.), *Les Universités Européennes du XVIe au XVIIIe siècle. Histoire sociale des populations étudiantes*, 2 vols. (Paris: Éditions de l'École des Hautes Études en Sciences Sociales, 1986–9), 2:25–486; see 354–6.

[8] Willem Frijhoff, "Surplus ou deficit? Hypothèses sur le nombre réel des étudiants en Allemagne à l'époque moderne (1576–1815)," *Francia: Forschungen zur westeuropäischen Geschichte*, 7 (1980), 173–218; see also Frijhoff, "Grandeurs des nombres et misères des réalités: la courbe de Franz Eulenburg et le débat sur le nombre d'intellectuels en Allemagne, 1576–1815," in Julia et al. (eds.), *Les Universités* 1:23–63. The work on Spain is Mariano Peset and Maria F. Mancebo, "La population des universités espagnoles au XVIIIe siècle," in ibid., 1:187–204; on Scotland, A. Chitnis, *The Scottish Enlightenment: A Social History* (London: Croom Helm, 1976), p. 133.

Studies by John Gascoigne on Cambridge and by Stone on Oxford show that these institutions, still serving essentially for those seeking orders, became increasingly identified with a restricted portion of society: the Anglican landowning gentry. At eighteenth-century Oxford, for example, enrollments of peers and barons held steady at about 5 percent; esquires increased from about 10 percent in the early century to 40 percent in 1810; gentlemen made up about 25 percent to 30 percent or more; and plebeians declined, from about 30 percent in 1686, to 10 percent in 1785–6 to only 1 percent in 1810. Part of the decline in plebeians is an illusion, as many of them began styling themselves gentlemen. But there was an absolute decline, due to increasing costs, tougher entrance requirements, and pressures to limit enrollments of plebs, including a tightening job market, thanks to elites penetrating professions once populated by plebs.[9]

Lawrence Brockliss's study of the University of Paris reveals the same trends. During the early modern era, culminating in the eighteenth century, the poor declined in numbers, at least in the faculty of arts. The number of nobles held fairly constant at 10 to 12 percent, though most of these were *noblesse de robe*. The decline of theology students in the French provinces gives evidence for a decline in poor students, as French provincial universities had had small arts faculties to begin with, and poor students seldom made their way into the law and medical faculties. Although no quantitative study exists for the Germanies, evidence also suggests a marginalization of the poor and their subjection to quotas governed by scholarship systems that tended to push them into the clergy.[10] Here we find the emergence of a quasi-caste of middle-class professionals whose sons were most numerous in the student body.

Monika Richarz has studied the case of the Jews, so we may make some observations here. Since the late fifteenth century, Jews might study at Italian universities with explicit papal privileges. By the late seventeenth century, Dutch universities became important sites open to Jewish students. Given the Anglican cast of Oxbridge, Jews were still proscribed from fully matriculating in the eighteenth century; about the situation at the more secularized Scottish universities, I have no information. Taking now the Germanies as a barometer of general trends from Western to Eastern Europe, virtually no Jewish students were at Catholic universities in the eighteenth century, although they began appearing by 1678 at Protestant universities. Counting

---

[9] See Gascoigne, *Cambridge*, pp. 20, 23; Stone, "The Size," pp. 37–46, 93; also Lawrence Stone and Jeanne F. Stone, *An Open Elite? England 1540–1800* (Oxford: Clarendon Press, 1984), pp. 262–6.

[10] On France, see Laurence Brockliss, "Patterns of Attendance at the University of Paris, 1400–1800," in Julia et al., *Les Universités*, 2:487–526, at 503–10; also Julia and Revel, "Les étudiants," p. 349; on the Germanies, Anthony La Vopa, *Grace, Talent, and Merit: Poor Students, Clerical Careers, and Professional Ideology in Eighteenth Century Germany* (Cambridge University Press, 1988), especially chap. 1; also William Clark, *The Hero of Knowledge (Homo Academicus Germanicus)* (Berkeley: University of California Press, forthcoming), chap. 4.

only the nine universities most visited by Jewish students, Richarz found 470 registrations from 1678 to 1800, all in medicine.[11]

If we can generalize to all eighteenth-century Europe, while a small but growing number of Jewish students appeared in parts of Europe, poor students declined in absolute numbers and were steered ever more into the clergy. Despite eighteenth-century sentiments regarding egalitarianism and meritocracy, the route to many or most occupations in science became more governed by social class, and the system of knowledge, as we shall see more later, fell ever more into the hands of a caste of landowners, gentlemen, and professionals.

The last considerations concern social class mobility; now we look at geographical mobility. The trend seems to be toward provincialism in attendance patterns and a restriction of academic peregrination to the aristocratic "grand tour." Of the Germanies, Eulenburg himself noted a drop in enrollments. Frijhoff's general deflation of Eulenburg's pre-1700 figures took up this point. Frijhoff argued that the greater geographical mobility of students pre-1700 resulted in an inflated body count in Eulenburg's figures: one must factor out the multiple enrollments of the peregrinators pre-1700. In a study of Dutch students, he found as well a big drop in peregrinations post-1650. Following the lead of Kagan on Castile, Peset and Mancebo surmised a rise in provincialism in student enrollments for Spain. And the analyses of Julia and Revel showed the same trend toward provincialism in the French universities.[12] Save a few exceptions (for example, Edinburgh, Göttingen, Leiden), eighteenth-century universities seem to have become not more but rather less cosmopolitan in the composition of their student bodies.

The matter of specialization points in a different direction in a few places. Except for Jesuit institutions, enrollments in arts (and sciences) faculties generally declined in early modern Europe. After 1750, the trend reversed at some universities, and the modern notion of the "major" in specific subjects in arts and sciences emerged: the mathematics major, the philology major, and so on. In the Germanies, the seminar system of teaching and the doctorate in philosophy appeared in the second half of the century. Such institutions and practices recast some students into active but elite members of the community, producing knowledge, and tending toward disciplinary specialization.[13]

The course of the century thus finds the population of European students

[11] See Monika Richarz, *Der Eintritt der Juden in die akademischen Berufe. Jüdische Studenten und Akademiker in Deutschland 1678–1848* (Tübingen: Mohr, 1974), especially pp. 13, 26–30, 41, 46, 227–9.

[12] Eulenburg, *Die Frequenz*, p. 136–7; Frijhoff, "Surplus"; Frijhoff, "Grandeurs"; Frijhoff, "Université et marché de l'emploi dans la République des Provinces-Unies," in Julia et al. (eds.), *Les Universités* 1:205–43 at 210; Peset and Mancebo, "La population," p. 196; Julia and Revel, "Les étudiants," pp. 303–35.

[13] On the Germanies, see William Clark, "On the Dialectical Origins of the Research Seminar," *History of Science*, 27 (1989), 111–54; Clark, "On the Ironic Specimen of the Doctor of Philosophy," *Science in Context*, 5/1 (1992), 97–137; on Scotland, see John R. R. Christie, "The Origins and Development of the Scottish Scientific Community, 1680–1760," *History of Science*, 12 (1974), 122–41, at 135–6.

declining in some places and faculties, and growing in others, with absolute numbers perhaps flat, if not falling. The student body appears less egalitarian, less cosmopolitan, more class-conscious and provincial. And in some places, an elite is forming that tends toward disciplinary specialization. It remains to be seen whether the student body reflects the community at large.

JESUITS

Steven Harris is the great contemporary prosopographer of Jesuit science.[14] The Jesuits resisted the modern bifurcation of the self into private versus public or professional parts: giving themselves wholly to the Society of Jesus, Jesuits had no private life, in the modern bourgeois sense. Strangely, that makes them perfect subjects for a prosopography in the spirit of a Poggendorffian lexicon, since Jesuit scientists can be reduced to their professional biographies. Two other characteristics made Jesuits vanguards of modernity: meritocracy and mobility. Let us first remark on how large their shadow loomed over Europe.

Until they were suppressed in France in 1762 and in Spain in 1767 and then abolished by the Pope in 1773 (although they continued in Russia), the Jesuits, if not essentially running the educational systems of Catholic countries, dominated them. By 1700 the Jesuits had more than seven hundred institutions of higher learning, with more than two hundred in Central Europe alone. They also, for example, operated about twenty-five astronomical observatories by 1773 and tended to fund physics cabinets at their institutions to a significantly greater extent than did Protestant states in the eighteenth century. Despite their efforts at accommodation with Protestants and others, the Jesuits oversaw a rival academic community until 1773.

We still do not know enough about Jesuit recruitment and advancement. In the early modern era, the Jesuits were often accused of caring mostly for the wealthy, so one cannot simply exculpate them from the tendency of marginalizing poor students. Still it is conceivable that, more than Protestant states, they steered scholarship students into the academic track of the Society. Meritocracy, as an oligarchy of talent, seems to have emerged in Europe through the Jesuits.[15] As we shall see, meritocracy was definitely not the essential value and character of other early modern groups devoted to knowledge, such as university professors. Given their notion of promotion through proven talent, the Jesuits may have recruited and advanced members in the Society irrespective of social class.

[14] Steven Harris, "Transposing the Merton Thesis: Apostolic Spirituality and the Establishment of the Jesuit Scientific Tradition," *Science in Context,* 3/1 (1989), 29–65; see also John Heilbron, *Elements of Early Modern Physics* (Berkeley: University of California Press, 1982), pp. 93–106.
[15] See David Knowles, *From Pachomius to Ignatius: A Study in the Constitutional History of the Religious Orders* (Oxford: Oxford University Press, 1966), p. 64.

Along with their meritocracy, the extreme mobility of Jesuits stands out. Jesuits as academics were frequently moved. At many institutions, a stay of only about five years or so was not atypical. A Jesuit might, for example, teach for five years at the University of Bamberg, then be moved by the Society to the University of Würzburg for five years or so, then, for the truly talented, moved to headquarters in Rome, and finally, for the worthy, perhaps on to Beijing or elsewhere.[16] This mobility served to break any tendencies toward national or provincial loyalties. The Jesuit was loyal to no particular faculty or university or academy. Jesuit science was cosmopolitan and international in that sense.

However, the Jesuits officially rejected an apparent tendency, as we shall see, among Protestant scholars and scientists: disciplinary specialization. Jesuit professors instead usually rotated through disciplines. Not unlike the British system of regenting, the Jesuit system typically had a professor teaching the same group of students for three years or so, moving with the students from discipline to discipline. In a Smithian sense, this resistance to a division of labor entailed resistance to creation of specific personae: the Jesuit was loyal to the Society and only secondarily or not at all to a discipline or international community of scholars.

But here two qualifications must be made. For a few disciplines, especially for mathematics, the Jesuits encouraged specialization for some scholars at some institutions. Taking the University of Würzburg as a good barometer, from 1700 to 1773, of its twelve mathematics professors, seven taught for three years or less, four were allowed to teach from seven to ten years, and one professor taught mathematics for about twenty years.[17] The best minds were often moved to Rome to pursue scientific work. Indeed, in advance of Protestant systems, the Jesuits set up sabbaticals: proven scholars obtained leave from teaching for two to six years to pursue and publish academic work.

For the period 1600 to 1773, Harris ascertained sixteen hundred Jesuits publishing on scientific subjects with a core group of two hundred who published seven or more items. Harris located a boom in Jesuit publications from 1740 to 1773. Jesuits still published much on their speciality, Aristotelian natural philosophy, but also as much on astronomy, mathematics, and modern natural and experimental philosophy. Harris and Heilbron see some

---

[16] On Jesuit mobility through institutions and disciplines, see L. W. B. Brockliss, *French Higher Education in the Seventeenth and Eighteenth Centuries: A Cultural History* (Oxford: Clarendon Press, 1987), pp. 46–7; Heilbron, *The Elements*, pp. 96–7; and, for example, Fritz Krafft, "Jesuiten als Lehrer an Gymnasium und Universität Mainz und ihre Lehrfächer . . . 1551–1773," *Beiträge zur Geschichte der Universität Mainz*, 2 (1977), 259–350.

[17] See Karl-Heinz Logermann, *Personalbibliographien von Professoren der philosophischen Fakultät der Alma Mater Julia Wirceburgensis . . . 1582–1803* (Diss. Med.: Erlangen-Nuremberg, 1970), p. 9; also Winfried Stosiek, *Die Personalbibliographien der Professoren der aristotelischen Physik in der philosophischen Fakultät der Alma Mater Julia Wirceburgensis von 1582–1773* (Diss. Med.: Erlangen-Nuremberg, 1972); Gudrun Uhlenbrock, *Personalbibliographien von Professoren der philosophischen Fakultät der Alma Mater Julia Wirceburgensis von 1582 bis 1803* (Diss. Med.: Erlangen-Nuremberg, 1973).

reconciliation of Jesuit science with "Protestant" science at the level of intellectual history.

But at the level of social history, we see two sorts of scientific communities in eighteenth-century Europe. Prosopographically, we can speak of a population of Jesuit scientists and their transnational but closed community, set against a population of Protestant scientists who espoused a cosmopolitan ideology of science but were actually essentially embedded in national or provincial communities. Such an opposition is overly simple, especially in view of other Catholics, atheists, vagabonds, and sundry sorts dwelling in the interstices. But the prosopography nonetheless suggests two essential and disjoint communities up to 1773.

EUROPEAN NATIONAL AND PROVINCIAL COMMUNITIES OF SCIENCE

A work by John Gascoigne allows some observations here.[18] As a good prosopographer, he delimited his population: Europeans and Americans in the *Dictionary of Scientific Biography* with birthdates from 1660 to 1760, producing 614 names. With all such delimitations, an element of the arbitrary enters as, in this case, one is drawing from a work dealing with essentially only the "most significant" individuals. Gascoigne's use of these data could, however, probably be generalized, with some caveats. Three of his results are of particular interest to us: the national affiliation of scientists, the seeming tendency to specialization during the century, and the displacement of the center of production outside the universities.

Gascoigne's results show that, of this population, more than 70 percent were born in three lands: France (30 percent), Great Britain (26 percent), and the Austro-German provinces (16 percent). Of the 188 (31 percent) who were professors or held similar teaching positions, Gascoigne found the following for the nationality of the final position held: 26 percent Austro-German, 19 percent French, 17 percent Italian, 15 percent British, with all other lands at 6 percent or less. The surprise in this list is France, which traditionally has not been associated with a vibrant university tradition in the early modern era. The lead of the Austro-German professoriate accords with received wisdom on the academic culture there; the placement of the Italian over the British professoriate also accords with received views.

Taking the professoriate as middle-of-the-road, if not conservative, in respect of institutional innovation, Gascoigne's results on specialization seem unobjectionable. He shows the rise during the century of specialized chairs or slots for specific natural sciences. By 1800, if we do not find the scientific

[18] John Gascoigne, "The Eighteenth Century Scientific Community: A Prosopographical Study," *Social Studies of Science*, 25 (1995), 575–81.

communities of each land divided into transnational subcommunities of mathematicians, physicists, and so on, we do find the body of natural philosophers and natural historians falling into the institutional hands of an academic division of labor.

Of the entire population of 614, more than 70 percent were educated at universities or like institutions. This indicates that having been a student remained the surest means of becoming a (significant) scientist. Gascoigne found, however, that 69 percent of his 614 scientists did not serve as professors or in a like capacity. Many questions arise here that the brevity of his article did not let him address. One would like to know, for instance, whether a higher percentage of the Jesuit population served as professors; indeed, one would like to know what percent of the 614 were Jesuits. In any case, Gascoigne's figures suggest, again according with received wisdom, that the productive center of science did not lie in the hands of the eighteenth-century professoriate, except in the Austro-German lands (and perhaps neglecting the Jesuits). We have a scientific community in which nonteaching "academicians" and "amateurs" played a big role.

From our prosopographical perspective, by "amateur" we mean one who pursues science as an avocation – that is, usually without remuneration – as opposed to a vocation – that is, usually with remuneration. In this sense we could say that someone might pursue science in one forum as an amateur and in another as a professional. As we shall see, few individuals in the eighteenth century seem to have been able to make a living solely by pursuing science: being a "scientist" was not a profession or vocation in the modern sense. My use of the terms "amateur" and "professional," like my use of "scientific community," is thus anachronistic in part, as later sections below will show. Nonetheless, I shall use them here.

Let us now make a technical distinction between a society and an academy. I shall use "academy" to refer only to institutions, such as the Académie des Sciences in Paris, for which at least the "ordinary" members received remuneration. To be an academician in this sense was a vocation. I shall use "society" to refer to institutions, such as the Royal Society in London, for which ordinary members received no (significant) remuneration. To be a member of a society was an avocation, and, in view of such a forum, we could say that all members appeared there in the persona of the amateur or lover of science. This nonpejorative, context-dependent sense of "amateur" helps illuminate the egalitarian ideology, insofar as it existed, in eighteenth-century societies. And, as the work of James McClellan shows, the eighteenth was the century of the scientific society.[19] In our sense, it was the Golden Age of the scientific amateur. Many people pursued science but, even of those remunerated, almost none could make a living.

---

[19] James McClellan, *Science Reorganized: Scientific Societies in the Eighteenth Century* (New York: Columbia Unversity Press, 1985).

Gascoigne's results on national affiliation may be skewed by some bias in his source. Nonetheless, I shall restrict attention to the three leading national regions as illuminated by his article: France, Great Britain, and the Austro-German lands. It would be nice to include other regions, such as Scandinavia or the Balkans or North America; but lack of space and especially of knowledge precludes me from so doing. In any case, sufficient diversity obtains between the three "leading" lands to warrant belief in some generality for the sections that follow.

FRANCE

Before turning to amateurs and academicians, let us quickly look at the non-Jesuit French professoriate. Laurence Brockliss's work offers an institutional overview of French higher education.[20] The Jesuitical heritage facilitated an early bureaucratization of appointments: a meritocratic system. Secular professorships in colleges and lower faculties of arts and sciences were officially filled by advertising the position and then testing the applicants via an exam, the *concours*. A faculty board determined the results and voted on the appointment. Except for the law faculty, this method was pretty much the rule. But the prosopographer would like to know whether de facto castes or dynasties emerged: to what extent did modernizing methods, such as state exams, break the hold of dynasties and classes over occupations? Did examination replace patronage and nepotism but favor the same old faces?

Professional faculties – theology, law, and medicine – could pay decent salaries; but salaries in colleges or arts faculties were in general too modest to support a lifetime occupation. A ten-year stint as college or arts and sciences professor was about the maximum, so turnover was great. The low salary would help explain why the non-Jesuit French professoriate did not form part of the core group for the pursuit of science there (if that was the case, as it seems to be).

Let us turn now to amateurs and confine ourselves to Daniel Roche's monumental study of French provincial "academies," most of which were societies in our sense. Roche uncovered about six thousand society members, of whom, roughly put, 37 percent were nobles, 20 percent higher clergy, and 43 percent commoners (*roturiers*). But these commoners were not so common. And, as the societies adopted a policy of restricted membership, with a hierarchy of honorary, ordinary, and associate members, more needs to be said. The nobles constituted 71 percent of honorary members in the provinces. For ordinary members, the majority, 61 percent, were still noble, whereas 37 percent were bourgeois. A majority of the associates, at the third level in the

---

[20] Brockliss, *French Higher Education*, especially pp. 39–48.

hierarchy, was drawn from the bourgeoisie (55 percent), primarily from the ranks of landowners, bureaucrats, physicians, lawyers, clerics, professors (18 percent), and other professionals and gentlemen. Indeed, more than three-fourths of the bourgeois group bore one of the three "black robes": theology, law, and medicine. Fewer then 4 percent of bourgeois members hailed from the ranks of merchants, manufacturers, and craftguilds. "The academy [society] is a phenomenon of an elite culture, of which the members, moreover, have a clear awareness."[21]

Roche exhibits this important population of amateurs of science as embedded in the patriciate and professional community. Here "who the guys were" were those who ran the town. And they were essentially "guys." Unlike the salon, to which we shall return, the society served as a homocentered site. It was, moreover, a site for certifying the worth of middle-aged men, as the great majority entered between their thirties and fifties. Roche notes that the pursuit of science as avocation still resonated with notions of the liberal arts, of *otium* versus *negotium:* leisure versus business. The tie of leisure and nobility facilitated the notion of an aristocracy of knowledge, insulated from mercantile cares. This, as he notes, made the inclusion of merchants and manufacturers especially problematic in France.

Roche sees the eighteenth-century society as a midpoint between polymathic Renaissance societies or cults and modern specialized scientific disciplinary communities, even though such societies remained polymathic in scope. Indeed, nonspecialization was part of the egalitarian ideology of such groups, insofar as it existed. Only very late in the century, if at all, did the society provide impetus toward a division of labor in science. The prosopography of such amateurs indicates the tangled web of aristocratic and egalitarian motifs in the nascent scientific community, at once provincial and international. Although replicating in a more-or-less uniform mold across France, thus offering some sense of a national community of savants, French societies, given Roche's prosopography, remained reflections of local or provincial society. In this regard, Roche returns to the ever-present specter of Paris over the French provinces.

"A professional bureaucrat could no longer be confused with the cultural polymath. . . . His position was conveniently linked to his functional role in the state, rather than the economic fruits of his labor. The existence of an academy of specialists [in Paris] once again reinforced his [the academician's] profoundly elitist values." Hahn stresses the nature and role of the academician as bureaucrat and expert thanks to his salary. In the eighteenth century, the distinction of the amateur from the professional emerged most clearly

---

[21] Daniel Roche, *Le siècle des lumières en province: Académies et académiciens provinciaux, 1680–1789* (Civilisations et Sociétes 62), 2 vols. (Paris: École des Hautes Études en Sciences Sociales, 1978), 1:90; see also 1:52–4, 197–8, 207–8, and 2:passim.

among academicians, and, by then-contemporary lights, the Parisians were *la crème de la crème*. Works by Roche, Hahn, and David Sturdy support this observation.[22]

Entry to the Parisian Académie des Sciences was via nomination by an academician, and then appointment by the king, who at times imposed his will. Social origin supposedly did not matter; but the low-born as well as most from manufacturing and mercantile backgrounds were essentially excluded. It would be nice to know how international recruitment was, but I know of no such figures for the century. Roche found that about 45 percent of the academicians came from the nobility. Of the "honorary" members, 80 percent were noble, while the third estate held 64 percent of the "pensioners" and "associates." Musing on the social composition of the other two great academies in Paris – l'Académie française and l'Académie des inscriptions – Roche noted that, while letters and history remained preserves of the nobility and higher clergy, the sciences were emerging as bourgeois. In a backhanded way, as he observed (as did Hahn earlier), academicians of science constituted themselves as an elite, an oligarchy, governed by meritocracy and specialization.[23]

Hahn also remarked that (as modern bureaucrats) academicians did not act like traditional occupational groups, such as craftguilds or academic faculties. Not only did academicians seldom intermarry, but they also seldom witnessed one another's weddings. In general, they did not socialize with one another outside the academy – and, indeed, as amateurs in societies – tended to do. Sturdy's work offers one great exception to this: until at least 1750, the incidence of nepotism rose. In our sense, except for this last matter, the academy was the social antithesis of the society. Hahn has further shown that academicians in Paris were not so well off. During the eighteenth century, the academy's budget did not keep pace with the increase in members. Salaries declined in absolute terms. Given the academy's hierarchy – adjuncts (two), associates (three), pensioners (three) – in each of its six specialized sections, achieving seniority meant that a new adjunct had to wait for five elders to pass on before becoming a pensioner, with the nice salary of 3,000 *livres,* instead of 1,800 to 1,200. Hahn noted that academicians did not form an occupational group in this further sense, since most of them were driven to make money elsewhere, as professors or administrators of institutes, or as military or technical advisers.

---

[22] The citation is from Roger Hahn, *The Anatomy of a Scientific Institution: The Paris Academy of Sciences, 1666–1803* (Berkeley: University of California Press, 1971), pp. 51–2; see also pp. 69–71, 80–1, 108; see also Hahn, "Scientific Research as an Occupation in Eighteenth-Century Paris," *Minerva,* 13 (1975), 501–13, especially 504–5. See also David Sturdy, *Science and Social Status: The Members of the Académie des Sciences, 1666–1750* (Bury St. Edmunds: Boydell Press, 1995).

[23] Roche, *Le siècle,* pp. 285–90; Sturdy, *Science,* pp. 376–8, 399–401, 414. The appendixes here at pp. 427–32, for 1702–50, list a good number of academicians of unidentified origins but, of those listed, show very few born outside France.

The spirit of research for the furtherance of the rational understanding of nature – which is my definition of scientific activity – neither coincided completely with the needs of the society of the *ancien régime,* nor was it encouraged on the scale required to create a professional class of scientists.[24]

If a professional class of scientists did not exist in eighteenth-century Paris, it existed nowhere.

GREAT BRITAIN

We noted earlier that the return of prosopography in the history of science emerged in works by Morrell, Shapin, and Thackray. These three authors concerned themselves essentially with the British context. And, in regard to the eighteenth century, the focus of their relevant works in the 1970s can be resolved to a tale of two cities: Edinburgh and Manchester. But let us first look very briefly at two other cities not unknown for learning.

Oxbridge remained essentially clerical. This produced a situation not unlike that for arts and science professors in the French secular universities: Oxbridge fellows as such did not tend to identify themselves with the production of science or learning. A fellowship rather provided a basis for making an extramural clerical career. The clerical cast of Oxbridge seems to have led modern scholars to focus on the politics instead of the political economy of knowledge. We know much about the politics – Whig versus Tory – of the universities and by implication many of their fellows; but social origins seem much neglected. As for the professors at Oxbridge, I know of no prosopographical studies on them and their social history.[25]

Let's now move to Scotland. Most agree that the Scottish professoriate comprised the core group in the eighteenth-century scientific community there. With the abolition of regenting at Edinburgh in 1708, a system of professorial chairs was introduced, and this system soon spread throughout Scotland. Academic aspirants, however, typically sought proficiency in multiple disciplines, thereby improving their chances for several chairs. Professors also often moved from chair to chair for larger salaries, fees, or even perhaps interest. Appointments to Edinburgh's chairs lay mostly in the hands of the Crown or Town Council. Morrell indicates, for Edinburgh at least, that a science professor was typically a native Scot, of at least middle-class origin, had studied at Edinburgh, and was often related to someone in the faculty. Nepotism was rife, with dynasties around the Gregories, Monros, and Stewarts. Politics and patronage also played a part. Shapin and Peter Jones have

---

[24] Roger Hahn, "Scientific Research," p. 512. See also Sturdy, *Science,* pp. 413–14.

[25] See Gascoigne, *Cambridge,* especially pp. 12, 15, 187–8; there is not the slightest attention paid to prosopography or even social class in L. S. Sutherland and L. G. Mitchell (eds.), *The Eighteenth Century,* vol. 5 of *The History of the University of Oxford* (Oxford: Clarendon Press, 1986).

stressed the role of the landowning elite as "patrons and partners," often antagonistic.[26]

At Edinburgh, omitting theology and law, four chairs existed pre-1700; fourteen more were created from 1708 to 1790. Other Scottish universities showed a similar pattern but fewer chairs. Taking the second half of the century and considering medicine, mathematics, astronomy, philosophy, and natural sciences, only forty individuals were professors at Edinburgh. Salaries ranged from £128 for botany to £113 for mathematics to zero for chemistry. Moral philosophy and natural history were well paid at £100, whereas Natural Philosophy had but £52. Such differences were an incentive to professors to change chairs.[27]

Given the system of chairs, institutional history might see the rise of specialization; but prosopography shows perpetuation of older practices. Publishing patterns might indicate a tendency to specialization and perhaps in accord with disciplines institutionalized by the chairs. Good Smithians usually, prosopographers, however, pay attention to the salary structure, and that put a brake on the creation of the modern scientist as professor, whose research tends to lie in the field of teaching. And one of the famous universities of the age amounted to a rather small community, bound by ties of blood not only spilt in faculty meetings. Scottish universities remained complex and interrelated moral communities, not unlike craftguilds. Here, as in traditional societies, the private life remained fused with the public or professional life.

Great Britain awaits its prosopographer of eighteenth-century societies. Except for Manchester and Edinburgh, few societies seem to have been studied. Michael Hunter's study of the Royal Society in London ends, alas, in 1700. He wrote, "[I]n statistical analysis of the Society's membership hitherto less attention has been paid to the occupations and social class of its supporters than to their political and religious affiliations . . . " As noted, the same seems to be the case for studies of Oxbridge. Gascoigne suggests that the eighteenth-century Royal Society was dominated by gentlemen and the landed classes. Given Hunter's results up to 1700, that seems reasonable; moreover, it shows the Royal Society of London to resemble French societies.[28]

Developing from the Medical and Philosophical Societies of the 1730s, the Royal Society of Edinburgh (RSE) of 1783 was a compromise between the

[26] See Chitnis, *The Scottish Enlightenment*, pp. 124, 132–5, 153–4; J. B. Morrell, "The University of Edinburgh in the Late Eighteenth Century: Its Scientific Eminence and Structure," *Isis*, 62 (1971), 158–71, at 160–4; Peter Jones, "The Scottish Professoriate and the Polite Academy, 1720–46," in Istvan Hont and Michael Ignatieff (eds.), *Wealth and Virtue: The Shaping of Political Economy in the Scottish Enlightenment* (Cambridge University Press, 1983), pp. 89–117, at pp. 91, 99, 111, 116–17; and Steven Shapin, "The Audience for Science in Eighteenth Century Edinburgh," *History of Science*, 12 (1974), 95–121; see also Christie, "The Origins."

[27] Morrell, "The University," pp. 162–5.

[28] Quotation from Michael Hunter, *The Royal Society and Its Fellows, 1660–1700: The Morphology of an Early Scientific Institution* (London: British Society for the History of Science, 1982–5; 2nd ed. 1994), p. 25; see also Gascoigne, *Cambridge*, p. 283.

Edinburgh professoriate and the landed literati. The founding fellows (about 150) included all professors of the University of Edinburgh and most from the other Scottish universities, along with a fair mixture of barons, ministers, clergy, lawyers, physicians, politicians, peers, and landed gentry. "[T]he RSE was bound to be at its inception very much an *ex officio* society, admission to its ranks being gained by status and not necessarily by intellectual achievement."[29] It was neither "a young man's society" nor, in view of its members, reflective of an espoused ideology of egalitarianism and meritocracy in the Republic of Letters. Shapin brought out the local and provincial dynamic that drove this supposed organ of a national and even international scientific community in the making. Like Roche's provincial French societies, Shapin's RSE was embedded, in view of its prosopography, in local and provincial culture, here that of Edinburgh and Scotland.

And what of our second city? By the end of the eighteenth century, as Shapin has noted, Britain may not have been Manchester, but it was on its way. In the 1780s and 1790s, "literary and philosophical" societies sprang up in British industrial centers, offering a new sort of society rather different from Roche's French provincial and Shapin's royal Scottish. Thackray has studied the Manchester Literary and Philosophical Society, founded in 1781. This society served to legitimate marginalized men – entrepreneurs and technicians – largely excluded from the society movement. In 1799–1803, nearly half of the Manchester society's twenty-six members were merchants and manufacturers, and only one was a gentlemen. "The new Manchester elite had little sympathy for honorable birth and hereditary wealth. The idea of a limited democracy of intellect and effort had greater appeal"; but natural knowledge here was to a striking extent "the private cultural property of a closely knit, continually intermarrying, almost dynastic elite . . . "[30] It was a new, ungentlemanly elite. Here, too, a prosopography of the amateurs of science embeds them in a local context and exhibits this seeming emblem of modernization as much like a traditional occupational group as was the Scottish professoriate.

THE AUSTRO-GERMAN LANDS

Let us begin with a story. In 1746 Freiherr Josef von Petrasch founded the "Society of the Incognito" in Olomouc. Since he did so with the consent of Empress Maria Theresa, it is unclear to whom the society was unknown. Three years later, Imperial Count von Haugwitz commenced laying plans for an academy of sciences in Vienna, now of notables. Enlisting the help of Petrasch,

---

[29] Steven Shapin, "Property, Patronage, and the Politics of Science: The Founding of the Royal Society of Edinburgh," *The British Journal for the History of Science,* 7 (1974), 1–41, at 37; see also Shapin, "The Audience," especially pp. 100, 110.

[30] Arnold Thackray, "Natural Knowledge in Cultural Context: The Manchester Model," *American Historical Review,* 79 (1974), 672–709, citations at 687 and 698.

he drew up a plan, on 5 January 1750, for an Austrian or Imperial Academy of Sciences in Vienna, to be modeled on the Parisian and Petersburg academies. The plan envisaged thirteen honorary members, selected per custom from the ranks of the nobles. There would be a president of the academy, definitely of blue blood. Under him would be two secretaries and the core of thirty ordinary and salaried academicians, sixteen of whom had to be pensioners paid a hefty salary. Next would follow ten adjuncts, whose remuneration was left open. Money should also be planned for four veterans and for about sixteen students attached to the academy. And about twenty to twenty-four corresponding members could be taken on. As opposed to this latter group, others had to be physically present in Vienna, Catholic and subjects of Habsburg lands.

Petrasch had a hard time figuring out where to find cash. A monopoly on the calendar or something else would bring in some funds, but probably not enough. That meant a subvention from the treasury would be needed. Supreme Treasurer and Imperial Count von Khevenhüller and other ministers objected here. They said that academicians tended to spend their time on projects of no use to the state. And were such an academy to be instituted, it must not compare adversely to the one in Berlin. Alas, this would mean buying expensive talent from abroad.

The matter of the academy was left "under review." Maria Theresa lent the virtual academy more *realitas* in 1775 when, on 25 January, she recalled to her subjects' minds that it was still on hers. Two new plans, it seems, had been presented to her in 1774. But despite the new plans of 1774 and the empress's mental state of 1775, cold cash did not simply sprout at night like mushrooms. As she remarked, on 25 November 1775, mindful of the only talent at court, "No way could I consider beginning an academy of sciences with [only] three ex-Jesuits and a single, even if valiant, Professor of Chemistry."[31]

From a prosopographical perspective, this virtual academy in Vienna has long been my favorite.[32] For the real ones – founded in Berlin (1700), Göttingen (1751), Erfurt (1754), Munich (1759), Mannheim (1763), and Prague (1784) – seem poor or paltry. Those in Munich and Prague were essentially societies in our sense, while those in Göttingen and Erfurt were actually university institutes. Only the academies in Berlin and Mannheim were academies of science in the sense of the Parisian. In any case, I know of no prosopographical studies of any of them, though lists of members have been published for some.[33]

---

[31] See Josef Feil, "Versuche zur Gründung einer Akademie der Wissenschaften unter Maria Theresa," *Jahrbuch für vaterländische Geschichte*, 1 (1861), 319–407, quotation at 382.

[32] Unless otherwise noted, this section is based on Clark, *The Hero*, and Clark, "On the Ministerial Archive of Academic Acts," *Science in Context*, 9/4 (1996), 421–86.

[33] See, for example, Georg Wegner (ed.), *Die Königliche Böhmische Gesellschaft der Wissenschaften, 1784–1884. Verzeichnis der Mitglieder* (Prague, 1884).

The Berlin academy was the most famous.[34] It is difficult to say how many members were salaried and, moreover, how many actually received their pay. Supposedly governed by meritocratic, republican, and democratic principles, elections to the academy seem to have been often rigged by an oligarchy within. From 1746 to 1786, the salad days of the academy, Friedrich the Great used French advisers (Maupertuis, d'Alembert, and Condorcet) in deciding whom to admit. Recruitment was international but, given the whims of the king and advisers, favored the French and Swiss while looking askance at Germans and Jews. Although nominated by a majority of the academy, Moses Mendelssohn was rejected by the king. Attempts to get Markus Hertz into the academy also failed. I think that no other Jews passed muster in the eighteenth century, and I doubt that practicing Catholics found favor either. The "Philosopher King" wanted a Prussian–Paris academy; he got a pale and poor imitation.

The academy of sciences in Munich was rather more a society.[35] Almost no one seems to have been paid until 1806, except for a few of the Protestants admitted. Although such admissions were undertaken to oppose the influence of the Jesuits in Bavaria, no Protestants were admitted at first. Of the twenty-five or so original ordinary and associate members, twelve were Benedictines, four of them professors in Salzburg, and another twelve were canons. Regular and high secular clergy formed the staff here, to whose ranks a healthy number of laymen with a "von" later appeared. The wanting prosopography of this group would doubtless reveal a Bavarian version of the English Royal Society.

As I know of no prosopographical studies of Austro-German societies, I shall look at Karl Hufbauer's study of the German chemical community.[36] Hufbauer sets at stage center Lorenz Crell's *Chemisches Journal . . .* (later *Chemische Annalen*), whose first issue appeared in 1778. The journal was, in effect, a society reduced to its corresponding members. That shifted the center of gravity from a local site but not yet to an international community. Hufbauer argues that, through this journal, German subscribers began to think of themselves as chemists in a professional sense and as German chemists to boot. The latter emerged most poignantly after 1789 in the face of a common foe: the new "French" chemistry of Lavoisier. Hufbauer also presents a center-margin analysis. Of the 146 contributors to the journal from 1784 to 1789, he finds a core group of 8. Hufbauer shows that this sort of scientific community,

---

[34] The standard source is still Adolf Harnack, *Geschichte der Königlich Preussischen Akademie der Wissenschaften zu Berlin* (Berlin: Reichsdruckerei, 1900), 4 pts. in 3 vols., with lists of members (1700–1812) in 1/1:242–4, 465–81; 1/2:645–54.

[35] See Ludwig Hammermayer, *Gründungs- und Frühgeschichte der Bayerischen Akademie der Wissenschaften* (Münchener Historische Studien, Abt. Bayerische Geschichte, 4) (Kallmünz: Lassleben, 1959).

[36] Karl Hufbauer, *The Formation of the German Chemical Community (1720–1795)* (Berkeley: University of California Press, 1982).

unlike the provincial society, was a bourgeois phenomenon, as about 90 percent of Crell's German subscribers were middle class, and Protestant as well.
It would be nice to know whether greater numbers of nobles subscribed to
journals of physics or natural history, but I know of no such studies.

I turn now to the scientific center in the Austro-German lands: the professoriate. From 1700 to 1789, the Austro-German lands had about forty-five
institutions with the status of universities. After the Reformation, a system
of disciplinary chairs or ordinary professorships developed. Small salaries led
to pluralism at many places, so that, for instance, the University of Altdorf in
1750 had only four arts and sciences professors, who held all the relevant chairs
among them. Ideally an academic began as lecturer (*Adjunkt* or *Dozent*) and
then, perhaps, became an extraordinary and finally, for the fortunate, an ordinary professor or chairholder, the only ones with a guaranteed salary. Since
the various salaries of chairs were set by statute, professors had to switch chairs
and faculties to gain higher salaries. Despite institutional appearances, the
salary and promotion structure worked against specialization in the division
of academic labor, as we have seen in the case of Scotland, whose academic
culture much resembled the German.

As a consequence of the Catholic-Reformation, Austrian arts and sciences
faculties and most German Catholic ones had fallen to the Jesuits, until their
suppression in 1773. After 1784 in the Austrian lands, chairs were supposedly
filled by examination, along the model of the French *concours*. As in France,
we see movement toward a bureaucratic meritocracy in the aftermath of a
Jesuitical past. German Catholic lands tended to fall in line with the Protestant ones after 1773.

German Protestant faculties, when they had a say about appointments,
weighed collegial and personal matters as much as, if not more than, impersonal and disciplinary. It counted for much if one was a graduate of the university. After 1648, religious confession supposedly did not matter but really
did. The professoriate remained a nationally endogamous body, if not an intramurally incestuous one. Nepotism worked everywhere. Its extent remains
partially hidden by a prosopographical failure to record women's maiden names.
Two interesting family trees of academic dynasties at Tübingen have been
published. From the sixteenth to the eighteenth centuries, in the Burckhardt-
Bardili family, ten professors' daughters can be found, and every one married
a professor at Tübingen. And in the Gmelin family in the eighteenth and nineteenth centuries, of eleven professors' daughters, nine married professors.[37]

The German academic cast(e) was bound with a larger one, more or less
analogous to that of Roche's French provincial societies. At the small University of Rinteln, professors formed an endogamous group with regional minis-

---

[37] In H. Decker-Hauff et al. (eds.), *500 Jahre Eberhard-Karls-Universität Tübingen: Beiträge zur Geschichte
der Universität Tübingen*, 3 vols. (Tübingen, 1977), 3:138–9, 168–9.

ters, pastors, and bureaucrats. Of the 171 professors altogether during Rinteln's history (1621–1809), 68 had clear blood or marital relations, and deeper relations have not yet been probed. In a prosopography of the professors at the middle-sized University of Marburg from 1653 to 1806, Hermann Niebuhr found the same pattern: with local ministers, bureaucrats, pastors, and others of this ilk, Marburg professors constituted a near-caste. No academics with farmers as fathers were to be found, and only 7.4 percent had fathers from craftguilds. Niebuhr found, moreover, that a full one-third of the professoriate in 1806 could trace its lineage by blood or by marriage all the way back to 1653. Famous cases of low-born boys in the German professoriate exist, but they seem few. There were no women, and the few Jews, initially all to teach "Oriental" languages, seem to have had to convert.[38]

During the mid- to late century, some Protestant states sought to rationalize appointments. Sovereigns had acquired not only the right to confirm faculty appointments but also the ability to make their own. Enlightening sovereigns in Berlin and Hanover, for example, endeavored to break the nepotistic bent of faculties, at least officially. After mid-eighteenth century, service and merit – with the latter usually demonstrated by publications and even offers from other universities – were to be the future keys to academic offices. The Hanoverian University of Göttingen proved trend-setting here, although much of its faculty turned out to be interrelated.[39]

Before this rationalization of academic life, German faculties, like the Scottish, behaved in the manner of traditional occupational groups: as complex moral communities. Some have seen "modernization" in the transformation of occupational groups from complex moral communities, in which public and private life were fused, into mere workforces in which professional life clove itself from the private sphere.[40] The formation of modern bureaucracies lay essentially in this transformation. In this sense, their traditional nepostistic bent aside, the University of Göttingen and the Parisian academy of sciences moved at the forefront of bureaucratic modernism and set the antithesis to the Jesuits' as well as to women's view of the academic and scientific community.

---

[38] See Hermann Niebuhr, *Zur Sozialgeschichte der Marburger Professoren, 1653–1806* (Quellen und Forschungen zur hessischen Geschichte, 44) (Darmstadt/Marburg: Hessische Historische Kommission, 1983); on Rinteln, see Gerhard Schormann, *Academia Ernestina. Die schaumburgische Universität zu Rinteln an der Weser (1610/21–1810)* (Academia Marburgensis. Beiträge zur Geschichte der Philipps-Universität Marburg, 4) (Marburg: Elwert, 1982), pp. 198–200; on Basel, Gießen and Marburg, see also Friedrich W. Euler, "Entstehung und Entwicklung deutscher Gelehrtengeschlechter," in Helmuth Rössler and Günther Franz (eds.), *Universität und Gelehrtenstand 1400–1800*, (Deutsche Führungsschichten in der Neuzeit, 4) (Limburg/Lahn: Stärke, 1970), pp. 183–232; Richarz, *Der Eintritt*, pp. 22–3.

[39] On such issues, see Clark, "The Ministerial."

[40] On the transformation of occupational groups, see Marc Raeff, *The Well-Ordered Police State: Social and Institutional Change through Law in the Germanies and Russia, 1600–1800* (New Haven, CT: Yale University Press, 1983).

WOMEN

In the eighteenth century, with students and Jesuits, women constituted an extraordinary group and sort of person in the nascent scientific community. The consideration of women will lead nicely to the next section, on the problem of delimiting the eighteenth-century scientific community itself. Here we rely much on a work by Londa Schiebinger.[41]

In Chapter 2 of *The Mind Has No Sex?* Schiebinger looks at "noble networks" in science. Until the late eighteenth century, well-born women played a role in science, as authors, translators, correspondents, patrons, and founders of academies. In Paris and those parts of Europe under its cultural sway, the salon emerged as a key site of the enlightened intellectual community, which we can only with difficulty prise apart from the scientific community. Unlike the society, the salon was a heterosocial site; it was also at first managed by well-born women.[42] Such salons could be found only, however, in those few parts of Europe in which significant numbers of aristocrats lived together in cities instead of in the country.

As the century wore on, an *embourgeoisement* took place. While maintaining its aura, the salon became detached from the nobility. In the second half of the century, Jews had become "salonfähig," able to be received in a salon. Deborah Hertz has studied the case of Berlin, 1780 to 1806, where Jewish woman not only participated in but also managed salons. Hertz constructed a collective biography of 417 intellectuals, of whom she found 100 participating in salons. Of this number, 38 were noble, 42 middle-class gentiles, and 20 Jewish. Only in the last group did the number of women (12) exceed that of men (8). Except for the entry by Jews, the social composition of the males in salons matches that of Roche's French societies. Most Berlin salon males were nobles, gentry, professors, and officials; only 4 percent were merchants.[43]

Dena Goodman has written, "Enlightenment salons were working spaces . . . which took play as their model."[44] In moving from noble networks to salons, we see the influence of the aristocratic ethos of leisure, stressed by Roche as still central in the provincial societies. Salons cultivated knowledge within the framework of the ancient notion of liberal arts, antithetical to mercantile values. More than the societies, salons (like the Jesuits) resisted the bourgeois separation of public and private. Indeed, as Hertz relates, the salon was a site of real sociability, from which friendships, affairs, and even marriages

---

[41] Londa Schiebinger, *The Mind Has No Sex? Women in the Origins of Modern Science* (Cambridge, MA: Harvard University Press, 1989); also see her article in this volume.

[42] Schiebinger, *The Mind*, pp. 30–2, 37–65, 153; also Dena Goodman, *The Republic of Letters: A Cultural History of the French Enlightenment* (Ithaca, NY: Cornell University Press, 1994), especially pp. 73–89.

[43] See Deborah Hertz, *Jewish High Society in Old Regime Berlin* (New Haven, CT: Yale University Press, 1988), especially pp. 20, 114–18; also Richarz, *Der Eintritt*, p. 7.

[44] Goodman, *The Republic*, p. 74.

resulted. To that extent, salons resembled complex moral groups, in the manner of faculties and guilds. The salon, as a place of working play or busy leisure, fused home and workplace, something that was typical for the nobility.

In Chapter 3 of *The Mind Has No Sex?* Schiebinger looks at women at the lower end of the social scale: the craft tradition. Because the early modern artisanal class, like the nobility, did not really separate home and workplace, gender roles were more fluid there. In many places, women might be full members of guilds and, more important, run their shop or craft, in the absence of a husband. Engraving, computation, and observation were essential crafts or skills underlying the new science, and women in the eighteenth century can be shown to have participated with such skills. In the case of astronomy, Schiebinger shows how the craft tradition of computation and observation went seamlessly into the theoretical tradition.

In Chapter 9 of *The Mind Has No Sex?* Schiebinger returns to considerations of Chapters 2 and 3 and brings us to the fruits of the eighteenth century: "Two developments – the privatization of the family and the professionalization of science – changed women's fortunes in science."[45] These developments essentially removed women or rendered them invisible assistants in the now private sphere of the home. As in the case of poor students, the eighteenth century led to the marginalization of women. By 1800, science had come into the hands of gentlemen, professionals, and middle-class men, and they separated home and workplace, public and private. This was the new scientific community of the modern era.

THE SCIENTIFIC COMMUNITY OF THE EIGHTEENTH CENTURY

Did the scientific community exist in the eighteenth century? And, prosopographically, can we say anything about it in general? Or is the very notion an anachronism improperly applied to early modern science? Prosopographers of the 1970s offered this very method, and its concept of the "scientific community," as a salve against anachronistic or Whig history. In the meantime, one must wonder whether a Whig sociology and social history simply tried to replace a Whig intellectual history. This would give good grounds for the cultural historian of science to look askance at the "barbarous" methods of the modern prosopographer.

But let us play this game until its end. Given that prosopographers tended to speak about the scientific community, let us turn the fragmentary nature their work, and thus of our knowledge, into the nature of the community itself. Taken as a whole, the eighteenth-century community of science was at

---

[45] Schiebinger, *The Mind,* p. 245.

least an ideological entity: call it the Republic of Letters. Cosmopolitanism and impartiality were two of its essential and ideal characteristics.[46] But when we look at particular or local instantiations, our prosopography finds a plethora of provincial and other interested groups.

In contrast to the enlightened cosmopolitan ideology associated with the Republic of Letters, our fragmentary prosopography has turned up localism, provincialism, and nationalism. The most cosmopolitan group we could find was the Jesuits. French provincial societies behaved as such. We have found that British societies in Edinburgh and Manchester were mired in local and provincial contexts of politics and patronage. Even Crell's chemical journal, a vehicle of disciplinary self-consciousness, was nationalistic. The Scottish and German professoriate, the academic avant garde then, we have seen as essentially endogamous groups, intramurally, locally, provincially or at least nationally. The case of Immanuel Kant, a self-styled cosmopolitan or "Welt-bürger" who never left his home town, we can take as emblematic. Even most students seem to have given up their wandering ways and stayed in the provinces. Except for the Jesuits, the academicians remain our best cosmopolitans. Yet, as the case of Berlin shows, chauvinistic policies of recruitment seem to have been the order of the day, at least in some places.

In contrast to egalitarian and meritocratic ideologies associated with the Republic of Letters, we have uncovered rather more a network of class and caste boundaries. The Jesuits again appear here to have been exceptions, as they perhaps moved personnel up the ranks and around the globe by merit. Moreover, we have speculated that the Jesuitical legacy in France and Austria was bound to the emergence of their systems of academic appointments based on examination. Among other groups and in other parts of Europe, we have found social class and caste asserting themselves. For one reason or another, the poor became marginalized in the student body. As in the case of the provincial societies, the enlightened student body looked patrician and professional, noble and gentlemanly, in origin. The emerging bourgeoisie perhaps thought in terms of merit and equality, but it acted in terms of influence and relations, as our tale of two cities showed. The Scottish and German professoriate thought as much, if not more, in terms of personal relations and monetary interests in academic appointments and advancements.

In contrast to modern professionalized and specialized disciplinary communities of scientists, we find many groups to be amateur and polymathic. The system of chairs at Scottish and German universities institutionalized specialized disciplines; but our prosopography shows resistance to a Smithian division of academic labor. The Jesuits resisted that as a matter of policy: one

---

[46] See Lorraine Daston, "The Ideal and the Reality of the Republic of Letters in the Enlightenment," *Science in Context*, 4/2 (1991), 367–86; also William Clark, Jan Golinski, and Simon Schaffer, "Introduction," in Clark, Golinski, and Schaffer (eds.), *The Sciences in Enlightened Europe* (Chicago: University of Chicago Press, 1999).

was loyal to the Society and not to some abstract international community or subcommunity of science. Protestant professors trained for multiple disciplines and chair-hopped for higher salaries. French provincial amateurs opposed disciplinary specialization, in part from egalitarian sentiments. Salons and societies upheld the antimercantile values of the liberal arts: the pursuit of science inhabited the sphere of aristocratic leisure, no doubt underlying part of its claim, in the next century, to be disinterested. Crell's chemical journal, reducing a society to its corresponding members, points to the next century, to the new persona and identity of the specialized scientist. Perhaps the few academicians spread throughout Europe embodied that persona, but I doubt it. In any case, there was no professional community of scientists.

Most groups pursuing science in the eighteenth century behaved like traditional complex moral communities, as we have called them. Had space and knowledge been available to consider other groups, such as artisans, engineers, and technicians, our prosopography might have turned up further evidence for this. In particular, like nobles, Jesuits, and enlightened women, artisans did not tend to separate their lives and selves into public and private parts. Thus, we see the force of Schiebinger's resolution of the century in regard to women and in general: the privatization of the family and the professionalization of science. The complex social spheres of the upper-class salon and the lower-class shop would no longer serve as suitable sites for middle-class science. Home and workplace became severed as private and public. Bourgeois personae came forth to fill these spaces. Like the new bureaucrat, the professional scientist occupied an official space in which the private self could be suppressed.

ENLIGHTENED PROSOPOGRAPHY

As a coda to the foregoing analysis, let us finally consider the eighteenth-century roots of our prosopography and inquire about its problems today. The lineaments of our prosopography lie in the genre of the academic éloge, the concept of population and the rise of "Statistik."

Roche's study of the éloge offers a basis.[47] Like the *Gelehrten-Lexicon,* the éloge, or funeral oration for scholars, may be taken as characteristic of the eighteenth century, albeit not unique to it. Bernard le Bovier de Fontenelle's éloges for members of the Paris Académie des Sciences established the genre for the eighteenth century and, indeed, did much to cast the persona of the modern scientist. Unlike the earlier, more rhetorical and panegyrical éloges, the new genre reflected historical interests. In imitation of the Parisian academy of sciences, perpetual secretaries of other scientific academies and societies

[47] Roche, *Le siècle,* 1:166–81.

typically kept detailed records on the lives of the members. They constructed the collective biography of the institutions for us.

Extrapolating from a typical case, an enlightened, scientific éloge wove itself around place and date of birth; full name; names of parents; condition and status of family and its relations; education, especially universities and professors visited; age and manner of début in the learned world; services and charges; travels; wedded state; tastes; objects of study; possession of a cabinet of curiosities and its character; possession of a library and its extent; works written, precisely listed and evaluated, if possible; reputation, friendships, and correspondences with relevant scholars; memberships in societies and academies; major private and public life events; character; lifestyle; health; cause of death; estate; and stature in the Republic of Letters.

Such éloges show similarities with the ancient "Lives of Philosophers." Reflecting the Christian tradition of the "Lives of the Saints," these éloges, beyond panegyrical and historical aspects, exhibit hagiographical ones as well, secularized in the spirit of the Enlightenment. To the prosopography or *curriculum vitae* of the list, the hagiographical moment bestows an ethic and ideology on the new subjects of science, "a saintly and sagacious life, divested of passion, the mastery of self authenticating the new saint."[48] The lives of these new saints and sages are not split into public and private parts by the enlightened academic éloge. In this sense, the éloge remains aristocratic and traditional. The scientific subject is still embodied as a complex, moral persona, whose virtue inheres in mastery of specific aspects of the self.

The éloge thus perpetuated the ancient, albeit now enlightened, tradition of collective biography. If we count the collective biographies of elites or small groups as prosopography in the broad sense, then prosopography in the narrow sense, the statistical study of populations or larger groups, if not springing full-born from the Enlightenment, nonetheless can trace its provenance therefrom. The nineteenth-century apotheosis of the professional "man of science," borne by a *Statistik* or *curriculum vitae* stripped of most aspects of private life, as reflected in Poggendorff's *Biographisch-literarisches Handwörterbuch zur Geschichte der exacten Wissenschaften,* finds its forebearer in eighteenth century "Lexica of the Learned." Their bent is bourgeois and even liberal, an effect of an egalitarianism bound up with statistics.

The eighteenth century, or the long Enlightenment generally, witnessed the birth of a sort of statistics or, rather, the emergence of the notion of population as a human group subject to quantifiable regularities and even social laws. In the 1660s "political arithmetic" arose in England as an attempt to quantify aspects of the social body, especially regarding population. In the eighteenth century, British "political economists," French "physiocrats," and German "cameralists," with their "police science" (*Policey-Wissenschaft*) and

---

[48] Ibid., 1:177.

Statistik, could argue that aspects of the social sphere constituted causal-functional ensembles and even self-regulating systems that were effects of populations and were independent of the plans, interests, and calculations of individuals, groups, and governments. Social practices and structures might have a functional value for society as a whole, or for classes or groups, without anyone having conceived such practices or structures.

For Adam Smith the division of labor was one such practice. As we heard from him, eighteenth-century political economists could conceive that social practices, such as the division of labor, might create new sorts of personae – philosophers versus street porters – instead of arising from them. Social identities are as much, if not more, an effect rather than a cause of social structures and practices. The eighteenth century laid the bases for a political economy of the subjects of science: our (statistical) prosopography, whose subject is the archetypal eighteenth-century middle-class man: *homo oeconomicus.*[49]

Progeny of the double-edged sword of the Enlightenment, our prosopography springs in part from British political economy and German police science, from political arithmetic and *Statistik.* Prosopography's aporias are part and parcel of those of the Enlightenment itself, the age of both the liberal and the bureaucratic state. One might thus view the recent hesitation about prosopography, this method deemed "barbarous" by some of its own practitioners, as a hesitation about the legacy of the eighteenth century itself: liberalism, materialism, and positivism. And mindful of Romanticism's critique of the Enlightenment, it is not without irony that prosopographers may contemplate the recent turn in the history of science from prosopography and social history toward "cultural" history. Is there anything more bound with Romanticism than our cultural history and its construction of the scientific identity?

[49] See Stone, "Prosopography," p. 59. See in general Louis Dumont, *From Mandeville to Marx: The Genesis and Triumph of Economic Ideology* (Chicago: University of Chicago Press, 1977), chaps. 3–6; Keith Baker, *Condorcet: From Natural Philosophy to Social Mathematics* (Chicago: University of Chicago Press, 1975); Michel Foucault, *Power/Knowledge: Selected Interviews and Other Writings 1972–1977* (New York: Pantheon, 1980), eds. and trans. Colin Gordon et al., pp. 146–82; Otto Mayr, *Authority, Liberty and Automatic Machinery in Early Modern Europe* (Cambridge, MA: MIT Press, 1986), part 2; Theodore Porter, *The Rise of Statistical Thinking 1820–1900* (Princeton: Princeton University Press, 1986), chap. 1; Lorraine Daston, "Rational Individuals versus Laws of Society: From Probability to Statistics," in Lorenz Krüger et al. (eds.), *The Probabilistic Revolution* (Cambridge, MA: MIT Press, 1987), 2 vols., 1:295–304; and Daston, *Classical Probability in the Enlightenment* (Princeton, NJ: Princeton University Press, 1988).

Part II

DISCIPLINES

10

CLASSIFYING THE SCIENCES

Richard Yeo

Since Plato and Aristotle, philosophers of the Western tradition have placed a premium on the organization of knowledge. When knowledge is ordered, subdivided, and controlled we speak of trees, fields, maps, and bodies – metaphors suggesting definite structures and relationships. When knowledge is regarded as chaotic, overwhelming, or undifferentiated, we speak of labyrinths, mazes, or oceans – still perhaps implying that an order exists but acknowledging that it is not yet visible. The ancient philosophers endorsed the first, and positive, side of this dichotomy in two related ways: first, by privileging logically demonstrable, or at least systematically organized, bodies of knowledge as *scientia* or science, distinguishing them from other forms of knowledge, such as opinion, craft, or technical skills (*techne*); second, by seeking to demonstrate how the various sciences are related, in some rational manner, to one another in an overarching classification of knowledge. These maps or charts indicated appropriate paths of education and learning. Schemes of this kind were produced by the scholastic thinkers of the Middle Ages and they informed, and were themselves reinforced by, the pedagogy and curricula of the universities through to the Renaissance and beyond.[1] To travel one of these paths was to master the "encyclopedy," the circle of sciences.

By the eighteenth century there had been significant changes in the social and cultural conditions that supported these earlier classifications of knowledge. For example, the universities were no longer the only avenue to knowledge, especially to information about science and the useful arts. But for at least the first half of the century, the terminology in which the sciences were discussed was still close to that of scholastic philosophy. Exposure to the formal language of textbooks, dictionaries, and scientific lectures of this period

[1] James A. Weisheipl, "The Nature, Scope and Classification of the Sciences," in David C. Lindberg (ed.), *Science in the Middle Ages* (Chicago: University of Chicago Press, 1978), pp. 461–82.

The writing of this chapter was supported by an Australian Research Council Grant, and by Griffith University. I also thank Jennifer Tannoch-Bland for research assistance.

can be a disturbing experience for the uninitiated. Words such as "Physicks" (and its apparent double, "Physick"), "Physiology," "Pneumaticks," "Pneumatology," "Phytology," "Somatology," and "Aerology" regularly occur in works apparently addressed not only to scholars but also to the reading public. At the same time, as the editors of *The Ferment of Knowledge* insist, far from being a stagnant period after the excitement of the Scientific Revolution, this century saw the consolidation of inquiry into the phenomena of electricity, magnetism, and heat, the revolution in chemistry, historical theories of the solar system, and the appearance of new subjects such as geology, biology, and psychology.[2] But it is precisely these developments, which appear reassuringly "modern," that make it crucial to resist any easy importation of later disciplinary categories into the discussion of eighteenth-century science. It is helpful to see these advances in two ways: first, the increasing success of the physico-mathematical sciences on the Newtonian model, such as astronomy, mechanics, and optics; second, the accumulation of empirical observations of the kind Francis Bacon (1561–1626) had called for in relatively new areas of inquiry, such as electricity, magnetism, physiology, and mineralogy, and in the taxonomy of the plant and animal kingdoms. This explosion of knowledge – by no means confined to the natural sciences – strained the old terminology and some of the classifications it embodied. It made new maps of knowledge necessary, while at the same time making them difficult to draw.

There can be no doubt, however, that the exercise appealed to a range of thinkers. Consider the prospect of the French philosopher, Antoine-Louis-Claude Destutt de Tracy (1754–1836), sitting in a prison at les Carmes in July 1794. Only a few days before his expected trial and possible death by guillotine, he struggled to work out a classification that would show the unity of the sciences.[3] This episode might be taken as an appropriate coda for a century that has often been seen as manifesting a passion for classification and universal systems. Yet it is also significant that Destutt de Tracy concluded that if there was a universal science, or a unity of the sciences, it rested on physiology rather than mathematics – thus inverting the position of thinkers such as Descartes and other progenitors of the Enlightenment movement.

Two points can be drawn from these observations: the first, and well-known one, is that the ancient quest for the unity of the sciences continued in the eighteenth century; the second is that there was only limited agreement about how the natural or physical sciences should be classified; moreover, the prospect of achieving a consensus was complicated, and diminished, by the end of the century as new scientific disciplines – such as Destutt de Tracy's favourite, physiology – emerged as largely autonomous fields of inquiry. When historians

---

[2] G. S. Rousseau and Roy Porter (eds.), *The Ferment of Knowledge: Studies in the Historiography of Eighteenth-Century Science* (Cambridge University Press, 1980), p. 2.

[3] Emmet Kennedy, "Destutt de Tracy and the Unity of the Sciences," *Studies on Voltaire and the Eighteenth Century,* 171 (1977), 223–39. Tracy survived and soon became a member of the new national Institut.

of the Enlightenment attempt to epitomize its intellectual character, this issue inevitably appears. Norman Hampson, in *A Cultural History of the Enlightenment,* suggested that the eighteenth century "regarded knowledge as a whole, rather than as a collection of separated parts." But this remark sits somewhat uncomfortably beside Thomas Hankins' comment in his *Science and the Enlightenment:* "The creation of the new scientific disciplines was probably the most important contribution of the Enlightenment to the modernization of science, and one that we might easily overlook."[4] These two attempts at generalization reflect the complexity of the question they imply: how did eighteenth-century thinkers perceive the relationships between the various sciences? How did they draw their maps of knowledge?

In logical terms, classification of knowledge involves assumptions about the demarcation of sciences from one another as discrete categories as well as views about relationships between various sciences, perhaps revealing an underlying unity. Classification implies division. But since the ancients there have been different, often coexistent, stresses on unification and division in classifying knowledge. The Aristotelian tradition divided the sciences into speculative or theoretical; practical; and artistic or productive, and within these, distinguished clearly between sciences in terms of subject matter and method. John Locke (1632–1704) followed a version of this in his *Essay Concerning Human Understanding* (1690), assigning the "sciences" to three groups: physics, ethics, and logic. But in reviewing this work in *New Essays Concerning Human Understanding* (1704), Gottfried Wilhelm Leibniz (1646–1716) argued that such divisions were arbitrary. Earlier, in 1679, he had remarked, "It does not make much difference how you divide the sciences, for they are one continuous body, like the ocean."[5] Thus, the conviction of unity did not necessarily require discrete categories, and it is thus not surprising to find an emphasis on either unity or diversity in eighteenth-century writers and in the work of historians studying them.

Whether the sciences were conceived philosophically as ultimately one or many, people of the eighteenth century did not share our modern sense of the scope and boundaries of scientific subjects. They certainly did not recognize the closely differentiated array of disciplines, often marked by special journals

[4] Norman Hampson, *A Cultural History of the Enlightenment* (New York: Pantheon Books, 1968), p. 86; Thomas Hankins, *Science and the Enlightenment* (Cambridge University Press, 1985), p. 11.
[5] Gottfried W. Leibniz, *Philosophical Writings* (London: J. M. Dent, 1995), ed. G. H. R. Parkinson, trans. M. Morris and G. H. R. Parkinson, p. 6. On the Aristotelian tradition, see James A. Weisheipl, "Classification of the Sciences in Medieval Thought," *Medieval Studies,* 27 (1965), 54–90, at 58–68; Charles Schmitt, *Aristotle and the Renaissance* (Cambridge, MA: Harvard University Press, 1983); William A. Wallace, "Traditional Natural Philosophy," in Quentin Skinner and Eckhard Kessler (eds.), *The Cambridge History of Renaissance Philosophy* (Cambridge University Press, 1988), pp. 210–35. For the problems seen by seventeenth-century thinkers, see Lorraine Daston, "Classifications of Knowledge in the Age of Louis XIV," in David L. Rubin (ed.), *Sun King: The Ascendancy of French Culture during the Reign of Louis XIV* (London: Associated University Presses, 1992), pp. 207–20. More generally, see Robert McRae, *The Problem of the Unity of the Sciences: Bacon to Kant* (Toronto: University of Toronto Press, 1961).

and institutions, that began to emerge in the early nineteenth century. Even the names of some modern disciplines, such as biology and geology, did not exist in the early part of the century; and of course, other names, such as "Physics," rather than denoting the set of subjects recognized today, usually referred to the entire study of causes in nature.[6] Aristotle called this "natural philosophy" and gave it higher status than mathematics, which he regarded as a subject dealing in abstract concepts that must be adjudicated by those searching for real causes in nature – namely, by the natural philosopher. The term still carried some connotations from its original meaning – a search for qualitative explanations based on the essential nature of bodies. For example, in a German encyclopedia (initiated in 1732 by Johann Zedler), the entry on "*Natur-Lehre*" recommended that physics be confined to the study of material objects but conceded that some people preferred the older view of it as also encompassing the properties of spiritual entities. But the dominant trend was a strengthening of the nexus forged between mathematics and natural philosophy during the preceding century and culminating in the work of Isaac Newton (1642–1727). This upset the Aristotelian subordination of mathematics to natural philosophy.[7] Another consequence of this nexus was the generally lower status accorded to nonquantitative studies of nature that did not boast the experimental method and mathematical formulation of the new natural philosophy. These observational and taxonomic studies were collectively called natural history, and in the century from 1660 to 1760 they represented at least 19 percent of the research activity within science, even though they accounted for only 4 percent of university chairs in science, the majority of which were in the established fields of mathematics, medicine, and natural philosophy.[8]

In an important essay on the array of sciences bequeathed to the eighteenth century from earlier periods, Thomas Kuhn distinguished between classical (mathematical) and experimental (Baconian) sciences. The former, he suggested, consisted of an uncontroversial "natural cluster" of five sciences – astronomy, harmonics, mathematics, optics, and statics (or mechanics) – those named by Aristotle as "the more physical parts of mathematics." Although practitioners of these sciences acknowledged a role for experiments, Kuhn contended that these were of a limited kind and were often "thought exper-

---

[6] See Benjamin Martin, *The Philosophical Grammar*, 2nd ed., 1738 [1st ed. 1735], (London: J. Noon), part 4, for use of "Geology"; but here it embraced not only the "terraqueous globe" but also vegetation and animal bodies. See Roy Porter, *The Making of Geology: Earth Science in Britain 1660–1815* (Cambridge University Press, 1977)

[7] *Grosses vollständiges Universal Lexicon*, 64 vols. (Halle: J. H. Zedler, 1732–50), vol. 23, column 1149. On mathematics and natural philosophy, see Peter Dear, *Discipline and Experience: The Mathematical Way in the Scientific Revolution* (Chicago: University of Chicago Press, 1996), pp. 35–8, 161–8; John Henry, *The Scientific Revolution and the Origins of Modern Science* (London: Macmillan, 1997), pp. 18–21.

[8] John Gascoigne, "The Eighteenth-Century Scientific Community: A Prospographical Study," *Social Studies of Science*, 25 (1995), 575–81, at 577–8.

iments" used as a jumping-off point for mathematical theories or, if actually performed, usually served to demonstrate a conclusion known in advance. In contrast, for the "Baconian" sciences of the seventeenth century, experimentation was preeminent, directed toward seeing "how nature would behave under previously unobserved, often previously nonexistent, circumstances." For Kuhn, this second category of sciences embraced a range of empirical inquiries, some of which were already commonly identified with named sciences, such as chemistry, whereas others were phenomena for new systematic investigation, such as electricity, magnetism, and heat. The Baconian sciences were associated with a new set of instruments for use in making and registering observations: microscopes, thermometers, barometers, air pumps, detectors of electric charges. Unlike the classical sciences, these fields were not marked by "a body of consistent theory," although their practitioners began systematically to concentrate their research around well-defined phenomena.[9]

Kuhn's analysis is useful in underlining the fact that the "sciences" of the eighteenth century were not all of one piece. In fact, it was the classical/ mathematical disciplines – to use his typology – that unproblematically qualified as sciences in the older sense of *scientia*, a meaning still endorsed by Samuel Johnson (1709–1784) in his *Dictionary of the English Language* of 1755. Thus, in Kuhn's account there was a body of mature, relatively stable sciences and another, more diffuse, group of subjects that pursued the Baconian program of collection, observation, and experiment but were not yet marked by strong consensus around a dominant theory. This view also allows a distinction between significant advances within physico-mathematical sciences, such as the wave theory of light in optics, and the consolidation of new areas of inquiry, such as those in physiology and geology.

What has been said so far indicates some of the issues historians of science have identified while trying to capture eighteenth-century assumptions. But how did contemporaries perceive the sciences? Did the appearance of what historians now see as new fields of inquiry, or significant advances within established subjects, lead to any reconfiguration of accepted maps of knowledge? We can answer only if we have a picture of how eighteenth-century thinkers regarded natural knowledge and how they placed it in relation to other parts of knowledge.

CLASSIFICATION IN PRACTICE

Where did thinking about classification of knowledge take place? The anecdote about Destutt de Tracy suggests that the ancient philosophical practice

---

[9] Thomas S. Kuhn, "Mathematical versus Experimental Traditions in the Development of Physical Science," in *The Essential Tension: Selected Studies in Scientific Tradition* (Chicago: University of Chicago Press, 1977), pp. 31–65, quotations at pp. 37, 47. See also Dear, *Discipline and Experience,* 168–79.

of discerning relationships – logical links, order of study, hierarchies of prestige – among subjects was alive and well at the end of the century. But there seems to be agreement that, by contrast with the centuries that preceded and followed, the contribution of the eighteenth century to the philosophical tradition of classifying the sciences was minor.[10] The writers who addressed this topic largely followed the earlier work of Bacon (1561–1626), Thomas Hobbes (1588–1679), Locke, or Leibniz, who in turn were either in agreement or dispute with Aristotle. This tradition was stronger in Germany than in France or Britain. Christian Wolff (1679–1754) and Immanuel Kant (1724–1804) saw it as important, but it is probably fair to say that no major philosopher made classification of the sciences his dominant preoccupation in the style of nineteenth-century writers such as Auguste Comte and Herbert Spencer; no natural philosopher of this period devoted such attention to this exercise as the French savant Andre-Marie Ampère (1775–1836). Indeed, the widespread distrust of *"esprit de système"* – associated with Aristotelian scholasticism and its metaphysical systems, and with Cartesianism – also led many writers to question the value of grand schemes of classification. And the variety of such schemes on offer began to encourage statements about their relative and arbitrary character.[11] Nevertheless, in spite of this skepticism, there were other practical imperatives that kept classification alive as an issue in a number of situations.

According to one commonplace image of the period, such an interest in classifying the sciences is to be expected. Noticing the passion for taxonomy in natural history – most obviously associated with Carl Linnaeus's (1707–1778) *Systema Naturae* (1735) and *Philosophia Botanica* (1751) – some writers have seen a drive to classify as indicative of a pervading thought style. In this perspective, both Linnaeus' works on natural history and encyclopedism are seen as contemporaneous and parallel classificatory projects – quests to name and order the world, both the Book of Nature and the circle of human learning.[12] Indeed, it is true that both projects assumed that this could be done by summarizing knowledge in textual form in a manner that was, in principle, universally accessible. In fact, George-Louis Leclerc comte de Buffon's (1707–

[10] See Robert Flint, *Philosophy as Scientia Scientiarum and a History of Classification of the Sciences* (Edinburgh: Blackwood, 1904); R. G. A. Dolby, "Classification of the Sciences: The Nineteenth Century Tradition," in Roy. F. Ellen and David Reason (eds.), *Classifications in Their Social Context* (London: Academic Press, 1979), pp. 167–93; Nicholas Fisher, "The Classification of the Sciences," in R.C. Olby, G. N. Cantor, J. R. R. Christie and M. J. S. Hodge (eds.), *Companion to the History of Modern Science* (London: Routledge, 1989), pp. 853–68.

[11] G. Tonelli, "The Problem of the Classification of the Sciences in Kant's Time," *Rivista critica di storia della filosofia*, 30 (1975), 244–94, at 265; Ernst Cassirer, *The Philosophy of the Enlightenment*, trans. F. Koelln and J. Pettegrove (Princeton, NJ: Princeton University Press, 1979), p. vii.

[12] Gunnar Broberg, "The Broken Circle," in Tore Frängsmyr, J. L. Heilbron, and Robin E. Rider (eds.), *The Quantifying Spirit in the 18th Century* (Berkeley: University of California Press, 1990), pp. 45–71, at pp. 45–6; see also Michel Foucault, *The Order of Things: An Archaeology of the Human Sciences* (London: Tavistock, 1970), pp. 125–65.

1788) *Histoire Naturelle,* published from 1749 (with supplements by collaborators), grew to forty-four volumes by 1804, making it far larger than most
encyclopedias. Both the systems of natural history and the compilations of
knowledge in encyclopedias were conceived and explained as places of display –
cabinets, museums, libraries, compendia – through which a larger, external
universe could be sampled and understood. Writing about the *Encyclopédie,*
Bernard Groethuysen captured this capacity of the *philosophes* to survey man's
intellectual estate as if it were a newly discovered land:

> The objects which we have assembled and which are found sometimes in a
> certain part of the island, sometimes in another, it is we who have collected
> and put them in the order which suits us, placing them in such and such a
> room of this universal museum which our *Encyclopédie* represents.[13]

The suggestion here is that the *philosophes* were audacious enough to arrange
knowledge as they liked, rather than following any traditional system. As we
shall see, this was indeed an issue, and one that also had its analogy in the
debate over natural versus artificial taxonomic systems in natural history. In
both cases, the question of whether classification was arbitrary was heightened by the problem of fitting expanding information into fixed categories of
a nomenclature.

In more recent scholarship, the eighteenth century has been seen as the
starting point of some phenomena that have reached their peak, or crisis
point, in our own time. Some historians have argued that leisure, consumerism, and information were significant issues in modern Western society
before the late twentieth century, and they regard the eighteenth century as
a watershed.[14] Since the 1970s, the notion of an information revolution has
been common in discussions of contemporary cultural crises. But it is also
possible to speak of an "information explosion" in the eighteenth century,
one that was associated with the massive circulation of printed material encouraged by an increasingly literate audience and the energies of print capitalism. As early as 1680 Leibniz confessed anxiety about the "horrible mass
of books which keeps on growing," so that it would soon be a disgrace rather
than an honor to be an author. Peter Burke has suggested that a pressing concern with ordering and managing this information was reflected in three areas:
the role of journals as filters of information; the practical need for cataloging

---

[13] Bernard Groethuysen, cited in Herbert Dieckmann, "The Concept of Knowledge in the *Encyclopédie,*"
in Herbert Dieckmann, Harry Levin, and Helmut Motekat (eds.), *Essays in Comparative Literature*
(St. Louis, MO: Washington University Studies, 1961), pp. 73–107, at pp. 84–5.

[14] For this literature, see Neil McKendrick, John Brewer, and J. H. Plumb, *The Birth of a Consumer So-
ciety: The Commercialization of Eighteenth-Century England* (London: Europe Publications, 1982); John
Brewer and Roy Porter (eds.), *Consumption and the World of Goods* (London: Routledge, 1993);
John Brewer and Ann Bermingham, *The Consumption of Culture, 1600–1800: Image, Object, Text*
(New York: Routledge, 1995).

of libraries; and the attempts at comprehensive summaries of knowledge in encyclopedias.[15] Although journals, libraries, and encyclopedias predate the eighteenth century, it is important to note that in this period they became more explicitly linked to the problems of organizing and selecting knowledge, by this time seen not merely as abstract philosophical issues but also as practical problems for all educated readers.

If we take the example of encyclopedias, we can say that there *was* something distinctive about the issue of the classification of knowledge in the eighteenth century. This era saw the emergence of the modern form of dictionaries and encyclopaedias – in vernacular languages – that sought to present the circle of sciences, both ancient and modern, to a readership wide enough to support the massive commercial investment they required. Beginning with the English dictionaries of arts and sciences by John Harris (1667?-1719) (*Lexicon Technicum,* 2 vols, 1704 and 1710) and Ephraim Chambers (1680?-1740) (*Cyclopaedia,* 2 vols, 1728), this reached a climax in the French *Encyclopédie* (1751) – the symbolic text of the Enlightenment – and concluded with the *Encyclopaedia Britannica* (3 vols, 1768–71), which started as three volumes but reached eighteen by 1797.[16] This was a period in which the various branches of knowledge were laid out on paper in a manner supposedly accessible to people outside the formal university system. For this reason, it is possible that the task of classifying sciences, placing them in relation to one another and choosing the most relevant for particular purposes, was made more public than it had ever been. Whereas a twelfth-century encyclopedist, Hugh of St. Victor (c. 1096–1141), advised his readers to learn everything because nothing was superfluous, the editors of the eighteenth-century works admitted that the full compass of arts and sciences could not be embraced by individual minds.[17] From this followed the need to select, but in doing so, to recognise the sector of the circle in which one was moving, to appreciate which subjects lay near by.

---

[15] Gottfried W. Leibniz, "Precepts for Advancing the Sciences and Arts," in Philip P. Wiener (ed.), *Leibniz: Selections* (New York: Scribner's, 1951), pp. 29–30; Theodore Roszak, *The Cult of Information: The Folklore of the Computer and the True Art of Thinking* (New York: Pantheon, 1986); Peter Burke, "Reflections on the History of Information in Early Modern Europe," *Scientiarum Historia,* 17 (1991), 65–73.

[16] The *Encyclopédie,* edited by Jean Le Rond d'Alembert (1717–1783) and Denis Diderot (1713–1784), began as a translation of Chambers' *Cyclopaedia,* projected first as four and then twelve volumes when it began to appear in 1751 but eventually becoming a dramatically larger work comprising seventeen volumes of text and eleven of plates by its completion in 1772. This was followed by four supplemental volumes of text, one supplemental volume of plates, and two supplemental volumes of index, 1776–80.

[17] The editors of the *Encyclopédie* explained that their project was unthinkable without the participation of many contributors. See Jean Le Rond d'Alembert, *Preliminary Discourse to the Encyclopedia of Diderot,* translated with an introduction by Richard N. Schwab (Chicago: University of Chicago Press, 1995), p. 3. On Hugh of St. Victor, see Pierre Speziali, "Classification of the Sciences," in Philip Wiener (ed.), *The Dictionary of the History of Ideas: Studies of Selected Pivotal Ideas,* 5 vols. (New York: Scribner's, 1968–74), vol. 1, p. 464.

MAPS OF SCIENCES IN ENCYCLOPEDIAS

I will focus the rest of this chapter on encyclopedias as a manageable way of approaching a number of questions about the way contemporaries regarded their intellectual landscape. Was there any consensus about the major divisions of knowledge? Where did natural knowledge – the object of what we now call "sciences" – lie on these maps of knowledge? Did perceptions on these matters undergo significant shifts by the end of the eighteenth century? This approach to the topic might seem paradoxical, because unlike earlier encyclopedic works, the major encyclopedias of the eighteenth century were alphabetically, rather than systematically, arranged. This format was in keeping with their titles (or in some cases subtitles): "*dictionaries* of arts and sciences." How, then, can they tell us anything about contemporary views on the organization of knowledge and the place of the various sciences within it? The answer, in part, is that editors regarded alphabetical arrangement as compatible with a classification of the sciences and even with a pedagogic order for reading the encyclopedia. In the prefaces to the leading publications, considerable rhetoric was invested in showing that an awareness of the relationships between the various fields of knowledge did inform the work in spite of its alphabetical listing of terms and concepts.[18]

Encyclopedic works of the Middle Ages and the Renaissance were topically, if not always systematically, arranged. Coherence could give way to miscellany, but rarely to alphabetical presentation. The order of their exposition of subjects was usually governed by some overarching pattern, such as the cosmological chain of being with the Divinity as its apex, the seven liberal arts, or the hierarchy of faculties in the university. Other schemes were also possible: in the fourteenth century, Domenicus Bandinus (c. 1335–1418) compiled an encyclopedic work, *Fons memorabilium universi*, which was divided into five parts to reflect the five wounds of Christ.[19] This power of theology was, of course, precisely what Enlightenment encyclopedists resisted, yet they did not dismiss the importance of classification, in spite of their departure from the traditional format of encyclopedias.

Some modern commentators are inclined to celebrate the advantages of strict alphabetical order more stridently than these eighteenth-century compilers. For example, Charles Porset, echoing the sentiment of Roland Barthes, writes, "As the zero degree of taxonomy, alphabetical order authorises all reading strategies; in this respect it could be considered an emblem of the

---

[18] See Richard Yeo, "Reading Encyclopedias: Science and the Organization of Knowledge in British Dictionaries of Arts and Sciences, 1730–1850," *Isis*, 82 (1991), 24–49.
[19] Lynn Thorndike, *A History of Magic and Experimental Science*, 8 vols. (New York: Columbia University Press, 1923–58), vol. 3, p. 560. For Renaissance works, see Neil Kenny, *The Palace of Secrets: Beroalde de Verville and Renaissance Conceptions of Knowledge* (Oxford: Clarendon Press, 1991).

Enlightenment."[20] Avoiding the hierarchies of systems, the alphabet is thus seen as egalitarian, reducing all subjects to the same level. In support of this view, we might add that, in principle, alphabetical arrangement allows indefinite expansion of content without the pressure to display connections or renegotiate categories. Commenting on advances during the Scientific Revolution, the economic historian Sir George Clark remarked that alphabetical ordering of information is not merely a matter of convenience and ready reference but rather reflects a situation "when knowledge is growing in many directions, and not in the framework of an accepted interpretation of the whole."[21] The early dictionaries of arts and sciences aimed to record and summarize data and doctrines from a wide range of intellectual territory – from Aristotelianism to Newtonianism, from gardening to heraldry. Given this, it is certainly fair to say that an alphabetical listing of short entries on terms avoided the need for synthesis, or the explicit placing of subjects in a philosophical taxonomy. Undoubtedly, Chambers and Diderot appreciated some of these advantages. Indeed Diderot's comment in the prospectus of 1750 to the *Encyclopédie* suggests that ease of access was an issue:

> We believe we have had good reason to follow alphabetical order in this work
> . . . If we treated each science separately and followed it with a discussion
> conforming to the order of ideas, rather than that of words, then the form
> of this work would have been even less convenient for the majority of our
> readers, who would have been able to find nothing without difficulty.[22]

Nevertheless, both Chambers's *Cyclopaedia* and the *Encyclopédie* carried charts of knowledge with supporting commentary, arguing that they allowed the careful reader to find the virtues of an encyclopedia within the pages of an alphabetical dictionary. "Former Lexicographers," wrote Chambers, "have not attempted any thing like Structure in their Works; nor seem to have been aware that a Dictionary was in some measure capable of the Advantages of a continued Discourse."[23] His diagrammatic display of the sciences was accompanied by a list of the terms belonging to each major subject so that, with cross-references, the reader could reconstitute a science that had been scattered alphabetically throughout the work. Similarly, in his *Preliminary Discourse* d'Alembert made it clear that the *Encyclopédie* was not just a dictionary:

> As an *Encyclopedia,* it is to set forth as well as possible the order and connec-
> tion of the parts of human knowledge. As a *Reasoned Dictionary of the Sciences,*

---

[20] Charles Porset, cited by Broberg, "The Broken Circle," p. 49.

[21] George Clark, *Science and Social Welfare in the Age of Newton* (Oxford: Clarendon Press, 1937, 2nd ed. 1970), p. 143.

[22] Diderot, cited in Cynthia J. Koepp, "The Alphabetical Order: Work in Diderot's *Encyclopédie*," in Steven Laurence Kaplan and Cynthia J. Koepp (eds.), *Work in France: Representations, Meaning, Organization, and Practice* (Ithaca, NY: Cornell University Press, 1986), pp. 229–57, at p. 237.

[23] Ephraim Chambers, *Cyclopaedia: or, an Universal Dictionary of Arts and Sciences,* 2 vols. (London: J. and J. Knapton, J. Darby, D. Midwinter et al., 1728), vol. 1, p. i.

Arts, and Trades, it is to contain the general principles that form the basis of each science and each art, liberal or mechanical, and the most essential facts that make up the body and substance of each.[24]

Before discussing these charts or maps of knowledge it is important to recognize that, even without them, the new dictionaries of arts and sciences were informed by certain assumptions about the division of knowledge. For a start, the category "arts and sciences," although a large one, excluded history, biography, and geography. These subjects were the province of a separate genre of reference work: the historical dictionary. The leading examples were Louis Moreri's (1643–80) *Grand Dictionnaire Historique, ou mélange curieux de l'histoire sacrée et profane,* first published in Lyon in 1674, and issued in English translation in 1694 as *The Great Historical, Geographical and Poetical Dictionary;* and Pierre Bayle's (1647–1706) famous *Dictionnaire historique et critique* of 1697. These works – and others that followed them in the eighteenth century, such as the *Biographia Britannica* (1747–66; 2nd ed., 1778–93), were concerned with the lives of notable figures rather than with explications of the arts and sciences.[25] Another important feature of the dictionaries of arts and sciences is that although they broke down information into short entries on scientific and technical terms, they nevertheless operated with larger categories, such as natural history and natural philosophy, that entailed distinctive groupings of subjects. Furthermore, some of them appealed to a unity or circle of arts and sciences (as implied by the word "encyclopedia") and advised that a methodical course of study could be conducted on the basis of these single works.

This suggests that in spite of their affirmation of the quick and easy consultation allowed by alphabetical arrangement, these scientific dictionaries or encyclopedias deferred to contemporary convictions about the importance of system and order in learning. The pedagogic message carried by the influential works of Isaac Watts (1674–1748) is worth noticing here. His *Logick; or, the right use of reason* appeared in 1726, with a second edition in 1728, and a supplement to it was published in 1741 as *The Improvement of the Mind.* In a section of this second work dealing with the sciences, Watts announced

> The best way to learn any Science, is to begin with a regular System, or a short and plain Scheme of that Science. . . . Systems are necessary to give an entire and comprehensive View of the several Parts of any Science, which may have a mutual Influence toward the Explication or Proof of each other: Whereas if a Man deals always and only in Essays and Discourses on particular Parts of a Science, he will never obtain a distinct and just Idea of the whole.[26]

---

[24] D'Alembert, *Preliminary Discourse,* p. 4.
[25] For this distinction, see Richard Yeo, "Alphabetical Lives: Scientific Biography in Historical Dictionaries and Encyclopaedias," in Michael Shortland and Richard Yeo (eds.), *Telling Lives in Science: Essays on Scientific Biography* (Cambridge University Press, 1996), pp. 139–69.
[26] Isaac Watts, *The Improvement of the Mind,* 3rd ed. (London: T. Longman and J. Buckland, 1743), p. 316.

At least a nodding approval of this position is found in unexpected places. Periodicals, such as the *Universal Magazine of Knowledge and Pleasure,* that professed to cover the arts and sciences as well as other subjects avowed that their successive issues consolidated into a "whole body of arts and sciences." This is also apparent in the textbooks on natural and experimental philosophy by writers such as John Theophilus Desaguliers (1683–1744) and Benjamin Martin (1704–1782), which perhaps offer a closer comparison with the scientific dictionaries. In *The General Magazine of Arts and Sciences* (1755), Martin worked to insert his product in a competitive market. Other magazines, he argued, gave no coherent coverage of these subjects; what they did supply they did "only by Peace-meal [sic], in Bits and Scraps, disjointed and mangled, without Order or Connection, and therefore of no Use to any one."[27] Even if not treating the whole circle of sciences, Martin needed to rely on the notion that there were recognizable parts in order to sell his works, including those on the Newtonian sciences, as a course of study more methodical than that offered by periodicals and perhaps by the alphabetical encyclopedias that may have been his unmentioned target.

The eighteenth-century scientific dictionaries covered a wider range of subjects than particular scientific textbooks. The charts or maps of knowledge in Chambers and the *Encyclopédie* were meant to display this range and also to help the reader see relationships between subjects. Chambers's claim that the *Cyclopaedia* promoted coherent understanding of sciences, in spite of the fragmentation wrought by the alphabet, can be understood as deference to the views espoused by Watts. One might also note that as works needing subscriptions from members of the educated elite – scholars, gentleman, clerics – the dictionaries of arts and sciences were in no position to violate openly respected educational opinion, even though part of their content was knowledge that fell outside the university curriculum. In this context, some continuing obeisance to the systematic bent of the encyclopedic tradition made good commercial sense. But it is clear that the charts of knowledge were more than mere rhetoric: when the *Encyclopaedia Britannica* (from 1768) decided not to have one, it made a special point of attacking the assumptions behind such charts and their role in a modern encyclopedia. This was significant, too, because the French editors made so much of the English Lord Chancellor's "Division of Human Learning" outlined in *The Advancement of Learning* in 1605. Partly because of the influence of the *Encyclopédie,* the division of the sciences given by Bacon became commonplace during the second half of the century. It therefore requires some discussion here.

---

[27] *Universal Magazine of Knowledge and Pleasure,* 1 (1747), preface, p. ii; Benjamin Martin, *The General Magazine of Arts and Sciences* (London: W. Owen, 1755), p. iii. See also Martin, *A Course of Lectures in Natural and Experimental Philosophy* (Reading, 1743). On the popularization of science, see Larry Stewart, *The Rise of Public Science: Rhetoric, Technology, and Natural Philosophy in Newtonian Britain, 1660–1750* (Cambridge University Press, 1992).

BACONIAN DIVISION OF THE SCIENCES

In "A Description of the Intellectual Globe" (written in 1612), Bacon said, "I adopt the division of human learning which corresponds to the three faculties of the understanding." By this he meant that different intellectual territories – History, Poetry, and Philosophy – depended, respectively, on Memory, Imagination, and Reason. History included natural history, geography, and political, ecclesiastical, and civil history, as well as the mechanical arts and crafts. Poetry covered the written and visual works of imagination, such as drama, painting, music, and sculpture. Philosophy, the largest group, contained "all arts and sciences," or, in Bacon's words, "whatever has been from the occurrence of individual objects collected and digested by the mind into general notions."[28]

This was a version of the classification given earlier in *The Advancement of Learning,* a work he later issued in Latin as *De Dignitate et Augmentis Scientarum* in 1623. The text of the Latin edition was reorganized on principles advocated by Petrus Ramus (1515–72), showing an argument proceeding from general to more specific propositions, and examples, by means of branching dichotomies. Thus, although Bacon did not include a chart, his division of the sciences was easily put into this form – as seen in many philosophical and pedagogic works of the sixteenth and seventeenth centuries.[29] Bacon's use of the tree metaphor also matched this approach, because it allowed him to say that the divisions between the sciences had a common point of origin and resembled "branches of a tree, that meet in a stem." This implied that there was a single "universal science," or *Philosophia Prima,* from which all other sciences derived. But this reference to unity was followed by a set of divisions. The Sciences were classed under Natural Philosophy and had two parts: the "inquisition of causes, and the production of effects." The former, or natural sciences, then divided into Physics and Metaphysic. Physics dealt with what was "inherent in matter, and therefore transitory"; Metaphysic with that "which is abstracted and fixed." Or, to put this in Aristotelian terms, Physics concerned efficient causes; Metaphysic formal and final causes.[30] Bacon also introduced the term "Mixed Mathematics" to denote subjects such as optics,

---

[28] Francis Bacon, "A Description of the Intellectual Globe," in *The Works of Francis Bacon,* collected and edited by James Spedding, Robert Leslie Ellis, and Douglas Denon Heath, 14 vols. (London: Longman, 1857–74; reprinted Stuttgart-Bad Cannstatt: F. Frommann Verlag, 1961–3), vol. 5, pp. 503–4.

[29] See Graham Rees (assisted by Christopher Upton), *Francis Bacon's Natural Philosophy: A New Source* (Chalfont St. Giles: British Society for the History of Science, 1984), p. 19, n. 45; Joseph S. Freedman, "Diffusion of the Writings of Petrus Ramus in Central Europe, c. 1570–c. 1630," *Renaissance Quarterly,* 46 (1993), 98–152, especially 103–5. I thank Marta Fattori and Graham Rees for advice (personal communications) about the absence of illustrations of the division of knowledge in editions of *De Augmentis* before the mid-eighteenth century.

[30] Bacon, *Advancement of Learning,* in *Works,* vol. 3, pp. 346, 351–4. For the version in *De Augmentis, Works,* vol. 4, p. 337.

astronomy, harmonics, and mechanics, as well as cosmography, music, and architecture, thus expanding Aristotle's category of "*scientia media.*"[31]

Bacon's scheme was novel, and deliberately so, because it departed from the traditional divisions of the sciences by subject area. Instead, he classified in terms of the mental faculty operating in the acquisition of three different branches of knowledge, yet still maintained that there were links between all branches of learning. His "Division of Human Learning" was a reference point for classifications of knowledge during the eighteenth century – not because it established a fixed and agreed system, but rather because it made distinctions while at the same time setting off debates about them.

Within what we now call science, Bacon drew a major dichotomy between natural philosophy and natural history. The former, located under the faculty of Reason and part of philosophy, embraced all the mathematical and physical sciences – disciplines that eighteenth-century writers recognized as the Newtonian sciences. In contrast, natural history belonged to Memory and was charged with producing adequate descriptions (histories), collections and taxonomies of minerals, plants, animals, and, significantly, accounts of the manual crafts and machines. Yet, at the same time, Bacon challenged the subordination of natural history to natural philosophy – that is, in the sense of mere facts compared with universals. He contended that the particular observations and "facts" of natural history were more secure and certain than many of the so-called demonstrations and axioms of the rival systems of natural philosophy on display in his own day.[32] Bacon's work thus became the framework for debates about the relations between the sciences in which some of his own divisions were adopted more rigidly than he intended before being abandoned by the end of the century. We should keep this in mind while discussing the classification of sciences in the major encyclopedias of the period.

HARRIS'S *LEXICON TECHNICUM*

The examples of Harris, Chambers, and Diderot and d'Alembert offer the chance to consider how classification of the sciences worked in three significant dictionaries of arts and sciences. In all three, alphabetical arrangement displaced pedagogic order, but each acknowledged the need for consideration of the larger subjects, which they reduced to numerous short entries (and in some cases, longer articles) on terms. However, the issues of classification and its display in the form of charts were handled differently in these three works.

[31] Bacon, *Works,* vol. 3, 360–1; Gary Brown, "The Evolution of the Term 'Mixed Mathematics,'" *Journal of the History of Ideas,* 52 (1991), 81–102, at 82–3; Sachiko Kusukawa, "Bacon's Classification of Knowledge," in Markku Peltonen (ed.), *The Cambridge Companion to Bacon* (Cambridge University Press, 1996), pp. 47–74, at pp. 49, 60.
[32] Lorraine Daston, "Baconian Facts, Academic Civility, and the Prehistory of Objectivity," *Annals of Scholarship,* 8 (1991), 337–63; Dear, *Discipline and Experience,* chap. 1.

Harris did not have a chart or map; Chambers used a chart of knowledge based on the branching dichotomies similar to those found in scholastic treatises and in the Ramist pedagogic texts from the sixteenth century; Diderot and d'Alembert revived Bacon's tripartite division of sciences by reference to mental faculty. Did these compilers agree on the main divisions within the sciences? How did they use charts of knowledge to indicate the relations between sciences?

Harris declared that the *Lexicon Technicum* was "a Dictionary not only of bare *Words* but *Things,*" or an explication of how "*Technical* Words" were used in the "*Liberal Sciences*" and some of the practical arts associated with them, such as navigation, ship building, the construction of mathematical and geometrical instruments, and also air pumps. A review in the *Philosophical Transactions* of the Royal Society of London endorsed this description, saying that "the design of this Dictionary is different from that of most others," and then, almost in Harris's own words, explained that it gave not only the "terms used in every Art and Science, but likewise the Arts and Sciences themselves." In the first volume Harris apologized for not being able to supply "at the End of the Book, a particular *Alphabet* for each *Art* and *Science* by it self."[33] But when the next volume appeared in 1710 (again covering the whole alphabet, but with new, and supplementary, entries) this list, accounting for the contents of both volumes, was appended.

Although he certainly did not present this "Index" as a grand scheme of classification, it did cluster the particular terms treated in the work under what Harris presumably regarded as recognizable subjects. Since there was no pagination in the *Lexicon,* this "Alphabetical Index" did not give page (or even volume) references to particular topics; rather, it listed the terms treated in the dictionary under twenty headings (or "Heads," in contemporary usage). This format allowed Harris to display a large number of subjects without having to place them in a hierarchy or delineate any relations between them. It also avoided the problem of naming some as arts and others as sciences. The list began with "Navigation" and ended with "Astronomy" and included headings for "Mathematical and Philosophical Instruments," "Fortification," "Dialling," "Anatomy," "Law," and "Heraldry." Within some of the headings, Harris's grouping of terms has a rough-and-ready look about it: some terms such as "Acids," "Earth," "Stones," and "Vegetables" appear under more than one heading. Given this, it is prudent not to exaggerate the evidence it provides; but the fact that it was done at all is revealing, since it suggests that Harris felt unable to let the alphabet stand without comment. As such, this Index offers some indication of how a dictionary maker and member of the Royal Society perceived the major areas of science.

---

[33] "An Account of a Book," *Philosophical Transactions,* vol. 24, no. 292, 1704, 1699–1702, at 1699; John Harris, *Lexicon Technicum; or an Universal Dictionary of Arts and Sciences* (London: Brown, Goodwin et al., 1704), vol. 1, "The Preface," no pagination.

There were three main headings covering natural knowledge: (1) "Natural Philosophy and Physicks"; (2) "Chymystry"; (3) "Botany, Natural History and Meteorology." Two of these – natural philosophy and natural history – were mentioned in the Introduction as large categories under which some new material had been incorporated in the second volume. There were also separate headings for "Mechanicks, Staticks," "Opticks and Perspective," and "Astronomy and the Doctrine of the Spheres." In part, this was because the *Lexicon* had so many entries from these sciences, reflecting Harris's interest – he was known as "Technical Harris" – but also because of their well-established status as disciplines of mixed mathematics. But it is clear from the definition of "Natural Philosophy" in the work itself that these subjects fell under that category, whereas "Geometry" and "Arithmetic and Algebra" – pure mathematics – did not. The contrast between the sciences under "Natural Philosophy" and those of "Natural History" (as described in the entry in Volume Two) was clear. The former were part of Newtonian philosophy, whereas the latter were mainly descriptive histories of the natural world – of earth, water, air, metals, minerals, fossils, and the beasts, birds, and fishes that inhabit the globe. The entry for natural history thus defined it as Bacon did, although in the Introduction to the second volume Harris advertised that he now also included schemes by which plants and animals "are ranged and distributed into their proper Orders."[34]

The only other physical science with a heading of its own in the *Lexicon* is chemistry. This reflected its position as a subject in which there were chairs at universities and specialist textbooks. Given Harris's emphasis on the physico-mathematical sciences it is not surprising that his treatment of this subject was fairly restricted, largely amounting to a definition or description of the names of chemical substances and techniques of analysis, collated under "Chymystry" in the Index and drawn from specialist chemical dictionaries cited in the Preface. This matches the humble definition of the subject, given in the first volume, as an "Art" aiming to "separate the Purer Parts of any mix'd Body from the more Gross and Impure."

CHAMBERS'S *CYCLOPAEDIA*

In his *Cyclopaedia* Chambers acknowledged Harris but claimed to go beyond previous dictionaries of arts and sciences by providing the option of a systematic reading of an alphabetical dictionary. Significantly, to allow such a methodical use of its content, he offered a diagrammatic chart of knowledge that portrayed the relationships of the sciences.[35] Chambers referred to this

---

[34] Harris, *Lexicon Technicum* (London: Brown, Goodwin et al., 1710), vol. 2, "Introduction," no pagination. The "Alphabetical Index" is at the end of this volume.

[35] On one context for this, see Richard Yeo, "Ephraim Chambers's *Cyclopaedia* and the Tradition of Commonplaces," *Journal of the History of Ideas*, 57 (1996), 157–75.

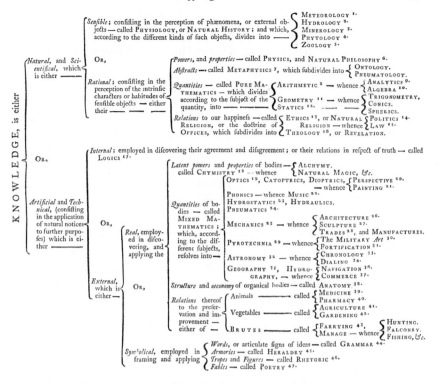

Figure 10.1 The "View of Knowledge" in the Preface of Chambers's *Cyclopaedia* (London, 1728). This appeared in all editions. The Wellcome Institute Library, London.

illustration, not as tree or map, but as a "View of Knowledge." Knowledge was categorized as either "Natural and Scientifical" or "Artificial and Technical" and then separated into further subdivisions, as in "method of dichotomies" of the Ramist kind (see Figure 10.1). After the first division, scientific (versus technical) knowledge of nature was divided into "sensible" or "rational," distinguishing between, say, meteorology and geometry. On the other hand, knowledge acquired for technical purposes was classed as either "internal" (logic) or, more frequently, "external," such as all the arts and crafts but also sciences such as optics, hydrostatics, pneumatics, mechanics, and chemistry. Clearly, then, this is not a simple contrast between arts and sciences, a distinction Chambers confessed to be unsure about. Indeed, the chart juxtaposes certain arts with particular branches of the mixed sciences: thus, Mechanics is linked with Architecture, Sculpture, Trades and Manufactures; Optics with Painting and Perspective; Astronomy with Chronology and Dialling. Chambers said that "the precise notion of an Art and Science, and their just, adequate Distinction, are not yet well fixed."[36]

[36] Chambers, *Cyclopaedia*, vol. 1, pp. i–v, for discussion of the chart. On the art/science relationship,

What relationships between the various sciences did this "View" convey? Chambers did not follow Bacon's classification by mental faculties, which allocated both the mechanical arts and natural history to Memory. Admittedly, in the Preface he did speak of different sciences deriving from the senses, reason, or imagination – thus apparently referring to mental faculties – but his classification is not a psychological one; in fact, he appears to have regarded the major divisions of arts and sciences as only conventional labels.[37] Moreover, his classification by dichotomies placed *both* natural history and natural philosophy on the "scientifical" branch, unexpectedly separating natural philosophy from the disciplines of mixed mathematics, which are located on the "artificial" or "technical" branch. Nevertheless, like Bacon and Harris, Chambers distinguished between the two large categories of natural history and natural philosophy: the former as "Sensible" and the latter as "Rational." And in the body of the work he made it clear that natural philosophy, or the version of it pursued by Newton – namely, "experimental philosophy" – was "scientific" in a way that studies in natural history were not: "In Effect, *Experiments,* within these 50 or 60 Years, are come into such Vogue, that nothing will pass in Philosophy, but what is founded on Experiment, etc. So that the new Philosophy is almost altogether *Experimental.*"[38]

In both Harris and Chambers, chemistry is the third man, neither conclusively under natural history nor conclusively under natural philosophy. Chambers's entry on this subject is more detailed than Harris's, describing chemistry as an art of analysis: "separating the several Substances whereof mix'd Bodies are compos'd." Chemistry was still defined as an art, and not a science, in Samuel Johnson's *Dictionary* of 1755 which, like Chambers, cited the definition given by Hermann Boerhaave (1668–1738), head of the medical faculty at the university of Leiden and professor of chemistry there from 1718 to 1729.[39] (Chambers and Peter Shaw (1694–1763) translated and published *A New Method of Chemistry* in 1727, based on Boerhaave's lectures.) But even though it was clearly recognized as a distinct subject, Harris and Chambers were unsure about its relation to the two large categories that informed their classification. This is highlighted by the fact that the list of terms under "Natural Philosophy" in the Index of the *Lexicon* included terms that, later in the century, would fall uncontentiously under chemistry – terms such as "Acidity," "Air," "Condensation," "Fermentation," "Phosphorus," "Spring of the Air," "Sulphur" and "Vapours." A few of these also occur under the heading for "Chymystry," but their presence in two places requires comment.

---

see the long discussion in the Preface at vol. 1, pp. vii–xvi, and "Science," vol. 2 (no pagination in body of the work).

[37] Ibid., vol. 1, p. ii.; also Yeo, "Reading Encyclopedias," pp. 28–9; Fisher, "The Classification of the Sciences," p. 861.

[38] Chambers, *Cyclopaedia,* vol. 1, "Experimental Philosophy."

[39] Chambers, *Cyclopaedia,* "Chymistry," vol. 1. See J. R. Partington, "Chemistry through the Eighteenth Century," in Alan Ferguson (ed.), *Natural Philosophy Through the 18th Century and Allied Topics* (London: Taylor and Francis, 1972), pp. 47–66 at p. 48.

In both these scientific dictionaries, the category of natural philosophy (or "physicks") operates in a different way from the most advanced disciplines usually accepted as part of it. In the *Lexicon* (Volume Two), the entry for "Physicks," or natural philosophy, confirms that the sciences of astronomy and optics, most illustriously pursued by Newton, certainly belong to this category. The entry is mainly a list of books that "will give the Reader a true and useful knowledge of Nature"; it begins with the *Principia*. But this bibliography is not confined to the so called Newtonian sciences, and it includes John Woodward (1665–1728) and William Whiston (1667–1752) on the history and theory of the earth. The heading of natural philosophy in the Index also goes beyond the mixed mathematical sciences in its list of terms – a curious catalog including not only the chemical terms mentioned above, but also some that seem to belong elsewhere, such as "Animals," "Earth," "Stones," "Vegetables," "Zoography." Some of these terms, as we might expect, also occur under "Botany, Natural History." Similarly, in his explication of "Physics, or Natural Philosophy" in the large footnotes accompanying the chart, Chambers does not mention the terms from the most obvious sciences – namely, those of mixed mathematics – because he gives them their own headings. Rather, this note shows the province of natural philosophy by listing terms pertaining to the "Powers" and "Properties" of nature such as attraction, elasticity, cohesion, electricity, and magnetism. Thus, natural philosophy functions as a general label for inquiries into the principles and causes of natural phenomena as well as a heading for a number of recognized disciplines. But it was not confined to the "classical sciences," as defined by Kuhn. Instead, some of what Kuhn called Baconian sciences were seen as legitimate, if undeveloped, parts of natural philosophy and its search for causes of phenomena in nature. This is why the German philosopher Christian Wolff, in *Preliminary Discourse on Philosophy in General* (1728), could consider a subject such as meteorology (classed as natural history in the English works) as natural philosophy, provided that it searched for causes of phenomena such as rain, rainbows, and lightning. Other forces and powers of nature, such as electricity and magnetism, thus came under this heading even though they had not been successfully explained on mechanical principles.[40] Chemistry, in particular, was seen as pressing its claim to be a science of causes and powers. Harris included the term "Acids" under natural philosophy and, in the Introduction to the volume of 1710, advertised the insertion of an unpublished paper, "De Acido," by Newton. He supplied a translation of this, noting how it made good Newton's suggestion in the *Optics* that attractive forces between small particles of matter could be understood in terms of laws of matter and motion.

---

[40] Christian Wolff, *Preliminary Discourse on Philosophy in General*, translated, with an introduction and notes, by Richard J. Blackwell (New York: Bobbs-Merrill, 1963), p. 42. See also Patricia Fara, *Sympathetic Attractions: Magnetic Practices, Beliefs, and Symbolism in Eighteenth-Century England* (Princeton, NJ: Princeton University Press, 1996), pp. 142–3.

Indeed, the last sentence of Chambers's entry gave an optimistic gloss on this story: "Dr. Friend [sic] has reduc'd *Chymistry* to *Newtonianism,* and accounted for the Reasons of the Operations on Mechanical Principles."[41]

It could be said that the main concern of Harris and Chambers was not the sophisticated mapping of the relation *between* sciences but rather was the listing of cognate terms under certain sciences. Chambers built on Harris's Index by showing the arts and sciences on a chart, but the main contribution of his work was the use of cross-references between the terms of *each* science. Nevertheless, both compilers assumed a larger classification as the foundation of their comments on the sciences. Whereas Harris's *Lexicon* had no map or chart and Chambers did not use Bacon's division by mental faculty, these two English dictionaries were informed by the contrast between natural philosophy and natural history. In fact, they may have adopted it more completely than Bacon, who always regarded the data of natural history as the "primary matter" on which the causal inquiries of natural philosophy were built.[42]

THE *ENCYCLOPÉDIE*

Diderot and d'Alembert acknowledged that Chambers had sought to sketch the relationships between the various sciences; but they claimed that this needed more attention and made much of rediscovering Bacon's contribution. The famous frontispiece of the *Encyclopédie* by Charles-Nicolas Cochin (1715–1790) was not prepared until 1764, but it expressed the message of both the prospectus (1750) and *Preliminary Discourse* (1751). It shows three figures. Reason, the most prominent, is lifting the veil from Truth (with the help of Philosophy); Memory and Imagination, each accompanied by its respective sciences and arts, are situated, respectively, to the right and left of Truth.[43] At the end of the *Preliminary Discourse,* the diagram (see Figure 10.2) depicting the Baconian system – which the editors usually referred to as an "encyclopedic tree" – made it clear that Reason controlled the largest number of arts and sciences. This point was graphically underscored later in the engraving of a tree of knowledge in the frontispiece to volume one of the supplementary index in 1780. This was a large folding sheet (39 by 24 inches) with tree and branches engraved by Robert Benard.[44] Here the trunk of Reason

---

[41] Harris, *Lexicon,* 1710, vol. 2, Introduction, for the paper by Newton; Chambers, *Cyclopaedia,* "Chymistry," vol. 1. This is a reference to the Oxford academic John Freind. On the point that chemistry was a "core subject at a time when physics [in the modern sense] had hardly achieved that status," see Maurice Crosland, *In the Shadow of Lavoisier:* The Annales de Chemie *and the Establishment of a New Science* (Oxford: British Society for the History of Science, 1994), p. 156.

[42] Bacon, *Works,* vol. 3, p. 356; Kusukawa, "Bacon's Classification of Knowledge," p. 53.

[43] Georges May, "Observations on an Allegory: The Frontispiece of the *Encyclopédie,*" *Diderot Studies,* 16 (1973), 159–74, at 162–4.

[44] For the tree image, see d'Alembert, *Preliminary Discourse,* p. 159; but he also used a "map" metaphor

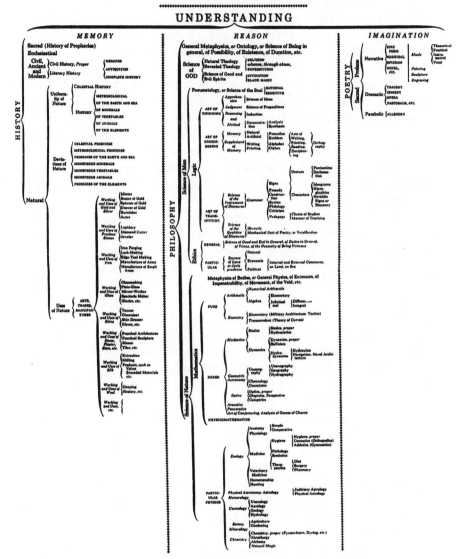

Figure 10.2 The classification of knowledge, influenced by Bacon, given in d'Alembert's *Preliminary Discourse* (1751). Source: *Preliminary Discourse to the Encyclopedia of Diderot,* translated by Richard N. Schwab, with the collaboration of Walter E. Rex; with an introduction and notes by Richard N. Schwab (Chicago: University of Chicago Press, 1995), pp. 144–5.

overwhelms the two main branches of Memory and Imagination; in fact, the sub-branch for Mathematics, shooting off from the main trunk of Reason, is itself more luxuriant than either of these.

Diderot and d'Alembert did not passively repeat Bacon's classification; they transformed his concept of Philosophy – the foundational trunk of all the sciences – into the Enlightenment torch of Reason. Robert Darnton suggests that Diderot and d'Alembert took "enormous risks" in undoing "the old order of knowledge" in this way: that is, by replacing Theology with Reason or Philosophy and excluding all knowledge without an empirical base, rather than allowing a separate tree, as Bacon did, for Divine Knowledge.[45] But for the purpose of this chapter, there is another issue: what did their use of the Baconian scheme entail for the way the *sciences* were classified? Apart from the restoration of the three mental faculties, did their chart of knowledge present a different arrangement of the sciences from that of Chambers or Harris?

The French editors stressed that "all encyclopedic trees necessarily resemble one another" in terms of the kinds of arts and sciences they included; the differences concerned the order and arrangement of the various branches. As d'Alembert put it, "One finds virtually the same names of the sciences in the tree of Chambers as in ours; yet nothing could be more different."[46] This is an admission that, like Harris and Chambers, they also worked with the categories of natural history and natural philosophy. The former, under Memory (where Bacon put it), included descriptions of the uniformities and deviations of nature together with the uses of nature exemplified in all the practical arts. Natural philosophy was not named as such, but all the sciences that Harris and Chambers placed in this category were now under "Science of Nature," which belonged to Philosophy and, of course, to the faculty of Reason. But Diderot and d'Alembert also began to undermine the earlier qualitative distinction between these two large categories. The membership of "Particular Physics" – the main grouping of physical sciences (apart from mixed mathematics) – included subjects such as zoology, meteorology, botany, mineralogy, and geology. Earlier mapmakers, such as Chambers, grouped these under Bacon's heading of natural history. Significantly, then, these subjects were now released from the lowly domain of Memory. Instead, they joined chemistry, which was also now indisputably a member of these sciences of nature, although it was also singled out as "the imitator and rival of nature," and the article on "natural history" declared that chemistry started where

---

(pp. 47–50). On the engraving, see Robert Shackleton, "The Encyclopaedic Spirit," in Paul J. Korshin and Robert R. Allen (eds.), *Greene Centennial Studies: Essays Presented to Donald Greene* (Charlottesville: University of Virginia Press, 1984), pp. 377–90, at pp. 386–7.

[45] Robert Darnton, "Philosophers Trim the Tree of Knowledge," in *The Great Cat Massacre and Other Essays in French Cultural History* (London: Penguin, 1985), pp. 185–207, at p. 187. See also Cassirer, *Philosophy of Enlightenment*, p. vii.

[46] D'Alembert, *Preliminary Discourse*, p. 159; see also pp. 151–5.

natural history ends.[47] Thus, under "Reason" in the *Encyclopédie,* there was now a continuum from pure and mixed mathematics to the experimental and observational sciences rather than a qualitative break between natural philosophy and natural history.

At one level, this reflected the rising status of the natural history disciplines. Diderot was an active defender of the organic sciences against the authority of mathematics, a campaign also assisted by Buffon in the introductory Discourse to his *Histoire Naturelle* (1749), where he insisted that the natural history disciplines must generalize. In 1777, Kant distinguished between *Naturbeschreibung* (description) and *Naturgeschichte* (historical development), thus opening the possibility of a study of *historical* causation in nature – thus, still distinct from natural philosophy but not by being limited to description and taxonomy.[48] In Britain, the shift to this more theoretical agenda for natural history was slower to appear. The *Encyclopaedia Britannica* maintained the distinction between natural history and natural philosophy, stating that only the latter had "universal laws of nature" as its province. This position continued in the third edition in the short entry for "Natural Philosophy"; but there was also the acknowledgment here that the data provided by natural history was the basis for more theoretical and causal speculations. James Hutton (1726–1797) put this more positively than some of his fellow Scots in 1794, arguing that "natural history and natural philosophy should proceed together with mutual advantage."[49] At another level, however, this more relaxed attitude to earlier distinctions between these two large categories reflected the demise of systematic classification of the sciences in encyclopedias of the late eighteenth century.

THE DEMISE OF MAPS OF
KNOWLEDGE IN ENCYCLOPEDIAS

In 1778, in the first volume of the *Deutsche Enzyclopädie,* its editors attacked Diderot and d'Alembert for their fixation on general principles and "the overgrown forest of a connected system" that they sought to convey. By way of contrast, this German work offered to "leave both the effort of frantically looking for a connection among materials and sciences that are barely or not at all connected and the honor of the task, to the compilers of the French

---

[47] Ibid., p. 155; "Histoire Naturelle," in *Encyclopédie; ou Dictionnaire Raisonne des Sciences, des Arts, et des Metiers* (Paris: Briasson et al., 1751–65), vol. 8, pp. 225–30, at p. 228.

[48] Cited in John Lyon and Phillip Sloan (eds.), *From Natural History to the History of Nature* (Notre Dame, IN: University of Notre Dame Press, 1981), p. 2. For Buffon' s "Initial Discourse," see pp. 97–128.

[49] "Physics," *Encyclopaedia Britannica,* 2nd ed. (Edinburgh: Bell and Macfarquhar, 1778–84), vol. 7, p. 6171; "Natural Philosophy," *Encyclopaedia Britannica,* 3rd ed., 18 vols. (Edinburgh: Bell and Macfarquhar, 1788–97), vol. 12, p. 670; James Hutton, *An Investigation of Principles of Knowledge* (Edinburgh: A. Strahan, 1794), vol. 3, p. 38.

Encyclopédie."[50] As it turned out, this restraint did not help the editors of this encyclopedia complete their project: it terminated at the letter *K* in 1804. But the doubts voiced here about classification of knowledge were already apparent in the *Encyclopédie*. At the end of the *Preliminary Discourse*, d'Alembert confessed that "our readers" might not be much interested in disquisitions on trees of knowledge. Although the discussion of Bacon's scheme undoubtedly brought the issue of classifying sciences to a wider public, the mixed metaphors of maps, charts, and trees that pervade the text may have contributed to doubts about this exercise. In spite of their comments on the importance of such classification, Diderot and d'Alembert made it clear that they regarded all systems of this kind as arbitrary and relative. The entry on "Philosophie" (published in 1765) referred to the non-Baconian system of Christian Wolff.[51]

With the publication of the *Encyclopaedia Britannica* between 1768 and 1771 there was a major encyclopedia with no map of knowledge.[52] The first two editions criticized Chambers's approach for fragmenting sciences into short entries on terms and proclaimed their "new plan": larger treatises on the major subjects, although still in alphabetical order, and short entries as satellites to the long treatises. By the third edition, starting in 1788, there was a frontal assault on the notion that a chart or map could assist the reader in reconstituting sciences that had been scattered by the alphabet. Acknowledging Chambers's efforts, the Scottish editors declared that his work "was still a book of shreds and patches, rather than a scientific dictionary of arts and sciences." Indeed, they went further, invoking the authority of Thomas Reid (1710–1796) to spurn all systematic classification of the sciences as presumptuous, as trying to "contract the whole furniture of the human mind in to the compass of a nutshell." They even included a copy of Chambers's chart, introduced by this note: "To be convinced of the truth of this assertion, one needs but cast his eye over the author's table of arrangement."[53] Significantly, the mere sight of this chart is taken here as an argument against it, and the *Cyclopaedia* is branded as a miscellany in spite of its attempt to provide a path through the various sciences. By the last quarter of the century, most encyclopedias had abdicated responsibility for any systematic classification of the sciences they covered. The emphasis was now on coherence at the level of

---

[50] Willi Goetschel, Catriona MacLeod, and Emery Snyder, "The Deutsche Encyclopädie and Encyclopedism in Eighteenth-Century Germany," in Clorinda Donato and Robert M. Maniquis (eds.), *The Encyclopédie and the Age of Revolution* (Boston: G. K. Hall, 1992), pp. 55–61, at p. 58.

[51] D'Alembert, *Preliminary Discourse*, p. 164; Tonelli, "The Problem of Classification," p. 264 referring to "Philosophie," in the *Encyclopédie*, vol. 12, pp. 509–15. See McCrae, *The Problem of Unity*, pp. 8, 109–122, on the ambiguous attitude of d'Alembert and Diderot to the classification of the sciences.

[52] See Yeo, "Reading Encyclopaedias," pp. 29–34. Some minor English works emulated the *Encyclopédie* by carrying a Baconian map of knowledge; but their prefaces stressed the difficulty of defending any single version of such grand taxonomies. See, for example, [John Barrow], *A Supplement to the New and Universal Dictionary* (London: printed for the Proprietors, 1754), preface, pp. 9, 14. The first volume of 1751 had a "Synopsis" of arts and sciences "arranged in their proper order," but no chart.

[53] *Encyclopaedia Britannica*, 3rd ed., 1788–97, vol. 1, pp. vii–viii.

disciplines, expounded in extensive treatises. Indeed, from 1782 the *Encyclopédie Methodique* (the successor to the *Encyclopédie*) was really a dictionary of dictionaries, so that, in the words of a reviewer, "every science will have its dictionary, or system, apart."[54] The *Britannica* continued this format, devoting large treatises to the major disciplines but placing them alphabetically within its volumes.

Soon after the decision to supply large treatises on each science there was recruitment of expert contributors. The third edition of the *Britannica* proudly announced its use of respected writers for various branches of science, a feature that became even more prominent in the *Supplement* of 1801, when John Robison (1739–1805) did almost all the physical sciences and Thomas Thomson (1773–1852) took over chemistry. This was linked with the need to keep abreast of the most recent advances. Thomson explained what this meant in the case of chemistry:

> So rapid has this progress been, that though the article Chemistry in the *Encyclopaedia Britannica* was written only about ten years ago, the language and reasoning of chemistry have been so greatly improved, and the number of facts have accumulated so much, that we find ourselves under the necessity of tracing over again the very elements of the science.[55]

The treatises on disciplines, written by experts, became more specialized: the cross-references from these articles were mainly to shorter entries on the cognate terms of a particular science, and not to other sciences. The boundaries between disciplines were often sharpened as contributors sought to codify the agreed data and principles of their own subject and as editors worked to allocate subject matter to different contributors.[56] In the case of topics such as heat, magnetism, and electricity, this could lead to artificially clean demarcations. But this drawing of boundaries did not renew consideration of the relations between various sciences under a broader natural philosophy. In fact, the entries on this term (and on natural history) were usually short, giving a historical gloss on its earlier meaning but then referring to the separate articles on the physical sciences, such as mechanics, hydrostatics, optics, and astronomy, and the new subjects forming around studies of magnetism and electricity. By the late eighteenth century, encyclopedias were also carefully registering the identity of new organic sciences that did not fit the old category of natural history. The *Britannica* explained that physiology "is a Greek word, which, in strict etymology, signifies that which discourses of

54 "Proposals for Publishing a Methodical Cyclopaedia," *Monthly Review,* 66 (1782), pp. 514–18, at p. 514.
55 Thomas Thomson, "Chemistry," in George Gleig (ed.), *Supplement to the Third Edition of the Encyclopaedia Britannica* (Edinburgh: T. Bonar, 1801), vol. 1, p. 210. On the use of specialists, see vol. 1, p. v.
56 Yeo, "Reading Encyclopedias," pp. 43–7; for the *Encyclopédie Methodique,* see Robert Darnton, *The Business of Enlightenment: A Publishing History of the Encyclopédie, 1775–1800* (Cambridge, MA: Harvard University Press, 1979), pp. 422, 451.

nature: but in its common use, it is restricted to that branch of physical science which treats of the different functions and properties of living bodies." This was a definite dismissal of the more general sense this term had earlier in the century, when it was still given by Harris, Chambers, and Martin as equivalent to physics or natural philosophy. It was now explicitly defined as a specialist discipline with a distinct identity: "We choose here to mark precisely the bounds of physiology, because we have always been led to imagine that it would be extremely fortunate for science that all its divisions were accurately defined, that each were restricted in its own sphere."[57]

CONCLUSION

With the collapse of the main categories of natural history and natural philosophy – which had been central to most classification of the sciences – encyclopedias abandoned any attempt to show how the various scientific subjects related to one another. The "circle of sciences" was no longer a path that readers were expected to follow. This did not mean that distinctions between the sciences became unimportant: as the example of physiology suggests, scientists were possibly becoming more concerned with marking out the boundaries of their specialist disciplines than earlier natural philosophers had been. Indeed, specialization stimulated the elaborate classification schemes of Comte, Ampère, Spencer, and others in the nineteenth century. But at the level of the public communication of science in encyclopedias, the emphasis on coherence was at the level of increasingly autonomous disciplines rather than on the position of these on a map or chart of sciences. It is a telling point that when the *Britannica* used the word "systems" on its title page, it referred to its treatises on particular sciences, and not to grand doctrines of natural philosophy or to the classification schemes that once prefaced earlier encyclopedias.

[57] "Natural Philosophy," *Encyclopaedia Britannica*, 3rd ed., vol. 12, pp. 670–1; "Physiology," vol. 14, p. 665.

II

PHILOSOPHY OF SCIENCE

Rob Iliffe

In recent years philosophy of science and the history of philosophy of science have been subjected to a number of critiques by scholars from areas such as sociology of science and history of science. The following is a litany of some of their complaints. Philosophers of science (it is argued) do not deal with the practical engagement with the world that is the central part of scientific activity, and their view of the nature and function of scientific theory is fanciful and biased ("theory" is seen as prior to, and more historically significant than "practice"). Historians of philosophy anachronistically decide what constituted important problems in the past, selecting for study the works of great men whose doctrines they wrench from their historical contexts. They then misinterpret and present the corpus of an individual's published writings as if it were coherent across various projects and over lengthy periods of time. Philosophers are taken to be in dialog with the timeless problems of their ancestors, and the "progressive," pure aspects of scientific work are divorced from other areas of an individual's intellectual output, such as theology and economics, which are seen as inferior productions. In dealing with the legacy of Newton, "Newtonians" merely develop and never radically challenge powerful suggestions that are inherent within the public texts of the Master, whereas "anti-Newtonians" are lumped together, whatever their doctrines, and whatever the traditions within which they write. As one corollary of Newtonocentrism, historians have tended to argue that all decent examples of exact science in the eighteenth century are the result of successfully grappling with problems laid out or "hinted" at in Newton's works.[1]

---

[1] See L. Laudan, "Theories of Scientific Method from Newton to Kant," *History of Science*, 7 (1968), 1–61; G. Buchdahl, *Metaphysics and the Philosophy of Science: The Classical Origins, Descartes to Kant* (Oxford: Basil Blackwell, 1969); for matter theory, see A. Thackray, *Atoms and Powers* (Cambridge University Press, 1970); R. E. Schofield, *Mechanism and Materialism: British Natural Philosophy in an Age of Reason* (Princeton, NJ: Princeton University Press, 1970); and P. Heimann and J. E. McGuire, "Newtonian Forces and Lockean Powers," *Historical Studies in the Physical Sciences*, 3 (1971), 233–306. Among many critiques of the underlying assumptions of these projects, see D. Bloor, *Knowledge and Social Imagery* (London: Routledge, 1976); S. Schaffer, "Natural Philosophy," in G. S. Rousseau

267

Although on the whole these criticisms have some force, their effect has been to grossly curtail historical discussions of the concepts and theories of the period. This surely constitutes something of an overreaction. In this article I assume that "philosophy of science" includes epistemology, methodology, ontology, and metaphysics, and I argue that what we now see as "extra-scientific" issues were intimately linked to the contemporary investigation of the natural world. I begin with a brief history of relevant developments before Newton's *Principia* and then show how the formulation of metaphysical questions and matter theories was related to theological and political contexts. Due partly to new research in light, heat, and electricity, natural philosophers increasingly devised dynamicist theories of matter in which forces and powers were considered to be immanent within matter, often with influence on space surrounding the matter itself. I then look at some ways in which the authority of Newton and Bacon was invoked in the Enlightenment, and I follow recent analyses in arguing that methodologies do not and could never describe the actual process of scientific inquiry, but instead have historically served a number of different functions. I conclude by briefly drawing attention to the way the exact sciences came to "bracket off" certain questions and styles of inquiry as unfruitful or illegitimate.

APPROACHES TO NATURAL PHILOSOPHY IN THE SEVENTEENTH CENTURY

Traditionally there was a deep-seated division between natural philosophy and mathematics that was enshrined in the relative status of the respective professors in universities. Medieval philosophy, drawing from the analysis in Aristotle's *Posterior Analytics,* dealt with qualitative, causal accounts ("explanations") of the nature of changes ("motions") of phenomena that were couched in terms of the four types of causes and the four Aristotelian elements. By means of the syllogism one "demonstrated" from observed effects to the sole and necessary cause, moving then to show how the effect was necessarily the result of the cause. The building blocks of the scholastic approach to nature were "experiences" that were universal and evident to every rational person; knowledge of these was knowledge of things that *had* to happen in nature, and not of singular events that by themselves were indications of no underlying regularity. The self-evidence of these elements was made problematic in the early seventeenth century by the advent of "contrived experiences," or what we would now call experiments. These were unnatural situations inaccessible

and R. Porter (eds.), *The Ferment of Knowledge: Studies in the Historiography of Eighteenth Century Knowledge* (Cambridge University Press, 1980), pp. 55–91; G. N. Cantor, "The Eighteenth Century Problem," *History of Science,* 20 (1982), 44–63; J. Schuster and R. Yeo (eds.), *The Politics and Rhetoric of Scientific Method* (Dordrecht: Reidel, 1986); J. Rée, *Philosophical Tales: An Essay on Philosophy and Literature* (London: Methuen, 1987).

to all but a few privileged individuals, and the natural knowledge gained from such situations was validated by authors claiming that they had created these experiences many times and in front of many people.[2]

On the other hand, despite being part of the university quadrivium (music, geometry, astronomy, and arithmetic), mathematics had traditionally had a much lower status than philosophy, pertaining as it did to *measurable* external dimensions such as size, quantity, and duration rather than to the essences of things. The difference between the two activities is perhaps best seen in astronomy, although Aristotle himself was unsure whether to class it as a branch of mathematics or physics (natural philosophy). The celestial sphere was unchanging and so not susceptible to explanations of an elemental type; moreover, there were no conventional means of determining which of a number of different "hypotheses" about the "real" motions of the heavens was true. Instead, following Ptolemy, most astronomers held that the business of their discipline was to "save the phenomena" and create geometric representations of the heavens (such as epicycles) without asserting their physical truth. For that reason astronomy would never be a proper "science" in the Aristotelian sense, since one could never specify the one necessarily true cause of the observed motions.[3]

One of the most innovative attempts to find certain knowledge about the natural world was pioneered by René Descartes. In the midst of a skeptical crisis about the possibility of true knowledge about religion and the natural world, Descartes aimed for a mathematically certain understanding of objects in his *Rules for the Direction of the Mind* of the late 1620s, although his epistemological project of achieving certain knowledge by means of introspection was to prove a barren resource for natural philosophers in the following century. In the *Discourse on Method* of 1637 he still promoted the value of deducing effects from first principles known a priori, but in the final part of the work he claimed that one needed to find "experiences" that would decide between different possible explanations of the world. Descartes's claim that the essence of matter was extension gave rise to a plenist ontology in which the motions of the heavens were explained by massive vortexes, terrestrial gravitation being accounted for by the pressure of the surrounding vortex. Whatever the plausibility of the specific hypothesis, this style of explanation was prominent for about a century in France, although its status as an absolutely certain "deduction" rather than as a plausible model was obviously dubious. The Cartesian attitude to the structure of the physical world and to the nature of scientific explanation remained extremely influential in France at the

[2] P. Dear, *Discipline and Experience: The Mathematical Way in the Scientific Revolution* (Chicago: University of Chicago Press, 1995).

[3] N. Jardine, *The Birth of the History and Philosophy of Science: Kepler's* A Defence of Tycho against Ursus with Essays on Its Provenance and Significance (Cambridge University Press, 1986), pp. 211–57, especially pp. 144–5; R. Westman, "The Astronomer's Role in the Sixteenth Century: A Preliminary Study," *History of Science*, 18 (1980), 105–47.

beginning of the eighteenth century, despite the fact that the majority of intellectuals were well aware of drawbacks in the details of Descartes's schemes.[4]

Although Descartes and others vigorously promoted the mechanical philosophy and it enjoyed widespread support throughout the seventeenth and eighteenth centuries, it was often accompanied by the view that we could not divine the *precise* mechanical workings of the cosmos a priori. For many contemporaries, the increasingly popular view that the essences of things were unknowable and even unintelligible tended to give rise to a version of nominalism – namely, that what could be inferred from our knowledge of observable entities could be given a name, without a commitment to its ontological underpinnings. On this view, although a causal account might be attainable in the long term, for the moment one would have to make do with generalizations that went no further than what was warranted by phenomena. A version of causal nescience (the position that one should eschew reference to unknown causes) was adopted by Galileo in the *Dialogue* of 1632 and the *Two New Sciences* of 1638. However, this derived from a different tradition to that of Robert Boyle, who also eschewed references to unexperienced "causes" later in the century. The probabilism adopted by a number of early seventeenth-century natural philosophers, and Boyle's view that our knowledge of the outside world could only be *morally* certain, was very different from Galileo's belief in a mathematically certain Aristotelian science. Galileo believed from early in his career that a true science should be conclusively demonstrable, and he suggested in the *Dialogue* that the ability of the Copernican system to explain the tides meant that it was the true and necessary cause of the phenomena of the world.[5]

In appealing to a mathematical conception of nature, Galileo drew from the so-called mixed mathematical sciences such as music and optics, which had succeeded in creating numerical representations of aspects of the real world. Galileo's mathematical approach constituted a new science of motion, disdaining reference to the unseen microworld in order to account for uniform acceleration. On the Second Day of the *Dialogue*, Galileo's mouthpiece Salviati rebuked the Aristotelian Simplicio for speaking of an entity, "gravity," as if it were the cause of uniform acceleration. Rather, *whatever it is* it should be *called* gravity, since we know only its name. In the *Two New Sciences*, Simplicio and Sagredo debated how acceleration might be explained in terms of

[4] E. McMullin, "Conceptions of Science," in D. C. Lindberg and R. S. Westman (eds.), *Reappraisals of the Scientific Revolution* (Cambridge University Press, 1994), pp. 32–5; D. Garber, *Descartes' Metaphysical Physics* (Chicago: University of Chicago Press, 1992); E. Aiton, *The Vortex Theory of Planetary Motions* (London: Macdonald, 1972).

[5] S. Shapin and S. Schaffer, *Leviathan and the Air-Pump: Hobbes, Boyle and the Experimental Life* (Princeton, NJ: Princeton University Press, 1985); Dear, *Discipline and Experience*. For a seminal account of the differences between the various approaches mentioned here, see T. S. Kuhn, "Mathematical Versus Experimental Traditions in the Development of Physical Science," in Kuhn, *The Essential Tension: Selected Studies in Scientific Tradition and Change* (Chicago: University of Chicago Press, 1977), pp. 31–65, especially pp. 41–59.

the relative strength of impressed force and weight. At this point Salviati violently interjected that this was not the time "to enter into the investigation of the cause of the acceleration of natural motion, concerning which various philosophers have produced various opinions . . . such fantasies, and others like them, would have to be resolved, with little gain." Such a dismissive attitude toward causal explanations was clearly troublesome to both Descartes and his Jesuit contemporaries, for whom this way of proceeding was unphilosophical and groundless, and an identical attitude colored many of the responses of Continental Europeans to Newton's *Principia Mathematica* of 1687.[6]

A sophisticated experimental approach was to prove equally important for natural philosophers in the early eighteenth century. In various works composed as part of his great reformist project *The Great Instauration,* Francis Bacon had earlier attempted to provide a new tool or method to produce certain knowledge of what he called the nature of "forms." Although veridical sense perception was basically built in to Aristotelian epistemology, Bacon distrusted bare senses as media of knowledge. He called for a collectivist natural history of the world, moving inductively from observational experience and history of single facts by means of artfully contrived experiments to the top of a pyramid where stood "metaphysic." His inquisitorial, interventionist approach to Nature (he was Lord Chancellor) was admired in the late seventeenth century by Robert Boyle, who likewise lauded the sort of knowledge gained from practical craft traditions and developed a more sophisticated experimental philosophy in the second half of the seventeenth century. Although he too distrusted the 'big systems' of Aristotle and others, he advocated a more skeptical attitude toward the possibility of attaining certain knowledge and argued that at best one could only have a concurrence of probabilities in support of a knowledge claim. Indeed, his experimental philosophy was based on the acquisition of experimental "singulars," created at a certain time and place and in principle replicable by readers of his texts.[7] Both advocates of experiment associated speculation with contention, although Boyle also had a detailed conception of the heuristic role of hypotheses in philosophy. Boyle criticized hasty speculation and attacked mathematical natural philosophers such as Blaise Pascal who described contrived "experiences" that were schematic representations of evident universal truths rather than descriptions of true and replicable experiments performed at specific historical moments. Despite the

---

[6] Galileo, *Two New Sciences: Including Centers of Gravity and Force of Percussion,* translated with an introduction by S. Drake (Madison: University of Wisconsin Press, 1974), pp. 157–9; Dear, *Discipline and Experience.*

[7] McMullin, "Conceptions of Science," pp. 27–92, especially pp. 45–54; R.-M. Sargent, *Diffident Naturalist: Robert Boyle and the Philosophy of Experiment* (Chicago: University of Chicago Press, 1995). For a comprehensive account of the way that Bacon's work was used in the middle of the seventeenth century, see C. Webster, *The Great Instauration: Science, Medicine and Reform, 1626–1660* (London: Duckworth, 1975).

popularity of Boyle's "style" of natural philosophy, Newton's successful math-ematization of the laws of motion and Universal Gravitation pointed to a much more ambitious and potent mode of natural philosophy at the start of the eighteenth century.[8]

THE HERITAGE OF NEWTON

Newton's approach can be seen as combining various aspects of the Galilean and Boylean styles of natural philosophy, and his *ex cathedra* statements about scientific method functioned as the scientific Ten Commandments of the fol-lowing two centuries. From the start of his tenure as Lucasian Professor of Mathematics in 1669, he assailed the probabilism of his contemporaries and argued that the science of colors could be made as certain as any other part of optics. In his theory of light and colors presented to the Royal Society in early 1672, he claimed mathematical certainty for his theory of the hetero-geneity of white light, demonstrated by means of a "crucial experiment." Lambasting all the alternative contemporary philosophies of science and es-pecially the appeal to "hypotheses," Newton prescribed a rigid methodology to the rest of the philosophical community based on experiment and then induction to general mathematical relationships or laws of nature.[9]

Unlike Galileo, Newton explicitly derided efforts to base natural philoso-phy on a multiplicity of experiences, nor did the pretence of narrating a Boy-lean, historical story of discovery at the start of his 1672 paper last for any length of time, even in the article itself. In the *Principia Mathematica* of 1687 he bequeathed a series of "Rules of Reasoning in Philosophy." The first two put limits on the sorts of causes that could be invoked, and the third gave a warrant for moving inductively from experienced qualities such as hardness and gravitation to all bodies whatever. The fourth expressed the view that one should prefer principles gained by induction from phenomena and that these should be taken as true until other, stronger, inductively derived evidence was brought to bear. Newton appealed to the method of "resolution" (or "analy-sis") and "composition," which in its full version finally made it into "Query" 31 in the 1717 edition. The first part of this approach included the making of experiments and observations, proceeding from "Compounds" to "Ingredi-ents" and then down to basic "forces," and in general to the most basic causes of all; the second part assumed these as principles and then explained the effects from them. For Newton, "hypotheses" might gain people a name but

[8] Sargent, *Diffident Naturalist*, pp. 26–61. For the cultural contexts of Boyle's distrust of claims to ab-solute (syllogistic or mathematical) certainty, see P. B. Wood, "Methodology and Apologetics: Thomas Sprat's *History of the Royal Society,*" *British Journal for the History of Science*, 12 (1980), 1–26.

[9] McMullin, "Conceptions of Science," p. 61; Newton to Oldenburg, 6 February 1671/2, in H. W. Turn-bull et al. (eds.), *The Correspondence of Isaac Newton*, 7 vols. (Cambridge University Press, 1959–81), vol. 1, especially pp. 92–107, pp. 96–7.

they were little better than a "Romance," and he carefully bracketed off natural phenomena covered by the laws of motion and universal gravitation from those (such as electricity and magnetism) whose mathematical description had not yet been given. As with Galileo, there was a strident provisionalism within his approach to natural philosophy: "To explain all nature is too difficult a task for any one man or any one age. 'Tis much better to do a little with certainty & leave the rest for others that come after you."[10]

Contemporaries saw a clear division between the phenomenalist statements with which he referred to gravity and even God, and the accounts in the Queries that, although stated in a hypothetical language, left little doubt about what Newton 'really thought'. Although he had attempted to account for gravity in terms of an ethereal mechanism in 1675, he generally professed ignorance about the cause of mutual attraction, saying in the General Scholium to the Second Edition of the *Principia* (of 1713) that to do so would constitute a "hypothesis": "to us it is enough that gravity does really exist, and act according to the laws which we have explained, and abundantly serves to account for all the motions of the celestial bodies." Nor, as he wrote to Richard Bentley in early 1693, should he ascribe the view that gravity was "essential & inherent to matter" to Newton: "That gravity should be innate inherent and essential to matter so that one body may act upon another at a distance through a vacuum without the mediation of any thing else by and through which their action or force may be conveyed from one to another is to me so great an absurdity that I believe no man who has in philosophical matters any competent faculty of thinking can ever fall into it." This conclusive rejection of action at a distance as a true cause of gravitation was extraordinarily influential in conditioning interpretations of Newton's "real" opinions on the subject when it was published in 1756.[11]

The General Scholium was crafted partly in response to a letter sent by Leibniz to Nicholas Hartsoeker and published in the weekly journal *Memoirs of Literature* of 5 May 1712. In it Leibniz had reflected adversely on the idea that all bodies attract each other by a law of nature that God implanted at the creation, calling this a "continual miracle" and "a fiction invented to support an ill grounded opinion." According to Newton the argument made by Leibniz about gravity could be made about even basic qualities such as hardness, namely that they "must go for unreasonable occult qualitys unless they can be explained mechanically." No more easily could inertia, extension,

---

[10] Newton, *Opticks* (New York: Dover, 1979, from fourth edition of 1730), pp. 401–6; Newton, *Principia: vol. 1, The Motion of Bodies; vol. 2, The System of the World*, trans. A. Motte in 1729 and revised by F. Cajori, (London: University of California Press, 1962), 2:398–405 and 547; J. E. McGuire, "Newton's Principles of Philosophy: An Intended Preface for the 1704 *Opticks* and a Related Draft Fragment," *British Journal for History of Science*, 5 (1970–1), 178–85, especially 185; Newton to Cotes, 28 March 1713, *Newton Correspondence*, vol. 5, p. 397.

[11] Newton, *Principia*, p. 547; Newton to Bentley, 17 January and 25 February 1692/3, in *Newton Correspondence*, vol. 3, pp. 240 and 254.

duration, and mobility of particles be explained mechanically, and yet no one took *them* for fictions or occult qualities. The same was true for gravity: "to understand this without knowing the cause of gravity, is as good a progress in philosophy as to understand the frame of a clock & the dependence of the wheels upon one another without knowing the cause of the gravity of the weight which moves the machine is in the philosophy of clockwork." The issue raised by Leibniz resurfaced in the correspondence between himself and Newton's protégé Samuel Clarke not long after the appearance of the General Scholium, and this, the most widely cited of all eighteenth-century philosophical controversies, was profoundly significant in bringing Newton's metaphysical views to the attention of Continental European intellectuals.[12]

Newton had seemed to argue definitively in the *Principia* that most of space was completely empty and that individual particles of matter experienced attraction in proportion to their mass. The second book, on motion in resisting media, raised a series of issues about whether planetary motion in a vortex would be consistent with Kepler's third law and whether the interaction of different vortexes would compromise their integrity so as to disturb the regularity of their motion and the entities supposedly borne by them. Nevertheless, Newton elsewhere posited an ether that would account for the phenomena of light, magnetism, and even gravity itself, and some of his previously unpublished work in this area trickled out into the public sphere in the middle of the eighteenth century. At the the end of the General Scholium he referred to a "most subtle spirit" existing in all gross bodies by which attractions were performed at very short distances, and this was described as being "electric and elastic" in the 1726 (third) edition published just before his death. In Query 18 of the 1717 *Opticks* he invoked the "Vibrations of a much subtiler Medium than Air" to account for the motion of heat and light, drawing attention to its elasticity in the process. In Query 21 and a revamped Query 28 he invoked the differential density of this ether to account for the gravitation of planetary bodies toward each other. In the former he conjectured that the ether could "contain Particles which endeavour to recede from one another (for I do not know what this Aether is)" and which were possibly even smaller than those of light, and in the final Query 31 he stated that "the small Particles of Bodies [have] certain Powers, Virtues, or Forces, by which they act at a distance." Few took the speculative form of these statements and the phenomenalist refrains seriously, and their sheer diversity meant that there was virtually no doctrine that could not be found somewhere in the Newtonian textual corpus.[13]

---

[12] Newton to the editor of the *Memoirs of Literature*, late 1712/early 1713, *Newton Correspondence*, vol. 5, pp. 299–300; H. G. Alexander (ed), *The Leibniz-Clarke Correspondence* (Manchester: Manchester University Press, 1956); G. Buchdahl, "Gravity and Intelligibility: Newton to Kant," in R. Butts and J. Davis (eds.), *The Methodological Heritage of Newton* (Oxford: Basil Blackwell, 1970), pp. 74–102. Leibniz had already attacked the doctrine of universal gravitation in his *Theodicy* of 1710.

[13] Newton, *Opticks*, pp. 349, 352, 363, 376.

In addition to stating that Newtonian gravitation was an occult quality and a permanent miracle, Leibniz argued for a plenum on the grounds that a contact mechanism was needed to explain gravity, and at a metaphysical level he argued that the more of something that existed, the more perfect the world was. For different reasons, many of the first Continental European readers of the *Principia* took it to be a majestic work of pure mechanics, and one reviewer, despite the overtly empirical content of Book Three, called for Newton to spend as much effort on explaining the physical world as he had in constructing his mathematical cosmos. Leibniz criticized Newton's notions of absolute space and time and proferred a different, relational view in which time and space were ideal things, the latter being "an order of things, observed as existing together." Leibniz's notion of a pre-established harmony was connected to a theology that stressed God's omniscience and His rational action; Newton and Clarke held that that derogated from the free will of both humans and God, and made humans believe that they could find out the reasons according to which God acted. Leibniz caricatured Newton's God as a faulty watchmaker who had to intervene continuously to mend his wretched contraption, whereas Clarke charged that Leibniz's God, having done it all at the Creation, now had nothing more to do with it.[14]

METAPHYSICS, THEOLOGY, AND MATTER THEORY

The nature of matter was perhaps the most keenly contested issue in eighteenth-century natural philosophy, and ontological positions were deeply entangled in epistemological, metaphysical, and confessional commitments. Mirroring debates over gravitation, the issues were initially centered on whether activity was essential and inherent to matter or whether it was superadded to particles that were essentially lifeless and passive. The former raised the materialist and atheistic specter of what Ralph Cudworth had called "hylozoism" in his *True Intellectual System* of 1678. There were two versions of the first view; the first, that the mere organization of matter gave rise to emergent properties such as consciousness and that there was thus no need to posit a dualism of spirit and matter; the second, that matter essentially contained within it some "force" or "active principle" that gave rise to impenetrability and other qualities. The notion that matter was essentially lifeless but had had other properties superadded to it was the position publicly adopted by Newton and many others. Systems premised on this view allowed a providentialist conception of God's activity in the world, but there was then the difficulty that the building blocks of the universe seemed to enjoy an existence independent of God. Debates on these points were extensive in theology, philosophy, and natural philosophy throughout the eighteenth century, and many referred to Locke's argument

---

[14] Alexander, *The Leibniz-Clarke Correspondence*, pp. 63, 72; Aiton, *Vortex Theory*, pp. 125–35.

in the *Essay* of 1690 that thinking matter – God could have made the power of thought and activity in general essential to certain kinds of matter – was not inconceivable.[15]

Although there were a number of British critiques of "activity" and of the nature of "causation," Continental European intellectuals subjected these issues to very different analyses. Leibniz's notion of a preestablished harmony attempted to bypass what he took to be the apparent absurdity of the voluntarist and providentialist account of God's relationship to His creation by suggesting that relations between cause and effect had been "programmed" into the world at the beginning of time. At the same time, Malebranche accounted for activity in the world by asserting that there were no real efficient causes in the world; only God could be the source of causal activity. Finally, the French were particularly worried by Locke's critique of innate ideas and by his doctrine that the mind was a blank slate on which were inscribed the experiences of a lifetime. This implied that the self might be no more than an effect of these physical influences on the mind, and those who thought that Locke had said that matter could think (he had merely stated that it was not inconceivable) had other textual items in his *Essay* to support their case. In the climate of the early eighteenth century, such criticisms were intimately allied to worries that Locke's philosophy might give solace to the atheist or unitarian.[16]

In Britain, orthodox theologians had to defend themselves in the early eighteenth century against anti-trinitarians and "deistic" writers such as Anthony Collins and John Toland. Toland claimed in *Letters to Serena* (1704) that matter was intrinsically active and possessed what he called *autokinesy* and later claimed Locke and Newton as supporters of his thesis. Even Samuel Clarke, with Newton presumably checking every word, held that Leibniz was not amiss in believing that "Mr Locke doubted whether the Soul was immaterial or no," although, according to Clarke, Locke had so far been followed "only by some Materialists, Enemies of the *Mathematical Principles of Philosophy*." Worst of all, the anti-trinitarian Newton was lumped by his critics – many of whom expertly detected anti-trinitarianism in the General Scholium – with the materialists he despised. In broader Augustan and Enlightenment cultures his philosophy enjoyed a vast reservoir of support and his system, in describing the rational and divine order of the cosmos, seemed to some to offer a blueprint for rational government on Earth.[17]

---

[15] J. Yolton, *Thinking Matter: Materialism in Eighteenth Century Britain* (Oxford: Basil Blackwell, 1984); J. Locke, *An Essay Concerning Human Understanding,* ed. P. Nidditch (Oxford: Clarendon Press, 1975), p. IV.3.6.

[16] J. Yolton, *Locke and French Materialism* (Oxford: Clarendon Press, 1991).

[17] Yolton, *Thinking Matter,* pp. 3–4, 14–24; Alexander, *Leibniz-Clarke Correspondence,* p. 12 and cf. 190; compare with Newton's "apology" for having called Locke a "Hobbist" and "for representing that you struck at the root of morality in a principle you laid down in your book of Ideas and designed to pursue in another book" during his breakdown of 1693; *Newton Correspondence,* vol. 3, p. 280.

Other writers explicitly aligned themselves against what they took to be the baneful hegemony of the Newtonian philosophy and drew from various sources to construct ontologies in which force and activity were immanent within matter. In defending orthodox Anglicanism in *The Principles of Natural Philosophy* of 1712 against the "Arianism" of Clarke and William Whiston and the "Papism" of Cartesianism and Newtonianism, Robert Greene denied that extension and solidity constituted the essence of matter and used Lockean arguments to show that we have as little idea of these as we do of the "essence" of matter. In his *Principle of Expansive and Contractive Forces* of 1727 he denied the existence of a vacuum and claimed that "matter" was not corpuscular but was rather constituted from certain innate forces that could be classed as either expansive or contractive. Independently, John Hutchinson affirmed the literal truth of the Bible and attacked both action at a distance and the antitrinitarianism of the General Scholium in his *Moses's Principia* of 1724–7. Hutchinson worried that positing the existence of some entity such as 'force' in the world introduced an immaterial agent that many people might equate with God Himself; this would make God the soul of the world. In fact, God had initially created particles of gross matter as well as finer particles that constituted light; and to combinations of the latter He had given motion. Hutchinson saw a correspondence between the Holy Trinity and the three different modifications in which this subtler form of matter could appear, namely fire, light, and spirit (air). In the second half of the century his cyclical cosmos was a potent resource for a group of Tory theologians and natural philosophers including Alexander Catcott senior and George Horne.[18]

The study of light, electricity, magnetism, and heat was one of the most fertile grounds for investigating the general nature of matter, and again Newton's work could be cited for any one of a number of projectile, fluid, vibration or wave theories. The notion (implied by Query 21) that light could be considered as corpuscular, with forces acting between light particles and other bodies, suffered from a number of drawbacks, as critics such as Benjamin Franklin and Leonhard Euler were not slow to point out. The idea that light initially came from the Sun implied that given any normal construal of its output, its power would soon be wasted. Particles from different sources would constantly be impacting with each other, creating a buzzing confusion; particles, however small, traveling at the kinds of speeds understood to be possessed by light would have an extraordinary force that had not been detected. Although experimental work in the 1750s by John Michell suggested that light did indeed possess mass, a number of individuals had already developed nonprojectile theories. The most significant of these was Euler, who

[18] Yolton, *Thinking Matter,* pp. 102–3; Greene, *The Principles of Natural Philosophy* (Cambridge, 1712); idem, *The Principles of the Philosophy of Expansive and Contractive Forces* (Cambridge, 1727); G. N. Cantor, *Optics after Newton: Theories of Light in Britain and Ireland, 1704–1840* (Manchester: Manchester University Press, 1983), pp. 97–102; C. B. Wilde, "Hutchinsonianism, Natural Philosophy and Religious Controversy in Eighteenth Century Britain," *History of Science,* 18 (1980), 1–24.

argued that light had to be transmitted through an elastic ether whose frequency of vibration was, by analogy with sound, equated with color. He also remained a steadfast adherent to the necessity of vortexes to account for planetary motion, despite the vogue for Newtonian attraction prominent in Europe in the 1740s.[19]

Although not necessarily constituting a coherent "tradition," other writers developed dynamicist accounts of matter in which particles were reduced to the constituent forces that gave rise to gravitation and impenetrability. Michell and Roger Boscovich independently developed theories in which matter was composed of centers of force around nonextended points, and, along with David Hartley's associationist psychology, these works were important for the materialist philosophy propounded by Joseph Priestley. A voluntarist and immaterialist until early 1774, Priestley had initially believed in a system not unlike that found in the General Scholium, in which an immaterialist God existed alongside "sluggish and inert" matter. Thereafter, his commitment to a monism, in which the difference between spirit and matter disappeared, was connected to his political and religious views as a radical dissenter. "Matter" was the subject of attractive or repulsive powers and was the source of life and not death, "spirit" was merely a more rarified form of matter, and Nature was "a plenum of intensive powers extended in space." Priestley argued that a reason accessible to all could progressively discover the truth of Christianity as much as it could uncover the laws governing the natural world. His histories of electricity (1767) and light (1772) showed how moral and intellectual progress had been made possible through individual discovery, and readers were invited to follow the actions and trains of thought of people such as Franklin. Newton was accordingly criticized for writing in a way that made it difficult to replicate his process of discovery.[20]

By the middle of the century, British natural philosophers and theologians became increasingly comfortable with the notion that some kind of power or force was immanent in matter, and vitalist accounts involving an organic substance that was intrinsically sensitive or "irritable" gained a great deal of support in medicine. In France, the Cartesian system remained a significant resource for theories of matter, cosmology, and human behaviour until the middle of the century. Nevertheless, it was eclipsed by other approaches because of the poverty and trivial nature of its explanations and by its association with the materialism that was part of the burgeoning clandestine literature.

[19] Cantor, *Optics,* pp. 10–12, 54–6, 117–21; C. Wilson, "Euler on Action at a Distance and Fundamental Equations in Continuum Mechanics," in P. Harman and A. Shapiro (eds.), *"The Investigation of Difficult Things": Essays on the History of the Exact Sciences* (Cambridge University Press, 1992), pp. 399–420, especially pp. 400–1.

[20] Cantor, *Optics,* pp. 71–2; Yolton, *Thinking Matter,* pp. 109–12; Schaffer, "Natural Philosophy," pp. 63–4; J. McEvoy and J. E. McGuire, "God and Nature: Priestley's Way of Rational Dissent," *Historical Studies in the Physical Sciences,* 6 (1975), 325–404; McEvoy, "Electricity, Knowledge and the Nature of Progress in Priestley's Thought," *British Journal for History of Science,* 12 (1979), 1–30.

One of the most important projects anywhere at the end of the century appealed to its own brand of Newtonianism. Pierre Simon de Laplace constructed a physics in which cohesion, capillary action, and chemical reactions were explained in terms of central forces that were either attractive or repulsive. Light, electricity, magnetism, and heat were conceived as being imponderable fluids composed of mutually repulsive particles which were nevertheless attracted by ponderable matter. Notoriously, he told Napoleon that there was no need for the "hypothesis" of God in his system.[21]

Elsewhere, the work of Kant provided a central metaphysical underpinning for the doctrines of *naturphilosophie* made famous by Friedrich Schelling. Kant dealt with the notion of space in his first work, *Thoughts on the True Estimation of Living Forces* (1747), and in his *First Ground of the Distinction of Regions in Space* of 1766 he introduced the argument that space could not be merely relational; if the contents of the universe consisted only of one hand and nothing else it would still be either a left hand or a right hand. Since this "handedness" was not configured with respect to another existing entity it would have to be with respect to an independent containing space. However, following the arguments of skeptics such as Hume he had come to believe by the early 1770s that the absolute space posited by Newton was "pertaining to the world of fable," and he developed the view that space and time were "forms" of sensible intuition, presupposed in our experience of the phenomenal world. In the *Critique of Pure Reason* of 1781 he argued that appearances were structured by concepts that imposed a "rule" on them and made them objects for intuition. These concepts were the "categories" of the understanding that Kant thought could be used to put the principles of Newtonian natural science on an a priori footing; ultimately, mind prescribed laws of nature to experience.[22]

In his *Metaphysical Foundations of Natural Science* of 1786, Kant attempted to display the conditions of possibility for an entire natural science. The central term appearing in Kant's system was that of matter conceived as made up of opposed attracting and repelling forces in equilibrium. Attraction, conceptualized in texts such as the General Scholium under an ontology of material particles with an immaterial force, was unintelligible; action at a distance could become intelligible only if matter was thought of as essentially composed of forces acting throughout space. Arguing that attraction had to be seen as essential to mass if gravitational attraction was to be seen as "proportional"

---

[21] R. Fox, "The Rise and Fall of Laplacian Physics," *Historical Studies in the Physical Sciences*, 4 (1974), 89–136; see also A. Vartanian, "Trembley's Polyp, La Mettrie and Eighteenth Century French Materialism," in P. P. Wiener and A. Noland (eds.), *Roots of Scientific Thought* (New York: Knopf, 1957), pp. 497–516; and S. Roe, *Matter, Life and Generation: Eighteenth-Century Embryology and the Haller-Wolff Debate* (Cambridge University Press, 1981).

[22] Alexander, *Leibniz-Clarke Correspondence*, pp. xlvi–xlviii; Kant, *Prolegomena to any Future Metaphysics that will be able to present itself as a Science*, ed. P. Gray-Lucas (Manchester: Manchester University Press, 1978), pp. 38–9, 82–9; Buchdahl, *Metaphysics*, pp. 574–615.

to mass, Kant's a priori conception of attraction had the status of a construct that served to "order" dynamical phenomena and "enlarge the field of action for the natural philosopher." Whether or not it actually existed was a matter that required empirical investigation. Likewise, Kant thought that other concepts could be shown to be useful for and constitutive of a more general natural science than Newtonian physics, although research would be required to determine whether such entities could be found in nature. In earlier works on cosmology Kant was strongly committed to the existence of a universal ether but treated it hypothetically in the *Metaphysical Foundations.* However, it played a central role in the analysis of science that he developed in the *Opus postumum* composed between 1790 and 1803.[23]

Drawing from Kant, the system of *naturphilosophie* developed by Schelling in 1798 posited the existence of a polarized force and stressed the unity and interconnectedness of apparently unrelated phenomena. Different powers manifested themselves in the three different realms of organic, universal, and inorganic nature, the last two being composed of, respectively, parallel categories of light, electricity, and the cause of magnetism; and chemical process, electrical process, and magnetism. His system was based on opposed forces, so that light, electricity, and magnetism were various manifestations of an underlying and basic "polar force" or "dualism." Schelling's work was significant for the more empirically inclined Johann Wilhelm Ritter, who held that the two principles underlying chemical, electrical, and magnetic activity were contained in the basic force, namely light. *Naturphilosophie* was also an important resource for Hans Oersted, who in 1799 produced two works on Kant's theory that matter was composed of attractive and repulsive basic forces. Having read Schelling more closely, Oersted extended these forces to include light, electricity, magnetism, and chemistry, and *naturphilosophie* was arguably a crucial influence on his momentous discovery of the effect of a voltaic pile on a magnetized needle in 1820.[24]

METHODOLOGY

The heroes of the scientific revolution claimed to be able to extract the essence of progress in natural philosophy and to pass it on to posterity in the form

---

[23] Kant, *Metaphysical Foundations of Natural Science,* in Kant, *Philosophy of Material Nature,* trans. J. W. Ellington (orig. 1786; Indianapolis: Hackett, 1985); Buchdahl, "Gravity and Intelligibility," pp. 90–9; M. Friedman, *Kant and the Exact Sciences* (Cambridge, MA: Harvard University Press, 1992), pp. 96–164; D. C. Barnaby, "The Early Reception of Kant's *Metaphysical Foundations of Natural Science,*" in R. S. Woolhouse (ed.), *Metaphysics and Philosophy of Science in the Seventeenth and Eighteenth Centuries: Essays in Honour of Gerd Buchdahl* (London: Kluwer, 1988), pp. 281–306.

[24] F. W. J. Schelling, *Ideas for a Philosophy of Nature,* 2nd ed., trans. E. E. Harris and P. Heath (Cambridge University Press, 1988, orig. 1803). See K. Caneva, "Physics and *naturphilosophie:* A Reconnaissance," *History of Science,* 25 (1997), 35–106, especially 39–41, 43, 45, 48–9; R. C. Stauffer, "Speculation and Experiment in the Background of Oersted's Discovery of Electromagnetism," *Isis,* 48 (1957), 33–50.

of a methodology. However, the function of such prescriptions is unclear. Paul Feyerabend pointed out that in his portrayal of the discovery of different degrees of refrangibility and his description of the crucial experiment, Newton had described an event that could never have taken place in terms that presupposed the truth of his theory. This was likely to be true for all methodological pronouncements. Moreover, since Newton retained the notebook of his researches, we now know that his route to the theory of the heterogeneity of white light was indeed very different from the way he described it. These considerations, allied to the demise of faith in the existence of a single scientific method, have led historians to reconsider the role of methodology in two ways. First, methodologies are essentially mythical and do not represent the way discoveries were made, and like the "discovery stories" themselves must be seen as serving a specific function in positioning the work with respect to other philosophies. Second, whatever this initial function, methodologies are extremely adaptable and are easily transformed to serve the interests of later writers. For example, naive Baconianism, wrenched from any "managerial" context it may have had for the conservative Lord Chancellor, was a resource both for egalitarians in the English Commonwealth and for revolutionaries after 1789. In the latter case, the supposed "misuse" of the philosophies of such men as Bacon and Newton proved too much for British opponents of the French Revolution such as John Robison.[25]

As Newtonian doctrines gained favor in Britain and France, other inquiries such as those into the "science of man," medicine, and even religion attempted to give their investigations the same epistemological status as Newtonian mechanics. For a few decades after 1740, the Newtonian "method," if not the doctrine of "attraction," was temporarily dominant across Europe. As with ontology, this "method" could be read in a number of ways. French savants such as d'Alembert saw the mathematical analysis of the *Principia* as the epitome of rational investigation, whereas Jean-Théosophile Desaguliers and the Dutch Newtonians Wilhelm 's Gravesande and Pieter van Musschenbroek – who extolled the experimental "Newtonian" method in their textbooks – promoted demonstration devices which physically realized Newtonian principles before the eyes of a large audience. The Newtonian system was particularly prominent in Scottish universities, and, as Paul Wood points out elsewhere in this volume (Chapter 34), both David Hume and Thomas Reid (among others) appealed to Newton's "method" in compiling

---

[25] P. K. Feyerabend, "Classical Empiricism," in Butts and Davis (eds.), *The Methodological Heritage*, pp. 150–70; J. Schuster, "Cartesian Method as Mythic Speech: A Diachronic and Structural Analysis," in Schuster and Yeo (eds.), *Politics and Rhetoric*, pp. 33–96; J. Martin, "Sauvages's Nosology: Medical Enlightenment in Montpellier," in A. Cunningham and R. French (eds.), *The Medical Enlightenment of the Eighteenth Century* (Cambridge University Press, 1990), pp. 111–37; R. Yeo, "An Idol of the Marketplace: Baconianism in Nineteenth Century Britain," *History of Science*, 23 (1985), 251–98; J. B. Morrell, "Professors Robison and Playfair and the 'Theophobia Gallica,'" *Notes and Record of the Royal Society*, 26 (1971), 43–63.

their moral philosophies, although there was a great difference between their approaches.[26]

Not all the features of the brand of "Newtonianism" that made a fetish of mathematical demonstration were intellectually palatable. In Britain, Hans Sloane's tenure as President of the Royal Society immediately after the death of Newton in 1727 raised the prominence of natural historians, and the high status of natural history culminated in the Presidency of the Society of Joseph Banks between 1778 and 1820. These men appealed to the authority of inductivism and argued that mathematics was an inappropriate structure for the organization of the plethora of new facts in botany and zoology. Although Denis Diderot prophesied an end to mathematical advances and announced that chemistry, electricity, and natural history were to be the next great human enterprises, it was his mathematophile co-editor of the *Encyclopédie,* Jean d'Alembert, who claimed in his *Preliminary Discourse* that Bacon was "the greatest, the most universal and the most eloquent of the philosophers." In his *Essay on the Origin of Human Knowledge* of 1740 Condillac cited Bacon as the first to point out that knowledge had its origin in sensory experience, whereas later in the century many *philosophes* transformed his epistemology into an attack on social elites and in particular on clerics. Nevertheless, Hume rated Bacon's importance in the history of science far below that of Galileo and attributed the opposite view to English partisanship.[27]

Perhaps the most sustained attack on English methodological hegemony was mounted by Johann Wolfgang Goethe and the *naturphilosophen* at the end of the century. Goethe argued that Nature was alive and possessed a wholeness in each of its parts and that a proper investigation of a number of different experimental situations would reveal the primordial type expressed in each single fact. Whereas the Baconian approach "foolishly exhausted itself" in a multiplicity of single facts before any induction could take place, Newton had falsely claimed to be able to derive a theory from a single experiment. Goethe worked in the early 1790s on a work titled *Contributions to Optics* in which he argued that Newton was a "tyrant" who had "enslaved" nations such as his own, and although Goethe preferred reform to revolution, he spoke of razing the Newtonian Bastille. German Newtonians were compared to hated Catholic priests, defending a canon whose obscurantist language

---

[26] R. Porter, "Medical Science and Human Science in the Enlightenment," in C. Fox et al. (eds.), *Inventing Human Science: Eighteenth Century Domains,* (Berkeley: University of California Press, 1995), pp. 53–87, 53–9; L. Stewart, *The Rise of Public Science* (Cambridge University Press, 1993); P. Brunet, *Les Physiciens Hollondais et la Méthode Expérimentale en France au XVIIIe Siècle* (Paris, 1926); L. Laudan, "Thomas Reid and the Newtonian Turn of British Methodological Thought," in Butts and Davis (eds.), *Methodological Heritage,* pp. 103–31; and P. Wood, "Reid on Hypotheses and the Ether: A Reassessment," in M. Dalgarno and E. Matthews (eds.), *The Philosophy of Thomas Reid* (Dordrecht: Kluwer, 1989), pp. 433–46.

[27] Yeo, "Idol of the Marketplace," pp. 253–4, 255, 256, 259; E. Cassirer, *The Philosophy of the Enlightenment* (orig. 1932; Princeton, NJ: Princeton University Press, 1979), pp. 73–4; d'Alembert, *Preliminary Discourse to the* Encyclopedia, trans. R. N. Schwab (orig. 1751; New York, 1963), p. 74.

made it impenetrable to outsiders, and the crucial experiment was a procedure whereby "the researcher tortured nature on the rack in order to elicit
a confession which the investigator had already anticipated." Newton was
classed with the "illuminati" of the secretive Masonic lodges that Goethe believed had been partly responsible for the French Revolution, and he called
for a much wider audience for natural philosophy: "The phenomena must
once and for all be brought out of the gloomy-empirical-dogmatic torture
chamber and presented before the jury of common sense." Newton's experiment was artificial, and only the genius who brought knowledge to the wider
public merited the respect of posterity. Although Goethe later disagreed with
the tenets of the *naturphilosophen,* its main exponent, Friedrich Schelling,
similarly condemned "the blind and mindless type of natural research, which
has generally established itself since the corruption of philosophy by Bacon
and of physics by Boyle and Newton."[28]

CONCLUSION

Despite still prevalent preconceptions about the barrenness of the period,
natural philosophy and rational mechanics were developed to a remarkable
extent in the eighteenth century. By 1800, there were serious efforts to eluci-
date the basic concepts of, and quantify, heat theory, electricity, and chemistry, and Euler and then Lagrange had formulated rational mechanics or
"analysis" into something like modern classical physics. Many questions that
had been an integral part of, and even constitutive of, subjects at the end of
the preceding century were now consigned to metaphysics, and disciplinary
divisions had come to resemble their present form. However, as the case of
Newton shows well, those who condemned metaphysics and the use of hypotheses were themselves necessarily committed to a metaphysics, and the
development of what we now call physics and mathematics constantly raised
novel philosophical issues. For example, advances in these fields raised problems associated with the foundations of the fluxional (Newtonian) and differential (Leibnizian) calculuses; the status of Newton's laws of motion; the
nature of force (the vis viva debate and the issue of whether force should be
treated as continuous or as an infinite number of small "impacts"); and the
introduction of new concepts and principles, such as Euler's development of
"pressure" and "stress," and d'Alembert's Principle of Least Action. Construed
as philosophy of physics, these issues were still the subject of the thoughts
and writing of practicing natural philosophers at the end of the century, and

---

[28] F. Burwick, *The Damnation of Newton: Goethe's Colour Theory and Romantic Reception* (Berlin: Walter de Gruyter, 1986); D. Sepper, *Goethe contra Newton: Polemics and the Project for a New Science of
Color* (Cambridge University Press, 1987); M. W. Jackson, "A Spectrum of Belief: Goethe's 'Republic' Versus Newtonian 'Despotism,'" *Social Studies of Science,* 24 (1994), 673–701, especially 682–4;
Schelling, *Philosophy of Nature,* p. 52.

few had developed such a pragmatic attitude toward them that they could afford to ignore them completely.[29]

However, if such concerns were still an integral part of the exact sciences, the relevance of epistemology is less obvious. Hume's "post-sceptical" arguments were made in a wide range of writings in history, politics, moral philosophy, and religion, and although he was *au fait* with many contemporary works in natural philosophy, his arsenal of skeptical challenges was extremely influential, not in natural philosophy but in Scottish and French moral philosophy and theology. Elsewhere, sophisticated critiques of the metaphysical and epistemological foundations of mathematics and physics were partly answered by some natural philosophers, although the success of the new techniques of analysis meant that a pragmatic attitude to the foundations of such tools was dominant. For example, Bishop George Berkeley's attacks on Newton's calculus implied that its foundations were based on a method that was no more "rational" than belief in a God, and Berkeley thought that the doctrine of absolute space came close to committing the heresy that space was God. These criticisms of the Newtonian hegemony were taken seriously by supporters of Newton such as James Jurin, but only in the following century were Berkeley's criticisms of the foundations of the fluxional calculus generally recognized as serious contributions to the topic. In some areas, work in "philosophy" had a powerful effect on the practice of natural philosophy: Kant's influence on Schelling and then, via him, on Oersted is a case in point, although the precise nature of this transmission is still difficult to ascertain. More representative was Locke's recognition in the early 1690s that in the stark demonstrations in his *Principia,* Newton had achieved a level of demonstrative rigor that was probably inimitable in spheres outside natural philosophy. Thereafter the practitioners of other disciplines, gazing on the success of mathematical physics with envious eyes, could only hope to mimic them.[30]

---

[29] For "analysis" see C. Truesdell, "A Program toward Rediscovering the Rational Mechanics of the Age of Reason," *Archive for the History of Exact Sciences,* 1 (1960–2), 1–36; T. Hankins, *Science and the Enlightenment* (Cambridge University Press, 1985), pp. 15–33; for "force" see C. Wilson, "D'Alembert Versus Euler on the Precession of the Equinoxes and the Mechanics of Rigid Bodies," *Archive for the History of the Exact Sciences,* 37 (1987), 233–73; and R. S. Westfall, *Force in Newton's Physics* (Cambridge University Press, 1971).

[30] See D. Norton (ed.), *Cambridge Companion to Hume* (Cambridge University Press, 1993), pp. 1–25, especially p. 5; J. Force, "Hume's Interest in Newton and Science," *Hume Studies,* 13 (1987), 166–216; M. Barfoot, "Hume and the Culture of Science in the Early Eighteenth Century," *Oxford Studies in the History of Philosophy* (Oxford: Clarendon Press, 1990), pp. 151–90; and D. Jesseph (ed. and trans.), *George Berkeley: De Motu and The Analyst: A Modern Edition with Introductions and a Commentary* (Dordrecht: Kluwer, 1992).

12

IDEAS OF NATURE
Natural Philosophy

John Gascoigne

The eighteenth century inherited a long tradition deriving from Greek antiquity that maintained that Nature could be understood by the exercise of reason. This belief underlay centuries of university practice in which natural phenomena had been explained by the use of logical deduction from first principles largely, although not exclusively, derived from the philosophy of Aristotle.[1] The long shadow cast by such an entrenched intellectual position was still evident for much of the eighteenth century in the links that remained between natural philosophy and the larger philosophical enterprise of explaining the fundamental purposes that underlay the works of God and humankind. At the beginning of the eighteenth century, natural philosophy remained a branch of philosophy along with metaphysics, logic, and moral philosophy.

But it was to be one of the striking features of the century that, as it progressed, natural philosophy more and more was loosened from such traditional moorings and began to assume an independent stance. Indeed, whereas once natural philosophy had deferred to metaphysics, natural philosophy increasingly assumed the status of the defining form of philosophy, which moral philosophy attempted to emulate and which called into question the worth of metaphysical inquiry. By 1771, for example, the *Encyclopaedia Britannica* could justify the study of moral philosophy on the grounds that it resembled natural philosophy in that it, too, "appealed to nature or fact; depended on observation; and built its reasonings on plain uncontroverted experiments."[2]

Not only did the eighteenth century see natural philosophy assume increasing independence from its philosophical origins, but also, as natural philosophy grew in scale and complexity, so, too, it began to give birth to separate disciplines. For most of the century the long-established belief that

---

[1] On the pluralistic nature of traditional scholastic natural philosophy see Charles Schmitt, "Towards a Reassessment of Renaissance Aristotelianism," *History of Science*, 11 (1973), 159–93.

[2] *Encylopaedia Britannica* (Edinburgh, 1771), 3 vols., "Moral philosophy," vol. 2, p. 270.

all knowledge was part of an underlying unity[3] helped to sustain the enterprise of explaining the workings of Nature in terms of a basic set of laws and procedures. But, by the end of the century, the domain of natural philosophy was beginning to be circumscribed. The methods of chemistry were sufficiently different to carve out a new principality that was substantially distinct from the overarching field of natural philosophy. Natural history, once regarded as the relatively lowly endeavor of collecting raw data from which the natural philosopher would distill fundamental laws, had gained a higher dignity with the growth of sophisticated forms of classification. These, in turn, helped to promote the division of natural history into the subdivisions of botany, zoology, and geology. As the domain of natural philosophy was restricted, the term "physics," which had traditionally been a synonym for "natural philosophy," began to take on the narrower connotation of the study of inanimate nature by means of experiment.[4] Overall, then, the eighteenth century saw the transition from natural philosophy as a branch of philosophy to the beginnings of an array of scientific disciplines that largely undermined the assumption of a unified view of Nature on which the enterprise of natural philosophy had traditionally been based.

When the eighteenth century dawned, different brands of natural philosophy competed for dominance in the intellectual vacuum created by the breakdown of the old Aristotelian-based scholastic order under the impact of that reconceptualization of the nature of the cosmos and of the Earth's place that we label the Scientific Revolution. In France and much of Continental Europe, scholastic natural philosophy had been replaced by or, to varying degrees, amalgamated with Cartesianism,[5] which, reassuringly, shared with scholasticism an integrated philosophical system in which natural philosophy could be deduced from a consistent set of metaphysical premises. And, for all Descartes's love of mathematics, his system of natural philosophy proceeded principally by logical deduction which again made it more palatable to scholastic-trained minds. Nonetheless, the intrusion of the foreign body of Cartesianism into the traditional philosophical corpus of learning increasingly separated natural philosophy from other forms of philosophy.

Like Cartesianism, the system of thought based on the work of Leibniz and later codified by Christian Wolff also integrated natural philosophy closely

---

[3] Thomas Hankins, *Jean d'Alembert, Science and the Enlightenment* (Oxford: Clarendon Press, 1970), p. 104.

[4] John Heilbron, "Experimental Natural Philosophy," in G. S. Rousseau and Roy Porter (eds.), *The Ferment of Knowledge: Studies in the Historiography of Eighteenth-Century Science* (Cambridge University Press, 1980), p. 362.

[5] On the attempts to merge scholasticism and Cartesianism, see Paul Dibon, *La philosophie néerlandaise au siècle d'or:* vol. 1, *L'Enseignement philosophique dans les universités à l'epoque précartesienne, 1575–1650* (Paris: Elsevier, 1954–), p. 253, and Michael Heyd, *Between Orthodoxy and the Enlightenment: Jean-Robert Chouet and the Introduction of Cartesian Science in the Academy of Geneva* (The Hague: M. Nijhoff, 1982), pp. 108, 134.

into a larger philosophical enterprise that drew on traditional scholastic origins. Leibniz also shared with Descartes the ambition of providing a thoroughgoing model of the workings of Nature. Although he rejected Descartes's basic premise that matter was inert, being characterized only by extension, and argued for an inherent force within matter, nonetheless, he, like Descartes, sought to construct a philosophically consistent model of the workings of the universe in terms of matter and motion. Although it had far fewer followers than Cartesianism, Leibniz's work was influential in the German-speaking lands and colored the work of a number of French natural philosophers. Some, such as the important Swiss natural philosopher Daniel Bernoulli (1700–1782), integrated aspects of Leibniz's work into a fundamentally Cartesian mold.[6] One tangible indication of the extent of Leibniz's influence on the Continent was the adoption of his notation for calculus rather than that of Newton.

In Britain, Leibniz's work had little impact, and, by the beginning of the eighteenth century, Cartesianism was beginning to be displaced by a system of natural philosophy that diverged still further from the mental habits of scholasticism: that of Newton. In contrast to that of Descartes, Newton's system of natural philosophy was not closely linked to a more overarching philosophical schema, nor was it deductive in the same manner as the scholastic and Cartesian systems. As Newton emphasized in his choice of title – *Principia Mathematica Naturalis Philosophiae,* in contrast to Descartes's *Principia Philosophiae* – his was a system based not on logic but on mathematics, thus clearly distinguishing the field of natural philosophy from other branches of philosophy.

The clash between these two systems of natural philosophy was to be a major theme of European intellectual life in the first half of the eighteenth century. Along with issues of national rivalry, the competition between the two chief world systems reflected differing conceptions of the scope and extent of natural philosophy. For the Cartesians their system offered a consistent model of Nature based on the thoroughgoing mechanical principles of particles in motion – principles that could be extended to all natural phenomena. In their eyes the Newtonian system, with its invocation of principles, such as gravitation, that had no clear mechanical basis, represented a return to the occult qualities of the old scholastic order. For the Newtonians, on the other hand, Cartesianism was a "Philosophical Romance" in its reliance on verbal explanations in contrast to the rigorous mathematical treatment on which they prided themselves. As Newton himself wrote in a draft of one of the queries of the *Opticks* directed at his Cartesian opponents, "But if without deriving the properties of things from Phaenomena you feign Hypotheses & think by them to explain all nature, you may make a plausible systeme of Philosophy

[6] C. Iltis, "The Decline of Cartesianism in Mechanics," *Isis,* 64 (1973), 356–73, especially 357.

for getting your self a name, but your systeme will be little better than a Romance."[7]

As Guerlac has stressed,[8] Newton's work represented a radical break from the traditions of natural philosophy in that it largely abandoned the attempt to construct a model of Nature based on philosophically consistent premises. In their different ways Descartes and Leibniz carried on the same mission as Aristotle; for they, too, sought to develop a system of physics or natural philosophy (the two terms being synonymous) that explained the fundamental causes of the workings of Nature. By contrast, Newton sought more limited ends: a mathematical model of the visible world that did not attempt to offer the same pictorial depiction of all natural phenomena that underlay other systems of natural philosophy. As one French popularizer of Newton wrote in 1743: "Descartes had the ambition of fabricating a world. Newton did not have the slightest desire in this regard."[9] Such different goals help to explain the slow acceptance of Newton's work on the Continent since it did not fulfill the traditional conception of the role of natural philosophy. It also helps to explain why, as Newton's work became more and more an established part of the intellectual terrain, it was particularly corrosive of the traditional association between natural philosophy and philosophy more generally.

Ultimately, it was Newton's work that commanded the greatest scientific authority throughout most of Europe by around the middle of the century.[10] But even though in the field of celestial mechanics and, to a lesser degree, terrestrial mechanics Newton largely reigned supreme, in other areas of science – notably those based on experiment – his shadow fell more lightly. In France the experimental sciences proceeded without any firm commitment to Newtonian concepts.[11] The equivocal attitude of Leonhard Euler (1707–1783) toward Newton's theory of light – ranging from an overtly anti-Newtonian rejection of the particle theory of light to a subsequent debt to elements of Newtonian mechanics in formulating his wave theory of light[12] – is another instance of the partial and provisional character of the Newtonian hegemony of the second half of the eighteenth century.

Nonetheless, the prestige increasingly accorded to Newton's achievement

---

[7] Henry Guerlac, "The Background to Dalton's Atomic Theory," in Guerlac, *Essays and Papers in the History of Modern Science* (Baltimore, MD: Johns Hopkins University Press, 1977), p. 220.

[8] Henry Guerlac, "Where the Statue Stood: Divergent Loyalties to Newton in the Eighteenth Century," and "Newton and the Method of Analysis," in *Essays and Papers*, pp. 133–4, 140, 210.

[9] Louis-Betrand Castel, *Le vrai système de physique generale de M. Isaac Newton, exposé et parallèle avec celui de Descartes* (Paris, 1743), p. 18, cited in Aram Vartanian, *Diderot and Descartes: A Study of Scientific Naturalism in the Enlightenment* (Princeton, NJ: Princeton University Press, 1953), p. 82.

[10] L. W. B. Brockliss, *French Higher Education in the Seventeenth and Eighteenth Centuries: A Cultural History* (Oxford: Clarendon Press, 1987), p. 360; and Tore Frängsmyr, "Swedish Science in the Eighteenth Century," *History of Science*, 12 (1974), 29–42, especially 35.

[11] Rod Home, "Out of a Newtonian Straitjacket: Alternative Approaches to Eighteenth-Century Physical Science," in *Studies in the Eighteenth Century,* vol. IV: *Papers Presented at the Fourth David Nichol Smith Memorial Seminar* (Canberra: Australian National University Press, 1979), pp. 239–40.

[12] Rod Home, "Leonhard Euler's 'Anti-Newtonian' Theory of Light," *Annals of Science*, 45 (1988), 521–33.

helped to ensure that mathematics became a distinguishing feature of natural philosophy. As this became increasingly pervasive so, too, natural philosophy came to be more sharply distinguished from other forms of philosophy. Increasingly, too, natural philosophy came to be distinguished by its recourse to experiments. Although this owed much to Newton's example, it was an aspect of natural philosophy that derived from a number of sources including the craft tradition and the often philosophically agnostic practice of Dutch experimentalists such as 's Gravesande and Musschenbroek.[13] "Newtonianism" was, then, a coat of many colors as it was tinctured by differing national intellectual traditions and as varying materials deriving from the rich and varied Newtonian corpus were given greater or lesser prominence. But, Protean though it may have been, Newtonianism set the intellectual boundaries within which much of the activity of eighteenth-century natural philosophy was conducted. We turn, then, to examine the way in which Newtonianism came to be established as the dominant form of natural philosophy first within Newton's native Britain and then throughout Europe more generally.

THE ESTABLISHMENT OF NEWTONIANISM WITHIN BRITAIN

The forbidding quality of the *Principia,* bristling with recondite mathematical exposition, reflected the character of its author – a man wary of controversy who wished to distance himself from the contentious world of natural philosophy. Indeed, Newton was reported by one of his contemporaries "to have made the *Principia* abstruse to avoid being baited by the little smatterers in mathematics."[14] Even within his own University of Cambridge, Newton's *Principia* was, at first, greeted with stunned incomprehension: "After Sir Isaac printed his *Principia,* as he passed by the students at Cambridge said there goes the man who has writt a book that neither he nor any one else understands."[15] There was, then, nothing foreordained about the intellectual reverence, verging on idolatry, that Newton's work was eventually accorded. The mighty vessel of the *Principia* required humbler intellectual tugs to tow it out of harbor. Important among those who performed such service in the immediate aftermath of its publication was a group of Scottish natural philosophers, of whom the main figure was David Gregory (1659–1708), professor of mathematics at the University of Edinburgh from 1683 to 1691 and, subsequently, Savilian professor of astronomy at Oxford from 1691 to 1708. At both Edinburgh and Oxford Gregory provided students with an introduction

---

[13] Pierre Brunet, *Les physicians hollandais et la méthode expérimentale en France au XVIIIe siècle* (Paris: Libraire Scientifique Albert Blanchard, 1926), p. 99.
[14] William Derham's recollections of Newton, King's College, Cambridge Keynes MS 130.
[15] Martin Folkes's recollections of Newton, ibid.

to Newton's chief cosmological conclusions, and this task was taken further by Gregory's student from Edinburgh, John Keill (1671–1721), who followed him to Oxford. There, as John Desaguliers (1683–1744), Newton's ardent disciple and active promoter of experiments, put it, Keill was the "first who publickly taught *Natural Philosophy* by *Experiments* in a mathematical manner."[16] Desaguliers's praise underlines the extent to which Newtonian natural philosophy was regarded as being characterized by a reliance on mathematics and experiment, thereby distinguishing itself from Cartesianism and, still more, from scholasticism.

Gregory and Keill were Episcopal refugees from Presbyterian Scotland who were naturally at home in the High Church Tory atmosphere of Oxford.[17] But Newtonianism did not flourish at Oxford, where natural philosophy as a whole was accorded a relatively lowly place in an intellectual environment dominated by the task of defending the Established Church by recourse to the traditional weapons of Aristotelian logic and knowledge of the Church Fathers. Although Cambridge had at first been slow to embrace Newton's work, it was there that it became an established part of the intellectual currency of the English elite.[18] The path of the *Principia* from an intellectual curiosity to an integral part of the curriculum owed much to the prevailing religious and political climate in the wake of the Revolution of 1688, which deposed the Catholic James II for the securely Protestant William and Mary; the enormity of justifying the deposition of the king and the appointment of a successor at the behest of Parliament, rather than according to the dictates of hereditary descent, deeply divided English churchmen and colored their attitude to the claims of reason. Those who regarded the Revolution Settlement in Church and State as consistent with rational principles tended to emphasize forms of theology that accorded reason a significant role. In particular, they turned to forms of natural theology that drew heavily on natural philosophy to illustrate how the hand of the Creator could be discerned in the Book of Nature as well as the Book of Scriptures. By contrast, those still wedded to a more traditional order in Church and State tended to be wary of natural theology and the forms of natural philosophy associated with it as distractions from a theology based on Divine Revelation, on which the claims of the Church and, with it, a divine right monarchy were based.

As Margaret Jacob's work[19] has illustrated such debates helped to create the climate in which Newton's work moved beyond a small circle of fellow specialists to become part of the general intellectual tenor of the age. Inter-

---

[16] William Strong, "Newtonian Explications of Natural Philosophy," *Journal of the History of Ideas*, 18 (1957), 49–83, especially 53.

[17] Anita Guerrini, "The Tory Newtonians: Gregory, Pitcairne and Their Circle," *Journal of British Studies*, 25 (1986), 288–311.

[18] John Gascoigne, *Cambridge in the Age of the Enlightenment: Science, Religion and Politics from the Restoration to the French Revolution* (Cambridge University Press, 1989), pp. 142–84.

[19] Margaret Jacob, *The Newtonians and the English Revolution 1689–1720* (Hassocks, Sussex: Harvester Press, 1976).

est in Newtonian natural philosophy, as much as any other system of ideas, was heightened by the extent to which it could be used in the controversies that dominated elite culture. For those who wished to emphasize the claims of reason in settling disputes about the character of Church and State, Newton's work provided a novel illustration of the way in which the human mind could unravel the secrets of Nature and, in doing so, provide instances of the hand of the Creator at work.[20] Thus, when the Boyle lectures "for proving the Christian Religion, against notorious Infidels" were established in 1691, the first lecturer chosen, the Cambridge classicist Richard Bentley (1662–1742), used the *Principia,* with Newton's encouragement, as a means of demonstrating the Argument from Design. Thus established, the tradition of linking natural philosophy and, in particular, Newtonianism, with the defense of religion was maintained by a number of Boyle lecturers. Appointments to such posts were largely controlled by those in the Church favorable to the Revolutionary Settlement, as, too, were some of the key posts at Cambridge, where Richard Bentley and his political and religious allies succeeded in institutionalizing Newtonian natural philosophy.

It was under Bentley's direction at Cambridge, too, that the second edition of the *Principia* was produced in 1713 with Roger Cotes (1682–1716), the first Plumean professor of experimental philosophy, as its editor. Cotes's labors helped to provide not only more copies of this now rare work (of which only three or four hundred copies had been printed[21]) but also to make it more comprehensible. In particular, his preface played a major role in asserting the merits of Newtonian natural philosophy as against that of Descartes both within Britain and on the Continent.[22] Cotes was one of a generation of Cambridge teachers of natural philosophy that included William Whiston, John and Samuel Clarke, and Nicholas Saunderson, who produced works that made Newton's work more accessible to undergraduates and to a larger public. Once established as part of the regular round of undergraduate studies at Cambridge, however, the Newtonian heritage there became rather ossified and formalized, with little attempt to develop it further. It was symptomatic, for example, that when, in 1726, there was a demand for a third edition of the *Prinicipia* it should have been produced not at Cambridge but by the Leiden-trained London physician Henry Pemberton (1694–1771). The Scottish universities, too, began to reassert their early lead in disseminating Newton's work; the most influential British Newtonian textbook of the mid-eighteenth century – Colin Maclaurin's *An Account of Sir Isaac Newton's Philosophy* (1748) – was the work of an Edinburgh professor of mathematics.

---

[20] On the early linkage between natural theology and Newton's work, see also Hélène Metzger, *Attraction universelle et religion naturelle chez quelques commentateurs anglais de Newton* (Paris: Hermann & Cie, 1938).

[21] I. Bernard Cohen, *Introduction to Newton's* Principia (Cambridge University Press, 1971), p. 138.

[22] Pierre Brunet, *L'Introduction des théories de Newton en France au XVIIIe siècle* (Paris: Librairie Scientifique Albert Blanchard, 1931), p. 66.

Thanks to the work of university teachers in both Cambridge and Scotland and to the way in which theologians, following the lead of the Boyle lecturers, incorporated elements of Newtonian natural philosophy into widely disseminated texts of natural theology Newton's work became closely associated with the established intellectual order in Church and State. It was an association furthered still more by the accolades that Newton himself received both for his intellectual achievement and his impeccable Whig credentials: Warden (1696) and Master (1699) of the Mint, a knighthood (1705), and President of the Royal Society (1703), an office he held until his death in 1727. Newton's chief philosophical spokesman, the Cambridge-trained cleric Samuel Clarke (1675–1729), was the confidant of Caroline, Princess of Wales, consort of the future George II. At her instigation, Clarke defended Newton's work against the philosophical aspersions of Leibniz. The great German metaphysician had alleged that "Sir Isaac Newton, and his followers, have also a very odd opinion concerning the work of God. According to their doctrine, God Almighty wants to wind up his watch from time to time . . . he had not, it seems, sufficient foresight to make it a perpetual motion."[23] Significantly, when defending Newton, Clarke used analogies between the role of God in the universe and the status of a king governing a kingdom, a comparison that, again, underlined the way in which in Britain Newton's work was largely appropriated by those defending the established order.[24]

One consequence of this identification of Newton with the defense of the established order in Church and State was that those who were out of sympathy with the status quo were inclined to look for alternative systems of natural philosophy. The foundation for much of the religious apologetic erected on Newton's system was its relegation of matter to the status of passive, homogeneous particles. The active forces needed to animate and sustain the cosmic order – such as the force of gravity, for which Newton had no mechanical explanation – were therefore regarded by Newton's early clerical apostles (with the characteristically cautious support of the master himself) as manifestations of God's involvement in sustaining His Creation. Much of the opposition to Newtonian natural philosophy centered, then, on the status of matter and the question of whether, indeed, it was as inert and lacking in purpose and direction as the Newtonian apologists claimed. From the radical republican and deist left, John Toland (1670–1722) turned to the works of that arch-heretic, the pantheist Spinoza, to develop a theory of matter contradicting that of Newton by arguing that motion is inherent in matter[25] – thus undermining the dualism that had been basic to the thought of Newtonian natural theologians.

[23] H. G. Alexander (ed.), *The Leibniz-Clarke Correspondence* (Manchester: Manchester University Press, 1956), pp. 11–12.
[24] Steven Shapin, "Of Gods and Kings: Natural Philosophy and Politics in the Leibniz-Clarke Disputes," *Isis*, 72 (1981), 187–215.
[25] Jacob, *The Newtonians,* pp. 235–9.

But Newton also attracted criticism from the right – from those who were disenchanted with the Revolutionary Settlement and its undermining of the hereditary principle in the State and the role of the priesthood in the Church. Although Cambridge University came increasingly under the sway of both the Whigs and the Newtonians, it produced a few who opposed the advancing political and intellectual tide. One such was the Tory sympathizer, Robert Greene (1678?-1730), who, in his massive tome *Principles of the Philosophy of the Expansive and Contractive Forces* (1727), set out (as he put it in his preface) to replace the mechanical philosophy (which he regarded as "the Product of Popish Countries") with "a Philosophy which is truly English." In a university and an age when theological debate colored all else, this was closely linked with his attack on those theologians, such as Newton's clerical popularizers, who, as he urged in an earlier work, were "too fond of what they call Rational, who put too great a stress upon their Reasonings from Nature, when so little of it is understood by us."[26] For Greene the key difference between his system and that of Newton was that he rejected "the Principles of a Similar and Homogeneous Matter" and viewed matter both as heterogeneous and as an active force. Matter, as his title suggested, could be "distinguished into the Expansive and Contractive Forces."[27]

Newton's theory of matter also came under attack from the Hutchinsonians, a group of religiously conservative theologians largely linked with Oxford who, like Greene, viewed Newton as an ally of a religious and political establishment with which they were out of sympathy. The instigator of the school, John Hutchinson (1674–1737), had attempted to construct a system of natural philosophy on thoroughgoing Biblical principles, arguing that the Christian Trinity was reflected in the universe in the three principles of fire, light, and air, which control the world through mechanical laws. He sought to establish these principles as an alternative to the Newtonian conceptions of force and gravitation which, since they lacked a mechanical explanation, seemed to imply God's direct involvement in the world.[28] This line of argument had been used by Newton's latitudinarian clerical popularizers to defend the Church and the political regime with which it was linked, but, in Hutchinson's view, it led to the heresy of pantheism by suggesting that God was in some sense part of His own Creation.[29] But the Hutchinsonians remained a

---

[26] Robert Greene, *The Principles of Natural Philosophy* . . . (Cambridge, 1712), preface.

[27] Robert Greene, *Principles of the Philosophy of Expansive and Contractive Forces* (Cambridge, 1727), preface, p. 409; Robert Schofield, *Mechanism and Materialism: British Natural Philosophy in the Age of Reason* (Princeton, NJ: Princeton University Press, 1970), p. 119.

[28] Geoffrey Cantor, "Revelation and the Cyclical Cosmos of John Hutchinson," in L. Jordanova and Roy Porter (eds.), *Images of the Earth: Essays in the History of the Environmental Sciences* (Chalfont St. Giles: British Society for the History of Science, Monograph Series no. 1, 1979), pp. 9–10; and Christopher Wilde, "Matter and Spirit as Natural Symbols in Eighteenth-Century Britain," *British Journal for the History of Science*, 15 (1982), 99–131, especially 106.

[29] C.B. Wilde, "Hutchinsonianism, Natural Philosophy and Religious Controversy in Eighteenth-century Britain," *History of Science*, 18 (1980), 1–24, especially 3–6.

small and increasingly embittered minority, because not only was their natural philosophy largely ignored but also their prospects for advancement in a
Church more and more dominated by those in sympathy with the principles
of the Revolution Settlement receded.

Although figures such as Greene or the Hutchinsonians had little direct
influence, the dualism between inert matter and active principles, which had
been fundamental to the work of Newton and his early theological and scientific popularizers, began to erode over the course of the century. By the time
of James Hutton (1726–1797) it was possible to use the Newtonian heritage as
part of his theory of the Earth to construct a view of Nature in which matter
was self-regulating, not requiring divine action to generate or sustain action.[30]
Similarly, Joseph Priestley (1733–1804) used Newtonian concepts to dissolve
Newtonian dualism by developing a view of matter characterised by extension and active forces.[31] As the century advanced, too, the diverse character
of "Newtonianism" became more evident, with different schools focusing on
varying aspects of the master's work. The very name, Newton, had become
too powerful a talisman for any but the occasional eccentric to put himself
outside the Newtonian fold, but this did not prevent considerable pluralism
as varying approaches to the study of Nature breathed an atmosphere permeated by Newtonian principles.[32] Thus, for example, even though Newton's
Opticks favored a particle theory of light this did not prevent some eighteenth-
century natural philosophers from continuing to speculate about wave theories of light while still claiming broad allegiance to a Newtonian conception
of the study of Nature.

And the imprint of Newton's own work bore much more heavily on some
areas than others. The *Principia* had drawn together Galilean conceptions of
terrestrial mechanics and a Keplerian understanding of celestial mechanics
into a powerful synthesis that became more impregnable as the century progressed. This further development of his work owed more to late eighteenth-
century French natural philosophers such as Laplace and Lagrange than to
Newton's British followers, who were perhaps too much in awe of the master to attempt to improve on him or who lacked the career opportunities for
uninterrupted scientific study made possible by the French absolutist and
revolutionary states. There were, then, core areas of the Newtonian heritage
on which few trespassed, but there were also whole areas of natural philosophy
on which Newton had touched lightly or not at all. This applied particularly
to the experimental sciences, which, in many cases, had received little more

---

[30] P. M. Heimann, "Voluntarianism and Immanence: Conceptions of Nature in Eighteenth-Century
British Thought," *Journal of the History of Ideas*, 39 (1978), 271–83, especially 281–2; and P. M. Heimann,
"'Nature is a Perpetual Worker': Newton's Aether and Eighteenth-century Natural Philosophy," *Am-
bix*, 20 (1973), 1–25, especially 24.

[31] P. Heimann and J. E. McGuire, "Newtonian Forces and Lockean Powers: Concepts of Matter in Eighteenth-Century Thought," *Historical Studies in the Physical Sciences*, 3 (1971), 233–306, especially 279.

[32] Geoffrey Cantor, *Optics after Newton: Theories of Light in Britain and Ireland, 1704–1840* (Manchester: Manchester University Press, 1983), pp. 11, 300.

than cursory, although suggestive, treatment in the queries to the *Opticks*. The growth of experiment was, indeed, to be one of the most striking features of eighteenth-century natural philosophy, owing much to the concomitant growth in sophistication of scientific instruments particularly in the last two decades of the century.[33] Along with mathematical rigor which was a more unambiguous aspect of Newton's influence, the increasing association between the study of Nature and the use of experiment served more and more to distinguish natural philosophy sharply from philosophy as traditionally understood and to lay the basis for its claims to be a privileged form of knowledge. The use of experiment also helped to create a public following for natural philosophy as traveling lecturers gained an exiguous living by demonstrating to the curious the way in which Nature could be controlled and made predictable by the use of experiment.[34]

THE DIFFUSION OF NEWTON'S
WORK ON THE CONTINENT

An increasing diversity of schools of natural philosophy operating within a framework broadly defined as Newtonian developed over the course of the eighteenth century in Britain. This diversity was still more evident on the Continent where earlier allegiances, such as that to Descartes or Leibniz, merged with and reshaped the later Newtonian tide. Christian Wolff (1679–1754), for example, drew on elements of Newton's work in his *Preliminary Discourse on Philosophy in General* (1728), which was written while Wolff was at the University of Marburg, but these elements became part of a mosaic that also owed much to traditional scholastic, Cartesian, and, above all, Leibnizian influences. For Wolff, Newton's work was an interesting exercise in mathematics, but it was not truly a work of natural philosophy since it lacked the philosophical breadth and depth that Wolff found in Leibniz and attempted to provide in his own work.[35] Wolff's *Preliminary Discourse* indicates, then, the persistence in the early eighteenth century of the long-established view that natural philosophy should be drawn into a larger synthesis that would include all branches of philosophy. Thus, using Aristotelian language he defined "that part of philosophy which treats of bodies" as "physics" (the traditional synonym for natural philosophy) and argued that "it is clear that metaphysics must precede physics, if the latter is to be developed demonstratively." Even this view that philosophy – including the study of Nature – was a deductive system reflects

---

[33] John Heilbron, *Electricity in the 17th and 18th Centuries: A Study of Early Modern Physics* (Berkeley: University of California Press, 1979), p. 78.

[34] Simon Schaffer, "Natural Philosophy and Public Spectacle in the Eighteenth Century," *History of Science*, 21 (1983), 1–43, especially 5.

[35] R. Calinger, "The Newtonian-Wolffian Controversy (1740–59)," *Journal of the History of Ideas*, 30 (1969), 319–30, especially 320.

the long shadow of a scholastic past, wherein a common logical approach to all philosophical problems gave unity to the different areas of philosophy.

Wolff was sufficiently in sympathy with modern science to acknowledge the importance of mathematics, but this, strictly speaking, fell outside the domain of natural philosophy as it had in the Aristotelian tradition. Mathematics became for Wolff, then, part of a system of deduction as did the use of experimental evidence to establish "the principles from which the reason can be given for what occurs in the nature of things." But for all the assault on final causes by major figures of the Scientific Revolution, Wolff still ascribed a limited role to natural philosophy: "Physics demonstrates the efficient causes of natural things, while teleology demonstrates their final causes."[36] It is a remark that underlines both the continuing adherence to the view that natural philosophy should form part of an overarching philosophical schema and the eclectic nature of much of eighteenth-century natural philosophy as elements of the old and the new were yoked together.

The strength of such views was less tenacious in medical faculties which were not subject to such a weight of philosophical tradition and which, by the nature of their subject matter, were more exposed to a range of empirical evidence that was corrosive of philosophical consistency. Such an academic environment helps to explain the wide scientific sympathies of Herman Boerhaave (1688–1738), one of the greatest scientific teachers of the century and one of the first outside Britain to incorporate Newton's work into his lectures. From 1701, as a member of the medical faculty of Leiden, Boerhaave attempted to base his chemical and medical investigations on a corpuscular theory of matter that, in the eclectic manner of many Continental natural philosophers, he owed in part to Descartes but more particularly to Boyle and Newton. Newton's influence, too, was apparent in the extent to which Boerhaave framed his explanations of the workings of the body in mechanical terms. In his lectures on chemistry (an important part of the medical curriculum) Boerhaave displayed the love of experiment that was to be a feature of Dutch science in the eighteenth century. He made full use of the best available instruments, using, for example, the latest balances constructed by Fahrenheit – an instance of the way in which developments in instrument-making fostered closer empirical investigation and, with it, the possibility of a more extensive use of mathematics. This emphasis on experimentation owed something to Newton's example (especially in the *Opticks*),[37] but, as well as owing much to local traditions of instrument-makers and mathematical practitioners, it also drew on earlier English influences such as Bacon and Boyle.[38]

---

[36] Christian Wolff, *Preliminary Discourse on Philosophy in General,* ed. and trans. Richard Blackwell (Indianapolis: Bobbs-Merrill, 1963), pp. 35, 49, 51, 54.

[37] I. Bernard Cohen, *Franklin and Newton: An Inquiry into Speculative Newtonian Experimental Science and Franklin's Work in Electricity as an Example Thereof* (Philadelphia: American Philosophical Society, 1956), p. 223.

[38] Schofield, *Mechanism and Materialism,* p. 136.

Boerhaave's student Petrus van Musschenbroek (1692–1761), although a graduate in medicine, helped to establish Newtonian natural philosophy within the philosophy curriculum. In 1717, two years after he graduated, he traveled to England and there met both Newton and his devoted disciple Desaguliers, whose public lectures based on experimental demonstrations helped bring Newton's work to a larger audience. Such influences helped nurture the experimental strain that Musschenbroek, a member of a well-known family of instrument-makers, had derived from Boerhaave. Musschenbroek's work indicates the close association between the British Newtonian tradition and the Netherlands, reflecting the long commercial and religious links between the two nations. But a Newtonian veneer was often placed on an active tradition of experimentalism that owed much to the Dutch universities' fostering of professional training in areas such as medicine and applied mathematics. As in Britain, such experimentation diversified and broadened the understanding of the domain of natural philosophy beyond the terrain of those areas that belonged directly to the Newtonian oeuvre. In his lectures on natural philosophy – given while professor of natural philosophy and mathematics at Utrecht from 1723 to 1740 and published in 1734 under the title *Elementa physicae* – Musschenbroek viewed physics as one branch of the broader study of philosophy, "the knowledge of all things both divine and human . . . which may be known by the understanding, the senses, reason, or by any other way." Along with physics, which "considers the space of the whole universe, and all bodies contained in it," Musschenbroek, like Wolff, included under the banner of philosophy "*teleology*, which investigates the ends, for the sake of which all things in the universe have their existence" together with other traditional branches of the philosophical canon such as metaphysics, logic, and moral philosophy. For Musschenbroek, like Aristotle, "motion is the principal object of Physicks"; whereas Aristotle would have considered the study of motion in qualitative terms, however, Musschenbroek laid greater emphasis on the quantitative possibilities experimentation made possible. He also emphasized the link between theory and practice and, in Baconian fashion, described it as a study that "discovers and improves the conveniences of human life."[39] His work indicates the extent to which the emphasis on experiment was serving to distance natural philosophy from philosophy more generally, although Musschenbroek still retained some conception of the way in which natural philosophy should engage with the larger philosophical enterprise.

In 1739 Musschenbroek returned to Leiden, the university of Boerhaave and of his long-standing scientific ally, Willem 's Gravesande (1688–1742). 's Gravesande and Musschenbroek taught together there, and, when the former died, Musschenbroek carried on his work in promoting experimental physics. Both men took the view that the inductive evidence obtained by the

---

[39] Petrus van Musschenbroek, *The Elements of Natural Philosophy*, translated from the Latin by John Colson (London, 1744), pp. 1, 2, 5, 9.

use of experiment could serve as the basis for deductive reasoning using mathematics in the manner exemplified by Newton.[40] But such mathematical reasoning must ultimately be confirmed by experiment, a position reflected in the title of 's Gravesande's published lectures given while professor at Leiden from 1717: *Mathematical Elements of Natural Philosophy, Confirmed by Experiments or, an Introduction to the Newtonian Physics* (the original Latin edition of 1720 being promptly published in translation by Desaguliers in 1720–1).

Nonetheless, 's Gravesande goes further than Musschenbroek in regarding natural philosophy as being distinguished from other areas of philosophy by the use of both experiment and mathematics. Indeed he defines natural philosophy in a preface colored by anti-Cartesian polemic as being "placed among those parts of Mathematics, whose Object is Quantity in general." Experiment and mathematics, he emphasized, played complementary roles in natural philosophy for "we are to discover the Laws of Nature by the Phaenomena, then by Induction prove them to be general Laws; all the rest is to be handled Mathematically."[41] It was a position that reflects 's Gravesande's well-developed Newtonian connections, which dated from his visit to England in 1715 as part of a Dutch embassy and resulted in his election to the Royal Society. Along with an introduction to Newton himself, the visit led to a continuing association with Newton's disciples John Keill and Desaguliers, from whom he learned the pedagogical value of using experiments to provide proof of scientific principles. 's Gravesande conceived his experimental program with its eschewal of speculative theorizing as being in the spirit of Newton's famous declaration, 'I frame no hypotheses.'[42] The influence of Newton became even more manifest in those sections of his lectures dealing with such core Newtonian concepts as gravitation and the laws of motion.

The conception of Newtonianism embedded in the work of Boerhaave, Musschenbroek, and 's Gravesande was, in some senses, a negative one: the rejection of the great Cartesian ambition of explaining all the phenomena of Nature by recourse to philosophical argument. Instead, the Dutch experimentalists regarded Newton – and especially his rejection of speculative hypotheses – as providing justification for their claim that the scope of natural philosophy should be defined largely by the conduct of experiments. By doing so they largely excluded the realm of living things, thus narrowing traditional natural philosophy to an area more closely approximating what the nineteenth century came to regard as "physics." But although such a position might be adopted to counter the sweeping claims of Cartesian mechanists, the Dutch experimentalists could not function without some philosophically based conception of the workings of Nature and naturally, like Newton, they largely

---

[40] Brunet, *Les physicians hollandais*, p. 100.

[41] Willem 's Gravesande, *Mathematical Elements of Natural Philosophy, Confirmed by Experiments*, trans. John Desaguliers, 2 vols. (London, 1747), 6th ed., pp. viii, xvi.

[42] Cohen, *Franklin and Newton*, p. 238.

turned to the dominant corpuscularian model. Their "Newtonianism" could also include an openness to other systems of natural philosophy; 's Gravesande, for example, published an article in defense of a Leibnizian conception of force. Nonetheless, they did share with Newton the radical position, which undermined the traditional association between natural philosophy and philosophy more generally, that natural philosophy could no longer be understood as a quest to understand fundamental causes. Like Newton, they argued that the task of the natural philosopher was both humbler and more circumscribed: the description of natural phenomena through mathematics or recourse to experiments. Ultimately they, like Newton, were prepared to make the painful acknowledgment that the causes of such basic phenomena as gravitation were unknowable.[43]

The Dutch example helped to give greater currency in France both to the use of experiment and to the study of Newton's work. 's Gravesande's textbook on Newtonian natural philosophy, for example, was widely disseminated in France.[44] There the introduction of new systems of natural philosophy had to combat a well-entrenched Cartesian regime that had largely institutionalized itself in what was then the largest scientific establishment in Europe. But within Cartesianism, as within Newtonianism, there were many mansions, some of which were more sympathetic than others to new movements in natural philosophy, including Newtonianism. The strain of idealist Cartesian philosophy developed by the philosopher-theologian Malebranche, for example, was less likely to be critical of Newton's lack of a complete mechanical explanation for phenomena such as gravity. That may help to account for the fact that Malebranchists, such as Jean-Pierre de Molières (1677–1742) and Jean-Jacques d'Ortous de Mairan (1678–1771), played a part in the early dissemination of Newton's work by attempting to reconcile Newtonianism and Cartesianism.[45] But even among Malebranche's disciples Newton tended to be regarded as an interesting geometer and experimenter rather than as someone who offered an alternative system of natural philosophy that, in true rationalist fashion, could be integrated into an overarching philosophical schema.[46]

But when Newton died in 1727 the éloge by Bernard de Fontenelle, Perpetual Secretary of the Academy of Sciences, revealed how far Newton was still largely viewed through Cartesian spectacles. While praising Newton, Fontenelle, as a loyal Cartesian, also endeavored to point out such inadequacies within his work as the idea of attraction.[47] Within a few years, however, a small group of Newtonian sympathizers had emerged within the Academy;

---

[43] Edward Ruestow, *Physics at Seventeenth and Eighteenth-Century Leiden: Philosophy and the New Science in the University* (The Hague: M. Nijhoff, 1973), pp. 121–4, 130–1, 134, 137.

[44] Brunet, *L'introduction*, p. 97.

[45] Henry Guerlac, *Newton on the Continent* (Ithaca, NY: Cornell University Press), p. 73; and Thomas Hankins, "The Influence of Malebranche on the Science of Mechanics during the Eighteenth Century," *Journal of the History of Ideas*, 28 (1967), 193–210, especially 195.

[46] A. R. Hall, "Newton in France: A New View," *History of Science*, 13 (1975), 233–50, especially 247.

[47] Brunet, *L'introduction*, p. 150.

they were led by Maupertuis and Clairaut[48] (the former had been directly introduced to Newton's work while in England in 1728).[49] Both men were involved in the Academy's expedition to Peru and Lapland in 1735 and 1736 to measure degrees of latitude close to the Equator and the North Pole. The aim of these expeditions was to test the Newtonian claim that the earth was not a perfect sphere and the experimental confirmation of the Newtonian position greatly enhanced Newton's prestige within France. However, the introduction of Newton's work was part of a more general influx of scientific and mathematical ideas into France – not only from Britain but elsewhere in Europe – made possible by the greater openness of the Academy to foreigners from 1720.[50]

Outside such specialist circles, however, educated French society had to wait until Voltaire published *Élements de la Philosophie de Newton* in 1738 for a readily comprehensible French introduction to Newton's work – or, at least, to those philosophical and theological aspects of it that Voltaire could use as part of his assault on the great philosophical system-builders, Descartes and Leibniz. It also formed part of Voltaire's more fundamental attack on a Church establishment that had, after some initial opposition, become closely associated with Cartesianism.[51] This helps to explain why Voltaire portrayed Newtonianism and Cartesianism as being sharply polarized – as being, in his words, "like the rallying cries of two parties."[52] Such polemic, however, served to obscure the way in which many French natural philosophers maintained a foot in both camps.

In Italy, too, the advocacy of Newtonianism over Cartesianism was linked to a critical view of traditional learning. There, however, as Vincenzo Ferrone has shown, the divisions were within Catholicism as the more avant-garde took advantage of the waning of Counter-Reformation militancy at the beginning of the eighteenth century to attempt to construct new syntheses between Christianity and modern learning. One of the fruits of this intellectual thaw was Francesco Algarotti's *Il Newtonianismo Per le Dame* (*Newtonianism for the Ladies*) (1737), which had its origins in a dispute with Cartesians over the nature of experimental evidence; Algarotti, in patriotic fashion, portrayed the empirical, antimetaphysical character of Newtonianism as being in the tradition of Galileo.[53] The title of the work is a reminder of the im-

---

[48] Robert Schofield, "An Evolutionary Taxonomy of Eighteenth-Century Newtonianisms," *Studies in Eighteenth-Century Culture*, 7 (1978), 175–92, especially 180.

[49] E. J. Aiton, *The Vortex Theory of Planetary Motions* (London: Macdonald, 1972), p. 201.

[50] J. Greenberg, "Mathematical Physics in Eighteenth-Century France," *Isis*, 77 (1986), 59–78, especially 75.

[51] Vartanian, *Diderot and Descartes*, p. 39.

[52] Colm Kiernan, "Science and the Enlightenment in Eighteenth-Century France," *Studies in Voltaire and the Eighteenth Century*, 59 (1968), 43.

[53] Vincenzo Ferrone, *The Intellectual Roots of the Italian Enlightenment: Newtonian Science, Religion, and Politics in the Early Eighteenth Century*, trans. Sue Brotherton (Atlantic Highlands, NJ: Humanities Press International, 1995), p. 96.

portance of women in the dissemination of new learning, including science – a role that was particularly marked in Italy.[54] However, increasing fears about the radicalism of the *philosophes* led, in the second half of the century, to Newtonianism becoming less associated with the intellectually novel as it was absorbed into more traditional forms of Catholic, scientifically based apologetic.[55]

In France as in Italy women were among the early promoters of Newton's work. As well as drawing on the example of his friend Francesco Algarotti, Voltaire's popularization of Newton greatly benefited from the collaboration of his mistress, Madame du Châtelet (1706–1749). Du Châtelet, who was first introduced to mathematical studies by Maupertuis, also produced a translation of the *Principia* published posthumously in 1756 – a work that owed much to the scientific advice of Clairaut, as did the more directly accessible summary account of the Newtonian worldview published in an accompanying volume.

But in France the ghost of Descartes continued to lurk behind the outward forms of Newtonianism that increasingly dominated the public scientific arena for, as d'Alembert wrote in 1743, the Cartesians were "a sect that in truth is much weakened today."[56] D'Alembert regarded himself as a Newtonian since he accepted the law of universal gravitation and rejected the Cartesian plenum with its dismissal of the idea of a vacuum. Nonetheless, for d'Alembert, as for Descartes, mathematical deduction took priority over evidence gained from experimentation, and he shared Descartes's ambition of deducing the sciences from first principles.[57] In his Preliminary Discourse to the great *Encyclopédie* (1754), d'Alembert dismissed experimental physics as differing "from the physico-mathematical sciences in that it is properly only a systematic collection of experiments and observations." And, for all d'Alembert's rejection of Cartesian system-building, in his map of knowledge "the science of nature" was still clearly a branch of philosophy along with its traditional bedfellows of logic and ethics. D'Alembert had sufficiently distanced himself from his Cartesian past to omit metaphysics as a separate branch of philosophy, although in developing the traditional scholastic distinction between "general" and "particular" physics he described the former as "the metaphysics of bodies" whereas the latter was a discipline that "studies the bodies in themselves and whose sole object is individual things."[58]

In France, then, as in much of the rest of Europe, Newtonianism, over the course of the century, came to assume ever greater importance, both for

[54] Paula Findlen, "Sciences as a Career in Enlightenment Italy: The Strategies of Laura Bassi," *Isis*, 84 (1993), 441–69; and Findlen, "Translating the New Science: Women and the Circulation of Knowledge in Enlightenment Italy," *Configurations*, 3 (1995), 167–206.

[55] Ferrone, *The Intellectual Roots*, p. 277.   [56] Guerlac, "Where the Statue Stood," p. 131.

[57] Hankins, *Jean d'Alembert*, pp. 169, 235.

[58] Jean d'Alembert, *Preliminary Discourse to the Encyclopedia of Diderot* (Indianapolis: Bobbs-Merrill, 1963), pp. 24, 54.

the mechanics explicated in the *Principia* and the experimental program sketched in the *Opticks*. However, the Newtonian heritage – which was itself diverse and amenable to different interpretations and emphases – was only one, albeit one of the most important, that shaped the practice of Continental natural philosophy. In France, as in the Netherlands, the experimental tradition had native roots that were often independent of Newton. Thus, the experiments of Abbé Jean-Antoine Nollet (1700–1770) – whose appointment to the first chair of experimental physics at the University of Paris in 1753 marked the increasing attention being accorded to experiment in France – were not dependent on any particular body of philosophy.[59] Indeed, if anything, his scientific outlook can be described as being anti-Newtonian.[60] It was a position that his visit to the Netherlands and contact with the Dutch experimentists served to strengthen despite their own Newtonian sympathies.[61] In France, then, the Newtonian tide was slower to inundate experimental physics than rational mechanics or astronomy.[62]

It is not that even such fields as mechanics were totally subdued by the English physicist's principles. In France, as the example of d'Alembert suggests, the fascination for the Cartesian project of constructing a system of natural philosophy by a process of rigorous deduction never entirely faded. It lingered, for example, in the majestic system of mathematical reasoning embodied in the *Méchanique analytique* (1788) of d'Alembert's close friend Joseph Lagrange (1736–1813). It was a work that sought to take an essentially Newtonian system of mechanics to a new level of mathematical purity by reducing its subject matter to a set of general formulas from which could be deduced all the equations necessary to solve any given problem. Similarly, the closely related *Méchanique céleste* (1799–1805 with supplements 1823–5) of Pierre-Simon Laplace (1749–1827), another of d'Alembert's protégés, eradicated many of the anomalies in the Newtonian explanation of the movement of the heavenly bodies by demonstrating with rigorous mathematical clarity that they reflected more fundamental basic laws. Indeed Laplace's ultimate ambition was to deduce the whole of celestial mechanics from the principle of universal gravitation.[63] His work highlights the one enduring feature of the Newtonian heritage in the face of the diverse shades it adopted in different times and places over the course of the century: the conviction that all phenomena, both celestial and terrestrial, could be explained by a uniform set of laws.[64]

---

[59] Jean Torlais, "La Physique Expérimentale," in René Taton (ed.), *Enseignement et diffusion des sciences en France au XVIIIe siècle* (Paris: Hermann, 1964), pp. 623, 627.

[60] Rod Home, "The Notion of Experimental Physics in Early Eighteenth-Century France," in Joseph Pitt (ed.), *Change and Progress in Modern Science* (Dordrecht: Reidel, 1985), p. 110.

[61] Brunet, *Les physicians hollandais*, pp. 109, 129.

[62] Roderick Home, "Newtonianism and the Theory of the Magnet," *History of Science*, 15 (1977), 252–66, especially 254.

[63] Charles Coulston Gillispie, *Pierre-Simon Laplace 1749–1827: A Life in Exact Science* (Princeton, NJ: Princeton University Press, 1997), p. 30.

[64] Roger Hahn, "New Considerations on the Physical Sciences of the Enlightenment Era," *Studies in Voltaire and the Eighteenth Century*, 264 (1989), 790.

CONCLUSION

Experiment and the increasingly sophisticated use of mathematics (both fostered by the growing accuracy of eighteenth-century scientific instrumentation) widened the divide between natural philosophy and philosophy more generally. It also gave credence to the increasingly assertive claims of the natural philosophers that theirs was a particularly privileged form of knowledge that avoided the deference to textual authorities and sophistic wordplay that had overshadowed the study of philosophy for much of its history. But as the territory of natural philosophy grew larger, so, too, did the tendency for some areas to secede and claim a separate identity. As chemistry and natural history established separate domains, natural philosophy began to lose that sense of a unified body of knowledge about all aspects of Nature that had long been one of its distinctive features.

The eighteenth century also saw the increasing use of the term "experimental philosophy" to describe those areas of natural philosophy that were concerned with subjects, such as electricity, magnetism, and optics that lent themselves more readily to experimental than mathematical treatment.[65] It was a category that defined in embryonic form the development of physics as the nineteenth century understood it:[66] a circumscribed discipline rather than, in the Aristotelian sense, the study of motion in all its forms.[67] As early as 1743 it was possible for one French reviewer to claim, "Apart from a few general principles . . . the entire study of physics today reduces to the study of experimental physics."[68]

As natural philosophy's aspirations to be an overarching view of Nature were weakened so, too, the different disciplines began to develop their own institutional forms with separate university chairs and, by the early nineteenth century, specialist professional associations outside the purview of the traditional academies such as the Royal Society or the Academy of Sciences, which had regarded themselves as being responsible for the study of all aspects of Nature. By the end of the eighteenth century the map of Nature was beginning to be divided into separate principalities, with tariff walls in the form of specialist training or the mastery of a particular body of knowledge. An early nineteenth-century *Dictionary of Arts and Sciences* underlined the way in which natural philosophy had become ever more diverse: "Natural philosophy is, however, obviously rather a system or aggregate of several branches

---

[65] Anders Lundgren, "The Changing Role of Numbers in Eighteenth-Century Chemistry," in Tore Frängsmyr, J. L. Heilbron, and Robin E. Rider (eds.), *The Quantifying Spirit in the 18th Century* (Berkeley: University of California Press, 1990), p. 256–8; and Maurice Daumas, "Precision of Measurement and Physical and Chemical Research in the Eighteenth Century," in A. C. Crombie (ed.), *Scientific Change* (London: Heinemann, 1963), pp. 418–30.

[66] David Miller, "The Revival of the Physical Sciences in Britain, 1815–1840," *Osiris,* 2nd series, 21 (1986), 107–34, especially 132.

[67] Thomas Hankins, *Science and the Enlightenment* (Cambridge University Press, 1985), p. 46; and D. S. L. Cardwell, "Science, Technology and Industry," in *The Ferment of Knowledge,* p. 458.

[68] Heilbron, *Electricity in the 17th and 18th Centuries,* p. 15.

of knowledge, than a simple and uniform science."[69] For, over the course of the eighteenth century, natural philosophy had largely broken loose from its traditional links with a wider body of philosophy, but in gaining independence it had produced offspring of its own that proved impatient of its basic claim to provide a unified understanding of the workings of Nature.

[69] George Gregory, *A Dictionary of Arts and Sciences*, 2 vols. (London, 1806–7), vol. 2, p. 255.

13

MATHEMATICS

Craig Fraser

Considered broadly, mathematical activity in the eighteenth century was characterized by a strong emphasis on analysis and mechanics. The great advances occurred in the development of calculus-related parts of mathematics and in the detailed elaboration of the program of inertial mechanics founded during the Scientific Revolution. There were other mathematical developments of note – in the theory of equations, number theory, probability and statistics, and geometry – but none of them reached anything like the depth and scope attained in analysis and mechanics.

The close relationship between mathematics and mechanics had a basis that extended deep into Enlightenment thought. In the Preliminary Discourse to the famous French *Encyclopédie*, Jean d'Alembert distinguished between "pure" mathematics (geometry, arithmetic, algebra, calculus) and "mixed" mathematics (mechanics, geometrical astronomy, optics, art of conjecturing). He classified mathematics more generally as a "science of nature" and separated it from logic, a "science of man." An internalized and critical spirit of inquiry, associated with the invention of new mathematical structures (for example, non-commutative algebra, non-Euclidean geometry, logic, set theory), represents characteristics of modern mathematics that would emerge only in the next century.

Although there were several notable British mathematicians of the period – Abraham De Moivre, James Stirling, Brook Taylor, and Colin Maclaurin among them – the major lines of mathematical production occurred on the Continent, a trend that intensified as the century developed.[1] Leadership was provided by a relatively small number of energetic figures: Jakob, Johann, and Daniel Bernoulli, Jakob Hermann, Leonhard Euler, Alexis Clairaut, Jean d'Alembert, Johann Heinrich Lambert, Joseph Louis Lagrange, Adrien Marie Legendre, and Pierre Simon Laplace. Research was coordinated by national

---

[1] For a study of British mathematics in the eighteenth century, see Niccolò Guicciardini, *The Development of Newtonian Calculus in Britain, 1700–1800* (Cambridge University Press, 1989).

and regional scientific academies, of which the most important were the academies of Paris, Berlin, and St. Petersburg. Roger Hahn has noted that the eighteenth-century academy allowed "the coupling of relative doctrinal freedom on scientific questions with their rigorous evaluations by professional peers," an important characteristic of modern professional science.[2] The academic system tended to promote a strongly individualistic approach to research. A determined individual such as Euler or Lagrange could emphasize a given program of research through his own work, the publications of the academy, and the setting of the prize competitions.

Although the academy as a social institution was inherently centralized and elitist, the writings of the academicians were more discursive, expository, and inclusive than would be the case in the specialized research journals of later science. The democratization of science that occurred in the nineteenth century, with the opening of scientific careers to a wide segment of society, was accompanied intellectually within each field by a rather narrow and proprietary specialization that was foreign to the spirit of inquiry in the age of Enlightenment. In comparing Euler's writings with those of a hundred or a hundred and fifty years later one is struck by the change in the way in which the audience is conceived from, in the first case, anyone in principle who is curious about mathematics to, in the second, a group of specialists who have already undergone considerable initiation and concerning whose knowledge many assumptions may be tacitly accepted.

This essay is devoted to major developmental trends in advanced theoretical mathematics during the eighteenth century. It is important nevertheless to call attention to the spread of mathematical methods and mentalities in a range of more practical subjects and pursuits. In navigation, experimental physics, engineering, botany, demography, government, and insurance there was an increasing emphasis on quantification and rational method. In the burgeoning industrial arts, instrument-makers achieved new levels of precision measurement. In French engineering schools, sophisticated mathematics – including the calculus – was introduced for the first time into the teaching curriculum, a practice that would be widely followed in later education. The operational, algebraic character of advanced theoretical analysis was reflected at a wider level in a pronounced instrumentalist understanding of the uses and nature of mathematics. In an overview of Enlightenment quantitative science John Heilbron writes as follows:

> [In the later eighteenth century] analysis and algebra, which, in contrast to geometry had an instrumentalist bias, became the exemplar of the mathematical method. . . . This instrumentalism was a key ingredient of the quantifying spirit after 1760. . . . Most of the leading proponents of the Standard

[2] Roger Hahn, *The Anatomy of a Scientific Institution: The Paris Academy of Science, 1666–1803* (Berkeley: University of California Press, 1971), p. 313.

Model [i.e. Laplacian molecular physics] . . . made clear that they understood it in an instrumentalist sense. . . . They found themselves in agreement with the epistemologies of Hume and Kant, and perhaps also with Condillac's teaching that clear and simple language, not intuitions of truth, conduces to the advancement of science.[3]

The rational "quantifying spirit" of the Enlightenment would find a lasting and pervasive legacy in the adoption at the end of the century in France of the metric system, a development that took place under the direct supervision of prominent mathematical scientists of the time.[4]

THE CENTURY OF ANALYSIS

Euler and Lagrange were leading and representative practitioners of analytical mathematics in the eighteenth century. Together they dominated the subject from 1740 until early into the next century. Their writings, and more particularly their extensive contributions to analysis, defined advanced mathematical activity. What is fundamental to an understanding of the intellectual fabric of mathematics of the period is the distinctive conception of algebraic analysis that guided their work. They conceived of the metaphysics of the calculus in a way that is significantly different from our outlook today. Although we tend to take the modern foundation for granted, the older approach of algebraic analysis was based on a different point of view, a different conception of how generality is achieved in mathematics, and a rather different understanding of the relationship of analysis to geometry and physics. The interest of the eighteenth-century work lies in considerable part in providing an example of an alternative conceptual framework, one with great historical integrity and cohesion.[5]

LEONHARD EULER

Euler became established as a mathematician of note during the decade of the 1730s. He was a young man in his twenties, a member of the St. Petersburg

---

[3] Heilbron's remarks are contained in his introduction to the volume, T. Frängsmyr et al. (eds.), *The Quantifying Spirit in the 18th Century* (Berkeley: University of California Press, 1990), pp. 3, 5.

[4] For studies of quantitative applied science in the eighteenth century, see H. Gray Funkhouser, "Historical Development of the Graphical Representation of Statistical Data," *Osiris*, 3 (1937), 269–404, and Laura Tilling, "Early Experimental Graphs," *British Journal for the History of Science*, 8 (1975), 193–213.

[5] Although the emphasis of the present essay is on calculus-related parts of mathematics, a concern for symbolic methods was also evident in such subjects as the theory of equations and number theory. See L. Novy, *Origins of Modern Algebra* (Leiden: Noordhoff International Publishing, 1973). Progress in formal mathematics was evident in probability and statistics; see Stephen M. Stigler, *The History of Statistics: The Measurement of Uncertainty before 1900* (Cambridge, MA: Harvard University Press, 1986).

Academy of Sciences and a colleague of Hermann, Daniel Bernoulli, and Christian Goldbach. Euler's interest in analysis is evident in writings from this period, including his major treatise of 1736 on particle dynamics, *Mechanica sive Motus Scientia Analytice Exposita*. Although the theme of analysis was well established at the time, there was in his work something new: the beginning of an explicit awareness of the distinction between analytical and geometrical methods and an emphasis on the desirability of the former in proving theorems of the calculus.

Euler's program of analysis would be launched in a series of comprehensive treatises on different branches of the calculus and celestial dynamics published between 1744 and 1766. During this period he was mathematics director of the Berlin Academy of Sciences. His capacity for calculation and tremendous output later led François Arago to confer on him the title of "Analysis Incarnate." In the last part of his career Euler returned to St. Petersburg where he continued to carry out research and to publish. In 1735 Euler lost the sight of his right eye, and shortly after his arrival in St. Petersburg, he lost the sight of his remaining eye. Despite working in conditions of near blindness he was able with the assistance of his family and servants to remain productive mathematically up to his death in 1783.

GRAPHICAL METHODS AND THE FUNCTION CONCEPT

The geometrical curve was an object of intensive mathematical and physical interest throughout the seventeenth and early eighteenth centuries. The study of the relations that subsist between the lengths of plane curves gave rise in 1718 in the writings of Count C. G. Fagnano to a theory of elliptic integrals. In the calculus of variations, a branch of mathematics pioneered by Jakob and Johann Bernoulli, classes of curves constituted the primary object of study; the goal of each problem was the selection of a curve from among a class of curves that rendered a given integral quantity a maximum or minimum. In analytical dynamics attention was concentrated on determining the relation between trajectories of particles moving in space and the forces that act on them. In the theory of elasticity, researchers studied the shape of static equilibrium assumed by an elastic lamina under various loadings, as well as the configurations of a vibrating string.

The curve also played a fundamental and very different role in the conceptual foundation of the calculus. By representing the relationship between two related variable magnitudes of a problem by means of a graphical curve the various mathematical methods that had been developed for the geometrical analysis of curves could be brought to bear on the problem. Graphical procedures had been employed by Galileo Galilei in his *Discorsi* of 1637 to relate the speed of a falling body to the time of its descent. They had become common in mathematical treatises by the late seventeenth century. Christiaan Huygens in his *Horologium Oscillatorium* (1673) and Isaac Barrow in his *Lec-*

tiones Geometricae (1670) represented quadrature relationships in this way. In his very first published paper in the calculus Gottfried Leibniz (1684) derived the optical law of refraction from the principle that light follows the path of least time. He considered two related magnitudes: the distance of the point of contact of the light ray along the interface, and the time of transit that corresponds to this distance. He represented this relationship graphically by means of a curve and proceeded to apply the differential algorithm, introduced earlier in the paper for the analysis of curves, to obtain the desired law. In his *Principia Mathematica* (1687) Isaac Newton investigated the inverse problem of central-force particle motion. In Propositions XXXIX and XLI of Book One he graphed the force as a function of the projection of position on the orbital axis and analyzed the resulting curve to arrive at expressions for the particle's trajectory. Jakob Bernoulli employed graphical methods throughout his researches of the 1690s. In his study of the elastica, the relation between the restoring force and the distance along the lamina was superimposed in graphical form on the diagram of the actual physical system.

Graphical methods played a role in the early calculus that would later be filled by the function concept. This point of view was formalized to some extent by Pierre Varignon in a 1706 memoir devoted to the study of spiral curves given in terms of polar variables.[6] Varignon considered a fixed reference circle ABYA with center C (Figure 13.1). A "courbe génératrice" HHV is given; a point H on this curve is specified by the perpendicular ordinate GH, where G is a point on the axis xCX of the circle. The line CX is conceived as a ruler that rotates with center C in a clockwise direction tracing out a spiral OEZAEK. Consider a point E on the spiral. With center C draw the arc EG. Let c = the circumference of the reference circle ABYA, x = arc AMB, CA = a, CE = y, GH = z and AD = b a constant line. The arc x is defined by the proportion c:x = b:z. Varignon wrote what he called the "équation générale de spirals à l'infini" as cz = bx. By substituting the value for z given by the nature of the generating curve into this equation, the character of the spiral was revealed. Depending on whether the generating curve was a parabola, a hyperbola, a logarithm, a circle, and so on, the corresponding spiral was called parabolic, hyperbolic, logarithmic, circular, and so on.

In Varignon's paper the equation of the spiral was formulated a priori in terms of Cartesian coordinates in the associated "generating curve." The latter embodied in graphical form the functional relationship between the polar variables and acted as a standard model to which this relationship can be referred.

From the very beginning of his mathematical career in the 1730s, Euler

---

[6] Pierre Varignon, "Nouvelle formation de spirales beaucoup plus différentes entr'elles que tout ce qu'on peut imaginer d'autres courbes quelconques à l'infini; avec les touchantes, les quadratures, les déroulemens, & les longueurs de quelques-unes de ces spirales qu'on donne seulement ici pour éxemples de cette formation générale," *Histoire de l'Académie royale des sciences avec les mémoires de mathématique et de physique tirés des registres de cette Académie 1704* (Paris, 1706), pp. 69–131.

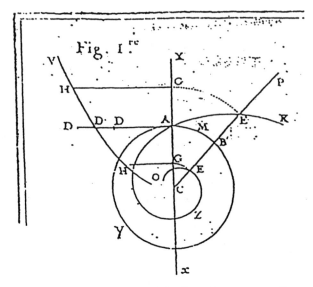

Figure 13.1. Varignon and the "Courbe génératrice."

imparted a new direction to the calculus by clearly emphasizing the importance of separating analysis from geometry. His program was evident in 1744 in his major treatise on the calculus of variations, *Methodus Inveniendi Lineas Curvas.*[7] A typical problem of the early calculus involved the determination of a magnitude associated in a specified way with a curve. To find the tangent to a curve at a point, it was necessary to determine the length of the subtangent there; to find the maximum or minimum of a curve, one needed to calculate the value of the abscissa that corresponded to an infinite subtangent; to find the area under a curve, it was necessary to calculate an integral; to determine the curvature at a point, one had to calculate the radius of curvature. The calculus of variations extended this paradigm to classes of curves.[8] In the fundamental problem of the *Methodus Inveniendi* it is required to select that curve from among a class of curves that makes a given magnitude expressing some property a maximum or minimum.

Near the beginning of his treatise (p. 13) Euler noted that a purely analytical interpretation of the theory is possible. Instead of seeking the curve that renders the given integral quantity an extremum, one seeks that "equation"

[7] Euler, *Methodus inveniendi lineas curvas maximi minimive proprietate gaudentes sive solution problematis isoperimetrici lattisimo sensu accepti* (Lausanne, 1744). Reprinted in Euler's *Opera Omnia,* Ser. 1, V. 24.

[8] For historical studies of Euler's calculus of variations, see Herman H. Goldstine, *A History of the Calculus of Variations from the 17th through the 19th Century* (New York: Springer-Verlag, 1980), chap. 3; and Craig Fraser, "The Origins of Euler's Variational Calculus," *Archive for History of Exact Sciences,* 47 (1994), 103–41; and Fraser, "The Background to and Early Emergence of Euler's Analysis," in M. Otte and M. Panza (eds.), *Analysis and Synthesis in Mathematics History and Philosophy,* Boston Studies in the Philosophy of Science, vol. 196 (Dordrecht: Kluwer, 1997).

between x and y that, among all such equations, renders the quantity a maximum or minimum. He wrote, "In this way questions in the doctrine of curved lines may be referred back to pure analysis. Conversely, if questions of this type in pure analysis be proposed, they may be referred to and solved by means of the doctrine of curved lines."

Euler's derivation of the basic equations and principles of the calculus of variations was formulated in terms of the detailed study of the properties of geometrical curves. Nevertheless, in Chapter 4 of his book he showed that a purely analytical interpretation of the theory was possible. He observed that "the method presented earlier may be applied widely to the determination of equations between the coordinates of a curve which render any given expression $\int Zdx$ a maximum or a minimum. Indeed it may be extended to any two variables, whether they involve an arbitrary curve, or are considered purely in analytical abstraction." He illustrated this claim by solving several examples using variables other than the usual rectangular Cartesian coordinates. In the first example he employed polar coordinates to find the curve of shortest length between two points. He was completely comfortable with these coordinates; gone was the Cartesian "generating curve" that Varignon had employed in his investigation of 1706 to introduce general polar curves. In the second example Euler displayed a further level of abstraction, employing variables that were not even coordinate variables in the usual sense.

A range of non-Cartesian coordinate systems had been employed in earlier mathematics but never with the same theoretical import as in Euler's variational analysis. Here one had a fully developed mathematical process, centered on the consideration of a given analytically expressed magnitude, in which a general equational form was seen to be valid independent of the particular interpretation conferred upon the variables of the problem.

Euler had succeeded in showing that the basic subject matter of the calculus – what in some ultimate sense the calculus is "about" – could be conceived independently of geometry in terms of abstract relations between continuously variable magnitudes. To develop this point of view systematically it was necessary to introduce formal concepts and principles. To do this Euler turned to the concept of a function, a concept that had appeared in earlier eighteenth-century work and that he made central in his mid-century treatises on the calculus.[9] His *Introductio in Analysin Infinitorum* of 1748 contained an explicit definition: "A function of a variable quantity is an analytical expression composed in any way from the variable and from numbers or constant quantities" (p. 4). Although he sometimes considered a more

---

[9] Carl Boyer observes that for Euler "analysis was not the application of algebra to geometry; it was a subject in its own right – the study of variables and functions – and graphs were but visual aids in this connection. . . . It now dealt with continuous variability based on the function concept . . . only with Euler did it [analysis] take on the status of conscious program." (*History of analytic geometry;* originally published as Numbers Six and Seven of The Scripta Mathematica Studies; republished 1988 by The Scholar's Bookshelf; the quoted passage appears on page 190 of the latter edition.)

general notion of a function, for example in the discussion of the solution to the problem of the vibrating string, the *Introductio* furnished the operative fundamental definition for eighteenth-century work in analysis.[10]

A notable example of Euler's functional approach is provided by his introduction of the sine and cosine functions. Tables of chords had existed since Ptolemy in antiquity, and the relations between sines and cosines were commonly used in navigation and mathematical astronomy. With the advent of the calculus, trigonometric relations were expressed in terms of geometrical infinitesimal elements contained in a standard reference circle. Euler, by contrast, defined the sine and cosine functions as formulas involving variables that were given independently of geometrical constructions or dimensional considerations. He also derived the standard power series for the trigonometrical functions, using multiangle formulas and techniques he had employed earlier in the treatise to obtain the exponential series. Although these expansions were not new, they had been derived by analytical principles: a function that was a solution to a definite differential equation had been expanded to yield the given series.[11]

DIFFERENTIATION

In the original Leibnizian calculus, the concept of differentiation possessed a dual character: algebraic/algorithmic on the one hand, and geometric on the other. The algebra comprised a set of rules that governed the use of the symbol d and was based on two postulates: $d(x + y) = dx + dy$ and $d(xy) = ydx + xdy$. Accompanying these rules there was also an order principle, according to which higher-order differentials in a given equation were to be neglected with respect to differentials of a lower order.

The differentials that appeared in a given problem could also be understood in another way: as the differences of values of a variable quantity at successive points in the geometrical configuration. The differential dx was set equal to the difference of the value of x at two consecutive points infinitely close together; higher-order differentials were set equal to the difference of successive lower-order differentials. Euclidean geometry was used to analyze the properties of the curve in terms of these differentials.

A good illustration of the dual character of differentiation is provided by

---

[10] For studies dealing with the history of the function concept, see Ivor Grattan-Guinness, *The Development of the Foundations of Mathematical Analysis from Euler to Riemann* (Cambridge, MA: MIT Press, 1970); A. P. Youschkevitch, "The Concept of the Function up to the Middle of the 19th Century," *Archive for History of Exact Sciences*, 16 (1976), 37–85; and Steven Engelsman, "D'Alembert et les Équations aux Dérivées Partielles," *Dix-Huitième Siècle*, 16 (1984), 27–37. In the secondary literature there has tended to be something of a historiographical divide. Authors such as Truesdell, Demidov, and Youschkevitch have emphasized Euler's modernism, whereas Grattan-Guinness and Fraser have in a less Whiggish vein called attention to historically particular features of his thought.

[11] For a historical account, see Victor J. Katz, "Calculus of the Trigonometric Functions," *Historia Mathematica*, 14 (1987), 311–24.

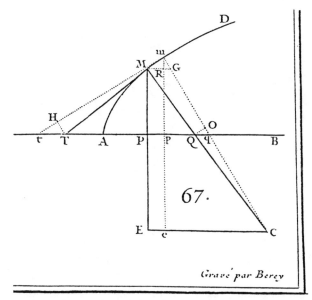

Figure 13.2. L'Hôpital and the center of curvature.

the derivations of the formula for the radius of curvature of a curve given by the Marquis de l'Hôpital in his textbook *Analyse des infiniment petits, pour l'intelligence des lignes courbes* (1696; second edition 1716). This formula was used in analytic geometry to calculate the evolute to a curve, that is the locus formed by the center of the radius of curvature. In mechanics it was known that the restoring force on an element of a stretched elastic string is proportional to the curvature (the reciprocal of the radius of curvature) of the string at the point where the element is located. The expression for the radius could be used to derive a differential equation to describe the string's motion.[12]

The first derivation of the formula that we shall consider was taken by l'Hôpital from a textbook published by Johann Bernoulli in 1691. Assume M is any point on the curve AMD (Figure 13.2). Let m be a point on the curve infinitely close to M. The normals to the curve at M and m intersect at the center of curvature C. The distance MC is the radius of curvature. Suppose $AP = x$ and $PM = y$ are the abscissa and ordinate of M. The lines MR and Rm parallel to AP and PM are the infinitesimal increments dx and dy of x and y. L'Hôpital calculated that $PQ = ydy / dx$. Let Q and q be the intersections of the normals MC and mC and the axis of the abscissae. L'Hôpital

---

[12] A detailed historical account of the theory of differentials from Leibniz to Euler, including a description of the calculation of the radius of curvature, is contained in Henk J. Bos, "Differentials, Higher-Order Differentials and the Derivative in the Leibnizian Calculus," *Archive for History of Exact Sciences,* (1974), 1–90.

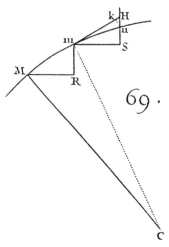

Figure 13.3. L'Hôpital and second-order differentials.

supposed that the quantity dx is constant, a step that corresponds from a modern perspective to the assumption that x is the independent variable in the problem. Since Qq = d(AQ) = dx + d(PQ), he applied the differential algorithm and obtained the expression Qq = dx +(dy² + yddy) / dx. Using similar triangles he proceeded to calculate the radius MC and obtained the formula

$$MC = \frac{(dx^2 + dy^2)\sqrt{dx^2 + dy^2}}{-dxddy} \tag{1}$$

In a subsequent derivation l'Hôpital employed a different procedure, calculating the second differentials directly in terms of the elements of the geometrical configuration. Consider again the portion of the curve AMD containing Mm (Figure 13.3). Let n be a point on the curve infinitely close to m. L'Hôpital conceived of the portion Mmn as composed of the polygonal segments Mm and mn. The second differential of y, ddy, is given as ddy = nS − mR = nS − HS = −Hn. By means of similar triangles he arrived at an estimate for the radius of curvature that reduced to formula (1).

Another illustration of the dual character of differentiation is provided in mathematical dynamics in the calculation of differential equations of motion connecting the force to the spatial coordinates of a moving particle. The usual procedure during the period involved the comparison of the dynamical system at three successive instants in time. The second differentials appearing in the equations of motion were calculated in terms of the second differences arising in these configurations. In the 1740s and the 1750s, in the writings of Euler and d'Alembert, the second differentials were calculated directly in terms of the differential algorithmic procedures of the calculus.[13] This method, as-

[13] Both methods of calculating second differentials were employed by d'Alembert in his *Traité de Dy-*

sociated today with the differential-equation form of Newton's second law, soon became standard in classical mechanics.

In his mid-century treatises Euler, as part of his program of separating analysis from geometry, made the algebraic conception of differentiation fundamental. In so doing he made the concept of the *algorithm* primary in his understanding of the foundations of the calculus. Some of the issues that arise in this shift in viewpoint are illustrated by his theory of differential expressions set forth in Chapters 8 and 9 of the first part of his 1755 *Institutiones Calculi*. Consider any formula containing dx, ddx, dy, ddy, ... Because these quantities are no longer interpreted geometrically the meaning of the formula is unclear; its value will depend on whether dx or dy is held constant, an assumption that is not evident in the algebra. For example, the quantity ddy / dx^2 is zero if dy is constant; if dx is constant its value will vary according to the functional relation between x and y. Conversely, certain expressions, such as (dyddx − dxddy) / dx^3, may be shown to be invariant regardless of which variable is taken to be independent.

Euler's solution to the problem of indeterminacy in differential expressions was to introduce notation that made clear the relations of dependency among the variables. He did so by eliminating higher-order differentials as such, replacing them instead with differential coefficients. Rather than write ddy / dx^2 (dx constant) we define the differential coefficients p and q by the relations dy = pdx and dp = qdx; ddy / dx^2 then becomes simply q. Euler provided rules and examples that showed how more complicated expressions can be reduced to ones containing only variables and differential coefficients. In addition to bringing order to the calculus, this emphasis on the differential coefficient was conceptually important in identifying the derivative as an independent object of mathematical study.[14]

INTEGRATION

Leibniz had regarded the integral as a kind of infinite summation carried out with reference to a sequence of values of one of the variables of the problem. He denoted integration using an elongated "S," which stood for the first letter of the Latin word "summa" for sum. Thus, the area under the curve $y = x^2$ was expressed as $\int x^2 dx$, where the limits of integration were understood to be given.

A significant modification of Leibniz's conception was introduced in the early 1690s by Johann Bernoulli, who replaced the concept of an integral as a sum with the quite different concept of the integral as an antiderivative.[15]

---

namique of 1744. See Craig Fraser, "D'Alembert's Principle: The Original Formulation and Application in Jean D'Alembert's *Traité de Dynamique* (1743)," *Centaurus*, 28 (1985), 31–61, 145–59.

[14] For a more detailed description of Euler's theory, see Bos, "Differentials, Higher-Order Differentials."

[15] Bernoulli's definition was contained in his *Die erste Integralrechnung*, a selection of his writings from

Taking the d-operation as logically primary, Bernoulli defined integration as the operational inverse of differentiation. The integral $\int x^2 dx$ was by definition equal to $x^3 / 3$, because the differential of the latter expression is equal to $x^2 dx$.

In his mid-century writings on analysis Euler adopted Johann Bernoulli's notion of the integral as an antiderivative, a point of view that Euler made fundamental in his two-volume *Institutiones Calculi Integralis* of 1768. It is clear that Euler held to this conception from a very early stage of his career. In the 1730s he had investigated the problem of determining orthogonal trajectories to families of curves, a subject that had been broached by Leibniz forty years earlier.[16] The latter had considered integrands consisting of expressions involving both a variable x and a parameter t. Leibniz showed that the partial derivative with respect to t of the integral is equal to the integral of the partial derivative of the expression itself with respect to t:

$$\frac{\partial}{\partial t} \int f(x,\ t)\,dx = \int \frac{\partial}{\partial t} f(x,\ t)\,dx \qquad (2)$$

To establish this result, known in modern calculus as Leibniz's rule, Leibniz used the fact that the differential of a sum of infinitesimal elements is equal to the sum of the differentials of each of the elements. In his studies of orthogonal trajectories Euler provided a quite different proof of the same result, a proof that rested on his understanding of the integral as an antiderivative.[17] To carry out the derivation Euler first established a preliminary theorem, showing that if f is a function of the two variables x and t then the second partial derivative of f is independent of the order of differentiation:

$$\frac{\partial}{\partial t} \frac{\partial}{\partial x} f(x,\ t) = \frac{\partial}{\partial x} \frac{\partial}{\partial t} f(x,\ t) \qquad (3)$$

With this result and his definition of the integral as an antiderivative Euler was able to deduce Leibniz's rule directly:

$$\frac{\partial}{\partial t} \int f(x,\ t)\,dx = \int \frac{\partial}{\partial x} \left(\frac{\partial}{\partial t} \int f(x,\ t)\,dx \right) dx =$$

$$= \int \frac{\partial}{\partial t} \left(\frac{\partial}{\partial x} \int f(x,\ t)\,dx \right) dx = \int \frac{\partial}{\partial t} f(x,\ t)\,dx \qquad (4)$$

the years 1691 and 1692 published in 1914, p. 3. See Carl Boyer, *A History of the Calculus and Its Conceptual Development* (New York: Dover Publications, Inc., 1959; originally published by Hafner Publishing Company in 1949 under the title "The Concepts of the Calculus, A Critical and Historical Discussion of the Derivative and the Integral"), pp. 278–9.

[16] For a historical survey of this subject, see Steven B. Engelsman, *Families of Curves and the Origins of Partial Differentiation* (Amsterdam: North-Holland, 1984).

[17] The derivation is contained in Euler, "De infinitis curvis eiusdem generis seu methodus inveniendi aequationes pro infinitis curvis eiusdem generis," *Commentarii Academiae Scientiarum Petropolitanae* 7 1734–1735 (1740), 174–89, 180–9. (Pages 190–9 were incorrectly numbered as 180–9.) In Euler's *Opera* Ser. 1, V. 22, pp. 36–56.

In his later writings Euler followed the pattern here, obtaining first equation (3) and then proceeding to derive Leibniz's rule; the proof rested at base on Euler's concept of the integral as an antiderivative. In his *Institutiones Calculi Integralis* he expounded in some detail on his operational understanding of the integral. Integration understood as the inverse of differentiation was analogous to subtraction as the inverse of addition, division as the inverse of multiplication, and the taking of roots as the inverse of the taking of powers. When it is not possible to express the inverse of a given expression Xdx in terms of known algebraic functions, then it follows that the resulting integral must be transcendental. The situation is analogous to the one with respect to the three inverse algebraic operations. When subtraction leads to numbers that are not positive then we arrive at negative numbers; when division results in nonintegral numbers we arrive at fractions; when the taking of roots leads to nonintegral numbers then we arrive at radicals.

The definition of integration as the operational inverse of differentiation was widely adopted in late eighteenth-century mathematics. By taking integrals one obtained new functional objects, and by applying functional inversion to these objects one obtained a further class of functions. The domain of analysis was thereby enlarged greatly. In an early memoir on elliptic integrals, Lagrange had observed that the investigation of the integrability of rational polynomials opened "a vast field to the researches of the analysts."[18] It should be noted that in this conception a given transcendental integral and its various properties were understood to be a consequence of the algebraic nature of the differential process. In particular, the various considerations of existence that are so fundamental in modern theories of integration did not arise at all.

THEOREMS OF ANALYSIS

A fundamental difference between eighteenth-century and modern analysis is the absence in the former of what is known today as the mean value theorem or the law of the mean. This result, a basic part of the classical arithmetic foundation of the calculus, is used in theorem-proving to localize a given property or relation at a definite value of the numerical continuum. The proposition is established by showing its validity at each value of this continuum.

Euler's viewpoint was quite different. A relation between variables was regarded by him as a primitive of the theory; it was not further conceptualized in terms of the numerical continuum of values assumed by each variable. This notion of a primitive abstract relation in large part defined his approach to

---

[18] Lagrange, "Sur l'intégration de quelques équations différentielles dont les indéterminées sont séparées, mais dont chaque membre en particulier n'est point intégrable," *Miscellanea Taurinensia*, 4; in Lagrange's *Oeuvres de Lagrange* 2, pp. 5–33. The quote is on p. 33 of *Oeuvres* 2.

analysis, distinguishing his point of view both from that of the early pioneers, who made the geometrical curve the basic object of study, and that of the nineteenth-century researchers, for whom the numerical continuum constituted a fundamental object of study.

Euler's proof in 1740 of (3), the theorem on the equality of mixed partial differentials, was analytical in a formal, nongeometrical sense. He was motivated to develop such a proof by a belief that a geometrical demonstration would be "drawn from an alien source."[19] He considered a quantity z that is a function of the variables x and a. He expressed the relevant differentials in terms of differential coefficients and showed by a suitable rearrangement of terms that the two partial differentials are equal. In modern real-variable analysis, Euler's argument is reformulated using the law of the mean and a limit argument. Suppose $z = z(x,a)$ and its first and second partial derivatives are defined and continuous on a rectangular region in the x-a plane. The law of the mean is used to obtain expressions for the relevant partial derivatives, which by rearrangement and a limit argument are shown to be equal.[20]

This example is typical of eighteenth-century calculus theorems and their counterparts in modern analysis. (Other examples are the fundamental theorem of the calculus, the theorem on the change of variables in multiple integrals, and the fundamental lemma of the calculus of variations.) The law of the mean introduces a distinguished value, localizing at a particular number the analytical relation or property in question. The result is then deduced using conditions of continuity and differentiability by means of a limit argument. In Euler's formulation, by contrast, there was no consideration of distinguished or individual values as such. Euler believed that the essential element in the demonstration was its generality, which was guaranteed by a formal analytical or algebraic identity. Thus, the key step in his proof rested on an algebraic identity that ensured the validity of the result.

ANALYTICAL PHILOSOPHY

Although the leading analysts of the eighteenth century did not formulate an explicit mathematical philosophy, implicit philosophical attitudes were evident in their handling of issues such as generality and the relationship of pure and applied mathematics. For Enlightenment mathematicians, each part of mathematics was understood to be given in some objective sense; its range of application and certainty derived from this objective nature and were not consequences of the particular method or set of concepts adopted by the mathematician. The generality of mathematics was a consequence of the general

---

[19] See Engelsman, *Families of Curves*, p. 129.
[20] Euler's derivation is studied in more detail in Craig Fraser, "The Calculus as Algebraic Analysis: Some Observations on Mathematical Analysis in the 18th Century," *Archive for History of Exact Sciences*, 39 (1989), 317–35, especially 319–21.

character of its objects, whether these be formulas of algebra or diagrams of geometry.

In the writings of the analysts the original problem of the calculus – to describe change along a curve – gave way to the study of formulas and relations. An analytic equation implied the existence of a relation that remained valid as the variables changed continuously in magnitude. Analytic algorithms and transformations presupposed a correspondence between local and global change, the basic consideration in the application of the calculus to the curve. The rules and procedures of the calculus were assumed to be generally valid. In a memoir published in 1751 Euler considered the rule $d(\log x) = dx/x$.[21] He rejected an earlier suggestion of Leibniz that this rule was only valid for positive real values of x with the following observation:

> For, as this [differential] calculus concerns variable quantities, that is, quantities considered in general, if it were not generally true that $d.lx = dx / x$, whatever value we give to x, either positive, negative or even imaginary, we would never be able to make use of this rule, the truth of the differential calculus being founded on the generality of the rules it contains. (pp. 143–4)

Eighteenth-century confidence in formal mathematics was almost unlimited. One historian has noted, "Sometimes it seems to have been assumed that if one could just write down something which was symbolically coherent, the truth of the statement was guaranteed," and another has commented on Euler's "naive faith in the infallibility of formulas and the results of manipulations upon them."[22] Functionality and operational efficacy were valued over deduction and logical verification. A belief in symbolic methods was supported by more general philosophical thinking about exact science. The writings of Nicolas Malebranche and his school had stressed the value of an arithmetical/ algebraic approach to mathematics. Somewhat later, Étienne Condillac emphasized the importance of a well-constructed language in rational investigation, and he cited algebra as the paradigm of what could be achieved in this direction.[23]

That the problems of geometry and mechanics should conform to treatment by pure analysis was something that eighteenth-century authors accepted as a matter of philosophical principle. Sergei Demidov, writing of the failure

[21] Euler, "De la controverse entre Mrs. Leibniz et Bernoulli sur les logarithmes des nombres negatifs et imaginaires, *Mémoires de l'académie des sciences de Berlin 5 (1749)*, (1751), 139–171; in the *Opera Omnia* Ser. 1, V. 17, 195–232.

[22] See Judith V. Grabiner, "Is Mathematical Truth Time-Dependent?" *American Mathematical Monthly*, 81 (1974), 354–65, especially 356, and Rudolph E. Langer, "Fourier Series, The Evolution and Genesis of a Theory," *American Mathematical Monthly*, 54, pt. 2 (1947), 1–86, especially 17.

[23] Malebranche's mathematical philosophy is discussed in Craig Fraser, "Lagrange's Analytical Mathematics, Its Cartesian Origins and Reception in Comte's Positive Philosophy," *Studies in the History and Philosophy of Science*, 21 (1990), 243–56. An account of Condillac's thought is contained in Robert McRae, "Condillac: The Abridgement of All Knowledge in 'The Same is the Same,'" in *The Problem of the Unity of the Sciences: Bacon to Kant* (Toronto: University of Toronto Press, 1961), pp. 89–106.

of Euler and d'Alembert to understand each other's point of view in the discussion of the wave equation, observes:

> A cause no less important of this incomprehension rests, in our opinion, on the understanding of the notion of a solution of a mathematical problem. For d'Alembert as for Euler the notion of such a solution does not depend on the way in which it is defined . . . rather the solution represents a certain reality endowed with properties that are independent of the method of defining the solution. To reveal these properties diverse methods are acceptable, including the physical reasonings employed by d'Alembert and Euler.[24]

A biographer of d'Alembert has noted his insistence on "the elementary truth that the scientist must always accept the essential 'giveness' of the situation in which he finds himself."[25] The sense of logical freedom that developed in later mathematics – expressed, for example, in Richard Dedekind's famous statement of 1888 that numbers are free creations of the human mind and the belief that the essence of mathematics consists in its autonomous conceptual development – reflects aspects of the modern subject that were quite absent in the eighteenth century.

JOSEPH LOUIS LAGRANGE

Lagrange's professional career was an exceptionally long one, spanning from 1754, when he was eighteen, to his death in 1813. From his birth in 1736 until 1766 he lived in Turin, participating in the founding of the Turin Society in 1757 and then becoming one of its active members; from 1766 to 1787 he was mathematics director of the Berlin Academy of Sciences; from 1787 to his death he lived in France as a pensionnaire of the Paris Academy of Sciences.

Although Lagrange's analytical tendencies were apparent from the very beginning of his career, his distinctive mathematical style became consolidated only in the period 1770 to 1776, when he was in his late thirties and comfortably settled at the Berlin Academy. In these years the value of analysis became an explicit theme in his writings for the Academy on a range of subjects in pure and applied mathematics.[26] In a memoir of 1771 on Kepler's problem he distinguished three approaches to its solution: one involving numerical approximation, a second using geometrical or mechanical constructions, and a third that is algebraic, employing analytical expressions. The last he cited for its "continual and indispensable use in the theory of celestial bodies." In a paper the next year on the tautochrone, a problem first investigated geo-

---

[24] Sergei Demidov, "Création et développement de la théorie des équations différentielles aux dérivées partielles dans les travaux de J. d'Alembert," *Revue d'Histoire des Sciences* 35/1 (1982), 3–42, especially 37.

[25] R. G. Grimsley, *Jean d'Alembert (1717–1783)* (Oxford: Clarendon Press, 1963), p. 248.

[26] For references to the publications of Lagrange cited in this section, see René Taton, "Inventaire chronologique de l'oeuvre de Lagrange," *Revue d'Histoire des Sciences,* 26 (1974), 3–36.

metrically by Huygens, Lagrange took as a starting point "analytical solutions" that had been advanced by Johann Bernoulli and Euler. In 1775 several memoirs appeared in which the value of analysis is promoted. In his paper on the attraction of a spheroid Lagrange attempted to show that the method of "algebraic analysis" provides a more direct and general solution than the "synthetic" or geometrical approach followed by Maclaurin. (This appears, incidentally, to be the first explicit appearance in his writing of the term "algebraic analysis.") In his study of the rotation of a solid Lagrange advanced an alternative to the mechanical treatment of d'Alembert and Euler, one that was "purely analytic," whose merit consisted "solely in the analysis" that it employed, and which contained "different rather remarkable artifices of calculation." In a memoir on triangular pyramids Lagrange noted that his "solutions are purely analytic and can even be understood without figures"; he observed that independent of their actual utility they "show with how much facility and success the algebraic method can be employed in questions that would seem to lie deepest within the province of Geometry properly considered, and to be the least susceptible to treatment by calculation."

The theme of analysis recurs in Lagrange's writings of the late 1770s and 1780s. In a 1777 study of cubic equations he described a method due to Thomas Harriot that avoided the geometrical constructions that had been used by mathematicians to investigate expressions for roots. In a memoir submitted to the Paris Academy in 1778 on the subject of planetary perturbations Lagrange offered a method for transforming the equations of motion that would "take the place of the synthetic methods proposed until now for simplifying the calculation of perturbations in regions beyond the orbit" and that "has at the same time the advantage of conserving uniformity in the march of the calculus." In 1780 he published a memoir on a theorem of Johann Lambert's in particle dynamics. The result in question had been demonstrated synthetically, and Lagrange expressed concern that it might be regarded as one of "the small number [of theorems] in which geometric analysis seems to be superior to algebraic analysis." His purpose was to present a simple and direct analytical proof. In a study in 1781 of projection maps he offered a "research, equally interesting for the analytic artifices that it requires as well as for its utility in the perfection of geographical maps." In the preface to his famous *Traité de la méchanique analitique,* completed around 1783, he announced that in it "no figures would be found," that all would be "reduced to the uniform and general progress of analysis." In a memoir of 1788 he discussed successes and difficulties in treating analytically the various subjects of Newton's *Principia Mathematica* and offered a new analysis of the problem of the propagation of sound.

Directness, uniformity, and generality were qualities that Lagrange associated with analysis; he sometimes also mentioned simplicity. Analysis was cited not simply for the results to which it led but also for the methods that it offered. In the writings discussed earlier he was affirming the value of analysis in situations in which an alternative geometrical or mechanical treatment existed; it was the possibility of this alternative that led him to explicitly assert his own methodological preferences. One should also note the sheer preponderance of pure analysis in his work of the 1770s and 1780s in such topics as the theory of equations, diophantine arithmetic, number theory, probability, and the calculus, subjects in which explicit questions of approach or methodology did not arise.

THEORY OF ANALYTICAL FUNCTIONS

By the end of the century a more critical attitude began to develop both within mathematics and within general intellectual culture. As early as 1734 Bishop George Berkeley in his work *The Analyst* had called attention to what he perceived as logical weaknesses in the reasonings of the calculus arising from the employment of infinitely small quantities. Although his critique was somewhat lacking in mathematical cogency, It stImulated writers in Britain and the Continent to explain more carefully the basic rules of the calculus. In the 1780s a growing interest in the foundations of analysis was reflected in the decision of the academies of Berlin and St. Petersburg to devote prize competitions to the metaphysics of the calculus and the nature of the infinite. In philosophy, Immanuel Kant's *Kritik der reinen Vernunft* (1788) set forth a penetrating study of mathematical knowledge and initiated a new critical conceptual movement in the exact sciences.

The most detailed attempt to provide a systematic foundation of the calculus was contained in two treatises by Lagrange published at the end of the century: the *Théorie des fonctions analytiques* (1797) and *Leçons sur le calcul des fonctions* (1801; rev. ed. 1806). The full title of the first work explains its purpose: "Theory of analytical functions containing the principles of the differential calculus disengaged from all consideration of infinitesimals, vanishing limits or fluxions and reduced to the algebraic analysis of finite quantities." Lagrange's goal was to develop an algebraic basis for the calculus that made no reference to infinitely small magnitudes or intuitive geometrical notions. In a treatise on numerical equations published in 1798 he set forth clearly his conception of algebra:

> [Algebra's] object is not to find particular values of the quantities that are sought, but the system of operations to be performed on the given quantities in order to derive from them the values of the quantities that are sought. The tableau of these operations represented by algebraic characters is what in algebra is called a *formula,* and when one quantity depends on other quantities, in such a way that it can be expressed by a formula which contains these quantities, we say then that it is a *function* of these same quantities.[27]

---

[27] Lagrange, *Traité de la résolution des équations numériques de tous les degrés* (Paris, 1798). The second edition was published in 1808 and was reprinted as Lagrange's *Oeuvres* 8. The quoted passage appears on pp. 14–15 of the latter volume.

Lagrange used the term "algebraic analysis" to designate the part of mathematics that results when algebra is enlarged to include calculus-related methods and functions. The central object here was the concept of an analytical function. Such a function y = f(x) is given by a single analytical expression that is constructed from variables and constants using the operations of analysis. The relation between y and x is indicated by the series of operations schematized in f(x). The latter possesses a well-defined, unchanging algebraic form that distinguishes it from other functions and determines its properties.

The idea behind Lagrange's theory was to take any function f(x) and expand it in a Taylor power series:

$$f(x + i) = f(x) + pi + qi^2 + ri^3 + si^4 + \cdots \qquad (5)$$

The "derived function" or derivative f'(x) of f(x) is defined to be the coefficient p(x) of the linear term in this expansion. f'(x) is a new function of x with a well-defined algebraic form; it is different from but related to the form of the original function f(x). Note that this conception is very different from that of the modern calculus, in which the derivative of f(x) is defined at each real value of x by a limit process. In the modern calculus the relationship of the derivative to its parent function is specified in terms of correspondences that are defined in a definite way on the numerical continuum.

Lagrange's understanding of derived functions was revealed in his discussion in the eighteenth lesson of the method of finite increments. This method was of historical interest in the background to his program. Brook Taylor's original derivation in 1715 of Taylor's theorem was based on a passage to the limit of an interpolation formula involving finite increments. Lagrange wished to distinguish clearly between an approach to the foundation of the calculus that uses finite increments and his own quite different theory of derived functions. In taking finite increments, he noted, one considers the difference $f(x_{n+1}) - f(x_n)$ of the same function f(x) at two successive values of the independent argument. In the differential calculus the object Lagrange referred to as the derived function was traditionally obtained by letting $dx = x_{n+1} - x_n$ be infinitesimal, setting $dy = f(x_{n+1}) - f(x_n)$, dividing dy by dx, and neglecting infinitesimal quantities in the resulting reduced expression for dy/dx. Although this process leads to the same result as Lagrange's theory, the connection it presumes between the method of finite increments and the calculus obscures a more fundamental difference between these subjects: in taking $\Delta y = f(x_{n+1}) - f(x_n)$ we are dealing with one and the same function f(x); in taking the derived function we are passing to a new function f'(x) with a new algebraic form. Lagrange explained this point as follows:

> The passage from the finite to the infinite requires always a sort of leap, more or less forced, which breaks the law of continuity and changes the form of functions. (*Leçons* 1806, p. 270)
>
> In the supposed passage from the finite to the infinitely small, functions actually change in nature, and . . . dy/dx, which is used in the differential

Calculus, is essentially a different function from the function y, whereas as long as the difference dx has any given value, as small as we may wish, this quantity is only the difference of two functions of the same form; from this we see that, if the passage from the finite to the infinitely small may be admitted as a mechanical means of calculation, it is unable to make known the nature of differential equations, which consists in the relations they furnish between primitive functions and their derivatives. (*Leçons* 1806, p. 279)

Lagrange's *Théorie* and *Leçons*, written when he was in his sixties, were notable for their success in developing the entire differential and integral calculus on the basis of the concept of an analytical function.[28] They contained several quite important technical advances. Lagrange introduced inequality methods to obtain numerical estimates of the values of functions, thereby providing a source of techniques that Augustin Cauchy was later able to use in his arithmetical development of the calculus. Another significant contribution was contained in Lagrange's exposition of the calculus of variations. To obtain the variational equations he modeled the derivation after an earlier argument in the theory of integrability. Although his derivation never quite achieved acceptance among later researchers, it remains historically noteworthy as an example of advanced reasoning in algebraic analysis. Lagrange also introduced the multiplier rule in both the calculus and the calculus of variations, a powerful method that allows one to solve a range of problems in the theory of constrained optimization.[29]

REFLECTIONS ON ALGEBRAIC ANALYSIS

It is important to appreciate the distinctive philosophical character of eighteenth-century algebraic analysis, understood within the larger historical and intellectual evolution of mathematical analysis. The algebraic calculus of Euler and Lagrange was rooted in the formal study of functional equations, algorithms, and operations on variables. The values that these variables received, their numerical or geometrical interpretation, was logically of secondary concern. Such a conception, strongly operational and instrumentalist in

---

[28] For a more detailed study of these works, see J. L. Ovaert, "La thèse de Lagrange et la transformation de l'analyse," in Christian Houzel et al. (eds.), *Philosophie et Calcul de l'Infini* (Paris: Francois Maspero, 1976), pp. 122–157; Judith V. Grabiner, *The Origins of Cauchy's Rigorous Calculus* (Cambridge, MA: MIT Press, 1981); and Craig Fraser, "Joseph Louis Lagrange's Algebraic Vision of the Calculus," *Historia Mathematica*, 14 (1987), 38–53. Grabiner and Fraser emphasize somewhat different aspects of the historiography. Grabiner calls attention to the origins of Cauchy's technical methods in Lagrange's writings and makes the concept of rigor central to understanding Cauchy's achievement. Fraser is concerned with highlighting the conceptual differences between the viewpoints of Lagrange and Cauchy and sees the latter's central accomplishment as having made the numerical continuum a fundamental object of concern.

[29] For a discussion of Lagrange's calculus of variations, see Craig Fraser, "J. L. Lagrange's Changing Approach to the Foundations of the Calculus of Variations," *Archive for History of Exact Sciences*, 32 (1985), 151–91, and Fraser, "Isoperimetric Problems in the Variational Calculus of Euler and Lagrange," *Historia Mathematica*, 19 (1992), 4–23.

character, should be contrasted with the geometrical approach of the early calculus, which relied heavily on diagrammatic representations and intuitions of spatial continuity. The geometrical emphasis of the early calculus conditioned how the subject was understood, allowing it to be experienced intellectually as an interpreted, meaningful body of mathematics.

Lagrange's algebraic analysis should also be contrasted with the much more conceptual and intensional mode of reasoning that was characteristic of classical real analysis, the field that developed in the nineteenth century and became the foundation of the modern subject. Although real analysis is logically independent of geometry, it continues to posit objects – defined using the concept of arithmetical continuity – that constitute its subject matter and define its point of view as a mathematical theory. A proposition about a function defined on some interval of real numbers under specified conditions of differentiability has a geometrical interpretation implicit in its very formulation. On a foundational level the algorithmic character of differentiation in real analysis is irrelevant to a conceptual understanding of this process; in algebraic analysis, by contrast, the notion of algorithm is fundamental to the whole approach.[30]

ROBERT WOODHOUSE AND GEORGE PEACOCK

The algebraic program of Enlightenment mathematics was taken up and extended by several English figures of the early nineteenth century.[31] Although these researches fall somewhat outside the period of this essay, they are worthy of note here as a direct continuation of what was primarily an eighteenth-century development. The appeal of algebraic analysis to the English was due in considerable part to a reaction against the prevalent geometric synthetic spirit of British mathematics. In his 1802 memoir "On the Independence of the analytical and geometrical Methods of Investigation; and on the Advantages to be derived from their Separation" Cambridge fellow Robert Woodhouse recommended the removal from analysis of all notation of geometrical

---

[30] For a more detailed discussion of this subject, see Fraser, "The Calculus as Algebraic Analysis," and Marco Panza, "Concept of Function, between Quantity and Form, in the 18th Century," in H. Niels Jahnke et al. (eds.), *History of Mathematics and Education: Ideas and Experiences* (Göttingen: Vandenhoeck & Ruprecht, 1996), pp. 241–74. For a social-intellectual study that deals with the place of algebraic analysis in nineteenth-century German mathematics, see H. Niels Jahnke, *Mathematik und Bildung in der Humboldtschen Reform,* volume 8 of the series *Studien zur Wissenschafts-, Sozial- und Bildungsgeschichte der Mathematik,"* eds. Michael Otte, Ivo Schneider, and Hans-Georg Steiner (Göttingen: Vandenhoeck & Ruprecht, 1990).

[31] For historical studies of this subject, see Joan L. Richards, "The Art and the Science of British Algebra: A Study in the Perception of Mathematical Truth," *Historia Mathematica,* 7 (1980), 343–65; Helena Pycior, "George Peacock and the British Origins of Symbolical Algebra," *Historia Mathematica,* 8 (1981), 23–45; and Menachem Fisch, "'The Emergency Which Has Arrived': The Problematic History of Nineteenth-Century British Algebra – a Programmatic Outline," *British Journal for the History of Science,* 27 (1994), 247–76.

origin. He urged, for example, that instead of writing sin x, a term whose etymology involved graphical associations, we employ the expression $(2\sqrt{-1})^{-1}(e^{x\sqrt{-1}} - e^{-x\sqrt{-1}})$. He also began to move toward a more careful explanation of the symbols of formal analysis. Thus he wrote the following concerning the symbol "=":

> It is true that its signification entirely depends on definition; but, if the definition given of it in elementary treatises be adhered to, I believe it will be impossible to show the justness and legitimacy of most mathematical processes. It scarcely ever denotes numerical equality. In its general and extended meaning, it denotes the result of certain operations. (p. 103)

Woodhouse illustrated this point with the inverse sine series

$$z = x + \frac{x^3}{3\cdot 2} + \frac{3x^5}{5\cdot 8} + \cdots,$$

in which

> nothing is affirmed concerning a numerical equality; and all that is to be understood is, that

$$z = x + \frac{x^3}{3\cdot 2} + \frac{3x^5}{5\cdot 8} + etc.,$$

> is the result of a certain operation performed [on the series for sin x]

$$x = z - \frac{z^3}{1\cdot 2\cdot 3} + \frac{z^5}{1\cdot 2\cdot 3\cdot 4\cdot 5} - etc.$$

Woodhouse's formal viewpoint was developed into a complete theoretical system by another Cambridge mathematician, George Peacock. In his "Report on the Recent Progress and Present State of certain Branches of Analysis," which was delivered to the British Association for the Advancement of Science in 1833, Peacock defined analytical science to include algebra, the application of algebra to geometry, the differential and integral calculus, and the theory of series. The first part of the report was devoted to an outline of his theory of algebra, which he based on something that he called the principle of the permanence of equivalent forms. An equivalent form is any relation that expresses the result of an operation of algebra: $(a + b)c = ac + bc$, $a^n \cdot a^m = a^{n+m}$, and so on. The principle of equivalent forms asserts as follows:

> Whatever equivalent form is discoverable in arithmetical algebra considered as the science of suggestion, when the symbols are general in their form, though specific in their value, will continue to be an equivalent form when the symbols are general in their nature as well as their form. (p. 199)

Because the relation $a^n \cdot a^m = a^{n+m}$ is an equivalent form when n and m are integers, it is by the principle also an equivalent form as a purely symbolic relation. Peacock regarded this fact as justification for extending the range of validity of $a^n \cdot a^m = a^{n+m}$ to non-integral values of n and m. In other branches of analysis – for example the theory of infinite series – the principle plays a similar role. Thus, because the relation $1 / (1 - x) = 1 + x + x^2 + x^3 + \ldots$ is valid for $x < 1$, it possesses by virtue of its form a general symbolical validity. The relation therefore remains valid, or at least meaningful, when $x > 1$, although in this case it is no longer interpretable in the usual sense in arithmetic.

The principle of equivalent forms is the formal statement of the idea contained in Euler's assertion of the universal validity of the relation $d(\log x) = dx / x$. In Peacock's system of analysis the principle had a dual purpose. It made legitimate the use of general symbolic relations and allowed one to assume an extended domain of validity for the variables contained in these relations. In addition, it ensured that the algebraic relations have at least a partial interpretation in arithmetic, and it thereby restricted the proliferation of purely abstract symbolical systems.

CONCLUSION

Eighteenth-century analysis achieved a theoretical completeness and sophistication not attained by other parts of mathematics. From a historiographical viewpoint, algebraic analysis provides an interesting example of a mature mathematical paradigm that would be replaced by a quite different paradigm in the later development of the subject. The transition from Euler and Lagrange to Cauchy and Weierstrass constituted a profound intellectual transformation in conceptual thought. The sort of relativism of viewpoint documented by Thomas Kuhn in the history of the physical sciences is also present in mathematics, albeit at a more purely conceptual level.[32] The case of mathematics is even in some important respects more striking, because the point of view embodied in the older paradigm retains a certain intellectual interest and validity not found in quite the same way in the discarded theories of older physics.

---

[32] For discussion of the relevance of Kuhn's ideas to mathematics, see Donald Gillies (ed.), *Revolutions in Mathematics* (New York: Oxford University Press, 1992).

14

ASTRONOMY AND COSMOLOGY

Curtis Wilson

During the eighteenth century the astronomy of the solar system became, in the words of William Whewell, "the queen of the sciences . . . the only perfect science . . . in which particulars are completely subjugated to generals, effects to causes."[1] The striking theoretical advances Whewell refers to were the work of Continental mathematicians, members of scientific academies in Paris, Berlin, and St. Petersburg, who, through elaborating the algorithms of the Leibnizian differential and integral calculus, elicited the consequences of Newtonian gravitation. Meanwhile, instrument-makers, chiefly British, so refined telescopes and graduated arcs that observational precision kept pace with theoretical prediction. Observatories, the chief of them nationally funded, took pride in contributing not only to the navigational needs of their nations' navies and merchant marines but also to the supranational goal of a perfected astronomy.

Ancillary to planetary and lunar astronomy was the construction of star catalogs: it was in relation to star positions that the positions of planets and the Moon were determined. The puzzle of apparent systematic motions of the stars was unraveled by James Bradley between 1729 and 1748 – a sine qua non for an astronomy precise to arcseconds. Meanwhile, a few thinkers speculated as to the large-scale structure of the universe. If gravitation were universal, why did not the stars collapse into one another? Was the cosmos a stable structure or in the process of change? Toward the end of the century, dynamical arguments and observational evidence were brought to bear on these questions and led to a new vision of an evolving stellar world.

---

[1] William Whewell, address, in *Report of the Third Meeting of the British Association for the Advancement of Science* (London, 1833), p. xiii.

THE ASTRONOMY OF THE SOLAR SYSTEM IN 1700: NEWTON'S FIRST EFFORTS TO DERIVE PRECISE ASTRONOMICAL PREDICTIONS

In 1700 planetary astronomy was scarcely yet touched by Newton's mathematical discoveries. The ellipticity of the planetary orbits (Kepler's first law) had come to be widely accepted, although it was never verified empirically with precision. Kepler's so-called second law – the equable sweeping out of area by the Sun-planet vector – was not so much an empirical law as a consequence of Keplerian dynamics; it had been generally rejected early because of mathematical difficulties it entailed. Substitute rules were proposed, but as Nicholas Mercator (ca. 1619–1687) pointed out in 1670, any viable substitute must closely approximate the Keplerian area rule.[2] In the *Principia,* Newton presented the areal rule as logically equivalent to a central force, and the ellipse with Sun in the focus as derivable from an inverse-square central force.

Kepler's third law, unlike the first two, stood on its own feet as an empirical law: the squares of the planetary periods varied as the cubes of their mean solar distances. It, too, Newton now showed, was a consequence of the inverse-square law. In Kepler's world it had been a "harmony" evincing the Creator's penchant for mathematical pattern. The first to make practical use of it had been Jeremiah Horrocks (1618?–1641) in the late 1630s. From observations of Venus and Mars he found that the horizontal solar parallax – the maximum angle subtended by the Earth's radius as seen from the Sun – needed to be reduced from Kepler's value of 1′ to about 15″ (the correct value is 8.8″). But then Horrocks discovered that, by using Kepler's third law to calculate the mean solar distances of these planets from their periods, he could bring prediction and observation into close agreement. This procedure was followed by Thomas Streete (1622–1689) in his *Astronomia Carolina* (1661); his planetary tables, reissued several times and still in use in the first decades of the eighteenth century, proved superior for the planets from Mercury to Mars.[3] Newton in his *Principia* also advocated the Horrocksian procedure.

Astronomy in 1700 owed to Horrocks another major innovation, his lunar theory. A modification of Kepler's lunar theory, it hypothesized an elliptical orbit with oscillating apsidal line and eccentricity. (The apsidal line was the ellipse's major axis, or line from perigee to apogee; the eccentricity was the Earth's distance from the ellipse's center, compared to the semimajor axis.)

---

[2] For a more detailed account of the developments mentioned in this and the next two paragraphs, see Curtis Wilson, "Predictive Astronomy in the Century after Kepler," in René Taton and Curtis Wilson (eds.), *Planetary Astronomy from the Renaissance to the Rise of Astrophysics, Part A: Tycho Brahe to Newton* (Cambridge University Press, 1989), pp. 161–206.

[3] See Curtis Wilson, "Horrocks, Harmonies, and the Exactitude of Kepler's Third Law," in *Science and History: Studies in Honor of Edward Rosen* (*Studia Copernicana 16;* Warsaw: The Polish Academy of Sciences Press, 1978), pp. 235–59.

The period of oscillation was a little less than seven months, the time for the Sun to move out of and back into alignment with the Moon's apsidal line. An account of this theory was first published in an appendix to Horrocks's *Opera posthuma* (editions in 1672, 1673, and 1678), with constants and tables supplied by John Flamsteed (1646–1719); a revised version was published in Flamsteed's *Doctrine of the Sphere* (1680).[4] Comparing Horrocks's lunar theory with other theories of the time – all of them complicated epicyclic mechanisms derivative from Tycho's theory – Flamsteed found Horrocks's far superior, especially as tested by micrometer measurements of the Moon's diameter. For Newton, Horrocks's theory had the advantage that, unlike other theories of the time, it lent itself to a dynamical explanation on Newtonian principles. In Corollaries 7–9 of Proposition I.66 of the *Principia*, he gave a qualitative explanation for the Horrocksian oscillations in apsidal line and eccentricity but did not seek to deduce their magnitudes.

Could this theory of the lunar motions be rendered quantitatively predictive? The first edition of the *Principia* did not attempt it. It set forth a wide-ranging argument in support of universal gravitation – not a *demonstratio* in Newton's sense of a strict deduction, but posing for the following century a vast program of inquiry.[5] Presumptively, an accurate lunar theory was derivable from universal gravitation; but how?

On 1 September 1694 Newton visited the Greenwich Observatory, and Flamsteed there showed him a record of discrepancies between lunar observations made at Greenwich and Flamsteed's version of the Horrocksian theory; the error at times reached nearly a third of a degree. At this date, let us recall, an accurate lunar theory, along with an accurate catalog of stars whereby the Moon's changing position could be determined, looked to be the most likely means for reliably finding the longitude at sea; it was with a view to providing these *desiderata* that Flamsteed had been appointed the King's Astronomical Observator in 1675. To find the longitude to within a degree would require a lunar theory accurate to 2 arcminutes; the discrepancies Flamsteed had found thus presented a serious challenge. Newton undertook to rectify the theory on the basis of an extensive set of Flamsteed's lunar observations. (Only a few of these are now identifiable, but Newton's complaints of Flamsteed as a supplier of observations seem petulant.)[6]

---

[4] On Horrocks's theory, see S. B. Gaythorpe, "Jeremiah Horrocks and his 'New Theory of the Moon,'" *Journal of the British Astronomical Association*, 67 (1957), 134–44; and Curtis Wilson, "On the Origin of Horrocks's Lunar Theory," *Journal for the History of Astronomy*, 18 (1987), 77–94.

[5] On the question as to whether Newton was able to *deduce* universal gravitation from the phenomena or was in the end forced to *hypothesize* it, see Howard Stein, "'From the Phenomena of Motions to the Forces of Nature': Hypothesis or Deduction?" in Arthur Fine, Micky Forbes, and Linda Wessels (eds.), *Proceedings of the 1990 Biennial Meeting of the Philosophy of Science Association* (1991), vol. 2, pp. 209–22.

[6] Nick Kollerstrom and Bernard D. Yallop, "Flamsteed's Lunar Data, 1692–95, Sent to Newton," *Journal for the History of Astronomy*, 26 (1995), 237–46.

The new theory that Newton elaborated was the Horrocksian theory with improved numerical parameters, plus several added terms. How Newton arrived at some of these added terms is uncertain. An account of the new theory was first published in Latin, in David Gregory's *Astronomiae Physicae & Geometriae Elementa* (Oxford, 1702), and then in English as *A New and most Accurate Theory of the Moon's Motion; Whereby all her Irregularities may be solved, and her Place truly calculated to Two Minutes* (London, 1702). No dynamical explanations were supplied; only numerical parameters and the mode of computing the Moon's position were included. The theory is not as accurate as advertised in the English title, and one of its terms carries the wrong sign, an error corrected later. Various versions of the theory were current during the first half of the eighteenth century. A version due to Flamsteed was published by Pierre-Charles Le Monnier (1715–1799) in his *Institutions astronomiques* of 1746; d'Alembert found the Le Monnier tables accurate to about 5′.

Edmond Halley (1656–1742) became Flamsteed's successor as Astronomer Royal in 1720 and during the following two decades carried out a comparison of lunar observations with positions derived from his own version of the Newtonian theory. Halley's hypothesis was that the discrepancies found – some as high as 8′ or 9′ – would repeat in each Saros cycle of some 18 years, when the Sun and the Moon return to nearly the same configuration. In fact, Halley's table of discrepancies could have been used for a fairly accurate solution to the longitude problem.[7] But it was published only after Halley's death, in 1749, and the idea was not pursued.

As for the planets, Newton in the first edition of the *Principia* assumed that the motions of the four inner ones would be predictable on the basis of the Keplerian rules. Streete and Mercator, contradicting Kepler, had claimed that the apsides of these planets (the points of their orbits farthest from the Sun) were at rest with respect to the stars; Newton in the *Principia* spoke of the "quiescence" of the aphelia but allowed that because of planetary and cometary perturbations the apsides would not be altogether immovable. In the case of the remaining two of the known planets – Jupiter and Saturn – Newton had been able to compute their masses relative to the Sun and so knew that their mutual perturbations should be detectable; but he had no viable way of computing them. In Proposition 13 of Book III he suggested putting a focus of Saturn's elliptical orbit in the common center of gravity of Jupiter and the Sun. In a letter to Flamsteed in the 1690s, and later in Proposition 13 of Book III of the second edition of the *Principia,* he stated that Saturn was subject to a Horrocksian-style oscillation of apse and eccentricity. Flamsteed into the 1710s attempted to find a 59-year or other period in the irregularities of these two planets but at last gave up in despair. Halley, seeking to reconcile both ancient

---

[7] See Nicholas Kollerstrom, "The Achievement of Newton's 'Theory of the Moon's Motion' of 1702," Ph.D. dissertation, University of London, 1994, chap. 13.

and modern observations with his *Tabulae astronomicae* (published in 1749), proposed a uniform acceleration in the motion of Jupiter and a uniform deceleration in the motion of Saturn over the centuries since ancient times.

THE FIGURE OF THE EARTH

In Proposition III.19 of his *Principia,* Newton supposed that the Earth had originally been a homogeneous fluid mass, acted on by universal gravitation and the centrifugal force due to its rotation, and he concluded without proof that it formed an ellipsoid of revolution. On these suppositions he found that the equatorial radius exceeded the polar radius by 1/229th part of the polar radius. In III.20, assuming that all cylinders from surface to center would counterbalance one another, he argued that the effective weight of a body would increase toward the poles, varying as the square of the sine of the latitude.

Christiaan Huygens (1629–1693) in his *Discours de la cause de la pesanteur* (1690), started from the assumption that gravitation was due to ethereal pressure and that each particle of the Earth was impelled toward the center of gravity of the Earth's mass with a force varying inversely as the square of its distance from that center. He found that the polar radius would be shorter than the equatorial radius by 1/578th of the latter.

Both the Earth's oblate form and the centrifugal force of its rotation must reduce the effective gravity at the Equator as compared with the effective gravity to the north or south. In 1672 Jean Richer (1630–1696), on an expedition sent out by the Paris Academy of Sciences to French Guiana, had found that a pendulum regulated to beat seconds in Paris required shortening at Cayenne (5° N. Lat.) if it was still to beat seconds, a result confirmed in a second expedition to the tropics in 1682.

Another consequence of the oblate shape was that a degree of latitude – a north-south distance on the Earth's surface over which the altitude of the celestial pole changed by 1° – would be lengthened as one went from the Equator toward either pole. Jean Picard (1620–1682) in the 1670s had undertaken to measure a degree of latitude along the meridian running through the Paris Observatory, and this measure was extended north and south by Giovanni Domenico Cassini (1625–1712) and his son Jacques (1677–1756) in the period 1700–18. The completed measurement showed a slight *decrease* in the length of the degree toward the north, and Jacques Cassini proclaimed that the Earth was a prolate spheroid, elongated toward the poles. This conclusion was challenged – orally by the astronomer Joseph-Nicolas Delisle in Paris and in a published article by Giovanni Poleni in Padua: the claimed variation fell within the range of likely observational error.[8]

---

[8] See John Greenberg, "Geodesy in Paris in the 1730s and the Paduan Connection," *Historical Studies in the Physical Sciences,* 13 (1983), 239–60; and Greenberg, *The Problem of the Earth's Shape from Newton to Clairaut* (Cambridge University Press, 1995), pp. 79, 118–19.

P. L. Moreau de Maupertuis (1698–1759), who assumed and defended New-tonian principles in his *Discours sur les différentes figures des astres* (1732), found Newton's derivation of the Earth's shape from dynamical principles obscure: the assumptions and argumentation were questionable. In 1733 he proposed to the Paris Academy that the question of the Earth's shape be investigated by geodetic measurements. The Academy sent out two expeditions: one to Peru (then including present-day Ecuador) to measure the length of a degree of latitude at the Equator, the other to Lapland to measure the length of a degree at the Arctic Circle. The Peruvian expedition set out in 1735, but its members returned only in the 1740s, and the first account of the equatorial measure, by Pierre Bouguer (1698–1758), appeared in 1749. The Lapland ex-pedition, led by Maupertuis, set out in 1736 and returned in 1737; Mauper-tuis's account of it appeared in 1738. The Lapland measure, compared with the earlier measures in France, confirmed the oblateness of the Earth but left the degree of flattening uncertain.[9]

Alexis-Claude Clairaut (1713–1765), who had been a member of the Lap-land expedition, in his *Théorie de la figure de la terre* (1743) addressed the mathematical problem with new mathematics: the theory of partial differ-ential equations developed by Alexis Fontaine des Bertins (1705–1771).[10] Clairaut showed that, assuming the Earth was made of ellipsoidal strata, the density of the strata diminishing from center to surface, and their ellipticity increasing, the overall flattening, contrary to Newton's assertion, would be less than in the homogeneous case. The trouble with this conclusion in the 1740s was that the measures in France and Lapland supported a flattening greater than in the homogeneous case (1/178, say, rather than 1/230). In the 1780s and 1790s, Pierre-Simon Laplace (1749–1827), using mathematical results obtained by A.-M. Legendre (1752–1833), and applying potential theory and statistical tests, labored to remove or account for the anomaly. From the seven measures of meridian degrees available to him in 1799, he found the most probable flattening to be 1/312, and the probable error in the Lapland measure 336 me-ters – an error he thought unbelievably large. But a remeasurement by Jons Svanberg in 1801–3 found a large error in the Maupertuis measure, possibly due to a deviation of the plumb line caused by local variation in the Earth's density. In later geodesy the method of least squares would allow an ellipsoidal shape with a flattening of about 1/300 to be chosen as the ideal from which local deviations were measured.

[9] On the Lapland and Peruvian expeditions, see Seymour L. Chapin, "The Shape of the Earth," in René Taton and Curtis Wilson (eds.), *Planetary Astronomy from the Renaissance to the Rise of Astrophysics, Part B: The Eighteenth and Nineteenth Centuries* (Cambridge University Press, 1995), pp. 22–34; Mary Terrall, "Representing the Earth's Shape: The Polemics Surrounding Maupertuis's Expedition to Lapland," *Isis*, 83 (1992), 218–37; Rob Iliffe, "'Aplatisseur du monde et de Cassini': Maupertuis, Pre-cision Measurement, and the Shape of the Earth in the 1730s," *History of Science*, 31 (1993), 335–75; Flo-rence Tristram, *Le procès des étoiles* (Paris: Éditions Seghers, 1979) (on the Peruvian expedition).
[10] See Greenberg, *The Problem of the Earth's Shape*, passim.

THE FIRST ANALYTICAL FORMULATION OF
THE PERTURBATIONAL PROBLEM: EULER

In the period of his dispute with the Leibnizians over priority in the invention of the calculus (the decade beginning in 1712), Newton put forward the claim that he had originally developed the argument of the *Principia* in the form of a fluxional analysis and then translated it into the language of traditional geometry. His manuscripts fail to support this claim.[11] The *Principia* indeed contains results that Newton can have obtained only by a symbolic calculus (for instance, in propositions I.40, II.35, III.26). But the *Principia's* central argument appears to have been worked out in just the way the *Principia* presents it: by means of a geometry augmented by Newton's doctrine of "first and last ratios" (ratios of infinitesimal increments or decrements).

In the early years of the eighteenth century a number of Continental mathematicians undertook to restate and solve problems of Newtonian mechanics in terms of the Leibnizian calculus. Beginning in 1700, Pierre Varignon (1654–1722) applied the Leibnizian algorithms to questions of orbital motion. When Leibniz asked him to grapple with the three-body problem, however, Varignon found himself stumped unless the third gravitating body was assumed to be immobile. In 1710 Jacob Hermann (1678–1733) and Johann Bernoulli (1667–1748) published solutions of "the inverse problem of central forces," proving that an inverse-square central force implied conic-section orbits. (Of this proposition, which is crucial to the central argument of the *Principia,* Newton gave no proof in the first edition, and in the editions of 1713 and 1726 went no further than to sketch the steps of a possible proof.)[12]

The first extended treatise applying Leibnizian algorithms to problems of Newtonian mechanics was Leonhard Euler's *Mechanica* of 1736. Euler (1707–1783) here addressed numerous problems of motion under the action of forces, proceeding algebraically and without appeal to geometrical or mechanical intuition. But the three-body or perturbational problem, which Newton had treated qualitatively in Proposition I.66, was notably absent. Why?

The Cartesian theory of vortexes was still widely accepted on the Continent in the 1730s, and Euler was one of its adherents. Still, in the 1760s he would be defending, after most others had abandoned it, the Cartesian principle that all forces must be forces of contact.[13] Thus, in the 1730s and 1740s Euler

[11] See D. T. Whiteside, "The Mathematical Principles Underlying Newton's *Principia,*" *Journal for the History of Astronomy,* 1 (1970), 118–20.

[12] See Bruce Pourciau, "On Newton's Proof That Inverse-Square Orbits Must Be Conics," *Annals of Science,* 48 (1991), 159–72; and Pourciau, "Newton's Solution of the One-Body Problem," *Archive for History of Exact Sciences,* 44 (1992), 125–46. We should add that Newton in Proposition 41 of Book I shows *in principle* how the problem of deriving the orbits corresponding to different centripetal force laws can be solved; he does not apply the proposition explicitly to inverse-square orbits.

[13] See Curtis Wilson, "Euler on action-at-a-distance and Fundamental Equations in Continuum Mechanics," in P. M. Harman and Alan E. Shapiro (eds.), *The Investigation of Difficult Things* (Cambridge University Press, 1992), pp. 399–420.

was doubtful that the inverse-square law was accurate to all distances, as Newton had assumed. Nevertheless, like his teacher Johann Bernoulli, Euler realized that many of the Newtonian propositions would have to be accommodated, either as approximations or as exact truths, in any final theory.[14] To test the truth of the Newtonian theory, it was necessary to elicit its consequences in detail. The obstacle to an algebraic-style exploration of the perturbational problem as posed under the inverse-square law was technical rather than philosophical: how to formulate the problem algebraically. Beginning in 1739, Euler at last found out how.

What was needed was a worked-out calculus of the trigonometrical functions.[15] Newton had known how to calculate derivatives and integrals of sines and cosines, but in the few places in the *Principia* where he utilized such operations (for instance, in Proposition III.26), the presentation was ostensibly geometrical and left the underlying trigonometrical calculus unarticulated. Roger Cotes in his posthumously published *Harmonia mensurarum* of 1722, gave the derivatives of the sine, tangent, and secant, but this promising start was not followed up. Only in 1739 did Euler codify the trigonometrical calculus as a means to the solution of linear differential equations with constant coefficients. Earlier, sines and cosines had been treated primarily as lines in diagrams; Euler was the first to treat them as ratios, with a consistent notation indicating their functional dependence on the central angles. He developed the identities equating powers of sines and cosines to sines and cosines of the multiple angles. These rules would play a key role in the Leibnizian-style approach to the perturbational problem. Euler showed how the same transformations could be expressed in terms of exponentials with complex powers, a procedure later utilized by d'Alembert.

By 1742, according to Euler's later account, he was at work on the perturbations of the Moon's motion. He was here assuming the accuracy of the inverse-square law. His first lunar tables were published in his *Opuscula varii argumenti* of 1746 but without exposition or explanation of the calculations by which they had been derived.[16]

Euler's detailed procedure for attacking the perturbational problem at last appeared in print in 1749 in his *Recherches sur la question des inégalités de Saturne et de Jupiter,* the prize-winning essay in the Paris Academy's contest of 1748. The topic for this contest had been chosen at a meeting of the prize commission in March 1746, when Le Monnier presented evidence that the motions of Jupiter and Saturn were subject to detectable inequalities. These, Le Monnier assumed, were due to each planet perturbing the other in accordance with the inverse-square law.

[14] See E. J. Aiton, *The Vortex Theory of Planetary Motions* (London: MacDonald and New York: American Elsevier Inc., 1972).

[15] See Victor J. Katz, "The Calculus of the Trigonometric Functions," *Historia Mathematica*, 14 (1987), 311–24.

[16] *Leonhardi Euleri Opera Omnia*, ser. II, v. 23, p. 2.

From the moment that the prize problem was set, two members of the prize commission – Alexis-Claude Clairaut (1713–1765) and Jean le Rond d'Alembert (1717–1783) – each unbeknownst to the other, launched their own Leibnizian-style assaults on the three-body problem. The rules prohibited members of the Academy from participating in the contest, but as Madame du Châtelet put it in a letter, "M. Clairaut and M. D'Alembert are after the system of the world, understandably they do not wish to be forestalled by the essays for the prize." By late 1746 both Clairaut and d'Alembert had derived the necessary differential equations. To establish priority, Clairaut deposited his derivation in a sealed envelope with the Secretary of the Paris Academy, whereas d'Alembert sent his derivation to the Berlin Academy. In their further work, they proceeded to apply their equations to the motions of the Moon, with results we shall report in a later section.

The lunar problem differs in an important respect from the planetary problem. The Sun-Earth distance is enormous relative to the Moon-Earth distance and is nearly constant; consequently, the varying distance between the Sun and the Moon, on which the perturbations depend, can be approximated by a series of a very few terms. In the case of Saturn perturbed by Jupiter, the distance between perturbed and perturbing planet varies by a factor of more than 5, and no manageable way of expressing this variation as a functional dependence on heliocentric angle was available before Euler showed how. His solution was the invention of trigonometrical series, with various devices for computing the coefficients of successive terms. Without this invention, the study of planetary perturbations would have been restricted to the method of numerical integration, tiresomely laborious in a pre-electronic age.

The chief task that Euler set himself in the *Recherches* of 1748 was to determine from Newton's theory the perturbations of Saturn due to Jupiter. The calculations had to be approximative. Because of algebraic errors, Euler obtained mistaken coefficients for some of the sinusoidal terms, one being off by 11 arcminutes. He carried the calculation only to the order of the first power in the eccentricities; the resulting theory therefore lacked terms of higher order, some of which Laplace would later show to be largest of all. Among the terms that Euler derived was an "arc de cercle," or term proportional to the time, which must eventually engulf all other terms; it was a mistake that arose from Euler's failure to allow for the precession of Jupiter's apse.

With all its flaws, Euler's theory was a major advance; much that came after it would be refinement and development of its ideas. One of its important conceptual advances was a way of calculating *secular* variations in the constants specifying the shape and orientation of a planetary orbit. In 1748 Euler introduced this procedure only for the planet's latitudinal variations. In the absence of perturbation, the inclination of the planet's orbital plane to a fixed plane of reference, and the position of the line of intersection of these two planes, would be constant, and the latitude of the planet (the heliocentric angle by which it departs from the plane of reference) is given by a simple formula, with the planet's longitude as independent variable. Perturbation, however, causes both the "constants" to vary, although at a rate exceedingly slow compared with the planet's longitude. Euler derived differential formulas for these slow variations. In a later essay, a memoir that won the prize in the Paris Academy's contest of 1752, he undertook to derive the secular variations of the aphelia and eccentricities for Jupiter and Saturn. The idea would eventually be applied to all six orbital elements and developed systematically by Lagrange.[17]

In comparing his theory with observations, Euler took a long stride toward introducing statistics into astronomy. In this comparison, the orbital elements of Saturn (corresponding to the arbitrary constants in the solution of the differential equations) had to be determined empirically; Euler used Jacques Cassini's values. But any such values unavoidably incorporated effects of perturbation due to Jupiter and therefore needed to be corrected. The observations to be accommodated were those of Saturn's heliocentric longitude, which is observable when Saturn is in opposition to the Sun, on average every 54 weeks. For each such recorded observation in the period from 1582 to 1745, Euler formed an equation in which the longitude as derived from Cassini's orbital elements, with differential corrections added, and the effects of perturbation also included, was set equal to the observed longitude. In the "equations of condition" thus formed, there were eight unknowns. To solve for them, Euler summed equations together – so as to maximize the coefficient of one unknown and minimize the coefficients of the others – and then neglected small terms. (The method of least squares was still a half century away.) His procedure permitted him to adjust the mean motion and to correct the largest error in the coefficients he had earlier derived from Newtonian gravitation.

The net outcome was that the adjusted theory fitted the observations with an average error of 3.3 arcminutes. Euler attributed the error to inaccuracy in the observations and in the inverse-square law, especially at large distances. The truth was that Euler's theory contained mistaken terms and lacked higher-order terms that were essential to a greater empirical success. His introduction of equations of condition, however, was important. Tobias Mayer (1723–1762) used them in 1750 to obtain a precise description of the libration of the Moon; Laplace in the 1780s was following Euler and Mayer when he used them to fit a greatly improved theory of Saturn to observations. The use of multiple equations of condition became *de rigueur* in the 1790s, years before the method of least squares was generally adopted.[18]

---

[17] A more detailed account of Euler's essay of 1748 is in Curtis Wilson, "The Great Inequality of Jupiter and Saturn: From Kepler to Laplace," *Archive for History of Exact Sciences*, 33 (1985), 15–290, especially 69–82, 90–6.

[18] On Euler's use of equations of condition in his essay of 1748, see Wilson, "The Great Inequality," pp. 221–7.

STAR POSITIONS AND PHYSICAL THEORY:
BRADLEY, D'ALEMBERT, AND EULER

As the filar micrometer came into use in the last third of the seventeenth century, astronomers began to detect motions of the stars relative to the standard equatorial frame of reference. Robert Hooke in the 1670s and Flamsteed
in the 1690s mistakenly claimed to have detected stellar parallax. James
Bradley (1693–1762) in the late 1720s at last correctly identified the chief apparent motion of the stars: the *aberration of light,* a displacement of each star
in the direction of the Earth's momentary motion about the Sun, amounting
at maximum to some 20″. Because the Earth's velocity has a finite ratio to the
velocity of light, the telescope must be canted slightly forward in the direction of the Earth's motion. Bradley's discovery was a confirmation of the fini-
tude of the speed of light and also of the Copernican hypothesis.

Having announced this discovery in the *Philosophical Transactions* for 1729,
Bradley went on to identify a further apparent motion in the stars: the *nutation.* This he explained as a wobble in the precessing axis of the Earth, caused
by the retrogradation of the Moon's orbit about the poles of the ecliptic, which
altered the direction of the Moon's net pull on the Earth's equatorial bulge.
Bradley delayed publishing this result until 1748 in order to trace the effects
of nutation over a full 18.6-year cycle of revolution of the Moon's nodes. The
effects were a ±9″ variation in the obliquity of the ecliptic and a ±17″ variation in the rate of precession of the equinoxes.[19]

These discoveries opened a new era in which observation could set itself
the goal of accuracy to arcseconds. In 1742 Bradley succeeded Halley as Astronomer Royal, and after fitting the Greenwich Observatory with new and
better-designed instruments, commenced in 1750 a series of observations that
would become the foundation of modern astrometry. The *Astronomiae fundamenta* of Nicolas-Louis de Lacaille, which appeared in 1757, was the first
extended publication of observational results taking aberration and nutation
into account; it gave the positions of 400 bright stars and 144 positions of the
Sun. Lacaille used these data as a basis for his *Tabulae solares* (1758), the first
such tables to incorporate not only aberration and nutation but also planetary perturbations.[20] The observational work of Tobias Mayer in Göttingen,
which was to lead to the first lunar theory accurate enough to permit determination of longitude at sea to within a degree (see the next section), likewise
depended on Bradley's discoveries.

Meanwhile, the nutation posed a question for the theorists: how could one
deduce it quantitatively from Newton's theory? Indeed, could the precession

---

[19] On Bradley's discovery of the aberration of light and the nutation, see S. P. Rigaud (ed.), *Miscellaneous Works and Correspondence of the Rev. James Bradley* (Oxford: Oxford University Press, 1832),
pp. xii–xxxv, lxi–lxx.

[20] See Curtis Wilson, "Perturbations and Solar Tables from Lacaille to Delambre," *Archive for History
of Exact Sciences,* 22 (1980), 168–88.

itself, on which the nutation was a superimposed wobble, be so deduced? Newton had correctly identified the source of the precession – the Moon's and Sun's attraction of the Earth's equatorial bulge – but his attempt at a deduction was deeply flawed. New principles were required for the derivation: a dynamics of rotating bodies. It was d'Alembert who first achieved a correct deduction of both precession and nutation, in his *Recherches sur la précession des equinoxes* (1749). Crucial to his deduction was the dynamical theorem called "d'Alembert's Principle," which treats the forces due to mutual actions and constraints in a system of masses as a system in equilibrium. Crucial also was the consideration of the equilibrium of the *moments* of the forces. Euler, stimulated by d'Alembert's success, went on in the next decade to formalize the dynamics of rigid bodies in terms of moments of inertia, torques, and angular accelerations.[21]

Still another motion of the "fixed" stars was deduced theoretically in these years. In his prize-winning memoir of 1748, Euler had shown that the perturbed planet's orbit would precess on the orbital plane of the perturbing planet. In the case of the Earth's orbit, this effect would produce a change in the obliquity of the ecliptic, the angle between the Earth's Equator and the plane of the Earth's orbit about the Sun. The question whether the obliquity had been decreasing since ancient times – Ptolemy's value for it was much larger than more recent determinations – had been in dispute. In a memoir completed in 1754 and published in 1756, Euler obtained a differential formula giving the approximate effect of planetary perturbations on the obliquity, and he found the rate of decrease to be $47.5''$ per century given the current configuration of orbital planes; his calculation assumed a conjectural relation to assign masses to Mercury, Venus, and Mars.[22] Attempts were made to determine the rate of diminution observationally. Consensus was reached only when Laplace derived an integral formula for the change, establishing its period and showing the eighteenth-century rate of decrease to be $47''$ per century.

THE LUNAR PROBLEM: CLAIRAUT, EULER, D'ALEMBERT, AND MAYER

In September 1747, Clairaut, as a member of the prize commission, read Euler's memoir for the contest of 1748. He had earlier concluded that only about half the motion of the lunar apse was derivable from Newton's inverse-square law, and he was delighted to find that Euler agreed with him. In mid-November Clairaut announced this result to the Academy of Sciences and proposed

---

[21] See Curtis Wilson, "D'Alembert *versus* Euler on the Precession of the Equinoxes and the Mechanics of Rigid Bodies," *Archive for History of Exact Sciences*, 37 (1987), 233–73.

[22] Leonhard Euler, "De la variation de la latitude des étoiles fixes et de l'obliquité de l'écliptique," *Mémoires de l'Académie des Sciences de Berlin*, 10 (1754, 1756), 296–336.

emending Newton's gravitational law by the addition of a small inverse-fourth-power term, so as to make the true apsidal motion derivable. The Count de Buffon objected vehemently, insisting that a law requiring two terms for its expression was metaphysically repugnant. The controversy continued through 1748.

In November 1747 Clairaut had carried his solution of the lunar problem only to a first-order approximation. A higher-order approximation was necessary if he was to achieve a theory accurate to arcminutes. Why did he (and Euler as well) reject the possibility that a further approximation could produce the missing half of the apsidal motion?

The answer may be as follows. Both Clairaut and Euler had begun by approximating the lunar orbit as a precessing ellipse. In respects other than the apsidal motion, this first-stage approximation seemed satisfactory in reducing the gap between observation and theory. No doubt, then, they began to imagine the orbit as, very nearly, a precessing ellipse. It seemed unlikely that the forward rotation of this ellipse could be doubled without destroying the rapprochement already obtained in other respects. Ergo, the full apsidal motion was not derivable from the inverse-square law.

But to picture the lunar orbit as a rotating ellipse is a mistake. Such a picture applies well enough to a planet such as Mars, where perturbations are small relative to the elliptic inequality. In the lunar case, the Sun severely distorts the orbit's shape, and any orbital shape assumed initially is at best an arbitrary starting point, to be corrected by deriving perturbations. At a time when measurement of the Moon's distance from the Earth was relatively imprecise, the apse was best conceived, not as the point of the orbit farthest from the Earth but rather as the fiducial point whence the anomalies were to be measured. Abstraction from geometrical thinking was requisite.

Clairaut at length undertook the second-order approximation, a tedious business. The first-order result, with numerical coefficients replaced by letters, was to be substituted back into an integral equation he had deduced from the original differential equations. In the result, small terms arising from the transverse perturbing force proved to have a large and hitherto unsuspected effect on the apsidal motion. Whereas Clairaut in the first approximation had found $1°30'11''$ for the apsidal motion, in the second approximation he obtained $3°\ 2'\ 6''$, as compared with the empirical value he accepted, $3°\ 4'11''$.

Euler, on hearing of the new result, went back over his own derivation but, as he wrote Clairaut in mid-July 1749, could find no error. When the St. Petersburg Academy initiated a series of prize contests, to begin with the year 1751, it chose the lunar problem as the first topic and designated Euler as chief judge in the contest. By March 1751 Euler was in possession of four of the submitted essays, including Clairaut's, the procedures of which he greatly admired and learned from. But he continued to seek a confirming derivation by his own very different route, assuming the empirical value of the apsidal motion but allowing for a deviation μ from the inverse-square law. At length he managed to show that, when the approximation was carried far enough, μ became effectively zero. His praise for Clairaut's discovery was unstinting: "It is only with this discovery that one can regard the law of attraction reciprocally proportional to the squares of the distances as solidly established; and on this depends the entire theory of astronomy" (letter to Clairaut of 29 June 1751).

The first to lay out the sequence of successive approximations in an ordered way was d'Alembert, who published his lunar theory only in 1754.[23] For the orbit postulated initially, he chose a circle concentric to the Earth as the most neutral assumption. He classed the small quantities involved in the calculation according to order of smallness and carried out a series of four increasingly refined approximations. The four resulting contributions to the apsidal motion were 1°30′39″, 1° 3′34″, 23′30″, 5′ 5″, which total 3° 2′48″.

The derivation of the Moon's apsidal motion was a theoretical, and not a practical, triumph. Tables computed from Clairaut's, Euler's, and d'Alembert's lunar theories led to predictions still in error by 3′ or more – too inaccurate to give the longitude at sea to within a degree. In 1714 the British Parliament had instituted a handsome prize for a method of determining the longitude at sea: £20,000 if accurate to within ½°, half as much if accurate to within 1°. The first to achieve lunar tables accurate enough for the second prize was Tobias Mayer. In deriving the lunar inequalities theoretically, Mayer applied what he had learned from Euler's prize essay of 1748 along with ingenious procedures of his own. He designed his tables to be easy and expeditious to use. The superior accuracy of his tables arose above all from the care with which he fitted his theory to observations. For this purpose he employed Halley's Saros cycle and undoubtedly also the method of "equations of condition" he had learned from Euler. By these means Mayer determined, along with the empirical constants of the theory, many coefficients whose accurate computation from the theory he judged to be prohibitively laborious. (All later lunar tables until the 1820s were similarly dependent on empirical fitting for the determination of some perturbational coefficients.) Finally, Mayer proposed using occultation of stars for finding the Moon's position observationally, and for this purpose he redetermined the positions of many zodiacal stars.[24]

In 1765 Parliament awarded £3,000 to Mayer's heirs, and £10,000 to John Harrison for a chronometer that gave the longitude to ½°. Nevil Maskelyne made Mayer's tables the basis of the *Nautical Almanac,* the first annual edition

---

[23] Jean d'Alembert, *Recherches sur differens points importans du système du monde, Premiere Partie* (Paris, 1754).

[24] For a more complete account of Mayer's tables, see Eric G. Forbes and Curtis Wilson, "The Solar Tables of Lacaille and the Lunar Tables of Mayer," in Taton and Wilson (eds.), *Planetary Astronomy, Part B,* pp. 62–8.

of which appeared in 1766. But seamen adopted the chronometers, which were an order of magnitude more accurate than the lunar method, as soon as they became affordable, in the 1780s and later.[25]

THE RETURN OF HALLEY'S COMET IN 1759

At the end of Book III of the *Principia,* Newton argued that comets move in conic sections having the Sun in a focus, and he showed how to determine from observations the elements of a parabola (perihelion, nodes, and orbital inclination) in which a comet might move. The elements thus determined would differ little from the corresponding elements for an elliptical orbit. If the comet moved in an ellipse, we could look for its return.

Edmund Halley, in his *Synopsis astronomiae cometicae* of 1705 (republished in an expanded version in his *Tabulae astronomicae* of 1749), calculated the elements of twenty-four comets observed since 1337. The elements of three of them – those of 1531, 1607, and 1682 – were almost identical, with the perihelion advancing and node receding a degree or less between apparitions. Halley proposed that these three were really the same. The periods from 1531 to 1607 and from 1607 to 1682 differed by more than a year, but he thought this difference might be explained as due to Jupiter's perturbing influence. A slight change in speed near perihelion could have a major effect on the date of the comet's return. The next return, he predicted (without supporting argument), would be in late 1758 or early 1759.[26]

Beginning in June 1757, Clairaut set out to compute the date of the anticipated return. His idea was to calculate by how many days perturbation had hastened or delayed the comet's return in the interval from 1607 to 1682 and then to do the same for the following period; the difference, added to the 1607–1682 interval, would give the interval from 1682 to the following return. To test the procedure, he applied it first to the two successive intervals from 1531 to 1607 and from 1607 to 1682. He took into account only perturbations due to Jupiter and Saturn. The computation was horrendous; most of the integrals had to be evaluated by numerical integration. He was assisted by the astronomer Joseph-Jérôme Lefrançais de Lalande (1732–1807) and Mme Nicole-Reine Étable de Labrière Lepaute – *la savante calculatrice,* as Clairaut called her.

In November 1758, to avoid being anticipated by the comet, Clairaut made a preliminary presentation to the Academy of Sciences. By his calculation (still to be completed and refined), perturbation shortened the period from 1607 to 1682, as compared with the preceding period, by 436 days; the actual

[25] See Dava Sobel, *Longitude* (New York: Walker Publishing, 1995), pp. 162–4.
[26] Craig B. Waff, "Predicting the Mid-Eighteenth-Century Return of Halley's Comet," in Taton and Wilson (eds.), *Planetary Astronomy, Part B,* pp. 69–82.

difference was 469 days, so the error of the calculation was 33 days. In the period from 1682 to 1759, as compared with the preceding period, perturbation had lengthened the period by about 618 days. Clairaut concluded that perihelion should occur in mid-April, give or take a month.[27]

The returning comet was first glimpsed by J. G. Palitzsch (1723–1788) in Prohlis, Saxony, on 25 December 1758 and then by Charles Messier (1730–1817) in Paris on 21 January 1759. The news became general only after perihelion and the comet's reemergence from the Sun's rays on 1 April. Perihelion proved to have been on 13 March. A verbal battle now broke out as to whether the error of Clairaut's calculation had been small or large. The larger fact was that Clairaut had turned a vague prediction into a precise one, which was verified to sufficient closeness to leave Newton's theory alive and vigorous.

THE TRANSITS OF VENUS OF 1761 AND 1769

For a precise, predictive astronomy, the mean horizontal solar parallax of the Sun – the angle subtended by the Earth's radius as seen from the center of the Sun – is a critical constant. Until the 1680s, solar theory (the theory of the Earth's motion) was based on meridian observations of the Sun's altitude, which had to be corrected downward for refraction and upward for parallax. The two effects were here inseparably mixed, and the traditional overestimation of parallax introduced into the Sun's (or Earth's) orbit an exaggerated eccentricity, which was then reflected in the theories of all the other planets. Flamsteed showed how to avoid this difficulty by determining the maximum equation of center from measurements of the Sun's right ascension.

An exact value of solar parallax remained a *desideratum* not only for the evaluation of celestial distances in terms of terrestrial units but also because of the role of solar parallax in determining the Earth's mass; the latter was needed in calculating the perturbations caused by the Earth in the motions of the other planets. The value of the Earth's mass is a function of the cube of the solar parallax; thus, Newton's choice of 10.55″ as the value of the solar parallax in the third edition of the *Principia* (the true value being 8.8″) exaggerated the Earth's mass by a factor of 1.7. Values of solar parallax proposed during the first half of the eighteenth century varied between 9″ and 15″.

In 1678 Edmond Halley had already asserted in print that the only sure way of determining solar parallax was by observation of one of Venus's transits across the face of the Sun from widely separated spots on the Earth's surface. At such times Venus is as close as it gets to the Earth, and timings of its entrance into, passage across, and exit from the Sun's disk by observers widely separated on the Earth's surface suffice to fix the parallax of Venus and hence

---

[27] On the errors involved in these calculations, see Curtis Wilson, "Clairaut's Calculation of the Eighteenth-century Return of Halley's Comet," *Journal for the History of Astronomy*, 24 (1993), 1–15.

(from the known solar distances of Venus and the Earth in Astronomical Units) the solar parallax. Halley again advocated this method in the *Philosophical Transactions* in 1691 and 1716. The next transits of Venus, he pointed out, would occur in June 1761 and 1769. After Halley's death in 1742, the championing of a worldwide effort to observe these transits was taken up by others, especially Joseph-Nicolas Delisle (1688–1768) and Lalande.

Both the British and the French organized expeditions for the observation of the transit of 1761; it was observed from some 120 spots on the Earth's surface. The resulting calculated values of the solar parallax varied from 8.28″ to 10.6″, a disappointingly wide range. Preparations for the transit of 1769 were more extensive and thorough; the British sent out 69 observers (including Captain Cook to Tahiti), the French 34, the Russians 13; 19 observers in the British colonies of North America observed the transit; 151 observations were attempted in all. The resulting values of solar parallax formed a narrower range than in the earlier transit, from about 8.4″ to 9.0″. A consensus, undoubtedly influenced by subjective factors, formed around the value 8.6″. This consensus continued to hold through the first half of the nineteenth century until arguments from other data suggested the need for an upward correction. Astronomers then returned to the original data and, with ever more refined statistical techniques, teased from them a slightly larger value.[28]

SECULAR AND LONG-TERM INEQUALITIES

"Secular inequalities" are sometimes described as inequalities that always increase with time, hence as noncyclical. Such was the acceleration that Halley, by comparing ancient and modern observations, believed he had found in the mean motion of the Moon. By similar comparisons he was led to assert an acceleration in the mean motion of Jupiter and a deceleration in the mean motion of Saturn. These trends, continued indefinitely, would lead to the dissolution of the planetary system.

In the work of Lagrange and Laplace, the term "secular inequalities" came to have a wider meaning: secular inequalities were distinguished from inequalities called "periodic." The latter are oscillations that run through their cycles once or more as the perturbing and perturbed planets go from conjunction to conjunction. One can define them by saying that they depend on the positions of the planets in their orbits. The secular inequalities, by contrast, depend on the relations of the orbits themselves; they lead to changed orbital shapes and orientations. They can be cyclical, but, if so, their periods

---

[28] For an account of the efforts made to observe the transits of 1761 and 1769, see Harry Woolf, *The Transits of Venus: A Study of Eighteenth-Century Science* (Princeton, NJ: Princeton University Press, 1959). On technical aspects, see Albert Van Helden, "Measuring Solar Parallax: The Venus Transits of 1761 and 1769 and Their Nineteenth-Century Sequels," in Taton and Wilson (eds.), *Planetary Astronomy, Part B*, pp. 153–68.

are of the order of tens of thousands of years, far longer than the periods of the planets.

An important question was whether the mean motions of the planets are subject to secular change, cyclical or unidirectional. Euler in his prize essay of 1752 (published in 1769), found (by faulty algebra) that both Jupiter and Saturn were undergoing acceleration. Lagrange, in a derivation published in 1766, found an acceleration for Jupiter and a deceleration for Saturn. At age twenty, Laplace, his curiosity piqued by the discrepancy between Euler's and Lagrange's results, undertook his own derivation and found that, to a high order of approximation, the mean motions of the two planets were immune to secular change. The apparent acceleration of Jupiter and deceleration of Saturn, he suggested, might be due to the gravitational action of comets. The memoir giving these results was submitted to the Paris Academy of Sciences before Laplace was elected to membership in March 1773 but was published only in 1776.[29]

In October 1774 Lagrange sent to the Paris Academy a memoir on the secular variations of the nodes and orbital inclinations of the planets. By a change of variables, he reduced the equations of the problem to first-order linear differential equations with constant coefficients, hence soluble without approximation. In the case of the largest planets – Jupiter and Saturn – he showed that the variations were oscillatory and bounded. Laplace, on reading the memoir, saw that the procedure was applicable to the aphelia and orbital eccentricities and so applied it in a memoir that appeared in 1775.[30]

To his eager young rival Lagrange offered to cede the entire topic of secular variations. But Laplace was stymied by the unaccountable acceleration and deceleration in the motions of Jupiter and Saturn. For the better part of a decade he turned his attention away from planetary perturbations to other topics: attractions of spheroids, tides, precession, nutation.

Lagrange, more concerned than Laplace with elegance and generality in the derivations, continued the inquiry, and during the period 1775–85 achieved seminal results. First (in 1779), he published a new and more general proof of the immunity of the mean motions to secular change. The proof assumed incommensurability of the mean motions (both Lagrange and Laplace took this for granted, although it is unverifiable). It was carried out by means of a function that Laplace would later dub "the perturbing function," a function from which the perturbations could be obtained by partial differentiation. In a second memoir of 1779, Lagrange used this function to derive the known integrals of motion for a system of gravitationally interacting bodies.[31]

In a lengthy treatise appearing in two parts in 1783–4, Lagrange gave a

---

[29] Curtis Wilson, "The Great Inequality of Jupiter and Saturn," *Archive for History of Exact Sciences*, 33 (1985), 150–68; Bruno Morando, "Laplace," in Taton and Wilson (eds.), *Planetary Astronomy, Part B*, pp. 131–5.

[30] Wilson, "The Great Inequality," pp. 168–92; Morando, "Laplace," pp. 135–6.

[31] Wilson, "The Great Inequality," pp. 198–210.

systematic derivation of the secular inequalities by the variation of orbital elements. He also sought to establish the stability of the system by showing that the orbital inclinations and eccentricities were confined within narrow bounds. His argument presupposed conjectured values for the masses of Mercury, Venus, and Mars, and he acknowledged that a general proof independent of the masses was desirable.[32]

When in 1784 or 1785 Laplace once more addressed the problem of planetary perturbations, he took his start from the Lagrangian innovations just mentioned. One further Lagrangian memoir may have triggered his reentry into the fray: a treatise on the periodic inequalities, of which the "Première Partie" was published in 1785. Here Lagrange introduced the sixth orbital element, the epoch, as a parameter subject to perturbational variation. After deriving its first-order variations, he appended a paragraph on how the large terms among the higher-order variations, proportional to powers and products of the eccentricities and inclinations, might be located expeditiously. It was an old theme mentioned in the earlier writings of Euler and Lagrange: the terms to be looked for were those that would be greatly enlarged by the double integration that the differential equations of celestial mechanics require. Lagrange's new formulation was more explicit. Each perturbation depends on the sine or cosine of a linear combination of the mean motions of the perturbed and perturbing planets. If the linear combination, θ, changes very slowly relative to the mean longitudinal motion, p, of the perturbed planet, then $\theta = \nu p$, where ν is very small relative to 1. On being twice integrated, the term will have $(1/\nu^2)$, which is very large, as a factor in its coefficient.[33]

Laplace's resolution of the anomaly in the motions of Jupiter and Saturn was presented to the Paris Academy of Sciences in November 1785. From the conservation of *forces vives,* Laplace showed that the apparent acceleration of Jupiter and deceleration of Saturn found by Halley were in just the ratio to be expected if they arose from mutual gravitational interaction. From a memoir of 1775 by J. H. Lambert (1728–77) it emerged that relative to the 1640s Jupiter's acceleration and Saturn's deceleration were both decreasing: the anomaly looked to be reversing, and so could be periodic!

If so, Laplace knew what he needed to find: a linear combination of the two rates of mean motion, n for Jupiter and n' for Saturn, that was very small relative to either n or n'. But since antiquity it had been known that 5 cycles of Jupiter are very nearly equal to 2 cycles of Saturn. Numerically, $5n' - 2n \approx n/74 \approx n'/30$. A sinusoidal term with the argument $(5n' - 2n)t$ would have a period of nearly 900 years. The coefficient of such a term (because $5 - 2 = 3$) is proportional to a cube or product of three dimensions of the orbital eccentricities and inclinations – a small factor. But because $(5n' - 2n)/n$ and $(5n' - 2n)/n'$ are small relative to 1, the same term will be enlarged after the integrations by a large factor.

---

[32] Ibid., pp. 210–15.   [33] Ibid., pp. 215–19.

In computing the complete coefficient in each case, Laplace used the perturbing function to "sharpshoot," picking out the terms that would be large and ignoring others. The total coefficient for Saturn proved to be about 49', that for Jupiter about 20'. In a detailed memoir of 1786, Laplace showed how his new theory accounted satisfactorily for the observations of Jupiter and Saturn from antiquity to his own day.

In the earlier memoir of November 1785, Laplace presented two other results of cosmological import. He supplied a proof of the stability of the solar system that ostensibly did not require knowledge of the planetary masses; in effect, from the conservation of angular momentum, he showed that certain sums involving the squares of the eccentricities and tangents of the orbital inclinations are constant.[34] Unfortunately for the proof, as Leverrier was later to show, it failed to take account of the large difference in order of magnitude between the masses of the planets from Mercury to Mars and those farther out – Jupiter, Saturn, Uranus. This difference invalidates the proof.[35]

Laplace's second result had to do with the first three of the four large satellites of Jupiter, whose mean motions (call them n, n', and n'') had been found to be such that $n - n' = 2(n' - n'')$, whence $n - 3n' + 2n'' = 0$. If the mean motions of these satellites were initially not too far from this resonant relation, Laplace showed, they would be pulled into exact accordance with it by the gravitational interactions of the satellites.[36] We know today that these mean motions evolve owing to tidal interactions with Jupiter; the resonance, once arrived at, remains stable in accordance with Laplace's demonstration.

Toward the end of 1787 Laplace announced the resolution of the last remaining major anomaly in the theoretical astronomy of the time: the apparent secular acceleration in the motion of the Moon. This, he argued, was the result of an indirect perturbation, the diminution of the Earth's orbital eccentricity in the present age, leading to a tiny decrease in the radial component of the Sun's perturbing force on the Moon. The effect would be reversed when the Earth's orbital eccentricity in a later age started to increase again, as the theory of planetary perturbations predicted. In 1854 John Couch Adams showed Laplace's derivation to be partially in error: only half the apparent secular acceleration had been accounted for. The rest of it would eventually be attributed to frictional slowing of the Earth's rotation due to tidal interaction with the Moon – an effect that Laplace had earlier dismissed as negligible. (In fact, the slowing of the Earth's rotation is such as to lead to a considerably greater apparent angular acceleration of the Moon, but it is counteracted by the Moon's rising, through tidal interaction with the Earth, into an ever higher orbit.) Laplace's triumphs of the 1780s supported the view that the

[34] Ibid., pp. 231–9.
[35] Jacques Laskar, "The Stability of the Solar System from Laplace to the Present," in Taton and Wilson (eds.), *Planetary Astronomy, Part B*, pp. 242–5.
[36] Morando, "Laplace," pp. 141–2; Robert Grant, *History of Physical Astronomy* (New York: Johnson Reprint, 1966), pp. 91–6.

solar system was quite stable, subject only to self-compensating oscillations. His drive toward explanations was to lead him in 1796 to propose that the system had *evolved* into its present stable state.

COSMOLOGY AND THE NEBULAR HYPOTHESIS

By the late seventeenth century, change and variability in the world of the stars was undeniable. Tycho in 1572 and Kepler in 1604 had observed supernovas. By 1640 J. P. Holwarda had discovered that the star Mira Ceti was variable; in 1667 Ismael Boulliau established the period between successive maxima as 333 days.[37] As yet, however, there was no reliable evidence that the *positions* of the stars underwent change.

In the *Principia*, Prop.III.14, Corollaries 1 and 2, Newton asserted that the stars are immovable and, as shown by their lack of detectable annual parallax, are at such great distances from the solar system as not to interact with it detectably. In late 1692 Richard Bentley (1662–1742), while preparing a sermon on the evidences of Christianity (one of a series endowed by Robert Boyle), addressed to Newton the following query: if, to begin with, matter was spread uniformly through a finite space, what would happen if it were allowed to move freely under the action of gravity? Newton replied that it would coalesce in the center, but if the space were infinite the number of clumps would be infinite; thus might the stars and Sun have been formed. But, countered Bentley, in such an infinite spread of matter, would not the gravitational pulls on any particle be equal in all directions? No: such an exact equilibrium, Newton explained, would be as unlikely as the standing upright of the sharpest needle on its point upon a looking glass. But then would not the stars also be subject to unbalanced gravitational forces? Yes. Thus, Newton and Bentley came to agree that the fixity of the "fixt stars" was a miracle, preserved only by divine power.

In a draft for a new Proposition III.15, Newton sought to connect the hypothesis of an approximately uniform distribution of the stars with observational fact. Since Ptolemy the stars had been classified according to "magnitude," first-magnitude stars being the brightest and sixth-magnitude stars just detectable by unaided eye. Following James Gregory, Newton assumed all stars to be of roughly the same intrinsic luminosity. The brightest stars would thus be the nearest, and if these stars were at equal distances from the Sun and one another, there would be twelve or thirteen of them. At the double distance there should be four times as many, at the triple distance nine times, and so on. Supposing magnitude to translate into distance, Newton expected these

---

[37] On the detection of variability among the stars from the sixteenth through the eighteenth centuries, see Michael Hoskin, *Stellar Astronomy: Historical Studies* (Chalfont St. Giles, Bucks, England: Science History Publications, 1982), pp. 22–55.

numbers to agree with the numbers of stars of different magnitudes in the star catalogs. The agreement was close for magnitudes 1 and 2, but thereafter the star counts increased more rapidly than the successive square numbers. As we now know, the psychophysiology of vision was involved (Fechner's law).[38]

In his published writings Newton gave only hints of his thoughts on stellar distribution. But David Gregory mentioned them in his published lectures, and Edmond Halley expanded on them in two papers given in 1721.[39] Here Halley adverted to the paradox later called "Olber's": on Newton's assumptions, the farther stars would transmit as much light to us as the nearer ones, so the whole surface of the night sky should be golden, and Halley's attempts at an explanation were partly wrong, partly obscure. But in 1744 J.-P. L. De Chéseaux pointed out that a slight scattering or absorption of light traveling from the stars would explain the paradox.

The Newtonian universe, kept safe from change only by miracle, was due for an assault. In 1750 the autodidact William Wright (1711–1786) of Durham published his *An original theory or new hypothesis of the universe,* in which he sought to reconcile astronomical theory and fact with theology.[40] Halley in 1718 had reported a change in latitude since antiquity of three zodiacal stars, and with this much warrant, Wright hypothesized that the stars of our system, distributed in either a spherical shell or a ring, were orbiting about a divine center to which they were gravitationally attracted. A terrestrial observer, his line of sight in a plane tangent to the sphere or ring, would see the myriads of stars forming the Milky Way. Here gravitation made for stability. In later, unpublished manuscripts, Wright adopted an evolutionary cosmology.[41]

Immanuel Kant, reading of Wright's *Original theory* in a Hamburg journal in 1751, took Wright to be explaining the Milky Way as the effect of viewing a disk-shaped galaxy along the plane of the disk. In his *Allgemeine Naturgeschichte* of 1755 he hypothesized that the universe contained a hierarchy of such "island-worlds": planets orbiting about suns, suns forming galaxies, galaxies forming clusters of galaxies, under the rule of gravity. The universe, originally chaotic, had been progressively organized by gravitational action and would undergo continuing evolution into order as well as beyond order into disorder, and then into order again. Kant's essay, of little influence during the eighteenth century, was an anticipation of later theory.[42]

A head-on challenge to Newton's notion of a uniform distribution of

---

[38] See Hoskin, *Stellar Astronomy,* pp. 71–95.

[39] Ibid., pp. 95–100.   [40] Ibid., pp. 101–16.

[41] On Wright's later speculations, see Simon Schaffer, "The Phoenix of Nature: Fire and Evolutionary Cosmology in Wright and Kant," *Journal for the History of Astronomy,* 9 (1978), 180–200. Another nonevolutionary cosmology, depending on gravitation for stability in a hierarchy of systems, was that sketched by J. H. Lambert in his *Cosmologische Briefe* of 1762; see Hoskin, *Stellar Astronomy,* pp. 117–23.

[42] On the cosmology of Kant's *Allgemeine Naturgeschichte* of 1755, see Hoskin, *Stellar Astronomy,* pp. 68–9, and Schaffer, "The Phoenix of Nature."

stars came in 1767 from John Michell, who argued that the actual frequency of star clusters was improbable unless many of the stars in apparent clusters were actually neighboring in three-dimensional space. Michell's conclusion, although faultily argued (it is in fact validly deducible), was widely accepted.[43]

During the 1770s William Herschel (1738–1822), a musician by profession, designed and constructed the first telescopes suitable for exploring the stellar world – reflectors of wide aperture and hence of unprecedented light-gathering power. In 1779 he began a search for double stars; these, if merely "optical" (collocated only in appearance), might by changes in relative position evince annual parallax. In the course of this search he discovered in 1781 a "comet," which proved to be the seventh planet, later named Uranus. The discovery brought membership in the Royal Society and a pension from the King. From 1783 onward, in a survey of nebulous objects, Herschel showed many of them to be resolvable into stars and hence at immense distances – island universes. He envisaged a "natural history" of the stars in which they progressively clustered under the action of gravity. In late 1790 he convinced himself that true nebulosity – actual cloudiness unresolvable into stars – exists and so moved toward his final cosmogonical view, according to which a series of types from true nebulosity to complete resolvability into stars constituted a temporal, evolutionary sequence. In 1803 and 1804 Herschel was able to confirm that some of his double stars were true binaries, orbiting about one another – the first sure evidence that gravitation was operative among the stars.[44]

Laplace, in Note 7 at the end of his *Exposition du Système du Monde* (1796), proposed the origination of the solar system from a hot, rotating, fluid disk – a nebula. The disk as it cooled would contract, leaving behind rings that would then agglomerate into rotating planets. By this proposal Laplace intended to account for the confinement of planets and satellites to nearly the same plane, the near circularity of their orbits, and their orbital revolution and axial rotation in a single sense, counterclockwise as seen from the ecliptic's north pole. The Nebular Hypothesis remains today, having survived many ups and downs, the most widely accepted theory as to the origin of the outer, giant planets.

Thus, with Herschel and Laplace, the earlier view of the universe as a stable clockwork, created once and for all in its final form, was replaced by a picture of stable forms arising by evolution in accordance with natural law.[45]

[43] See Oscar Sheynin, "The Introduction of Statistical Reasoning into Astronomy from Newton to Poincaré," in Taton and Wilson (eds.), *Planetary Astronomy, Part B*, pp. 193–5.

[44] On Herschel, see J. A. Bennett, "'On the Power of Penetrating into Space': The Telescopes of William Herschel," *Journal for the History of Astronomy,* 7 (1976), 75–108; Simon Schaffer, "Uranus and the Establishment of Herschel's Astronomy," *Journal for the History of Astronomy,* 12 (1981), 11–26; Hoskin, *Stellar Astronomy,* pp. 125–42.

[45] See Morando, "Laplace," 131–50; Ronald L. Numbers, *Creation by Natural Law: Laplace's Nebular Hypothesis in American Thought* (Seattle: University of Washington Press, 1977); Stephen G. Brush, *Nebulous Earth* (Cambridge University Press, 1996), pp. 14–137.

CONCLUSION: THE LAPLACIAN
SYNTHESIS IN THE 1790s AND LATER

In the midst of the Terror, in August 1793, the National Convention abolished the Paris Académie des Sciences. A product of the *ancien régime,* it was viewed as elitist. Prominent members of it, friends of Laplace such as Bochart de Saron and Lavoisier, were guillotined. Laplace himself, in December 1793, was removed from the Commission on Weights and Measures on suspicion of lacking "republican virtues and the hatred of kings." He took himself off to Melun, southeast of Paris, and there began the writing of his *Exposition du Système du Monde.*

In simple language, without a single equation, Laplace's *Exposition* set forth the picture of the celestial world at which mathematical astronomy, according to Laplace, had arrived. The chief known celestial phenomena, so Laplace claimed, had been accounted for on the basis of the single principle of universal gravitation. The main thrust of the *Exposition* was to assert the beautiful equilibration of the celestial system, which Laplace saw as analogous to the adaptation of forms in organic nature.

Astronomy's achievement, Laplace tells us more than once, was a triumph of *Analyse* – mathematical and mechanical analysis. The meaning was partly Newtonian. "For the whole difficulty of Philosophy," Newton had said, "seems to turn on this, that from the phenomena of motion we should investigate the forces of Nature, then from these forces demonstrate the remaining phenomena" (*Principia, Praefatio ad Lectorem*). Newton's method was analytical in that principles were to be drawn from observations and experiments by an analytical or resolutive process. From the principles so established (corrigibly, with "moral" rather than "mathematical certainty"), the consequences could then be demonstrated mathematically. Newton's "new way of inquiry"[46] thus avoided resorting to hypotheses and metaphysical commitments, such as the Cartesian commitment to forces of contact. This sense of "analysis," of which Newton had provided the most impressive exemplification, had come to permeate Enlightenment thought; Laplace made it his own.

But for Laplace the term "Analyse" also meant Leibnizian algorithmic analysis, the differential and integral calculus as set forth by Leibniz and developed by Jacob and Johann Bernoulli, Euler, Lagrange, and others. Misled by Newton's own statements, Laplace supposed that Newton's geometrical presentation of his findings had everywhere been preceded by a prior symbolic derivation. The error was perspectival; it seemed impossible that Newton could have obtained his results in the absence of the algorithmic processes by which these results were now so easily derived. In any case, it was the language of Leibnizian calculus in which Laplace's *Mécanique Céleste* was couched, its first four massive volumes (1798–1805) so tersely argued that

---

[46] William Harper and George E. Smith, "Newton's New Way of Inquiry," in Jarrett Leplin (ed.), *The Creation of Ideas in Physics* (Boston: Kluwer, 1995), pp. 113–66.

extensive commentary from Nathaniel Bowditch and others was required to unpack the meaning and logic of its derivations.

What a contrast was Laplace's *Exposition,* with its simple language and clear, unified vision of a beautifully equilibrated and stable celestial world! In some sort it was a response to the chaos of the Terror and the challenge of the Jacobins. The extreme Jacobins such as Robespierre were Rousseauians, for whom the model sciences were the life sciences – sciences accessible to ordinary, nonmathematical observers, revealing a life world with which humans could feel akin. In his *Exposition,* Laplace sought to make the results of mathematical astronomy accessible to a popular audience; the universe, in the picture that he projected of it, was friendly to human life. In certain respects, his picture was premature and tendentiously drawn.

With the Thermidorean reaction of the summer of 1795, the regime dropped its quarrel with mathematical physics for practical reasons. The nation was at war, and astronomy was necessary to the navigation of its navy and mercantile fleet. In June 1795 the Bureau des Longitudes was established, to calculate and publish ephemerides, to direct the Paris Observatoire, and to perfect the theories of celestial mechanics. Among the first members named to it were Lagrange and Laplace. The role of science in the nation's life was reaffirmed. Toward the end of 1795 the old Académie des Sciences was reconstituted as part of a newly created Institut de France, and Lagrange and Laplace were named as members of its mathematical class.

The triumphant picture of mathematical astronomy that Laplace had projected would be dominant in France for some years to come; and over the Bureau des Longitudes he would exercise a de facto rule. Some of his theoretical conclusions were hasty, driven by his desire to wrap things up; and his domination of the Bureau des Longitudes was in some respects harmful. The cosmic evolution that he proposed was artifically confined to the past. His proof of the stability of the solar system was flawed. A number of his approximative moves in deriving perturbations would have to be abandoned in the further development of the science. Rigorous celestial mechanics would find its foundations not in Laplace's *Mécanique Céleste* but in the *Mécanique Analytique* of Lagrange (1788).

In Laplace's domination of the Bureau des Longitudes, all was sacrificed to the verification of his planetary and lunar theories, and the determination of star positions suffered as a result. Delambre, at the end of his life, complained that no new star catalog had emerged from the Paris Academy for more than a century. For this he blamed the dynasty of the Cassinis at the Paris Observatoire up to the Revolution, and the dominant influence there of Laplace after 1795: Never, he wrote in a posthumous note, should a geometer (that is, a mathematician) be put at the head of an observatory.[47] But star catalogs

---

[47] Jean-Baptiste Joseph Delambre, *Histoire de l'astronomie au dix-huitième siècle* (Paris: 1827), p. 291; G. Bigourdan, "Le Bureau des Longitudes: Son Histoire et ses Travaux de l'origine (1795) à ce jour," France, Bureau des Longitudes, *Annuaire* (1931), A1–A72.

were the foundation of all accuracy in astronomy. It was the Greenwich transit observations, accumulated with regularity and by uniform procedures, that in the end would prove most important as a basis for the testing of theory.

These criticisms of Laplace's influence should not obscure the importance of his accomplishment for later astronomy. Through his discoveries in celestial mechanics, he showed that a complete explanation of celestial phenomena in terms of Newtonian gravitation was conceivable and indeed near at hand. Astronomy, both observational and mathematical, would derive from that accomplishment the impulse for its later drive to perfection.

15

MECHANICS AND EXPERIMENTAL PHYSICS

R. W. Home

Older-style histories of science that depicted the growth of science as a gradual accretion of new knowledge, and that devoted much attention to identifying when discoveries were made and by whom, allocated little space to the physics of the eighteenth century. Although some interesting discoveries, especially in relation to electricity, were acknowledged, the period was generally presented as a fallow one compared with the periods of dramatic advance in physical understanding that preceded and followed it. More recently, as historians have adopted a less restricted view of their task, eighteenth-century physics has come to be seen in a more favorable light: as the period when physics became a field recognizably like the one we know today.

Physics as traditionally understood was not an experimental science, and neither was its subject matter the same as it is today. Consistent with the meaning of the Greek word φυσις from which it drew its name, physics was taught in universities throughout Europe as "natural philosophy," that is, as the part of the standard undergraduate course in philosophy dealing with "nature" in general. The primary concern was with broad principles rather than particular natural effects, and above all with the nature of body and the conditions determining natural change. Everywhere for several centuries the Aristotelian treatises *Physica, De caelo, De generatione et corruptione, Meteorologica,* and *De anima* were the standard texts, and in many places they were still being used at the start of our period, notwithstanding the dramatic changes in intellectual outlook that had occurred during the preceding century and a half.

Those changes were, however, beginning to have an effect in some parts of western Europe. In some of the more progressive universities, non-Aristotelian ideas had begun to infiltrate the curriculum. In addition, new institutions had begun to emerge that provided a home for self-consciously non-Aristotelian approaches to understanding nature. Everywhere, physics was distinguished from natural history on the traditional ground that the latter merely provided descriptions of phenomena, whereas physics dealt with causes. Yet "physics" continued to embrace not only the topics it includes today but also questions that we now assign to the chemical, biological, and human sciences. During the eighteenth century, the scope of the subject changed to become more like that of today.

Along with new ideas about nature's workings came a change of emphasis in the teaching of physics, with less attention being paid to traditional questions concerning "the nature, essence and properties appropriate to all bodies" – *physica generalis,* as it was called – and more being paid to "the examination and discussion of the particular bodies or corporeal individuals this universe contains" – that is, to *physica particularis.*[1] The change was already apparent in the enormously popular *Traité de physique* of Jacques Rohault (1620–1672), first published in 1671, in which Cartesian physics was expounded, without the metaphysical underpinnings that Descartes himself had stressed, as an experimental science chiefly distinguished by explanations based on corpuscular mechanisms in a universal ether. The new emphasis became universal in textbooks in the early decades of the eighteenth century.

Within the traditional categories of knowledge, physics, the science concerned with matter and change, had been carefully distinguished from the eternal verities of mathematics. The so-called subordinate sciences – astronomy, geometrical optics, harmonics – in which mathematical arguments were developed, starting from premises grounded in physics, occupied an intermediate position. It was in these mathematized sciences inherited from the ancients and in mechanics, successfully mathematized by Galileo, that the major advances occurred during the seventeenth-century Scientific Revolution.[2] Progress was less dramatic in physics itself. At a methodological level, however, outside the universities, the empirical approach made ground. Eventually, this began to make inroads even within university teaching programs.

The best known of the new institutions – the Académie Royale des Sciences in Paris and the Royal Society in London – were both established in the 1660s. Others were later founded all over Europe in imitation of them. Whereas the universities saw their task as transmitting received knowledge, these new organizations were committed to advancing it – that is, to promoting new research. In Paris, one of the two meetings held each week was devoted to mathematical investigations and the other to experimental inquiries, whereas in London the weekly meetings were largely given over to reports of experiments. Members of both institutions adopted a skeptical outlook toward theory. Jesuits and Cartesians alike were excluded from the Paris Academy in its early years because they were considered too committed to their particular intellectual systems and insufficiently open-minded in the pursuit of truth; Hobbes was excluded from the Royal Society for the same reason. At both the Royal Society and the Paris Academy, the "systematic spirit" was abandoned.

---

[1] *Mémoires de Trévoux,* 57 (1756), 582.
[2] T. S. Kuhn, "Mathematical versus Experimental Traditions in the Development of Physical Science," *Journal of Interdisciplinary History,* 7 (1976), 1–31.

As Fontenelle (1657–1757) wrote in 1699, "Systematic Physics must refrain from building its edifice until Experimental Physics is able to furnish it with the necessary materials."[3]

In France, where the lines of demarcation were particularly clear, one can distinguish three separate groups in the late seventeenth century that were concerned with "physics," as traditionally defined. First, there were the university professors. Their method was verbal and expository; their theory, Aristotelian; their objective, knowledge that was certain; their primary concern, the elucidation of causes. Members of the *physicien* group at the Académie Royale des Sciences were committed to the discovery of new knowledge by means of experiment and were inclined to mechanistic theories to explain their observations, but they were skeptical about all theory and less and less concerned as time passed with the discovery of causes. And there were the Cartesians, whose objective was, as with the professors, certainty and a knowledge of causes, but whose method was experimental and whose theory was programmatically mechanistic. The separations among these groups began to break down in the 1690s. Elsewhere, it was never so sharply defined, and in many parts of Europe Cartesianism never found a significant foothold, whereas, in some places, alternatives did (for example, Leibnizianism in Germany). The situation in France brings out clearly, however, the issues involved. In effect, our concern in this chapter is with the regrouping of forces that occurred during the eighteenth century, a regrouping that led to the reconstitution of "physics" as a recognizably modern science.

The Parisian Academy was reorganized in 1699, one consequence being that Cartesians were admitted as members for the first time. In time, this somewhat diluted the skepticism that marked the Academy's early years. Yet the Academy remained an exclusive and, intellectually speaking, an inward-looking body. Although its members did some important work, except for Malebranche they had surprisingly little impact on intellectual patterns in French society more generally. In particular, their empirical outlook had little influence outside the Academy's walls. Empiricism had to fight its battles anew at the more popular level of university and public lecture, with little assistance from the King's academicians.

A tradition of public lecturing in which scientific principles were illustrated by dramatic, specially designed experimental demonstrations was established in Paris by Rohault in the 1660s. A similar course of experimental demonstrations was introduced into the University of Paris in the 1690s in conjunction with the Cartesian invasion of the curriculum. The lecturer was Pierre Polinière (1671–1734), who was commissioned to mount a course of demonstrations to illustrate the principles that the Cartesian professor Guillaume Dagoumer (d. 1745) was presenting in his lectures. Polinière's demonstrations

---

[3] Quoted by Roger Hahn, *The Anatomy of a Scientific Institution: The Paris Academy of Sciences, 1666–1803* (Berkeley: University of California Press, 1971), p. 33.

proved extremely popular, and he continued to give the course several times each year. He also offered courses for the general public, and these, too, were most successful, the young King Louis XV himself attending in 1722. In due course Polinière published his lectures, the first edition appearing in 1709. His work represented the first major infiltration of an experimental outlook into the French educational system. Polinière himself tells us that, after his book was published, the example of the Parisian masters in utilizing his work was imitated elsewhere. Successive editions of his book reveal Polinière's own growing confidence in experiment as a route to knowledge: at first this was said to be merely supplementary to the theoretical principles being expounded, but in the third and subsequent editions it became the only dependable way of arriving at a true physics.[4]

The Royal Society of London was a more open institution than the Paris Academy, and its advocacy of an experiment-based approach to natural knowledge spilled over more easily to a wider public. At the Society's meetings, what had initially been envisaged as collective experimental investigations quickly became demonstrations of experiments that had already been tried in private, the implications of the experiments rather than the doing of them becoming the main subject of discussion at the meetings. Courses of experiments were introduced at Oxford in 1704 by John Keill (1671–1721) and at Cambridge soon afterward. Rohault's *Traité de physique* in a Latin edition by Samuel Clarke, decorated with Newton-inspired footnotes that controverted Rohault's Cartesian explanations, became a popular text. In London, James Hodgson began presenting courses of public lectures with experimental demonstrations in 1705, lecturing in association with the instrument-maker Francis Hauksbee (d. 1713), the Royal Society's "curator of experiments." For Hauksbee, as for many other instrument-makers who followed his example, the lectures were a form of advertising for the instruments he made, and he soon began offering his own, independent course of lectures. Other, rival courses proliferated and became a feature of polite society in Britain for more than a century. Hauksbee's own lectures were taken over following his death by the man who was to become the most successful lecturer of them all, John Theophilus Desaguliers (1683–1744), who also succeeded Hauksbee as curator of experiments at the Royal Society.[5]

In the Netherlands, experimental demonstrations were part of the physics course at Leiden from the 1670s.[6] In 1715, Willem 's Gravesande (1688–1742) visited London and attended various lectures on "experimental philosophy."

---

[4] R. W. Home, *Electricity and Experimental Physics in Eighteenth-Century Europe* (Aldershot: Variorum, 1992), chap. 7; Geoffrey V. Sutton, *Science for a Polite Society: Gender, Culture, and the Demonstration of Enlightenment* (Boulder, CO: Westview Press, 1995), chap. 6.

[5] Alan Q. Morton and Jane A. Wess, *Public and Private Science: The King George III Collection* (Oxford: Oxford University Press, 1993), pp. 41–50.

[6] Edward G. Ruestow, *Physics at Seventeenth and Eighteenth-Century Leiden: Philosophy and the New Science in the University* (The Hague: M. Nijhoff, 1973), pp. 96ff.

Two years later, he was appointed to a chair at Leiden and mounted his own course of lectures illustrated by experimental demonstrations. These were published under the title *Physices elementa mathematica experimentis confirmata, sive introductio ad philosophiam newtonianum* (1720–1). The work was a huge success, going through several Latin and English editions. 's Gravesande's student Pieter van Musschenbroek (1692–1761), who succeeded him in the chair at Leiden after many years at Utrecht, likewise produced an enormously successful textbook, numerous editions of steadily increasing size being published in various languages between 1726 and 1769. During the 1730s and 1740s, courses of experimental lectures also became common in Italy, Germany, and elsewhere.

The Dutch texts played a major part in redefining the scope of "physics" by omitting the botany, zoology, anatomy, and physiology that had traditionally been included. In France, they helped shape the thinking of Jean Antoine Nollet (1700–1770), who took over Polinière's role and for more than three decades was the dominant figure in French experimental physics, offering courses of public lectures with experimental demonstrations that attracted royal patronage and became a feature of Parisian social life.[7] Nollet's *Leçons de physique expérimentale,* published in six volumes between 1743 and 1748 and reprinted many times, confirmed the new, narrower definition of the field. In 1739 Nollet was elected, as Polinière never was, a member of the Paris Academy of Sciences, and in 1753 the king created a new chair for him in *physique expérimentale* at the Collège de Navarre. It became a model for similar chairs established in colleges and universities throughout Europe.

The courses of lectures, whoever delivered them and wherever they were delivered, quickly settled into a pattern established by the first generation of textbooks, in which various aspects of physics – the laws of motion, the principles of simple machines, statics, hydrostatics, pneumatics, heat, light, sound, magnetism, electricity, the system of the Sun and planets – were expounded by means of striking and ingenious demonstrations. Lecturers invested large sums in their apparatus. Models of machines and of the solar system were displayed and the principles on which they worked enunciated. Experiments with an air pump were always a highlight, as indeed they had been ever since the pioneering work of Robert Boyle (1627–1691) in the 1660s; they were soon joined by equally spectacular experiments using frictional electrical machines. Especially in larger centers such as London and Paris, there was considerable competition between lecturers, leading to constant efforts to keep courses up-to-date and to devise ever more striking demonstrations. Generally speaking, lecturers claimed only to be illustrating already established principles. If pressed, however, most would also have claimed that these principles could be deduced from experiments of the kind they presented. Some, including both

---

[7] Pierre Brunet, *Les physiciens hollandais et la méthode expérimentale en France au XVIIIe siècle* (Paris: Blanchard, 1926).

Polinière in his later years and Nollet, insisted that the principles of physics were (or should be) straightforward generalizations from experience: "The aim of experimental physics," Nollet wrote, "is to know the phenomena of nature, and to show the causes of these by proofs of fact."[8]

During the seventeenth century, older ways of explaining natural phenomena were displaced by the new "corpuscular" or "mechanical" philosophy, according to which all natural changes were to be understood in terms of the motions and impacts of particles, whether of the gross matter involved or of the subtle ether in which some scholars, and above all those who adopted Descartes's mode of explanation, supposed all ordinary matter was immersed. Following Newton's work, it became a matter of dispute as to whether motion and impact sufficed to account for all natural phenomena, as most seventeenth-century mechanists had supposed. Newton, in his *Principia mathematica philosophiae naturalis* (1687), constructed a science of motion in which changes were ascribed to the actions of forces that might or might not be caused by impacts; most famously, he had explained the motions of the heavens on the basis of a universal gravitational force acting between particles of matter, the cause of which he professed not to know but which, he concluded, was not mechanical. In his *Opticks* (1704) and especially in new material that he added to the 1706 edition of this, Newton suggested that many other phenomena might also find their explanation in terms of forces acting at a distance in some unexplained way between separated corpuscles. The proper method for science, he declared, was to proceed from an investigation of the phenomena of nature to a discovery of the forces that acted in the world, and only afterward to worry about how these forces might be caused.

Many who have written about eighteenth-century science have depicted it as a struggle for supremacy between Cartesian mechanisms and Newtonian ideas, with Newtonianism gradually triumphing, at first in Britain around the beginning of the century, then in Holland a decade or two later, and eventually in France (and, by implication, in the rest of Europe) some time in the late 1730s or early 1740s. For the remainder of the century, we are told, Newtonianism reigned supreme.[9] Initially constructed as an account of the triumph of Newtonian celestial mechanics, this schema was later expanded to include an experimental Newtonianism inspired by Newton's *Opticks*. Neither the magnitude of Newton's achievements, mathematical and experimental, nor their impact on those who came after him can be denied. Yet, as various historians have noted, defining a whole century of science on this basis creates

---

[8] Jean Antoine Nollet, *Oratio habita . . . cum primum Physicae Experimentalis cursum professor a Rege institutus auspicaretur* (Paris, 1753), p. 9.

[9] Pierre Brunet, *L'introduction des théories de Newton en France au XVIIIe siècle* (Paris: Blanchard, 1931); I. Bernard Cohen, *Franklin and Newton: An Inquiry into Speculative Newtonian Experimental Science and Franklin's Work in Electricity as an Example Thereof* (Philadelphia: American Philosophical Society, 1956); Robert E. Schofield, *Mechanism and Materialism: British Natural Philosophy in an Age of Reason* (Princeton, NJ: Princeton University Press, 1970).

many problems. It ignores the significance of other schools of thought such as, in particular, that deriving from Leibniz. It suggests a much more linear history of science than the original sources reveal. It implies that scientists throughout the eighteenth century continued to be preoccupied with the issues that concerned people in Newton's day, whereas in fact most of these issues had long since been either settled or agreed to lie beyond the bounds of scientific inquiry. Above all, it has proved impossible to characterize "Newtonianism" in a way that adequately embraces the variety of approaches adopted by eighteenth-century scientists to whom the term has been applied. Inevitably, later work was influenced by familiarity with Newton's extraordinary achievements. Yet instead of worrying about past disputes, eighteenth-century scientists looked forward to the resolution of new and quite different sets of scientific problems.

MECHANICS

Among Newton's achievements, his reconstruction of the science of motion reigns supreme. In Book I of the *Principia,* he presented a systematic analysis of the motions of point masses under the action of forces, focusing particularly on "the two principal cases of attractions" – namely, oscillatory motions brought about by forces varying directly as the distance between two bodies, and motions in conical orbits brought about by forces varying inversely as the square of the distance. In Book II, he discussed bodies moving in resisting media, culminating in a demonstration that bodies carried around in Cartesian-style vortexes would not obey Kepler's laws. Finally, in Book III he applied the propositions demonstrated in the earlier parts of the work to the heavenly motions. Having concluded that a universal gravitational force operated in the world, he proceeded on the basis of the theory of perturbations he had developed to account, quantitatively and in detail, for various observed deviations from perfect ellipticity in the lunar and planetary motions. He also applied his analysis to the motions of comets, to the shape of the Earth and the precession of the equinoxes, and to the theory of tides. Almost incidentally, he provided an explanation of Boyle's law based on the assumption that air was composed of mutually repelling particles, and a theory that accounted for the sine law of refraction of light by assuming that rays consisted of streams of corpuscles that, at the surface of a refracting body, were acted on by a force normal to the surface. He also formulated a pioneering analysis of the propagation of vibrations through a medium, thus making a major contribution to the theory of sound.

At the heart of Newton's discussion lay his notion of impressed force as that which caused changes in the motions of bodies. For both Descartes and Leibniz, on the other hand, "force" was what a body possessed by virtue of its motion, the power that it had of producing mechanical effects – something that was not, in fact, a well-formed concept in Newton's mechanics. Descartes had argued that this should be measured by the quantity of motion, mv, the product of the quantity of matter involved and its speed. He also maintained that the total quantity of motion in the universe was constant. Leibniz, however, had criticized Descartes, arguing that it was vis viva, the quantity mv^2, that was the proper measure of the force, and the quantity that was conserved in nature. In the 1720s, this dispute broke out again, with various of Newton's supporters, such as Samuel Clarke (1675–1729) and Colin Maclaurin (1698–1746), joining forces with Cartesians, such as Fontenelle and Jean-Jacques Dortous de Mairan (1678–1771), in opposing the Leibnizian position. The battle lines were not clear-cut, however, with both the Newtonian 's Gravesande and the Cartesian Johann Bernoulli (1667–1748) siding with Leibniz's cause. Some, most notably 's Gravesande, tried to resolve the question experimentally, whereas others, such as Jean d'Alembert (1717–1783) and Roger Boscovich (1711–1787), declared that it was a matter of words. Eventually, around mid-century, the dispute died out without ever being explicitly resolved. In retrospect, we can see that it involved a disagreement over the nature of matter – if the ultimate particles of which matter is composed are completely inelastic hard atoms, vis viva cannot be conserved – but also depended on the fact that, despite Newton's work, clear distinctions among and definitions of various mechanical quantities such as impulse, momentum, work, power, and, above all, force, had not yet been established.

Eighteenth-century mathematicians made major advances in clarifying these distinctions and establishing the principles of mechanics on a more definite basis. They also developed principles that could be brought to bear on a much wider range of problems than Newton had treated, including the motions of extended bodies and fluid media. Whereas d'Alembert, in his *Traité de dynamique* (1743), sought to banish the concept of force altogether, Leonhard Euler (1707–1783), who in his *Mechanica* (1736) rendered much of Newton's work on the dynamics of a particle into the new mathematical language of the calculus, sought to define the concept more precisely. In 1750 Euler succeeded, announcing, as "a new principle of mechanics" that applied to all mechanical systems, the relationship commonly known today (despite Newton's not having stated it in this form) as "Newton's second law of motion" or the principle of linear momentum, namely (for force P in the x-direction and mass M)

$$M \frac{d^2 x}{dt^2} = P$$

and similar equations for the other two spatial axes. Recognizing the full generality of the principle, Euler went on to derive what are now called "Euler's equations" for the motion of rigid bodies.

A second important principle of rational mechanics that was established and developed to full generality during this period was the principle of moment of momentum. Having its origins in work by Jakob Bernoulli (1654–1705), his nephew Daniel Bernoulli (1700–1782), and Euler, it was eventually proclaimed to be an independent axiom of mechanics by Euler in 1775.[10]

The concept of internal pressure in a fluid was another that was similarly clarified and generalized. Newton, in Book II of the *Principia,* developed a few propositions in hydrostatics but went no further; Johann Bernoulli, in his *Hydraulica* (1738), expressed the idea clearly in relation to fluids in tubes; d'Alembert in 1749 and then Euler much more elegantly in 1755 generalized it to a fluid occupying any part of space and, on this basis and using the principle of linear momentum, constructed a comprehensive theory of hydrodynamics.

Eighteenth-century mechanics built on a number of other principles as well. These included the so-called principle of virtual velocities for solving problems of equilibria; "d'Alembert's principle," which effectively reduced difficult dynamical problems to more familiar statical ones; and the seemingly teleological "principle of least action" announced in 1744 by Maupertuis (1698–1759) and subsequently the subject of violent polemics in the Berlin Academy of Sciences, of which Maupertuis was President. All these principles were brought together in the set of fundamental equations describing the motions of systems of bodies announced by Joseph-Louis Lagrange (1736–1813) in his grand work of synthesis, *Mécanique analytique,* published in 1788.

A striking feature of almost all these eighteenth-century developments in mechanics was their remoteness from experimental inquiry. Although physical questions usually provided the starting point for research, most of the work was driven by purely mathematical considerations, above all by the desire to construct a comprehensive system of truly *rational* mechanics.[11] In many cases, the simplifications that had to be introduced to render problems mathematically tractable left any solutions obtained completely inapplicable to the real-world situation: as Daniel Bernoulli complained of d'Alembert's research on the winds, "After one has read his paper, one knows no more about the winds than before."[12] The mathematical discussions of the shape of the Earth by Alexis Clairaut (1713–1765) and others in the 1730s and 1740s made effectively no contact with the measurements made by members of the Paris Academy during their famous expeditions to Lapland and Peru in these same years. The controversy that developed over the solution of the wave equation for a vibrating string, eventually involving d'Alembert, Euler, Daniel Bernoulli, Lagrange, and Pierre-Simon Laplace (1749–1827), almost immediately left the

---

[10] Clifford A. Truesdell, "Whence the Law of Moment of Momentum?" in Truesdell, *Essays in the History of Mechanics* (Berlin: Springer, 1968), pp. 239–71.

[11] H. J. M. Bos, "Mathematics and Rational Mechanics," in G. S. Rousseau and Roy Porter (eds.), *The Ferment of Knowledge: Studies in the Historiography of Eighteenth-Century Science* (Cambridge University Press, 1980), pp. 327–55.

[12] Quoted by Thomas L. Hankins, *Jean d'Alembert: Science and the Enlightenment* (Oxford: Clarendon Press, 1970), p. 3.

physical problem behind in favor of arguments over the nature of a mathematical function. Mechanics in the eighteenth century was a branch of mathematics, and, with few exceptions, the old separation between mathematics and physics remained firmly in place.

EXPERIMENTAL PHYSICS

Newton's *Opticks* was his second great bequest to the eighteenth century. The work was chiefly concerned with phenomena relating to color, and it took as its starting point Newton's celebrated analysis of white light into its constituent colored rays by refracting it through a prism. It also included an extensive account of Newton's pioneering investigations into the colors of thin films, thick plates, and natural bodies and, more briefly, into double refraction and "inflection" (that is, diffraction). At the end Newton included, in increasing numbers from one edition to the next, so-called Queries in which he set out, in the form of questions, many of his underlying theoretical assumptions.

Newton's experiments with prisms and, more particularly, the conclusions he drew from them caused a flurry of controversy when they were first published in the 1670s. In France his ideas were actually rejected for a generation because his experiments could not be replicated, but in the 1710s they came to be fully accepted. Thereafter, almost everyone agreed – Goethe being a notable exception[13] – that white light was a mixture of permanently existing colored rays of differing refrangibilities.

Newton's research on periodic phenomena associated with light began with his investigation of the colors of thin films ("Newton's rings") in the 1660s. Drawing a bold analogy, Newton concluded that the colors of natural bodies arise from the corpuscles of which they are composed interacting with light in exactly the same way as thin films do. Which rays would be reflected and which absorbed would depend solely on the sizes and densities of the corpuscles involved, and not at all on their chemical composition; indeed, the theory seemed to provide a way of determining the sizes of the corpuscles of which various bodies are composed from the colors the bodies display. In most cases, colors would, if analyzed with a prism, prove to be compound as a result of bands of different order overlapping; Newton's measurements on thin films, far in advance of anything achieved by anyone else for another hundred years, gave him precise quantitative control over this phenomenon. Underpinning the whole scheme was Newton's conception that light consists of streams of rapidly moving corpuscles. From the outset, Newton assumed that the phenomena with which he was dealing resulted from light corpuscles exciting

---

[13] Dennis Sepper, *Goethe contra Newton: Polemics and the Project for a New Science of Color* (Cambridge University Press, 1988).

vibrations in a subtle matter or ether that occupied the spaces between the corpuscles of bodies. However, when he came to compose his *Opticks,* he replaced
all references to vibrations by his phenomenologically based but mysterious
concept of fits of easy transmission and easy reflection of light. His thus
stripping his account of the physical ideas that had underpinned the research
rendered it much less comprehensible than it might have been. For much of
the eighteenth century, very little further work was done on the subject: people
seem simply to have stood in awe of Newton's achievement.[14]

Geometrical optics based on the rectilinear propagation of light rays and
the law of reflection originated in the ancient world. Following the announcement by Descartes in 1637 of the sine law of refraction, a mathematical treatment of refracted rays and hence of lens systems also became possible. From
a physical point of view, however, the perfection of lens systems was limited
by chromatic dispersion. Newton from his experiments concluded that any
deviation of light was accompanied by a dispersion and hence that no lens
system was possible that was free of chromatic aberration. Euler, however,
argued in 1747 that this could not be correct since the operation of the human
eye proved otherwise, and he set out to design a compound lens that would
be achromatic. Subsequent developments cast an interesting light on the respective predilections of eighteenth-century mathematicians and experimentalists. Euler's argument was opposed by the London optician John Dollond
(1706–1761) on the basis of Newton's reported results. Later, however, Dollond
did his own experiments, found that Newton had made significant errors,
and succeeded in making compound lenses that were acceptably achromatic
and that quickly found a ready market among astronomers. Far from being
pleased by this, however, Euler rejected Dollond's claims, maintaining that
any improvements he had made could not be due to his lenses being achromatic. The problem was that Dollond's measurements seemed to show that
there was no relationship between the refractive indexes of different colored
rays in one medium and those in another, and that the refractive indexes had
to be measured individually in each case; whereas Euler, in a manner characteristic of eighteenth-century mathematical physicists, maintained that there
had to be a mathematical law covering the situation.[15]

Most accounts of eighteenth-century optics, while noting occasional instances of experimental work such as the photometric investigations of Pierre
Bouger (1729) and Johann Heinrich Lambert (1760), concentrate on discussions of the nature of light. Newton, as indicated, adhered to a corpuscular or
projectile theory, and in some of the Queries in his *Opticks* he presented arguments in favor of this theory and against the theory set out by Christiaan

---

[14] Alan E. Shapiro, *Fits, Passions, and Paroxysms: Physics, Method, and Chemistry and Newton's Theory
of Colored Bodies and Fits of Easy Reflection* (Cambridge University Press, 1993).
[15] Keith Hutchison, "Idiosyncrasy, Achromatic Lenses, and Early Romanticism," *Centaurus,* 34 (1991),
125–71.

Huygens (1629–1695), according to which light consisted of impulses transmitted through an ether. Newton's conclusion was widely accepted in France and Britain during the eighteenth century but much less so east of the Rhine, where Euler's adherence to the transmission theory won many followers. The predominance of the projectile theory has often been ascribed to the weight of Newton's authority rather than to the theory's scientific merits. Yet Newton's arguments were, in the context of the time, very strong. If light were "a pression or motion" propagated through a medium, he said, shadows would not form: "Pression or Motion cannot be propagated in a Fluid in right Lines, beyond an Obstacle which stops part of the Motion, but will bend and spread every way into the quiescent Medium which lies beyond the Obstacle." Double refraction in Iceland crystal was also a problem: Huygens, despite his elegant geometrical account of the unusually refracted ray, had been unable to explain what happened when he tried the refraction in two successive pieces of crystal, whereas Newton's notion that the light corpuscles might have an in-built polarity provided at least a basis for an explanation. The differently colored rays of differing refrangibility could, Newton indicated, be accounted for on the assumption that "the Rays of Light be Bodies of different Sizes"; he was, however, dissimulating here, because in the explanation of refraction outlined in the *Principia,* the sizes of the corpuscles should not affect the amount by which they were refracted. Yet on the other side, Huygens had confessed himself unable to offer any explanation at all for color, the reason being that his was a theory of transmission of impulses and not of regularly repeating waves. A few years later, Nicolas Malebranche (1638–1715) proposed that what was propagated were waves and suggested by analogy with sound that the different colors of light arose from pressure vibrations of different frequencies. Euler, in proposing that light consisted of displacement waves in an elastic ether, followed a similar line. In the absence, however, of a principle of interference – this was first proposed by Thomas Young (1773–1829) in 1800 – Euler was unable to offer any explanation of the fits of easy transmission and easy reflection of light.

Euler made a valiant effort to rebut Newton's claim that, on any transmission theory, light would bend into the shadow. Euler's argument was illicit, for it depended on the idea that, whereas waves were propagated in all directions from a source, individual pulses progressed linearly in the directions in which they were first emitted. That this was incorrect was not readily apparent at the time, however, for deep conceptual difficulties still confronted those (including Euler) who were trying to construct an adequate theory of wave motions in elastic media. These difficulties were not satisfactorily resolved until the work of Augustin Fresnel (1788–1827) in 1816–19, in which he employed the principle of interference to construct a complete mathematical theory of diffraction that also overcame Newton's objection regarding the formation of shadows. Until then, the status of Euler's argument vis-à-vis Newton's was uncertain. Meanwhile, other arguments that Euler put forward seemed to have considerable weight. If light rays consist of streams of material corpuscles, he said, either the mass of an emitting body should in time be detectably diminished, or the density of the rays must be improbably small – an argument that he backed up with some convincing figures regarding the rate of emission of light from the Sun and the effect this should have on its mass. Euler also asked how two rays could intersect at such incredible speed without disturbing each other's motion and what structure transparent bodies could possibly have that permitted light to pass freely through them in all directions.

Although there were attempts to resolve such arguments experimentally,[16] the disagreement depended mostly on conceptual issues. As noted, although majority eighteenth-century opinion sided with Newton, Euler's views also attracted a following. In the 1780s, however, support for the wave theory evaporated in the face of convincing new experimental evidence that light was material. These experiments came from the realm of chemistry. They showed that light was an essential ingredient in certain chemical reactions – in particular, photosynthesis and the blackening of silver salts – and led Antoine-Laurent Lavoisier (1743–1794) to include *lumière* in the famous table of simple substances or elements set out in his *Traité élémentaire de chimie* in 1789. Only with the work of Fresnel did the wave theory again come into its own.[17]

Throughout his life, Newton, like most other seventeenth-century scientists, believed that ordinary matter was suffused with a subtle matter or matters that were instrumental in bringing about many observed effects. In a section added to the second (1713) edition of his *Principia,* Newton alluded to this "spirit" as the cause of the cohesion of bodies; the attracting and repelling power of electricity; the emission, reflection, refraction, inflection, and heating effect of light; and the functioning of the nerves. He also drafted new Queries describing this substance for the next (1717) edition of the *Opticks,* but these were never published. He did, however, include other Queries suggesting the existence of a universal ether that took over some (but not all) of the explanatory functions previously attributed to the subtle matter and that, extending throughout the universe, might also explain gravity.[18] Being the cause of gravity, it was itself imponderable. Few of Newton's readers distinguished between the subtle matter and this ether: they simply drew from his work the idea that, in addition to ordinary perceptible matter, the world includes an all-pervading invisible subtle matter.

Adaptations of this idea became a feature of eighteenth-century physics. Newton's writings were not the only source, however, on which people drew.

---

[16] Cf. Casper Hakfoort, "Nicolas Béguelin and His Search for a Crucial Experiment on the Nature of Light (1772)," *Annals of Science,* 39 (1982), 297–310.

[17] Casper Hakfoort, *Optics in the Age of Euler: Conceptions of the Nature of Light, 1700–1795* (Cambridge University Press, 1995); Geoffrey Cantor, *Optics after Newton: Theories of Light in Britain and Ireland, 1704–1840* (Manchester: Manchester University Press, 1983).

[18] Home, *Electricity,* chap. 2.

Descartes had also invoked a universal ether, motions within which underpinned all the explanations he offered for natural phenomena. During the first decades of the eighteenth century, there were various attempts to improve Descartes's theory of celestial vortexes.[19] At a more mundane level, his explanation of magnetism in terms of smaller vortexes of a special kind of subtle matter, passing axially through a magnet and then returning through the external air, was adopted even by Newton and continued to be widely accepted until at least the late 1750s.[20]

Equally influential was the work of the Dutchman Herman Boerhaave (1668–1738), who in his *Elementa chemiae* (1724) argued that the world is filled with an all-pervading, material, elastic Fire, the presence and activity of which was revealed by the expansion of bodies – including the mercury in the thermometers developed by Daniel Fahrenheit (1686–1736) that detected and measured it. Opposing this expansive power, Boerhaave said, is a contractile power inherent to ordinary matter, which causes bodies to contract when cooled; bodies at a fixed temperature maintain a constant volume because the two powers are in equilibrium.

In Boerhaave's account, it was unclear whether it was the activity or the mere presence of Fire that caused the expansion recorded by a thermometer. Others denied the materiality of fire altogether, preferring the idea, deriving from Newton and ultimately from Francis Bacon, that fire and the heat accompanying it were merely manifestations of an increased motion among particles of ordinary matter. The materialist view gained support from the work of Joseph Black (1728–1799) in the early 1760s on changes of state from solid to liquid and liquid to vapor. Black showed that a determinate amount of heat disappeared in converting ice to water, or water to steam. It appeared, however, that this heat could be retrieved by reversing the process. Black concluded that the heat had not been destroyed but had merely become undetectable by his thermometer; it had become "latent" heat. By observing temperature changes in mixtures, Black also arrived at the notion that different substances had different capacities for heat, that is, that a given quantity of heat would increase the temperature of different substances by different amounts.[21]

Black's work on heat influenced James Watt (1736–1819) as Watt developed his greatly improved steam engine in the 1760s. It also coincided with an explosion of interest among chemists (to which Black also contributed) in gases. Inspired by Black's ideas, Lavoisier, who with Laplace carried out a famous series of calorimetric experiments in the early 1780s, developed the notion that vaporization was a chemical combination between the substance being vaporized and the matter of heat, or "caloric" as Lavoisier called it. The amount

[19] E. J. Aiton, *The Vortex Theory of Planetary Motions* (London: Macdonald, 1972).

[20] Home, *Electricity*, chap. 4.

[21] Douglas McKie and Niels H. de V. Heathcote, *The Discovery of Specific and Latent Heats* (London: Edward Arnold, 1935).

of caloric required – the latent heat – depended on the chemical affinity between the particular substance concerned and caloric. By extension, it was assumed that any gas was a compound of some substance with caloric and that when a gas took part in a chemical reaction, caloric would be set free. In particular, when, in accordance with Lavoisier's new theory of combustion, oxygen combined with some combustible substance, the caloric present in the oxygen gas would be released. Just as Lavoisier used the balance to keep track of the different ponderable species involved in a chemical reaction, so he sought with the calorimeter to keep track of the caloric.[22]

The notion of a subtle matter, whether Fire or something else, associated with ordinary matter, likewise provided the early context in which the phenomena of frictional electricity were understood. Confined at first to the attraction, known since ancient times, that rubbed amber or glass exerted on nearby light objects, "electricity" came during the eighteenth century to refer instead to a cluster of surprising and remarkable effects. Those writing on the subject in the late years of the seventeenth century were unanimous that the electrical attraction was due to the agitation, brought about by the rubbing, of a subtle matter located in the pores within ordinary bodies, an agitation that caused this matter to be ejected into the surrounding space. Since what was being explained involved an inward rather than an outward motion, however, it was clear that something else must be involved as well. Opinion differed on what this was and also on the nature of the subtle matter.

In the first years of the eighteenth century, Francis Hauksbee showed, in spectacular experiments with a rubbed glass globe mounted on a spindle, that electrification was linked to the emission of light – whether in the form of glow discharges inside an evacuated globe or of sparks drawn from one filled with air. He also showed that the traditional electrical attraction was normally followed by a repulsion and that glass was apparently transparent to the electrical action since a rubbed piece of glass or sulphur brought near a glass globe affected threads hanging inside it; and he devised experiments that seemed to demonstrate the direction of flow of the subtle matter in the space surrounding an electrified body. Indeed, in Hauksbee's experiments it seemed that the electrical effluvia could be felt, seen, and heard.

The story became more complicated when Stephen Gray (1666–1736) in 1731 announced his discovery that the attracting power of electricity could be transmitted over great distances, provided that the conducting line was made of an appropriate material and was suitably supported. His experiments, repeated and systematized by Charles-François Dufay (1698–1739), led to a distinction being drawn between "electrics," which could be electrified by friction but would not transmit the electrical action, and "non-electrics," which could not be electrified by friction but would transmit the effect. Dufay also

[22] Henry Guerlac, "Chemistry as a Branch of Physics: Laplace's Collaboration with Lavoisier," *Historical Studies in the Physical Sciences*, 7 (1976), 193–276.

announced a further distinction: one between the electricity acquired by substances such as glass and that acquired by resinous substances. Pieces of paper attracted to rubbed glass and repelled after making contact would subsequently also be repelled by other pieces of rubbed glass but would be attracted by pieces of rubbed resin, *e contra*.

Meanwhile, Hauksbee's rotating-globe electrical machine and the experiments he had performed with it became a popular feature of the rising tide of public lectures on experimental physics. Strengthened and improved, the machine became in the early 1740s a source of increasingly powerful effects. The sparks that were obtained were used to ignite alcohol and other inflammable substances; sparks were even drawn from a block of ice! Many people, including Nollet, who had taken over Dufay's mantle as France's leading "electrician," were convinced by this that the subtle matter involved was either Boerhaave's all-pervading fluid Fire or something very similar to it. Nollet developed an elaborate account that attracted widespread support. He envisaged streams of agitated fiery matter streaming out from electrified bodies while other streams moved inward to replace that which had left. Bodies were attracted or repelled depending on whether they were located where the inward or outward moving stream was stronger; when the streams were sufficiently concentrated, the particles composing them would collide head-on, splitting open their surrounding envelopes of sulphureous matter and releasing the active fiery matter within. In accordance with his general approach to theory construction, Nollet insisted that the basic premises of this theory, far from being hypothetical, were straightforward matters of fact.

In 1746, Musschenbroek announced to the world the discovery of "the Leyden experiment," in which a bottle filled with water and then electrified delivered a terrible shock if contact were made simultaneously with the water and the outer, conducting surface of the bottle. The experiment caused a sensation, with enthusiasts everywhere rushing to confirm the report for themselves. So spectacular was the effect that accounting for it at once became the primary objective of any theory of electricity. And unfortunately for Nollet, as the nature of the phenomenon was clarified, it also gradually became clear that he was unable to provide a coherent explanation. In the meantime, Benjamin Franklin (1706–1790) offered an alternative account that dealt comparatively successfully with the Leyden experiment while at first ignoring traditional concerns about explaining the attractions and repulsions. There were fierce debates between the two opposing camps throughout the 1750s and lingering support for Nollet thereafter; but in time ideas based on Franklin's became generally accepted.

Franklin's theory also began with the notion that ordinary matter was suffused with an elastic subtle matter. He did not, however, identify it with either Boerhaave's Fire or the Newtonian ether but rather assumed it to be a specifically electric fluid. He supposed that any sample of ordinary matter contained a quantity of electric fluid natural to it and that the process of electrification amounted to a transfer of subtle matter between rubber and thing rubbed, so that one finished with more than its natural quantity and the other with less. In Franklin's terminology, one was electrified "plus" and the other "minus." Whereas in Nollet's theory, degree of electrification was the fundamental quantity, Franklin's yielded the notion of a body being "charged" either positively or negatively. These two possibilities he identified with Dufay's vitreous and resinous electricity, respectively. Noting the ability of pointed conductors to discharge nearby charged objects, Franklin conceived an experiment, first successfully performed near Paris in May 1752, to demonstrate that thunderclouds were electrified and that lightning was an electrical discharge. His conclusion that erecting pointed conductors ("lightning rods") on buildings could protect them from lightning strikes was hailed as a triumph of reason over nature.

Franklin was less successful in providing a coherent dynamical basis for his theory or in explaining what came to be known as "electrostatic induction." In 1759, however, Franz Aepinus (1724–1802) published a fully consistent version of Franklin's theory. He abandoned Franklin's notion that charged bodies were surrounded by "atmospheres" of electric fluid, the interactions of which with other atmospheres or with ordinary bodies gave rise to the observed electrical motions, in favor of a fully fledged action-at-a-distance account. To render the theory consistent, however, Aepinus assumed not only that the particles of electric fluid repelled each other while being attracted by particles of ordinary matter, as Franklin had done, but also that the particles of ordinary matter mutually repelled one another. Many people rejected this notion as being in conflict with Newtonian gravitation – something Aepinus himself denied – and supposed instead that there were actually *two* electric fluids that normally neutralized each other but that became separated in the process of electrification; on this theory, the additional repulsive force that Aepinus invoked was attributed to the second electric fluid rather than to ordinary matter. Operationally, however, there was no way to distinguish between the two-fluid theory and the one-fluid alternative.[23]

In the same work Aepinus constructed a theory of magnetism analogous to his modified version of Franklin's theory of electricity. He supposed the existence of a subtle magnetic fluid, the redistribution of which gave rise to positive and negative magnetic "charges" – which Aepinus identified with north and south magnetic poles – and thus to the various phenomena associated with magnets. The theory worked well and fairly quickly displaced the traditional magnetic-vortex theory deriving from Descartes. However, in this case, too, for the same reason as in the electric case, many of those who ac-

[23] *Aepinus's Essay on the Theory of Electricity and Magnetism*, ed. R. W. Home, trans. P. J. Connor (Princeton, NJ: Princeton University Press, 1979); J. L. Heilbron, *Electricity in the 17th and 18th Centuries: A Study of Early Modern Physics* (Berkeley: University of California Press, 1979).

cepted the basic ideas behind the theory supposed that there were two magnetic fluids and not one.

TOWARD A QUANTIFIED PHYSICS

Physics in the eighteenth century was not the mathematized science we know today. Although mechanics became almost entirely mathematical, the mathematics was often remote from any real-world situation. Meanwhile, most experimental investigations were wholly qualitative, and the theories developed to account for the experimental observations were often expressed too vaguely to sustain a reformulation in mathematical terms. There were also social barriers to be overcome, since mathematicians and experimentalists mostly constituted separate communities that had little to do with each other. After about 1760, however, more experimentalists undertook quantitative work. Conversely, around the end of the century, some of the powerful analytical tools developed by the mathematicians began to be applied successfully to problems arising from experimental physics. The social barriers between the two groups also began dissolving, especially in France, where the newly founded École Polytechnique provided an institutional base for the first generation of mathematical physicists when it emerged in the early 1800s.

Transforming a qualitative experimental subject into a fully mathematized one was by no means straightforward. Many experimental investigators actively opposed the use of mathematics in physics. Nollet regarded imprecision as inseparable from experimental physics. "It is dangerous," he said, "for a Physicist to develop too great a taste for Geometry," while conversely he stressed the need for "geometrical exactitude not to be disdainful of lowering itself to the *à peu près.*" Such attitudes persisted. Even in the early decades of the next century, many German physics professors still discounted mathematical description on the ground that mathematics tended to lead physicists away from their proper field of study.[24]

Other experimentalists were convinced by Newton's example of the importance of discovering quantitative empirical laws but found the task beyond them. Musschenbroek, for example, was familiar with Newton's achievements regarding the law of gravity, and he devoted much effort in the 1720s to determining the law governing the force acting between two magnets. Eventually, however, Musschenbroek was forced to admit defeat: "I can only conclude," he wrote, "that there is no proportion between forces and distances." In hindsight, we can see that he was measuring the wrong thing – namely, the force between one entire (spherical) magnet and another, rather than, as Charles-

---

[24] Kenneth L. Caneva, "From Galvanism to Electrodynamics: The Transformation of German Physics and Its Social Context," *Historical Studies in the Physical Sciences,* 9 (1978), 63–159.

Augustin Coulomb (1736–1806) did in the 1780s, that between two effectively isolated magnetic poles; but Musschenbroek had no suitable theory to guide him.

The expanding trade in scientific instruments played a significant role in encouraging the rise of experimental physics in general and also in facilitating quantitative experimentation. Driven by the demands of astronomy, navigation, and geodesy, the London instrument trade in particular achieved new levels of precision in the construction of linear and angular scales and of the instruments to which they were attached. New techniques were linked to and often lay behind advances in physics, and they gave the subject an increasingly practical and utilitarian cast. Dollond's work on achromatic lenses has already been mentioned. As another example, in the 1740s, Gowin Knight (1713–1772) devised a method of making artificial magnets that were stronger and more permanent than naturally occurring ones. These dramatically improved the reliability of mariners' compasses, and Knight's achievement also stimulated new magnetic research. Much improved declinometers and dip circles came on the market, and enthusiasts began systematically recording the daily variations in the Earth's magnetic field. Improved barometers and thermometers found immediate application in meteorological recording. We have already noted the link between the development of reliable thermometers and advances in the science of heat – advances that themselves clarified the distinction between temperature, which thermometers measured, and quantity of heat. The development of calorimetry required that attention be paid to heat losses from the apparatus, leading to the invention of the ice calorimeter by Lavoisier and Laplace. In the case of electricity, the first attempts at quantitative work using an electroscope date from the 1740s, but the quantity people wished to measure, "degree of electrification," was not well connected theoretically to the measurements taken. Later in the century, with the concept of "charge" established, greatly improved electrometers were devised; thanks to the incorporation of the "condenser" invented by Alessandro Volta (1745–1827), they were capable of detecting minute amounts of electrification. These instruments played a key role in Volta's research leading to his discovery in 1800 of the electric "pile" as a source of continuous electric current. It long remained unclear, however, what they measured: charge or (by analogy with air) the pressure or tension in the electric fluid. Volta decided that they measured tension and, recognizing that different bodies had different "capacities" for electricity, guessed that the quantity of electricity on a charged body depended on the linear relationship

$$\text{quantity} = \text{capacity} \times \text{tension}$$

More generally, how to establish a mathematical relationship between experimental variables remained most unclear. The increasing precision allowed by the new instruments highlighted the variations between measurements, but few eighteenth-century physicists concerned themselves with how to handle these. The idea of "experimental error," the graphing of data, and the application of statistical procedures were developments of the nineteenth rather than the eighteenth century. Meanwhile, eighteenth-century workers in reporting quantitative data often presented long strings of digits uncritically when in fact these were merely products of their numerical computations, or they announced general conclusions on the basis of astonishingly small bodies of evidence. Coulomb, for example, in his famous 1785 determination of the law of force between electrically charged bodies, presented only three sets of experimental data, one of which did not even fit his proposed inverse-square law very well!

Other problems, too, confronted the would-be mathematizer of physical theory. Had anyone tried, it would have been impossible to translate the circulating-vortex theory of magnetism that was widely accepted during the first half of the eighteenth century into a mathematical theory, since the then available mathematical hydrodynamics would not have been up to the task. Similarly, Euler's attempts to develop a mathematically expressed wave theory of light were severely constrained by the undeveloped state of the theory of waves in fluid media. Often, a qualitatively expressed theory was not in a fit state to be mathematized. Franklin's theory of electricity, for instance, despite its success in rendering the Leyden experiment comprehensible, was insufficiently coherent to sustain mathematization. Only when its basic principles had been "cleaned up" and rendered mutually consistent by Aepinus, sometimes in ways far removed from Franklin's own conceptions, did it become a candidate for mathematical treatment. Yet even then, because Aepinus could not prove that the law of force between charges was inverse-square in form, he failed to advance beyond a semimathematical formulation of the theory. Henry Cavendish (1731–1810) in 1771 took the process somewhat further, but the development of a fully mathematical theory had to await Coulomb's work and, more particularly, that of Siméon-Denis Poisson (1781–1840) in the early years of the following century.

There was thus a set of interlocking problems – experimental, conceptual, and mathematical – that lay in the way of a mathematically expressed but experimentally based physics. Few eighteenth-century scientists were equipped to tackle these problems. Prevailing social patterns meant that few experimenters had more than the most rudimentary mathematical understanding, and few mathematicians had a sense of the constraints within which they had to work if their analyses were to become a part of physics. The few individuals who did bridge the gap – Aepinus, for example – found it difficult to find an audience for their work, with experimentalists finding the mathematics beyond them and the mathematicians finding it so elementary as to be devoid of interest. In Aepinus's case, this led to his work not having its full impact until a generation after it was published. Even at the end of the century, although both rational mechanics and experimental physics had established generally accepted norms of practice, they remained largely separate activities. Physics as a science in which these two realms of practice were successfully integrated was a nineteenth-century invention.

16

CHEMISTRY

Jan Golinski

Writing in 1855, of the period now known as the Enlightenment, the Scottish Whig Henry Brougham commented that "the science of chemistry [was] almost entirely . . . the growth of this remarkable era." One hundred years later, the British historian Herbert Butterfield, renowned for his critique of the Whig interpretation of history, issued a much more negative judgment of the chemistry of the Enlightenment. In his *Origins of Modern Science* (1949), Butterfield notoriously relegated eighteenth-century chemistry to a kind of limbo, where it was awaiting its "postponed scientific revolution," which arrived only in the last two decades of the century with the work of Antoine-Laurent Lavoisier (1743–1794). Enlightenment chemistry had been "immature," hindered by philosophical confusions and the absence of an adequate intellectual framework. The difference of opinion between Brougham and Butterfield has an intriguing connection with their divergent political outlooks. Whereas the Whig writer saw a lengthy period of gradual progress, culminating in Lavoisier's individual accomplishments, the anti-Whig historian saw the French chemist as the first person with true insight into the fundamental ideas of the science, a beacon in an otherwise dark landscape of confusion and error.[1]

The perspectives of Whiggism and anti-Whiggism have continued to dominate much of the historical writing on the sciences of the eighteenth century, not least chemistry. Whiggish historians have looked to catalog specific and permanent factual discoveries – steadily accumulating positive knowledge – such as findings of new gases, mineral species, and salts. Butterfield's anti-Whiggism reflected the approach of Alexandre Koyré and, before him, the tradition of philosophical history derived from Immanuel Kant, which searched for organizing intellectual schemes, worldviews, *Weltanschauungen*, or

[1] Henry Brougham, *Lives of the Philosophers of the Time of George III* (*Works of Henry Lord Brougham*, 1) (Edinburgh: Adam & Charles Black, 1872), p. xxi; Herbert Butterfield, *The Origins of Modern Science* (London: Bell, 1949), p. 191.

paradigms. These have rarely been found before Lavoisier, whose accomplishment has usually been seen as the provision of a previously absent theoretical framework for chemistry. In the twentieth century, various formulations have been given of the essence of Lavoisier's theoretical achievement, emphasizing different aspects of his work but frequently reiterating his self-representation as one who had broken decisively with previous chemical tradition. Some scholars have identified the key step as a new theory of combustion, in which atmospheric oxygen was accorded its true role and the fictional principle of combustion, "phlogiston," was discarded. Alternatively, Lavoisier has been hailed for his novel understanding of chemical composition and his pragmatic definition of an element as the product of the best available methods of analysis. Or again, it is the recognition of the gaseous state of matter that has been taken to be the decisive innovation – the realization that substances could be made into gases by the addition of heat without changing their chemical nature. Or, finally, it has been claimed that the crucial development was the insistence on the conservation of matter as it undergoes chemical change, a law discovered by use of the precision balance and represented formally by chemical equations.

All these readings of Lavoisier's accomplishment stress its theoretical character as the laying of an intellectual foundation for subsequent chemical science. Although recent historians have been careful to distance themselves from Butterfield's pronouncement, assertions that Lavoisier's work represented the revolutionary first steps in the constitution of chemistry as a true "science" are sometimes still made. The "origin myth" created by Lavoisier himself and his immediate disciples has long retained its hold, with the consequence that the work of his predecessors is relegated to the shadows.[2] The anti-Whiggish outlook that identifies Lavoisier's achievement as a conceptual revolution has made it difficult to discern the lineaments of the chemistry that came before him.

In the essay that follows, I try to avoid the extremes of Whiggism and anti-Whiggism. I shall not present the chemistry of the Enlightenment simply as a process of accumulating factual data. We shall see that chemical discoveries were not always neutral facts but could rather be tokens in disputed theories – crucial to some interpretive schemes and ignored by others. But we shall also see that chemistry existed as a discipline well before Lavoisier gave it the theoretical framework familiar to modern eyes. Chemistry in the eighteenth century was a body of practical techniques, instruments, and materials, which were organized in written texts and oral lectures. Students were taught ways of assimilating new information and were given a clear sense of the history of the subject. Chemistry, however, was a discipline without entirely rigid boundaries; it engaged in productive exchanges of concepts and experimental phe-

---

[2] Bernadette Bensaude-Vincent, *Lavoisier: Mémoires d'une révolution* (Paris: Flammarion, 1993), especially pp. 363–92.

nomena with neighboring sciences, especially natural philosophy and natural history. Chemists applied various theoretical schemes to the interpretation of such phenomena, sometimes disagreeing about whether they were properly the business of chemistry at all. This diversity assumed critical importance as novel experimental discoveries were produced and their implications for chemical theory explored.

I shall therefore begin with an exploration of the identity of chemistry in the eighteenth century, mentioning how the subject was organized and taught, the social locations in which it was undertaken, and the material, instrumental, and discursive resources that were used to pursue it. Against this background, we can assess the importance of the philosophy of matter, whether mechanistic (invoking specific corpuscular shapes and motions) or Newtonian (invoking specific forces of attraction between particles). These theories of matter can be compared with other available means of conceptually organizing chemical information, such as the schemes for classifying the composition of salts and the popular tables of "affinities" between substances. A survey of chemists' ways of organizing the properties of the substances they encountered prepares us for the arrival on the scene of novel and anomalous chemical entities around the middle of the eighteenth century. Arguments about how to understand the new gases and "imponderables" (heat, light, and phlogiston) will be seen to have led directly to Lavoisier's self-proclaimed "revolution." His achievement will thus appear, not as the goal of a teleological progress toward modern science but rather as a brilliantly creative response to the need to fulfill chemists' task of organizing information about substances, their properties and behavior, in the face of unsettling new phenomena. The new system was communicated by Lavoisier and his allies by reworking the traditions of teaching and laboratory practice bequeathed to them by their chemical predecessors. New instruments, new experimental methods, new textbooks, and a new language were the tools of the new chemistry, which, notwithstanding its revolutionary rhetoric, preserved more than a few signs of its historical inheritance.

DISCIPLINE AND ENLIGHTENMENT

Lavoisier's revolution unfolded against the backdrop of a lengthy *ancien régime* of chemical practice. The beginnings of chemistry as an organized discipline, in textbooks and lectures, have been traced to the end of the sixteenth century. In terms of the overall organization of the contents of the discipline, there was a substantial degree of continuity from this period to the late eighteenth century. In its instrumental resources, chemistry, it has also been said, experienced a *longue durée* of relative stability. Basic laboratory equipment – glassware, crucibles, and furnaces – remained largely unchanged for at least a hundred and fifty years before the middle of the eighteenth century.[3] The advent of novel experimental phenomena concerning heat and gases, and moves toward more precise measurements of chemical quantities, introduced significant changes in laboratory practice in the late eighteenth century. Only then did the *ancien régime* of chemical practice begin to break down.

Owen Hannaway has persuasively argued that the modern tradition of the chemical textbook was launched by the *Alchemia* (Frankfurt, 1597) of Andreas Libavius (c.1540–1616). Libavius, a Lutheran schoolmaster, took the ancient textual form of a collection of recipes for chemical preparations and systematically organized them under the headings of the operations involved. He adopted the methods of the Humanist pedagogues: defining the subject matter of the discipline, dividing the definition and defining each part in turn, and so on. Presenting the whole subject in a series of dichotomies, represented in the form of branching tree diagrams, Libavius asserted the autonomy of chemistry as a discipline and its primacy over various practical arts. This kind of pedagogical exposition of chemistry was a way of striking against what Libavius saw as the obscurantism and impious mysticism of the late sixteenth-century followers of the Swiss alchemist and physician Paracelsus.[4]

A proclaimed adherence to systematic method and a declared abhorrence of what was seen as the willful obscurity of alchemical writings remained prevalent aspects of chemical textbooks in the eighteenth century. Other formal features of these texts were also derived in principle from Libavius, albeit subjected to some degree of reorganization as time went on. It was routine to introduce details of preparative procedures with a discussion of the apparatus of the typical laboratory. Chemical operations, such as distillation, sublimation, filtration, and dissolution, would be listed and categorized. The practice of beginning with a definition of chemistry, which would then be unpacked in the course of the exposition, persisted, for example, in the *Elementa Chemiae* (1732) of Herman Boerhaave (1668–1738). One feature that some eighteenth-century teachers added to the traditional outline of the chemical text was an introductory review of the history of the subject. Boerhaave, at Leiden, and William Cullen (1710–1790), at Glasgow and Edinburgh, were among those who did this. The historical introduction, which owed something to other Enlightenment exercises in conjectural history, consolidated the message of the integrity and continuity of the discipline.[5]

In pursuing their careers in universities, Boerhaave and Cullen took advantage of one of the important institutional niches gained by chemistry in the course of its *ancien régime*. German universities took the lead in establish-

[3] Frederic Lawrence Holmes, *Eighteenth-Century Chemistry as an Investigative Enterprise* (Berkeley: Office for History of Science and Technology, University of California, 1989), pp. 17–20.

[4] Owen Hannaway, *The Chemists and the Word: The Didactic Origins of Chemistry* (Baltimore, MD: Johns Hopkins University Press, 1985).

[5] John Christie, "Historiography of Chemistry in the 18th Century: Herman Boerhaave and William Cullen," *Ambix*, 41 (1994), 4–19.

ing positions in chemistry in the early seventeenth century. However, the academic profile of the subject remained dependent on the demand for medical education. Teachers were frequently unsalaried or underpaid, and they relied on collecting fees from medical students, physicians, apothecaries, and anyone else who was interested. While chemistry flourished at Leiden and Edinburgh, because of the highly successful medical schools at those universities, there were many decades when it languished at Oxford and Cambridge.

Universities were not by any means the only places where chemists found employment for their skills in the eighteenth century. In Germany and Scandinavia, there were openings at mining and administrative academies. In Paris, the Académie Royale des Sciences nurtured a distinguished tradition of chemical research by its salaried academicians throughout the century. Lecturers were also appointed at the Jardin du Roi and elsewhere in the French capital to give courses to members of the public: Diderot and Rousseau were among those who attended lectures at the Jardin. In the second half of the century, the foundation of provincial academies and local learned societies in many European countries provided further opportunities for chemists to lecture and pursue their research. In England, the public scientific lecturer (who might also be an author) became a feature of the expanding commercial market for education and leisure. Peter Shaw (1694–1763) was the first chemist to explore this kind of occupation in the early 1730s; by the 1770s he had several imitators in London and the provinces.[6]

It was in these circumstances of relations with a large and heterogeneous audience in various institutional settings that chemistry acquired the profile of an Enlightenment science. Its perceived utility was the key to this, but utility was understood as more extensive and solid the more securely founded were the scientific or "philosophical" credentials of the discipline. Enlightenment chemists thus further developed Libavius's claim that chemistry was an autonomous science by virtue of organizing and providing foundations for many of the practical arts. The relationship with medicine was the closest and most venerable, a long-lasting legacy of Paracelsus and his followers, who had pioneered the use of chemically prepared drugs. By the eighteenth century, chemical medicines had an accepted place in the pharmacopeia, even if their apothecary advocates were still regarded with suspicion by some physicians. For this reason, an up-to-date medical education would include lectures on chemistry. In Germany and Scandinavia, the links between chemistry and the arts of mineralogy and metallurgy were also traceable to the Renaissance and were substantially developed during the eighteenth century as new mineral resources were exploited. Entirely new areas of chemical technology were also opened up. In Scotland, Cullen and other chemists participated in local

[6] Jan Golinski, *Science as Public Culture: Chemistry and Enlightenment in Britain, 1760–1820* (Cambridge University Press, 1992), pp. 52–63; Karl Hufbauer, *The Formation of the German Chemical Community, 1720–1795* (Berkeley: University of California Press, 1982).

societies devoted to national economic improvement, working on applications of chemistry to dyeing and bleaching, the manufacture of salts, and the use of agricultural fertilizers. Hopes for further progress in these arts, chemists declared, should be invested in the science that was fundamental to them all.[7]

By forging social connections with practitioners and patrons of the arts and by the concrete work of experimental research, eighteenth-century chemists positioned their science at a pivot point in the relationship between natural knowledge and power over the material world, a relationship that was emerging as a characteristic of the age. It is easier for historians to read the traces of this enterprise in what eighteenth-century chemists *said* than in what they *did.* Their writings have survived, but the traces of their actions have been obscured by time. Very little is known about the oral traditions and those of tacit knowledge that sustained the discipline in the course of its *longue durée.* Textbooks rarely seem to have sufficed to train a chemist, who also had to see and feel to learn – and indeed to smell, taste, and hear, since cultivation of all the senses was understood to be a vital part of the chemist's formation. The chemist, according to one writer, needed "his thermometer at the tips of his fingers and his clock in his head."[8] Embodied skills such as this have left few traces, even fewer than the remains of the material apparatus that displaced some of them in the eighteenth century. By the end of the century, real thermometers and clocks were regular items of laboratory equipment, along with other devices for measurement: barometers, eudiometers, calorimeters, gasometers, and, most important, balances. The end of chemistry's *ancien régime was* marked by a shift from reliance on the senses and on informal estimates of quantities to a culture of increasingly precise measurement with a range of refined instrumentation.[9] The subject was reconfigured by disciplinary means that extended well beyond revisions in the textbooks. "Discipline" now acquired a meaning that embraced the training of chemists in the manual skills demanded by the new equipment of the laboratory. Academic chemistry had begun as an outgrowth of Humanist pedagogy; by the beginning of the nineteenth century, like other sciences in the era of the Industrial Revolution, it was being inculcated through a regimen of laboratory training.

---

[7] Archibald Clow and Nan L. Clow, *The Chemical Revolution: A Contribution to Social Technology* (London: Batchworth, 1952); Arthur L. Donovan, *Philosophical Chemistry in the Scottish Enlightenment: The Doctrines and Discoveries of William Cullen and Joseph Black* (Edinburgh: Edinburgh University Press, 1975), pp. 34–92.

[8] G. F. Venel, quoted in Isabelle Stengers, "L'affinité ambiguë: le rêve newtonien de la chimie du XVIIIᵉ siècle," in Michel Serres (ed.), *Éléments d'histoire des sciences* (Paris: Bordas, 1989), pp. 297–319, quotation at p. 309.

[9] Lissa Roberts, "The Death of the Sensuous Chemist: The 'New' Chemistry and the Transformation of Sensuous Technology," *Studies in History and Philosophy of Science*, 26 (1995), 503–29; Trevor H. Levere, "Practice, Apparatus, and the Growth of Eighteenth-Century Chemistry," paper delivered at the Dibner Institute for the History of Science and Technology, MIT, Cambridge, MA, 28 November 1995.

THE PHILOSOPHY OF MATTER

Butterfield claimed that chemistry had waited until the end of the eighteenth century to achieve a coherent framework of fundamental ideas. Yet an earlier historian, Hélène Metzger, had already demonstrated the importance of philosophies of matter in chemistry from the early seventeenth century. Metzger's pioneering studies, especially *Les doctrines chimiques en France* (1923) and *Newton, Stahl, Boerhaave et la doctrine chimique* (1930), reconstructed the development of the matter theories associated with the mechanical philosophy, with the doctrines of Newton, and with the works of the Halle professor Georg Ernst Stahl (1660–1734). Metzger's studies continue to command respect, and her lead in the exploration of philosophical theories of matter has been followed by subsequent scholars, but historians have also reconsidered the question of the relationship between matter theory and other domains of chemical thought and practice. The historiography to which Metzger was committed tended to prejudge this issue by assuming the primacy of fundamental philosophical ideas in the development of the sciences, whereas recent research has disclosed a more problematic and ambivalent influence.[10]

The mechanical philosophy of the seventeenth century drew upon the ancient concept of atoms to make several contributions to the theory of chemistry, but these had little impact on practice and remained relatively marginal to the tradition of chemical writing. A mechanistic ontology was applied to account for properties of acids in the *Cours de chymie* (1675) of Nicholas Lemery (1645–1715). Lemery proposed that the acrid taste and corrosive action of acids could be explained by their sharp-pointed particles which were capable of penetrating into the pores of other bodies. But these speculations occupied only a small portion of the "reasonings" he added to descriptions of chemical operations and they did not substantially affect the largely traditional contents of his text. Similarly, Robert Boyle's attempts to rationalize chemical operations in terms of the shapes and textures of particles of matter were advanced cautiously in some of his experimental essays but were little attended to by other chemical writers. Boyle earned more recognition for his elegant dialog *The Sceptical Chymist* (1661), where he argued against the existence of elements or principles that would retain a consistent chemical identity through analysis by fire. These arguments were reiterated by others, including Lemery, and contributed to diminishing the authority of the traditionally identified chemical elements. A contemporary judged that

---

[10] Hélène Metzger, *Les doctrines chimiques en France du début du XVIIe à la fin du XVIIIe siècle* (Paris: Presses Universitaires, 1923); Metzger, *Newton, Stahl, Boerhaave et la doctrine chimique* (Paris: Alcan, 1930); John R. R. Christie, "Hélène Metzger et l'historiographie de la chimie du XVIIIe siècle," and Evan M. Melhado, "Metzger, Kuhn, and Eighteenth-Century Disciplinary History," in Gad Freudenthal (ed.), *Études sur / Studies on Hélène Metzger* (Leiden: E. J. Brill, 1990), pp. 99–108, 111–34.

Boyle had "not so much laid a new Foundation of Chemistry, as he has thrown down the Old."[11]

In the early eighteenth century, the mechanical philosophy was succeeded as a source of chemical matter theory by the ideas of Newton. Again, an ambitious program for the reconstruction of chemistry was announced, but it failed to match the pragmatic aims of most chemists. The influence of Newton's natural philosophy on chemistry was more subtle and indirect than the early "Newtonians" hoped for. The first chemical writers to identify themselves with the Newtonian philosophy were John Keill (1671–1721) and John Freind (1675–1728), both working in Oxford in the first decade of the century. Keill's 1708 paper in the *Philosophical Transactions* of the Royal Society followed closely on Newton's own remarks about chemical phenomena in the twenty-third Query of the Latin translation of *Opticks* in 1706 (subsequently revised as Query 31 of the 1717 edition of the text). There, Newton had applied to a variety of chemical phenomena the notion of microscopic forces, which would be analogous to the gravitational force but act on a much smaller scale. The notion was especially relevant to displacement reactions, for example, when copper was added to a solution of a silver salt in acid, and silver precipitated. Such a reaction was to be explained by saying that the specific force of attraction between copper and the acid is stronger than that between silver and the acid so that the silver is displaced from combination. Newton noted that the metals could be arranged in a consistent order of their relative attraction for the acid in question and that similar orders could be constructed for other kinds of reactions. Although the phenomenon was not new to many chemists, Newton's discussion made it plausible that it could be explained in terms of a microscopic analogue of the force of gravity.[12]

This was the point developed by Keill and Freind. Keill's paper presented a series of putative axioms for explaining chemical phenomena on the grounds of specific attractions, which would in turn be explained by such factors as the relative density, shapes, and textures of the particles of bodies. Freind's text, *Praelectiones Chymicae* (1709), which was derived from lectures delivered in the basement laboratory of the Ashmolean Museum in Oxford, reproduced these axioms and went on to apply them to generate explanations for various

---

[11] Metzger, *Doctrines chimiques*, pp. 281–338; T. S. Kuhn, "Robert Boyle and Structural Chemistry in the Seventeenth Century," *Isis*, 43 (1952), 12–36; Antonio Clericuzio, "Carneades and the Chemists: A Study of *The Sceptical Chymist* and Its Impact on Seventeenth-Century Chemistry," in Michael Hunter (ed.), *Robert Boyle Reconsidered* (Cambridge University Press, 1994), pp. 79–90; John Freind, *Chymical Lectures* (London: J. W. for Christian Bowyer, 1729), p. 4.

[12] Isaac Newton, *Opticks* (New York: Dover Publications, 1952), pp. 381ff.; John Keill, "Joannis Keill . . . In qua Leges Attractiones Aliaque Physices Principia Traduntur," *Philosophical Transactions of the Royal Society*, 26 (no. 315) (1708), 97–110; Keill, "De Operationum Chymicarum Ratione Mechanica," trans. A. Guerrini and J. Shackelford, *Ambix*, 36 (1989), 138–52; Arnold Thackray, *Atoms and Powers: An Essay on Newtonian Matter-Theory and the Development of Chemistry* (Cambridge, MA: Harvard University Press, 1970), pp. 8–82; Anita Guerrini, "Chemistry Teaching at Oxford and Cambridge, circa 1700," in P. Rattansi and A. Clericuzio (eds.), *Alchemy and Chemistry in the 16th and 17th Centuries* (Dordrecht: Kluwer, 1994), pp. 183–99.

chemical operations. His exposition failed, however, to encompass the familiar chemical attributes of bodies. Attractive powers seemed to bear no relation to the properties that chemists had identified with particular substances. For this reason, Freind's text was occasionally cited by natural philosophers but rarely by chemists in the following decades. A few writers continued to link the relative attractions or "affinities" among chemical substances to the Newtonian concept of forces, but, as we shall see, the ordering of chemical affinities had an important function in eighteenth-century chemistry quite independently of Newton's ideas.[13]

As well as discussing attractive forces in his influential Query, Newton also mentioned the possibility of forces of repulsion. He noted how "fermentation" – by which he meant any process that released "air" from solids or liquids – "seems unintelligible, by feigning the particles of Air to be springy and ramous, or rolled up like hoops, or by any other means than by a repulsive power." These remarks assumed considerable importance in the origins of pneumatic chemistry, in which the processes of release of aerial fluids, and the reverse processes of "fixing" air in solids or liquids, were central. Stephen Hales (1677–1761), vicar of Teddington in Middlesex, investigated these phenomena in his *Vegetable Staticks* (1727), claiming that they demonstrated the existence of repulsive forces between air particles. Hales provided the conceptual vocabulary in which interactions between aerial fluids and solid or liquid substances could be explicated: The repulsive forces responsible for the expansion of air could be overcome by sufficiently strong attraction by particles of more ponderous matter, in which case the air would be "fixed," he claimed. Hales also developed the instrumentation for studying these processes. He used the "pneumatic trough" to collect and measure samples of air given off in chemical reactions by leading it into vessels filled with water and held upside-down over a water-filled basin. In the light of the subsequent differentiation of aerial fluids into chemically distinct gases, Hales has sometimes been criticized for failing to discriminate between the airs he manipulated; he continued to assume that airy fluids were essentially one kind of entity, albeit sometimes contaminated by mixtures of other substances. He was, in fact, much less interested in the chemical differentiation of airs than in their role in a providential economy of nature sustained by a balance of attractive and repulsive forces. It was in these terms, for example, that he understood the reduction in the volume of air surrounding a burning body as due to the release of sulfurous or acidic particles, which were strongly attractive and so reduced the elasticity of the air they contaminated. In ascribing a crucial providential role to air – "this noble and important element, endued with a most active principle [by] the all-wise Providence of the great Author of nature" – Hales was using it to address the problem of divine action

---

[13] John Freind, *Praelectiones Chymicae* (London: J. Bowyer, 1709).

in the world, a theme of central importance in eighteenth-century natural philosophy.[14]

AFFINITIES AND COMPOSITION

In what became Query 31 of the *Opticks,* Newton pointed out that chemical substances could be arranged consistently in order of the strength of their attraction for a certain other substance. His idea seemed to have come to fruition in 1718, when Etienne François Geoffroy (1672–1731) presented to the Paris Academy a "Table of the different relationships observed between different substances" (Figure 16.1). The sixteen columns of the table were headed with symbols for the different acids, alkalis, and metals. Below each symbol were arranged the symbols for those substances that could form combinations with them, in descending order of strength of the combination. Geoffroy carefully avoided use of the terms "attraction," with its specifically Newtonian connotations, and "affinity," which could invoke alchemical notions of occult sympathies. By referring simply to *rapports* (relationships), he tried to maintain neutrality on the theoretical issues that divided Newtonians from the Cartesians who still prevailed in French science – a precaution that did not, in fact, prevent suspicions that Geoffroy was covertly representing the Newtonian outlook.[15]

Geoffroy's was the first of a large number of such tabulations that appeared in the course of the eighteenth century. Affinity tables, as they came to be called, became larger and more elaborate, summarizing more information about reactions and combinations. In 1775, the Swedish chemist Torbern Bergman (1735–1784) presented a table in two parts (for wet and dry reactions), with thirty-four columns and as many as twenty-seven substances listed in each column. Some historians have read the prevalence of these tables as an indicator of the influence of the Newtonian philosophy of matter on chemical thinking in the eighteenth century. Accounts of a "Newtonian tradition" of chemistry have referred to the tables as an important thread of that tradition. Others, however, have argued that affinity tables should be understood in relation to their uses in chemists' research and teaching, reflecting not a specifically Newtonian tradition but intrinsically chemical ways of thinking about such issues as combination and reactions. The latter reading seems more in line with the attitude recommended by Bernard de Fontenelle, perpetual secretary of the Academy, when Geoffroy's table was first published. Fontenelle wrote: "It is here that sympathies and attractions would become

---

[14] Newton, *Opticks,* p. 396; Stephen Hales, *Vegetable Staticks* (London: Scientific Book Guild, 1961), pp. 176–7; Arthur Quinn, "Repulsive Force in England, 1706–1744," *Historical Studies in the Physical Sciences,* 13 (1982), 109–28.

[15] Thackray, *Atoms and Powers,* pp. 85–95; Stengers, "L'affinité ambiguë," pp. 300–2.

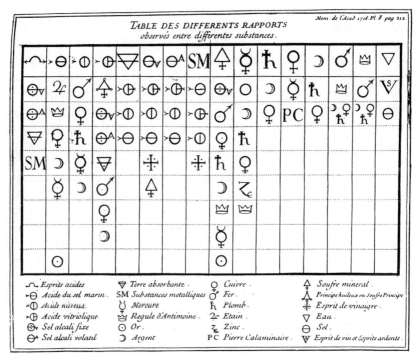

Figure 16.1. The "Table of different relationships observed between different substances," submitted to the Paris Academy of Sciences by E. F. Geoffroy in 1718. Geoffroy used traditional symbols for the acids, alkalis, and metals. The order of the symbols in each column of the table shows (in descending order) the relative strength of combination of each substance with that at the top of the column. Geoffroy deliberately avoided the terms "affinity" or "attraction" in labeling his table. (From *Mémories de l'Académie Royale des Sciences*, 1718, p. 212.)

appropriate, if there were such things. However, leaving as unknown that which is unknown, and holding to certain facts, all chemical experiments prove that a particular body has more disposition to unite with one body than with another, and that this disposition has different degrees."[16] Talk of sympathies and attractions was beside the point, Fontenelle suggested. The table should be valued as a means of ordering information about chemical operations, thereby easing learning and directing research.

[16] Fontenelle, quoted in Ursula Klein, "The Chemical Workshop Tradition and the Experimental Practice: Discontinuities within Continuities," *Science in Context*, 9 (1996), 251–87, quotation at p. 276 (translation slightly modified). On subsequent tables, see Stengers, "L'affinité ambiguë"; A. M. Duncan, "Some Theoretical Aspects of Eighteenth-Century Tables of Affinity," *Annals of Science*, 18 (1962), 177–94, 217–32; Lissa Roberts, "Setting the Table: The Disciplinary Development of Eighteenth-Century Chemistry as Read through the Changing Structure of Its Tables," in Peter Dear (ed.), *The Literary Structure of Scientific Argument: Historical Studies* (Philadelphia: University of Pennsylvania Press, 1991), pp. 99–132. J. W. Goethe was to make use of the notion of differential

Fontenelle's remark directs attention away from the possible connections between affinities and the theories of natural philosophy and toward chemical operations themselves and how they were conceived. Ursula Klein has demonstrated that the operations recorded in Geoffroy's table are all to be found in books of metallurgical and pharmaceutical chemistry in the seventeenth century.[17] Particularly important were the operations by which salts were formed. Frederic L. Holmes has documented the importance of research on analysis and synthesis of the so-called middle salts – those formed by combining an acid and an alkali – in the Academy in the two decades before the table appeared. Particularly important contributions to this research were made by Geoffroy himself and by Wilhelm Homberg (1652–1715).[18] In 1702, Homberg had distinguished three classes of middle salts: those formed by acids in combination with, respectively, a fixed alkali, an alkaline earth, and a metal; in addition, he recorded a class of ammoniac salts. Although he did not explicitly reject either the ancient notion of elements or the ontology of corpuscles, Homberg's understanding of middle salts utilized a more pragmatic and operational concept of chemical compounds and the processes by which they were composed and decomposed. Geoffroy's table reflected the notion that chemistry was concerned with the combination and separation of chemically identifiable entities by operations that were always in principle reversible. As he put it in a paper of 1704, "What completely assures us that we have succeeded in investigating the composition of bodies is, having reduced mixta into the simplest substances that chemistry can provide, we can recompose them by reuniting these same substances."[19]

Eighteenth-century affinity tables have been discussed by Klein and Holmes as part of a largely autonomous chemical practice that was substantially independent of the matter theory handed down by natural philosophy. Chemists worked with ideas about composition and chemical processes that were adapted to the materials they dealt with and the operations they carried out. Chemical operations were conceived as essentially reversible combinations and separations of parts that were ascribed stable identities in terms of their chemical properties. Other historians have linked this conceptual outlook specifically with the influence of Stahl. The German chemist provided his contemporaries with a popular vocabulary for labeling chemical substances and changes, which distinguished them from bodies and operations considered from a purely physical point of view. He defined chemistry as concerned specifically with

chemical affinities as a metaphor for human attraction in his novel *Elective Affinities* (1809). See Jeremy Adler, "Goethe's Use of Chemical Theory in His *Elective Affinities*," in Andrew Cunningham and Nicholas Jardine (eds.), *Romanticism and the Sciences* (Cambridge University Press, 1990), pp. 263–79.

[17] Klein, "Chemical Workshop Tradition."

[18] Holmes, *Eighteenth-Century Chemistry*, pp. 33–55; Holmes, "The Communal Context for Etienne-François Geoffroy's 'Table des rapports,'" *Science in Context*, 9 (1996), 289–311.

[19] Geoffroy, quoted in Klein, "Chemical Workshop Tradition," p. 272.

bodies considered as "mixts" or compounds – that is to say, from the point of view of their chemical constitution rather than their physical "aggregation." Mechanics was concerned with taking bodies apart into their homogeneous physical components, whereas chemistry, in Stahl's view, was concerned with a more intimate kind of composition, in which bodies were found to be constituted of heterogeneous substances that did not share the properties of the compound in which they occurred. Stahl's ontology designated a way of studying the objects in the world that ascribed to chemists distinctive and independent skills of analysis and synthesis.[20]

Stahl's support for the autonomy of chemistry was one reason for the popularity of his doctrines among European chemists in the mid-eighteenth century. In France, his ideas were introduced in lectures given at the Jardin du Roi between 1742 and 1768 by Guillaume-François Rouelle (1703–1770). Rouelle's influential lectures reiterated the Stahlian insistence on a realm of specifically chemical entities and processes, the domain of an autonomous discipline of chemistry. The same line was taken in the article "Chymie," published by Gabriel François Venel (1723–1775) in the *Encyclopédie* of Diderot and d'Alembert in 1753. Venel urged his readers to reject philosophical hypotheses about the nature of chemical composition and not to assume that the destiny of chemistry lay in its reduction to the principles of natural philosophy. Chemists could take pride, he maintained, in their ability to comprehend chemical processes without resorting to uncertain physical hypotheses. On these grounds, they could assert the authority of their discipline over the practices of the chemical arts.[21]

Along with his often-echoed assertions of the independence of chemistry from physical theory, Stahl also advocated somewhat more debatable doctrines. He identified three different earthy principles in the composition of mixts: a vitrifiable earth, a mercurial or metallic one, and the sulfurous principle or "phlogiston." The last was the principle of flammability, which was present in all combustible matter and was released as light and heat by burning bodies or by metals in the course of corrosion or calcination. Phlogiston was given a particularly important role in the work of French chemists following a crucial reinterpretation of the doctrine by Rouelle. In his lectures, Rouelle identified the principle not with the earths but with the ancient element "fire," which he characterized as a physical agent (or "instrument") *and* a chemical element. Fire, in other words, was both a cause of chemical changes and a participant in them, capable of entering into the composition of substances. Rouelle ascribed the same duality of role to the other classical elements, most importantly including air, which, as Hales had shown, was

---

[20] G. E. Stahl, *Philosophical Principles of Universal Chemistry,* trans. Peter Shaw (London: Osborn and Longman, 1730); Metzger, *Newton, Stahl, Boerhaave,* pp. 93–188.

[21] Rhoda Rappaport, "G. F. Rouelle: An Eighteenth-Century Chemist and Teacher," *Chymia,* 6 (1960), 68–101; Stengers, "L'affinité ambiguë," pp. 309–11.

capable of entering into chemical combination. Air could exist free in the atmosphere or fixed in chemical combination – for example, in aerated mineral waters. Fire, similarly, could act as a physical instrument, rarefying bodies, or it could enter into chemical combination as phlogiston in metals or combustible matter.[22]

This kind of account was found appealing by many French chemists. It was adopted from Rouelle's lectures by Venel, for example, in his articles in the *Encyclopédie,* and by Pierre Joseph Macquer (1718–1784), author of the widely read *Dictionnaire de chymie* (1766). When Lavoisier, who had attended Rouelle's lectures himself, launched his assault on the phlogiston theory in the 1780s, it was this prevailing version that he attacked. It owed its appeal to the broad sweep of its explanatory capabilities and to its resonance with chemists' claims to theoretical autonomy. The realization that metallic calcination was the same process as combustion was a dramatic accomplishment, recognized even by Whiggish historians as a positive achievement of the phlogiston theory, and it consolidated the claims of chemists to authority over metallurgical practices. As a distinctively chemical entity, phlogiston was a strategic tool in the campaign to establish the boundaries of the discipline and its credentials as an Enlightenment science.

GASES AND IMPONDERABLES

Notwithstanding the efforts of chemists to secure the credentials of their discipline, chemistry did not in fact operate independently from other sciences in the eighteenth century. The study of gases connected the interests of chemists with those of natural philosophers and medical men who were exploring such issues as respiration and the healthiness of different kinds of air. Phlogiston was given a remarkably broad application in many domains of Enlightenment science. As one of a group of "imponderable" (weightless) fluids, it was invoked in connection with phenomena as diverse as static electricity, nervous impulses, and terrestrial heat. Thus, chemists found themselves sharing a crucial explanatory concept with many other scientific practitioners; far from working in a cultural vacuum, they found they inhabited a climate of ideas dense with distinctive gases and imponderables such as phlogiston.

Following the lead of Hales, pneumatic chemistry flourished particularly in Britain. Two fairly distinct paths were explored. The first concerned the role of heat in physical transitions among solid, liquid, and gaseous states. In Scotland, beginning in the 1740s, an attempt was made to connect these phenomena with chemical reactions understood in terms of affinities. In this

---

[22] Rhoda Rappaport, "Rouelle and Stahl: The Phlogistic Revolution in France," *Chymia,* 7 (1961), 73–102; Martin Fichman, "French Stahlism and Chemical Studies of Air, 1750–1770," *Ambix,* 18 (1971), 94–122.

tradition, phlogiston appeared as a weightless agent of physical and chemical change, rather than as a participant in chemical composition; in fact, it assumed some of the functions of the "ether," a subtle and imponderable fluid freighted with the weight of Newton's authority by virtue of its mention in the Queries of the *Opticks*. The second line of research, pursued largely by English natural philosophers, considered the chemical characteristics of different gases, which were initially still regarded as different species of air. In the hands of Joseph Priestley (1733–1804), a number of new gases were produced and distinguished by the degree to which they appeared to contain phlogiston. In the early 1780s, Priestley and others went so far as to identify "inflammable air" with pure phlogiston. For them, phlogiston was not a subtle fluid but a regular factor in chemical composition.

It was Cullen who initiated Scottish research on heat in relation to physical and chemical change. In doing so, he made use of an influential doctrine advanced by Boerhaave, who had taught his students at Leiden that the four ancient elements should be regarded, not as components of matter responsible for its properties but rather as "instruments" of all kinds of physical and chemical change. Earth was the "matrix" of certain transformations, water was the solvent that permitted others to occur, air was the medium of combustion and respiration, and fire was the great instrument of activity in the cosmos.[23] Fire, for Boerhaave, was an imponderable material fluid, capable of passing into or out of normal weighty matter; it was the prime agent of chemical change, but it did not itself participate in chemical combination. Cullen identified Boerhaave's fire with Newton's ether, characterized as imponderable, expandable, subtle, and repelled by normal matter – properties routinely ascribed to the ether since Newton's brief description had been elaborated upon in the early 1740s.[24] All substances, Cullen suggested, were composed of normal attractive matter permeated by a cloud of repulsive ether. The relative densities of matter and ether would determine the state of aggregation of the body: if attractions exceeded repulsions, the body would be solid; if they were approximately in balance, it would be liquid; if repulsions overwhelmed attractions, the body would become a vapor. A change of physical state of a body could thus be related to the addition or subtraction of ether, that is, heat or fire.

Cullen extended this perspective to try to relate exchanges of heat to chemical transformations. He explored reactions that involved the production or absorption of heat: the addition of water to dehydrated salts, for example, which releases heat, or the evaporation of volatile liquids, which absorbs it. He was obliged to admit failure, however, to explain chemical reactions in all

[23] Rosaleen Love, "Herman Boerhaave and the Element-Instrument Concept of Fire," *Annals of Science*, 31 (1974), 547–59.

[24] J. R. R. Christie, "Ether and the Science of Chemistry, 1740–1790," in G. N. Cantor and M. J. S. Hodge (eds.), *Conceptions of Ether: Studies in the History of Ether Theories 1740–1900* (Cambridge University Press, 1981), pp. 85–110.

their qualitative variation. Although states of aggregation could be explained in terms of a balance of ethereal fluid and ponderable matter, it was not possible to account for the differing strengths of attractions between chemical substances on the same basis. The theory had limited success in the realm of chemical change, but it enjoyed its most triumphant application in the discoveries of specific heat capacities and latent heats by Cullen's pupil Joseph Black (1728–1799). Black showed that different bodies had different capacities to absorb heat to produce a measured change in temperature, a finding that was clearly rooted in the conception of heat as a material fluid absorbed to a greater or lesser degree by normal matter. Similarly, Black's revelation of the latent heats of fusion and evaporation, which were required to change the state of a body but not revealed by the thermometer, reflected the influence of Cullen's framing of the problem of heat and aggregation.[25] The significance of this work for chemistry was that it pointed toward an understanding of the gaseous state: a gas came to be seen not as a variety of air but as a state that all bodies could attain, given sufficient heat. Lavoisier, who knew of the work of Black and his Scottish colleague William Irvine, was to turn this insight to telling effect in his research on heat and gases, the first avenue to be explored in his revolutionary remaking of chemical theory.

Black also played a significant part in the second line of development of pneumatic chemistry, investigating the chemical identities of different airs or gases. His *Experiments upon Magnesia Alba* (1756) scrutinized the air given off by heating "magnesia alba" (magnesium carbonate). Using Hales's terms, Black labeled this vapor "fixed air" and showed that its release diminished the weight of the salt. Two other findings were crucially important. First, the fixed air proved to have different properties from normal atmospheric air: it turned lime-water milky and did not support combustion or respiration. Second, after being deprived of its fixed air, the magnesia alba was also found to have lost its alkalinity; apparently the air contributed to its chemical properties when present in the compound. These observations, established by an impressive series of careful experiments, indicated how gases were to assume the status of chemical entities. Black had shown, at least in the case of fixed air, that they could be characterized by tests of their chemical identity and that their effects on the properties of the bodies in which they were compounded could be ascertained.[26]

Further exploration of the chemical identities of gases was largely the work of English researchers. In 1766, Henry Cavendish (1731–1810) distinguished Black's fixed air from the "inflammable air" given off by metals dissolving in acids. Priestley followed by systematically producing and distinguishing nu-

[25] Douglas McKie and Niels H. de V. Heathcote, *The Discovery of Specific and Latent Heats* (London: Edward Arnold, 1935); Donovan, *Philosophical Chemistry*, pp. 222–49.
[26] Donovan, *Philosophical Chemistry*, pp. 183–221.

merous new airs, of which he published detailed descriptions in his *Experiments and Observations on Different Kinds of Air* (3 vols., 1774–7). With quite modest equipment, skillfully used, Priestley was able to identify (among others) "fixed air," "inflammable air," "nitrous air," "marine acid air," "alkaline air," and "phlogisticated air" (to give them their modern names: carbon dioxide, hydrogen, nitric oxide, hydrochloric acid, ammonia, and nitrogen). To Priestley, the significant difference among them were their degrees of phlogistication, that is, the amount of phlogiston they contained. He developed a diagnostic test for this, the nitrous air test, in which the gas to be tested was mixed with nitrous air and the product showed a diminution in volume as part of it was absorbed by water. The diminution appeared to Priestley to be proportional to the "purity" of the test air; he used the procedure to support his theory that in respiration the purity of air was reduced by discharge of harmful phlogiston from the body into the atmosphere. Priestley's test became the basis for instruments of various designs, collectively known as "eudiometers," with which investigators toured sites throughout Europe to make assessments of the healthiness of the air. Heavily phlogisticated air, such as that found in marshes or overcrowded urban areas, was regarded as unhealthy. At the other end of the spectrum, Priestley's most exciting new discovery, prepared by heating the red calx of mercury with a burning glass, was named "dephlogisticated air." This air, which Lavoisier was to construe quite differently as "oxygen," appeared to Priestley as the purest and most suitable air for respiration.[27]

The Swedish chemist Carl Wilhelm Scheele (1742–1786), who preceded Priestley in the isolation of this gas, had already interpreted it in the light of the phlogiston theory, as *Feuerluft*, the agent responsible for producing fire when united with phlogiston in combustion. The difficulties of assigning priority for the discovery of oxygen in these circumstances have confused – or delighted – historians. It seems sufficient for our purposes to note the significance of the fact that the new gas was understood by Priestley and Scheele in terms of the theory of phlogiston. Gases entered the domain of chemistry, acquiring identities as chemical beings, in connection with the theory by which chemistry had proclaimed its autonomy in the eighteenth century. In the event, however, the phlogistic appropriation of the new gases was a short-lived affair. Lavoisier was to conceptualize their nature in different terms, using resources from the tradition that had linked them to the study of heat. After his work, it turned out that chemistry could survive, and indeed flourish, without phlogiston at all.

[27] John G. McEvoy, "Joseph Priestley, 'Aerial Philosopher': Metaphysics and Methodology in Priestley's Thought," *Ambix*, 25 (1978), 1–55, 93–116, 153–75; *Ambix*, 26 (1979), 16–38; Simon Schaffer, "Measuring Virtue: Eudiometry, Enlightenment and Pneumatic Medicine," in Andrew Cunningham and Roger French (eds.), *The Medical Enlightenment of the Eighteenth Century* (Cambridge University Press, 1990), pp. 281–318.

THE MAKING OF A REVOLUTION

Lavoisier's interest in the nature of gases can be traced to notes he composed in 1766, in which he suggested that air might be a compound of a certain chemical basis with the matter of fire. It has been suggested that his sources included papers by the Berlin academician J. T. Eller and an essay on "Expansibilité," in the *Encyclopédie*, by the French *philosophe* A. R. J. Turgot. Turgot's model enabled Lavoisier to conceive of different gases as distinct chemical entities that owed their gaseous form to a temporary combination with the material fluid of fire or heat, which he was to name "caloric."[28] A physical model of the gaseous state appealed to Lavoisier, whose research extended beyond chemistry into many of the other fields of physical science, but it was consistently linked with an emphasis on chemical combination. Lavoisier always viewed caloric as a chemical entity, capable of being exchanged and combined in chemical reactions.

In the early 1770s, Lavoisier put this model to work in connection with the processes of combustion and calcination. He viewed as particularly significant the determination, by Louis Bernard Guyton de Morveau (1737–1816), that metals gained weight as they underwent calcination. Guyton had measured an increase in weight that had been noted on occasion before but not consistently quantified. For Lavoisier, the weight gain was telling evidence that combustion and calcination were processes in which air was fixed by solids, releasing its caloric in the form of light and heat. The flames characteristic of combustion were therefore signs not of phlogiston but of caloric. This new understanding of combustion and calcination had implications for the reverse process, the reduction of calxes to their metallic bases. Lavoisier experimented with the reduction of lead calx (litharge) to the metal by heating with charcoal. Traditionally, the charcoal had been regarded as a source of the phlogiston necessary to form the metal; Lavoisier found it difficult at first to account for its function, since he saw reduction as the release of fixed air from the calx. A critically important step was taken in the wake of Priestley's visit to Paris in October 1774, when he described to Lavoisier his experiments on the reduction of the red calx of mercury. This reduction was of interest because it could be performed without charcoal, in which case it yielded the fascinating new gas that Priestley was shortly to name dephlogisticated air. Lavoisier repeated these experiments, finding that the air released by the reduction of mercury calx was the only part of the atmosphere absorbed in calcination or combustion. This "purest part of the air" also turned out to be the part consumed in respiration by humans and animals; it was thus also named "eminently respirable air." Finally, this gas was given the name

---

[28] R. J. Morris, "Lavoisier and the Caloric Theory," *British Journal for the History of Science*, 6 (1972), 1–38; Robert Siegfried, "Lavoisier's View of the Gaseous State and Its Early Application to Pneumatic Chemistry," *Isis*, 63 (1972), 59–78.

"oxygen" after its role as generator of acids in such processes as the combustion of sulfur and phosphorus.[29]

By 1777, Lavoisier had achieved an understanding of combustion and calcination, and of the role of oxygen in the formation of acids, that enabled him to glimpse a comprehensive alternative to the phlogiston theory. He launched his first outright attack, declaring phlogiston a purely imaginary entity. At first, few chemists were convinced. Some, such as Guyton and Claude Louis Berthollet (1748–1822), accepted that a portion of air was fixed in combustion, but they continued to believe that phlogiston was also released in the process. Lavoisier won them over with his work of the early 1780s, in which he introduced methods of precision measurement to determine quantities of substances in reactions. This new direction to his work followed a period of collaboration with the mathematician and physicist Pierre Simon de Laplace (1749–1827), with whom Lavoisier developed the ice calorimeter in 1782–3 to measure the heat released by processes of combustion and respiration. Precision measurement using the balance was applied particularly to experiments on the analysis and synthesis of water, following Cavendish's finding that ignition of inflammable air yielded a small quantity of water. In 1785, in Paris, Lavoisier staged public demonstrations of the synthesis of water from oxygen and the gas he named "hydrogen" and its analysis into these constituents. To show that quantities were conserved in the course of the reactions, precise measurements were taken of the weights of reactants and products. Lavoisier claimed a "demonstrative proof" had been accomplished of the compound nature of water and the role of oxygen in combustion.[30]

These demonstrations were of considerable importance in securing assent to the new theory among prominent chemists, initially in France and subsequently throughout Europe. Lavoisier had been privately expressing his ambition to make a "revolution" in the science since 1773; ten years later, the chemist Antoine François de Fourcroy (1755–1809) acknowledged his success by using the same term. Lavoisier recruited as allies Fourcroy, Berthollet, and Guyton, who collaborated with him to prepare a new system of chemical nomenclature, which was published as *Méthode de nomenclature chimique* (1787).[31] In this system, hydrogen and oxygen appeared as elements, as did

---

[29] Henry Guerlac, *Lavoisier – The Crucial Year: The Background and Origin of his First Experiments on Combustion in 1772* (Ithaca, NY: Cornell University Press, 1961); Maurice P. Crosland, "Lavoisier's Theory of Acidity," *Isis,* 64 (1973), 306–25.

[30] Carleton Perrin, "The Triumph of the Antiphlogistians," in Harry Woolf (ed.), *The Analytic Spirit: Essays in the History of Science in Honor of Henry Guerlac* (Ithaca, NY: Cornell University Press, 1981), pp. 40–63; Henry Guerlac, "Chemistry as a Branch of Physics: Laplace's Collaboration with Lavoisier," *Historical Studies in the Physical Sciences,* 7 (1976), 193–276; Maurice Daumas and Denis I. Duveen, "Lavoisier's Relatively Unknown Large-Scale Decomposition and Synthesis of Water, February 27 and 28, 1785," *Chymia,* 5 (1959), 113–29.

[31] Maurice P. Crosland, *Historical Studies in the Language of Chemistry* (London: Heinemann, 1962), pp. 168–92; Trevor H. Levere, "Lavoisier: Language, Instruments, and the Chemical Revolution," in T. Levere and W. R. Shea (eds.), *Nature, Experiment, and the Sciences* (Dordrecht: Kluwer, 1990),

the various metals and simple non-metallic substances such as carbon and sulfur. Compounds were to be named to reflect their makeup according to the new chemistry, with different degrees of oxidation indicated by different suffixes, as in "sulfite" and "sulfate." Chemists had long been calling for reform of the language of their science to eliminate what were recognized as anachronisms and ambiguities. Lavoisier and his allies answered this call in a particular way: following the lead of the philosopher Etienne Bonnot de Condillac (1715–1780), they forged a scientific language designed to reflect nature directly rather than to follow the conventions among chemists. To speak the new language was henceforth to adopt the new theory. In 1789, the new system was further codified in Lavoisier's textbook, *Traité élémentaire de chimie,* which allowed for students to be trained in the new theory and in the use of the apparatus, including the calorimeter and the balance, by which it had been achieved. Again, Lavoisier presented his revolutionary new doctrines in a form adopted from chemical tradition, reworking the genre of the textbook that had been a standard feature of chemical education since Libavius. In a revolutionary gesture of rupture from the past, he dropped the standard historical introduction. Students were now to be taught that chemistry had effectively begun with Lavoisier.[32]

Historians have argued about the balance of the old and the new in Lavoisier's system. Consideration of the reactions of his contemporaries may help us discern the continuities and the discontinuities between his work and the prior traditions of chemistry. The most striking discontinuities occurred, broadly speaking, in the realm of methods. Lavoisier, in collaboration with Laplace and other physicists, introduced methods of precision measurement that had never previously been put to such telling use in chemistry. The balance, in particular, was Lavoisier's instrument of choice for accurate measurement, sometimes used in conjunction with other apparatus such as the calorimeter or the gasometer and linked to accountancy procedures for keeping track of the quantities involved in reactions. He had balances of almost unrivaled precision constructed by the best instrument-makers in Paris, leading to criticism that the experiments he performed with them could not be readily repeated by others without such resources.[33] Priestley denounced Lavoisier for the expense and exclusivity of his apparatus, connecting his choice of instrumentation with his remaking of the language of the science. Both tactics appeared to Priestley as illegitimate impositions on the community of chemists – brute displays of power rather than reasonable attempts to persuade. But the steady success of Lavoisier's attempts to convince chemists encouraged adoption of

pp. 207–23; B. Bensaude-Vincent and F. Abbri (eds.), *Lavoisier in European Context: Negotiating a New Language for Chemistry* (Canton, MA: Science History Publications, 1995).

[32] Bensaude-Vincent, *Lavoisier,* pp. 285–312.

[33] Ibid., pp. 197–230; Jan Golinski, "'The Nicety of Experiment': Precision of Measurement and Precision of Reasoning in Late Eighteenth-Century Chemistry," in M. Norton Wise (ed.), *The Values of Precision* (Princeton, NJ: Princeton University Press, 1995), pp. 72–91.

the apparatus and methods by which conversion was frequently achieved, as, for example, when the experiments of analysis and synthesis of water were repeated in the Netherlands.[34] Thereafter, the new chemistry was taught by textbooks and, increasingly, by laboratory training in the use of the new instrumentation. Chemistry was thus prepared to play a central role in the significant growth of laboratory sciences characteristic of the early nineteenth century. Measurements of weights of reactants and quantities of gases, for example, became standard procedure in chemical laboratories; in the hands of John Dalton and others, both turned out to be of great theoretical importance in the years after Lavoisier's revolution.

Although some of the methods of the revolution were borrowed from other physical sciences, it would be wrong to present Lavoisier's overall accomplishment as anything like a takeover of chemistry by physics. He had almost nothing to say in the *Traité* about the philosophy of matter, of the Newtonian or any other kind. Instead, his focus was consistently on the kinds of issues that had long concerned chemists: the nature of chemical elements and compounds and the course of chemical reactions. Although he viewed the gaseous state as a physical, rather than a chemical, condition, he ascribed a chemical role to caloric, which he continued to list among the elements. One area of traditional chemical theory that Lavoisier did not address was the study of affinities, a lacuna in his system of which he was well aware. His work did, however, reflect the substantially increased knowledge of chemical composition that eighteenth-century chemists had discovered and represented in affinity tables. Lavoisier was happy to replace his senses with instruments when possible, but he continued to have at his fingertips and in his head the plentiful information – about metals, calxes, acids, alkalis, salts, and the new gases – that his chemical predecessors had gathered.

In some respects, then, Lavoisier's revolution attested to the disciplinary maturity that chemistry had already attained, particularly in its Enlightenment role as philosophical foundation for many of the practical arts. This accounts for the quite ready acceptance of the new theories in Scandinavia and Germany, the regions where knowledge of chemical composition had made the greatest strides, especially in connection with mineralogy and metallurgy, and where the pragmatic approach to the identification of elements and compounds was already well entrenched.[35] From this point of view, the antiphlogistic account of combustion, calcination, and acidification was of secondary importance, merely displacing a few terms in the previous accounts of these processes. Many German chemists appear to have accepted the new theory

---

[34] T. H. Levere, "Martinus van Marum and the Introduction of Lavoisier's Chemistry in the Netherlands," in R. J. Forbes, E. LeFebvre, and J. G. Bruijn (eds.), *Martinus van Marum: Life and Work*, 6 vols. (Haarlem: Tjeenk, Willink, and Zoon, 1969–76), 1:158–286.

[35] Theodore M. Porter, "The Promotion of Mining and the Advancement of Science: The Chemical Revolution of Mineralogy," *Annals of Science*, 38 (1981), 543–70; Evan M. Melhado, "Mineralogy and the Autonomy of Chemistry around 1800," *Lychnos* (1990), 229–62.

quite readily, in a debate largely focused on just these reactions.[36] In other contexts, however, the dislodging of phlogiston was not so easily accomplished; and this is a sign of how chemistry, notwithstanding its maturity, remained a discipline with somewhat permeable boundaries. Scottish chemists and natural philosophers – for example, the geologist James Hutton (1726–1797) – were invoking phlogiston in connection with phenomena such as the earth's heat; for some of them it was too useful a notion to be given up.[37] In the first decade of the nineteenth century, some chemists in Germany and England resorted to phlogistic explanations for the effects of electric currents passing through solutions of salts. Even Humphry Davy dabbled with the idea of reviving a version of the phlogiston theory to connect chemical phenomena with those of heat, electricity, and light.[38] For all of Lavoisier's success in answering the questions of chemical composition, his victory over phlogiston deprived chemists of a valuable resource for making interdisciplinary connections of this kind.

Lavoisier's revolution will probably continue to cast its light backward on the eighteenth century. We can draw from it a salutary lesson on the inadequacy of Whiggish views of scientific development. The weight gain of metals in the course of calcination, for example, was scarcely a significant "fact" until Guyton and Lavoisier gave it a cogent reinterpretation. On the other hand, however, the anti-Whiggish belief that Lavoisier created a science *de novo* by providing a new conceptual scheme must also be abandoned. His new concepts and methods reshaped the discipline, but they did so by exploiting knowledge chemists already had about substances and reactions. His instrumentation and techniques of precision measurement were striking innovations, and Lavoisier very deliberately presented his system as a radically new one, especially in the *Traité*. But in its central focus on chemical composition and processes, the new system fundamentally reaffirmed the autonomy to which chemistry had already laid claim. In this sense, the new chemistry was the fulfillment of the old, not as the outcome of a teleological process but rather as a theoretical system created to provide an understanding of what were recognized as intrinsically chemical phenomena. Lavoisier transformed chemistry, but he did so by appropriating and reshaping the traditions consolidated in the course of its *ancien régime*.

---

[36] Hufbauer, *German Chemical Community*, pp. 118–44.
[37] Douglas Allchin, "James Hutton and Phlogiston," *Annals of Science,* 51 (1994), 615–35.
[38] David Knight, *Humphry Davy: Science and Power* (Oxford: Basil Blackwell, 1992), p. 68.

17

THE LIFE SCIENCES

Shirley A. Roe

For much of the eighteenth century, the biological world was seen as a very ordered place. Plants and animals yielded to Linnean classification. Physiological functioning was envisioned in mechanistic terms. And the generation of new animals and plants proceeded from preformed germs that had existed since the creation. All this order arose from God, who had created and organized the world for humans to understand and thereby to appreciate His handiwork and lead moral lives. Even the seemingly disordered, such as monsters and wonders of Nature, were generally brought under the paradigm of order.

All this was to be challenged by mid-century. Mechanistic physiology, based on the analogy of living organisms with machines, was to be considerably broadened by the introduction of Newtonian forces into physiology. The clear borders between the animal and plant kingdoms, and even between the plant and mineral worlds, were to be called into question by new experimental evidence. And the comfortable synthesis of mechanism with reproduction from preexisting germs was to encounter serious opposition from new theories of gradual development that raised the specter of materialism.

Some scholars have characterized eighteenth-century life sciences as a move from mechanism to vitalism, from a view that living phenomena could be explained by matter and motion alone to one that argued that living organisms possess some special force or principle that makes them distinct from dead matter.[1] Yet, as Thomas Hankins has pointed out, creating such absolute dichotomies can be misleading because it ignores the "middle ground."[2] In the late seventeenth and early eighteenth centuries, one does find mechanical explanations dominating physiology. Hankins claims, in fact, that by 1670

---

[1] See, for example, Theodore M. Brown, "From Mechanism to Vitalism in Eighteenth-Century English Physiology," *Journal of the History of Biology*, 7 (1974), 179–216; and Brown, *The Mechanical Philosophy and the "Animal Oeconomy"* (New York: Arno Press, 1981).

[2] Thomas L. Hankins, *Science and the Enlightenment* (Cambridge University Press, 1985), p. 120.

all major physiologists were mechanists. One has only to think of Giovanni Borelli's (1608–1679) *De motu animalium* (1676), with its hydraulic model of muscular contraction, or René Descartes's (1596–1650) *Traité de l'homme* (1664), in which sensations cause nerve fibers to open little doors in the brain so that particulate animal spirits can travel out to the muscles to produce contraction and movement.

These strictly mechanical explanations, based on the motion of material particles alone, began to fall away as the century progressed. Particularly as the ideas of Isaac Newton (1642–1727) became more widely known, one finds more and more that attractive forces are called upon in physiology and generation theory. But it would be a mistake to label all such forces as vitalistic, if what one means by this is a force that does not operate on the same principles as forces in physics do. Here is where we come to Hankins's middle ground. Several naturalists saw themselves as following in Newton's footsteps in using attractive forces, modeled on gravity, to explain living phenomena, but few of them would have labeled themselves as "vitalists." There were, of course, different varieties of Newtonianism that flourished in the early and mid-eighteenth century. There is the Newton of the *Principia* (1687), in which gravity is explained as a universal force whose operation can be understood solely as a function of mass and distance and whose existence is known from phenomena but whose cause is not to be speculated on. But then there is the Newton of the *Opticks*, where in the 1706 edition short-range attractive forces are introduced particularly to explain chemical phenomena and in the 1718 edition the immaterial (or at least nonmaterial) ether is thrown into the mix. The addition of forces to matter opened up a whole new explanatory horizon for naturalists. Some thinkers saw these forces as being purely mechnical, even though their causes were unknown. Others saw them as immaterial and necessary to life. And still others would, later in the century, see them as purely material, with no cause or origin other than matter itself.

The relationship between matter and activity was one of the burning issues of the eighteenth century, particularly in the life sciences. Was matter inherently passive, with its animation coming from an initial "push" by God (à la Descartes) or from mechanical forces added to matter by God? Or could matter itself possess active principles; could it be inherently active? What made this question a significant one in this period were its implications with regard to God and His role in the world. One of the dangers of the mechanical philosophy, recognized by Newton in his critique of Descartes's mechanism, was the possibility that a wholly mechanical universe could have formed by chance from the interaction of matter and motion. So somewhere a necessary role for God needed to be found. One answer was to argue that God could be known and understood by studying the intricacies of His creation. This approach can be seen in the flowering of natural theology in the late seventeenth and early eighteenth centuries. John Ray's (c. 1628–1705) *The Wisdom of God*

Manifested in the Works of the Creation (1704), for example, powerfully illustrates the richness of natural history for binding God to Nature. But God's existence could also be proven if one claimed that matter itself is purely passive and requires the addition of active forces or principles or spirits to produce the phenomena of the world. Much of the work in physiology and in generation theory in the eighteenth century was done with an eye on this question and with an awareness of the implications of attributing too much to matter. And by the middle of the century the question of active matter became the defining issue in the emergence of biological materialism and in the counterefforts among naturalists to retain a role for God in living phenomena.

This essay will focus on the two principal areas where questions of mechanism, vitalism, and materialism arose: physiology and theory of generation. In both areas we can see a shift from simple mechanism to more-complex versions, the emergence of more-vitalist approaches, and the challenge of materialism after mid-century. In all this we must remain aware of the extraordinary importance placed on the implications of various biological theories for the existence and role of God and for the moral basis of society.

Before turning to our two topics, let us briefly consider one other historiographical issue regarding the life sciences in the eighteenth century. Some scholars, most notably Michel Foucault and François Jacob, have argued that writing a history of "biology" in the eighteenth century is impossible because, as Foucault put it, "life [as we now think of it] does not exist: only living beings."[3] What he meant was that, in this period, the passion for natural history was to classify and order all natural objects. "Life" was an observable phenomenon, simply a characteristic of a class of natural beings. Thus, Foucault claimed, "the naturalist is the man concerned with the structure of the visible world and its denomination according to characters. Not with life."[4] Jacob has argued similarly that in the "Classical period" (the seventeenth and eighteenth centuries) naturalists were concerned with explaining living beings according to the same laws of mechanics governing inanimate objects. Living beings were looked upon as combinations of visible elements, these resting on an arrangement of particles united through mechanical action or forces. The concept of an organism that possess "life," Jacob claimed, did not arise until the early nineteenth century, when the notion developed that the organism contains within itself the principles of formation and regulation. Physiology and the study of reproduction then began to be founded on a necessary distinction between living (or organic) and nonliving (or inorganic) beings. In the eighteenth century, Jacob maintained, "There were not

---

[3] Michel Foucault, *The Order of Things: An Archaeology of the Human Sciences* (New York: Random House, 1973), p. 160.
[4] Ibid., p. 161.

yet functions necessary to life; there were simply organs which function. The aim of physiology was to recognize their machinery and mechanisms."[5]

Although it is perhaps overdramatic to say that life did not exist in the eighteenth century, Foucault and Jacob are pointing to an important characteristic of eighteenth-century life science.[6] It would be a mistake to characterize the mechanists of the eighteenth century as reductionists, as striving to explain the phenomena of living organisms solely in terms of physics. The point is that reductionism would have been a concept difficult to think of in this period because the kind of distinction on which it rests was simply not made. Life as a category of existence having a completely different character from existence in the inorganic realm was not a basic premise. This is not to say that animists and vitalists did not object to the overuse of mechanism in explaining the phenomena of living beings or that those materialists who wished to place life in matter itself did not imbue matter with qualities mechanists would have had little use for.[7] But there is a difference between the organism of the nineteenth century and the organized being of the eighteenth. It may be significant that the famous French *Encyclopédie* of Denis Diderot (1713–1784) and Jean Le Rond d'Alembert (1717–1783) does not contain an entry for "organisme," only one for "organisation": "the arrangement of parts that make up the animal body. . . . The organization of the solid parts happens by mechanical movements." And for "vie" the *Encyclopédie* states, "Life is the opposite of death . . . I define it as a continual movement of solids and fluids throughout the animal body."[8] Let us keep these issues in mind as we explore eighteenth-century physiology and generation theory.

THE RISE OF NEWTONIAN PHYSIOLOGY

To speak of "a" Newtonian physiology would be misleading. In the eighteenth century there were a variety of "Newtonianisms" and a variety of individuals who interpreted what being Newtonian meant in very different ways. In physiology we can identify three principal ways in which individuals either identified themselves as being Newtonian or invoked the image and

---

[5] François Jacob, *The Logic of Life: A History of Heredity,* trans. Betty E. Spillmann (New York: Pantheon Books, 1973), p. 34; see also pp. 32–4, 74–5, 88–9.

[6] See also Jacques Roger, "The Living World," in G. S. Rousseau and Roy Porter (eds.), *The Ferment of Knowledge: Studies in the Historiography of Eighteenth-Century Science* (Cambridge University Press, 1980), pp. 255–83.

[7] One cannot deny some continuity between early nineteenth-century physiologists who stressed the difference between life and nonlife and eighteenth-century vitalists. See François Duchesneau, *La physiologie des lumières: empirisme, modèles et théories* (The Hague: M. Nijhoff, 1982), pp. 442, 465.

[8] Denis Diderot and Jean Le Rond d'Alembert (eds.), *Encyclopédie, ou dictionnaire raisonné des sciences, des arts et des métiers,* 17 vols. (Paris: Birasson, David, Le Breton, Durand and Neufschâtel: S. Fauche, 1751–67), 11:629; 17:249. The article on "organisation" is unsigned; that on "vie" is by Louis de Jaucourt.

sanction of Newton in their work. Of course, individuals at different times and in different places and social settings might have widely different reasons for claiming the endorsement of Newtonianism. But there are some common threads that we can explore.

One group of Newtonian physiologists identified primarily with the Newton of the *Principia,* especially with Newton's reliance on mathematical reasoning and attractive forces. Theodore Brown and Anita Guerrini have identified a group of Newtonian physiologists clustered around Archibald Pitcairne (1652–1713) and David Gregory (1659–1708), who were active in the last decade of the seventeenth century and first decade of the eighteenth.[9] Pitcairne, Gregory, and their students relied on atomistic matter and short-range attractive forces to explain physiological phenomena. James Keill's (1673–1719) *Account of Animal Secretion,* published in 1708, was the high point of this group's efforts. Keill's account was based on attractive forces operating among a few basic types of particles in animal blood. His work bears strong resemblance to that of previous mechanists, with attraction added to replace mechanical devices in separating and cohering particles.

Both Brown and Guerrini see this early Newtonian group dissolving after Keill. But they differ in their views on the subsequent career of Newtonianism in physiology. Brown points out that in the 1720s, Pitcairne's and others' iatromechanical views received a great deal of criticism. Much of the criticism rested on what was seen as a lack of an experimental foundation for these theories. This did not constitute a rejection of Newtonism, for some of the critics claimed that the Newtonian physiologists had not been Newtonian enough because they had not relied on experiments. Brown also points to the rise to preeminence of Herman Boerhaave (1668–1738), the renowned professor of medicine in Leyden who identified himself as a Newtonian. Boerhaave's influence dovetailed with the expansion of Newtonianism that occurred after the appearance of Newton's *Opticks,* which clearly sanctioned the experimental method. Boerhaave's own emphasis on experiments and on the fiber as the basic building block of the animal body soon were reflected in a new group of physiologists in Britain such as George Cheyne (1671–1743) and Richard Mead (1673–1754). It is in this new emphasis on experiment that I think we can identify a second variety of Newtonianism in eighteenth-century physiology. Both through Newton's *Opticks,* particularly the Queries he continued to revise and add to through the years, and Boerhaave's works on physiology, which appeared in English beginning in the 1710s, an observational and experimental tradition arose, particularly in British physiology, that was self-consciously Newtonian.

---

[9] Brown, "From Mechanism to Vitalism"; Brown, *The Mechanical Philosophy;* Anita Guerrini, "James Keill, George Cheyne, and Newtonian Physiology, 1690–1740," *Journal of the History of Biology,* 18 (1985), 247–66; Guerrini, "The Tory Newtonians: Gregory, Pitcairne, and Their Circle," *Journal of British Studies,* 25 (1986), 288–311; and Guerrini, "Archibald Pitcairne and Newtonian Medicine," *Medical History,* 31 (1987), 70–83.

Guerrini has argued that another strand of Newtonianism stemmed from the *Opticks,* a strand which is exemplified in the later work of Cheyne. She points to the 1718 edition of Newton's *Opticks,* in which Newton suggested that attractive phenomena might be explained on the basis of a "subtle spirit," the ether. She sees Cheyne as relying on this ether-attraction equation in his introduction of a "spiritual" dimension in physiological explanation and in his view that a "self-active and self-motive Principle" needed to be added to the mechanical body for it to function as an animal.[10]

This third category of Newtonianism, particularly if we broaden it a bit, brings in the whole question of how attractive forces can be added to matter to explain the phenomena of living function. Are these forces material, non-material, spiritual? Within the mechanistic view they must be given to passive matter by God, but if some kind of ether is the material agent of God's activity, then how close is one coming to materialism? On the other hand, if activity in matter, particularly in living organisms, is based on something nonmaterial, then is one then bordering on animism? We know that Newton himself struggled with these questions in trying to explain the force of gravity.[11] It is also evident that such questions arose among those Newtonians who worried about how to explain the material activity that relying on forces entailed.

On the Continent, Cartesian mechanism remained dominant far longer than it did in Britain. Although Pierre-Louis Moreau de Maupertuis (1698–1759) and a few others introduced Newtonianism into France in the 1730s, Jacques Roger has pointed out that the impact of Newtonianism in the life sciences was not really evident until after 1745.[12] In the German areas, Friedrich Hoffmann (1660–1742), whose *Fundamenta medicinae* appeared in 1695, supported a mechanical foundation for physiology, although this was challenged by his colleague at the University of Halle, Georg Friedrich Stahl (1660–1734).[13] Newtonian inroads begin to be made by Boerhaave's physiology, although primarily in his sanctioning of the experimental method.

One of the principal ways that Boerhaave's physiology became well known was through an edition of his teachings prepared by his student Albrecht von Haller (1708–1777), the *Praelectiones academicae in proprias institutiones rei*

---

[10] Guerrini, "James Keill," pp. 260–5.

[11] This is especially evident in Newton's unpublished papers. See Betty Jo Teeter Dobbs, *The Janus Faces of Genius: The Role of Alchemy in Newton's Thought* (Cambridge University Press, 1991), and Betty Jo Teeter Dobbs and Margaret C. Jacob, *Newton and the Culture of Newtonianism* (Atlantic Heights, NJ: Humanities Press, 1995), chap. 1.

[12] Jacques Roger, *Les sciences de la vie dans la pensée française du XVIIIe siècle: la génération des animaux de Descartes à l'Encyclopédie* (Paris: Armand Colin, 1963), p. 250. Recently translated, except for the last chapter, as *The Life Sciences in Eighteenth-Century French Thought,* ed. Keith R. Benson, trans. Robert Ellrich (Stanford, CA: Stanford University Press, 1997).

[13] On Hoffmann and Stahl, see Duchesneau, *Physiologie,* chaps. 1–2; Lester S. King, "Stahl and Hoffmann: A Study in Eighteenth-Century Animism," *Journal of the History of Medicine,* 19 (1964), 118–30; and King, "Medicine in 1695: Friedrich Hoffmann's *Fundamenta Medicinae,*" *Bulletin of the History of Medicine,* 43 (1969), 17–29.

medicae (1739). Haller, who became one of the century's most important physiologists, was also a strong proponent of the Newtonian experimental method. Haller argued that the naturalist must explain, through mechanical forces, how sensation, motion, digestion, growth, and reproduction take place in the animal body. Yet he was cautious about applying mechanical laws in too simple a manner to physiological processes, because the organism is much more complex than a physical machine. His goal was to create a distinct "animal mechanics," in which the laws that govern physiology operate in the same manner as physical laws even though they may not be the same laws. Thus, forces may operate in organisms that are not found elsewhere.[14]

Haller is best known for demonstrating the existence of just such a force, that of "irritability." Haller performed a set of experiments on dogs and other animals, in which he exposed various parts of a live animal's body and irritated it by touch or chemicals. He found that muscles reacted by contracting, thereby demonstrating that they possess irritability. Haller separated sensation from irritability, arguing that the nerves possess "sensibility" and thereby the ability to transmit sensations to the brain.[15]

Haller's Newtonianism emerged not only in his championing of an experimental approach to physiology but also in his refusal to elaborate on the cause of irritability. As he argued,

> What therefore should hinder us from granting irritibility to be that property of the animal gluten in the muscular fiber, such that upon being touched and provoked it contracts, to which moreover it is unnecessary to assign any cause, just as no probable cause of attraction or gravity is assigned to matter. It is a physical cause, hidden in the intimate fabric, and discovered through experiments, which are evidence enough for demonstrating its existence, [but] which are too coarse to investigate further its cause in the fabric.[16]

One can almost hear in one's head when reading this passage Newton's famous declaration in the General Scholium to the *Principia* that he would not feign any hypotheses concerning the cause of gravity because it was enough to know that it exists from the phenomena.

The Newtonian sanctioning of forces that can be deduced from experiments without needing further explanation became quite popular in the mid-eighteenth century. It was even used on both sides of an argument. For example, one of Haller's opponents, Robert Whytt (1714–1766) also invoked the image of Newton. Whytt argued that both muscular movement and

---

[14] On Haller's physiology, see especially Shirley A. Roe, "*Anatomia animata:* The Newtonian Physiology of Albrecht von Haller," in Everett Mendelsohn (ed.), *Transformation and Tradition in the Sciences* (Cambridge University Press, 1984), pp. 273–300; and Duchesneau, *Physiologie,* chap. 5.

[15] Albrecht von Haller, "De partibus corporis humani sensilibus et irritabilibus," *Commentarii Societatis Regiae Scientiarum Gottingensis,* 2 (1752), 114–58. See also "A Dissertation on the Sensible and Irritable Parts of Animals," ed. and with an introduction by Owsei Temkin, *Bulletin of the History of Medicine,* 4 (1936), 651–99.

[16] Haller, "De partibus," p. 154.

sensation were caused by a "sentient principle" residing in the nerves. This sentient principle is immaterial and is the agent of the soul in the body.[17] We shall return to animism later, but what is important to note here is that Whytt thought all forces are immaterial, even the force of gravity. So when he argued that he did not have to explain exactly how the sentient principle operates because we accept gravity without knowing how it operates, Whytt thought of gravity in a very different light than did Haller. But the sanctioning of unexplained forces by Newton's example played a significant role in the ways in which Newtonianism emerged in different kinds of physiology in this period.

The application of Newtonian forces to physiology faced another problem, which Haller experienced as well. This was the specter of materialism. Why could not Haller's unexplained force of irritability be a property of matter and nothing else? Haller in a sense wanted to have it both ways; irritability was not immaterial, nor was it material in the sense of being possessed by matter on its own. Rather, it is a mechanical force added to passive matter by God. But not everyone saw it this way. The infamous materialist Julian Offray de la Mettrie (1709–1751) was quick to seize upon irritability, especially as evidenced in muscles that move after having been removed from the body, as proof that there is no need to postulate anything other than matter to explain living phenomena. As he proclaimed in *L'Homme machine* (1748), "Let us then conclude boldly that man is a machine and that there is in the whole universe only one diversely modified substance."[18] Haller was quick to rebut what he saw as La Mettrie's misuse of irritability and, in particular, La Mettrie's claim to an intellectual kinship with him.[19]

ANIMISM, VITALISM, AND THE REJECTION OF MECHANISM

A significant strand of antimechanism also ran through the eighteenth century, first in the animism propounded by Stahl and then later in the school of vitalism that arose in Montpellier, France. Further vitalist ideas can be found later in the work of naturalists such as John Hunter (1728–1793) and Johann Friedrich Blumenbach (1752–1840). Of course, the lines between mechanism, vitalism, and materialism can get quite blurred. It is fairly easy to classify someone as a vitalist who argues that there is something immaterial that exists in living organisms and not in matter. At times, this has led to the erroneous belief that anyone who thought that living organisms possess special

[17] See R. K. French, *Robert Whytt, the Soul, and Medicine* (London: Wellcome Institute of the History of Medicine, 1969); and Duchesneau, *Physiologie,* chap. 6.

[18] Julian Offray de La Mettrie, *Machine Man and Other Writings,* trans. and ed. Ann Thomson (Cambridge University Press, 1996), p. 39. See also Kathleen Wellman, *La Mettrie: Medicine, Philosophy, and Enlightenment* (Durham, NC: Duke University Press, 1992).

[19] See Roe, "*Anatomia animata,*" pp. 282–4.

forces should be called a vitalist. Thus, even someone like Haller has been classified as a vitalist. But, as we have seen, since he argued that the forces that act in living organisms are the same kind as those that operate in matter in general, he should be classified among the mechanists. Another problem sometimes arises in distinguishing between materialism and vitalism. A materialist believes that all the properties of life are contained in matter. But does this then make matter vital somehow, even though these properties emerge only in certain situations?

Stahl is one of the clearest cases of someone who can be labeled a vitalist (at least of the "animist" variety). Although Stahl recognized that mechanical principles operate in the organism, he felt that they were insufficient to account for physiological phenomena. Instead he argued, especially in his *Theoria medica vera* (1708), that a conscious, rational soul, or anima, governs vital functions. Life, he argued, is the conservation of the organism against dissolution. Matter itself could not accomplish this without the immaterial anima as the directing agent.[20]

Stahl is particularly remembered for his critiques of the mechanistic theories of Hoffmann and Boerhaave and for his influence on the group of vitalists who became active at the medical school at Montpellier later in the century. Stahl's ideas were introduced there in the 1730s through François Boissier de Sauvages's (1706–1767) medical lectures. By mid-century, physicians at Montpellier had developed their own ideas on the singularity of life, expressed especially by Théophile de Bordeu (1722–1776) and Paul-Joseph Barthez (1734–1806).[21] These physicians believed that in living organisms there exists a power or principle different in kind from other forces found in nature. In his *Recherches anatomiques sur la position des glandes et sur leur action* (1752), Bordeu argued that the glands operate in the body not by a mechanistic process of compression but rather by nervous action and some kind of force acting in the body. Although Bordeu referred frequently to this force he did not attempt to explain its action, noting only that we know that it exists from its actions in the body. Bordeu viewed the living body as a harmonious whole made up of separate "departments." Using what became his well-known metaphor of a hive of bees, Bordeu claimed that all the organs of the body have separate "lives" that coordinate and function together in the whole.

Barthez, who succeeded Bordeu at Montpellier, developed the most influential synthesis of vitalist thought to come out of the medical school. In his *Nouveaux élémens de la science de l'homme* (1778) Barthez argued that life is governed by a "vital principle" that is essentially unexplainable but not

---

[20] See Duchesneau, *Physiologie*, chap. 1; and Lester S. King, *The Philosophy of Medicine: The Early Eighteenth Century* (Cambridge, MA: Harvard University Press, 1978), pp. 143–51. See also note 13.

[21] On Montpellier vitalism, see Elizabeth A. Williams, *The Physical and the Moral: Anthropology, Physiology, and Philosophical Medicine in France, 1750–1850* (Cambridge University Press, 1994), chap. 1; Duchesneau, *Physiologie*, chap. 9; and Elizabeth Haigh, "Vitalism, the Soul, and Sensibility: The Physiology of Théopile de Bordeu," *Journal of the History of Medicine*, 31 (1976), 30–41.

unknowable. Barthez clearly distanced himself from Stahl and the idea of a rational soul operating in animal functions. Rather, Barthez saw the vital principle as an "abstraction" from observable phenomena. Invoking our now familiar sanctioning of Newton, Barthez likened the unknowability of the vital principle to gravitation.

Other vitalists emerged in different settings at approximately the same time, most notably John Hunter in Britain, whose *Lectures of the Principles of Surgery* were delivered in 1786–1787, and Johann Friedrich Blumenbach in Germany, whose *Ueber den Bildungstrieb und das Zeugungsgeschäfte* appeared in 1781.[22] Both individuals postulated the existence of vital forces or principles, and both of them saw life as irreducible to physical explanation. An interesting question to ask is why we see vitalist theories arising in France, Britain, Germany, and elsewhere in the latter half of the eighteenth century. To fully answer this question is beyond the scope of this essay. Furthermore, one must explore the local contexts to understand the motivations behind vitalist theories being suggested by any one of these naturalists. But we can also see a similar progression – from simple mechanism to force mechanics to the introduction of vital principles or vital matter – over the century in the area of generation theory. Let us turn now to that topic.

MECHANISTIC PREFORMATION

"Do you say that beasts are machines just as watches are?" asked Bernard de Fontenelle (1657–1757) in 1683. "Put a male dog-machine and a female dog-machine side by side, and eventually a third little machine will be the result, whereas two watches will lie side by side all their lives without ever producing a third watch."[23] In generation theory, mechanism, particularly of the matter and motion variety, reached its outer limits of explanatory capability, at least if one wanted to explain generation through epigenesis, as a gradual, mechanical development of an organism from unorganized particles. This is what Descartes attempted to do, and the theory of preformation arose largely as a response to the problems inherent in mechanistic epigenesis. The idea of preformation was that God had created, at one time, all of the organisms that would ever populate the earth and encased them within one another as tiny germs. The theory of preformation put forward in the late seventeenth century was thus often called "the preexistence of germs." To understand eighteenth-century theories of generation, we must first look at developments in the late seventeenth century.

---

[22] See François Duchesneau, "Vitalism in Late Eighteenth-Century Physiology: The Cases of Barthez, Blumenbach and John Hunter," in W. F. Bynum and Roy Porter (eds.), *William Hunter and the Eighteenth-Century Medical World* (Cambridge University Press, 1985), pp. 259–95; and Duchesneau, *Physiologie*, chap. 8.

[23] Bernard de Fontenelle, *Lettres galantes*, in *Oeuvres* (Paris: Libraires Associés, 1766), 1:312.

Descartes had hoped to include a mechanical theory of generation in his *Traité de l'homme,* where he used corpuscular mechanics to explain a variety of physiological phenomena such as digestion, movement, and sensation. Yet the key to generation eluded him for another decade, and it was not until shortly before his death in 1650 that he finally worked out his explanation for the mechanical formation of animal embryos from the mixing of semen from both parents, through a fermentation of particles. His resulting treatise, *De la formation de l'animal,* appeared posthumously with his earlier physiological work in 1664. Descartes's mechanistic explanation of generation based solely on matter in motion was for him the capstone of his mechanistic view of life.

Descartes's theory, however, was seen as both insufficient and disturbing. Nicolas Malebranche (1638–1715), who first formulated a theory of preexistence of germs in 1674, was responding directly to Descartes when he wrote:

> The rough sketch given by this philosopher may help us to understand how the laws of motion are sufficient to bring about the gradual growth of the parts of an animal. But that these laws should form them and link them together is something no one will ever prove. Apparently M. Descartes recognized this himself, for he did not press his ingenious conjectures very far.[24]

Malebranche's fullest discussion of preformation, in his *Entretiens sur la métaphysique et sur la religion* (1688), was in the form of a dialogue among a teacher (Malebranche), a student, and a priest. As a Cartesian in physics, Malebranche believed that God's role in the physical universe was limited to imparting to the material world the initial motion that then would be communicated from body to body for as long as the world existed. The question regarding generation was, could this communication of motion from one particle of matter to another be sufficient to create a new organism? Malebranche answered no, claiming rather that all the parts of the fly are in the grub, only awaiting their development. "At the time of the Creation," Malebranche argued with regard to God, "he constructed animals and plants for all future generations; he established the laws of motion that were necessary to make them grow. Now he rests, for he does nothing other than follow these laws."[25] As to the question of how all future organisms of each species could possibly be contained in their first representative created by God, Malebranche argued that because matter is infinitely divisible (another Cartesian position) this is at least conceivable. But what was not conceivable, he maintained, was that the laws of the communication of motion themselves could create new organisms at each generation. Thus in preexistence Malebranche combined mechanistic physics with the Cartesian view of God's initial involvement in our world. After creating the matter out of which it would form

---

[24] Nicholas Malebranche, *Oeuvres complètes,* ed. André Robinet, 20 vols. (Paris, Librairie Philosophique J. Vrin, 1958–), 12:264.

[25] Ibid., pp. 252, 253.

and imparting to it the initial necessary motion, God ceased to be directly involved in nature.

Malebranche was well aware of the microscopical research being done by his contemporaries, and he referred to the work of Marcello Malpighi (1628–1694), Jan Swammerdam (1637–1680), and Nehemiah Grew (1641–1712) to support his ideas on preexistence. Swammerdam had included some brief remarks in favor of preexistence of germs in a Dutch work of 1669, and these views became more widely available in Latin in 1672.[26] Malpighi's observations on chick development, published in 1673, gave clear evidence that the rudiments of the embryo could be seen in a fertilized egg that had not yet been incubated (but not in an unfertilized egg). Even so, Malpighi's observations, along with Swammerdam's, were often cited as evidence for the preexistence of the tiny preformed organism in the chick egg or the frog's egg, or in the plant seed.[27]

Yet the fact that these observations were immediately taken up by those making preformationist claims indicates that the concept of preexistence was not one that grew from observational evidence alone. It is clear, as several scholars have pointed out, that preformation through preexistence was a theory that responded more to philosophical than to observational needs. Roger was the first to explore the ties between preexistence of germs and mechanism and to point out that because the mechanistic view of the universe rested on passive matter and a noninterventionist God, preexistence made very good sense: "All of nature, in becoming mechanized, loses all spontaneity, becoming pure passivity in the hands of God, of the God who created it and is now content to conserve its existence and maintain its movement. At the limit, nothing could appear in nature that had not come from the creation of all things."[28]

The preformation-mechanism synthesis was very successful. Although there was some early opposition among a few mechanists, preexistence became the dominant theory by 1705 and remained so until mid-century. Among its supporters one can cite a number of figures who did not agree in many other aspects of their philosophies, including, for example, Boerhaave, Gottfried Wilhelm Leibniz (1646–1716), Fontenelle, and René Antoine Ferchault de Réaumur (1683–1757). (Some believed in preexistence in the male spermatozoa, but most were ovists.)[29] The notion that God must be fundamentally involved in each instance of reproduction, but only from His initial creative

[26] On Swammerdam's work and that of Antoni van Leeuwenhoek, who observed spermatozoa with the microscope, see Edward G. Ruestow, *The Microscope in the Dutch Republic: The Shaping of a Discovery* (Cambridge University Press, 1996).

[27] See Roger, *Sciences de la vie*, pp. 334–53; and Shirley A. Roe, *Matter, Life, and Generation: Eighteenth-Century Embryology and the Haller-Wolff Debate* (Cambridge University Press, 1981), pp. 2–9.

[28] Roger, *Sciences de la vie*, pp. 328–9.

[29] For a recent discussion of ovism versus animalculism, see Clara Pinto-Correia, *The Ovary of Eve: Egg and Sperm and Preformation* (Chicago: University of Chicago Press, 1997).

act and the mechanical laws that He had established, had widespread appeal. It rested on mechanistic science while avoiding atheism.

ORGANISMS AT THE BORDERS

One of the first indications that the neat synthesis of mechanism and preformation was about to be challenged was new evidence that emerged in the early 1740s showing that not all organisms may fit so well into the clear animal-plant distinction. The most widely discussed of these was the freshwater polyp, rediscovered by Abraham Trembley (1700–1784) in 1740. (The polyp had been first described by Antoni van Leeuwenhoek [1632–1723] in 1702 but not studied in depth.) The polyp (today called the hydra) exhibits both animal-like and plantlike properties. In locomotion, feeding, and response to stimuli it resembles an animal. Yet in addition to being green, it also resembles plants in its methods of multiplying. Its normal method of reproduction is by budding, but it also can regenerate from cut-off pieces. Both were contrary to all known methods of animal reproduction, and both challenged the idea of preexistence of germs.[30]

Other much-discussed organisms were the "plant worm" and the "vegetable fly," both produced by fungi that attack dead caterpillars or insects, sending out plantlike stalks and thus giving the appearance of an animal transforming itself into a plant. Plants displaying animal characteristics included the mimosa, or sensitive plant, and the tremella, an alga that seems to move spontaneously. First described by Michel Adanson (1727–1806) in the 1760s, the tremella was quickly seen as another link between the animal and plant kingdoms.

Many scholars in the eighteenth century were proponents of the chain of being, the idea that one could arrange all plants and animals on a scale of increasing complexity, stretching from the lowliest lichen up to human beings. The discovery of new organisms that linked the animal and plant kingdoms was hailed by naturalists such as Charles Bonnet (1720–1793) as confirmation of the existence of such a chain. As Bonnet remarked, "Nature descends by degrees, from Man to the Polyp, from the Polyp to the Sensitive plant, from the Sensitive plant to the Truffle, etc. The superior Species always adhere by some character to the inferior Species; these latter to Species more inferior still."[31] Although some people argued that the new discoveries showed that

---

[30] On the freshwater polyp, see Virginia P. Dawson, *Nature's Enigma: the Problem of the Polyp in the Letters of Bonnet, Trembley and Réaumur* (Philadelphia, PA: American Philosophical Society, 1987), and Aram Vartanian, "Trembley's Polyp, La Mettrie, and Eighteenth-Century French Materialism," *Journal of the History of Ideas,* 11 (1950), 259–86.

[31] Charles Bonnet, *Contemplation de la nature,* in *Oeuvres d'histoire naturelle et de philosophie* (Neuchâtel: Samuel Fauche, 1779–83), 8:514. See also Dawson, *Nature's Enigma,* pp. 167–76. The classic study is Arthur O. Lovejoy, *The Great Chain of Being* (Cambridge, MA: Harvard University Press, 1936). See

nature is far more nuanced and complex than our simple divisions into king-doms and species would suggest, others continued to claim that nature's order is evident in the hierarchy of organisms.

Microscopic organisms also presented new challenges at mid-century to the ordered view of living things. First described by Leeuwenhoek in the 1670s, microorganisms received considerable attention beginning in the late 1740s, and they emerged as one of the principal battlegrounds for materialism de-bates. Why did microscopic organisms become a source of controversy? Why could one not simply assume that they were just very little animals, as the name most commonly used for them, "animalcule," implies? One of the problems arose from where microorganisms were most commonly found. Most observers mixed up infusions, steeping various sorts of organic materials, usually from plants, in water. After several days, the water was found to be teeming with moving microscopic beings. One of the questions asked about these tiny creatures was what they were – animals, plants, or some sort of borderline or-ganism? What were the essential characteristics an organism needed to possess in order to be called an animal? And, more important, how had these micro-scopic organisms arisen in these infusions? Did they come from eggs or seeds already in the infusion or dropped in from the air or by insects? Or, more problematically, could they have arisen from the infused material itself, from some kind of spontaneous generation?

For the first half of the eighteenth century, the animal nature of micro-organisms was not called into question. Louis Joblot (1645–1723); for example, described observations he had made in 1716 on covered and uncovered infu-sions of boiled hay. These convinced him that microscopic organisms arise, like all others, from eggs. Henry Baker's (1698–1774) popular volume, *The Microscope Made Easy* (1743), expressed as well the widely held view that micro-organisms are simply little animals that reproduce from eggs.

All this was to be called into question with the work of John Turberville Needham (1713–1781), whose first book, *An Account of Some New Microscop-ical Discoveries* (1745), established him as a skilled observer of the microscopic world. Three years later, Needham published the results of a series of obser-vations on infusions of seeds, pulverized wheat, and mutton broth. He sealed and heated the mutton broth infusion but still found it teeming with micro-organisms after opening it several days later. In the wheat infusions he found filaments that seemed to release moving globules that then turned into fila-ments again. Enthralled, he wrote, "I own I cannot but wonder to this Day at what I saw; and tho' I have now seen them so often, I still look upon them with new Surprize." He thought that perhaps he was observing the conver-sion of plants into animal and back into plants.[32] Needham also thought he

---

also William F. Bynum, "The Great Chain of Being after Forty Years: An Appraisal," *History of Sci-ence*, 13 (1975), 1–28.
[32] John Turberville Needham, "A Summary of some late Observations upon the Generation, Compo-

was witnessing the activity of a vegetative force that, once released in the infusion, operated to produce new organisms from dead matter. Vegetating, active matter, not preexisting germs, produced these new organisms.

GENERATION THROUGH NEWTONIAN FORCES

Needham was not alone at mid-century in believing that vegetative forces, or forces of some kind, are at work in the generation of new organisms. With the broadening of mechanism that Newtonianism allowed, it became possible to conceive of generation through epigenesis based on matter and forces. Maupertuis, for example, argued in his *Vénus physique* (published anonymously in 1745) that since attractive forces are evident in physics and chemistry they might also play a role in generation. "Why should not a cohesive force, if it exists in Nature," he claimed, "have a role in the formation of animal bodies?"[33] In his theory, these attractive forces acted when seminal fluids from both parents mixed together. Yet Maupertuis also realized that attraction alone, acting on passive matter, would be insufficient to form an organized living creature. Over the next few years, he developed the idea that somehow the particles making up a living organism "remember" their former locations and instinctively unite in the reproductive matter. Although he did not develop these ideas further, Maupertuis's dilemma brings up a question that was key in all attempts to explain development gradually by epigenesis. What accounts for the resulting organization in a complex living being? If preexisting germs do not exist, how does matter "know" how to develop into an organism?

Maupertuis's ideas played a major role in influencing the thoughts of another – and perhaps the most significant – challenger of preexistence at mid-century: Georges Louis Leclerc, comte de Buffon (1707–1788).[34] Buffon began writing an account of his theory of generation in the mid-1740s, apparently after having several talks with Maupertuis on the subject. Buffon's theory was based on a distinction between two kinds of matter: organic and brute. When organisms eat, their digestive systems separate out the "organic molecules" and send them to various parts of the organism's body. Each part then sends

---

sition, and Decomposition of Animal and Vegetable Substances; Communicated in a Letter to Martin Folkes Esq.; President of the Royal Society," *Philosophical Transactions*, 45 (1748), 615–66; quotation on 647. See also Shirley A. Roe, "John Turberville Needham and the Generation of Living Organisms," *Isis*, 74 (1983), 159–84; and Renato G. Mazzolini and Shirley A. Roe (eds.), *Science Against the Unbelievers: The Correspondence of Bonnet and Needham, 1760–1780* (Studies on Voltaire and the Eighteenth Century, 243) (Oxford: The Voltaire Foundation, 1986), pp. 10–23.

[33] Pierre-Louis Moreau de Maupertuis, *The Earthly Venus*, trans. Simone Brangier Boas (The Sources of Science, no. 29) (New York: Johnson Reprint Corporation, 1966), p. 56. See also Michael H. Hoffheimer, "Maupertuis and the Eighteenth-Century Critique of Preexistence," *Journal of the History of Biology*, 15 (1982), 119–44; and Mary Terrall, "Salon, Academy, and Boudoir: Generation and Desire in Maupertuis's Science of Life," *Isis*, 87 (1996), 217–29.

[34] For more on Buffon, see especially Jacques Roger, *Buffon*, trans. Sarah Lucille Bonnefoi (Ithaca, NY: Cornell University Press, 1997).

off representative particles to become seminal fluid in the reproductive organs. This notion was part of Maupertuis's theory as well. But rather than invoke some kind of "memory" in these particles, as Maupertuis did, Buffon thought the seminal fluids from both parents form into the new organism through an "internal mould" (*moule intérieur*) and the action of "penetrating forces." Like Maupertuis, Buffon thought a Newtonian-style attractive force must be involved in the coalescing of these organic particles into an organism. Buffon was adamant in his opposition to generation from preexistence of germs, arguing that such an idea "is not only admitting that we do not know how it is accomplished but also renouncing all desire to conceive of it."[35] Yet one can argue that Buffon was not a true epigenesist either, because he thought that the "internal mould" rather quickly organized the seminal fluids from both parents into an organism.[36] Yet his clear rejection of preexistence of germs and his willingness to locate life in a particular kind of matter set his views apart from those of his contemporaries. Furthermore, Buffon unequivocally promoted active matter as the basis of life; "Living and animation," he boldly proclaimed, "instead of being a metaphysical degree of being, is a physical property of matter."[37]

Buffon's theory of generation was published in 1749 in the second volume of his *Histoire naturelle*. This contained not only his theory as formulated in the mid-1740s but also the results of series of observations carried out jointly with Needham, who arrived in Paris in 1746 full of enthusiasm for observing the microscopic world. The two set out to examine the "animalcules" found in seminal fluid (even searching for them in female "seminal fluid") and the microscopic beings swimming around in infusions. These joint observations served, in Buffon's view, to confirm fully the ideas he had already formulated about generation. And for Needham, they formed the basis for his theory of generation based on active, vegetative forces.

Prior to joining forces with Buffon, Needham had made some startling observations that, like those he made on infusions, seemed to provide evidence for active matter. He became interested in tiny whitish fibers that one finds among grains of blighted wheat. He found that when he added water to them they appeared to come to life as tiny worms – or, as he called them, "eels" – that moved in a twisting motion for several hours. Furthermore, he found that the same thing happened even two years later in a sample of blighted grains he had saved. Coupled with Trembley's observations on regeneration in the polyp – that no matter how many pieces one cut a polyp into they all seemed to be able to grow into a new polyp – Needham's witnessing of re-

---

[35] Georges Louis Leclerc, comte de Buffon, *Histoire naturelle, générale et particulière, avec la description du cabinet du roy* (Paris: L'Imprimerie Royale, 1749–89), 2:28. See also Roger, *Sciences de la vie*, pp. 527–84.

[36] Roger, *Buffon*, pp. 137–8. See also Peter J. Bowler, "Bonnet and Buffon: Theories of Generation and the Problem of Species," *Journal of the History of Biology*, 6 (1973), 259–81.

[37] Buffon, *Histoire naturelle*, 2:17.

vivification added observational evidence that called into question the simple picture of generation from preexistent germs.

THE RESURGENCE OF PREEXISTENCE THEORIES

Immediately upon their publications, the theories of Buffon and of Needham were controversial. Both theories rested on notions of active matter, and both utilized evidence from the microscopic world. As a result, over the next decade or two, preexistence theories reached their most fully articulated and observationally based form. In the work of Haller, Bonnet, and Lazzaro Spallanzani (1729–1799), preexistence of germs became the linchpin in a whole new line of defense against the tide of materialism rising in France. All three scholars were profoundly affected by encountering the theories of Buffon and Needham, and all three spent a good part of their professional lives trying to make sure that God remained in control of the generative process and thereby of the world of living (including human) creatures.

Haller, Bonnet, and Spallanzani voiced three principal concerns about the new theories of generation. First, they did not see how forces could, on their own, be responsible for the generation of complex, seemingly designed, living organisms. Second, they questioned, and in the case of Spallanzani tried to disprove, the experimental observations made by Needham and Buffon. And third, and perhaps most importantly, they worried about the implications of active matter – not just for understanding living phenomena but even more for religion and morality.

Haller, in a critique he wrote of Buffon's theory in 1751, raised the question of how forces could be responsible for generation. "I do not find in all of nature," he remarked, "the force that would be sufficiently wise to join together the single parts of the millions and millions of vessels, nerves, fibers, and bones of a body according to an eternal plan. . . . M. Buffon needs here a force that seeks, that chooses, that has a purpose, that against all the laws of blind combination always and infallibly casts the same throw."[38] As we saw earlier in Haller's work on irritability, his idea of how forces operate in nature was very clear to him. After having been given to matter by God, all forces operate, like gravity, on a physical basis and manifest themselves automatically. But they can possess no self-guiding abilities and thus cannot be responsible for the organization of a new living organism.

To bolster his arguments for preexistence, Haller made a series of observations on chick embryos – observations that confirmed to him that one could find evidence that all parts of the chick exist essentially from the earliest stages and that development proceeds via the heart beating and sending

[38] Albrecht von Haller, *Réflexions sur le système de la génération, de M. de Buffon* (Geneva: Barrillot, 1751), p. 42.

fluids through the rudimentary organism. Shortly after publishing these observations Haller became involved in a controversy with yet another epigenesist, Casper Friedrich Wolff (1734–1794). Wolff proposed his own force-based theory in 1759, arguing that a *vis essentialis* guided the gradual formation of the embryo. Haller and Wolff argued back and forth for a number of years, both on the issue of forces and on specific observations on chick development.[39]

Charles Bonnet announced his own critique of the views of Buffon and Needham shortly after learning of Haller's observations on chick eggs. In his *Considérations sur les corps organisés* (1762), Bonnet raised similar objections to Haller's, claiming that "there is in nature no true generation; rather what we improperly call generation [is] the beginning of a development that renders visible what previously we could not perceive."[40] At this point neither Bonnet nor Haller was able to counter the microscopical observations made by Buffon and Needham. But all this was to change when, in 1765, Spallanzani announced in his *Saggio di osservazioni microscopiche concernenti il sistema della generazione dei signori di Needham e Buffon* (1765) the results of a series of microscopical observations he had made on infusions. Spallanzani claimed that he had proof that all microscopic beings are true animals, that there is no such thing as microscopic animals turning into plants and vice versa, and that all animalcules in infusions generate from eggs and not from decomposing matter. These observations were just what Bonnet and Haller were looking for to bolster their campaign against the implicit materialism of Buffon's and Needham's theories of generation.

THE RISE OF MATERIALISM

Haller's and Bonnet's worries were not unfounded, for Needham's and Buffon's ideas and observations found a receptive audience among materialists such as Diderot and Paul-Henri Thiry d'Holbach (1723–1789). (La Mettrie died unexpectedly in 1751, well before the impact of materialism was felt in generation theory.) Diderot's interest in living organisms was piqued in 1749 when he first encountered the ideas of Buffon. In his own *Pensées sur l'interpretation de la nature* (1753), Diderot began questioning Buffon's distinction between organic particles and brute matter, asking whether there were really any differences between living and dead matter other than organization and self-movement. He remarked, for example, in a letter, that by eating food, organisms grow, so that, as he put it, "something dead put alongside something

---

[39] See Roe, *Matter, Life, and Generation*, chaps. 3–4.
[40] Charles Bonnet, *Considérations sur les corps organisés, où l'on traite de leur origine, de leur développement, de leur réproduction, etc.*, 2 vols. (Amsterdam: M. M. Rey, 1762), 1:169. See also Mazzolini and Roe, *Science Against the Unbelievers*, pp. 23–52.

living began to live." Yet, he continued, "you might as well say that if I put a dead man in your arms he would come back to life."[41] Within a few years, Diderot rejected Buffon's distinction, arguing that something he called "sensibility" is a property of all matter – inert in brute matter but rendered active in living organisms.

The culmination of Diderot's materialist thinking was his witty and provocative dialog, the *Rêve de d'Alembert,* written in 1769 but not published in his lifetime. Diderot's interlocutors are fictional figures whose names are the same as some of his contemporaries; d'Alembert, fellow *philosophe* and former co-editor with Diderot of the *Encyclopédie;* Mademoiselle de l'Espinasse, a salonièrre; Dr. Bordeu, the vitalist physician from Montpellier; and Diderot himself. In the second part of the *Rêve,* Diderot portrayed the fictional d'Alembert in the midst of a dream in which, at one point, he is looking through a microscope at an infusion, observing the ceaseless activity in the microscopic world. "In Needham's drop of water everything happens and passes away in the twinkling of an eye," d'Alembert reports. "You have two great phenomena," he continues, "the passage from the state of inertia to the state of sensibility, and spontaneous generation; let them suffice for you: draw from them the correct conclusions."[42] For Diderot, Needham's microscopical observations provided the model for a world based on ceaseless activity and change rather than preordained stability.

Needham's observations found their way into d'Holbach's *Système de la nature* (1770) as well. Although d'Holbach's materialism was more chemical than biological in nature, he argued, like Diderot, that matter is fundamentally active, requiring only the proper circumstances for this to become manifest. D'Holbach cited Needham's work as providing evidence that inanimate matter can become living matter and vice versa. "Would the production of a man independent from the ordinary means," he queried provocatively, "be more marvelous than that of an insect from flour and water?"[43] Diderot and d'Holbach were part of a group of *philosophes* who dined together regularly at d'Holbach's house, where their discussions ranged over all of the more radical issues of the day.[44]

Although Haller and Bonnet opposed d'Holbach's views, the theory of preexistence that they had placed so much faith in to preserve God and

---

[41] Denis Diderot, *Correspondance,* ed. Georges Roth (Paris: Editions de Minuit, 1955–70), 2:283; letter of 15 October 1759.

[42] Denis Diderot, *Oeuvres philosophiques,* ed. Paul Vernière (Paris: Editions Garnier Frères, 1964), pp. 299, 303. On Diderot's materialism, see Roger, *Sciences de la vie,* pp. 585–682; and John Furbank, *Diderot: A Critical Biography* (New York: Knopf, 1992), pp. 324–42.

[43] [Paul-Henri Thiry d'Holbach], *Système de la nature, ou des loix du monde physique et du monde moral, par M. Mirabaud* (London [Amsterdam], 1770), p. 23 n. 5. On d'Holbach's materialism, see Roger, *Sciences de la vie,* pp. 678–82; and Shirley A. Roe, "Metaphysics and Materialism: Needham's Response to d'Holbach," *Studies on Voltaire and the Eighteenth Century,* 284 (1991), 309–42.

[44] See Alan Charles Kors, *D'Holbach's Coterie: An Enlightenment in Paris* (Princeton, NJ: Princeton University Press, 1976).

mechanism fell out of favor with their deaths. Epigenesis replaced it, particularly in the German context, where Blumenbach's *Bildungstrieb* ("building force") opened the door to a teleological epigenesis that was a far cry from the mechanical epigenesis of Descartes or the developmental epigenesis of Wolff. German naturalists went on to build a philosophy of nature on the notion that organization and development are built into Nature, and are not something requiring explanation.[45]

CONCLUSION

Around 1800, the word "biology" began to be used independently by Jean-Baptiste Lamarck (1744–1829) and Marie-François-Xavier Bichat (1771–1802) in France, and by Karl Friedrich Burdach (1776–1847) and Gottfried Reinhold Treviranus (1776–1836) in Germany. Although each of them meant something slightly different by the term, all four saw a need to unify the life sciences in distinction to sciences dealing with the nonliving world. Lamarck attempted to define life in physical terms but as a separate level of nature. In some ways he carried on the materialism of the *philosophes* by opposing any notion of a vital principle separate from matter. But in other ways he wanted to move beyond the supposition that matter contains in itself all the properties of the organic world. "Life," as he put it, "is an order and state of things in the parts of every body that possesses it."[46]

One can overgeneralize about the "thought of a century." Yet we can say that many of the central questions that occupied naturalists in the eighteenth century when confronting the phenomena of living organisms were quite different from those that would occupy their counterparts only decades later. The burning concerns about proving or disproving the existence of God, about passive or active matter, about Newtonian forces or vital principles – all these were to become nonissues. As Jacques Roger has remarked, "At the beginning of the century everybody spoke of God; at the end everybody spoke of nature, and they probably thought that great progress had been made, without realizing that they were substituting a new ideology for the old one."[47] It is in fact the ideology of the eighteenth century and how it led naturalists to view and understand living organisms that has occupied us in this essay.

---

[45] See Roe, *Matter, Life, and Generation,* chap. 6.
[46] Jean-Baptiste Lamarck, *Recherches sur l'organisation des corps vivans* (Paris: Maillard, 1802), p. 71.
[47] Roger, "Living World," p. 283.

18

THE EARTH SCIENCES

Rhoda Rappaport

As defined in the eighteenth century, "natural history" meant description (then a synonym for "history") and classification of everything in nature, from the cosmos to the insect. Understandably, then, few naturalists attempted surveys or syntheses of so shapeless a range of subjects. One of the few, Carl Linnaeus, tried to chart the order in all realms of nature in a series of taxonomic works devoted to the animal, vegetable, and mineral kingdoms. Another, Georges-Louis Leclerc, comte de Buffon, criticized taxonomies as incapable of accurately depicting nature in all its variety; by omitting botany, Buffon's *Histoire naturelle* narrowed its focus in one respect while broadening it in others, as the author included the origin of the solar system, the history of the earth, and a treatment of animals that went beyond anatomy into such matters as environments and heredity.

Some naturalists contented themselves with producing compendia of "curiosities," and others tried to give unity to these collections by indicating their aim of revealing, in John Ray's famous title, *The Wisdom of God Manifested in the Works of the Creation* (1691). Many sought a degree of completeness by selecting either a geographical or a topical focus. As examples of the former, one can cite the long British tradition of local histories that effectively began with Robert Plot's *Oxfordshire* (1677) and eventually included one literary classic, Gilbert White's *Natural History and Antiquities of Selborne* (1789). Although studies of this kind seem to have been less common outside Britain, a striking feature of almost all such works was the attention given to human artifacts, chiefly those of antiquity, and often to such topics as language, customs, and migrations. When naturalists bothered to explain these choices, they indicated their preference for the factual and hence the unbiased aspects of human culture; these merited the same descriptive treatment accorded to nature.

Other writers tried to be exhaustive by selecting clusters of related topics. As a small sample of those who included "natural history" in their titles, John Woodward studied rocks and fossils (1695); Dezallier d'Argenville, what he called "two of the main parts of natural history," rocks and shells (1742); Vitaliano Donati, chiefly the flora and fauna of the Adriatic Sea (1750); Rudolf Erich Raspe, "new islands born from the sea" (1763); and John Williams, "the mineral kingdom" (1789).[1] To such authors, as to the local historians, the common enterprise was description, even if causal explanations inevitably entered into all these works.

Since there could be a natural history of virtually anything – as David Hume showed by producing a controversial *Natural History of Religion* (1757) – to write a history of this pervasive genre is not feasible. It may seem anachronistic here to single out the earth sciences – or the life sciences (see Chapter 17 in this volume) – but one unifying theme, recognized at that time, was in fact called "the theory of the earth." This phrase received wide currency thanks to Thomas Burnet, whose *Sacred Theory of the Earth* (1681) aroused debate in Britain and on the Continent. Burnet's phrase would be employed again, for a very different synthesis, in the first volume of Buffon's *Histoire naturelle* (1749). After mid-century, such large syntheses, dubbed "systems," were generally repudiated, the most admired geologists being those "known for their travels, field observations, and caution against overeager generalization or systematizing."[2] At the same time, wishing to "get good information and organize it in the right way," men such as Horace-Bénédict de Saussure and Déodat Dolomieu continued to say, in the 1790s, that their aim was to contribute to an improved "theory of the earth."[3]

By that decade, "geology" and "geognosy" were coming into use, along with such older terms as "mineralogy" and "physical geography," the latter now being redefined as fields subordinate to geology. This new/old science of 1800 relied on physical laws, chemical analysis, and historical reconstruction. A century earlier, Burnet's allies had been Cartesian physics and ancient historical texts, sacred and profane. The result was as much a history (in the modern meaning of the word) as a theory, in that Burnet transformed the hypothetical cosmogony and geogony of René Descartes into a sequence of irreversible events, documented by both nature and Scripture. More than Nicolaus Steno's

---

[1] Quotations are from the title pages of the several works. The diversity of subjects can be glimpsed in, but is not exhausted by, Nicholas Jardine et al. (eds.), *Cultures of Natural History* (Cambridge University Press, 1996). On p. 129 of this volume, Daniel Roche quotes the definition of natural history in the Diderot *Encyclopédie;* see also *Encyclopaedia Britannica* (1771), s.v. Natural History.

[2] This quotation and the next are from Kenneth L. Taylor, "The Beginnings of a French Geological Identity," *Histoire et nature,* 19–20 (1981–2), 75, 79.

[3] Dolomieu, "Discours sur l'étude de la géologie," *Journal de physique, de chimie et d'histoire naturelle,* 2 (1794), 270–1. Horace-Bénédict de Saussure, "Agenda," appended to the last volume of his *Voyages dans les Alpes* (1779–96), and also published in *Journal des mines,* 4 (1796), 1–70. A recent synthesis is provided by Gabriel Gohau, *Les Sciences de la terre aux XVIIe et XVIIIe siècles* (Paris: Albin Michel, 1990). A synthesis with a different focus is by Martin J. S. Rudwick, "The Shape and Meaning of Earth History," in David C. Lindberg and Ronald L. Numbers (eds.), *God and Nature* (Berkeley: University of California Press, 1986), chap. 12. For Britain in particular, see Gordon L. [Herries] Davies, *The Earth in Decay* (New York: American Elsevier, 1969), and Roy Porter, *The Making of Geology* (Cambridge University Press, 1977). For a valuable reference tool, see François Ellenberger, *Histoire de la géologie,* vol. 2 (Paris: Lavoisier, 1994).

famous *Prodromus* (1669), Burnet can be said to have set the agenda for the next decades: how to reconstruct the earth's past and how to combine natural evidence with human records.[4]

FOSSILS AND THE FLOOD

In Burnet's day and for a time thereafter, the process of reconstruction entailed finding answers to two key questions: what are fossils (a subject ignored by Burnet himself), and what role should be attributed to Noah's Flood? Debate about fossils centered primarily on marine shells – rather than, for example, petrified wood – which not only were hard to identify but also could be found far from modern seas and far above or below sea level. Men such as Steno, Paolo Boccone, Robert Hooke, and Agostino Scilla argued cogently for the view that such fossils were remains of true animals. Others disagreed because they wondered why some fossils had no living analogues, why some forms seemed to be clustered chiefly within certain strata, and how supposedly marine creatures had been transported to burial sites on land. To such men as Filippo Buonanni, Edward Lhwyd, Martin Lister, and Joseph Pitton de Tournefort, most marine fossils had been produced in the rocks themselves, whether by a "plastic power" or by a process of growth from seeds or eggs. In general, these objects were *lusus naturae,* or nature's playful way of imitating genuine organisms.[5]

These issues were vigorously debated from the 1660s until about 1710 or 1720. No decisive discovery or idea could "refute" a concept such as plastic power or the growth of organisms within rocks, but even Edward Lhwyd admitted that his own theory of seeds seemed bizarre. In a period of increasing allegiance to the view that nature is a "machine," one detects growing aversion to essentially mysterious and "unnatural" processes. French academicians, for example, paid little attention to Tournefort's seeds, whereas German naturalist Johann Jacob Baier would in 1708 declare his own dislike of "conjuring up some agent distinct from God, which . . . directs and modifies corporeal creatures, sometimes toying idly with them, often fashioning absurdities and

---

[4] Jacques Roger, "La Théorie de la terre au XVIIe siècle," *Revue d'histoire des sciences,* 26 (1973), 23–48. Other valuable studies of Burnet and his era include Marjorie Hope Nicolson, *Mountain Gloom and Mountain Glory* (Ithaca, NY: Cornell University Press, 1959); Mirella Pasini, *Thomas Burnet: Una storia del mondo tra ragione, mito e rivelazione* (Florence: La Nuova Italia, 1981); Roy Porter, *Making of Geology,* chap. 3; Rhoda Rappaport, *When Geologists Were Historians, 1665–1750* (Ithaca, NY: Cornell University Press, 1997), chap. 5; and Paolo Rossi, *The Dark Abyss of Time,* trans. Lydia G. Cochrane (Chicago: University of Chicago Press, 1984), chaps. 7–11. That Steno was associated chiefly with the origin of fossils is implicit in Victor A. Eyles, "The Influence of Nicolaus Steno on the Development of Geological Science in Britain," *Acta Historica Scientiarum naturalium medicinalium,* 15 (1958), 167–88.

[5] Strictly speaking, fossil animals and plants growing within the rocks were not *lusus,* but the organisms had never lived in their natural environments. The best analysis is in Martin J. S. Rudwick, *The Meaning of Fossils* (London: Macdonald, 1972), chap. 2; see also chap. 1 for use of the word "fossils."

monstrosities."[6] Significant in Baier's text was his retention of the category, not the concept, of *lusus naturae*. Here he placed various forms that awaited explanation – such as the suspiciously symmetrical shapes (round, conical, star-like) that might be crystals or concretions rather than organisms. For some decades thereafter, Baier's cautious solution applied to belemnites in particular, as naturalists struggled to interpret these baffling fossils.[7]

Robert Hooke believed that the organic origin of fossil shells would prove unacceptable unless coupled with an explanation of their transport from the sea to their various burial sites. He thus proposed that the earth had had a history of localized earthquakes, which were responsible for the elevation and depression of tracts of land. (These upheavals, he added, had perhaps destroyed those fossil species without living analogues.) Hooke's theory aroused almost no enthusiasm in the Royal Society of London, some of the Fellows arguing that human records should confirm the frequency of such striking events; without adequate confirmation of this kind, the theory would remain mere speculation. Eventually, Hooke himself admitted that Fellows were finding attractive an alternative that ancient writings did confirm: the possibility that the Flood had deposited marine fossils.[8]

Flood geology can be dated from the publication of *An Essay toward a Natural History of the Earth* (1695) by John Woodward, one of Burnet's critics. Although writers such as Hooke and Boccone had not ignored the Flood, they and their successors generally regarded that event as only one of a series of episodes responsible for the long succession of sedimentary strata. Woodward thought otherwise, and his book played at least three vital roles. First, as an expert fossilist, he examined at length and in convincing fashion the evidence for considering marine fossils to be organic in origin. Second, he argued that he had solved the problems of their transport and deposition by attributing them to the Flood. (In fact, he held that very little geological change was post-diluvial, the earth's major landforms being products of the Flood and of up-heavals immediately following the retreat of the waters.) Third, by presenting the Flood as miraculous in cause and natural in effects, he induced readers to ask whether miracles could properly enter into natural philosophy. Ironically, Burnet had been attacked for virtually excluding the divine from nature, and Woodward was perceived by critics and even by some disciples as going too far in the opposite direction.[9]

---

[6] Quoted from Baier, *Oryktographia norica* (Nuremberg, 1708), in *The Lying Stones of Dr. Johann Bartholomew Adam Beringer*, trans. and ed. Melvin E. Jahn and Daniel J. Woolf (Berkeley: University of California Press, 1963), pp. 181–2. Also, Rappaport, *When Geologists Were Historians*, pp. 129–35.

[7] Baier, *Oryktographia*, pp. 30–1, and the rehearsal of interpretations of belemnites by Jean-Etienne Guettard, *Mémoires sur différentes parties de la physique, de l'histoire naturelle, des sciences et des arts* (Paris, 1768–86), vol. 6, pp. 215–96.

[8] Rhoda Rappaport, "Hooke on Earthquakes: Lectures, Strategy and Audience," *The British Journal for the History of Science*, 19 (1986), 129–46. For a different approach to Hooke, see Yushi Ito, "Hooke's Cyclic Theory of the Earth in the Context of Seventeenth-Century England," *British Journal for the History of Science*, 21 (1988), 295–314.

[9] The best study is Joseph M. Levine, *Dr. Woodward's Shield: History, Science, and Satire in Augustan England* (Berkeley: University of California Press, 1977). See also Victor A. Eyles, "Woodward," *Dictionary of Scientific Biography*, XIV, pp. 500–3.

Some Woodwardians tried to substitute natural causes for miracle – hence the proposal by William Whiston in his *New Theory of the Earth* (1696) that a comet passing near the earth had caused the Flood. Others, including Whiston, argued that Woodward had underestimated the turbulence of the Flood and its aftermath, because sedimentary strata and their enclosed fossils had rarely been deposited in the order of their specific gravities, as Woodwardian theory required. A worried Swiss naturalist, Louis Bourguet, would in 1729 propose another solution to the latter problem, suggesting that the Flood had been a series of successive sedimentary events. To disciples, then, it seemed clear that the Flood had played a major role but that Woodward's theory needed improvement. To critics, the difficulties called for rejection of the theory. In Paris, for example, academicians such as Réaumur and Antoine de Jussieu found their observations to be incompatible with diluvialism. So, too, did Antonio Vallisneri and Anton Lazzaro Moro, who examined Woodward's views in detail and firmly rejected all use of the Flood. To combine miracle with science would, in Vallisneri's words, merely produce "an indigestible mixture of science and morality."[10] The latter viewpoint would be shared by Buffon.

Controversy about the Flood continued throughout the century, with some naturalists remarking that diluvialism constituted one school of thought in conflict with a serious rival, namely, the view that wherever marine fossils are found, there had occurred "a long presence" of the sea.[11] It should be noted that members of the latter school usually did not deny that the Flood had taken place, the historical evidence consisting not only of Genesis but also of flood legends gathered from the Americas, Asia, and ancient Greece. For decades, historical and geological evidence would produce no consensus about the Flood: Was it universal? Was it literally a flood or, perhaps, a giant earthquake? Did it have detectable natural effects? Was it one of a series of events, unusual in being recorded historically? Was it perhaps a recent event, coming late in a lengthened history of the earth? No historian has yet analyzed these issues as they appear in writings especially of the post-1750 period, although it remains part of folklore in the history of geology that James Hutton liberated geology from the constraints of biblicism.[12]

BUFFON'S SYNTHESIS AT MID-CENTURY

Topics other than fossils and diluvialism did not bulk large in geological writings until after mid-century. Here it seems appropriate and convenient to use

[10] Antonio Vallisneri, *De' Corpi marini, che su' monti si trovano,* 2nd ed. (Venice, 1728), p. 65. Disciples and critics are discussed in Rappaport, *When Geologists Were Historians,* chap. 5.
[11] Louis Bourguet, *Lettres philosophiques* (Amsterdam, 1729), pp. 177–80; Elie Bertrand, *Recueil de divers traités sur l'histoire naturelle de la terre et des fossiles* (Avignon, 1766), pp. 32–3.
[12] E.g., Victor A. Eyles, "Hutton," *Dictionary of Scientific Biography,* VI, pp. 577–88. Some interpretations of the Flood are treated in Rappaport, "Geology and Orthodoxy: The Case of Noah's Flood in Eighteenth-Century Thought," *British Journal for the History of Science,* 11 (1978), 1–18.

that widely read work of synthesis and originality, Buffon's "History and theory of the earth" (1749), to examine both the state of knowledge at that time and the provocations Buffon offered his readers.

Buffon manifested little interest in fossils (until later in his career), being able to rehearse briefly the older dispute about their origins. Nor did he pause to analyze the few earlier discussions of fossil populations peculiar to various locales or the failure of French scientists to find any fossils at the highest elevations of the Cordilleras.[13] Taking as the task of all scientists the search for large patterns in nature, Buffon used the worldwide occurrence of fossil mollusks to argue for an implicitly long history of marine sedimentation. His own minimal observations, in conjunction with other material, showed that such accumulations could not be the work of the Flood. The present-day activity of the sea, as Buffon knew, was visible in the constant alteration of shorelines, but he also argued that in the past marine currents had built (and were still building) landforms on the sea floor. The latter argument left him with a question he could not answer: how had submarine landforms emerged from the sea? With a graceful shrug, he allowed that he had good evidence for the building of mountains on the sea floor – his critics would find this dubious – but no way at all to explain their subsequent elevation.[14]

Like most of his predecessors, Buffon could find no mechanism for uplift. Some had been studying such topics as the existence of heat within the earth and the possible role of volcanoes and earthquakes in shaping the earth's relief. By 1749, consensus seems to have been reached in two areas. For one, volcanoes – and earthquakes were commonly thought to be associated with them – were superficial phenomena, the sulfureous odors detectable during eruptions and the sulfur compounds in volcanic ejecta both showing that what had taken place was the burning of bitumens located in the relatively recent crust. For another, however, it seemed fairly clear that the earth did possess some internal heat even in nonvolcanic regions, as shown by the heat gradient in mines. What kind of heat, its extent, causes, and depth, baffled scientists for the entire century.[15] For Buffon, the earth's heat had existed at the time of the planet's formation, but he ignored (until years later) the possible persistence of residual heat. Volcanoes were thus "accidental," and were not "general" fea-

---

[13] Buffon, *Histoire naturelle*, vol. 1 (Paris, 1749), Articles 8, 12. In Antoine de Jussieu's examination of fossil ferns in Lyonnais, the author posited marine currents transporting the plants from the West Indies to France; Buffon's theory required normal movement of the sea in the opposite direction. As for the Cordilleras, Buffon's theory had all mountains originating on the ocean floor; marine fossils should thus have been found even at the highest elevations.

[14] The best analysis is by Jacques Roger, *Buffon: Un philosophe au Jardin du roi* (Paris: Fayard, 1989), chaps. 5–7. For the evidence of successive sedimentation drawn from the geographer Varenius, as used by Buffon and his predecessors, see Rappaport, *When Geologists Were Historians*, pp. 164–5, 173–4, 245–6, and Claudine Cohen, *Le Destin du mammouth* (Paris: Seuil, 1994), pp. 73–4.

[15] The early experiment by Nicolas Lemery (published in 1703) to simulate an eruption, using sulfur and iron filings, was still being cited by the *Encyclopaedia Britannica* (1771), s.v. Vulcano. The best mid-century summary is by Jean-Jacques Dortous de Mairan, *Dissertation sur la glace* (Paris, 1749). See the later discussion for these issues in Hutton's day.

tures of the earth's history; in effect, they were mountains that had happened to become inflamed.

Buffon leveled at his predecessors two important methodological criticisms. First, he consistently ignored human (ancient) texts, sacred and profane, insisting that modern observations were more reliable. His second point, integral to the first, was emphasis on a method later dubbed "actualism": processes observed in the present are our only guide to those of the past. Earlier writers, to be sure, had had much to say about the inviolable laws of nature, but Buffon considered himself to be more rigorous in the application of this principle. At the same time, he had to admit that uniformity of *effects* depends on the conditions under which physical laws operate. As he put it, the same causes now at work might, under earlier conditions, have worked more rapidly or with more striking effects.[16]

NEW APPROACHES AT MID-CENTURY

Geology after about 1750 was in some respects markedly different from what had gone before, and one is tempted to attribute much to Buffon. Twenty years later, Genevan Charles Bonnet could still say that to read Buffon was to understand what was known and what remained to be known.[17] Certainly, an old problem, now made glaringly obvious, was the need to explain the elevation of land masses. At the same time, relatively few scholars seem to have objected to Buffon's refusal to use ancient texts, and it may be that he persuaded some readers to consider that the geological past long antedated the appearance of humankind. In these decades, however, new discoveries and methods also meant that Buffon's text was in some respects becoming outdated. During the 1750s and 1760s, discovery of many hitherto unsuspected volcanic sites raised once again the question of the role of heat. In the same years, geologists outside Germany and Sweden began to recognize that they had neglected the relevance of chemistry to the reconstruction of the past. Above all, post-1750 geologists resurrected the older charge against Cartesianism, now applied to Buffon: system-building was premature, large syntheses being based on slender foundations. To signal his own priorities, Nicolas Desmarest in 1757 outlined in detail the observations required of naturalists seeking to understand the "physical geography" of the earth. Abraham Gottlob Werner's *External Characters of Minerals* (1774) was essentially a guide useful for fieldwork, as

---

[16] Roger, *Buffon*, p. 148. Reflections on the problem of knowing past conditions are in Johann Friedrich Henckel, *Pyritologie*, trans. d'Holbach (Paris, 1760), vol. 2, pp. 410–11, including further commentary by one of Henckel's pupils. A thoughtful analysis can be found in the introductory section of R. Hooykaas, *Natural Law and Divine Miracle: The Principle of Uniformity in Geology, Biology and Theology* (Leiden: Brill, 1959).

[17] Letters dated 1769, in *The Correspondence between Albrecht von Haller and Charles Bonnet*, ed. Otto Sonntag (Bern: Huber, 1983), pp. 819–27.

Werner enumerated those properties allowing for accurate identification of minerals and rocks. By 1794, Déodat Dolomieu could announce to his students that the laboratory and the museum, albeit vital adjuncts, could not substitute for travels, hammer in hand.[18] Such programs did not mean that geologists avoided theorizing, but their efforts in this respect became self-consciously tentative, as even the most famous of syntheses, Wernerian neptunism, was also intended as a field guide, the schema to be modified on the basis of local observations.

Desmarest, Werner, and Dolomieu also represent a new breed of investigator. In earlier decades, and to a lesser extent in later ones, writers on geological topics might be characterized as chiefly independent scholars and amateurs. However bookish, these men did pay at least some attention to fieldwork, as is evident in articles by members of the Paris Academy of Sciences and in John Woodward's *Brief Instructions for Making Observations in all Parts of the World* (1696). Nonetheless, it was common practice to gather specimens (or get access to those in a museum) and to examine them with, at best, erratic attention to the places where they had been found. After mid-century, and occasionally earlier, geologists tended to be practical men, associated with schools of mines, subsidized by various governments, supported by individual patrons, to see what they could do to find new natural resources or to improve existing technologies. To contrast Buffon with Desmarest may here be instructive. As *intendant* of the Jardin du roi, Buffon had institutional support for his writings, but that support consisted wholly of his post as administrator, not a professor at the Jardin and not a person expected to write the kinds of books he produced. Desmarest, on the other hand, had state support for his travels in Auvergne – support not for the study of volcanoes but for the extensive travel required of an "inspector of manufactures." Some French scientists with interests in mining also received subsidies, so the belated founding in Paris of a school of mines (1783) should not be taken to mean a lack of an institutional center for research pertinent to geology. As in France, Italian support came not from specialized institutions but from reform-minded bureaucrats in the various small states, and sometimes from individual patrons. In the latter category was Sir John Strange, one of a small number of British diplomats who developed an interest in geology. (Most famous among these was Sir William Hamilton in Naples.) Strange not only conducted

---

[18] Desmarest, "Géographie physique," in Diderot, *Encyclopédie*, vol. 7 (1757), pp. 613–26. (References to the *Encyclopédie*, in this and later notes, are all to the original folio edition.) Victor A. Eyles, "Abraham Gottlob Werner (1749–1817) and His Position in the History of the Mineralogical and Geological Sciences," *History of Science*, 3 (1964), 105–7. Dolomieu, "Discours sur l'étude"; here, as elsewhere, his model was Saussure's travels. See Kenneth L. Taylor, "Desmarest" and "Dolomieu," *Dictionary of Scientific Biography*, IV, pp. 70–3, 149–53, and Alexander M. Ospovat, "Werner," ibid., XIV, pp. 256–64. Valuable articles by both Ospovat and Taylor are in Cecil J. Schneer (ed.), *Toward a History of Geology* (Cambridge, MA: MIT Press, 1969), pp. 242–56, 339–56. Probably the best study of Werner consists of the introduction and notes to Werner, *Short Classification and Description of the Various Rocks*, trans. with introduction and notes by Ospovat (New York: Hafner, 1971).

his own research on volcanoes and Roman antiquities but also financed some of the travels of Alberto Fortis, friend and disciple of Giovanni Arduino. In Sweden and some of the German states, support came in more organized fashion in the form of schools of mines and governmental boards of mines. In all these cases, travel and observation were part of the job. In this younger generation of geologists, antipathy toward system-building and recognition of the need for more observation were combined with increasing opportunities to conduct the requisite fieldwork.[19]

One of the quasi-novelties after 1750 was the classification of formations, which was developed almost simultaneously by three men: Giovanni Arduino, Johann Gottlob Lehmann, and Guillaume-François Rouelle. In general, earlier writers had been aware of distinctions between "older" and "younger" parts of the earth's crust, the older perhaps igneous (Leibniz, Buffon), or simply dating from the Creation, or a product of those chemical processes sometimes said to have been employed during the Creation. The younger parts were marine sediments. For Arduino, Lehmann, and Rouelle, this scheme was too vague. Lehmann (1756) thus specified a sequence of events and corresponding formations· granitic masses, visible at high elevations in the Hartz mountains, were the oldest, dating from the Creation; on the flanks of these mountains were strata containing marine fossils, and attributable to the Flood; quite recent were alluvial terrains and volcanic rocks. Arduino's scheme (1760) is broadly similar, with two striking exceptions: his oldest rocks are "vitrescent," and he has no reference at all to Creation and Flood. For Rouelle (1750s, 1760s), the oldest masses were formed by aqueous crystallization, and what would later be called the Coal Measures he thought constituted a sort of "intermediate" formation.[20] The three men did not agree on all points, Lehmann being unique in his use of the Flood, Arduino in his employment of subterranean heat, and Rouelle in his focus on chemical crystallization – and these issues would be debated for the rest of the century. What they shared, however, was the conviction that they were describing universal structural relationships and sequences corresponding to periods of time. Within the larger scheme, both Arduino and Lehmann drew sections of the strata observed in

[19] This pattern is gleaned from studies of individuals, such as those cited in n. 18. Also, Ezio Vaccari, "Mining and Knowledge of the Earth in Eighteenth-Century Italy," *Annals of Science*, 57 (2000), 163–80. For Sir John Strange and his circle, see Ciancio, cited in n. 28. Britain seems to have been an exception, as most geologists continued to be independent scholars and amateurs.

[20] Arduino, "Due lettere . . . ," *Nuova Raccolta d'Opuscoli scientifici e filologici*, 6 (1760), clviii–clxix; also, Ezio Vaccari, *Giovanni Arduino (1714–1795)* (Florence: Leo S. Olschki, 1993), pp. 149–68; Lehmann, *Versuch einer Geschichte von Flötz-Gebürgen* (1756), in Lehmann, *Traités de physique, d'histoire naturelle, de minéralogie et de métallurgie*, trans. d'Holbach (Paris, 1759), vol. 3, pp. 212–341; also, Bruno von Freyberg, *Johann Gottlob Lehmann (1719–1767)* (Erlangen: University of Erlangen, 1955); "Cours de chymie de M. Rouelle," Bibliothèque de la ville de Bordeaux, MSS 564–5, fols. 559–87, 663–73 (the volumes of this manuscript being continuously paginated). Desmarest, in his *Géographie physique*, vol. 1 (Paris, 1794), discusses at length Rouelle on the Coal Measures (pp. 421–4), but he also suggests (p. 413) another "intermediate" formation consisting in part of debris from the oldest rocks and resembling one of Arduino's categories.

specific localities, but how they thought the detailed variations from one locale to another should be explained remains unclear.[21]

This mid-century pattern of classification became commonplace, as one
writer after another referred to the earth's crust as comprised of primitive formations (mainly granites), secondary (marine sediments), and relatively recent
(alluvial terrains). On the whole, volcanoes were judged not to be primitive,
but just where they belonged, their role in the earth's past, and their relationship to the earth's internal heat would become matters of dispute. From
explorers of the Alps and the Pyrenees to students of Derbyshire, this scheme
provided the framework for analysis of local terrains, as both John Whitehurst
and Johann Jakob Ferber (who visited Whitehurst in 1769) commented on
the "universal" order of strata characteristic of all of Derbyshire, interrupted
now and then by "accidental" deposits.[22] The same general framework can
be found in the teaching and writing of Werner, but further developed in
some respects, as he gave some attention to defining the "formation" or the
subset of rocks that seemed to comprise a unit in certain ways: a more or
less coherent lithology, distinct from neighboring assemblages, and thus presumably attributable to one time and mode of origin. Werner also outlined
a chronological sequence of formations that he expected to have general validity on a global scale, even while proper attention had to be given to local
variations.[23]

THE ROLES OF FIRE AND WATER IN EARTH SCIENCE

If geologists could agree on the large pattern, causal explanations provoked
debate. Traditionally, the issues have been summed up as dispute between
neptunists and vulcanists, or between those who gave water the larger geological role and those who favored heat or fire. Furthermore, the basalt controversy has customarily been used as a sort of microcosm for these issues,
neptunists reputedly denying the igneous origin of basalt as advocated by
vulcanists. Such terms and labels are in fact misleading, for many geologists

---

[21] Arduino's section of 1740 was published in 1744 by G. G. Spada; for this and his later sketches, see
Vaccari, *Giovanni Arduino*, pp. 34–5 and illustrations. There is a useful discussion of Arduino and
Lehmann in John C. Greene, *The Death of Adam* (1959; reprint New York: New American Library,
1961), pp. 67–72, 76–8. See also Chapter 25, by Charlotte Klonk, in this volume, and Martin J. S.
Rudwick, "The Emergence of a Visual Language for Geological Science, 1760–1840," *History of
Science*, 14 (1976), 149–95.

[22] John Whitehurst, *An Inquiry into the Original State and Formation of the Earth*, 2nd ed. (London,
1786), chap. 16; Johann Jakob Ferber, *Versuch einer Oryktographie von Derbyshire* (1776), translated
in John Pinkerton, *A General Collection of the Best and Most Interesting Voyages and Travels* (London,
1808–14), vol. 2, especially pp. 469–74.

[23] Ospovat in Werner, *Short Classification*, pp. 17–24; and Rachel Laudan, *From Mineralogy to Geology*
(Chicago: University of Chicago Press, 1987), chap. 5. For other examples of the primitive-secondary-
recent framework, see Numa Broc, *Les Montagnes vues par les géographes et les naturalistes de langue
française au XVIIIe siècle* (Paris: Bibliothèque nationale, 1969), pt. 2, chap. 1.

agreed that all or most basalts were volcanic, that volcanoes were more common than their predecessors had suspected, and that, nonetheless, the basic structure of the earth revealed predominantly aqueous agencies. The fundamental issue, therefore, was whether heat or fire existed deep within the "bowels" of the earth or was confined to the more superficial crust.

As indicated earlier, scientists before 1750 had given attention to subterranean heat and had concluded only that it existed; its nature and depth remained unknown. Even the rare writers, such as Leibniz and Buffon, who argued for the earth's igneous origin found little or no use for such heat in the subsequent history of the globe. When A. L. Moro argued in 1740 that volcanic eruptions were responsible for the elevation of all land masses, few found his work impressive and many dismissed his theory as a mere "system," in conflict with the obvious fact that marine sediments showed no signs of the operations of fire.[24] More than a dozen years later, these issues were once again on the agenda, chiefly because of Jean-Etienne Guettard's discovery, published in 1756, that the *puys* of Auvergne were volcanic. (One year earlier, in fact, the great Lisbon earthquake had also drawn attention to subterranean forces.) Many would visit Auvergne in the next decades, and, more significantly, geologists began to look for and to find volcanic rocks in areas of Germany and Italy that had no recorded history of eruptions. Interest in active volcanoes also increased in the aftermath of eruptions of Vesuvius in the 1760s and 1770s and after the great Calabrian earthquake of 1783. Desmarest provided a further stimulus by his description of the prismatic, columnar basalts of Auvergne; because of their association with lava flows, these columns, he concluded, were volcanic products. Other geologists, finding similar basalts, often without recognizable lavas or cones, sometimes followed Desmarest in claiming that here, too, was evidence of ancient volcanic activity.[25]

The so-called basalt controversy was thus related to the larger issue acknowledged by Sir William Hamilton (who had found no columnar basalt near Vesuvius and Etna) when in 1776 he expressed the hope that "subterraneous fires will be allowed, to have had a greater share in the formation of mountains, islands, and even great tracts of land, than has hitherto been suspected." Ever cautious, he immediately added:

---

[24] Contrast Rappaport, *When Geologists Were Historians*, pp. 224–6, with Rose Thomasian, "Moro," *Dictionary of Scientific Biography*, IX, pp. 531–4.

[25] François Ellenberger, "Précisions nouvelles sur la découverte des volcans de France: Guettard, ses prédécesseurs, ses émules clermontois," *Histoire et nature*, 12–13 (1978), 3–42; and Ellenberger, *Histoire de la géologie*, pp. 230–3. Thomas D. Kendrick, *The Lisbon Earthquake* (Philadelphia: J. B. Lippincott [c. 1955]). Augusto Placanica, *Il filosofo e la catastrofe: Un terremoto del Settecento* (Turin: Giulio Einaudi, 1985). Worthwhile discussions are in Carozzi's introduction to Rudolf Erich Raspe, *An Introduction to the Natural History of the Terrestrial Sphere*, trans. and ed. A. N. Iversen and A. V. Carozzi (New York: Hafner, 1970), pp. xxxviii–lii, and in Rev. William Hamilton, *Letters concerning the Northern Coast of the County of Antrim* (Dublin, 1786), letters 10–11. The latter appeared in German translation in 1787, French in 1790. See also Otfried Wagenbreth, "Abraham Gottlob Werner und der Höhepunkt des Neptunistenstreits um 1790," *Freiberger Forschungshefte*, ser. D, 11 (1955), 183–241.

Such as have attributed the formation of all Mountains, to the operation of water alone, are certainly not founded in their system, and perhaps the same may be as truly applied, to those who have insisted, that every Mountain has been formed by explosion from subterraneous fires.[26]

By that date, a few writers had begun to extend the role of heat in the earth's history, and others would follow. As early as 1760, for example, Arduino had proposed that the vitrified rocks of primitive formations showed sufficient analogy to volcanic products to support the idea of the earth's igneous origin, and he would later turn this tentative statement into a more detailed argument. At the same time, he admitted to a correspondent his hesitation in putting emphasis on igneous causes because this seemed so odd as to be "absurd."[27]

Having read Buffon, Arduino in 1760 had sought a subterranean force capable of elevating land. So too, did his friends and disciples – including Alberto Fortis, Johann Jakob Ferber, and Sir John Strange – but they could not wholly agree that volcanic fire stemmed from a central source within the earth; indeed, it seemed as likely that there had been a long intervening period separating igneous origins from the later ignition of volcanoes. Similarly, Ferber's friend Whitehurst thought he had found such a force, which was manifested at the earth's surface by earthquakes and volcanoes; when he concluded that even continents had been elevated by subterranean explosions, he also admitted that the nature and depth of subterranean fires remained unknown. In a comparable vein, Peter Simon Pallas, explorer of Russia and Siberia, suggested that subterranean fires had perhaps been ignited soon after the formation of primitive granites and had thereafter been responsible for the elevation of land.[28]

These writers knew that they faced strong evidence against their own views, and it is worth noting that this evidence was well known before Werner produced his "neptunist" synthesis. If Desmarest's conclusions about the columnar basalts of Auvergne were shared by later observers, what of regions in Hesse and Saxony, and in County Antrim, Ireland, where such columns were not associated with any visible cones or even with rocks of known volcanic origin? In Ireland, the Rev. William Hamilton argued that the columns

---

[26] Sir William Hamilton, *Campi Phlegraei* (Naples, 1776), p. 6; text in both English and French. Sir William is not to be confused with the Irish cleric of the same name (n. 25).

[27] Arduino, "Due lettere," pp. clx–clxi, clxxvi–clxxvii, and Arduino, "Saggio fisico-mineralogico di Lythogonia, e orognosia," *Atti dell'Accademia delle Scienze di Siena*, 5 (1774), especially 243–4, 254–6, 299. The latter text was included in a volume of Arduino's articles (1775), translated into German in 1778. His letter of 1773 is quoted by Vaccari, *Giovanni Arduino*, p. 168 n. 140.

[28] For the circle around Arduino, see Luca Ciancio, *Autopsie della terra: Illuminismo e geologia in Alberto Fortis (1741–1803)* (Florence: Leo S. Olschki, 1995), chap. 2; Whitehurst, *An Inquiry*, chap. 9 and p. 115, and discussion in Davies, *The Earth in Decay*, pp. 130–3; Albert V. Carozzi and Marguerite Carozzi, "Pallas' Theory of the Earth in German (1778) Translation and Reevaluation Reaction by a Contemporary: H.-B. de Saussure," *Archives des sciences*, vol. 44, fasc. 1 (1991). The Carozzis interpret Pallas' uncertainties and hesitations to mean that he thought even granites to be igneous in origin, p. 103; but contrast the text by Pallas, pp. 20–1.

of the Giant's Causeway did resemble those of known volcanic sites, and the regularity of shape could have been produced by fusion followed by slow cooling. To most chemists, however, with only occasional dissent, the cooling of igneous melts produced glass, not crystals – and certainly not the spectacular prismatic formations of the Giant's Causeway.[29]

The same chemical objection applied to primitive granites. When Pallas argued for the existence of subterranean fires at some depth, he hesitated about granite, apparently leaning in the direction of aqueous crystallization. The latter conclusion seemed entirely obvious to Saussure, as it did also to Wernerians and to vulcanologists, Desmarest and Dolomieu. Even before intense debate began in the 1770s, the Baron d'Holbach, very knowledgeable in matters of chemistry and geology, thought it evident that the earth's granitic core had been produced by crystallization in water. To d'Holbach and others, igneous origins were advocated by system-builders, as in the volcanic system of Moro or Arduino and the cosmological system of Buffon. That system-building was a key issue seemed clear also to J.-B. Romé de l'Isle, who in 1779 attacked the physics and the hypothetical reasoning of Buffon and Dortous de Mairan, reaching the conventional conclusion that volcanoes are products of the burning of bitumens.[30]

Lending support to the chemical argument were two common perceptions about subterranean fire. First, if it was akin to ordinary fire, then it needed both fuel and an air supply. Since both air and a suitable fuel could be found only in the relatively superficial crust, it seemed obvious that mountains had existed long before becoming inflamed and transformed into volcanoes. Second, and more generally, fire was equated with destruction and disorder. Well aware of this, Sir William Hamilton hoped that his own studies would persuade readers that volcanic regions were not merely "torn to pieces by subterraneous fires" but that such fires should be seen "in a CREATIVE rather than a DESTRUCTIVE light."[31] Under these circumstances, it is hardly

---

[29] Hamilton, *Letters*, pp. 146–50, 160–1, 163–4. For Hesse and Saxony, see Carozzi in Raspe, *Terrestrial Sphere*, p. xlvi. Valuable discussions of the chemical issues are in Laudan, *From Mineralogy to Geology*, especially pp. 63–5, and Cyril Stanley Smith, "Porcelain and Plutonism," in Schneer, *Toward a History of Geology*, especially pp. 321, 326–8, for the occasional crystals produced in igneous melts.

[30] Romé de l'Isle, *L'Action du feu central bannie de la surface du globe* (Stockholm, 1779); the target here was Buffon's mature work, *Les Epoques de la nature* (1778), which began with the earth's igneous origin and examined at length the effects on climate and organisms of a long period of cooling. D'Holbach, "Caillou," in Diderot, *Encyclopédie*, vol. 2 (1751), pp. 533–6; in articles such as "Crystal" and "Crystallisation," vol. 4 (1754), pp. 523–4, 529, d'Holbach cited the chemical studies of Rouelle. Albert V. Carozzi and John K. Newman, "Dialogic Irony: An Unusual Manuscript of Horace-Bénédict de Saussure on Mountain Building," *Archives des sciences*, 43 (1990), especially 248–50. Dolomieu seems to have changed his mind in his very last publication, reporting on his travels in Auvergne; here he remarked, with obvious surprise, that volcanic fire seemed to have its source *beneath* the primitive granite. Dolomieu, "Rapport fait à l'Institut national . . . sur ses voyages de l'an V et de l'an VI," *Journal des mines*, 41–42 (1798), 393–8.

[31] Hamilton, *Campi Phlegraei*, pp. 3, 13. See also Ellenberger, "Précisions nouvelles," pp. 22–3, and Kenneth L. Taylor, "Nicolas Desmarest and Italian Geology," in Gaetano Giglia et al. (eds.), *Rocks, Fossils and History* (Florence: Festina Lente, 1995), p. 98.

surprising that James Hutton's theory of the earth initially aroused little enthusiasm. What did he mean by "fire or heat," a phrase he used often but could
not explain? If Hutton likened heat to a force, readers still wanted to know
what the thing was: something resembling ordinary fire, or animal heat, or
fermentation, or kinetic heat. Critics made known these and other questions
soon after the first publication of Hutton's views (1788), and he could not provide answers in the expanded version of 1795. In the latter text, Hutton took
the high road earlier followed by Buffon: if he could not explain heat and
Buffon the elevation of land, both men insisted that the facts were sufficiently
clear to warrant their own interpretations of the earth's past.[32]

As implied earlier, the famous "universal ocean" of Wernerian neptunism
did not originate with Werner. Crystalline primitive rocks were products of
that ocean, as were subsequent secondary formations. The sea being obviously more mobile than land, its level had apparently fluctuated, its content
had varied repeatedly (as shown by the diverse materials deposited), and, at
least to some writers, sea basins had seemingly changed location from time
to time. As early as 1708 and in later publications, Antonio Vallisneri remarked
that his observations in the Apennines led him to believe that marine deposits
had been laid down by fluctuating levels of the sea; near the end of the century, Antoine-Laurent Lavoisier would produce a different kind of analysis but
a somewhat comparable conclusion, namely, that the boundaries of the sea
had changed repeatedly, as shown by alternations of pelagic and littoral strata.
Neither Vallisneri nor Lavoisier could explain the sea's behavior, but in the intervening decades evidence had accumulated about changing sea levels.[33]

On the one hand, Italian and Dutch engineers had long wrestled with
coastal problems apparently stemming from a rising sea. (This phenomenon
at Venice was familiar to Vallisneri in Padua.) On the other hand, conflicting evidence came from Sweden, where the Baltic Sea seemed to be steadily
retreating from Scandinavian shores. When Anders Celsius in 1743 published
his report on the Baltic, he relied in part on recent memory – former fishing
grounds now showing large boulders, harbors no longer able to accommodate
deep-draft ships – and in part on records kept by his colleagues for about fifty
years. The latter information allowed him to attempt a calculation of the height

---

[32] Controversy about heat receives little attention in Dennis R. Dean, *James Hutton and the History of Geology* (Ithaca, NY: Cornell University Press, 1992), but see Rachel Laudan, "The Problem of Consolidation in the Huttonian Tradition," *Lychnos* (1977–8), 195–206, and Patsy A. Gerstner, "The Reaction to James Hutton's Use of Heat as a Geological Agent," *British Journal for the History of Science*, 5 (1971), 353–62. Especially valuable is Jacques Roger, "Le Feu et l'histoire: James Hutton et la naissance de la géologie," in *Approches des lumières: Mélanges offerts à Jean* Fabre (Paris: Klincksieck, 1974), pp. 415–29, reprinted in Roger, *Pour une histoire des sciences à part entière*, ed. Claude Blanckaert (Paris: Albin Michel, 1995), pp. 155–69.

[33] Anonymous review (by Vallisneri) of Woodward, in *La Galleria di Minerva*, 6 (1708), 17, and Lavoisier, "Observations générales sur les couches modernes horizontales," *Mémoires de l'Académie royale des sciences*, 1789 (1793), 351–71. See also Kenneth L. Taylor, "The *Epoques de la nature* and Geology during Buffon's Later Years," in Jean Gayon (ed.), *Buffon 88: Actes du Colloque international pour le bicentenaire de la mort de Buffon* (Paris: Vrin, 1992), p. 378.

of the sea some two thousand years earlier. (He carefully indicated that his calculations were based on the assumption of a uniform rate of diminution.) These results provoked debate in the Academy of Stockholm, where critics could point to the contrary evidence published by Italian and Dutch scientists.[34]

The dissemination of Swedish research needs further historical investigation, but it is likely that the findings of Celsius were seen as harmonizing with the established view that the sea had formerly occupied the high elevations where marine sediments could be found. In Germany, the Baltic evidence was being used by Lehmann as early as 1756, and it would later be integrated into the Wernerian version of the universal ocean.[35] In an intelligent and flexible synthesis of available information and interpretation, Wernerian theory began with the aqueous crystallization of granites followed by a series of events occurring in a sea alternately tranquil and turbulent, diminishing in level but with occasional resurgences. Behavior of the sea was correlated with an orderly sequence of the various materials deposited. Perhaps most strikingly, one no longer needed to seek an explanation for the elevation of land, as continents were merely being *exposed* by the retreat of the ocean. To be sure, one still could not explain the sea's behavior – calm and agitated, rising and falling – but few geologists expected immediate solutions to such problems, being aware of earlier speculations about waters retreating into hypothetical caverns within the earth. If it was hard to avoid wondering about causes, geologists had learned that observation and patience would eventually fill gaps in the theory of the earth.

FOSSILS, TIME, AND CHANGE

Geologists after 1750 did continue to discuss two topics prominent in earlier decades: fossils and the Flood. Neither has attracted the sustained attention of historians of science, except for the dramatic case of mammoths, culminating in the researches of Georges Cuvier. The next paragraphs cannot fill this considerable void but will suggest that these topics entered into a new relationship with one another and with efforts toward a theory of the earth.

After the debate on fossil origins had died down, collectors continued to amass specimens, and taxonomies were produced by such men as Carl Linnaeus, John Hill, E. Mendes da Costa, and A.-J. Dezallier d'Argenville. On

---

[34] Tore Frängsmyr, *Geologi och skapelsetro* (Stockholm: Almqvist & Wiksell, 1969), pp. 199–206, and Frängsmyr, in Frängsmyr (ed.), *Linnaeus: The Man and His Work* (Berkeley: University of California Press, 1983). See also Rappaport, *When Geologists Were Historians*, pp. 226–34. Lengthy analyses are in Desmarest, *Géographie physique*, pp. 133–50, 307–18, and more briefly in d'Holbach, "Mer," in Diderot, *Encyclopédie*, vol. 10 (1765), pp. 359–61. The debate in Sweden became more urgent after the publication of *Telliamed* (1748), in which the diminution of the sea was associated with eternal cosmological cycles; the author, Benoît de Maillet (d. 1738), was unaware of earlier Swedish research.

[35] Lehmann, *Traités de physique*, vol. 3, p. 201. For Werner, see n. 18.

the whole, most attention went to marine forms arranged in standard zoological categories: univalves, bivalves, and so on. Before 1700, literature on
ammonites and other forms had raised the question of whether such creatures
were extinct, and this in turn prompted reflections on God's design, wisdom,
and beneficence. That a shift in priorities occurred after mid-century is suggested by the German naturalist who remarked that extinct species had probably served whatever purpose God intended and had then vanished.[36] With
or without such larger concerns, extinction was hard to prove, as naturalists
realized that marine forms, found as fossils, might still be alive in inaccessible
ocean depths.

It is easier in retrospect than it was in the eighteenth century to see that
extinction had vital implications for the earth's history: did different forms
of life characterize different periods of the past? Without quite posing the
question in this way, various writers did comment on the way fossil shells
seemed to be grouped or segregated in colonies. In mid-century, Rouelle
apparently planned (but never wrote) a study of the distribution of fossil mollusks. In 1774, Arduino would remark briefly that fossils "on the whole differ
from stratum to stratum," and comparable statements can be found scattered
elsewhere. That the subject merited study was indicated by Johann Ernst
Immanuel Walch who declared that one needed to know "the different fossils
found in each stratum."[37] Apparently Werner, himself no fossilist, taught his
students that fossils could be used to characterize and to reveal the relative
ages of sedimentary strata.[38] Efforts to do so schematically, rather than in the
detail proposed by Walch or advocated by Werner, were being produced by
such men as Johann Friedrich Blumenbach, Jean-Louis Giraud Soulavie, and
François-Xavier Burtin, each arguing that different "epochs" of the past had
distinct assemblages of fossils.[39]

---

[36] Johann Friedrich Esper, *Description des zoolithes nouvellement découvertes d'animaux quadrupèdes in-
connus et des cavernes qui les renferment,* trans. J. F. Isenflamm (Nuremberg, 1774), p. 76. For earlier
discussions of extinction, see Rudwick, *The Meaning of Fossils,* pp. 64–5.

[37] Walch, *Recueil des monumens des catastrophes que le globe terrestre a essuiées,* trans. from the German
(Nuremberg, 1768–78), vol. 1, p. 99. Arduino, "Saggio fisico-mineralogico," p. 256. For Rouelle,
see d'Holbach, "Fossile," in Diderot, *Encyclopédie,* vol. 7 (1757), p. 211. Other examples are in François
Ellenberger and Gabriel Gohau, "A l'aurore de la stratigraphie paléontologique: Jean-André De Luc,
son influence sur Cuvier," *Revue d'histoire des sciences,* 34 (1981), 236–8, 239, 247, 256.

[38] Martin Guntau, "The Beginning of Lithostratigraphic and Biostratigraphic Thinking in Germany,"
in Giglia, *Rocks, Fossils and History,* p. 152, citing one of Werner's students writing in 1804.

[39] Walter Baron, "Blumenbach," *Dictionary of Scientific Biography,* II, pp. 203–5. Burtin, "Réponse à la
question physique, proposée par la Société de Teyler, sur les révolutions générales, qu'a subies la surface de la terre, et sur l'ancienneté de notre globe," *Verhandelingen, uitgegeeven door Teyler's tweede
genootschap,* 8 (1790); text in both French and Dutch. Soulavie, *Histoire naturelle de la France méridionale* (Paris, 1780–4), vol. 1. When in 1785 the Academy of St. Petersburg offered a prize for a classification of rocks according to their properties, modes of origin, and periods of formation, Soulavie
was one of the contestants; he dealt at length with rocks, devoting his last pages to four epochs for
fossils. Soulavie, "Les Classes naturelles des minéraux et les époques de la nature correspondantes à
chaque Classe," in *Mémoires présentés à l'Académie Impériale des Sciences* (St. Petersburg, 1786). Enough
was published on this subject to make it unnecessary for historians to wonder whether Georges
Cuvier and Alexandre Brongniart were familiar with the unpublished work of William Smith; for a

These and other writers assumed that many fossils represented extinct species, but persuasive evidence for extinction would ultimately come from Cuvier who in 1796 began to publish his studies of the bones of great fossil quadrupeds. As Cuvier knew from the start, his research would at last show that past epochs had had their distinctive forms of life; and this, he expected, would have important implications for the theory of the earth.[40]

The quadrupeds made famous by Cuvier, and that made Cuvier's fame, had not been ignored by his predecessors, and there exists a considerable earlier literature on "elephant" bones in particular.[41] Such anatomical studies of the elephant as existed were consulted by naturalists who nonetheless reached no consensus about whether the fossils belonged to still-living species or to some extinct relative. Even more controversial was the question of how such exotic creatures had been transported from presumably tropical locations to burial sites in Britain, Italy, Germany, Russia, and North America. Various explanations had been canvassed before 1750, and two of these – climatic changes in the past and transport by the Flood – remained current in the later decades. Such use of the Flood merits attention, for it had long been evident that "elephant" bones were relatively recent deposits, in alluvial terrains or barely consolidated sediments, with organic materials often still detectable in the fossils. To earlier writers, the antediluvian part of the earth's history could in general be measured by using Biblical chronology. After mid-century, however, it became increasingly clear that the accumulation of secondary formations required considerable extension of antediluvial (and prehuman) time. If few denied that the Flood had occurred, the event became a recent one in a lengthened history of the earth. Now and then one encounters objections even to this diminished role for the Flood, as when Barthélemy Faujas de St.-Fond, professor of geology at the Museum of Natural History in Paris, wondered why, given the progress of geological science, there still existed efforts to harmonize Genesis with geology; these efforts he identified with the British and with such odd people as Jean-André Deluc.[42]

Faujas also remarked on "a diluvial flood . . . which was not unique" in its kind, thus indicating his view that the earth's past was punctuated by floods as well as other violent events. Much has been written about the "catastrophism" of the early nineteenth century (a subject that could use further study), commonly defined as advocacy of worldwide upheavals, violent in nature, their

---

detailed and inconclusive study of the latter kind, see Joan M. Eyles, "William Smith, Sir Joseph Banks and the French Geologists," in Alwyne Wheeler and James H. Price (eds.), *From Linnaeus to Darwin: Commentaries on the History of Biology and Geology* (London: Society for the History of Natural History, 1985), pp. 37–50.

[40] Rudwick, *The Meaning of Fossils,* chap. 3.

[41] See Cohen, *Le Destin du mammouth,* chaps. 2–5; Rappaport, *When Geologists Were Historians,* chap. 4, and especially Greene, *The Death of Adam,* chap. 4; and George Gaylord Simpson, "The Beginnings of Vertebrate Paleontology in North America," *Proceedings of the American Philosophical Society,* 86 (1942), 130–88.

[42] Faujas de St.-Fond, *Essai de géologie,* vol. 1 (Paris, 1803), pp. 19–20.

causes unknown and even unknowable.[43] This characterization – one is tempted to say caricature – needs reexamination, for it is dubious when applied to the late eighteenth century. A key problem turns out to be linguistic, as geologists often referred to "revolutions" in the earth's past. What that word meant depends on the contexts used by each author; thus, when Alberto Fortis, cleric and disciple of Arduino, announced in 1802 that the earth had undergone "revolutions," he explained that he referred to transformations "by slow and regular causes . . . even if this requires a thousand million years."[44] In addition to the "slow and regular," no geologist denied the importance of violence, for such episodes were among the "actual" causes within their experience: floods, rapid erosion, landslides, earthquakes, volcanic eruptions. Causes of particular phenomena might be uncertain but not unknowable. Although worldwide upheavals were occasionally suggested by men such as Blumenbach and Whitehurst, all were based on naturalistic examples, extended in scale. Perhaps most commonly, violence seemed necessary to explain mountain-building, the disturbances of strata formerly horizontal, and the excavation of valleys. In commenting on valleys, Dolomieu insisted, as a proper empiricist, that what modern rivers could not do – and a weak force would not have greater effects, merely given more time – clearly required a more powerful agent, namely, a rapid retreat of waters that had excavated these terrains.[45]

Dolomieu and most of his contemporaries, including diluvialist Jean-André Deluc, show no signs of being hampered by a short timescale, supposedly inducing them to seek rapid agents of geological change. On the contrary, they had no difficulty imagining a long, prehuman history of the earth. On the whole, the eternalism detected by critics of James Hutton proved as unacceptable as the short time of Biblical chronology. But to say this much leaves open a vast field of choice between the eternal and six thousand years. Some few writers did attempt calculations or estimates of the age of the earth, but it proved impossible to find a suitable, reliable, invariant, and agreed-upon natural chronometer.[46] In short, geologists lacked neither time nor a

[43] Ibid., pp. 20–1. For example, William F. Bynum et al. (eds.), *Dictionary of the History of Science* (Princeton, NJ: Princeton University Press, 1981), s.v. Catastrophism. Contrast Rudwick, "Shape and Meaning," p. 311.

[44] Quoted in Ciancio, *Autopsie della terra,* p. 275 n. 110. Careful attention to contexts can be found in Albert V. Carozzi, "Une nouvelle interprétation du soi-disant catastrophisme de Cuvier," *Archives des sciences,* 24 (1971), 367–77, and Martin J. S. Rudwick, *Georges Cuvier, Fossil Bones, and Geological Catastrophes* (Chicago: University of Chicago Press, 1997), pp. 173–83. Compare the thoughtful analysis of Cuvier by Rudwick, *The Meaning of Fossils,* chap. 3, with the conventional one by Cohen, *Le Destin du mammouth,* chap. 6. See also François Ellenberger, "Etude du terme Révolution," *Documents pour l'histoire du vocabulaire scientifique,* 9 (1989), 69–90; much more briefly, Ellenberger, *Histoire de la géologie,* pp. 63–5.

[45] Broc, *Les Montagnes,* p. 145, and pt. 2, chap. 3. A similar view can be found in at least one of Hutton's disciples, Sir James Hall. See Victor A. Eyles, "Hall," *Dictionary of Scientific Biography,* VI, pp. 53–6.

[46] In addition to Broc, *Les Montagnes,* see Ellenberger, *Histoire de la géologie,* pp. 35–9; Rudwick, "Shape and Meaning," and the discussion of chronometers by Davies, *The Earth in Decay,* pp. 99–103. For time and chronometers before 1750, see Rappaport, *When Geologians Were Historians,* pp. 189–99.

commitment to naturalism. One might suggest that devotion to empiricism limited their outlook, since they could not imagine that small, incremental, even "insensible" changes visible to or deducible by the modern observer could ultimately produce major alterations of landforms.[47]

These concluding statements, and much of this chapter, differ markedly from older interpretations of "premodern" geology. In the traditional view, the fact that early geologists often reached conclusions manifestly wrong by modern standards could be explained in one of two ways: they had not been sufficiently empirical, or they had subordinated their observations to prevailing religious beliefs. The first point seems to be based on the assumption that science is really quite easy, requiring only that one observe properly in order to reach correct conclusions. The second point stems from the modern view that science and the Bible are inherently different and thus incompatible. In the past, however, the common assumption was that two truths, the scientific and the Biblical, could not contradict each other. Rather than assume an inevitable conflict, eighteenth-century geologists either tried to harmonize the two truths (usually by reinterpreting Genesis, not geology) or ignored Genesis as irrelevant to their concerns.[48]

Although the older interpretation has proved tenacious, scholarly research in the past two or three decades has done much to revise this tradition, in part by a return to a more careful reading of the original documents. In addition, historians have increasingly adopted a more sophisticated approach to the nature of the sciences. Instead of pure empiricism leading inevitably to correct conclusions, the sciences may be better understood as entailing legitimate disagreements about how to interpret evidence and, in fact, how to select the evidence deemed most important. Such considerations make the study of the eighteenth century especially rewarding, because geologists of that period had become remarkably self-conscious about methodology, the nature of knowledge, and the problems peculiar to their own discipline.

Standard studies of timescales are wholly inadequate for the eighteenth century; see, e.g., Francis C. Haber, *The Age of the World: Moses to Darwin* (Baltimore, MD: Johns Hopkins University Press, 1959), and Stephen Toulmin and June Goodfield, *The Discovery of Time* (London: Hutchinson, 1965).

[47] The final sentence is prompted in part by remarks in Stephen Jay Gould, *Ever since Darwin* (New York: W. W. Norton, 1979), pp. 147–52 ("Uniformity and Catastrophe"), and Gould's *Hen's Teeth and Horse's Toes* (New York: W. W. Norton, 1983), pp. 94–106 ("The Stinkstones of Oeningen").

[48] The essential work is by John H. Brooke, *Science and Religion: Some Historical Perspectives* (Cambridge University Press, 1991). For a spirited review of historiography, see Mott T. Greene, "History of Geology," *Osiris*, ser. 2, 1 (1985), 97–116. Informative also is Claude Blanckaert's introduction to Roger, *Pour une histoire*.

19

THE HUMAN SCIENCES

Richard Olson

Historians have long seen the search for a viable "science of man [*sic*]" as a central feature of eighteenth-century intellectual life. David Hume's (1711–1776) desire to be "the Newton of the moral Sciences" and his insistence in 1740 that "'tis at least worthwhile to try if the Sciences of man will not admit of the same accuracy which several parts of natural philosophy are found susceptible of"[1] have been taken to represent the views of a huge number of intellectuals throughout the century and across all nations of Europe and North America. Moreover, the centrality of the human sciences to the Enlightenment project is acknowledged not only by those sympathetic to the goals of that project and fundamentally optimistic about its liberating consequences[2] but also by those who have found the goals misdirected and the consequences fundamentally destructive.[3]

The issue of how to portray the relationships between such twentieth-century professional disciplines as anthropology, economics, geography, history, linguistics, psychology, or sociology and various eighteenth-century attempts to establish human sciences is both extremely complex and a matter of intense debate.[4] Eighteenth-century authors and readers often thought in

---

[1] David Hume, *An Abstract of A Treatise of Human Nature* (London, 1740).

[2] See Ernst Cassirer, *The Philosophy of the Enlightenment* (Boston: Beacon Press, 1964, from 1933 German original), and Peter Gay, *The Enlightenment: An Interpretation,* 2 vols. (New York: Vintage, 1966–9).

[3] See, for example, Lester Crocker, *An Age of Crisis: Man and World in Eighteenth Century France* (Baltimore, MD: Johns Hopkins University Press, 1959), which views the undermining of traditional religiously grounded morality as disastrous, and Theodor W. Adorno and Max Horkheimer, *Dialectic of Enlightenment* (New York: Herder and Herder, 1972), which sees the new focus on science and reason as tyrannical in its own right. On the negative social consequences of the application of "reason" to madness, see Michel Foucault, *Madness and Civilization: A History of Insanity in the Age of Reason* (New York: Pantheon, 1965).

[4] Michel Foucault has denied that any true science of man could exist in the eighteenth century. See Foucault, *The Order of Things: An Archeology of the Human Sciences* (London: Tavistock, 1970), p. 309. For the argument that the presumption of disciplinary continuities is largely misleading and that there is little to be gained by talking about such disciplines as psychology in the eighteenth century, see, for example, Roger Smith, "Does the History of Psychology Have a Subject?" *History of the Human Sciences,* 1 (1988), 147–77, and Graham Richards, *Mental Machinery: The Origins and Consequences of*

terms of categories that differ from those in use today. Thus, for example, the phrases "the natural history of man" and "philosophical history" were frequently used to include many topics now included in anthropology, linguistics, and sociology, along with some that now belong to political science and aesthetics. At the same time, "anthropology" was used in German-speaking regions to cover physiology as well as topics from the first three twentieth-century disciplines. In what follows we will try to keep distinct the categories of eighteenth-century actors from those of modern vintage.

NOTIONS OF "SCIENCE" IN THE HUMAN SCIENCES

When Hume wrote *A Treatise of Human Nature* (1739), he subtitled it *An Attempt to introduce the Experimental Method of Reasoning into Moral Subjects.* In doing so he typified two major features of eighteenth-century trends in the studies of humans. First is the emphasis on experiment, or more properly, on observation; as Hume was careful to point out, attempts to manipulate human subjects would almost certainly distort the operation of natural principles. As a consequence, insisted Hume:

> We must glean up our experiments in this science from a cautious observation of human life, and take them as they appear in the common course of the world, by men's behavior in company, in affairs, and in their pleasures. Where experiments of this kind are judiciously collected and compared, we may hope to establish on them a science, which will not be inferior in certainty, and will be much superior in utility to any other of Human comprehension.[5]

Most important seventeenth-century attempts to found human sciences, especially those of Thomas Hobbes (1588–1679), Benedictus Spinoza (1632–1677), and Gottfried Wilhelm Leibniz (1646–1716), had incorporated strong rationalist tendencies. Each of these thinkers purported to be able to deduce the most desirable features of civil society from definitions of human nature. In the cases of Hobbes and Spinoza in particular, the systems of morality and society grounded in these definitions were appalling to most contemporaries because they seemed to lead in peculiarly self-centered, secular, and atheistic directions. One consequence was the creation of an empiricist, antimetaphysical

---

Psychological Ideas, 1600–1850 (Baltimore, MD: Johns Hopkins University Press, 1992), pp. 1–11. For the counterargument – that given a reasonable amount of caution, it makes eminent sense to see modern sciences such as anthropology and psychology as the continuation (with modifications) of traditions that existed in the eighteenth century – see Loren Graham, Wolf Lepenies, and Peter Weingart (eds.), *The Functions and Uses of Disciplinary Histories* (Dordrecht: D. Reidel, 1983). This issue is also addressed in several of the essays in C. Fox, R. Porter, and R. Wokler (eds.), *Inventing Human Science: Eighteenth-Century Domains* (Berkeley: University of California Press, 1995).

[5] David Hume, *A Treatise of Human Nature,* edited with an introduction by Ernest C. Mossner (New York: Penguin, 1969, from 1739 original), p. 46.

backlash that shaped much eighteenth-century discourse about human nature and human institutions.

Perhaps the most important example of this self-conscious backlash was *Traité des systèmes* (1749) by Etienne Bonnet Condillac (1715–1780). Defining a system as "a disposition of the different parts of an art or a science in an order in which they mutually support one another and in which the last are explained by the first,"[6] Condillac explicitly attacked the metaphysical systems of Spinoza and Leibniz, arguing that all systems grounded in abstract principles or definitions are fundamentally misguided. On the other hand, systems grounded firmly in facts established by experience, such as Newton's system of the world, represented the pinnacle of scientific knowledge. Between these two extremes were hypothetical systems based on provisionally held suppositions. When such systems were used for heuristic purposes – to propose new experiments or observational tests of the supposition, as both John Locke (1632–1704) and Newton had suggested,[7] then they might be tremendously valuable; but if used uncritically as explanatory principles, hypotheses could be almost as dangerous and misleading as metaphysical principles.

Among those thinkers who developed the human sciences, some, such as David Hartley (1705–1557) and Jean-Jacques Rousseau (1712–1778), argued that the phenomena associated with human actions and interactions were so intricate and extensive that there was no hope of inducing principles directly from experience; thus, they viewed the use of provisional hypotheses as essential.[8] In his *Discourse on the Origin and Foundations of Inequality among Men* (1755), Rousseau argued that one should proceed as follows:

> Begin by setting all facts aside, for they do not affect the question. The researches that can be undertaken concerning this subject must not be taken for historical truths, but only for hypothetical and conditional reasoning better suited to clarify the nature of things than to show their true origin, like those our physicists make every day concerning the formation of the world.[9]

Others, such as Charles Louis Secondat, Baron de Montesquieu (1689–1755) and Adam Ferguson (1723–1816), were so disturbed by the use of untestable hypotheses by authors such as Hobbes and Rousseau that they argued against the use of hypotheses entirely.[10]

---

[6] Etienne Bonnet de Condillac, *Oeuvres philosophiques de Condillac*, ed. Georges Le Roy, 3 vols. (Paris: Presses Universitaires de France, 1947–51), 1:121.

[7] On the importance of the methodological suggestions of Newton and Locke as well as the multiple interpretations of their works by those developing the human sciences in the eighteenth century, see especially Sergio Moravia, "The Enlightenment and the Sciences of Man," *History of Science*, 17 (1980), 247–68.

[8] See David Hartley, *Observations On Man, His Frame, His Duty, and His Expectations* (London: Thomas Tegg and Son, 1834, 6th ed.; 1st ed., 1749), pp. 4–5.

[9] Jean Jacques Rousseau, *The First and Second Discourses*, ed. Roger D. Masters (New York: St. Martins, 1964), p. 103.

[10] See Adam Ferguson, *An Essay On The History of Civil Society*, ed. Duncan Forbes (Edinburgh: Edinburgh University Press, 1966), pp. 2–6.

A second issue related to the complexity of human phenomena – an issue that divided empiricist students of humanity – was also emphasized by Condillac. According to Condillac all legitimate knowledge must be formulated through a process, often called analysis, in which the complexity and initial chaos of sensations and thoughts is brought into order by isolating or abstracting its salient and simpler features and then recombining them into a whole that is "understood" to be the simple sum of its parts.[11] Among those concerned with the human sciences prior to about 1796, few disagreed; but toward the end of the eighteenth century the general presumption of analyticity came under attack by a group of Parisian thinkers, the Idéologues, who had either been trained in medicine at Montpelier or had studied with graduates of Montpelier, where a new anti-iatromechanical, provitalist, Hippocratic revival was underway. These scholars, led by Pierre Cabanis (1757–1808), whose *Rapports du Physique et du Moral de l'Homme* (Paris, 1796) sought to ground morality in physiological psychology, argued that the complexity of human life and interactions derives from the fact that different factors *interact* in unpredictable ways, so that human phenomena simply cannot be understood as the sum of the effects of a set of isolatable simple causes.

Condillac did suggest one critical nonempirical criterion for evaluating scientific explanatory systems. Borrowing from Jean d'Alembert's *Treatise on Dynamics* of 1743, Condillac argued that "a system is the more perfect as the principles are fewer in number: it is even to be hoped that they could be reduced to one."[12] He used this principle in his own psychological theorizing to reduce Locke's account of knowledge acquisition, which depended on both sensation and reflection, to a system based on sensation alone. More important, the focus on simplicity was openly appropriated from Condillac in the discussions of many other proponents of the human sciences, including Claude Adrien Helvétius (1715–1771), Denis Diderot (1713–1784), Julien Offray La Mettrie (1709–1751), and Adam Smith (1723–1790).

Not all eighteenth-century attempts at human sciences embraced either the empirical emphasis or the focus on simplicity advocated by Condillac. A small number of French scholars in particular, who seem to have been influenced by the science of rational mechanics, persisted in believing that human institutions might be derived directly from the definition of "man" without recourse to observations. This view was particularly prevalent among the mid-century group of political economists known as physiocrats, who railed

---

[11] See, for example, Etienne de Condillac, *La Logique,* trans. W. R. Albury (New York: Abaris Books, 1980, from 1778 original), pp. 63–87.

[12] See note 6. D'Alembert's formulation of this principle, repeated in his *Preliminary Discourse to the Encyclopedia of Diderot,* translated and with an introduction by Richard Schwab (Indianapolis: Bobbs-Merrill, 1963), p. 22, is as follows: "The more one reduces the number of principles of a science, the more one gives them scope, and since the object of a science is necessarily fixed, the principles applied to that science will be so much more fertile as they are fewer in number."

against the tyranny of the past. According to one of their spokesmen, Mercier de la Rivière (1719–1792):

> I do not cast my eye on any particular nation or sect. I seek to describe things as they must *essentially* be, without considering what they have been, or in what country they may have been. . . . By examining and reasoning we arrive at knowing the truth self-evidently, and with all the practical consequences which result from it. Examples which appear to contrast with these consequences prove nothing.[13]

Although a number of major eighteenth-century political economists, including David Hume, the Abbé Ferdinando Galiani (1728–1787), and Adam Smith, rejected this rational mechanics model for political economy in favor of a more empiricist approach, David Ricardo (1772–1823) revived the style in the early nineteenth century, and it has remained the dominant style everywhere through the late twentieth century. A similar, although slightly less virulent version of this perspective informed the writings of Anne-Marie de Condorcet (1743–1794), who insisted that because all humans were, by definition, capable of reason, they therefore deserved equal treatment, regardless of sex, race, or religion.[14]

Perhaps the most interesting eccentric views about the "scientific" nature of human science were those of Giambattista Vico (1688–1744), author of *Principles of a New Science Concerning the Nature of Nations* (1725), who hearkened back to the fifteenth-century views of legal humanists in insisting that humans can have scientific knowledge only of that which they have created themselves. This methodological perspective severely limited the impact of Vico's work during most of the eighteenth century. But it seemed to be particularly compatible with Kantian and Neo-Kantian scientific perspectives; so it was revived and embraced at the very end of the century in Germany by Johann Herder (1744–1803), among others.

NOTIONS OF "HUMAN" IN THE HUMAN SCIENCES

With rare exceptions, students of "man" shared a few basic assumptions regarding what it meant to be human. First, even those who took a deterministic view of human actions argued that people act or should act as if they were capable of making choices. Thus, for practical purposes, consistent determinism was not part of the eighteenth-century human sciences. Second, no one denied that self-preservation and the search for individual happiness

---

[13] From Mercier de la Rivière, *L'ordre naturel et essentiel des sociétés politiques* (1767), cited in Terence Hutchison, *Before Adam Smith: The Emergence of Political Economy* (Oxford: Basil Blackwell, 1988), p. 293.

[14] Anne Marie Condorcet, "On the Admission of Women to the Rights of Citizenship," in *Condorcet, Selected Writings*, ed. Keith Baker (Indianapolis: Bobbs-Merrill Co., 1976), p. 98.

played central roles in human behavior, although a number of important authors insisted on the existence of independent feelings of sociability, benevolence, or sympathy as well. Third, almost every student of human actions and institutions continued to use the dualistic categories of the physical and the moral. For most thinkers, these two categories were ontologically separate, the residue of Cartesian matter and spirit. Yet it seemed clear to all of them that physical circumstances had a strong bearing on moral choices, so the answers to many questions regarding human nature were formulated in terms of the relationships between *l'homme physique* and *l'homme moral*. Even materialists such as La Mettrie, Helvétius, and Cabanis, although they denied the separate existence of *l'esprit* (spirit), continued to use the linguistic dichotomy between the physical and moral; and the goals of their human sciences, like those of the dualists, were overtly moral.

Although they focused attention on moral issues, however, the human sciences treated those issues in a nontraditional way, insisting on secular understandings of what had long been understood as the primary domain of revealed religion. Prior to the eighteenth century, the Bible had been widely accepted as the primary source of moral guidance in Judeo-Christian Europe. Some intellectuals, including d' Holbach and Helvétius, turned away from this source of morality because they viewed all religions as impositions on an ignorant and emotionally needy populace by a clerical elite whose primary goal was to accumulate power and wealth. Others continued to remain deeply committed to Christianity but argued that God acted in things human, as in nonhuman nature, through the mechanism of natural laws. Thus, Gershom Carmichael, the orthodox Presbyterian professor of moral philosophy at Glasgow, wrote in 1727 that moral philosophy is nothing but "the demonstration of the duties of man and citizen from knowledge of the nature of things and the circumstances of human life."[15]

Finally, even though the tendency through the century was to view the passions as increasingly important in shaping human actions, the capacity to reason continued to be seen as uniquely human. Even at the end of the century, Mary Wollstonecraft (1759–97) could write, "In what does man's preeminence over the brute creation consist? The answer is as clear as that a half is less that a whole; in Reason."[16] Almost no author would have disputed this claim, even though many of them would have seen the extent of the powers of human reason as severely limited; and a very small number, including Vico and Herder, would have seen formal reasoning as a historical accretion rather than as a universal characteristic of humans.

---

[15] Cited in James Moore and Michael Silverthorne, "Gershom Carmichael and the Natural Jurisprudence Tradition in Eighteenth-Century Scotland," in Istvan Hont and Michael Ignatieff (eds.), *Wealth and Virtue: The Shaping of Political Economy in the Scottish Enlightenment* (Cambridge University Press, 1983), p. 76.

[16] Mary Wollstonecraft, *A Vindication of the Rights of Woman* (New York: W. W. Norton, 1975), p. 12, especially note 2.

THE RESERVOIR OF HUMAN "EXPERIMENTS":
HISTORY AND TRAVEL ACCOUNTS

Nineteenth-century social scientists looked down on the human scientists of the eighteenth century because of their failure to carry out experiments or observations under sufficiently controlled conditions and because those who theorized about human nature and institutions frequently got their data secondhand, from travel accounts and historical literature.

Even eighteenth-century scholar François Catrou recognized that even though he might have taken the greatest of care in researching and composing the four volumes of his *Histoire romaine, depuis la fondation de Rome* (twenty-one volumes from 1725 to 1737) that dealt with the Punic wars, his extremely negative view of Carthage and his admiration for Roman virtue were inevitably shaped by the fact that all the available sources were by Roman and Greco-Roman historians and that no sources from the Carthaginian side remained. Furthermore, it was well understood that subsequent interpretations of antiquity added their own filters to the information that was offered in firsthand accounts. Thus, wrote Adam Ferguson, all historical accounts

> are made to bear the stamp of the times through which they have passed in the form of tradition, not of the ages to which their pretended descriptions relate. The information they bring is not like the light reflected from a mirror, which delineates the object from which it originally came; but, like rays that come broken and dispersed from an opaque or unpolished surface, only give the colors and features of the body from which they were last reflected.[17]

Perhaps even more important, it was also well recognized that reports of non-European cultures were shaped by the interests and assumptions of European reporters. From the mid-sixteenth-century, for example, almost all accounts of Native American and Pacific Island inhabitants tended to treat them as either noble and unspoiled or ignorant, vicious, and cruel. This dichotomy had been established during the 1550s as scholars fought over the treatment of indigenous populations by Spanish Conquistadors. The tradition of the ignoble savage was continued most extensively by ships' captains and settlement leaders who feared the consequences when sailors and community members went native, threatening the success of their projects.[18] The image of the noble savage, on the other hand, was perpetuated and promoted by Dutch scholars who supported the revolt against Spanish rule. It was intensified in writings such as the 1703 *Supplément aux Voyages du Baron Lahontan ou l'on trouvé des dialogues curieux entre l'auteur et un sauvage de bon sens*

---

[17] Adam Ferguson, *An Essay on the History of Civil Society* (London: Transaction Books, 1980 reprint of 1767 original), p. 76.
[18] See especially B. W. Sheehan, *Savagism and Civility: Indians and Englishmen in Colonial Virginia* (Cambridge University Press, 1980), chap. 1.

qui à voyagé written by Louis Armond de Lom d'Arce, baron de La Hontan (1666–1715). This work expressed the admiration held by a French soldier who had lived and fought among the Canadian Indians for twenty years for their values and ways of life in comparison with what he viewed as the corrupt lifestyles and institutions of the French. It was appropriated by Jean-Jacques Rousseau in formulating his eccentric but influential *A Discourse on the Origin and Foundations of Inequality Among Men* (1755). And it was reintroduced as a self-conscious device from Rousseau into the travel literature genre by Georg Forster, whose account of Captain James Cook's second voyage, *A Voyage Around the World* (1777), is counted among the best pieces of eighteenth-century ethnographic reporting.[19]

In spite of all their limitations and the tendency of European observers to impose their own "presentist" categories of analysis on their interpretations of distant others, by the eighteenth century, both the historical narratives and the travel accounts – which had been accumulating since the simultaneous initiation of European voyages of exploration and the humanistic revival of interest in antiquity – contained huge masses of information that had not previously been available. Amid the credulous and the accidentally or intentionally distorted accounts for which readers showed a voracious appetite, there were many serious, self-aware, and respectful, although not worshipful, descriptions and discussions of other cultures from which discerning philosophical historians drew much of their "experimental" knowledge.[20]

If European categories were sometimes imposed upon others, immersion in non-European cultures also initiated a reevaluation of traditional European categories and assumptions about human institutions. In his *Origin of the Distinction of Ranks: Or an Enquiry into the Circumstances which give rise to Influence and Authority in the Different Members of Society* (1779, third edition of the 1771 original), John Millar (1735–1801), for example, drew heavily from the ethnographic accounts of the Iroquois Nations in the *Histoire et Description Générale de la Nouvelle France* (1744) by Pierre-François Xavier de Charlevoix (1682–1761) and the *Moeurs des sauvages américains comparées aux moeurs des premiers temps* (1724) by Joseph-François Lafitau (1681–1746) in challenging both the notion that monogamous marriage is a "natural" and ubiquitous institution and the notion that all "governing" structures are inevitably

---

[19] Among the best of the many accounts of the noble savage tradition is that in Urs Bitterli, *Die "Wilden" und die "Zivilisierten": Grundzuge einer Geistes-und Kulturgeschichte der europaisch uberseeischen Begegnung* (Munich, 1976). Unusual in its complexity, it emphasizes the late eighteenth-century return of literary images of the noble savage into travel journals such as that of Forster. See also Bitterli's *Cultures in Conflict* (Stanford, CA: Stanford University Press, 1986), especially chaps. 3 and 7.

[20] P. J. Marshall and Glyndwr Williams, *The Great Map of Mankind: Perceptions of New Worlds in the Age of Enlightenment* (Cambridge, MA: Harvard University Press, 1982), offers an excellent evaluation of travel literature available during the eighteenth century in England. Michelle Duchet, *Anthropologie et Histoire au siècle des lumiérs: Buffon, Voltaire, Helvétius, Diderot* (Paris: François Maspero, 1971), contains an extensive annotated list of French sources.

patriarchal.[21] Similarly, between 1550 and 1750 the European concept of "Liberty" was transformed almost unconsciously, largely as a result of its use to describe relationships within Native American cultures.[22] It began as a term that had been defined in connection with special class privileges to engage in certain activities in its original Roman context; but by the mid-eighteenth century it had become identified with universal rights to absence of interference.

Origin of the Distinction of Ranks also suggests that the more-sophisticated human scientists of the eighteenth century were considerably more critical in their use of travel narratives than their nineteenth-century critics were inclined to admit. Millar, for example, insisted that no factual claim be accepted unless it met three conditions: it had to be confirmed by another independent observer separated from the first by a significant period of time and coming from a different national and religious background, so that biases and fictive claims could be controlled; it had to be about an issue regarding which the observers could be assumed to have no prior theoretical expectations; and it had to be explicable as an illustration of some general system of thought with wide applicability. If all these conditions were met, he argued, "the evidence becomes as complete as the nature of the thing will admit."[23]

LEGAL LOCALISM, MORAL PHILOSOPHY, AND PHILOSOPHICAL HISTORY: THE TRIUMPH OF ENVIRONMENTALISM AND THE STADIAL THEORY OF SOCIAL CHANGE

If one looks at the backgrounds of those figures who were major philosophical historians, a vastly disproportionate number were associated in one way or another with legal studies. Montesquieu, Vico, Millar, Henry Home, Lord Kames (1696–1782), and James Burnett, Lord Monboddo (1714–1799) were all trained in the law and practiced as lawyers or judges. Hume and Adam Ferguson served as legal librarians, and Francis Hutcheson and Adam Smith lectured on jurisprudence and modeled their moral philosophy courses on Samuel Pufendorf's *On the Duties of Man and Citizen according to Natural Law* of 1673.

Donald Kelley has argued that this fact is a direct consequence of early modern legal conflicts that emerged as part of the growth of centralized nation-states.[24] In connection with Renaissance humanist legal studies a strong

---

[21] John Millar, *The Origin of the Distinction of Ranks* (1779 ed.), reprinted in William C. Lehman, *John Millar of Glasgow, 1735–1801* (Cambridge University Press, 1960), pp. 184–200.

[22] William Brandon, *New Worlds for Old: Reports from the New World and their Effect on the Development of Social Thought in Europe, 1500–1800* (Athens: Ohio University Press, 1986).

[23] Millar, *Origin of Ranks,* pp. 180–1.

[24] Donald R. Kelley, *The Human Measure: Social Thought in the Western Legal Tradition* (Cambridge, MA: Harvard University Press, 1990), passim.

Romanist tradition developed within university culture. According to this tradition, Roman law, especially as it was codified by Emperor Justinian, was universally valid, and its source lay in natural law, or the very nature of humans. On the other hand, throughout Europe there were locally varying common law traditions. Princes and kings seeking to consolidate power wanted both to appropriate the authority to establish their own laws and to institute a sense of national identity by emphasizing the uniqueness and aptness of local legal systems. This nationalism led to an important justificatory literature.[25]

By the beginning of the eighteenth century, the question of just exactly *why* there should be so much local variation in human customs and laws had become one of the central puzzles for legal scholars and for moral philosophers. This was an especially critical issue in regions such as Scotland and southern Italy, which saw the rapid growth of commercial-cosmopolitan centers such as Edinburgh, Glasgow, and Naples in regions that had large rural precommercial populations and where the local common law traditions were not well suited to commercial activities.[26] Similarly, it was true in nations, such as France, where there was major conflict between local authorities, who sought to retain some autonomy, and the central monarch, who claimed a virtually unlimited authority to legislate for the nation. It was also true toward the end of the century in Germany, where patriotic scholars were battling French intellectual and political hegemony.

The first major eighteenth-century attempt to confront these issues was that of the Neapolitan, Giambattista Vico in his *Principi de una scienza nuova d'intorno alla natura delle natzione* (first edition, 1725, expanded in 1744). Vico viewed the historical development of societies as the working out of God's providential plans through humans' creation of their own languages and institutions. He was, however, certain that humans seldom anticipated all the providentially ordained consequences of their decisions; thus he articulated a principle – since known as the principle of unintended consequences – that became central to the arguments of virtually all philosophical historians and to conservative thinkers throughout the next 250 years.

Vico also argued that the growth of the human individual, from infancy through young adulthood to full maturity, provided the fundamental pattern for the development of civil societies. Just as the human infant is incapable of the same kind of rationality as the adult, societies are not fully rational in their early stages, when most critical institutions such as religion, marriage, and burial are established. Religion is established by anthropomorphizing natural entities; proto-legal customs are formed as humans project their desire

---

[25] See, for example, Jean Bodin, *The Six Bookes of the Commonweal,* ed. Kenneth D. McRae (Cambridge, MA: Harvard University Press, 1972).

[26] See especially David Lieberman, "The Legal Needs of a Commercial Society: The Jurisprudence of Lord Kames," in Hont and Ignatieff (eds.), *Wealth and Virtue,* pp. 203–34.

for vengeance against those who injure them onto powerful divinities and use divine oracles to pronounce judgments; and the values of the society are expressed in myths and fables. During the adolescent stage of societies, values are incorporated into stories of heroes to be emulated, and justice remains largely uncodified and within the domain of the personal. Only in the final, adult stage can the values of a society be incorporated into a systematic moral philosophy and can justice be formulated in terms of an abstract set of principles.

Although all aspects of society in a particular stage are consistent with one another, it makes no sense to evaluate the mores and institutions of a society in one stage in terms of the expectations and presuppositions of another. Societies at the same stage of development can be assumed to share some characteristics; so, for example, the early history of European cultures can be illuminated by considering contemporary American ones. But even societies at the same stage of development must be shaped by local physical conditions and linguistic developments; thus each society is unique and should be understood on its own terms. Even human nature is different from one stage to another; so there are no universal standards by which to judge the institutional, moral, or even aesthetic preferences of a society.

Ideen zur Philosophie der Geschichte der Menschkeit (1784–91), by Johann Herder (1744–1803), incorporated many of Vico's views into a system that posited the progress of humanity through the successive flowering of different cultures, or *Völker,* only then did the emphasis on the uniqueness of each people, the organic growth of cultures through various life stages, and the critical role of language as the unique shaper of each culture reenter the human sciences to become cornerstones of the German *Geisteswissenschaften.* With respect to law codes in particular, J. S. Putter (1725–1809), long-term professor of law at Göttingen, and his student Gustav Hugo initiated a historical school of law that, although it apparently developed independently of Vico's ideas, shared his emphasis on local reason and the fit of law to particular stages in cultural development.[27]

More immediately important than Vico's *New Science* was Montesquieu's *Spirit of the Laws* (1749). As a young president of the Bordeaux *parlement,* Montesquieu began a long career criticizing the central monarchy and defending local privilege and custom in his *Lettres Persanes* (1721), which exploited the interest in travel literature by purporting to be a series of observations on Parisian customs and institutions by two visiting Persian diplomats. In 1734 he turned to ancient European history, giving it a philosophical twist by focusing on causal relationships rather than narrative in his *Considerations of the Greatness of the Romans and their Decline.* In 1749 he published his masterpiece, *The Spirit of the Laws,* which sought to provide a more comprehensive

---

[27] See Kelley, *The Human Measure,* pp. 239–42. Also see Peter Reil, *The German Enlightenment and the Rise of Historicism* (Berkeley: University of California Press, 1975).

understanding of why different laws and customs existed in different places than had ever been offered before.

Insisting, contrary to Vico, that human nature was constant over time and space and that there were universal physical and moral laws that account for human interactions, Montesquieu argued that behaviors nonetheless vary widely because the preexisting conditions, or "general spirits," of different nations shape the way in which the universal laws operate, much as the initial conditions and boundary conditions produce radically different-looking solutions to physical problems governed by the same physical laws.

From classical political theory, Montesquieu argued that the governing structures under which people live (republican, monarchical, or despotic) and their corresponding dominant principles (virtue, honor, or fear) influence the customs and laws they are capable of living under. From Jean Bodin (1530?–1596), Montesquieu borrowed the idea that physical environment plays a major role in suiting people to a particular set of laws; although he replaced the old humoral basis for this claim with a theory grounded in the physiological arguments of John Arbuthnot (1667–1735), whose 1733 *An Essay Concerning the Effects of Air on Human Bodies* suggested that different temperaments dominated in different regions because cold temperatures cause tissues to contract and to respond more slowly to stimuli. Other factors, such as religion, the quality of the soil, and population density received Montesquieu's consideration; but perhaps most important and innovative was his discussion of the relationship between laws and "the manner in which the several nations procure their subsistence."[28]

Montesquieu developed a fourfold classification of societies into hunting, pastoral, agricultural, and commercial, and he argued that the laws would be radically different in nations according to which economic activity predominated. Laws would be very simple in hunting societies because there was little private property to protect; they would be slightly more complex in herding nations; more complicated in agricultural nations in which private land ownership emerged; and most complex in commercial nations in which the variety of forms of property was greatly increased. Although Montesquieu was convinced that commercial activities bred peace among nations because of the need among traders to establish relations of trust and cooperation, he was the first of many to suggest that peace among nations was bought at the unfortunate cost of increasing competition and lessening social bonds within local communities.

One likely reason that Montesquieu's emphasis on subsistence became important was that it corresponded with changing social circumstances in reorienting our theoretical understanding of the relationship between political and economic activities. For classical political theorists, including Aristotle

---

[28] Montesquieu, *The Spirit of the Laws*, trans. Thomas Nugent, 2 vols. in 1 (New York: Hafner, 1949), vol. 1, p. 275.

and Machiavelli, productive and reproductive activities were understood as somehow less significant than "public" political activity, whether it was represented by deliberation about law, administering the state, or taking part in military activity. For virtually all the developers of Montesquieu's ideas, on the other hand, this valuation is inverted, and governments and political life are seen as serving broader social and economic interests that matched those of a growing bourgeoisie throughout Europe.

Montesquieu's fourfold taxonomy of societies was temporalized and turned into a theory of the progress of societies – from the earliest and most primitive hunting stage through pastoral and agricultural stages into the commercial stage – by Anne Robert Turgot (1727–1781) in France and by the whole school of Scottish philosophical historians including Ferguson, Smith, Millar, and Kames, each of whom offered his own modifications of Montesquieu's basic themes.[29]

Among the most interesting of this group was Adam Ferguson, whose *An Essay on the History of Civil Society* (1767) initiated a revision of traditional ideas regarding the functions of conflict in society. Ferguson argued that social and legal progress emerge all but exclusively from conflict between parties and classes. Moreover, he insisted that community solidarity depends in large measure on the perception of hostile outside enemies and that humans have such a taste for competition that when there are no military activities to allow them to exert themselves, they make up competitive games to take their place.

The only kind of competition that Ferguson saw as destructive was the privatizing economic competition that developed in commercial societies. Ferguson was at one with Montesquieu and his earlier Scottish colleagues, such as Francis Hutcheson (1694–1746), in believing that social passions offer the greatest scope for human happiness and that private ones are more often the source of anxiety, jealousy, fear, and envy.

Although they were sometimes ambivalent about the consequences of the development of societies from one stage to the next, Adam Smith and his student John Millar had no question that what we now call economic considerations, growing out of self-interest, were the foundation on which all human institutions – familial, social, and formally legal – were built and that all other aspects of society thus had to change as patterns of economic activity did. In his lectures on jurisprudence delivered from about 1750 to 1765, Smith explored a broad range of factors dependent on the stages of economic development.[30] In *Origin of the Distinction of Ranks* (1771), Millar focused attention on the relation of two particular issues to the four-stage theory: the roles of women and of slaves.[31] In both cases Millar argued that exploitative

[29] Perhaps the best brief account of the rise of the stadial theory of social development is Ronald Meek, *Social Science and the Ignoble Savage* (Cambridge University Press, 1976).

[30] See Adam Smith, *Lectures on Jurisprudence*, eds. R. L. Meek et al. (Oxford: Clarendon Press, 1978).

[31] See William C. Lehman, *John Millar of Glasgow: 1735–1801: His Life and Thought and His Contribu-*

and oppressive arrangements had developed naturally and appropriately in earlier societies but that they were inappropriate to the emerging commercial society of lowland Scotland.

Henry Home, Lord Kames, with whom Millar had lived as a law student, produced one of the most comprehensive, eclectic, and eccentric philosophical histories – in his eight-volume *Sketches of a History of Man* (1774) – when he was nearly ninety. Although far less coherent than that of Smith or Millar, Kames's work seems to have had a substantially wider audience, most likely because his religious and social conservatism were less disturbing and because his intense Scottish nationalism and antipathy to Native American cultures had substantial local appeal. This work was carried to America by Scottish educators, where it was widely embraced by those of European background, who found in it a rationalization for their sense of superiority.

In France, at least two self-consciously anti-Montesquieu traditions of philosophical history emerged. One, associated initially with Claude Adrien Helvétius and with some political economists, saw Montesquieu's tendency to justify practices simply by their existence as fundamentally perverse. Since most institutions emerged before humans had the knowledge and wisdom to design them well, the history of human institutions read more as a history of mistakes entered into from ignorance than a history of desirable rational arrangements. For Helvétius, one of the clearest examples of this phenomenon occurred as a consequence of the growth of money-based economies that used durable goods as mediums of exchange. In his *Treatise on Man* of 1774, Helvétius argued that the convenience of using long-lasting and easily transportable commodities for exchange was, unfortunately, accompanied by the ease of hoarding and of creating huge distinctions of wealth which eventuated in the exploitation of the many by the few and led to open class warfare. No money-based economy could avoid the establishment of some divergences of wealth. No society that also allowed for the legal passage of unlimited property to a single heir could avoid the amplification of initially small distinctions of wealth into huge and destructive ones.[32]

The second French anti-Montesquieu tradition of philosophical history was initiated by Jean Jacques Rousseau in his *Discourse on the Sciences and the Arts* (1749) and *Discourse on the Origin and Foundation of Inequality Among Men* (1755). Drawing very heavily from the noble savage literature to form his picture of natural man, Rousseau argued that virtually every feature associated with increasing "civilization" – refinements in knowledge and the arts, the multiplication of forms and amounts of wealth, cosmopolitanism, and so on –

---

tions to Sociological Analysis (Cambridge University Press, 1961), and Paul Bowler, "John Millar, The Four-Stage Theory, and Women's Position in Society," *History of Political Economy*, 16 (1984), 619–38.

[32] Claude Adrien Helvétius, *A Treatise on Man*, trans. W. Hooper (New York: Burt Franklin, 1969), especially pp. 103–27. For an excellent general evaluation of Helvétius's works, see D. W. Smith, *Helvétius: A Study in Persecution* (Oxford: Oxford University Press, 1965).

tended to the corruption rather than to the improvement of morals and to the destructive creation of artificial inequalities among people.

RACE AND THE PLACE OF HUMANS IN THE NATURAL ORDER: THE BACKGROUND TO PHYSICAL ANTHROPOLOGY

During the second half of the eighteenth century, two new sets of issues emerged among those who were trying to formulate a natural history or philosophical history of man. The first of these had to do with the relationship between humans and "orangutans," a term used to describe chimpanzees and apes as well as what we now call orangutans. The second issue had to do with the characteristics and origins of the different "races" of humans. Before midcentury the traditional Judeo-Christian notion that humans were radically distinct from all other creatures because of their immortal souls, by virtue of which they alone were made in God's likeness and by virtue of which they alone could exhibit moral choices, was seldom challenged. Nor were there many serious scholars who questioned the descent of all present humans from Adam and Eve. But in the second half of the century increasing numbers of scholars were either irreligious or antireligious; and even religiously orthodox scientists often insisted on completely naturalistic accounts of humans. At the same time, increasing evidence was amassed suggesting the close anatomical and physiological similarities between apes and humans as well as a growing range of anatomical differences among groups of humans.

The first explicit inclusion of humans in a comprehensive classification of natural organisms appeared in the first (*1735*) edition of the *Systema naturae sive regna tria naturae* of Carl Linnaeus (1707–1778). There, the medical student and taxonomist included the genus, Homo, with a single species, *sapiens,* having four varieties (*europeanus albus, americanus rubescens, asiaticus fucus,* and *africanus niger*) under the order Anthropomophora. As increasing information came in, especially from South America and the South Pacific, Linnaeus's organization of the genus Homo became increasingly complex. By the tenth edition of the *Systema* in 1758, the new order of primates had been introduced; the genus Simia had been much expanded; two new varieties, *Homo sapiens ferus,* "wild man," and *Homo sapiens monstrosus* (including Hottentots and Patagonians) had been added; and an entirely new species of Homo, Homo *Troglodytes* (including orangutans), had been introduced, implying the possibility of polygenetic origins of humans. Thus, when the learned Scottish jurist Lord Monboddo argued in his *On the Origin and Progress of Language* (six volumes, 1773–1792) that it was the capacity for language that distinguished humans from other animals and that orangutans had been shown to have vocal cords capable of producing sounds varying in pitch and loudness, he accepted the conclusion that there was no legitimate reason to deny that orangutans were indeed human, although of a precivilized sort.

Monboddo's colleague Lord Kames was inclined to accept the traditional monogenetic source of humans; but he was careful to point out that accumulating evidence on the geographical distribution of groups of humans was consistent with the alternative hypothesis of independent origins of different races of humans in different places.

Most students of humanity rejected Monboddo's suggestions, but they tended to accept morphological criteria for deciding the issue. Thus, for example, in his *De generis humani varietate nativa* (1775), Johann Friedrich Blumenbach (1752–1840) emphasized the absence of an "intermaxillary" bone in humans and their upright posture to distinguish between apes and humans; and Peter Camper conducted anatomies on orangutan vocal organs, emphasizing their differences from those of humans. On the other hand, Blumenbach turned Kames's rejected suggestion regarding the separateness of different races into the foundation of a racial taxonomy that was immensely influential during the nineteenth century.

Offering a radical alternative to the Linnean-morphological-approach to species in general was George Louis le Clerc, comte du Buffon (1707–1788). Buffon argued that the term "species" should be reserved in natural history for collections of organisms that are reproductively connected with one another over both time and space. Especially in volumes two and three of his *Histoire naturelle des animaux* (1749–67), he argued for the monogenetic origins of all humanity; but he turned monogenism into a strongly Eurocentric doctrine by arguing that humans had originated in the Eastern Mediterranean and, like other species, had degenerated as a consequence of environmental differences when they moved away from their place of origin and as world climates changed over time.[33] Buffon was at times puzzled about the relationship between humans and apes, suggesting as a possibility that the apes were extreme examples of human degeneracy.

ENRICHING THE STATE AND ITS CITIZENS: CAMERALISM AND POLITICAL ECONOMY

The term "political economy" was coined by Antoine de Montchretien (1575–1621) around 1615, but it came into prominence only after the 1767 publication of Sir James Steuart's (1713–1780) *Inquiry into the Principles of Political Economy:*

---

[33] For Buffon's anthropology, see *Buffon: De l'homme,* ed. Michelle Duchet (Paris: Maspero, 1971). For Blumenbach's, see *The Anthropological Treatises of Blumenbach,* ed. Thomas Bendyshe (London: Lonman, Green, Longman, Roberts, and Green, 1865).

Being an Essay on the Science of Domestic Policy in Free Nations, In Which Are Particularly Considered, Population, Agriculture, Trade, Industry, Money, Coin, Interest, Circulation, Banks, Exchange, Public Credit, and Taxes. The term was used almost immediately by Adam Smith, among others, to identify any work that focused, although not often exclusively, on the revenue of both the people and the state. Steuart's usage was unusually broad, for he had composed the *Inquiry* while in exile at Tübingen and under the influence of the cameral sciences favored by German-speaking authors, who were unwilling to isolate economic issues from issues associated with general administration, public health and safety, political autonomy, the perceived quality of life, and even national character. Most non-German political economists, on the other hand, tended to exclude issues that did not have an immediately "economic" content. In what follows, I will use "political economy" to identify the more narrowly construed discussions favored by French, English, Dutch, Scottish, and Italian authors, using "cameral sciences" to identify the broader approach favored in the German states, Austria, Scandinavia, and Russia.

One critical feature of the difference between cameral science and political economy was a consequence of the medical training of many of the most important seventeenth-century founders of the disciplines. The German physician-cameralists, such as Johann Becher (1635–1682) had been trained in anti-Greek alchemical and Paracelsian medicine. According to Paracelsus it was the physician's task to improve on Nature by proactively intervening in the life of a patient to see that such things as diet, sanitation, and even working conditions were improved to promote well-being. By the same token, the cameralist physician to the body politic advocated a broad range of state interventions, including the central planning and regulating of social and economic affairs, to enhance the well-being of the nation. The medical founders of political economy, such as William Petty (1632–1687) and John Locke, on the other hand, were trained in the Hippocratic/Galenic tradition, enriched by the iatromechanist approaches of William Harvey and the Cartesians. According to this version of medical theory, Nature was essentially self-regulating and self-perfecting. Illness or disease occurred because some pathological entity was present, impeding the natural processes. The primary function of the physician was simply to remove the impediment and stand aside. Although he was a clergyman rather than a physician, Josiah Tucker (1713–1799) expressed the laissez-faire implications of this perspective for eighteenth-century Anglo-French political economy particularly well in his *Elements of Commerce* (1755):

> Hence, therefore, the physician to the body politic may learn to imitate the conduct of the physician to the body natural, in removing those disorders which a bad habit, or a wrong treatment hath brought upon the constitution; and then to leave the rest to nature, who best can do her own work. For after the constitution is restored to the use and exercise of its proper fac-

ulties and natural powers, it would be wrong to multiply laws relating to commerce as it would be to be forever prescribing physic.[34]

Seventeenth-century political economy and cameral science had been produced all but exclusively as advice to government officials (usually heads of state, but in the British case, to Parliament as well) by court officials or persons seeking patronage. They thus tended to center on ways in which princes or the governments for which they stood could enrich themselves; and they tended to be uncritical of established authorities, seeking gently to sway them into developing policies intended to increase the wealth of citizens because it was accepted that the wealth of a prince was dependent on the well-being and wealth of his subjects. Throughout the eighteenth century the cameral sciences continued to be formulated as friendly advice to paternalistic princely rulers; but political economy took on a much more critical character in Western Europe. Most works of political economy still offered policy advice to governments; but virtually all of them did so in the name of the general good; and they often reflected the interests of particular groups. Moreover, as the century went on, both among cameral scientists and among political economists, increasing numbers of works sought to provide comprehensive theories or systems of the functioning of commercial economies in order to provide a general framework for the formation of specific policies.

Among early eighteenth-century political economists, Pierre de Boisguilbert (1646–1714), a farmer who also had legal training, stands out as particularly impressive. Boisguilbert was deeply distressed by the French crown's taxation policies as well as by export and price controls on grain. Together, these policies seemed to be driving increasing numbers of farmers into bankruptcy and to a consequent rapid decline in both private and public revenues throughout France. In a series of private and public tracts, including *Détail de la France* (1695), *Factum de la France* (1705), and *Dissertation de la nature des richesses* (1707), Boisguilbert provided compelling arguments that any exchange between two uncoerced parties would inevitably lead to the benefit of both, so that a free and unregulated market would ensure that both farmers and consumers benefited. Thus, he argued against price controls. If the government insisted on setting maximum grain prices in years of poor crops, it should likewise provide price supports in years of overproduction. Furthermore, Boisguilbert emphasized the centrality of consumption for the economy, arguing that consumption is increased by circulation of money and that circulation is increased by putting more money into the hands of the poor, who spend their incomes faster than do the wealthy. Thus, he urged tax policies that were progressive as well as policies that encouraged the creation of increased productive capacities.

Ernst Ludwig Carl (1682–1743) was the first major eighteenth-century

---

[34] Cited in Terence Hutchison, *Before Adam Smith* (Oxford: Basil Blackwell, 1988), p. 231.

cameral scientist. After studying law and the cameral sciences at Halle, Carl
was sent to Paris as an agent of the Margraves of Bayreuth and Ansbach. There
he met up with the work of Boisguilbert and studied French manufacturing
policy. In 1722 and 1723 Carl published his three-volume *Traité de la Richesse
des Princes et de leurs États* which incorporated some western political economy
into the cameralist framework. Admitting the self-interested motives of economic actors and the potential for natural market regulation of economic
exchanges, Carl nonetheless argued that through shortsightedness and ignorance, most individuals subverted the market process, creating a need for state
regulation. Among the newly emerging issues Carl addressed were the importance of the division of labor for increasing productivity and the idea that
each nation has a comparative advantage in producing some goods for exchange. As a consequence, international trade is not a zero-sum game; but it
can be carried out to the advantage of all participants.

After Carl, there was very little originality in cameralist works, although
Johann Heinrich Gottlob von Justi (1717–1771), author of numerous works including the widely popular *Staatswirtschaft* (1755) and *System des Finanzwesens*
(1766), was inclined to warn his bureaucrats-in-training of how easy it was to
simply raise taxes rather than to control spending; thus, he tried to emphasize the responsibility of the governors to those governed to a greater extent
than most prior cameralists. Joseph von Sonnenfels, Professor of Cameral
Science at Vienna, proved to be such an effective promoter of cameralist ideas
that his *Grundsatze der Polizei, Handlung und Finanzwissenschaft* (1763) continued to be the leading textbook of cameral science for nearly a century.

International trade and banking were the chief interests of Richard Cantillon (c. 1690–1734), an Irish-born Parisian international banker, whose comprehensive and immensely influential *Essai sur la nature du commerce en générale*
circulated in manuscript for decades before it was finally published in 1755.
Although primarily a critical attack on John Law's policies, which had led to
the creation and collapse of the Mississippi Company, Cantillon's work was
among the most comprehensive economic systems prior to that of Adam
Smith. Dividing the costs of production into labor, rents, and profits on capital, Cantillon argued that producers would produce only enough of a commodity to satisfy a demand that would maintain a price that would oscillate
around the cost of production. In addition, he particularly focused on the
importance of entrepreneurship and the rewards that came from taking risks.
He explained the causes of inflation, analyzed exchange rates, and showed that
productive capacity was the ultimate source of wealth; as a result, precious
metals moved rapidly away from nations that acquired them by mining into
the hands of those who produced finished goods.

Between 1756 and 1774, French political economy was dominated by a group
of men who identified their movement as physiocracy, the rule of Nature, as
opposed to monarchy, aristocracy, or democracy, the rule of the one, the few,
or the many. Led by the farm-bred, surgically trained, and autocratic François

Quesnay (1694–1774), who had come to an interest in political economy only in his sixties, the physiocrats sponsored their own journals, *Journal de l'Agriculture, du Commerce, et des Finances* (1765–6) and *Éphémérides du Citoyen* (1768–72) under the editorship of Pierre Samuel Du Pont de Nemours (1739–1817). In general, they adopted Boisguilbert's special concern with agriculture and free market exchanges along with Cantillon's emphasis on investment. Thus, they pushed to ensure an annual net profit from agriculture that would allow for continued investment in capital improvements and hence, productivity; and they expected to achieve this situation by deregulating grain prices and exports. Moreover, they argued for tax policies that would minimize impediments to production. Among their most important technical developments – a mathematical model of circulation of money in an economy – was the *Tableau Économique* (1758).

Among the nonphysiocratic political economists in France during the second half of the century, the most prominent was probably Anne Robert Jacques Turgot (1727–1781), a career administrator who served as finance minister of France, initiating a brief period of unregulated grain trade. Unfortunately, his experiment coincided with several years of very poor crops, and the public outcry against high bread prices forced the king to accept his resignation. Turgot's major general work, *Reflections on the Formation and Distribution of Riches* (1769–79), slightly modified the physiocratic obsession with agriculture, analyzed various forms of capital, and explored how relatively stable exchange values of commodities were established by communities of buyers and sellers. Although of relatively little immediate impact on political economy, the analyses of subjective preference in establishing what economists now call individual utility functions by Etienne Bonnot de Condillac in his *Le commerce et le gouvernement* (1776) have received much attention by twentieth-century utility theorists.

Italy produced several significant liberal political economists, including the Milanese friends Caesar Beccaria (1738–1794) and Petro Veri, whose mathematical treatment of the relationship between utility, scarcity, and price expresses one of the first mathematical "laws" of political economy. But the greatest of eighteenth-century Italian political economists was probably a Neapolitan, Ferdinando Galiani (1728–1787). As a twenty-two-year-old, Galiani wrote an excellent treatise, *Della moneta,* in the liberal tradition. But twenty years later, after living for some time as a Neapolitan diplomat in Paris, he wrote perhaps the most scathing attack ever against the physiocrats, appropriating the historical perspectives of both Vico and Montesquieu to argue that the effects of the operation of economic laws depend critically on local conditions, including the form of government under which people live and their customs. Thus, for example, in *Dialogs sur le commerce des blés* (1770), Galiani argued that even though deregulating the grain trade would, if accomplished, eventuate in equilibrium prices that would benefit everyone, it should not be attempted in France for at least two reasons. First, freedom of trade was inconsistent with monarchical government because it would inevitably lead to both higher costs of living and a net transfer of wealth to the peasantry, undermining the inequalities of wealth and status that support monarchical government and increasing pressures in favor of republicanism. Second, but equally important, time may be a critical factor, especially in connection with essential foodstuffs. Although an equilibrium price might eventually be established, it could (as it in fact did a few years later) take so long to do so that masses of poorer people, fearing starvation, would rebel against intolerably high prices in the meantime.

Galiani was not alone in trying to link historical and economic issues. Many political economists, including Cantillon and Turgot, had serious interests in philosophical history, and virtually all philosophical historians were concerned centrally with problems of subsistence. But it was in Scotland and particularly in Adam Smith's *An Inquiry into the Nature and Causes of the Wealth of Nations* (1776) that philosophical history and political economy were fused into a comprehensive synthesis so compelling that it made virtually all preceding work in political economy obsolete. Starting with a historical analysis of the role of the division of labor in increasing productivity and the need to increase markets to maximize the benefits of the division of labor, Smith proceeded to a full discussion of the role of markets in setting prices and establishing a distribution of economic resources that would maximize wealth. He went on to argue that only in growing economies with labor shortages will wages be above the subsistence level. He covered all the traditional topics of political economy, drawing heavily from the writings of Turgot and incorporating his own extensive historical evidence; he also included extensive critiques of cameralist, physiocratic, and earlier scholastic discussions of moral economy.

QUANTIFICATION IN THE HUMAN SCIENCES

With rare exceptions, political economists sought mathematical regularities in the phenomena they studied. Some, including the followers of John Graunt and William Petty as well as those of such cameralists as Johanne Becher, argued that both public and private decisions would be improved by being guided by the use of quantifiable information regarding "the lands and hands of the territory . . . according to all their intrinsic and accidental differences."[35]

Inventories of persons and economic resources had been widely used for establishing tax liabilities throughout Europe since Roman times; but advocates of "political arithmetic" and "statistics" sought to collect more information and to use it in more extensive ways. As early as 1693, Edmund Halley, for example, used birth and death data from Breslau to illustrate how to cal-

[35] John Graunt, *Natural and Political Observations . . . Made Upon the Bills of Mortality,* ed. Walter Wilcox (Baltimore, MD: Johns Hopkins University Press, 1939, from 1662 original), p. 78.

culate life expectancies and the prices to be charged for single life annuities or tontines (which were widely used by governments to generate income, much as lotteries are in the twentieth century).[36] Abraham de Moivre (1667–1754) soon figured out how to calculate the cost of multiple life annuities. The use of quantitative data for public resource management was pioneered by the German cameralists, who developed techniques in forestry management for estimating the volume of wood in large areas through sampling and so on.[37]

Sweden was the first nation to establish an effective statistical office when it established an Office of Tables in 1749 through efforts led by Andre Berch, professor of economics at Uppsala University and author of *Politisk Arithmetica* (1746).[38] In spite of the enthusiasm of some advocates, however, the effective use of social and vital statistics was virtually impossible everywhere well into the nineteenth century. Although one might be able to calculate life expectancies based on the total population of a specific locale, such a population did not reflect the population of those likely to purchase insurance; thus, insurance companies quite reasonably ignored the theoretical calculations of actuaries until the mid-nineteenth century. Moreover, gathering accurate data was nearly impossible both because relatively weak central governments were unable to enforce uniform procedures for collecting information and because information was deliberately withheld or misreported by nearly everyone out of fear that it might be used by central authorities against their interests.[39]

Finally, the eighteenth century saw some of the first attempts to supply probability theory to a social issue when Condorcet attempted to explore the conditions under which the majority decisions of representative assemblies should be accepted in his *Essai sur l'application de l'analyse pluralité des voix* of 1786. Again, however, the practical application of mathematical theory to public policy was far in the future.

SENSATIONALIST/ASSOCIATIONIST PSYCHOLOGY, UTILITY, AND POLITICAL SCIENCE

When Condorcet drafted his *Projet de décret sur l'organization générale de l'instruction publique* for the National Assembly in 1792, he argued that there

---

[36] Edmund Halley, "An Estimate of the Degrees of Mortality Drawn from Curious Tables of Births and Funerals at the City of Breslau, with an Attempt to Ascertain the Price of Annuities upon Lives," *Philosophical Translations of the Royal Society in London,* 17 (1693), 596–610.

[37] See Henry Lowood, "The Calculating Forester: Quantification, Cameral Science, and the Emergence of Forestry Management in Germany," in J. L. Heilbron and Robin E. Rider (eds.), *The Quantifying Spirit in the Eighteenth Century* (Berkeley: University of California Press, 1990).

[38] See August Johannes Hjet, *Det svenska tabrlltrerkets uppokomst, organisation och tidigare verk samhet* (Helsingfors: O. W. Backmann, 1900).

[39] On the eighteenth-century resistance to governmental attempts to collect statistical information, see Peter Buck, "People Who Counted: Political Arithmetic in the Eighteenth Century," *Isis,* 73 (1982), 28–45.

should be three lecturers on the social or human sciences at each *lycée*. One would be responsible for philosophical history, one for political economy, and the third and most important would combine the analysis of sensations and ideas, scientific method, morality, and "the general principles of political institutions."[40] In the late twentieth century the functions of this third lecturer would be spread across departments of psychology, philosophy, and political science, members of which would probably deny any connection with one another. But during the eighteenth century they were frequently linked because many thinkers argued that the structures of governments should be suited to their function, which was to meet the needs and wants of the citizens or, in terms used by eighteenth-century theorists, to increase their happiness and allay their fears. If governments were to truly serve these functions, it was first necessary to determine precisely what made people happy and fearful and how they might act so that the needs of the general population, rather than those of a small segment, were met – that is, to address issues of psychology and morality. Finally, through the use of scientific methods, one could determine how to organize societies so that people were made or allowed to act to serve the general good.

This pattern – of moving in political discourse from an analysis of human perceptions, desires, and aversions, through an analysis of moral "rights" and duties, to a prescription for social and political arrangements – had been well established during the seventeenth century, especially in the writings of Hobbes and Spinoza. Virtually all those who approached the psychology-morality-civil society complex of issues in this way during the eighteenth century shared one other general feature: they began from and refined the empiricist perspective that had been explored in John Locke's *Essay Concerning Human Understanding* (1690) and in the writings of Pierre Gassendi (1592–1655) rather than the "rationalist" perspective which had informed the works of Hobbes and Spinoza.

David Hume's difficult *A Treatise of Human Nature* and its more popular expositions in *An Enquiry Concerning Human Understanding* (1762), *An Enquiry Concerning the Principles of Morals* (1751), and *Essays, Moral, Political, and Literary* (two volumes, 1741–42) were among the most important attempts to formulate an empirically based psychology, morality, and politics. One of Hume's chief arguments was that reason plays a vastly more limited role and our passions play a much greater role in motivating our actions than most moralists and political theorists had admitted. Humans link ideas much more often through their association, which is largely a matter of habit, than through logical connection, which is a product of intentional rationality. Furthermore, the human inventory of passions is much more complex and extensive than had been realized. In particular, humans are driven not only by self-

---

[40] See Keith Michael Baker, *Condorcet: From Natural Philosophy to Social Mathematics* (Chicago: University of Chicago Press, 1975), Appendix A, p. 389.

interest but also by a variety of social passions grounded in sexual attraction and attachment to children but extended by habitual associations to family and community members. As a consequence, if institutions were to be effectively devised to accommodate human desires and aversions, they would have to be much more complicated than the primitive and simplistic psychological assumptions of earlier theorists had supposed. Indeed, Hume was so skeptical of the human ability to anticipate all the complex consequences of any new institution that he urged extreme caution in political innovations.

Much more optimistic about our ability to change human behavior by manipulating experiences so as to produce desirable associations and about our ability to design institutions that will promote human happiness was David Hartley (1705–1757). His *Observations on Man, His Frame, His Duty, and his Expectations* (1749) became a kind of holy book for radical reformers in both Britain and America. According to Hartley, as we mature, we naturally develop increasingly benevolent passions, which seek the welfare of others; so if we can simply avoid the pathological development of artificially great divergences of wealth and status, there will be a natural accommodation between the desires of each individual and the well-being of the entire community. Joseph Priestley, James Mill, William Godwin, Benjamin Rush, and even the early nineteenth-century socialist Robert Owen, were all self-styled followers of Hartley, although the moderate Anglican clergyman would probably have been appalled at how far they pushed the egalitarian and democratic implications of his work.[41]

More moderate in his political views, and more concerned with the need to force individuals to act in such a way as to benefit society as a whole, was Jeremey Bentham (1748–1832), who popularized the term "utility" and the notion that the aim of all governments should be the greatest good for the greatest number. Beginning in his *An Introduction to the Principles of Morals and Legislation* (1789), Bentham published a series of tracts developing his "Utilitarian" philosophy and his "calculus of felicity," which was intended to offer a way of quantifying the relative desirability of various policies based on their differential abilities to promote happiness and reduce pain and anxiety. During the nineteenth century and in part through its development by John Stuart Mill in England and Etienne Dumont in France, Utilitarianism became a popular political movement. Its members were instrumental in legal reform, health and sanitation reform, and in the extension of voting rights through the English Reform Bill of 1832.

In France, Condillac played much the same role that Hartley did in England. Although he was not a political reformer, Condillac's psychological writings and discussions of scientific method in his *Essai sur l'origine des connaissances humaine* (1746), *Traité des systèmes* (1749), *Traité de sensations* (1754), *Traité des*

---

[41] See Isaac Kramnick, "Eighteenth-Century Science and Radical Social Theory: The Case of Joseph Priestley's Scientific Liberalism," *Journal of British Studies,* 25 (1986), 15ff.

Animaux (1755), *Le commerce et le gouvernment* (1776), and *La Logique: ou les premiers développmens de l'art de penser* (1782) provided the starting place for a huge number of attempts at educational, economic, social, and political reforms through the period of the French Revolution.

Claude-Adrien Helvétius was among the most influential and original of those who developed the social and political implications of Condillac's ideas. A member of a wealthy family who increased his fortune by early successes as a tax farmer, Helvétius "retired" at age 42 to become a savant and experimental farmer. His *De l'esprit* (1758) and *De l'homme* (published posthumously in 1774) began by arguing, contra Montesquieu, that most existing human institutions, because they had been founded before the principles of sensationalist psychology were recognized, were based on false understandings of human nature and were thus the causes of untold suffering. Self-interest, including the desire to be admired and to exercise authority over others, drives all human actions. But most people come to identify their own interests with those of a group of people with whom they share status and functions. In this way actions come to be shaped by class interests; and in societies where large differences in wealth and authority have come to exist, the clergy, the wealthy, and the governing elites recognize that it is in their interest to keep the mass of human beings ignorant and poor. Immense wealth and power are thus concentrated in the hands of a few, while everyone else toils in misery. The trick is to reverse this situation and create institutions that reward actions that serve the general good. In fact, it was Helvétius who insisted that the goal of government should be to establish the greatest good for the greatest number, a notion later adopted by Bentham with acknowledgment of his debt to Helvétius.

In 1758 Helvétius hoped that educational reform spearheaded by the scientific intelligentsia could succeed in producing peaceful reform; but the hostile clerical and governmental reaction to *De l'esprit* convinced him that progressive change could probably not be achieved short of violent revolution. Thus, Helvétius became one of the first theorists to advocate the creation of an egalitarian and classless society through overthrow of the present authorities.

After Helvétius's death, his wife continued the salon that she and her husband had initiated. It was in this environment that Condorcet developed his educational ideas, his proposals to extend the vote to all citizens regardless of sex or race, and his applications of probability to social issues. Similarly, it was at the salon of Mme. Helvétius (and Mme. Condorcet after her husband's death in 1794) that the Idéologues – Pierre Cabanis, Antoine Destutt de Tracy (1758–1836), and Jean-Baptiste Say (1767–1832) – began their careers as social reformers and social theorists.

On the Continent outside of France, Helvétius's ideas were particularly central in the development of the utilitarian arguments of Cesre Beccaria, whose *Dei Delitti e della Pene* (1764) initiated a period of penal reform throughout Europe.

One of the final eighteenth-century movements to derive its foundations from the psychological theorizing of Condillac's followers was the early feminism of Catherine Macaulay and Mary Wollstonecraft, who, like Helvétius, saw many cultural practices grounded in mere custom as antithetical to the dictates of an egalitarian associationist psychology. Thus, in her *Letters on Education* (1790), Macaulay wrote as follows:

> It ought to be the first care of education to teach virtue on immutable principles, and to avoid that confusion which must arise from confounding the laws and customs of society with those obligations that are founded on correct principles of equity . . . There can be but one rule of moral excellence for beings made of the same materials, organized after the same manner, and subject to similar laws of nature. . . . [It follows] that all of those vices and imperfections which have been generally regarded as inseparable from the female character, do not in any manner proceed from sexual causes, but are entirely the effects of situation and education.[42]

The reaction of many observers to the radical political agendas reflected in feminism and the French revolution was to turn violently against the psychologically based theories that seemed to provide their rationales and to reassert the importance of historically oriented theories. Thus, the very end of the eighteenth century and the early nineteenth century saw a resurgence of philosophical history, keyed by the antirevolutionary sentiments of authors such as Edmund Burke, whose *Reflections on the Revolution in France* (1790) signaled the beginnings of the reactionary historicist trends.

GENERAL EVALUATION OF EIGHTEENTH-CENTURY HUMAN SCIENCES

During the nineteenth century, most of the subject matters discussed by eighteenth-century thinkers under the general category of sciences of man [sic] – or our category of the human sciences – were reorganized under the twofold influences of Comtian Positivism and the professionalization of academic disciplines. When that happened, there was a general downplaying of the scientific significance of the work done by almost all the figures discussed in this essay. For reasons that we cannot explore here, Comte was adamant in insisting that the introspective methods underlying sensationalist and associationist psychology were unscientific and misleading; so as psychology became professionalized in Germany by Wilhelm Wundt and others under Positivist influences, it was recast as a physiological discipline, and eighteenth-century discussions were relegated to the "metaphysical" prehistory of the discipline. As sociology and anthropology became professionalized in the later nineteenth

---

[42] Catherine Macaulay, *Letters on Education* (New York: Garland Publishing, 1974, reprinted from 1790 London original), pp. 201–4.

century, the works of Vico, Montesquieu, and the Scottish school of philo-sophical history, as well as those of Herder and the German historians of jurisprudence, were appreciated for the issues they raised, but their authors were often condemned for being "armchair philosophers" who based their speculations on the uncontrolled and often credulous tales of travelers and ancient historians rather than real scientists who grounded their discoveries in carefully controlled and extensive fieldwork. In political economy, virtually all nineteenth-century professionalizers continued to see Adam Smith's *Wealth of Nations* as the foundational text of their discipline; but Smith's portrayal of physiocracy and cameralism as unscientific and politically destructive special pleading also served as a barrier to interest in earlier political economy among nineteenth-century practitioners.

The upshot of all these factors was that nineteenth-century practitioners of the human and social sciences virtually stopped reading and giving serious consideration to their eighteenth-century predecessors, taking literally the Comtian notion that knowledge of human interactions in society attained the status of positive knowledge only in the nineteenth century. In this percep-tion, they were cutting themselves off from their roots in a way that has per-sisted into the late twentieth century.

20

THE MEDICAL SCIENCES

Thomas H. Broman

Mention the term "medical science" to someone, and it is likely to evoke an image of white-coated scientists working at a laboratory bench. In the mind of a more historically informed listener, the term might produce a more specific image – of Louis Pasteur gazing at a test tube, of Xavier Bichat bending over one of his corpses in the Hotel-Dieu, or even of William Harvey ligating a vein – but the general meaning would remain largely the same, because for us the association between "medical science" and "experiment" is a powerful one. Yet for all its pervasiveness, this association is misleading when we consider the medical sciences in the eighteenth century. An image far more appropriate than the laboratory would be the simple podium or lectern, for the medical sciences were understood by eighteenth-century physicians primarily as a body of theoretical doctrines that formed one part of the university medical curriculum. The medical sciences, especially the subjects of physiology and pathology, furnished the bridge between medical knowledge proper and the domain of natural philosophy. And natural philosophy attempted in turn to provide a comprehensive theoretical knowledge of the elemental makeup of the world and the motions of matter.[1] Therefore, insofar as physiology and pathology explained the composition and actions of the living body in its healthy and diseased states and rendered those explanations in terms consistent with natural philosophy, they legitimated medicine's claim to the status of scientific knowledge.

A recognition of the doctrinal and pedagogical role of medical science is essential to an understanding of it, and it carries two important consequences. First, it highlights the role of universities in constituting and certifying scientific knowledge, a function that was particularly significant for a university-

[1] On the contours of academic natural philosophy in the eighteenth century, see L. W. B. Brockliss, *French Higher Education in the Seventeenth and Eighteenth Centuries* (Oxford: Oxford University Press, 1987), and William Clark, "German Physics Textbooks in the *Goethezeit*," *History of Science*, 35 (1997), 219–39, 295–363. Clark's article presents a comprehensive survey of the evolution of academic natural philosophy in Germany from the mid-eighteenth century to the 1820s.

based profession such as medicine. Of course, universities in the eighteenth century did not possess exclusive regency over the domain of scientific knowledge; the academies of science that began appearing during the second half of the seventeenth century also played a major role in its adjudication, and arguments have also been advanced for the role of salons, coffeehouses, and other institutions. But because medicine was a university subject and medical faculties had the right to examine and grant licenses to physicians and other healers, what those faculties taught effectively defined medical science as a prescribed body of doctrine.[2]

A second consequence of treating medical science in its pedagogical setting is that it renders visible a spectrum of scholarly work that today looks to us quaintly bookish, even scholastic in the most prejudicial sense of the word. When an eighteenth-century medical professor wanted to produce a work of scholarship, he (they were all male) might write a textbook on pathology or general therapeutics, a history of venereal disease from antiquity to modern times, or a dissertation on some theoretical issue that would subsequently be defended by a medical student in formal disputation. Needless to say, these works would often be written in Latin, which remained the language of choice for academic writing until well after mid-century, especially in Central Europe and Italy. Even a renowned experimentalist such as Albrecht von Haller (1708–1777) produced mountains of scholarly writings that had no bearing whatsoever on experiment. In short, the range of scholarship comprehended under the rubric "medical science" was far broader in the eighteenth century than would subsequently be the case.

With this context in mind, this article will present a survey of the medical sciences as they evolved over the course of the eighteenth century. I will begin with a brief discussion of the structure of medical education in 1700, with particular emphasis on the central role of the "institutes of medicine" (*institutiones medicae*) in offering students an introductory overview of medical theory. Then I will discuss in greater detail developments in the two core subjects of medical theory: physiology and pathology. As will rapidly become evident, the stories of how physiology and pathology evolved over the course of the eighteenth century are quite different and have little direct bearing on each other. As a result, one can detect a fragmentation in medical theory over the course of the century that was being remarked upon by contemporaries in the 1790s. I will close with a brief discussion of the situation in medical theory at the end of the century.

---

[2] The prescriptions defining acceptable medical doctrine reach back to the earliest history of universities. As early as 1309, Pope Clement V dictated to medical students seeking a license from the University of Montpellier a knowledge of specific writings from Galen, Hippocrates, and Avicenna, among others. See Hastings Rashdall, in F. M. Powicke and Ab. B. Emden (eds.), *The Universities of Europe in the Middle Ages*, vol. 2 (Oxford: Oxford University Press, 1936), p. 127. For a general introduction to the origins and early history of university medical education, see Nancy G. Siraisi, *Medieval and Early Renaissance Medicine* (Chicago: University of Chicago Press, 1990), pp. 55–77.

Before turning to those matters, however, a word about the national coverage of this article is in order. The equation of medical science with university medical curricula works nicely in parts of Europe where the universities maintained their traditional function of educating the medical elite, as in Scandinavia, France, German-speaking Central Europe, Switzerland, Italy, and the Netherlands. Its relevance to Great Britain, however, is much more problematic. The two medieval English universities of Oxford and Cambridge did continue to grant a handful of medical degrees during this period, but their hold on medical education was challenged first by the Scottish universities, especially Edinburgh, and also by a system of nonuniversity training that emerged in London. Whereas the curriculum at Edinburgh was relatively consistent with the Continental pattern we will be examining, that of London looked substantially different. Medical education there featured a variety of private lecture courses and considerable opportunities for "walking the wards" in the great metropolitan hospitals to observe surgeons and physicians doing their charitable work.[3] Medical theory in London did not differ markedly from what was being taught elsewhere, but the clinical orientation of the London schools meant that the introduction to medical theory given students there was tailored to bedside practice. As a result of the different pedagogical and institutional setting, the problem of how medicine should be understood as a science took on different contours in England. For this reason, the following description focuses most closely on medical science on the Continent.

THE SHAPE OF MEDICAL EDUCATION

No matter how progressive a medical faculty believed itself to be, medical education in the eighteenth century was largely, if not exclusively, a matter of lecture and explanatory comment. But whereas professors had formerly built their courses around a canon of ancient medical texts, by the opening of the eighteenth century standard practice called for them to use a textbook as the centerpiece of the course. Lectures consisted of reading aloud passages from the textbook, which were then embroidered with illustrative and explanatory comments. For this reason, textbooks often were written in an aphoristic style in numbered paragraphs, a format that facilitated cross-referencing and perhaps gave inattentive students an opportunity to locate their place in the lecture. We can see this method at work, for example, in two well-known textbooks of the period: Friedrich Hoffmann's (1660–1742) *Fundamenta medicinae* (1695) and Herman Boerhaave's (1668–1738) *Institutiones medicae* (1708). Something

---

[3] London was a major center of medical and surgical training. By mid-century, some one hundred students each year were signing hospital registers and paying fees for the privilege of accompanying the consulting staff on their rounds. Susan C. Lawrence, *Charitable Knowledge: Hospital Pupils and Practitioners in Eighteenth-Century London* (Cambridge University Press, 1996), p. 111.

like what students might have heard in Boerhaave's course can be surmised
from the edition of Boerhaave's lectures published by Albrecht von Haller as
the *Praelectiones academicae in proprias institutiones rei medicae* (Academic lectures on the particular institutes of medicine, 1739–44). The *Praelectiones* consists of short series of numbered paragraphs from the *Institutiones,* followed
by sections containing Boerhaave's extensive explanatory comments on each
paragraph.[4]

At many universities, such a course in the *institutiones* constituted a student's introduction to medical theory, and it invariably covered five topics:
physiology, pathology, semiotics (the interpretation of symptoms), therapeutics, and dietetics (the rules of preserving health). To the modern eye, this may
seem an odd mixture of topics for a course of medical theory, but the composition of the *institutiones* had both historical and intellectual justifications.
Historically, as Nancy Siraisi has shown, courses in the *institutiones* were
descendants of sixteenth-century medical courses based on translations of the
Canon of Avicenna, an early eleventh-century Muslim scholar.[5] Intellectually, the specific topics covered in the *institutiones* were more intimately connected than mere historical juxtaposition might suggest. One significant thread
running through them was the close attention they devoted to a person's interactions with the external environment. These interactions, which had been
known since antiquity as the "nonnaturals," were grouped into six general
categories under the headings of air, food and drink, motion and rest, excre-
tions and retentions, sleep and wakefulness, and psychological affects.[6] From
the perspective of pathology, the nonnaturals figured into an understanding
of the causes of disease. A sudden chill, for example, or overindulgence in spicy
foods, or too much vigorous dancing – a particular danger for young ladies,
in the eyes of more than one eighteenth-century physician – could be the
specific trigger for the onset of illness. In dietetics, proper regulation of the
nonnaturals was the key to maintaining health, and some of the eighteenth century's most famous guides to healthy living, such as George Cheyne's (1671–
1743) *Essay of Health and Long Life* (1724), Samuel August Tissot's (1728–1797)
Avis au peuple sur sa santé (1761), and Christoph Wilhelm Hufeland's (1762–
1836) *Die Kunst das menschliche Leben zu verlängern* (1797), were plainly
organized around the general conceptual pattern provided by the nonnaturals.

---

[4] The *Praelectiones* also featured footnotes added by Haller at the bottom of each page in the commentary sections. These contain extensive bibliographic references from ancient and modern sources – references that make the *Praelectiones* an outstanding research tool for the modern scholar – along with
Haller's own comments on Boerhaave's doctrines.

[5] Nancy G. Siraisi, *Avicenna in Renaissance Italy* (Princeton, NJ: Princeton University Press, 1987). See
pp. 101–2 for comments on the *Canon*'s influence on later textbooks of medical theory.

[6] L. J. Rather, "The 'Six Things Non-Natural,'" *Clio Medica,* 3 (1968), 333–47. In the terminology of
Galenists, the "naturals" referred to those things pertaining to the basic structure and functions of
life: the elements, the temperaments, the humors, the spirits and natural heat, the organs, the faculties
and functions, and generation. The nonnaturals, therefore, can be thought of as agents capable of disrupting the body's disposition "by nature" toward health. See Siraisi, *Medieval and Early Renaissance
Medicine,* p. 101.

Likewise in therapeutics, regulation of the nonnaturals played a major role in the treatment of illness, alongside routine therapies such as phlebotomy and administration of drugs.

Typically, the *institutiones* were a staple offering in the first year of a three- or four-year curriculum. Other courses in the first year might include botany, anatomy, and chemistry – three specialized subjects whose profile in medical teaching had expanded in the seventeenth century and would continue growing throughout the eighteenth. At the next level of the curriculum, a student might take courses in general and special pathology, materia medica (which taught the various kinds of medications and their effects on the body), and perhaps therapeutics. Finally, at the third and final stage of the curriculum, students would typically take courses in general and special therapeutics, the method of writing prescriptions, surgery, and finally clinical practice, in which they might have an opportunity to participate in the care of patients. Needless to say, there was considerable variation in this basic pattern, and even to label it a "structure" is somewhat misleading, because at many universities students were not required to take the courses in a particular sequence, although they might have been advised to do so.[7]

Even more variable than the subjects in the curriculum was the method of instruction. Students in chemistry courses did not necessarily have a chance to perform chemical experiments themselves. Botany students probably had more opportunities for studying their specimens firsthand, although not necessarily in a botanical garden. Students in anatomy courses, meanwhile, had the most difficult time of all, because the supply of cadavers for dissection was limited and remained so throughout the century. Despite frequent calls by professors for more opportunities for student dissections and despite repeated efforts by governments to make more cadavers available, it was by no means a common occurrence for a physician trained in a university to have studied anatomy through hands-on dissection.[8]

At the opening of the eighteenth century, therefore – and indeed until its very end – the centerpiece of medical education remained the spoken and written word. As in the case of anatomy, this situation in part reflected shortcomings that were felt by the professors themselves – but only in part: it also reflected a culture that valued the collation and criticism of medical writings both ancient and modern as fundamental to scholarship. Medical theory

---

[7] For descriptions of the medical curriculum in the eighteenth century, see Brockliss, *French Higher Education,* pp. 391–400; Thomas H. Broman, *The Transformation of German Academic Medicine, 1750–1820* (Cambridge University Press, 1996), pp. 28–9; and Lisa Rosner, *Medical Education in the Age of Improvement: Edinburgh Students and Apprentices, 1760–1826* (Edinburgh: Edinburgh University Press, 1991), pp. 44–61.

[8] Broman, *Transformation of German Academic Medicine,* p. 29(n). Brockliss and Jones describe how "dissecting riots" broke out in Montpellier and Lyons over the provision of the bodies of poor inmates of hospitals for student dissection. See Laurence Brockliss and Colin Jones, *The Medical World of Early Modern France* (Oxford: Oxford University Press, 1997), pp. 713–14. Similar difficulties beset anatomical teachers in London as well. See Lawrence, *Charitable Knowledge,* pp. 194–200.

formed the foundation for the physician's social identity just as much as did bedside practice. Indeed, insofar as physicians were only one group among a diverse crowd of surgeons, apothecaries, midwives, barbers, bathkeepers, charletans, itinerant drug peddlers, and others who undertook to offer health care in this period, their credentials as learned gentlemen were essential in distinguishing themselves from others lower down the social hierarchy. Yet if the role of medical theory as a badge of professional identity was relatively stable, the contents of that theory were not. Over the course of the century, the teaching of the *institutiones medicae* would gradually disappear, replaced by individual courses in physiology and pathology. And as we shall see later, this separation in textbook subjects mirrored a growing theoretical separation between the core subjects of physiology and pathology.[9]

PHYSIOLOGY

If medical science could be properly considered a branch of natural philosophy, it was physiology's unique task to connect that branch to the tree of knowledge. The term "physiology" itself was a sixteenth-century neologism; if Galen had heard the word, he would have understood it as a broader study of nature. Galen did write extensively on what we would call "physiology," of course, but he never produced a systematic treatise on the functions of living bodies. However, synthesis was much more the order of the day in the academic culture of the sixteenth century, when Galen's writings first became widely available in Latin translation, along with other synthetic treatments of Galenic physiology, such as Avicenna's *Canon.* As a result, Galen's writings on physiology became the basis for Galen*ism,* a natural philosophy of living beings.

The same impulse toward a systematic comprehension of the living body continued to shape academic medical treatises throughout the seventeenth century. A declaration in 1695 by Friedrich Hoffmann that "as far as medicine uses the principles of physics it can properly be called a science" only echoed similar statements made by a multitude of predecessors – and indeed by Galen himself in *On the Natural Faculties,* in which Galen credited Hippocrates with joining philosophy and medicine.[10] Yet Hoffmann's conception of physiology deviated considerably from the theory of his predecessors. Whereas the early seventeenth-century Wittenberg professor Daniel Sennert (1572–1637)

---

[9] This change can be seen both in published lecture catalogs produced by various universities and by inspection of the titles of medical textbooks. In the generations following Boerhaave and Hoffmann, it became increasingly uncommon to write medical theory in the form of an *institutiones medicae* and much more the practice to produce separate textbooks of physiology and pathology.

[10] Friedrich Hoffmann, *Fundamenta medicinae,* trans. and intro. Lester S. King (New York: American Elsevier, 1971), p. 6; Galen, *On the Natural Faculties,* trans. A. J. Brock, Loeb Classical Library vol. 71 (Cambridge, MA: Harvard University Press, 1991), p. 9.

had employed a standard schema when he defined three principal faculties – the nutritive, the augmentative, and the generative – and four secondary faculties of attraction, retention, concoction and expulsion,[11] by the later seventeenth century Galenic physiology had begun to come apart. One line of criticism emanated from the experimental work of William Harvey (1578–1657) on the heartbeat and circulation of the blood, which undermined Galen's claims for the function of specific organs such as the heart and the liver. A second serious challenge to Galenism emerged from the promulgation of corpuscular natural philosophy in the works of René Descartes, Pierre Gassendi, and Robert Boyle. For these scholars, the proper subject matter of natural philosophy was the motion of matter and not the secondary qualities that give objects their form. To the extent that they considered form at all, natural philosophers such as Boyle attempted to explain it as arising from corpuscular motion.[12]

It took a generation or two for physicians to adjust their own ideas to conform to the new mechanical natural philosophy, but by 1700 physiology had become thoroughly "mechanized." Hoffmann proclaimed the new dogma forthrightly in his *Fundamenta medicinae:* "Medicine is the art of properly utilizing physico-mechanical principles, in order to conserve the health of man or to restore it if lost." And just so that there would be no mistaking his position, he specified what a mechanical explanation would consist of: "Size, shape, motion, and rest are entire basic states of simple bodies. From these, therefore, the reasons for all natural phenomena and effects are to be sought."[13] Boerhaave, whose teachings would soon transform the University of Leiden into Europe's leading medical school, also presented the human body as a vast mechanical contrivance. In one well-known passage from his *Institutiones medicae,* he characterized the body in the following terms:

> The solid parts are either membranous pipes, or vessels including the fluids; or else instruments made up of these, and more solid fibers, so formed and connected, that each of them is capable of performing a particular action by the structure, whenever they shall be put into motion we find some of them resemble pillars, props, cross-beams, fences, coverings, some like axes, wedges, levers, and pulleys; others like cords, presses, or bellows; and others again like sieves, strainers, pipes, conduits, and receivers; and the faculty of performing various motions by these instruments, is called their function; which are all performed by mechanical laws, and by them only are intelligible.[14]

---

[11] Daniel Sennert, *Epitome institutionum medicinae* (Amsterdam, 1644), pp. 26–7. For a good survey of the principles undergirding Galenic medical theory, see Lester S. King, *The Philosophy of Medicine: The Early Eighteenth Century* (Cambridge, MA: Harvard University Press, 1978), pp. 41–63.

[12] Norma Emerton, *The Scientific Reinterpretation of Form in the Seventeenth Century* (Ithaca, NY: Cornell University Press, 1984). For an excellent discussion of form and matter in Aristotelian and Galenic natural philosophy, see the articles collected in C. H. Lüthy and W. R. Newman (eds.), "The Fate of Hylomorphism: 'Matter' and 'Form' in Early Modern Science," *Early Science and Medicine,* 3, 1 (1997).

[13] Hoffmann, *Fundamenta medicinae,* pp. 5, 7.

[14] Herman Boerhaave, *Institutiones medicae* (Leiden, 1730), § 40, pp. 12–13. The translation is taken from *Dr. Boerhaave's Academical Lectures on the Theory of Physick,* 6 vols. (London, 1742–6), vol. 1, p. 81.

Boerhaave and Hoffmann were only two among a host of medical professors in 1700 who advocated the reformulation of physiological theory along the lines of the mechanical philosophy.[15] From a superficial perusal of their writings, a reader might easily conclude that they subscribed to the all-sufficiency of mechanical principles for explaining vital phenomena. But in fact the situation was more complex. Shortly after writing that matter is differentiated only by shape and size and that its motion is the "most universal principle of things and the efficient cause of all forms," Hoffmann introduced the rich lexicon used by chemists to describe the body's different temperaments. The blood, for example, contains particles that Hoffmann characterized as "earthy, branching, watery, salinous, volatile, fixed, alkali, or sulfurous."[16] The role of bile in digestion, he wrote elsewhere, is that it "corrects sourness, dissolves oily particles of the foods so that they can be intimately mixed with water and constitute the chyle, removes the earthy viscidity, and through its saline-sulfurous spicules stimulates the intestine to excretion."[17] In such passages one sees an eclectic mix of terms indicative of mechanical processes – those little "spicules" delivering their pinpricks to the intestine, for example – with a catalog of chemical qualities that are not immediately reducible to the mechanics of matter in motion. Boerhaave also described digestion as a largely chemical process.[18]

By granting digestion a place in the animal economy as a chemical phenomenon, Hoffmann and Boerhaave were not tacitly admitting the inconsequence of the mechanical philosophy for physiology. Although they were convinced that mechanical principles were the foundation of all changes observed in Nature, they realized that certain kinds of vital phenomena were not immediately explicable in mechanical terms. Nor is there anything surprising about this: the task of physiology, after all, was to explain vital phenomena and not to link itself seamlessly to physics. Within physiology's own explanatory domain, it made sense to offer the best accounts possible of, say, the action of the stomach and the glands, without worrying unduly over the ultimate causes of their actions. What Hoffmann and Boerhaave were particularly at pains to avoid was the attribution of vital processes such as digestion to what they regarded as the occult "faculties" of the body's organs so favored by their predecessors. For them, a proper explanation of what goes on in the body consisted of how such processes could arise from the motions of the body's constituent particles.

All the same, the introduction of the mechanical philosophy into physiology posed difficulties for medical theorists, for it soon became apparent that the doctrine of matter in motion could not very well account for the mani-

---

[15] See François Duchesneau, *La physiologie des lumières*, Archives internationales d'histoire des idées, vol. 95 (The Hague: M. Nijhoff, 1982), pp. 32–64; King, *The Philosophy of Medicine*, pp. 95–124; and Brockliss, *French Higher Education*, pp. 405–8.

[16] Hoffmann, *Fundamenta medicinae*, p. 11.   [17] Ibid., p. 22.

[18] Boerhaave, *Institutiones medicae*, § 76–89.

fest *functions* of living beings. Not only were the chemical transformations that accompany digestion and other vital processes not obviously explicable in terms of the mechanical philosophy but also at another level entirely, there was a goal-directedness and coordination of such activities that mechanistic models were unable to confront. Such functions had been comprehended in Galenic physiology far more easily, because in it an organ's form had determined its properties and its functions. To take but one case, Galen had argued that the kidneys must possess an attractive faculty or power for urine (a faculty determined by the kidneys' form), because otherwise it would be inexplicable how urine can be filtered out of the blood without the blood also losing its serum and other liquid components. For our purposes, it is noteworthy that in developing this explanation Galen specifically refuted the Epicurean doctrine that such attraction occurs via "the rebounds and entanglements of atoms." What remains, therefore, could be understood only as a specific attractive faculty.[19]

It was precisely on this ground – the inadequacy of the mechanical philosophy for understanding living beings – that Georg Ernst Stahl (1659–1734), Hoffmann's colleague at the Prussian University of Halle, planted his flag. In several polemical writings on the subject, Stahl argued against the assimilation of vital to physical phenomena. The actions of living bodies, he claimed, are devoted principally to the prevention of corruption, and it is this manifestly teleological activity that distinguishes them from nonliving bodies.[20] The agent responsible for preserving living bodies from decay, Stahl claimed, is the *anima* (soul), a term that since antiquity had been used to designate the seat or source of vital processes. Stahl insisted that the anima is immaterial but nonetheless real, and motion is the means by which the anima communicates with the body and guides its functions.[21]

Stahl's doctrine preserved the distinctiveness of vital processes against reduction to purely mechanical models on the one hand, and his insistence on the nonmateriality of the anima fended off the possibility of imputing vital powers to matter itself, a doctrine that Stahl and many of his contemporaries found unacceptable on religious grounds. Unfortunately, as the philosopher Gottfried Wilhelm Leibniz (1646–1716) and numerous others pointed out, Stahl's own explanation was literally unintelligible, because it depended on

---

[19] Galen, *On the Natural Faculties*, pp. 91–3. To avoid any possible misunderstanding, let it be specified that "form" here refers to an entity's formal cause (the cause that makes something what it is and none other) and not to its spatial conformation, the sense by which we would understand the term.

[20] Georg Ernst Stahl, *Paraenesis ad aliena a medica doctrina arcendum* (Halle, 1706).

[21] On Stahl's critique of the mechanical philosophy, see Duchesneau, *La physiologie des lumières*, pp. 6–23, and King, *The Philosophy of Medicine*, pp. 143–51. For a broader assessment of Stahl, see Johanna Geyer-Kordesch, "Die 'Theoria Medica Vera' und Georg Ernst Stahls Verhältnis zur Aufklärung," in Wolfram Kaiser and Arina Völker (eds.), *Georg Ernst Stahl (1659–1734): Wissenschaftliche Beiträge der Martin-Luther-Universität Halle-Wittenberg*, 66 (E73) (Halle, 1985), pp. 89–98. For a comparison of Stahl's chemical theories with those of Boerhaave, an arena in which differences over the mechanical philosophy also played a role, see Hélène Metzger, *Newton, Stahl, Boerhaave et la Doctrine Chimique* (Paris: F. Alcan, 1930).

the immaterial anima being somehow directly responsible for moving matter. It was not the possibility of a connection between the soul and the body that critics rejected; voluntary muscular motion, after all, was only one among many phenomena suggestive of some kind of link between them. Rather, it was Stahl's claim that the non-material anima constituted an explanation for the vital movements of bodies that his opponents could not accept.[22]

The argument over the adequacy of mechanical explanations in physiology, which revolved in part around the question of what the aims of physiological science should be, continued with undiminished vigor during the century. One line of thinking followed Boerhaave and Hoffmann in implicitly rejecting the traditional causal framework of physiological theory – the attribution of vital processes to an organ's faculties, for example – and replacing it with an account based on the principle of matter in motion. Of particular importance here was the introduction of the idea of "force" into such explanations as a way of accounting for cause-and-effect relationships. The inspiration for this move was Isaac Newton's perceived success in describing planetary motion by means of gravitation, a force that Newton refused to characterize other than through its effects on matter. Independently, and at nearly the same time as Newton, the physician John Locke (1632–1704) was making the same point about our idea of "power" in his *Essay Concerning Human Understanding* (1690). Powers, Locke argued, are not intelligible as real things in themselves; instead, they are the relations that we observe among objects whereby one object appears to effect changes in another.[23]

By far the most influential application of this thinking to medical theory came in the work of Albrecht von Haller (1708–1777), a Swiss physician and professor at the University of Göttingen. Following the model established by Newton, Haller attempted to describe vital phenomena as the effects of powers or forces in living matter and to correlate those forces with anatomical structure.[24] His most famous work in this vein was read to the Royal Scientific Society in Göttingen in 1752 and published the next year in the society's journal as "De partibus corporis humani sensibilis et irritabilibus" (On the sensible and irritable parts of the human body). In this paper, Haller identified two basic vital forces: irritability, located in the muscles, and sensibility, located in the nerves. Consistent with his Lockean epistemology, Haller identified the presence of these forces by the regular effects obtained from certain

---

[22] On Leibniz's criticism of Stahl, see Karl E. Rothschuh, *Physiologie: Der Wandel ihrer Konzepte, Probleme und Methoden vom 16. bis 19. Jahrhundert* (Freiburg, 1968), pp. 155–6; and Duchesneau, *La physiologie des lumières*, pp. 87–102, especially 96.

[23] John Locke, *An Essay Concerning Human Understanding*, Bk. II, chap. xxi (Repr. ed. New York: Dover Publications, 1959).

[24] For a general introduction to Haller's thinking and especially to his epistemology, see Shirley A. Roe, "*Anatomia animata*: The Newtonian Physiology of Albrecht von Haller," in Everett Mendelsohn (ed.), *Transformation and Tradition in the Sciences: Essays in Honor of I. Bernard Cohen* (Cambridge University Press, 1984), pp. 273–300; and Richard Toellner, *Albrecht von Haller: Über die Einheit im Denken des letzten Universalgelehrten: Sudhoffs Archiv*, Beihefte, Heft 10 (Wiesbaden, 1971).

experimental manipulations. Irritability was said by Haller to be present in any part of the body "which becomes shorter upon being touched," by poking with a needle, for example, or applying alcohol or caustic chemicals to it. Sensibility, by contrast, presented a more complicated case. "I call that a sensible part of the human body," Haller wrote, "which upon being touched transmits the impression of it to the soul; and in brutes, in whom the existence of a soul is not so clear, I call those parts sensible, the irritation of which occasions evident signs of pain and disquiet in the animal."[25]

By insisting that irritability was uniquely the property of muscles and that sensibility belonged exclusively to nerves, Haller attempted to combat Stahl's claim that muscular contraction depended directly on some immaterial cause, such as the anima. Haller was especially eager, too, to avoid becoming involved in speculations regarding the ultimate source of these phenomena, but in this he was betrayed by his own experimental method. In contrast to irritability, sensibility could be demonstrated only by the registration of the irritation in the subject as "evident signs of pain and disquiet." Consequently, sensibility was not strictly a vital phenomenon, because its demonstration required assumption of a consciousness in the experimental subject that could sensibly register the pain. This fact would ensnare Haller in multiple controversies with his contemporaries. One such dispute pitted Haller against the Edinburgh professor Robert Whytt (1714–1764), whose own theory of muscular motion posited a "sentient principle" resident in the nerves and distributed throughout the body. According to Whytt, it was this sentient principle that perceived external stimuli, even if such perceptions never came to consciousness in the brain, and prompted muscles to react. Thus for Whytt all muscular motion, whether voluntary or not, depended on the soul, and Haller's results with sensibility were fully in accord with Whytt's theory.[26]

For his part, Haller refused to accept the idea that the soul is coextensive with the body. He defended himself against Whytt's criticisms by insisting that irritability was innate to muscles, a *vis insita,* and completely separated from sensibility, not subordinate to it, as Whytt believed. This opened up Haller to materialist interpretations of his work by those claiming that his experimental demonstration of irritability proved that matter was self-moving. Such possibilities were brought unpleasantly to his attention by Julien Offray de La Mettrie (1709–1751), like Haller an erstwhile student of Boerhaave. In his *L'homme machine* (1747), La Mettrie, claiming to have been inspired by Haller's early comments about irritability, described human thought and the soul as nothing more than the products of organized matter. As if this were

---

[25] Albrecht von Haller, "A Dissertation on the Sensible and Irritable Parts of Animals," reprinted in Shirley A. Roe (ed.), *The Natural Philosophy of Albrecht von Haller* (New York: Arno Press, 1981), pp. 658–9. It is not clear how "corpori humani" in the original Latin title became "animals" in the English translation.

[26] On the controversies between Whytt and Haller, see R. K. French, *Robert Whytt, the Soul, and Medicine* (London: Wellcome Institute of the History of Medicine, 1969), pp. 63–76.

not reason enough for the deeply pious Haller to take offense, La Mettrie dedicated his scandalous work to him, a bow toward Haller that barely attempted to conceal something of a smirk as well.[27]

Haller's appropriation of an experimental approach to physiology was in part an attempt to avoid the intrusion of metaphysical issues into what he considered natural-philosophical questions. For him, arguments about the source or ontological status of forces such as irritability were pointless or, in the hands of a La Mettrie, potentially blasphemous. Yet Haller's desire to avoid such causal entanglements encountered opposition from another group of scholars who took up Stahl's line of thought if not his specific doctrines. For these physicians, causal issues, especially the final causes evident in vital functions, could not so easily be left aside. The most influential center for this vitalist physiology was the University of Montpellier in southeastern France. Its vanguard arrived in the person of François Boissier de Sauvages (1706–1767), whose lectures and academic writings in the latter 1730s began invoking the presence of an anima in the body as a way of explaining why living bodies do in fact move, whereas the bodies of recently deceased beings are incapable of such motion. It has been claimed that Boissier de Sauvages's commitment to the real presence of an anima appears to have been less thoroughgoing than Stahl's had been, and it may have done little more than perform the same linking function between empirical phenomena that "force" did for Newton and Haller or "power" for Locke.[28] Yet even if Boissier de Sauvages was less rigorously committed than Stahl to the physiological role of the anima, his choice of the term could scarcely have been a neutral one in the 1730s. It indicates that, whatever Boissier de Sauvages thought the anima actually was, he intended it to compensate for the mechanical philosophy's perceived deficiencies in the explanation of living phenomena.

Boissier de Sauvages's introduction of vitalist physiology into the Montpellier curriculum was embraced by Théophile de Bordeu (1722–1776) in his *Recherches anatomiques sur la position des glandes et leur action* (1752). Bordeu's choice of glandular function for developing his physiology was an apt one, for glands were widely believed to exercise specific functions within the overall animal economy. Bordeu's anatomical studies discounted one idea held by mechanists, that glands produced their humors as the result of being squeezed by muscles. Instead, he argued that each gland was endowed with a specific sensitivity, which when stimulated prompted the production of the gland's

---

[27] On the medical context of La Mettrie's work, see Kathleen Wellman, *La Mettrie: Medicine, Philosophy, and Enlightenment* (Durham, NC: Duke University Press, 1992).

[28] This position is taken in Julian Martin, "Sauvages's Nosology: Medical Enlightenment in Montpellier," in Andrew Cunningham and Roger French (eds.), *The Medical Enlightenment of the Eighteenth Century* (Cambridge University Press, 1990), pp. 111–38. See also Elizabeth Haigh, *Xavier Bichat and the Medical Theory of the Eighteenth Century*, Medical History, suppl. 4 (London, 1984), p. 31. For a contrasting interpretation of Sauvages, see Roger French, "Sickness and the Soul: Stahl, Hoffmann and Sauvages on Pathology," in Cunningham and French (eds.), *The Medical Enlightenment*, pp. 88–110.

particular humor. It was, as Brockliss and Jones have pointed out recently, a view of organic function not very far removed from the Galenic doctrine of organs as endowed with specific faculties. Bordeu's doctrine of specific vital functions encountered criticism from Paul-Joseph Barthez (1734–1806), who nonetheless shared Bordeu's disdain for mechanical explanations of vital functions. Barthez, who was numbered among the horde of eighteenth-century scholars who styled themselves the "Newton" of their discipline, attributed vital actions and their coordination to what he called a vital principle that extends over the body but is not identical with matter itself.[29]

At one level, of course, the disagreements between Haller and his opponents amounted to the oft-cited battle between "mechanism" and "vitalism" that has long been a standard story in histories of eighteenth-century medical science. At the same time, we should note that this dispute covered a more fundamental disagreement over the very nature of physiology as a medical science. On one side of the question stood those physicians whose thinking tended to assimilate physiology into natural philosophy. We have seen how Hoffmann and Boerhaave attempted as far as possible to explain vital phenomena as examples of corpuscular mechanics, whereas Haller's physiology incorporated a mechanics of forces based on Newton's model of gravity. Although his approach to the explanation of vital phenomena differed from that of his predecessors, Haller's adoption of experimental methods did have the same effect of making physiology a branch of natural philosophy. By the end of the century, the number of scholars conducting experiments on vital phenomena had swelled considerably, comprising a sizable group that included physicians such as Luigi Galvani (1737–1798) and Joseph Black (1728–1799), along with nonphysicians such as the chemist Antoine Laurent Lavoisier (1743–1794) and the naturalist Alexander von Humboldt (1769–1859).

What remained unclear in the wake of this program for the experimental study of vital phenomena, however, was how physiology could continue providing the theoretical foundations for a unified *medical* science. The dominant explanatory models available in natural philosophy after 1700 appeared to leave little place for the kinds of goal-directed actions that were so characteristic of living beings. Ultimately, this is what Stahl objected to in the mechanical philosophy, and it also motivated the objections of Whytt and the Montpellier physicians to Haller's work. It is important to recall that our subject here is medical science and not "biology" *avant la lettre*. Of course, there was a domain of life science in the eighteenth century, which is discussed in Chapter 17 of this volume by Shirley Roe. But the topic of greatest concern to physicians in a medical context was not *life* so much as *health* and its correlate, *illness*. And even in the eyes of a mechanist such as Boerhaave, health was understood in terms of functional coordination as "that faculty of the body

---

[29] Brockliss and Jones, *The Medical World of Early Modern France,* pp. 425, 427–30; and Haigh, *Xavier Bichat,* pp. 31–42.

in which all its parts are duly enabled to perform their respective offices with perfection."[30] The axis defined by health and illness is surely one reason that the goal-directedness of vital actions remained a matter of central concern to many physicians and that the experimental study of vital phenomena undertaken by Haller and others increasingly placed physiology in an ambiguous relationship with the rest of medical science.

PATHOLOGY

Since we have just circumscribed the central concerns of medical theory on the axis of health and illness, it might be supposed that physiology and pathology, the theory of illness, were similarly paired on opposite ends of the axis. Although this was true to a certain extent, their relationship was by no means a straightforward one. One reason for this is that, as the quotation just given from Boerhaave suggests, the healthy body was considered by eighteenth-century physicians to be the natural condition, requiring no particular explanation. Illness, by contrast, represented a falling away from the ideal of health, and consequently the theory of illness was called upon to offer a causal account of instances of illness. Epistemological issues of causality, which could be avoided more or less successfully in physiology, could not so easily be written out of pathology. Second, we must not lose sight of the social milieu in which physicians worked. Doctors did not confront illness as an abstract theoretical puzzle; instead, they encountered it at the bedside in the sufferings of their patients. Whereas physiological phenomena such as muscular contraction and the circulation of the blood and other fluids could be described more or less successfully with mechanical models drawn from natural philosophy, the phenomena of central concern in pathology, diseases such as smallpox, pleurisy, or apoplexy, derived from concrete disturbances in the body and were recognizable by their symptoms. Pathology thus had not only to explain causally how a condition such as pleurisy could arise, but it also had to teach doctors how to recognize pleurisy and to interpret the underlying meaning of its symptoms. In this latter task, pathology shared a problem domain with semiotics, another branch of the traditional *institutiones medicae*.

The problems with which a science of pathology had to contend, therefore, were to a considerable degree separable from those dealt with by physiology. As we have seen, the adoption of the mechanical philosophy had a number of important consequences for physiology, not only in terms of its specific doctrines but also in terms of its goals and methods. The impact of the new natural philosophy on pathology was less direct. Whereas the specific expla-

---

[30] Boerhaave, *Institutiones medicae*, § 695.

nations offered for the occurrence of illnesses often displayed the direct influence of mechanical models, especially in the early part of the century, the task of pathology as a whole – the explanation of illness as a phenomenon fundamentally different from the state of health – remained virtually unchanged.

One circumstance that made pathological theory accessible to the mechanical philosophy was the long-standing practice of dividing illnesses between the solids and the humors. Both categories readily adapted themselves to mechanical explanations. For example, Boerhaave described three general categories of afflictions in the body's solids. First, there could be problems such as excessive rigidity or laxness with the most basic fibers that constitute the body's solid parts. Second, organs and other solids themselves are susceptible to a host of problems, ranging from displaced position and obstructions of an organ's passages to improper shape or motion. Third, changes in structure could cause alterations in a solid part's function, producing what Boerhaave described as organic illnesses (*morbi organici*).[31] The body's fluids, too, were susceptible to a range of problems derived from changes in either quantity or chemical quality. Too much or too little of a particular fluid could have severe consequences, as could reduction in the circulation of a fluid such as the blood, which if stagnant could begin to ferment in an unhealthy manner. So important were the fluids in Hoffmann's pathology that he declared that "in every disease the motion of the blood and of the fluid parts is either diminished or increased."[32] As had been true of their physiology, in pathology, too, Boerhaave and Hoffmann displayed a ready eclecticism when discussing fermentations and other changes in the humors as chemical phenomena.[33]

Even though a catalog of pathological changes inside the body may have seemed to Hoffmann and Boerhaave the surest vehicle for demonstrating the mechanical philosophy's significance for pathology, such an approach did not illuminate the problem of how diseases arise in particular circumstances. Nor did it provide much guidance to the semiotic interpretation of symptoms as a means of identifying diseases. Both aspects are evident in Hoffmann's discussion of scurvy. Scurvy, he wrote in the *Fundamenta medicinae*, "is nothing but the maximum impurity and irregularity of the lymph and blood." Yet his very next statement characterized the disease slightly differently, claiming that the symptoms of scurvy "are derived mostly from the spasmodic contraction of the nerves."[34] Now, the obvious question that one could put to Hoffmann is, what exactly is this "scurvy"? Does it subsist in the first-mentioned impurities in the lymph and blood, or in the spasmodic contractions of the nerves, or indeed in the visible symptoms themselves? In favor of the latter possibility,

---

[31] Ibid., § 699–712.   [32] Hoffmann, *Fundamenta medicinae*, p. 41.

[33] For example, Hoffmann characterized leprosy as a "scabies" arising from "obstruction and from profound destruction of fleshy tubules by an acrid steaming serum." Ibid., pp. 63–4.

[34] Ibid., p. 57.

one could plausibly argue that scurvy is nothing but a symptomatic predicate; that is, only a particular collection of visible signs allows any given illness to be designated as "scurvy."[35]

The problem illuminated by this example was well appreciated by eighteenth-century physicians, because it touched on the vexing question of what exactly the cause of disease was. The assessment of causation involved a combination of bodily dispositions, general environmental factors, and specific triggering circumstances in the production of an ailment. This explanatory structure received a full treatment in the *Institutiones pathologiae medicinalis* (1758) of Hieronymus David Gaub (1705–1780), probably the most prominent pathology textbook to appear during the second half of the century. Gaub's discussion of causation focused special attention on the role of the "nonnaturals" in producing illnesses. Taking up each one in turn, he discussed the influences (*potentia*) exerted by the air, food and drink, bodily movement or inactivity, and the other nonnaturals. Yet the nonnaturals alone could not induce sickness in someone, for it was also necessary that the person's body be made susceptible in some way to the external morbific influence. Accordingly, Gaub continued with a discussion of the internal predispositions (*seminia*) that, in combination with an external triggering circumstance, can lead to illness.[36] However, even this combination of an external trigger and an internal predisposition did not close the causal chain, because Gaub, like many of his contemporaries, recognized that something else was required: a specific bodily affliction that would be the product of remote causes described earlier and would itself then furnish the necessary and sufficient cause for the appearance of the symptoms of illness in the patient. If anything was deemed worthy of being called the disease, this "proximal cause," as physicians named it, was it.[37]

With the proximal cause thus defined, Gaub then reviewed much of the same ground covered by Boerhaave and Hoffmann in cataloging the possible pathological changes undergone by the body's solid parts and humors. In most places, he evidently saw little reason for venturing far from the mechanical and chemical models put to work by his predecessors. But Gaub also introduced a significant new distinction between illnesses in the body's solid parts and illnesses arising in those solid parts that are animated. These latter parts, he noted, are distinguished from other solids by their possession of what Gaub labeled a vital force (*vis vitalis*). Taking a position similar to Robert Whytt's, Gaub described the life force as having two components: an ability to perceive stimuli and an ability to react to them. It was, he wrote, a force unlike any other hitherto discovered.[38]

---

[35] Semiotics in the eighteenth century is a complicated and largely unexplored field, for which Foucault's *Birth of the Clinic* is an unreliable guide. For a good discussion of some of the relevant issues, see Volker Hess, "Spelling Sickness: The *Aufschreibesystem* of Medical Semiotics in the Eighteenth Century," in Cay-Rüdiger Prüll (ed.), *Traditions of Pathology in Western Europe* (forthcoming).

[36] Hieronymous David Gaub, *Institutiones pathologiae medicinalis,* 2nd ed. (Edinburgh, 1762), § 419–605.

[37] Ibid., § 60.   [38] Ibid., § 169–72, 186.

Gaub's vague indication of a role for vital forces in pathology was given a much fuller treatment by the Edinburgh medical professor William Cullen (1710–1790). Cullen's theory of pathology gave pride of place to the nervous system in the genesis of diseases, and he defined the large majority of diseases as involving some kind of disruption in the stimulus-response mechanism argued over so vigorously by Haller and Whytt. So wedded, in fact, was Cullen to Whytt's view of the subordination of muscular irritability to nervous sensibility that he argued against the anatomical distinctiveness of the nervous and muscular systems. It was pointless, Cullen claimed, to describe the nerves and muscles as separate organ systems, because their functional integration was so complete that they became a single system.[39]

Despite some of his anatomical pronouncements and his insistence that disease was first and foremost a disruption in the actions of the nervous system, Cullen was no crude systematizer. Like Gaub, Cullen believed that the process leading to disease was a complex combination of remote and proximal causes, and he shared Gaub's conviction that the establishment of firm causal connections in the case of particular diseases was a difficult endeavor, requiring the painstaking collection and comparison of exact observations from individual case histories.[40] One widely recognized approach was to associate the symptoms of a disease with pathological changes found inside the body upon autopsy, should the illness lead to the patient's death. This was the method deployed in the eighteenth century's most famous work of pathological anatomy, *De sedibus et causis morborum* (The Seats and Causes of Diseases, 1761), by the Padua medical professor Giovanni Battista Morgagni (1682–1771). The goal of his work, as Morgagni plainly stated in the treatise's preface and five dedicatory letters, was to extend and correct an earlier collection of observations in pathological anatomy, the *Sepulchretum anatomicum* (1679) of Théophile Bonet (1620–1689). First, Morgagni intended to add to this collection the results of his own anatomical observations and the unpublished autopsies performed by his teacher at Bologna, Antonio Maria Valsalva (1666–1723).[41] More significantly, Morgagni considered it of utmost importance to make this knowledge accessible to practitioners. In his view, one of the chief faults of the *Sepulchretum* was the way it divided a single case among several different sections of the work so a reader might encounter the same case history

[39] On Cullen, see A. Doig et al. (eds.), *William Cullen and the Eighteenth-Century Medical World* (Edinburgh: Edinburgh University Press, 1993); and W. F. Bynum, "Cullen and the Study of Fevers in Britain, 1760–1820," in W. F. Bynum and V. Nutton (eds.), *Theories of Fever from Antiquity to the Enlightenment, Medical History,* suppl. 1 (London, 1981), pp. 135–47. For a description of the medical environment in Edinburgh, see Christopher Lawrence, "Ornate Physicians and Learned Artisans: Edinburgh Medical Men, 1726–1776," in W. F. Bynum and Roy Porter (eds.), *William Hunter and the Eighteenth-Century Medical World* (Cambridge University Press, 1985), pp. 153–76.

[40] See Gaub's comments on this point in *Institutiones pathologiae medicinalis,* § 44–9.

[41] Giovanni Battista Morgagni, *The Seats and Causes of Disease Investigated by Anatomy,* vol. 1, trans. Benjamin Alexander, repr. ed. with an intro. and trans. of five letters by Paul Klemperer (New York: Hafner, 1960), p. xxiii.

over and over. Worse still, the system of cross-referencing employed in the *Sepulchretum* was both vague and filled with errors, so that a reader in search of information about similar cases faced the tedious prospect of digging through pages of irrelevant material to find the desired references. These flaws Morgagni sought to correct with the compilation of no fewer than four indexes for his own work, in which a reader could, for example, find all cases containing the description of a particular symptom or, conversely, all cases associated with a particular pathological finding after dissection. As a result, he wrote,

> if any physician [should] observe a singular, or any other symptom in a pa-
> tient, and desire to know what internal injury is wont to correspond to that
> symptom; or if any anatomist [should] find any particular morbid appearance
> in the dissection of the body, and should wish to know what symptom has
> preceded an injury of this kind in other bodies; the physician, by inspecting
> the first of these indexes, the anatomist by inspecting the second, will imme-
> diately find the observation which contains both (If both have been observed
> by us).[42]

This was no mere prefatory bluster, for *De sedibus et causis morborum* is a brilliant and wonderfully comprehensive work, something that perhaps ex-plains why it took Morgagni so long to publish it (it finally appeared when he was seventy-nine years old). Morgagni divided the treatise first into the major parts of the body, beginning, as was customary, with the head ("Dis-orders of the Head," "Disorders of the Thorax," etc.), and then within each major division by individual symptoms or groups of symptoms. So in the diseases of the thorax, one reads of individual chapters with such titles as "Of Respiration being injured from Aneurisms [*sic*] of the Heart, or Aorta, within the Thorax," or "Of Pain in the Breast, Sides, and Back."[43] As these chapter headings suggest, Morgagni mostly tried to avoid associating his postmortem findings with specific *diseases*. Instead, he sought to present the morbid phe-nomena as correlated with particular *symptoms*. The individual case histories narrated in the chapters also follow this method to a great extent, describing the progression of symptoms without naming diseases.

Yet the connection between the "seat," or proximal cause, of a disease and the production of a particular set of symptoms was not as easily maintained as all this might imply. In the first place, as Morgagni himself conceded, any

---

[42] Ibid., p. xxx.

[43] It is worth noting that individual chapters in *The Seats and Causes of Diseases* were written in an epistolary genre. By doing so, Morgagni remarked in the preface, he was following the model of "ancient
and modern physicians" as well as the "greatest anatomists" of more modern times. One suspects
that what made this model suitable was its appropriateness for the narration of individual case
histories. Morgagni wanted these case histories above all to constitute the core of his collection.
Ibid., pp. xxvi–xxvii. The centrality of case histories in the creation of medical knowledge in the
early modern era has been scarcely recognized by historians up to now. One exception is Johanna
Geyer-Kordesch, "Medizinische Fallbeschreibungen und ihre Bedeutung in der Wissensreform des 17.
und 18. Jahrhunderts," *Medizin, Geschichte und Gesellschaft*, 9 (1990), 7–19.

given illness could often be the result of divergent, even opposite, causes.[44] Second, Morgagni's method demanded that certain symptoms be singled out as the most characteristic in a given case, with others relegated to a secondary status. For example, at one point he remarked that difficulty in breathing is often present with other symptoms in a wide variety of illnesses, although not all such cases would properly belong under the heading of "Disorders of Respiration," even were postmortem autopsies to reveal significant damage in the lungs.[45] The conclusion one could draw from such musings is that, when faced with a collection of symptoms deriving from a variety of pathological changes inside the body, Morgagni was forced to decide, if only implicitly, what in fact this or that patient had been sick with as a prerequisite to placing the case in one of his anatomical/symptomatic categories.

The point here is not to charge Morgagni with inconsistency but rather to suggest that the "modernity" of Morgagni's pathological anatomy, which has long been contrasted to the foggy speculativeness of eighteenth-century pathological theory, was less dramatic than might appear. As Morgagni himself attested, he was only working a field made ready by a host of predecessors. More important, there was no real difference between Morgagni's conception of pathology and Gaub's or Cullen's. Morgagni's understanding of how diseases were to be interpreted at the bedside ultimately depended on the same kinds of judgments and categories that Gaub and Cullen described in their textbooks. But whereas Gaub and Cullen taught the formal categories by which illness was to be understood, Morgagni's treatise offered the content of that understanding. In the context of eighteenth-century medical science, however, both were clearly needed in pursuit of a fully elaborated theory of pathology. Only in the markedly changed theoretical circumstances of nineteenth-century pathological anatomy could Morgagni appear to be a man ahead of his times.

CONCLUSION: THE MEDICAL SCIENCES IN THE 1790s

As I hope the preceding discussion has shown, the two core medical sciences of physiology and pathology each experienced significant development during the eighteenth century, although it could scarcely be claimed that they evolved in concert with each other. That the two doctrines could take such independent paths can be understood on the basis of the particular doctrinal positions occupied by each science. Physiology mediated the relationship between natural philosophy and medical science. As long as physicians found it expedient to claim the status of a natural science for their discipline, something they had done for centuries, it was incumbent on them to make medical theory

---

[44] Ibid., Letter of dedication to Johann Friedrich Schreiber.
[45] See Morgagni's comments on this point in ibid., Letter XV, article 3, p. 359.

account for its phenomena in terms of the dominant natural philosophy. The doctrinal role of pathology, by contrast, was far more complex. In the first place, pathological theory was charged with explaining the relationship between health and illness, the basic phenomena that defined the domain of medical theory. In this respect, pathology could and did find an anchor in physiology. Yet eighteenth-century pathology was not merely "the physiology of illness," because pathology also occupied the crucial doctrinal position between medical theory and a physician's experience of illness at the bedside. It bears repeating here that for most physicians in the eighteenth century illness was not an abstract theoretical category. Instead, once one moved beyond the most basic formal definition of illness (such as Boerhaave's, given earlier), physicians grasped the problem of pathology as one of first identifying and then explaining the occurrence of cases of illness in individual patients.

Alongside these developments in the relationship between physiology and pathology, it must also be noted that the cultural milieu in which medical science could be articulated was also changing in ways that would significantly affect academic medicine. Although a complete assessment of this setting lies well beyond the constraints of this article, one such development might be briefly indicated. First, the eighteenth century witnessed the appearance of a host of new institutions of sociability, such as masonic lodges, salons, coffeehouses, and reading societies, as well as an explosive growth in print media such as newspapers and periodicals. Together, these institutions constituted what is commonly labeled the "public sphere," in which members of civil society recognized themselves as members of the public. One of the most distinctive products of the public sphere was the distinctively modern institution of "criticism," a kind of discourse in which private individuals sought to speak both for and to the public by defining objective standards of reason and taste. Criticism took a number of forms, ranging from political confrontation with the state to the definition of esthetic canons in literature and the visual arts. Most important, critical discourse in the eighteenth century called for knowledge of all kinds to demonstrate its utility in social practice. For medicine, the consequence of these developments would be the appearance of calls to forge a tighter and more intimate link between medical theory and bedside practice than had been previously believed necessary or desirable.[46]

By the end of the century, the relationship between physiology and pathol-

[46] The fountainhead for the explosion of recent works on the public sphere is Jürgen Habermas, *The Structural Transformation of the Public Sphere*, trans. Thomas Burger with Frederick Lawrence (Cambridge, MA: MIT Press, 1989). For recent general commentaries on the public sphere, see Margaret Jacob, "The Mental Landscape of the Public Sphere: A European Perspective," *Eighteenth-Century Studies*, 28 (1994), 95–113; Anthony J. La Vopa, "Conceiving a Public: Ideas and Society in Eighteenth-Century Europe," *The Journal of Modern History*, 44 (1992), 79–116; and Dena Goodman, "Public Sphere and Private Life: Toward a Synthesis of Current Historiographical Approaches to the Old Regime," *History and Theory*, 31 (1992): 1–20. Among writings by historians of science that deal with the public sphere, see Paul Wood, "Science, the Universities, and the Public Sphere in Eighteenth-Century Scotland," *History of Universities*, 14 (1994), 99–135; and Thomas Broman, "The Habermasian Public Sphere and 'Science *in* the Enlightenment,'" *History of Science*, 36 (1998), 123–49.

ogy, and more generally the status of medical science, was being subjected to critical scrutiny from a number of directions. It might usefully serve as a conclusion to this essay to describe briefly two prominent themes in such discussions. The first was the erosion of certain assumptions that had undergirded the doctrinal role of theory in medical teaching. When one examines Boerhaave's and Hoffmann's textbooks in medical theory, one sees no concern voiced over the circumstances and methods through which their knowledge is obtained. The organs operate in such a manner, they are susceptible to these illnesses – the basis of these claims is not explicitly adjudicated. It is not that Boerhaave and Hoffmann were oblivious to such considerations but rather that such discussions were not germane to the exposition of medical theory. During the following decades, however, such matters would become very germane indeed. Haller's experimental work and the opposition it encountered from Whytt and the Montpellier physicians centered in part on how one might define the phenomena of relevance to physiology. The entire debate over the nature of forces in physiology, a debate that paralleled a similar one in natural philosophy, concerned the reality of the forces thus uncovered by experiment. That forces such as gravity, electricity, and Haller's irritability could be called forth and empirically demonstrated was beyond question for most scholars. But what exactly *were* these forces? In the critical environment of the eighteenth century, this seemingly metaphysical question was transformed into an epistemological one: how do we come to know the effects of such forces, whatever their ontological status? Along the same lines, the concern voiced by Gaub and Morgagni over the problem of how one can identify the causes acting in particular illnesses is conspicuous in comparison with the natural philosophy articulated by Boerhaave, Hoffmann, and Stahl at the beginning of the century.

Toward the end of the century these epistemological issues began to occupy center stage. Writers such a Jean-Georges Cabanis (1757–1808), in *Du degré de la certitude de la médecine* (1788), and Johann Christian Reil (1759–1813), in *Von der Lebenskraft* (1795), produced influential manifestos that redefined medical theory as first and foremost an epistemological problem. For both Cabanis and Reil, medical theory could not advance without possession of a proper method for acquiring medical knowledge. Characteristically, both Cabanis and Reil pointed to analysis as the key to a more secure medical theory, the kind of analysis represented in the philosophy of Etienne Bonnot de Condillac (1714–1780) and put into scientific practice by the French chemists led by Lavoisier. Reil in particular held high expectations for the promise held out by chemical analysis to become the basis for a unified medical science.[47]

A second kind of reaction consisted of attempts to synthesize a unified medical theory from the new developments in physiology and pathology. One

[47] On Cabanis, see John E. Lesch, *Science and Medicine in France: The Emergence of Experimental Physiology, 1790–1855* (Cambridge, MA: Harvard University Press, 1984). For a discussion of Reil, see Broman, *Transformation of German Academic Medicine*, pp. 86–8.

prominent attempt in this direction was made by John Brown (1735?–1788), an Edinburgh-trained physician whose *Elementa medicinae* (1780) defined all illnesses as resulting from either a general systemic over- or understimulation. Adopting the view of life as the product of an external stimulus acting on irritable matter, Brown insisted that illness came about as the result of an imbalance between stimulus and irritability, and the proper therapy involved taking measures to restore the balance, most often by altering the stimuli acting on the patient. Brown's theory found adherents in Edinburgh, Vienna, and Pavia, and it created a tremendous uproar in Germany. A variant even took root in the New World, in the teachings of the prominent American physician Benjamin Rush (1745–1813).[48]

Whereas Brown's theory described illness as a dynamic imbalance of forces and virtually ignored pathological anatomy, the French anatomist Xavier Bichat (1771–1801) attempted to synthesize physiology and pathology in a different and perhaps more traditional way: by focusing on the functional integrity of body parts. In a series of publications during his short career, Bichat examined the steps by which death occurs and defined two varieties of life force: the organic, which resides mainly in the heart, and the animal, which is seated in the brain. Yet what made Bichat's mature work distinctive was his location of vital processes not for the most part in intact organs, but rather in the twenty-one different tissues from which the body was built up. As he described them in his *Anatomie générale* (1801), these tissues perform distinctive functions in the body – secretion, excretion, absorption, contraction, and so on. Illness then is comprehended readily as the disruption of these individual functions.[49]

As different as they were, each of these medical reformers attempted to define a new kind of medical theory and not just to revise the doctrines of existing theory. The temptation is great at this juncture to see in these reforms the anticipations of what would come in the nineteenth century, which witnessed the establishment of a laboratory-based experimental physiology, the development of physiological chemistry, and the synthesis of physiology and pathology into a physiological pathology based on the cell theory. But I think that seeing it this way reads too much into their thinking. Instead, it seems more appropriate to interpret the reforms of Cabanis and the others as their recognition that medical theory could no longer continue exercising its traditional doctrinal and pedagogical function as the link between medicine and natural philosophy. Henceforth – and this was already evident in the writers we have just been examining – the adequacy of medical theory would be judged against its ability to serve as the grounding for medical practice.

[48] For further discussion of Brown's medical theories, see W. F. Bynum and Roy Porter (eds.), *Brunonianism in Britain and Europe: Medical History,* suppl. number 8 (London, 1988); Guenter B. Risse, "The History of John Brown's Medical System in Germany During the Years 1790–1806" (Ph.D. Diss., University of Chicago, 1971); and Thomas Henkelmann, *Zur Geschichte der pathophysiologischen Denkens: John Brown und sein System der Medizin* (Berlin: Springer-Verlag, 1981).

[49] Haigh, *Xavier Bichat,* and Lesch, *Science and Medicine in France.*

21

MARGINALIZED PRACTICES

Patricia Fara

In August 1783, three eminent men of science made the thirty-mile journey from London to Guildford to watch one of their colleagues, James Price, fulfill the alchemists' ancient dream of extracting gold from mercury. This distinguished chemist, a wealthy Oxford graduate who had been elected to the Royal Society when only twenty-nine years old, had already publicly demonstrated his alchemical skills and had published a book advertising his successful transmutations. Concerned to preserve the Royal Society's reputation, its President, Joseph Banks, instructed Price to repeat his experiments before an expert audience. But instead of the process of lucrative creation they had been promised, Banks's delegates witnessed only one of self-destruction, as Price swallowed a glass of laurel water and died in front of their eyes.

Price was pushed into making this ultimate sacrifice in the cause of Enlightenment rationality. Some of his critical peers were preoccupied less with the validity of his claims than with the threat his activities posed to the status of established institutions. One of Banks's confidants, Charles Blagden, articulated this interest in guarding the proprieties of scientific behavior rather than monitoring its results:

> Was ever any country more completely disgraced than ours has been by the conduct of the University. For, granting that Price has made the discovery held out in his book, should it not have been said to him that the man who having hit upon an improvement in science, keeps it from the world deserves rather to be excluded from the Society of learned men than to be adorned with extraordinary academic honours?[1]

Although Price constructed a singularly dramatic scenario for excluding himself from the realm of legitimated science, this incident does highlight several of the characteristics exhibited by practices, such as alchemy, astrology,

---

[1] Blagden's letter to Banks of 6 August 1782, quoted at p. 111 of H. Charles Cameron, "The Last of the Alchemists," *Notes and Records of the Royal Society of London*, 9 (1951), 109–14.

and animal magnetism, that were accredited to varying degrees but came to be marginalized. Enlightenment rhetoricians frequently proclaimed that the reforming power of reason had eradicated older traditions or superstitious beliefs, but they survived in various guises throughout the century, even among the well-educated classes. Encyclopedists' projects to redraw maps of knowledge entailed cultural transformations as well as epistemological ones, and elite practitioners policed the newly delineated scientific territories to determine not only what types of knowledge they should encompass but also who should be allowed to enter them. Subsequent historians have endorsed these polemic visions of scientific progress, bracketing together a wide range of disparate practices united chiefly by their common exclusion from the modern domains of recognized science.

Two other chapters in this book – those by Roger Cooter and Mary Fissell (Chapter 6) and Richard Yeo (Chapter 10) – consider more generally how some practices became legitimated sciences while others were banished from the map of knowledge. In this chapter, I explore some characteristics of these excluded practices. I shall briefly survey rhetorical attitudes toward discredited systems of belief and then discuss five examples with very different historical trajectories – animal magnetism, physiognomy, astrology, alchemy, and Hutchinsonianism – to illustrate the features that they share as well as the contrasts between them.

RHETORICS OF ENLIGHTENMENT

"The time will come," wrote the Marquis de Condorcet, "when the sun will shine only on free men who have no master but their reason."[2] Thus one of France's leading *philosophes* articulated two major motifs of Enlightenment rhetoric: the primacy of rational thought and the power of illumination. Throughout the eighteenth century, philosophical writers often used images of light to underpin their claims that a reasoned approach toward the natural world – including human civilizations – would lead to political as well as intellectual liberation. As part of their quest to improve society, they sought to demarcate legitimated areas of rational study and eliminate traditional practices based on belief. The compilers of the *Encyclopédie*, the most ambitious rationalizing project of the period, redrew the tree of knowledge to delineate new academic disciplines and construct boundaries between the known and the unknowable, thus excluding sacred learning as well as magic, superstition, and the occult.[3]

Although specific objectives varied, many writers employed similar strategies

---

[2] Quoted in Dorinda Outram, *The Enlightenment* (Cambridge University Press, 1995), p. 1.
[3] Robert Darnton, *The Great Cat Massacre and Other Episodes in French Cultural History* (Harmondsworth: Penguin, 1985), pp. 185–207.

to promote their own ideas, achievements, and status by denigrating those of their opponents. Influential authors such as Pierre Bayle and François-Marie Voltaire constructed progressive historical narratives asserting that modern enlightened philosophers had eradicated the erroneous beliefs afflicting previous generations. They systematically ridiculed older works as well as contemporary ones, deriding them for purveying magic or superstition, often inviting their readers to relish the folly of particularly mockable quotations. Although the English were rather circumspect in their attacks on religion, many Continental writers decried the Church for sustaining spiritual faith, thus encouraging belief in demons and magic.

Until the past twenty years, most accounts of the eighteenth century were strongly colored by these claims of Enlightenment rhetoricians that the light of reason was dispelling the dark clouds of superstitious error and ignorance. Focusing largely on France, scholars analyzed eighteenth-century texts to show how Enlightened philosophers successfully discriminated between magic and science, eliminated charlatans trading on popular ignorance, and coolly debated the status of miracles. Steeped in their own cultural values prizing rationality, historians endorsed modern celebrations of scientific achievement by patronizingly categorizing as pseudosciences belief systems such as astrology and physiognomy that, although now discredited, attracted many respected adherents in the past.

Nowadays, scholars no longer perceive a unitary and uniform Enlightenment but are examining the differences between the beliefs and activities of groups that were socially and geographically separated. Although historians of science formerly glossed the eighteenth century as an inactive interlude separating the innovations of Isaac Newton from the great scientific advances of the Victorian era, recent studies portray a vital period of epistemological and cultural conflict, during which polemicists eliminated rivals to the increasingly unchallengeable Newtonian orthodoxy and established men of science as elite producers of knowledge about the natural world. In contrast with earlier interpretations of a pan-European explosion of rationality, increasing emphasis is being placed on the distinguishing features of individual national groups and on the practices, institutions, and local interests underpinning changes in intellectual attitudes. As part of this reappraisal, much interest has focused on retrieving information about what is often referred to as "the dark underbelly of the Enlightenment" to provide less-biased accounts of practices such as Freemasonry, Mesmerism, and magic that, although suppressed, provided the essential "Other" against which polemicists characterized the period.[4]

---

[4] Outram, *Enlightenment*, pp. 1–13; Roy Porter and Mikuláš Teich (eds.), *The Enlightenment in National Context* (Cambridge University Press, 1981); Jan Golinski, *Science as Public Culture: Chemistry and Enlightenment in Britain, 1760–1820* (Cambridge University Press, 1992); Larry Stewart, *The Rise of Public Science: Rhetoric, Technology, and Natural Philosophy in Newtonian Britain, 1660–1750* (Cambridge University Press, 1992); Patricia Fara, *Sympathetic Attractions: Magnetic Practices, Beliefs, and Symbolism in Eighteenth-Century England* (Princeton, NJ: Princeton University Press, 1996), pp. 11–30, 208–14.

At the same time as Enlightenment rationality was being reevaluated, sociologists of science were breaking down the philosophical barriers between
the sciences and the so-called pseudosciences and those between orthodox
and quack medical practices. By meticulously re-creating the circumstances
leading to the rejection of practices such as Mesmerism or phrenology, they
demonstrated the primacy of cultural criteria in establishing boundaries that
reinforced the prestige of legitimated science and medicine. Such studies repudiated Whiggish, celebratory accounts by depicting science not as successful progress toward true knowledge but rather as a social practice sharing many
features with other activities and belief systems. Interpreting rhetorical claims
with more cynical eyes showed how enlightened philosophers were promoting their own role as society's guardians by advertising the power of reason.
By setting themselves up as protectors of "the ignorant masses" – a common
phrase – they ensured the perpetuation of an educated elite. Legitimating
some practices and marginalizing others entailed drawing social as well as
epistemological boundaries.[5]

These historiographical trends have led to some fine analyses of particular
practices among what are now variously named the fringe, alternative, or non-
normal sciences of the eighteenth century. However, even this more sympathetic labeling entails a retrospective value judgement, since it implies not only
that such sciences were perceived as peripheral at the time but also that they
held some essential characteristics in common that distinguished them from
their more respectable competitors. By clustering an array of diverse practices
under a single umbrella, even this apparently charitable nomenclature danger-
ously invites misunderstanding because it conceals several fundamental distinctions. Important among these are the absence of a shared epistemological
basis between disparate practices, differences in the histories of their marginalization, and contrasts in beliefs and attitudes relating to a single practice at
various places and times – in other words, the extent to which such practices
are culturally situated. The significance of considering the social processes
contributing to their eventual exclusion is too easily disguised by grouping
these practices together.

Although the title I have chosen – "Marginalized Practices" – still depends
on hindsight to determine what this retrospectively constructed category
should include, it does emphasize the social nature of the transformations
leading to their reduced status, and it also indicates that this status changed
over time. At various periods in the past, astrology, physiognomy, and alchemy
were all held in high esteem as valuable sources of knowledge about the world
and could not necessarily be distinguished from belief systems that we cur-

---

[5] Roger Cooter, "Deploying 'Pseudoscience': Then and Now," in Marsha Hanen, Margaret Osler, and
Robert Weyant (eds.), *Science, Pseudo-Science and Society* (Ontario: Wilfred Laurier University Press,
1980), pp. 237–72; Roy Porter, "Before the Fringe: 'Quackery' and the Eighteenth-Century Medical
Market," in Roger Cooter (ed.), *Studies in the History of Alternative Medicine* (London: Macmillan,
1988), pp. 1–27.

rently perceive to have laid the foundations for modern science. Conversely, innovations that we now regard as significant were often criticized. For example, after the French Revolution, many English Tories were scathingly critical about the chemical experiments of men with radical leanings, such as Joseph Priestley and Humphry Davy, who are now esteemed as founding heroes. In his anonymous "The Birth of Wonders!", Sceptic's succession of imaginary babies included Mesmeria, an infant who leapt into a frog being eaten by Luigi Galvani, and Antiphlogiston: he thus indiscriminately satirized sciences we view as historically central – Lavoisierian chemistry, galvanic electricity, pneumatic chemistry – alongside Mesmerism, subsequently marginalized.[6] Newtonian rhetoricians so successfully suppressed some contemporary groups of opposition – notably the High Church followers of John Hutchinson – that they have even been excluded from catalogs of the marginalized, since this supposedly historical category is determined by modern definitions of alternative sciences. My insistence on the intrinsic Whiggishness of retrospectively bracketing together an assortment of practices questions the validity of the very topic on which I am writing; on the other hand, appreciating the social processes of marginalization greatly enhances our understanding of the legitimated sciences.

Although the natural world gradually replaced the Bible as the source of knowledge, establishing the supremacy of reason entailed major changes in attitudes toward obtaining, judging, and presenting facts of nature and the evidence supporting them. Contrary to the Enlightenment-inspired rhetorics still propagated by some modern historians, there was no simple or sudden transition from a magical world to one ruled by the force of reason. Broad transformations are now seen as being far slower and less uniform than previously claimed, and because of the great religious and political differences between eighteenth-century communities, there were significant regional variations. Rather than simply banish magic and eliminate occult powers, the new mechanical sciences redefined them to naturalize the world of spirits; the role of the supernatural was diminished by recategorizing many prodigious phenomena as natural events. The new scientific societies proclaimed their commitment to rejecting secrecy by substituting public knowledge, but this ideology of openness conflicted with their advertised status as specialized searchers after truth.[7]

To demonstrate their mastery over the world of nature, natural philosophers captured and dominated polite audiences by displaying their expertise in

---

[6] Anon., *The Sceptic* (Retford: for West & Hughes, 1800), pp. 1–11.

[7] Key texts include Simon Schaffer, "Godly Men and Mechanical Philosophers: Souls and Spirits in Restoration Philosophy," *Science in Context*, 1 (1987), 55–85; Lorraine Daston, "Marvelous Facts and Miraculous Evidence in Early Modern Europe,'" in James Chandler, Arnold I. Davidson, and Harry Harootunian (eds.), *Questions of Evidence: Proof, Practice, and Persuasion across the Disciplines* (Chicago: University of Chicago Press, 1994), pp. 243–74; William Eamon, *Science and the Secrets of Nature: Books of Secrets in Medieval and Early Modern Culture* (Princeton, NJ: Princeton University Press, 1994).

experimental performances. Popular philosophical entertainers flourished by directing overt appeals to this suppressed fascination with the occult, and London's leading instrument-makers marketed sophisticated magic tricks under the guise of rational entertainment. Claims by natural philosophers of their intellectual control over nature were inseparable from their bids for authority within society. For instance, Establishment practitioners sought to outlaw competitors by condemning them as quacks, but they were, like lecturers, similarly engaged in commercialized activities.[8] Social negotiations were central to the marginalizing processes entailed in constructing epistemological boundaries.

Eighteenth-century educators favored cartographical metaphors, purveying misleading visions of clean delineations between adjacent territories to match the allegedly sharp separation of murky bygone eras from the brightness of Enlightenment reason. But the boundaries between past and present, between the traditional and the modern, between popular and elite, were blurred. Like Enlightenment polemicists, many historians simplistically distinguish science from magic, ignoring not only the tangled relationships between hermeticism, occultism, and magic but also the different types of magic and their relevance to the foundations of science. In his 1727 tract *A System of Magick,* Daniel Defoe was not just being satirical but was also reiterating older distinctions when he classed astronomy and philosophy as natural magic, differentiated from artificial magic (such as using charms and spells) and diabolical magic (summoning up evil spirits).[9]

Although rationalizing philosophers repeatedly declared that superstitious customs had been eradicated, traditional practices were not eliminated but rather concealed. As Hester Thrale observed in 1790, "Superstition is said to be driven out of the World – no such Thing, 'tis only driven out of Books & Talk."[10] Elite French writers regretted persistent beliefs in demons and witchcraft, whereas English dictionary and encyclopedia compilers retained detailed discussions of supernatural phenomena. Private diaries and letters reveal a confidence in religious therapeutics, astrology, and witchcraft whose public declaration was hindered by ecclesiastical and political disputes. The Royal Society officially declined to enter into public debates about "any doubtfull Matter," but "in their private and separate Capacitys" several fellows clan-

[8] Simon Schaffer, "Natural Philosophy and Public Spectacle in the Eighteenth Century," *History of Science,* 21 (1983), 1–43; Patricia Fara, "'A treasure of hidden vertues': The Attraction of Magnetic Marketing," *British Journal for the History of Science,* 28 (1995), 5–35; Roy Porter, "The Language of Quackery in England, 1660–1800," in Peter Burke and Roy Porter (eds.), *The Social History of Language* (Cambridge University Press, 1987), pp. 73–103; Barbara M. Stafford, *Artful Science: Enlightenment Entertainment and the Eclipse of Visual Education* (Cambridge: MIT Press, 1994).

[9] Daniel Defoe, *A System of Magick: or, a History of the Black Art* (London, 1727), p. 49; Brian P. Copenhaver, "Natural Magic, Hermetism, and Occultism in Early Modern Science," in David C. Lindberg and Robert S. Westman (eds.), *Reappraisals of the Scientific Revolution* (Cambridge University Press, 1990), pp. 261–301.

[10] Hester Lynch Thrale, *Thraliana: The Diary of Mrs Hester Lynch Thrale, 1776–1809,* ed. Katherine C. Balderston, 2 vols. (Oxford: Oxford University Press, 1942), 2:786.

destinely requested and compiled information about topics such as "the gentleman who has ye Gift of Second Sight."[11]

After more than two centuries of suppression, it is hard to retrieve evidence of marginalized practices, and there are very few studies in English dealing with countries other than England and France.[12] The secondary literature is patchy and reflects a modern fascination with the bizarre as much as the preoccupations of the period. Far more has been written about the mesmeric merchants of magnetic medicine than the eminent enthusiasts promoting the widespread application of electrical treatments,[13] and there are virtually no studies in English of rhabdomancy (divination) or chiromancy (palmistry).[14] Similarly, historians have recently become fascinated by the Masonic networks perpetuating ancient knowledge of alchemy, Paracelsian medicine, and hermetic beliefs; ironically, some researchers have so overinterpreted the scanty material on Freemasonry that their conclusions mirror Enlightenment paranoia about clandestine international organizations.[15]

The isolated studies of specific characters and episodes that have been completed reveal the great variety of marginalized practices, thus confirming the importance of studying them contextually. Practices such as animal magnetism and astrology were conducted very differently in France, England, and Germany, because their reception depended on local attitudes towards, for

[11] Kay S. Wilkins, "Attitudes to Witchcraft and Demonic Possession in France during the Eighteenth Century," *Journal of European Studies*, 3 (1973), 348–62; Arthur Hughes, "Sciences in English Encyclopædias, 1704–1875–I," *Annals of Science*, 7 (1951), 340–70; Michael MacDonald, "Religion, Social Change, and Psychological Healing in England, 1600–1800," *Studies in Church History*, 19 (1982), 101–25. Letter from Henry Baker to Archibald Blair of 10 February 1749, Correspondence of Henry Baker, 8 vols. (John Rylands Library, Manchester University, MS/9), 4:67; this letter formed part of an extended correspondence.

[12] For references to foreign literature, see Gloria Flaherty, "The Non-Normal Sciences: Survivals of Renaissance Thought in the Eighteenth Century," in Christopher Fox, Roy Porter, and Robert Wokler (eds.), *Inventing Human Science: Eighteenth-Century Domains* (Berkeley: University of California Press, 1995), pp. 271–91. For an influential study on America, see Herbert Leventhal, *In the Shadow of the Enlightenment: Occultism and Renaissance Science in Eighteenth-Century America* (New York: New York University Press, 1976).

[13] Despite its prevalence and its importance for understanding later debates about life (in *Frankenstein*, for instance), the literature on electrical medicine remains sparse: Margaret Rowbottom and Charles Susskind, *Electricity and Medicine: History of their Interaction* (San Francisco: San Francisco Press, 1984), pp. 1–54; Geoffrey Sutton, "Electric Medicine and Mesmerism," *Isis*, 72 (1981), 375–92; pp. 68–78 of Simon Schaffer, "Self Evidence," in James Chandler, Arnold I. Davidson, and Harry Harootnian (eds.), *Questions of Evidence: Proof, Practice, and Persuasion across the Disciplines* (Chicago: University of Chicago Press, 1994), pp. 56–91.

[14] But for French rhabdomancy, see Luca Ciancio, "La Resistibile Ascesa della Rabdomanzia: Pierre Thouvenel e la 'Guerra di Dieci Anni,'" *Intersezione*, 12 (1992), 267–90, and for the science of manual gestures, see James R. Knowlson, "The Idea of Gesture as a Universal Language in the XVIIth and XVIIIth Centuries," *Journal of the History of Ideas*, 26 (1965), 495–508.

[15] Clarke Garrett, *Respectable Folly: Millenarians and the French Revolution in France and England* (Baltimore, MD: Johns Hopkins University Press, 1975); John M. Roberts, *The Mythology of the Secret Societies* (London: Secker & Warburg, 1972); M. Keith Schuchard, "Freemasons, Secret Societies, and the Continuity of the Occult Traditions in English Literature" (Ph.D. thesis, University of Texas at Austin, 1975); and Margaret C. Jacob, *The Radical Enlightenment: Pantheists, Freemasons and Republicans* (London: Allen & Unwin, 1981) are good sources for relevant primary material, although their interpretations are contested.

instance, religion, politics, and the role of women. In this chapter, I shall concentrate on five examples of marginalized practices to illustrate how social processes played vital roles in their exclusion. I shall discuss specific situations because only by considering local episodes is it possible to demonstrate how processes of marginalization are culturally fashioned; however, I have chosen these five examples to present a wide range of both analytical approaches and historical transformations.

ANIMAL MAGNETISM

Animal magnetism, the therapeutic technique often called Mesmerism after its initiator, Franz Mesmer (1734–1815), has been the major discredited science to be treated sympathetically by recent historians. Animal magnetizers promoted various techniques but generally claimed to effect cures for chronic ailments by their ability to redistribute the magnetic or nervous fluid circulating through a patient's body, frequently inducing a trancelike state. After a few years of lucrative popularity, Mesmer himself was outlawed by a governmental inquiry, but his followers developed and propagated his ideas, which enjoyed recurrent but diverse bouts of popularity throughout Europe and America.

The subject of numerous studies, animal magnetism exemplifies the close alliance between political interests and decisions governing which types of knowledge become validated.[16] In his pathbreaking interpretation, Robert Darnton depicted French Mesmerism as a scientific vogue akin to ballooning that offered a new secular faith for democratic reformers favoring superstition over Enlightenment rationality. Through his detailed analyses of Parisian and provincial practices and propaganda, he argued that Mesmerism was a medical fashion enlisted by radicals as a vehicle for communicating their ideas and mobilizing public discontent. Historians have started to explore the introduction of animal magnetism into other countries; their work corroborates the politicized nature of Mesmerism by showing the contrasts between processes of marginalization in different cultural situations.[17] Here

---

[16] Key studies include Robert Darnton, *Mesmerism and the End of the Enlightenment in France* (Cambridge, MA: Harvard University Press, 1968) and Alan Gauld, *A History of Hypnotism* (Cambridge University Press, 1992), pp. 1–123. For a fuller discussion of the literature, see pp. 127–31 of Patricia Fara, "An Attractive Therapy: Animal Magnetism in Eighteenth-Century England," *History of Science* 33 (1995), 127–77, on which this account is based.

[17] For instance, Heinrich Feldt, "Der Begriff der Kraft im Mesmerismus: Die Entwicklung des physikalischen Kraftbegriffes seit der Renaissance und sein Einfluβ auf die Medizin des 18. Jahrhunderts" (Ph.D. dissertation, University of Bonn, 1990); Joost Vijselaar, "The Reception of Animal Magnetism in the Netherlands," in Leonie de Goei and Joost Vijselaar (eds.), *Proceedings of the 1st European Congress on the History of Psychiatry and Mental Health Care* (Rotterdam: Erasmus Publishing, 1993), pp. 32–8; James E. McClellan, *Colonialism and Science: Saint Domingue in the Old Régime* (Baltimore, MD: John Hopkins University Press, 1992), pp. 163–200; Fara, "An Attractive Therapy," and Fara, *Sympathetic Attractions*, pp. 193–207.

I shall compare the fate of animal magnetizers in late Enlightenment France and England to illustrate the importance of studying marginalized practices contextually.

In France the state played a far greater role in scientific and medical innovation, both through its financial backing for research projects and through its control of how therapies were marketed. In the early 1780s, two investigators commissioned by the supervisory Société Royale de Médecine reported their inquiries into the claims of magnetic therapists, including Mesmer, who had transferred his medical practice from Vienna to Paris in 1778. They praised the curative, calming value of wearing magnets fashioned into different shapes and tied to painful parts of the body, and they recommended further research.

Mesmer himself moved from this literal application of magnets as therapeutic devices to a more metaphorical vision of a universal magnetic fluid.[18] The central feature attracting wealthy patrons to his fashionable music-filled salon was the *baquet*, a large oaken tub filled with magnetic materials, magnetized water, and aromatic herbs. Clients, predominantly women, absorbed its healing magnetic powers by holding protruding iron bars and wrapping ropes around afflicted limbs. Mesmer also treated patients individually, passing his hands around them while gazing intently into their eyes to achieve healing crises resembling fits or trances. These somnambulistic episodes prompted frequent accusations of sexual misconduct, despite Mesmer's theoretical explanations that he was redirecting the flow of universal magnetic fluid through the body. Mesmer basked in lucrative popularity for several years, although he quarreled with former adherents – notably Charles Deslon – who became his rivals. In 1784, an official inquiry nominally headed by Benjamin Franklin discredited Mesmer by publicly ridiculing his therapeutic claims, although unpublished documents reveal greater concern with sexual impropriety.

Mesmer fled to London but shortly returned to continental Europe after failing to rouse support among men of science who had been swayed by the verdict of their eminent Parisian colleagues. Darnton led the way in showing how Mesmeric practices subsequently spread throughout Revolutionary France, often purveyed through Masonic networks of men fascinated by the occult. In regional centers, radical promoters developed their own versions of animal magnetism as they sought a democratic yet scientific medicine, attracting faithful enthusiasts very different from the rich metropolitan clientèle swooning under Mesmer's powerful gaze.

Although animal magnetism flourished throughout France well into the nineteenth century, in England it gained only a brief spell of metropolitan

---

[18] Robert G. Weyant, "Protoscience, Pseudoscience, Metaphors and Animal Magnetism," in Marsha Hanen, Margaret Osler, and Robert Weyant (eds.), *Science, Pseudo-science and Society* (Ontario: Wilfrid Laurier University Press, 1980), pp. 77–114.

popularity. In contrast with France, because of the close alliance between natural philosophy and private commercial ventures, English philosophical entrepreneurs were more concerned with the practical value of magnets for navigation than abstract research into their properties; similarly, medical practitioners openly competed for patients in a pluralistic market unrestricted by government controls. It was only in 1784 that English animal magnetizers launched their new therapy – not, as in the earlier situation in Paris, into a favorable milieu of customers impressed by existing magnetic cures but rather among skeptics deterred by an official inquiry denouncing Mesmer as a charlatan.

In London, rival animal magnetizers adapted French methods to capture various sectors of the market for novel therapies. For instance, in Covent Garden, John Bell sought to re-create the magnetic money-laden ambiance of Mesmer's salon with its oaken *baquet,* but he was eclipsed by John de Mainauduc, a trained surgeon and midwife, who astutely located himself in fashionable Bloomsbury. Turning the Parisian condemnation to his advantage, de Mainauduc prospered by disclaiming any affiliation with Mesmer and physical magnets. His lectures and treatments, which reflected contemporary medical insistence on restoring natural circulation and equilibrium, attracted leisured members of the aristocracy, newly wealthy Quaker industrialists, and a small artistic group engaged in occult activities.

Numerous critics denigrated de Mainauduc and his less famous competitors, but, unlike in France, radical enthusiasts showed little interest in adopting their activities. Instead, as "Animal Magnetism" – the title of a successful theatrical farce – came to symbolize popular gullibility, diverse polemicists adopted it as a vehicle for attacking their opponents in a variety of debates. Enlightenment rationalists advertised the elite's responsibility to protect credulous women and the uneducated masses, different religious sects accused one another of harboring magnetizers, and Whigs and Tories alike published magnetic satires on the political opposition. Because it had been imported from France, animal magnetism's condemnatory import in ideological controversies intensified after the Revolution, when a flurry of panicky pamphlets characterized it as a subversive practice contributing to social unrest, materialist philosophies, and frivolous foreign behavior – "a diabolical practice" originating "in that *antichristian Empire of Atheism,* France."[19] Ostensibly banished, the English animal magnetizers nevertheless had articulated current preoccupations with the relationship between the mind and the body, and Romantic authors imbued magnetic imagery with new connotations of personal attraction.

These contrasting studies of France and England illustrate the value of a close analysis of marginalized practices within an immediate context. By re-

---

[19] *Supernatural Magazine* (1809), p. 8.

creating how they were perceived at the time, historians not only achieve a deeper understanding of sciences that once enjoyed credibility but also gain new insight into a broad range of contemporary attitudes.

PHYSIOGNOMY

In their promotional literature, several animal magnetizers established historical precedents to authenticate their therapies by claiming continuity with earlier healers; they also benefited from the traditional associations of magnets with sexual attraction and medical remedies. Nevertheless, Mesmer's practices were essentially new: his *baquet* and other magnetic equipment relied on recent techniques for permanently magnetizing steel, and he derived his theories from Newtonian models of gravitational attraction. In contrast, physiognomy – the science of judging people's characters from their appearances – was an ancient system that was given a new prominence in the last third of the century by a Swiss pastor, Johann Caspar Lavater (1741–1801).[20] Lavater's colleagues recognized him as a gifted physiognomist long before he published his own ideas, which built on the work of his predecessors and articulated contemporary concerns to discern an individual's true character. Unlike Mesmer, Lavater's influence spread through his publications rather than his activities; his work was extremely well known and generally commanded considerable respect, although attitudes toward his physiognomy varied in different countries and changed over time. Although we may now bracket physiognomy with Mesmerism as a discredited or even laughable belief, many eighteenth-century writers referred to it in all seriousness as a useful science with a long history, one to which Lavater had made an important contribution. For instance, historians often cite Robert Southey's lengthy denunciation of animal magnetism, but when this Enlightenment rationalist visited Lisbon, he surveyed the galley slaves "with a physiognomic eye to see if they differed from the rest of the people."[21]

Physiognomic perception is fundamental to human interaction: we all read people's faces to infer their moods and personalities, and we even invest

---

[20] The most comprehensive account is Graeme Tytler, *Physiognomy in the European Novel: Faces and Fortunes* (Princeton, NJ: Princeton University Press, 1982), pp. 3–165. See also John Graham, *Lavater's Essays on Physiognomy: A Study in the History of Ideas* (Berne: Peter Lang, 1979); Roy Porter, "Making Faces: Physiognomy and Fashion in Eighteenth-Century England," *Etudes Anglaises*, 38 (1985), 385–96; Michael Shortland, "The Body in Question: Some Perceptions, Problems and Perspectives of the Body in Relation to Character c. 1750–1850," 2 vols. (Ph.D. thesis, University of Leeds, 1984); E. H. Gombrich, "On Physiognomic Perception," in *Meditations on a Hobby Horse and other Essays on the Theory of Art* (London: Phaidon, 1963), pp. 45–55.

[21] Robert Southey, *Letters from England,* ed. J. Simmons (London: Cresset Press, 1951), pp. 304–19, and a letter to his brother of 23 May 1800, in *The Life and Correspondence of Robert Southey,* ed. Charles Southey, 6 vols. (London: Longman, Brown, Green, and Longmans), 2:68–74 (quotation at p. 73).

animals or clouds with anthropomorphic qualities. Although not the earliest, Aristotle's treatise relating physiognomical features to moral dispositions influenced many subsequent attempts to provide a systematic basis for intuitive interpretations. From the eleventh century, European physiognomy was closely bonded with medicine and astrology, but writers increasingly stressed its artistic relevance. Interest increased and altered during the eighteenth century: several medical men published detailed studies of physiognomy based on new anatomical and experimental approaches to the human body; and esthetic theorists compared the roles of the rational and the imaginative approaches to works of art and explored the relationships between physical and moral characteristics. Particularly in France, novelists drew on physiognomic principles to portray two categories of fictional character: idealized heroines and grotesque villains and eccentrics.

Lavater viewed physiognomy as a theological study designed – as the full title of his voluminous *Essays on Physiognomy* (1775–8) spells out – to help people understand and love one another.[22] Through formulating precise rules for studying bodies and faces, he sought to aid the practicing physiognomist's instinctive perceptions of moral and intellectual attributes. His books included analyses of famous portraits as well as silhouettes and drawings of people, animals, and individual features that illustrated particular types. For example, he judged that Denis Diderot's forehead revealed not only his intelligence but also his gentleness and lack of enterprise, whereas Wilhelm von Humboldt's betrayed his stubbornness. The flexibility of an elephant's trunk indicated the animal's prudence, the bridge of Satan's nose marked violence, and an eye reflecting pride and courage might be marred by a weak lower lid. Lavater focused on the face at rest, searching for a person's essential physiognomic characteristics, teaching that expressions – more properly the subject of pathognomy – betrayed only temporary passions. In his physiognomic search for inner reality, Lavater contended that each member of God's creation is unique and that each part of an individual, down to the very nerves and blood, contains the character of the whole.

Lavaterian physiognomy was immediately highly successful throughout Europe, although it subsequently became more fashionable in France and England than in the German-speaking countries. Between 1775 and 1810, fifty-five different editions and translations of Lavater's *Essays* were produced; his close friend Henry Fuseli collaborated with William Blake to illustrate one of the English editions. It was handsomely produced and included fine engravings for wealthy purchasers or members of the new book clubs, and the countless abridgements and reviews meant that any literate person must have been familiar with Lavater's ideas. One journal reported, "A servant would . . . scarcely be hired but the description and engravings of Lavater had been

---

[22] Johann Caspar Lavater, *Physiognomische Fragmente zur Beförderung der Menschenkenntniss und Menschenliebe*, 4 vols. (Leipzig, 1775–8).

consulted in *careful* comparison with the lines and features of the young man's or woman's countenance."[23]

But although Lavater's supporters wrote favorable reviews, critics accused him of being insufficiently systematic and mocked his claims to have converted physiognomy into a science. The harshest satirist was the natural philosopher Georg Lichtenberg, who, placing a greater reliance on pathognomy than physiognomy, argued that actions reveal more about character than appearances. The prominent Tory bluestocking Hannah More, ever attentive to the moral welfare of the uneducated, complained to her friend Horace Walpole that "In vain do we boast . . . that philosophy had broken down all the strongholds of prejudice, ignorance, and superstition; and yet, at this very time . . . Lavater's physiognomy books sell at fifteen guineas a set."[24]

Despite the mixed nature of the reactions to Lavater's work, novelists developed more subtle physiognomical portraits of their fictional characters, reflecting Romantic preoccupations with self-knowledge and authenticity. An increasing number of texts on physiognomy and pathognomy appeared during the first half of the nineteenth century, many of them associated with the rising science of phrenology, which used skull shape to determine character. Although many modern historians belittle physiognomy as a pseudoscience, at the end of the eighteenth century it was not merely a popular fad but also the subject of intense academic debate about the promises it held for future progress.

ASTROLOGY

Lavater claimed to be improving an ancient system by formulating its principles more precisely, but he wanted to dissociate physiognomy from the equally long-rooted practice of astrology. In contrast with physiognomy, polemicists had already marginalized astrology by the beginning of the eighteenth century, although versions survive even today outside elite legitimated discourses. Nowadays, philosophers and scientists cite astrology as a typical pseudoscience that inevitably became discredited through its failure to produce meaningful predictions, whereas historians focus on displaying the cultural pressures affecting transformations in astrological practices.

Like other marginalized practices, astrology has no fixed definition, but the term broadly refers to systems that focus on interpreting the human or terrestrial significance of the stars. To illustrate how adherence to such beliefs is culturally situated, I shall discuss the example of England, which is the

---

[23] From the *Gentleman's Magazine*, quoted at p. 61 of Graham, *Lavater's Essays*, with an erroneous reference.

[24] Letter to Horace Walpole of September 1788, reproduced in W. S. Lewis, *The Yale Edition of Horace Walpole's Correspondence*, 48 vols. (London: Oxford University Press, 1937–83), 31:279–81 (quotation at p. 280).

subject of the best analyses showing how transformations of this ancient science were embedded within political and social changes.[25]

English astrology reached its peak during the upheavals of the mid-seventeenth century. Rare celestial events were viewed as supernatural manifestations, and politically interested astrologers offered knowledge of the future in almanacs and other cheap books. But after the Restoration, astrological practices altered to accommodate a double suppression: politicians took measures to control this radical threat, and astronomizing philosophers sought to displace astrologers' authority and assume their powerful public role. After a brief period of crisis and unsuccessful attempts at reform, astrological practices remained stable throughout the eighteenth century. There were three basic types of astrology, and they differed in their approach, the audiences to whom they appealed, and their historical pattern.

First, at a popular level, astrologers thrived by articulating the astral beliefs firmly retained by rural laborers and urban artisans. With their lives closely governed by the seasons, these people sought reassurance about unusual starry wonders as well as more routine information about the weather and their health. Constantly attacked for purveying superstition to the ignorant, popular astrologers based their predictions on relatively simple analyses of the moon and the sun. At mid-century their almanacs were probably being read by around one-third of the population; by far the most successful – well into the next century – was *Moore's Vox Stellarum*, with a print run of more than ten times that of the *Gentleman's Magazine*.

Judicial astrology entailed far more complex interpretative and astronomical abilities, and it provided horoscopes (maps of planetary positions) drawn up for specific occasions. In addition to the complex mathematical calculations required to plot astral locations, judicial astrologers had to master an intricate body of doctrine concerning plantary influences. Practitioners included gentlemanly antiquarians such as William Stukeley but more typically were autodidactic surveyors or mathematicians, living predominantly in the Midlands area. John Cannon, for instance, was an excise-man who engaged in an ambitious program of learned reading but who also obsessively consumed astrological works, copying out texts and complicated diagrams. Like many of his contemporaries, he felt that the natural and the human worlds were closely bonded and that philosophical astronomers were incapable of explaining the dramatic celestial events sent as messages from God. Thus religious prophets viewed a spectacular aurora borealis "as a *bloody flag*, hung out by divine resentment, over a guilty world."[26] Educated provincial men

---

[25] The most comprehensive account is Patrick Curry, *Prophecy and Power: Astrology in Early Modern England* (Cambridge: Polity Press, 1989). See also Simon Schaffer, "Newton's Comets and the Transformation of Astrology," in Patrick Curry (ed.), *Astrology Science and Society: Historical Essays* (Woodbridge: Boydell Press, 1987), pp. 219–43.

[26] John Money, "Teaching in the Market-Place, or 'Caesar Adsum Jam Forte: Pompey Aderat': The Retailing of Knowledge in Provincial England during the Eighteenth Century, in John Brewer and Roy Porter (eds.), *Consumption and the World of Goods* (London: Routledge, 1993), pp. 335–77; James Hervey, *Meditations and Contemplations*, 2 vols. (Paisley, 1774), 2:54.

such as Cannon were relatively isolated from the higher echelons of society, but as metropolitan values spread throughout the country, traditional judicial astrology declined. Instead, men like Francis Barrett and Ebenezer Sibley launched a new style of publication that initiated the Victorian middle-class taste for an astrological amalgam of science and magic.[27]

The third type, philosophical or cosmological astrology, relied on current astronomical measurements and theories but was rooted in theological natural philosophy. Producers of natural knowledge reformulated older astrological discussions about the structure, functioning, and governance of the universe in the vocabulary of Newtonian philosophy. By constructing laws describing the behavior of comets and other unusual phenomena, they claimed to pre-dict those events that had formerly been predictions of disaster. Nevertheless, they also discussed the influence on people's lives and health of planets and comets, speculating about whether or not they were inhabited. Legitimated as natural philosophy, these astrological ideas circulated among wider audi-ences through the expanding numbers of books and lectures. For example, the Newtonian arguments of Richard Mead that the sun and the moon af-fect the earth's atmosphere, and thus explain periodic phenomena such as epilepsy and menstruation, were frequently republished and provided Mesmer with the basis of his magnetic theories.[28]

This analysis provides a continuous history for English astrology by show-ing how transformations in its practice were closely tied to local social changes and the emergence of a middle class that initially embraced patrician ide-ological ideals. The necessity of contextualizing the study of marginalized practices is underlined by comparing France, with its strong, centralized re-ligious and political control. There, astrology was similarly appropriated by natural philosophy but rapidly diminished in the later seventeenth century at more popular levels; during the reign of the Enlightenment *philosophes,* instead of a middling judicial astrology sustaining a subsequent revival, as-trology reappeared as a topic of historical inquiry.[29]

ALCHEMY

The most important alchemist living in the eighteenth century was Isaac New-ton. This simple statement, which only fifty years ago would have seemed

[27] Francis Barrett, *The Magus, or the Celestial Intelligencer; Being a Complete System of Occult Philosophy* (London: Lackington, Allen & Co, 1801), pp. 3–11, 142–76; Allen G. Debus, "Scientific Truth and Occult Tradition: The Medical World of Ebenezer Sibly (1751–1799)," *Medical History,* 26 (1982), 259–78, especially 265–8.

[28] Michael J. Crowe, *The Extraterrestrial Life Debate 1750–1900* (Cambridge University Press, 1986); Frank A. Pattie, "Mesmer's Medical Dissertation and Its Debt to Mead's *De Imperio Sollis ac Lunae,*" *Journal of the History of Medicine,* 11 (1956), 275–87.

[29] Jacques E. Halbronn, "The Revealing Process of Translation and Criticism in the History of Astrol-ogy," in Patrick Curry (ed.), *Astrology, Science, and Society: Historical Essays* (Woodbridge: Boydell Press, 1987), pp. 197–217.

almost blasphemous, not only articulates how historians have drastically reappraised their opinions of Newton but also illustrates the way in which sensitive studies of marginalized practices can enhance our understanding of scientific history. Enlightenment rhetoricians denigrated alchemy as occult superstition, the preserve of mystical eccentrics living in a bygone age, and cynically joked about the "political alchymists" who sought to transform paper into gold by suspending cash payments against banknotes. Nevertheless, scholars now perceive alchemy as fundamental to the thought of one of science's major heroes.

Although they did not realize it, England's leading natural philosophers of the eighteenth century were developing theories formulated within an alchemical context. This paradoxical centrality of a tradition relegated to the peripheries of legitimated science underlines the falsity of demarcating scientific practices with hard epistemological boundaries. Like astrology and animal magnetism, alchemy aroused fierce antagonism among its critics because of its ambiguous relationship with more orthodox experimentation. The Midlands painter Joseph Wright encapsulated this problematic status of Enlightenment science in his 1771 painting, *The Alchymist, in search of the Philosopher's Stone, discovers Phosphorus and prays for the Successful Conclusion of his Operation, as was the Custom of the Ancient Chymical Astrologers.* Kneeling in front of a bright flask of phosphorus irradiating his crowded laboratory, the hopeful alchemist raises his eyes to heaven for guidance. As in Wright's other pictures of experimental demonstrations and blacksmiths' forges, the centrality of the manmade light recalls religious imagery, but this chemical luminescence hovers between secular illumination and divine transience. Hidden inside his Platonic cave, the self-seeking philosopher ignores God's natural light of the moon glimmering through the Gothic window, preferring to seek progress by artificial processes toward an unrealizable goal.[30]

Promoters, detractors, and historians of alchemy have interpreted the term in different ways. Its diverse material aims included the transmutation of base metals into silver and gold, the production of pearls and other precious objects, and the concoction of medical remedies, particularly the elixir alleged to prolong life; furthermore, some commentators viewed alchemy as the endeavor to transform the imperfect human soul into a spiritual entity. Alchemists were united in their belief that all the visible forms of matter are based on a single essential substance: hence the possibility of changing lead into gold, or healing a human body with a mineral or vegetable derivative. A single vegetative force governed development – the animate growth of seeds and embryos as well as the production of minerals and metals; similarly, human bodies and souls were related to those of metals, and alchemical processes were subject to astrological influences. Practitioners saw them-

---

[30] Ronald Paulson, *Emblem and Expression: Meaning in English Art of the Eighteenth Century* (London: Thames and Hudson, 1975), pp. 184–203.

selves as the inheritors of a divinely revealed art whose secrets should not be divulged.

Newton scoured the alchemical literature throughout most of his working life, compiling voluminous notes and copying whole treatises by hand. He engaged in intensive periods of experimentation and wrote copious reports of his findings. He sought evidence of the alchemists' universal animating spirit, through which he believed God constantly molded the universe: for him, gravity, alchemy, and God were intimately linked through his etherial speculations. Newton's alchemical pursuits were not ancilliary to his natural philosophy but rather formed an essential part of his religious endeavor to study God's activities from as many aspects as possible.[31]

Newton's alchemical works were never published, and any suggestion of his interest in such matters was systematically repressed. Concealing Newton's alchemical activities contributed to widening the divide between marginalized practices and the increasingly inviolable edifice of Newtonian science. This Enlightenment dismissal means that it is now hard to find much surviving evidence of alchemical practitioners in England. As a mid-century translator reported, "The Number of Operators in that Way, has of late Years mightily encreased . . . though they endeavour to conceal themselves . . . to avoid that Ridicule, which generally attends the Professors of the Occult Sciences." Editors who puffed new translations of alchemic texts as being of historical interest may have been catering to such covert enthusiasts.[32]

Historians have managed, however, to retrieve substantial amounts of information about a few individuals nearer the end of the century, suggesting local continuities augmented by foreign contacts. In the Germanic countries, traveling alchemists had traditionally been integrated in court culture, promising financial gain in exchange for princely patronage of their projects. Practitioners increasingly concealed their activities, but Johann Semler, a theology professor in Halle, prospered by publishing popular books on demonic possession, hermetic medicine, and alchemical transmutation. Europe's most celebrated Enlightenment alchemist was the Comte de Saint-Germain, a charismatic and knowledgeable polymath who traveled widely and was acclaimed in the highest social circles, including the court of Louis XV. Probably a Sephardic Jew, he concealed his origins while encouraging rumors that he had discovered the elixir of life and was hundreds of years old. In addition to his alchemical expertise, he was a skilled chemist and musician and was employed in several missions of diplomatic espionage. Chaim Schmul Falck became

---

[31] Betty Jo Teeter Dobbs, *The Janus Faces of Genius: The Role of Alchemy in Newton's Thought* (Cambridge University Press, 1991); for references to the literature, see pp. 1–4.

[32] Quotation from unpaginated preface of Joannes Henricus Cohausen, *Hermippus Redivivus: or, the Sage's Triumph over Old Age and the Grave* (London: for J. Nourse, 1744); Camillus Leonardus, *The Mirror of Stones: in which the Nature, Generation, Properties, Virtues and Various Species of more than 200 Different Jewels, Precious and Rare Stones, are distinctly Described* (London: for J. Freeman, 1750), pp. vii–xiv.

London's most famous practicing alchemist. A Jewish emigré, he had astounded German noblemen with his ability to cure illnesses and perform apparently miraculous feats. Known as Doctor Falkon, this wealthy recluse lived modestly and enjoyed an international reputation for his alchemical and cabalistic expertise.[33]

James Price – the unfortunate "Paracelsus of Guildford," as Banks sarcastically dubbed him – was not the only alchemical experimenter at the Royal Society: the eminent chemist Peter Woulfe blamed the failure of his own assiduous search for the elixir of life on his incomplete moral preparation. Both men appear to have belonged to a small circle of occult enthusiasts that flourished in London toward the end of the century and whose members included William Blake, Richard Cosway, and Philip de Loutherbourg. The surviving evidence of their activities is often ambiguous, and it is difficult to know how many of them actually practiced alchemy, although skeptics did circulate rumors that de Loutherbourg's wife had jealously smashed his crucible. As well as their cabalistic, alchemic, and Mesmeric interests, these men were particularly fascinated by the doctrines of the Swedish mystical theologian Emanuel Swedenborg (1688–1772). After thirty years as a conventional and well-respected natural philosopher in Sweden, Swedenborg experienced a spiritual conversion and dedicated his life to expounding his metaphysical religion. His influential books vividly portrayed his heavenly visions and expounded a Neoplatonic cosmology that, relying on scriptural interpretation, conceived the material world as a perpetual divine emanation. Swedenborgian activity in London centered on the New Jerusalem Church, organized by Wesleyan preachers five years after his death. Some of his English followers were also involved in international Masonic networks such as the Avignon Society, a group of mystical enthusiasts that attracted wealthy gentlemen from all over Europe. Prestigious members included Count Tadeusz Grabianka of Poland, then the center of Jewish mysticism, and General Charles Rainsford – governor of Gibraltar, Fellow of the Royal Society and Banks' own cousin – who crammed notebooks with alchemical lore in three languages, copying out diagrams and experimental details.[34]

Some entrepreneurial writers ensured that alchemical ideas reached wider audiences. Barrett's *The Magus* remains the most famous book on the occult since the end of the seventeenth century. An exotically packaged compendium of ancient knowledge, it deliberately appealed to those with a predilection for

[33] Pamela H. Smith, *The Business of Alchemy: Science and Culture in the Holy Roman Empire* (Princeton, NJ: Princeton University Press, 1994); Raphael Patai, *The Jewish Alchemists: A History and Source Book* (Princeton, NJ: Princeton University Press, 1994), pp. 455–79.

[34] Garrett, *Respectable Folly*, pp. 97–120; Schuchard, "Freemasons," pp. 230–475; N. L. Danilewicz, "'The King of the New Israel': Thaddeus Grabianka (1740–1807)," *Oxford Slavonic Papers*, 1 (1968), 49–73; Clarke Garrett, "Swedenborg and the Mystical Enlightenment in Late Eighteenth-Century England," *Journal of the History of Ideas*, 45 (1984), 67–81. Wellcome Institute, London, MSS 4032–9.

arcane mysteries. After a historical review of alchemy's venerable tradition, Barrett embedded his detailed instructions for increasing a quantity of gold within moral exhortations prescribing the appropriate behavior essential for the successful alchemist. In contrast, Sibly was a qualified physician who wrote several texts blending the latest scientific and medical theories with older hermetic views of a harmonious vitalistic universe, including alchemical notions. Although ignored by his establishment contemporaries, Sibly's books were repeatedly published well into the nineteenth century. He thus perpetuated at a popular level the interest in alchemy shared by more elite neoplatonic authors such as Thomas Taylor, who influenced the Romantic poets.[35]

HUTCHINSONIANISM

During the eighteenth century, Newton's supporters consolidated his iconic status so successfully that, until recently, historians ignored both his alchemical activities and the voices of his critics. Natural philosophers embracing diverse theoretical positions judged it advantageous to label themselves "Newtonian," and writers employed the vocabulary of religious and military warfare to describe the conflicts between self-styled Newtonian "disciples" and opposing philosophical sects such as Cartesians. In England, many of these opponents were High Church Tories, in contrast with the general, but by no means exclusive, alignment of Newtonians with Whig Latitudinarianism. Bishop Berkeley, for instance, is now famous as an idealist philosopher, but during his lifetime he was renowned for his attacks on Newton's calculus and his religious imprecations against materialism. In England, the most aggressive anti-Newtonians were the Hutchinsonians, a small but vocal group effectively marginalized by polemicists constructing an unassailable Newtonian ideology.[36] Retrieving information about such dissidents yields a more comprehensive picture of the social negotiations underlying the consolidation of a Newtonian orthodoxy that concealed the variations embedded within it. Since Hutchinsonianism was important in England and Scotland, studying

---

[35] Barrett, *Magus*, especially pp. 51–70; Ron Heisler, "Behind 'The Magus': Francis Barrett, Magical Balloonist," *The Pentacle*, 1(4) (1985), 53–7; Debus, "Sibley," pp. 263–5.

[36] Important literature includes Geoffrey Cantor, "Revelation and the Cyclical Cosmos of John Hutchinson," in Ludmilla Jordanova and Roy Porter (eds.), *Images of the Earth: Essays in the History of the Environmental Sciences* (St Giles: British Society for the History of Science, 1978), pp. 3–22, and "Light and Enlightenment: An Exploration of Mid-Eighteenth-Century Modes of Discourse," in David Lindberg and Geoffrey Cantor (eds.), *The Discourse of Light from the Middle Ages to the Enlightenment* (Los Angeles: University of California Press, 1985), pp. 67–106; Michael Neve and Roy Porter, "Alexander Catcott: Glory and Geology," *British Journal for the History of Science*, 10 (1977), 37–60; Christopher B. Wilde, "Hutchinsonianism, Natural Philosophy and Religious Controversy in Eighteenth-century Britain," *History of Science*, 18 (1980), 1–24, and "Matter and Spirit as Natural Symbols in Eighteenth-century British Natural Philosophy," *British Journal for the History of Science*, 15 (1982), 99–131. For further references, see Fara, *Sympathetic Attractions*, pp. 24–30, 210–12.

the processes contributing to the Hutchinsonians' marginalization suggests the possibility of parallel analyses to investigate how other anti-Newtonian groups were suppressed in Britain and abroad.[37]

Although extreme in his views, John Hutchinson (1674–1737) provides a valuable antidote to glib accounts of Newtonian rationality triumphantly sweeping across an Enlightened nation. The confrontational title of his 1724 *Moses's Principia* reveals Hutchinson's conviction that natural truth is to be found not in the mathematical book of Newton – that "Cobweb of Circles and Lines to catch Flies in" – but rather in the holy scriptures dictated by God.[38] Hutchinson devoted immense effort to restoring an uncorrupted Hebrew version of the Bible, since he believed that deciphering the original God-given texts formed the essential route to knowledge about the natural world. Like the Behmenists, a pietist group centered around William Law and John Byrom, Hutchinsonians believed that words resonated with intertwined material and spiritual meanings, so that, for example, the Biblical word for "gravity" also signified the glory of God. For them, the language of the scriptures bonded the physical, human, and spiritual worlds in a complex metaphorical web, so that natural philosophical and theological inquiries were inextricably linked.

Hutchinson envisaged the universe as a large machine driven from the sun by a perpetual circulation of three forms of a subtle fluid – fire, light, and spirit – analogous to the holy Trinity. Initially set in motion by God, these three manifestations of the universal divine fluid operated on ordinary matter by direct mechanical contact. Hutchinson's objections to Newton's cosmology were theologically based. Most fundamentally, Hutchinson felt that the principle of gravitational attraction acting through a vacuum gave an agency to inert matter, thus reducing the distinction between God's spirituality and the passive product of His creation. He accused Newton of contradicting Biblical authority and further limiting God's power by equating Him with space and requiring Him to be constantly active in maintaining the equilibrium of the universe. In contrast, the Hutchinsonian cosmos was a self-sustaining plenum in which movement was effected by impulse. Whereas Newton's universe was a directional one in which the quantity of motion was constantly decreasing, Hutchinson envisaged a closed system that conserved the total amount of matter and motion. Like Berkeley and other critics, Hutchinson denied that

---

[37] For Jesuit and Cartesian resistance in France, see François de Dainville, "L'Enseignement Scientifique dans les Collèges Jésuites," in René Taton (ed.), *Enseignement et Diffusion des Sciences en France au XVIIIe Siècle* (Paris: Hermann, 1964), pp. 27–65, and James Evans, "Fraud and Illusion in the Anti-Newtonian Rear Guard," *Isis*, 87 (1996), 74–107. For English Behmenism, see Simon Schaffer, "The Consuming Flame: Electrical Showmen and Tory Mystics in the World of Goods," in John Brewer and Roy Porter (eds.), *Consumption and the World of Goods in the Eighteenth Century* (London: Routledge, 1992), pp. 489–526.

[38] John Hutchinson, *The Philosophical and Theological Works of the Late Truly Learned John Hutchinson*, 12 vols. (London: for J. Hodges, 1748–9), 5:222.

abstract mathematical reasoning could yield useful knowledge about the divinely created world, and he argued that Newton's methodological procedure was inverted: instead of inferring God's nature from observations of the physical world, philosophers should learn about nature by studying God's holy texts.

A small coterie of Hutchinson's supporters published his books and clandestinely circulated his ideas through correspondence networks, and, after his death, the movement gathered strength at the predominantly Tory University of Oxford. Prominent members included George Horne, later vice chancellor of the University and Bishop of Norwich, and William Jones of Nayland, subsequently an influential theological writer, a pamphlet ally of Edmund Burke, and author of two books of natural philosophy. For most of their lives, these men denied allegiance to Hutchinson, by then denigrated as an obscure religious eccentric. Jones provided the most systematic exposition of Hutchinsonian antagonism toward Newton's cosmology, seeking to undermine the Newtonian hegemony by emphasizing, "An experiment in nature, like a text in the Bible, is capable of different interpretations, according to the preconceptions of the interpreter."[39] Focusing on the theological problems raised by attraction through a vacuum, he insisted – as did many of his non-Hutchinsonian contemporaries – that making matter active opened the door to materialism and atheism. In London, Jones's teachings remained influential among Tory High Church affiliates well into the nineteenth century, and the Hutchinsonian leanings of the Scottish Episcopalian clergy influenced the entries on natural philosophy in early editions of the *Encyclopaedia Britannica*.

Neither Hutchinsonian nor Newtonian natural philosophy was unitary or static. For example, Jones highlighted the internal differences between famous Newtonian adherents by juxtaposing their completely contradictory assertions about whether attraction was a cause or an effect. During the second half of the century, exegetes paid increasing attention to Newton's ethereal speculations, and writers of a Hutchinsonian inclination imbued Newtonian ethers of light, heat, and electricity with theological significance. Newton's disciples made overt attacks on his opus increasingly impossible, yet didactic writers such as Adam Walker and George Adams eclectically incorporated aspects of Hutchinsonianism within their teachings while nominally conforming to the prevailing Newtonian orthodoxy. Thus in processes paralleling those in which legitimated astronomy absorbed philosophical astrology, Newtonian rhetoricians marginalized Hutchinsonian practitioners even though their contemporaries were accommodating transformed versions of Hutchinsonian cosmology within the corpus of Newtonian thought.

[39] William Jones, *Physiological Disquisitions: or, Discourses on the Natural Philosophy of the Elements* (London: J. Rivington, 1781), p. 148.

CONCLUSION

In many respects, these five marginalized practices – animal magnetism, physiognomy, astrology, alchemy, and Hutchinsonianism – bear little resemblance to one another. I have brought them together in this essay to illustrate some of the cultural processes that contributed to the new disciplinary maps of knowledge being drawn up at the end of the eighteenth century. As I have shown, judgments about these practices were affected by their political or religious implications as well as by their epistemological value: to us, Mesmer's universal magnetic fluid appears no more intrinsically ridiculous than contemporary explanatory mechanisms, such as phlogiston or electrical atmospheres, that were widely accepted.

The major characteristic shared by these discredited belief systems is their exclusion from modern science. Observers at the time viewed the sciences rather differently from the way we see them. Writers pointed to physiognomy, for instance, as an ancient practice that had been systematically codified during the Enlightenment and would become increasingly scientific in the future. No one could have foreseen that the Mesmerism still being practiced on the Continent would be reimported into England during the nineteenth century or that astrological beliefs would survive at a popular level. Conversely, some experimental activities that seemed to have been effectively marginalized turned out to have important consequences. At his Pneumatic Institute in Bristol, Thomas Beddoes invested his research into the effects of gases with hopes of radical reform similar to those held by the French Mesmerists, but Beddoes was satirized as a foolish and revolutionary enthusiast. Although the gas we now call nitrous oxide provided material for public entertainment, its anaesthetic value was not appreciated until the 1840s.[40] Some of Beddoes's English contemporaries would have been surprised to learn that his historical fate proved very different from that of de Mainauduc.

One important attribute held in common by these diverse marginalized practices is their strong impact on science and literature: although the practitioners may have been ostracized, the effects of these practices resonated throughout society. Polemicists successfully suppressed criticisms of the Newtonian edifice, but, like any dominant system, the new sciences were shaped by their advocates' responses to perceived threats. Although the Hutchinsonians could be facilely dismissed as religious cranks, the questions they posed about the concept of attraction threatened the very core of Newtonian philosophy, which its adherents constantly modified as a protective strategy. Poets such as the Hutchinsonian Christopher Smart and the Behmenist Henry Brooke deliberately infused their verse with anti-Newtonian sentiments. Their essentialist views of language were inimical to scientific writers seeking to strip their prose of metaphorical allusions, but these views enriched the poetry of

---

[40] Golinski, *Science as Public Culture*, pp. 166–87.

Romantic authors such as William Blake and Percy Bysshe Shelley. Mesmer and de Mainauduc were derided as quacks, but their magnetic imagery permanently affected the English language; like writers of the period, we still refer to the "magnetic attraction" that draws lovers together.

Sites of controversy are historically the most rewarding to study, and the ferocity with which some practitioners were marginalized indicates the centrality of the issues they were addressing. The animal magnetizers, for instance, dangerously straddled distinctions between the genius and the madman, between experimental philosophers and theatrical performers, or between the clinician's penetrating gaze and the charlatan's mesmerizing stare. Lavater's treatises were controversial precisely because a physiognomical approach was seen to be valuable in deciphering the true nature of an individual concealed behind the mask of social convention. The chemist Joseph Black apparently felt so insecure about his discipline's historical antecedents that he related Price's sad demise to his students as a cautionary tale to warn them against alchemical temptations. The arcane pursuer of esoteric knowledge, the Faustus perpetually transmuted to suit new audiences, remains an important mythical figure; the enduring appeal of Mary Shelley's *Frankenstein* offers the most obvious evidence of the alchemist's continued symbolic significance. Like Enlightenment rationalists, modern scientists still fear the power of the practitioners they dismiss so contemptuously: in 1975, more than two hundred eminent experts felt the need to condemn publicly a research study whose conclusions endorsed astrological character assessments.

Three hundred years after Newton's birth, the economist John Maynard Keynes reminded a shocked audience that since history is continuous, the predecessors of the Enlightenment should be regarded with respect:

> Newton was not the first of the age of reason. He was the last of the magicians, the last of the Babylonians and Sumerians, the last great mind which looked out on the visible and intellectual world with the same eyes as those who began to build our intellectual inheritance rather less than 10,000 years ago.[41]

Inheriting the progressive views of eighteenth-century rationalists, we too readily relegate ancient traditions as well as short-lived fashions to anecdotal status. Examining marginalized practices is valuable because of the illumination they cast on how beliefs of the past affected the legitimated sciences and other cultural activities. Although we retrospectively tend to bracket together these heterogeneous belief systems as discredited alternatives to modern orthodoxy, during the eighteenth century they made important formative contributions to the definition and consolidation of the new scientific disciplines.

---

[41] Quoted at p. 258 of Richard Yeo, "Genius, Method and Morality: Images of Newton in Britain, 1760–1860," *Science in Context*, 2 (1988), 257–84.

Part III

SPECIAL THEMES

22

EIGHTEENTH-CENTURY SCIENTIFIC INSTRUMENTS AND THEIR MAKERS

G. L'E. Turner

"The diffusion of a general knowledge and of a taste for science, over all classes of men, in every nation of Europe, or of European origin, seems to be the characteristic feature of the present age." So wrote James Keir (1735–1820), the pioneer industrial chemist, in the preface to his *The First Part of a Dictionary of Chemistry* of 1789.[1] There can be no question that the study of the material world – then described as experimental natural philosophy – seriously impinged on the popular consciousness for the first time in the course of the eighteenth century. This was achieved by means of a remarkable social and educational phenomenon: the lecture demonstration.

Science today is understood to be the sphere of activity of the "scientist," a term that was first coined in the 1830s by William Whewell (1794–1866), author of *The History of the Inductive Sciences*. The coinage marks a transition between the mainly amateur natural philosopher and the professional scientist. This is not, of course, to say that science was not studied, and used professionally, centuries earlier in Europe. What was missing in the classical Greek approach to the natural world was the use of experiment. Ideas were tested by reason alone, following the authority of Aristotle, which was broadly accepted throughout the Middle Ages. For example, Aristotle denied the possibility of a vacuum because he reasoned that bodies would move with infinite velocity, a theory that could not then be checked by experiment.

With the Italian Renaissance of the fifteenth century came the impulse to question traditional concepts and to put accepted ideas to the test. At the same time the great voyages of discovery, and the worldwide expansion of trade, required astronomers to address the practical problems of deep sea navigation. They, and the navigators and surveyors who sailed to undiscovered parts of the world and mapped them, were the earliest professional scientists, and it was their need for instruments that gave a strong impetus to the craft of the scientific instrument-maker. In turn, these skills were also put to use

[1] J. K. [James Keir], *The First Part of a Dictionary of Chemistry &c* (Birmingham, 1789), p. iii.

G. L' E. Turner

by the experimental philosophers of the seventeenth and eighteenth centuries, who were engaged in a process of continuous inquiry into the operation of the natural world.

The intellectual force behind experimental science was that of Francis Bacon (1561–1626), who, in his *Novum Organum* (1620), argued that scientific truth must have its basis in the real world, for "neither the naked hand nor the understanding left to itself, can do much; the work is accomplished by instruments and helps." Bacon enunciated the philosophy of ceaseless inquiry into natural phenomena, constantly tested by experiment, and resulting in practical benefit to the community. This was the basis of the activity of the Royal Society of London, founded on 28 November 1660, with the very Baconian motto *Nullius in Verba*.

The founding of the Royal Society of London was anticipated by three years in Italy, where the Medici court at Florence was remarkable for its patronage of learning, particularly the new natural philosophy. In 1657, Prince Leopold of Tuscany (1617–1675) formed the Accademia del Cimento, where organized experiments were carried out, an account of which has come down to us under the title *Saggi di Naturali Experienze fatte nell' Accademia del Cimento*.[2] The society had, however, only a short life of ten years, being disbanded in 1667. Much of its apparatus, however, survived, first in a museum established in 1775 and later in the collection of the Museo di Storia della Scienza in Florence.[3] In France, as in England, the origin of the Académie royale des Sciences was in informal meetings of men of science that were given official status in 1666, when King Louis XIV granted pensions to the members and provided a fund for the purpose of acquiring instruments and carrying out experiments.[4]

This, then, was the background to the emergence of experimental science in the seventeenth century. The last decades of that century saw the new science beginning to be taught in universities in both England and the Netherlands. Science chairs were established at Oxford and Cambridge Universities, and both were influenced by original members of the Royal Society – such men as Isaac Newton, who held the Lucasian Chair of Mathematics at Cambridge; John Wallis, Savilian Professor of Geometry, Oxford; and Thomas Millington, Sedleian Professor of Natural Philosophy, Oxford.[5] The vacuum research of Robert Boyle (1627–1691) at Oxford, assisted by Robert Hooke (1636–1703), who was Curator of Experiments to the Royal Society, attracted international interest. It was after a visit to England in 1674 that Burchardus

[2] W. E. Knowles Middleton, *The Experimenters: A Study of the Accademia del Cimento* (Baltimore, MD: Johns Hopkins University Press, 1971).

[3] Mara Miniati, *Museo di Storia della Scienza: Catalogo* (Florence: Istituto e Museo di Storia della Scienza, 1991), pp. 132–47.

[4] Roger Hahn, *The Anatomy of a Scientific Institution: The Paris Academy of Sciences, 1666–1803* (Berkeley: University of California Press, 1971).

[5] G. L' E. Turner, "The Physical Sciences," in L. S. Sutherland and L. G. Mitchell (eds.), *The History of the University of Oxford, V: The Eighteenth Century* (Oxford: Clarendon Press, 1986), pp. 659–81.

de Volder (1643–1709), professor of physics at Leiden University, had a *Theatrum Physicum* built, where he used an air-pump for class demonstrations.[6]

John Desaguliers (1683–1744), who lectured at Oxford for two years before setting up in London in 1713, tells us, "The first who publikly taught Natural Philosophy by Experiments in a mathematical Manner" in an English university was John Keill (1671–1721), who came from Edinburgh to Oxford in 1694.[7] He gave his course from 1700 to 1709, when he left for New England, but he was back in 1712 as the Savilian Professor of Astronomy. During this break, lectures on experimental philosophy were given by Desaguliers. At Cambridge University, similar lectures were being delivered from 1707 by William Whiston (1667–1752), Newton's successor as Lucasian Professor. Whiston was, in fact, the father of the popular lecture demonstration; when he was dismissed from Cambridge in 1710 for heresy, he went to London to continue his lecturing. He and his successors were able to place great emphasis on the use of apparatus by cooperating with an instrument-maker. Whiston's partnership with Francis Hauksbee senior (c. 1666–1713) was particularly significant, since Hauksbee was a Fellow of the Royal Society, had lectured on his own account, and produced his lectures in published form under the title *Physico-Mechanical Experiments* (1709), later translated into Italian, Dutch, and French. Working as a team, however, Whiston lectured while Hauksbee carried out the experiments, and Whiston published the illustrated text of his series of lectures in 1714. Another notable lecture demonstrator in London at the same time was John Theophilus Desaguliers whose career began in Oxford, at Hart Hall, but who also moved to the capital and became well known for his lectures, course books, and translations of French and Dutch scientific texts.

A younger exponent of the lecture demonstration was Stephen Charles Triboudet Demainbray (1710–1782), who provides a good example of the strong links between England and Holland in the field of experimental philosophy. He studied the subject under Willem 's Gravesande at Leiden University before being appointed in 1754 tutor to the Prince of Wales, later King George III, and subsequently to the King's children. His collection of demonstration apparatus is now part of the King George III collection, newly displayed at the Science Museum, London, in 1993. Demainbray also lectured in Dublin and traveled extensively in France, delivering his course in Toulouse, Montpellier, Lyons, and Paris.[8]

Willem Jacob 's Gravesande (1688–1742) visited England as part of a state deputation in 1715, and there he met Newton, who helped him to an appointment as professor at Leiden University two years later. During his stay

---

[6] E. G. Ruestow, *Physics at the 17th- and 18th-Century Leiden* (The Hague: M. Nijhoff, 1973), pp. 96–8.
[7] Turner, "Physical Sciences," pp. 671–2.
[8] A. Q. Morton and J. A. Wess, *Public & Private Science: The King George III Collection* (Oxford: Oxford University Press and the Science Museum, 1993). For the career of Demainbray, see chap. 4.

in London, 's Gravesande was elected a Fellow of the Royal Society and had the opportunity to attend the Hauksbee-Whiston demonstrations. He was one of a group of notable teachers of science at Leiden, including Boerhaave and Gobius, men who helped to make that University famous throughout Europe. As was the case with William Whiston, 's Gravesande's success as a lecture demonstrator owed much to his collaboration with an instrument-maker, Jan van Musschenbroek (1687–1748), whose father, Samuel, had made the air-pump used by De Volder. 's Gravesande, following the now familiar pattern, published the text of his lectures, and the result was numerous orders for instruments from Jan van Musschenbroek. Petrus (1692–1761), Jan's younger brother, studied under 's Gravesande at Leiden University, took a doctorate in 1719, and eventually occupied the chair of natural philosophy and mathematics at Utrecht University from 1723 to 1740, moving to a professorship at Leiden University from the latter date until his death. He was 's Gravesande's successor, and his lecture notes, collected into ever larger volumes, were widely used and were translated from Latin into all the main European languages.[9]

The art of the lecture demonstration reached its zenith in the *cours de physique* of Jean-Antoine Nollet (1700–1770), performed with some 350 different instruments.[10] Nollet was a peasant boy whose village curé recognized his intelligence and recommended him for the Church. He duly went to Paris to study theology but instead devoted himself to science, joining in 1728 the Société des Arts under the patronage of the Comte de Clermont. Through this short-lived society, Nollet was able to meet members of the Académie des Sciences, who helped his career. Also, perhaps even more significantly, he met the mathematician Pierre Polinière (1671–1734), a successful public lecturer on natural philosophy from whom Nollet inherited both an example and an audience. Two leading academicians – Charles-François Dufay (1698–1739) and René-Antoine de Réaumur (1683–1757) – used Nollet's assistance in their scientific investigations and enabled him to visit both England and Holland, where he met Desaguliers and 's Gravesande.[11] On his return from Holland in 1735, Nollet decided to follow the career of scientific lecturer, and he made, and trained workmen to make, the apparatus he needed (Figure 22.1). Throughout much of his career he continued to supervise, on behalf of individual collectors and institutions, the making of pieces of demonstration apparatus. His success as a lecturer was phenomenal, and the expanded syllabus of his lectures, *Leçons de Physique Expérimentale*, which appeared in six

[9] Peter de Clercq, *At the Sign of the Oriental Lamp: The Musschenbroek Workshop in Leiden 1660–1750* (Rotterdam: Erasmus, 1997); de Clercq, *The Leiden Cabinet of Physics: A Descriptive Catalogue* (Leiden: Museum Boerhaave, 1997).

[10] René Taton, *Enseignement et diffusion des sciences en France au XVIIIe siècle*. Histoire de la pensée XI (Paris: Hermann, 1964), pp. 619–45.

[11] John L. Heilbron, "Nollet, Jean-Antoine," *Dictionary of Scientific Biography*, X, 145–8.

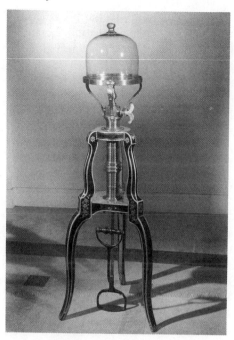

Figure 22.1. An air pump made for Jean-Antoine Nollet (1700–1770), the successful French scientific lecturer and deviser of demonstation apparatus. The instrument is decorated in the ornate, black-and-gold style that is typical of Nollet.

volumes between 1743 and 1748, was equally successful and often reprinted and translated. In 1743, he lectured before the royal family at Versailles and later was offered the newly created chair of physics at the Collège de Navarre. More than simply a popularizer of science, Nollet made a serious contribution to the study of electricity, but it is as the most skilled exponent of the lecture demonstration that he is best remembered.

Nollet made his reputation lecturing to educated, and mainly aristocratic, audiences. There is much evidence that the lecture demonstration, and its domestic imitations, became the vogue in polite society. Sophie v. La Roche (1731–1791), a German aristocrat and novelist, visited London in 1786 and recorded her travels in a diary. A typical entry ends, "Our evening passed at physical experiments, which most certainly form part of divine service, showing us as they do the inner qualities of being, and so leading a sensitive soul to increased and rational reverence for its Creator."[12] This view of experimental philosophy as a means toward a deeper understanding of the majesty of God is typical of the eighteenth century.

---

[12] *Sophie in London 1786 being the Diary of Sophie v. la Roche. Translated from the German with an Introductory Essay by Clare Williams* (London, 1933), p. 136.

From the middle years of the century, the lecture demonstration began to reach a much wider spectrum of the population, certainly in the British Isles and in Holland. One of the first lecturers in England to transport his apparatus by road in order to lecture in town after town was Benjamin Martin (1704–1782), who, from 1740, used Reading and Bath as centers from which he made tours into the west country and as far north as Chester.[13] Adam Walker (1731–1821) traveled the north of England, using Manchester as a base, and had "philosophic apparatus" whose extent is shown by an advertisement in the *York Courant* of 1772. As well as astronomical apparatus and optical instruments, it included "All the mechanical powers, with working Models of various Cranes, Pumps, Water-Mills, Pile-Drivers, Engines, the Centrifugal Machine, and a working Fire-Engine for draining Mines, of the latest construction." Audiences for lecture demonstrations ranged from royalty and the nobility to the most humble citizens. James Watt (1736–1819), the pioneer of steam power, came from an artisan background but had read 's Gravesande's lecture course before he was fifteen, and he later studied the works of Desaguliers on the Savery and Newcomen model engines. The same was true of many of the engineers and business entrepreneurs who laid the foundations of the Industrial Revolution.[14]

The teaching of science by this method became extensively institutionalized as the century progressed, in universities, as might be expected, but also in learned societies, which supplied adult education in science for the benefit of citizens. Haarlem in the Netherlands provides a good example of such activity. In 1776, Martinus van Marum (1750–1837) was appointed by the town council of Haarlem to lecture on philosophy and mathematics. Making use of the collection of apparatus that he gradually amassed in the Teyler's Museum, he continued to give public lectures well into the nineteenth century.[15]

Individuals, too, were inspired to acquire scientific instruments for their own use by attending lecture demonstrations. Many of them, of course, were aristocrats, and the extensive collections of some of them have found their way into science museums. Other collections are known to us only from the surviving auction catalogs, such as those of John Stuart (third Earl of Bute and a close friend of George III)[16] and of the French aristocrat, Bonnier de la Mosson (1702–1744). Bonnier was, like Nollet, a member of the Société des Arts, and he established in his Paris house, the Hotel du Lude, a remarkable cabinet that embraced physics, mechanics, chemistry, pharmacy, and wood-,

[13] John R. Millburn, *Benjamin Martin: Author, Instrument-maker, and "Country Showman"* (Leiden: Noordhoff, 1976).

[14] A. E. Musson and E. Robinson, *Science and Technology in the Industrial Revolution* (Manchester: Manchester University Press, 1969).

[15] G. L' E. Turner and T. H. Levere, *Martinus van Marum Life and Work, Volume IV: Van Marum's Scientific Instruments in Teyler's Museum* (Leiden: Noordhoff, 1973).

[16] G. L' E. Turner, "The Auction Sales of the Earl of Bute's Instruments, 1793," *Annals of Science*, 23 (1967), 213–42.

ivory-, and metal-working equipment, as well as the more usual natural history specimens. His house was later demolished and the collection dispersed, but a set of architect's drawings, dated 1739 and 1740, a contemporary description, and a sale catalog, together with a painting of the cabinet that survived the demolition, provide a unique record of what the collection included and how it was arranged.[17] We also know that the purchase of instruments for the home extended right through the social scale. Henry Baker, the best-selling author of *The Microscope Made Easy* (1742), whose house was in Fleet Street, London, advised many friends living in the provinces on how to acquire optical instruments. The microscope and telescope, the electrical machine, and the air-pump gradually became sources of improving entertainment in many homes.[18]

Our knowledge of what the experimental philosophy apparatus of the eighteenth century consisted of, and how it was arranged, rests on more exact and detailed evidence than pictorial or literary information. A number of eighteenth-century cabinets of instruments have survived more or less intact to the present day. These, as one would expect, contained the apparatus of institutions set up for the purpose of teaching science but are otherwise varied in the location and style. Perhaps the most comprehensive example, since the majority of the instruments are still in their original cases in the splendid Oval Room designed to house them (Figure 22.2), is Teyler's Museum, Haarlem, the apparatus having been acquired, as already mentioned, by Martinus van Marum. Van Marum bought extensively from auction sales, revealing how many private collectors of instruments existed at that time. He also ordered instruments from leading makers in London and other European centers, building his collection over many years. What is exceptional about his cabinet is that most of the documentation concerned with his purchases has been preserved.

Another well-documented collection is that belonging to Harvard University in the United States. Harvard College received, in 1727, the endowment of the first scientific chair in America and the gift of five chests of philosophical apparatus, which was housed in the Philosophy Chamber of the Old Harvard Hall. All were destroyed in a disastrous fire in 1764. The then professor of natural philosophy, John Winthrop (1714–1779), was given the task of rebuilding the collection of scientific instruments. He did this, buying largely from London, notably from Benjamin Martin (who became a successful instrument retailer after his career as lecture demonstrator),[19] over a period of fifteen years. The Harvard cabinet was cataloged in this century by David Wheatland,[20] who also acquired, and later sold to the David M.

[17] C. R. Hill, "The Cabinet of Bonnier de la Mosson (1702–1744)," *Annals of Science*, 43 (1986), 147–74.
[18] G. L' E. Turner, "Henry Baker, F.R.S., Founder of the Bakerian Lecture," *Notes and Records of the Royal Society of London*, 29 (1974), 53–79.
[19] Millburn, *Benjamin Martin*.
[20] D. P. Wheatland, *The Apparatus of Science at Harvard 1765–1800: Collection of Historical Scientific Instruments, Harvard University* (Cambridge, MA: Harvard University Press, 1968).

Figure 22.2. The Oval Room in the Teyler Museum, Haarlem, the Netherlands. The museum was founded by the will of the Haarlem silk merchant Pieter Teyler van der Hulst (1702–1778). The oval museum hall was built behind Teyler's house by Leendert Viervant and was completed in 1782, the first part of what was to become a complex of buildings to house a scientific cabinet, mineralogical and paleonological collections, and works of art. Copyright Teylers Museum, Haarlem, The Netherlands.

Stewart Museum in Montreal, Canada, twenty-three pieces of eighteenth-century philosophical apparatus that had come originally from the Académie de Dijon and are made to the design of Nollet.

Other cabinets of apparatus still in their original setting are to be found at the Benedictine seminary housed in the magnificent baroque monastery of Kremsmunster, Austria; and in the University of Coimbra, Portugal. The latter collection owes its existence to the Marques de Pombal (1699–1782), who, while ambassador to the Court of St. James between 1739 and 1745, had ample opportunity to attend lecture demonstrations in London. In 1772,

Pombal transferred to Coimbra University the "professors, machines, and instruments of astronomy and experimental physics" belonging to the Colegio dos Nobres in Lisbon, in the charge of the professor at the Colegio, an Italian from Padua, Giovanni Antonio della Bella. It was he who was largely responsible for amassing the cabinet, following the usual practice of having some pieces locally made and buying others from the London trade.[21]

These, however, are only some of the cabinets that have remained virtually intact. Characteristic of them is the arrangement of the instruments in cases around the walls of a large room used for lecturing, each case often bearing the name of the category of instrument it contains. The rooms and the style of the instruments are appropriate to the period and of the prevailing style of architecture. The instruments of the Nollet type are particularly recognizable and have been described as finely finished in red and black *vernis Martin*, decorated with gilding. The rooms at Kremsmunster are in the baroque style. At Teyler's Museum in Haarlem, the elegant Oval Room, virtually unchanged from when it was first built, has arched, glass-fronted cases all around the walls for the apparatus, with bookcases above to hold the library, accessible from a gallery, and flat showcases through the center of the room, for mineralogical specimens.

All over Europe, however, there exist, already preserved in museums or even now being discovered, examples of eighteenth-century philosophical apparatus that originated in universities, colleges, learned societies, or the collections of individuals during the period appropriately named the Enlightenment. The George III collection in the Science Museum, London, has already been mentioned. The Deutsches Museum in Munich houses the collection of the Bavarian Academy, much of it made by the great Augsburg instrument-maker Georg Friedrich Brander (Figure 22.3).[22] There is a quantity of eighteenth-century demonstration apparatus in the Museo di Storia della Scienza in Florence. At the present time, all over Italy, scientific instruments used for teaching are being discovered in universities and schools. In Denmark, there was an interesting variation in the provenance of a cabinet. The apparatus amassed for his personal use in the last decades of the eighteenth century by Adam Wilhelm Hauch (1755–1838), soldier, administrator, and amateur of science, eventually found its way to Soroe Academy, an old-established school, where it was used for science teaching.[23] The cabinet's history has been studied, and the instruments restored, by a science teacher at the school in recent years.

[21] G. L' E. Turner, "Apparatus of Science in the Eighteenth Century," *Separata da Revista da Universidade de Coimbra*, 45 (1977).

[22] Alto Brachner (ed.), *G. F. Brander 1713–1783: Wissenschaftliche Instrumente aus seiner Werkstatt* (Munich: Deutsches Museum, 1983).

[23] A. W. Hauch, *Det Physiske Cabinet eller Beskrivelse over de til Experimental-Physiken henhörende vigtigste instrumenter tilligmed brugen deraf*, 2 vols. (Copenhagen, 1836, 1838).

G. F. Brander

1713–1783

Wissenschaftliche Instrumente aus seiner Werkstatt

Eröffnung der Ausstellung am 27. 9. 1983

Deutsches Museum

Figure 22.3. The German instrument-maker Georg Friedrich Brander (1713–1783) of Augsburg produced a wide range of instruments in his large workshop. Though not an innovator, he was a fine craftsman, and many of his instruments survive in museums, notably the Deutsches Museum in Munich, some of whose Brander pieces are illustrated here. Courtesy of the Deutsches Museum.

THE ROLE OF APPARATUS IN LECTURES

The lecture demonstration offered a new field of study for the educated and a source of excitement and wonder for the curious – similar to that provided by fireworks displays. But there was a practical benefit from the lecture demonstration. As David Brewster pointed out in the Preface to the second edition of the *Lectures* of James Ferguson (1710–1776), another highly successful practitioner, "We must attribute [to itinerant lecturers] the general diffusion of scientific knowledge among the practical mechanics of this country, which has, in great measure, banished those antiquated prejudices, and erroneous maxims of construction, that perpetually mislead the unlettered artist [artisan]."[24] The lecture demonstration proved beyond all question that the best way to teach science to those with no, or little, basic knowledge of it is by *showing* how it works.

What, then, were the topics dealt with in these lecture demonstration courses? They remained remarkably consistent throughout the eighteenth century, comprising mechanics, magnetism, astronomy, hydrostatics, pneumatics, heat, optics, electricity, and chemistry. Mechanics included the classical machines – levers, pulleys, balances – and some large set-pieces to show the parallelogram of forces, the rebound and trajectory of balls, cycloidal motion, and centrifugal forces. Some practitioners developed composite pieces of apparatus for ease of display, a good example of which is 's Gravesande's table of forces, to show the equilibrium of bodies. Also included under mechanics would often be models of practical devices that operate under the laws of mechanics – for example, the capstan, crane, pile driver, and varieties of mill. Magnetism was a popular wonder, and lodestones were held to possess magical properties, so examples of different types of magnets would be included, as well as the practical employment of magnetic attraction in the compass. If astronomy were included in the course, globes would be used, as would the astronomical telescope as well as, certainly in the latter part of the century, mechanical devices for displaying the movements of the heavenly bodies, such as the planetarium.

Hydrostatics and hydraulics attracted much practical attention in the eighteenth century because of the taste for fountains and elaborate garden designs using water. Demonstration pieces here would include the classical Hero's fountain, specific gravity experiments, suction and force pumps, and capillary attraction. The air-pump was one of the best known of all eighteenth-century demonstration instruments and would be used in the lecture on pneumatics, with many accessories for showing the effects it could produce. The effects of heat would be shown in measuring devices such as the thermometer and hygrometer and in instruments revealing the power of steam, notably the engine devised by Newcomen.

---

[24] John R. Millburn, *Wheelwright of the Heavens: The Life & Work of James Ferguson, F.R.S.* (London: Vade-Mecum Press, 1988). For the printing history of *Lectures on Select Subjects,* see pp. 285–7.

Optical instruments became the most popular instruments of the century for use in the home, few of which did not have a microscope or a telescope, to reveal the worlds of the very small and the very distant. Also for demonstration at an optics lecture would be distorting mirrors, prisms, and lenses. The eighteenth century was the age of static electricity, and the electrical machine equaled the air-pump and the microscope for popularity. Martinus van Marum had built in Teyler's Museum a giant electrostatic generator, with glass disks five feet six inches in diameter, but the main market was for small, portable machines. Chemical experiments were concerned mostly with the combustion of oils, phosphorus, and carbon and the oxidation of mercury.

The lecturers of the eighteenth century laid down a remarkably consistent pattern of design in demonstration apparatus that lasted through two and a half centuries. This is most obvious in mechanics, a study that goes back to classical times, but it is also clear in hydrostatics, hydraulics, and pneumatics, and is present, if not so extensive, in optics. Virtually all the demonstrations in mechanics used by Whiston at the premises of Francis Hauksbee in 1714, which are illustrated in six engraved plates in his *Course*, can be identified in the pages of the *Catalogue of Scientific Apparatus* issued by J. J. Griffin and Sons Ltd. in 1912. The same designs, although modernized and later made in plastic, appear in school textbooks of the 1960s and in educational toy catalogs today. From the predominantly middle-aged audience of the eighteenth century, through the undergraduates of the nineteenth century and the secondary school pupils of the twentieth century, the process of teaching the basic elements of science by demonstration is now continuing in primary schools and in the playroom.

What occurred in Europe during the eighteenth century was, quite simply, an outburst of interest in the working of the natural world, as revealed by means of experimental apparatus. One of Abbé Nollet's successful publications was a three-volume work titled *L'Art des Experiences, ou Avis aux Amateurs de la physique, sur le Choix, la Construction, et L'Usage des Instruments*. This is a practical guide on choosing, making, and acquiring apparatus for the uninitiated. The collections of instruments of experimental philosophy that have survived from the eighteenth century represent only a tithe of the huge quantity of such apparatus sold to private owners and institutions and constantly exchanged between individuals and at auction sales. This was the immediate effect of the immensely popular lecture demonstration. Its long-term legacy was to lay the foundations of the Industrial Revolution of the nineteenth century and of our own science-dominated age.

INSTRUMENTS IN SCIENTIFIC RESEARCH

This is not to suggest, however, that serious scientific research was lacking in the eighteenth century. Henry Baker (1698–1774) not only wrote popular books on the microscope but also read papers on microscopy to the Royal Society and received its Copley Medal for his work on the morphology of crystals.[25] Jean André de Luc (1727–1817), a Swiss who became tutor to the British royal family, contributed to meteorology, and Benjamin Wilson (1721–1788) studied static electricity, demonstrating at the Pantheon in London before King George III and arguing publicly with Benjamin Franklin on the subject of lightning conductors. This was essentially empirical research, but its impor- tance should not be underestimated. The studies of Wilson, and many others, on static electricity gave rise to galvanic electricity, the voltaic electric pile, and the experiments with current electricity eagerly taken up from 1800 on- ward. Another example of far-reaching empiricism was the development by John Dollond (1706–1761) of achromatic lens systems for the telescope. Tech- nical difficulties and the lack of any theory about the formation of the opti- cal image prevented his son, Peter Dollond, from extending the telescopic breakthrough to the microscope, but Dollond's achievement, together with the popular interest in microscopy, certainly accelerated improvements to the performance of objective lenses.

The patenting in 1758 of Dollond's achromatic lens combinations for the telescope was of particular importance because in the eighteenth century, positional astronomy had state patronage, which was spurred by the urgent economic need to improve ocean navigation. Prize money was on offer to anyone who could solve the problem of finding the longitude on board ship when out of sight of land; this feat was finally achieved by John Harrison with his portable marine chronometer, of 1761. State observatories had been built in England and France during the seventeenth century (Figure 22.4), and in the next century, astronomical activity across Europe and in North America was stimulated by the transits of the planet Venus across the Sun in 1761 and 1769. The improved accuracy of instruments made it possible to calculate the measure of the solar system from this phenomenon. At Harvard College, John Winthrop organized an expedition to observe the transit in Newfoundland in 1761, and Captain Cook led an expedition to the Southern hemisphere in 1769. The Russian Imperial Academy ordered a range of in- struments from London makers in 1768.

Equipping the many observing stations was a task that fell largely upon the London precision instrument trade. In a study of the eighteenth-century transits, Professor Harry Woolf listed the observing stations and, whenever possible, the telescopes used.[26] At the 1761 Transit, he records thirty-seven reflectors, three achromats, and sixty-seven unspecified; for the 1769 Transit, forty-nine reflectors, twenty-two achromats, and fifty-seven unspecified. To judge from the focal lengths given, most of those unspecified were refractors,

[25] Turner, "Henry Baker," p. 63.
[26] Harry Woolf, *The Transits of Venus: A Study of Eighteenth-Century Science* (Princeton, NJ: Princeton University Press, 1959).

Figure 22.4. The observer's room of the Radcliffe Observatory, Oxford, illustrated in Ackermann's *History of Oxford*. In 1763, Thomas Hornsby (1733–1810) succeeded the great astronomer James Bradley as Savilian Professor of Astronomy and was concerned that the University had no observatory. He petitioned the Radcliffe trustees for funds to build and equip an observatory, his proposals were accepted, and building began in 1773. The unusual design, modeled on the Tower of the Winds in Athens, was by James Wyatt, but the work was carried out by Henry Keene. The astronomical instruments were supplied in 1774 by John Bird. Those shown here are now in the Museum of the History of Science, Oxford, with the exception of the 10-ft Herschel reflector. Bodleian Library, University of Oxford.

probably nonachromatic. For achromatic refracting telescopes, Peter Dollond, inheriting his father's patent, was the prime supplier. For reflectors, which use polished metal mirrors rather than lenses, James Short (1710–1768), who described himself in trade directories as an "Optician Solely for Reflecting Telescopes," was the supplier chosen by many astronomers, to judge from the numbers of 24-inch and 18-inch telescopes he made in the 1760s. Short gave each instrument he made a serial number, and this has made it possible to put a figure on his production: 1,370 over thirty-five years.[27]

Unlike the telescope, the microscope was used throughout the eighteenth century largely for recreational purposes because of the aberrations that marred the image. The quality of glass available, the small size of the lenses needed, and the difficulty of identifying and communicating what was seen were factors that limited the instrument's scientific performance. It was not until the nineteenth century that the microscope achieved its full potential, following the work on lens systems of Joseph Jackson Lister. Nevertheless, the microscope was hugely popular. Simple microscopes for field use by naturalists, and compound models, their stands improved by the use of brass, for home observation, were made and sold in large numbers. What serious work was done was mainly in the fields of mineralogy, classification of plants and insects, and zoology.

The study of chemistry, as of natural philosophy, received an important impetus from the University of Leiden, where Herman Boerhaave (1668–1738) taught medical students in a course that included the preparation of drugs in a laboratory. Throughout the eighteenth century, there was much preoccupation with gases, the nature of "air," and the processes of combustion and oxidation. The century produced great chemists, Joseph Black (1728–1799) and Antoine Lavoisier (1743–1794), to name only two. The apparatus used by chemists for their research was, however, simpler and, being generally made of glass, less permanent than that used by physicists. Some chemical apparatus used for teaching has survived in, for example, the Teyler Museum and the Playfair Collection in the National Museums of Scotland,[28] but pre-nineteenth-century apparatus is rare and was generally made not by instrument-makers but by glass-making firms or mechanics attached to the laboratory. The exception is the chemical balance, which came into its own in the eighteenth century. Joseph Black, at Glasgow and then at Edinburgh, first carried out reasoned sets of chemical experiments in which the balance was used at every stage.[29] Lavoisier, in Paris, also stressed the importance of quantitative studies and the need for sensitive balances in the laboratory. These balances were made by the instrument-makers whom we shall be considering.

METHODS, MATERIALS, AND MAKERS

Scientific instrument-making developed from the art of engraving the metal plates used for printing. This manual skill was supplemented by a good measure

[27] G. L' E. Turner, "James Short, F.R.S., and his Contribution to the Construction of Reflecting Telescopes," *Notes and Records of the Royal Society of London,* 24 (1969), 91–108.

[28] R. G. W. Anderson, *The Playfair Collection and the Teaching of Chemistry at the University of Edinburgh 1713–1858* (Edinburgh: National Museums of Scotland, 1978).

[29] On Joseph Black, see Anderson, *Playfair Collection,* chap. 2, which has extensive notes. For an introduction to the balance, see John T. Stock, *Development of the Chemical Balance* (London: HMSO for the Science Museum, 1969).

of mathematical knowledge, and the two came together in Flanders in the sixteenth century, where they were notably practiced by the great mapmaker Gerard Mercator. A contemporary of Mercator left Louvain to settle in London, where he called himself Thomas Gemini. He made his name by engraving the plates for a fine edition of the classic work of Vesalius, whose *De humani corporis fabrica* was published at Basle in 1543. For this work, Gemini received a pension from Henry VIII, and he also engraved maps and made mathematical instruments. Gemini was the first among a group of instrument-makers who flourished in the reign of Queen Elizabeth I.[30] Many of these makers, employed in what was a new craft, found their place in the London guild structure by becoming free of the Grocers' Company, one of the Twelve Great Livery Companies.[31] By this means several master/apprentice dynasties were established that continued through the seventeenth and eighteenth centuries and into the nineteenth.[32] Many of the makers referred to later, including Thomas Heath, the Adams family over three generations, and the Troughtons, were free of the Grocers' Company.

A leading London maker who spanned the turn of the seventeenth and eighteenth centuries was Edmund Culpeper (1660–1738).[33] Culpeper was apprenticed to Walter Hayes, whose premises in Moorfields he took over just before 1700. A range of objects is still to be found bearing his signature – Edmund Culpeper Fecit – and among the more unusual are an engraved design for a dessert trencher and a memorial brass. His conventional products still extant in museums include rules, sectors, sundials, backstaffs, and surveying instruments. Culpeper began making small pocket microscopes about 1700, judging from a dated example, and many of these so-called Wilson screw-barrel microscopes have Culpeper's name engraved on them. Another type of microscope, a large tripod form customarily referred to as the Culpeper type, is thought to have been first produced in 1725. The attribution rests on the trade card, bearing Culpeper's sign of the crossed daggers, and his name as the engraver of the printing plate. It is possible that Culpeper decided to extend the range of articles that he sold and so took on first the Wilson-type microscopes that are made of turned ivory or brass and whose stands involve flat, folding feet constructed in the manner of rules or sectors. It is more likely, however, that Culpeper bought from other specialist tradesmen the large tripod microscopes that do not depend on any delicate workmanship apart from the lenses.

---

[30] G. L' E. Turner, *Elizabethan Instrument Makers: The Origins of the London Trade in Precision Instrument Making* (Oxford: Oxford University Press, 2000).

[31] J. Brown, *Mathematical Instrument-Makers in the Grocers' Company 1688–1800* (London: Science Museum, 1979).

[32] Gloria Clifton, *Directory of British Scientific Instrument Makers 1550–1851* (London: Philip Wilson, 1995). See also D. J. Bryden, *Scottish Instrument Makers 1600–1900* (Edinburgh: National Museums of Scotland, 1973).

[33] G. L' E. Turner, "Micrographia Historica: The Study of the History of the Microscope," *Proceedings of the Royal Microscopical Society*, 7 (1972), 120–49. Reprinted in G. L' E. Turner, *Essays on the History of the Microscope* (Oxford: Senecio Publishing, 1980).

Another craftsman who chose to diversify was George Lindsay (fl. 1728–1776) of the Strand, who was a watchmaker to King George III. Lindsay issued a proposal for a portable microscope in 1728 and patented it in 1743. It was said of him that he employed all the skill of the watchmaking trade to produce a microscope that would pack away into the space of a snuffbox.

At around the time of Culpeper's death in 1738, wood and leather became old-fashioned, and brass was increasingly used to make the bodies and stands of microscopes and telescopes and the frames of octants. Under the guidance of Henry Baker, John Cuff (1708–1772) made, in about 1743, a radical change in the design of the microscope, using an all-brass construction except for the wooden box foot.[34] It was said of Cuff that he was one of the best workmen of his trade in London and could make a microscope in two weeks. Nevertheless, by 1750 he had become bankrupt. Some said it was because he was too honest, whereas others said that he was too slow in producing the goods ordered.

The London instrument-making trade was, by 1700, a complex interlocking of specialist makers and retailers. A shopkeeper would sell his own products and would accommodate his customers by providing a comprehensive stock drawn from the rest of London. There were some who made only for the trade, such as Jack Dunnell (fl. 1680) who made vellum tubes for telescopes and microscopes, and John Morgan (fl. 1740), a brass worker. Others were only retailers, a leading eighteenth-century example being Benjamin Martin, whose early lecturing activity has already been referred to and who later ran a successful business in London selling scientific instruments of all kinds. In 1768 his shop in Fleet Street was visited by Jean Bernouilli, a Swiss-born astronomer from the Royal Observatory in Berlin. He described in his *Lettres astronomiques,* published in 1771, his travels to Germany, France and England, and he considered that Martin's shop was one of the best equipped and notable for the lecture demonstrations given there by Martin.[35] The catalog of the auction sale of Martin's effects after his death in 1782 shows the comprehensive range of instruments he sold: spectacles, opera glasses, optical toys, telescopes, microscopes, instruments for surveying and navigation, sundials, drawing instruments, air-pumps, electrical machines, planetaria, clocks, barometers, thermometers, gunners' gauges, and so on. The range of his stock and his efficient publicity make it unsurprising that he was chosen to supply Harvard College with replacement teaching apparatus after the fire of 1764.

Benjamin Martin went bankrupt in his later years, possibly through over-diversification. James Short, a Scot who confined himself to making reflecting telescopes with consummate skill, died worth a fortune. He concentrated on the careful polishing and matching of the speculum metal mirrors of the

---

[34] Turner, "Henry Baker," pp. 63–4.
[35] Jean Bernouilli, *Lettres Astronomiques où l'on donne une idée de l'état actuel de l'astronomie practique dans plusieurs villes de l'Europe* (Berlin, 1771), pp. 72–4.

telescopes, and the excellence of his instruments led to his election to Fellowship of the Royal Society at the age of twenty-six and to his being a candidate for the office of Astronomer Royal in 1764. Short died a few months before Bernouilli arrived in London, but the latter attended the auction sale of Short's stock, to which he devoted an entire letter. It is interesting to note that Short asked, and had no difficulty in getting, more than twice the usual price for his telescopes. For a 24-inch reflector with rackwork, Martin asked fourteen guineas, Henry Pyefinch sixteen guineas, George Adams twenty guineas, and Short thirty-five guineas.[36] With such a range of quality suppliers, it is not surprising that London was the world's marketplace for precision instruments.

Bernouilli, an astronomer, was specially interested in the mural quadrants and sectors made by the specialists, Graham and Bird. George Graham, who had a workshop in Fleet Street from 1713 until his death in 1751, was originally a clockmaker and had been trained by the great Thomas Tompion. Graham turned his skill to astronomical and observatory instruments, doing much to improve their accuracy.[37] He is known to have sold transit and zenith instruments and astronomical clocks. He was made a Fellow of the Royal Society in 1720. John Bird, a younger man, worked with both Graham and Jonathan Sisson and made the first dividing engine to allow for variations caused by changes of temperature. Bird supplied instruments to the Royal Greenwich Observatory and other observatories across Europe[38] and worked for the Board of Longitude. Sisson, too, was singled out for mention by Bernouilli, as was Jesse Ramsden. Born in 1735, Ramsden was a leading London instrument-maker and retailer throughout the second half of the eighteenth century, with premises in Piccadilly. He became a Fellow of the Royal Society in 1786 and was an author and inventor as well as a businessman. He had close links with other instrument-makers and was married to Sarah, sister of Peter Dollond.

Other European scholars kept diaries of their travels and included descriptions of visits to London instrument-makers.[39] Marc Auguste Pictet, professor of experimental philosophy at Geneva, visited London in 1801 and wrote enthusiastically of the excellence of the dividing engine and other instruments by Edward Troughton FRS (1753–1835).[40] Here we meet another instrument-making family, for Edward and his brother, John, were trained by their uncle, also John Troughton (1716–1788), and were in business together until

[36] Turner, "James Short," p. 101.
[37] J. A. Bennett, *The Divided Circle: A History of Instruments for Astronomy, Navigation, and Surveying* (Oxford: Phaidon-Christie's, 1987). See chap. 7 on observatories.
[38] V. L. Chenakal, "The Astronomical Instruments of the Seventeenth and Eighteenth Centuries in the Museums of the U.S.S.R.," *Vistas in Astronomy*, 9 (1968), 53–77.
[39] G. L' E. Turner, "The London Trade in Scientific Instrument-Making in the Eighteenth Century," *Vistas in Astronomy*, 20 (1976), 173–82.
[40] Marc Auguste Pictet, *Voyage de trois mois en Angleterre, en Ecosse, et en Irlande pendant l'Eté de l'an IX (1801 v. st.)* (Geneva, 1802).

1804.[41] Edward continued the firm after his brother's retirement, and the firm of Troughton & Simms operated throughout the nineteenth century and well into the twentieth century as Cooke, Troughton & Simms. The Danish astronomer Thomas Bugge visited London in 1777 and kept a diary illustrated with his own drawings.[42] He bought instruments costing more than £88 from the firm of Nairne & Blunt. Edward Nairne (1726–1806) had premises in Cornhill from 1749 that he took over from Matthew Loft, and he numbered Jesse Ramsden among his workmen. Nairne received his FRS in 1776 and a royal appointment to King George III in 1785, and he was in partnership with Thomas Blunt between 1774 and 1793. Nairne & Blunt advertised the full range of scientific instruments and described themselves as "Optical, Mathematical and Philosophical Instrument Makers."

One other London instrument-making family must be mentioned, as indeed they are by the traveling diarists: the Adamses.[43] George Adams the elder was born in 1709, learned his craft as apprentice to Thomas Heath, and ran a well-known business at the sign of Tycho Brahe's Head in Fleet Street. Adams knew all about the advantages of advertising by writing and lecturing and was particularly well known for his globes, although he sold the full range of instruments. He was succeeded by his elder son, also George, who lived only until the age of 45; the younger son, Dudley, took over, and he continued the highly successful business until 1817 (Figure 22.5). George Adams Jr., who succeeded to his father's royal appointment, supplied many instruments to the Dutchman Martinus van Marum, who was engaged, during the final decades of the eighteenth century, in building an extensive collection of instruments for teaching and research at the Teyler's Foundation in Haarlem.

It becomes clear from this brief account of only some of the leading London instrument-makers of the eighteenth century that these were men of intelligence and education, capable of writing about their work, and willing and able to innovate. They were recognized as distinguished members of the scientific community and of society in general. The enthusiasm with which European visitors to London described the shops, the range and quality of the instruments, and the knowledge and ability of their owners confirms this assessment. It may well have been the case, however, that in the second half of the century, London was so well supplied with precision instrument-makers that a young man at the end of his apprenticeship might look abroad for business opportunities. One man who trained in London but decided to emigrate to Amsterdam was John Cuthbertson (1743–1821).[44] The move brought

[41] A. W. Skempton and J. Brown, "John and Edward Troughton, Mathematical Instrument Makers," *Notes and Records of the Royal Society of London,* 27 (1973), 233–62.

[42] Thomas Bugge, travel diary August–December 1777 (Copenhagen, Kongelige Bibliotek, MS Ny kgl. Saml. 377e).

[43] John R. Millburn, *Adams of Fleet Street, Instrument Makers to King George III* (Aldershot: Ashgate, 2000).

[44] W. D. Hackmann, *John and Jonathan Cuthbertson: The Invention and Development of the Eighteenth-Century Plate Electrical Machine* (Leiden: Museum Boerhaave, 1973).

Figure 22.5. A trade card of Dudley Adams (1762–1830), younger son of George Adams Senior, who took over the family business at the premises at "Tycho Brahe's Head" in Fleet Street, London, on the early death of this elder brother, George, in 1795. Science Museum/Science & Society Picture Library.

success, for he was considered to be the best maker of electrical machines and air pumps in Europe. He made the largest of all plate electrical machines: that in the Teyler Museum in Haarlem. Much of his business depended on the patronage of the Teyler Foundation, and following a disagreement about the design of friction pads for electrical machines with his patron, Van Marum,

Cuthbertson returned to London in 1793 and opened a shop there. His brother, Jonathan, also started an instrument-making business in Holland, at Rotterdam, and never returned to England.

THE INSTRUMENT TRADE IN
EUROPE AND NORTH AMERICA

To break into the London instrument trade if one were a foreigner and lacked local backing was not easy. Jacob Bernhard Haas (1753–1828) was born at Biberach, in southern Germany, became a skilled instrument-maker, and settled in London. He found needed capital and someone willing to help him in the person of Johann Heinrich Hurter, a Swiss painter who achieved success at the English court, and the two men established what was described as "a manufactory of philosophical instruments." Haas made several pieces, including a fine chemical balance, for Van Marum, and for three years, from 1793 to 1795, was in formal partnership with Hurter, signing his instruments with both names. In 1791, Haas tried unsuccessfully to obtain the post of mechanic to the Teyler Foundation, explaining in his letter how difficult it was to succeed in London. He finally went to work in Lisbon as head of the instrument-making workshop of the Portuguese Admiralty.

Holland was a center for instrument-making second only to London,[45] due in considerable measure to the influence of Leiden University, where lectures in experimental philosophy were established from 1675. The instrument-maker who supplied much of the apparatus for these lectures was Jan van Musschenbroek (1687–1748); the Musschenbroek workshop has been studied in detail by Peter de Clercq.[46] The business, at the sign of the Oriental Lamp, was established in the Rapenburg, Leiden's great university street, by Jan's uncle, Samuel, in the 1660s. It was continued by Johan van Musschenbroek, Samuel's younger brother, and then by Johan's son of the same name, generally known as Jan. Although the Musschenbroeks had a reputation for the high quality of their instruments, there is no evidence that their workshop was a large one. Indeed, apart from the family, only one man seems to have been employed. This was Anthony Rinssen, who, when he set up his own business, advertised that he had been Jan's assistant for twenty years. This concentration on a bespoke (custom) trade by a single skilled craftsman, with few assistants, was the norm in the first half of the eighteenth century, but that changed after 1750, when the great London instrument firms, and also those in Germany and France, employed many skilled artisans, sometimes as many

---

[45] For a preliminary listing, see Maria Rooseboom, *Bijdrage tot de Geschiedenis der Instrumentmakerskunst in de noordelijke Nederlanded tot omstreeks 1840* (Leiden: Rijksmuseum voor de Geschiedenis der Natuurwetenschappen, 1950).

[46] de Clercq, *At the Sign of the Oriental Lamp.*

as fifty. Jan Paauw (c. 1723–1803), also of Leiden, was among the instrument makers whom Thomas Bugge met on his tour of 1777, when he visited Holland as well as England. Paauw was a graduate of Leiden University and took the degree of doctor of philosophy at the University of Franeker. In the second half of the eighteenth century there were few Dutch collections of scientific instruments, public or private, that did not contain apparatus from his workshop. In Amsterdam, Bugge met the instrument-makers Adam Steitz and Jan van Deijl, the latter certainly because of his reputation as a maker of optical instruments. Jan and his son, Harmanus van Deijl (1738–1809), were in business together, and Harmanus is credited with being the first to produce commercially an achromatic microscope, which was described in 1807.[47] The instrument followed the conventional pattern of the day and had two sizes of objective with focal lengths of 20 and 30mm. They were composed of a plano-convex flint-glass lens and a bi-convex crown-glass lens; the plane surface of the first turned toward the object to be viewed. The achievement was in making the required lenses in a size small enough for use in the microscope. Another notable Dutch instrument-making family, which existed through three generations was that of Jan van der Bildt (1709–1791) of Franeker. Van der Bilt lived and worked there all his life, achieving a reputation for the construction of telescopes.

Germany did not become a united national state until the end of the nineteenth century. The numerous separate principalities in the eighteenth century did not create the political, economic, and social conditions for many precision instrument workshops to flourish. Of those that did exist, the most notable was that of Georg Friedrich Brander (1713–1783) of Augsburg.[48] Born in Regensburg, Brander studied mathematics at the University of Altdorf and in 1737 founded his workshop, which produced every type of philosophical, mathematical, and optical instrument. His inventive skill was in the construction of glass micrometers for use in microscopy and astronomy, for which he designed his own ruling machine, starting production in 1761. The design of his microscopes was, however, much influenced by those of the London makers, and his interest in microscopes was stimulated by the publication in German of Henry Baker's book *Employment for the Microscope,* in 1753. Brander supplied many of the instruments for the Bavarian Academy, now housed in the Deutsches Museum in Munich. Another notable German instrument-maker of the eighteenth century was Johann Christian Breithaupt (1736–1799), of Kassel.[49] In Vienna, a successful business was established by Johann Christoph Voigtlander (1732–1797). German instrument-making produced its genius in the person of Joseph von Fraunhofer (1787–1826), whose inno-

---

[47] Turner and Levere, *Van Marum*, pp. 301–2.   [48] Brachner, *Brander.*
[49] L. von Mackensen, *Feinmechanik aus Kassel: 225 Jahre F. W. Breithaupt & Sohn, Festschrift und Ausstellungsbegleiter* (Kassel: Georg Wenderoth Verlag, 1987).

vations in glass production made possible the German triumphs with optical instruments during the nineteenth century.[50]

In France, too, the economic and social climate was less favorable than in England to innovation in precision instrument-making until the last decades of the eighteenth century.[51] There were, however, some notable makers. Jacques and Pierre Lemaire, father and son, were active from 1720 to 1770, making a variety of mathematical and navigational instruments. The most important workshop of this period was that of Claude Langlois, who was in business from 1730. He became engineer to the *Académie des sciences* and had premises in the Louvre. He made instruments for the Paris Observatory, including two six-foot mural quadrants, and quadrants and sectors for expeditions to Peru and Lapland. Langlois's business after his death was taken over by his nephew, Canivet, who continued to make astronomical instruments of high quality. In the middle of the century, optical instruments were made in France that were remarkable for their fine and elaborate workmanship, but they were not innovative; they have been described as "salon pieces." The best-known makers were Claude Paris, who produced reflecting telescopes like those of the London maker, Scarlett; and Claude Siméon Passemant (1702–1769), an astute self-publicist and a fine craftsman. Passemant adopted John Cuff's design for the compound microscope but produced many variants.

In the last quarter of the eighteenth century, a determined effort was made to promote French science, and two important workshops emerged: those of Lenoir and Fortin. Etienne Lenoir (1744–1832) established a reputation as the best maker of navigational and observatory instruments.[52] He made the prototype Borda reflecting circle, and his instruments were used to measure the meridian. Throughout the Revolution and into the empire, "Citizen Lenoir" presided over the most important precision instrument workshop in Paris and was able to pass it on to his son in the new century. Nicolas Fortin (1750–1831) made his reputation as instrument-maker to Lavoisier, for whom he made balances, thermometers, and apparatus for oil combustion. He also worked for Gay-Lussac, and the jury of the 1819 Paris Exhibition commented, "Devoting his energies especially to the construction of physical instruments, Fortin, by his talent, made possible the work of the French physicists which has revolutionized the science of physics and created modern chemistry." Fortin's name is generally associated today with the type of barometer that he invented.

---

[50] Alto Brachner, *Mit den Wellen des Lichts: Ursprünge und Entwicklung der Optik im süddeutschen Raum* (Munich: Olzog Verlag, 1987).

[51] Maurice Daumas, *Les instruments scientifiques aux XVIIᵉ et XVIIIᵉ siècles* (Paris: Presses universitaires de France, 1953. English translation by Mary Holbrook, London: Batsford, 1972).

[52] A. J. Turner, *From Pleasure and Profit to Science and Security: Etienne Lenoir and the Transformation of Precision Instrument-Making in France 1760–1830* (Cambridge, MA: Whipple Museum of the History of Science, 1989).

Although the vast majority of scientific apparatus used for experiment and teaching in the American colonies during the eighteenth century was imported from Europe – the re-equipping of Harvard College by Benjamin Martin has been referred to – mathematical instruments for the practical purposes of surveying and navigation were being produced by American makers. Because it was readily available, many of the surveying instruments were made of wood, and not of brass as they would have been in Europe at the time. There were some immigrant makers, such as John Dabney, who came to Boston from London, and Anthony Lamb, also apprenticed in London, who settled in New York.[53] But it was not until the nineteenth century that a flood of immigrants from Europe came to the United States in search of business opportunity.[54]

Two of the most important mathematical practitioners in North America were the Rittenhouse brothers of Philadelphia: David (1732–1796) and Benjamin (1740–c. 1820). David Rittenhouse used instruments of his own design and construction to survey the boundary between Pennsylvania and Delaware in 1763, and in 1770 he built and equipped the first American astronomical observatory in Philadelphia. Benjamin was superintendent of the government's gunlock factory and made clocks and surveying instruments. Another notable surveyor and maker of surveying instruments was Andrew Ellicott (1754–1820), who, in 1801, was offered the position of Surveyor General of the United States by President Jefferson. Ellicott employed as his assistant in survey work Benjamin Banneker (1734–1806), a free Negro who was self-educated and achieved fame as the producer of astronomical almanacs. Other notable American instrument-makers were James Wilson of Vermont, the first native globe maker, and Joseph Pope of Boston, who constructed a superb orrery for Harvard University.

A SCIENTIFIC COLLABORATION

Because science was popularized for the first time in the eighteenth century and because for some people it was a source of entertainment, it is often assumed that little serious scientific work was done. This was not the case. The popularization was part of a general educational process. People seek education for a variety of reasons that include curiosity and distraction as well as the serious desire to learn. It must also be remembered that the gulf between amateur and professional did not then exist as it does today. Many of those who were interested in natural philosophy – a more contemporary term than

---

[53] Silvio A. Bedini, *At the Sign of the Compass and Quadrant: The Life and Times of Anthony Lamb,* Transactions of the American Philosophical Society, 74, pt. 1 (1984).

[54] Silvio A. Bedini, *Early American Scientific Instruments and Their Makers,* United States National Museum Bulletin 231 (Washington, DC, 1964. Reprinted, with addenda to Bibliography, Rancho Cordova, CA: Landmark Enterprises, 1986).

"science" – did not earn their livelihood by its study, so it is assumed that they were unlikely to produce a serious addition to knowledge. Scientific knowledge grew during the century, and the foundations were laid for the huge technological advances of the Industrial Revolution. The fact that the new science had a broad base of popular interest achieved two results. First, the instrument-making trade was so well supported that it was capable of producing new research instruments. Second, the pressure of popular interest stimulated patronage for science from governments, learned societies, and wealthy individuals, providing the power to drive the engine of scientific research in this very scientific century.

23

PRINT AND PUBLIC SCIENCE

Adrian Johns

CULTURES OF PRINT AT THE
ONSET OF ENLIGHTENMENT

During the eighteenth century, natural knowledge became the focus, the vehicle, and the archetype of public enlightenment. This chapter describes some of the most important conditions underpinning that development. Its central subject is a distinctive realm of print that matured toward the end of the seventeenth century and lasted until the first quarter of the nineteenth – a realm differing in important respects from anything that had existed before. The chapter explains its principal characteristics, showing how they came about and why in the end they proved unstable. It outlines how printed materials were made, circulated, and put to use. From there it proceeds to explain how the features of this realm affected the creation and distribution of knowledge. The materials created by printers and booksellers – not only books themselves but also new objects such as periodicals – substantially changed the construction and representation of knowledge. The chapter's major claims in this regard are of a general character. They are certainly applicable to what we would now call science; but they also extend far beyond that, and encompass knowledge of many other kinds.

The world of the book in the eighteenth century was simultaneously uniform and various. On the one hand, the régimes of custom and regulation guiding the conduct of printing and publishing in most countries rested, to a greater or lesser extent, on similar mechanisms of guilds, licensing, patronage, and privileges. In France, for example, Louis XIV's reign saw the establishment of a comprehensive system of press regulation based on these foundations that would last until the revolution a century later. Similar systems were likewise developed in Spain, Austria, Sweden, and Denmark. The German lands embraced them too, as did the Dutch. Broadly parallel institutional forms existed in all these nations. But on the other hand, all these parallel mechanisms were legally and politically distinct. The specific decisions reached by any particular régime were of no force beyond the borders of its jurisdiction. A printer familiar with the conventions in one region would probably recognize the practical customs reigning in another, but the accumulated archive of past decisions on which the book trades depended would be completely unknown to him or her. One consequence was that unauthorised printing in a territory's neighboring regions – a perfectly legal activity – became a persistent source of complaint in many regions of Europe. Such complaints had considerable effects on the perceived character of print.[1]

The situation in English-speaking countries was significantly different. In early eighteenth-century England, contemporaries perceived that a profound change was occurring. After sustained criticism from radical Whigs such as John Locke, the Restoration regime's Press Act was allowed to lapse in 1695 for what proved to be the last time. This had three major effects: it ended prepublication oversight by agents of the government or established church; it removed all statutory sanction against the unauthorized printing of works already claimed by others; and it permitted the unregulated expansion of the printing trade into the provinces. In all three respects it reduced the chances of creating a comprehensive system like that of most Continental powers, based on a guild (in this case the Stationers' Company), licensers, and royal privileges. During the same years, the printing industry in America expanded rapidly from a very small base to become a large and influential trade spread across the colonies – a trade that was too diverse for effective regulation by government. The pressures created by such changes would call forth redefinitions of print culture itself, from which the rest of Europe could not stand aloof for long. With them would come revisions of readership and authorship. And at the beginning of the nineteenth century the character of printed communication would change radically once more, with the invention and eventual introduction of steam-powered industrial presses.

According to some of our most commonly held assumptions, these statements would seem to make little sense. We think we know what we mean by "print culture," and we can identify it by reference to what we are sure we know about print itself. The press creates large numbers of identical texts: that is its very definition. The characteristics of a print culture derive from this essentially technological cause. They include a reduction in the cost of books and an increase in their availability. Above all, the mechanical repetition central

[1] M. Treadwell, "The History of the Book in Eighteenth-Century England, Ireland, and America," *Eighteenth-Century Life*, 16 (1992), 110–35; J. Popkin, "Print Culture in the Netherlands on the Eve of the Revolution," in M. C. Jacob and W. W. Mijnhardt (eds.), *The Dutch Republic in the Eighteenth Century: Decline, Enlightenment, and Revolution* (Ithaca, NY: Cornell University Press, 1992), pp. 273–91; J.-P. Lavandier, *Le Livre au Temps de Marie-Thérèse* (Berne: P. Lang, 1993); S. G. Lindberg, "The Scandinavian Book Trade in the Eighteenth Century," in G. Barber and B. Fabian (eds.), *Buch und Buchhandel in Europa im Achtzehnten Jahrhundert* (Hamburg: Hauswedell, 1981), pp. 225–48; D. M. Thomas, *The Royal Company of Printers and Booksellers of Spain, 1763–1794* (Troy, NY: Whitston, 1984); L. Domergue, "Les Freins à la Diffusion des Idées Nouvelles," in B. Bennassar (ed.), *L'Espagne: de l'Immobilisme à l'Essor* (Paris: CNRS, 1989), pp. 145–55.

to the practice of printing means that printed texts display a uniform fidelity inconceivable in manuscript reproduction. In all these respects, print culture seems at root a matter of machinery. Accordingly, there can only ever be one such culture, manifested to greater or lesser extents in different times and places.[2]

If that were true, then talk of new and diverse print cultures would be virtually oxymoronic. Yet the assumption that print culture is straightforwardly derivative of the press is not beyond challenge. It seems clear enough that the press, like other technological devices, can be subjected to different practical uses, and that as a result it can generate a variety of cultural consequences. This seems to have been the case in early modern Europe. A region divided by profound rifts in politics, geography, and religious allegiance produced a corresponding variety of regimes of print, ranging from the relative laxity of London to the strict regulation of Madrid or Naples. The production, circulation, consumption, and reading of printed materials differed across those regimes. What resulted, certainly by the late seventeenth century if not long before, was a number of distinct and local print cultures. And we can even argue that the articulation of some of the most characteristic concepts of Enlightenment depended on this being so. During the eighteenth century, attempts to secure civil collaboration across those different cultures combined and conflicted powerfully with developing arguments within them. It was this process that gave rise, not so much to a harmonious "public sphere" as to contemporaries' strenuous assertions that such a sphere *must* exist. That is, it provoked their insistence that there must be one true culture of print and that that culture was the only one capable of supporting true Enlightenment.[3]

The major initial components of the process can be clearly identified in the example of England, where the refusal to follow Continental regulatory norms laid bare their principal advantages and drawbacks. The Press Act had first been passed in 1662, shortly after the Restoration of Charles II. Charles's and James II's governments had prized it as a valuable aid in the struggle against sedition. With its final lapse ended a regime for overseeing printed publications that had persevered, on and off, for more than a century. The Act had also enshrined in law the central protocols of the Stationers' Company. This Company, chartered in the mid-sixteenth century, was in many ways representative of the craft organizations or guilds to be found in most major European centers. It purported to embrace all men and women involved in the book trades, and over a long period it had successfully created a complex

---

[2] The definitive argument along these lines is E. L. Eisenstein's *The Printing Press as an Agent of Change: Communications and Cultural Transformations in Early Modern Europe*, 2 vols. (Cambridge University Press, 1979).

[3] For a comparative survey of international trade networks that shows the interaction of booksellers as far apart as Lisbon and Warsaw (with a notable gap in Spain), see G. Barber, "Who Were the Booksellers of the Enlightenment?" in Barber and Fabian (eds.), *Buch und Buchhandel*, pp. 211–24.

set of practices for maintaining an ordered realm of print. It had its own court, its own maxims of judgment, its own searchers licensed to enter any premises where book-trade practices were believed to be conducted, and even, at one time, its own prison cells in which to intern offenders against Company conventions. The Company regulated the size, number, and constitution of printing houses – and in doing so it affected the size, number, and constitution of their products. In short, it created, defined, and exemplified a print culture. With the end of governmental oversight of the press in 1695, what contemporaries perceived to be ending was not just the practice of licensing but this culture in its entirety. But only in Britain did it end. In other European countries, including Holland, similar corporate and governmental cooperation persevered.

The resulting British experience highlighted the centrality to craft culture, in English and Continental companies alike, of the concept of *propriety*. The word now means civility, and in fact this is not an inappropriate allusion. But in eighteenth-century English it was far more commonly employed as a synonym for *property*. And this is what it meant in the context of printing and bookselling. In the book trade, it had long been accepted that a "register" recorded what one may, with briefly permissible anachronism, call literary property. But in fact this was scarcely property by modern standards at all. True, it was supposed to persist indefinitely, and it could be mortgaged, alienated, and inherited. But control over its transfer continued to reside within the craft community, and explicit legal recognition remained lacking. It was effectively an artifact of craft civility, bolstered by the company's alliance with the crown.[4] This was where the lapse of the Press Act ended up mattering most. Seeking to identify for every published book an individual who could be deemed responsible for its content, Parliament had in 1662 decreed that all books must be not only licensed but also registered. In doing so it had provided legal recognition for craft propriety, albeit rather by default. In 1695, however, with licensing disappeared this legal recognition of craft propriety. The unauthorized reprinting of registered titles became unpunishable. A kind of anarchy seemed to be in prospect. Not everyone saw it that way, of course: Locke resented the existing property conventions as monopolistic and intellectually hobbling. But some new practice securing probity in print would have to be forged. The major result – one achieved after a long struggle lasting most of the eighteenth century – was a very different regulating concept, based not on custom but on authorial labour and its legal recognition. In Britain, first, this was reflected in the neologism of "copyright." Other countries persevered longer with proprietorial concepts resting on guild and royal prerogative, but by the end of the century similar regimes were being mooted in most European regions. Natural philosophical work bore a close

---

[4] Compare the representation of similar conventions in the Netherlands: P. G. Hoftijzer, "'A Sickle unto thy Neighbour's Corn': Book Piracy in the Dutch Republic," *Quaerendo*, 27 (1997), 3–18, especially 6–9.

and interesting relation to this debate. That relation is one of the more consequential elements of the history of science during the period.

This chapter tells the story, then, of a period that has been called the typographical *ancien régime*.[5] In the French context, we can identify this period as running from the creation of a comprehensive press regime by Colbert in the 1660s, through its wholesale (and, for the printed book, disastrous) destruction by the Jacobins in the 1790s, to the reconstruction of publishing as an industry in the early nineteenth century.[6] In Germany and the Habsburg Empire, it similarly extends to the elaboration of literary property regimes in the early 1800s. Dutch laws against piracy also took effect from the end of the century. In England, the term is if anything even more apt but in a different sense. The period can be said to extend from the lapse of the Press Act in 1695, which effectively ruled out the development of a regime of print on a Continental model, to the consolidation of industrial publishing, which began around 1810 and was well advanced by the 1830s. During this period, the machinery of the printing press remained much as it had for the previous 150 years; the cultural changes experienced by contemporaries were, however, remarkable. What kinds of books were made, how they were represented, what it was to become an author – all were very different in 1800 from what they had been in 1660.

PROPERTY AND PIRACY IN THE PRODUCTION OF ENLIGHTENMENT

In the sixteenth and seventeenth centuries, printing was predominantly a metropolitan enterprise. Overseen by the local guild – that of St. Luke in the Netherlands, for example, of St. Jacques in Paris, or St. John in Madrid – was a varying number of small, normally domestic printing houses; an unknown number of "private" presses scrabbled for subsistence by producing piracies and radical pamphlets. In fact, it was rather a hand-to-mouth economy even for the licensed master printers. The quotidian survival of most printing houses could be secured only by producing tickets, bills, and pamphlets, and even those printers engaged in larger projects such as atlases or bibles would think little of interrupting work to print such ephemera for ready cash. The exception was in the Low Countries, where printing houses were sometimes much larger. This had long been the center of the European trade. Premises here could afford to dedicate presses to long-term enterprises, and their rel-

---

[5] R. Chartier, "l'Ancien Régime Typographique: Réflexions sur Quelques Travaux Récents," *Annales E.S.C.*, 36 (1981), 191–209.

[6] An appropriate template, vol. 2 of Roger Chartier and Henri-Jean Martin, *Histoire de l'Édition Française*, 4 vols. (2nd ed.; Paris: Fayard, 1989–91), covers the period 1660–1830. For the origins of the system, see H.-J. Martin, *The French Book: Religion, Absolutism, and Readership, 1585–1715*, trans. P. Saenger and N. Saenger (Baltimore, MD: Johns Hopkins University Press, 1996), pp. 31–53.

atively easy access to a huge Continental market meant that they could employ economies of scale. Dutch wholesaling booksellers such as the Wetsteins would complain to their London counterparts that English books, in particular, were far too expensive to import; so they would happily reprint their own versions against the futile objections of the London Stationers. Amsterdam thus maintained its role as the center of European learned publishing, but at the same time could be construed by others as a centre of "piracy," too. And it was here that the new social identities on which a transnational learned community depended – the editor, the international publisher, and something very like the literary agent – first came into being.[7]

Most European countries maintained that the state had a legitimate duty to regulate such a potent enterprise. They developed a variety of systems of licensing to achieve this end. All books to be published had first to be vetted by a representative of the state or church. Only Britain discarded such a regime before 1700, and even there printers and booksellers spent years trying to re-institute it because of its role in limiting the size of the trade and guaranteeing its political safety. But licensing systems, like those of craft propriety, extended no further than the boundaries of any given legislation. Organizations like the Société Typographique de Neuchâtel, in Switzerland, could thus print with impunity books banned in France, and earn large profits by transporting them across the border and across the French regions. The activities of such organizations helped to define some of the major characteristics of Enlightenment. And even within particular jurisdictions, enforcement varied widely: in France, the chief licenser, Malesherbes, actually connived at the circulation of the nominally illicit *Encyclopédie*.

Another major similarity between different realms was the perceived prevalence of "piracy." By this term contemporaries meant more than just verbatim reprinting. It referred to a number of distinct practices, all of which impinged directly on the role of print as a reliable medium for the communication of information. If a printer were engaged to print one thousand copies of a work, for example, then simply by printing another two hundred he could expect to garner substantial extra profit. This was certainly represented as a real hazard by such figures as John Flamsteed, who accused Isaac Newton of collaborating with bookseller Awnsham Churchill to produce supernumerary copies of Flamsteed's *Historia Coelestis*. And the Royal Society took care expressly to forbid its "printers" (in fact, they were booksellers) from practicing supernumerary printing. Other claims of piracy focused on imitation, translation, or abridgment. Again, the Royal Society stipulated that its printers not reprint any work "in epitome." Translation was altogether less easy to restrain, since it often took place in foreign countries and thus under different legal regimes. The Society found that several important publications, including the

---

[7] A. Goldgar, *Impolite Learning: Conduct and Community in the Republic of Letters, 1680–1750* (New Haven, CT: Yale University Press, 1995).

Philosophical Transactions, were imperiled by unauthorized translations published in Geneva and the Netherlands. (The same fate was visited on the *Journal des Sçavans.*) Finally, hacks embraced flexible authorial personae including those of living rivals. Henry Fielding, like the Scriblerians and William King, issued imitations of the *Transactions,* and, in an age of unauthorized translations, could persuade some readers to believe them genuine. There was little any academy could do to prevent this altogether. By 1731, the Royal Society in London was even offering a reward to anyone who could name the author of printed reports purporting to reveal Society proceedings that had never in fact occurred. And the Society was repeatedly traduced in almanacs, despite trade grandees' own efforts to track down offenders. Faced with this multiplicity of hazards, the virtuosi found themselves forced repeatedly to reconsider their protocols throughout the century.[8]

A piracy could be anything from a backstreet robbery to a sophisticated international enterprise. It could not easily be countered by a writer acting in isolation.[9] An academy needing to use print might do better, however, and institutions such as the Royal Society of London and the Académie Royale des Sciences in Paris pioneered attempts to do so. Their initiatives – the appointment of privileged printers whose conduct was strictly delimited, the development of periodical publications, the articulation of protocols for editing, and the articulation of polite reading conventions – formed the model for those adopted by other academies in countries spread across Europe. They succeeded by enrolling the printers as allies even as they redefined their civility: in Sweden, for example, the Royal Academy of Sciences' privileged printer became the most powerful individual in the country's book trade.[10] It is notable that these were actions that learned societies *had* to take, and that they had real effects on the fortunes of participants in the book trades.

The world of print that emerged in the early eighteenth century was thus only potentially a reliable medium for learned communication. In actuality, the authenticity of printed materials was extremely fragile. The major printers and booksellers themselves saw the need to bolster it. Across Europe, guild conventions were losing their efficacy – and occasionally, as in revolutionary France, even their legal recognition. Copyowners reasonably feared that their valuable investments would be rendered worthless. They responded in two

---

[8] Richard Savage, quoted in R. Straus, *The Unspeakable Curll* (London: Chapman and Hall, 1927), p. 43; British Library, ms. Add. 4441, fols. 2ʳ-5ᵛ, 20ʳ, 120ʳ-121ᵛ; Royal Society, Ms. Dom. 5, No. 23; N.B. Eales, "A Satire on the Royal Society, dated 1753, attributed to Henry Fielding," *Notes and Records of the Royal Society,* 23 (1968), 65–7; Archive of the Stationers' Company, London: Supplementary Documents, Box D, Envelope 15 (a list of faults in Thomas Carnan's almanacs, 1776, including "Reflections on the Royal Society," "scurrilities of all kinds" against it, a "long and illiberal Reflecion" on the Society, and an account of a comet that constitutes "a nasty ludicrous performance," reflecting on the Society and the Astronomer Royal).

[9] See especially Hoftijzer, "'A Sickle unto thy Neighbour's Corn,'" and F. Moureau (ed.), *Les Presses Grises: La Contrefaçon du Livre (XVIe-XIXe siècles)* (Paris: Aux Amateurs de Livres, 1988).

[10] Lindberg, "Scandinavian Book Trade," pp. 230–1.

principal ways. One was to propose repeated statutory measures in an effort to formulate an acceptable legal environment for literary commerce. Their efforts in this regard are described in a later section. Their other strategy was to take matters into their own hands, moving to replace communal propriety with new collective practices. This they did by perfecting a social and economic mechanism that had already been developing in a somewhat inchoate fashion since the 1670s: the semipermanent alliance between printers and booksellers, with rules committing them to oppose transgressions upon their "property."

Such an alliance was typically composed of about fifteen booksellers who would act together to buy up an edition and distribute it wholesale. Their "conger," as it was sometimes called, although more an expedient understanding than a close conglomeration, served two principal purposes. First, it guaranteed a basic reward to the undertaker of the publication, who would not have to venture large amounts of capital in advance on the uncertain prospect of retail sales in a piratical realm. The alliance in this way obviated much of the necessity driving printers to pursue only short-term projects. It encouraged the production of large, even multivolume works at a time when the only reasonable alternative, subscription publishing, had begun to stall thanks to a poor reputation fostered by repeated failures. Like subscriptions, congers appeared first in England, where they congealed around Samuel Smith, printer to the Royal Society, to protect his expensive pharmacopeias. But from that origin the strategy grew to offer a collective solution to the specter of piratical anarchy in general. It militated against piracy not only by underwriting publications but also by implicitly threatening to blacklist suspect printers and booksellers. Such exclusion could be severely damaging, since individuals identified as outsiders probably sustained a lasting blot to their credit. What began as ad hoc alliances consequently developed into major powers in the trade. Their private auctions, at which editions would be divided among invited participants, formed the prototype for a new trade civility. In fact, the alliances and their customs effectively grew to constitute an informal, self-selecting successor to the guild protocols that had reigned over preceding generations. They did not always succeed – Dutch attempts, in particular, proved abortive – but some success was better than none.[11]

The number of antipiracy alliances multiplied in the first quarter of the eighteenth century. Before long they were engaging not only in distribution but also in publishing itself. In this way, the collective solution they offered could be applied not just to the de facto defense of particular editions, but to the de jure protection of copy. They thus soon began to auction among their

---

[11] N. Hodgson and C. Blagden, *The Notebook of Thomas Bennet and Henry Clements (1686–1719), with Some Aspects of Book Trade Practice* (Oxford: Bibliographical Society, 1956), pp. 67–100; J. Feather, *A History of British Publishing* (London: Routledge, 1988), pp. 67–77; Hoftijzer, "'A Sickle unto thy Neighbour's Corn,'" p. 17.

members not only material volumes but also copies and even shares in copies. The auctions took place at taverns and coffeehouses. They were termed "trade sales," and as they proliferated, a kind of private share market developed. Booksellers could trade, say, a forty-eighth share in Thomas Burnet's Anglo-Cartesian *Sacred Theory of the Earth*. These shares were then capable of further dealing, but only at succeeding closed auctions. The result was a system that preserved the protected status of copy by transforming it. It relied on social exclusion rather than corporate custom. Persisting throughout the eighteenth century, the trade-sales network lasted well into the nineteenth.[12] But piracy proved tenacious, too. It would remain a routine hazard in the Germany of Goethe and the London of the 1820s–40s, as radical scientists found to their cost. There was always a delicate balance between the assertion of property and the threat of its infringement.[13]

These collective measures exemplified attempts within a range of print cultures to provide for a reliable commerce in knowledge in the absence of traditional corporate oversight. The adoption of such restrictive practices served moral and epistemic ends, as well as the economic ones probably at the forefront of booksellers' own minds. In the eighteenth century, however, an even more difficult challenge arose. This challenge was not so easy to defeat by social exclusion, for the simple reason that its sources were already excluded. This was the challenge of unauthorized reprints made outside the judicial realm in which the original producer lived. Since both conventional craft civilities and state privileges extended no further than the political bounds of the realm, reprints beyond those bounds were in no sense illicit. But they were resented, by both book-trade personnel and authors. And they too had epistemic as well as economic consequences.

At first, the main threat to copies came from local printers and booksellers. The "pirate kings" of Augustan England, for example, were London figures such as Henry Hills Jr. and Edmund Curll. But in the 1730s printers in Edinburgh, Glasgow, and Dublin, profiting from a new political stability in both Scotland and Ireland, began to look beyond their regional markets and see an opportunity to exploit English readerships. They would eye the London market for promising titles, reprint them, and import them back into the English capital. Books like Burnet's *Sacred Theory* remained among the top ten piratical favorites for decades. In both Ireland and Scotland, lawyers were prepared to argue that the precedents and statutes defending English copies like Burnet's were of no efficacy in what were after all different legal systems.[14] And in all likelihood they were right; the mere reprinting of English titles was indeed

[12] T. Belanger, "Booksellers' Trade Sales 1718–1768," *The Library*, 5th s., 30 (1975), 281–302.

[13] M. Woodmansee, *The Author, Art, and the Market: Rereading the History of Aesthetics* (New York: Columbia University Press, 1994); A. Desmond, *The Politics of Evolution: Morphology, Medicine and Reform in Radical London* (Chicago: University of Chicago Press, 1990), pp. 15, 44, 74, 120–1, 163, 204, 230–3, 237–9, 246–8, 338–9, 412–14.

[14] R. C. Cole, *Irish Booksellers and English Writers, 1740–1800* (London: Mansell, 1986), pp. 1–21.

legal. Importation back into England was not, however, and men such as Andrew Millar, Thomas Birch's bookseller, believed that it threatened calamity for the commerce of letters. The extent of the danger was brought home with exquisite rudeness when Edinburgh dealer Alexander Donaldson opened a shop in London itself, openly intending to sell his reprints there.

The problem now facing the London booksellers was representative of the European trade as a whole. Geography – in the form of territorial boundaries and brute physical obstacles such as seas and mountains – played a major role in determining the limits of different print cultures. The Dutch, for example, were proficient in reproducing English works, and could do so with impunity; Isaac Newton was both a sufferer and a closet exploiter of their industry. Robert Darnton has shown how immensely profitable the reprinting of Diderot and d'Alembert's *Encyclopédie* could be, if pursued in a city enjoying mountainous and unpoliceable land routes into southern France. Sometimes such "piracies" became prestigious national projects, especially in the Italian states. This also applied to the three-hundred-odd German principalities. German governments regularly encouraged the unauthorized reprinting of neighbors' publications on mercantilist grounds. In the region of the Holy Roman Empire, imperial privileges aspired to protect titles from piracy everywhere, but they were often granted to these unauthorized reprints rather than to their originals. In any case, they were patchily obeyed. The enterprise of the Dublin and Edinburgh publishers was thus one that was recognizable across the European nations. This was the practical reality underlying enlightened representations of literary internationalism. Those representations came into being partly in response to a realm of the book riven by national and juridical boundaries the transcendence of which was often identified, not as cosmopolitanism, but as piracy.[15]

Polite leisure fueled such cross-border infiltration. The book trades expanded vigorously as pleasure became a highly profitable commercial opportunity.[16] Printers and booksellers soon established themselves in most large provincial towns, concentrating at first on newspapers and advertising – a business they did not quite invent but were the first fully to exploit. By mid-century their distribution networks had become sophisticated, rapid, and all but universal. Although the provinces could not yet seriously challenge the metropolitan monopoly on copyownership, those networks became the basis on which publishers built something resembling a national and even international literary culture. With diverse audiences now accessible, the enterprising could create and exploit new markets. And they did so with extraordinary energy,

[15] R. Darnton, *The Business of Enlightenment: A Publishing History of the* Encyclopédie, *1775–1800* (Cambridge, MA: Harvard University Press, 1979), pp. 19–20, 34–5; Woodmansee, *The Author, Art, and the Market*, p. 46.

[16] P. Langford, *A Polite and Commercial People: England 1727–1783* (Oxford: Oxford University Press, 1989), pp. 90–9; J. Brewer, *The Pleasures of the Imagination: English Culture in the Eighteenth Century* (London: HarperCollins, 1997), pp. 125–97.

identifying readers ranging from the Italian women who pored over Algarotti's *Newtonianism for the Ladies* to the English children who learned their own Newtonianism from Tom Telescope.[17] Edition sizes increased, too, encouraged by the expanding readership and protected by the informal structures of trade sale and conger. Consolidation soon began to replace the small-scale domestic printing houses of the seventeenth century with larger enterprises. John Watts's premises, employing fifty men by the 1720s, would have been unthinkable in London a generation earlier. It was more akin to Dutch houses than to its own predecessors.[18]

None of this expansion was inevitable. Nor, even when it had occurred, was it irreversible. As any observer who had lived through the period of the South Sea Bubble knew, commercial achievements could be discomfitingly ephemeral. In a realm of print uneasily balanced between piracy and property, its prospects always remained insecure. Constant work was needed to reinforce them. That work centered on the establishment of credit. In effect, the need of natural philosophers and gentlemen for transnational communication coincided with a parallel need among printers and publishers for a commercial civility. A bookseller held his "Property," contemporaries knew, only "at the Curtesy of the designing Pirate." The latter must be defeated, or else "we must never expect to see again a beautiful edition of a book."[19]

Writers and publishers put the idiomatic style first developed in experimental philosophy to fresh use in this realm. The matter of fact modestly reported, once the centerpiece of virtuoso civility, now became both a vehicle of profit and a source of pleasure. Print in the eighteenth century was built by accrediting "facts" of all kinds. Facts were profitable things, whether one worked in London or Paris, Neuchâtel or Edinburgh. The press distributed professedly factual reports between diverse readerships, and knowledge was discriminated from error in terms of their reception. A telling indication of the motivations behind this is gleaned from the printers who based other sales on their credit-building. Respectable John Newbery and Jacobite pirate William Rayner both marketed medical elixirs, for example; the principal ingredient of Newbery's seems to have been boiled dog.[20] In France, too, the

---

[17] J. Secord, "Newton in the Nursery: Tom Telescope and the Philosophy of Tops and Balls, 1761–1838," *History of Science*, 23 (1985), 127–51; C. Welsh, *A Bookseller of the Last Century* (London: Griffith et al., 1885), pp. 89–117; K. Shevelow, *Women and Print Culture: The Construction of Femininity in the Early Periodical* (London: Routledge, 1989); P. Findlen, "Translating the New Science: Women and the Circulation of Knowledge in Enlightenment Italy," *Configurations*, 2 (1995), 167–206.

[18] B. Franklin, *Autobiography*, ed. J. A. L. Lemay and P. M. Zall (New York: W. W. Norton, 1986), pp. 36–7.

[19] R. Campbell, *The London Tradesmen* (London: T. Gardner, 1747), pp. 133–4.

[20] J. Secord, "Extraordinary Experiment: Electricity and the Creation of Life in Victorian England," in D. Gooding, T. Pinch, and S. Schaffer (eds.), *The Uses of Experiment: Studies in the Natural Sciences* (Cambridge University Press, 1989), pp. 337–83 (Secord in fact refers to the possibilities generated by the steam press); Welsh, *A Bookseller of the Last Century*, pp. 21–9, 36; M. Harris, *London Newspapers in the Age of Walpole: A Study of the Origins of the Modern English Press* (London: Associated University Presses, 1987), pp. 91–8.

medical advertising promulgated by such entrepreneurs linked the highest and lowest of cultures. One could even buy "reason pills" to help in the promotion of Enlightenment. Provincial newspapers depended on readers' evaluations of such advertisements, perhaps even more than on that of their reports. And no realm of medicine, faculty or "quack," lay beyond the credit-building strategies that they pioneered. Beyond medicine, city and Continental publishers' vast financial investments in the great projects of the age, such as Diderot and d'Alembert's *Encyclopédie,* depended all the more on such investments of faith.[21] Booksellers filled their shops with travelers' tales, every one of them asserting – with all due modestly – its plausibility in conveying reliably witnessed matters of fact. Histories, too, were much in demand, and they appropriated similar rhetorical conventions to argue for their often wildly conflicting accounts of the past. And if you were executed, your autobiography and last words would appear almost before you reached the scaffold, buttressed with witnesses' statements proving the genuineness of the text. Finally, novels – the hybrid product of factual rhetoric with fanciful content – flourished on the back of this fashion. A pinnacle of sorts was reached when a Deist minister and printer's man named Robert Nixon tried to prove his credibility as a chemist and as a pamphleteer in one blow, by combining the efficacy of experimental demonstration with the power of print. Nixon walked into the palace of Westminster and exploded a bundle of his own polemical tracts with a homemade bomb. In the seventeenth century, matters of fact had been largely the province of gentlemen, lawyers, and philosophers. Now, in the hands of Nixon's community of printers, booksellers, editors, and hacks, they were extended to reshape the nature of print.[22]

It would be easy to claim that this appropriation of the fact constituted substantial progress toward public knowledge and away from superstition. And many people did make that claim, loudly and repeatedly. But not all. Credibility could all too easily shade into credulity; in several instances, fictional narratives were mistaken for relations of actual events. English readers thought that Defoe's *Journal of the Plague Year* was an authentic record, their French counterparts that Hennepin's spurious exploration stories told of a real expedition. One could soon pay to see Robinson Crusoe's shirt in a London coffeehouse. Consequently, the rhetoric of moral certainty did not fool everyone. At the end of the century Edmund Burke pointedly observed that the republic of letters, although it strenuously ridiculed the giants and fairies

---

[21] Campbell, *London Tradesman,* pp. 133–4; C. Jones, "The Great Chain of Buying: Medical Advertisement, the Bourgeois Public Sphere, and the Origins of the French Revolution," *American Historical Review,* 101 (1996), 13–40; G. Feyel, "Médecins, Empiriques, et Charlatans dans la Presse Provinciale à la Fin du XVIIIe Siècle," in *Le Corps et la Santé: Actes du 110e Congrès Nationale des Sociétés Savantes* (Paris: CTHS, 1985), pp. 79–100; S. Schaffer, "Defoe's Natural Philosophy and the Worlds of Credit," in J. Christie and S. Shuttleworth (eds.), *Nature Transfigured: Science and Literature, 1700–1900* (Manchester: Manchester University Press, 1989), pp. 1–44.
[22] Harris, *London Newspapers in the Age of Walpole,* pp. 96–7.

of earlier ages, had in effect fostered its own dangerous gullibility. Its practitioners were obsessed with "the marvelous in life, in manners, in characters." One might add that even supposedly obsolescent marvels such as ghosts, witches, and astrologers did not in fact disappear; they returned in newly polite rhetorical dress. Accreditation was no more problematic than it had ever been. "Even in their incredulity," Burke argued, the adepts of public reason "discover[ed] an implicit faith."[23]

The people who manufactured this vast outpouring of facts had not really existed before. Although there had been individuals condemned for "dulness" as early as the mid-seventeenth century, they had never before been present in considerable numbers, let alone as a distinct and identifiable type. Throughout the eighteenth century they were condemned and ridiculed. These were the hacks of Pope's *Dunciad*, who resided in the real Grub Street of north London and were paid by booksellers to produce poetry and prose by the line. The usual accusation against them was that they had reduced writing to a "mechanick" trade. Bereft of inspiration, their reason, inasmuch as they showed any, must be critical and not creative. But such individuals were not always as slavishly unoriginal as highbrow legend insists, nor as impoverished. After all, *The Dunciad* included Fellows of the Royal Society among the hacks, and the distinction between even Pope and the likes of Edmund Curll was one that had to be fought for and not simply observed (Pope himself fought for it and lost). By the 1750s, one of the most successful of the duces, the notorious "Sir" John Hill, was making some £1,500 a year from his botanical works.[24] Superfluous items from Hans Sloane's natural history collection ended up displayed in a coffeehouse. It is possible to argue that the establishment of Newtonianism, like that of Enlightenment itself, was as much a matter of hacks and duces as of Hauksbees and Desaguliers.[25]

Duces' reason, where it was recognized as reason at all, was most often

[23] E. Burke, *Reflections on the Revolution in France,* ed. J. G. A. Pocock (Indianapolis: Hackett, 1987), p. 150; M. McKeon, *The Origins of the English Novel* (Baltimore, MD: Johns Hopkins University Press, 1987), pp. 45–128; P. G. Adams, *Travelers and Travel Liars 1660–1800* (Berkeley: University of California Press, 1962), pp. 48–9; J. A. Champion, *The Pillars of Priestcraft Shaken: The Church of England and Its Enemies, 1660–1730* (Cambridge University Press, 1992), pp. 25–52; J. M. Levine, *The Battle of the Books: History and Literature in the Augustan Age* (Ithaca, NY: Cornell University Press, 1991), pp. 267–413; M. Harris, "Trials and Criminal Biographies: A Case Study in Distribution," in R. Myers and M. Harris (eds.), *Sale and Distribution of Books from 1700* (Oxford: Oxford Polytechnic Press, 1982), pp. 1–36. The fortunes of magic and astrology in the public sphere deserve sustained study: for suggestive approaches, see P. Curry, *Prophecy and Power: Astrology in Early Modern England* (Cambridge: Polity, 1989), pp. 95–117, 153–68, and I. Bostridge, *Witchcraft and its Transformations, c.1650–c.1750* (Oxford: Clarendon Press, 1997), especially pp. 233–43.

[24] P. Rogers, *Grub Street: Studies in a Subculture* (London: Methuen, 1972), pp. 175–217 and passim; Harris, *London Newspapers,* p. 142; A. S. Collins, *Authorship in the Days of Johnson* (London: Robert Holden, 1927), pp. 26–7.

[25] Rogers, *Grub Street,* p. 271 (for a contemporary casting Newton himself as a hack); P. and R. Wallis, *Newton and Newtoniana, 1672–1975* (Folkstone: Dawson, 1977). L. Stewart, *The Rise of Public Science: Rhetoric, Technology, and Natural Philosophy in Newtonian Britain, 1660–1790* (Cambridge University Press, 1992), pp. 143–6, sets the scene for this study.

characterized as derivative and critical. But it could take positive form, too. And it might prove radical, as Lynn Hunt and Robert Darnton have shown in their studies of pornography – another new genre, and one particularly reliant on a rhetoric of witnessed facts. Among the *livres philosophiques* that flooded the French book trade of the Enlightenment, works by Voltaire and Rousseau were vastly outnumbered by copies of *Venus in her Cloister* and gross sexual libels directed against the court. Darnton has revealed the extent of their circulation, which often originated beyond the French borders in Switzerland or the Low Countries. He has highlighted the centrality of hack writers to the public culture of the *ancien régime* and in particular the materialist philosophy explicit in their pornographic books. Darnton argues that their authors were driven by resentment at their inability to rise in *ancien régime* society and that their increasingly vitriolic prose set the stage for regicide.[26]

The cultures of print that developed in the eighteenth century were thus highly competitive and consciously riven. As such, they gave rise to problems that impinged directly on learned communication and at the same time drew resources from such communication to address those problems. The trade fostered literary conventions manifesting accreditation, not least because the credit of printed materials was always under threat.[27] In this sense, the hacks pioneered public awareness of the literary conventions of the new science. Yet those same hacks perpetrated a culture in which gentility and authorship seemed all but incompatible. What emerged from this conflicted realm was something that looked to contemporaries like a broad, harmonious, and reasonable culture. The texts that did circulate were apparently routinely accredited as knowledge. The press alone is not enough to explain why. The question should therefore be this: How was this widespread accreditation achieved?

Collective action was the best bet. The Royal Society – under Newton's presidency a respected element of the establishment – could hope to use its size, status, and legal right to restrain its printers in order to support the reputation of its books and of the *Philosophical Transactions*. In London, at least, it found some success. By the 1720s, although many lampoons of the virtuosi continued to appear, none really threatened its continuance as those of the Restoration wits perhaps had. But there still seemed little that a lone gentleman could do in the face of such a competitive and discrediting print culture, even if that gentleman happened to be an FRS. No citizen of the republic of letters could be guaranteed immunity, and no truth was so transcendent

---

[26] R. Darnton, *The Literary Underground of the Old Regime* (Cambridge, MA: Harvard University Press, 1982), pp. 199–201; R. Darnton, *The Forbidden Best-Sellers of Pre-Revolutionary France* (New York: W.W. Norton, 1995), pp. 169–246; L. Hunt (ed.), *The Invention of Pornography: Obscenity and the Origins of Modernity, 1500–1800* (New York: Zone, 1993). For the continuing history of radical pornographers after 1790, see also I. McCalman, *Radical Underworld: Prophets, Revolutionaries, and Pornographers in London, 1795–1840* (Cambridge University Press, 1988), pp. 31, 204–31.

[27] An example is the footnote, as elegantly discussed by Anthony Grafton in *The Footnote: A Curious History* (Cambridge, MA: Harvard University Press, 1997).

that it could be sure to survive immersion untarnished. The Dutch physician Herman Boerhaave was a notable victim.[28] Desaguliers, too, found that his *Lectures of Experimental Philosophy* were pirated, in an edition claiming to have been "approved" by Desaguliers himself. He never really managed to erase this piracy. The implications extended to Desaguliers's own reputation, which he feared might be compromised by allegations of plagiarism. He responded by personally signing every legitimate copy. In America, Benjamin Franklin adopted the technique of "nature printing" to much the same end. Nature printing was a technique for creating a trace on paper by the direct use of plants or other natural bodies, without the intervention of artist or craftsman. By eliminating the engraver – or rather, as one of Franklin's friends put it, by replacing him with "the Greatest and best Engraver in the Universe" – it promised to produce images of untarnished veracity. Franklin himself put that veracity to powerful use. Making impressions from leaves, each of which was unique, he found that he could guarantee the immunity of his imprints from piracy, since no pirate could reproduce his template. The technique prevented the piracy of one kind of printed paper in particular: cash. By eliminating the threat of counterfeiting, it played a central part in the establishment of paper money in America. So successful was it that historians today do not know how Franklin's technique worked.[29]

READING AND THE REDEFINITION OF REASON

Historians often talk of a new social practice that appeared in the late seventeenth or early eighteenth century and transformed the creation, adjudication, and consequences of claims to knowledge. This practice is identified with the development of a "public sphere." The proclaimed emergence of this "sphere" is taken to mark a decisive break with previous political and epistemic forms. Briefly, in seventeenth-century European states the royal court was represented as the concentration of power and knowledge. It embraced a localized notion of a "public" as the subject of a display of authority, to be "imprinted" by the court's glory. All of society might be present there on particular occasions, but only in its regimented "orders." The court was thus the sole source of advancement and of judgment on matters of law, politics, philosophy, and the arts. By the end of the seventeenth century, print

---

[28] *Gentleman's Magazine*, 2 (1732), 1099–1100; K. Maslen and J. Lancaster (eds.), *The Bowyer Ledgers* (London: Bibliographical Society, 1991), iv, p. 227 (no. 2968).

[29] J. T. Desaguliers, *A Course of Experimental Philosophy*, 2 vols. (London: J. Senex etc., 1734–44), I, sigs. c4ᵛ, C2ᵛ; Desaguliers, *Lectures of Experimental Philosophy* (London: W. Mears etc., 1719), sig. A3ʳ; Desaguliers, *A System of Experimental Philosophy, Prov'd by Mechanicks* (London: B. Creake etc., 1719); R. Cave and G. Wakeman, *Typographia Naturalis: A History of Nature Printing* (Wymondham: Brewhouse Press, 1967), pp. 12–13.

was making possible a competing source of authority in the shape of an urban reading public.[30]

The nature and impact of that new source of authority may be seen represented most vividly in Condorcet's *Sketch for a Historical Picture of the Progress of the Human Mind* (1793). Condorcet's polemic – which became something of a manifesto for the Jacobins, although they hounded its writer to his death – declared the pivotal role of print in hurling Europe into enlightenment and, before long, revolution. Condorcet presented the first full account of the press as a unique force for cultural transformation. He articulated what he called the "revolution that the discovery of printing must bring about." That revolution had occurred first in natural science; it must soon extend to every aspect of human life. The press had enabled readers to obtain any book they wanted. As "knowledge became the subject of a brisk and universal trade," so it had achieved a new certainty. This laid the foundations for continued progress. The reading public became "a new sort of tribunal" – a virtual polity, superseding boundaries of rank or nation. Independent of prodigal displays mounted at Versailles, the court of public opinion "no longer allowed the same tyrannical empire to be exercised over men's passions but ensured a more certain and more durable power over their minds." It took even politics to be its province. The result – according to its own propagandist, an inevitable one – was the political revolution of 1789.[31]

Condorcet's representation identified most of the important characteristics of the "public sphere": its independence from the court, its identification with a reading public, and its assumption of a judicial role over all aspects of human knowledge and morality, including not only science, but also the arts, religion, and politics. Print, according to this pronouncement, had the power to create a distributed tribunal exercising disinterested reason. Printed objects crossed civic and national boundaries (often becoming piracies in the process, we may note), and remained identical as they did so. They embodied comprehensible reporting styles, and as a result all reasonable readers could concur in their meanings. Unlike the elaborate imagery of baroque courts, which was now identified with Catholic and arbitrary rule, nothing was hidden from

---

[30] The historiography of the "public sphere" derives in large part from J. Habermas's *The Structural Transformation of the Public Sphere,* trans. T. Burger (Cambridge: Polity, 1989; originally published in 1962). Habermas's work is now largely superseded by more recent interpretations; see R. Chartier, *The Cultural Origins of the French Revolution* (Durham, NC: Duke University Press, 1991), pp. 20–37 for a critical survey. For American developments see M. Warner, *The Letters of the Republic: Publication and the Public Sphere in Eighteenth-Century America* (Cambridge, MA: Harvard University Press, 1990), and C. Armbruster (ed.), *Publishing and Readership in Revolutionary France and America* (Westport: Greenwood, 1993).

[31] J. A. Nicolas de Caritat, Marquis de Condorcet, *Sketch for a Historical Picture of the Progress of the Human Mind,* trans. J. Barraclough, ed. S. Hampshire (New York: Noonday Press, 1955), pp. 98, 99–123, 167–9, 175–6; Chartier, *Cultural Origins,* pp. 32–3, 65–6; R. Chartier, *Forms and Meanings: Texts, Performances, and Audiences from Codex to Computer* (Philadelphia: University of Pennsylvania Press, 1995), pp. 8–13.

the view of the onlooker. A new community was constituted of those readers. Invisible to one another, all were in principle equal.[32] But it was not just print that underwrote Condorcet's public tribunal. The consolidation of distributed readers came about by virtue not only of the intrinsic power of the press but also of that power's artful deployment. Two elements in this process deserve particular attention: periodical publication and reading.

A new form of publishing provided the essential foundation for belief in printed knowledge.[33] Across Europe, periodicals linked readers in cities otherwise separated by geography, ideology, or confession. Their regularity reinforced the apparent bonds among such readers and allowed for quick retaliation in the event of error or infringement. The readers of such periodicals, rather than of folio treatises, constituted the public conceived by Condorcet. This public was no longer a passive recipient of spectacle but an active participant: to be a fully fledged citizen, one must contribute to periodicals as well as read them. From their earliest exemplars, among which Bayle's *Nouvelles de la République des Lettres* was perhaps the most influential, the journals explicitly aimed to unite this "republic." To do so, their creators fostered new personae and practices. The editor, the corrector, and peer review were all their ideas. At the same time, however, they rendered most of the hackwork of journal production socially invisible, preserving thereby the fiction of a polite correspondence untainted by commercial interest. The polite rhetoric of experimental philosophy once more served this purpose. Editors, printers, publishers, and *colporteurs* were to be as invisible as the "laborants" so firmly relegated to the back of the experimentalist's stage. By contrast, courtly displays were repeatedly criticized by unmasking the labor of "juggling" on which they rested. This asymmetry was essential to securing the reading public as the tribunal of reason.[34]

It is important to recognize that the fortunes of periodicals were by no means unambiguously rosy. All of them encountered economic and cultural problems related to their core purpose, not the least of which was "piracy" perpetrated outside their home governments' jurisdictions. In this sense the professedly borderless republic of letters always rested on practices that very much de-

---

[32] F. Waquet, "L'Espace de la République des Lettres," in Waquet and H. Bots (eds.), *Commercium Litterarum: La Communication dans la République des Lettres, 1600–1750* (Amsterdam: APA-Holland University Press, 1994), pp. 175–89; Chartier, *Cultural Origins*, pp. 20–7; B. Stafford, *Artful Science: Enlightenment Entertainment and the Eclipse of Visual Education* (Cambridge, MA: MIT Press, 1994), pp. 8–23.

[33] For surveys of scientific periodicals, see A. A. Manten, "Development of European Scientific Journal Publishing Before 1850," in A. J. Meadows (ed.), *Development of Science Publishing in Europe* (Amsterdam: Elsevier, 1980), pp. 1–41, and D. Kronick, *A History of Scientific and Technical Periodicals: The Origins and Development of the Scientific and Technical Press, 1665–1790* (Metuchen, NJ: Scarecrow Press, 1976); for medical periodicals see also W. R. Lefanu, *British Periodicals of Medicine 1640–1899* (Oxford: Wellcome Unit, 1984).

[34] Goldgar, *Impolite Learning*, pp. 70–4 and passim; S. Shapin, *A Social History of Truth: Civility and Science in Seventeenth-Century England* (Chicago: University of Chicago Press, 1994), pp. 361–9, 378–407; Stafford, *Artful Science*, pp. 76–88.

pended on traditional political boundaries, whether for protection by laws or from them. Periodicity certainly helped in the effort to destroy competitors' credit, since it required relatively low capitalization and facilitated rapid responses. But it was not always enough, and producers were sometimes forced to employ methods not readily represented as elements of public reason, including brute force. Again, the representation of a public sphere rested in practice on "negotiations" that must remain inexplicit. When the Royal Society adopted the *Philosophical Transactions* as its own in 1752, it faced real difficulties in articulating its relation – epistemically distanced yet commercially and editorially managerial – with the journal. These were matters best left obscure.[35]

Not even a major player in the Republic of Letters – a Pierre Bayle, say, or a Fontenelle – could extend control so far as to affect the *reading* of periodicals. Reading was a practical skill, and like other practical skills it changed with respect to its setting. With periodical publishing and new distribution networks came new sites, and therefore new practices, of reading. The coffeehouse was a prime example. Coffeehouses could be found across European capitals and in most provincial towns. Their clientèle was widely regarded as a representative microcosm not only of the "political" nation but also of society itself. Women and children could be found here, as well as men of all "sorts." What they encountered was not just an unprecedented degree of social mixture, but one of real consequence for the circulation of knowledge. Even William Whiston, Newton's fallen successor as Lucasian professor, could be heard lecturing at a coffeehouse.[36] At the same time, however, one could turn to another coffee-drinker and have a conversation about the newspapers scattered around the coffeehouse – another practice in which civility and printed facts rested on each other. Or he could discuss with a third the publishing of his poems and the alleged piracy of his letters.[37] It was as such a place that the coffeehouse produced what contemporaries recognized as a new kind of reading. Coffeehouse readers were boisterous, skeptical to the point of Hobbesian atheism, critical, witty, and above all voracious. And the character of their reading went a long way toward dictating the character of printed output, since folio Latin treatises on predestination were hardly appropriate fare for the coffeehouse. For perhaps two generations, coffee and print depended on each other, not least economically.[38]

The coffeehouse was only one among a number of new sites for reading. At these places, novel practices of consumption transformed the implications

---

[35] BL ms. Add. 4441, fols. 21ʳ-24ʳ.

[36] Stewart, *Rise of Public Science*, pp. 101–82; S. Schaffer, "Natural Philosophy and Public Spectacle," *History of Science*, 21 (1983), 1–43, especially 3–15.

[37] For Pope's intricate negotiations with booksellers, see D. Foxon, *Pope and the Early Eighteenth-Century Book Trade*, ed. J. McLaverty (Oxford: Clarendon Press, 1991), pp. 23–46.

[38] Harris, *London Newspapers*, pp. 30–1; A. Ellis, *The Penny Universities: A History of the Coffee-Houses* (London: Secker and Warburg, 1956), pp. 223–5.

even of traditional texts. It was in these different appropriations, as much as in the printed objects themselves, that innovative implications lay. The *Encyclopédie* itself, for example, was purchased predominantly by elites not given to revolutionary zeal. But reading rooms made expensive books such as this available to the less wealthy, too. In Paris one did not even have to visit a *cabinet de lecture;* it was even possible to rent books – and parts of books – by the hour. The mere existence of the *Encyclopédie* explains little without this novel form of consumption.

In short, with the multiplicity of new reading spaces (coffeehouse, salon, home, private library, and *cabinet*) and new reading materials (increased print runs, new genres, and smaller and cheaper formats) came new reading practices. These were reckoned to be predominantly critical and witty, as befitted the world of hackery. Readers applied them "extensively" to a large number of short texts, rather than "intensively" to a small number of long ones. But at the same time they could be pronouncedly private, creating a secluded space in which individuals could act out their parts as agents of a "public" reason independent of any interest. Immanuel Kant described this practice most pithily in his essay "What is Enlightenment?" and it has been claimed recently that a "revolution" in reading practices along these lines took place in mid-eighteenth-century Germany to spur his reflections. One need not endorse this argument unequivocally – reading aloud remained in many ways as important as silent reading, for example – in order to concur with the general sentiment.[39]

Sentiment, in fact, was the core of the issue for many people in the eighteenth century. The practice of reading had not only a history but also a philosophy – and a natural philosophy. And this natural philosophy had to be understood if the practice were to be made the keystone of a new, public rationality. The scandalized representation of women readers rendered hysterical by their consumption of novels was a commonplace example, founded as it was on physiological considerations of female nature. In a sense, the representation was a very old one; in the seventeenth century, men had been able to explain away female authorship in terms of the body's susceptibility to the passions, and their arguments drew on humoral theories of the body dating back to antiquity.[40] But work in neurology and anatomy in the later seventeenth

[39] I. Kant, "What is Enlightenment?" in Kant, *Political Writings,* trans. H. B. Nisbet, ed. H. Reiss (Cambridge University Press, 1970), pp. 54–60; Chartier, *Cultural Origins,* pp. 67–91; R. Chartier, "Leisure and Sociability: Reading Aloud in Early Modern Europe," in S. Zimmerman and R. F. E. Weissman (eds.), *Urban Life in the Renaissance* (Newark: University of Delaware Press, 1989), pp. 103–20.

[40] For the longevity of such representations, and their problematic relation to women's actual experiences of reading, see N. Tadmor, "'In the even my wife read to me': Women, Reading and Household Life in the Eighteenth Century," in J. Raven, H. Small, and N. Tadmor (eds.), *The Practice and Representation of Reading in England* (Cambridge University Press, 1996), pp. 162–74, and P. H. Pawlowicz, "Reading Women: Text and Image in Eighteenth-Century England," in A. Bermingham and J. Brewer (eds.), *The Consumption of Culture 1600–1800: Image, Object, Text* (London: Routledge, 1995), pp. 42–53.

and eighteenth centuries provided this kind of representation with new authority. John Locke's educational advice, which was extraordinarily influential for the raising of children of both sexes throughout the century, rested on such work. It proposed ways to establish "habits" in the child's body capable of countering the ill effects of passionate reading. In mid-century, David Hartley's association psychology provided still more resources to explain reading experiences as processes of embodied reason – and to explain away the same experiences as passionate should they prove inconvenient. Hartley's mechanisms, first outlined in his *Observations on Man* (1749), were still being used for this purpose well into the nineteenth century.[41] There is a complex and consequential history of such arguments about the physiology and psychology of reading, and it is only beginning to be recovered.

In conclusion, then, eighteenth-century Europeans understood public reason to have come into being largely as a result of a confluence of several factors surrounding the press. Natural philosophy was one foundation: physiological explanations underpinned what readers thought to be happening to them when they read their periodicals. The rhetorical tropes developed by experimental philosophers bolstered the credit of their reports. But natural knowledge was to be not only the foundation of this new tribunal but also its immediate subject. In an enlightened society, men such as Joseph Priestley thought, claims to such knowledge should be adjudicated largely by the dispersed and disinterested laity. This was an important reason why Priestley objected to what he saw as the restrictive obscurity of Lavoisier's neologisms. In this regard, his public epistemology represented the apogee of Enlightenment.[42]

AUTHORSHIP, GENIUS,
AND THE END OF ENLIGHTENMENT

The eighteenth century began with the greatest single author in the history of science about to become president of the Royal Society. Isaac Newton took great care to ensure that he secured this unequaled authorial status. He closely monitored the editing, printing, and publishing of the second edition of his *Principia,* and skillfully manipulated the backstage practices of the international

[41] J. Locke, *Some Thoughts Concerning Education* (London: A. and J. Churchill, 1693), pp. 32–7, 46, 63–5, 175–90; G. S. Rousseau, "Nerves, Spirits, and Fibres: Towards Defining the Origins of Sensibility," *Studies in the Eighteenth Century,* 3 (1976), 137–57; A. Manguel, *A History of Reading* (London: HarperCollins, 1996), pp. 28–39; G. J. Barker-Benfield, *The Culture of Sensibility: Sex and Society in Eighteenth-Century Britain* (Chicago: University of Chicago Press, 1992), pp. 1–36; R. W. F. Kroll, *The Material Word: Literate Culture in the Restoration and Early Eighteenth Century* (Baltimore, MD: Johns Hopkins University Press, 1991), pp. 183–238; R. Darnton, "Readers Respond to Rousseau: The Fabrication of Romantic Sensitivity," in Darnton, *The Great Cat Massacre and Other Episodes in French Cultural History* (New York: Basic Books, 1984), pp. 215–56.
[42] J. Golinski, *Science as Public Culture: Chemistry and Enlightenment in Britain, 1760–1820* (Cambridge University Press, 1992), pp. 50–90.

book trade that supported the overt civility of the republic of letters. At the same time he deviously employed the registration protocols of the Society to destroy claims of rivals such as Leibniz, and he allied himself with Edmond Halley and bookseller Awnsham Churchill to overcome the domestic opposition of Astronomer Royal John Flamsteed.[43] In all these ways Newton can be seen to have pioneered strategies and representations that during the eighteenth and early nineteenth centuries became central to the creation of scientific authorship.

The eighteenth century is perhaps the most significant and well-studied period in the history of authorship in general, if not scientific authorship in particular. Between the contesting of traditional guild and privilege regimes starting around 1700 and the establishment of the steam press a century later, the modern author seems to have come into being. The transformation began with a shift in the physical location of debates over print. Hitherto, such debates had been pursued under guild auspices, and had accordingly drawn on craft customs and civilities that were relatively autonomous from wider public concerns. In the eighteenth century, as craft organizations decreased in power, these disputes moved to political authorities and courts of law. It became appropriate for the first time to argue at the level of high principle. The particular principle in question was that of property. At first, copyowning booksellers largely succeeded in defending a strong concept of perpetual property against allegedly piratical rivals. The courts accepted their argument for a fundamental property right in creative works, infringement of which remained punishable by law. But the rivals continued their efforts and constructed powerful counterarguments. Both sides seized on the cultural resources they found around them. Three in particular stood out as useful. They derived from prior considerations in philosophy and natural science.[44]

First, natural philosophy and technology furnished exemplary authors for all sides. Aristotle, Gassendi, Descartes, and Newton were the key authorial archetypes for more than one combatant.[45] The question was, what did this prove? Many were prepared to aver that a Newton did not write for cash. Proprietorial authorship was thus alien to true creativity. "Glory is the reward of Science," as Lord Camden famously pronounced in the 1770s, "and those who deserve it, scorn all meaner Views." The real Newton's behind-the-scenes maneuvering was systematically obliterated in such representations. In the case of earlier writers, among them Robert Boyle, the repute of authorial identity

---

[43] R. C. Iliffe, "'Is He Like Other Men?' The Meaning of the *Principia Mathematica,* and the Author as Idol," in G. MacLean (ed.), *Culture and Society in the Stuart Restoration: Literature, Drama, History* (Cambridge University Press, 1995), pp. 159–76; Johns, *Nature of the Book,* pp. 543–621. See also L. Stewart, "Seeing through the Scholium: Religion and Reading Newton in the Eighteenth Century," *History of Science,* 24 (1996), 123–65.

[44] For a perceptive discussion of the historiography of authorship, see R. Chartier, "Figures of the Author," in Chartier, *The Order of Books: Readers, Authors, and Libraries in Europe between the Fourteenth and Eighteenth Centuries,* trans. L. G. Cochrane (Cambridge: Polity, 1994), pp. 25–59.

[45] *A Letter to the Society of Booksellers* (London: J. Millan, 1738), p. 27.

was actively crafted by eighteenth-century editors such as Peter Shaw and Thomas Birch, just as it was deployed to bolster the very print culture that permitted their republication.[46]

The second important resource was John Locke's theory of property, refracted through the efforts of various eighteenth-century editors, abridgers, translators, and interpreters. Copyowners freely appropriated Lockean arguments, relating them to intellectual labor as Locke himself had to physical. The "marks" of labor could be discerned in literary works as much as in land, their would-be proprietors claimed.[47] That meant that the writer of a book gained a legitimate property in it by virtue of the labor expended. That property was perpetual and could be alienated. In other words, the copies "owned" by elite metropolitan booksellers should be considered true properties, and provincial rivals condemned as pirates. "Authors have ever had a Property in their Works founded upon the same fundamental Maxims by which Property was originally settled," as one pamphlet argued. "The Invention of Printing did not destroy this Property of Authors, nor alter it in any Respect, but by rendering it more easy to be invaded." The new notion of "copyright," which ordained protection for a limited temporal span, was thus unsatisfactory and must be considered at most a statutory appendage to the fundamental common-law right. It has been claimed that in making this argument the booksellers were the first to enunciate a strong concept of creative and proprietorial authorship.[48]

So-called pirates, on the other hand, argued for the temporality of copyright by constructing an analogy to craft inventions. There was no perpetual property generated by the crafting of inventions, even though "a great deal of *mental Labour* is often bestowed upon *Mechanical* Inventions, as well as upon *Literary* Productions." They were customarily protected by means of royal patents. Patents also had their equivalent in the book trade: one could gain a monarch's "privilege" for exclusive publication of a work for a given period. In France and Germany these *privilèges* were for most of the century the major foundation of literary propriety. But patents provided temporary protection only, and they were discretionary. They acknowledged no *right* on the part of the writer or producer, rather taking the status of royal gifts to such individuals. Wealthy booksellers thus preferred the argument from Lockean principles, which promised a secure property right independent of royal whim. Their rivals constructed sophisticated analogies to argue that printed texts should be considered as inventions. In insisting upon his "copy," a possessive

[46] J. Golinski, "Peter Shaw: Chemistry and Communication in Augustan England," *Ambix*, 30 (1983), 19–29. Compare M. de Grazia, *Shakespeare Verbatim: The Reproduction of Authenticity and the 1790 Apparatus* (Oxford: Oxford University Press, 1991).

[47] *Speeches or Arguments of the Judges of the Court of King's Bench* (Leith: W. Coke, 1771), p. 49; W. Blackstone, *Commentaries on the Laws of England*, 15th ed. (London: A. Strahan, 1809), vol. 2, pp. 405–7.

[48] M. Rose, *Authors and Owners: The Invention of Copyright* (Cambridge, MA: Harvard University Press, 1993); *The Case of Authors and Proprietors of Books* (n.p., n.d.), p. 1.

bookseller was said to be acting in the same way as the "Inventor of any small Mechanical Instrument." By the 1760s, judges were thus building verdicts by contrasting Lockean philosophy with Harrison's chronometer. It should be the same, concluded one, "whether the Case be *mechanical,* or *Literary;* whether it be an *Epic Poem,* or an *Orrery.*"[49]

This argument had rather profound philosophical import. True knowledge was, of course, based in Creation. In that case it could hardly be restricted to one redactor. "Invention and labour," one judge remarked, "cannot change the nature of things, or establish a right, where no private right can possibly exist." Truth could not legitimately be parceled out. "The Inventor of the Air-Pump had certainly a Property in the *Machine* which He formed," the same judge splendidly opined, "but did He thereby gain a Property in the *Air?*"[50] In the same spirit, Enlightenment principles could lead one to question whether such a thing as intellectual property were conceivable at all. After all, Boyle's theories were in a sense his literary creations, and if it was absurd to assert a property in Boyle's law, then it might be equally ridiculous to propose one in any other idea. In France, it was again the Marquis de Condorcet who expressed the ultimate objective of such an argument. Condorcet recommended that the literature of Enlightenment should get rid of all piratical transgressions by abolishing the object of their violation. Condorcet's ideal realm of print would simply define piracy out of existence by eliminating textual properties themselves. This would have dramatic practical implications. Readers would select works by subject matter only and not by authorial name. Printed texts would be organized entirely into topical periodicals. The author would be dead almost before having a chance to be born.

After 1789, the revolutionary regime briefly attempted to put Condorcet's theory into practice. The result was calamitous. With "freedom of the press" came liberation from the protective structure of the publishing guild, from the accumulated capital of *ancien régime* literature (especially in the hitherto bankable domains of law, religion, and royal politics), and from security of literary property in the future. With all but the most rapidly produced works subject to unauthorized reprinting, the Parisian book trade first turned into a pamphleteering industry and then virtually collapsed altogether. Bankruptcy overtook its major participants, and book production shrank to almost zero. Before long, even Condorcet realized that the consequences of unfettered deregulation were going to be disastrous. He urged the reinstitution of a limited authorial copyright. Put into effect in 1793, his suggestion now formed the origin of the modern copyright system in France. But the French book still took years to recover from its taste of unmitigated Enlightenment.[51]

---

[49] *Speeches or Arguments,* p. 30; J. Burrow, *The Question Concerning Literary Property* (London: W. Strahan and M. Woodfall, 1773), pp. 3, 42–4, 70, 101; C. MacLeod, *Inventing the Industrial Revolution: The English Patent System, 1660–1800* (Cambridge University Press, 1988), pp. 196–200.

[50] *Speeches or Arguments,* p. 50; see also Rose, *Authors and Owners,* p. 87, for a similar sentiment with respect to Newton.

[51] C. Hesse, *Publishing and Cultural Politics in Revolutionary Paris, 1789–1810* (Berkeley: University of

Opponents of deregulation across western Europe had predicted this outcome and developed arguments to defend against it. They reached fruition first in Germany. The central strategy was to allege a distinction between the "form" of a composition, the truth it contained, and its particular material manifestation. A property might then be identified in the first of these, which was taken to be the product of a particular author or artist's personal creativity. Proponents of this argument constructed an almost paradoxical alliance of concepts. On the one hand, they strenuously insisted on the irreducible individuality of the author – and before long began to call this individuality "genius." On the other, they asserted the commercial, pecuniary application of that high ideal in the grubby world of print. The argument culminated at the hands of Kant, Herder, Fichte, and Goethe, who extended it to a radical redefinition of creativity itself. Their arguments took shape in response to a failed attempt to establish on Enlightenment grounds a "German Republic of Letters" peopled by authors liberated from the control of booksellers. In reaction they forged a Romantic concept of the author that would transform images of scientific discovery and genius. That concept hailed the author as a unique soul, whose work involved not the rule-mongering of stale reason but rather processes that were essentially creative. In effect, those creative processes reiterated the fecundity in Nature itself. Reading was therefore an encounter, not with a reason that was in principle common to all, but rather with an irreducibly foreign spirit.

The brutal conflicts of the Enlightenment here gave way to what might be termed a *Naturphilosophie* of authorship. Applied to the memory of Isaac Newton, for example, such representations would result in oblivion for the practical labor he had dedicated to print and in the retrospective creation of an archetypal scientific genius. Newton became and remained a national hero. Even the hostility that Romantic writers sometimes directed at Bacon, Locke, and Rousseau effectively confirmed their iconic status, albeit as symbols of a sterile reason to which they were steadfastly opposed. In choosing him to blow the last trump, it could be said, William Blake revealed himself to be of Newton's party without knowing it.[52] We now tend to think of the authorship of which Newton is the exemplary case as a category established during

---

California Press, 1991), pp. 102–14; Hesse, "Economic Upheavals in Publishing," in R. Darnton and D. Roche (eds.), *Revolution in Print: The Press in France 1775–1800* (Berkeley: University of California Press, 1989), pp. 69–97; J. D. Popkin, *Revolutionary News: The Press in France 1789–1799* (Durham, NC: Duke University Press, 1990), p. 183. For scientific publishing in this difficult setting, see J. Dhombres, "Books: Reshaping Science," in Darnton and Roche (eds.), *Revolution in Print*, pp. 177–202.

52 Woodmansee, *The Author, Art, and the Market*, pp. 35–55. For Romantic creativity and organicism in the sciences, see N. Jardine, "*Naturphilosophie* and the Kingdoms of Nature," in N. Jardine, J. Secord, and E. Spary (eds.), *Cultures of Natural History* (Cambridge University Press, 1996), pp. 230–45, especially pp. 231–4. For a survey of the reading of chemical texts in Germany at this time, see K. Hufbauer, *The Formation of the German Chemical Community (1720–1795)* (Berkeley: University of California Press, 1982). For reconstructions of Newton, see R. Yeo, "Genius, Method, and Morality: Images of Newton in Britain, 1760–1860," *Science in Context*, 2 (1988), 257–84. For Blake's Newton, see *Europe: A Prophecy* (1794), Plate 13, lines 4–5 (among many other places).

the Enlightenment and as one derived from Enlightenment principles. It is a notion that comfortingly identifies the origins of modern authorial conventions with those of public reason, and perhaps of science itself. But the French experience implied that the book trade's experience of unmitigated Enlightenment could in fact be a disaster for the author, who tended to be economically throttled where he was not physically guillotined. Like the scientist, the modern scientific author is a later and more compromised creation than we might suppose.

24

SCIENTIFIC ILLUSTRATION IN THE EIGHTEENTH CENTURY

Brian J. Ford

Illustration emerges from complex and diverse motives. The portrayal of an objective reality may seem to lie at its heart, but there may be other, subtle factors at work. Preconception, for example, guides many an illustrator's hand. A wish to project known realities onto nascent concepts distorts reality in its own ways, and the process of transmuting the subtle realism of Nature into an engraver's line imposes constraints and conventions of its own.

There is a general principle in artwork, often unrecognized: the culture of each era dictates its own arbitrary realities. Our experience of this is largely intuitive, but it explains why we can relate a specific image (a saint from a thirteenth-century psalter or the countenance of the Statue of Liberty) more easily to the time it was produced than to the identity of the artist or the name of the subject. In just this way, a scientific illustration is a mirror of contemporaneous preoccupations and a clue to current prejudice. It is more than a didactic symbol. Some illustrations create, and then perpetuate, icons that transcend reality and provide a synthesized convention that passes from one generation of books to the next. These icons are created for textbooks, and they populate their pages as decorative features that do little to reveal reality.[1]

Early in the century, François Legaut's *Voyages et Aventures* (1708) featured a rhinoceros with a second horn projecting forward from its brow. This structure is never found in life. Why should it be featured in an eighteenth-century illustrated textbook? The first published study of a rhinoceros (made by Albrecht Dürer in 1515), although powerful and otherwise realistic, boasts a small secondary horn on the shoulders, which projects forward. The image was repeatedly plagiarized and – with each generation of copying – this imaginary forward-projecting second horn increased in size. By the time it was included in Legaut's book, the imaginary horn was equal in size to the real one.

---

[1] Discussions on the relationship between reality and interpretation are found in Brian J. Ford, *Images of Science: A History of Scientific Illustration* (New York: Oxford University Press, 1992); see also the author's "Images Imperfect, the Legacy of Scientific Illustration," *Yearbook of Science and the Future* (Chicago: Encyclopedia Britannica, 1996), pp. 134–57.

For the technician, keen to capture scientific imagery, some notion of "realism" is the declared aim. Yet cultural interpretations of this term vary considerably. European conventions, for example, aimed at conveying the right proportions, the correct number of scales, the alignment of petals in precise order. Eastern illustrators, on the other hand, portrayed nature with less absolute accuracy, but with more style and panache. A Japanese illustrator, faced with the precision of a picture from a Western encyclopedia of ichthyology, observed, "Yes, but I don't find it appetising."

The art of representing three-dimensional, subtle realism on a flat sheet with lines and bands of color took time to mature. Each new convention bestowed its terms of reference on the illustrators of the generation that followed, and they nurtured the paradigm, improved it, honed and refined it for their own contemporaneous audience. Representational illustration became a hallmark of eighteenth century reference books, and we can glimpse in earlier centuries flashes of the inspirational realism on which this tradition was founded.

Earlier examples are the photo-real *Viola odorata* portrayed for Jacopo Filipo of Padua in the 1390s, followed by the brilliant rendering of *Mandragora autumnalis* by Giacomo Ligozzi around 1480, and the clump of turf of Albrecht Dürer's *Das Große Raßenstück* of 1503.[2] There remained no method of duplicating these vivid images, which could not find publication as scientific illustrations in mass-produced works. Most earlier published scientific illustrations were of poor quality, and the crude woodcut of a horse skeleton by Ferrari in 1560 (as an example) shows little more than a half-decayed and distorted corpse with little attempt at realism. Others were far more attentive to detail. Carlo Ruini produced well-observed studies of the horse skeleton as early as 1598, and his work was still being used as a reference by Snape in 1683.

As the eighteenth century dawned, new philosophies were emerging in a cascade of intellectual renewal. Locke and Spinoza, Leeuwenhoek and Leibniz, Descartes and Newton were publishing revelations in a continuing stream. The techniques of graphical representation began to be more fully understood, and accepted conventions could now be applied to the panoply of natural philosophy. After centuries of haphazard block-making and essentially crude drawing, an era of representational scientific illustration was suddenly to emerge. Now that artists were aware of the way to apply their craft, and natural philosophers were identifying the realities that surrounded them, lucid illustrations suddenly began to become commonplace. A flowering banana, for example, was portrayed, for the first time in scientific history, as clearly as you would expect to find it in a present-day textbook or a botanical home page on the World Wide Web. The horse skeleton was beautifully embodied

---

[2] Reference is made to Hastings Hours manuscript (< 1480) (London: British Library, Additional 4787 f49]), and to Filipo Jacopo, *Viola odorata* (1390) (London: British Library, Egerton 2020 [94]). The original of Dürer is preserved as Albrecht Dürer, *Das Große Raßenstück* (1503) (Vienna: Albertina collection), and the Giacomo Ligozzi, *Mandragora autumnalis* (c. 1480), is at Florence (Uffizi Gallery, Gabinetto Disegnis).

in engravings of the highest quality, transcending anything that had gone before, and better than most illustrations available today. Majestic images of the heavens and detailed drawings of the intricate communities of ponds and hedgerows were suddenly laid before an enthusiastic audience. As an era of science began to mature, the illustration of its findings came to a state of representational perfection that sometimes exceeded the standards in modern reference sources.

ILLUSTRATION BEFORE THE EIGHTEENTH CENTURY: A TRADITION OF OBSCURANTISM

Representational realism as a routine had begun to appear in previous centuries, although not usually in academic texts. The many herbals, for example, frequently featured hideously distorted and exaggerated versions of medicinal plants. To the tyro, their value as a guide to identification was slight; herbalists, of course, knew perfectly well what was portrayed. To seek the best early portrayals of nature we turn to the religious artist. The most accessible source is the Books of Hours perfected by the Flemish illustrators of the fifteenth and sixteenth centuries. One example, the *Hasting Hours,* painted before 1480, is a perfect small tortoiseshell butterfly sitting on a primrose. The religious painters were content to set down images of nature as an aid to worship and had no need to distort or falsify.

Why, then, do we find unrecognizable images of plants in the specialist herbals? The reason lies in a wish to dignify the trivial through obscurantism. Herbalists did not wish to have the public *au fait* with their art, and the purpose of the imagery was to keep outsiders at bay. In modern science we find the equivalent when false-color transmission electron micrographs are used to decorate articles for the public and when complex terminologies are used to designate the simplest of concepts. There is a phenomenon, for example, in which patients clench their molars and rub them together while asleep. It is known to doctors as tooth-grinding – but only amongst themselves. Once the topic is likely to receive a wider audience, it becomes transmuted into "temperomandibular joint syndrome." In just the same way, a white cell remains a white cell among the scientists in a hematology laboratory. If an outsider comes into the room, these cells become "polymorphonuclear granulocytes," and they remain polymorphonuclear granulocytes until the visitor has left.

A desire to keep specialisms special – and to prevent nonacademics from gaining undue insights into fields that authorities like to keep for themselves – has often guided the *rapporteurs* of scientific progress. In the field of scientific illustration we find resonances of the same ideal. The modern world is replete with images of twisting spirals of nucleic acid, seductive impressions of black holes in the deep recesses of outer space, meaningless vistas of integrated circuits animated in a choreographed sequence in a television commercial –

images designed to impress the public with a sense of unattainable complexity rather than to illuminate a simple and assimilable truth. Atoms are still illustrated as though they were composed of billiard balls, even though this concept has not been applicable since the postquantum era of the 1930s.

A RESPITE OF REALISM

In the eighteenth century viewers enjoyed a respite from this kind of practice. The clumsy caution of the earliest illustrators was maturing into a full appreciation of the wonder of reality, and the present-day tendency to distort and impress was yet to arise. In this singular century we saw the new currency of scientific honesty beginning to emerge. It bequeathed to us a legacy of vivid and striking pictures, capturing for the first time the extent of human discovery. Science began to emerge as a recognizable discipline, and its illustrators served it well. Great voyages of discovery were bringing back collections of natural history, plant specimens, and geological samples. Explorers were using artists to record their finds and were servicing collectors on their return. The era of scientific exploration was under way.

When the century was born, work on a great illustrated work was only beginning. Maria Sibylle Merian (1647–1717) had just returned to Amsterdam from Surinam, laden with drawings and collections of plant and insect material from that South American colony. She came from a family with long-established connections to this field. Her father, Mathäus Merian the Elder, had inherited the *Florilegium Novum* of his father-in-law, Johann Theodor de Bry, and published a new edition of this work in 1641. After his death a few years later, Maria's mother married Jacob Marrell, a painter of floral arrangements, and Maria in turn married Johann Graff, one of her stepfather's pupils from Nuremberg. Some years afterward she left her husband for religious reasons, and in 1698 travelled to Surinam to study, collect, and draw. The results appeared in a beautifully illustrated work, *Metamorphosibus Insectorum Surinamensium* (1705); an enlarged second edition, containing additional plates made by her daughter Joanna, appeared in 1714. These editions are illustrated with fine engravings, each painstakingly hand-colored, and provide an interesting exemplar of the transitional period between the partly imaginary corpus of earlier illustrations and the move toward representationalism. The illustrations have an exaggerated quality, with salient features distorted to suit the view of the engraver. Some of the aspects (and several of the species) are imaginary, and some have anatomical details in the wrong order. Her cashew nut, for instance, is drawn upside-down in relation to its stem.[3]

---

[3] See Johann Theodor de Bry, *Florilegium Novum* (Frankfurt, 1611); Maria Sibylle Merian, *Metamorphosibus Insectorum Surinamensium* (Amsterdam, 1705), was published in a second edition, with additional plates, in 1714. Rudolph Ackerman published *Thirty Studies from Nature* (Munich, 1812), an example of publishing in which teams of colorists were employed to embellish engraved plates.

Joseph Pitton de Tournefort (1656–1708), Professor of Botany at the Jardin Royal in Paris, another important figure in the early part of the century, first published his *Élémens de Botanique* in Paris in 1694. It became well known for its much-revised English edition, *The Compleat Herbal,* published in two volumes (1719 and 1730). Tournefort cataloged 8,846 vascular plants in this book, which was illustrated with some five hundred engravings on copper. Whereas Maria Sibylle Merian had explored the New World, de Tournefort took leave from his post in Paris and set off to explore the Middle East. He was accompanied by Andreas Gundelscheimer – a physician, close friend, and naturalist – and a noted apothecary and illustrator, Claude Aubriet. They set sail from Marseilles in 1700 and headed for Crete, where they botanized for three months. After exploring the islands of the Aegean they traveled to Turkey, spending time in Armenian and Kurdish communities en route to Georgia, discovering many species new to Western botanists. Tournefort published an account of the journey and their findings in a book illustrated by Aubriet: the *Voyage au Levant* of 1718. The engravings are recognizable but are unsubtle and done without grace. As in the case of Sybille Merian's publications of the same era, there are detectable resonances of the herbals and other illustrated books of an earlier age. However, in the *Élémens de Botanique* Tournefort had produced the first comprehensive treatise on botany, and the work of later taxonomists (Linnaeus, for example) should be considered in conjunction with this work.[4] Aubriet himself illustrated several works on botany. His studies were stylized, too, but the vigor of his vernacular enhanced, rather than detracted from, the vivid realism with which he portrayed his subjects.

The great work in England was published by John Ray (1627–1705), the son of a blacksmith-cum-herbalist from Essex. With the support of Francis Willughby (1635–1672), Ray toured Europe between 1663 and 1665. Many of Ray's other publications appeared during the seventeenth century, but his greatest work appeared in the eighteenth. Publication of the great *Historia generalis plantarum* was completed in 1704. Although a greater work than Joseph Pitton de Tournefort's *Élémens de Botanique* (Ray describes 18,600 plant species in a work of almost 3,000 pages), it was unillustrated. There was a growing belief that the student of high science needed no pictures to enlighten the mind. This concept persisted well into the twentieth century. Today's senior British botanists know Clapham, Tutin, and Warburg's *Excursion Flora of the British Isles,* for example, which contains no illustration of any of the species it describes.

A close friend of John Ray's, Mark Catesby (1683–1749) – a self-taught artist – was inspired to visit the Americas to document what had recently been discovered in the territories still new to natural philosophy. Catesby visited the Carolinas, Virginia, and the Bahamas and published a *Natural*

---

[4] Joseph Pitton de Tournefort published *Voyage au Levant,* (Paris, 1718) and *The compleat Herbal* (London, 1719, 1730). The quality of engraving was not always high, but the large number of species described set in train the work by taxonomists such as Linnaeus.

history of Carolina, Florida and the Bahama islands in 1731–43. Catesby had an urge to draw everything he could see, and his books were well illustrated. There are resonances of an earlier era, for some of his illustrations lack a certain realism. Many are imbued with a caricature-like quality that is not truly representational but reflects his self-taught status. In a later era, cartoon-like resonances of Catesby's work can be detected in some of the drawings by Edward Lear, in which birds are given almost anthropomorphic expressions. For all his stylistic oddities, Catesby is rightly celebrated as the pioneering great naturalist to work in America.

George Edwards (1694–1773), who traveled widely in Europe as an apprentice, undertook to publish much of Catesby's output, and he wrote a number of published natural history books. The *Natural history of birds* was published as a three-volume set (1743–50) and was subsequently produced in a French edition. Edwards's greatest work was published in four volumes as *The Natural history of uncommon Birds, and some other rarer undescribed Animals*. It appeared between 1743 and 1751 and was followed by *Gleanings of Natural History* (three volumes, 1758–64).

The vigor of Catesby's illustrations can be compared with those of Griffith Hughes, whose *Natural History of Barbados* (1750) fulfills the need of the book collector rather than the naturalist. The marine invertebrates Hughes figures are faithfully portrayed and represent study in the field, but each one is set with mathematical precision on a plate faithfully dedicated to its particular sponsor. The result is a dull book, pandering to the wealthy patrons but doing little to entice readers with the exuberance of discovery in the New World.

Catesby deserves his reputation as an innovative naturalist of the Americas, but he was not the first natural philosopher ever to illustrate the species of the New World. Charles Plumier (1666–1706) had visited the Americas between 1689 and 1695. He published several illustrated works, including the *Nova Plantarum Americanarum Genera* of 1703 and the *Traité des Fougères de l'Amerique,* published in 1705. Both books feature detailed illustrations of the genera Plumier studied, and the 170 plates on American ferns (engraved on copper by Plumier from his own specimen drawings) became a standard work (Figure 24.1). Here too the images are slightly stylized, but they come close to the modern ideal of a line drawing. Plumier died on a journey to Peru in his fortieth year, and his final book was published posthumously. This was *Plantum Americanum,* edited by Johannes Burmann. Plumier left a legacy of inspiration, too, for it was his late seventeenth-century travels to the New World that inspired Maria Sibylle Merian to set out and explore for herself.

Catesby was followed by William Bartram (1739–1823), who was the first great naturalist born in America, and a fine artist. He has also been called "America's first ecologist" (a title perhaps more correctly bestowed upon Henry Chandler Cowles [1869–1939] whose publication, *Vegetation of the Sand Dunes of Lake Michigan,* was published in 1899). Bartram, the son of King George III's

Figure 24.1. Plumier's study of American ferns (1705). Botanical illustration, and the flora of the Americas, came to the fore in the eighteenth century. As exemplified by this Jamaican holly fern, copper engraving brought a clarity of line to the illustrations in Charles Plumier's *Traité des Fougères de L'Amérique,* published in 1705. Artwork © Brian J. Ford.

botanist in America, certainly understood that there were subtle interactions between plant and animal communities, and his great work was published as *Travels through North and South Carolina, Georgia, East and West Florida, etc,* in 1791.

Carl Linnaeus (1707–1778) of Råshult, Sweden, was the great systematist who brought an enduring order to the taxonomy of the natural world. The most innovative of his pioneering books was *Systema naturæ regnum vegetabile,* first published in 1735, but his *Genera plantarum* (1737) soon followed and was itself succeeded in 1751 by the *Philosophia Botanica* and in 1753 by *Species Plantarum,* in which the binomial convention appeared for the first time. Linnaeus is commemorated as the person who popularized the system of nomenclature that we have inherited. It is interesting to note that the drive towards this simplicity was motivated not by the desire for scientific clarity but by the need to economize on paper. Linnaeus felt that too much space was devoted to the lengthy Latinized plant descriptions that were then current. By reducing the description to genus and species (one word for each) he reduced his printing costs. His work gave a considerable impetus to the development of illustrated botanical books. Pulls of many of the plates used in his own books were used as wallpaper in Linnaeus's country home at Hammarby and survive there in their original condition.

One of those inspired by the output of Linnaeus was Carl Peter Thunberg (1743–1828), who set off from Sweden in 1700 for a journey through Japan, Java, the Cape, and Ceylon. He documented about two thousand new species of plants, and they appear in his 293 publications on natural history and medicine. His many books were influential for generations of botanists who followed.[5] Michel Adanson (1727–1806), of Aix-en-Provence, France, was one of them. Adanson explored Senegal in 1749 collecting specimens. His *Histoire naturelle du Sénégal* (1757) was published in Paris, appearing in English in volume 16 of J. Pinkerton's *General collection of . . . voyages* (1814, London). Adanson's extensive two-volume *Familles des Plantes* (1763, 1764) was published in Paris. The engravings, taken from specimens collected on the way, are vivid and realistic. Interestingly, his plates on conchology show specimen shells with the apex downward, a convention followed by French publishers at variance with traditions elsewhere in European illustration.

During the eighteenth century, the flora and fauna of Southeast Asia remained unknown to the West until they were recorded by Georg Eberhardt Rumpf (1627–1702). His works were published posthumously. The *D'Amboinsche Rariteitkamer* (1705) contains more than sixty plates featuring mollusks and crustaceans, and his seven-volume *Herbarium Amboinense* (1741–55) is also richly illustrated with discoveries new to science.

FROM WOOD TO METAL ENGRAVING

The traditional use of woodcut in illustration survived into the eighteenth century (Figure 24.2), but its eclipse was marked when H. L. Duhamel de Monceau published his *Traité des arbres et des arbustes* (1755). All the illustrations were old woodcut blocks. As this book was published, the young Georg Dionysius Ehret (1708–1770) was already demonstrating the use to which fine and detailed copper engraving could be put. Although a vogue for exploration and discovery provided a hallmark for eighteenth-century biologists,

---

[5] The principal botanical volumes are as follows: Carl Peter Thunberg, *Flora Japonica* (Leipzig, 1784); *Sera uti Europa, Africa, Asia förätad ären 1770–79* (Uppsala, 1788–93); *Flora Capensis* (Uppsala, 1807); *Prodromus plantarum Capensium [etc]* (Uppsala, 1794–1800).

Figure 24.2. The end of wood-cuts: Rundbeck (1701). Wood-cuts survived into the eighteenth century, marking an historic connection with the dawn of printed scientific illustration. This late example, of *Crinum,* was published by Olof Rundbeck at Uppsala in the *Campi Elysii* of 1701. Only the second of twelve planned volumes was produced; a fire destroyed the wood blocks for the eleven others in 1702. Artwork © Brian J. Ford.

Ehret did not travel the globe. He spent all his life in Europe, but he produced some of the most memorable and accomplished illustrations in the history of botany. Born in Heidelberg, Germany, Ehret was the son of a gardener. His father died young, but not before he had painstakingly taught the young Ehret to draw from nature. Ehret's early years were not a success. He worked in Karlsruhe as a gardener for Karl III of Baden who was impressed by his Ehret's art and invited him to paint some botanical studies. Ehret's status as a favorite caused problems with his fellow gardeners, and by 1726 matters had become so strained that he moved to Vienna. An apothecary named J. W. Weinman commissioned Ehret to produce one thousand botanical studies, but after he had produced the first half Weinman felt dissatisfied, paid him half a year's salary, and the contract was terminated. Some of the Ehret engravings appeared in Weinman's *Phytanthozoa Iconographia.* However, this book also contains many imaginary organisms and thus in terms of context, if not technology, has more in common with a work from a previous century.

While working for a Regensburg banker, Ehret was befriended by Johann Ambrosius Beuer, a trainee apothecary and keen amateur botanist, who introduced him to his uncle. This was a crucial event for the young Ehret. The uncle was Christophe Jacob Trew, of Nuremberg, who became Ehret's most successful patron and an enduring friend. Trew encouraged Ehret in his work. A collection of six hundred of Ehret's botanical watercolours was sold in 1732 for 200 thalers (Trew had been paid only 20 thalers for his contract with Weinman). The collection, known as *Herbarium Vivum Pictum,* may be the otherwise unidentified collection of watercolor studies now in the library of the Earl of Derby at Knowsley Hall, England.[6] It was Ehret's first commercial success.

Ehret spent time during the following years travelling in Europe, visiting botanic gardens and studying as widely as he could. His many flower paintings were sent back to Trew, who added them to his collections and cared for them. Ehret's pictures were always identified by name of specimen and often with details such as where and how the plants were grown. Ehret visited England, where he became friendly with George Clifford, a Netherlands banker whose gardening assistant and physician was the young Linnaeus. Linnaeus published a description of the rare plants in Clifford's gardens, and it was illustrated with twenty plates by Ehret. The plates were engraved by Jan Wandelaar, and the finished product appeared under the title *Hortus Cliffordianus* (1737). The beautiful illustrations marked the move toward scientific accuracy in delineation of detail. Since Linnaeus considered the sexual characteristics of angiosperms crucial to classification, Ehret featured many of these details, painstakingly dissected out, in his illustrations. At the Chelsea Physic Garden, he was befriended by the curator and horticulturist Philip Miller, whose sister-in-law Ehret married. Miller had kept notes on many remarkable plants at the Garden, and published *Figures of the most Beautiful, Useful and Uncommon Plants* between 1755 and 1760. The illustrations were all by Ehret.

[6] Reference may be made to Christophe Jacob Trew, *Hortus Nitidissimus* (Nuremberg, 1750–92); *Plantae Selectae* (Nuremberg, 1750–73).

By this time Ehret was beginning to engrave his own copper plates. He set himself a project that was published as *Plantae et Papiliones Rariores* (1748–59). Although the plants are confidently engraved, the butterflies are not identified by the artist and serve merely as embellishments to his floral compositions. Some of the plates show dissected details of floral structures, and each book was painstakingly hand-colored. Meanwhile Trew was compiling his own books for publication, using engravings of Ehret's fine watercolours as illustrations. Trew published the ten-part *Plantae Selectae* between 1750 and 1773, and a book on garden plants, *Hortus Nitidissimus*, between 1750 and 1792. Ehret's illustrations were included in other books; for example, a fine engraving of *Pinus pinaster* illustrates the *Description of the genus Pinus* (1803–24), written by Aylmer Bourke Lambert. Ehret worked for a year unsuccessfully as curator of the Oxford Botanic Garden but quarreled with Humphrey Sibthorp, Professor of Botany. Ehret was then given patronage by the Duchess of Portland, a keen horticulturist and collector, who asked Ehret to train her daughters in botanical illustration. Though Ehret did not produce any major illustrations devoted to his work, he served as a great teacher and an inspiration to the botanical illustrators who were to follow. In the next few years, botanical illustration was to reach the height of observational accuracy.

Humphrey Sibthorp himself visited Vienna in 1785 to study the manuscripts that recorded the ancient teachings of Dioscorides. Here he became friendly with Ferdinand Bauer, a young and promising botanical painter, and the two embarked on one of the greatest floral books ever published. Together they toured Greece in 1786 and 1794, the findings of which were immortalized in the *Flora Graeca*, which Sibthorp wrote with J. E. Smith. The illustrations are of high quality, in the form of copper engravings watercolored by hand, and by the time the finished work appeared (1806–40) it represented the best of eighteenth-century botanical illustration. Ehret, nevertheless, remained the finest plant illustrator in Europe until Franz Bauer (1758–1840) came to Kew Gardens at the invitation of Sir Joseph Banks. The two Bauer brothers, Franz and Ferdinand, became a formidable pair of talents, and they brought illustration to a pitch of perfection that has not been exceeded.

The most famous botanical artist of the period was the Belgian Pierre-Joseph Redouté (1759–1840), who was born at St. Hubert in the Ardennes. His greatest influence, as a young man, was Charles Louis L'Héritier de Brutelle (1746–1800), who commissioned Redouté to illustrate his *Stirpes Novae aut Minus Cognitae* (1764–85). Some of the plates were printed in color *à la poupée*. In this technique, colored inks are used on differing parts of the engraved plate, producing a full-color printed image. The engravers began working on plants collected from the Spanish colonies when, in 1786, they were instructed to return them to the Spanish Ambassador in Paris. Realizing what a valuable property they held, Redouté and L'Héritier had the collection shipped urgently to London, where they were cared for by Sir Joseph Banks. Redouté was thus introduced to many members of English scientific society. (L'Héritier, who

returned to Paris, was murdered in the streets in 1800.) Redouté gained an
additional reputation through his role as protege of Josephine Bonaparte, and
he illustrated books by Augustin de Candolle and Philippe la Peyrouse. As
the century closed, he was working on his *Liliacées,* published in Paris between
1802 and 1816, which some observers have claimed as the greatest illustrated
work in the history of botanical science.

EARLY TECHNICAL PROBLEMS

The production of botanical illustrations was a labor of considerable magnitude. The drawings themselves took time, but the painstaking rendering of
each anatomical feature as fine scratches on a copper sheet proved to be an
arduous task. As in sculpture, the removal or indentation of the raw material
had to be done with care. Mistakes during the engraving process were not easy
to repair.

Robert Thornton (1768–1837) planned to publish *A New illustration of the
sexual system of Carolus Linnaeus.* The first part appeared in 1799, and Thornton found he could no longer bear the cost of production. He appealed to
friends in Parliament, who prepared an Act that would allow Thornton to
hold a national lottery to pay for publication. Although further portions
of the great work appeared up to 1807, the lottery was insufficient to pay for
the project, and Thornton ended his career as a publisher of scientific illustrations a ruined man.

Other projects never came close to completion. Sir Joseph Banks had embarked on the grandiose publication of a folio work, the *Florilegium.* Massive copper plates were engraved of the new plant species from Australasia,
the intention being to print them in full color *à la poupée.* The cost proved to
be unbearable, and although some pages were published inked with black ink
only, the great set of engraved plates was wrapped in paper and stored away.
They were rediscovered, in nearly perfect condition, in storage in London and
were published by Editions Alecto in 1990. The editors had the plates meticulously polished and plated, and they were then inked with contemporaneous
pigments and the pages produced with a nineteenth-century printing press.
The original engraving had been carried out in 1770, so it took more than
two centuries for the final work to appear. Prolonged delays in publication
are not unique in the field of scientific illustration. Eustachio's *Tabulae Anatomicae Viri* took 162 years from completion of the engraved plates to publication
of the completed work (1552–1716).

ACKNOWLEDGED AND UNACKNOWLEDGED REUSE

Human fascination with the dynamics of the animal world, an organic rival to
human existence, had for centuries given rise to countless fanciful descriptions

of demons, dragons, and denizens of the deep. To illustrators unaccustomed to unfamiliar forms of animal life, the stories brought back by travelers from far-off lands were hard to interpret as images on paper. Thus, the mysterious one-horned rhinoceros was portrayed as a unicorn, and the first images of elephants looked like hogs with an elongated snout. Plagiarism was rife.

The process of rendering a three-dimensional animal as an illustration is more complex than the portrayal of a plant: unlike herbarium specimens, collections of pressed animals are not available. For this reason, images were often reengraved from published illustrations. Some eighteenth-century authors acknowledged their sources (for example, Henry Baker paid homage to van Leeuwenhoek, whose images Baker copied in his own books), but most others protested that the work of others was all their own and continued to plagiarize the work of earlier illustrators since that was far easier than creating an image *de novo*. It has been observed that the exaggerated protestation of unprecedented novelty and originality in the introduction to an illustrated book is usually a sure sign that the images were derivative and not original. The publisher who feels the need to insist that the reader should never imagine he has borrowed someone else's inspiration must, on deconstruction, have felt the need to disclaim responsibility. In some cases the results are amusing, as in the example of the great auk, which was vividly portrayed by the Danish naturalist and collector Ole Worm in the *Museum Wormianum* of 1655. His own great auk was a house-trained pet that he regularly took for walks. Naturally, it wore a collar of silk, and the engraver who portrayed it in the great book figured the silken band around the bird's neck. For centuries thereafter, the great auk was shown with a pale collar around its neck, as though this were a feature of the plumage of that species. Rarely was plagiarism so fittingly traced to its origins. In the modern era, published images are often used by scientific illustrators as points of reference for a new piece of artwork. The use of a photograph, or a published plate, as a reference is now being recognized in the commercial world of scientific publishing. The fees for the use of a picture as a reference are becoming comparable to standard reproduction charges.

Early eighteenth-century zoological illustration was marred by a host of books containing mythical creatures like those from two or three centuries earlier. François Valentijn published *Oud en Nieuw Oost-Indien* between 1724 and 1726, filling the book with extraordinary images of sea creatures allegedly drawn from life. These did little to celebrate the wondrous realities of nature, for they were grotesque versions of distorted and unreal creatures already published by Louis Renard in his *Poissons, Écrevisses et Crabes* (1719). The book, published in Amsterdam and dedicated to the King of England, is filled with hand-colored illustrations of a lurid and unreal nature. They have always been dismissed as figments of a vivid imagination. However, a scholarly investigation by Theodore W. Pietsch, published in 1995, has shown how most of the figures can be related to existing species. In this analysis, the comical images published by Renard were based on real studies after all. The artist imposed such artificialities on the drawings brought back from afar as to make the images grossly distorted and, at first, unrecognizable. Pietsch may have done much to rehabilitate an illustrator previously held to have invented more than he observed.[7]

ZOOLOGY: A NEW REALISM

In the field of ichthyology, which was one of economic importance as well as a burgeoning field of pure scientific study, the trend toward representational accuracy makes the eighteenth century the crucial period of refinement. Some of the most attractive illustrations of fish appeared in Japan at this time, where natural history illustrations were beginning to acquire increasing importance. The Eastern cultures embodied more cultural resonances than the strictly representational illustrators of the West. The spiritual elevation of the eel, for instance, gave these fish the status of objects of desire in China and Japan, whereas they were traditionally reviled in the West; the national iconographies reflect these distinctions. Examples surveyed by Aramata (1989) clearly exemplify the difference.[8]

In Europe between 1785 and 1794, Carl von Meidinger published *Icones Piscium,* with vivid hand-coloured engravings. The technique of illustrating fish species reached its heights with perhaps the most attractive and appealing illustrated work on fish ever to be published. Compiled by Marcus Elieser Bloch (1723–1799) and published under the title *Icthyologie, ou Histoire Naturelle Générale et Particulière* (1785–97), it is a work of stunning beauty and impressive size. The large folio pages are decorated with striking and confident engravings of fish species from around the world. The images are hand-colored, and silver paint is frequently used to convey a sense of realism to the fish.

Elsewhere in zoology, the refinement of illustration was being harnessed to record the new realism that was seen in science. Butterflies, always a popular subject (frequently seen in religious paintings and Books of Hours), were well represented by the engraver's art. Eleazar Albin (1713–1759) prepared many beautiful studies which were published as hand-colored engravings. As was common at the time, the name of the sponsor of each plate was featured as a dedication prominently displayed below each image. The *Natural History of English Insects* (1720) featured a series of well-observed studies. It was followed by many others, including a small pocket book, *Natural History of English Song*

---

[7] Published as Theodore W. Pietsch, *Fishes, Crayfishes and Crabs* (Baltimore, MD: Johns Hopkins University Press, 1995).

[8] Hiroshi Aramata, *Fish of the World* (Tokyo: Heribonsha, 1989), includes some artwork from Eastern sources. Conventional histories rely upon a tradition of Eurocentrism, which this wide-ranging book serves to correct.

Birds (1737) which contained plain engraved plates that the owners would embellish with coloring of their own. Albin was not above plagiarism. The frontispiece of his *Natural History of Spiders, and other Curious Insects* (1736) shows a fine engraving of the author himself, seated on horseback and surrounded by arthropods of various shapes and sizes. The most prominent feature is an "original" study of a mite, which is in reality copied line by line from Robert Hooke's celebrated *Micrographia,* which had originally appeared in 1665. Butterflies and moths were meticulously portrayed in *The Aurelian* (1766), by Moses Harris, who seems to have provided the inspiration for a book by Jacob L'Admiral titled *Nauwkeurige Waarnemingen omtrent de Veranderingen van Veele Insekten* (1774), in which the illustrations are somewhat less reliable. Images of great charm, and in hand-painted plates of Technicolor vividness, were published by J. C. Sepp in *Beschouwing der Wonderen Gods* (1762–1860). Symmetrical patterns of butterflies and moths appear in Benjamin Wilkes's *Twelve New Designs of English Butterflies,* published in 1742. The plates in most surviving copies are printed in black ink, although some copies of the rare hand-colored edition are also in library holdings.

By the turn of the century, some of the most beautiful books on the lepidoptera were in print. James Edward Smith published his *Natural History of the Rarer Lepidopterous Insects of Georgia, from the Observations of John Abbott* in 1797, and the wondrous variety of insect life was featured in books such as Dru Drury's *Illustrations of Natural History, wherein are Exhibited Figures of Exotic Insects* (1770–82), which shows how appealing hand-colored copper-engraved images can be. The peak of perfection in insect illustration must be Edward Donovan's *Epitome of the Natural History of the Insects of India.* It was engraved and richly colored in the closing years of the century and was released to an adoring public in 1800.

Roesel von Rosenhof, originally a painter of miniatures (and a good microscopist), published *Historia Naturalis Ranarum* in 1759, filled with richly detailed hand-coloured illustrations of amphibia and occasional reptiles in their natural habitats. Although the drawings have a certain stylistic boldness, they are of unmistakable realism. The frontispiece, showing salamanders and frogs clustered around an engraved plaque, is one of the most vivid and memorable in the history of scientific illustration. The shells of the mollusks became popular objects for collectors, and many books were published on onchology, corals, and the like. The year 1742 marked the publication of two works with striking folio plates: a book by Antoine-Joseph Dezallier d'Argenville titled *Le Lithologie et la Conchyliologie* (1742) and Niccolo Gualtieri's *Index Testarum Conchyliorum.* John Ellis published *Essay towards a Natural History of the Corallines* (1755), followed by *History of the Zoophytes* (1786), both illustrated with engraved plates. Among the finest books on conchology were those with hand-colored plates. Thomas Martyn published *The Universal Conchologist* in four volumes between 1743 and 1778, and they contain so many remarkable images that few copies survive with all 160 plates intact.[9] The largest colored illustrations in any book on shells appear in the *Choix de Coquillages et des Crustaces* (1758), the images here being markedly more accurate than those in the *Neues Systematisches Conchylien-Cabinet* (1769–95) by F. H. W. Martini and J. H. Chemintz. These are such impressive and magnificent tomes that they must have acted as a stimulus to the endeavors of the natural philosopher. Here was color publishing available on a grand scale for the first time in history. The results are as vivid as anything available today, but this was in an era when the public had never before seen anything so beautiful and captivating. Within a generation, realism and vivid color had become widespread.

Not all biological illustration required vivid colors to convey its message. The development of animal anatomy during the eighteenth century was perfectly paralleled by a maturation in the engraving technique of the scientific illustrators. The exemplar is George Stubbs (1724–1806), whose painstaking dissections and diligent studies raised the anatomy of the horse to a peak of perfection that is lyrical in its beauty and impressive in its accuracy. Stubbs was born in Liverpool, the son of a leather-dresser, who encouraged the young George to study the anatomy of the animal carcasses he saw. George Stubbs took an apprenticeship at Knowsley Hall, engraving pictures from the Earl of Derby's collections, but found it uncongenial and resolved to study on his own rather than through formal channels. He therefore set out to "study from nature herself, and consult and study her only," and in this frame of mind became a painter in Leeds, mostly painting portraits. He studied anatomy under a surgeon in York and began giving lectures on the subject to medical students. In 1754 he visited Italy but soon returned to set up home in Lincolnshire. There he resolved to complete a major undertaking, later published as *The Anatomy of the Horse* in 1776 (Figure 24.3).

Stubbs lived and worked in a deserted Lincolnshire farmhouse with his partner, Mary Spencer, who was euphemistically described on different occasions as his "aunt" and his "niece." He worked continually, painstakingly removing the hide, then the muscle layers, and finally the sinews and on down to the bones. The odor of decay was heavy and oppressive, bringing complaints from neighbours many miles downwind. Stubbs, a man of great physical strength, used to carry the cadaver of a horse up several flights of stairs to his attic dissection room. The specimen was suspended on wires and ropes in a lifelike attitude while Stubbs worked his way down through the layers and meticulously recorded each detail of what he observed. He made each engraved plate himself. In some of the illustrations he portrayed the specimen as seen, adding a second plate in which an outline diagram of great delicacy of line bore the annotations by which the figure was interpreted. His engravings run in sequences, so that, by turning the pages, the reader is presented with a time-lapse voyage of discovery from the surface layers to the internal

[9] Published as Thomas Martyn, *The Universal Conchologist*, 4 vols. (London, 1743–78).

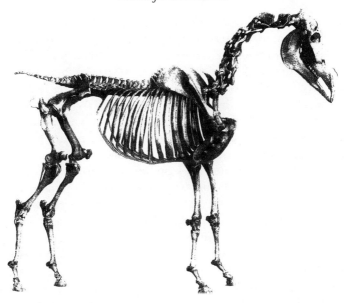

Figure 24.3 The horse skeleton by Stubbs (1776). When George Stubbs completed his work on *The Anatomy of the Horse* in 1776, it marked the first definitive account of equine bone and musculature. He dissected cadavers, suspended from the rafters of his isolated Lincolnshire farmhouse, diligently recording the results. Artwork © Brian J. Ford.

anatomy, stage by stage. The book, when published, was enough to recommend Stubbs to a wider audience. Joshua Reynolds was among his patrons at that time. Stubbs spent the rest of his life engaged in the painting of pictures, mostly of famous horses.

NEW STUDIES IN HUMAN ANATOMY

As the eighteenth century began, the ancient teachings of the much-revered anatomists still lingered. They had been much modified by Andreas Vesalius (1514–1564) whose *De Humani Corporis Fabrica* (1543), published when the author was only 28 years old, contained six hundred woodcut illustrations. It remained a popular source in the early eighteenth century. Curiously, one of the great eighteenth-century illustrated works on human anatomy was written by a contemporary of Vesalius. This was the *Tabulae Anatomicae Viri* of Bartolomeo Eustachio (1524–74). It is rich in striking copper-plate engravings. This work was completed by 1552 but was not published until 1714 when the plates were found in storage in Rome; the gap of more than one and a half centuries was exceeded only by the delay of two centuries in the publication of the *Florilegium*. Its appearance almost coincided with the publication of

William Cheselden's *Anatomy of the Humane Body* (1712). This sudden flurry of new publications served to stimulate research, and within the next decades the standards of anatomical illustration increased dramatically. The studies in Cheselden's first work were extended and improved upon in his *Osteographia* (1733), and this became a standard work of reference. The images are good, although lacking in fine detail. Some of them show the human body in action, occasionally as two forms in combat. They were inspiring images in their time and were plagiarized by Sir Charles Bell a century later when he came to compile his own work on skeletal anatomy.

One seminal new illustrated work on human anatomy was the *Tabulae Sceleti* (1747) by Bernhard Albinus, which was replete with strikingly stylized engravings with a three-dimensional appearance, each dissected human body standing in a scene of classical splendor. Some of the muscular figures were portrayed in a magnificent garden setting or against a work of sculpture; some were set in juxtaposition with a grazing rhinoceros, and, when the skeletal subject was reversed to show the opposing aspect, the rhinoceros also turned its back. This work was followed by William Smellie's fine studies. His *Sett of Anatomical Tables* (1754) takes a further step toward photo-realism. The drawings were made from his dissections by Jan van Rymsdyk and engraved on copper by Grignion. Van Rymsdyk used his years with Smellie as a training period for his greatest creative period as an illustrator, for he went on to work for the celebrated anatomist William Hunter (1718–1783). William and his younger brother John (1728–1793) revolutionized the documentation of medicine in many respects. William Hunter studied medicine at the University of Glasgow, graduating in 1750, and moved to London where he was joined by his younger brother. The latter, who did not attend medical school, went to London in 1748, and continued his studies by assisting his brother in the dissecting room. He studied surgery under Cheselden (*supra*), and his dexterity as a surgeon became renowned. He was subsequently elected Master of Anatomy at Surgeon's Hall.

William Hunter, meanwhile, specialized in obstetrics from 1756 and began to establish the science as a branch of formal medicine rather than the concern only of the midwife. With Jan van Rymsdyk at his side, Hunter made detailed studies of the human uterus in pregnancy and childbirth. He published three works, the results of his obstetrical anatomical studies appearing in *The Anatomy of the Human Gravid Uterus* (1774). This is a marvelous work, replete with rich and precise illustrations with a luminous quality of vivid realism. His work was followed, at the turn of the century, by the *Museum Anatomicum* of Eduard Sandifort (1793–1835). This fine book, which features bold – if stylized – engravings of human anatomy was not superseded until Henry Gray published his celebrated *Anatomy* in 1858. To this day, *Gray's Anatomy* is a standard teaching text, and reproductions of some of the original plates have appeared in editions published as the late twentieth century merged into the twenty-first.

A NEW VIEW: MICROSCOPY

The impact of the magnified image left its mark on eighteenth-century scientific illustration, although not to the extent one might imagine. In the late seventeenth century, Antony van Leeuwenhoek (1632–1723) had introduced the concept of a microbial universe to the world of natural philosophy, and many of his letters to the Royal Society of London were published in volume form during the early part of the eighteenth century. The published illustrations are testimony to the technical limitations of the era, for they convey only a crude impression of the vital quality that exists in the original red crayon drawings Leeuwenhoek regularly sent to London. Interestingly, Leeuwenhoek never drew; he employed a limner to capture images of his observations and directed the artist as to how to finish the study. On occasion, he records, he had to tell the appointed draftsmen to hurry up with the work, for they tended to spend time looking in wonderment at the new sights Leeuwenhoek's home-made microscopes revealed.[10]

His young compatriot Jan Swammerdam (1637–1680) lived a tortured and turbulent life, which ended at the age of forty-three when he took up a fanatical religious asceticism. Swammerdam studied insects in minute detail, using injections of mercury to emphasize the course of vessels within the dissected insect body and recording his observations in illustrations made with meticulous accuracy. After his death the papers eventually found their way into the hands of Herman Boerhaave. He wrote a biography of this lost genius, the true founder of anatomical illustration in the arthropod world, and had the illustrations published at his own expense in a grand folio volume. The book appeared as *Bybel der Natuure* in two volumes (1737–38) and became a standard reference work in its field. Although it took more than sixty years to appear in print, the book was still considered advanced in its time. The engravings reveal the extent to which a gifted observer could discern detail using the simplest of optical apparatus.

One of the books to popularize microscopy after Leeuwenhoek's time was Henry Baker's *The Microscope Made Easy* (1743). Baker refers to Leeuwenhoek

---

[10] Most of Leeuwenhoek's volumes appeared during the late sixteenth century, but those appearing after 1700 are as follows: Antony van Leeuwenhoek, *Sevende Vervolg der Brieven* (Delft, 1702); *Arcana Naturae Detecta,* third edition (Lugduni Batavorum, 1708); *Arcana Naturae Detecta,* fourth edition (Lugduni Batavorum, 1708); *Continuato Arcanorum Naturae . . .,* reprint of 1697 edition (Lugduni Batavorum, 1722); *Continuato Epistolarum,* third edition (Lugduni Batavorum, 1715); *Send-Brieven, zoo aan de Hoog-edele Heeren van de Koninklyke Societet te Londen* (Delft, 1718); *Brieven seu Werken No 19* (Delft, 1718); *Epistolae ad Societatem Regiam Anglicam* (Lugduni Batavorum, 1719); *Epistolae Physiologicae Super compluribus Naturae Arcanis,* reissue of 1718 edition (Lugduni Batavorum, 1719); *Anatomia Seu Interiora Rerum,* retranslation of 1687 edition (Lugduni Batavorum, 1722); *Omnia Opera, seu Arcana Naturae* (Lugduni Batavorum, 1722); *Continuato Epistolarum,* fourth edition (Lugduni Batavorum, 1730).

Reference can also be made to the author's *Single Lens: The Story of the Simple Microscope* (London: Heinemann, 1981); and *The Leeuwenhoek Legacy* (London: Farrand and Bristol, 1991).

and reproduces several of the Leeuwenhoek figures, in a reengraved and somewhat degraded form. Baker shows particular interest in Leeuwenhoek's observations of *Hydra viridis*, devoting a book published in 1743 to this interesting organism. Baker did little to advance our knowledge of these freshwater polyps, however, and the scientific study of *Hydra* was not further advanced until it came to the attention of Abraham Trembley (1710–1784), a Swiss-born teacher who used *Hydra* as a topic for the teaching of children in the Netherlands during his work as a private tutor. Trembley published illustrations of his experiments in the form of stipple engravings that, notwithstanding a certain artificiality of line, convey a powerful impression of the organisms in their living state (Figure 24.4). Baker's hydroids, by comparison, look moribund and distorted.

The aspects of Trembley's work of interest to the student of the history of scientific illustration concern the great delicacy of line employed in producing his plates. *Hydra* has tenuous tentacles, and they unravel into the surrounding aquatic environment as fine, undulating structures. This is all well conveyed in the plates. The remarkable analysis by Lenhoff and Lenhoff (1986) includes a full facsimile of the original, complete with folding plates where appropriate.[11] The transparency of a glass vessel (Plate 3 in the 1986 work) is carefully conveyed through the use of the finest engraved lines in the plate. In the Trembley plates, note should be made of the judicious use of stipple in the engraver's technique. This allows the most subtly graded shading, but only if done by a master of the craft. Plate 5 shows stipple used to convey an impression of a living *Hydra*. Rarely has it been used to greater effect. The illustrations for *The Universal Conchologist* were also printed from stippled plates. Great subtlety of texture can be conveyed by this exacting technique.

Glassware came to be more frequently portrayed in engravings as chemical experiments came to the fore. The research that led to Joseph Priestley's discovery of the gas Antoine Lavoisier called "oxygen" is clearly illustrated in his *Experiments and Observations on Different Kinds of Air* (1744–77). The glassware is neatly engraved in copper plates, although comparatively little attention is paid to the need to convey the refractive clarity of the glass (Figure 24.5). In France, when the essential nature of oxygen was recognized by Lavoisier, it was his wife who prepared the drawings that illustrated his published papers. She was Marie-Anne Paulze, a diligent technical artist who paid attention to the correct assembly of apparatus and recorded it faithfully in her scientific illustrations. The engraver, D. Lizars, conveys the details of the experiments in plates that, although sometimes distorted in the interests of clarity, make the nature of the experiments plain.

---

[11] See Abraham Trembley, *Mémoires d'un genre des polypes d'eau douce* (Geneva, 1744). Reproduced in annotated facsimile by S. Lenhoff & H. Lenhoff, *Hydra and the Birth of Experimental Biology* (Pacific Grove, CA: Boxwood, 1986).

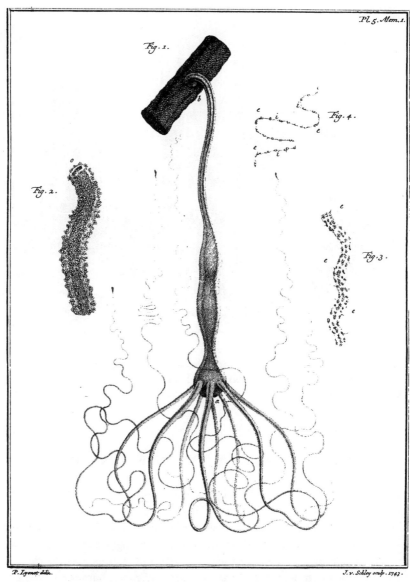

Figure 24.4. Trembley's study of *Hydra* (1744). Abraham Trembley's striking study of *Hydra* was published in his *Mémoires . . . de Polypes d'eau Douce* of 1774. This work describes his regeneration and transplant experiments, each illustrated with vivid engravings. Artwork © Brian J. Ford.

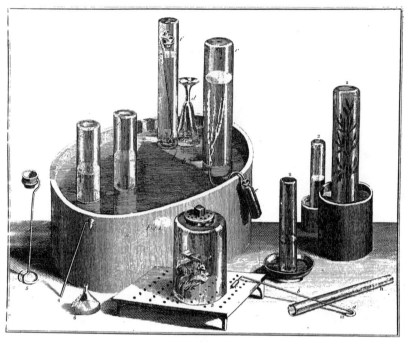

Figure 24.5. Joseph Priestley and oxygen (1774). Oxygen attracted the attention of Lavoisier and Priestley. This engraving is from Priestley's *Experiments and Observations on Different kinds of Air,* published in London 1774–7. Priestley worked also on ammonia, sulphur dioxide, and the generation of acids. Artwork © Brian J. Ford.

NEW TECHNOLOGY FOR A NEW CENTURY

The move from woodcuts to metal plate engraving brought with it a significant increase in the details it was possible to convey. In the chemical and physical sciences, as in astronomy and mathematics, line diagrams were the stock in trade of the scientific illustrator, and copper-engraved plates provided the ideal medium for graphs and maps, charts and circuits. Geological science, perhaps the most vivid of the new earth sciences, harnessed the new techniques and exploited them to the full. Jean Étienne Guettard (1715–1786) recognized that several key features of the geology of France were volcanic in origin. His concept was extended by Nicholas Desmarest (1725–1815), and the resulting impetus attracted great interest in the newly expanding science of geology. Meanwhile, an English canal-builder, William Smith, was making copious notes as he traveled the length and breadth of Britain, and from these notes he compiled the most detailed geological maps of the era. His work, published early in the nineteenth century, took the form of colored engraved sheets and laid the groundwork on which today's geological maps are based.[12]

By the end of the century, the standard of scientific illustration had reached a level rarely exceeded in present-day volumes. Earlier woodcuts had been progressively superseded by higher-quality engravings on copper, some of them inked with different pigments *à la poupée,* thus allowing the publisher to mass-produce engraved plates in color for the first time. Volumes rich in hand-colored illustrations became available, and by the end of the century lithography had appeared.

The lithographic printing process was invented by Alois Senefelder (1771–1834), at the age of twenty-six, through pure chance. In 1796 Senefelder, an unsuccessful copper engraver, jotted down a shopping list in wax crayon on a piece of smooth Bavarian limestone. It occurred to him that the wax might resist an etchant with which he could remove surrounding stone. As his experiments proceeded, he realized that the wax itself would attract ink, which would not mark the stone if it was moistened with water. Eventually, Senefelder's process was based on the production of a wax image on a stone surface, which was wetted with water prior to printing. When a thin coating of oil-soluble ink was rolled across the image, the ink adhered to the wax while being repelled by the water. A sheet of paper applied to the inked surface would thus pick up an ink copy of the original wax image. The concept was soon applied to scientific illustration, the first lithographic work of botanical illustration appearing in 1812. The first zoological book, on ornithology, appeared in 1818.

Senefelder's definitive account, a *Comprehensive Course in Lithography,* was also published in 1818. His work in the late eighteenth century led to present-day lithographic printing, in which a polymer sheet is used in place of a limestone block. The Bavarian royal family awarded Senefelder a life pension for his work, and his inspiration is extant in almost every illustrated science book of the modern era. Thus, in terms of scientific illustrators' methods, the eighteenth century bridges the gap between the earliest woodcuts and the foundations of today's printing technology.[13]

---

[12] See Jean Étienne Guettard, *Atlas et Description Minéralogiques de la France* (Paris, 1780); see also Nicholas Desmarest, *Histoire de l'Académie Royale des Sciences* (Paris, 1774); and William Smith, *A Delineation of the Strata of England and Wales with part of Scotland* (London, 1815).

[13] The work that set out the principles of lithography was Alois Senefelder, *Vollständiges Lehrbuch der Steindruckerey* (Munich, 1818), and the first illustrated work in zoology to be produced by lithographic printing was a modest volume: Karl Schmidt, *Beschreibung der Vögel* (Munich, 1818).

25

SCIENCE, ART, AND THE REPRESENTATION OF THE NATURAL WORLD

Charlotte Klonk

No interested person in the eighteenth century would have accepted the recent insistence by a professor of biology and publicizer of science that art and science are entirely separate and different disciplines.[1] Nobody was astonished in 1760, when the Society for the Encouragement of the Arts, Manufactures and Commerce, founded to promote scientific inventions for the benefit of the national economy, staged the first public art exhibition in Great Britain. In an age when politicians, scientists, and artists met in coffeehouses and clubs and debated subjects ranging from natural history to political scandals, the boundaries between art and science were not clearly marked. In eighteenth-century Europe, botany was not only a cutting-edge science but also the favorite subject matter for the drawings of aristocratic ladies; those who collected and classified shells were also the arbiters of Rococo taste, and the foremost anatomists were employed by art academies to lecture to their students.[2] Each of these areas would be a suitable subject for a chapter dedicated to science and art in the eighteenth century. So, too, might the way that some artists represented the activities of experimental scientists and technologists. Yet it was not such experimental investigators but rather the "natural historians" – descriptive explorers of the natural world – who were the central fig-

[1] Lewis Wolpert in the *Independent on Sunday,* 15 December 1996, p. 46; 23 February 1997, p. 41; and 9 March 1997, p. 41.
[2] For a survey of botanical treatises, see Wilfrid Blunt, *The Art of Botanical Illustration* (London: Collins, 1950), and Gill Saunders, *Picturing Plants: An Analytical History of Botanical Illustrations* (Berkeley: University of California Press, 1995). For the Rococo taste and shells, see Andrew McClellan, "Watteau's Dealer: Gersaint and the Marketing of Art in Eighteenth-Century Paris," *Art Bulletin,* 78 (1996), 432–53. For anatomy and art, see Martin Kemp, *Dr William Hunter at the Royal Academy of Arts* (Glasgow: University of Glasgow Press, 1975); Barbara Maria Stafford, *Body Criticism* (Cambridge, MA, MIT Press, 1991); Michel Lemire, "Fortunes et infortunes de l'anatomie et des préparations anatomiques, naturelles et artificielles," in Jean Clair (ed.), *L'âme au corps: arts et sciences 1793–1993* (Paris: Gallimard/Electra, 1993), pp. 70–101; Jean-François Debord, "De l'anatomie artistique à la morphologie," ibid., pp. 102–17.

I am very grateful to Roy Porter, Rhoda Rappaport, and Michael Rosen for their careful reading of this chapter and insightful comments.

ures in the age of the Cult of Nature, and it is their concerns that intersect most clearly with those of artists in the eighteenth century. It is on the changing perceptions of landscape as they developed among artists and natural historians that this chapter will concentrate.

THE ARCHIVE OF NATURE

On 10 December 1777 Johann Wolfgang von Goethe (1749–1832) climbed the Brocken mountain in the Harz. The sun was shining brightly when he reached the summit. Overwhelmed by what he experienced, he turned to the book of Job to express his emotions. "What is man that thou art mindful of him?" he wrote in his diary.[3] Seven years later, Goethe returned to this climb in the little essay "On Granite." He writes, "Sitting on a high and bare mountain peak and surveying a wide area, I can tell myself: Here you are resting directly on a foundation that reaches to the deepest parts of the earth; no newer layer, no piled-up, no washed-together ruins come between you and the firm ground of the original world . . . these peaks are before all life and after all life."[4]

Goethe's assertion here is unintelligible except in relation to contemporary geological research, and the work of his countryman, Johann Gottlob Lehmann (1719–1769), in particular. In an essay on the history of mountains, published in 1756, Lehmann, a mining administrator and doctor, distinguished between primitive mountains – which, he claimed, originated at the Creation – and secondary mountains, which he believed to have resulted from the Mosaic Deluge. According to Lehmann, primitive mountains, usually granitic, form the highest elevations and have steep slopes; secondary mountains, usually composed mainly of limestone, are much lower and slope gently against the steeper rise of the first class of mountains.[5] Lehmann saw this structure displayed particularly clearly in the Brocken, with its core of granite, surrounded by transitional rocks of limestone and greywacke, and followed by the secondary *Flötz* rocks,[6] and it was undoubtedly the geological interest of this formation that induced Goethe to undertake his winter expedition in the Harz.[7]

---

[3] Johann Wolfgang von Goethe, *Gesamtausgabe der Werke und Schriften*, 22 vols. (Stuttgart: J. G. Cotta, 1949–63), vol. 20 (1960), "Schriften zur Geologie und Mineralogie," ed. Helmut Hölder und Eugen Wolf, "Aus dem Tagebuch," p. 107.

[4] Ibid., "Über den Granit," p. 323.

[5] Johann Gottlob Lehmann, *Versuch einer Geschichte von Flötz-Gebürgen* (Berlin: Gottlieb August Lange, 1756). I am using the French translation by baron d'Holbach, "Essai d'une histoire naturelle de couches de la terre," which appeared as the third volume in 1759 in *Traités de physique, d'histoire naturelle, de mineralogie et de métallurgie*, trans. Paul Henri Tiry d'Holbach, 3 vols. (Paris: J.-T. Hérissant, 1759–69).

[6] Ibid., vol. 3, p. 222.

[7] Lehmann was not alone in classifying mountains into primary and secondary. Many contemporary authors across Europe, from Giovanni Arduino in Italy to Torbern Bergman in Sweden, were investigating mountains and coming to similar conclusions. See Gabriel Gohau, *A History of Geology*, rev. and trans. by Albert V. Carozzi and Marguerite Carozzi (New Brunswick, NJ: Rutgers University Press, 1990). Also Martin Guntau's "The Natural History of the Earth," in N. Jardine, J. A. Secord,

Before the end of the seventeenth century few people would have shared this enthusiasm. Mountains were generally regarded as a blemish on the earth, a manifestation of the wrath of God consequent on the Fall of Man. By the middle of the eighteenth century, however, not only had mountains come to be celebrated as examples of the glory of Creation but also they were seen as providing the most important evidence for the earth's continuing history.[8] As one eighteenth-century writer, Peter Simon Pallas, a German natural historian and explorer of the mountains in Russia, explained, "mountains . . . offer the most ancient chronicle of our globe . . . They are the archives of nature."[9] The geologists' understanding of mountains as giving direct and palpable evidence of the ancient history of the earth quickly captured the imagination of the broader European public; geology became, in the words of one writer, the favorite science of the day.[10]

An indication of the interest taken by the larger educated public in geological research is the many sumptuous publications produced by geologists. These publications were illustrated with views that drew on the traditions of landscape art familiar to the audience. Not only did the geologists borrow artistic conventions for depicting nature, but also their own interests increasingly came to inform landscape art itself.[11] The problem was how to depict what

and E. C. Spary (eds.), *Cultures of Natural History* (Cambridge University Press, 1996), pp. 211–29. A more detailed recent account of these authors and the history of geology in Europe in the eighteenth century is Rachel Laudan, *Mineralogy to Geology: The Foundations of a Science, 1650–1830* (Chicago: University of Chicago Press, 1987).

[8] The history of geology has many ramifications, such as the social and political conditions of each country in which the research was carried out, the debates over the role of fossils, and the interests of state mining bureaucracies, training schools, and private landowners, which supplied a utilitarian motive for the study of minerals. Good surveys are Gohau, *A History of Geology*, and, with extensive further bibliographical references, Laudan, *Mineralogy to Geology*. For Britain in particular, see Roy Porter, *The Making of Geology: The Foundations of a Science, 1650–1830* (Cambridge University Press, 1977). On the debate regarding the role of fossils, see Martin J. S. Rudwick, *The Meaning of Fossils: Episodes in the History of Palaeontology* (New York: Science History Publications, 1972), and, with a particular emphasis on visual representations, Rudwick, *Scenes From Deep Time: Early Pictorial Representations of the Prehistoric World* (Chicago: University of Chicago Press, 1992).

[9] Peter Simon Pallas, *Betrachtungen über die Beschaffenheit der Gebürge* (Frankfurt: n.p., 1778), p. 48.

[10] A. C. von Ferber's preface to Giovanni Arduino, *Sammlung einiger mineralogisch-chymisch-metallurgisch- und oryktographischer Abhandlungen* (Dresden: Waltherische Hofbuchhandlung, 1778), p. ii. It should be pointed out that I am using the words "geology" and "geologists" anachronistically, although in accordance with most late twentieth-century writers on the subject. The study of the earth, which is now understood by the word "geology," came under many descriptions in the eighteenth century, such as cosmology, mineralogy, natural history, and lithology. It was only at the end of the eighteenth century that the word "geology" came into use more commonly, assuming a distinct identity separate from, for example, mineralogy or biology (Gohau, *A History of Geology*, pp. 2–5).

[11] One art historian who has particularly looked at the illustrations that accompanied the work of naturalist travelers is Barbara Maria Stafford, in her *Voyage into Substance: Art, Science, Nature, and the Illustrated Travel Account, 1760–1840* (Cambridge, MA: MIT Press, 1984). Whereas Stafford explicitly excludes from her discussion a high proportion of landscape art, because it did not match the empiricism of the naturalists, Timothy F. Mitchell, *Art and Science in German Landscape Painting, 1770–1840* (New York: Oxford University Press, 1993), hardly discusses imagery produced by the latter. A ground-breaking article that specifically looked at the visual imagery that geologists used in the eighteenth century to transmit their research to a wider audience is Martin J. S. Rudwick, "The Emergence of a Visual Language for Geological Science, 1760–1840," *History of Science*, 14 (1976),

Goethe had brilliantly described in his essay "On Granite": to find a form of representation capable of evoking an awareness of nature's long history. Although many geologists experimented with depictions of geological sections and schematic diagrams, there was, as yet, no universally accepted set of conventions for theoretical sections or the other more abstract, formalized, and theory-laden modes of representation that later evolved.[12] The great majority of pictorial representations by geologists are – certainly during the second half of the eighteenth century – simply landscape views.

HISTORY PAINTING AND COSMOGONIES

On the face of it, the established traditions of landscape art would not appear to have been particularly promising for the naturalists' purposes. The most influential traditional account is to be found in Sir Joshua Reynolds's (1723–1792) *Discourses on Art*. In the *Discourses*, delivered at the Royal Academy between 1769 and 1790, Reynolds presented an essentially Aristotelian view. According to Reynolds, even the most beautiful natural forms are inherently deficient and hence are unsuitable material for serious art as they are immediately given. Artistic beauty comes from an ideal distillation of nature's particular, empirical forms. Since the Renaissance, landscape art had been counted among the inferior branches of art, but Reynolds argued that the same distinction between higher and lower forms was applicable within landscape art itself. Works could be classed as higher or lower according to the degree to which the artist departed from nature as directly observed and so made his or her subject striking to the imagination. In the broader hierarchy of genres, history painting was considered to involve the greatest degree of imagination, the greatest involvement of the intellect. In Reynolds's opinion, landscape was at its highest when it came closest to "history painting" (the name given to mythological or religious subjects in which the human figure plays a predominant role). This was epitomized, according to Reynolds, in the work of the seventeenth-century masters Salvator Rosa, Peter Paul Rubens, Claude Lorrain and Nicolas Poussin.[13] The idealized Roman landscapes of Claude, Poussin, and Rosa, as well as Rubens's rhythmically enclosed visions of nature, are suffused with the aura of their mythological and historical subjects, however reduced in size the figures themselves may be in the paintings. The ideal of beauty in these landscapes embodied the Aristotelian notion of an

149–95. This article has stimulated further recent research by art historians into the pictorial representation of landscape used in scientific publications. See Susanne B. Keller, "Sections and Views: Visual Representation in Eighteenth-Century Earthquake Studies," *British Journal for the History of Science*, 31 (1998), 129–59; and Charlotte Klonk, *Science and the Perception of Nature: British Landscape Art in the Late Eighteenth and Early Nineteenth Centuries* (New Haven, CT: Yale University Press, 1996).

[12] Rudwick, "The Emergence of a Visual Language," p. 152.

[13] Sir Joshua Reynolds, *Discourses on Art*, ed. Robert R. Wark, 2nd ed. (New Haven, CT: Yale University Press, 1975), pp. 237–8.

eternally stable nature. Yet, as far as science was concerned, this notion had become defunct by the beginning of the eighteenth century.

The early modern period saw many cosmogonies that accepted that the earth had not remained unchanged since its creation. These cosmogonies relied heavily on hypothetical conjectures and on one privileged piece of historical information: the testimony derived from the first five chapters of Genesis – in particular, the account of the Creation and the Flood. Typical of the theories that appeared at the end of the seventeenth century was Thomas Burnet's *Telluris Theoria Sacra* (Sacred Theory of the Earth) (see Rhoda Rappaport, Chapter 18 in this volume).[14] Burnet (1635–1715) gave a physical interpretation of Biblical claims and in this way attempted to make scriptural events rationally intelligible. He explained the formation of mountains and plains as the outcome of a process of collapse, an event that he related to the surging of waters during the Deluge. Burnet's frontispiece to his second volume (Figure 25.1) gives a simultaneous representation of the history of the earth, as conceived from his Christian perspective. It shows Christ – above whose halo appears the famous phrase from the Book of Revelations: "I am Alpha and Omega" – standing astride the first and last phases of the earth's development, each represented by a distinctively different-looking globe, starting with the dark primeval Earth and the smooth paradisiacal state without mountains or seas. The paradisiacal earth is followed by a globe covered by the agitated waters of the Deluge. The fourth image shows the present state of the earth, with mountainous continents surrounded by the oceans, and the fifth is the anticipated conflagration in which the earth will be consumed by fire, to be followed by the New Earth, similar in aspect to the paradisiacal state, and, finally, the New Heavens.[15]

In this kind of history, which treats the Bible as its empirical guide, there is no conflict between the conventions and traditions of high art and those of scientific representation. The image is historical, even if human beings are absent, because it is based on testimony about the history of the earth.

What is more surprising is that the writer who most vehemently attacked the Christian conceptions of authors such as Burnet chose to illustrate his own account by recourse to the same pictorial traditions. Georges-Louis Leclerc, comte de Buffon (1707–1788) became famous in 1749 when the first volumes of his *Histoire naturelle* were published. By the time that the last volume appeared, posthumously, in 1804, it had reached its forty-fourth volume. In the first volume, Buffon presented his theory of the earth's history, supporting his account with a range of empirical evidence that he terms "proofs" (see Rhoda Rappaport, Chapter 18 of this volume).[16] Buffon restricted explanation

---

[14] Thomas Burnet, *Telluris Theoria Sacra*, 2 vols. (London: G. Kettilby, 1681, 1689).

[15] For a discussion of this image and others in Burnet's publication, see Stephen Jay Gould, *Time's Arrow, Time's Cycle*, 2nd ed. (London: Penguin, 1988), pp. 21–59.

[16] Georges-Louis Leclerc, comte de Buffon, *Histoire naturelle, générale et particulière*, vol. 1 (Paris: Imprimerie Royale, 1749).

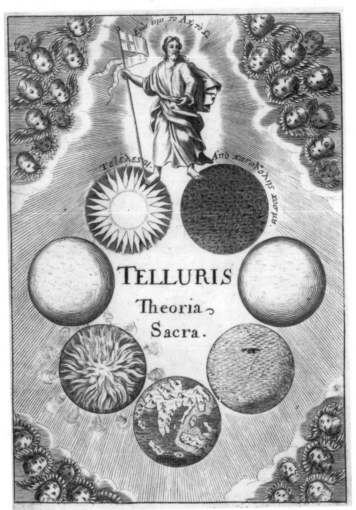

Figure 25.1. Frontispiece, engraving, from Thomas Burnet, *Telluris Theoria Sacra,* vol. 2 (London: G. Kettilby, 1689), Bodleian Library, University of Oxford, 4° C.49 Th.

to causal processes that could still be observed. On this assumption, the magnitude of the effects to be observed in natural phenomena required, he argued, a much longer time span than the six thousand to eight thousand years of the Christian account.[17] Although, at this stage of his career, Buffon (like

---

[17] Ibid., p. 79. Buffon was not the first to break away from the Biblical account. His countryman René Descartes made no effort to reconcile with Genesis the mechanical explanation of the formation of the earth that he gave in 1644, and, earlier in the eighteenth century, Benoît de Maillet came to the conclusion that mountains were the result not of one event, such as the Creation or the Deluge, but rather of successive phases in the earth's history.

most other eighteenth-century writers) believed that mountains were basically the result of the retreat of the ocean that had once covered the whole earth, he did not believe that the required consolidation of suspended material could have been possible within the time span given in the Bible, as argued by earlier writers. For contemporary mountains to have arisen within the period allowed in the Bible would have required miracles or the operation of physical laws then unknown.

Although Buffon did much to redirect his contemporaries' energies toward field observation, he himself did not refrain from offering his own hypothetical account of the earth's formation, albeit one that did not follow the story of Genesis. In the third section of the first volume of the *Histoire naturelle,* titled "Proofs of the Theory of the Earth," Buffon gives an account of the formation of the planets and the earth. He suggests that a comet brushed against the sun and detached a stream of molten matter, which then split into the six planets. The comet's detachment of matter from the sun is illustrated in a magnificent image (Figure 25.2).[18] Instead of Buffon's comet simply communicating motion to a quantity of matter, we see, surprisingly, God the Father, surrounded by angels, swirling dramatically into the picture from the top right. This God, in his movement and gesture, is clearly inspired by the most famous of all depictions of the Creation, Michelangelo's first three scenes in the Sistine Chapel. It is likely, of course, that the artists employed by Buffon produced their designs in the tradition of Biblical illustrations of the Creation already familiar to them. For the scene of the creation of the planets Buffon employed a little-known Irish artist, Nicholas Blakey, who worked in Paris. But the image is too close to Buffon's description of the detachment of matter by a comet not to have been executed under his supervision. In Buffon's illustration the six planets have just been formed but are still fireballs surrounded by cloud. God's pointed finger suggests that it was He who directly caused the planets to detach themselves from the sun, rather than the force of impulsion that Buffon goes to great lengths to calculate.

The *Histoire naturelle* was also illustrated with vignettes that open each chapter and that are, like the images for the *Theory of the Earth,* conventional in conception. The principal artist employed for these was Jacques de Sève, a painter who mainly illustrated works by classical authors. The vignettes are mostly allegorical depictions, a genre that, of course, already had a long tradition both in book illustration and in painting. The chapter "Natural History of Man" in the second volume of the *Histoire naturelle* is headed by a vignette (Figure 25.3) showing a shepherd proudly commanding a paradisiacally peaceful tableau of a cow, a horse, a stag, a lion, some goats, a boar, and a dog, all placed in front of a generalized mountainous landscape.[19] Behind the shep-

---

[18] This image was placed facing the earlier general introduction to the "History and Theory of the Earth" (ibid., opposite p. 65).

[19] Buffon, *Histoire naturelle,* vol. 2, p. 429.

Figure 25.2. Nicholas Blakey, illustration for "De la formation des Planètes," engraving (St. Gessard), from Georges-Louis Leclerc, comte de Buffon, *Histoire naturelle, générale et particulière*, vol. 1 (Paris: Imprimerie Royale, 1749), opposite p. 65, Bodleian Library, University of Oxford, RSL. 1996 d. 433/1.

herd (possibly Adam counting the animals) a classically draped seminude female sits dictating to an angel. The female figure is an allegory; her many breasts identify her as Natura (Mother Earth). Allegorical representations of this kind traditionally evoked a conception of nature as eternal and unchanging. Yet, just as the religious imagery chosen by Buffon to illustrate the *Theory of the Earth* contradicted the secular explanation to be found in it, so, too, it was just this idea of nature as permanent and unhistorical that the *Histoire naturelle* challenged.

Allegorical representations remained popular throughout the eighteenth

Figure 25.3. Jacques de Sève, vignette for "Histoire naturelle de l'homme," engraving (F. A. Aveline), from Georges-Louis Leclerc, comte de Buffon, *Histoire naturelle, générale et particlière,* vol. 2 (Paris: Imprimerie Royale, 1749), p. 429, Bodleian Library, University of Oxford, RSL. 1996 d. 433/2.

century as frontispieces to natural history publications as well as on the walls of art exhibitions. Buffon's arch rival, Carolus Linnaeus (1707–1778), for example, included the multibreasted Natura, or Diana *polymastos,* in half length on a pedestal featuring flora and fauna and set in the middle of a landscape among elks and goats as a frontispiece to his *Fauna Suecica* of 1746. At this stage, this was simply another representation of the generalized idea of nature.[20] A decade later, in the tenth edition of his *Systema naturae,* however, it became significant for Linnaeus's work: the allegorical figure of Mother Nature with her many breasts gained significance in his classification of those animals (including humans) having hair, three earbones, and a four-chambered heart, as "mammalia" (although, in fact, breasts were the least significant common characteristic of the group).[21]

NATURE'S LONG HISTORY AND THE EMERGENCE OF THE SUBLIME

The images so far discussed had a significant tradition in book illustrations and elsewhere. With their religious or allegorical subject matter, they might be

[20] Carolus Linnaeus, *Fauna Suecica* (Stockholm: L. Salvius, 1746).
[21] Londa Schiebinger, *Nature's Body: Sexual Politics and the Making of Modern Science* (London: Pandora, 1993), chap. 2.

included within Reynolds's esteemed category of history painting. One early eighteenth-century naturalist, however, went further than any other toward bringing together geological discoveries and the established tradition of landscape art. In so doing he came to modify that tradition considerably. Johann Jacob Scheuchzer (1672–1733), a doctor of medicine and professor of mathematics at Zurich, is today known mostly for his contribution to the debate about fossils and, in particular, for his discovery of *Homo diluvii testis* – the Man who Witnessed the Flood.[22] Like most of his contemporaries, Scheuchzer did not conceive of his scientific work as antithetical to the teachings of theology, and he also produced a commentary on the Bible magnificently illustrated with scenes ranging from Genesis to the Apocalypse.[23] In the introduction, he explains that, as a teacher of mathematics and physics, he had attempted to explain the Holy Bible according to the principles of recent philosophy and science. But, in looking at illustrated Bibles, he had found that all the plates were "either only concerned with the histories [those scenes in which the human figure plays a predominant role], or stemmed almost entirely from the self-fabricated imagination of the artist or engraver and are not correctly composed and wrongly conceived."[24] Scheuchzer engaged the imperial court engraver, Johann Andreas Pfeffel (1674–1748) of Augsburg, under whose direction eighteen engravers worked on the publication. Scheuchzer also employed (and closely supervised) two artists: the local painter Melchior Füßli (1677–1736), who did the views, and Johann Daniel Preißler (1666–1737) from Nuremberg, who was responsible for the ornate baroque frames that surround the scenes. These frames are particularly remarkable. In many cases, they are used to convey some of the scientific knowledge with which Scheuchzer wished to substantiate the account given in the Bible, even if, as in some cases, this did not immediately fit the main narrative illustrated in the image.[25]

Scheuchzer's *Physica sacra,* as he came to call his Bible commentary, opens with the first two days of the Creation, the initial darkness and chaos being followed by the creation of light and the separation of heaven and earth.[26]

---

[22] Melvin E. Jahn, "Scheuchzer: Homo Diluvii Testis," in Cecil J. Schneer (ed.), *Toward a History of Geology* (Cambridge, MA: MIT Press, 1969), pp. 193–213.

[23] Johann Jakob Scheuchzer, *Kupfer-Bibel, in welcher die Physica Sacra, oder geheiligte Natur-Wissenschafft derer in Heiliger Schrift vorkommenden natürlichen Sachen deutlich erklärt und bewährt von Johann Jakob Scheuchzer,* 4 vols. (Augsburg: C. U. Wagner, 1731–5). The work first appeared in German and Latin and a year later also in French.

[24] Scheuchzer, *Kupfer-Bibel,* vol. I, preface [p. 2].

[25] See, for example, ibid., plate XXIII, opposite p. 29. The main scene shows Adam sitting in Paradise just after his creation. This is surrounded by a frame displaying an extraordinary life-size illustration of the development of the human fetus from the human egg "the size of an anis corn" (no. I) to a four-month-old embryo (no. XI). The frame shows not only the human fetal development, but also its cyclicity (the embryos holding the eggs) as well as conveying a baroque *memento mori* message (the embryos are skeletons, and one wipes its tears). Yet, at this stage in the Creation as illustrated in Scheuchzer's "Picture-Bible," the *Physica sacra,* there is no Eve, without whom this cycle could not begin.

[26] This first section of Scheuchzer's publication is admirably discussed and amply illustrated in Martin J. Rudwick, *Scenes from Deep Time,* pp. 4–17.

These illustrations still fall within the tradition of depictions of the globe as seen from outer space. On the Third Day, however, when the mountains and rivers and oceans are created, Scheuchzer introduces the first of his landscape views (Figure 25.4). We see a bare mountainous view through which a river meanders into the sea in the background. From the rock in the foreground to the left, some water still runs off into the river, an illustration of God's command that the waters that had covered the earth should gather together in the seas and let the dry land appear. The rocks appear rugged and have bizarre outlines, similar to those to be found in the fictional mountain scenes of the sixteenth-century German artist Albrecht Altdorfer. The latter were explicit fantasies that existed beyond actual landscapes, human concerns, or the passage of time.[27] Scheuchzer's landscape, although still a place without a recognizable topography, is very much a landscape that already has and will continue to have a history. And it is presented accordingly: its elaborate engraved baroque frame marks the image as seeking a status equivalent to that of history painting. As yet, however, this is nature without human beings or vegetation and is quite uninhabitable, and this is marked in the image by the abrupt way in which the scene confronts the viewer. In contrast to the later landscapes in the *Physica sacra,* the viewer of the picture is given no imaginary standpoint within the scene.

After illustrating the Creation, taking the opportunity to discuss botany and zoology, Scheuchzer moves on to discuss life in the Garden of Eden and after the Fall. He then comes to the Deluge, the event that most naturalists at that time took to be responsible for the way in which mountains then appeared. In the text Scheuchzer gives an account of the current physical explanations. He agrees with the theory that the mountains were formed under the surface of the Deluge, to be revealed after the latter's retreat. But Scheuchzer had also observed that not all strata in the Alps were disposed in even horizontal layers. These irregularities led him to speculate that God, after the Deluge, had displaced a number of formerly horizontal beds and raised them above others.[28] Scheuchzer gives an illustration of this phenomenon in a plate showing three views of the Alps around Lake Urn, the eastern part of the Lake Vierwaldstätter (Figure 25.5). In this plate, the artist, Füßli, adapts the existing topographical tradition to the depiction of natural phenomena. Topographical art existed alongside the more stylized modes of landscape representation that Reynolds deemed to be of higher value. Topographical art concentrated on the delineation or description of features of a particular locality, often covering a wide terrain as in Scheuchzer's illustrations. The central object of interest in topographical depictions was human settlement or activity. In Scheuchzer's illustrations, however, the single recognizable feature of the image is the dramatic formations of the mountains.

[27] Christopher S. Wood, *Albrecht Altdorfer and the Origins of Landscape* (London: Reaktion, 1993).
[28] Scheuchzer, *Kupfer-Bibel,* vol. 1, p. 64.

TAB. VI.

GENESIS Cap. I. v. 9. 10.

Opus tertiæ Diei.

I. Buch Mosis Cap. I. v. 9. 10.

Drittes Tagwerck.

Figure 25.4. Melchior Füßli and Johann Daniel Preißler, *Genesis Cap. I. v. 9. 10. Opus Tertiae Diei*, engraving (J. A. Corvinus), from Johann Jakob Scheuchzer, *Kupfer-Bibel, in welcher die Physica Sacra, oder geheiligte Natur-Wissenschafft derer in Heiliger Schrift vorkommenden natürlichen Sachen deutlich erklärt und bewährt von Johann Jakob Scheuchzer,* vol. 1 (Augsburg und Ulm: Christian Ulrich Wagner, 1731), plate VI, by permission of the British Library, London, 9. g. 1.

Figure 25.5. Melchior Füßli, *Genesis Cap. VII. v. 21. 22. 23. Cataclysmi Reliquia,* engraving (J. A. Corvinus), from Johann Jakob Scheuchzer, *Kupfer-Bibel, in welcher
die Physica Sacra, oder geheiligte Natur-Wissenschafft derer in Heiliger Schrift vorkommenden natürlichen Sachen deutlich erklärt und bewährt von Johann Jakob Scheuchzer,*
vol. 1 (Augsburg und Ulm: Christian Ulrich Wagner, 1731), plate XLVI, by permission of the British Library, London, 9. g. 1.

A seventeenth-century Dutch artist, Jan Hackaert, had already produced
topographical depictions that featured Alpine scenery, probably as a result of a
commission from merchants in Amsterdam wanting to develop the trading
route into Italy. But Scheuchzer was the main influence in leading artists to
concentrate on the mountains. Felix Meyer, usually accepted as the founder

of Swiss landscape painting, was one of the artists to provide sketches illustrating Scheuchzer's books on the Alps.[29] Scheuchzer started publication of the *Description of the Natural Histories of Switzerland* in 1706. The third part, published in 1708, included an account of a tour through the Alps he had undertaken in 1705.[30] Not only did Scheuchzer provide many geological and mineralogical observations, but he also gave an account of the Alps as the positive result of divine purpose, rather than, as Burnet still believed, being the ruins of the world as it was before the Deluge. Around the turn of the eighteenth century, to restore a sense of divine purpose, wisdom, and beneficence more and more writers drew attention to the physical function of mountains as part of the natural water cycle, pointing to the many springs and rivers that rise in them. However, the images in the *Description of the Natural Histories of Switzerland* add a further element; they reflect awe at the work of the great Master Builder.[31] Not only is God wise and good, the message is, but also His work is superhuman and thus awe- and terror-inspiring.

The image *Planten Bruck* (Figure 25.6) was drawn by the artist Melchior Füßli who seems to have accompanied Scheuchzer on his tours through the Alps. This view shows two male figures dressed in contemporary clothes and standing on a little plateau in the foreground in front of a deep gorge. The stream known as the Sandbach, the main source of the river Linth at the edge of the Glarner Alps, flows through the gorge. The figures are dwarfed by the steep elevation of the dark cliffs on either side, an effect that is magnified by the fact that we cannot see the tops of the mountains and so gain an idea of their scale. Further up the mountains, a bridge, which gives the plate its name, spans the two sides of the gorge. On the bridge there is a cart drawn by an ox. Two further tiny figures appear on a path to the right leading to the bridge. The bridge itself is fragile-looking in comparison with the solidity of the mountains on either side. This image of overwhelming mountains is one of the first pictorial renderings of the esthetic of the sublime in which awe – an emotion of mingled terror and exultation, thought to be appropriate to God alone – was transferred to the natural realm. When Scheuchzer writes that the mountains display an immensity exceeding our perceptual capacity but that our mind "is able to comprehend," he is anticipating the same conception that Immanuel Kant would later refine and systematize in his chapter on the sublime in *The Critique of Judgement*.[32]

From now on, the sublime was to become the standard esthetic mode

---

[29] For an introductory account of the development of Swiss landscape art and the role of natural scientists, see Peter Wegmann, "Felix Meyer und Caspar Wolf: Anfänge der malerischen Entdeckung der Alpen," *Gesnerus*, 49 (1992), 323–39.

[30] Johann Jakob Scheuchzer, *Beschreibung der Natur-Geschichten des Schweizerlands*, 3 parts (Zürich: author, 1706–8).

[31] Ibid, p. 176.

[32] Immanuel Kant, *Die Kritik der Urteilskraft*, 1790, ed. Wilhelm Weischedel (Wiesbaden: Suhrkamp, 1957), pp. 164–91.

Figure 25.6. Melchior Füßli, *Planten Bruck,* engraving, from Johann Jakob Scheuchzer, *Beschreibung der Natur-Geschichten des Schweizerlands,* part 3 (Zurich: author, 1708), opposite p. 27, by permission of the British Library, London, 444. b. 19.

within which mountains were perceived.[33] Instead of drawing the curtains of their coaches while traveling through the Alps on their way to Italy, travelers on their Grand Tour would actively seek out such sights.[34] The grand and vast character of the mountains, with their dreadful precipices and irregular

[33] The classical work on the emergence of the sublime is Marjorie Hope Nicolson, *Mountain Gloom and Mountain Glory: The Development of the Aesthetics of the Infinite* (Ithaca, NY: Cornell University Press, 1959). She does not, however, deal with the sublime as a pictorial category.

[34] For an account of the esthetic appreciation of mountains and early tourism in the Alps, see Monika

features, was the principal key to their sublime character, but there was also a temporal dimension: the mountains were believed to be the residues of the first ages of history.

After Scheuchzer, many naturalists explored the Alps and described their age and sublimity both in their texts and, frequently, also in the accompanying illustrations. Artists all over Europe soon adopted this esthetic for the depiction of mountain scenes whose market had become assured. In depicting the immensity of physical size, the sublime analogously suggests the huge temporal gulf that separates the modern observer from the formative events in the earth's history. Scheuchzer believed that all mountains were essentially of the same age; they all came into being at the time when the Deluge retreated. Lehmann, on the other hand, believed that mountains had a more varied history, consisting of three distinct stages. The first and second stages stemmed from the Creation and Deluge, and the third embodied minor later alterations of the earth's crust. However, as in the case of Buffon, a view of the history of the earth as developing over a much longer period began to spread in the eighteenth century. The mineralogist Abraham Gottlob Werner (1749–1817) believed that the earth had undergone multiple transformations over millions of years since the Creation. Despite important differences about when and how rock structures were formed (see Rhoda Rappaport, Chapter 18 in this volume), the authors who took this view agreed on two things: that rocks had an aqueous origin and that the massive forces that had created them were now exhausted and no longer in operation. It comes as no surprise that Horace-Bénédict de Saussure (1740–1799), probably the most famous pioneer of alpine research, uses the same style of depiction as his predecessors. Saussure's account of the history of the earth, although it involved more complex and extensive changes than Scheuchzer's, still referred to events happening at a period far remote from the present.

Saussure captured the imagination of the European public by climbing Europe's highest summit, the dangerous Mount Blanc, in 1787.[35] Saussure was usually accompanied on his tours by the artist Marc Théodore Bourrit (1735–1815). On the occasion of the ascent of Mont Blanc, Bourrit set out with the group but, too weak to carry on, he had to return to Chamonix, and so Saussure's famous assault remained unillustrated. A year later, Saussure undertook another tour into the Mont Blanc massif, this time to the Giant's Needle, where he and his son, accompanied by guides, spent seventeen days. Here Saussure carried out the extensive scientific experiments that, because of the shortage of time, he had been unable to do on Mont Blanc. Bourrit was again absent,

Wagner, "Das Gletschererlebnis - Visuelle Naturaneignung im frühen Tourismus," in Götz Groáklaus and Ernst Oldemeyer (eds.), *Natur als Gegenwelt: Beiträge zur Kulturgeschichte der Natur* (Karlsruhe: von Loeper Verlag, 1983), pp. 235–63.

[35] The account appeared in the fourth volume of Saussure's *Voyages dans les Alpes*, whose publication stretched from 1779 until 1796. Horace-Bénédict de Saussure, *Voyages dans les Alpes*, 4 vols. (Neuchâtel: S. Fauche et al., 1779–96).

Figure 25.7. Théodore de Saussure, *Vue de l'aiguille du Géant, prise du côté de l'Ouest,* engraving, from Horace-Bénédict de Saussure, *Voyage dans les Alpes,* vol. 4 (Neuchâtel: Louis Fauche-Borel, 1796) plate III, Bodleian Library, University of Oxford, 4°BS. 391.

but this time Saussure's son, Théodore, provided the illustrations that accompany the account. One image (Figure 25.7) shows a view of the Giant's Needle from the west across the vast white expanse of the Entrèves glacier. This is set off by a stretch of dark rock in the foreground on which Saussure and his entourage had set up camp. The camp consisted of two tents and a little stone hut, which provided not only a place to sleep at night and protection against the intense sun during the day but also laboratory space (Saussure carried with him a hygrometer and thermometer, a barometer, a graphometer, an electrometer, and a magnetometer). On the left, a pole is visible from which Saussure suspended the hygrometer and thermometer for some of his experiments, and, between the tents, a figure is shown collecting water melted from snow, the group's only source of liquid. Although the picture is precise in its rendering of the location of the camp and the vertical structures of the rock formations around it (mentioned particularly by Saussure), the image emphasizes the sublime character of the spectacle – something that Saussure himself did not fail to note, despite being busy with his scientific observations.[36] The picture is marked with lettering, giving the viewer an orientation in an otherwise overpowering expanse. The emptiness of the scene is emphasized by the tininess of the figures on the glacier. Saussure's depiction is more than an attempt to convey the topography of the area. It embodies

[36] Ibid., vol. 4, p. 220.

an esthetic appreciation of the Alps whose sublime emptiness conveys the vastness of the mountains, in both their perceived spatial and imagined temporal scale.

The sublime is by no means, however, the only esthetic mode employed by Saussure. The beautiful and the topographical style also appear. In his preface, Saussure praises the countryside of Switzerland for embodying all the genres of landscape, from terrible mountains to the smiling plains with their "mild image of spring in a fortunate climate."[37] "Smiling plains" framed within a prospect of hills and valleys was the ideal of landscape beauty found by Reynolds and others in the pictures of Claude Lorrain.[38] Saussure did, however, initiate one new pictorial tradition that would flourish in the early nineteenth century. In the first volume of his *Tour in the Alps,* he included a "*Vue Circulaire*" (Figure 25.8), taken from the summit of the glacier of Buet by Bourrit. It was Saussure's idea to give a 360-degree view by drawing concentric circles on graph paper, and Bourrit's picture shows apparent perspectival distortions in the near foreground. This is characteristic of the fish-eye depiction that results from Saussure's technique.[39] All-round topographical views became immensely popular in the numerous panoramas that followed Robert Barker's first purpose-built rotunda, opened in Leicester Square, London, in 1793. In this case the visitor moved about as he or she took in a series of perspectival renderings, displayed at a distance in order to avoid the foreground distortions of Saussure's technique. At the same time, prints of fish-eye views started to circulate widely.

Saussure's impressive folio volumes were clearly intended to reach a wide public. That public would have understood what his use of different esthetics implied: the topographical for a more descriptive account, the sublime for the ancient history of the mountains and the beautiful for the eternally recurring nature of life on earth. Nor was this esthetic division restricted to the visual. Witness Goethe again in his essay "On Granite." A descriptive account of the ancient uses of granite and its characteristics is followed by his sublime experience on the Brocken. But then Goethe's viewpoint changes. He describes hunger and thirst making themselves felt, and he turns his gaze down to the prospect of fertile fields underneath him and envies "the inhabitants of these verdant, resourceful plains who . . . rake up the dust of their ancestors,

---

[37] Ibid, vol. 1, pp. viii–ix.
[38] A less evocative and more descriptive style was used to illustrate the important geological observations such as the S formation, found, for example, at the Cascade of Arpenaz near Faucigny (ibid., vol. 1, plate IV), which Saussure explained, after dismissing subterranean fires, as due to "*refoulement*" (sideways push) (ibid., vol. 4, p. 183). At this point, however, Saussure stopped short. As Gohau has argued, he had clearly broken the Wernerian heritage of aqueous explanation of all rock formations by recognizing compression "but seemed incapable of proceeding any further" (Gohau, *A History of Geology,* p. 117).
[39] Bourrit did, however, include a sideways view of the base of the glacier. At least, the image is shaded as if it had been taken from the front. Saussure duly notes that he could not remember having seen this, ibid., vol. 1, p. 512.

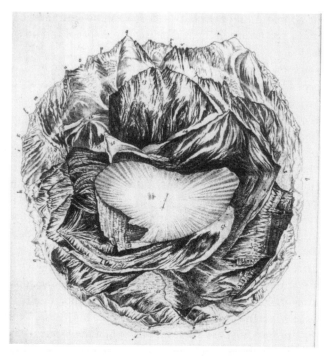

Figure 25.8. Marc Théodore Bourrit, *Vue Circulaire des Montagnes qu'on découvre du sommet du Glacier de Buet,* engraving, from Horace-Bénédict de Saussure, *Voyage dans les Alpes,* vol. 1 (Neuchâtel: Samuel Fauch, 1779), plate 8, Bodleian Library, University of Oxford, 4° BS. 388.

and calmly satisfy the small need of their days in a small circle."[40] The ideal beauty celebrated by Poussin and Claude Lorrain, the vision of an eternal cycle of human life and nature, did not lose its appeal with the rise of the sublime.

In the latter half of the eighteenth century the scope of the sublime esthetic came to be extended from the experience of high mountains to include other, less ancient but equally overwhelming, phenomena, such as violent sea storms, earthquakes, and volcanoes. This was mainly due to the publication of Burke's *A Philosophical Enquiry into the Origins of our Ideas of the Sublime and Beautiful,* which appeared in 1757.[41] Burke was the first to explore systematically the effect of untamed nature on the human mind. For Burke, it was not just mountains that produced the effect of the sublime, but whatever is terrible as it affects the sense of sight. When a massive earthquake shook Calabria

[40] Goethe, "Über den Granit," p. 324.
[41] Edmund Burke, *A Philosophical Enquiry into the Origin of our Ideas into the Sublime and Beautiful,* ed. J. T. Boulton (London: Routledge, 1958).

and the Sicilian trading city of Messina, the Neapolitan Royal Academy of Sciences immediately sent a scientific expedition to investigate the devastated landscape. A year later a sumptuous folio publication appeared whose engravings carefully recorded the devastation left behind. Its ruling pictorial formula was that of the sublime.[42] Mount Vesuvius, another violent natural feature, had attracted much attention following its major eruption of 1631. Most seventeenth-century artists depicted the volcano in a way that associated the event with the emblems of divine intervention.[43] A Jesuit, Athanasius Kircher, included a picture of Mount Vesuvius erupting in his *Mundus subterraneus* of 1655. Kircher believed that volcanoes were an intrinsic part of God's design for the earth, an outlet for the central fire that powers the earth's mechanism and keeps it "alive" like an animal's body.[44] Vesuvius was almost continuously active between about 1750 and the early nineteenth century. At this period, it was treated by artists, such as the French painters Claude-Joseph Vernet and Pierre-Jacques Volaire, as a sublime natural spectacle. By the time Joseph Wright of Derby (1734–1797) visited Naples in 1774 and witnessed an eruption of Vesuvius, there already existed a considerable market for the kind of paintings that he went on to produce over the next twenty years: pictures that, through the dramatic contrast between the cool light of the moon and the sizzling red heat of the exploding mountain, are overwhelming to the eye and thus, in Burke's account, sublime.

It is likely that Wright's interest in Vesuvius was stimulated by the communications sent to the Royal Society in London by Sir William Hamilton (1730–1803), the British envoy to Naples. Hamilton's descriptions were published in the *Philosophical Transactions* from 1768 onward.[45] His accounts were intended to promote the acceptance of the major role played by volcanoes in shaping the earth. Hamilton hoped to change attitudes toward volcanoes from sublime spectacles to equally sublime pieces of evidence for the great age of the earth. Although they were not remnants of the first ages of the earth, for Hamilton, volcanoes were just as important for deciphering the earth's varied past. He believed that no traces of the original crust of the earth had survived and that its present appearance was entirely the result of action by rain, sea, earthquakes, and volcanoes.[46] Two close friends of Wright – John Whitehurst and Erasmus Darwin, both of whom had a keen interest in

[42] Keller, "Sections and Views," pp. 150–9.

[43] Alexandra R. Murphy, *Visions of Vesuvius* (Boston: Museum of Fine Arts, 1978).

[44] Athanasius Kircher, *Mundus subterraneus,* vol. 1 (Amsterdam: J. Janssony & E. Wyerstraet, 1665), between pp. 14 and 15.

[45] The first two letters by Hamilton were received with enthusiasm by the Royal Society and were published in 1768 and 1769 ("An Account of the Last Eruption of Vesuvius," *Philosophical Transactions of the Royal Society,* 57 (1768), 192–200, and 58 (1769), 1–12). Hamilton continued to communicate his observations to the Royal Society until 1795.

[46] John Thackray, "'The Modern Pliny': Hamilton and Vesuvius," in Ian Jenkins and Kim Sloan (eds.), *Vases & Volcanoes: Sir William Hamilton and his Collection* (London: British Museum, 1996), p. 69.

geology – were members of the Royal Society and may very well have drawn his attention to Hamilton's accounts. When he saw the eruption in Naples, Wright regretted that Whitehurst was not there to see it with him.[47]

In 1776 Hamilton published his studies of volcanoes in Sicily in what has rightly been called "one of the most lavish books of the eighteenth century."[48] The text of *Campi Phlegraei: Observations on the Volcanos of the Two Sicilies* was a reprint of six of Hamilton's communications to the Royal Society. They were accompanied, however, by fifty-four magnificent, hand-colored engravings, and a further five illustrations were issued as a supplement in 1779.[49] The artist employed for the illustrations was the Neapolitan Pietro Fabris (active 1756–1784). Fabris worked under Hamilton's close supervision. The images in *Campi Phlegraei* range from topographical views to close-up depictions of different specimens of volcanic matter; from sublime night views of Vesuvius's eruption to beautiful vistas of the coast of Sicily. At times, the focus is more descriptive. One illustration, for instance, shows the changes that took place in the crater of Vesuvius during an eruption in 1769; Hamilton was almost obsessive in his attempt to register every moment of an eruption and every trace left behind afterward. Even though Fabris' images clearly follow the contemporary pictorial formulas of the beautiful and the sublime, this is not at the expense of precise observation. Hamilton was personally responsible for directing the attention of several landscape artists (in particular the German Jakob Philipp Hackert) toward a closer study of natural phenomena in their paintings than had hitherto been the case.[50] The view of the big eruption of Vesuvius in 1767 (Figure 25.9) displays the sublime spectacle of light and color, and yet at the same time attention is paid to the particular way the lava flows. But, as in the depictions of the Alps, the categories of the sublime and the beautiful fulfill more than a narrowly esthetic role.[51] The depiction of the sublime force of the lava flows indicates Hamilton's conviction that successive violent events through the earth's long history were capable of entirely reshaping its appearance.

In the 1750s, at the same time as Vesuvius was starting its extended period of activity, other geologists were finding evidence of volcanic activity in regions where no active volcano was apparent (see Rhoda Rappaport, Chapter 18 in this volume). Jean-Étienne Guettard was the first person to identify the

---

[47] David Fraser, "Joseph Wright of Derby and the Lunar Society," in Judy Egerton (ed.), *Wright of Derby* (London: Tate Gallery, 1990), p. 15.

[48] John Thackray, "The Modern Pliny," p. 70.

[49] Sir William Hamilton, *Campi Phlegraei: Observations on the Volcanos of the Two Sicilies* (Naples: n.p., 1776), and *Supplement to the Campi Phlegraei* (Naples: n.p., 1779).

[50] Mark A. Cheetham, "The Taste for Phenomena: Mount Vesuvius and Transformations in Late 18th-Century European Landscape Depiction," *Wallraf-Richartz Jahrbuch*, 45 (1984), 137.

[51] Plate XVII in *Campi Phlegraei* is an example of an image that is reminiscent of Claude Lorrain's beautiful arcadia. Hamilton uses the classical ideal of beauty in the pictures to make the same point as do his references to classical authors: Strabo, Pliny, Justin, and others all confirm the antiquity of volcanic activity.

Figure 25.9. Pietro Fabris, *View of the great eruption of Vesuvius from the mole of Naples in the night of the 20th Oct. 1767*, hand-colored engraving (P. Fabris), from Sir William Hamilton, *Campi Phlegræi: Observations on the Volcanos of the Two Sicilies* (Naples: n.p., 1776), plate VI, by permission of the British Library, London, 33. h. 6.

mountains of the Auvergne as extinct volcanoes. But it was Nicholas Desmarest (1725–1815) who, in 1765, after intensive research in the same area, first identified basalt as being of volcanic origin.[52] Desmarest's claim was striking in two respects. In the first place, Hamilton had not observed the production of prismatic basalt in any of the active volcanoes that he studied. On the other hand, basalt had been found among sedimentary strata in areas where no volcanoes had been observed. If Desmarest's thesis was true, it meant that volcanic activity had been much more widespread than previously assumed. Two naturalists – Rudolf Erich Raspe (1737–1794) and Barthélemie Faujas de Saint-Fond (1741–1819) – produced further research in Germany and France to confirm Desmarest's claim. Both believed that volcanic activity was to be considered an ancient geological phenomenon, and thus it comes as no surprise that they illustrated their work with images that were sublime in character.[53]

The extraordinary shape of polygonal basalt columns had been noted before their association with extinct volcanoes caused them to be seen as part of the ancient history of the earth and, hence, suitable subjects for sublime depiction. The most famous site of basalt columns was the Giant's Causeway in County Antrim on the north coast of Ireland. The first investigation into its striking arrangement of vertical columns appeared in the Royal Society's *Philosophical Transactions* in 1693.[54] This article, by Sir Richard Bulkeley, was essentially descriptive, drawing attention to a phenomenon that was regarded as a freak of nature. It did, however, stimulate further interest, which lasted well into the eighteenth century.[55] The Rev. Samuel Foley quickly replied to Bulkeley's account and published his observations together with notes by Thomas Molyneux (1661–1733) in the *Philosophical Transactions* a year later.[56] This was accompanied by the first illustrations of the Causeway by Christopher Cole, a map inset in a kind of topographical view.[57] As the article makes clear, the artist himself realized that the view was a hybrid between a map and

---

[52] This was, however, not published until 1774 and 1777: Nicolas Desmarest, "Mémoire sur l'origine et la nature du basalte à grandes colonnes polygones," *Histoire de l'Académie Royale des Sciences, Année 1771*, Paris (1774), 705–75; "Mémoire sur le basalte, troisième partie," *Histoire de l'Académie Royale des Sciences, Année 1773*, Paris (1777), 599–670.

[53] Rudolf Erich Raspe, "A Letter . . . containing a short Account of some Basalt Hills in Hassia," *Philosophical Transactions of the Royal Society*, 61 (1771), 580–3 (on Raspe's theory, see Albert V. Carozzi, "Rudolf Erich Raspe and The Basalt Controversy," *Studies in Romanticism*, 8 (1969), 235–50). Bartélemi Faujas de Saint-Fond, *Recherches sur les Volcans Éteints du Vivarais et du Velay* (Grenoble: J. Cuchet et al., 1778) (see, for example, plate II, opposite p. 271).

[54] Richard Bulkeley, "Part of a Letter Concerning the Giants Causeway," *Philosophical Transactions of the Royal Society*, 17 (1693), 708–10.

[55] Sergei Ivanovich Tomkeieff, "The Basalt Lavas of the Giant's Causeway District of Northern Ireland," *Bulletin Volcanologique*, 6 (1940), 89–95.

[56] Sam Foley, "An Account of the Giants Causway [sic] in the North of Ireland . . . ; and Notes thereupon, serving for farther Illustration thereof by T. Molyneux," *Philosophical Transactions of the Royal Society*, 18 (1694), 169–82.

[57] The early visual representations of the Giant's Causeway are discussed in Martyn Anglesea and John Preston, "A Philosophical Landscape: Susanna Drury and the Giant's Causeway," *Art History*, 3 (1980), 253–73.

a view.[58] Molyneux continued to study the area, with particular attention to the superposed segments of the basalt and their jointing. He was the first to identify the rock as "Lapis Basaltes." Dissatisfied with the limited topographical and geological information derivable from Cole's designs, Molyneux proposed to members of the Dublin Society that another artist should be commissioned to produce a second picture. The result was a much more typical topographical view by Edwin Sandys. Sandys's view was published in the *Philosophical Transactions of the Royal Society* in 1697, a year before the publication of Molyneux's own further research.[59] Although this view (Figure 25.10) is more precise in its depiction of the topography of the countryside and the jointing of the basalt, it still contains pictorial exaggeration of the ruggedness of the coast as well as the rock formation. An especially striking inaccuracy is that the pillars in the hills to the left are shown with tree tops.[60]

This fact should alert us to the errors that can occur in the depiction of a phenomenon to which a naturalist wants particularly to draw attention. Although most educated people in the eighteenth century were taught drawing, the complexities of a topographical view usually exceeded their own skill and so the naturalist would have had to commission a professional artist. The resulting drawing would then be given to an engraver, who could make significant alterations to the initial design, particularly with regard to the texture of the formation. Sandys's drawing would have been engraved in London, and it is quite likely that this misinterpretation of the columns as trees was the engraver's doing.[61] Almost all the naturalists who provided illustrations to accompany their work emphasized in their introductions what care was given to the supervision of the plates.[62]

How treacherous it could be to base empirical conclusions solely on such visual representations is shown in the case of Nicholas Desmarest. Desmarest, who had not himself been to Ireland, saw engravings by François Vivarès of the Giant's Causeway. Desmarest judged that the hillside of Aird Snout, as represented in one of the engravings, consisted of volcanic cones like those of the extinct volcanoes in the Auvergne.[63] From this he concluded that the basalt of the Causeway, like that of the Auvergne, was of volcanic origin. This conclusion remained in dispute until the end of the eighteenth century. Until his death, Abraham Werner upheld the view that basalt was a sedimentary rock. In 1745 Richard Pocock had given an account of the Giant's Causeway as the result of the repeated precipitation of particles suspended in water or

[58] Foley, "An Account of the Giants Causway," p. 175.

[59] Thomas Molyneux, "A Letter . . . Containing some additional Observations on the Giants Causway [sic] in Ireland," *Philosophical Transactions of the Royal Society*, 20 (1698), 209–23. The map appeared in *Philosophical Transactions of the Royal Society*, 19 (1697), opposite p. 777.

[60] Anglesea and John, "A Philosophical Landscape," pp. 255–6.

[61] This point is made in Anglesea and John, "A Philosophical Landscape," p. 257.

[62] See, for example, Faujas de Saint-Fond, *Recherches*, p. xviii; Hamilton, *Campi Phlegraei*, p. 5; Saussure, *Voyages*, vol. 1, p. xviii.

[63] Anglesea and John, "A Philosophical Landscape," p. 261.

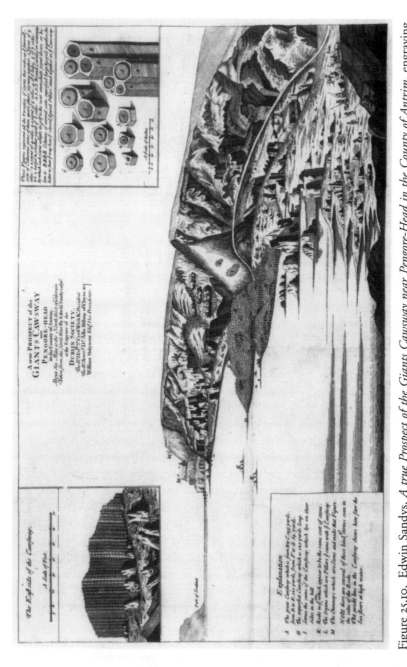

Figure 25.10. Edwin Sandys, *A true Prospect of the Giants Causway near Pengore-Head in the County of Antrim*, engraving, from *Philosophical Transactions of the Royal Society*, 19 (1697), opposite p. 777, Bodleian Library, University of Oxford, AA 75 Med.

mud. Although Desmarest proved to be right, the observations on which he based his conclusion were wrong. There is no volcanic cone in the hills behind the causeway. The prints Desmarest saw were engraved by Vivarès from two drawings done by Susanna Drury from Dublin, who published them in 1743–4. They show views of the Giant's Causeway from the west (Figure 25.11) and from the east, with features marked by letters. Desmarest's error notwithstanding, these images are remarkable for their detail and accuracy. The legend makes it clear that Susanna Drury was familiar with the debate regarding the nature of the stone and the importance of its jointing.[64] The images are also remarkable achievements in the art of topography for this early date. Drury, who was probably trained by a Dutch artist,[65] was evidently influenced by the Dutch tradition of topography. It was only somewhat later that this tradition really came to flourish in England as a consequence of the increased need for military surveys.[66] But topographical depictions, however much they gave precise geological information regarding one particular locale, were not capable of conveying what was of particular concern to eighteenth-century naturalists: a sense of the vast extent of the history of the earth.

BEYOND THE IMMEDIATELY OBSERVABLE: GEOLOGICAL SECTIONS AND DIAGRAMS

Desmarest himself commissioned a map from two royal engineers to depict his understanding of the successive periods of volcanic activity in the Auvergne.[67] Using five different seminaturalistic symbols, he indicated ancient volcanic flows, ancient massifs of molten lava, modern flows, flows with isolated older prismatic basalt, and those with basalt balls. Other writers had already experimented with showing their findings in geological maps, but they tended to confine themselves to indicating with "spot-symbols" the distribution of outcrops.[68] Rarely did they attempt to show the extent of the underlying strata or their assumed temporal succession. An exception is the *Petrographical Map* appended by Johann Friedrich Wilhelm Charpentier (1738–1805) to his publication on the mineralogical geography of the Electorate of Saxony of 1778 (Figure 25.12).[69] Charpentier used washes of color to indicate eight

[64] Ibid., p. 261.   [65] Ibid., pp. 262–3.

[66] Michael Charlesworth, "Thomas Sandby Climbs the Hoober Stand: The Politics of Panoramic Drawing in Eighteenth-Century Britain," *Art History,* 19 (1996), 247–66.

[67] Desmarest, "Mémoires," 1774, opposite p. 774. For the development of the geological map, see V.A. Eyles, "Mineralogical Maps as Forerunners of Modern Geological Maps," *The Cartographic Journal,* 9 (1972), 133–5; Rudwick, "The Emergence of a Visual Language," pp. 159–64; Endre Dudich (ed.), *Contributions to the History of Geological Mapping: Proceedings of the Xth INHIGEO Symposium* (Budapest: Akad'emiai Kiad'o, 1984).

[68] Rudwick, "The Emergence of a Visual Language," p. 160.

[69] Johann Friedrich Wilhelm Charpentier, *Mineralogische Geographie der Chursächsischen Lande* (Leipzig: S. L. Crusius, 1778).

THE WEST PROSPECT OF THE GIANTS CAUSWAY IN THE COUNTY OF ANTRIM IN THE KINGDOM OF IRELAND

Figure 25,II. Susanna Drury, *The West Prospect of the Giant's Causway in the County of Antrim in the Kingdom of Ireland*, 1743/4, engraving (F. Vivarès), paper size c.69 x 44 cm, © The British Museum.

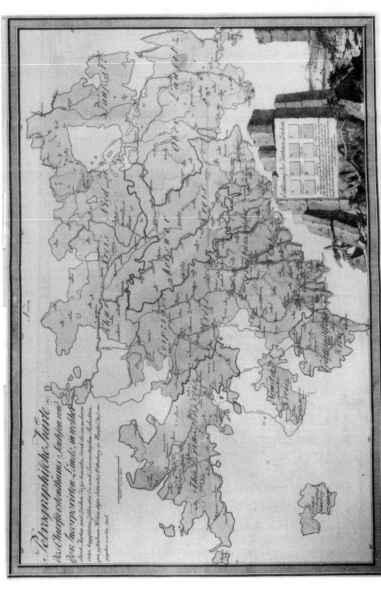

Figure 25.12. Johann Friedrich Wilhelm Charpentier, *Petrographische Karte des Churfürstentums Sachsen und der Incorporierten Lande*, hand-colored engraving, from Johann Friedrich Wilhelm Charpentier, *Mineralogische Geographie der Chursächsischen Lande* (Leipzig: S.L. Crusius, 1778), plate I, Bodleian Library, University of Oxford, 4° I. 407.

different types of rock together with the more conventional spot-symbols for other outcrops. This alone, of course, does not imply anything about the temporal succession of the strata. But instead of representing hills in abbreviated form, sideways on and almost all the same size, as had previously been common in maps, Charpentier depicts the mountains almost from above and in relief. This he achieved by hatching the slopes. The higher the slopes, the darker the hatching appears.[70] In this way, Charpentier made it possible for the informed reader to draw conclusions, at least tentatively, about the presumed ages of the rock formations. The colors pink and violet, which denote granite and gneiss, respectively, for example, are generally found on Charpentier's map in the areas with the highest mountains. Granite, as we have seen, was widely believed to be the oldest rock, forming the highest mountains with the steepest slopes. Charpentier endorsed this view and also added gneiss to the constituents of ancient mountains. By contrast, the colors to be found on the map in the lower and more level regions reveal that these parts of the country were almost entirely covered by more recent, secondary, Flötz rocks.[71] A map such as this one, which goes beyond marking the evidence of outcrops, already incorporates theoretical assumptions that go beyond immediate visual evidence. As Rudwick has pointed out, in using color to show "an extrapolation of surface outcrops between actual scattered exposures of the rocks," Charpentier's map is an early example of the increasingly abstract visual representations used by geologists to communicate their research.[72] After the turn of the century, geological maps would also contain three-dimensional extrapolations of the succession of underground strata as well as Charpentier's extrapolation of surface outcrops.

Naturalists in the eighteenth century did experiment with three-dimensional representations of the succession of strata, but there was no attempt to incorporate this theoretical assumption into maps. Lavoisier's vertical sections to indicate the stratigraphic arrangement of the earth in particular areas covered by the map were printed in the margins of the *Atlas et Description minéralogiques de la France* (1780) which he prepared with Guettard and Antoine Monnet.[73] Sections of strata had been depicted and published in the seventeenth century, but these were general diagrams to illustrate cosmological or cosmogonical theories. In the eighteenth century, sections came to be representations of ob-

---

[70] The watercolor artist Paul Sandby has been credited with introducing this innovation in mapmaking while working for the Board of Ordnance in Scotland (Jessica Christian, "Paul Sandby and the Military Survey of Scotland," in Nicholas Alfrey and Stephen Daniels (eds.), *Mapping the Landscape: Essays on Art and Cartography* (Nottingham: University Art Gallery, 1990), pp. 18–22. How to shade mountains to indicate height remained controversial throughout the nineteenth century; see Ulla Ehrenvärd, "Color in Cartography: A Historical Survey," in David Woodward (ed.), *Art and Cartography: Six Historical Essays* (Chicago: University of Chicago Press, 1987), pp. 123–46.

[71] Charpentier, *Mineralogische Geographie*, pp. 386–8.

[72] Rudwick, "The Emergence of a Visual Language," p. 162.

[73] Rhoda Rappaport, "The Geological Atlas of Guettard, Lavoisier, and Monnet: Conflicting Views of the Nature of Geology," in Cecil J. Schneer (ed.), *Toward a History of Geology*, pp. 272–87.

servations made at some particular locale. John Strachey's section of a coal field in Somerset is an early example, but it does not carry temporal connotations.[74] One of the earliest horizontal sections used to show the temporal sequence of different types of rocks is Lehmann's illustration of stratification on the southern edge of the Harz mountains, published first in 1756 (Figure 25.13).[75] The sequence is established by extending downward the outcrops found in the area. Lehmann had established earlier that the elevation of rock formations decreased the younger they are. In this plate, the formation that is most vertical, numbered 31 in the plate, is identified as belonging to the class of primitive mountains. According to Lehmann, the less-inclined secondary formation starts at number 19 and ends in a nearly horizontal bed at number 1. This depiction is, of course, deeply theory-laden. The cutaway representation of the earth beneath the surface is a considerable reconstruction and extension beyond what is immediately observable – or, indeed, for the most part, observable at all. Lehmann still attempts to mediate the unobservable part of his depiction, the section, by showing the countryside above in abbreviated form with mountains and stylized trees. By the turn of the nineteenth century, even this nod toward naturalistic representation would be more or less eliminated as sections became an essential tool for geologists in communicating even more complicated structural configurations, with causal and temporal implications for the history of the earth.[76] Although the history of strata is implicit in geological sections, it would be necessary for the viewer to have an independent grasp of the underlying geological theory for that to become apparent.

An unusual and ingenious solution to the depiction of geological history was found by J. L. Giraud Soulavie (1752–1813). Soulavie, who pioneered the use of fossils in the dating of rocks, produced a comprehensive regional study of the area between Auvergne and Montpellier in his *Natural History of Southern France* in 1780 and 1784.[77] Each volume is illustrated with as many as five plates, some representations of specimens in isolation, some sections, and some depictions of basalt formations that follow contemporary conventions. But in the fourth volume, in which he is concerned with the chronology of volcanic activity, Soulavie introduces an illustration (Figure 25.14) divided into two halves: one shows what he took to be the traces left behind by primeval volcanoes, and the second shows the traces of the second and third phases. Although it is not immediately apparent (since the images lack the foreground framing to give a standpoint to viewers), the images are views of formations observed by Soulavie in two specific places in the south of France. The first one shows two granitic mountains in the Vivarais divided by a steep

---

[74] John Strachey, "A Curious Description of the Strata Observ'd in the Coal-Mines of Mendip in Somersetshire," *Philosophical Transactions of the Royal Society*, 30 (1719), plate II, opposite p. 947.

[75] Lehmann, "Essai," p. 324.

[76] Rudwick, "The Emergence of a Visual Language," pp. 164–72.

[77] J. L. Giraud Soulavie, *Histoire naturelle de la France méridionale*, 7 vols. (Nîmes: C. Belle, 1780–4).

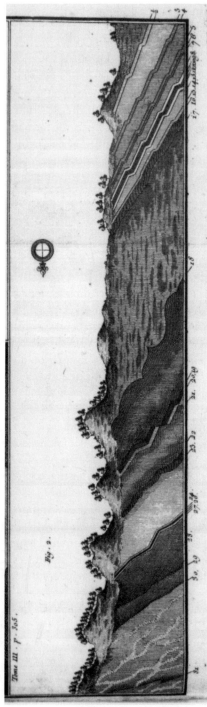

Figure 25.13. Section of stratification on the southern edge of the Harz mountains, engraving, from Johann Gottlob Lehmann, *Traités de physique, d'histoire naturelle, de minéralogie et de métallurgie*, vol. 3: "Essai d'une historie naturelle de couches de la terre" (Paris: J.-T. Hérissant, 1759), plate IV, fig. 2, Bodleian Library, University of Oxford, Vet. E5 f. 59.

Figure 25.14. *Volcan de la Première Époque, Volcan de la Seconde & Troisième Époque,* engraving, from J. L. Giraud Soulavie, *Histoire Naturelle de la France méridionale,* vol. 4 (Nîmes: C. Belle, 1718), plate 1, by permission of the British Library, 955. h. 11.

valley. In the text, Soulavie invites us to imagine the peaks as if they were joined together into a single mass. This, he says, would be the true image of that "primordial time" when the granitic mass was deposited at the bottom of the universal ocean and water currents had not yet worked to produce valleys.[78] A volcanic vein, which can be seen in the mountain to the right, is,

---

[78] Ibid., vol. 4, p. 33.

according to Soulavie, all that is left of the first phase of volcanic eruption. Soulavie's history of the earth is a history of the destruction of volcanic features by erosive forces in the course of time. The fewer the traces of this activity that remain, the more ancient is the volcanic activity. The second image shows four volcanic mountains in the Ardèche along the river Dorne. It illustrates the remains of the second and third periods, characterized by the retreat of the universal ocean and the erosive activity of fresh water. Each mountain is topped by a bed of a conglomerate of limestone, granite, and basalt, an aqueous deposit. The next traces of volcanic activity are the crowns of basalt on top of this bed. As the lines in the illustration indicate, these basalt formations were once a single mass before water carved the valleys between them.[79]

And so the volcanic history of the earth continues – altogether, Soulavie distinguishes six periods – leaving more traces of volcanic action each time until the most recent stage, when almost nothing has been destroyed. Soulavie's illustrations of the first three periods are remarkable because they combine the depiction of what can be seen today with a clear indication of the chronology of its formation. Of all the types of illustration discussed here, these are the most adequate visual equivalent to the verbal aspirations of eighteenth-century naturalists: to be able to read mountains as historians read ancient chronicles and from them to write the history of the world. In contrast to Saussure's depictions of the Alps, for example, Soulavie found a way of showing successive stages in the history of the earth without having either to evoke the sublime esthetic to indicate past time or to abandon immediately visible phenomena in favor of the abstractions of geological maps and strata depictions. However, the history of the earth as written by Soulavie, like that of most of his contemporaries, was of a planet with a past but not a future.[80] Like most naturalists of the eighteenth century, he assumed that ancient forces, such as the universal ocean, were more powerful than anything existing today and that therefore the main formation of the earth had been completed. Soulavie's natural history of the South of France appeared shortly before the French Revolution. Although, like Buffon, he clashed with the Church because he rejected the Christian picture of the earth's history, his account restricted dramatic change to the earth's geological past.

By the early nineteenth century, however, naturalists came to challenge this view. Just as the French Revolution had undermined belief in a stable, traditional social order throughout Europe, so geologists abandoned the view that dramatic upheavals took place only long in the past and came to understand the present as one stage in the continuing development of the earth. Furthermore, rather than there being a universal history for the whole world, each region was now believed to have its own unique history. The sublime and the beautiful, with their implications of vast temporal distances or timeless har-

---

[79] Ibid., vol. 4, pp. 104–6, 122–6.   [80] Gohau, *A History of Geology*, p. 107.

monious cycles, were no longer adequate forms of visual representation. Geologists and artists either adopted more theory-laden representations or produced imaginary reconstructions of worlds long lost.[81]

It was only during the nineteenth century that the dualistic view of the relationship between art and science, which is the subject of so much debate today, came to be widely held. In the eighteenth century, art and science were commonly understood as forms of knowledge with a shared procedure: using generalizations drawn directly from repeated observations. At that stage their boundaries were fluid. They could complement each other, influence each other, and generally support each other's aims. During the first half of the nineteenth century, the prestige of science as the most reliable source of knowledge about the ultimate nature of reality greatly increased, although this was resisted by writers in the German idealist tradition who claimed that art was an essential means to gain knowledge about the world. After the middle of the century, developments in the sciences came to emphasize the creative, constructive aspects of the scientific imagination and so opened the way for many of the vehemently held positions that today inform the argument about art and science.[82]

There were (and are) those who, like the professor of biology mentioned at the beginning of this chapter, see science as uniquely capable of forming a developing, objective view of the world. In this view, "science is about progress, art about change."[83] In the other camp, there are those who, because science is confined to empirical reality, attempt to transcend science in favor of religion, metaphysics, and art as more fundamental forms of human understanding. More frequently, nowadays, we find claims that the act of creation is the same in science and in art. As one author puts it, in each case its "symbols are meant as guides to a physical reality hidden beyond appearances."[84] In the eighteenth century, by contrast, what brought art and science together was not an endeavor to "express the inner beauties of nature that lay beyond the immediate senses"[85] but rather their common quest for knowledge of nature *within* the realm of the senses. When this common ground had disappeared, there would never again be a single esthetic within which all three constituencies – scientists, artists, and the broader public – could represent the long development of the earth.

[81] Some examples are discussed in Rudwick, *Scenes from Deep Time.*

[82] Maurice Mandelbaum, *History, Man, & Reason* (Baltimore: John Hopkins University Press, 1971), pp. 19–20.

[83] Lewis Wolpert in the *Independent on Sunday,* 23 February 1997, p. 41.

[84] Arthur I. Miller in the *Independent on Sunday,* 23 February 1997, p. 41 (printed in reply to Wolpert).

[85] Ibid., p. 40, and also Arthur I. Miller, *Insights of Genius: Imagery and Creativity in Science and Art* (New York: Springer Verlag, 1996).

26

SCIENCE AND
VOYAGES OF DISCOVERY

Rob Iliffe

Natural history and geographical knowledge were transformed in the eighteenth century by means of the systematic analysis of virtually all the accessible parts of the planet. From the 1760s onward, the nature of voyages with a broadly scientific goal underwent a rapid evolution. Although some degree of international cooperation was necessary to achieve this change, the increasing mastery of the Pacific was overshadowed by vigorous competition in the same area among the major European powers. Recently there has been an explosion of interest in this development, in particular among scholars located on the Pacific Rim, and the great voyages of the late eighteenth century have been linked to a number of political, imperial, and commercial contexts. Ostensibly scientific missions were usually accompanied by a set of instructions regarding the discovery of either the Northwest Passage, which was supposed to offer a northern entrance into the Pacific, or of *terra australis incognita,* an area that since classical times had been posited as necessary to "balance" the putative excess of land in the Northern hemisphere. In this chapter I survey the major explorations of the century and analyze their broad achievements in a diversity of scientific fields such as ethnography, botany, cartography, and zoology. I argue that the scientific motives behind these forays were usually bound up with, and often inextricably part of, the strategic concerns of governments in Britain, France, Russia, and Spain.[1]

[1] For the significance of mapping in the period, see J. Brotton, *Trading Territories: Mapping in the Early Modern World* (London: Reaktion, 1997). In general, see V. T. Harlow, *The Founding of the Second British Empire, 1763–93,* 2 vols. (London: Longmans, 1952–64); M. Steven, *Trade, Tactics and Territory: Britain in the Pacific, 1783–1823* (Melbourne: Melbourne University Press, 1983); A. Frost, "Science for Political Purposes: European Exploration of the Pacific Ocean, 1764–1806," in R. MacLeod and P. F. Rehbock (eds.), *Nature in Its Greatest Extent: Western Science in the Pacific* (Honolulu: University of Hawaii Press, 1988), pp. 27–44; D. Baugh, "Seapower and Science: The Motives for Pacific Exploration," in D. Howse (ed.), *Background to Discovery: Pacific Exploration from Dampier to Cook*

For comments on a previous version of this chapter, I would like to thank Elsbeth Heaman, Roy Porter, Roger Cooter, Chris Lawrence, David Edgerton, Pascal Brioist, Simon Schaffer, Janet Browne, and Larry Stewart.

Two sets of voyages epitomize the change in the scale of operations that took place during the century. At the beginning of the period, the British Admiralty equipped Edmond Halley in the *Paramore* between 1698 and 1700 to note variations of the magnetic compass and to find the longitudes of various locations on African and South American ports. Nevertheless, he failed to reach the Pacific, a feat that had been achieved by the buccaneer William Dampier a decade earlier. Dampier's book *A New Voyage Round the World* of 1697, with its important observations on ethnography and natural history in the South Pacific and especially New Holland (Australia), prompted support from the Royal Society and convinced the Royal Navy to make him a civilian captain of an expedition to find *terra australis* in 1699. On the *Roebuck,* which sank at Ascension Island in February 1701, Dampier enjoyed even worse relations with his crew than did the unfortunate Halley, although Dampier's expedition did chart some of the coastline of New Holland and he found an island he called New Britain off the coast of New Guinea. At this time all such enterprises were prey to the vagaries of scurvy, a situation not helped by the glaring inability of sailors to determine longitude to a tolerable degree of accuracy while at sea.[2]

A century after the *Roebuck*'s demise, the very names of the vessels indicated a new approach to exploration. In 1801 the British sent out the *Investigator,* commanded by Matthew Flinders, to establish a presence on the west and south of Australia before a French expedition – commanded by Nicolas Baudin in the corvettes *Le Géographe* and *Le Naturaliste* and already on its way – did the same. Despite the strategic imperial function of each undertaking, both missions were overtly "scientific." Baudin had a team of "philosophical travellers" who had been specially primed with lengthy instructions on physical anthropology and craniometry by Georges Cuvier and on ethnography and anthropology by Joseph-Marie Degérando. Both of them, like Baudin, were members of the newly formed Société des Observateurs de l'Homme, and the expedition's entourage boasted a total of seven of its members. This veritable traveling academy was more than matched by the *Investigator,* whose planning and equipping were organized by Joseph Banks, President of the Royal Society between 1778 and 1820. The best scientific instruments and clocks were ordered, and alongside the most up-to-date charts, Banks included

(Berkeley: University of California Press, 1990), pp. 1–55; G. Williams, "The Achievement of the English Voyages, 1650–1800," in Howse, *Background,* pp. 56–80; P. Petitjean, C. Jami, and A. Moulin (eds.), *Science and Empires: Historical Studies about Scientific Development and European Expansion* (Dordrecht: Reidel, 1992); D. Turnbull, *Mapping the World in the Mind: An Investigation of the Unwritten Knowledge of the Micronesian Navigators* (Geelong: Deakin University Press, 1991).
[2] N. Thrower (ed.), *The Three Voyages of Edmond Halley in the "Paramore," 1698–1701* (London: Hakluyt Society, 1981); Commander D. W. Waters, "Captain Edmond Halley, F.R.S., the Royal Navy and the Practice of Navigation," in N. J. Thrower (ed.), *Standing on the Shoulders of Giants: A Longer View of Newton and Halley* (Berkeley: University of California Press, 1990), pp. 171–202; J. C. Shipman, *William Dampier: Seaman-Scientist* (Lawrence: University of Kansas Libraries, 1962); G. Williams, *The Great South Sea: English Voyages and Encounters 1570–1750* (New Haven, CT: Yale University Press, 1997), pp. 106–30.

his own *pièce de résistance:* a greenhouse that would shelter plants against insects, rats, and seawater. With a broad range of scientific goals, these expeditions had efficacious antiscorbutics and could find their longitude accurately by two different methods. Both were concerned with the *detail* of exploration, and it was arguably this "more minute examination" that would transform discovery into commercial and imperial advantage.[3]

Voyages of discovery unlocked the potential for new sources of wealth and imperial expansion, and they captivated an audience back in Europe that was reassessing its own values by means of philosophical, literary, and ethnographic accounts of the "nature" of humankind. Deist challenges to traditional religion forced a rethinking of the truth of revealed religion and in particular of Christianity, and in the second half of the century, critiques of civilization such as those mounted by Rousseau increasingly raised questions about the depravities of the modern world. From the end of the seventeenth century, print culture spawned astonishing numbers of collections of travels and voyages that owed much to the genre of the Grand Tour narrative. Such tales made heroes of George Anson, James Cook, and Louis Antoine de Bougainville, and the plethora of *Voyages* provided numerous resources for fantasy and cultural self-assessment. In turn, contemporary literature colored the expectations both of the travelers to the South Seas and of the artists who illustrated the *Voyages* that inevitably resulted from them. Eighteenth-century fiction exhausted the possibilities of the exotic, from early castaway narratives to the self-discovery and redemption of Crusoe in 1719, and then to a host of "Robinsonades," which, like *Robinson Crusoe* itself, became extremely popular in France. Nevertheless, Edenic depictions of Noble Savages in the 1760s and 1770s were transformed into more jaundiced views of non-Europeans, hastening the appearance of missionaries in the South Seas at the end of the 1790s.[4]

[3] J. Dunmore, *French Explorers in the Pacific,* 2 vols. (Oxford: Clarendon Press, 1965–9), vol. 2, pp. 9–40; M. Hughes, "Tall Tales or True Stories? Baudin, Péron and the Tasmanians, 1802," in R. MacLeod and P. F. Rehbock (eds.), *Nature in its Greatest Extent: Western Science in the Pacific* (Honolulu: University of Hawaii Press, 1988), 65–86, 67–71; R. Jones, "Images of natural man," in J. Bonnemains, E. Forsyth, and B. Smith (eds.), *Baudin in Australian Waters . . .* (Oxford: Oxford University Press, 1988), pp. 35–64; J.-M. Dégerando, *Considérations sur les diverses méthodes à suivre dans l'observation des peuples sauvages* (1800), trans. F. C. T. Moore as *The Observation of Savage Peoples* (Berkeley: University of California Press, 1969); D. MacKay, *In the Wake of Cook: Exploration, Science and Empire, 1780–1801* (London: Croom Helm, 1981), pp. 3–6.

[4] M. N. Bourguet, "L'explorateur," in M. Vovelle (ed.), *L'Homme des Lumières* (Paris: Seuil, 1996); G. R. Crone and R. A. Skelton, "English Collections of Voyages and Travels, 1625–1846," in E. Lynam (ed.), *Richard Hakluyt and His Successors* (London: Hakluyt Society, 1946), pp. 63–140; P. J. Marshall and G. Williams, *The Great Map of Mankind: British Perceptions of the World in the Age of Enlightenment* (London: Dent, 1982); N. Rennie, *Far-fetched Facts: The Literature of Travel and the Idea of the South Seas* (Oxford: Oxford University Press, 1995); B. M. Stafford, *Voyage into Substance: Art, Science, Nature, and the Illustrated Travel Account, 1760–1840* (Cambridge, MA: MIT Press, 1984); B. Smith, *European Vision and the South Pacific: A Study in the History of Art and Ideas,* 2nd ed. (New Haven, CT: Yale University Press, 1985).

THE BACKGROUND TO SCIENTIFIC VOYAGES

For a number of reasons "voyages of discovery" might be connected to what can loosely be called "scientific expeditions," pioneering examples of which were sponsored by the Académie Royale des Sciences in the seventeenth century to procure information useful for navigation. Under the protection of Colbert and Louis XIV, the Académie actively promoted three voyages between 1668 and 1670 that were undertaken with the explicit intention of testing the feasibility of using clocks or Jupiter's satellites to determine longitude, the perennial problem of the sailor. While cartography prospered in Paris through the work of Gian Domenico Cassini and Jean Picard, Jean Richer traveled to Cayenne in 1672 and made measurements on a pendulum that were used by Newton in his *Principia Mathematica* of 1687 to demonstrate that the earth was flattened at the poles.[5] The Royal Society of London published advice in the early numbers of the *Philosophical Transactions* for sailors and gentlemen travelers to make observations in ethnography and natural history and to report back to both the Society and the Admiralty. This influenced the Narborough expedition of 1669–71, which was supposed to report in detail on the coastlines, minerals, and flora and fauna of the South Pacific, although strategically the affair was disappointing. The *Phil. Trans.* regularly published news from all over the known world, and in 1694 Tancred Robinson, a secretary of the Society, remarked anonymously in his introduction to an *Account of Several Late Voyages and Discoveries* that journals kept at sea should be more detailed and "'tis to be lamented, that the English nation have not sent along with their Navigators some skilful Painters, Naturalists and Mechanists."[6]

Any voyage had to respect the current political climate which might prevent a vessel from continuing or even being given fresh water if it was forced into an unfriendly port. With a sizable presence in Indonesia, the Dutch made it difficult to reach the Pacific via the Cape of Good Hope; to Spain the sea west of the Americas was *mare clausum,* and the latter guarded entry around Cape Horn via the Straits of Magellan. Attempting to enter the Pacific from the north, a number of efforts were made in the early eighteenth century to find the celebrated passage that was supposed to exist between the Atlantic and the Pacific. Attention focused on Hudson Bay and the poorly charted coastal region between the Hudson Bay Company post at Fort Churchill and Southampton Island in the north of the Bay. Because of the height of the tides in Ross Welcome Sound, Luke Foxe had suggested in the 1630s that there might

---

[5] S. Chapin, "The Men from across La Manche: French Voyages, 1660–1790," in D. Howse (ed.), *Background*, pp. 81–127; J. W. Olmstead, "The Voyage of Jean Richer to Acadia in 1670: A Study in the Relations of Science and Navigation under Colbert," *Proceedings of the American Philosophical Society,* 104 (1960), 612–34; Williams, *Great South Sea,* pp. 115–16.

[6] Williams, *Great South Sea,* pp. 115–16; M. Deacon, *Scientists and the Sea, 1650–1900: A Study of Marine Science,* 2nd ed. (Aldershot: Ashgate, 1997).

be a passage to the west of Southampton Island. Despite a disastrous expedition led by James Knight in 1719, efforts to find a passage were promoted by Arthur Dobbs, who, spurred on by the presence of whales in the Sound, suspected the Company of making an inadequate effort to locate the passage. However, further expeditions, including that of Christopher Middleton in 1741–2, failed to find the mythical route that was supposed to exist, and they ended with a substantial loss of life. Nevertheless, a reward offered by an Act of Parliament of 1745 gave further encouragement to private ventures to find a navigable passage, and interest in the possibility of such a route continued in the 1750s and indeed into the nineteenth century.[7]

THE IMPORTANCE OF VENUS

International cooperation on expeditions had been prominent in the 1730s and 1740s, when the French had worked with Spanish and Swedish personnel near the Equator and in Lapland to determine the nature of the shape of the Earth. As a result of these efforts it became accepted by the end of the 1730s that the Earth was flattened at the poles (that is, an oblate spheroid), as Newton and Christian Huygens had argued. Developing still further the models of scientific cooperation provided by these expeditions, astronomers all over the inhabited parts of the globe prepared for an even more ambitious undertaking in the midst of the Seven Years' War (1756–63). This produced a concerted effort to observe the first of two transits of Venus across the Sun in order to find the mean distance of the Earth from the Sun. As with the cartographic and geodesic measurements of the 1730s and 1740s, observations had to be taken from different parts of the globe in order to determine this value more accurately. Prominent in organizing these expeditions was Joseph-Nicolas Delisle, who had been based in Russia between 1725 and 1747. In 1753 he helped to coordinate international observers to measure a transit of Mercury, and useful experience was gained for the 1761 transit. He produced a *mappemonde* that outlined the precise locations on the globe from which various moments of the 1761 transit could be seen, and he sent it to a number of scientific academies across Europe.[8]

Supported generously by the Crown, French astronomers were dispatched to all parts of the planet. Alexandre-Gui Pingré left France at the beginning of

[7] O. H. K. Spate, *The Pacific since Magellan*, vol. 1: *The Spanish Lake*, vol. 2: *Monopolists and Freebooters* (Canberra: Australian National University Press, 1979, 1983); P. Brioist, *Espaces Maritimes au XVIIIe Siècle* (Paris: Atlande, 1997); G. Williams, *The British Search for the Northwest Passage in the Eighteenth Century* (London: Longmans, 1962), pp. 17–25, 46–72. Middleton, who had published a number of papers on the topic in the *Philosophical Transactions* was made a Fellow of the Royal Society and received the Copley Medal for his investigation of magnetic attraction.

[8] R. Iliffe, "'Aplatisseur du monde et de Cassini': Maupertuis, Precision Instruments and the Shape of the Earth in the 1730s," *History of Science*, 31 (1993), 335–75; H. Woolf, *The Transits of Venus: A Study of Eighteenth Century Science* (Princeton, NJ: Princeton University Press, 1959), pp. 31–4, 48–9.

1761 accompanied by an assistant, Denis Thuillier, who had been instructed by the Comte de Buffon to make some collections in natural history. After some initial problems the astronomers arrived at the Isle Rodrigue in the Mascarenes; although their observations of the transit itself on 6 June were affected by cloud cover, they managed to make accurate assessments of the island's flora and fauna as well as its precise location. After a difficult journey from St. Petersburg, Jean-Baptiste Chappe d'Auteroche had a clear view of the transit from Tobolsk in Siberia. However, drawing from Montesquieu's analysis of the influence of climate on physique and morals, he upset Catherine II with some ill-judged remarks about despotic government and the coarse bodies and unrefined minds of people of Northern Europe. The trip of Guillaume-Hyacinthe-Jean-Baptiste le Gentil de la Galaisière was doomed to failure since the British were blockading his Indian Ocean destination (Pondichery) even as he left France, and he was actually at sea when the transit occurred. Using the Isle de France as a base, he remained in the area for a decade and made important astronomical and cultural observations in the Philippines, Madagascar, and, in particular, India.[9]

Nevil Maskelyne oversaw the organization of instruments for the 1761 transit expeditions organized by the Royal Society to Bencoolen in Sumatra and to the island of St. Helena in the South Atlantic. Maskelyne was chosen as chief observer for the St. Helena expedition, and Charles Mason was made principal observer of the Bencoolen voyage with Jeremiah Dixon as an assistant. Responding to an overtly nationalistic appeal from the Society, the Crown made the unprecedentedly generous award of £800 for each. Maskelyne made a number of important measurements of latitude and longitude, although Mason and Dixon experienced tragedy and then farce on their way to Bencoolen. Their ship, the *Seahorse,* was engaged by the *le Grand* in the Channel with the loss of eleven lives, and Mason in particular was convinced to go on to Bencoolen only when the Royal Society threatened legal action, warning that failure to complete their mission would probably result in their "utter ruin" and "an indelible stain on their character." By the time they reached the Cape of Good Hope, news had arrived that the French had taken Bencoolen. Nevertheless, the observers had a clear view of the transit at Cape Town and made an accurate determination of its longitude. Overall, the results of the 1761 expeditions were inconclusive, not least because there was disagreement about the precise moment at which Venus passed the rim of the Sun in its ingress and egress (the points at which the planet's extremities were first and last seen to touch the rim of the Sun).[10]

Despite the importance of Joseph Banks's botanical interests in assessing

[9] Woolf, *Transits of Venus,* pp. 57–61, 66–8, 97–130; F. Marguet, *Histoire de la Longitude à la Mer au XVIIIe siècle, en France* (Paris: A. Challamet, 1917).

[10] Woolf, *Transits of Venus,* pp. 73–96, 130–4; Howse, *Maskelyne,* pp. 18–38; T. D. Cope and H. W. Robinson, "Charles Mason, Jeremiah Dixon and the Royal Society," *Notes and Records of the Royal Society* 9 (1951), 55–78.

the legacy of the voyage of the *Endeavour*, the ostensible scientific aim of the voyage was to make observations of the 1769 transit of Venus in Tahiti. Other expeditions to observe this transit also tested cures for scurvy and the two basic methods for determining longitude – namely, by lunar distance and chronometer. Although Sweden and Russia sponsored an impressive number of observations, the most significant expeditions were those mounted by Britain and France. Le Gentil went to Pondichery, where he had full cooperation from the British Governor but not from the heavens, and he missed the relevant moment because of cloud cover. With two Dollond achromatic refracting telescopes, Chappe d'Auteroche went to San José del Cabo on the southern tip of Baja California and managed to observe the transit on 3 June before he tragically died soon afterward, along with three-quarters of the disease-ridden town. Pingré went to Cap-François, Saint-Domingue, where his efforts to observe the external and internal moments of contact at ingress were successful. The Royal Society organized four expeditions, sending William Wales to Fort Churchill, Mason to Donegal, Dixon to Hammerfest (an island off the coast of Norway), and Maskelyne's assistant William Bayly on the same voyage to North Cape (eight miles northeast of Hammerfest). The fourth involved sending Charles Green to Tahiti on board the *Endeavour* under Lieutenant James Cook. Although there was still difficulty in obtaining exact accounts of the locations at which observations were made, and disagreement over precise moments of ingress and egress, the range of values was much smaller than that obtained from the 1761 expeditions and gave a figure for the mean distance of the Earth from the Sun that was much closer to the modern value.[11]

IMPERIAL VOYAGING

With the Seven Years' War concluded in its favor, the Admiralty took note of the advice of its First Lord, George Anson, that the Falkland Islands be used as a way station into the Pacific, and it launched an expedition under Commander John Byron in 1764. Byron was to survey the Falklands and then explore New Albion – the American coastline north of San Francisco so named by Francis Drake – for the strait of Juan de Fuca, believed by some to be the Pacific entrance of the Northwest Passage. Instead, having accomplished the Atlantic part of the mission and having rounded Cape Horn, Byron turned west for the Solomons but sailed too far north to find the legendary islands. Not long after his return to England in 1766, his ship, the *Dolphin*, was refitted for another foray around the Horn; the new captain, Samuel Wallis, was given secret instructions to search for land between New Zealand and Cape Horn in more southerly latitudes than Byron had looked. Leaving Plymouth in August 1766, Wallis was accompanied by the *Swallow* under the

[11] Woolf, *Transits of Venus*, pp. 154–70, 190–5.

leadership of Philip Carteret, first lieutenant on Byron's ship. The *Dolphin* was stocked with the latest antiscorbutics and had on board a purser who was able to calculate longitude according to the methods laid down by Maskelyne. The two ships parted company after a fraught trip around the Horn, and the ill-equipped *Swallow* limped back to England in May 1769, exactly a year after Wallis's return. Wallis's voyage was made famous by the discovery of Tahiti in June 1767, although it left behind a legacy of venereal disease that was later blamed on the French. Its most significant impact with respect to the first Cook expedition was the belief of the crew that they had sighted the northern tip of *terra australis*.[12]

Having founded a settlement on the Falkland Islands (Les Malouines) in 1764, Louis Antoine de Bougainville was sent back in the *Boudeuse* at the end of 1766 to formally hand it over to the Spanish, whence it became Las Malvinas. The mission was prompted by Byron's voyage and by Charles de Brosses's book, which made the existence of *terra australis* more likely; a further goal of the mission was to find a base in the South Pacific that could serve as the foundation of imperial expansion. Believing by early 1768 that he had sailed too far north to find *terra australis,* Bougainville methodically dismissed the existence of many lands posited by previous sailors and confessed proudly that he was "a voyager and a sailor, that is to say, a liar and an imbecile in the eyes of that class of slothful and arrogant writers, who speculate the livelong day . . . in the penumbra of their study, thus impertinently submitting nature to their imaginations." In early April they reached Tahiti, where Bougainville and his naturalist, Philibert Commerson, enthused over the innocence of the people while the Tahitians marveled at the fact that Commerson's traveling companion, Jean Bart, was actually a woman, Jeanne Baret. Describing the land as *Nouvelle Cythère* after the home of Aphrodite, Bougainville rhapsodized about the therapeutic qualities of the climate and took a Tahitian, Ahutoru, back to France. This practice was copied on Cook's second voyage by Tobias Furneaux of the *Adventure* when he took another Tahitian, Omai, back to England. In June Bougainville reached the Great Barrier Reef whereupon he sailed north through the Solomons, arriving in Batavia after much hardship. Passing Carteret in February 1769, he reached France in March of the same year.[13]

[12] Williams, *Great South Sea,* pp. 214–58; Frost, *Science for Political Purposes,* pp. 27–31; J. C. Beaglehole, *The Exploration of the Pacific,* 3rd ed. (Stanford, CA: Stanford University Press, 1966), pp. 196–213; W. Eisler (ed.), *The Furthest Shore: Images of Terra Australis from the Middle Ages to Captain Cook* (Cambridge University Press, 1995).

[13] Beaglehole, *Exploration of the Pacific,* pp. 213–28; Stafford, *Voyage into Substance,* pp. 48–9; L-A. de Bougainville, *A Voyage around the World,* trans. J. R. Forster (London, 1772); J. E. Martin-Allanic, *Bougainville,* 2 vols. (Paris: Presses Universitaires de France, 1964); E. Taillemite *Bougainville et ses Compagnons autour du Monde, 1766–1769,* 2 vols. (Paris: Impr. Nationale, 1977); M. Mollat and E. Taillemite (eds.), *L'Importance de l'exploration maritime aux siècle des Lumières (à propos du voyage de Bougainville)* (Paris: CNRS, 1982); C. de Brosses, *Histoire des Navigations aux Terres Australes* (Paris, 1756); E. H. McCormick, *Omai, Pacific Envoy* (Auckland: Oxford University Press, 1977).

Commerson was dropped off at Mauritius on the homeward trip and, although he made significant investigations of the local flora and fauna, he died before he could organize the collections made on the voyage. On Mauritius he continued the work of a number of botanists, many of whom were surgeon-naturalists who, since the early part of the century, had searched for material that could form a part of private collections of the Jardin du Roi. Attention turned in the 1740s and 1750s to the role of climate and to the effects of deforestation on moisture in the atmosphere. When Commerson alighted at Mauritius in November 1768, the results of a vigorous policy of deforestation had already been noted by Pierre Poivre, now commissaire-général-ordonnateur of the island and a key figure in early attempts to develop an ecological understanding of natural habitats. Commerson's report on the nature of the island, and of the South Pacific ambience in general, appeared in the *Mercure de France* of February 1769. This text, along with Rousseau's *Julie, ou La Nouvelle Héloïse* and Bougainville's *Voyage,* was among the most influential writings for forming French public opinion about the beauties of island or garden paradises.[14]

TERRA AUSTRALIS: COOK'S FIRST TWO VOYAGES

Before Cook's first voyage in 1768, natural history was the central reason for long-distance travel, with the Jardin du Roi and the Chelsea Physic Garden functioning as important centers for the reception of exotic materials. From the 1730s, a large number of expeditions was organized for his students by Carl von Linné (Linnaeus), whose binomial system of classification was increasingly being adopted by botanists. A number of these protégés, such as Daniel Solander, Herman Spöring, and Anders Sparrman, were also involved in the Cook voyages, although this extensive travel did not result in the import substitutions of staples that Linnaeus wished would benefit his native Sweden. In Britain, thanks to the protection of men such as Sir Hans Sloane, natural history and especially botany had become highly fashionable by the middle of the century. A great deal of this interest was connected to the vogue for stocking gardens with exotic flowers, and a series of networks was created that facilitated the movement of mineralogical and organic merchandise from the corners of the empire back to London. These networks involved collectors in London such as Peter Collinson, John Fothergill, and John Ellis, and various employees of the East India and West India Companies. They

---

[14] R. Grove, *Green Imperialism: Colonial Expansion, Tropical Island Edens and the Origins of Environmentalism, 1600–1860* (Cambridge University Press, 1996), pp. 158–65, 179–90, 216–41; J. Roger, *Buffon: un Philosophe au Jardin du Roi* (Paris: Fayard, 1989); Poivre, *Travels of a Philosopher; or, Observations on the Manners of Various Nations in Africa and Asia* (Glasgow, 1770); W. Stearn, "Botanical Exploration to the Time of Linnaeus," *Proceedings of the Linnean Society of London,* 169 (1958), 173–96; Y. Laissus, "Les voyageurs naturalistes du Jardin du roi et du Muséum d'Histoire naturelle: essai de portrait-robot," *Revue d'histoire des sciences,* 34 (1981), 259–317. See also J.-J. Rousseau, *Letters on the Elements of Botany addressed to a Lady,* trans. T. Martyn (London, 1782).

also centered on sites such as the Chelsea Physic Garden – whose chief gardener in the middle of the century was Philip Miller – and Kew Gardens, unofficially run by Joseph Banks after 1772.[15]

Natural history was to constitute one of the major scientific purposes of the voyage of *Endeavour,* although the mission had a number of other goals. The Royal Society initially favored Alexander Dalrymple (an expert on what was then known of the South Seas) as captain, but as a civilian Dalrymple ruled himself out by refusing to go either as a passenger or "in any other capacity than having the total management of the ship." Not long before departure they also had to take into account the news of the existence of Tahiti and the supposed *terra australis* that had been conveyed by Wallis. The King generously authorized £4,000 for the undertaking on 24 March 1768, and Lieutenant James Cook was appointed to the command of the vessel. The decision to go to Tahiti (King George's Island) was made on 9 June, and Cook and Charles Green were appointed official observers. The Council of the Society also told its Secretary to "request that Mr B &c may be permitted to go the voyge [sic] & consequently be receivd on board the Ship with their Baggage." The man in question, Joseph Banks, equipped *Endeavour* lavishly for research in natural history with "all sorts of machines for catching and preserving insects" and "many cases of bottles with ground stoppers, of several sizes to preserve animals in spirits." He and his companions were to play a pivotal role in the expedition's success.[16]

During the Seven Years' War, Cook had surveyed the St. Lawrence River in preparation for the assault on Quebec, and in the first half of the 1760s he had charted the coasts of Nova Scotia and Newfoundland, making a number of impressive measurements of various positions by dint of his skill in astronomy. Although he received "hints" from the President of the Royal Society, the Earl of Morton, on issues such as how to deal with natives, his main instructions for the *Endeavour* mission came from the Admiralty. In an attached packet, Cook also had a list of "secret" instructions that urged him to search for the conceivably massive tract of land to the south of the path taken by the *Dolphin.* From Tahiti he was to go to 40°S and search for *terra australis* between that position and 35°S until hitting either Van Dieman's Land (Tasmania) or New Zealand. If the supposed "Large Continent" were discovered he was to make

[15] L. Koerner, "Purposes of Linnaean Travel: A Preliminary Research Report," in D. D. Miller and H. D. Reill (eds.), *Visions of Empire: Voyages, Botany and Representations of Nature* (Cambridge University Press, 1996), pp. 117–37; H. J. Braunholtz, *Sir Hans Sloane and Ethnology* (London: British Museum, 1970); H. B. Carter, *Sir Joseph Banks 1743–1820* (London: British Museum, 1988); J. Gascoigne, *Joseph Banks and the English Enlightenment: Useful Knowledge and Polite Culture* (Cambridge University Press, 1994), pp. 76–94, 109–10; D. E. Allen, *The Naturalist in Britain: A Social History* (Harmondsworth: Penguin, 1978).

[16] J. C. Beaglehole (ed.), *The Voyage of the Endeavour 1768–1771* (Cambridge University Press, 1968), pp. 512–14; H. T. Fry, "Alexander Dalrymple and Captain Cook: The Creative Interplay of Two Careers," in R. Fisher and H. Johnston (eds.), *Captain James Cook and his Times* (Seattle: University of Washington Press, 1979), pp. 41–58.

all sorts of observations of the local people and flora and fauna, as well as the local minerals, which he was to take back to London along with seeds and grains. Failing that, he was to chart the coastline of New Zealand, enjoining the crew not to breathe a word of what they had seen.[17]

A veteran of a surveying expedition to Newfoundland led by his friend Constantine Phipps in 1766, Banks was connected to the London collectors and a growing number of natural historians. He distrusted armchair speculation and system-building and had a penchant for collecting both artificial and natural objects. Nevertheless, although he believed that there was a major difference between the virtuoso collector and the serious natural historian, he viewed the voyage as a Grand Tour unparalleled in its extent. His hand-picked companions included Daniel Solander, whom he knew from the British Museum, and Herman Spöring; they were to be naturalist and assistant naturalist respectively. Although Cook now had some accomplished draftsmen on board, Banks allowed himself the luxury of two men who were highly skilled in pictorial representation. They were Sydney Parkinson, who had already worked for both Banks and the naturalist Thomas Pennant, and Alexander Buchan, who was to die only a matter of days after arriving at Tahiti. Their drawings, copied by engravers for John Hawkesworth's edition of Cook's *Voyage* in 1773, were profoundly influential in shaping the European visual perception of the South Seas, but, as with many writings on the subject, published etchings of non-Europeans and their surroundings tended to depict them according to neoclassical conventions.[18]

Leaving at the end of August 1768, Cook arrived at Tahiti on 13 April 1769. This gave time for Green to supervise the construction of an observatory (at Point Venus) and for Banks and others to collect flora and fauna and to sample the generous hospitality of the locals. Like Bougainville, Banks described the Tahitians and their culture in classical terms, whereas Cook composed a less florid account. Of seminal importance was the decision to take a Tahitian, Tupaia, with them on the remainder of their journey, for it was he who would prove that Tahitians shared a common language with a range of peoples encompassing a much greater expanse of the globe than had previously been imagined. Completing the observations and failing to sight *terra australis,* Cook turned west to New Zealand, where he discovered that the Maoris and Tupaia could communicate with each other. The existence of a widely dispersed family of languages, implying a single root culture, was one of Cook's greatest discoveries. Later he compiled a comparative vocabulary of Tahitian and Maori

---

[17] Beaglehole, *Endeavour Journal,* 514–19, pp. cclxxix–cclxxxi, cclxxxii–cclxxxiv.

[18] Gascoigne, *Joseph Banks and the English Enlightenment,* pp. 76–8, 112–15; Smith, *European Visions of the Pacific,* pp. 11–12, 16–18; M. Bowen, *Empiricism and Geographical Thought: From Francis Bacon to Alexander von Humboldt* (Cambridge University Press, 1981); A. M. Lysaght, "Banks's Artists and His Endeavour Collection," in T. C. Mitchell (ed.), *Captain Cook and the South Pacific* (Canberra: National Library of Australia, 1979), pp. 9–80; R. A. Rauschenberg, "Daniel Carl Solander: Naturalist on the *Endeavour,*" *Transactions of the American Philosophical Society,* 58 (1968), 1–66; D. J. Carr (ed.), *Sydney Parkinson, Artist of Cook's Endeavour Voyage* (London: Croom Helm, 1983).

and was extremely impressed by Maori carving, which he thought "very little inferior [to] work of the like kind done by common ship carvers in England." Cook made an extremely accurate chart of the North and South Island in six months, and at the end of March 1770 he sailed west again to New Holland, where he encountered aborigines and allowed Banks and Solander to botanize. However, he nearly came to grief on the wrong side of the Barrier Reef. Having charted the coast of New Holland and claimed much of the territory for the King, Cook proceeded to test Torres's claim to have sailed all the way along the south coast of New Guinea. With a chart of Torres's route given by Dalrymple to Banks, Cook successfully negotiated the strait and thence sailed to Batavia for repairs. The expedition greatly increased the sum total of scientific knowledge about the areas covered, and Linnaeus called the material brought back to Europe "a matchless and truly astonishing collection, such as has never been seen before, nor may ever be seen again."[19]

The voyage prompted a number of strategic undertakings by the French and the Spanish to forestall British presence in the Pacific. In turn, the British organized two new voyages. One of these was determined by the belief that ice could not form in saltwater and that there might be a sea free of ice near the North Pole that could serve as a Northwest Passage. The second voyage was to test the hypothesis that *terra australis* might lurk even farther south than other voyages had explored. The northern expedition, led by Constantine Phipps, reached 80°N in the summer of 1773 before it was almost scuppered by ice and was forced to return to England. For the southern trip the newly promoted Cook was naturally chosen as leader, to be accompanied by Banks. This time Cook requested and received the use of two vessels – *Resolution* and *Adventure* (under Tobias Furneaux) – and for the former Banks organized an even larger team than on *Endeavour*. His entrouage included four portrait painters and draftsmen along with Solander and Joseph Priestley, but the latter's religion proved to be too unorthodox for some and James Lind was chosen instead. However, all these preparations came to naught when Cook became concerned by the structural changes that Banks was proposing to make to *Resolution*. Banks accordingly withdrew, instead making a consolation jaunt to Iceland. The Admiralty now appointed Anders Sparrman, Johann Reinhold Forster, and Forster's son Georg as naturalists, with William Hodges as painter, and it selected William Wales and William Bayly as astronomers.[20]

Cook's extraordinary journey in the Southern Ocean began in July 1772.

---

[19] Beaglehole, *Exploration of the Pacific*, pp. 231–60; B. Finney, "James Cook and the European Discovery of Polynesia," in R. Fisher and H. Johnston (eds.), *From Maps to Metaphors: The Pacific World of George Vancouver* (Vancouver, BC: UBC Press, 1993), pp. 19–30; W. Shawcross, "The Cambridge University Collection of Maori Artefacts, Made on Captain Cook's First Voyage," *J. Polynesian Society*, 17 (1970), 305–48. Solander later teamed with John Ellis to produce a pioneering work in zoology: J. Ellis and D. Solander, *Natural History of Many Curious and uncommon Zoophytes . . .* (London, 1786), and P. Cornelius, "Ellis and Solander's *zoophytes* 1786: Six Unpublished Plates and Other Aspects," *Bulletin of the British Museum (Natural History), (Historical Series)* 16 (1988), 17–87.

[20] Williams, *Northwest Passage*, pp. 159–66; Smith, *European Visions of the Pacific*, pp. 53–4.

Going eastward around the Cape of Good Hope, the expedition was the first to go south of the Antarctic circle, in January 1773, and, having braved icebergs and extraordinary meteorological conditions, *Resolution* reached New Zealand at the end of March. From here Cook went due east until July, whereupon he sailed north to Tahiti. In October he arrived back in New Zealand, having established the position of a number of islands on the way. Here he stocked up on antiscorbutics and, like the naturalists, made further observations on Maori culture. From New Zealand he set out on the perilously high latitudes of the oceans between New Zealand and Cape Horn; discovering no land in the vicinity, he decided to head north to Easter Island, which was sighted in March 1774. Finding that the islanders spoke a language similar to those of the Tahitians and the Maoris, Cook returned to Tahiti via the Marquesas. He left Tahiti in May and spent time charting New Hebrides and New Caledonia before returning to New Zealand; from here he made the long journey eastward at approximately 55°S, which took him to Cape Horn and then all the way to the Cape of Good Hope.[21]

Cook arrived back as a great hero in England at the end of July 1775, having determined that no large continent existed in temperate southern latitudes in either the Pacific or the Atlantic. In many ways, the scientific importance of the second expedition was greater than that of the first: the two methods for determining longitude were successful, Cook again lost no sailor from scurvy, and the Forsters made exceptionally important ethnological observations of non-Europeans. The elder Forster had taught at the Warrington Academy in the late 1760s and had then earned his living by translating texts such as Bougainville's *Voyages* into English. His wide reading was important in determining his more speculative assessments of the influence of climate upon morals, which made up a large bulk of his *Observations*. The Forsters had difficulty obtaining plates from the Admiralty to publish in their books, but the texts exerted a deep influence, especially on the Continent. Johann Forster explicitly set out to describe "nature in its greatest extent," and in his *Ideas for a Philosophy of the History of Man* (1784–91) Johann Herder described him as the "Ulysses" of the Pacific, praising his work on "philosophico-physical geography." Georg Forster, who was one of the first to recognize and articulate the baneful if inevitable effects of Europeans upon other peoples, had important discussions with Alexander von Humboldt before the latter toured Spanish America in 1799. The painter, Hodges, worked closely with Wales and Cook and made accurate drawings of various coastlines for the captain, and in appealing to the art establishment based around the Royal Academy, he made striking efforts to depict the unusual light effects that were experienced in the Antarctic. Wales and Bayly published their results after their return,

---

[21] Beaglehole, *Exploration of the Pacific*, pp. 261–84; see also A. Gurney, *Below the Convergence: Voyages towards Antarctica, 1699–1839* (London: Pimlico, 1998), pp. 86–185.

and, although Wales disliked the Forsters, he shared with them an interest in meteorological phenomena.[22]

THE NORTHWEST PASSAGE: COOK'S FINAL VOYAGE

Travels in search of a Northwest Passage were still being keenly promoted by lobbyists such as Daines Barrington, and in 1775 a new Act of Parliament was passed offering a reward of £20,000 to anyone who found it. The Admiralty accordingly made plans for a journey to the northwest coast of America and chose the *Discovery* to partner *Resolution,* which was refitted at Woolwich. In early 1776 Cook retired, accepting a post as captain at Greenwich Hospital but with a proviso that his retirement could be canceled if any work that required his special talents should arise. However, doubtless spurred on by the Act, he was soon convinced that he should command the Pacific voyage, and Charles Clerke was appointed captain of the accompanying ship *Discovery.* Journeying via the Cape of Good Hope, Cook was to search for some islands apparently discovered by previous voyagers in the South Pacific, and having deposited Omai in Tahiti he was to speedily move up the northwest coast of America from 45°N until reaching 65°N, after which a careful and detailed survey of the coastline was to take place. This latitude was chosen because since Samuel Hearne had reached the Arctic Ocean from Fort Churchill in the early 1770s without crossing a saltwater strait, it was now known that there was no passage between Hudson Bay and the Pacific. In that case the best choice for a passage appeared to be one from the Pacific into the Arctic Ocean. A Russian map by J. von Stählin of the St. Petersburg Academy of Sciences had just been published, and it depicted Alaska as a large island.[23]

The Russians were perhaps the best placed to mount exploratory missions into the North Pacific, but, due in the main to the largely impassable terrain and the vast distances involved, the colonization of the Kamchatka peninsula (on the far east of Siberia) did not take place until the early eighteenth century. Under Ivan III the Russians had moved eastward from Moscow in the

[22] Smith, *European Visions of the Pacific,* pp. 55–85; M. Hoare (ed.), *The* Resolution *Journal of Johann Reinhold Forster 1772–1775* (London: Hakluyt Society, 1982); M. Hoare, *The Tactless Philosopher: Johann Reinhold Forster, 1729–98* (Melbourne: Hawthorn Press, 1976); G. Forster, *A Voyage Round the World . . . during the Years 1772, 1773, 1774 and 1775,* 2 vols. (London, 1777); J. R. Forster, *Observations made during a Voyage around the World* (London, 1778); J. G. Herder, *Outlines of a Philosophy of the History of Man,* trans. T. Churchill (London, 1800); H. West, "The Limits of Enlightenment Anthropology: Georg Forster and the Tahitians," *History of European Ideas,* 10 (1989), 147–60; W. Wales and W. Bayly, *Original Astronomical Observations made in the course of a Voyage towards the South Pole* (London, 1777).

[23] Williams, *Northwest Passage,* pp. 173–4, 184–92; Williams, "Myth and Reality: James Cook and the Theoretical Geography of North-West America," in Fisher and Johnston (eds.), *Captain James Cook,* pp. 59–76, especially pp. 66–71; H. R. Wagner, *The Cartography of the Northwest Coast to the Year 1800* (Berkeley: University of California Press, 1937).

late sixteenth century, discovering that Siberia had navigable rivers and large reserves of sable, or "soft gold." After their conquests in the Amur Valley were turned over to the Manchu under the Treaty of Nerchinsk in 1689, many Russians – including the entrepreneur fur traders (*promyshlenniki*) – immediately moved northeast to the difficult but profitable land of the Kamchatka peninsula. Because they repeatedly depleted and extinguished commercially valuable fauna as they went, there was clearly an economic need to search the lands to the far east and to exploit whatever might lie on the northwest coast of the American continent.[24]

The main purpose of the famous Kamchatka Expeditions of 1725–30 and 1733–43 (the first under the Dane Vitus Bering and the second under Bering and Aleksei Il'ich Chirikov) was commercial, since Semon Dezhnev had proved in 1648 that Asia and America were separated. Although the first expedition was not deemed to be a success, the second, prompted by the geographer Ivan Kirilov (who saw it as a way of opening trade to China and Japan) was a massive affair comprising as many as two thousand men and taking ten years. Although it had beneficial long-term trading effects, the naturalist who accompanied the second expedition, Georg Steller, was given only a matter of hours to make some observations. The expedition sighted and charted various parts of the Alaskan coastline and brought back sea otter furs, which prompted the *promyshlenniks* to cross the Aleutian islands and begin hunting on the American mainland. At the same time as the second Bering expedition, another one traveled to the west coast of America. In 1732 Mikhail Gwosdev depicted what is now Cape Prince of Wales, although the map was not made public, and the relationship between the coastlines of Gwosdev and Bering remained unclear until Cook's third voyage.[25]

A number of Russian leaders, beginning with Peter the Great in the early eighteenth century, were keen to procure the services of talented foreign personnel to help with cartography and natural history in the massive empire. For example, J.-N Delisle helped to train many of the astronomers who went on the large number of overland expeditions that were organized in the next decades, with varying sorts of scientific purpose. Scientific travel was often generously supported by the Crown, and in 1767 Catherine the Great authorized the acquisition of twenty-one telescopes for six separate expeditions

[24] J. R. Gibson, "A Notable Absence: the Lateness and Lameness of Russian Discovery and Exploration in the North Pacific, 1639–1803," in Fisher and Johnston (eds.), *Maps to Metaphors*, pp. 85–103, 88–90; G. V. Lantzeff and R. A. Pierce, *Eastward to Empire: Exploration and Conquest on the Russian Open Frontier, to 1750* (Montreal: McGill-Queen's University Press, 1973).

[25] Gibson, *A Notable Absence*, pp. 95–6, 102; Williams, *Myth and Reality* pp. 60–5; G. W. Steller, *Journal of a Voyage with Bering, 1741–1742*, trans., M. A. Engel and D. W. Frost (Stanford, CA: Stanford University Press, 1988), p. 75; R. H. Fisher, *Bering's Voyages: Whither and Why* (London: C. Hurst, 1977); E. G. Kushnarev, *Bering's Search for the Strait: The First Kamchatka Expedition, 1725–1730*, trans. and ed. E. A. P. Crownhart-Vaughan (Portland: Oregon Historical Society, 1990); S. Krasheninnikov, *Explorations of Kamchatka, 1733–1741*, trans. E. A. P. Crownhart-Vaughan (Portland: Oregon Historical Society, 1972); and C. L. Urness, *Bering's First Expedition: A Re-examination Based on Eighteenth Century Books, Maps and Manuscripts* (New York: Garland, 1987).

in connection with the observations of the 1769 transit of Venus, sending letters to a number of foreign academies inviting astronomers to observe on Russian soil. With regard to other areas of science, Peter Simon Pallas made many important observations in Russian natural history and maintained a correspondence with a number of other naturalists such as Thomas Pennant.[26]

Given the difficulty of the overland route, some observers had argued that there would be a number of advantages from reaching Kamchatka by means of a circumnavigation, but the next major expedition in the area followed the same overland route to the east coast. Despatched to chart the Aleutians, this effort set off two years late in 1766, and although it surveyed the tip of the Alaskan peninsula and the Shumagin Islands, all four ships under Pyotr Krenitsyn and Mikhail Levashov were wrecked off the Kamchatkan coast. The *promyshlenniks* had a virtual monopoly of the fur trade until the 1780s and knew parts of the Alaskan coastline extremely well, but news of Cook's third voyage and the forthcoming La Pérouse expedition prompted Pallas to recommend a geographical and astronomical expedition to investigate the northern coasts of Russia. This lasted between 1785 and 1794, and although the ship was commanded by one of Cook's crew, Joseph Billings, its results were deemed to be relatively unsuccessful. The substantial Russian achievements in charting the Kamchatkan, Aleutian, and Alaskan waters remained largely unpublished until William Coxe's book on the subject in 1780. This secrecy was a result of a wish to preserve the fur trade monopoly and to disguise the true state of Russian presence in the east, although the Russians' knowledge of the coastlines in the area was patchy.[27]

Cook sailed in July 1776 just over a week after the American Declaration of Independence; ominously, neither the French nor the Spanish believed this time that the major function of this expedition was any other than strategic. With James King as second lieutenant, William Anderson as surgeon-naturalist, and John Webber as draftsman, Cook completed the first part of his voyage and en route to America made contact with a group of isles he named the Sandwich Islands (Hawai'i) in January 1778. He stopped long enough to allow himself and others to make some observations, and again he marveled at the fact that the people spoke a language closely related to Tahitian. On 7 March the northwest coast of America was sighted at 44° 33'. Coasting northward and unwittingly following in the tracks of recent Spanish voyages, Cook believed that he had disposed of the supposed straits to the Atlantic. However, detailed surveying was not to take place until higher latitudes, and in fact he mistook a number of islands for mainland. He landed at Nootka Sound,

---

[26] Woolf, *Transits of Venus*, pp. 23–49, 179–80; J. R. Masterton and H. Brower, *Bering's Successors, 1745–1780: Contributions of Peter Simon Pallas to the History of Russian Exploration toward Alaska* (Seattle: University of Washington Press, 1948); and C. L. Urness (ed.), *A Naturalist in Russia: Letters from Peter Simon Pallas to Thomas Pennant* (Minneapolis: University of Minnesota Press, 1967).

[27] Gibson, "A Notable Absence," pp. 91–4, 102; W. Coxe, *Account of the Russian Discoveries between Asia and America* (London, 1780).

and repairs on *Resolution* were carried out until April. Cook continued north and then west along the southern coast of Alaska, filling in the areas that were represented only by dotted lines on older maps. As he traveled, he came to share Bougainville's frustration with speculative geographers as the existence of a navigable strait to the Arctic Ocean became increasingly unlikely. Not for the first time he was to fulfill a mission by proving the nonexistence of the geographical object he had set out to find.[28]

Cook continued up the coast until he reached 70° 44'N and encountered impenetrable walls of ice; not only was there no strait to the Arctic from southern Alaska, but a journey through the Bering Strait and across the Arctic Ocean also appeared to be impossible. *Resolution* traversed the Bering Strait to the Asian coast and then turned south, and after meeting Russian traders at Unalaska Cook headed for the Sandwich Islands. Cook was killed in a skirmish in Kealakekua Bay in January 1779, and the ship left Hawai'i only in March, moving north again to Kamchatka. Heading back along the coast of Asia, the two vessels arrived in England in October 1780. In the realms of botany and zoology, the results of the Cook voyages were astonishing, and a number of people, including Cook himself, continued to collect for Banks on the last two voyages. Using the very latest equipment, Cook's surveys ensured that there would be fewer squiggles on maps and thus fewer calls for attempts to find the Northwest Passage and *terra australis*.[29]

IMPLICATIONS OF COOK'S VOYAGES: LONGITUDE AND SCURVY

Although latitude could be measured reasonably accurately, the inability to locate one's longitude (the distance to the west or to the east of a given or "prime" meridian) was the central source of navigational inaccuracy. A favored method was to sail as far as one could to one side of a target and then, having reached the desired latitude, to sail in the supposed direction of the destination. Complex astronomical techniques for determining longitude by means of the satellites of Jupiter had always proved unworkable at sea given the conventional equipment available to the navigator, and a more precise determination awaited the construction of a more suitable instrument or timepiece. With the latter, one could in principle locate one's own position if one knew

---

[28] Beaglehole, *Exploration of the Pacific*, pp. 290–300; Williams, *Northwest Passage*, pp. 179–83.

[29] Beaglehole, *Exploration of the Pacific*, pp. 301–15; Williams, *Northwest Passage*, pp. 203–11; W. Stearn, "The Botanical Results of Captain Cook's Three Voyages," *Pacific Studies* 1 (1978), 147–62; P. J. P. Whitehead, "Zoological Specimens from Captain Cook's Voyages," *Journal of the Society for the Bibliography of Natural History* 5 (1969), 161–201; S. P. Dance, "The Cook Voyages and Conchology," *Journal of Conchology* 26 (1971), 354–79; A. Kaeppler, *"Artificial Curiosities": Being an Exposition of Native Manufactures Collected on the Three Voyages of Captain James Cook* (Honolulu: Bishop Museum, 1978); H. Friis, *The Pacific Basin: A History of Its Geographical Exploration* (New York: American Geographical Society, 1967).

the time at the prime meridian, since longitude is also given by the simultaneous difference between local time (which could be found relatively easily) and the time at the prime meridian.[30] As an incentive to develop new instruments for this purpose, a Longitude Act of 1714 offered a large reward for an instrument that could accurately determine longitude at sea, and a Board of Longitude was appointed to oversee the business. An added condition – that the technique be easy for use at sea – was brought closer to realization with the invention of the double-reflection quadrant by John Hadley and the American Thomas Godfrey in 1731 and its transformation into the sextant twenty-five years later. Compared with the quadrant, the sextant doubled the number of days each month in which relevant observations were possible. In addition, one would be able in principle to work out what Greenwich time was at any moment if one knew the motions of the moon in advance, by measuring the angular distances between the moon and the sun or stars (the lunar-distance method). Or one could find out Greenwich time by the simpler method of having a timepiece that kept time to relevant accuracy.[31]

In 1757 Tobias Mayer, using equations provided by Leonhard Euler, sent the Board a set of tables that could be used for the lunar-distance method. The Astronomer Royal James Bradley compared them with his own observations at Greenwich and found them sufficiently accurate to determine the moon's place to 75 arcseconds and thus in principle to within a sufficient degree of precision to win the prize. The Seven Years' War made trial at sea difficult, and the tables were first tested properly when Nevil Maskelyne went to St. Helena to observe the transit of Venus. When he returned from St. Helena he published a description of the lunar-distance method in his *British Mariner's Guide*. Mayer's tables, being more accurate than those of Nicolas-Louis Lacaille utilized in the *Connaissance des Temps* (an earlier and less ambitious version of the *Almanac*), were the basis of Maskelyne's *Nautical Almanac and Astronomical Ephemeris for the Year 1767* and the associated *Tables,* which were published at the start of 1767. The Board decided the the *Almanac* should be published three years in advance, an undertaking that required the work of two full-time "computers." Despite the short-term success of the method, Cook ran out of sheets of the *Almanac* on all three voyages, and the future of longitude determination lay with marine chronometers.[32]

The development of chronometers in Britain into a relatively cheap, accurate, and usable device was due to the work of a number of men, in particular John Harrison. Pioneering techniques for coping with problems of lubrication and for compensating for changes in temperature and barometric pressure, Harrison worked on a number of designs from the 1730s. The third timepiece designated with the prefix "H" was not completed until 1757, eight

[30] D. Howse, "Navigation and Astronomy in the Voyages," in Howse (ed.), *Background,* pp. 160–83.
[31] Howse, *Navigation,* pp. 168–9.
[32] D. Howse, *Maskelyne: The Seaman's Astronomer* (Cambridge University Press, 1989), pp. 14–15, 85–96.

years after he had been recognized by the Royal Society with the award of the
Copley Medal. Harrison showed the next and ultimately prize-winning timepiece, H.4, to the Board of Longitude in 1760, and, after some problems, a
trial of the machine in Barbados in 1764 was a complete success. Further difficulties ensued with Maskelyne and the Board before Harrison could receive
his reward, the most serious being the need to make a complete disclosure of
the mechanism to a small selection of watchmakers and scholars. Although
this was done in August 1765, the Harrisons remained at loggerheads with
various members of the Board, especially Maskelyne, until the second half of
the £20,000 prize was granted in 1773.[33]

One of the horologers present at the demonstration, Larcum Kendall, took
two and a half years to complete his copy of H.4, and it was this device (K.1)
that went with Cook on his second voyage. The watch had maintained a constant going "rate" (that is, the amount of seconds by which a timepiece gains
or loses per day) of between 9 and 13 seconds since April 1773, and when this
was taken into account it was remarkably accurate; Cook called it his "trusty
friend" and "never-failing guide." Three of John Arnold's chronometers went
with K.1. on Cook's *Resolution*, whereas Thomas Earnshaw produced the standard design used in the construction of the marine chronometer for the next
150 years. Two of his timepieces would later accompany George Vancouver
on the *Discovery*, along with K.3 and two Arnold chronometers.[34] The French
had long been developing their own version of the marine chronometer, and
Pierre LeRoy's "A" and "S," and Ferdinand Berthoud's No. 6 and No. 8 timepieces were tested on a number of ships between 1768 and 1773, with Berthoud
receiving most of the awards. Although some manufacturers such as Earnshaw
pioneered the standardization of manufacture for chronometers and built large
numbers of them, LeRoy's complex and innovative watches were never built
in large quantitites. On the other hand, in 1792 Berthoud remarked that more
than fifty of his own timepieces had been used in eighty voyages.[35]

It was generally recognized in the late sixteenth century that oranges and

---

[33] Howse, *Maskelyne*, pp. 40–1, 46–7, 50–2; H. Quill, *John Harrison: The Man Who Found Longitude*
(London: John Baker, 1966); R. T. Gould, *The Mariner's Chronometer: Its History and Development*
(London: J. D. Potter, 1923); W. J. H. Andrewes (ed.), *The Quest for Longitude* (Cambridge, MA:
Collection of Historical Scientific Instruments, 1996); W. J. H. Andrewes, "Even Newton Could Be
Wrong: The Story of Harrison's First Three Sea Clocks," in Andrewes (ed.), *Quest for Longitude*,
pp. 189–234, especially pp. 195–6, 206–7, 211.

[34] Quill, *Harrison;* Gould, *Chronometer*, p. 258; J. Betts, "Arnold and Earnshaw: The Practical Solution," in Andrewes (ed.), *Quest for Longitude*, pp. 312–28, especially pp. 320, 326–8; D. Howse, "The
Principal Scientific Instruments Taken on Captain Cook's Voyages of Exploration, 1768–1780,"
Mariner's Mirror, 65 (1979), 119–35; A. C. Davies, "Vancouver's Chronometers," in Fisher and Johnston (eds.), *Maps to Metaphors*, pp. 70–84.

[35] Gould, *Chronometer*, p. 83; C. Cardinal, "Ferdinand Berthoud and Pierre Le Roy: Judgement in
the Twentieth Century of a Quarrel Dating from the Eighteenth Century," in Andrewes (ed.), *Quest
for Longitude*, pp. 281–92, especially pp. 287–8, 292; J. Le Boy, "Pierre le Roy et les horloges marines,"
in C. Cardinal and J-C Sabrier (eds.), *La Dynastie des Le Roy, Horlogers du Roi* (Tours: Musée de
Beaux-Arts, 1987), pp. 43–50; C. Cardinal, *Ferdinand Berthoud, 1727–1807: Horloger Mécanicien du
Roi et de la Marine* (La Chaux-de-Fonds, Switzerland: Musée International d'Horlogerie, 1984).

lemons could restore the health of sailors on long voyages, especially combined with dry clothes and clean ship, although citrus fruits were more often than not accompanied by copious amounts of cider. In the eighteenth-century British Royal Navy, scurvy caused more losses than enemy action, and until 1750 there was little progress in relating the onset of the disease to a lack of specific foodstuffs. In 1746 James Lind, who had joined the Navy as a surgeon's mate in 1739, tried an experiment aboard a ship where scurvy had broken out in which six groups of two men with the disease were given different remedies, including oranges and lemons. Those given a diet of oranges and lemons did best, although this information was not seen as conclusive by the Navy.[36]

Lind argued in his treatise of 1753 that scurvy, being unknown in dry places, was caused by moisture that clogged the skin's pores and made the air unfit for breathing. Rather than the ventilation of ships, exercise and some raw onion and garlic would steel the mariner against the vagaries of climate, whereas the best cure was a change of air. However, in the 1772 edition of Lind's work, he was more inclined to recommend regular exercise and a diet "of easy digestion." Significantly for Cook's voyages, the President of the Royal Society, Sir John Pringle, argued that fever among troops and prisoners resulted from air vitiated by filth and sweat, and he recommended the use of "antiseptic" substances such as vinegar, lemon juice, tobacco smoke, and "fixed air" (carbon dioxide). Following this, David MacBride urged the consumption of wort, a malt preparation, as it contained a large amount of fixed air. This was found to be revolting during trials in the early 1760s, but Pringle nevertheless urged that it be tried on long voyages. It was for his attempts to impregnate water with fixed air that Joseph Priestley was awarded the Copley Medal by the Royal Society in 1773, and Cook took some of this specially prepared liquid on the second voyage.[37]

When he took command of *Endeavour* in 1768, Cook was determined that the voyage would constitute a rigorous trial of both new and familiar cures. Although nobody died from the disease on the voyage, he had a hand-picked crew and the vessel was never away from land for more than seventeen weeks. On board were various preventives for scurvy, including sauerkraut, which the sailors initially despised but then ate when they saw Cook and his officers doing the same. Along with this, wort and rob (the boiled essence of oranges and lemons) became favored foods on *Endeavour*, although Joseph Banks continued to praise the antiscorbutic virtues of fixed air and fresh vegetables (despite having "flown" to lemon juice when he thought he had contracted the disease). On his second voyage Banks guarded particularly against the

[36] K. J. Carpenter, *The History of Scurvy and Vitamin C* (Cambridge University Press, 1986), pp. 15–21.
[37] Carpenter, *History of Scurvy*, pp. 51–60, 64–6, 70–1; C. Lawrence, "Disciplining Disease: Scurvy, the Navy, and Imperial Expansion, 1750–1825," in Miller and Reill, *Visions of Empire*, pp. 80–106, especially pp. 86–7; James Lind, *A Treatise of the Scurvy* (Edinburgh, 1753; 3rd ed. London, 1772).

putrefaction of air that could result from dirty bedding and clothing, and he was awarded the Copley Medal in 1776 for a paper on scurvy in which he praised wort and sauerkraut but also the value of proper discipline. Not least because of repeated ignorance of what actually caused scurvy, lemon juice was adopted on a large scale by the British Admiralty only in the late 1790s, and other nations' navies continued to suffer from the disease well into the following century.[38]

AFTER COOK

The Cook expeditions provided a template for the way in which all countries organized similar undertakings before 1800. Of these, the most ambitious was probably that mounted by the French under the leadership of Jean-François Galoup de La Pérouse, which sailed from France at the start of August 1785. This expedition was designed to investigate the American and Asian coastlines and to explore the potential for fur trading and whaling – as well as to assess the degree of Russian, American, and British participation in these activities. However, it was also probably the most extensive scientific expedition mounted to that time, and La Pérouse's *Boussole*, accompanied by the *Astrolabe*, possessed a large number of astronomers, skilled draftsmen and naturalists. La Pérouse was given detailed instructions on how to create a "descriptive catalogue" of "natural curiosities" and on how to collect ethnographic information: "he will order the garments, arms, ornaments, utensils, tools, musical instruments, and everything used by the different people he shall visit to be collected and classed; and each article to be ticketed, and marked with numbers corresponding to that assigned in the catalogue." Traversing most of the Pacific Rim, the expedition tragically foundered on the reef of Vanikoro (off the Santa Cruz islands) in early 1788. However, La Pérouse managed to send back information as he went, and his voyage was influential in reversing the positive view of non-Europeans that was standard at the time – influenced no doubt by the massacre of the captain of the *Astrolabe* and eleven others at Samoa.[39]

After the Revolution in 1789 the French mounted another spectacular voy-

---

[38] Carpenter, *History of Scurvy*, pp. 77–83; Lawrence, *Disciplining Disease*, pp. 87–8; J. Watt, "Medical Aspects and Consequences of Cook's Voyage," in Fisher and Johnston (eds.), *Captain James Cook*, pp. 129–58, especially pp. 130–5. Of *Endeavour*'s company, forty-one of the original ninety-eight died before Cook returned, the majority dying of malaria and dysentery in Batavia and en route from Batavia to the Cape of Good Hope in early 1771.

[39] Dunmore, *French Explorers*, vol. 1, pp. 250–82; Beaglehole, *Exploration of the Pacific*, pp. 318–19; Smith, *European Visions*, pp. 137–54, especially p. 139; M. L. A. Milet-Mureau, *A Voyage around the World . . . by J.F.G. de La Pérouse*, ed. J. Johnston, 4 vols. (London, 1798); C. Gaziello, *L'Expédition de La Pérouse, 1785–88: Réplique Française aux Voyages de Cook* (Paris: CTHS, 1984); J. Dunmore and M. de Brossard, *La Voyage de Lapérouse, 1785–88: Récit et Documents originaux* (Paris: Imprimerie Nationale, 1985).

age to the Pacific, this time commanded by Bruni d'Entrecasteaux. The two ships – *Recherche* and *Espérance* (under Huon de Kermadec) – went in search of La Pérouse, and, although they did not attain this objective, they made a number of significant discoveries and cartographic corrections in the Australasian region, particularly in charting the correct geography of the Solomons. Their investigation of the southeast corner of Tasmania and of the southwest corner of Australia paved the way for the more extensive charting of the Australian coastline by Baudin, although before that (in 1797) Bass demonstrated that Tasmania and Australia were in fact separate. The undertaking ended when the two captains died within a few months of each other in 1793 and other members of the team were captured and imprisoned at Java – anticipating the fate of Matthew Flinders a decade later when he was detained in Mauritius by the French for seven years. The account of the expedition by *Recherche*'s naturalist, Jacques Julian de Labillardière, offered yet another view of non-Europeans, this time depicting them as stoical, rational, and capable of civilization.[40]

The Spanish invested more money in botanical expeditions over the last few decades of the century than any other nation, but it was the British who were best able to collate and make use of the vast amounts of information (Spanish material largely excepted) pouring back into Europe from the peripheries. Joseph Banks played a pivotal role in organizing the British expeditions that followed Cook, and he built up a series of vast networks that allowed local analysis, minerals, and organic material itself to be sent back to London. As President of the Royal Society, Banks had extremely close connections with powerful institutions such as the Admiralty and the Board of Trade, and he supplied them with information that might prove strategically useful just as they helped serve his real interests in natural history. Banks took it upon himself to continue the style of the explorations begun by Cook, and in 1780 James King (who had taken over command of *Discovery* after Clerke died) told Banks that he looked upon him "as the common Centre of we discoverers."[41]

Aided by his membership in organizations such as the African Association, Banks promoted travels such as those by Mungo Park into the interior of Africa, and he was keenly interested in acquiring useful information that might arise from the Macartney expedition to China in the early 1790s. He also had contact with well over a hundred collectors from China to South America and Africa, and with the help of powerful patrons such as the Duchess of Portland

[40] Dunmore, *French Explorers*, vol. 1, pp. 283–314; J. J. H. de Labillardière, *Voyage in Search of La Pérouse, Performed by Order of the Constituent Assembly during the Years 1791, 1792, 1793 and 1794* (London, 1800).
[41] D. P. Miller, "Joseph Banks, Empire, and 'Centres of Calculation' in Late Hanoverian London," in Miller and Reill, *Visions of Empire*, pp. 21–37, especially p. 29; D. MacKay, "A Presiding Genius of Exploration: Banks, Cook and Empire, 1767–1805," in Fisher and Johnston (eds.), *Captain James Cook*, pp. 21–40.

he sponsored more than twenty individuals on specific collecting missions. These men sent back observations and analysis not merely of flora and fauna of interest to natural history but also of various flowers and crops that might either flourish as exotics in Britain or be transplanted to colonial outposts. Plant specimens were analyzed at Kew, and other "curiosities," reflecting Banks's continuing gentlemanly interest in antiquities and later in anthropology, were kept at his houses in New Burlington St. and (later) Soho Square.[42]

Banks played crucial roles in a number of ventures, such as the colonization of New South Wales and attempts to grow staples there; the efforts to transplant breadfruit from Tahiti to the West Indies to feed slaves; and the voyage led by George Vancouver to survey the northwest coast of America and to consolidate British presence. All these endeavors explicitly entangled scientific knowledge with imperial power and commercial advantage. A number of reasons, such as its strategic benefits and the fear that La Pérouse was going to install a colony in New Zealand, prompted the British to take a serious interest in the imperial potential of New South Wales. The Governor of the colony, Arthur Phillip, was keen to develop the production of cotton, cochineal, and coffee and turned to Banks for advice; Philip reciprocated by sending botanical and zoological exotica back to London. Plans to take breadfruit to West Indian plantations were stalled by the American War of Independence, but when this ended in 1783 the need to bolster the plantations was more pressing than ever. Banks was responsible for the appointment of William Bligh as commander of the *Bounty*, and after the notorious events on that voyage, Banks made the complex botanical arrangements for Bligh's ensuing trip in the *Providence*. This succeeded where the *Bounty*, despite enjoying the specialist skills of the gardener David Nelson, had conspicuously failed. Banks's expertise was crucial in ensuring the success of the breadfruit transplantation, although this did not mean that the foodstuff was readily incorporated into the local diet.[43]

The voyage led by George Vancouver between 1791 and 1795 was the last great exploratory mission launched by the British in the eighteenth century. Its context was both imperial and commercial, and Vancouver's mission had a complex prehistory that was related to the rapid expansion of whalers and fur traders into the Pacific. Initially headed for the South Atlantic, the route to be followed by *Discovery* was suddenly changed in the wake of the Nootka

[42] D. MacKay, "Agents of Empire: the Banksian Collectors and Evaluation of New Lands," in Miller and Reill (eds.), *Visions of Empire*, pp. 38–54, especially pp. 39, 45–6, 49–50; Gascoigne, *Joseph Banks*, pp. 80–2, 149–57; S. Schaffer, "Visions of Empire: Afterword," in Miller and Reill (eds.), *Visions of Empire*, pp. 335–52, especially p. 345.

[43] D. Mackay, *In the Wake of Cook*, pp. 123–40; C. A. Bayly, *Imperial Meridian: The British Empire and the World, 1730–1830* (London: Longman, 1982); A. Frost, *Convicts and Empire: A Naval Question, 1776–1811* (Oxford: Oxford University Press, 1980). Nelson's plants were all thrown overboard during the mutiny, and he died before the *Bounty* returned to England; see also J. Browne, "A Science of Empire: British Biogeography before Darwin," *Revue d'Histoire des Sciences* 45 (1992), 453–75, especially 465.

Sound incident and in the light of new fears about U.S. activity near what is now Vancouver Island. A veteran of the second and third Cook expeditions, Vancouver was made commander, with Banks's choice, Archibald Menzies, as naturalist, although the extent of scientific representation on board was relatively small. Vancouver accurately surveyed the coastlines between 30°N and 60°N (that is, south of where Cook had made his detailed survey) and showed that there was no Northwest Passage hiding behind the large islands. Menzies was given a detailed set of instructions by Banks for observing, collecting, and preserving various specimens he might encounter on his passage, and he was told to pay special attention to the suitability of a location for settlement. Although Vancouver fell out badly with Menzies even before they set out and his rather dull book of the voyage appeared when the vogue for such publications was on the wane, his meticulous enterprise fulfilled the multifunctional roles that had long been the norm for all such travels and paved the way for the voyages of Flinders and his successors.[44]

SPANISH VOYAGES

Like the Russians, the Spanish had a policy of keeping the bulk of their results secret, and rumors of real or false Spanish discoveries on the northwest coast of America galvanized a number of voyages in the eighteenth century. By the beginning of the century, these voyages had amassed a great deal of information relating to the South Pacific and the west coast of South America and New Spain. For example, Luis Vaez de Torres had passed between Papua New Guinea and Australia through the straits named after him in 1606, although this fact long remained secret, leaving a number of geographers for more than a century and a half to believe that Papua New Guinea was actually the northernmost point of *terra australis*. In the eighteenth century, Spanish claims to the Pacific and its coastlines looked increasingly fragile as the French, Russians, and British sought to construct colonies and trading presences in the South Seas and on the coast of North America.[45]

Spaniards accompanied a number of French expeditions to South America in the first half of the eighteenth century, the most significant being the presence of Antonio de Ulloa and Jorge Juan on the journey to Peru between 1735 and 1744. Juan and Ulloa produced a best-selling public version of their extraordinary experiences on the ill-fated voyage, and they also wrote "Noticias secretas de América" for the eyes of the king only. However, serious support for Spanish botanical exploration began in the reign of Carlos III (1759–88).

---

[44] Mackay, *In the Wake of Cook*, pp. 57–116; J. C. H. King, "Vancouver's Ethnography: A Preliminary Description of Five Inventories from the Voyage of 1791–95," *J. Hist. Collections*, 6 (1994), 35–58.

[45] Beaglehole, *Exploration of the Pacific*, pp. 98–103; W. L. Cook, *Flood Tide of Empire: Spain and the Pacific Northwest, 1543–1819* (New Haven, CT: Yale University Press, 1973).

The monarch sought to promote a Spanish Enlightenment with a strong emphasis on natural knowledge and its practical benefits, and he sponsored the building of a Royal Botanical Garden, a Museum of Natural Science, a Royal Academy of Medicine, and an Astronomical Observatory. With his backing, two naval officers accompanied Chappe d'Auteroche in observing the transit of Venus in 1769, and in 1777 the king supported the expedition through Chile and Peru of the botanists Hipólito Ruiz and José Antonio Pavón and the French naturalist Joseph Dombey, an expedition that lasted until 1788.[46]

When the Spanish heard in 1774 of possible extensions by the Russians from Kamchatka to the American continent, the Viceroy of New Spain, Antonio María Bucareli y Ursúa, was ordered to send exploratory teams up the Pacific coast to determine the extent of Russian involvement and to take formal possession of the coast. The Spanish were more interested in turning the natives to Christianity than were the early French or British expeditions, and there was a strategic interest in a realistic portrayal of peoples who might be significant allies against other European powers. Bucareli chose Juan Peréz, an experienced seaman, to command the first expedition, and the latter sailed in the *Santiago*. Unable to land, Peréz made significant observations of the communities on the northwest coast of America with his second officer, Esteban José Martínez. Since no formal claims of possession had been made, Bucareli sent off another voyage the following year, under Bruno de Hezeta. In the accompanying schooner, the *Sonora*, Juan Francisco de la Bodega y Quadra landed at 57° 2' and took possession of the land for Spain in sight of Mount Edgecombe. When he learned of Cook's intended voyage in 1776, the Minister of the Indies ordered a new expedition; this was led by Ignacio Arteaga on the *Princesa*, accompanied by Bodega on the *Favorita*, and left San Blas in February 1779. It produced detailed cartographic and ethnological accounts of the region around Bucareli Bay on the west coast of Prince of Wales Island.[47]

When news arrived in the mid-1780s of substantial Russian presence as far south as Nootka Sound, the Spanish again responded. This time Martínez was given command of an expedition, which sailed in March 1788, and with the threat of U.S. involvement in the region, yet another undertaking under Martinéz was ordered for 1789 to bolster Spanish claims to sovereignty of the coast. It was Martínez who put James Colnett in irons in July 1789 and sent him, along with his *Argonaut*, down to San Blas, setting in motion the so-called "Nootka Sound incident" that paved the way for negotiations between the Spanish and the British to discuss the sovereignty of the northwest coast

[46] I. H. W. Engstrand, *Spanish Scientists in the New World: The Eighteenth Century Expeditions* (Seattle: University of Washington Press, 1981), pp. 6–8; V. von Hagen, *South America Called Them* (New York: Knopf, 1945), p. 300; A. R. Steele, *Flowers for the King: The Expedition of Ruiz and Pavón and the Flora of Peru* (Durham, NC: Duke University Press, 1964).

[47] C. I. Archer, "The Spanish Reaction to Cook's Third Voyage," in Fisher and Johnston (eds.), *Captain James Cook*, pp. 99–119, especially pp. 100–9. Hezeta discovered the Columbia River and named it the Entrada de Hezeta, although the American Robert Gray renamed it in 1792.

of America. Despite – or because of – their territorial goals, all the Spanish expeditions in the 1770s, 1780s, and 1790s placed a high level of importance on the need to acquire accurate information on mineralogy, meteorology, and ethnography, and the remarkable voyage of Alejandro Malaspina was explicitly devised to rival the scientific achievements of the LaPérouse and Cook expeditions.[48] Having completed a circumnavigation between 1786 and 1788, Malaspina and José Bustamante y Guerra submitted a plan to the Spanish Minister of Marine, Antonio Valdés, for "a Scientific and Political Voyage around the World." Malaspina requested two botanists or naturalists and two artists and was granted the command of the *Descubierta;* Bustamante was given charge of the *Atrevida*. Malaspina selected Lieutenant Antonio Pineda y Ramírez of the Royal Spanish Army as the main natural historian, to be assisted by Luis Née, who had a great deal of experience working for the Royal Botanical Garden. Meanwhile, the newly crowned Carlos IV recommended that the expedition take advantage of the availability of the botanist and naturalist Tadeo Haënke (who became the first person to describe the redwood tree in a European publication).[49]

During 1790 the ships moved slowly up the west coast of South America until they arrived off the coast of Panama, where they went on separate routes. The *Atrevida,* with Arcadio Pineda, Née, the artist José Guío, and the physician-naturalist Pedro María González, sailed on to Acapulco and then on to San Blas. Joining the *Descubierta* in Acapulco in April 1791, Malaspina announced that he had a new brief from Carlos IV to find the Northwest Passage. Accordingly, the ships set sail in May, leaving behind Antonio Pineda and others to explore the local flora and fauna; the personnel at sea had been increased by the addition of the artists José Cardero and Tomás de Suría.[50] Once Malaspina had left, two botanists and an artist from the Royal Scientific Expedition to New Spain (1785–1803), led by Martin de Sessé, accompanied Bodega y Quadra, whose ultimate goal was to sort out territorial issues with Vancouver. The most significant naturalist on board was José Moziño, who made by far the most detailed contemporary linguistic, ethnographic, and historical study of the Nootka Indians. In the meantime the recently returned *Descubierta* and *Atrevida* left Acapulco in December 1791 to explore the Pacific Islands. Leaving the Philippines early in 1792, the expedition toured various sites on the Pacific Rim and finally reached Cadiz in February 1794. Some members enjoyed success: Née collected more than ten thousand plants on his tour and spent a great deal of time ordering his observations in Madrid.

[48] Archer, "Spanish Reaction," pp. 109–12, 114; Cook, *Flood Tide,* pp. 146–99.

[49] Engstrand, *Spanish Scientists,* pp. 44–9; D. C. Cutter, "The Return of Malaspina: Spain's Great Scientific Expedition to the Pacific, 1789–1794," *American West,* 15 (1978), 4–19; M. D. H. Rodriguez, *La Expedición Malaspina 1789–1794* (Madrid: Ministerio de Cultura, 1984); V. G. Claverán, *La Expedición Malaspina en Nueva España (1789–1794)* (Mexico: El Colegio de Mexico, 1988).

[50] Engstrand, *Spanish Scientists,* pp. 50–76; T. Vaughan, E. A. P. Crownhart-Vaughan, and M. P. de Iglesias, *Voyages of Enlightenment: Malaspina on the Northwest Coast 1791/1792* (Portland: Oregon Historical Society, 1976).

Malaspina was more unfortunate, being first compromised at court and then sentenced to ten years' imprisonment.[51]

CONCLUSION

In alliance with imperial and commercial interests, scientific travel extended the bounds of European empires and brought home the effects of European expansion both on the natural world and on fellow human beings. As reading publics became sated with depictions of Others, the culture of collecting that accorded value to items on the grounds of their exotic value was increasingly disparaged. From the 1790s onward, commanders of voyages had specific instructions to make detailed assessments of coastlines, and naturalists were to collect botanical and zoological specimens for *analysis*. The same "analytic" approach applied also to the study of non-Europeans. The collection of ethnographies revealed novel patterns and differences and facilitated the appearance of a value-laden ethnology and anthropology. As indigenous peoples slowly recovered from the ravages of European diseases, they were beset by a scientific racism allied to craniometry. However, it also became clear that the planet was not an inexhaustible resource and that it would require careful management if it was not soon to be ravaged.

By the end of the eighteenth century, navigators and naturalists had at their disposal instruments that were undreamed of a hundred years earlier, capable of measuring phenomena of which their forbears were equally unaware. With a grasp of detail it was now possible to begin the systematic investigation of regional similarities and differences over a planet much diminished in size, and various forces could be linked to form a general science of terrestrial phenomena. Ambitious efforts such as those of Alexander von Humboldt to reveal the "cooperation of physical forces" and hence to display the underlying unity of Nature promised to link all corners of the Earth in a "global physics." Humboldt's narrative of his five-year odyssey to South America depicted what Mary Louise Pratt has called "a dramatic, extraordinary nature, a spectacle capable of overwhelming human knowledge and understanding," and his experiences were made widely available in his popular *Ansichten der Natur*. A new science was necessary to capture the sublime magnificence of such a phenomenon, and Humboldt's study of "vegetation" linked previously disparate areas of research such as botany and geography to form what he called

[51] Engstrand, *Spanish Scientists*, pp. 104–9, 111–18, 123; H. W. Rickett, "The Royal Botanical Expedition to New Spain," *Chronica Botanica*, 11 (1947), 1–81; I. H. Wilson, ed. and trans., *Noticias de Nutka: An Account of Nootka Sound in 1792 by José Mariano Moziño* (Seattle: University of Washington Press, 1970); R. McVaugh, *Botanical Results of the Sessé and Moziño Expedition (1787–1803)* (Ann Arbor: University Herbarium, University of Michigan, 1977), pp. 97–195; X. Lozoya, *Plantas y Luces en Mexico: La Real Expedición Científica a Nueva España, (1787–1803)* (Barcelona: Serbal, 1984).

"earth history." Mapping was central to this enterprise, and an understanding both of historical geology and of zoological regionalization were prolegomena to the transformation of the analysis of the history of the earth and its inhabitants in the following century.[52]

52 See also M. Nicolson, "Alexander von Humboldt, Humboldtian Science and the Origins of the Study of Vegetation," *History of Science*, 25 (1987), 167–92; M. Dettelbach, "Global Physics and Aesthetic Empire: Humboldt's Physical Portrait of the Tropics," in Miller and Reill (eds.), *Visions of Empire*, pp. 258–92; and M. Louise Pratt, *Imperial Eyes: Travel Writing and Transculturation* (London: Routledge, 1992), pp. 111–43, especially p. 120.

Part IV

NON-WESTERN TRADITIONS

27

ISLAM

Emilie Savage-Smith

The eighteenth century was a period characterized by confrontations, exchanges, and misunderstandings between Europe and Islam. The European commercial and military expansion that began in the sixteenth and seventeenth centuries continued throughout the eighteenth century, and in its wake some early modern European technologies and scientific ideas were introduced into the Middle East. These concepts and technologies coexisted, sometimes uneasily, with medieval Islamic practices. It was a period of ambivalence among Islamic rulers as well as scholars as to the relevance or acceptability of Western science and technology.

The Napoleonic Expedition of 1798 symbolizes the organized introduction of European science, medicine, and technology into the Near East, for engineers and scientists accompanying the expedition to Egypt methodically introduced the latest European ideas while at the same time recording the indigenous technologies they encountered.[1] Prior to that, the introduction of European scientific ideas was sporadic, and occasionally there was a lengthy time lag before their introduction into the Ottoman, Safavid, and Mughal worlds. After its introduction, the integration of a new technology into the culture occurred (if at all) only after a considerable interval of time during which there were social and sometimes ideological adaptations.

Historians have given relatively little attention to scientific, medical, and technological activities in the eighteenth century in the Islamic world. The sources for this period are fragmentary and difficult to interpret. Relatively few treatises written in Arabic, Persian, or Turkish during this period have been studied by scholars, and historians are largely dependent on records of European travelers, diplomats, and missionaries – accounts that are often superficial or prejudiced.[2]

---

[1] Charles C. Gillispie and Michel Dewachter (eds.), *Monuments of Egypt, the Napoleonic Editions: The Complete Archaeological Plates from "La description de l'Égypte,"* 2 vols. (Princeton, NJ: Princeton Architectural Press, 1987; reprint of 1809 edition).

[2] Ahmad Gunny, *Images of Islam in Eighteenth-Century Writings* (London: Grey Seal, 1996), pp. 9–36.

The Islamic world of the eighteenth century can be viewed as consisting of four parts: (1) India, where European courts had established a presence at the Mughal court since the seventeenth century, (2) Persia, ruled by the Safavids until the demise of the regime in the early eighteenth century, with many European contacts through envoys from courts and religious missionaries, (3) the Anatolian peninsula, which was the administrative center of the Ottoman Empire, and (4) Syria, Egypt, Iraq, and North Africa, parts of which were under Ottoman rule. In the seventeenth century the great trading companies were established – notably, the East India Company, the (Dutch) Ooost-Indische Compagnie, and the Compagnie Française des Indes – and they established trading stations, or "factories," at the major ports and near the courts in India, Persia, and Turkey and within the Ottoman provinces, at places such as Aleppo. These mercantile companies played major roles in the exchange of information and technologies between the Middle East and Europe. Before 1710 Persia or India was the preferred destination for travelers, traders, and missionaries from Europe, but after that time Persia was less frequently visited and it was to Syria and Egypt that most European travelers were attracted. The Arabian peninsula had relatively little European contact, although in 1760 Carsten Niebuhr led a Danish expedition to Saudi Arabia, which he described in *Description de l'Arabie*.[3] For the purposes of this essay, attention will focus on the Ottoman Empire and on the Safavid Empire (and the later regime of Nādir Shāh and his successors in Persia) rather than the Mughal Empire in India, for India is dealt with elsewhere in this volume (Chapter 28).

French merchants and diplomatic missions were especially influential in the Middle East in the eighteenth century. They supplied gunnery experts to Persia, as well as many artisans, such as jewelers and clockmakers.[4] For the Ottomans as well, the French were major suppliers of technologies and craftsmen, and it was to France that the Ottoman court turned for military support in an attempt to form alliances against the Russians and Austrians. So great was the influence of France on the Near East that all Europeans were referred to as "Franks" (*farangī* or *al-ifranj*).

Under the Ottoman Sultan Ahmed III, who ruled from 1703 to 1730, contacts with Europe were openly encouraged and new ideas and technologies imported. From 1718 to 1730 the Grand Vizier serving Sultan Ahmed III was İbrahim Paşa, who was particularly notable for his patronage of learning. During this twelve-year period, İbrahim Paşa commissioned twenty-five scholars to translate Arabic and Persian historical writings into Turkish, established

---

[3] *The Arabian Journey: Danish Connections with the Islamic World over a Thousand Years* (exhibition catalogue) (Århus: Prehistoric Museum Moesgård, 1996), pp. 57–65.

[4] H. J. J. Winter, "Persian Science in Safavid Times," in Peter Jackson and Laurence Lockhart (eds.), *The Cambridge History of Iran*, vol. 6: *The Timurid and Safavid Periods* (Cambridge University Press, 1986), pp. 581–609.

five public libraries (which included scientific works among their collections), encouraged the establishment of the first Turkish printing press, and permitted the practice of the chemical medicine developed in Europe by followers of Paracelsus. This period of purposeful contact with Europe came to an abrupt end in 1730 through a revolution in which Ahmed III and his vizier died.

In 1720–1 Sultan Ahmed III sent Yirmisekiz Çelebi Mehmed Efendi on a mission to France, ostensibly to inform the French that the Ottoman state would permit them to repair the Church of the Holy Sepulchre in Jerusalem, although archival documents suggest that the true purpose was to observe the military and technological innovations in France and report on those suitable for introduction into the Ottoman Empire. As he traveled to Paris, he noted the locks, canals, bridges, tunnels, and other technologies that he saw in the French countryside. In the city he was taken to the observatory, where he described many of the instruments; to a mirror factory, where he saw concave burning mirrors "as big as one of our large dining trays of Damascus metalwork"; to the museum of natural history, where he was especially interested in the wax anatomical models; to a large glass greenhouse (apparently a new concept to him); and to the royal library, where he was astonished (according to the French accounts) by the large number of Turkish and Arabic manuscripts in the collection. At the observatory he was presented with corrections of astronomical tables made by Gian Domenico Cassini and hand-copied by Cassini's son Jacques, who was then director of the Paris observatory, and at the natural history museum he was given two wax anatomical models: one of an animal and one of a human male. Mehmed Efendi was seriously interested in scientific and military matters, and his astronomical descriptions make up a large portion of his embassy report.[5]

A less experienced attitude was shown by the ambassador Mustafa Hattı Efendi, who was sent in 1748 to Vienna:

> At the emperor's command we were invited to the Observatory, to see some of the strange devices and wonderful objects kept there. We accepted the invitation a few days later, and went to a seven- or eight-storey building. On the top floor, with a pierced ceiling, we saw the astronomical instruments and the large and small telescopes for the sun, moon, and stars.
>
> One of the contrivances shown to us was as follows. There were two adjoining rooms. In one there was a wheel, and on that wheel were two large, spherical, crystal balls. To these were attached a hollow cylinder, narrower than a reed, from which a long chain ran into the other room. When the wheel was turned, a fiery wind ran along the chain into the other room, where it surged up from the ground and, if any man touched it, that wind struck his finger and jarred his whole body. What is still more wonderful, is that if the man who touched it held another by the hand, and he another, and so formed

[5] Fatma Müge Göçek, *East Encounters West: France and the Ottoman Empire in the Eighteenth Century* (Oxford: Oxford University Press, 1987), pp. 4, 57–61, 142–4.

a ring of twenty or thirty persons, each of them would feel the same shock in finger and body as the first one. We tried this ourselves. Since they did not give any intelligible reply to our questions, and since the whole thing is merely a plaything, we did not think it worthwhile to seek further information about it.

Another contrivance which they showed us consisted of two copper cups, each placed on a chair, about three ells apart. When a fire was lit in one of them, it produced such an effect on the other, despite the distance, that it exploded as if seven or eight muskets had been discharged.

The third contrivance consisted of small glass bottles which we saw them strike against stone and wood without breaking them. Then they put fragments of flint in the bottles, whereupon these finger-thick bottles, which had withstood the impact of stone, dissolved like flour. When we asked the meaning of this, they said that when glass was cooled in cold water straight from the fire, it became like this. We ascribe this preposterous answer to their Frankish trickery.

Another contrivance consisted of a box, with a mirror inside and two wooden handles outside. When the handles were turned, rolls of paper in the box revealed in stages, each depicting various kinds of gardens, palaces, and other fantasies painted on them.

After the display of these toys, a robe of honour was presented to the astronomer and money given to the servants of the Observatory.[6]

This response may reflect the fact that the astronomer and staff of the observatory in Vienna were themselves viewing many of the innovations as amusements and were conducting shows of marvels and wonders, perhaps particularly for foreigners. In Europe, electrical demonstrations and experiments were, of course, held before a wide range of audiences, often for educational purposes.[7] The accounts of such demonstrations before viewers from the Near East, however, suggest that in these instances little education or information was transmitted.

The practice among Europeans of amusing spectators with scientific technology can be seen later in the century in the account given by the Baron de Tott, a French military officer who advised the Ottoman court on the establishment of a naval school of engineering in Istanbul in 1773. As part of the entertainments for a birthday celebration in Istanbul for Sultan Mustafa III (reg. 1757–74), Baron de Tott set up an apparatus that gave electrical shocks to everyone. He gave the following account:

---

[6] Bernard Lewis, *The Muslim Discovery of Europe* (New York: W. W. Norton, 1982), pp. 221–38, quotation at pp. 231–2.

[7] See, for example, Simon Schaffer, "The Consuming Flame: Electrical Showmen and Tory Mystics in the World of Goods," in John Brewer and Roy Porter (eds.), *Consumption and the World of Goods* (London: Routledge, 1993), pp. 489–526; and John L. Heilbron, *Electricity in the Seventeenth and Eighteenth Centuries: A Study of Early Modern Physics* (Berkeley: University of California Press, 1979), pp. 134–66.

I had prepared some Electrical-Experiments, which I proposed to shew him as a kind of chamber Fire-works, that might amuse us for the rest of the evening. So great was the effect of the Electrical-Phoenomena at first, that I had much difficulty in erasing the suspicion of Magic, which began to take root in their minds, and to which every new experiment gave additional strength . . . The next day the City resounded with the Miracles I had performed.[8]

If this was the customary approach to the latest technologies by the European demonstrators addressing Near Eastern audiences, it is scarcely surprising that the observers of such shows would consider the technology to be amusements, akin to fireworks displays. It is also possible that European presentations of these electrical demonstrations reflected in part a contemptuous attitude toward their audience - in this case Arabs and Turks – who, like children, could be amazed and entertained but were not deserving of a serious explanation or a discussion of the technologies' usefulness. In any case, this type of display did little to facilitate the serious transfer of new scientific and technological ideas.

MILITARY TECHNOLOGY AND CARTOGRAPHY

The earliest and most wholehearted adoption of a Western technology can be seen in the changing military equipment. In the sixteenth and seventeenth centuries, the Safavids, Ottomans, and Mughals eagerly adopted European firearms and field artillery following a number of humiliating defeats in which the inadequacy of their traditional military equipment had been painfully demonstrated. The extent of the integration of European firearms and cannon into the military was not, however, uniform across all the territories, nor was it achieved without some reluctance. There was a persistent prejudice against regiments using firearms in comparison with the cavalry, for firearms were clumsy and made an unwelcome noise. It was not really until the invention of flintlock that firearms could be used from horseback, and even then speed and mobility – two hallmarks of traditional warfare – were sacrificed.[9]

The Safavids did not use as much field and siege artillery (cannon) as did the Ottomans or the Mughals, possibly because the terrain was not conducive to its deployment, with few major waterways by which such heavy equipment

[8] François Baron de Tott, *Memoirs of Baron de Tott. Containing the State of the Turkish Empire and the Crimea, during the late War with Russia. With numerous Anecdotes, Facts and Observations, on the Manners and Customs of the Turks and Tartars* [trans. from the French], 4 parts in 2 volumes (London: Printed for G. G. J. and J. Robinson, Pater-Noster Row, 1785), vol. I, part 2, p. 86.

[9] Rudi Matthee, "Unwalled Cities and Restless Nomads: Firearms and Artillery in Safavid Iran," in Charles Melville (ed.), *Safavid Persia: The History and Politics of an Islamic Society* (London: I. B. Tauris in association with the Centre of Middle Eastern Studies, University of Cambridge, 1996), pp. 389–416.

could be transported. Although supervised by a French master gunner, the Safavid artillery was still at a disadvantage in the battle of Gulnabad in 1721 against the Afghan army and in the siege of Isfahan the next year. In the event, however, it was not the more effective use of firearms and artillery that brought the Safavid rule to an end, but rather the time-honored tactic of starvation. Cannon played a more prominent role in the Ottoman military strategy of the eighteenth century than in either Iran or India.

There was interest in cartographic and navigational innovations as well as military technology. In all these instances the adoption of European techniques was a slow and uneven process, with the traditional approaches continuing alongside the early modern European ones throughout the century. For example, the late medieval cartographic conventions evident in the Ottoman military maps of the sixteenth and seventeenth centuries continued to be used for many maps until the end of the nineteenth century, although a plan of the Battle of Prut in 1711 was drawn using contemporary European cartographic methods, and an increasing number of similar military maps were produced in the following decades. The Ottoman state produced the military maps, as well as maps of waterways and state-sponsored architectural projects, but other terrestrial mapping appears to have been a private enterprise and displays greater conservatism. For example, the planispheric world maps of the Dutch cartographer Joan Blaeu (d. 1673) were the source for the maps illustrating a particularly popular Turkish encyclopedia titled *Ma'rifetname* written in 1756–7 by Erzurumlu İbrahim Ḥaḳḳı. His encyclopedia is concerned primarily with cosmological matters, but it includes some early modern European ideas, such as the magnetic compass and the heliocentric theory of the universe, the latter presented in addition to the classical geocentric theory. For illustrations of world maps he turned to the planispheric maps from Blaeu's *Atlas Maior,* which had been translated into Turkish between 1675 and 1685. As a result, Ḥaḳḳı's map of the New World repeats Blaeu's depiction of California as an island even though by Ḥaḳḳı's day new European maps had corrected this error. His use of an outdated map indicates either the unavailability of contemporaneous planispheric maps or an attitude that the use of the latest knowledge was not necessary for an essentially literary work.[10]

In 1734 a school of engineering with emphasis on military science was established at Üsküdar, a suburb of Istanbul, and several Turkish versions of European texts on military and engineering topics were produced for use at the school. These included a treatise on military science by an Italian soldier, Count Montecuccoli.[11] Whereas in the Ottoman empire there was consid-

---

[10] Ahmet T. Karamustafa, "Military, Administrative, and Scholarly Maps and Plans," in J. B. Harley and David Woodward (eds.), *The History of Cartography: Volume Two, Book One: Cartography in the Traditional Islamic and South Asian Societies* (Chicago: University of Chicago Press, 1992), pp. 209–27; J. M. Rogers, *Empire of the Sultans: Ottoman Art from the Collection of Nasser D. Khalili* (Geneva: Musée d'art et d'histoire, 1995), pp. 121–3.

[11] Lewis, *The Muslim Discovery of Europe,* p. 235; Abdülhak Adnan-Adıvar, *La science chez les Turcs ottomans* (Paris: G.-P. Maisonneuve, 1939), pp. 142–4.

erable interest in recent European military, navigational, cartographic, and engineering innovations, in Persia available evidence suggests that there was not a comparable interest in European developments in any of these areas except military technology.

MECHANICAL CLOCKS AND WATCHES

A different pattern of reaction to Western technology is evident in the responses to the introduction of mechanical clocks. The first documented interest in the Ottoman Empire in mechanical clocks (as opposed to water clocks) occurred in 1531 when Sultan Süleyman I bought a gold ring with a watch on it. Thereafter, Europeans frequently gave clocks and watches as presents to the Sultan and also to local rulers and officials. In an agreement made in 1547 between Austria and the Ottoman Empire, Austria was to pay a yearly tribute of silver ornaments and clocks, in addition to a large sum of money, to deter Ottoman aggression. As a consequence of this agreement, clocks and watches were produced in Europe exclusively for an Ottoman market, and this market continued even after the cessation of the tribute. During the eighteenth century, French, Swiss, and English clockmakers competed for this market, and correspondence between European watchmakers at this time includes lists of clocks and watches intended for shipment to the Ottomans. The decoration of the timepieces was often adapted for that market, with Islamic dials or scenes from the Bosphorus or Mecca, and on occasion the European watchmakers engraved their signatures in Arabic script. The Topkapi Palace acquired many such clocks and watches during the eighteenth century, and miniature paintings sometimes depicted the ceremonial presentation to the Sultan of gifts that included clocks.[12]

Little information is available, however, on the use made of these clocks and watches. Available evidence suggests that their function was more ornamental than utilitarian, for they seem to have been mounted on any object in need of ornamentation, one preserved example being in the base of a gold birdcage. The clock's function as a precise measurer of the passage of time or the watches' ability to provide portable and private timekeeping (both highly valued in Europe) seem not to have been their primary value in the Ottoman context.

The reason for this was the use among Ottomans of the unequal hour for Islamic civil and religious timekeeping. This unit was calculated by dividing the period between sunset and sunrise, and that between sunrise and sunset, by twelve. For nonequatorial locations, on only two days of the year (the equinoxes) would the twelve hours of the day be equal in length to the twelve

---

[12] Göçek, *East Encounters West*, p. 105; Otto Kurz, *European Clocks and Watches in the Near East* [Studies of the Warburg Institute, 34] (London: The Warburg Institute, 1975), pp. 70–88.

hours of night. Not only do night and day lengthen and shorten as the seasons progress, but the progress of the seasons varies from one latitude to another, so unequal hours vary from place to place. For such a timekeeping system, the European clock and watch based on twenty-four hours of equal length would have been inappropriate (except for astronomical calculations).

There seem to have been no indigenous clockmakers or repairers in Istanbul. Palace records suggest that all repairs on the imported timepieces were made by foreign artisans residing in Constantinople. From the late sixteenth century through the eighteenth century, European watchmakers and goldsmiths, particularly from Geneva, would work for a few years in Constantinople after completing their apprenticeships and then return home having gained experience and money. This technological dependence on Europe prompted Voltaire, in 1771, to write in a letter to Frederick the Great (when discussing the need to create new markets for the products of some fifty religious refugees from Geneva, all of whom were watchmakers); "It is now sixty years since they [the Ottomans] have been importing watches from Geneva, and they are still not able to make one, or even regulate it."[13]

Regarding the availability and reception of European clocks and watches in the Ottoman provinces of Iraq, Egypt, and Syria, there is less information available than for the court at Istanbul, and virtually nothing is recorded about the role of mechanical clocks in Persia at this time.

THE PRINTING PRESS

A yet different response to new technology is evident in the reception of the printing press in the Near East, where its very late adoption has prompted some historians to speculate on its absence being a major factor in the very slow assimilation of European technology, science, and "modernism."[14] Printing with movable type was eventually adopted throughout the Middle East, but the transition from a scribal to a print culture was achieved only after a considerable period of time and with marked social adjustments and repercussions. The minority communities were the first to adopt the printing press in the Ottoman Empire. In the fifteenth century Jewish exiles from Spain and Portugal were allowed to set up a press in Istanbul but to print only in Latin or Hebrew characters. An edict of Sultan Bayezıd II in 1485, and another by Sultan Selım I in 1515, explicitly forbade Muslims from printing texts in Arabic script, although a *firmān* (a certificate of authority) granted by Sultan Murad III in 1588 permitted the import of printed Arabic texts from Europe,

[13] Quoted by Lewis, *The Muslim Discovery*, p. 234, and Kurz, *European Clocks*, p. 71.
[14] For example, see George Atiyeh, "The Book in the Modern Arab World," in Atiyeh (ed.), *The Book in the Islamic World: The Written Word and Communication in the Middle East* (Albany: State University of New York Press, 1995), p. 235.

where Arabic texts had been produced by the Medici Press in Rome since 1586 and even earlier at other locations in Italy. In 1567 an Armenian press was established in Istanbul by the priest Abgar Tibir of Tokat, who had learned printing in Venice, where he acquired the fonts. The first Greek press was set up in 1627 by Nicodemus Metaxas, who had graduated from Balliol College, Oxford, in 1622 and had started printing religious books in London. He went to Istanbul (with the Greek fonts) at the invitation of the Patriarch Cyril Lucaris, who wished him to use the press against Jesuit propaganda, but in 1628 Jesuit complaints forced the Janissaries to close the press. Within the confines of the Ottoman Empire, the first book printed in Arabic script was a Bible, in Arabic translation, printed in 1716 in Aleppo. In 1720 and 1721 Ottoman decrees reflect some social disturbances caused by the use of printed material by some Armenian priests to convert.

There was no Muslim Turkish press until that established by İbrahim Müteferrika (d. 1745), a Hungarian Unitarian who had converted to Islam. He composed a treatise for Sultan Ahmed III on the advantages of the printing press and subsequently obtained permission to set up the first press that could print Turkish (using Arabic script). The *firmān* granted to him by Ahmed III in 1727 states the following:

> By virtue of your having composed a learned tract about, and having expertise in, the various above-mentioned activities, you will see to the necessities and expenditures without loss of time, so that on a fortunate day this Western technique will be unveiled like a bride and will not again be hidden. It will be a reason for Muslims to say prayers for you and praise you to the end of time. Excepting books of religious law, Qur'ānic exegesis, the traditions of the Prophet, and theology, you asked the Padishah's permission in the aforementioned tract to print dictionaries, history books, medical books, astronomy and geography books, travelogues, and books about logic. . . . Copies will be printed of dictionaries, and books about logic, astronomy and similar subjects, and so that the printed books will be free from printing mistakes, the wise, respected and meritorious religious scholar specializing in Islamic Law, the excellent *Kazī* of Istanbul, Mevlana İshak, and Selaniki's *Kazi*, Mevlana Sahib, and Ghalata's *Kazi*, Mevlana Asad, may their merits be increased, and from the illustrious religious orders, the pillar of the righteous religious scholars, the *Sheykh* of the Kasim Paşa Mevlevikhane, Mevlana Musa, may his wisdom and knowledge increase, will oversee the proofreading. With the actual setting up of the press, the above-mentioned books in history, astronomy, geography, logic and so forth, after they pass the review of the learned scholars, shall become numerous. However, you will take special care to see that the copies remain free from error and depend on the noble learned men for this. Ordered in middle of [the month] of Dhū al-Qaddah in the year 1139 [end of June 1727] in Istanbul the protected.[15]

[15] Translation by Christopher M. Murphy, in Atiyeh (ed.), *The Book in the Islamic World*, pp. 284–5.

Thus, although Müteferriḳa was granted permission to establish a printing press, he was restricted to the publication of secular books, and even these had to be submitted for proofreading and approval by a panel of three legal authorities and one religious scholar. There seems also to have been concern that more social unrest would result from this inexpensive means of communication, a concern possibly resulting from the earlier trouble with the Armenians, Greeks, and Jesuits.[16]

Printed maps and geographical material were a particular interest of Müteferrika, some of which included early modern European ideas, though ones by then out of date. The press operated from only 1729 to 1743, when Müteferriḳa fell ill, and only seventeen books were produced. Müteferriḳa's heirs were able to obtain additional *firmān*s to permit printing to continue, but by 1797, when the press closed down, only seven additional books had been printed.

Outside the central Ottoman Empire, experimentation with printing was even slower. In Egypt, the printing press was not introduced until the Napoleonic expedition of 1798, when the fonts were supplied from France and Rome. In Persia, Carmelite friars in Isfahan tried to establish a press with Arabic fonts sent from Rome in 1629, but they seem never to have successfully issued a book. After 1648 the press was kept in storage in the Ooost-Indische Compagnie and then in 1669 returned to the Carmelites, where it remained unused. Early in the seventeenth century an Armenian press was set up in Isfahan; although it did issue some volumes, it also fell into disuse by the end of the century, and another Armenian press was not established until 1771. As for books in Persian (using Arabic script), some texts were imported from Europe (since 1639 they had been printed in Leiden), and toward the end of the eighteenth century they were imported from India, where the East India Company had begun printing Persian (as well as Arabic) books in Calcutta in the 1780s. But within Persia itself, a viable press issuing books in Persian was not functioning until 1817.

Although there were these intermittent experiments with printing, the production of books remained predominantly in the hands of scribes until well into the second half of the nineteenth century. Printing in Arabic script (which was employed to write Turkish and Persian as well) is a more difficult enterprise than printing in Greek, Latin, Hebrew, or Armenian, all of which have discrete block letters. Arabic script, on the other hand, is only cursive, with the letters interlocking and of varying sizes, thus presenting considerable problems to typesetters. Moreover, a printed form of Arabic script loses much of the calligraphic beauty of a hand-copied text, and the cost of equipment for printing is extremely high compared with that required for a man-

---

[16] Göçek, *East Encounters West*, pp. 108–15; H. A. R. Gibb and Harold Bowen, *Islamic Society and the West: Volume One: Islamic Society in the Eighteenth Century*, 2 parts (Oxford: Oxford University Press, 1950–7), part 2, p. 153.

uscript at a time when the cost of labor was very low. In addition there was a large, readily available, and highly respected workforce of trained scribes who would be displaced by the wholesale introduction of printing. When the Bolognese scholar Luigi Ferdinando Marsigli (d. 1730) visited Istanbul, for example, he said there were eighty thousand copyists in the city. Moreover, the *Qur'ān* is considered God's eternal word, and Arabic is venerated because it was the medium through which God's word was revealed, and for these reasons the religious scholars (the *'ulamā'*) initially opposed the use of metal equipment imported from Christendom to print Arabic texts, particularly religious and *Qur'ānic* ones. In addition, the *'ulamā'* may well have feared that mass-produced religious and legal texts might undercut their control of the educational and legal systems.[17] With these shortcomings and difficulties associated with printing by movable type, it is not surprising that the extensive use of such techniques would not have been thought as necessary or desirable in the Near East as in Europe. Lithography, on the other hand, was almost immediately adopted by all countries using Arabic script (India, Persia, the Ottoman Empire, and North Africa) following its invention by Alois Senefelder of Munich in 1796, for it produced inexpensive multiple copies and yet permitted all the calligraphic and esthetic features of a manuscript without the limitation of fonts. The role that printing, or the lack of it, might have played in the assimilation and dissemination of early modern European ideas and technologies in the Near East has yet to be rigorously examined.

ASTRONOMY

Astronomical instrumentation displays even greater conservatism at this time. Earlier contacts in the sixteenth and seventeenth centuries with European celestial cartography did not have much lasting impact. For example, the planispheric star maps based on the early modern European star maps printed about 1650 by Melchior Tavernier and engraved in 1654 on astrolabe plates by the instrument-maker Muhammad Mahdī of Yazd seem to have had no further influence on Islamic celestial cartography or instrument design. Similarly, the magnificent gilt-metal celestial and terrestrial globes produced in 1579 in the workshop of Gerard Mercator and presented to the Ottoman Sultan Murad III (reg. 1574–95) appear to have had no influence on globe design in the Ottoman empire or elsewhere in the Islamic world.[18] Nor did the European instruments, such as telescopes and microscopes, presented

---

[17] G. Oman, Günay Alpay Kut, W. Floor, and G. W. Shaw, "*Maṭba'a* [printing]," in C. Bosworth, E. van Donzel, et al. (eds.), *The Encyclopaedia of Islam*, 2nd ed., vol. 6 (Leiden: Brill, 1991), pp. 794–807; Atiyeh (ed.), *The Book in the Islamic World*, pp. 209–53, 283–92.

[18] F. Maddison and E. Savage-Smith, *Science, Tools, and Magic*, 2 parts (London/Oxford: Azimuth Editions and Oxford University Press, 1997), Part 1, pp. 173–5; Christie's London, *The Murad III Globes* (sale catalogue), 30 October 1991.

as embassy gifts have an impact on locally produced instrumentation in Persia or the Ottoman Empire. The celestial cartography and astronomical instrument design remained throughout the century steadfastly Ptolemaic and traditional.

Although few astrolabes were made in Europe after the seventeenth century, in Persia astrolabes continued to be produced in great numbers. This large output of astrolabes is probably more a reflection of the court's interest in astrology than a devotion to astronomical investigation – an interpretation supported by the invariable inclusion of astrological tables on these instruments. In the Ottoman Empire, on the other hand, relatively few astrolabes were produced at this time, but there are numerous eighteenth-century Ottoman quadrants, which were used for surveying, for measuring the angular elevation of a star or planet, and for other trigonometric calculations. Another instrument produced by eighteenth-century Ottoman instrument-makers was a portable instrument called a *dā'irat al-mu'addil,* which combined a sundial with a *qiblah*-compass, a device for determining the direction a worshipper must face during prayer. All these instruments were variations on medieval instrumentation and did not reflect current European astronomical instrumentation.[19]

As for early modern European astronomical treatises, a number of tables (such as those by Jacques Cassini published in France in 1740 and Joseph de Lalande printed in Paris in 1759) were translated into Turkish during the eighteenth century. These were generally devoid of theory, however, and there is no evidence that any of the recent European astronomical discoveries had any impact. The heliocentric theory was mentioned in some Ottoman writings alongside traditional Ptolemaic theories, but it appears to have been treated as a secondary technical hypothesis and given little attention, although in one instance it drew overt criticism. İbrahim Müteferrika, in a chapter supplementing his 1732 printing of a seventeenth-century geographical work, described in some detail the new astronomy and presented a history of the heliocentric theory based on the writings of Edmond Pourchot (d. 1734). In doing so, Müteferrika qualified the theory as unfortunate and invalid, and he urged scholars to contribute to astronomy by criticizing it and marshaling support for the geocentric interpretation.[20]

As for Persia, exhausted by wars and political turmoil in the first quarter of the eighteenth century, many scholars emigrated from there to the Mughal Court in Delhi. The Persian astronomical writings of this period display greater interest in astrology and a preference for late medieval astronomical treatises.

---

[19] Maddison and Savage-Smith, *Science, Tools, and Magic,* Part 1, pp. 248–59, 266, 277–9.
[20] Ekmeleddin İhsanoğlu, "Introduction of Western Science to the Ottoman World: A Case Study of Modern Astronomy (1660–1860)," in Ekmeleddin İhsanoğlu (ed.), *Transfer of Modern Science & Technology to the Muslim World: Proceedings of the International Symposium on "Modern Sciences and the Muslim World"* (Istanbul: Research Centre for Islamic History, Art and Culture, 1992), pp. 67–120.

MEDICINE

Early modern European medical ideas began to filter into the Middle East in the sixteenth and seventeenth centuries and continued in the eighteenth, but there was always a considerable delay between the European development and its transmission to the Middle East. This time lag is evident, for example, in regard to Vesalius's Latin treatise *De humani corporis fabrica,* printed in Basel in 1543. Present evidence suggests that it was not until the seventeenth century that his treatise was known in the Ottoman empire, and from Persia there are anonymous manuscripts of the late seventeenth to mid-nineteenth centuries that contain ink sketches of the skeletal and muscular figures, and some individual organs, derived from those in Vesalius's *Fabrica.* The treatise on botany and materia medica by Pietro-Andrea Mattioli, published in Italian in 1548, was translated into Turkish in 1770. The description of the circulation of the blood given by William Harvey (d. 1657) was not mentioned by Turkish writers until the end of the eighteenth century, even though in 1664 a Greek translator at the Ottoman court, Alexander Mavrocordato, had written a dissertation at the University of Bologna on the discovery.[21]

During the reign of Sultan Ahmed III, the chemical medicine that developed in Europe from the theories of Paracelsus acquired a following among some Muslim physicians in Istanbul. The concept of "chemical medicine" had initially been introduced to the Ottoman court of the seventeenth century through the writings of a court physician, a Syrian named Ṣāliḥ ibn Naṣr ibn Sallūm, who in 1655 translated into Arabic extracts of Latin treatises by Oswald Croll (d. 1609), professor of medicine at the University of Marburg, and Daniel Sennert (d. 1637), professor of medicine at Wittenberg. Both men were followers of Paracelsus (d. 1541), who employed mineral acids, inorganic salts, and alchemical procedures in the production of remedies. Many of the medicaments required distillation processes and plants that were indigenous to the New World, such as guaiacum and sarsaparilla. Ibn Sallūm's treatises not only reflected the new chemical medicine but also described for the first time in Arabic a number of "new" diseases, such as scurvy, anemia, chlorosis, the English sweat (a type of influenza), and plica polonica (an East European epidemic of matted and crusted hair caused by infestation with lice). Ibn Sallūm, however, drew his theoretical considerations of the causes and symptoms for the most part not from the Paracelsians but from late medieval Islamic writers of the thirteenth to sixteenth centuries, in particular the Arabic commentary written in 1283 by Quṭb al-Dīn al-Shīrāzī on the *Qānūn fī al-ṭibb (Canon of*

---

[21] Gül Russell, "'The Owl and the Pussy Cat': The Process of Cultural Transmission in Anatomical Illustration," and Ramazan Şeşen, "The Translator of the Belgrad Council Osman b. Abdülmennan & His Place in the Translation Activities," in İhsanoğlu (ed.), *Transfer of Modern Science & Technology to the Muslim World,* pp. 180–212, 371–84; Maddison and Savage-Smith, *Science, Tools, and Magic,* Part I, pp. 14–24; Adnan-Adıvar, *Le science chez les Turcs,* pp. 99–100; Albert Hourani, *Islam in European Thought* (Cambridge University Press, 1991), p. 140.

Medicine) by Ibn Sīnā, known to Europeans as Avicenna (d. 1037), and a medical compendium by the Syrian physician Dāʾūd al-Anṭākī (d. 1599), with the result that his treatises were pastiches of late medieval Islamic medical thinking alongside seventeenth-century European medical chemistry. These versions of Paracelsian tracts were subsequently used in the eighteenth century by Turkish medical and chemical writers such as Ömer Sinan al-İznikı and Ömer Şifai.[22]

The interest in Paracelsian medicine, however, met with some opposition, as illustrated by a decree issued in 1704 prohibiting the practice of the "new medicine." It referred to "certain pseudo-physicians of the Frankish community who abandoned the way of the old physicians and used certain medicaments known by the name of the new medicine (*ṭibb-i jedīd*)" and stated that Mehmed, a convert to Islam, and his partner, a European doctor, who had opened an office at Edirne, were consequently to be expelled from the city.[23] Whatever the motivation for this decree, it is evident that its effect was not widespread nor long-lasting, for a large number of Turkish treatises composed (or copied) throughout the eighteenth century concerned themselves, at least in part, with the new chemical medicine.

When plague befell Istanbul in the middle of eighteenth century, the Ottoman Sultan Mustafa III ordered that a Turkish translation be made of two treatises by the Dutch medical reformer Herman Boerhaave (d. 1738): the *Institutiones medicinae* and *Aphorismi de cognoscendis et curandis morbis,* published in 1708 and 1709 respectively. The Turkish version, completed in 1768 by the court physician Ṣubḥī-zāde ʿAbd al-ʿAzīz in collaboration with the Austrian interpreter Thomas von Herbert, attempted through explanations and glosses to harmonize European medicine with medieval medicine.[24]

In Persia, the unsettled political conditions at the end of the seventeenth century and beginning of the eighteenth prompted many physicians to move to the Mughal court, then in Delhi. One of the most prominent refugees to India was ʿAlavī Khān, who left Shiraz in 1699 to eventually become court physician in Delhi, returning to Persia for a while to act as physician to Nādir Shāh (reg. 1736–47). The mercantile companies in Persia sometimes sent physicians to various factories to attend to the health of the European merchants and missionaries and occasionally that of a ruler, for Safavid (and subsequent) rulers sometimes consulted European physicians. Nādir Shāh, for example,

---

[22] E. Savage-Smith, "Drug Therapy of Eye Diseases in Seventeenth-Century Islamic Medicine: The Influence of the 'New Chemistry' of the Paracelsians," in *Pharmacy in History,* 29 (1987), 3–28; Nil Sari and M. Bedizel Zülfikar, "The Paracelsian Influence on Ottoman Medicine in the Seventeenth and Eighteenth Centuries," in İhsanoğlu (ed.), *Transfer of Modern Science & Technology to the Muslim World,* pp. 157–79; Adnan-Adıvar, *La science chez les Turcs,* pp. 96–8, 128.

[23] Quoted by Lewis, *The Muslim Discovery,* p. 231. See also Sari and Zülfikar, "Paracelsian Influence," pp. 168–9; and G. A. Russell, "Physicians at the Ottoman Court," in *Medical History,* 34 (1990), 243–67, especially 265–6.

[24] C. E. Daniëls, "La Version orientale, arabe et turque, des deux premiers livres de Herman Boerhaave: Étude bibliographique," *Janus,* 17 (1912), 295–312.

sought a European medical adviser when his personal physician, 'Alavī Khān, left on pilgrimage to Mecca. For a while Nādir Shāh used the services of Fr. Damian of Lyons, a Capuchin friar, before requesting medical assistance from the English factory at Isfahan and, when that proved unsatisfactory, from the Dutch factory.[25] In North Africa a similar pattern occurred, wherein the Turkish governors in Tunisia consulted European doctors while also employing local court physicians.[26] The general population, however, was probably unaffected by these contacts with European medicine.

The nucleus of most Arabic, Turkish, and Persian medical compositions during the eighteenth century was medieval Islamic medicine, despite some acquaintance with European medical ideas and the occasional contact with European doctors. During the eighteenth century, complete translations of Avicenna's *Canon of Medicine* were made into both Turkish and Persian. Dietetics, drug remedies, and self-help manuals were the primary focus of medical writings, whereas in Arabic-speaking areas there seems to have been a renewed interest in didactic medical poetry concerned primarily with dietetics and drug lore.

European travelers of the eighteenth century frequently noted the failure of physicians in the Middle East to keep up with European developments in medicine or to maintain the best of the practices represented by the learned Arabic medieval medical compendia.[27] A particularly interesting account of the medical care in Syria (then under Ottoman rule) between 1742 and 1753 was given by Alexander Russell (d. 1768), who was physician to the English factory in Aleppo at that time. In his *The Natural History of Aleppo* he commented on the streets of Aleppo being narrow but well paved and "kept remarkably clean," also observing that "the people here have no notion of the benefit of exercise, either for the preservation of health, or curing diseases." In 1742, while describing an epidemic of smallpox, he noted that "inoculation is only practiced here among the Christians, and is not yet general even among them."[28] Of medical practice in general, he goes on to say

> Though the Turks are predestinarians, they are taught however to believe, that tho' God has afflicted mankind with disease; yet he has sent them also the remedies, and they are therefore to use the proper means for their recovery: so the practitioners in physic are here well esteemed, and very numerous. These are chiefly native Christians, and a few Jews. The Turks seldom made

[25] Cyril Elgood, *A Medical History of Persia* (Cambridge University Press, 1951), pp. 348–436; Elgood, *Safavid Medical Practice* (London: Luzac, 1970), pp. 97–105, 285–8.

[26] N. E. Gallagher, *Medicine and Power in Tunisia, 1780–1900* (Cambridge University Press, 1983), especially pp. 17–18.

[27] See Gunny, *Images of Islam*. For similar nineteenth-century reactions, see Rhoads Murphey, "Ottoman Medicine and Transculturalism from the Sixteenth through the Eighteenth Century," in *Bulletin of the History of Medicine*, 66 (1992), 376–403.

[28] Alexander Russell, *The Natural History of Aleppo, and Parts Adjacent* (London: A. Millar, 1756 [mispr. 1856]), pp. 5 and 194. See also Abraham Marcus, *The Middle East on the Eve of Modernity: Aleppo in the Eighteenth Century* (New York: Columbia University Press, 1989), pp. 252–68.

this their profession. Not one of the natives, however, of any sect, is allowed to practice without a licence from the *Hakeem Bashee* [chief physician]; but a few sequins are sufficient to procure this to the most ignorant; and such most of them are egregiously, for they have no colleges in which any branch of physic is taught: and as the present constitution of their government renders the dissection of human bodies impracticable, and that of brutes is a thing of which they never think, they have a very imperfect idea of the situation of the parts, or their functions.

Of the use of chemistry in medicine they are totally ignorant, but now and then one amongst them just acquires a smattering enough of alchemy to beggar his family by it. The books they have amongst them are some of the Arabian writers: Ebnsina [Avicenna] in particular, whose authority is indisputable with them. They have likewise some translations of Hippocrates, Galen, Dioscorides, and a few other ancient Greek writers. But their copies are in general miserably incorrect. Hence it may easily be seen, that the state of physic among the natives in the country, as well as every other science, is at a very low ebb, and that it is far from being in a way of improvement.

But ignorant as they are in regard to physic, they are great masters in temporizing, and know how to suit a plausible theory to the patient's way of thinking, in doing which they scruple not to quote the authority of Hippocrates, Galen, and Ebensina, in support of opinions the most ridiculous and absurd. It is from the pulse alone that they pretend, and are expected, to discover all diseases, and also pregnancy.

. . . What has been said with regard to practitioners in physic, relates solely to the natives; for the Europeans, of whom there are several, practice in their own way, and are greatly respected by the inhabitants; though, partly to save their money, and partly from a notion of their giving violent medicines, they seldom apply to them, till they have tried their own doctors to no purpose.[29]

According to Russell, the chemical medicine of the Paracelsians, which was discussed in a number of Turkish treatises, does not seem to have been current among practitioners in Aleppo, at that time part of the Ottoman Empire. Pierre-Charles Rouyer, a French army pharmacist who accompanied the Napoleonic expeditionary force to Egypt, commented that at the end of the century "the Egyptians, having become apathetic and indolent, have let a large number of their medications fall into disuse." From his report it is evident that some European drugs were available to Egyptians but that mercurial drugs were almost unknown. Of the eighty-one plant drugs listed by Rouyer as available in Cairene shops (excluding those run by and for Europeans), 90 percent were to be found in contemporary European pharmacopeias, and 23 percent were regularly exported to France, although Rouyer appears to have been unaware of this indebtedness of European pharmacology to Egypt and other areas of the Middle East.[30] Like most Europeans traveling to the Middle East at that time, he enthusiastically recorded the practices of a strange and exotic culture while concluding that all comparisons with European ways only demonstrated the excellences of Europe and the deficiencies of the foreign culture.

EUROPEAN INTEREST IN THE MIDDLE EAST

Expeditions, such as those led by Carsten Niebuhr to Saudi Arabia, included cartographers, natural scientists, and physicians, with the purpose of improving maps, finding supporting botanical evidence for the work of Carl von Linné (Linnaeus, d. 1778), recording medical conditions among the populations, and collecting medieval Arabic and Persian manuscripts for European libraries. Medieval Islamic medical literature greatly interested some physicians in eighteenth-century Europe. In 1725–6 the physician John Freind (d. 1728) published in two volumes his *History of Physick.* The second volume is devoted to medieval Arabic medicine, with considerable attention given to the tenth-century Spanish physician Abū al-Qāsim al-Zahrāwī (known in Latin as Albucasis) and to the Eastern physician Muḥammad ibn Zakarīyā' al-Rāzī (d. 925), known to Europe as Rhazes, drawing attention to the latter's treatise on smallpox and measles (although no version of it was then available in Latin). Freind himself knew no Arabic but used Latin material and versions made by Salomon Negri, a Damascene translator residing in London, and John Gagnier, an Oxford Arabist. In 1747 Richard Mead published, as an appendix to his *De variolis et morbillis liber,* a Latin translation of Rhazes's treatise on smallpox and measles that had been prepared by Salomon Negri and John Gagnier and revised by Thomas Hunt, Laudian Professor of Arabic at Oxford. A second translation into Latin, with Arabic text, was published in 1766 by a London apothecary and Arabist, John Channing, whose edition and Latin translation of the surgical writings of Albucasis was published in 1778. It was also to medieval Arabic sources that Georg Fuchs turned for his study of the occurrence and treatment of *Dracunculus medinensis,* the parasitic guinea worm, also called Medina worm or dragon worm, which was published in 1781.[31]

---

[30] J. Worth Estes and Laverne Kuhnke, "French Observations of Disease and Drug Use in Late Eighteenth-Century Cairo," *Journal of the History of Medicine,* 39 (1984), 121–52, quotation at p. 131.

[31] D. M. Dunlop, "Arabic Medicine in England," in *Journal of the History of Medicine,* 11 (1956), 166–82; E. Savage-Smith, "John Channing: Eighteenth-Century Apothecary and Arabist," in *Pharmacy in History,* 30 (1988), 63–80; Georg F. C. Fuchs, *Commentaria historico-medica de dracunculo Persarum sive vena medinensi Arabum* (Jena: widow of J. R. Croecker, 1781).

THE INTERMINGLING OF TRADITIONS

Although in the Islamic world of the eighteenth century there was exposure to some European innovations, nearly all of them were technologies, and even then the adoption was selective. The new scientific and medical philosophies developed in Europe were not assimilated and perhaps at times were purposely avoided. For example, when Paracelsian medicine was introduced through the versions prepared by Ibn Sallūm, the theoretical and philosophical ideas were omitted and only the chemical procedures and compound remedies were translated, first into Arabic and then into Turkish. Similarly, when the Ottoman court turned to the works of Herman Boerhaave, the treatises were selectively translated and presented in a way that would reconcile the material with the traditional medieval medicine, stripping it of much of the new scientific philosophy then developing in Leiden.

In Europe by the beginning of the eighteenth century a mechanical philosophy of nature dominated much of the learned discourse, and there was wide acceptance of the experimental and mathematical approach to nature, but none of these ideas played any significant role in seventeenth- or eighteenth-century scientific thinking in the Islamic world. The major European scientific and philosophical developments of the seventeenth century remained unknown to the Islamic world until the very end of the eighteenth century. The Copernican revolution is viewed by many to be a major transformation of man's conception of the universe and his position within it. Yet this philosophical shift, and its subsequent development by Galileo Galilei, Johann Kepler, and Isaac Newton, did not interest the eighteenth-century Islamic world, although the heliocentric theory was occasionally mentioned alongside the Ptolemaic one. Nor were the ideas of Robert Boyle or the philosophy of René Descartes discussed. Of the numerous anatomical discoveries of the seventeenth century, such as those by Thomas Willis, Francis Glisson, Marcello Malpighi, Robert Hooke, Giovanni Alfonso Borelli, and William Harvey, only the last seems to have been mentioned in eighteenth-century Turkish literature, and then only at the very end of the century.

As for eighteenth-century European thinkers, the morbid anatomy of Giovanni Battista Morgagni, the mechanical model proposed by Julien Offray de la Mettrie in his *L'Homme machine* published in 1747, the quantification and hemostatic experiments of Stephan Hales, and the developments in chemistry and physics by Henry Cavendish, Antoine Laurent Lavoisier, or Luigi Galvani – to give only some examples – all remained unknown to eighteenth-century Islamic scholars. Even smallpox inoculation was slow to be adopted, at least in Ottoman Syria (if the evidence given by Russell is reliable), although Lady Mary Wortley Montagu, wife of the Ambassador Extraordinary to the Ottoman court, had brought widespread publicity to the procedure in the 1720s by describing its use in Turkey. Furthermore, surgical innovations, such as the removal of a cataractous lens from the eye as pioneered by Jacques

Daviel, do not seem to have been imported into the medical world of the Middle East.

Various explanations have been offered for this lack of receptivity to contemporaneous European thinking in the eighteenth century.[32] Underlying the question is the tacit assumption that Western experimental science is always desirable and is always a means of obtaining the truth. Such an assumption would not be acceptable for the eighteenth-century Islamic world, which manifestly did not view the major philosophical shifts in science and medicine to be a desirable or necessary means of attaining truth. The notion of "science" defining truth was unacceptable in Muslim terms, for truth could be known and determined only by God. An educational system very different from that in Europe, centering on the *Qur'ān* and stressing memorization and recitation of texts; the unquestioned belief in the omnipotence of God and His possible intervention at any moment and at any level; and the precedence given in Islam to the ideas and practices expressed by the early Islamic community – all these were fundamental factors in determining the intellectual climate into which these foreign scientific ideas were introduced. The resulting respect for tradition and authority was not conducive to generating new ideas and certainly not ones that would be considered "truths" as defined by the experimental method or a mathematical model of nature.

An ambivalent attitude toward European technologies is evident in the sources so far examined for the eighteenth century, and a disinterest – and at times selective filtering out – of the philosophical issues can be discerned. There appears to have been concern for maintaining the social and religious norms of Islamic society. The selected European technologies and scientific ideas were intermingled with traditional practices and concepts in the writings of the learned, highly educated sector of the society. To what extent any of these newer ideas and techniques affected other segments of society is unknown.

By the end of the eighteenth century, it is evident that some eighteenth-century European ideas – such as a mechanical view of the human body or current anatomical atlases – could find a receptive audience in the Islamic world. The *Encyclopédie* edited by Denis Diderot and Jean d'Alembert and published in Paris between 1762 and 1772, for example, was the (unacknowledged) source for a treatise published in 1820 by the Ottoman physician and court historiographer 'Atā'ullāh Şānizāde (d. 1826). In his treatise, Şānizāde spoke of the human body as a machine and carefully copied anatomical plates that represented more current knowledge than those previously available, although he omitted the allegorical settings surrounding the anatomical figures in the European originals and presented the figures in complete

[32] For example, see Toby E. Huff, *The Rise of Early Modern Science: Islam, China, and the West* (Cambridge University Press, 1993), pp. 202–36, 322–59.

isolation.[33] In the nineteenth century, a fundamental change occurred in the teaching of science, technology, and medicine throughout the Middle East, for Western European ideas were introduced on a massive scale, and the Islamic world was drawn more and more into the orbit of Europe. Nonetheless, a conspicuous thread of traditional practices still coexists today alongside modern, basically European, science.

[33] Russell, "'The Owl and the Pussy Cat,'" pp. 200–12.

28

INDIA

Deepak Kumar

In the history of South Asia, the eighteenth century is unique in the sense that it saw the decline of precolonial systems as well as the inauguration of systematic colonization. This single century encompasses both the precolonial and the colonial phases. Although every historical period is a period of transition, the theme of transition is more applicable to the eighteenth century than to any other period in Indian history. In this century the mighty Mughals broke up, and this collapse has been explained in terms of religious differences, economic crises, and cultural failures. The crucial nature of the last factor has of late been emphasized: "It was this failure that tilted the economic balance in favour of Europe"; it was this failure again that sapped "the capacity to grapple with agrarian crises"; "even military weaknesses flowed from the intellectual stagnation that seem to have gripped the Eastern world."[1] Is "stagnation" the right description? Was it really an "age of decline"? What was the state of techno-scientific knowledge in this age of political turmoil? It is true that the Eastern knowledge *corpus* and its implements were no match for what was then happening in the West. But why? Was it because of some "structural fault" in the Indo-Islamic society or some built-in defect in its ideological framework? What was the size and composition of the intelligentsia? What were their economic interests and cultural predilections?[2] Many questions emerge for which only partial explanations can be attempted.

Eighteenth-century India inherited a long-lived tradition in both philosophical and material terms. Centuries before, Said al Andalusi (1029–1070), in his *Tabaqat al Uman* (probably the first work on the history of science in any language), referred to India as the first nation that cultivated the

[1] Athar Ali, "The Eighteenth Century – An Interpretation," *Indian Historical Review*, 5 (1979), 175–86.
[2] Irfan Habib, "Reason and Science in Medieval India," in D. N. Jha (ed.), *Society and Ideology in India: Essays in Honour of Professor R.S. Sharma* (New Delhi: Munshiram Manoharlal Publishers, 1996), pp. 163–74.

The academic help received from Professors Shireen Moosvi, S. R. Sarma, and I. G. Khan of the Aligarh Muslim University is gratefully acknowledged.

sciences.[3] Later, India adopted post-Ghazali Islam, which was marked by a bitter theological opposition to *falsafa* (philosophical rationalism). Knowledge in the Islamic framework was divided between *ilm-al-Adyan* and *ilm-al Dunya*.[4] Accordingly, Muslim scholars were divided into those who relied on *manqul* (traditional knowledge) and those who favored the touchstone of reason (*maqul*). The former, greater in number and more powerful, opposed Sultan Muhammad Tughlaq (1325–1351) when he tried to patronize *ilm-i-maqulat*. However, Mughal India was somewhat eclectic, and because there was no consolidated, systematic, and detailed curriculum, the channels of learning were not at all closed to *maqul* ideas.[5] Along with the debates within the Islamic framework, there were several attempts at cross-cultural fertilization. In the late fourteenth century, Mahendra Suri (an astronomer at the Court of Firoz Tughlaq and author of *Yantraraja*) had tried to introduce Arab and Persian astronomy into the Sanskrit *Siddhanta* tradition. This flow of astronomical ideas, as well as instruments, continued into the seventeenth century, providing the basic materials for those training in the Ptolemaic system.[6] Similarly, in 1337 a compendium of general medicine (*Majma'ah-i-Diya'i*) was compiled at the order of Sultan Muhammad Tughlaq on the basis of numerous Arabic, Zoroastrian, Persian, Buddhist, and Hindu works.[7] Later, in 1512, Mian Bhuwah prepared a manual of medicine (*Ma'din al-shifa-i-Sikandar-Shahi*) based on the Ayurvedic and Yunani traditions.[8] However, no real synthesis could emerge. The Sanskrit *tols* and Islamic *madrassas* continued to cling to their own distinct astronomical and medical systems. These schools did influence each other and occasionally came together under an enlightened ruler, only to fall apart.

It seems that during the late medieval period no comprehensive attempt was made to explore India's scientific heritage, much less to keep it abreast of the developments then taking place in the Western hemisphere. Unlike

---

[3] Andalusi wrote, "Among all nations India was known as the mine of wisdom and the fountain-head of justice. Although their colour belongs to the first grade of blackness, yet God the Exalted has kept them immune from evil character, base conduct and low nature. He (God) has thus exalted them (Indians) over many brown and white peoples. . . . They also obtained profound and abundant knowledge of the movements of the stars, the secrets of the celestial sphere and all other branches of mathematical sciences. Moreover, of all the peoples they are the most learned in the science of medicine and well-informed about the properites of drugs and the nature of composite elements." M. S. Khan, "Qadi Saiid al-Andalusi's Account of Ancient Indian Sciences and Culture," *Journal of the Pakistan Historical Society,* 45 (1997), 1–31.

[4] The Sufi nomenclature was *ilm Batin* (knowledge of self) and *ilm Zahir* (knowledge outside self).

[5] Rather, in the eastern regions of Awadh and Bihar, the subjects bearing on *maqulat* were compulsory. North West India was more orthodox, and here these subjects were optional. Muhammad Umar, *Islam in Northern India during the Eighteenth Century* (New Delhi: Munshiram Manoharlal Publishers, 1989), p. 272.

[6] G. G. Joseph, *The Crest of the Peacock: Non European Roots of Mathematics* (Harmondsworth: Penguin, 1990), pp. 347–8.

[7] A. Rahman (ed.), *Science and Technology in Medieval India: A Bibliography of Source Materials in Sanskrit, Arabic and Persian* (hereafter cited as *Bibliography*) (New Delhi: Indian National Science Academy, 1982), p. 55.

[8] C. A. Storey, *Persian Literature: A Bio-Bibliographical Survey,* vol. 2, part 2 (London: Royal Asiatic Society, 1971), p. 231.

Alberuni's *Kitbu-l-Hind,* Abul Fazl's *A'in-i-Akbari* (a classic on Mughal times) barely touches science. Alberuni could cite numerous Greek texts; Abul Fazl refers only to Aristotle and Ptolemy. He cites the tables of specific gravity from Alberuni but makes no attempt to verify Alberuni's calculations, which were made almost 550 years earlier (1030 A.D.).[9] It appears that scientific curiosity was in decline, and Abul Fazl admits it. But he shows great interest in technology, especially the smelting process and liquor distillation. Abul Fazl appreciated the importance of technological improvements for the state economy; socially he enunciated *Sulh-i-Kul* which emphasized tolerance and coexistence; and intellectually he was not at all dogmatic.[10] Yet he was unable to move beyond the classical theoreticians. The only Mughal noble who took a little more interest in modern astronomy, geography, and anatomy was Danishmand Khan. He employed the French physician François Bernier (1659–66), who translated for him the works of Gassendi and Descartes. Bernier even dissected sheep to explain Harvey's discovery of the circulation of blood. But the Indian followers of Galen (*Yunani Tibb*) remained unimpressed.[11] Similarly clocks and watches did not impress the Mughals, who, unlike the Ottomans and Manchus, refused these marvels a closer look. But items of military concern, such as artillery and shipbuilding, were favorably looked upon. Other items, such as mirrors, window panes, pumps, and pistols aroused interest, but no attempt was made to learn the techniques behind them.[12] So the scenario remains complex, with several gray areas. In the absence of a deeper understanding of the texts written in classical languages, and sometimes in the absence of an authentic source itself, it is difficult to say with precision why certain new scientific ideas did not germinate or find favor or why a new technique was ignored. A theory of decline does not explain everything. One thing, however, appears certain: Indians were not xenophobes.

THE THREE SHADES OF OPINION

On the basis of scattered, if not scanty, evidence, one can identify three major shades of opinion. To the first category belong the majority of contemporary European travelers and several subsequent British officials and scholars, who found everything in India "black and bleak." In sharp contrast, another set of opinions is quite enthusiastic about India's scientific credentials and

[9] Irfan Habib, "Reason and Science."

[10] When Abul Fazl was reminded by Shaikh Ahmed Sirhindi of Ghazzali's dictum that all sciences not found in the scriptures were useless, he is said to have replied, "Ghazzali spoke nonsense." Habib, "Reason and Science."

[11] François Bernier, *Travels in the Mughal Empire, A.D. 1656–68,* trans. A. Constable (rev. ed; New Delhi: S. Chand, 1968) p. 339; see also M. Athar Ali, *The Mughal Nobility under Aurangzeb* (Delhi: Asia Pub. House, 1966), p. 179.

[12] A. J. Qaisar, *The Indian Response to European Technology and Culture, 1498–1707* (Delhi: Oxford University Press, 1982), pp. 128–39.

potentialities in the precolonial period. A third set of opnions treads cautiously and offers guarded comments. From the accounts of the European travelers came the stories of the "oriental mind" and Indian resistance to innovation and change, and these were to become the obsession of European scholarship for generations to follow. These accounts do throw some light on the level of science and technology in precolonial India.[13] Astronomy, medicine, and the Indian textile and steel-making processes impressed travelers the most. Their accounts usually begin with feelings of surprise and admiration and end on a suspicious, even arrogant, note. For example, Indian astronomy was lauded as "a proof still more conspicuous of their extraordinary progress in science," and its accuracy was found to be on a par with that in modern Europe.[14] The Indian observatories were seen as "gigantic relics of the zeal in the pursuit of science manifested in former days." Then comes the indictment: "It is carried on by mechanical rules, without any idea of the principles upon which they depend . . . The instruments employed are rude in the extreme."[15] The standard criticism of Indian astronomy was as follows:

1. "It gives no theory, nor even any description of the celestial phenomena, but satisfied itself with the calculation of certain changes in the heavens, particularly of the eclipses of the sun and moon."[16]
2. The Indian astronomers were satisfied with their traditional systems; they did not bother to improve on, nor did they welcome any criticism of, the Puranic and Siddhantic systems.

Most of the travelers recorded that Indians had made remarkable progress in mathematics and astronomy in ancient times – progress that gradually fell from grace, particularly after the establishment of the Muslim rule. Later, this statement was uncritically accepted by several British and Indian historians. Thus grew the notion that science flourished only in ancient India and not in medieval India.[17] However, scholars such as Rahman, Dharampal and Ansari

---

[13] Satpal Sangwan has referred to numerous pamphlets, travelogues, and books published mostly during the eighteenth and nineteenth centuries, in *Science, Technology, and Colonization: An Indian Experience 1757–1857* (Delhi: Sage, 1991), pp. 167–87. Michael Adas uses several travelogues as evidence of "first encounters" in *Machines as the Measure of Men* (Ithaca, NY: Cornell University Press, 1989).

[14] W. Robertson, *An Historical Disquisition Concerning the Knowledge Which the Ancients Had of India* (London, 1791), pp. 302, 308, quoted in Sangwan, *Science Technology*, pp. 2–3.

[15] Hugh Murray, *Historical Account of Discoveries and Travels in Asia from the Earliest Ages to the Present Time* (Edinburgh, 1820), pp. 310–11, quoted in Sangwan, *Science Technology*.

[16] John Playfair, "Remarks on the Astronomy of the Brahmins," *Transactions of the Royal Society of Edinburgh*, II (1790), 135–92, reproduced in Dharmpal, *Indian Science and Technology in the Eighteenth Century* (Delhi: Impex India, 1971), pp. 12–13.

[17] For example, George Sarton wrote, "Hindu culture was stifled, if not stamped out, in many places by Muslim conquerors." George Sarton, *Introduction to History of Science* (London, 1947), vol. 2, p. 107. Later, a prestigious publication on the history of science in India records, "India had her period of glory in the Classical Age and made remarkable progress . . . even right up to the twelfth century A.D. Thereafter the creative endeavour showed signs of decay due largely to the traditional compulsion and political vicissitudes." D. M. Bose et al., (eds.), *Concise History of Science in India* (New Delhi: Indian National Science Academy, 1971), pp. 484–6.

have raised strong objections to this notion. In his *Bibliography of Source Materials in Sanskrit, Arabic and Persian,* Rahman argues that throughout the medieval period scientific and technological activity was both continuous and vigorous. Second, although the major contributions lie in the fields of astronomy, mathematics, and medicine, they cover a wide range of scientific and technological subjects. Third, as compared with contributions of a general nature, there are a large number of special treatises. The number of manuscripts listed in Rahman's bibliography is quite large. In the sphere of astronomy alone, in Persian, 411 manuscripts are said to have been compiled from the tenth to the nineteenth centuries, of which 32 belong to the eighteenth century; in Arabic, of 346 manuscripts, 22 were written in the eighteenth century. Sanskrit has the greatest number of manuscripts (2,136), of which 190 belong to the seventeenth century and 37 to the eighteenth century. As for the nature of these manuscripts, of the 32 Persian manuscripts written in the eighteenth century, 21 are of a general nature, two are commentaries, one is of a special nature, two are translations, and six are almanacs; of the 22 Arabic manuscripts, eight are of a special nature, six are commentaries, and eight are almanacs; and of the 37 Sanskrit manuscripts, three are of a general nature, 15 are special, eight are commentaries, two are translations, four are anthologies, and five are almanacs.[18] The list is impressive, but to determine whether they contain the seeds of modern science, or at least reflect the advance then achieved in science, would require further study.

The third set of opinions advocates neither an unqualified denunciation nor a naive (perhaps revivalist) appreciation of the precolonial science and technology. Writing in the 1930s, at the peak of the Indian national movement, B. K. Sarkar compared India and the West in terms of the following equations:[19]

$$\text{India in exact science (B. C. 600 – A.D. 1300)} \atop = \text{Europe in exact science (B. C. 600 – A.D. 1300)} \tag{1}$$

$$\text{Renaissance in India (1300–1600)} \atop = \text{Renaissance in Europe (1300–1600)} \tag{2}$$

$$\text{India in exact science (1600–1750)} \atop = \text{Europe in exact science (1300–1600)} \tag{3}$$

Thus it was during the seventeenth and eighteenth centuries, the post-Renaissance epoch (that of Descartes and Newton), that Europe began to outdistance India in the natural sciences. Dharampal also concedes that "it is possible that the various sciences and technologies were on a decline in India around 1750 and perhaps had been on a similar course for several centuries

[18] Rahman (ed.), *Bibliography,* pp. 11–16.
[19] B. K. Sarkar, *India in Exact Science: Old and New* (Calcutta, 1937), p. 7.

previously."[20] Irfan Habib does not accept any description of the precolonial technology as primitive but calls for "a wider study of the social constraints that prevented either an endogenous development of industrial technology comparable to that of modern Europe or, at least, a rapid absorption of European technology itself."[21] He argues that many mechanical principles frequently employed in modern machines were in use in Mughal India but adds that the range of their application was rather limited. Precolonial science and technology were definitely not primitive. A better description perhaps would be "proto-science and technology," clearly distinguishing it from the post-seventeenth-century modern scientific tradition based on the experimental method.[22]

ASTRONOMY

Perhaps the best example of proto-science in Mughal India can be found in the realm of astronomy, especially in the construction and use of astrolabes and celestial globes.[23] The astrolabe was probably introduced in India by Al Biruni, and between 1567 and 1683 a number of these instruments were produced by the family of Allahdad at Lahore. Of this family, Diya al-Din Muhammad was most prolific and versatile; he produced about thirty-two astrolabes and sixteen celestial globes with innovative designs. These instruments figure in several Mughal miniatures and testify to royal patronage.[24] This patronage, however, was motivated more by astrological than other considerations. Unfortunately, the Hindu astronomers did not make much use of these instruments even though many of them took notice of their worth. Padmanabha (about 1400) and Ramchandra Vajpeyin (1428) discussed astrolabes extensively. Later, a Jaina monk, Megharatna, used several Arabic and Persian technical terms. In 1621, Narsimha from Benaras refers to the celestial globe as *bhagola*

[20] Dharampal, *Indian Science and Technology,* p. 32.

[21] Irfan Habib, "Technological Changes and Society: 13th and 14th Centuries," Presidential Address, Medieval India Section, *Proceedings of the Indian History Congress,* Delhi, 1969, 139; Habib, "Technology and Barriers to Social Change in Mughal India," *Indian Historical Review,* 1–2 (1979), 152.

[22] R. A. L. H. Gunawardana, "Proto-Science and Technology in Pre-colonial South Asia," in S. A. I. Tirmizi (ed.), *Cultural Interaction in South Asia in Historical Perspective* (New Delhi: South Asia Books, 1993), pp. 178–208.

[23] The celelestial globe (*al-Kura*) consists of a spherical globe made usually of brass on which the celestial equator, ecliptic, tropics, and other circles are plotted. Upon this grid are marked the positions of about 1,020 fixed stars according to the coordinates given either by Ptolemy in his *Almagest* or by Ulugh Beg in his Tables. The astrolabe (*asturlab*), on the other hand, is a versatile instrument for observation and computation. Here the great circles and other circles, the star positions, etc., are drawn on a plane of two dimensions by a method called stereographic projection. It enables one to read the configurations without recourse to long and tedious computations. S. R. Sarma, "The Lahore Family of Astrolabists and Their Ouvrage," *Studies in History of Medicine and Science,* 13, 2 (1994), 205–24.

[24] In a miniature painted for Shah Jahan (1628–1658), his grandfather Humayun is shown surrounded by two angels, one holding a globe and the other a ring dial. These two instruments perhaps symbolized cosmic space and time. S. R. Sarma, "Astronomical Instruments in Mughal Miniatures," *Studien zur Indologic und Iranistik,* Reinbek (1992), 235–75.

but adds, "the stars known to the Muslims do not serve our purpose. Observation of unfamiliar stars would lead to misfortune." The Hindu astronomers
were interested in the coordinates of a very limited number of stars and not
in all the 1,018 stars marked on the Islamic globe.[25] In Akbar's time serious
attempts were made to bring the two closer. Several Sanskrit works were
translated into Persian, and Ulugh Begh's astronomical tables were translated
into Sanskrit. Yet there remained a cultural gap. It was at the begining of the
eighteenth century that there appeared a scholar-prince who tried to assimilate and synthesize the astronomical knowledge then available to him. He
was Sawai Jai Singh (1688–1743) of Amber.[26]

A LONE LIGHT

Jai Singh ascended the throne of Amber in 1699 and later emerged as a trusted
lieutenant of the Mughal king Muhammad Shah, who was beseiged with numerous rebellions and attacks. The Mughal empire was crumbling, and it was
a period of uncertainty and unrest. At the same time the European presence in
India had increased, and certain members of the Indian nobility had evinced
some interest in certain aspects of European ideas and artifacts. In the midst
of such politico-cultural turmoils, Jai Singh tried to do something different.
He wanted to explore why the time of different celestial phenomena, especially the eclipses of the sun and the moon, differed according to Siddhantic
and Greco-Arabic astronomy and did not often tally with actual occurrence.
He consulted a large number of almanacs, traditional scholars, and European
travelers. He was not satisfied with calculations done through astrolabes (brass
instruments) and thought that stone observatories, larger and fixed in one place,
would give more-accurate results. So he constructed large masonry observatories in Delhi, Jaipur, Mathura, and Varanasi. He was also presented with a
telescope by a French Jesuit.

Jai Singh is also credited with evolving a systematic scientific method. He
sent his scholars to Central and West Asia and invited European scholars to
his court; the results of his efforts were compiled in the *Zij-i-Muhammad Shahi*
(1728), which is considered to be the most important astronomical work of
medieval India. Several commentaries were later written on it. That same year
he sent a delegation to Lisbon led by a Jesuit priest, Emmanuel de Figuerado,
who in 1730 brought de la Hire's *Tabulae Astronomicae*. Although Jai Singh
was convinced of the reliability of his own data, he borrowed from de la Hire's
tables some refraction corrections and geographical coordinates. Later, in
1734, Jai Singh invited two French Jesuits – Claude Boudier (1686–1757) and

[25] S. R. Sarma, "From al-Kura to Bhagola: On the Dissemination of the Celestial Globe in India," *Studies in History of Medicine and Science*, 13, 1 (1994), 69–85.
[26] To date the best work on Jai Singh is V. N. Sharma, *Sawai Jai Singh and His Astronomy* (Delhi: Moti-
lal Banarsidass Publishers, 1995).

Francis Pons (1698–1752) – who confirmed the defects in Hire's tables. Jai Singh planned to send another scientific delegation to Europe, but death intervened in 1743.

Critics argue that his choice of Lisbon was not appropriate; he should have contacted astronomers in Paris and London. In addition, his obsession with masonry instruments (which was not the European tradition), accuracy, the calender, and so on is usually taken to mean that Jai Singh's outlook was medieval and limited to the Ptolemaic concept of the universe. In all probability he remained ignorant of the contents of the *Revolutionibus* and the *Principia* until the very end of his days. Some scholars, however, believe that although Jai Singh did not acknowledge Copernicus and Kepler explicitly, he may have known their theories. Sobirov, in his translation of the *Zij* from Persian to Russian, quotes Jai Singh as saying the following:

> The predecessors of astronomy, namely Hipparchus and Ptolemy and others, gave the principles of the movements of planets and description of the orbits of their movement but their description is far from the truth. The system of the world is in reality the movement of the planets occurring contrary to the descriptions given by the above-mentioned scientists. The orbits of the movement of the planets have a different form. Above all, it should be mentioned that the orbits have elliptical shape in one of the centres of which lies the sun.[27]

This is taken as his acceptance of the Copernican model, indicating his open-mindedness and true scientific spirit. Sobirov also credits Jai Singh with the full use of the telescope. The *Zij* says:

> As our artisans have constructed the telescope so excellent that with its aid we can see bright and luminous stars even about midday in the middle of the sky, by employing such powerful telescope, the new moon can be seen even before the time the astronomers have determined for its rays to begin emanating. And also after it has entered the prescribed limit of its invisibility, it still remains visible (through the telescope).[28]

Another important deviation that Jai Singh made from the traditional Greco-Arabic astronomy related to the so-called "fixed stars." Ptolemic astronomy puts the stars into two categories – the wandering stars and the fixed stars – the latter conceived as immovable. In the seventh section of his *Zij*, Jai Singh refutes this theory: "Those stars that are termed Fixed Stars in the terminology of astronomers are not stationary in reality. Nor do they move with one rate of velocity, but with different velocities."[29]

[27] G. Sobirov, "Samarkand Scientific School of Ulugh Begh," *Dushanbe,* 1975, quoted in A. Rahman, "Maharaja Sawai Jai Singh: Purposes and Contributions," paper presented at a seminar on Sawai Jai Singh, New Delhi, Oct. 1989.
[28] S. A. K. Gori, "The Impact of Modern European Astronomy on Raja Jai Singh," *Indian Journal of History of Science,* 15, 1 (1980), 55.
[29] Ibid., 56.

Rahman surmises that Jai Singh's aim was to bring about, through the application of science, a renaissance in India. Another scholar claims that the path to the final "reawakening" (the Scientific Revolution) was blocked by the onset of colonization.[30] These enthusiastic estimates are not, however, shared by several scholars. It is argued that Jai Singh was no theoretician and that he adhered to the old Ptolemaic concept. There is no doubt that he at least thought that the brass astrolabes were not accurate, and he was brilliant enough to devise new ways of measurement. But his obsession with finding the exact moment and with accuracy calls for some explanation. Was it motivated by astrological concern? Obsession with the exactness of time (for example, the time of *yagna* or marriage) has been an important feature of Indian social life, and Jai Singh was naturally part of it. Moreover, he was an intensely religious and ritual-minded person and had performed difficult Vedic *yagnas* (sacrifices) such as Vajpeya and Asvamedha. Was his *Zij* intended only as a means to compute accurate time and not as a treatise to show off his acumen or document his new findings? The telescope had come to India even before Jai Singh's birth. He was definitely aware of it, but it is doubtful that he made full use of it. In the absence of a chronometer, one could see the distant objects through a telescope but could not measure them. So despite his enthusiasm and efforts, Jai Singh may appear as a sort of historical anachronism who belonged intellectually to the medieval tradition of *Zij* astronomy but lived chronologically in the modern age of astronomy.[31]

This, however, is not to minimize Jai Singh's efforts. With a little more foresight and courage he could have transcended his cultural limits. It was not as if the Ptolemaic system was always blindly followed in India. Earlier, during Shah Jahan's time, Mulla Mahmud Jaunpuri had ventured to raise doubts about the system in his *Shams-e-Bazegha*. Later, in a commentary on Jai Singh's *Zij*, Mirza Khairullah Khan argued as follows:

> Whenever we calculate the different positions of the Sun and other planets in accordance with equations of the circle, they do not conform with the actually observed ones. On the contrary, when the equations are derived, taking the orbits elliptical and calculating the positions, they generally conform with observations. Hence the orbits must be elliptical.[32]

It is significant that this remark is based on observation alone. There is no evidence to suggest that Khairullah Khan had any knowledge of Kepler. It was only in the second half of the eighteenth century that a few tracts appeared, and they were either translations of a new European work or were composed under European supervision. For example, Abul-Khair Ghiyasuddin made a

[30] S. M. R. Ansari, "Zij-i-Muhammadshah: The Astronomical Tables of Jai Singh," paper presented at a seminar on Sawai Jai Singh, New Delhi, Oct. 1989.

[31] R. K. Kochar, "The Growth of Modern Astronomy in India," *Vistas in Astronomy*, 34 (1991), 72.

[32] W. H. Abdi, "Mulla Mahmud Jaunpuri's Theory of Moon-Spots," *Indian Journal of History of Science*, 22, 1 (1987), 47–50.

Persian translation of William Hunter's book on the Copernican system, and this was done under the supervision of Hunter himself. Jai Singh and his associates may not have known new astronomy, but they did not adopt the old one blindly. Their parameters, eclipse tables, and a number of subsidiary planetary tables differ from those of Ulugh Beg. They determined new parameters and new tables for the planets, the obliquity of the ecliptic, and the geographical coordinates of a number of localities in India. They did not, however, attempt to change Ulugh Beg's (that is, Ptolemy's) basic planetary models despite their contact with Europeans.[33]

A significant aspect of Jai Singh's reign is that he brought together a number of astronomers and scribes from different parts of the country and established a virtual colony of astronomers.[34] Notable among them were Jagannath Samrat, Kevalverma, Nayansukha, and Harilal. Jagannath had learned both Arabic and Persian. In 1727 he translated Tusi's Arabic version of Euclid's *Elements* and called it *Rekhaganita*. In 1732 he wrote *Samratsiddhanta* based on the Arabic recension of Ptolemy's *Almagest*. Similarly, Nayansukha did not simply render an Arabic text into Sanskrit literally but instead expanded those passages that he found particularly difficult. Kevalverma was rigid in following the *Surya Siddhanta* and even ignored the new parameters being worked out by Jagannath and Jai Singh.[35] Although Jagannath himself respected observation as *pramana* (proof), he would finally succumb to the *siddhantas* (canons) as "divine" authority. Jai Singh and his pundits just could not transcend the barriers.

MAQUL IN EDUCATION

The *madrassas* and *maktabs* in India had adopted *Silsilai Nizamiya* in conformity with educational practices throughout the Islamic world. The main subjects taught according to this system were grammar, rhetoric, philosophy, mathematics, theology, and law. Philosophy included physics and metaphypics (based on Aristotelian principles), and mathematics meant Ptolemaic astronomy, algebra, geometry, and arithmetic. The main feature of this general curriculum was the balance between scientific and humanistic studies, and in practice *maqul* seems to have received greater attention than *manqul*.[36] Akbar wanted the syllabi to include mathematics, medicine, agriculture,

---

[33] David Pingree, "Indian and Islamic Astronomy at Jayasimha's Court," in David King and G. Saliba (eds.), *From Deferent to Equant: A Volume of Studies in the History of Science in the Ancient and Medieval Near East in Honour of E.S. Kennedy* (New York: New York Academy of Sciences, 1987), pp. 313–27.

[34] Unfortunately, the concept of a scientific society or journal was to come a few decades later in 1784, when William Jones established the Asiatic Society in Calcutta.

[35] V. N. Sharma, "Sawai Jai Singh's Hindu Astronomers," *Indian Journal of History of Science*, 28, 2 (1993), 131–55.

[36] S. M. Jafar, *Education in Muslim India* (Peshawar City, India: S. Muhammad Sadiq Khan, 1936), p. 23.

geography, and even some Sanskrit texts such as Patanjali to balance the orthodox emphasis on Islamic studies. Even a deeply religious monarch such as Aurangzeb is said to have reproached his teachers for not teaching geography and subjects useful for administration and having wasted his youth "in the dry, unprofitable, and never-ending task of learning words!"[37]

The eighteenth century saw two great educationists in North India: Shah Waliullah, who taught at Madrasa Rahimiyya in Delhi until his death in 1762, and Mulla Nizamuddin Sahalwi, who taught at Firangi Mahal, Lucknow, until his death in 1748. The former was scholastic and orthodox. The topics of his writings range from the deep nuances of the Quranic words to how the sun in reality revolves around the earth! In contrast, Mulla Nizamuddin developed a course called *Dars-i Nizami,* which of course included *hadis* and *tafsir* (traditional studies) but put more emphasis on *mantiq* (logic) and *hikmat* (metaphysics). The number of books prescribed in these two representative schools on different subjects makes the difference clear.[38]

Subject	Number of Books Prescribed	
	Madrasa Rahimiyya, Delhi	Firangi Mahal, Lucknow
Grammar	2	12
Rhetoric	2	2
Philosophy	1	3
Logic	2	11
Theology	3	3
Jurisprudence	4	5
Astronomy and Mathematics	2	2
Medicine	1	0
Mysticism	5	1

Apart from the higher number of books, the *Dars-i Nizami* preferred different, sometimes new, texts.[39] But "new" knowledge was yet to enter. At best, it tried to bring together secular and theological education in the Greco-Arab tradition. This was in conformity with the Safavid and Mughal practices, in contrast to the Ottomans, who patronized *manqul* more. As greater numbers of Iranians settled in India under the Mughals, the Iranian skills in rational sciences were carried to the fertile grounds of India.[40] Here they were not

[37] I. G. Khan, "Rationalistic and Technical Content in the Eighteenth Century Education," paper presented at Indian History Congress, Bangalore, 1997. This paper forms the basis of this section.
[38] G. M. D. Sufi, *Al-Minhaj: Being the Evolution of Curriculum in the Muslim Educational Institutions of India* (Delhi: Idarah-i Adabiyat-i Delli, 1977), pp. 68–75.
[39] For details, see Muhammad Umar, *Islam in Northern India,* pp. 274–7.
[40] Francis Robinson, "Ottomans – Safavids – Mughals: Shared Knowledge and Connective Systems," *Journal of Islamic Studies,* 8, 2 (1997), 151–84. I am grateful to my friend S. Irfan Habib for this reference.

seriously attacked by the "purists," who later did so when faced with the "new threats" from expanding Europe.

Apart from the *madrasas,* private tuition was also in vogue and was popular with upper classes. A *riyazi* (mathematics) scholar could earn good money by producing horoscopes, calendars, or revenue estimates. Similarly, physicians were always valued by the nobility. Astronomers as well as physicians remained tradition-bound. They virtually ended where they began, always invoking the authority of Aristotle or Charak, Ibn Sina or Bhaskara. Numerous commentaries were written; they were not mere repetition, but none was trail-breaking. Pre-British India had no scientific society and no network of communications between experts. Individual brilliance operated under severe socio-cultural limitations. A parasitic nobility encouraged parasitic intellect. It took the surplus from the land but did nothing to introduce new tools or methods in craft or agricultural production. Of course, some new texts were written about fruit trees and cash crops, but their influence remained rather limited.[41] The lack of vernacular prose literature as a vehicle of the expression of knowledge prevented craftsmen from transmitting their experiences and problems.[42] They were unable to obtain any theoretical knowledge that could help them professionally. Yet, at least in name, useful knowledge and crafts were honored. In the eighteenth century a Bangash prince, Qaim Khan, is known to have excelled in crafting leather shoes and casting cannons. Another Pathan, Muhammad Hayat Khan, became an authority on arithmetic, algebra, and astronomy, including the Siddhantas. This century also saw mobility in terms of professions. The son of a noble could join the revenue service, whereas the son of a religious scholar might join the army. There does appear to have been a qualitative change in the post-Moghul nobility. A contemporary text (*Kitab Amoz-al Munshi,* 1782) says

> Every gentlemen should be taught the numerals, measuring of time, all the calendars, the harvests, names of the planets, the auspices days of the year, mathematics, ways of reading pulse, a few medicines, classifying people and animals, the imperial offices and ways of addressing the lower cadres in the army, about flattering women, etc.[43]

These nuggets of wisdom could be of no use when the European traders flexed political muscles!

MEDICINE: ITS TEXTS AND PRACTICES

Medicine has always been a significant part of the Indian heritage. Its major concern was how to prolong life and to preserve health and vitality as far as

---

[41] For example, Ahmad Ali, *Nakhlbandiya* (1790); Anon., *Risala-i-Zara't* (1785).

[42] Surendra Gopal, "Social Set-up and Science and Technology in India," *Indian Journal of History of Science,* 4 (1969), 52–7.

[43] Quoted in I. G. Khan, "Rationalistic and Technical Content."

possible. Curing illness, by itself, was not enough. The term *Ayurveda* meant "the science of (living to a ripe) age."[44] Although the ancient Indian practitioners realized that the body was controlled by natural law, their knowledge of human physiology was utterly inaccurate. Still, they were extremely good at therapeutics, as was recognized by the Islamic medical men who introduced the Galenic tradition. There gradually appeared a hybrid Muslim-Hindu system known as the *Tibb*. They differed in theory, but in practice both traditions seem to have interacted and borrowed from each other.

A fine example of this interaction is *Ma'din al-shifa-i-Sikandarshahi* (A.D. 1512), which was authored by Miyan Bhuwah.[45] He leaned heavily on the Sanskrit sources and even thought that the Greek system was not suitable for the Indian constitution and climate. From the Islamic side the concept of *arka* entered Ayurveda. Several Sanskrit medical texts were translated into Arabic and Persian, but instances of Islamic works being translated into Sanskrit are rare. The eighteenth century is significant because of the appearance of two Sanskrit texts – *Hikmatprakasa* and *Hikmatpradipa* – which refer to the Islamic system and use numerous Arabic and Persian medical terms.[46] The concept of individual case studies and hospitals (*bimaristans*) also came from the *unani* practitioners.[47] In 1595 Quli Shah had built a huge *Dar-us-Shifa* (House of Cures) in Hyderabad.[48] During the reign of Muhammad Shah (1719–1748) a large hospital was constructed in Delhi, and its annual expenditure was more than Rs. three hundred thousand. Numerous medical texts, mostly commentaries, were written during this century – for example, Akbar Arzani's *Tibb-i-Akbari* (1700), Jafar Yar Khan's *Talim-i-Ilaj* (1719–25), Madhava's *Ayurveda Prakasha* (1734), and *Bhaisajya Ratnavali* of Govind Das. A Christian Mughal, Dominic Gregory, wrote *Tuhafatul-Masiha* (1749), which, along with descriptions of diseases, anatomy, and surgery, contains important notes in Persian and Portuguese on alchemy and the properties of various plants, along with drawings of instruments and, interestingly, a horoscope.[49]

---

[44] According to Charaka (A. D. 100), Ayurveda involved

  1. general principles of medicine (*sustra-sthana*),
  2. pathology (*nidana-sthana*),
  3. diagnostics (*vimana-sthana*),
  4. physiology and anatomy (*sarira-sthana*),
  5. prognosis (*indriya-sthana*),
  6. therapeutics (*chikitsa-sthana*),
  7. pharmaceutics (*kalpa-sthana*), and
  8. means of assuring success in treatment (*siddhi-sthana*)

A. L. Basham, "The Practice of Medicine in Ancient and Medieval India," in C. Leslie (ed.), *Asian Medical Systems* (Berkeley: University of California Press, 1976), pp. 18–43.

[45] The manuscript was first printed by Nawal Kishore Press, Lucknow, in 1877.

[46] G. J. Meulenbeld, "The Many Faces of Ayurveda," *Journal of the European Ayurvedic Society,* 4 (1995), 1–9.

[47] S. H. Askari, "Medicines and Hospitals in Muslim India," *Journal of Bihar Research Society,* 43 (1957), 7–21.

[48] D. V. Subba Reddy, "Dar-us-Shifa Built by Sultan Muhammad Quli: The First Unani Teaching Hospital in Deccan," *Indian Journal of History of Medicine,* 2 (1957), 102–5.

[49] Rahman (ed.), *Bibliography,* p. 57.

An outstanding physician of this century, Mirza Alavi Khan, wrote seven texts, of which *Jami-ul-Jawami* is a masterpiece embodying all the branches of medicine then known in India.[50] Another great physician during the period of Shah Alam II (1759–1806) was Hakim Sharif Khan, who wrote ten important texts and enriched *unani* medicines with indigenous *ayurvedic* herbs.[51] Some works were unique and ahead of their time. For example, Nurul Haq's *Ainul-Hayat* (1691) is a rare Persian text on plague, and Pandit Mahadeva's *Rajsimhasudhasindhu* (1787) refers to cowpox and inoculation.[52]

A number of European physicians visited Mughal India. François Bernier, Niocolao Manucci, Garcia d' Orta, and John Ovington wrote extensively on Indian medical practices. The Western medical episteme was not radically different from that of Indian physicians; both were humoral, but their practices differed greatly. Neither of them was able to develop a comprehensive theory of disease causation, but there seems to be a general agreement that the Indian diseases were environmentally determined and should be treated by Indian methods. Europeans, however, continued to look at the Indian practices with curiosity and disdain.[53] They preferred blood-letting, whereas the *vaidyas* prescribed urine analysis and urine therapy. But in the use of drugs Europeans and Indians learned from each other, as the works of van Rheede, Sassetti, and d'Orta testify.[54] The Europeans introduced new plants in India that were gradually incorporated into the India pharmacopeia. They also brought venereal diseases, such as syphilis, which was noticed as early as the sixteenth century by Bhava Misra, a noted *vaidya* in Benaras, who called it *Firangi roga* (disease of the Europeans). Indian diseases received graphic description in Ovington's travelogue.[55] The best account of smallpox and the Indian method of "variolation" was given by J. Z. Holwell in 1767. To him this method, although quasi-religious, still appeared "rational enough and well-founded."[56] The travelers depicted Indian medical practices more as a craft – and one that was governed by caste rules and wrapped in superstition. Yet they could not help admiring the wonder called rhinoplasty (on which

---

[50] R. L. Verma and N. H. Keswani, "Unani Medicine in Medieval India: Its Teachers and Texts," in N. H. Keswani (ed.), *The Science of Medicine in Ancient and Medieval India* (New Delhi, 1974), pp. 127–42.

[51] Hakim Abdul Hameed, *Exchanges between India and Central Asia in the Field of Medicine* (New Delhi: Institute of History of Medicine and Medical Research, 1986), 41.

[52] Rahman (ed.), *Bibliography*, pp. 127, 165.

[53] A European traveler, Edward Ives (1755–7), thus writes of the Indian belief that "man was divided into two or three hundred thousand part; ten thousand of which were made up of veins; ten thousand of nerves; seventeen thousand of blood, and a certain number of bones, choler, lymph, etc. And all this was laid down without form or order, either of history, disease or treatment." Quoted in H. K. Kaul, *Travellers' India: An Anthology* (Delhi: Oxford University Press, 1979), p. 299.

[54] For details see John M. de Figueiredo, "Ayurvedic Medicine in Goa According to European Sources in the Sixteenth and Seventeenth Centuries," *Bulletin of History of Medicine*, 58, 2 (1984), 225–35.

[55] A. Neelmeghan, "Medical Notes in John Ovington's Travelogue," *Indian Journal of History of Medicine*, 7 (1962), 12–21.

[56] J. Z. Holwell, *An account of the manner of innoculating for the small pox in the East Indies* (London, 1767), p. 24.

modern plastic surgery is founded), nor could they deny the efficacy of Indian drugs. The Indians, for their part, did not completely insulate themselves from the "other" practices. As the interaction grew in the eighteenth century, the *vaidyas* even took to bleeding in a large number of cases. Yet while the European medical men were gradually moving, thanks to the works of Vesalius and Harvey, from a humoral to a chemical or mechanical view of the body, Indians remained faithful to their texts.[57]

TOOLS AND TECHNOLOGIES

As in astronomy and medicine, the state of Indian technology evoked a mixed response from foreign observers. Several of them were awestruck by the quality of Indian steel (called *wootz*) as well as Indian textiles. They were impressed by the end result, but they found the tools, the method, and the process clumsy, crude and defective. It is quite possible that they were unable to appreciate a treatise or a device that would appear "appropriate" only when viewed against the existing socio-economic context. Or was this response – part appreciation and part denunciation – a part of the process of hegemonization? Whatever the case, the need to place certain technological developments in a comprehensive historical context has led to interesting deductions. Dharampal, for example, argues as follows:

> Smallness or simplicity of construction, as of the iron and steel furnaces or of the drill-ploughs, was in fact due to social and political maturity as well as arising from understanding of the principles and processes involved. Instead of being crude, the processes and tools of eighteenth century India appear to have developed from a great deal of sophistication in theory and an acute sense of the aesthetic. . . . In the context of the value and aptitudes of Indian culture and social norms (and the consequent political structure and institutions) the sciences and technologies of India, instead of being in a state of atrophy, were in actuality usefully performing the tasks desired by Indian Society.[58]

There is no doubt that agricultural tools, irrigation methods, and certain crafts were "appropriate" and in tune with the existing capabilities and requirements, but the "sophistication" in theory to which Dharampal alludes is markedly absent. The variety of agricultural implements, the drill plough, the system of rice transplantation, the rotation of crops, and experiments in fruit crops speak of the rich experience of Indian peasants.[59] Similarly, local

---

[57] M. N. Pearson, "The Thin End of the Wedge: Medical Relativities as a Paradigm of Early Modern Indian-European Relations," *Modern Asian Studies*, 29, 1 (1995), 141–70.

[58] Dharampal, *Indian Science and Technology*, pp. 63, 65.

[59] S. Sangwan, "Level of Agricultural Technology in India 1757–1857," *Indian Journal of History of Science*, 26, 1 (1991), 79–101.

conditions determined irrigation methods, which involved community man-agement of water resources. These systems were the outcome of the experi-ences and collective wisdom of "practical" peasants. Yet they do not stand com-parison with practices in eighteenth-century Japan, where row cultivation was introduced, the number of plant varieties was increased using deliberate seed selection, and irrigation by treadmills and Dutch pumps was improved and extended.

After agriculture the most important sectors were textiles and steel manu-facture. Textiles involved the labor-intensive processes of starching, bleach-ing, dyeing, winding, warping, and weaving. The Europeans tried to imitate Indian dyeing techniques, without much success. But their growing com-mercial interest in Indian textiles led to the introduction of the filature sys-tem, drum warping, and the fly shuttle technique. These tools were used for mercantilist "penetration" or "intervention" by the European companies, and the Indian weavers gradually suffered impoverishment and virtual elimina-tion.[60] Later, a similar fate awaited the Indian steel producers, but in the eighteenth century this industry was considered a success story. Historians of metallurgy believe that the Indian iron smelters had acquired an advanced and precise knowledge about the production technology of iron and steel – their thermo-mechanical behaviour, heat treatment, and so on.[61] The result was a high-carbon ingot (*wootz*) that commanded respect in international markets. The Dutch carried a large amount of *wootz* from Masulipatnam to Batavia and Persia. So *wootz* was neither "handicraft" nor a "primitive tradi-tional" production, yet it remained localized at a time when Europe was fast moving toward mass production. The Indian smiths could not obtain high temperatures and opt for large furnaces because they did not know how to generate power except through the use of draft animals or charcoal.[62] Except in one or two places, water power remained unthought-of and untapped. This resulted in a high cost of production, and so naturally Indian peasants kept the use of iron to the bare minimum.

Similarly, mining itself was done on a small scale. It involved barely more than scratching the surface of the earth using crowbars and spades. Mining below the water level and haulage were simply out of the question. Curiously, although gunpowder was used for armament, it was never used for mining purposes.[63] But there did exist a flourishing metallurgical industry, which was run almost like a cottage industry. Slags of iron and steel and metals such as copper, zinc, lead, and, to a smaller extent, silver and cobalt, in parts or

[60] V. Ramaswamy, "South Indian Textiles: A Case for Proto-Industrialization?" in Deepak Kumar (ed.), *Science and Empire* (Delhi: Anamika Prakashan, 1991), pp. 41–56.

[61] B. Prakash, "Metallurgy of Iron and Steel Making and Blacksmithy in Ancient India," *Indian Journal of History of Science*, 26, 4 (1991), 351–71.

[62] H. C. Bhardwaj, "Development of Iron and Steel Technology in India during the Eighteenth and Nineteenth Centuries," *Indian Journal of History of Science*, 17, 2 (1982), 223–33.

[63] A. K. Ghose, "History of Mining in India, 1400–1800, and Technology Status," *Indian Journal of History of Science*, 15, 1 (1980), 25–9.

Rajasthan, Bihar, and Deccan bear testimony to this.[64] Zinc production in India preceded that in Europe.

In the realm of armaments, the finest example of Indian ingenuity lay in the use of "Bana" rockets by Hyder Ali and Tipu Sultan, who ruled Mysore during the last quarter of the eighteenth century and fought several wars against the British. These rockets were much more advanced than any the British had seen or known; the propellent was contained in tough iron tubes, which gave higher bursting pressures in the combustion chamber and hence higher thrust and longer range for the missile.[65] The rockets consisted of a tube (about 60 mm in diameter and 200 mm in length) fastened to a 3-m bamboo pole, with a range of 1–2 km. In the battle of Pellilur (1780) the British were defeated because their ammunition tumbrils were blasted by the Mysore rockets. In the last Anglo-Mysore war, Wellesley (later the hero of Waterloo) himself was shocked by the "rocket fire." Several rocket cases were sent to Britain for analysis, and these led to a great interest in rocketry in Europe. Under the supervision of William Congreve, scientific principles were applied and appropriate designs were made, tested, and evaluated. This the eighteenth-century Indians were unable to do.

REFLECTIONS

In the early eighteenth century, Ramchandrapant, an *amatya* (minister) of Kolhapur, wrote about the activities of the European traders and "factors." He called them *topikars* (hat-wearers) and recognized that their strength lay in "navy, guns and ammunition." His prompt advice was to avoid the *topikars,* "neither troubling them nor being troubled by them."[66] This was an early sign of withdrawal, of playing safe. But this attitude tempted the *topikars* to attempt conquest along with commerce, and their success was virtually ensured. But during the same period one finds Sawai Jai Singh inviting Jesuits to India and sharing astronomical knowledge with them. Even for earlier periods Indians cannot be held guilty of xenophobia. There were several areas in which interaction between the East and the West resulted in acceptance and improvement: shipbuilding, armaments, metallurgy, cloth printing, and architecture. "But as long as there was an alternative or appropriate indigenous technology which could serve the needs of Indians to a reasonable degree, the European counterpart was understandably passed over."[67]

Several important developments, however, such as mechanical clocks, the printing press, telescopes, coal, and so on remained mere curios. Since these

---

[64] R. D. Singh, "Development of Mining Technology during Nineteenth Century India," *Indian Journal of History of Science,* 17, 2 (1982), 206.

[65] R. Narasimha, *Rockets in Mysore and Britain, 1750–1850* (Bangalore, 1985) (mimeographed).

[66] Sarkar, *India in Exact Science,* pp. 8–9.   [67] Qaisar, *The Indian Response,* p. 139.

were not found culturally compatible, they did not attract the attention of the Indian nobility. In addition, neither the nobility nor the merchants would invest in the upgrading of technology. Tools remained the sole concern of the poor artisans who sought to compensate for this poverty of tools by the acquisition of individual skills – skills that are manifest in Dacca *muslin*, brilliant dyes, and *wootz*. Even this craft production, although superbly executed, did not stand on its own. It was heavily dependent on the agrarian system that, once under strain (as in the eighteenth century), triggered adverse chain reactions, leading to the fall of the Mughal rule.

Another important aspect that needs to be taken into account is the caste system, which has always been a unique feature of Indian society. P. C. Ray was the first historian of science who saw in the caste structure "something that made science a prey to creeping paralysis."[68] Caste led to the ruinous separation of theory from practice – of mental work from manual work. Ray wrote as follows:

> The intellectual portion of the community being thus withdrawn from active participation in the arts, – the how and why of phenomena – the co-ordination of cause and effect – were lost sight of – the spirit of enquiry gradually died out. Her [India's] soil was rendered morally unfit for the birth of a Boyle, a Descartes, or a Newton.[69]

In eighteenth-century India this paralysis was compounded by an enormous intellectual (cultural) failure on the part of the ruling class. Jai Singh had attracted several scholars to his court, but he never thought of establishing an institution that would continue and improve on his work. It was a curious situation. On the one hand, one finds Mushibullah al-Bihari writing *Risalah Juz 'la Yatajazza,* an Arabic treatise on the indivisible atom, and two other texts on motion and time (1700); on the other hand is Walih Musawi (1700–1770) writing *Murgh-namah* (on cock fighting) and *Kabutar-namah* (on pigeons).[70] As the British strengthened their grip at the end of the eighteenth century, the Indians did not continue this withdrawal. As interaction with the West grew, Indians did try to look out and look within. For example, in 1790 Mir Hussain Isfahani wrote *Risalah-i-Hai'at-i-Angrezi,* a Persian text on European astronomy.[71] Many commentaries were written during this period; although they did not entail a paradigmatic change, neither were they slavish. In fact, composing commentaries was considered a civilized form of making progress.[72] In several instances (especially in medicine) these commentaries explain scientific knowledge in terms of its own rationality and

---

[68] Debiprasad Chattopadhyay, *History of Science and Technology in Ancient India: The Beginnings* (Calcutta: South Asia Books, 1986), p. 10.
[69] P. C. Ray, *History of Hindu Chemistry,* vol. 2 (London: Williams and Norgate, 1909), p. 195.
[70] Rahman (ed.), *Bibliography,* p. 494; Storey (ed.), *Persian Literature,* p. 410.
[71] Rahman (ed.), *Bibliography,* p. 333.
[72] Frits Staal, *Concepts of Science in Europe and Asia* (Leiden: IIAS, 1993), p. 26.

logic, but in the final analysis when the validity of certain knowledge was put to test, the sacred texts were always the standard measure. More than three hundred years before P. C. Ray, Abul Fazl had mourned "the blowing of the heavy wind of *taqlid* (tradition) and the dimming of the lamp of wisdom. . . . The door of "how" and "why" has been closed; and questioning and enquiry have been deemed fruitless and tantamount to paganism."[73]

Had this illustrious historian lived in the mid-eighteenth century, he would have perhaps been more harsh.

[73] Quoted in Irfan Habib, "Capacity of Technological Change in Mughal India," in A. Roy and S. K. Bagchi (eds.), *Technology in Ancient and Medieval India* (Delhi: Sundeep Prakashan, 1986), pp. 12–13.

29

CHINA

Frank Dikötter

Historians of science and technology have not identified the eighteenth century as one of the most significant periods in Chinese history. The ambitious
examination of the world of science and civilization in China by Joseph
Needham is explicitly confined to the period up to the end of the sixteenth
century, and other works, examining the contributions of the Jesuits, stress
the importance of the seventeenth century. The more conservative atmosphere of the mid-Qing (c. 1720–1820), marked by the orthodox neo-Confucianism promoted by the Manchu rulers, stands in contrast to the more open
intellectual climate of the late Ming (c. 1550–1644) and early Qing (c.
1644–1720). By the early eighteenth century, Jesuits were limited both by the
relatively obsolete nature of their knowledge and by their closer integration
at court level. Outside the imperial capital at Beijing, the most important
trends in eighteenth-century scholarship were marked by a shift away from
an interest in Jesuit science toward a rediscovery of ancient knowledge. In the
Yangzi Delta, followers of evidential scholarship (*kaozhengxue*), or philological "search for evidence," were concerned with precise scholarship and practical matters, but they generally appropriated Jesuit science in efforts to "rediscover" their own presumed scientific tradition rather than attempting to
contribute new knowledge to mathematics and astronomy.

JESUIT SCIENCE

If the seventeenth century was a significant period of cultural interaction
between Jesuit missionaries and Confucian scholars, little further scientific
knowledge was transmitted during the eighteenth century. Not only were the
Jesuits mainly interested in using science as a way of achieving religious aims,
but also the Church's injunction in 1616 against the teaching of heliocentric
astronomy, as well as other aspects of science, severely limited the nature of
their knowledge. As a result, they continued to promote the obsolete cosmol-

ogy of Ptolemy and Tycho Brahe well into the eighteenth century. Only in 1761 did the Jesuit Michel Benoist explain Copernician cosmography to the Emperor Qianlong: not until the end of the century was his work translated and circulated among a number of thinkers in China.[1] Jesuit influence was also restricted by the very nature of imperial interest in science. Manchu rulers acted as patrons of European science but generally kept Jesuit scientists and technicians confined to the court. As a consequence, their work – either in the form of publications in such fields as cartography, mathematics, astronomy, armaments, and medicine, or as court-related projects – could not have a widespread audience among scholars outside the inner city of Beijing. The work accomplished by the Jesuits was often unimpressive and inadequate, remained far behind contemporary discoveries in Europe, or was caught up in the mechanical trivia they were forced to produce for court diversion. As Jonathan Spence underlines, the Qing palaces filled up during the eighteenth century with all kinds of European bric-a-brac, while court favorites hoarded the hundreds of clocks and watches that could instead have contributed to spreading the new technologies they represented.[2] Even the vaster architectural projects, such as the grandiose summer palaces built by the middle of the century according to Jesuit design, remained restricted to imperial use. The political entanglements of the Jesuits with different factions at court as well as the demands of the papal legate in 1708 also contributed to the perception of Christian practices as "deviant" and "heterodox" (*xie*). This trend continued after the dissolution of the order in 1773, as both Christian and Protestant missionaries were suspected of collaboration with foreign forces. The strict control of the capital by imperial sponsors, the confinement of the royal family in the inner city, and the concomitant absence of aristocratic estates that might have provided scholars with the means to pursue their activities are other factors that were unhelpful in the development of mechanical or experimental sciences.

Not only were contacts between European missionaries and Chinese scholars limited, but also the emperor Kangxi (r. 1662–1722) considered Jesuit science to be a tool of government only. The transmission of science went directly via the emperor, who acted as a sponsor of a number of specific projects that could bolster his authority and legitimacy.[3] In the capital, the Imperial Astronomical Bureau (*Qintianjian*) represented the imperial control of an institution that had revolved around the calendar, meant to ensure the cosmological correspondence between Heaven and Earth, in which the Emperor was considered to be the intermediary link. Research conducted in this

---

[1] Nathan Sivin, *Science in Ancient China* (Aldershot: Variorum, 1995), pp. 1–53.

[2] Jonathan D. Spence, "The Dialogue of Chinese Science," in Spence (ed.), *Chinese Roundabout: Essays in History and Culture* (New York: W. W. Norton, 1992), p. 151.

[3] Catherine Jami, "L'empereur Kangxi (1662–1722) et la diffusion des sciences occidentales en Chine," in Isabelle Ang and Pierre-Etienne Will (eds.), *Nombres, astres, plantes et visceres: Sept essais sur l'histoire des sciences et des techniques en Asie orientale* (Paris: Institut des Hautes Études Chinoises, 1994), pp. 193–210.

institution had clear political implications, and Jesuit science was appropriated within this institutional context dominated by a concern for political legitimation. Under imperial sponsorship, a number of projects and institutions were set up to promote astronomical knowledge. The Bureau itself had a long and venerable pedigree and was in charge of editing reference works and teaching young astronomer-mathematicians under Jesuit supervision.[4] Under the patronage of Kangxi, it also compiled two major texts published as a voluminous encyclopedia (*Lüli yuanyuan*) in 1723. A Library for the Education of Children (*Mengyangzhai*) was further established by Kangxi, as was an Academy of Mathematics (*Suanxue guan*) in 1713.

The Jesuits also initiated and supervised a number of cartographic projects. They often relied on Chinese cartographic traditions, including local gazetteers, to provide detailed maps with statistical, physical, economic, and geographic information. After the arrival of French Jesuits in China in 1687, when Paris had become the center of cartography in Europe, they strengthened their position at court and produced maps in the service of the emperor. By the first decade of the eighteenth century, the Kangxi emperor had become acutely aware of the need for a reliable geographical representation of China because the empire had been rapidly expanding. Under imperial sponsorship, the Jesuits were given responsibility for a general survey of the empire in 1708. The most accurate image of the Qing empire at the time, it was first published in 1717 as the *Huangyu quanlan tu* (General Atlas of the Empire) and was reprinted many times until the end of the nineteenth century. A second Jesuit survey, carried out between 1756 and 1759, was later authorized by the Qianlong emperor (r. 1736–1799). Completed in 1769 with a special edition by Michel Benoist, it included for the first time data on the strategically sensitive outer regions of the empire.[5]

Artillery, especially cannons, was also of special interest to Qing emperors, although in this field too the most important Jesuit contributions were made in the seventeenth century. The designs developed at the cannon foundry of Ferdinand Verbiest until his death in 1688, for instance, were still used at the time of the Opium War in 1839. Despite the sharp decline of Jesuit influence in Beijing in the eighteenth century, European missionaries at the Qianlong court carried out a range of technical activities, including the supervision of glass-making, the construction of furnaces, the building of complex hydraulic machinery, and even the development of electro-convulsive shock therapies for nervous illness.[6]

[4] Jonathan Porter, "Bureaucracy and Science in Early Modern China: The Imperial Astronomical Bureau in the Ch'ing Period," *Journal of Oriental Studies*, 13 (1980), 61–76.

[5] Theodore N. Foss, "A Western Interpretation of China: Jesuit Cartography," in C. E. Ronan and Bonnie B. C. Oh (eds.), *East Meets West: The Jesuits in China, 1582–1773* (Chicago: Loyola University Press, 1988), pp. 209–51.

[6] Joanna Waley-Cohen, "China and Western Technology in the Late Eighteenth Century," *American Historical Review*, 98, 1 (1993), 1533.

In addition to the Jesuit presence in China, a small number of Dutch, Russian, and British traders or envoys also contributed to maintaining a degree of contact between Europe and China. The most significant attempt to gain a footing in China was the Macartney expedition to China in 1797. Later described as "a tedious and painful employment" by the British envoy, the mission concluded a century of limited exchange between Europe and China. Part of a new wave of explorations by the end of the eighteenth century, specialists who accompanied the Macartney mission were expected to measure, record, tabulate, and collect "facts" based on contemporary ideas of "scientific" exactness. In a spirit of scientific exchange, the British envoys also presented technological instruments and scientific knowledge at court to favorably impress the emperor and facilitate diplomatic relations and commercial exchanges. Twenty objects were offered by the British, including a planetarium and a reflecting telescope built by William Herschel. Although the presents may have impressed scholars in China to a greater extent than has been previously acknowledged, they did little to change an atmosphere of imperial control over foreign science in the capital. Imperial aspirations to universal authority by the Qianlong emperor dictated both a proclaimed disinterest in foreign technology and the strict control of access to scientific information. Later misrepresented in Europe as a sign of the "arrogant and insupportable pretension" of the Chinese, the Qing court's dismissal of the British gifts has been endowed with symbolic significance by some historians. Mobilized in narratives of a "clash of civilizations," it has been understood as an emblem of profound cultural differences imagined between an "immobile" and "stagnant" China in its confrontation with a more dynamic Europe, although readily available evidence indicates that domestic politics rather than any "mental attitude" was the primary reason for a public denial of interest in technological advances.[7] In contrast to leaders in Japan, treated elsewhere in this volume (Chapter 30), the emperor did not encourage scholars to examine European works, principally for reasons of political legitimacy.

EVIDENTIAL SCHOLARSHIP

Beijing and the Yangzi Delta were the two principal geographical locations marked by a concentration of wealth and an interest in scientific knowledge in the eighteenth century. Although scholarship in these two places was informed by different motivations and embedded in divergent philosophical currents, both referred to the same Confucian body of knowledge.[8] Within

---

[7] Ibid., 1525–44.
[8] Catherine Jami, "Learning Mathematical Sciences in the Late Ming and Early Qing," in Benjamin A. Elman and Alexander Woodside (eds.), *Education and Society in Late Imperial China, 1600–1900* (Berkeley: University of California Press, 1994), pp. 223–56.

the scholarly community of the Yangzi Delta, however, the scientific contributions of the Jesuits generally encouraged a return to the Classics rather than the introduction of new knowledge based on European sources.

The Yangzi Delta had harbored anti-Manchu scholars since the foundation of the Qing dynasty in 1644. Many scholars who had held office under the Ming refused to serve the new dynasty, including influential thinkers with a sustained interest in early Jesuit science such as Fang Yizhi (1611–1671), Wang Fuzhi (1619–1692), and Gu Yanwu (1613–1682). The new rulers imposed a strict interpretation of the neo-Confucianism propounded centuries before by Zhu Xi, an orthodox approach that limited the range and nature of texts used in the civil service examination.[9] Evidential scholarship (*kaozhengxue*), or the philological "search for evidence," flourished in the early Qing in reaction against neo-Confucianism. Supported in the Yangzi Delta, where the most powerful official and private patrons could be found, evidential scholars blamed the downfall of the Ming on the sullied nature of Confucian developments, seen to have become corrupted with Buddhist and Taoist influences since the Song. Many scholars sought to reconstruct what was thought to be the authentic Confucian vision of social order through philological examination of ancient texts, and they openly rejected Zhu Xi's method of interpreting the Classics. Evidential scholars relied on patronage of leading officials and only rarely entered the civil service, which was premised on an acceptance and knowledge of the Zhu Xi tradition.

Encouraged by the Jesuits' introduction of aspects of exact sciences, the evidential research movement was also characterized by a concern with precise scholarship and practical matters (*jingshi*). Although inquiries into natural phenomena remained ancillary to philosophical concerns, important research began to be conducted, and the institutions required for precise scholarship were gradually established. In a shift away from numerological explanations toward empirical induction, the development of mathematics and astronomy in Confucian discourse gradually transformed intellectual life in the eighteenth century.

The influence of Jesuit science on the development of evidential scholarship has been well attested, and claims about a "revolution in scholarly discourse" have been made.[10] The interest in philological studies can also be portrayed in a less positive light, as an initial interest in natural philosophy during the seventeenth century was abandoned in favor of more philologically focused studies, leading to what has been characterized as "indifference" in science.[11] However one might wish to characterize the rise of evidential scholarship,

---

[9] Susan Naquin and Evelyn S. Rawski, *Chinese Society in the Eighteenth Century* (New Haven, CT: Yale University Press, 1987).

[10] B. A. Elman, *From Philosophy to Philology: Intellectual and Social Aspects of Change in Late Imperial China* (Cambridge, MA: Harvard University Press, 1984), p. 84.

[11] Willard Peterson, "From Interest to Indifference: Fang I-chih and Western Learning," *Ch'ing-shih wen-t'i*, 3, 5 (Nov. 1976), 72–85.

compilation projects, initiated both by local patrons and by the government, became a major feature of the eighteenth century. The government-sponsored Complete Library of the Four Treasuries (*Siku quanshu,* 1772–82), the most important project of the eighteenth century, undertook to collect for a reprint the best editions of books and manuscripts considered to be the most important in China.[12] Under the direction of Dai Zhen (1724–77), one of the most distinguished scholars versed in these new intellectual trends, fresh methods in evidential research were used in the work of the compilers. The Four Treasuries project, moreover, gave many scholars the opportunity to gather and examine ancient texts on mathematics and science, and it enabled them to relate these texts to contemporary issues. The *Chouren zhuan* (Biographies of astronomer mathematicians) compiled between 1795 and 1799 under Ruan Yuan (1764–1849), for instance, attempted to present the technological aspects of astronomical and mathematical knowledge received from Europe in an indigenous context.

In efforts to rediscover the complexity of contributions to knowledge by the first Confucian scholars, Ruan Yuan, Qian Daxin (1728–1804), Wang Mingsheng (1722–1798) and other evidential scholars attempted to restore the glory of ancient knowledge in mathematics and astronomy: the study of "Western methods" became a means of reasserting an indigenous heritage. They claimed that "Western methods" had originally been invented in ancient China before being adopted in the West and that the use of such methods in calculating time was not intrinsically opposed to indigenous ways. This approach heavily influenced mathematical and astronomical knowledge. The Kangxi emperor, for instance, was personally interested in astronomy and mathematics and openly favored the work of Mei Wending (1633–1721), who posited an indigenous origin to European astronomy. Mei Wending compiled a work on the origins of the calendar that was included in the *Lixiang kaocheng* (Verification of astronomical observations and calculations), designed in 1722 and published two years later. This work was considered to be superior to the *Xiyang xinfa lishu* (Astronomical treatise based on Western new methods), printed for the first time in 1646 under the supervision of Jesuits. Mei Wending's contribution, moreover, was still part of a cosmological approach in which abnormalities in the measure of time were considered to be part of a cosmic order.[13] Likewise, other empirical treatises on astronomy throughout the first half of the eighteenth century continued to be openly linked to the past achievements of indigenous astronomy. Chen Yuanyao, to take another example, attempted to revive the astronomical work of Du Yu (222–284 B. C.) in his scholarship.

[12] Kent Guy, *The Emperor's Four Treasuries: Scholars and the State in the Late Ch'ien-lung Era* (Cambridge, MA: Harvard University Press, 1987).

[13] Jean-Claude Martzloff, *Recherches sur l'oeuvre mathématique de Mei Wending (1633–1721)* (Paris: Còllege de France, Institut des Hautes Études Chinoises, 1981); Martzloff, *Histoire des mathématiques chinoises* (Paris: Masson, 1988).

The shortcomings of Jesuit science were also used in the second half of the century as evidence of the superiority of ancient Confucian ways.[14] Under the reign of the Qianlong emperor, the most inconsistent aspects of Jesuit astronomical works were publicly denounced. After the failure to predict the appearance of an eclipse in 1730, a revised astronomical study titled *Lixiang kaocheng houbian* (Sequel to the Verification of astronomical observations and calculations), compiled by Ignatius Kögler, Andrea Pereira and the Mongol Minggantu, appeared in 1738. It introduced a number of refinements, including Kepler's eclipse and new observations by Cassini and Flamsted, although it did not adhere to the heliocentric theory. A number of Chinese scholars incorporated into their own conception of astronomy some of the ideas of Tycho Brahe and Johannes Kepler. Sheng Bai'er (fl. ca. 1756), an expert in astronomy and trigonometry, critically compared Brahe with Ptolemy in his *Shangshu shitian* (An explanation of astronomy in the Classics of History, 1749–53).

Ruan Yuan and Qian Daxin carried on the work of their seventeenth-century predecessors by criticizing numerological derivations of astronomical constants and calendrical periods. Although their work could be interpreted as evidence of the relative decline in correlative constructions and cosmological analogies, it should also be pointed out that several noted scholars continued to think in terms of cosmological numerologies. Jiang Yong (1681–1762), a major eighteenth-century classical scholar and major contributor to mathematical knowledge, applied correlative numerology in his study on harmonics. Generally, however, resistance against the more extravagant systems of correspondence propounded by the neo-Confucian cosmologist Shao Yong (1011–1077) was a significant theme in eighteenth-century China.

Scholars also continued to critically examine geometrical cosmographies, favoring irregular lines of demarcation in cartography and astronomy. Ruan Yuan thus opposed the use of a solar calendar, which might have eliminated the need for intercalation, on the basis that it was impractical and unnatural as well as in disagreement with the meaning of the Classics. In their predilection for naturally formed lines of demarcation as opposed to sharply defined boundaries, some scholars even questioned the existence of fixed temporal divisions on a cosmic scale. Jiang Yong, mentioned earlier, denied that a point of cosmic origin could ever be calculated. The critique of spatial, temporal, and cosmological boundaries emphasized astronomical anomalies and insisted on a lack of accord between prediction and phenomenon. In their speculations on astronomical irregularities, mid-Qing scholars were generally far less imaginative than their predecessors.

For many evidential scholars, "science" was studied not for the discovery of natural laws but as a means to make moral statements about the political

---

[14] Harriet T. Zurndorfer, "Comment la science et la technologie se vendaient à la Chine au XVIIIe siècle: Essai d'analyse interne," *Études Chinoises*, 7, 2 (1988), 59–90.

order. Zhang Xuecheng (1738–1801), one of the major eighteenth-century scholars interested in these astronomical anomalies, wrote an essay titled "Analogy of Heaven" (*Tianyu*): he compared the processes by which astronomical systems degenerate and are reformed to human society. Eighteenth-century scholars, contrary to their contemporaries in Europe, did not posit the existence of a uniform and predictable order in the physical universe. As John Henderson observes, the "rejection of traditional cosmology might well have inhibited the development of modern science in China."[15] Similarly, geometry and trigonometry, as deductive systems based on proofs and demonstrations, were virtually absent in the eighteenth century. Neither did optics, which contributed so much to the development of the telescope and the microscope in Europe, undergo any fundamental developments.

Some unorthodox ideas did emerge in demography. Hong Liangji (1746–1809), a scholar versed in evidential research, developed a vision of overpopulation in a short essay published in 1793, five years before Thomas Malthus' *Essay on the principle of population*. Hong Liangji compared the unlimited increase in the population with the limited increase of the means of subsistence and found a correlation between demographic growth and economic decline. Although he proposed a number of measures to alleviate population pressure, such as the full cultivation of available land, reclamation of wasteland, a reduction in taxes, prohibition of luxury, equalization of wealth, and the opening of more granaries, he did not develop any systematic demographic theory comparable to that of his European counterparts. His essay was part of a series of politically controversial texts that led to his banishment to the frontiers of the empire. As Toby Huff has underlined, science thrived most in Europe, where individuals enjoyed a variety of "neutral zones" and public spaces free from political and religious control.[16] No comparable intellectual autonomy and scientific interest existed in the mid-Qing.

MEDICINE

Despite restrictions on medical work from both the Vatican and the Qing emperors, many Jesuits worked as physicians and apothecaries, although their medical works left few traces in the eighteenth century. The most notable exception was the work of Dominique Parennin (1665–1741), who compiled on imperial order eight volumes on anatomy in the Manchu language, complete with ninety hand drawings of human organs.[17] He added a ninth volume

---

[15] John B. Henderson, *The Development and Decline of Chinese Cosmology* (New York: Columbia University Press, 1984), p. 256.
[16] See Toby E. Huff, *The Rise of Early Modern Science: Islam, China, and the West* (Cambridge University Press, 1993).
[17] See F. R. Lee and J. B. Saunders, *The Manchu Anatomy and Its Historical Origin* (Taipei: Li Ming Cultural Enterprise, 1981).

on chemistry, toxicology, and pharmacology, but his work was never printed, presumably because of court intrigues. Influences from non-European countries on medical knowledge are less well known, although one could point to the example of Liu Zhi (1660–1730), who localized different mental functions in the brain in a syncretic work inspired mainly by Arabic medical science.[18] Although in imperial China the brain was not generally thought to be an organ, the pharmacist Zhao Xuemin (ca. 1719–1805) also wrote that "memory is housed in the brain."[19]

Beyond anatomical knowledge, however, a huge discursive explosion can be found in medical specializations such as women's health, childbirth, smallpox, and typhoid fever. In the seventeenth and eighteenth centuries, commercial companies such as the Xin'an printing house in Anhui province thrived on the publication of medical books, a trend that was indicative of the influence and importance of medical activities during the Qing. Reflective of common interests between the reading public, medical experts, local elites, and publishing houses, many of these books of vulgarization were printed with the financial assistance of local elites, often including rich merchants. Many of these publications were only of local significance and are no longer readily available, although their mere quantity indicates a remarkable dissemination of medical knowledge in late imperial China. Some popular treatises, on the other hand, were frequently reprinted. The *Dashengpian* (Book on successful childbirth) of 1715, one of the most widely circulated booklets on reproductive health, was reprinted more than a dozen times in the eighteenth century. In response to the gradual decline of government intervention in medical matters, privately sponsored efforts at medical relief by local notables, acting as philanthropists, also developed under the Qing. Growing involvement in medical aid and organized charity characterized the local elites of the Yangzi Delta. Under their guidance, local institutions such as dispensaries and infirmaries continued to develop in the eighteenth century.[20]

Distributed by a flourishing print culture that thrived on changing economic and social conditions as well as higher rates of literacy,[21] medical publications catered to a broad readership in the urbanized centers of the coastal region. The widespread compilation of encyclopedias, initiated by the Qing to rally support from scholars, also expanded the circulation of medical works. One example is the influential *Yizong jiajian* (Golden Mirror of Medicine), compiled by Wu Qian, a member of the Imperial Academy of Medicine in

---

[18] Zhu Yongxin, "Historical Contributions of Chinese Scholars to the Study of the Human Brain," *Brain and Cognition*, 11, 3 (Sept. 1989), 133–8.

[19] Quoted in Zhang Binglun, "Renti jiepou shenglixue de fazhan" (The development of human physiological anatomy), in Gou Cuihua, Wang Zichun, Xu Weishu et al. (eds.), *Zhongguo gudai shengwuxue shi* (The history of ancient biology in China) (Beijing: Kexue chubanshe, 1989), p. 181.

[20] Angela Ki Che Leung, "Organized Medicine in Ming-Qing China: State and Private Medical Institutions in the Lower Yangzi Region," *Late Imperial China*, 8, 1 (June 1987), 134–66.

[21] Evelyn S. Rawski, *Education and Popular Literacy in Ch'ing China* (Ann Arbor: University of Michigan Press, 1979).

the second half of eighteenth century. Medical knowledge, moreover, transcended the confines of educated culture: family encyclopedias and cheap handbooks made medical knowledge available to a much larger section of the reading public, including women. The blurring of social distinctions by increased economic prosperity, the growth of a culture of conspicuous consumption, greater social mobility, and a heightened competition over status all had an impact on the diffusion of medical specializations throughout the late imperial period.

A number of scholars criticized the abandonment of ancient recipes contained in the Classics of medicine and opposed the spread of alternative recipes in the Qing. Huang Yuanyu, an eighteenth-century medical writer who lost an eye as the consequence of incorrect treatment, was virulent in his denunciation of new trends in medicine. He was scathingly critical of authors who prescribed ingredients that were not mentioned in the Classics. Like other eighteenth- century scholars interested in evidential scholarship, Xu Dachun (1693–1771) also promoted a return to a strict interpretation of the Classics. He proscribed the use of tonics and recommended only consumption of the five grains after an illness.[22] As part of the cultural reorientations that permanently shattered the foundations of orthodox neo-Confucianism – notably, evidential scholarship (*kaozhengxue*) and a movement in favour of a return to antiquity (*fugu*) – many commentaries on ancient medical texts appeared, often written by medical writers from the Yangzi Delta who were critical of medical theories that had flourished since the Song. Many of these commentators favored a return to the most ancient medical texts in order to reconstruct the classical tradition.

Concern with the philological examination of ancient medical texts generally prevailed over the development of new hypotheses or the discovery of new knowledge. In general, evidential scholarship in the Yangzi Delta, which was so dominant in defining the nature of knowledge in eighteenth-century China, favored a return to the Classics rather than the development of science, from mathematics on the one hand to medicine on the other.

---

[22] Paul U. Unschuld, *Medicine in China: A History of Ideas* (Berkeley: University of California Press, 1985).

30

JAPAN

Shigeru Nakayama

The eighteenth century was one of Western recognition of Japan against the Chinese background. During that period, Japanese thinkers became critical of the Chinese scholarship with which they had struggled to keep pace in the previous century; for the first time, Japanese intellectuals from the extreme eastern regions of Asia began to compare Chinese scholarship with the infiltrating Western science. It is extremely interesting to see what happens to a paradigm from one culture – and the scholarly traditions that have evolved around it – when it is introduced into another. In the following pages we shall examine the impact of this transplantation, mainly on three disciplines: mathematics, astronomy, and medicine.[1]

The Jesuits had been evangelizing in Japan since the mid-sixteenth century. Eventually, the Japanese government, considering Christianity a threat to the cohesiveness and integrity of Japanese culture, successfully banned all Westerners from the country with the exception of Protestant Dutch traders,[2] who were restricted to the port of Nagasaki. This ban, which remained in effect until the mid-nineteenth century, was reinforced with bans on Jesuit writings in Chinese in the 1630s and further intensified in the 1680s.

The beginning of the eighteenth century was thus the nadir of access to information on all things Western. Throughout the eighteenth century, a gradual relaxation of the ban brought an awareness of East-West comparisons based on limited sources of available information.[3]

---

[1] Shigeru Nakayama, *Characteristics of Scientific Development in Japan* (New Delhi: The Center for the Study of Science, Technology, and Development, 1977).

[2] C. R. Boxer, *Jan Compagnie in Japan, 1600–1817* (The Hague: M. Nijhoff, 1936).

[3] Masayoshi Sugimoto and David L. Swain, *Science and Culture in Traditional Japan* (Rutland, VT: C. E. Tuttle, 1989).

SCIENCE AS AN OCCUPATION

Peace prevailed throughout the century in Japan. The economy and demographics remained stable, and the class hierarchy was tightly maintained. The samurai – around 6 percent of a total population of approximately 25–27 million – were at the top of the class structure, and their sons learned the orthodox Confucian classics in clan schools. Commoners were much less literate until village schools evolved toward the end of the eighteenth century with the development of the rural economy.

The samurai class held hereditary stipends that only firstborn males could inherit. Families without sons had to adopt heirs if they were to continue. Younger sons, on the other hand, had to find their own livelihood. Most of them were adopted by other families, and others found occupations outside the old rigid structure in medicine or Confucian scholarship.

There were no clear-cut scientific occupations. Astronomy and medicine were esteemed but offered opportunities only to a few talented men. These fields offered opportunities outside the conventional structure, allowing some to take advantage of a social mobility that was not otherwise available. But attempts were continually made to subordinate men in these fields to the hereditary tradition that governed the rest of Japanese life. It was expected, for instance, that the son of a doctor would eventually be registered as a doctor, regardless of how little aptitude or motivation he might have. The shogunal and fief governments needed talented professionals, however, and governmental authorities often resolved the conflict by advising a professional family to adopt a gifted youngster.

Elsewhere in the Chinese cultural domain, including Korea and Vietnam, scientific professionals were tightly bound to central government institutions through civil service examinations. However, in Japan after the tenth century, as the Chinese-type court bureaucracy atrophied and military power became dominant, these examinations disappeared.

Even during the peaceful Tokugawa period (1603–1869), the shogunal government had no power to impose its recruiting policy on the fief government. During the eighteenth century, the shogunate discussed reviving the examinations, but this was carried out only tentatively in the last decade of the century by testing candidates in Confucian studies from the lower samurai class.

However, an egalitarian examination system was not possible within a hereditary structure. In practice, those who passed with the highest grades received only a prize, varying with their family status, but it did not bring a permanent increase in social status.

Medical examinations began some years before those in Confucian studies. The shogunate, in need of several hundred doctors, gave written as well as oral examinations. They, too, were not intended to change the social status

of graduates but to encourage the sons of medical families to study diligently
rather than simply claim their sinecures.

To find experts to fill posts in technical fields such as astronomy (there were
only ten or twenty such posts), personal references were sufficient.

THE BAN ON WESTERN SCIENTIFIC KNOWLEDGE

During the eighteenth century, woodcut printing flourished. A set of printing blocks could produce around two hundred clear copies, and, therefore,
publishing a book with a corresponding potential readership was commercially feasible. By the end of the century, a popular culture of reading evolved
to the extent that more than ten thousand copies of a bestseller might be in
circulation. During the eighteenth century the center of publishing moved
from Kamigata (Kyoto and Osaka) to Edo (Tokyo). Most academic works were
written in classical Chinese, whereas popular works used a native style that
combined Chinese characters and phonetic kana.

Even as the book trade grew, an official ban severely restricted knowledge of
the West. In the 1630s the government banned imports of Sino-Jesuit writings,
thus depriving the reading public of information on Western science. In 1685
the censors defaced and destroyed what previously had been two important
sources of European cosmology: *Huan yu ch'uan,* written by the Portuguese
Jesuit Francisco Furtado (now preserved at the Bibliotheque Nationale, Paris)
and the sequel to a Chinese work with Jesuit influence, *T'ien ching huo-wen*
(Queries on the Heavens), which will be discussed later in this chapter.

A collection of treatises by Matteo Ricci, *T'ien-hsueh chu hand,* had previously been available in Japan. It consisted of two parts: *li,* catechetical and
theological, and *ch'i,* scientific. The latter does not appear to have been strictly
censored. The world map compiled by Ricci, and some popular astronomical books that reflected some Western influence such as *T'ien ching huo wen*
(Tenkei Wakumon), had escaped the attention of the censors at the port of
Nagasaki. These documents, as they spread among Japanese intellectuals, influenced their worldview as well as their cosmology.

Furtado's book *Huan yu ch'uan* is a popular treatise on Western cosmology
and cosmography, with the first half covering Christian theology and the second half devoted to scientific matters, as in Ricci's collection and in contemporary popular books. This format implied that the two parts, written by
the same author, were distinct but inseparable. When the authorities became
aware of this, public access to scientific and overtly religions works was forbidden, and at the beginning of the eighteenth century the Sino-Jesuit treatises were still heavily censored.

The Jesuits, who arrived in China in the seventeenth century, challenged
traditional astronomy with their superior parameters and methods of calcu-

lation, which became apparent in competitions to predict solar eclipses. It was clear that in astronomy – the foremost subject of a traditional exact science – the criterion of quantitative precision transcended East and West, and observation of celestial phenomena precluded human manipulation. Accordingly, Jesuit astronomers took over the Astronomical Bureau in 1644, won a decisive prediction contest, and quickly carried out a calendar reform that established a largely Western system, the Shih-hsien li. Jesuit control over the Bureau, although challenged several times, survived until the end of the empire.

The Japanese learned about this Chinese reform from imported annual almanacs. Because of the ban on Jesuit works since the 1630s, not enough information could be obtained to reform the Japanese system. Shibukawa Harumi (1639–1715), the first genuine reformer, judged from the crude values of Western parameters given in the popular *T'ien ching huo-wen* that the Shih-hsien calendar was no improvement on its predecessors. Shibukawa followed the great Shou-shih computational system of 1279, which had not been significantly improved before the arrival of the Jesuits.

T'ien ching huo-wen provided Japanese intellectual circles with a standard pre-Copernican cosmological picture, but its sequel was banned at Nagasaki for the official reason that its contents were occult and therefore unhealthy. This volume has long been unknown in China. I had an opportunity to look at the copy in the Seikado Library in Tokyo, a modern acquisition from a Chinese private collector. It contained nothing fantastic. Its history of Chinese calendrical astronomy clearly stated that the Western Jesuits carried out the recent Shih-hsien reform. If contemporary Japanese read this book, the news may have caused great concern and jeopardized Shibukawa's native Jokyo reform, which was based on a purely Chinese model.

We still do not know why the sequel was banned. It is my guess that under the seclusion policy the government and its "Confucian" censors feared that it would convince intellectuals that Western astronomy was superior to traditional Chinese astronomy and perhaps would lead eventually to the belief that Christianity was superior – something that was, after all, the ultimate aim of the Jesuits astronomical activity in China.

In the early part of the eighteenth century, members of the scientific elite in the shogunate consulting bodies began to suspect that the Chinese approach to calendar-making had been replaced by that of the West. Because of this suspicion, in 1720 Tokugawa Yoshimune, the eighth shogun (1684–1751), who was himself eager to collect Western knowledge, ordered specialists in calendrical astronomy and mathematics to carefully examine the banned books that were stored in the shogunal library. Nakane Genkei (1661–1723), a private scholar not previously allowed to see Shibukawa's Jokyo calendar, at Yoshimune's request read the Sino-Jesuit astronomical writings and concluded that they would be useful for the next calendar reform.

The shogunate adopted his recommendations and encouraged elite scholars to study foreign languages, particularly Dutch, which was the sole language used for trade with the West during the seclusion period.

This event, a watershed in the official recognition of Western science, would have shocked those who respected the Chinese model. Because the central government demanded a monopoly on information about the West, it never publicly announced the lifting of the ban. People were still wary of becoming involved with anything related to Western learning. However, for the remainder of the eighteenth century a number of intellectuals perceived that the policy was not being rigorously enforced, and they copied and circulated Sino-Jesuit works. To escape the censors' notice, they often put false titles on the front pages and avoided direct quotation.

In 1726 an incomplete set of the *Li-suan ch'an-shu* (Complete works on calendrical mathematics, 1723), by the great mathematician Mei Wenting, reached Japan. This work was influenced by the Jesuits, but because Mei's work was purely scientific and technical it was much safer to disseminate in Japan. Mei's treatises convinced the Japanese that Western astronomy was superior, and that brought about a further moderation of the ban.

TRANSLATIONS OF WESTERN WORKS

Japanese intellectuals could read classical Chinese writings without difficulty, and Sino-Jesuit publications were likely to be widely read if available; hence the ban. However, because no astronomer had mastered European languages, there seemed little need to ban these publications.

On the other hand, at the port of Nagasaki there were about fifty official interpreters of Dutch, twenty-three of them hereditary. They had enough linguistic knowledge to communicate verbally with Dutch traders and ships' doctors. In the mid-eighteenth century, with the relaxation of the seclusion policy, these interpreters began to study Dutch books and undertake translations. They usually worked at the request of feudal lords who rewarded them well. They translated the few books they received from their foreign contacts. These materials – initially they were generally maps, seamen's almanacs, and other navigators' essentials – stimulated curiosity about Western countries. Eventually, Dutch merchants were asked to import various books by way of Batavia (Indonesia), but these were expensive and limited in subject matter.

The official interpreters were bound by their official duties, and their translations were never intended for publication. Doctors, on the other hand, as intellectuals and men of culture, were more open-minded, independent of government institutions, and eager to publish books that would improve medical practice. Their knowledge of Dutch was greatly inferior to that of the Nagasaki interpreters, but, nevertheless, a few of them obtained Western medical books and began to translate them. The first publication was *Kaitai Shin-*

sho (New Book of Anatomy) in 1773. A pioneer of the project, Sugita Gen-paku (1733–1817), managed to obtain official approval in advance for the first published translation of a Western book, and, as the title of his memoir states, this was viewed as the dawning of Dutch learning in Japan. This breakthrough encouraged other physicians and intellectuals to learn Dutch and to investigate Western science. By the turn of the century, a group of physicians and intellectuals formed a society with the aim of exchanging information on Western science.[4]

THE INDEPENDENT TRADITION OF MATHEMATICS

The Western world first became aware of the independent tradition of Japanese mathematics (*wasan*) through the publication in 1914 of *A History of Japanese Mathematics* by Eugene Smith and Mikami Yoshio. Japanese mathematicians had built this tradition on the basis of late seventeenth-century Chinese mathematics and had developed it independently in the eighteenth century. Despite the influence of Western culture on other aspects of life, Japanese algorithms and the Japanese style of writing equations are quite distinct. A number of its characteristic problems did not exist, or appeared later, in the history of Western mathematics.

Reckoners used the abacus or counting rods on a grid. When symbolic algebra appeared after Seki Takakazu (d. 1708), symbolic (as opposed to merely numerical) written calculation (on paper) became possible. *Wasan* has been compared to Newton's and Leibniz's differential and integral calculus; but Seki and his immediate successor, Takebe Katahiro (1664–1739), leading figures in the tradition, showed little interest in solving mechanical problems, the calculation of the area of a circle being a more typical preoccupation. Pure mathematics, in other words, was pursued as a hobby and was not associated with physical science.

In the seventeenth century, mathematicians established the tradition of *wasan* by the practice of "bequeathed problems." Anyone who solved a difficult problem for the first time would write a treatise and would then bequeath a new problem; whoever solved this problem would bequeath another, and so on. This ongoing competition gave the tradition its momentum. Emphasis was placed on problem solving by unusual means and on presenting problems for which solutions were unknown or perhaps did not exist. Mathematicians increasingly valued complexity and emphasized the transformation of systems of simultaneous equations into higher-degree equations.

This ostentation for its own sake prompted Seki Takakazu to introduce an important innovation: the *tenzan* algebra. This was a system for expressing

---

[4] Donald Keene, *The Japanese Discovery of Europe, 1720–1830,* revised edition (Stanford, CA: Stanford University Press, 1969).

unknowns, previously solved only via numerical equations, in symbols. He also developed a theory of equations that recognized imaginary and negative roots (*daijutsu bengi no ho, byodai meichi no ho*) but rejected them as "sick solutions" – thereby ruling out a theory of imaginary numbers. In *wasan,* as in Chinese arithmetic, problems were presented in the form of questions and answers and often omitted the method of derivation. Therefore, they did not encourage investigation of the basic nature of equations.

Seki was exceptional in investigating the general nature of equations. His orthodox standard for posing and solving problems fits the Kuhnian defini- tion of "paradigm." Once this paradigm was established, "normal science" could follow.

Before the time of Seki and Takebe, mathematicians were concerned with the practical problems of calendrical astronomy: surveying and so forth. The paradigmatic approach ruled out application. Subparadigms appeared in the course of development – for instance, Takebe's *enri* (circle theory) calculus. Ajima Naonobu applied this method not only to circles but also to curves and curved surfaces in general. Wada Yasushi furthered the development of math- ematical analysis by compiling tables of definite integrals and applying them to the mathematically infinite and infinitesimal, and so on. All these innova- tions remained, however, within the *wasan* tradition. There was little discussion of fundamental theories, and practitioners continued to solve increasingly complicated geometrical figures by algebraic means.

Another stimulus for amateur mathematicians came from a somewhat dif- ferent source. Artists and poets had established a tradition of offering their masterpieces, painted on wooden plaques, for display in the public gallery of Shinto shrines. Mathematicians followed suit, exhibiting a tablet with both problem and answer displayed, usually accompanied by an elegant geomet- rical diagram for public entertainment. Amateurs seeking acclaim often spent a great deal of money on such pursuits. Thus, mathematicians valued playful competition rather than basic research or application, and intuitive break- throughs for elaborate problems were regarded more highly than logical con- sistency or rigor. Indeed, when Euclid's *Elements* first came to Japan, *wasan* experts – noting its figures only – judged it rudimentary and unchallenging.

The Sino-Jesuit treatises had introduced trigonometry, and by the end of the century it was used by astronomers and surveyors. Although *wasan* mathematicians were capable of mastering it, they continued to use their traditional methods, valuing problems arising from their own tradition and not those posed by technical practice. The eccentric mathematical genius Ku- rushima Yoshihiro wrote, "In mathematics, it is more difficult to raise a prob- lem than to answer it. Only mathematicians incapable of inventing problems borrow them from other fields such as calendrical science."[5]

Because of their lack of interest in practical applications, the nucleus of

[5]Naonobu Ajima, *Seiyo Sanpo* (1779), preface.

Japanese mathematicians did not compete with Western mathematicians in solving practical problems until the middle of the nineteenth century. Unlike other Japanese intellectuals, *wasan* practitioners, as they moved from paradigm creation to the formation of a support group, advanced to new technical frontiers without the need to consult foreign authorities. In isolation they underwent vigorous growth in normal science, largely unaware of the developments in Chinese or Western mathematics. *Wasan* did not substantially influence scholarship in other fields or in cultural matters, although those practitioners who applied their mathematical skills to the practice of land surveying or calendrical calculation were well-versed in the algorithms, formula, and notations of traditional *wasan* mathematics.

MATHEMATICS AS AN OCCUPATION

Seki Takakazu issued a license to teach, and it was developed by his successors into a five-stage system of degrees. This system did not guarantee employment as a teacher but was primarily considered an honor, certifying that a certain level of mathematical mastery had been achieved. Without a solid occupational basis it is difficult to estimate how many people were engaged in *wasan* or to distinguish amateurs from professionals.

There is evidence that even peasants occupied themselves with *wasan* puzzles in the agricultural off-season. Mathematicians were primarily hobbyists, and the traditions existed only in the private sector. Although the shogunate attempted to maintain the occupational hereditary system, it did not consider mathematics worthy of perpetuation through this system.

Only a handful of leading mathematicians were able to support themselves. From the late eighteenth century, a number of them traveled from village to village, visiting amateur groups and enthusiasts and conducting problem-solving competitions, thereby following the practice of other arts such as *haiku* poetry.

PUBLICATION IN MATHEMATICS

Wasan mathematicians circulated their solutions to problems by copying by hand, although the more famous published theirs. Popular mathematical works for general readership were even more widely published. The printing blocks were often cut in an informal running style of calligraphy, which made it easy for the literate public to read. They were bestsellers by Japanese standards, some selling more than several thousand copies.

The literary world did not consider mathematics to be true scholarship, and, from the end of the seventeenth century, mathematics was often classified in book catalogs as a hobby on a par with flower arrangement or the tea ceremony. *Wasan* authors, in an attempt to increase the prestige of their books,

invited Confucian philosophers to write prefaces, which usually bore no relation to the technical content.[6]

ASTRONOMY WITHIN THE TRADITIONAL FRAMEWORK

The traditional Chinese approach ("calendrical astronomy") investigated the
apparent motion of the sun and moon to construct a method for generating
lunisolar calendars. The ultimate test was the precision with which a system
could predict solar eclipses. Successive reforms refined parameters for solar
and lunar motions by testing them on previous records of solar eclipses. Planetary phenomena attracted relatively little attention.[7]

Throughout the written history of Japan, the Chinese lunisolar calendar
had been accepted. From the late seventeenth century onward, the shogunate
adopted its own Jokyo system, merely revising the Chinese Shou-shih system
of the thirteenth century to incorporate the difference in latitude between
China and Japan.

An order of the Shogun Yoshimune in 1720 assigned responsibility for a
new calendar reform to Nishikawa Masayasu. He was the son of the noted
Nagasaki scholar Nishikawa Joken, Japan's foremost expert on the West.
Masayasu was not a professional astronomer, but, assisted by professionals,
he undertook a reform based on Sino-Jesuit writings. When Yoshimune died,
a family of court astronomers in Kyoto tried to restore the emperor's prerogative of issuing the calendar. This conservative backlash ignored Yoshimune's
goal of reform and their Horyaku system, issued in 1754, made matters worse.
Masayasu was subsequently dismissed, and his associates could only edit their
records for use by a future generation.

Real reform came a generation later in the Kansei calendar revision (1797).
It was undertaken by Asada Goryu (1734–1799), a physician and amateur
astronomer, and his followers, who had access to most of the Sino-Jesuit
treatises.

In astronomy, the new Western paradigm did not replace the traditional
Chinese one. Rather, new data and mathematical techniques were simply
incorporated into the old framework. This was also the case in China from
the seventeenth century onward, with the structure, style, and purpose of
Chinese calendrical astronomy unchanged. As Hsu Kuang-ch'i, the high official who had collaborated with the Jesuit Matteo Ricci on several projects,
remarked, "We melted down their materials and poured them into the [old]

---

[6] Shigeru Jochi, "The Influence of Chinese Mathematical Arts on Seki Kowa," unpublished Ph.D. dissertation (University of London, 1993).
[7] Shigeru Nakayama, *A History of Japanese Astronomy: Chinese Background and Western Influence* (Cambridge, MA: Harvard University Press, 1969).

Ta-T'ung mould."[8] Until the mid-nineteenth century, official Japanese astronomers adopted this attitude and even repeated Hsu's slogan in their treatises.

Throughout the eighteenth century, Japanese calendrical astronomy adopted the view that astronomical parameters varied in time. In 1684, the government adopted the Jokyo system, whose originator, Shibukawa Harumi (1639–1715), restored the variable tropical year length of the Chinese Shou-shih calendar. He reasoned that such a minute variation reflected high precision. In reality, however, it provided no gain in accuracy.

Ogiu Sorai (1666–1728), the most influential Confucian philosopher of his time, supported Shibukawa's notion on ideological grounds, commenting in his *Gakusoku Furoku* (supplement to School Rule), "Heaven and earth, sun and moon are living bodies. According to the Chinese calendrical technique, the length of the tropical year was greater in the past and will decrease in the future. As for me, I cannot comprehend events a million years ahead."[9] In Ogiu's dynamic view of nature, everything was subject to change and it was therefore impossible that ancient laws could still hold. Since the heavens were imbued with vital force, the length of the year could change freely, and thus constancy was not to be expected in the heavens. Indeed, only a dead universe could be governed by law and regularity. Since it was precisely the vital aspects of nature that interested Ogiu, he remained an agnostic in physical cosmology.

Lack of interest toward the search for regularities in nature prevailed in the School of Ancient Learning (*Kogaku*), of which Ogiu was the leader. Nature was observed in the light of social and ethical concerns. This moralistic, anthropocentric, and often anthropomorphic view of nature was common among Japanese Confucian intellectuals. Few of them imagined that mathematical astronomy was deserving of attention except to provide an accurate calendar. Hence, the official astronomers' recognition of Western superiority did not immediately influence conventional intellectuals.

The next calendar reform, the Horyaku (1755), replaced Shibukawa's notion of changeable parameters. The value of yearly change was much too large and the discrepancy between observation and calculation, as it increased, was bound eventually to become apparent. This secular variation, nevertheless, was again adopted without reflection in the system that was to follow, with a predictable growth in inaccuracy.

The traditional eclipse records used as benchmarks for astronomical parameters were not supplemented by Western observational records in Jesuit writings. Asada Goryu (1734–1799) collected all the available records, tradi-

---

[8] Shigeru Nakayama, *Kinsei Nihon no Kagaku Shiso* (Japanese Scientific Ideas in the Eighteenth and Nineteenth Centuries) (Kodansha Gakujutsu Bunko, 1993), p. 70.

[9] *Gakusoku Furoku* (Appendix to the Principles of Learning) (1727).

tional and Western, and tried to represent them all with a single formula of his own. He varied not only tropical year length but also other astronomical parameters in a twenty-six-thousand-year cycle of precession. His approach was purely numerical, and it was incorporated in the next Kansei calendar reform in 1798.

In Asada's time, knowledge of Western astronomy was still limited to Sino-Jesuit writings in Chinese, which made no mention of Copernican doctrines. Toward the turn of the nineteenth century, Asada's pupil Takahashi Yoshitoki (1764–1804) began to study a Dutch translation of Lalande's post-Newtonian *Astronomie*. His was a purely academic interest in the kinematics of planetary motions, although celestial mechanics were still beyond him.

Because calendrical astronomy was an official domain, advisers urged that the shogunate recognize Western superiority. However those bureaucrats employed in astronomy had no authority except in purely technical matters, and they neither intended nor had the power to speak publicly on the merits (or otherwise) of Western science. Their influence in nonastronomical fields was negligible.

An index of Western influence can be taken from the use of the Sino-Jesuit 360 degrees for coordinates as opposed to the traditional Chinese count of approximately 365.25 (the old degree, *tu,* was defined as one day's mean solar motion). Official astronomers working on the Kansei calendar reform (1798) first used the former, after which it spread gradually into general use.

ASTRONOMY AS AN OCCUPATION

Calendrical astronomy was a state monopoly. Issuing the annual ephemeris and reforming computational methods were purely a matter of prestige for the ruling government. In the eighteenth century, although dynastic legitimacy could not be removed from the imperial court in Kyoto, real political power lay entirely in the hands of the military dictator, the shogun, in Edo (Tokyo). Certain astronomical prerogatives were a hereditary right of the Tsuchimikado family in Kyoto, the imperial court astrologers, but only the shogun's astronomers had the actual power to reform the calendar and apply the science. There were eight shogunal families of astronomers, who, like their imperial predecessors, had become hereditary. They and their associates totaled between fifty and one hundred officials. The Tsuchimikado, other families of lower rank, and temporary associates brought the total in Kyoto to less than fifty. Hoping to restore their ancient authority, they successfully intervened at the time of Horyaku reform, but the Kyoto revival was short-lived.

The hereditary astronomers did not require talent or even much skill to calculate the annual ephemeris. They met the greater demands of the two eighteenth-century reforms by acquiescing in the appointment of – or even by adopting – well-qualified individuals. Some fief governments occasionally hired astronomers, usually because the ruling *daimyo* family was interested in the astrological prediction of natural disasters. Some remote areas such as Satsuma, one of the larger fiefs, appointed permanent astronomers and issued their own calendars, but these did not diverge significantly from Shogunal astronomical practice.

PUBLICATION IN ASTRONOMY

As a shogunal practice, astronomy was by no means accessible to the general public. The most important official treatise on the new calendrical system, the product of the calendar reform, was never published. Three manuscripts were submitted to the Shogunal Library, the Imperial Court Library, and the Library of the Ise Grand Shrines, to each of which only a few high-ranking officials had access. The government feared that criticism from the private sector would destroy public esteem for this particular governmental function.

As a result, those who wanted to learn computational astronomy could study only the past system, in particular the Shou-shi calendrical treatise of the late thirteenth century, the highest achievement in Chinese mathematical astronomy and the model for the Japanese calendar until the middle of the eighteenth century. Many illustrated guides and commentaries on this were complied and printed in Japan. The main source for cosmology was *T'ien ching huo wen,* the seventeenth-century treatise that incorporated some Jesuit elements. Again, many Japanese-illustrated versions and textbooks satisfied the intellectual needs of the day, and the needs of the general public were met with yearly almanacs printed and distributed by a network controlled by the hereditary imperial court astrologers.

INTRODUCTION OF
COPERNICANISM AND NEWTONIANISM

Because the hereditary astronomers' interest remained confined to the traditional model of calendrical science, the introduction of the core of modern Western astronomy was left in the hands of the official interpreters. The first to become involved, perhaps, was Motoki Ryoei (1735–1794), who invented his own system of transliteration from Dutch to Japanese phonetics, using Chinese characters. We know of no similar activity in China at the time.

Ryoei was interested in translating a history of Western astronomy to add to the margins of large navigational charts. However, he was concerned to learn that Galileo had been persecuted because of his writings on Copernical cosmology. Ryoei realized that this would be a delicate subject in Japan since it was related to the strictly proscribed Christianity, and his translation, drafted in 1771, omitted discussions of the trial of Galileo. However, he found Copernicanism important and interesting, and his later translations gradually revealed details of Copernicanism, a full translation being completed in 1793. Borrowing from Ryoei, Shiba Kokan (1738–1818), an illustrator and popularizer, published many books that disseminated the theory of heliocentricism in Japan.

Shizuki Tadao (1760–1806) was also born into a family of official translators. He left his inherited profession to concentrate on the translation of Western books and was the first Newtonian in East Asia to introduce such concepts as the molecule and force. Because traditional Japanese Confucian learning was not concerned with natural philosophy and was unaware of late Chinese writing on the discipline, in translating Newtonian concepts terminology had to be borrowed from the Ten Wings of the Book of Changes, Buddhist speculation, and neo-Confucian writings. Tadao translated the work of John Keill, the popularizer of Newton, adding a great many comments of his own, some quite original and going far beyond Keill. Not entirely satisfied with Newtonian laws, Tadao attempted to base them on traditional Yin-yang metaphysics. He tried to introduce the inverse-cube centrifugal force or quadruple of distance to explain such phenomena as chemical affinity and plant physiology. Tadao is also known for his nebular rotation view of the solar system – a similar idea later attributed to Kant and Laplace – although, essentially, he applied neo-Confucian cosmogony to the solar system.[10]

PHYSICIANS AS INTELLECTUAL CONNOISSEURS

In the seventeenth century, mainstream Chinese medicine dominated that of Japanese with the exception of surgery, which the Jesuits had introduced in the sixteenth century to meet the needs of endemic civil war.

In the eighteenth century, a new group became critical of scholarship of the physiology and pathology that had prevailed in China since the Chin and Yuan periods. They claimed to be returning to a simpler reasoning that more directly reflected the clinical practice of the ancient *Shang han lun* (Treatise on Cold Damage Disorders, between A. D. 196 and 220) but showed little interest in the more theoretical and speculative *Huangti Neiching* (Yellow Emperor's Inner Classics). This group called itself *Koiho* ("Back to Ancient Medicine School").[11]

The school preferred simple and drastic medical prescriptions as opposed to the great variety of Chinese formulas, some simple and some complex, some strong and some mild, some formed by theory and some by direct experience of drug action, all of which neutralized effects. The Koiho defined their goals in terms of utility, which made Chinese complexity seem more of

---

[10] Tadashi Yoshida, "The Rangaku of Shizuki Tadao: The Introduction of Western Science in Tokugawa Japan," unpublished Ph.D. dissertation (Princeton University, 1974).

[11] Yu Fujikawa, *Kurze Geschichte der Medizin in Japan* (Tokyo: Kaiserliches Unterrichtsministerium, 1911).

an impediment. Because they wished to tackle disease as directly as possible, they refused to view it as a microcosm. As Yoshimasu Todo (1701–1773), the foremost figure of this school, declared, "Yin and yang are the *ch'i* of the universe, and thus have nothing to do with medicine."[12]

The Koiho were materialists in the sense that they rejected abstraction, trusting only that which was tangible, and thus they developed abdominal palpation, which did not exist in China.[13]

FROM THE ENERGETIC TO THE
SOLIDIST VIEW OF THE HUMAN BODY

In Chinese and Japanese medicine, disease was attributed to an imbalance of *ch'i*, which circulated throughout heaven and earth and thus through the human body. It is now considered imponderable and incorporeal energy, but Ch'ing Chinese considered it the material basis of life. This view was close to that of Western humoralists, who believed disease resulted from an imbalance between the humors circulating through the body rather than to a pathological abnormality in a particular organ.[14]

Goto Gonzan (1659–1733), a precursor of the Koiho school, reduced traditional physiology and pathology to a simplistic scheme in which every disease originates in the stagnation of *ch'i* and in which the *ch'i* was a more materialistic concept than the accepted Japanese abstract and incorporeal matter. Goto's successors took a position much closer to that of the solidists than had previously been possible in Japan. Lacking abstract concepts, functional analysis lost its importance, and the Koiho physicians studied the physical organs for their own sake.[15]

In conventional *ch'i* physiology, dissection does not yield meaningful information, as the dead body contains no *ch'i*. Koiho physicians, on the other hand, showed a genuine interest in dissection and their organ-centered approach brought a recognition that the traditional anatomical charts were crude and inaccurate. In 1754 Yamawaki Toyo (1705–1762), a leader of the Koiho school, was the first to examine the corpse of a criminal for anatomical purposes. He questioned Chinese anatomical charts and wrote *Zo shi* (Chart of internal organs, 1759) on the basis of his findings. Yamawaki's achievement was, however, limited to challenging the old scheme of six yin and five yang organs (*wu-tsang liu-fu*) in preference to nine. No interest was shown in the

---

[12] "Idan," 1795, in *Kinsei Kagaku Shiso* (Ideas of Early-Modern Science), vol. 2 (Iwanami, Nihon Shiso Taikei, 1971), p. 540.

[13] Yu Fujikawa, *Japanese Medicine*, translated from the German by John Ruhrah, P. B. Hoeber (1934).

[14] C. Leslie (ed.), *Asian Medical Systems: A Comparative Study* (Berkeley: University of California Press, 1976).

[15] Norman Takeshi Ozaki, "Conceptual Changes in Japanese Medicine during the Tokugawa Period," unpublished Ph.D. dissertation (University of California, San Franciso, 1979).

investigation of the skull, which was regarded as a reservoir of medullary tissue.

In the East Asian tradition, it made no sense to ask in which organ thought took place because physicians did not think in solidist terms. They attributed every activity, mental or physical, to the fundamental agency of *ch'i*, which permeated the microcosm as well as the macrocosm and which circulated harmoniously in both. Thus, there was no reason to attribute thought as a function of the brain. Thought, imagination, and emotion were functions, not as a physical organ but as a bodywide system of energy circulation.

In 1771, however, Sugita Genpaku and his followers abandoned the Chinese physiological tradition, relying on Dutch anatomy charts,[16] although even they showed no interest in examining the contents of the skull after the decapitation of a criminal. Genpaku therefore found it difficult to translate Dutch writing on the brain, often resorting to guesswork and borrowing from Buddhist terminology in order to coin new words for sensory perception. His confusion created difficulties for successive generations of medical students, who used his writings as a base for the understanding of cerebral function.

Thus, the translation of Western anatomical books was significant not only as the beginning of Western learning in Japan but also for the introduction of the solidist school of thought into East Asian culture. However, it is unlikely that the average eighteenth-century Japanese doctor understood the function of the brain.[17]

Practitioners of Chinese medicine viewed disease holistically, and a given disease was not usually associated with a particular body location since pathological *ch'i* as well as life-sustaining *ch'i* usually affected the whole microcosm. For example, when doctors referred to a cardiac or hepatic dysfunction they were not referring to the physical organ but to a whole-body system of functions that the organ merely regulated and that the disease affected. They also treated the body holistically. To treat a headache, for example, needles were inserted into the foot. This approach did not require precise anatomical charts in the clinic. Unlike the physicians trained in the Chinese method, those surgeons cooperating with Sugita Genpaku discovered a remote ancestry in Western origins. This would explain their openness to a solidistic way of thinking.[18]

When the power of Western anatomical knowledge was first realized by the Japanese, it was naturally assumed that associated therapies would also be more effective, although there was no evidence for this belief. Indeed, therapeutically, there was very little choice between the systems of internal medicine that were evolved in the various advanced civilizations before the end of the

---

[16] Sugita Genpaku, *Dawn of Western Science in Japan* (Rangaku Kotohajime, 1815), translated by Rytz Matsumoto (Tokyo: The Hokuseido Press, 1969).
[17] John Z. Bowers, *Western Medical Pioneers in Feudal Japan* (Baltimore, MD: Johns Hopkins University Press, 1970).
[18] Harm Beukers et al. (eds.), *Red-Hair Medicine: Dutch-Japanese Medical Relations* (Amsterdam: Rodopi, 1991).

nineteenth century, although European medicine was more drastic and more likely to harm the patient than most.[19]

The traditionalists naturally objected to anatomy, a common response being that anatomy and dissection were irrelevant to the improvement of therapeutic practice. Other objections were based on traditional physiology. Sano Antei, in his *Hi Zoshi* (A Refutation of the Anatomical Charts, (1760), said, "What the *tsang* [the spheres of function and their associated viscera] truly signify is not a matter of morphology. They are constant containers that store vital energy with various function. Lacking that energy, the *tsang* became no more than empty containers."[20] In other words, the internal organs were characterized not by their morphology but by differences in function, defined by their proper *ch'i*, and, therefore, nothing could be learned by dissecting a cadaver, since its *ch'i* did not exist. Because they were based on dissection, the anatomical charts that had caught Toyo's imagination gave no indication of the dynamic functions of the body.

The same point emerges in another criticism of Antei: Yamawaki Toyo's anatomical charts did not demark the large and small intestines. Antei himself did not believe that they were morphologically dissimilar. A physiological difference followed. What made them different was that the large intestine was responsible for absorbing and excreting solid wastes, while the small intestine performed the fluid waste functions. This crucial difference would be undetectable in the dead body. Figure and appearance were significant only in terms of their relation to function. Antei, unlike the Koiho radicals, did not claim to be a pure empiricist. "The observation of two obvious facts is of much less value than groping speculation . . . even a child is as good an observer as an adult";[21] a scholar who did not investigate the connections between form and function was no better than a child.

In spite of such reactions by conventional physicians to the radical Koiho school, the solidist tradition that it had initiated paved the way for Western anatomy. Genpaku took up the study of anatomy because it seemed the most tangible and, therefore, the most comprehensible part of Dutch medicine. A solidist breakthrough resulted from this viewpoint and, at the turn of the century, physics and chemistry were studied by Genpaku's successors. The impact of anatomy challenged the energetics and its functional beliefs not only of medicine but also of natural philosophy, and eventually led to the wholesale introduction of modern Western science.[22]

[19] Mieko Mace, *L'anatomie occidentale et l'experience clinique dans la mecine japonaise du XVIe au XVIIIe siècle*, in Isabelle Ang and Pierre-Étienne Will (eds.), *Nombres, astres, plantes et viscères: sept essais sur l'histoire des sciences et des techniques en Asie orientale* (Paris: Collége de France, Institut des Hautes Études Chinoises, 1994).

[20] "Hi Soshi" in Koichi Uchiyama, "Nihon Seiri Gakushi" (History of Japanese Physiology), in *Meijizen Nihon Igakushi* (History of Japanese Medicine before Meiji Era), vol. 2 (1955), p. 122–3.

[21] Ibid.

[22] Wolfgang Michel, *Hermann Buschof – Das genau untersuchte und auserfundene Podagra, Vermittelst selbst sicher-eigenen Genaesung und erloesenden Huelff-Mittels* (Heidelberg: Haug Verlag, 1993).

THE MEDICAL PROFESSION AS AN OCCUPATION

Medical practitioners who began to take on the challenge of Western science constituted the largest scientific profession during the Tokugawa period. Medicine, unlike astronomy, was a private concern and not subject to any form of constraint in terms of response to new ideas. Because there was no public health program at the time, medical practice was essentially a relationship between physician and patient. Each community usually had a private physician or healer. The samurai class had its government doctors and fief doctors, and townspeople and peasants had their local practitioners. Medicine was not a profession, and practitioners did not form organizations or even common ties. They were not regulated by the central government and were not subject to the traditional expectation that physicians should be sons of physicians.[23]

Edo, as the seat of the shogunate, was a center of professional activity. The important schools of medicine were scattered as far as Nagasaki, where a tradition of Western surgery was maintained through access to Dutch interpreters. Osaka, for instance, was famous for its number of physicians, whose patients were mainly from the merchant class. This decentralization made medicine one of the few geographically mobile professions in Japan.

Unlike the medical profession in contemporary Europe, which was well established and able to develop through the universities, Japanese doctors remained marginal. Government appointments were not the only possible source of income for physicians. A doctor hired by a fief government had no more status than other petty intellectual officials (Confucian scholars, astronomers, or interpreters), but private practice could bring a much higher income.

Doctors – unlike official astronomers – were independent of government hierarchy, and, in the period shortly before the modernization of Japan, they were among those most receptive to liberal thought.

Medical practitioners were not licensed. Even those who were able to read medical classics could advertise themselves as physicians. Often, those hoping to qualify as Confucian scholars supported themselves by practicing medicine. The shogunal and fief governments appointed physicians, usually with small stipends, to take care of lords and samurai families. The government often encouraged physicians to adopt a talented young man to ensure a reliable supply of medical practitioners rather than bequeathing a first-born son. Toward the end of the century, as public living standards improved, towns and even small villages supported their own doctors, although most of the peasants found Omyoji (traditional diviners) adequate to meet their medical requirements.

Physicians were usually trained by apprenticeship. A young man wishing to embark on a medical career would become the pupil of a practitioner, living in his house for several years to gain "hands-on" experience. The ap-

[23] Erhard Rosner, *Medizingeschichte Japans* (Leiden: E. J. Brill, 1989).

prentice would then move from place to place to gain clinical experience, finally returning home to set up in practice. The more ambitious would seek medical training as far away as Edo, Kamigata, or Nagasaki (for Dutch medicine). It is difficult to estimate the number of practitioners trained in medicine, but I would estimate it to be in the region of several tens of thousands.

The end of the century saw the emergence of therapists – practicing even in villages. For example, the second son of a village chief, with aspirations of becoming a country doctor, would spend many years as an apprentice to neighboring practitioners and finally return to his native village. He would be expected not only to provide medical services but also to undertake educational and cultural duties.

Young men who showed academic promise but with no conventional prospects were often advised to study medicine to achieve a secure livelihood. Those who studied Dutch medicine as a means to a medical career often became experts in Western learning and, much later in the mid-nineteenth century, were to have a revolutionary influence on political affairs.

Among the intellectual professions of the Tokugawa period, it was only physicians who were able to achieve an independent position: they were able to view the world from new perspectives and thus bring modern (universal) science to Japan. However, their independence was bought at the cost of alienation from the true sources of power in Japan – the samurai governments. Their role was thus limited to that of connoisseurs of cultural novelty.[24]

From the late eighteenth century on, the Rangakusha (scholars of Dutch learning) were mainly free-lance physicians.[25] The more successful tended to live in cities, often with government appointment. In Edo, particularly, doctors met and exchanged information on Dutch learning. In 1794, they started their celebrations of the Western New Year and drank European wine. Some connoisseurs wrote entertainingly on curious aspects of Western culture, and their books became best sellers. Otsuki Gentaku (1757–1827), who published a heavily edited and revised version of the *Kaitai Shinsho,* founded a school of Dutch language in 1789. Most of his students were doctors employed in the public sector, but people of any social status could attend. His school was followed by other institutions of Dutch learning. Motivated by a taste for exoticism (novelty), these scholars were not hindered by feelings of inferiority toward Western science.[26] At the turn of the nineteenth century a few recognized that it provided something that was lacking in the Eastern tradition, namely the natural philosophy that generated modern science.[27]

[24] Takeo Nagayo, *History of Japanese Medicine in the Edo Era: Its Social and Cultural Backgrounds* (Nagoya: University of Nagoya Press, 1991).
[25] G. K. Goodman, *Japan: The Dutch Experience* (London: Althlone Press, 1986).
[26] Yoshio Kanamaru, "The Development of a Scientific Community in Pre-Modern Japan," unpublished Ph.D. dissertation (Columbia University, 1981).
[27] Togo Tsukahara, *Affiniti and Shinwa Ryoku: Introduction of Western Chemical Concepts in Early Nineteenth-Century Japan* (Amsterdam: Gieben, 1993).

MATERIA MEDICA

Materia medica, a practice ancillary to medicine, included the study of substances derived from plants, animals, and minerals, and writings on these subjects were indispensable to practitioners. The government often sponsored these voluminous writings, which formed a large pharmaceutical encyclopedia that followed the pattern of the Chinese classification of drugs – mainly according to symptoms – and provided a rough classification of sources. A Chinese treatise, *Pents'ao kang-mu* (Systematic materia medica, 1596, imported 1607), taxonomically arranged, provided a standard pattern in Japan as well as in China. An important Japanese concern was the comparison and identification with local species of animals and plants mentioned in the Chinese classics. This concern led to a dependence on actual observation rather than on the study of classical works. This not only furthered the trend toward morphological study but also introduced a new criterion of classification according to habitat and environment, such as distinguishing insects living on or in water, and fish living in fresh or sea water, as shown in *Yamato Honzo* (Japanese materia medica, 1708) by Kaibara Ekken (1630–1714).

Most scholars of materia medica were physicians, but in the latter half of the century their interest extended from conventional writings of materia medica toward encylopedic natural history, which added new species without proven medical properties, including materials imported from the West.

CONCLUSION

From the seventeenth century on, when Western knowledge began to produce claims distinct from Chinese learning, Japanese thinkers were critically attentive. The conviction that European technical knowledge was superior brought about a switch to the new model. Yoshimune and his astronomer mathematicians clearly recognized the superiority of Western over Chinese astronomy. As bureaucrats or technicians, their interest in Western science was limited to the precision of astronomical data and methods of calculation, and they did not jeopardize their hereditary posts by entering into the mechanistic philosophy of early modern Western science.

Professional interpreters in Nagasaki, well versed in the Dutch language, became acquainted with the concepts of Western science. They too, as hereditary officials, remained within the boundaries of their duty to translate faithfully. Neither official astronomers nor interpreters wrote for a general readership.

From the late eighteenth century on, many Dutch works on *Natuukunde* (the study of nature) found their way into Japan. They aroused the interest of independent scholars, who set about translating them even though their language skills were inferior to those of the Nagasaki interpreters. The majority of these "Dutch scholars" were medical practitioners who were not necessarily occupationally motivated and were thus able to indulge in dilettantism.[28] By the end of the century their interest extended to anything Western.

Astronomy was the first discipline to bring about the conviction of the superiority of Western learning. The idea that this was also the case in other fields of scientific endeavor first spread among the independent physicians. Although only a few realized the power of mechanistic Western science, and fewer still knew of the Enlightenment, many were professionally interested in Western medicine. Unfavorable comparisons were not usually drawn; the thinking was that East is East and West is West and that curious things were going on in the West that brought interesting comparisons with the Eastern tradition. It was believed that Japanese intellectuals could gain advantages from both.

Later in the century, Western aggression toward East Asia – especially that of Russia – was to become prominent. It was not yet on a scale that prompted a radical reevaluation of the need to change political and technological institutions.[29] This took place after the 1840s, when Western aggression increased, leaving the shogunate face-to-face with advanced Western military technology.

[28] Herman Heinrich Vianden, *Die Einfhrung der deutschen Medizin im Japan der Meiji-Zeit* (Dusseldorf: Triltsch Verlag, 1985).
[29] Shigeru Nakayama, "Japanese Scientific Thought," *Dictionary of Scientific Biography*, XV, pp. 728–58.

31

SPANISH AMERICA

From Baroque to Modern Colonial Science

Jorge Cañizares Esguerra

Formal systems of "non-Western science" created by the Aztec, Maya, and Inca seemed to have evaporated into thin air in the wake of the Spanish conquest.[1] The collapse of large indigenous polities and the disappearance of courts capable of sustaining elite knowledge appear to be the cause. Nancy Farriss has argued that the Maya in Yucatan lost the institutions that had kindled their taste for large cosmic riddles. Although the Maya elites did not disappear – and actually became important brokers in the operation of colonial labor systems – they lost interest in those theological and cosmological questions that had driven the astronomical and calendrical investigations of classic and post-classic Maya civilizations. As the Maya elites were left in charge of ever more simplified polities, their interest became narrowly parochial. Under Spanish colonial rule the former complex social structures of the Inca, Maya, and Aztecs gave way to simplified communities lacking all intermediate social tiers: gone were the indigenous pan-regional polities of the past whose courts had maintained large retinues of priests, scribes, and scholars – producers of elite pre-colonial non-Western knowledge. The new simplified native elite class embraced Catholic images, shrines, temples, and rituals, and those few religious leaders who kept native religions (and thus non-Western scientific traditions) alive went underground, losing the source of much of their prestige, which lay in maintaining communal cohesion through *public* sumptuous worship. By the eighteenth century indigenous systems of knowledge had transmuted into hybrid forms of folk Catholicism and had moved to the margins of Latin American societies.[2]

---

[1] I am limiting my remarks to elite systems of natural philosophy that disappeared at conquest. Indigenous scientific knowledge has survived to this day in the vernacular tradition.

[2] Nancy Farriss, *Maya Society under Colonial Rule* (Princeton, NJ: Princeton University Press, 1984), passim. Cf. Inga Clendinnen, *Ambivalent Conquests: Maya and Spaniard in Yucatan, 1517–1570* (Cambridge University Press, 1987), and James Lockhart, *The Nahuas After the Conquest: A Social and*

The author wishes to thank Marcos Cueto, Felipe Fernández-Armesto, Thomas Glick, Antonio Lafuente, and Roy Porter for the comments on previous drafts of this chapter.

This chapter does not deal with the hybrid forms of popular knowledge that developed at the margins of colonial Latin American societies, such as those produced by millions of African slaves who arrived in Spanish and Portuguese America over the course of four centuries. The history of science of colonial Latin America, by and large, does not belong in the "non-Western world." The scientific practices and ideas that became dominant were those brought by Europeans as they strove to create stable, viable colonial societies. Portugal, however, failed to introduce learned institutions in Brazil until the early nineteenth century, when, in the wake of Napoleon's invasion, the crown fled Lisbon to settle in Rio de Janeiro. This chapter focuses primarily on the viceroyalty of New Spain. In the eighteenth century, Mexico produced most of the wealth Spain derived from its colonies. Mexican elites were wealthy and cosmopolitan; the cathedral cloister in Mexico City, for example, inaugurated in 1734 one of earliest standing orchestras of the "Western world."[3] Eighteenth-century Mexico is ideally suited for a study of the connections of science to Baroque culture (one characterized by an emblematic view of nature), to colonialism, and to nationalism.

EARLY INSTITUTIONS

In the early eighteenth century, the Spanish crown appeared to be saddled with highly autonomous colonial societies. Although for some two centuries the New World had supplied Europe with silver, sugar, and dyes, Spanish America was hardly a colonial outpost solely serving the needs of the core. The loose political structure of the Spanish empire and the reduced markets of the metropolis had conspired to turn the colonies into semi-independent entities ruled by viceroys and *audiencias* (high courts) always observant of the needs of local elites.

Like any other early modern European society, Spanish America was built on corporate privileges and social estates. Yet unlike other European societies, Spanish America had social estates that overlapped with additional racial and cultural hierarchies: African blacks were slaves; "Indians" were treated as

---

Cultural History of the Indians of Central Mexico, Sixteenth through Eighteenth Centuries (Stanford, CA: Stanford University Press, 1992). Serge Gruzinski has described the rise and fall of a hybrid Indo-Christian culture in the central valley of Mexico in the mid-sixteenth century. Observant Franciscan friars, using techniques derived from Renaissance humanism, trained a cadre of native classicists who acted as cultural translators. Native and friar humanists produced polyglot texts (Latin, Spanish, and Nahua), including a monumental encyclopedia of Nahua lore, the *Florentine Codex*, and a Nahua herbal, the *Libellus de medicinalibus Indorum herbis* (or *Codex Badianus*). The *Libellus* introduced Nahua glyphs and esthetic conventions into the genre of European herbals. See Serge Gruzinski, *La colonisation de l'imaginaire. Sociétés indigènes et occidentalisation dans le Mexique espagnol xvie–xviiie siècle* (Paris: Gallimard, 1988), chap. 1, especially 76–100, and Barbara E. Mundy, *The Mapping of New Spain: Indigenous Cartography and the Maps of the Relaciones Geográficas* (Chicago: University of Chicago Press, 1996).

[3] Craig A. Russell, personal communication.

peasant commoners and regarded for legal purposes as childlike members of a separate "republic"; and Spaniards and their descendants enjoyed special privileges and thought of themselves as patricians. *Castas* (mixed bloods) lived in the interstices of the original three-tier system, and as they blurred the carefully policed colonial racial boundaries they gave way to complex social taxonomies and intermediate groups of commoners.

By the early eighteenth century Spanish America had developed a set of institutions that fostered scientific activities: universities and colleges, cloisters, private libraries, pharmacies, and viceregal and ecclesiastical courts. Spanish America boasted some twenty universities and dozens of religious colleges. Chartered on the model of the medieval University of Salamanca, colonial universities were institutions controlled by religious orders (mostly Dominicans and Jesuits) that trained theologians, lawyers, and a few physicians on rigid neoscholastic curricula.[4] The Jesuits developed powerful educational institutions of their own that catered to the needs of local elites. Their support for philosophical eclecticism allowed the followers of St. Ignatius Loyola to introduce their charges to some innovative European thought, including experimental philosophy. The Jesuits subordinated science to their apostolic mission and created vertically integrated and technically efficient economic systems (for example, haciendas and plantations) to support their colleges and missions. The operation of their pharmacies is a case in point. Luis Martín has argued that in the seventeenth and eighteenth centuries the Jesuits ran from their college of San Pablo in Lima a network of pharmacies in various colonial cities. Italian and, later, German brethren were charged with making the pharmacies profitable, and they set up labs and collected and exchanged plants in hopes of identifying new remedies for trade. According to Martín, the Jesuit pharmacy in Lima held a monopoly on quinine and bezoar stones in European markets throughout the seventeenth and early eighteenth centuries, and the profits helped the order to maintain missions and colleges.[5] The Jesuits also subordinated cartography and natural history to the strategic needs of the order, which sought to expand post-Tridentine Catholicism to the frontiers of the colonial Spanish empire. From 1628 to 1767, the year in which they were expelled from all Spanish possessions, the Jesuits were the officially appointed cosmographers of the Indies.[6] The order promoted coordinated astronomical observations and the writing of natural histories.[7]

---

[4] John Tate Lanning, *Academic Culture in the Spanish Colonies* (London: Oxford University Press, 1940).

[5] Luis Martín, *The Intellectual Conquest of Peru: The Jesuit College of San Pablo, 1568–1767* (New York: Fordham University Press, 1968), pp. 97–118.

[6] Philip III appointed the Jesuits in 1628 as sole official cosmographers, consolidating the teaching of astronomy at the Imperial College of Madrid of the Jesuits. See "Expediente sobre la asignación de la cátedra de cosmógrafo en el Colegio Imperial de Madrid," August through October of 1760. Archivo General de Indias, Seville (herafter AGI), Indiferente General 1520.

[7] On coordinated astronomical observation by Jesuit cartographers, see Christian Reiger, "Memorial del cosmógrafo mayor al Consejo de Indias sobre limitaciones que tiene el cosmógrafo para ejercer las expectativas puestas sobre él deacuerdo al título," June 30, 1761, AGI, Indiferente General 1520.

Viceregal courts and religious cloisters offered alternative patronage systems to that of the universities, allowing some innovative philosophers to emerge. Baroque polymaths kept cabinets of curiosities and alchemical laboratories, and they were summoned by patrons to do astronomical observations and maps, to cast horoscopes, to design machines for courtly and public entertainment, to tend to the sick, and to help design sacred and secular public buildings. Carlos Sigüenza y Góngora best typifies the Spanish American Baroque polymath. Holder of the chair of mathematics at the University of Mexico, Sigüenza taught medical astrology while also working as a censor for the Inquisition and as chaplain for a local hospital. He drew maps, helped coordinate the works to drain the lake on which Mexico City was built, and led a surveying expedition to the borderlands (to the bay of Pensacola in Florida). Sigüenza was an accomplished scholar who kept a cabinet of curiosities and a telescope and a microscope and who did not fear to engage in heated debates with European astronomers. In 1690 Sigüenza published his *Libra Astronómica* to take on the German Jesuit Eusebio Francisco Kino, former professor at the University of Ingolstadt and leading missionary in California, for having espoused antiquated theories on the origin of comets. As Sigüenza sought to prove that comets were not earth exhalations, harbingers of disease, and omens of evil as the German and other astrologers had long argued, he showed familiarity with the writings of Kepler, Galileo, and Descartes. Finally, Sigüenza spent a great part of his scholarly life seeking to clarify Biblical chronologies through a detailed study of Mesoamerican calendars and codices.[8]

More often than not, Baroque polymaths were summoned to participate in the rituals of power which facilitated the smooth reproduction of social structures.[9] *Cabildos* (city councils), cloisters, viceroys, and prelates supported savants to design emblems for triumphal arches and funeral pyres, to build machines to impress the public in religious processions, to write and deliver commemorative sermons, and to uncover hidden "signatures" in Nature and religious images. It is no wonder therefore that Neoplatonic and hermetic currents enjoyed wide currency among the learned. The Neoplatonic writings and theories of the seventeenth-century authors Juan Caramuel, Athanasius Kircher, and Gaspar Schott held a particularly strong and lasting influence over the imagination of most Baroque Spanish American scholars.[10] In fact Kircher himself exchanged letters with many a Mexican scholar, to whom he sent academic advice, books, religious images, and mechanical toys in exchange for

---

[8] Irving A. Leonard, *Don Carlos de Sigüenza y Góngora* (Berkeley: University of California Press, 1929), and Eliás Trabulse, "La obra científica de Don Carlos de Sigüenza y Góngora 1667–1700," in Antonio Lafuente and José Sala Catalá (eds.), *Ciencia colonial en America* (Madrid: Alianza Universidad, 1992), pp. 221–52. Sigüenza was not different from other European Baroque polymaths; see Gunnar Eriksson, *The Atlantic Vision: Olaus Rudbeck and Baroque Science* (Canton, MA: Science History Publications, 1994).

[9] José Antonio Maravall, *La cultura del barroco* (sixth edition, Madrid: Ariel, 1996), and Irving A. Leonard, *Baroque Times in Old Mexico* (Ann Arbor: University of Michigan Press, 1959).

[10] Octavio Paz, *Sor Juana or the Traps of Faith* (Cambridge, MA: Harvard University Press, 1988).

curiosities, patronage, and dozens of pieces of Mexican prime chocolate (which the German Jesuit seems to have loved).[11] Kircher was so grateful that he dedicated one of his treatises on magnetism to one of his Mexican correspondents, Alexandro Favián.[12] Sigüenza, again, typifies the Spanish American Baroque polymath collaborating in the theatricalization of power. He wrote commemorative pamphlets on civic and religious events and designed emblems for triumphal arches. Drawing on cosmic metaphors, he readily represented secular and religious authorities as "suns" and "planets" of a hierarchical politico-social order. Sigüenza maintained correspondence with Caramuel, and his will revealed that the tooth of a "giant," a collection of Mesoamerican codices, and the works of Kircher were among his dearest possessions. Sigüenza donated his body to be dissected.[13]

PATRIOTIC, NEOPLATONIC, AND EMBLEMATIC DIMENSIONS

Colonial Baroque scholarly traditions were strongly colored by patriotism. As David Brading has shown, by the seventeenth century "Creole patriotism" had penetrated most learned circles in the colonies. Creoles, those born in the colonies of Spanish descent, saw themselves as being discriminated against by first-generation Spanish migrants. Creoles took the newcomers to be lowly commoners whose activity in commerce and mining had given them access to landed wealth and false claims to patrician origins. Creoles were in turn perceived by peninsular Spaniards as idle dilettantes who, either by astral influences or by cultural proximity to the natives, had turned into degenerate Indians. Although the Crown had sought to privilege peninsulars in colonial appointments, Creoles took advantage of Spain's chronic fiscal crisis to buy offices, and thus they were kept out of only the highest colonial posts. Creoles secured positions in the church and gained power over local ecclesiastical institutions and religious orders. Creoles also thought of themselves as a local landed nobility. But since colonial landed elites were never allowed to turn into feudal lords or grandees (making land an unprofitable and insecure investment), Creoles tended to be downwardly mobile and resented the success of Spanish newcomers, who paradoxically aspired to gentility by marrying into Creole families. As learned clergymen, Creoles wrote patriotic treatises praising the glories of the land, themselves, and their ecclesiastical establish-

---

[11] This correspondence has been collected and edited by Ignacio Osorio Romero in *La luz imaginaria: Epistolario de Atanasio Kircher con los novohispanos* (México: Universidad Autónoma de México [hereafter UNAM], 1993).

[12] Athanasius Kircher, *Magneticum naturae regnum* (Rome: Ignacio de Lazaris, 1667). Osorio has also reproduced Kircher's Latin dedication as well as excerpts from Kircher's *Magneticum* with references to the work and contributions of Favián to the study of magnetism. See *Luz imaginaria*, pp. 111–28.

[13] "Testamento de Don Carlos de Sigüenza y Góngora," in Francisco Pérez de Salazar, *Biografía de D. Carlos de Sigüenza y Góngora* (México: Antigua Imprenta de Murguía, 1928), pp. 170–2.

ments (cloisters, temples, universities).[14] Natural philosophy in the colonies was also patriotic.

Steeped in Neoplatonic and hermetic doctrines, the Creole clergy was constantly looking in nature for underlying hidden signatures with patriotic significance. For them the body, the earth, and the cosmos constituted Baroque "theaters" (in that objects were reduced to a language of images) interlocked by micro- and macrocosmic analogies.[15] All objects held polysemic meanings, and the exegetical skills of the clergy helped discover their underlying import, revealing a cosmos suffused with providential designs that favored the colonies. For example, Creole scholars concluded that astral phenomena that in Europe would have caused natural disasters in the Indies were benign. In 1638 the Augustinian friar Antonio de la Calancha argued that eclipses in Asia, Africa, and Europe were ominous signs; should they happen under Aries, Leo, and Sagittarius they would set off horrible aerial visions, harmful comets, and devastating fires; under Gemini, Libra, and Aquarius, they would trigger famines and epidemics. According to Calancha, Peru, however, had so many new stars that the land was under the domain of entirely different and auspicious "Zodiac" signs. The five-star constellation of *Cruzero* (The Southern Cross), for example, was at the southern pole, and its crosslike form kept away demons responsible for stirring up the waters. It was for this reason, he argued, that the South Seas were calm and received the name of "Pacific" Ocean. Calancha, in fact, thought that Peru was blessed; God had chosen to protect it by giving Peru not only crosslike constellations but also crosslike fossils, stones, and plants.[16] In Mexico things were slightly more complex, for demons appeared to be so entrenched that they have even shaped the landscape. In the 1690s Cristobal de Guadalajara, the mathematician and cartographer from Puebla, realized after drawing a map of the rivers and lakes of the central valley of Mexico that the hydrographic contours of the valley represented the head, body, tail, horns, wings, and legs of a Satanic beast (Figure 31.1).

---

[14] David Brading, *The First America: The Spanish Monarchy, Creole Patriots, and the Liberal State, 1492–1867* (Cambridge University Press, 1991).

[15] For examples of micro- and macrocosmic analogies, see Didaco Osorio y Peralta, *Principia medicinae epitome, et totius humanis corporis fabrica seu ex microcosmi armonia divinium, germen* (Mexico: Heredes viduae Bernardo Calderon, 1685). On emblematic-religious reading of natural objects, see Antonio de la Calancha, *Coronica moralizada del orden de San Agustín en el Perú* (Barcelona, 1638), pp. 57–59 (plant and flower of "granadilla" that resemble the symbols of the Passion – nails, sponge and lance, wounds, bindings, and crown of thorns – and that therefore induce pain and general malaise); and Agustín de Vetancurt, *Teatro mexicano*, 2 vols. (México, 1698), pp. 1:22–3 (on precious stones that reveal their medicinal value through the symbolism of their colors: those white cure milk-related diseases; those red cure blood ilnesses; those green with black spots stop bilious-hepatic attacks ["hijadas"]; those green with red spots cure intestinal bleedings; and, finally, those green with white spots help dissipate kidney stones), p. 38 (on a spring that turns to be medicinal because it produces crosslike stones), p. 42 (on bananas whose cores look like a crucified Christ), p. 51 (on "tlahulitucan" trees whose crosslike leaves keep demons away). On patriotic astrology, see Jorge Cañizares Esguerra, "New World, New Stars: Patriotic Astrology and the Invention of Indian and Creole Bodies in Colonial Spanish America, 1600–1650," *American Historical Review*, 104 (February 1999), 33–68.

[16] *Coronica moralizada*, pp. 48–50, 58–9.

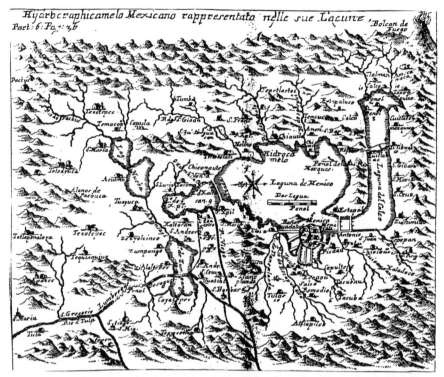

Figure 31.1. "Hydrographicamelo Mexicano reppresentato nelle sue Lacuna (hydro-camel represented in the lake of Mexico)." From Gemelli Careri, *Giro del Mundo* (Napoles, 1700). The map was first drawn in 1618 by Adrián Boot, a Flemish engineer hired to oversee the works leading to the drainage of the central valley of Mexico. The figural reading of the map, however, appears to be Guadalajara's. Along with the map, Gemelli published cabalistic readings of the names of the ten Aztec monarchs that proved that the added sum of their names was 666, the number of the beast.

To counteract the kingdom of darkness of the Aztecs, God had fortunately sent Spanish conquerors and the Virgin Mary to set the Indians free. The Virgin – who had appeared to a Nahua commoner, Juan Diego, in Tepeyac in 1531, leaving her image stamped on his cape – held the key to understanding Mexico's destiny. The image, a variation of a rather common European representation of the Immaculate Conception, had a Virgin standing on the moon; surrounded by sun rays while eclipsing the sun; wearing a blue, starred, heavenly shawl; and held up by an angel. In 1648, Miguel Sánchez argued that the description of the woman of the Apocalypse (Rev. 12:1–9) (pregnant with a would-be Messiah; clad with stars; persecuted by a multiheaded dragon, the devil, who wanted her to abort; protected by God, who sent an army of angels headed by the archangel Michael to destroy the dragon) was a prefiguration of Our Lady of Guadalupe, who had routed the kingdom of darkness of the

Aztecs. Sánchez offered interpretations of every detail of the image: the moon underneath the Virgin represented her power over the waters; the Virgin eclipsing the sun stood for a New World whose torrid zone was temperate and inhabitable; the twelve sun rays surrounding her head signified Cortés and the conquistadors who had defeated the dragon; and the stars on the Virgin's shawl were the forty-six good angels who had fought Satan's army (Sánchez used cabala to calculate the number of good angels).[17] Sánchez thought that the image was the most important icon in Christendom, and his interpretation inaugurated a literature of exegesis in which contemporary Mexicans appeared as God's new elected people.[18] Throughout the seventeenth and eighteenth centuries, Baroque Creole scholars debated the meanings of the image and reached conclusions such as that it had dominion over the sphere of water because the Virgin was standing over the moon, which, in turn, was related to tides, floods, and droughts.[19] Every time Mexico City was submerged in the waters (and that happened often), thousands of anguished citizens took to the streets to parade her image, and to their relief the waters always subsided.[20] If properly understood, religious images could be deployed at cardinal points of cities and towns to act as symbolic "fortifications" (*baluartes*), to fend off sublunar, evil "intelligences" capable of causing natural calamities (Figure 31.2).[21]

Latin American historians of science have not given proper attention to the emblematic and Neoplatonic dimensions of colonial science and have misread key colonial Baroque texts. Elías Trabulse, a prolific Mexican historian of science, has revealed the figure of the Mercederian Diego Rodríguez

---

[17] Miguel Sánchez, "Imagen de la Virgen María Madre de Dios de Guadalupe" [1648], *Testimonios históricos guadalupanos* (México: Fondo de Cultura Económica [hereafter FCE], 1982), p. 168 (on the crown of twelve stars); p. 219 (on the eclipsed sun); pp. 223–4 (on the moon); pp. 226–7 (on the forty-six stars).

[18] Brading, *The First America*, chap. 16, and Jacques Lafaye, *Quetzalcóatl et Guadalupe* (Paris: Editions Gallimard, 1974).

[19] Dozens of treatises appeared in the seventeenth and eighteenth centuries seeking to uncover the hidden meanings of the image. For example, some scholars thought that an eightlike figure on the robe of the Virgin signified that the image was the eighth wonder of the world. Others thought that the mark was in fact a Syrian-Chaldean character that, along with other "Oriental" symbols in the picture, indicated that St. Thomas had come to Mexico with the image in the first century A.D. after having preached the gospel in the Orient. See Mariano Fernández de Echeverría y Veytia, *Baluartes de México* (written ca. 1778; Méjico: Alejandro Valdés, 1820), p. 12; Miguel Cabrera, "Maravilla Americana" (1756), in *Testimonios históricos guadalupanos*, p. 519; Ignacio Borunda, "Clave general de geroglíficos Americanos" (c. 1792) in Nicolás León (ed.), *Biblioteca mexicana del siglo XVIII* (Boletín del Instituto Bibliográfico Mexicano, 7, Mexico, 1906), sección primera, tercera parte, pp. 276–7; Servando Teresa de Mier, "Sermón predicado en la Colegiata el 12 de diciembre de 1794," *Obras completas: El heterodoxo*, 4 vols. (México: UNAM), 1:249–50.

[20] Francisco de la Maza, *El guadalupanismo mexicano* (reprint 1953 edition; México: FCE y Secretaría de Educación Pública, 1984), pp. 43–5, 177.

[21] Francisco Florencia, "Estrella del norte de México" (1688), *Testimonios históricos guadalupanos*, pp. 394–5. See also Juan José de Eguiara y Eguren, "Panegírico de la Virgen de Guadalupe" (1756), ibid., p. 487; Cayetano de Cabrera y Quintero, *Escudo de armas de México: Celestial protección de la Nueva España y de casi todo el Nuevo Mundo* (México: Joseph Bernardo de Hogal, 1746); and Echeverría y Veytia, *Baluartes de México*.

Figure 31.2. Frontispiece to Cabrera y Quintero's *Escudo de Armas de México* (México, 1746). The image of Our Lady of Guadalupe is held up in the air by *putti* and acts as a shield protecting Mexico from negative heavenly influences, reportedly the underlying cause of the epidemics of *matlazahuatl* that hit Mexico between 1736 and 1738. The image of the Virgin acts as a "fortification" to keep evil intelligences away.

as a seminal Creole natural philosopher, the first holder of the chair of mathematics at the University of Mexico and the author of highly sophisticated mathematical treatises. The merit of Trabulse's work on Rodríguez is that it puts to rest the rather popular construct that Catholic colonial Spanish

America was intellectually barren, a land choked by the Inquisition. Trabulse has presented Rodríguez as a "modern" whose treatise on the comet of 1652 sought to do away with "superstitious" beliefs that held that comets were harbingers of evil.[22] Trabulse has insisted correctly that Rodríguez showed acquaintance with the works of Kepler, Galileo, and Descartes. But Rodríguez was a Creole patriot engaged in Neoplatonic and emblematic readings of Nature. In fact, Rodríguez's treatise shows that his rejection of comets as ominous signs stemmed from his repudiation of a common idea held by Europeans that the constellations of the New World were different from those of the Old World and caused biological degeneration; his patriotic belief that Mexico's skies were protected by the Immaculate Conception; and his conviction that "there is no sign [over the skies of the viceroyalty of New Spain] that, although shocking and surprising for the ignorant, does not serve the Queen of Heaven and [help] explain her glories."[23] The basic assumption that moved Rodríguez to deny that the comet of 1652 was a harbinger of disease and death was that its path through the Zodiac revealed its symbolic associations with the Immaculate Conception. The comet had moved through the constellations of Noah's dove and Medusa. The constellation of the dove stood for purity not unlike that of Mary, whose immaculate conception had spared her a post-lapsarian human nature; that of Medusa, on the other hand, represented the dragon that had sought to kill the pregnant Virgin. Rodríguez assumed therefore that the comet and the Virgin Mary were symbolically linked. Rodríguez also presupposed that since the image of the Immaculate Conception was a Virgin eclipsing the sun, eclipses as well as any other heavenly phenomena in Mexico, a land under the protection of Our Lady of Guadalupe, could be harbingers only of joyous news. According to Rodríguez, the comet of 1652 would make Mexican leaders wiser because it had passed across Mars, a planet that stood for wisdom. To be sure, his argument for the benignity of the comet was not sustained solely on the learned exegesis of nature and religious images. Based on the allegorical interpretation of classical mythology as physical events and on close knowledge of contemporary astronomy, Rodríguez sought to prove that comets were supralunar phenomena and therefore had no negative physical effects on the sublunar world. According to Rodríguez, the sub- and supralunar worlds were qualitatively different.

Emblematic readings of nature lasted well into the eighteenth century. In 1742, for example, the Augustinian friar Manuel Ignacio Farías sought to explain the cause of a bout of *matlazahuatl* (plague) that had killed thousands of indigenous peoples of Michoacán. Like Rodríguez, Farías thought that the image of the Immaculate Conception and heavenly phenomena in the

---

[22] Elías Trabulse, *La ciencia perdida: Fray Diego Rodríguez, un sabio del siglo XVII* (México: FCE, 1985).

[23] Diego Rodríguez, *Discurso etheorológico del nuevo cometa* (México: Viuda de Bernardo Calderón, 1652), fol. 4v. Trabulse has reproduced the more "modern" parts of Rodríguez's *Discurso etheorológico* in Trabulse (ed.), *Historia de la ciencia en México (edicion abreviada)* (México: FCE y Consejo Nacional de Ciencia y Tecnología, 1994), pp. 324–37.

Americas were symbolically interwoven and that the skies of Mexico were under the special patronage of the Virgin. Empirical evidence, however, proved
that the epidemics had occurred in the wake of a solar eclipse. If it was true that
the image of Our Lady of Guadalupe protected Mexico, the recent eclipse that
had been observed in Michoacán should have never triggered such natural
disaster, for the image showed the Virgin eclipsing the sun. Calancha and Rodríguez had already argued that eclipses in the New World could be harbingers only of joyous news. Thus, Farías thought that the recent event was an
anomaly that had been prefigured in the Bible. Interpreting words of King
David as predicting that those standing to the right of the Temple would never
be protected, Farías maintained that the indigenous peoples of northern Michoacán lived to the "right" of the shrine of Guadalupe in Valladolid (capital
of Michoacán), for the image of Our Lady of Guadalupe in the temple was
facing west to replicate in a larger scale the message of the painting (that the
image should "eclipse" the sun at sunrise to act as a protective shield for the
city). Given the logic of Farías's cosmology, those who "stood" to the "right"
of the image were the Indians of the north.[24]

These kinds of emblematic, sacred, and patriotic readings of natural phenomena were often used in medical treatises. Eighteenth-century Creole
doctors concluded, for example, that pulque, a liquor from the maguey plant,
was a panacea not simply because clinical evidence indicated so but also because maguey was the plant from which was obtained the fabric on which
the miraculous image of Our Lady of Guadalupe had been stamped; the Virgin had subtly let medical doctors know about the divine virtues of maguey.[25]
Many of these physicians thought that the virtue of plants could also be identified by studying Nahua etymologies. In 1746 Cayetano Francisco de Torres
argued that pulque was a universal remedy (*polychresto*) because it had symbolic-
material association with the image of Our Lady of Guadalupe and because
the Aztec name of the plant ("teometl") said so clearly, "divine plant."[26]

The idea that Nahuatl and Quechua, the tongues of the Inca and the Aztec,
were "Adamic" languages exerted an important attraction over the imagina-

---

[24] Manuel Ignacio Farías, *Eclypse del divino sol causado por la interposición de la immaculada luna* (México:
Maria de Rivera, 1742).

[25] Francisco Fuentes y Carrión, "Discurso sobre las virtudes del pulque" (1733), Biblioteca Nacional de
México (hereafter BNM), Ms. 1540.

[26] Cayetano Francisco de Torres, "Virtudes maravillosas del pulque, medicamento universal o polychresto" (1748), BNM, Ms. 23, fol. 1–16. Colonial physicians used etymology to identify medical
virtues. In a heated debate that took place in Mexico City in 1782 over the efficacy of live lizards for
curing tumors, the Protomédico (a learned physician charged by the crown with the regulation of
medical practice in the city) Joseph Giral Matienzo supported José Vicente García de la Vega. García
de la Vega had claimed that lizards could cure not only cancer but many other illnesses as well, including pulling out splinters by means of hidden sympathies. Giral Matienzo argued that García de
la Vega had understood that lizards were "robust" remedies, able to cure many diseases, because he
had identified the meaning of its "hieroglyph," namely, the hidden Latin etymology for lizard ("lac-
ertus," a synonym for "robur," robust). See Joseph Giral Matienzo, "Aprovación," in José Vicente
García de la Vega, *Discurso crítico sobre el uso de las lagartijas, como específico contra muchas enfermedades*
(México: Felipe de Zuñiga y Ontiveros, 1782), n.p.

tion of Creole scholars of Spanish America throughout the entire colonial period. In their efforts to claim for their homelands the status of independent kingdoms, endowed with prestigious genealogies and loosely affiliated with a universal Spanish monarchy, some Creoles converted the indigenous past into their own classical antiquity. Many Creole scholars therefore were willing to compare the languages of the Aztecs and Inca favorably to Greek and Latin. And some naturalists even claimed that Nahua and Quechua taxonomies pointed to the essences of plants.

The naturalist Francisco Hernández, commissioned by Philip II to identify the properties of plants in the New World, was perhaps the first to suggest, circa 1574, that Nahuatl, the language of the Aztecs, was an Adamic language. He insisted that, like the ancient Hebrews, Mesoamerican peoples had named things after their essences. He expressed surprise that nations he thought were barbarous could have developed such sophisticated languages.[27] In 1637, on the advice of the holder of the chair of Quechua (the Inca language), Alonso Huerta, the cloister of the University of Lima, turned down a proposal to open a new chair of medicine devoted to botanical studies on the grounds that physicians should rather study Quechua. Peruvian Indians, the cloister argued, had already discovered the property of plants and had named them after their virtues.[28] In the eighteenth century some Peruvian Creole naturalists insisted that new educational institutions devoted to the study of Nature in the colonies should educate students in Quechua,[29] and we know that important late eighteenth-century naturalists in Lima, such as Juan Tafalla and Francisco González Laguna, who were deeply involved in the Royal Botanical Expedition to Peru (1777–1808), also sought to promote the publication of Quechua grammars.[30] In Colombia and Mexico, prestigious Creole naturalists such as Francisco José de Caldas and José Antonio de Alzate openly praised the value of Quechua and Nahua taxonomies.[31] Finally, the Spaniard Martín de Sessé, head of the Botanical Expedition to Mexico (1787–1803), made it his priority to learn Nahuatl on the assumption that the language of the Aztecs was "an elegant language [in which the] names of plants signified their virtues."[32]

[27] Francisco Hernández, *Antiguedades de Nueva España* (Madrid: Historia 16, 1986), pp. 128, 147.

[28] Libro de Claustros, University of San Marcos, 1637, quoted in Hipólito Unanue, "Introducción a la descripción científica de las plantas del Perú," *Mercurio Peruano*, 2 (1791), 71.

[29] Letter of José Eusebio de Llano Zapata to the Marquis of Villa Orellana, ca. 1761 in Llano Zapata, *Memorias histórico-físicas-apologéticas de la América Meridional* (Lima: Imprenta y Librería de San Pedro, 1904), p. 595. Llano Zapata proposes the creation of a "College of Metallurgy" to train students in experimental philosophy, mathematics, geometry, mechanics, and natural history, as well as in Italian, French, German, Greek, Latin, and Quechua.

[30] Joseph Manuel Bermúdez, "Discurso sobre la utilidad e importancia de la lengua general del Perú," *Mercurio Peruano*, 9 (1793), 178–9.

[31] Francisco José de Caldas, "Prefación a la geografía de las plantas de Humboldt," *Obras completas* (Bogotá: Universidad Nacional de Colombia, 1966), and José Antonio de Alzate, *Linneo en México*, ed. Roberto Moreno (México: UNAM, 1989), p. 25.

[32] Letter of Sessé to Casimiro Gómez Ortega, 15 January 1785, reproduced in Xavier Lozoya, *Plantas y*

IN SERVICE TO CROWN AND COMMERCE

By the mid-eighteenth century, these Baroque, hierarchical yet fiercely patriotic societies were changed by forces outside their control. As Spain embarked on a project of economic and cultural renewal, the crown set out to revitalize Spanish commerce and to reform outdated colonial policies. This new set of mercantilist policies were designed to regain for Spain the status of Continental power and to prove to the rest of Europe that Spain was not intellectually barren, as Western European sarcasm had long maintained. To work, the program of reform needed to turn overseas territories into colonial dependencies, politically and economically subservient to the needs of the metropolis. It also required that science be enlisted in the service of the state and the new economy.[33] Science would provide the crown with new secular discourses of political legitimacy to undermine the power of the church.[34] Natural history, experimental philosophy, astronomy, and cartography would help the crown to exploit botanical and mineral resources and to regain control over loosely controlled frontiers and borderlands.[35] Under the aegis of Charles III and using a revitalized navy, the crown sent numerous expeditions staffed by Europeans to America to draw accurate maps of territories being lost to England, Portugal, France, and Russia.[36] It also sent botanical expeditions to help identify botanical resources to establish new commercial monopolies.[37] Much energy was wasted by numerous bureaucrats and naturalists bent on finding in tropical America cloves, cinnamon, and tea to challenge the monopoly held by Dutch and British merchants.[38] Spanish

---

luces en México: La real expedición científica a Nueva España (1787–1803) (Barcelona: Ediciones del Serbal, 1984), p. 30.

[33] Richard Herr, *The Eighteenth-Century Revolution in Spain* (Princeton, NJ: Princeton University Press, 1958).

[34] On the rise of a new cultural authority for science linked to the search for new forms of political legitimacy, see Dorinda Outram, *The Enlightenment* (Cambridge University Press, 1995), pp. 47–62, 96–113.

[35] This pattern of subordination of colonial science to mercantilist policies has also been identified in eighteenth-century Haiti by James E. McClellan III in *Colonialism and Science: Saint Domingue in the Old Regime* (Baltimore, MD: John Hopkins University Press, 1992).

[36] Horacio Capel, *Geografía y matemáticas en la España del siglo XVIII* (Barcelona: Oikus-Tau, 1982). The largest and most significant of these cartographic expeditions was that headed by Alexandro Malaspina; the expedition also included numerous naturalists and painters. See Iris H. W. Engstrand, *Spanish Scientists in the New World: The Eighteenth-Century Expeditions* (Seattle: University of Washington Press, 1981); Virginia González Claverán, *La expedición científica de Malaspina en Nueva España (1789–1794)* (México: El Colegio de México, 1988); and Juan Pimentel, *La física de la monarquía: Ciencia y política en el pensamiento colonial de Alejandro Malaspina* (Madrid: Ediciones Doce Calles, 1998).

[37] The literature on eighteenth-century botanical expeditions has witnessed an explosion. Some representative titles are Arthur Steele, *Flowers to the King* (Durham, NC: Duke University Press, 1964); Lozoya, *Plantas y luces en México*; and Marcelo Frías Nuñez, *Tras El Dorado vegetal: José Celestino Mutis y la Real Expedición Botánica del Nuevo Reino de Granada* (Sevilla: Diputación de Sevilla, 1994). For a survey of recent historiography, see Miguel Angel Puig Samper and Francisco Pelayo, "Las expediciones botánicas al nuevo mundo durante el siglo XVIII: Una aproximación histórico-bibliográfica," in *La Ilustración en América colonial*, pp. 55–65.

[38] Francisco Javier Puerto Sarmiento, *Ciencia de cámara: Casimiro Gómez Ortega (1741–1818) el científico cortesano* (Madrid: C.S.I.C., 1992), pp. 148–209; and Frías Nuñez, *Tras El Dorado vegetal*, pp. 159–250.

botanical expeditions in America also sought to tap the continent's vast and unknown pharmaceutical resources to find "a panacea for the diseases of the century."[39] In 1777, Casimiro Gómez Ortega, the architect of Spanish scientific expeditions to the New World, promised José de Gálvez, minister of the Indies, that "twelve naturalists . . . spread over our possessions will produce in their expeditionary pilgrimages a profit incomparably greater than could an army of 100,000 strong fighting to add a few provinces to the Spanish empire."[40] Finally, the crown also sent parties of Spanish and German chemists, geologists, and mining experts charged with improving the production of silver and mercury.[41]

Spain obtained access to the new European sciences by sending students abroad, by hiring foreign technicians and savants, and by creating new institutions of learning at home and in the colonies.[42] The crown sought to reform universities as a means to reign in the church. Many of the educational institutions of the Jesuits were dismantled when the order refused to embrace the new regalist and Jansenist principles of state-church relation and was forced out of all Spanish territories in 1767. The crown seized the momentum of the Jesuit expulsion to undermine the autonomy of all universities and to introduce reforms. Yet the universities did not become sites where the New Science set roots, for the cloisters understood correctly that the new learning was associated with attempts to undermine their corporate privileges and to terminate the clerical monopoly over learned institutions.[43] Eighteenth-century

---

[39] Puerto Sarmiento, *Ciencia de cámara*, p. 174. When compared to the colonial botanical agendas of other European powers the Spanish emphasis on the search for pharmaceuticals in the tropics appears exaggerated. On the eigtheeenth-century colonial botanical agendas of Britain and Sweden, see David Miller and Peter Hanns Reill (eds.), *Visions of Empire: Voyages, Botany, and Representations of Nature* (Cambridge University Press, 1996). On France, see McClellan, *Colonialism and Science*, pp. 111–15, 147–62.

[40] Quoted in Puerto Sarmiento, *Ciencia de cámara*, pp. 155–6.

[41] A. P. Whitaker, "The Elhuyar Mining Missions and the Enlightenment," *Hispanic American Historical Review*, 31 (1951), 557–85, and Modesto Bargalló, *La minería y la metalurgia en la América española durante la época colonial* (México: FCE, 1955).

[42] Italian and French court physicians in the entourage of the Bourbons also played an important role in Spain's scientific renewal; see J. Riera, "Médicos y cirujanos extranjeros de cámara en España del siglo XVIII," *Cuadernos de historia de la medicina española*, 14 (1975), 87–104. Italian and French doctors arrived in the colonies as court physicians for viceroys and prelates, introducing Newtonian and iatromechanical ideas early in the eighteenth century; see, for example, Federico Bottoni, *La evidencia de la circulación de la sangre* (Lima, 1723), reproduced in Alvar Martínez Vidal (ed.), *El nuevo sol de la medicina en la ciudad de los reyes* (Zaragoza: Comisión Aragonesa Quinto Centenario, 1990), and Juan Blas Beaumont, *Tratado de la agua mineral caliente de San Bartholomé* (México: Joseph Antonio de Hogal, 1772). Bottoni arrived in Spain with the entourage of Isabel de Farnecio, second wife of Philip V. Bottoni worked in Lima as court physician for two viceroys and for a Franciscan prelate, and he introduced Peruvian doctors to Harvey's theory of circulation. Juan Blas Beaumont, Latinist surgeon, holder of the chair of anatomy at the University of Mexico, and in the retinue of the archbishop Francisco Antonio de Lorenzana, was in all likelihood the son of Blas Beaumont, a French surgeon at the court of Philip V who contributed to the renovation of early eighteenth-century Spanish medicine.

[43] John Tate Lanning has sought to give Spanish American colonial universities their due by claiming that they played a key role in the eighteenth-century cultural renewal; see Lanning, *The Eighteenth-Century Enlightenment in the University of San Carlos de Guatemala* (Ithaca, NY: Cornell University

modern Spanish science, as Antonio Lafuente and José Luis Peset have maintained, grew under the patronage of the military.[44] The army and the navy
created a set of alternative establishments to the universities where state-of-
the art sciences were taught (including mechanical philosophy, Newtonian
physics and the calculus, and experimental philosophy). The military set up
hospitals in Spain and in the colonies for learned surgeons, mathematical
academies in Spain for gunners and engineers, and an observatory in Cadiz
for cartographers. The military also helped support a network of botanical
gardens to acclimatize tropical plants that might prove useful for new commercial ventures. But the crown also sponsored new institutions that were
independent from the military. Academies of art were created in Spain and
Mexico to train masons, architects, textile designers, and botanical illustrators
in neoclassical taste.[45] The Royal Academy of History (1738) in Madrid and the
Archive of the Indies (1784) in Seville were founded to bring critical methods
to bear not only on the study of historical documents but also on geography
and natural history (the Academy of History had as one of its most important
responsibilities the writing of "critical" natural histories). And medical academies, the Royal Botanical Garden (fl. 1774–88), and a revived *protomedicato*
worked in Spain and in the colonies to standardize the training of healers, to
undermine guilds of apothecaries and surgeons, to check quacks and midwives
(and "shamans" in the case of the colonies), and to improve the lowly status
of physicians. Like other European monarchies, the Spanish crown linked
mercantilism to medical reform in the hopes of augmenting the population,
and it avidly endorsed the massive use of smallpox immunization at home
and in the colonies. The crown also supported the development of a "public" sphere all over the empire and promoted salons and patriotic societies to
help disseminate a new utilitarian learning.[46]

Press, 1956). Yet the evidence against Lanning's thesis seems to be overwhelming; see, for example,
Enrique González, "El rechazo de la Universidad de México a las reformas ilustradas (1763–1777),"
in *Estudios de historia social y económica de América* (Alcalá), 7 (1991), 94–124; Marc Baldó, "La Ilustración en la Universidad de Córdoba y el Colegio de San Carlos de Buenos Aires (1767–1810)," ibid.,
31–54 ; Antonio E. Ten, "Ciencia y universidad en la América hispana: La Universidad de Lima," in
Ciencia colonial en América, pp. 162–91; Diana Soto Arango, "La enzeñanza ilustrada en las universidades de América colonial: Estudio historiográfico," in D. Soto Arango, Miguel Angel Puig Samper, and Luis Carlos Arboleda (eds.), *La Ilustración en América colonial* (Madrid: Doce Calles, CSIC,
and Colciencias, 1995), pp. 91–119. For a brief overview of the resistance of Spanish universities to
change, see Mariano and José Luis Peset, "La renovación universitaria," in Manuel Sellés, José Luis
Peset, and Antonio Lafuente (eds.), *Carlos III y la ciencia de la Ilustración* (Madrid: Alianza Universidad, 1992), pp. 143–55.

[44] Antonio Lafuente and José Luis Peset, "Las academias militares y la inversión en ciencia en la España
ilustrada (1750–1760)," *Dynamis*, 2 (1982), 193–209. See also Horacio Capel, Joan-Eugeni Sánchez,
and Omar Moncada, *De Palas a Minerva: La formación científica y la estructura institucional de los
ingenieros militares en el siglo XVIII* (Barcelona: CSIC y Ediciones Serbal, 1988).

[45] The Academy of Art of San Carlos was founded in Mexico in 1783 to teach artisans and masons, among
other things, optics and mathematics; see Thomas Brown, *La Academia de San Carlos en Nueva España*
(México: Septentas, 1976).

[46] For current scholarship on the new Spanish scientific institutions described in this paragraph, see
the articles in *Carlos III y la ciencia de la Ilustración*.

But the mercantilist project proved unsuccessful in the long run. The French Revolution sent chills down the spine of the Spanish monarchy, whose active involvement in the conformation of a public sphere in Spain was suddenly checked, freezing most initiatives for cultural renewal. Continuous war against England threw a monkey wrench in the efforts of the crown to reorganize colonial trade. The frustrated policies of economic and cultural renewal did not leave colonial science unscathed. Francisco Puerto Sarmiento has described how the institutions created by the crown to exploit colonial botanical resources slowly deteriorated, leaving Spain with little to show after some thirty years of official, unparalleled patronage of naturalists. Botanical gardens set up to acclimatize tropical plants were shut down or never used; the efforts to standardize the production of quinine to take advantage of Spain's secular monopoly on the febrifuge failed; production of cinnamon and cloves never succeeded; and most ironically, the patriotic evidence that could have at last proven to the rest of Europe that Spain was indeed "modern" (treatises of taxonomies, travel accounts, maps, and thousands of botanical and anthropological illustrations accumulated by dozens of scientific expeditions) remained, by and large, unpublished.[47] The institutions and expeditions created to increase the productivity of silver and mercury mines at Mexico and Peru failed to introduce significant technological changes; and although production of silver grew manifold, it resulted largely from the extensive use and overexploitation of manual labor.[48] Moreover, as Juan José Saldaña has demonstrated, the Colegio de Minería (college of mining), established in Mexico by the crown in 1792 to churn out bureaucrats and miners enlightened on subjects such as subterranean geometry, geology, and chemistry frequently was not funded; it remained a vibrant institution only because the local elites chose to keep it alive.[49] Medical reform in the colonies did not improve public health significantly, nor did reform elevate the social status of physicians.[50]

But despite all these failures, the cultural milieu of the colonies was forever changed in the wake of the Bourbon efforts at imperial renewal. A public sphere, for one, was initiated. Newspapers, periodicals, salons, cafes, and patriotic societies appeared everywhere, divulging new European thought and the gospel of utilitarian knowledge. Eventually this dynamic public sphere was to demand from the crown new forms of democratic political

---

[47] F. J. Puerto Sarmiento, *La ilusión quebrada: Botánica, sanidad y política científica en la España ilustrada* (Barcelona: Serbal, 1988).

[48] Kendall Brown, "La recepción de la tecnología minera española en las minas de Huancavelica, siglo XVIII," *Saberes andinos*, ed. Marcos Cueto (Lima: Instituto de Estudios Peruanos, 1995), pp. 59–90; Carlos Contreras and Guillermo Mira, "Transferencia de tecnología minera de Europa a los Andes," in *Mundialización de la ciencia*, pp. 235–49.

[49] Juan José Saldaña, "Ilustración, ciencia y técnica en América," in *La Ilustración en América colonial*, pp. 19–49, especially pp. 43–6.

[50] J. T. Lanning, in John Jay TePaske (ed.), *The Royal Protomedicato: The Regulation of the Medical Professions in the Spanish Empire* (Durham, NC: Duke University Press, 1985).

participation.[51] "Newtonianism" and mechanical philosophy permeated most public discourse. Even opponents of the New Science (who correctly perceived that this new learning was developing at the margin of clerical institutions, contributing to the secularization of society and to demise of scholastic theology) studied it.[52] The rhetoric of experimentation and the fad of collecting cabinets of experimental apparatus overtook the learned.[53] Desiderio de Osasunasco set up a most elaborate experimental plan in the 1780s to discover the cause of his own intolerance to chocolate, subjecting himself to systematic and painful self-experimentation.[54] The doctor Juan Manuel Venegas thought little of the natives, whom he considered ignorant savages, yet he wrote a medical treatise on indigenous herbal lore on the assumption that savages knew things through trial and error, that is, through hard-won experimentation.[55] Learned medical debates were conducted by exalting the authors' own experimental authority and by undermining the reliability and lack of experimental work of the authors' opponents. The language of metaphors and emblems of the Baroque lost its appeal. The works on geology by Francisco Xavier de Orrio, a mid-eighteenth-century Spanish Jesuit teaching at Zacatecas, and by Andrés Ibarra Salazán, a student at Mexico City's college of mining at the turn of the nineteenth century, typify this cultural change. Whereas for Orrio the Earth was a recently created organic macrocosm in which mercury was transmuted into gold and in which hidden sympathies made stones resemble flora and fauna, for Ibarra Salazán the Earth had a long

---

[51] On the public sphere in the colonies, see Jaime E. Rodríguez O, *La independencia de la América española* (México: FCE and Colegio de México, 1996), pp. 58–63 and passim; Renán Silva, *Prensa y revolución a finales del siglo XVIII* (Bogotá: Banco de la República, 1988); and Jean-Pierre Clément, "L'apparation de la presse periodique en Amérique espagnole: Le cas du 'Mercurio Peruano,'" in *L'Amérique espagnole à l'epoque des lumières: Tradition, innovation, représentation.* Colloque franco-espagnol du CNRS, 18–20 Septembre 1986 (Paris: Editions du CNRS, 1987), pp. 273–86.

[52] Francisco Ignacio Cígala, *Cartas al Ilmo, y Rmo P. Mro. F. Benito Gerónymo Feyjoó Montenegro. Carta Segunda* (México: Imprenta de la Biblioteca Mexicana, 1760). Cígala, with the approval of the dean of the University of Mexico, Juan José Eguiara y Eguren, and of a leading Mexican Jesuit, Francisco X. Lazcano, took Feijoó, a Benedictine friar largely responsible for the popularization of Newton and Descartes in early eighteenth-century Spain, to task for challenging Aristotle and scholastic theology. Cígala, Eguiara y Eguren, and Lazcano, however, were not blindly holding to the past. They chastised the moderns for claiming to have created a new philosophy when it had already been developed by the ancients. Cígala criticized Feijoó from a position of strength, revealing Feijoó's lack of understanding of the mechanics of air and of the writings of Boyle and Leibniz. For an unsympathetic reading of Cígala as throwback reactionary, see Pablo González Casanova, *Misoneismo y modernidad cristiana en el siglo XVIII* (México: Colegio de México, 1958), pp. 114–29. On Newton in colonial Spanish America, see Luis Carlos Arboleda, "Acerca del problema de la difusión científica en la periferia: el caso de la física newtoniana en la Nueva Granada," *Quipu* (Revista Latinoamericana de la Historia de la Ciencia y la Tecnología), 4 (1987), 7–32; and Celina A. Lértora Mendoza, "Introducción de las teorías newtonianas en el Río de la Plata," in *Mundialización de la ciencia,* pp. 307–23.

[53] The pages of *Gazetas de México* carry descriptions of at least ten cabinets in Mexico City by 1790. See *Gazetas,* August 1790, vol. 4, 16, pp. 152–4.

[54] Desiderio de Osasunasco, *Observaciones sobre la preparación y usos del chocolate* (México: Felipe de Zúñiga y Ontiveros, 1789).

[55] Juan Manuel Venegas, *Compendio de la medicina: o medicina práctica* (México: Felipe Zúñiga y Ontiveros, 1788).

history revealed in layers of rock and fossils that were documents that spoke about slow processes of geological change.[56]

TRAVELERS AND CULTURAL CHANGE

The remarkably different fates of two foreign scientific expeditions in Spanish America speak volumes about the profound changes in cultural outlook that were undergone in the second half of the eighteenth century by the Spanish American elites. From 1735 to 1745 an expedition headed by three French academicians – Pierre Bouguer, Louis Godin, and Charles-Marie de La Condamine – stayed in the Andes to measure on the Equator three degrees of a meridian's arch in order to settle the Newtonian-Cartesian debate over the shape of the globe. The expedition proved an unmitigated disaster for the French. The much-publicized expedition arrived back in Paris eight years too late to have any impact on the final resolution of the debate. Seeking to do away with charges of scholarly incompetence, La Condamine wrote an account of the travails of the expedition that is both painful and hilarious; more important, it shows that the French worked amid a hostile population. The expedition was not only besieged by accidents (such as instruments lost to strange meteorological phenomena and philosophers battered and killed by debilitating "tropical" fevers) but also doomed, for from the start the locals greeted it with open hostility. The Indians, portrayed by La Condamine as submissive and stupid, either destroyed, stole, or moved the signposts for trigonometric calculations built by the French on mountain summits. The Indians also systematically refused to work as scouts for the Europeans, and when they agreed to become guides they often fled and left the academicians stranded in the most rugged territories (Indian porters managed to get La Condamine's luggage lost twice). Blacks and *castas* were no more sympathetic. To the eyes of the French they were an unruly "plebe" who, in open defiance of European decorum, carried swords with which they stabbed a servant of the expedition. The French did not find in the white elites shelter from the hostility of Indians, black slaves, and mestizo plebeians. Although warmly welcomed by "enlightened" Jesuits and by a select handful of scholarly Creoles, the academicians often faced the rage of both imperial and provincial white authorities (Figure 31.3). For years the French had to fight in court charges of engaging in illegal trade and of building unauthorized commemorative monuments. The monuments had self-congratulatory inscriptions that left out the Spaniards. Finally, in the wake of a popular riot led by the provin-

---

[56] F. Xavier Alexo de Orrio, "Metalogía o physica de los metales," BNM, Ms. 1546; and Andrés Ibarra Salazán (AYS), "Tratado de las montañas y rocas" (written ca. 1810), BNM, Ms. 1510. Ibarra Salazán, however, compares geological layers to layers of tissue of the human body (see fol. 15v–16r).

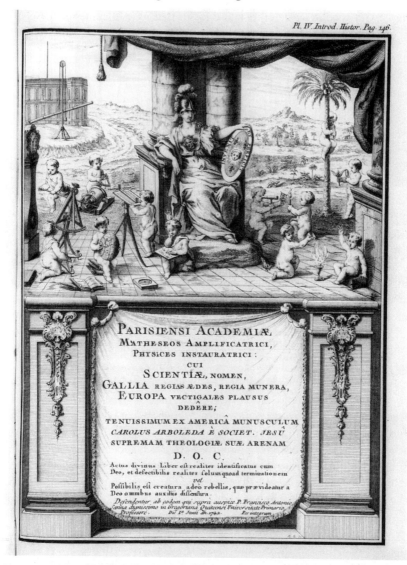

Figure 31.3. Frontispiece of a thesis defense dedicated to the French academicians Bouguer, La Condamine, and Godin at a Jesuit college in Quito- Ecuador, in June 1742. From Charles-Marie de La Condamine, *Journal du voyage fait par ordre du roi à l'Équateur* (Paris: Imprimerie Royale, 1751). The *putti* busy at measuring and gathering "matters of fact" with the help of all sorts of experimental apparatus might well signify a turning point in Creole sensibility. In the second half of the eighteenth century, discourses on experimental and mechanical philosophy, along with new secular learned institutions, arrived in the colonies. Reproduced with permission of the John Carter Brown Library, Brown University.

cial Creole elites of Cuenca in which the surgeon of the expedition, Jean Seniergues, was stoned and stabbed to death after being accused of promiscuity and deflowering a local beauty (and in which the rest of academicians were forced to flee for their lives), the French got involved in a trial against the Creole ring-leaders that lasted three years and led to no punishments.[57]

Some sixty years later (1799–1804), Alexander von Humboldt and Aimée Bonpland visited several Spanish American colonies. Throughout their journey these two philosophical voyagers were greeted as heroes. Upon their return to Paris, Humboldt published thirty volumes of observations and philosophical reflections, which, unlike La Condamine's, portrayed the Spanish American colonies most favorably. The imperial authorities and local literati not only embraced the Europeans warmly but, more important, gave the foreigners the results of forty years' worth of their own collective investigations. Humboldt's thirty volumes should be read not only as the product of a genius working in isolation but also as a summary of the Spanish American Enlightenment.[58]

A UNIFYING THEME

Changes in Creole cultural sensibilities were also reflected in the new scientific idioms they chose to express their age-old patriotic longings. Thomas F. Glick has shown that although the new scientific institutions created by the Bourbon in the colonies were, by and large, staffed and led by peninsulars, they contributed in the training of cadres of patriotic Creole natural philosophers. These Creole scientists spearheaded the wars of independence against Spain (1810–24) as they became aware of their status as colonized subjects much earlier than other sectors of the population. According to Glick, Creole

---

[57] Charles-Marie de La Condamine, *Journal du voyage fait par ordre du roi à l'Équateur* (Paris: Imprimerie Royale, 1751), passim. The events surrounding Seniergues's murder are exquisitely recounted by La Condamine in *Lettre a Madame***sur l'emeute populaire excitée en la ville de Cuenca au Pérou le 29 d'Aôut 1739* (Paris, 1746). Two other French expeditions had already visited Peru prior to La Condamine's, suggesting an early pattern of Spanish-French collaboration under the Bourbon that lasted throughout the eighteenth century. See Louis Feuillée, *Journal des observations physiques mathematiques et botaniques sur les côtes orientales de l'Amérique Méridionale et dans les Indes Occidentales, depuis l'année 1707 jusque en 1712*, 2 vols. (Paris: Pierre Giffart, 1714); Louis Feuillée, *Journal des observations physiques mathematiques et botaniques sur les côtes orientales de la Amerique Meridionale et aux Indes Occidentales, et dans un autre voyage fait par le meme ordre à la Nouvelle Espagne et aux isles de l'Amérique* (Paris: Jean Matiette, 1725); Amédée François Frézier, *Relation du voyage de la Mer du Sud aux côtes du Chily et Pérou fait pendant les années 1712, 1713 et 1714* (Paris, 1732). In *Joseph Dombey, médicin, naturaliste, archéologue, explorateur du Pérou, du Chili et du Brésil (1778–85)* (Paris: E. Guilmoto, 1905), E. T. Hamy recounts yet another French expedition to the Andes in the last quarter of the century. See also the expedition of Chappe d'Auteroche to Mexico in *Voyage en Californie pour l'observation du passage de Vénus sur le disque du soleil. le 3 juin 1769. Contenant les observations de ce phénomene et la description historique de la route de l'auteur à travers le Mexique* (Paris: Charles-Antoine Jombert, 1772).

[58] Alexander von Humboldt, *Voyage de Humboldt et Bonpland: Voyage aux régions équinoxiales du nouveau continent*, 30 vols. (Amsterdam: Theatrum Orbis Terrarum, 1970–3).

natural philosophers became "Newtonian" liberals who sought to create national sciences around the defense of Andean and Nahua taxonomies (threatened by the expansion of new Linnean botanical classifications that arrived in
the colonies along with tactless Spanish imperial scientists), the identification
and development of local materia medica distinct from European ones, and
resistance to European negative characterizations of the American climate.[59]

Creoles modified the scientific idioms in which they cast their proto-
nationalism. Whereas Baroque patriots had used astronomy and astrology to
exalt God's providential designs and had praised the mineral and pharmaceutical wonders of their land, late eighteenth-century scholars sought to tap
the agricultural potential of the colonies. They argued that each colony was
endowed by Providence so that it would become a leading commercial emporium in the world. Naturalists presented their local territories as micro-
cosms of the globe in which the multitude of ecological niches and endless
equatorial agricultural cycles made the lands capable of supplying every need
of the world's markets. These naturalists also assumed that the natural laws of
the Americas were different from those of Europe and that New World phenomena could be studied only by Creole scientists. José Antonio de Alzate,
a leading Mexican naturalist and editor of several periodicals, insisted that Mexico's rare natural productions undermined and upset all scientific hypotheses
devised by Europeans and sought to create a science that only Mexicans could
foster and interpret.[60] Peruvian Creole physicians took advantage of humoral
theory to create a form of medical nationalism that maintained that the
climate, bodies, and diseases in Peru were singular and thus that only Peruvian
physicians could identify and cure local diseases.[61]

To be sure, profound cultural differences separated the worlds of Baroque
and "Newtonian" Creole scientists. But patriotism remained a constant, unifying theme throughout the long eighteenth century.

---

[59] T. Glick, "Science and Independence in Latin America (with Special Reference to New Granada),"
Hispanic American Historical Review, 71 (1991), 307–34.

[60] Jorge Cañizares Esguerra, "Nation and Nature: Natural History and the Fashioning of Creole National Identity in Late Colonial Spanish America," Cultural Encounters in Atlantic Societies, 1500–
1800, International Seminar on the History of the Atlantic World. Working Paper Series (The Charles
Warren Center for Studies in American History, 1998).

[61] Ibid., and Jorge Cañizares Esguerra, "La utopía de Hipólito Unanue: Comercio, naturaleza y religión,"
in *Saberes andinos,* pp. 91–108.

Part V

RAMIFICATIONS AND IMPACTS

32

SCIENCE AND RELIGION

John Hedley Brooke

Commentaries on the Enlightenment often propose a highly schematic account of the changing relations between science and religion. Whereas the seventeenth century is credited with a notional "separation" of the sciences from religious control, the eighteenth is characterized by a more devastating form of secularization in which the methods and conclusions of the natural philosophers were turned against the authority of the established Churches. With carefully selected examples, this story can be attractive and plausible. Early in the seventeenth century, Francis Bacon (1561–1626) had warned against the mixing of biblical exegesis with natural philosophy, and, in France, René Descartes (1596–1650) had mechanized a universe no longer anthropocentric. Both men had devised stringent criteria that truth claims had to meet and both had rejected final causes from the explanation of natural phenomena. During the second half of the seventeenth century, enduring scientific societies had come into existence in both London and Paris, and within them religious disputation was banned. By the end of the century, Isaac Newton (1642–1727) had articulated his laws of motion and the law of universal gravitation, laws that to later generations would symbolize a universe characterized by order and regularity rather than divine caprice.

Newton is brought within the schema in other ways. If his *Principia* was a towering monument to the power of mathematical reasoning, his *Opticks* displayed the power of a rigorous experimental method.[1] Seemingly the stage was set for the displacement of theology, once the queen of the sciences, by more bracing sciences that promised an improvement of the world and a brighter destiny for humankind. Newton's shedding of light on light itself and John Locke's (1632–1704) line on the senses as ultimate conduits for the acquisition of knowledge helped to forge a new epistemology, in which

[1] Gerd Buchdahl, *The Image of Newton and Locke in the Age of Reason* (London: Sheed and Ward, 1961).

"vision was queen among the senses"[2] and observation, rather than innate ideas or revelation, defined the route to secure knowledge.

Such a schema can accommodate many features routinely associated with a new Age of Reason: confidence in human powers to transform the world and the extension of scientific methods to the study of those powers. It can accommodate the attacks on the Catholic Church and the pleas for greater religious tolerance advanced by Voltaire (1694–1778) and others denigrated as "deists" by their orthodox critics. From Protestant cultures, too, there are examples of revolt against the credal formulations of the past, such luminaries as David Hume (1711–1776) in Scotland and Joseph Priestley (1733–1804) in England reacting against the oppressive Calvinism of their formative years. As both dissenting minister and experimental philosopher, Priestley typifies an optimistic spirit, visible among the *philosophes,* that scientific knowledge coupled with enlightened programs of education would erase superstition from a world debilitated by otherworldy concerns. In radical clandestine literature, such as the *Testament* of the heretical French priest Jean Meslier (1664–1729), a secular critique of religion would be pushed to extremes: it was humans who made gods, not God who made humans; and a future life was a fiction foisted on people by ruling elites. In the defense of theology, traditional arguments based on miracles and fulfilled prophecy continued to be wheeled out, but a vogue among religious apologists for physico-theology – in which the argument for design was prominent – illustrates the incursion of new forms of rationalism into theology itself, reflecting allegations of irrationality it was obliged to refute.[3]

Here, however, a crack appears in this notional schema. While new forms of "natural religion" were indeed set up in opposition to both Catholic and Protestant creeds, the physico-theology of John Ray (1627–1705), Robert Boyle (1627–1691), and such successors as Richard Bentley (1662–1742), Samuel Clarke (1675–1729), and William Derham (1657–1735) was articulated in defense of a Christian theism against libertarian and atheistic opponents. In their writings the latest science would be appropriated to demonstrate what Ray had called the *Wisdom of God Manifested in the Works of Creation* (1691). Thus, Ray celebrated the greater elegance of the Copernican universe over that of Ptolemy; Bentley saw in Newton's gravitational forces evidence of nonmaterial agencies in nature; Clarke interpreted Newton's laws as a summary of the way God normally chose to act in the world, and Derham welcomed the expansion of the universe as an exhilarating escape from a theology that had been overly anthropocentric.[4]

[2] Peter Hulme and Ludmilla Jordanova, "Introduction," in Peter Hulme and Ludmilla Jordanova (eds.), *The Enlightenment and its Shadows* (London: Routledge, 1990), p. 4.

[3] For a fuller introduction and attendant qualifications, see John Hedley Brooke, *Science and Religion: Some Historical Perspectives* (Cambridge University Press, 1991), pp. 152–225; Dorinda Outram, *The Enlightenment* (Cambridge University Press, 1995), pp. 31–62.

[4] John Gascoigne, "From Bentley to the Victorians: The Rise and Fall of British Newtonian Natural

One effect of such argumentation was to give the sciences a higher profile. Consequently, in certain contexts, there could be a symbiotic relation between promoting the sciences and promoting a respectable religion. It is visible in the rhetoric of the Swedish taxonomist Carl Linnaeus (1707–1778) who elevated the scientist even as he elevated his Creator:

> If the Maker has furnished this globe . . . with the most admirable proofs of his wisdom and power; if this splendid theatre would be adorned in vain without a spectator; and if man . . . is alone capable of considering the wonderful economy of the whole; it follows that man is made for the purpose of studying the Creator's works.[5]

THE DIVERSITY OF NATURAL RELIGION

The prevalence of such discourse in eighteenth-century texts reflects the fact that, through their incorporation into natural religion, the sciences could be enlisted both to attack and defend a Christian theism. Natural religion itself had many meanings. For Voltaire it denoted an alternative to Catholic Christianity with its own simple and universal creed:

> When reason, freed from its chains, will teach the people that there is only one God, that this God is the universal father of all men, who are brothers; that these brothers must be good and just to one another, and that they must practise all the virtues; that God, being good and just, must reward virtue and punish crimes; surely men will be better for it, and less superstitious.[6]

But an appeal to natural religion could also be part of a defense of Christian orthodoxy – even the principal part, according to the Anglican Bishop of Durham, Joseph Butler (1692–1752), in his *Analogy of Religion* (1736). It could be the principal part perhaps, but never the sufficient part since the existence of a covenant between God and His people could never be established by the light of reason alone. In Britain, advocates of a natural religion would often say, as did William Wollaston (1660–1724), that "so far from undermining true revealed religion, . . . it rather paves the way for its reception."[7] Whereas deists such as Henry St. John, Viscount Bolingbroke (1678–1751), would say that "in natural religion the clergy are unnecessary,"[8] the clergy were often

---

Theology," *Science in Context*, 2 (1988), 219–56; Neal C. Gillespie, "Natural History, Natural Theology and Social Order: John Ray and the 'Newtonian Ideology,'" *Journal of the History of Biology*, 20 (1987), 1–49.

[5] C. Linnaeus, *Reflections on the Study of Nature* (1754), trans. J. E. Smith (1786), quoted in David Goodman, *Buffon's Natural History* (Milton Keynes: Open University Press, 1980), p. 18.

[6] Voltaire, "Sermon of the Fifty," in Peter Gay (ed.), *Deism: An Anthology* (Princeton, NJ: Van Nostrand, 1968), pp. 152–3.

[7] David Pailin, "What Is Natural Religion?" in Arvind Sharma (ed.), *The Sum of our Choices* (Atlanta: Scholars Press, 1996), pp. 85–119.

[8] Pailin, "Natural Religion," p. 92.

its advocates. Importantly, the more reflective advocates of a natural religion recognized that the relationship between natural and revealed theology was more complex than a simple contrast between the two might suggest. For example, could the attributes of the deity, ostensibly deduced from the natural order, have been known had they not first been inferred from revelation? This was an important question for Priestley, who defined "natural religion" as "all that can be demonstrated, or proved to be true by natural reason" even though "it was never, in fact, discovered by it; and even though it be probable that mankind would never have known it without the assistance of revelation."[9] The study of other nations taught Priestley the salutary lesson that those destitute of revelation had made little headway in their religious instruction.

RELATING THE SCIENCES TO RELIGION

The diversity of natural religion, in both scope and purpose, is not the only complication when one relates new cultures of science to the religious sensibilities of the eighteenth century. To speak at all of "relations between science and religion" already presupposes some distinction between the bodies of "science" and "religion" in their cognitive claims and/or their practices. Yet in the first two decades of the new century (and long afterward in certain contexts), theistic arguments were incorporated into scientific debate in ways that blurred such distinctions. The term "natural philosophy," used by Newton and his contemporaries, denoted a discipline broader in scope than the word "science" came to signify as a result of later specialization. Newton himself, in the second edition of his *Principia* (1713), wrote that "to discourse of [God] from the appearances of things does certainly belong to natural philosophy." Although one can analyze the mathematical reasoning of the *Principia* without engaging theological issues, the depth of Newton's own interests in alchemy and Biblical exegesis suggests that holding together his various intellectual projects was a preoccupation with the manner of divine activity in the world. His conceptions of absolute space and time were explicitly informed by his theology, and his confidence in a *universal* law of gravitation reflected the supposition of a single and omnipresent deity whose will had been imposed on the world.[10]

The debate that took place in the second decade of the eighteenth century between Newton's advocate Samuel Clarke and his detractor Gottfried Wilhelm Leibniz (1646–1716) shows a continuing fusion of empirical with metaphysical and theological elements. Whereas Newton had required a periodic

[9] Pailin, "Natural Religion," p. 95.
[10] Andrew Cunningham, "How the *Principia* Got Its Name," *History of Science,* 29 (1991), 377–92; Betty Jo Dobbs, *The Janus Faces of Genius* (Cambridge University Press, 1991); J. E. McGuire, "Newton on Place, Space, Time and God," *British Journal for the History of Science,* 11 (1978), 114–29; R. S. Westfall, *Force in Newton's Physics* (London: Macdonald, 1978), p. 340.

"reformation" of the solar system in order to correct destabilizing tendencies, Leibniz protested that this demeaned the deity, who was not to be reduced to a second-rate clockmaker. Whereas Newton's universe allowed for empty space, even within matter itself, Leibniz would have none of it:

> To admit a vacuum in nature is ascribing to God a very imperfect work. I lay it down as a principle, that every perfection, which God could impart to things without derogating from their other perfections, has actually been imparted to them. Now let us fancy a space wholly empty. God could have placed some matter in that space: Therefore there is no space wholly empty: Therefore all is full.[11]

In this politically charged debate Clarke championed Newton's view that what God had freely chosen to do in the world should be discovered by inspecting the world and not by legislating for "Him." In retaliation, Leibniz insisted that this world could not be *shown* to be the best of all possible worlds unless criteria for goodness could be established independently of what empirical methods then proved to be the case. Far from the seventeenth century having produced a separation of "science" from "theology," it had, in the words of one historian, produced an "unprecedented fusion."[12] There was, however, an important consequence. If, as in Newton's natural philosophy, traditional attributes of the deity – dominion, omnipotence, omnipresence – gained new and specific meanings through scientific redescription, through that very specificity they became more vulnerable. By the close of the eighteenth century, the God who, according to Newton, engineered the reform of the solar system through a resourceful use of comets, was embarrassed by the calculations of Pierre Simon de Laplace (1749–1827), according to which the system was *self*-stabilizing.[13]

In a letter to Thomas Burnet (c. 1635–1715), Newton had insisted that where there were natural causes at hand, God would use them to effect His purposes – a view that Burnet clearly shared, for in his own account of the *Sacred Theory of the Earth* (1684) he had explained how Noah's flood had erupted through the release of subterranean water when the Earth's crust cracked. To conservative religious critics this was a presumptuous reduction of miracle to mechanism; yet Burnet did not consider that he was detracting from divine Providence. He saw in the synchronization of the flood with the moral decay described in Genesis a powerful argument for divine prescience. The image in both Burnet and Newton is one of a deity controlling and working through "secondary" causes. This signals another point of fundamental importance

---

[11] G. W. Leibniz, "Fourth Paper . . . ," in H. G. Alexander (ed.), *The Leibniz Clarke Correspondence* (Manchester: Manchester University Press, 1956), p. 44.

[12] Amos Funkenstein, *Theology and the Scientific Imagination from the Middle Ages to the Seventeenth Century* (Princeton, NJ: Princeton University Press, 1986).

[13] Roger Hahn, "Laplace and the Mechanistic Universe," in David C. Lindberg and Ronald L. Numbers (eds.), *God and Nature* (Berkeley: University of California Press, 1986), pp. 256–76.

when one correlates "scientific" with "religious" beliefs. It was usually possible to describe events in terms both of natural (or "secondary") causes and of divine Providence. It was not a question of either/or, as it became for later polemicists. This means that the extension of scientific knowledge would not automatically lead to an expulsion of God from the world.[14] As in the seventeenth century, the disclosures that were most damaging to the Christian religion would come from encounters with other cultures rather than with scientific innovation. The former invited cultural relativism of the kind explored by Charles-Louis de Montesquieu in his *Lettres Persanes* (1721) and fueled heterodox speculation about multiple origins for the human races – even the possibility, mooted as early as 1655 by the French Calvinist Isaac La Peyrère (1596–1676), that there had been pre-Adamic progenitors.[15] By contrast, as long as scientific *laws* were still understood in terms of divine legislation, extending their province need not be profane.

If the debate between Leibniz and Clarke shows that different interpretations of nature could be informed by competing metaphysical and theological views, it is also true that one and the same scientific innovation could be given both sacred and secular readings. For this reason the "relations between science and religion" cannot be reduced to a simple pattern of religious retreat as the sciences advanced. The great diversity of interpretation to which Newton's science was susceptible provides a striking example. William Whiston (1667–1752), who succeeded Newton in the Lucasian Chair of Mathematics at Cambridge, identified the gravitational force with the interposition of God's "general, immechanical, immediate power." Among his opponents was Anthony Collins (1676–1729), for whom Newton's forces proved the inherent activity of matter. Whiston was no orthodox divine. Like Newton himself, he had strong Arian tendencies, denying that Christ had been of one substance with the Father. But Whiston was elated by Newton's provision of a science in which God's continuing dominion over nature was celebrated in opposition to the distant, spiritually aloof deity of the Cartesian mechanists.[16] It is the interpretative flexibility that strikes the modern reader. In his attempt to purge Christianity of all that was mysterious, the Irish emigré John Toland (1670–1722) insisted that Newton's science did not have to be interpreted as Newton prescribed. Why should the force of gravity and powers of self-movement not be essential attributes of matter?[17] Despite the attempt of

---

[14] John Hedley Brooke, "Science and Theology in the Enlightenment," in W. Mark Richardson and Wesley J. Wildman (eds.), *Religion and Science: History, Method, Dialogue* (New York: Routledge, 1996), pp. 7–27.

[15] Richard H. Popkin, *Isaac La Peyrère (1596–1676): His Life, Work and Influence* (Leiden: Brill, 1987).

[16] James E. Force, *William Whiston: Honest Newtonian* (Cambridge University Press, 1985); and Force, "Biblical Interpretation, Newton, and English Deism," in Arjo Vanderjagt (ed.), *Scepticism and Irreligion in the Seventeenth and Eighteenth Centuries* (Leiden: Brill, 1993), pp. 282–305.

[17] Margaret C. Jacob, "John Toland and the Newtonian Ideology," *Journal of the Warburg and Courtauld Institutes*, 32 (1969), 307–31.

Richard Bentley to dissociate Newton's atomism from the atheistic atomism of antiquity, a Lucretian reinterpretation of Newtonian science could always be pressed, as it was later in the century by the self-avowed atheist Baron d'Holbach (1723–1789). He would say that "matter moves by its own peculiar energies; that its motions are to be attributed to the force which is inherent in itself."[18] The inherent powers of gunpowder helped him, but he also enlisted Newton's philosophy. According to one's predispositions, innovative science could be appropriated for traditional theistic positions, for alternative systems of natural religion, and, less commonly, for an explosive atheism.[19]

The reference to predispositions is important because investment in what were seen as rationally superior forms of religion could be encouraged by considerations that had nothing directly to do with the sciences. The baleful effects of religious warfare were often uppermost in the minds of those who deplored the dogmatism of the Churches. That Christianity might destroy itself had been a concern of Robert Boyle as he reflected on the proliferation of Puritan sects during the Interregnum in mid-seventeenth-century England. A natural theology held out the attractive prospect of reestablishing common ground, of binding more closely those who would otherwise quarrel over the finer points of doctrine. A natural religion could grow from considerations internal to Christian theology itself. How were those to be judged who had never heard the gospel of Jesus Christ? With reference to Romans 1: 20 it was often said that all possessed the light of nature in sufficient degree to discern the power of a God to whom they were answerable. Christian apologists would tend to say that people would be judged according to the gifts they had been given. In his *Christianity as old as the Creation* (1730), often described as the Bible of the deists, Matthew Tindal (1657–1733) peddled the reductive formula that each person would be judged according to the use he or she had made of his or her reason. In another reduction Tindal asserted that the duties of a Christian and a good citizen were one and the same. In shaking off the shackles of "priestcraft" and "superstition," those who targeted the Catholic church would rail against mystifying doctrines such as transubstantiaton and dubious practices of confession that conferred power on a priesthood privy to secret knowledge. Another form of cynicism could take its toll. Looking back over the recent religious history of England, it was not difficult to argue that there had been a good deal of trimming as individuals had changed their beliefs in accord with political expediency. In seeking to find a more rational basis for religious belief, it would be observed, as had Locke, that

---

[18] Michael J. Buckley, *At the Origins of Modern Atheism* (New Haven, CT: Yale University Press, 1987), p. 280.

[19] Michael Hunter and David Wootton (eds.), *Atheism from the Reformation to the Enlightenment* (Oxford: Clarendon Press, 1992).

personal religious convictions were of a kind that could not be legislated and that, if reason were not allowed a determining role, all people would be slaves to the religious mores of the country of their birth. Critiques of Judaism and Christianity would be launched on moral grounds, as when Voltaire protested against the seemingly arbitrary and vengeful acts of the God of the Old Testament. Liberating concepts of what was truly "natural" could also be turned against a repressive sexual morality, as when Denis Diderot (1713–1784) celebrated the reported sexual freedom of the Tahitians. Moral considerations of a quite different kind could also inform attitudes to religious claims. Natural disasters could be disastrous for natural theology. How could the *beneficence* of the deity be inferred from a world in which ten thousand lives were destroyed by the Lisbon earthquake of 1755, which for Voltaire was so ruinous of the "best of all possible worlds"?

These are merely fragments of arguments used in different contexts to challenge conventional claims for religious authority. But they suffice to show that relating religious beliefs to scientific change can be artificial and reductive if mediating circumstances are ignored. The way in which scientific innovations were mobilized for religious and political purposes was also critically dependent on both national and local contexts.[20] The Leibniz/Clarke controversy was politically supercharged because it had local, national, and international dimensions. A priority dispute over the invention of the calculus had been a running sore for several years when, in November 1715, Leibniz held Newton and Locke responsible for a decline of natural religion in England. Jealousies raged because, with the Hanoverian succession to the English throne, Leibniz seemingly coveted the role of philosopher to the English court – a prospect not exactly relished by champions of Newton. The fact that Princess Caroline, through whom Leibniz conducted the correspondence, fell under the tutelage of Clarke added poignancy to exchanges in which the two antagonists accused each other of a disreputable deism. There had been an international dimension to the construction of Leibniz's philosophy of nature because he had sought a system that would be acceptable to both Catholics and Protestants and so would help achieve a reunification. In England both Locke and Newton had been vehemently anti-Catholic, Newton associating the papacy with the Antichrist. The voluntarist theology developed by Newton and Clarke, which stresssed the freedom of the divine will, could be used in England to justify the removal of a pro-Catholic king, James II, from the throne. But so great an emphasis on a Sovereign will was unpalatable to Leibniz because, in another context, it could be used to legitimize the pretensions to absolute power of an earthly sovereign – notably Louis XIV, whose expansionism Leibniz perceived as a threat to the German states. Even such broad-brush contextualization shows that

---

[20] Roy Porter and Mikuláš Teich (eds.), *The Enlightenment in National Context* (Cambridge University Press, 1981).

there was more to the Leibniz/Clarke controversy than a philosophical quarrel over the best way of relating science to religion.[21]

The importance of placing such debates in their local contexts is brought out by the fact that Paris was perhaps the only European center that would experience a wave of outright atheism in the 1750s.[22] British visitors, such as David Hume, Edward Gibbon (1737–1794), and Joseph Priestley, found the phenomenon surprising. Gibbon deplored "the intolerant zeal of the friends of d'Holbach and Helvétius, who preached the tenets of scepticism with the bigotry of dogmatists, and rashly pronounced that every man must be either an Atheist or a fool."[23] Priestley recorded that when at a French dinner party he had declared himself a believer, he had not been believed! The *philosophes,* in his view, had rejected a corrupt form of Christianity but had jettisoned too much. Part of his mission was to reeducate them in a rational Christianity.[24] Contrasts can also be drawn for *fin de siècle* attitudes, when the science of Laplace, with its exclusion of God from the solar system and its origins, chimed with the secular ethos of the Revolutionary period. By contrast, in Britain, revulsion against the terror meant that Laplacian science would often be stigmatized or had to be resacralized. The latter was not impossible because one could always argue (as Leibniz had) that a system requiring no intervention afforded the better testimony to divine prescience.

SCIENCE AND SECULARIZATION

If relations constructed between science and religion were more equivocal, mediated, and complex than is often supposed, it is nevertheless true that the elevation of the sciences in European and American cultures had subtle and indirect effects that were perceived as damaging to religious sensibilities. A striking example is provided by the appearance during the first half of the eighteenth century of metropolitan and itinerant lecturers who captured an audience for Newtonian science with spectacular demonstrations of nature's forces. Electrical sparks would fly in London's inns and coffeehouses as entrepreneurs such as Jean Theophilus Desaguliers (1683–1763) and Francis Hauksbee (1688–1763) turned experimental inquiry into popular entertainment. As part of the rhetoric of these public performances, lecturers would often claim to be demonstrating divine powers, or at least powers placed by God in nature. Theologically, such displays were therefore deeply ambiguous. They could be

---

[21] Steven Shapin, "Of Gods and Kings: Natural Philosophy and Politics in the Leibniz-Clarke Disputes," *Isis,* 72 (1981), 187–215.

[22] A. C. Kors, *D'Holbach's Coterie: An Enlightenment in Paris* (Princeton, NJ: Princeton University Press, 1976).

[23] Buckley, *Origins,* p. 255.

[24] John Hedley Brooke, "'A Sower Went Forth': Joseph Priestley and the Ministry of Reform," in A. Truman Schwarz and John G. McEvoy (eds.), *Motion Toward Perfection: The Achievement of Joseph Priestley* (Boston: Skinner House, 1990), pp. 21–56.

presented as reverential, as a form of natural theology; but in two respects they might be considered presumptuous. Here were men with their impressive apparatus controlling and manipulating forces that had once been a divine prerogative. And in so doing they were mirroring, even usurping, the privileged role of priest.[25] The ambiguity is visible in an announcement of lectures to be given in Virginia, in 1766, by one William Johnson, who advertised a "course of experiments, in that instructive and entertaining branch of natural philosophy, called electricity." His performance was to include a demonstration that lightning was electrical fire and, following Benjamin Franklin (1706–1790), that lightning conductors offered protection. The advertisement continued: "we have the utmost reason to bless God for a discovery so important." There was reference to Proverbs 22:3 and a final invocation of the God of Nature to bless the enterprise: "As the knowledge of nature tends to enlarge the human mind, and give us more exalted ideas of the God of Nature, it is presumed that this course will prove to many an agreeable and rational entertainment."[26]

However, the God of Nature was not always the God of more sensitive religious spirits. In Britain there was opposition to Newtonian science from High Churchmen, partly because Newton, Clarke, and Whiston were perceived as Arian heretics, partly for the very reason that those who peddled the new science in popular lectures were posing as a secular priesthood. Arguably "the audience for natural philosophy outstripped the capacity of religious authority to control the experimental medium."[27] High Church frustration was voiced by George Horne, Oxford don and future bishop of Norwich, who complained in 1753 of the "stupid admiration" shown to those whose experimental displays degraded "the philosopher into the mechanic." Horne's sarcasm concealed a real concern. That Newton had wanted only "a glass bubble, and a board with a hole in it, to describe all the wonders of light" might mean that "even women and children may hereafter commence sage philosophers, by blowing phlegm through a straw, or staring at soapy vehicles." No one was safe from the seductions of experiment.

If science popularizers played on spectacular visible effects, speculations about the invisible could also shift attention from the sacred to the secular. The invisible workings of the human mind provided a crucial locus for such a shift because, in some cases, new models of the mind already reflected detachment from a traditional theological vocabulary of soul, spirit, and free will. Where authority is freshly located within human beings, rather than in gods or in semidivine monarchs, "people themselves as sources of power become

---

[25] Simon Schaffer, "Natural Philosophy and Public Spectacle in the Eighteenth Century," *History of Science*, 21 (1983), 1–43; Larry Stewart, *The Rise of Public Science* (Cambridge University Press, 1992).

[26] I. Bernard Cohen, *Benjamin Franklin's Science* (Cambridge, MA: Harvard University Press, 1990), pp. 144–5.

[27] Larry Stewart, "Seeing Through the Scholium: Religion and Reading Newton in the Eighteenth Century," *History of Science*, 34 (1996), 123–65.

enormously more interesting."[28] The study of mental faculties could then open further possibilities for the secularization of knowledge through mechanistic theories that explained how the mind might operate through the association of ideas. Scientists were sometimes destructive despite themselves.[29] The physician David Hartley (1705–1757) affords an example. In his *Observations on Man, his Frame, his Duty, and his Expectations* (1749), Hartley wrote as a theologian, claiming that the design of the mind ensured that humanity would progress toward virtue and happiness. It did so because human conduct was guided by the attempt to maximize pleasure and to minimize the pain associated with actions and their consequences. But when he proceeded to correlate mental associations with patterns of vibration that, once stamped on the material substance of nerves and brain, could be reexcited, a thoroughly naturalistic account of mental phenomena became possible, which in other hands might displace his theological framework. That process began with Priestley's use of Hartley to buttress his own brand of determinism and rational dissent. It culminated in anachronistic accounts of Hartley by nineteenth- and twentieth-century psychologists who claimed him as a scientific, and not a religious, thinker.[30]

Although the extension of naturalistic explanation often went hand-in-glove with natural theology, a certain distancing of the Creator from the creation could easily result. In this respect suggestive contrasts can be drawn between late eighteenth-century attitudes and those of a hundred years earlier. Whereas Boyle had expressed a sense of immediate dependence on God, receiving "pregnant hints" even as he conducted chemical experiments, Priestley dismissed belief in a divine influence on the mind as vulgar superstition. Collapsing the matter/spirit dualism that had been expressed in extreme form by Descartes, Priestley lodged the powers formerly ascribed to spirit agencies in a more capacious concept of matter. In cosmology, esthetically pleasing features of the solar system (the fact that the planets orbited the sun in the same direction and almost the same plane), which Newton had ascribed to intelligent design, were, a hundred years later, subsumed by Laplace under a nebular hypothesis that required only the gradual cooling of a rotating solar envelope to generate the planets and their orbits.

One hundred years also separated the *Theory of the Earth* (1795) of James Hutton (1726–1797) from the *Sacred Theory of the Earth* propounded by Thomas Burnet. The word "sacred" was lost. Hutton's geological cycles were not intrinsically atheistic, and he frequently spoke of nature as if it were a designed system. But when he wrote that geology furnished no vestige of a beginning, no prospect of an end, his detractors complained that the deity could hardly

[28] Jordanova, *Enlightenment and Its Shadows*, p. 206.

[29] Peter Burke, "Religion and Secularisation," in *The New Cambridge Modern History*, vol. 13 (Cambridge University Press, 1979), pp. 293–317.

[30] Roger Smith, *The Fontana History of the Human Sciences* (London: Fontana, 1997), pp. 252–5.

be distanced further from the creation. In the life sciences, too, a hundred years saw a remarkable change. When John Ray had spoken of the wisdom of God in the works of creation, the word "creation" had referred to a world that had remained essentially unchanged since its inception. By the close of the eighteenth century, the earth had acquired a history, full of incident, which included the gradual modification of living things. In his *Epochs of Nature* (1778), Georges-Louis Leclerc, Comte de Buffon (1707–1788), no longer saw earth history and human history as coextensive. Speculative European philosophers, including Erasmus Darwin (1731–1802) in Britain and Jean-Baptiste Lamarck (1744–1829) in France, distanced Creator from creation through a new vocabulary. In the transformism of Lamarck, creatures ceased to be creatures and became "nature's products."[31] In Lamarck's interpretation of fossil shells, organic transformation was presented as an alternative to the admission of extinction; but once his rival Georges Cuvier (1769–1832) focused attention on extinct quadrupeds, there was to be yet another threat to the *balance* and *economy* of nature, so pervasive in clerical natural history and so conspicuous in Gilbert White's *The Natural History of Selborne* (1789).

In yet another respect, the sciences could become associated with secular trends without necessarily being a primary agent of secularization. Incorporated into systems of physico-theology they contributed to a form of religious apologia that invited its own refutation. It is not simply that arguments for design emphasised God the Creator at the expense of God the Redeemer, although this can hardly be denied. Rather, inappropriate and inflated claims were made in deducing the personality of the deity from impersonal forces. Rhetorical gestures to the effect that the design argument is the *only* theistic proof worthy of serious consideration would prove a liability when alternative accounts of the appearance of design became plausible; hence the claim that natural theology dug its own grave by positively inviting the atheistic response of Diderot and d'Holbach.[32] In his *Pensées Philosophiques* (1746) Diderot seemed to accept a voguish physico-theology; but the great architect of the *Encyclopédie* soon flipped from deism to an atheism in which the appearance of design in nature was illusory. In 1753 he published the speculation that, over millions of years, organic matter might have passed through an almost infinite number of organized states, the defective combinations falling by the wayside. The polarity between worldviews informed by natural theology and such atheistic visions was captured by Pierre Maupertuis (1698–1759) in his *Essai de Cosmologie* (1756). Allocating all contemporary philosophers to one of two sects, he characterized one group as wishing to subjugate

---

[31] Ludmilla Jordanova, "Nature's Powers: A Reading of Lamarck's Distinction Between Creation and Production," in James R. Moore (ed.), *History, Humanity and Evolution* (Cambridge University Press, 1989), pp. 71–98; Roy Porter, "Erasmus Darwin: Doctor of Evolution?" ibid., pp. 39–69.
[32] Buckley, *Origins*, chaps. 3–6.

nature to a purely material order, while the other, penetrating the Creator's intent in the minutiae of creation, saw divine power and goodness painted on the wings of butterflies and in every spider's web.[33] But if proofs of God's beneficence were suspended, as they were by the New England divine Jonathan Edwards (1703–1758), from so fragile a thread as a spider's web, the pious reading could easily snap if the piety were not firmly grounded in other forms of religious experience or instruction.

For mid-century French materialists such as Diderot and Julien Offray de la Mettrie (1709–1751), transport into materialism was eased by three empirical disclosures of the 1740s. One was evidence for the spontaneous generation of microorganisms, seemingly conjured from rotten corn by an English Catholic priest, John Turbeville Needham (1713–1781). Another was evidence produced by the Swiss naturalist Albrecht von Haller (1708–1777) that muscular tissue had inherent powers of motion independent of vital force or soul. Even when removed from the body it would automatically contract when pricked. The third revelation was sensational. A humble freshwater polyp, the hydra, when cut into pieces had the ability to regenerate itself. Abraham Trembley's (1710–1784) discovery was repeated all over Europe, conferring new credibility on the view that matter could organize and reorganize itself.

Such disclosures did not entail a materialist philosophy. Neither Needham nor von Haller went down that road. Indeed, as a priest, Needham fell victim to Voltaire's jibe that he had been faking a miracle. Trembley's polyp might call into question an indivisible animal or vegetable soul; but it was susceptible of a conservative interpretation. A missing link in the great chain of being, it corroborated a taxonomic ideal at one with belief in the plenitude of God's creation. Nevertheless, such discoveries provided potent symbols of *nature's* powers and were not lost on La Mettrie, who in his *L'Homme Machine* (1747) combined a materialist physiology with a secular philosophy in which religious beliefs were dispensable in the conduct of one's life.[34]

PROVIDENCE AND THE UTILITY OF SCIENCE

If a certain distancing of the Creator from creation could be a consequence (even an unintended one) of scientific innovation, such a process probably impinged most on public consciousness through claims for scientific utility. As cultures of "improvement" grew in various European towns, so did a rhetoric that invested in the sciences the promise of economic prosperity, a richer agriculture, better medicines, more-efficient industrial processes. The rhetoric

---

[33] Hahn, "Laplace," p. 265.
[34] Shirley A. Roe, *Matter, Life and Generation: Eighteenth-Century Embryology and the Haller-Wolff Debate* (Cambridge University Press, 1981); Aram Vartanian, "Trembley's Polyp, La Mettrie and Eighteenth-Century French Materialism," in P. Wiener and A. Noland (eds.), *Roots of Scientific Thought* (New York: Basic Books, 1957), pp. 497–516.

would reflect local circumstances. In Edinburgh the chemist William Cullen (1710–1790) courted Scottish landowners; through the Lunar Society of Birmingham, Priestley tapped the wealth of industrial entrepreneurs such as Matthew Boulton, James Watt, and Josiah Wedgwood. In both cases the utility of chemistry was underlined in distinctive relations of reciprocity.[35] Utopian visions would sometimes outstrip the ability of science to deliver. Not all of Priestley's gases had the curative properties of which he dreamed.[36] But there were cases in which ameliorative control over natural forces was achieved. The use of lightning conductors to protect church towers provides a sensitive indicator of public attitudes because here was progress through science in a context where the ringing of church bells had been the traditional method of warding off storms, a practice sadly more likely to attract than repel the fatal bolt.

For later rationalists the reluctance of the clergy to fit the new device was a paradigm case of religious obscurantism versus scientific vision. To enliven his account of the *warfare* between science and Christian theology, Andrew Dickson White would count the bellringers who had unnecessarily met their deaths. Recent scholarship has been more sensitive to the complexity of the issues. A conductor was fitted to St. Mark's cathedral in Venice in 1766 – belatedly in White's account, but no more than fourteen years after Franklin's invention. There was indeed resistance to the new technology both in Europe and America, but for diverse reasons. One stemmed from popular confusion between the use of ungrounded rods to attract lightning from clouds for experimental purposes and the allegedly protective use when the rods were earthed. The fear was that even a grounded rod might *attract* a strike that would otherwise have been avoided. There was even resistance from electrical connoisseurs, such as the Abbé Nollet (1700–1770), an active member of the Paris Academy of Sciences, who saw Franklin as a rival. Nollet warned that ringing bells only made matters worse; but, in a report of 1764, he would not endorse the use of conductors, believing them "more suitable to attract the fire of thunder to us than to preserve us from it." Franklin himself was angered by Nollet's stand but not surprised that unlearned men, "such as commonly compose our church vestries," should be hesitant.[37]

In their attacks on the Catholic Church, the *philosophes* often pitted science and reason against religion and superstition. The case of the lightning conductor shows that their rhetorical formula did not always fit. There are

[35] Jan Golinski, *Science as Public Culture: Chemistry and Enlightenment in Britain, 1760–1820* (Cambridge University Press, 1992), pp. 11–90. The distinctively different audience in France for the *Encyclopédie* is described by Robert Darnton, *The Business of Enlightenment* (Cambridge, MA: Harvard University Press, 1979), p. 526.

[36] Golinski, *Public Culture*, pp. 91–128; Roy Porter, *Doctor of Society: Thomas Beddoes and the Sick Trade in Late Enlightenment England* (London: Routledge, 1991).

[37] Cohen, *Franklin's Science*, pp. 118–58. For Nollet, see also Jean Torlais, *Un Physicien au Siècle des Lumières: L'Abbé Nollet, 1700–1770* (Paris: Sipuco, 1954), and R. W. Home, *Electricity and Experimental Physics in Eighteenth-Century Europe* (Aldershot: Variorum, 1992).

examples of popes (notably Benedict XIV from Bologna) and priests whose attempts to erect the device were thwarted by a deeply mistrustful populace. Franklin's friend Priestley would always place science and enlightened religion on the same side against popular superstition. Regional diversity is also important in charting the progress of Franklin's device. Reporting from London in 1773, Franklin was pleased that "some churches, the powder magazine at Purfleet, the queen's house in the park" were now protected. By contrast, in isolated regions of Catholic Europe, storms were still accompanied by the tolling of bells a hundred and more years later. A debate that occurred in Boston during the 1750s shows the importance of local circumstances, in this case an earthquake. In what became a famous quarrel with John Winthrop of Harvard, the Reverend Thomas Prince aired the discomfiting thought that, in discharging electricity to earth, the effect might be to increase the incidence of earthquakes in the region. Boston had more "points of iron" than anywhere else in New England and seemed to be "more dreadfully shaken." This was not as naive as it seems, since a correlation between lightning and earthquakes was congruous with contemporary science.[38]

For A. D. White, clerical resistance to lightning rods was rooted in the belief that it would be presumptuous to interfere with Providence. The situation was undoubtedly more complex. The issue of presumption did, however, arise because clearly there were beguiling questions concerning the relationship between divine and human control. The nineteenth-century atheist Richard Carlile would claim that belief in divine providence made any attempt to improve the world sacrilegious. Yet remedial practices, preeminently those of medicine, had long enjoyed the blessing of the Churches. Accordingly, when the issue of presumption was debated in Philadelphia in 1760, in what became the American Philosophical Society, it was medicine that quelled the doubts: "with what care we endeavour to guard against the bad effects of other elements, . . . to prevent and remove disorders of the body plagues and sickness of every sort, and this without any imputation of presumption; why then should it be imagined more presumptuous in the present case?"[39] For such ameliorists, as for Priestley and other apostles of "rational dissent," improvements in technology could be subsumed under Providence as fulfillment rather than interference.

RELIGION AND THE LIMITATIONS OF REASON

The utility just considered was a practical utility in which scientific knowledge extended human control. But in Europe's new scientific societies one would also hear claims for moral utility. Intensive study of nature would divert young minds from temptations of the flesh; it might excite awe and wonder; it might

---

[38] Cohen, *Franklin's Science*, pp. 145–54.   [39] Cited by Cohen, *Franklin's Science*, pp. 142–3.

even provide proofs of that divine Providence which welcomed human collaboration in the improvement of the world.[40] The use of physico-theology
to *prove* the existence and attributes of the deity was, however, a form of rationalism within theology itself that could not escape criticism. In different
ways David Hume and Immanuel Kant (1724–1804) exposed the limitations
of reason as currently employed in defending Christianity. Hume observed
that arguments for design and arguments based on the miraculous could not
reinforce each other since the one presupposed a determinate order in nature,
the other its violation. Nor could either argument separately serve as a foundation for faith. No testimony, argued Hume, was sufficient to establish a
miracle unless the testimony was such that its falsehood would be more miraculous than the fact it purported to establish. It was a question of weighing
probabilities. The uniformity of past experience, encoded in laws of nature,
was such as to create a high antecedent probability against a violation having
occurred – so high, in Hume's opinion, that it is always more probable that
those who report a miracle are deceived than that the event occurred as they
report it. Hume buttressed his critique with observations made by previous
deists and skeptics – how miracles abounded among the more barbarous nations and how human testimony was unreliable, especially when vested interests were involved. He allowed that reports of unusual occurrences (such as
total darkness over the earth for eight days) might be believed if there were
sufficient coherent testimony; but where assent could be given there was an
onus not to declare a miracle but to search for the natural causes that had made
the unusual possible.[41]

Such discussions brought religious belief and probability theory into the
same discourse. The critical question was, "what degree of extrinsic probability would counterbalance or outweigh the great intrinsic improbability
of a violation of the laws of nature?"[42] This question attracted French mathematicians such as Condorcet, Laplace, and Poisson as well as British philosophers. Laplace, like Hume, would insist on the "immense weight testimony
must carry in order to admit the suspension of natural laws; and how great
an abuse it would be to apply the ordinary rules of criticism to such cases."
Poisson even advised that one should doubt the testimony of one's own senses
in cases where natural laws appeared to be abrogated. Such accounts made it
increasingly difficult to affirm that reports of miracles, in sacred texts or in
more recent testimony, could provide rational foundation for religious belief.
In inimitable style, Hume concluded that only one miracle survived his analy-

---

[40] Arnold Thackray, "Natural Knowledge in Cultural Context: The Manchester Model," *American Historical Review,* 79 (1974), 672–709; Derek Orange, "Rational Dissent and Provincial Science: William Turner and the Newcastle Literary and Philosophical Society," in Ian Inkster and Jack Morrell
(eds.), *Metropolis and Province* (London: Hutchinson, 1983), pp. 205–30.

[41] J. C. A. Gaskin, *Hume's Philosophy of Religion,* 2nd ed. (London: Macmillan, 1988); J. Houston, *Reported Miracles: A Critique of Hume* (Cambridge University Press, 1994).

[42] Lorraine Daston, *Classical Probability in the Enlightenment* (Princeton, NJ: Princeton University Press,
1988), pp. 306–42.

sis: that anyone should still believe in miracles, for such a belief required the subversion of one's understanding.

There were rejoinders. Priestley conceded those cases where there was inadequate testimony or where (as in the Virgin birth) there could be no witnesses. But there were miracle stories in the New Testament where the event had allegedly been witnessed by large numbers. Such instances were not to be dismissed lightly. Priestley's fellow architect of Unitarianism, economist and preacher Richard Price (1723–1791), also rebuffed Hume. Price conceded that the uniformity of past experience created an ever-increasing probability that miracles would not occur in the future; but this was far from proving them impossible. In discussions of the miraculous much would depend on whether one was already committed to a theocentric position. For the evangelical reformer John Wesley (1703–1791), author of a compendium of popular science, it did not strain credulity to believe in a miracle-working God. A deity who had performed the supreme miracle of making a world from nothing would surely be able to perform lesser miracles. The crucial role of such presuppositions in theological debate was exposed both by Hume and Kant in their respective critiques of the argument for design.

It was through the design argument that some of the most intimate links between science and religion were forged. But did it not already assume, as we have just seen Wesley assume, the role of a Creator? Hume exposed the circularity in his posthumous *Dialogues Concerning Natural Religion* (1779). Even if the world could be shown to resemble a machine or some other human artifact, this would not *prove* the existence of a single transcendent Mind, because many minds can be involved in the design and construction of machinery. Through the character of Philo in the *Dialogues,* Hume exposed the frailty of analogical argument on which the proof of a Designer depended. The seeming order and purpose in nature that the physico-theologians captured with their mechanical analogues could, with equal propriety, be predicated of an animal or vegetable. And if the world, as Philo suggested, resembles an animal or vegetable *more* than a watch or a knitting loom, the cause of the world might be an egg or superseed rather than an intelligent Creator. Hume also insisted that a cause must always be proportioned to its effects, with the consequence that one could not properly infer the infinite attributes of a transcendent deity from the patterns of a finite world. Moreover, if one argued for the beneficence of the Creator on the strength of seemingly providential features of creation, for consistency it would surely be necessary to infer a maleficent deity from the high degree of misery. The world might remotely resemble a work of human intelligence, but the analogy was surely too weak to bear on human conduct.[43]

Hume had raised the objection that to explain the presence of order in the

---

[43] David Hume, *Dialogues Concerning Natural Religion, and The Natural History of Religion,* ed. J. C. A. Gaskin (Oxford: World's Classics, 1993).

world by postulating mental order in a Creator was to invite an infinite regress because the source of that mental order was left unexplained. The regress could be avoided only if mental order in a deity could be taken as self-explanatory. But surely this would be an a priori assumption. For similar reasons Kant also concluded that the design argument failed. It simply assumed that a self-existent Being could be established as the First Cause of the cosmos. In his *Critique of Pure Reason* (1781) Kant showed that rational proofs of such a Being were unattainable. Indeed one effect of Kant's critical writings was to disentangle threads that had bound science and theology together. Although he could still say that scientific investigation was possible only when nature was conceived as if its laws were the result of design, the emphasis fell on the "as if." That laws of nature were to be regarded as if they had been prescribed by a lawgiver was not sufficient to establish that they had. Only if no other explanation for the appearance of design could be found would the inference to a designer be secure. But one could not know that all other possibilities had been exhausted. Kant did not deny that natural science had metaphysical foundations. But it was possible to separate what he called the "metaphysics of corporeal nature" from general metaphysical issues concerning God, freedom, and immortality. The supreme deficiency of physico-theology was that, no matter how much ingenious artistry might be displayed in the physical world, it was powerless to demonstrate the *moral* wisdom that had to be predicated of God. To make the world morally coherent, it was necessary to postulate a rational and moral Being who, as creator and sustainer, has the necessary power to make happiness proportional to virtue. But this was a far cry from claiming that the objective existence of such a Being could be rationally demonstrated. In deobjectifying religion, Kant not only licensed a separation of scientific from religious discourse but also by dwelling on the postulates of individuals rather than an assured knowledge of the deity, ushered in new possibilities for agnosticism.[44]

THE LEGACY OF ENLIGHTENMENT CRITIQUES

Assessing the consequences of Enlightenment critiques of religion is bedeviled by long-term and short-term complications. A current "postmodern" focus on local rationalities, on the distinctiveness of local scientific and religious communities, is a far cry from Enlightenment projects designed to uncover a universal "reason" by which the absolute rationality of specific beliefs might be determined. And back in the eighteenth century Hume would not have been surprised by the survival both of popular religious belief and of natural

---

[44] John L. Mackie, *The Miracle of Theism: Arguments For and Against the Existence of God* (Oxford: Oxford University Press, 1982); John P. Clayton, "Gottesbeweise," in Gerhard Krause and Gerhard Müller (eds.), *Theologische Realenzyklopädie* (Berlin: De Gruyter, 1984), pp. 724–84.

theology. By his own insistence, justice, morality, politics, and religion were grounded not in reason but in habit and custom. It was this that made a sociological inquiry into the human condition both possible and urgent. But if religious beliefs were grounded in habit and custom, they were unlikely to be shaken by a reasoned critique. Scottish commonsense philosophers, notably Thomas Reid (1710–1796), argued that belief in an intelligent deity was an intuitive and ineradicable belief. Some, such as Dugald Stewart (1753–1828), even turned a Humean empiricism against Hume's own skepticism. Hume had interpreted causality in nature not as a form of necessity in which effects were bound by some hidden power to their causes but rather as an expression of an expectation in us that because particular causes and effects had been constantly conjoined in past experience, they would continue to be. In Stewart's reading, Hume's rejection of an invisible necessity in the linkages of nature was consistent with a voluntarist theology: "Mr. Hume's doctrine . . . keeps the Deity always in view, not only as the first, but as the constantly operating efficient cause in nature, and as the great connecting principle among all [phenomena]."[45]

Responses to Kant on causality were equally diverse. In his *Critique of Teleological Judgment* (1790) Kant insisted that the purposive causality found in living organisms could not be explained by analogy with a work of art. The formative power of an organism was inherent within the organism itself: living things were both cause and effect of themselves. Here was another reason to disentangle the physiologists' unavoidable references to teleology from a theological superstructure. In Germany, where Kant had his greatest impact, there were physiologists who clearly welcomed that liberation.[46] Goethe's vision of ideal morphological types from which living systems were derived was inspired both by Kant and by his own quest for beauty in both physical and spiritual realms.[47] Yet the Kantian legacy was richly ambivalent in that to expose the limits of reason in the context of theistic proofs might be read as making way for faith rather than destroying it. In his *Critique of Pure Reason* Kant himself had proposed that otherwise discredited proofs could still regulate the ideal of a Supreme Being to ensure that it remained an impeccable ideal. The traditional arguments did not work as proofs, but they were still useful in purifying the concept of God, ensuring that it was self-consistent.

It was therefore possible for a natural theology to survive as long as it moderated its claims. Indeed a resurgence rather than an eclipse of the design argument has been detected in the English-speaking world as the eighteenth century drew to a close. Whereas earlier in the century there were perceptions that the battle against atheism had been won, by the 1790s apologists had to

[45] Dugald Stewart, *Elements of the Philosophy of the Human Mind*, vol. 1, in Sir William Hamilton (ed.), *The Collected Works of Dugald Stewart* (Edinburgh: Constable, 1854), vol. 2, p. 479.
[46] Timothy Lenoir, *The Strategy of Life* (Dordrecht: Reidel, 1982), pp. 17–53.
[47] Nicholas Jardine, *The Scenes of Inquiry* (Oxford: Oxford University Press, 1991), pp. 37–43.

contend with the atheism of Diderot and d'Holbach, the skepticism of Hume, the naturalistic vista of earth history proposed by Buffon, a mechanism for the transformation of species mooted by Erasmus Darwin – and at a time when all eyes were on Paris.[48] The burning of Priestley's laboratory by a Birmingham mob furnishes a powerful symbol of conservative reaction to revolutionary terror – a reaction in which the sciences, as an expression of free thought, were easily blamed for the fomenting of revolutionary politics.[49] In William Paley's *Natural Theology* (1802) it is possible to see a response to this cumulative challenge. His preoccupation with the finesse of anatomical structures has also been seen as reflecting the new world of industrial machinery.[50] Far more editions of Paley than of Hume would be read in nineteenth-century Britain. In France, too, an authoritarian response to the terror was expressed through the reaffirmation of spiritual values. In Maine de Biran (1788–1824) there was a refocus on the inner life and free will of the human agent; in the romantic writings of the royalist Chateaubriand (1768–1848), worship of the deity was commended for its esthetic rewards; in those of the Catholic monarchist Joseph de Maistre (1754–1821), the orgies of the Revolution had been "satanic," and such fantasies as a pristine "state of nature" or Rousseau's "social contract" violated the fundamental truth that human society could be stabilized only through God-given laws.[51] In Romantic reactions against French materialism, science as well as religion could play its part. In London's newly established Royal Institution, Humphry Davy (1778–1829) showed how, from the same two elements nitrogen and oxygen, gases with radically different properties could be made: dosed with nitrous oxide one choked with laughter; on the brown fumes of nitrogen dioxide one would choke to death.[52]

To conclude with these conservative images would, however, be to skew a legacy that was of incalculable importance to those who continued to reconstruct nature in naturalistic terms. The impress of the Scottish Enlightenment has been seen in Charles Lyell's introduction to his *Principles of Geology* (1830–3), in which deference to Biblical authority was seen as obstructive to the science.[53] Nineteenth-century models of an evolving cosmos built on those of Buffon, Kant, and Laplace. Charles Darwin's "bulldog," Thomas Henry Huxley (1825–1895), would rediscover Hume; and Darwin's own experience

[48] Brooke, *Science and Religion*, pp. 209–25; David Burbridge, "William Paley, Erasmus Darwin, and the System of Appetencies: Natural Theology and Evolutionism in the Eighteenth Century," *Science and Christian Belief*, 10 (1998), 49–71.

[49] Maurice Crosland, "The Image of Science as a Threat: Burke versus Priestley and the 'Philosophic Revolution,'" *British Journal for the History of Science*, 20 (1987), 277–307.

[50] Neal C. Gillespie, "Divine Design and the Industrial Revolution," *Isis*, 81 (1990), 214–29.

[51] Jordanova, *Enlightenment and Its Shadows*, pp. 209–16.

[52] David Knight, *The Transcendental Part of Chemistry* (Folkestone: Dawson, 1978), pp. 61–84; Knight, *Humphry Davy: Science and Power* (Oxford: Blackwell, 1992), pp. 73–88.

[53] Rachel Laudan, *From Mineralogy to Geology* (Chicago: University of Chicago Press, 1987), pp. 202–3.

of the Brazilian jungle ("twiners entwining twiners . . . beautiful lepidoptera – Silence – hosannah") shows how a life in science could spark its own surrogate religion. In these and other respects the further articulation of the sciences would continue to challenge established religious verities. When combined with the *historical* criticism of Scripture regrounded in Hegel's grasp of the radically different and obsolete thought forms of a bygone era, the sanctity of the sacred texts was placed in even greater jeopardy.[54] And if a greater religious tolerance was finally won during the nineteenth century, it had been made possible by the creative engagement of this difficult issue in the campaigning literature of the eighteenth.[55] It was a difficult issue because, as Priestley had discovered, to lobby for the emancipation of Catholics in Britain was to incur the censure of one's fellow dissenters, who worried lest their own case be jeopardized. Priestley would eventually find solace in the America of Thomas Jefferson (1743–1826), whose bid to privatize confessional religions and to construct a rational religion for public consumption had an enduring constitutional legacy. Traces of this eighteenth-century project are still discernible in the academic study of religion whenever investigations into comparative religion are driven, albeit anachronistically, by the quest for a common core.[56]

[54] W. Neil, "The Criticism and Theological Uses of the Bible, 1700–1950," in S. L. Greenslade (ed.), *The Cambridge History of the Bible* (Cambridge University Press, 1975), pp. 238–93; Marilyn Chapin Massey, *Christ Unmasked: The Meaning of the "Life of Jesus" in German Politics* (Chapel Hill: University of North Carolina Press, 1983).

[55] Outram, *Enlightenment,* pp. 36–46.

[56] John P. Clayton, "Thomas Jefferson and the Study of Religion," an Inaugural Lecture (Lancaster: Lancaster University, 1992).

33

SCIENCE, CULTURE, AND THE IMAGINATION

Enlightenment Configurations

George S. Rousseau

The Arts and Sciences brighten'd *Europe's* face,
Learning did no more noble blood debase,
T'was honour's genuine stamp, and dignify'd the race.
(John Mawer, *The progress of language,
an essay . . .* , London, 1726)

Hence the fine arts become like the mechanical; genius is fettered by precedents; and the waving line of fancy exchanged for a perpetual round of
repetitions.
(William Rutherford, *A View of Antient History; including
the progress of literature and fine arts,* London, 1788–91)

A CENTURY OF CHANGE

Alexander Pope (1688–1744), reputedly the greatest English poet of his age and
a man whose satiric lash spared no target and whose panegyric pen captured
entire lives in a single couplet, exalted Isaac Newton this way in the widely
read *Epitaph Intended for Sir Isaac Newton In Westminster Abbey:*

Nature, and Nature's Laws lay hid in Night.
God said, *Let Newton be!* and All was *Light.*

These lines were widely quoted, paraphrased, and translated into every European language within a few years of Newton's death in 1727. Leibniz, Voltaire,
and most of the *philosophes* knew them by memory, as did the French and
the Italians. Goethe, that unparalleled Enlightenment man (enlightened in
almost all the senses in which this label was used in the eighteenth century),
imagined himself in Newton's place, and Byron composed variations on the
Pope couplet for poetic sport. One could fairly predict that the Newton whom
Pope epitomized as a mortal man, his couplet art transformed into an immortal – a veritable god. The analogy was this: God–Newton, Newton–light. In

case anyone missed the point, Pope reiterated it in another famous line in his philosophical poem *An Essay on Man* (1733), where he extended the analogy:

> Superior beings, when of late they saw
> A mortal Man unfold all Nature's law,
> Admir'd such wisdom in an earthly shape,
> And shew'd a NEWTON as we shew an Ape (2. 31–34).

A century later Charles Lamb (1775–1834), not the most celebrated commentator on the past but certainly one of the best read, performed an about-face and damned Newton at a dinner given by the painter Robert Haydon for the circle of Wordsworth and Keats. At Haydon's silver-laid table Lamb denigrated the worthless Newton as "a fellow, who believed nothing unless it was as clear as the three sides of a triangle," a crude charlatan who "had destroyed all the poetry of the rainbow by reducing it to prismatic colours."[1]

The shift was monumental. How can it have occurred in less than a century? This essay aims to answer the question and address the problems raised by the question without reducing either interrogation or repertoire of answers to neat patterns of forced explanation. More than anything, it aims to show that the entirety of knowledge and its rational and imaginative components – and not merely the individual progress of either poetry or science, art or truth – were at stake in the transformation. Lamb's heroes were the dramatists and poets, the great tradition from Shakespeare to Milton and Wordsworth, in whose imaginary company Lamb was forever complaining (to Coleridge) that he was a scientific ignoramus: "a whole Encyclopaedia behind the rest of the world."[2] He lamented that "Science has succeeded to Poetry no less in the little walks of Children than with Men," and wondered, "Is there no possibility of averting this sore evil?"[3]

Unlike Coleridge, Lamb judged "luddite science" to be the scourge of the future. It would deaden the imagination and wreck the arts, he claimed, especially poetry. It was literal, obtuse, transparent, and antithetical to the ways of imagination, with Locke and Newton its most terrifying, if successful, representatives.[4] A century earlier, Swift, like Lamb, had been the odd man out: if Pope and his contemporaries extolled Newton as a saintly presence,[5] Swift

---

[1] See Benjamin Robert Haydon, *The Autobiography and Memoirs of Benjamin Robert Haydon* (London: Humphrey Milford, 1927), p. 392.

[2] Charles Lamb, "The Old and the New Schoolmaster 1821," in Roy Park (ed.), *Lamb as Critic* (London: Routledge, 1980), p. 160.

[3] Charles Lamb, "Children's Books 1802," in Park (ed.), *Lamb as Critic*, p. 165.

[4] For the perception of eighteenth-century science in some of these categories, see Andrew Cunningham and Nicholas Jardine (eds.), *Romanticism and the Sciences* (Cambridge University Press, 1990); G. S. Rousseau, "Literature and Science: The State of the Field," *Isis*, 72 (1981), 406–24; John Christie and Sally Shuttleworth (eds.), *Nature Transfigured: Science and Literature, 1700–1900* (Manchester: Manchester University Press, 1989).

[5] See M. H. Nicolson, *Newton Demands the Muse* (Princeton, NJ: Princeton University Press, 1946); Henry Guerlac's study of the image of Newton, *Newton on the Continent* (Ithaca, NY: Cornell University Press, 1981); and more generally Walter Schatzberg, Ronald A. Waite, and Jonathan K. Johnson

had harbored grave doubts despite his never naming names (Newton). Swift's fierce satire in the "Voyage to Laputa" in *Gulliver's Travels* (1726), published one year before Newton's death, is silent on Newton and the Newtonians but makes plain that praise, let alone blind deification à la Pope, was unthinkable for Swift.[6] This is another inconsistency that must be taken into account from the outset. But there is a far more crucial distinction to be made whose confusion imperils our enterprise: the notion that the European "Romantics" despised science or were antiscientific in outlook and temperament.[7] They were not, despite many attempts to present them as such on both sides of the Channel. Furthermore, the differences among them are so great as to render impossible the consideration of them as a unit. They hardly constituted one mind or view, and the label itself – Romantics or Romanticism – is highly misleading.[8] They harbored, for example, a different notion of genius than did the generation of Pope and Voltaire, even if this difference in itself does not prevent their extolling scientific genius, as several of them did of the German natural philosophers.

This pursuit of genius produces little advance.[9] Furthermore, such exploration assumes that *both* literature and science were then stable, or even somewhat stable, categories, something that they were not and had not been since Bacon's proposed reforms; they were no more stable than any other broad Enlightenment labels forever brought under the stress of different thinkers and the stresses of their agendas and ideologies. For some of our contemporaries this last caveat is superfluous: all categories, they argue, are culturally constructed and must be filled in – reconstructed – to be historically valid. For others, these unstable categories must be driven to a desperate skeptical extreme, so far that we cannot meaningfully invoke their labels at all. The balanced and integrated approach surely lies somewhere in the middle. Otherwise, the historical enterprise itself would be practically invalidated.[10]

DOCTRINES OF OPTIMISM

During Alexander Pope's adult years (1714–44) diverse groups of anti-Newtonians flourished – those who themselves were opposed not merely to

---

(eds.), *The Relations of Literature and Science: An Annotated Bibliography of Scholarship, 1880–1980* (New York: Modern Language Association of America, 1987).

[6] See Marjorie Hope Nicolson's pioneering studies on Swift reprinted in *Science and Imagination* (Ithaca, NY: Cornell University Press, 1956).

[7] The case for the idea that the Romantics despised science has been made by Hans Eichner, "The Rise of Modern Science and the Genesis of Romanticism," *PMLA*, 97 (1982), 8–30.

[8] A. O. Lovejoy, *Essays in the History of Ideas* (Cambridge, MA: Harvard University Press, 1948).

[9] It is adroitly argued by Simon Schaffer in Cunningham and Jardine (eds.), *Romanticism and the Sciences*, pp. 82–98.

[10] For further discussion of the stability of these categories see G. S. Rousseau, "Discourses of the Nerve," in Frederick Amrine (ed.), *Literature and Science as Modes of Expression* (Dordrecht: Kluwer, 1989), pp. 29–60.

Newton's theories but also to what he symbolized – and it falsifies history to pretend they did not or that they were not broadly based; for the anti-Newtonians, like the Newtonians, came in different casts and colors, so to speak. Moreover, it took almost a generation from the 1680s and 1690s for Newton's theories to infiltrate the Low Countries and France, but there, too, all sorts of anti-Newtonians flourished.[11] In England John Hutchinson (1674–1737), the Yorkshire "physico-theologian" (not to be confused with Francis Hutcheson, the Glasgow moral philosopher), was one of them. But there were others who often called themselves "Hutchinsonians," which became a buzzword *contra* Newton.[12] The aim of the Hutchinsonian way of thinking was the creation of a system reconciling God with the physical evidence of Nature while rejecting Newton's work as "a cobweb of circles and lines to catch flies in."[13] Hutchinson and his followers reinterpreted the Old Testament in the light of new scientific theories, locating analogical and metaphorical references to the physical forces that the new thinking saw as responsible for the creation of the universe, all of which anticipated the Blakean revolt against Newton later in the century.

Also in England, writers as diverse as Horace Walpole, John Wesley, and Samuel Coleridge reveled in Hutchinson's anti-Newtonian philosophy. But their opposition, no matter how Hutchinsonian, amounted to a drop in the ocean. The dissemination of Newtonianism was widespread and quick, even among the unscientific and uneducated. Farther north, the Scots quickly converted from the old maths to Newtonianism, to the degree of incorporating his reforms into their curricula, and the same occurred on the Continent, from Amsterdam to Geneva and Vienna.

The anti-Newtonians (so far as they can collectively be generalized about) not only attacked the arrogance of science's quest – the belief that it could discover all the universe's laws – but also argued against its theological intentions and moral foundations. Awed by the mathematical proofs proposed by the scientists, anti-Newtonians emphasized the word – the ancient *logos* – and its role in the deity's revelations to mankind through reason and imagination. Their ideological program reveals much that the world of Pope and Voltaire aimed to champion, especially a scientific genius symbolic of the march forward toward nationhood and world presence. Newton fit the bill more than anyone else in English history, as had Huyghens in Holland and as Lavoisier would in France, exceeding Bacon, Sydenham, and all the other early members of the Royal Society. From the 1660s forward, when Dryden and Pepys were as enthusiastic as Pope and his generation would become about

---

[11] For the counter-movement see Margaret C. Jacob, *The Radical Enlightenment: Pantheists, Freemasons and Republicans in Early Modern Europe* (London: Allen & Unwin, 1981); Margaret C. Jacob (ed.), *The Dutch Republic in the Eighteenth Century* (Ithaca, NY: Cornell University Press, 1992).

[12] See Albert Kuhn, "Glory or Gravity: Hutchinson vs Newton," *Journal of the History of Ideas*, 22 (1961), 303–22; Brian Stock, *The Holy and the Demonic* (Princeton, NJ: Princeton University Press, 1983).

[13] See Kuhn, "Glory or Gravity," p. 307.

science's ability to transform modern society into the quasi-utopian state these writers imagined, this need to celebrate one figure as symbolic of science's prowess was self-evident.[14] A nation's power, whether that of England, France, or even the still-untamed Russia, seemed encased in its degree of progress; its progress seemed predicated on scientific achievement, which translated into practical technological advance and then in turn to wealth and prosperity. William Rutherford, educated in the best classical methods of the period and head of an academy at Uxbridge, eloquently stated the claim at the end of the century when introducing his treatise on *Progress of Literature and Fine Arts:* "It redounded to the praise of the antients, that a taste for the arts and sciences was frequently united with talents for public affairs; the road to business and the path to literature coincided; and the hero and the statesman joined to the cares of the commonwealth an elegant intercourse with the Muses."[15] Earlier, the anti-Newtonians proselytized and wrote but without chipping the block of science's optimistic might.

Nevertheless, however symbolic a cultural icon Newton became after his death in 1727, resistance was considerable, especially among imaginative and philosophical types. What we today somewhat anachronistically call "imaginative literature" – drama, poetry, novels, romance – was then still, it is true, largely realistic and not at loggerheads with the moral implications of Newtonian optical and mathematical theory vis-à-vis the godhead: that is, mimetic of a readily grasped external reality available to the human senses and captured in literary forms; not yet symbolic in any sense that clouded the reality of marriage and the family, nationhood, war and peace, and the advancement of science and knowledge. European literature as an institution still construed its task as conveying to audiences the concerns of *both* the private and the public spheres, whether through the imagery of Nature, landscape, or the body or through the workings of governments and their ministers. Readers craved up-to-the-minute information in their literature, which was not yet "escapist" to our modern degree, vicarious though some of its fictions were. Much of the literary triumph of the poetry of the period, more so than its prose didacticists, lay in conveying the fundamental essences of humanity itself, and the concerns of science were prominent in this endeavor. Even the drama abounded with references to the latest discovery of new aspects to the heavens or the mighty seas.

Poetry could accomplish this end without bombast or comic rhetorical inflation. Pope's *Rape of the Lock* (1712–14), perhaps the finest mock epic poem in any language, contains a fourth canto titled "The Cave of Spleen" that demonstrates complete familiarity with the (then) most recent theories of med-

---

[14] Marjorie Hope Nicolson, *Pepys' Diary and the New Science* (Charlottesville: University of Virginia Press, 1965).

[15] William Rutherford, *A View of Antient History; including the progress of literature and fine arts* (London, 1788–91), p. v. See also for a contemporary discussion, J. D. D. Anderson, *Progress of Arts and Sciences* (London, 1784).

ical hysteria. James Thomson's *Seasons* (1726–8), the most widely read English poem of the eighteenth century bar none (it was translated into many languages and was a best seller in Britain but globally could not compete with scriptures, psalms, and hymns), incorporated Newton's optical and gravitational theories. Samuel Garth's *Dispensary* (1699) – like *The Rape of the Lock*, another triumph in the mock heroic form – took medicine as its primary subject matter and wittily versified the battle of the apothecaries and surgeons over the role of "dispensing" medicines. Mark Akenside's *Pleasures of Imagination* (1744) lyricized the creative act and its afforded pleasures. In the same year John Armstrong – a trained doctor who, like many of his day, also published poetry – versified diet, death, and disease in his widely read *Art of Preserving Health*. All had absorbed some version of Newtonianism into the fabric of their poetry.

South of the Alps, the same could be claimed despite Roman Catholic resistance. Lodovico Antonio Muratori (fl. 1740s), a brilliant and erudite Italian poet contemporary with these British writers, was unabashed about his Newtonianism and openly professed to be applying Newton's theories to the dreamscapes of his poetry.[16] Farther north, other poets were even more explicitly didactic in their agendas, as when English doctor-poet Malcolm Flemyng published an extended epic poem in hexameters called *Neuropathia* (1740) about the intricate and microscopic peregrinations of the nerves, spirits, and fibers; it was an anatomical parallel to Erasmus Darwin's *Loves of the Plants* (1789), which performed a similar task for the sex lives of the microscopic botanical world just at the moment when the American colonies were declaring themselves free of the mother country. Such poetry varied enormously in range and quality but shared a post-Newtonian confidence that wondrous physiological fibers (Flemyng) pulsating on the inner highways of the bloodstream – a type of busy Enlightenment anatomical Internet – and lascivious botanical sex lives (Darwin) could sustain long epic poems. The biographically unknown Flemyng paved the way for a much greater poet, William Blake, to build upon his (Flemyng's) fibrous Newtonian vision of the human condition by subverting it. Blake replaced Flemyng's nervous positivism with a mysterious fibrous animism that awakened at birth and then developed through the equally fibrous cycles of love, marriage, senescence, and death – a view that the anatomical fibers, unseen by any microscope, encompassed the whole circle of human life but could never be dissected or plainly reduced in the Newtonian way.

A similar dissemination occurred on the Continent in the realms of literature and art, although nothing in German natural theological poetry

---

[16] See Muratori's *Book on Imagination and Dreams* (1741, trans. English 1747) and A. Andreoli, *Nel mondo di Lodovico Antonio Muratori* (Bologna: Il Mulino, 1972). Only the barest traces of Flemyng's life remain: whatever papers and manuscripts existed in the eighteenth century have disappeared, and nothing remains to be discovered about him anywhere.

approximated poems so didactic and extensive as Flemyng's. One wonders whether there was pietistic resistance in middle Europe to naturalize, and technically "scientize," these analogies; this is a bewildering lacuna in that no sturdy equivalents for the British literary tradition can readily be found so far as the internal effects of dream and reverie were concerned. But the moralization of the nerves (for example) was commonplace in the newly developed German literature of sensibility. Here, an old tradition of picaresque was grafted to a novel of education (*Bildungsroman*) whose psychology of character often built on the chain of nervous arousal, followed by sympathy based on this nervous sensibility, and culminating in the empathy that formed the true mark of the educated person. In more rarefied German literature – usually poetic, sublime, esthetic – a cult of *Einfühlung*, or sensitive and empathic affinity, lorded over literary and fictional figures. This is the route leading to much early German Romantic homoerotic verse and to the esthetics of Winckelmann, Goethe, and Kant.[17]

This Germanic literature (Austrian and Prussian as well) could not have developed without a prior theory of nervous anatomy on which the sensibility shapes human empathy.[18] Not much further south, Johann Georg Zimmermann (1728–1795), a distinguished Swiss physician-writer who published books on nervous disorders as well as poetry translated into most European languages, represented the norm of Enlightenment diversity rather than the exception. The difference lay in esthetic distinction: he wrote more perfectly than most. And his appointment as private physician to George III and Frederick the Great merely legitimated his biographical stature as a genuine product of the new "Enlightenment." Philosophically, Zimmermann began life as a Hallerian (whose life he wrote) in the belief that the fiber – the basic substance of the anatomical nerve – was the fundamental matter of life. This theory he extended to other realms, medical and nonmedical, historically to eras past and present, and by so doing he laid the foundation for the early nineteenth-century anthropological debates about the fibrous basis of civilized societies in history. Zimmermann also corresponded widely: with Herder, Blumenbach, Wieland, Russia's Empress Catherine, the French ideologues, and Parisian literati – in short, with everyone important, and he generated a strong case for national pride much to be cultivated among folk peoples such as the Swiss. He never weakened in his belief that experience counted for much more than reason in the repair of the pathological body restored to normalcy. He observed the remarkable events of the French Revolution from the eastern side

---

[17] For two different approaches to Goethe's literary-scientific enterprise, see Frederick Amrine, *Goethe in the History of Science* (New York: P. Lang, 1996); and Alice A. Kuzniar (ed.), *Outing Goethe and His Age* (Cambridge University Press, 1996).

[18] See C. Brunschwig, *Enlightenment and Romanticism in Eighteenth-Century Prussia* (Chicago: University of Chicago Press, 1974); Hans-Peter Schramm (ed.), *Johann Georg Zimmermann* (Weisbaden: Harrassowitz, 1998).

of the Jura mountains, where he often wandered alone like Wordsworth's leech-gatherer. But even before then Zimmermann had written his most famous treatise – *Solitude considered with respect to its dangerous influence upon the mind and heart* (1792) – a bible of Romanticism he published in German in 1784. It created a sensation in England and was prolifically illustrated there before the century was out. These diverse activities gave currency to the view that Zimmerman was known to be an eccentric amalgam of sentimentalism, melancholy, and enthusiasm, and this may explain why he, unlike Newton, appealed to Lamb. Yet there was nothing literal or logical about Zimmermann, the man or his poetry.

One aim of these writers was to explain the esthetic implications of the new scientific domain. There was then much less doubt than there is today about science's "edge of objectivity"[19] or its unique ability to sustain the progress necessary to create the perfection for which humankind yearned. In the Orient, in the Levant, in the Mediterranean crescent, the wisest sages dedicated themselves to this question: how to find the progress that would lead humankind to its perfection. Few writers explicitly invoked the word "progress," but it lurks under the skin of every page of their writing: the notion that things were getting better, each day fairer and sunnier than the last. It was not a mindless optimism but a rosy outlook. Writers did not debate science's metaphysical meanings; that was left to others. They would have agreed with the encyclopedists who derived their definitions from the traditions of *scientia* as knowledge that was accurate, communicable, predictable, and knowable through the rational faculty. Thus, at the beginning of the period they would have agreed with John Harris – author of the first truly general encyclopedic work in English to which readers could turn – that "science is Knowledge founded upon, or acquir'd, by clear, certain, and self-evident Principles";[20] and later, with the principally Scottish authors of the first edition of the *Encyclopaedia Britannica* (1768) that "science is any doctrine deduced from self-evident principles";[21] or at the end of the period, in 1819, with Londoner Abraham Rees, who repeated this dictum but added that science was everything that lay in opposition to art;[22] or one year later, with Robert Watt, a Scottish compiler and encyclopedist, who in his entry under science noted epigrammatically that whereas "science plans, art performs."[23] In all these finely tuned discriminations, the authority of John Locke (1632–

[19] Charles Gillespie, *The Edge of Objectivity: An Essay in the History of Scientific Ideas* (Princeton, NJ: Princeton University Press, 1960).

[20] John Harris, *Lexicon Technicum: An universal English dictionary of arts and sciences*, 2 vols. (London, 1736), entry on "science."

[21] Society of Gentlemen in Scotland, *Encyclopaedia Britannica*, 3 vols. (Edinburgh, 1771), entry on "science."

[22] Abraham Rees, *The Cyclopaedia; or, Universal Dictionary of Arts, Sciences, and Literature*, 45 vols. (1819–20), entry on "science."

[23] Robert Watt, *Bibliotheca Britannica*, 4 vols. (Edinburgh, 1820), vol. 4, entry on "science."

1704) counted for much as the voice of science or certainty, especially in contrast to the "sceptics" – David Hume and his Scottish colleagues – who doubted that science could guarantee the *certainty* it claimed.

As crucial in the mindsets implied by these configurations was the still integral unit of knowledge constituted by science and religion and morality. They had not yet been classified into the species, or disciplines, of science, theology, and philosophy that we take for granted. This may be one reason that the developing novel subsumed functions of their split into specialized discourses by providing the illusion that an organic unit had lingered despite apparent fragmentation.[24] Seeded in "Quixotic" fifteenth-century Spain and in England in the Elizabethan period, the novel was born in France (to the degree its origins can be grounded in a time and place)[25] in the seventeenth century in works such as Mme de La Fayette's (1634–1693) remarkable *Princess of Clèves* (1678). Her fiction anticipates the eighteenth century's sentiment and sensibility as well as its search for realism. Thematically, the early novel's origins lay in romance and in the centrality of erotic love. The old medieval romances had been its treasure trove, the source of many of its fictions. But equally important was its message that the true Godhead lay within the *self* rather than outside it in any external physical world. But as the French novel of sensibility developed after the appearance of the *Princess of Clèves*, its "nervous content" increased. The form gradually absorbed the new physiology and its assumptions about body and selfhood into its esthetic, so emphatically that a "nervous French novel" can also be posited in the eighteenth century, exemplary in Diderot's *Rêve*, apparent in the French imitators of *Tristram Shandy*, and culminating in the Marquis de Sade's exquisitely "nervous" *Justine* and his pornographic *Philosophy of the Bedroom*.[26] It may be folly to classify novels by their national stereotypes, as if they were not fictions but diseases (i.e., the French disease, Dutch disease, Italian disease, and so forth), but broad differences based on national culture then existed.

The French novel especially cultivated nervous sensibility because its philosophic concerns turned prominently to the springs of romantic intrigue and pangs of the erotic heart, whereas the English novel (which could be romantic and sentimental) nevertheless foregrounded marriage, a grass-roots morality, and the family. In all this prose fiction the *self within* eventually emerged as the novel's true subject matter no matter how ingeniously disguised or fancifully presented. Loosely speaking, one could say that the writer of prose fiction was a *scientist of the interior self* who penetrated the psyche's

[24] This point has been adroitly developed in M. McKeon, *The Origins of the English Novel, 1600–1740* (Baltimore, MD: Johns Hopkins University Press, 1987); and in his "The Origins of Interdisciplinary Studies," *Eighteenth Century Studies*, 28 (1994), 17–28.
[25] I am skeptical of the theory of the novel's origins in ancient Greece and Turkey developed by Margaret A. Doody in *The True Story of the Novel* (New Brunswick, NJ: Rutgers University Press, 1996).
[26] For the nervous French novel, see G. S. Rousseau, "Cultural History in a New Key: Towards a Semiotics of the Nerve," in Joan Pittock and Andrew Wear (eds.), *Cultural History* (London: Macmillan, 1991), pp. 25–81.

entrails: a Newton of the mind, as it were, dissecting human nature and its ulterior motives with the same precision as the natural philosopher charting the physical world.

For the novelist, language and its cadences formed the essential crucible of his or her métier. Nor were the poet's issues of originality and the burden of the past stressful: there was little anxiety of influence to eschew, given the novelty of the form. Whether as Oronooko, Robinson Crusoe, Lemuel Gulliver traveling in remote parts, or Moll Flanders in London and Essex; whether as Pamela Wilson or Clarissa Harlowe in familiar country houses, the new novelists, somewhat like the new anti-Newtonians of the early eighteenth century, were experts of explanation whose forte was the mystery of human nature and the riddle of character difference. Writers of fiction explored character in their quest to unlock the psychological complexities of the human head and heart; physical scientists, the material universe in search of its laws; and both were explorers calibrated to provide a curious audience with in-depth explanations.[27]

PARALLEL MENTAL UNIVERSES

If the new English fiction of Behn and Defoe, Richardson and Fielding, produced novels of morality, an opposite moral state – amorality – cannot be said to have consisted primarily of things scientific. That is, today science clearly exists in both realms – moral and amoral – as well as immoral. But in the world of Defoe and Diderot and their successors, science was coming into its own primarily as good or bad, so to speak, according to conditions of evidence and proof rather than as moral imperatives. Science had not yet assumed its categorical moral stances outside the religious domain, nor had government yet incorporated it in any form resembling the infrastructures of the nineteenth century. Moral and amoral science required the institutionalizations and national science policies of bureaucracies and big governments.

But there can be no doubt about the moral preoccupations of the developing novel: to show that character stood in relation to the crucial functions of manners and marriage in ways that virtually dictated the norms of society itself. Samuel Richardson (1689–1761) perfected the formula in *Clarissa* by penetrating into the psychological heartland of his characters: their innermost emotional fibers and essences. Fielding may have been a psychologist *manqué*, but he was well read in empirical psychology and construed his artistic feat as lying in the construction of a "new province of writing" through the exposure of personal affectation and hypocrisy. Smollett, despite his scurrilous satiric thrust and picaresque penchant, retained a deep, almost Presbyterian,

---

[27] For the "self-within," see Stephen Cox, *The Stranger Within* (Pittsburgh, PA: University of Pittsburgh Press, 1979).

moralistic strain throughout his novels. His medical training and profound
knowledge of human anatomy left him perfectly poised to dissect the passions in their moral contexts with a clinical eye.[28]

But if the novel was inherently moral, science *also* was. The notion of an
amoral science – amoral through professional and institutional corruption
and because truth, no matter how damaging, and not morality, was its self-
iterated criterion – had not yet developed. That came at the beginning of the
nineteenth century. However, like imaginative literature, still in the last stran-
gleholds of the neoclassical "rules" prescribing the old Aristotelian unities of
time and place, science was already perceived along lines of (approximately)
good and bad as true and false. Its corruptions and politics, as Swift and
Mandeville had shown, were rife. In both *The Fable of the Bees* and *Gulliver's
Travels* the discerning reader could find a comparison of the two realms: literature and science.[29] Swift had also argued that science's intrinsic morality
necessarily lay in the truthfulness of its findings but also in the honest intentions of those discovering its secrets and – par excellence – in the utility
of its agendas. But this complex definition was not remote from that of the
new novelists intent to prove the validity of their plots to an inquiring readership. The difference lay in the methods of proof used by writers and artists:
formalistic, ironic, rhetorical, witty.

Literature and science, broadly construed, were thus allied in this era in different ways, especially as the new form – the novel – matured and staked out
its truth claims. Fielding as psychologist paraded ridicule in his novels as the
most predictive test of human moral worth. But Smollett compared his prose
métier to a broad canvas showing the remarkably diverse picture of human
life. Richardson and Sterne turned inside to the interior nerve of human
truth: private, idiosyncratic, unpredictable, irrational, incommunicable, sexual. But even they differed: for Richardson the "self within" was ultimately
tragic, for Sterne ultimately risible and parodic of any great truths.

As the British novel developed at mid-century it altered its mooring from
these earlier anchors, moving toward sentimental emotion, gothic horror,
political allegory, and – in the case of Laurence Sterne (1713–1768), surely its
most original voice – into an inward stream of consciousness that charted the
mind's interior spaces and private associations. These interior mental zones
differed from the exterior Newtonian spaces cultivated and estheticized by the
pastoral and landscape poets.[30] Sterne's cock-and-bull narrative in *Tristram*

---

[28] Terry Castle has noted the ambiguities of scientific/medical readings of the emotions in *The Female Thermometer: Eighteenth-Century Culture and the Invention of the Uncanny* (Oxford: Oxford University Press, 1994).

[29] As well as in Mandeville's other extended work, *A Treatise of Hysterick and Hypochondriack Passions* (London, 1711), perhaps the premier dialogic example in the period of an overlap of the two realms, not to say two cultures, as they had not yet been bifurcated; see G. S. Rousseau, "Mandeville and Europe: Medicine and Philosophy," in *Mandeville Studies* (London: Oxford University Press, 1975), pp. 139–47.

[30] For discussion of the esthetics of Newtonian space, see G. S. Rousseau, "'To Thee, whose Temple is all Space': Varieties of Space in *The Dunciad*," *Modern Language Studies*, 9 (1979), 37–47.

Shandy (1759–68) actually suggested more than this: that the protagonist hero, Tristram Shandy, *is* his inner mental life, his external surroundings mere props wired up by his mental associations.[31] In dozens of full-length fictions by Sarah Fielding, Eliza Haywood, Charlotte Lennox, and other women in the 1750s, to Anne Radcliffe, Clara Reeve, Mary Hays, Elizabeth Inchbald, and Mary Wollstonecraft in the 1790s, novelists adumbrated the truths of the human head and heart in strategies fundamentally similar to those of scientists generating theories about physics or geology. Theory was necessary in both realms (science was impossible without its hypotheses). Theory was calculated to explain what was at stake in this pursuit of truth, whether to fictionalize the convolutions of gender and sex and race as did the novelists, or to theorize about the heavens, the seas, and the bowels of the Earth.

Aphra Behn's racial fictions of the Restoration, it is true, gave way a century later to William Godwin's class-based fantasies of moral justice and tales based on philosophical inquiry into the perils of conservative rule. And the familiar mores of daily life, especially the secrets of women and the home, loomed increasingly large in British novels as the eighteenth century wore on. But one thing remained relatively constant: that among the forms of literature, poetry was still summoned to versify the most technical of scientific areas. And after three generations the novelist remained the type of writer most proximate to the Enlightenment scientist. Differences existed, as we shall see, but the similarities must also be noted, and they were increasing as time progressed. The same case could not have been made for Shakespeare's or Galileo's world, where imaginative writers were less consciously explanatory and clarificatory than were the new novelists of the world of Newton and Priestley.

Quandaries about creation and the origins of life, routinely parodied in "shambolical" works such as Laurence Sterne's *Tristram Shandy* and John Hill's ephemeral *Lucina sine concubitu* (1750) – a satire on theories of parthenogenesis and immaculate conception – developed in tandem with mechanical theories and practical applications. Linnaeus, for example, a politically very conservative Swede, generated a bisexual botanical classification grounded in analogy and personification (the whole plant world, like the human, was metaphorized to lawful marriages of husbands and wives producing children, etc.) and construed his new "system" as advancing science and furthering human progress. The new botany would "progress society," as did all inventors and discoverers who crossed lines between the true arts and sciences. Exemplars include Benjamin Franklin (1706–1790), the brilliant colonial "Renaissance Man" and avid pursuer of electricity,[32] and, in middle Europe, the

---

[31] Erich Kahler was one of the first modern critics to notice this symmetry in *The Inward Turn of Narrative* (Princeton, NJ: Princeton University Press, 1973).

[32] See J. A. Leo Lemay and G. S. Rousseau, *The Renaissance Man in the Eighteenth Century* (Los Angeles: Papers of the Williams Andrew Clark Memorial Library, 1978); John L. Heilbron, *Electricity in the 17th and 18th Centuries: A Study of Early Modern Physics* (Berkeley: University of California Press, 1979).

prolific Protestant Swiss naturalist-doctor Samuel Tissot (1728–1797), who capitalized on medicine as economic commodity and pioneered a new medical anthropology interpreting human health as part of the workplace and local community. Another example is Erasmus Darwin (1731–1802; see later section), Charles's grandfather, who practiced both the arts and the sciences (*ars combinatoria*), in the elder Darwin's case practicing medicine, writing poetry, inventing mechanical things. Even Darwin's versification of the "*illicit* loves of plants" – not married husbands and wives, as in the taxonomies of Darwin's guide, Linnaeus, but rather conjugating unmarried lovers and polygamous libertines – was generated "in the service of" truth (science) and beauty (art). So too was much of eighteenth-century mimetic literature.

An equal share of beauty and truth was claimed to drive both literature and science, as each segued into the other, accompanied by the natural amalgam of theoretical leap and practical application that formed a cornerstone of Enlightenment science. One could even probe further and demonstrate that convergence was then a sine qua non for the best scientific advances. Surprise lies in the sheer number of applications made. Newton's and Locke's authority was such that wholesale imports of their theories were routinely made in all the arts and sciences: Newtonian physics applied to painting, musical composition, human morals, and the working of government, and Lockean psychology applied to social behavior and artistic creativity. The notion of an emerging "science" of poetry or painting, in an age when the allied arts were so filially construed, was not improbable. Direct applications are made in all ages, as in literary theory in our time: poststructuralist or deconstructionist. The difference *then* was their civil and utopian edge. More recent applications appear to be generated today in the name of detached truth rather than ethical or social advancement. Progress as utopian-based has largely fallen off in our century, except in medicine and technology, and even here there is abundant evidence that the "diseaseification of everything" has disadvantages.

OPTIMISM AND DOUBT

Yet such a cultural reconstruction of literature and science as I have been attempting here – that is, as one unit rather than separate disciplines or subjects developing individually – assumes that cultures are whole and organic, their individual tiles, so to speak, parts of one organic mosaic. Individual discourses or disciplines – the arts and sciences found in today's universities – had not yet become fragmented in our postmodern sense. Egypt was thought to have been the first great civilized nation in history precisely because it did not separate knowledge so artificially.[33] The optimistic culture I delineate here

---

[33] "The arts and sciences which arose in Egypt, contributed to the celebrity which this people enjoy. As they were the first civilised nation in the world, and had no examples to imitate, they are entitled

necessarily contained its individual parts: arts as well as speculative and applied sciences. As the Industrial Revolution developed at mid-century, much literature dedicated itself to celebrating its technological triumphs; this was not merely wonderment that such feats had been accomplished but also congratulation that society would be happier and more progressive as a result. In the 1750s Diderot had dialogically prognosticated the development in *Le Rêve d'Alembert* (*The Dream of d'Alembert*) when gazing into the future of the technology of the age.

A decade later, all the arts – painting no less than literature and music – were at least glancing at the miracle of progress within their realm: the true Newtonian legacy. Thinkers as diverse as the philosopher David Hume and the painter and theorist of esthetics Allan Ramsay, in their discourses – *Of the Standard of Taste* and *A Dialogue on Taste,* respectively – examined the proposition that taste progresses in conjunction with society. Ramsay in particular saw a close correlation between the two: "Good taste in poetry proceeds from good poetry, good poetry from good philosophy, and good philosophy from good government."[34] Progress and power were conjoined by the proclaimed certainty of science and were enabled by developing nationhood. Both induced a sense that their culture of optimism was justified rather than a chimera eventually to be aborted or disenfranchised by political change or revolution.

Yet doubt lingered. Had they still been alive, the Tory satirists, Swift and Pope most of all, would have condemned such confidence as naive and hypocritical cant. Their belief in progress and national power was more restrained, even gloomy; and their poetic legatees – Gray, Collins, all the mid-century lyric poets – concurred. Wordsworth's gladness-madness syndrome in "Resolution and Independence" ("We Poets in our youth begin in gladness; / But thereof come in the end despondency and madness") could be their collective epigraph. Such doubt began with their own lives and reasoned upward to generality. The same Pope who began optimistically as a young poet paraphrasing ancient pastorals ended his life in cultural despair, believing that "Universal Darkness buries All" as civilization is extinguished. Thomas Gray, an *erudito* in matters of natural history, had retreated into seclusion, broken and bitter after a temporary infatuation, Aschenbach-like, with a Swiss student named Victor von Bonstetten. Collins and Cowper and Smart all suffered their own afflictions: the one devoured by chronic manic depression, the other two by intermittent religious melancholy that invaded every aspect of their creative intelligence and even dictated the "inner voices" they heard. Their broken lives occluded their view of the world, which was hardly cheerful or steeped in any sense of an enduring global progress.

---

to the praise of genius and invention. Their particular situation urged their application to some arts, and favoured the cultivation of others. Agriculture, the mother of the arts, originated in Egypt" (Rutherford, *Antient History,* p. 37). See n. 15.

[34] Allan Ramsay, *A Dialogue on Taste,* 2nd ed. (London, 1762), p. 74.

None of these writers, long since canonical in our literary pantheons, was *contra* science; all of them partook of its wonderment and celebrated it. It was science's optimism they doubted: the sense that science could solve society's main problems or transform the individual self from the Miltonic hell it normally occupied. Not one was in any systematic sense a philosopher of Enlightenment science. Yet to the degree that they, like Swift, speculated about science's social contribution, they doubted its ability even in a pure and apolitical state to transform the lot of humankind. Transformation remained the rub. Philosophers since Hobbes and Malebranche in the seventeenth century had demonstrated that amelioration was contingent on knowledge of truth and that truth in any absolute sense was predicated on the human nature discovering that truth. Later, the authority of Locke, who extended some of Newton's optimism for discovering everything that could be known, endured. Still later, Mandeville, Hume, Fergusson, and Adam Smith claimed that human nature no less than other natural realms could be studied as a "science": the science of human nature. This was the subject of essays on man, such as Pope's "Leibniz," as well as countless essays and allegories about the human condition. A proleptic gaze instructs us that the modern social sciences – anthropology, psychology, sociology, political science – originated here.

More seminally, this developing science of human nature was virgin territory for systematic scientific thinkers such as Bernardino Ramazzini (an expert in what we would call the sociology of medicine) and Tissot (who brought sex and economics firmly into the medical sphere) and for imaginative poets and essayists. No single group could claim *human nature* as lying solely within its own province. Nevertheless, it formed the basis for the truest human science: a position that some (the Warburtonians and their opponents) claimed the pagan ancients themselves had struggled to define. Perhaps it was religion after all – Christian and pagan, civilized and savage – that lay at the base of human nature; this is one reason among several that much writing of the eighteenth century cannot be classified either as literature or science but merely as didactic or explanatory or moral.

Yet if the "sciences of man" were necessarily being founded on a *universal* human nature, they were not exactly congruent, and they harbored a profound sense of the Other.[35] Enlightenment thinkers were eager to prove that human beings, "savage" and civilized, were globally identical, their apparent distinctions arising from different religions, climates, and governments. Dissolve these and a base layer, a substratum, of human nature would be evident that was universally commensurate. This urge formed a cornerstone of the Enlightenment campaign to identify and define the Other not to measure its defects but rather to survey their differences in order to compare them – a sort of primitive comparative anthropology. The more the enterprise was extended,

---

[35] Bernard McGrane, *Beyond Anthropology: Society and the Other* (New York: Columbia University Press, 1989), and A. Pagden, *The Fall of Natural Man* (Cambridge University Press, 1986), chaps. 6 and 7.

the clearer it became that the mental nature of human beings – mind and its affections, sentiments, and emotions – still lay buried in darkness.

Nevertheless, in keeping with this culturally condoned proclivity to survey the progress of all subjects and then scientize them by claiming what could be known with certainty about them, the progress of the "science of mind" was construed as a *natural* history of the brain. Thomas Willis had recently expounded the intricate anatomy of the brain; now it remained to apply his findings within a newly developing psychology of sentiment and sensibility and their consequent mental states. Hence, Locke charted the realms of association, Hallerian irritability, Bonnetian attention, Baconian empiricism, and Cartesian rationalism. But the poets did not lag as they supplied their own theories of consciousness and conscience, not to be confused.[36] They also shared in the Lockean and (in Middle Europe) Bonnetian enterprise by peering deeply, and didactically, into matters of mind and mood. This activity was no less vigorous than their reading and explication of Newton's nonmathematical texts. Their original "research" was composed of both reading and speculation (as when they fantasized on the implications of Newtonianism), and even if they did not perform original research in the sense in which a science-based writer today, such as Don de Lillo or Pynchon, might, they hardly relied on the science of their grandfathers for the knowledge they invoked. They were assiduously up-to-the-minute.

FORMS OF REPRESENTATION

The human body represented in the arts was concurrently both liberated and restrained in the formation of the new science. As Foucault began to demonstrate in the 1960s, the body had always been the contested site of gender difference, sexual rivalry, and power base; the acutest feminist critics proved him right by documenting the vast record and especially by demonstrating that the female "breast" became the embattled anatomic and symbolic site of the new sexuality.[37] But as travel to distant parts enlarged human horizons and opened imaginary vistas, the body's compass was also extended. The artistic consequences were as varied as the streams feeding into them: new restraints on the body's license after a Restoration period of libertinism; a contrary tendency sublimating these restraints into the new pornography; an old academic anatomic tradition constricting these trends.

The artistic results were varied. Most novelists and poets harbored notions of the body that we would recognize as conventional – tall-short, fat-thin,

---

[36] Jean Hagstrum, "Towards a Profile of the Word *Conscious* in Eighteenth-Century Literature," in C. Fox (ed.), *Literature and Psychology in the Eighteenth Century* (New York: AMS, 1987), pp. 23–50.

[37] See Ruth Perry, "Colonizing the Breast: Sexuality and Maternity in Eighteenth-Century England," *Journal of the History of Sexuality*, 2 (1991), 204–34.

light-dark – with their attendant mythologies. The Christian body of the painters grew secularized, their saints and sinners less pale and frail than their medieval and Renaissance forebears. If portraiture applied principles of symmetry and contrast to human faces, the much larger realm of painting represented bodies derived from realistic landscapes in towns and cities rather than from symbolic zones. Even the caricaturists merely exaggerated these forms through grotesque size and competing colors. William Blake's (1757–1827) pictorial bodies stand apart from these traditions – not wholly or without precedent, but in resembling the Romantic bodies of Füseli and Casper David much more than those of Hogarth and his literary cousins: Fielding, Sterne, Smollett.

Bodies gone pathological became deranged and "mad" (Figure 33.1). Madness, even in its most anatomical version, was said to lie in the blood (as in George III's) or tissues (the composite of spirits, fibers, and nerves) rather than in a noncorporeal soul, mind, or even brain. But gradually the brain became the contested site. One would search far and wide then for artistic sketches of the demented brain's gray matter or the offending vital spirit's geometry and chemistry. Anatomists and physiologists alike descanted on the deranged animal spirits and magical fibers (Sterne even conjured a way to open *Tristram Shandy* on this note – surely the most brilliant opening of any novel of the century); but no one harbored any certain sense of their visible forms, and microscopes were still too weak for the likes of gray matter.

Even dedicated European Newtonian anatomists (especially in the Dutch school in Leiden: Bernhard Albinus, Petrus Camper, and Boerhaave's disciples) never drew pictures of these *interior* anatomic zones.[38] They were prolific on frames and skulls, muscles and ribs, but they rarely drew the interior body beyond the organs, let alone the body's unseen physiological mechanisms, healthy or pathological. One reason that the language used to describe this physiology is heavily metaphorical is that it is an attempt to give the invisible palpable physical form.[39] Hence, madness was loosely claimed to be *bodily;* but how and why remained sealed mysteries. A sense barely existed of social or situational dementia outside the realms of erotic and religious melancholy. Most post-Battie/post-Monro theories in the formative second half of the eighteenth century medicalized imagination and felt *au courant* doing so. All the pathways and constellations of brain (gray matter, nerves and fibers

---

[38] A point made by Stephen J. Gould, "The Analogistic Tradition from Anaximander to Bonnet," *Ontogeny and Phylogeny* (Cambridge, MA: Harvard University Press, 1977), pp. 13–32, and developed by Barbara Stafford in *Artful Science: Enlightenment Entertainment and the Eclipse of Visual Education* (Cambridge, MA: MIT Press, 1994).

[39] See Martin Kemp, "'The Mark of Truth': Looking and Learning in Some Anatomical Illustrations from the Renaissance and Eighteenth Century," in W. F. Bynum and Roy Porter (eds.), *Medicine and the Five Senses* (Cambridge University Press, 1993), pp. 85–121. For the type of illustrated anatomical book the poet-artist William Blake would have consulted, see Andrew Fyfe, *A compendious system of antomy. In six parts. Part I. Osteology. II. Of the muscles, etc. III. Of the abdomen. Part IV. Of the thorax. V. of the brain and nerves. IV of the senses* (London, 1790).

Figure 33.1. Robert Pine, portrait of a deranged or possessed woman, painted in the late eighteenth century and engraved by William Dickenson. Reprinted by kind permission of the Wellcome Museum and Photographic Collection, London. The original painting has not been located.

throughout the body, the nervous system) were medicalized, but the socioeconomic determinants were overlooked. A humanitarian sense existed, having evolved from ancient times: the notion that poverty dehumanized and deranged people. But for the most part, madness lay in the body's fibrous firmament. Few centuries have empowered their anatomic bodies to this degree. Anatomic illustration also thrived, but the imaginative pictorial arts had not yet tapped into these theories for their canvas possibilities, as they have in the MRI-enthralled twentieth century.

Even less was there consensus about the reasons for the "low spirits" and the "chronic melancholy" we call depression, although theories and therapies abounded. Young Robert Hume (not to be confused with Scottish philosopher David Hume) wrote to William Beckford (1760–1844), the millionaire nomad and squire of Fonthill Abbey in Wiltshire, when his young Portuguese lover

(Gregorio Franchi) lay dying in a London room: "The Dr. says that medicines are useless where the mind is disordered."[40] That was in 1817. A century earlier, the view had been less skeptical. Satellite states of anxiety, panic, and mood swing had not yet come into their own. The older seventeenth-century categories of melancholia as erotic, religious, sedentary, and so forth were refined and entirely medicalized, but the sustained version (chronic depression) was only just becoming psychologized. Several of the early psychiatrists suspected links between madness and creativity; indeed it was John Dryden, the poet, who a century earlier had claimed that "madness and wit were near allied."

These were random chants, so to speak, unintegrated into the whole choir. Still, the artists themselves possessed a surer sense than the doctors about their condition in an era when depressed artists surfaced in numbers, especially among poets and composers. The lists are long, although each case – from Collins and Cowper among the poets to composers Mozart and Schubert, who also had their feet firmly in the eighteenth century, Mozart entirely so – requires scrutiny based on its individual circumstances. Perhaps for the first time in Western cultural history, art that was adjudged too difficult, or complex, for the ordinary person, or even the connoisseur, was said to be "deranged." When Beethoven first presented his string quartets to his patron, Razumovsky, not far on the other side of the century (1806), they were said to be the work of a deranged man. However, Lord George Gordon Byron (1788–1824), very much of a piece with the eighteenth century and born just as Europe became embroiled in the capriccio of revolution and continental war, was perhaps the best example of the stakes involved in creative malady. Unstable almost from birth, Byron persuaded himself that he was born under the spirit of an evil psychological northern Scottish curse, and he proceeded to act out, so to speak, on the premise: failing in his personal and domestic relations (except perhaps among his homoerotic boyhood ones at school), falling into alleged incest with his half-sister, charged with the attempted sodomy and murder of his wife. The less stable he was, the more Byron seemed able to rise from his emotional ashes and create great poetry, as in his epic *Don Juan,* from the abyss of despair. He may be the greatest poet of rage in the language; if so, his emotional reservoirs lay mostly on the far side of reason.

Minds healthy and sick, in the medico-religious sense, were not the only natural sites for breeding extraordinary imagination. The physical landscape and its soil also were, and in virtually all the literary genres, not merely pastoral or the ballad. In literature, the poets were particularly responsive to Newtonian theories of light, which translated, for them, into color and a new sense of light and dark. James Thomson, the dreamy poet of the *Seasons* mentioned earlier, and Richard Jago, a landscape and graveyard poet, perfected this importation from Newtonian light theory into a "school" to the

---

[40] Audrey Williamson, *William Beckford: A Life* (London: Oliver Boyd, 1973), pp. 19–20.

degree that they were bound by the theory's components: the construction of light, color, and pictorial contrast. Thomson was particularly impressionistic, dreamily weaving prisms and rainbows into associated poetic images. Here the arrows of influence were clear enough: from Newtonian light theory to poetry and painting, and then to color that configured the rainbow anew. Thomson's translators in Europe make evident their sensitivity to these connections between scientific theory and poetic senses of the landscape. In other instances the direction of the influence was more equivocal, as in the earth sciences (geology especially) giving rise to an art of (in Matthew Arnold's phrase) "mountain gloom and mountain glory." Geology was a late-developing science compared with optics and physics; its secrets, requiring new technologies for exploration of ocean floors and beneath desert sands, translated into explorers' dreams of conquering places such as the vast geographical Sahara. The quest for the discovery of the longitude, which all manner of writers versified, was more pressing yet, as it had terrific implications for war.[41]

Besides, poets and painters alike had described their visions of the Earth's interiors long before the late Georgian geologists (Hutton, Werner, and company) postulated what the truth actually was. This facet of the Western imagination had been shaped by Scripture and travel writing as much as anything else. Conversely, explorers and discoverers set out to discover the truth because their minds had been galvanized about oceans, forests, and deserts. As the century wore on, each domain of nature claimed its artistic preserve: oceans in literary epic (think of the long tradition from the Icelandic sagas and Beowulf to the sea in *Don Juan* and, in its mock-heroic version, as the River Thames in Pope's *Rape of the Lock*); forests in prose romance and gothic fiction (as in the great forest scenes of Smollett's *Ferdinand Count Fathom* and Anne Radcliffe's *The Forest);* and endless deserts in fictions set deeply in Orientalism, as in Beckford's tales. Crags – alpine and bare, Swiss and Scottish – became the preserve of the ruminative wanderer and ponderous dreamer: the preserve of solitary leech gatherers meditating the meaning of their own existence.

These developments – naturalistic landscapes as well as those of minds and bodies – responded to the colossus of travel, especially the *scientific* Grand Tour about which our scholarship has remained largely silent. Gentlemen had routinely been traveling in numbers for more than a century. The difference now was that travel conditions (roads, inns, safety, health, and passports, as well as the new wealth underpinning all these) had improved to such a degree that more young gentlemen, encouraged by their parents, undertook it. Their new wealth also permitted them the time to collect and retrieve objects. Such amassed artifacts naturally altered everyone's horizon about the future of the beyond, especially when the objects were remote flora and fauna, minerals, gems, fossils, and other things buried in the earth. As cabinets were filled with

---

[41] See, for example, the exploratory yearning sense in Francis Moore, *Travels into Africa* (London, 1738).

gathered curiosities and natural objects classified, writers, especially novelists, felt obliged to describe these retrievals fictionally.

The popular imagination responded diversely to such luxuries in European nations where much of the population were struggling to feed themselves. The didactic and moral aspects of such retrievals were another matter. To the studious gentleman and (occasionally) gentlewoman, the collection of such remote information was obligatory for the progress of knowledge; for the creator, in words or pictures, it was the starting point for new pictorial compositions, as in Giovanni Paolo Pannini's reflexive *Modern Rome* (1757) and the geologically inspired *Rocks at Sandy Bay, Saint Helena* by William Daniel (1794).[42] The constant factor may have been the firing of the moral imagination in a sunny period bent on the belief that the future would be better than the present and much improved over the past. Viewed either way, the reports of real travelers continued to be one of the deepest inspirations sustained by the artistic imagination.

SCIENCE AND REVERIE

Not everyone was capable of sustaining such attention to minutiae and remote detail, nor did Europe's "mind doctors" tell them why. The scientific analysis of consciousness and conscience had not yet been studied. The latter was dissected not by a scientist of any type but by a writer who has been called the inventor of stream of consciousness long before James Joyce: Laurence Sterne.[43] The subconscious had not even been named, although strong cases have been made that the Greek and Elizabethan tragedians were its discoverers.[44] In brief, the point is that again and again it was imaginative literature (and painting) that was supplying scientific knowledge (that can be relied on for certainty and predictability) with its first images. In realms of reverie, which had not been medicalized in the way madness and melancholia had (discussed earlier), the only *analyses* were artistic. Even poets such as James Thomson (already mentioned) and symbolic painters such as Henry Füseli turned their attention to realms imaginary and symbolic despite keeping another eye or ear to influences Newtonian or Hartleyan. Still others, especially

[42] The recent historiography of the concept of *progress* has touched on the arts: see F. Teggart, *The Idea of Progress* (Berkeley: University of California Press, 1929); David Spadafora, *The Idea of Progress in Eighteenth-Century Britain* (New Haven, CT: Yale University Press, 1990); for the arts in the eighteenth century, see L. Lipking, *The Ordering of the Arts in the Eighteenth Century* (Princeton, NJ: Princeton University Press, 1970).

[43] See James E. Swearingen, *Reflexivity in* Tristram Shandy: *An Essay in Phenomenological Criticism* (New Haven, CT: Yale University Press, 1977).

[44] Bruno Snell, *The Discovery of the Mind: The Origins of European Thought* (Oxford: Basil Blackwell, 1953); and, more recently, Ruth Padel, *In and Out of Mind: Greek Images of the Tragic Self* (Princeton, NJ: Princeton University Press, 1992), which has profound implication for the "two cultures" in the ancient world.

Continental painters and travelers heading south over the Alps, were more blatantly dreamlike: free spirits unfettered by the determinants of time or space. Like Rousseau in his famous meditations and reveries, or Sade in erotic fantasies about the contortions of sexual gender in the *boudoir,* these artists ruminated about, and associated, all sorts of images psychologically. It was almost as if they were drugged.

Such extravagant deliriums, however, should not imply that the scientific image or some specific theory did not guide even them at some deep level. Christopher Smart's (1722–1771) "lamb," in whom he continued to "rejoice," never has his feet secure on any English pasture or meadow no matter how vividly described. Beckford (already mentioned) could double, if necessary, as an *English* Rousseau in another key, having titled his epistolary travels through Europe *Dreams, Waking Thoughts and Incidents* (1783). The Mediterranean steps and stones he wandered among are remarkably specific, down to particular paintings and window panes. But a larger purpose always lies in roaming through his exotica, no matter how concretely and discretely grounded. This extensive work, now destroyed except for a few leaves, revealed so much of Beckford's private persona that his family forced him to suppress it. Despite vast differences, these diverse dreamers took their place beside Blake-the-total-Londoner, who attached confabulated names to his alteregos and damned them with the "woof of Newton and Locke." Science (certain knowledge) and reverie (dream states) had not been at odds either north or south of the Alps; they merely existed in discrete empires of the same mind.

The philosophical point is worth restating: the modern history of science usually assumes that science endows artists with the ideas that they in turn copy, imitate, versify, or paint; but the reality is far more complex. Mark Akenside (1721–1770) serves as a case in point. He was a qualified medical doctor – he wrote a dissertation at Leiden in record time (six weeks) – and a polymathic savant. He practiced medicine in England afterward, was scientifically and philosophically erudite, and applied his didactic knowledge to the "operations of the imagination," especially in a long blank-verse poem called *Pleasures of Imagination* (1744).[45] This remarkable poem-of-ideas heralds evolution in the Lamarckean-Darwinian sense long before it was fashionable. Building on Locke's associationism and Newton's physics of nervous transmission, Akenside explained how pictures are formed in the mind and then "sublimed" into poetic images both pleasing and beautiful. He required both types of learning to perform this feat: one without the other would have been insufficient.

---

[45] John F. Norton edited and annotated Akenside's epic poem in a doctoral dissertation he never published; the only available editions are dated 1744 and 1825, and there is no modern annotated edition; see also George R. Potter, "Mark Akenside, Prophet of Evolution," *Modern Philology,* 24 (1927), 55–64.

But why artificially label or pigeonhole him as a poet *or* a psychologist? Apart from the odium of constricting labels, Akenside was clearly both, as were Dutch-born Bernard Mandeville (1670–1733), the doctor-poet, and Thomas Gray, the poet-naturalist-historian. In Akenside's generation, only David Hartley was able to achieve claims as a psychologist in both realms: medicine and literature. But Akenside and his lyrical contemporaries (especially poets from Collins and Cowper to Coleridge and Shelley) also provided high-level analyses – usually in their poetry but also in epistolary correspondence – of the nature of energy (not yet the entropy it became in the Victorian era). And by the end of the eighteenth century Blake transformed the much-enlarged concept of energy into the *métier* on which he would later build his vast mythology of Enitharmon: the symbol of spiritual beauty, who, as the representative of eternal female energy, is also the creator of all space.[46] Energy, quite apart from Blake's elevated view of it, was also celebrated as an inherent attribute of art, especially poetry. Hence, these elements – energy, light, the imagination itself – existed in geographical mind sets and national contexts. Northern literature, for example, was symbolically configured as young, imaginative, and energetic, corresponding to youth in the development of national literatures; southern, especially in the Mediterranean, old, mature, and restrained. These trends held profound implications for the further development of such concepts as mind and energy within literature.

These developments and overlaps were bound to raise profound questions about progress in an epoch when sunny optimism about the future abounded. They did. Progress naturally existed in many hues and shapes, light to dark, small to large; from personal advancement to national and global; and what was forward movement for a family or town may not have been so for a nation. Still, the new differentiations were primarily between two major types: social (which included the scientific) and those others associated with the advancement of the human condition itself (artistic, religious, moral). The latter depended much less on science, no matter how construed between the world of Newton and the three Darwins.

It would have been difficult then (and is now) to make any sustainable case that poetry had "improved" since the Greeks or Shakespeare; art, since the Renaissance or Leonardo da Vinci; philosophy, since Plato and Aristotle; the love of God and His kingdoms, since the days of primitive Christianity. Artists responded, in part, to these apparent tensions by their anxiety over the possibility of progress. American literary critic Harold Bloom, much influenced by Freud, has generated a theory along these lines in which the artist, especially "strong poets," creates by overcoming the Father Figure.[47] Other

---

[46] Stuart Peterfreund, "Organicism and the Birth of Energy," in F. Burwick (ed.), *Approaches to Organic Form: Boston Studies in the Philosophy of Science* (Boston: Reidel, 1987), pp. 105, 113–52; for the later tradition of energy, see Patrick Brantlinger (ed.), *Energy and Entropy: Science and Culture in Victorian Britain* (Bloomington: Indiana University Press, 1989).

[47] Harold Bloom, *The Anxiety of Influence: A Theory of Poetry* (New York: Oxford University Press, 1973).

artists (especially poets of the ilk of Coleridge and Wordsworth, who deeply believed that poetry had to reach "the Common Man" if it was to survive), were politically more reactive: as they watched the events in the Bastille, they summoned their strongest inspiration. Nevertheless, many writers and painters who created and composed out of a sense that progress in the arts actually existed, did not doubt that they themselves were proving the theory in their works.

PROGRESSES TO PERFECTION

Hence, we return to one of the themes of this chapter: the complex connections between certainty and creativity. Did the former enhance, or detract from, the latter? And if progress in the arts exists, why not perfection? Perfection, however, then carried spiritual baggage unknown to the more secular progress. Yet the step from progress to perfection (not the social perfection of the French ideologues, *perfectibilité*, but formal perfection in the arts) is not a Herculean task if one possesses the right frame of mind. Historically considered, perfection in the arts had never been attainable, not even within the great Aristotle's theoretical grasp in his esthetic pronouncements. The translation by English literary critic John Dryden of Bellori accounted for the reason in the Preface to Du Fresnoy's *De Arte Graphica*: "All things which are sublunary are subject to change, to deformity, and to decay."[48] Therefore, the *idea* of a *perfect* work of art, universally recognized and entirely unblemished, is outlawed even on epistemological and historical grounds as the necessary consequence of mutation.

Theory and practice differed on the reasons. The neoclassical piety (different from the practice) was that artistic perfection could be approached, but not reached, if the ancient rules were followed and, only occasionally, broken, as in Alexander Pope's famous dictum: "And snatch a Grace beyond the Reach of Art" (*An Essay on Criticism,* line 155). In practice this implied adherence to a theory of realism and representation and mandated an inflexible notion of literary phylogeny and the genres: tragedy, comedy, tragicomedy, pastoral, lyric, ode, and so forth (at the time called "the literary kinds" in English, and in French *la lois du genres*).

But European perfection then also bore another resonance – scientific and spiritual – echoing much esthetic theory and practice of the eighteenth century. Seventeenth-century perfection remained fundamentally spiritual: a temple in which to lodge the soul. From the metaphysical poets (Vaughan, Donne, Traherne, even Marvell) to the Cambridge Platonists; from Bunyan and William Law to the post-Civil-War Mystics and Quietists of the 1690s; the poet's or artist's soul bore a direct relation to the art it could conceive and

---

[48] John Dryden, *Essays of John Dryden,* ed. W. P. Ker, 2 vols. (Oxford: Clarendon Press, 1900), 2:117.

execute. Molière wrote his greatest plays in the same decade that John Milton, the blind bard, published *Paradise Lost* (1667–74): neither was composed by a conscious attempt to produce a "perfect" play or epic poem, but both were preoccupied with notions of spiritual perfection. Concurrently, Restoration knowledge in England was punctured by the memory of a civil revolution and the effects of a scientific revolution that replaced (among other substitutions in medicine and mathematics) soul with anatomical mind and brain. France and its Cartesians had contributed to the transposition but had no such sustained movement as occurred in England. This was the *other* revolution of the Oxford and London physiologists – the body as transformed by anatomist-doctor Thomas Willis and his Royal Society colleagues – that still commands awe among historians of science and medicine. But consult the other end of the era – perfection c. 1800 – and all is changed in the aftermath of the French Revolution. If the old cults of the perfection of the soul (spiritual, quietistic, chiliastic, even vegetarian) were not entirely washed away, the new ones were based on guaranteed personal rights and civic justice, much more so than the spiritual domains now vulgarly mocked as so much babble and prattle.[49]

The art inspired by these cults of perfection also differed: in both literature and painting, it was formally less recognizable and much less imitative of the old genres, conceived in new blended forms, addressed to an audience from below rather than above, so to speak, and realistically and earnestly concerned with moral justice and esthetic freedom. This last thread could describe the development of the novel and other prose forms from about the 1720s, diverse though they were. For all their varied wit and irony, the novelists of the period assumed the burden of moral justice and personal liberty, whether of a Man Friday or Tom Jones, a Pamela Wilson or Clarissa Harlowe, with the zeal of religious crusade. Furthermore, this time line, occurring approximately in the 1740s, was created by an esthetic theory and practice that were absorbing, and reacting to, a Lockean and Newtonian aftermath. Each influence in the arts, early and late, claimed to base itself on identifiable perfection of a sort. The difference was that perfection itself had radically changed, as the most general critic of the day, Samuel Johnson (1709–1784), pronounced over and again. His *Dictionary* cites the paradoxically multiple definition offered by English theologian Richard Hooker (1554–1600): "Man doth seek a triple *perfection:* first a sensual, consisting in those things which very life itself requireth, either as necessary supplements, or as ornaments thereof; then an intellectual, consisting in those things which none underneath man is capable of; lastly, a spiritual and divine, consisting in those things whereunto we tend by supernatural means here, but cannot here attain."[50]

---

[49] Nils Thune, *The Behmenists and the Philadelphians: A Contribution to the Study of English Mysticism in the 17th and 18th Centuries* (Uppsala: Almqvist & Wiksell, 1948).

[50] See the entry for "perfection" in S. Johnson, *A Dictionary of the English Language,* 2 vols. (London, 1755).

The skeptics denied all this. Hume and his followers in Scotland (including poets such as James Beattie, who applied these principles to the arts),[51] as well as other "self-doubting philosophers," claimed nothing in human nature had changed, least of all progress in the arts or any notion that advancement in science translated into esthetic improvement. Some observers argued that social progress itself was a type of illusion, given that old problems give way to new. That civic enlightenment was as equivocal as it was esthetic, even when treated by the artists: therefore, consult Hogarth's satires or Fielding's – the argument went – and you see that human nature remains a constant. Governments rise and fall, artists are more or less talented, but the essence of humankind remains the same over time.

Others, such as Clara Reeve (1729–1807) in England, were more optimistic[52] when affirming dialogically among her three representative philosophers – Hortensius, Sophronia, and Euphrasia – that when societies advance, so do their literatures: "*Euph*. As a country became civilized, their narrations were methodized, and moderated to probability – From the prose recitals sprung History, – from the war-songs Romance and Epic poetry."[53] These were challenges science rarely faced head on. The great Newton himself was preoccupied with calculations of the chronologies of ancient kingdoms; Locke, in his old age – he lived for fifteen years after publishing the revolutionary *Essay concerning human understanding* in 1689 – amused himself in alchemical pursuits.[54] Chronology and alchemy were then not indications of a rationalism or empiricism *manqué*. Alchemy was in decline, even in middle Europe – dedicated Paracelsus country – but hardly dead or eradicated as a wellspring of inspiration for aspiring poets and painters.[55] Study systematically the biographies of Enlightenment scientists, and you discover less direct confrontation with the big issues than among the philosophers and artist-writers.

Even less direct was the scientist's engagement with civic progress.[56] In

---

[51] J. Beattie, *Essays on the Nature and Immutability of Truth in Opposition to Sophistry and Scepticism* (Dublin, 1762).

[52] Clara Reeve, *The Progress of Romance, through times, countries, and manners; with remarks on the good and bad effects of it, on them respectively; in the course of evening conversations* (Colchester, 1785), p. iv: "In the following pages, I have endeavoured to trace the progress of this species of composition, through all its successive stages and variations, to point out its most striking effects and influence upon the manners, and to assist according to my best judgement, the reader's choice, amidst the almost infinite variety it affords, in a selection of such as are most worthy of a place in the libraries of readers of every class, who seek either for information or entertainment." For Reeve, see A. K. Mellor, "A Criticism of Their Own: Romantic Women Literary Critics," in John Beer (ed.), *Questioning Romanticism* (Baltimore, MD: Johns Hopkins University Press, 1995), pp. 29–48; Suzy Halimi, "La femme au foyer, vue par Clara Reeve," *Bulletin de la societé d'Études Anglo Americaines de XVIIe et XVIIIe Siécles*, 20 (1985), 153–66.

[53] Reeve, *Progress of Romance*, p. 14.

[54] For Locke's alchemy, see Michael Ayers, *Locke: A Biography*, 2 vols. (London: Routledge, 1991).

[55] See Robert Markley, *Fallen Languages: Crises of Representation in Newtonian England, 1660–1740* (Ithaca, NY: Cornell University Press, 1994), and Ernest Lee Tuveson, *Thrice Avatars of Hermes* (Lewisburg, PA: Bucknell University Press, 1986).

[56] For this tradition, see Paul Alkon, *Origins of Futuristic Fiction* (Athens: University of Georgia Press, 1987); W. Hirsch, "The Image of the Scientist in Science Fiction," *American Journal of Sociology*, 63

France and Germany it was virtually nonexistent, despite the French fascination for (what we would call) science fantasy.[57] By the end of the eighteenth
century some of the paid intellectuals at Catherine's Russian court were pon-
dering the matter, but it was the British traditions of commingled satire and
utopian fiction that genuinely took up this gauntlet. Science fiction and science fantasy, still in their infancy, were nevertheless developing forms then.
In Ireland, Swift's satiric future in *Gulliver's Travels* was grim, pervaded by the
sense that scientific agendas were sinking under the weight of political corruption and that human pride had not improved one iota since prelapsarian
times. Swift was hardly antiscientific as he gazed deeply into science's connection to politics, especially the way science became infected by corrupt
politicians tampering with its agendas.

No eighteenth-century memoirist of the future was more prescient of the
moral that science destroys those who misuse it than the now-unknown English writer Samuel Madden (1686–1765) in his *Memoirs of the Twentieth Century* (London, 1733). The work claimed to be published in six volumes, but
only one ever appeared; it contained the subtitle "Original Letters of State,
under George the Sixth: Relating to the most Important Event in Great-
Britain and Europe . . . from the Middle of the Eighteenth, to the End of the
Twentieth Century, and the World. Received and Revealed in the Year 1728;
and now published . . . In Six Volumes." Madden quickly repressed the work
for reasons now lost to time. Paul Alkon, the American scholar who first re-
suscitated it, demonstrated that it was not influenced by *Epigone,* the anonymous seventeenth-century futuristic French satire (the French led the way in
such science fantasy in the early modern period). Alkon also showed that it
combines dystopian, gothic, and romance elements. *Memoirs of the Twentieth
Century* is, then, a vital eighteenth-century link between *Gulliver's Travels* (1726)
and *Frankenstein* (1819).

Written within the epistolary framework, its past-tense narrative comments exclusively on the future, suggests a new esthetics, and feeds directly
into our Big Brother world as we approach 2000. Still, Alkon adjudges it
"failed satire:"[58] perhaps as the result of its obscure religious butts and other
incoherent targets, despite Alkon's corroboration of many of Madden's intuitions about the future. For example, Madden's persona claims that English
will be spoken in the twentieth century. And Madden uses German polymath Athenasius Kircher's wondrous tour of the solar universe in *Itinerarium
Extaticum* (1656). A guardian angel arrives one night in Madden's chamber,
where he composed (he claims) in 1728, and handed over documents from the
twentieth century, the last of which is dated 1 May, 1998. Always preoccupied

(1958), 507–12; Rosalyn D. Haynes, *From Faust to Strangelove: Representations of the Scientist in Western
    Literature* (Baltimore, MD: Johns Hopkins University Press, 1995).
[57] For the French tradition, see Arthur B. Evans, *Jules Verne Rediscovered: Didacticism and the Scientific
    Novel* (Westport, CT: Greenwood Press, 1988).
[58] Alkon, *Futuristic Fiction,* p. 93.

with negotiating between science and art on the eve of the millennium of 2001, Madden's fiction ambulates through several nations – England, France, Italy, Turkey, Russia during the 1990s – in search of "the infinite incredible verities in the world of science."[59]

Madden was succeeded anonymously in 1763 by French journalist Louis-Sébastien Mercier's *L'An 2440*, a kaleidoscopic tour of Paris in which a benign government is contrasted with corrupt regimes of the time.[60] Mercier's imaginary state remains blissfully at peace with its neighbors, no longer consumed by colonial dreams of wealth or slavery. If Madden and Mercier read *Gulliver's Travels* they absorbed much, especially about political corruption, and its opposites, in future governments. Science and technology, newly invigorated in the generation after Newton's death, may uncannily have encouraged pan-European wars, as well as elicited dreams of their cessation, to a degree unknown at the time of the Thirty Years' War a century earlier. As Alkon writes, Madden's vision is "a jingoistic fantasy of future warfare conducted by an improbably efficient ideal ruler in a twentieth century that is just like the eighteenth century except for the presence of a King George who does everything successfully and finally even conquers France."[61]

Nevertheless, the conundrum about the "future" in the eighteenth century was a topic too surcharged to neglect: within one year after Mercier's fiction protagonist observed the new life of Paris in 1771, Mercier's was translated into English as *Memoirs of the Year Two Thousand Five Hundred* (1772), paving the way for achievements such as *Nineteen Eighty Four*. These writers removed the old Erasmian utopias from the ineffectual realms of No Place to the influential arena of future possibilities set in specific sites. Other French writers at the end of the eighteenth century – especially Restif de La Bretonne, in *Les Posthumes* (1802) – portrayed a distant future marked by planetary transformation and biological evolution.[62] Big Brother and H. G. Wells had not yet been conceived, and even a menace such as Frankenstein could only be imagined in the aftermath of the 1791 cannibalization of society in the French megalopolis.[63] Here was the future unraveling before the French *citoyen's* eyes, and not in science fiction or fantasy. Whether on the banks of moonlit Lake Leman or in haunted Italian palaces such as Otranto on the remote Straits of Taranto, gothic fiction, with its horrific visions of haunted houses and lonely landscapes, dismembered and murdered fugitives and avenging villains, captured these terrors. Poets throughout Europe dealt with the sublime aspects.

[59] Quoted in Alkon, *Futuristic Fiction*, p. 111.
[60] The fullest account to date in English is found in Robert Darnton, *Forbidden Bestsellers of Pre-Revolutionary France* (Cambridge, MA: Harvard University Press, 1996), pp. 118–36, 226–32, and discussed in Alkon, *Futuristic Fiction*, pp. 4, 111–13.
[61] Alkon, *Futuristic Fiction*, p. 112.   [62] Ibid., p. 4.
[63] See Stephen Bann (ed.), *Frankenstein: Creation and Monstrosity* (London: Reaktion, 1994); Tim Marshall, *Murdering to Dissect: Grave-Robbing, Frankenstein, and the Anatomy Literature* (Manchester: Manchester University Press, 1995); Radu Florescu, *In Search of Frankenstein: Exploring the Myths Behind Mary Shelley's Monster* (London: Robson, 1996).

The more mundane future, in factories and cities, was left to social critics who were persuaded neither that society was improving nor that science would save the world.

THE IMAGINATIONS OF CONSUMERS

Sunny optimism about the prospects for human progress and social perfection, clouded by these dissenting strains, could not have developed without the consuming societies fueling the optimism (consuming in the economic rather than the medical sense) – that is, collecting rather than wasting away. People – highborn and low and throughout the middle social ranks – purchased, amassed, and collected goods and, most important, justified their newfound luxury in unprecedented ways despite the persistence of poverty.[64] Luxury in the emerging nations of Europe (France, Britain, Sweden, Austria, Russia) became the sine qua non of modern life in the second half of the eighteenth century, spilling over, in England, into the affluent Regency drawing rooms and new town crescents, whose grandeur intimated the goods that lay inside.

Luxury then was not merely a by-product of strong economic development but also a force and habit affecting the artistic life of the nation. And if we have learned anything in a generation of magnificent historical scholarship – I am thinking especially of the three volumes about patterns of consumption edited by John Brewer and his colleagues – it is the degree to which this human activity was becoming idealized among artists: in paintings, in poems, even in the dozens of new novels realistically describing everything from armoires to clothing, and dresses to silver.[65] The sciences are usually described as theoretical or technological, even in the early modern period. Yet they stretched their tentacles into the familiar domestic scene, where the improvement of everything was attributed to "sciences" abetting humankind's lot: the familiar Whig theory of history. For this reason, among others, even the schoolmasterish William Rutherford, mentioned earlier, pronounced that the arts and sciences were under a *single* umbrella, viewing them as fundamentally similar human practices, especially in their changing forms and under the weight of the new consuming passion.[66]

What did amassing and collecting have in common with scientific practices? European Enlightenment travelers to Holland knew the answer after

---

[64] John Sekora, *Luxury: The Concept in Western Thought, Eden to Smollett* (Baltimore, MD: Johns Hopkins University Press, 1977), remains the only modern study, but much work is in progress.

[65] John Brewer and Roy Porter (eds.), *Consumption and the World of Goods* (London: Routledge, 1993); John Brewer and Susan Staves (eds.), *Early Modern Conceptions of Property* (London: Routledge, 1995); Ann Bermingham and John Brewer (eds.), *The Consumption of Culture, 1600–1800* (London: Routledge, 1996).

[66] See Rutherford, *Antient History*.

their first visit, after seeing the treasures Dutch travelers brought back with them while accumulating their East Indies wealth.[67] Luxury throughout the fabric of life then spilled out into enriched scientific collections, cabinets, and museums of rarities, all of which piqued the popular imagination of novelists and artists alike. Today it is a commonplace that science could have not developed as it did without collection, retrieval, and classification. But the "collections" were also estheticized and idealized: the true site of their crossover into the literary-cultural realm, a primary aim of this essay. And the obsessive or excessive collector was eventually stereotyped as a figure of extremity – first lampooned (Pope) and then romanticized (Horace Walpole and Beckford) when retrieval was popularized as a solitary activity requiring sacrifice, loneliness, and exile. Nomadic poetry of the period begins innocently and ends by the time Romanticism is in full bloom, with ancient Coleridgean mariners stranded on silent seas and lonely Wordsworthian wanderers gathering leeches. Novelist Fanny Burney domesticated the topos, genderized and moralized it in *The Wanderer; Or, Female Difficulties* (1814), and firmly excavated the "science" from it.

Natural history bound many of these practices. No leisure activity then rivaled botany for its uncanny ability to tap into curiosity, particularly curiosity about the sexual undercurrents of the era, or to form such an intrinsic part of the female mind set that contemporary women brought to gardens in the countryside and the development of seaside resorts. This was truer of northern European countries than of southern – the result principally of the socioeconomic status of women.[68] The eighteenth century was the epoch in which the country ramble developed and was even perfected (in tandem with the many other "perfections" discussed earlier), as well as the era of the urban sprawl of the city walk, as in English city-poet John Gay's *Trivia: or the Art of Walking the Streets* (1716), the antidote to his earlier *Rural Sports* (1713).[69] The new botany developed in both these topographical locales, albeit differently in each setting, women roaming the countryside in search of flowers and leaves in the one while planting new herb gardens in the other. Botany for the ladies swelled into a passion rather than pastime because women began to comprehend its scientific basis (Linnaeus),[70] which they usually could not in the new astronomy, mathematics, or light theory (Newton). They were also

---

[67] For French collecting, which developed later than Dutch, see Marc E. Blanchard (ed.), *Writing the Museum: Diderot's Bodies in the Salons* (Lexington, KY: French Forum Pubs., 1984).

[68] See Olwen H. Hufton, *The Prospect Before Her: A History of Women in Western Europe, Vol. 1, 1500–1800* (London: Harper Collins, 1995), pp. 59–98.

[69] See Anne D. Wallace, *Walking, Literature, and English Culture: The Origins and Uses of Peripatetic in the Nineteenth Century* (Oxford: Clarendon Press, 1993), chap. 1 for the eighteenth century.

[70] For the context, see N. Jardine, J. A. Secord, and E. C. Spary (eds.), *Cultures of Natural History* (Cambridge University Press, 1996); Ann B. Shteir, *Cultivating Women, Cultivating Science: Flora's Daughters and Botany in England, 1760 to 1860* (Baltimore, MD: Johns Hopkins University Press, 1996); Londa Schiebinger, *Nature's Body: Sexual Politics and the Making of Modern Science* (London: Pandora, 1994); and Schiebinger, *The Mind Has No Sex? Women in the Origins of Modern Science* (Cambridge, MA: Harvard University Press, 1989).

fascinated by botany's reproductive dimension, in which they held a natural stake. Often these differences were determined by nothing more recondite, respectively, than levels of education in the natural sciences and the widespread female dread of death during pregnancy. Men, too, harbored natural stakes, of course, especially dynastic and economic anxieties for their own progeny. But their reproducing bodies were far less vulnerable than women's; hence their lesser curiosity about the intricacies of plant and flower reproduction.

Within natural history, botany was represented pictorially as well as discursively: idealized in drawings and captured in cartoons. The latter were thematically but not visually monochromatic (most cartoons were colored) and captured the male sense that women were now rivaling them as botanists. These caricatures of early female botanists expressed an anxiety about a pursuit that had remained a male prerogative all over Europe. The new difference visually was botany's realism and its discovered sexual substratum – not merely the Linnean bisexual revolution and its attendant taxonomic politics but the *visible* artistic gardens women designed and planted themselves: in England, Lady Luxborough's exotic jungle, Sarah Abbot's diverse herbarium, Marie Jackson's flower garden, Jenny Lawrence's award-winning horticultural exposé.[71] Similar gardens, if fewer in number, existed in Sweden, Denmark, and the Low Countries. Hence, a hitherto masculine botany in the time of Nehemiah Grew and Philip Gerard (1690–1720) was transformed within little more than one generation (c. 1740–1770) into a partly female one.

After Linnaeus prevailed in the pan-European debate about botanical nomenclature (1760–1780), women could see the sexual organs of the plants with their own unaided eyes: the vulva, the womb, the cavity. And what had been an interdiction in nomenclature before Linnaeus – the constriction on speaking or writing about the sexual organs of flora – would soon become popular poetic liberation after the Lichfield doctor with a luxurious imagination – Erasmus Darwin, grandfather of Charles of evolutionary fame, and an ambassador of culture in his own right – published his *Loves of the Plants* (1786). Hence, his contribution to women's anatomy, women's plants, women's loves: a set of analogies transforming both the professional and the amateur practice of botany by the end of the eighteenth century. But philosophically considered, was this development *art* or *science* or something else altogether? The substantive matter, and not the label, is what matters. Labels count for little in practice, except when the histories of complex relationships are at stake, as in the heartland of this essay.

It may be fanciful to raise the question. For the fact is that no other field had come under such remarkable gender stress: what had been a male province was not merely challenged by women now but also commandeered by them. This transformation of theoretical botany and its leisurely pursuits was a

---

[71] See Miles Hadfield, *A History of English Gardening*, 3rd ed. (London: Murray, 1979), p. 259; Shteir, *Cultivating Women*, p. 55 n. 69, p. 120.

process – theoretical and practical, scientific and artistic – occurring beneath society's social dermis, as it were, but nonetheless palpable. No other scientifically based endeavor of the period had such implications for gender's new yoking to popular science. Women, it is true, rarely collected fossils, butterflies, or Elgin marbles to this degree, in part because they did not routinely take the Grand Tour. Therefore, they were rarely drawn. However, on the few occasions when artists did portray them, they were humorously rather than realistically captured. The connections between gender and morality certainly extended to these leisure activities, making it evident how deeply it cut into the fabric of society then. If post-Newtonian science was crucial to the era in its moral dimension, as having terrific implications for the implementation of civil and humane societies, it was as vital for the transformation of received notions of gendered leisure activity.

But Uppsala-Linnaeus was no Lichfield-Darwin. Not only were they dissimilar, but also their agendas differed. Historically speaking, no one did more to bring sex and gender into botany *and* poetry than did Erasmus Darwin, whose human and scientific profile has altered in our generation as much as the botany of his *Loves of the Plants* transformed popular taste. Today some scholars think that Darwin was as important to the development of theories of evolution and natural selection as his grandson Charles. And there is no longer any doubt that he possessed an extraordinary intuition for such diverse endeavors as canal lock gates, steam airplanes, and new weather measurements.[72] Even so, it would be an error to view his "sexual science" or its intellectual underpinnings apart from their moorings. Theories of gender and sex c. 1750–90 were much spurred on by the new biology springing forward from anatomy and physiology; these, in turn, intermixed with broadly based artistic inspiration to spawn further inventions in gothic fiction, art (especially scientific illustration), didactic literature (the novels that abound with long quotations and passages lifted verbatim from scientific works), and early anthropological literature of the bizarre type found in English social philosopher Martin Madan's (1726–1790) *Thelypthora*, a weird if serious plea for the legalization of polygamy in Britain.[73]

Traffic between literature and science (in our modern sense) through natural history was then energized by a revived sexual imagination: curious, brave, but not so prurient or pornographic as some have suggested.[74] Prior decades of convergence of literature and science, from the peak of "sexual Hogarth" in the 1750s in England to the experimental anatomic work of the Hunter

---

[72] D. G. King-Hele, *Erasmus Darwin* (London: Macmillian, 1963); Erasmus Darwin, *The Letters of Erasmus Darwin*, ed. D. G. King-Hele (Cambridge University Press, 1991). The debate about the origins of the theory of evolution continues into our own generation, moving further back than Darwin and Maupertuis to previous centuries and even the ancients.

[73] Martin Madan, *Thelyphthora; or, A Treatise on Female Ruin*, 2 vols. (London, 1780).

[74] Peter Wagner has argued for a pornographic slant; see *Erotica and the Enlightenment* (Frankfurt/Main: P. Lang, 1990); Wagner, *Eros Revived: Erotica of the Enlightenment in England and America* (London: Secker and Warburg, 1988).

brothers at mid-century, must also constitute an intrinsic part of Darwin's broad mind set. It is insufficient to localize his imagination in the groves of illicit analogy: a type of post-Linnean Forest of Arden. Darwin drew on a larger sexual milieu – new stress between the genders, assaults on the old masculinity, a decaying institution of marriage with new hope for reform in divorce law, even the possibility of marital liberation and polygamous utopia – for his illicit loves of plants and the equally daring new evolution *contra* creationism in *Zoonomia* (1794). *Au fin* these are contexts that culminate beyond the French Revolution, in the next generation, as in Mary Shelley's wild invention of Frankenstein in 1818. Through their *et in arcadia ego* vision of sexuality unshackled from all fetters, they lent to Darwin *grandpère* a type of Blakean apocalyptic side.

But merely a version, for the differences between Darwin and Blake are as great as the similarities. If Darwin's polygamous imagination ran amok when describing the "unmarried amours" of plants, his contemporary's (William Blake's) fantasy life was more perpetually in the cosmic heavens, even when sketching the body's *corpora fabrica* in his illustrations of *Job* and *Thel* and *America*.[75] The two men were near contemporaries, born only twenty-five years apart, and Blake's affinities with Darwin are stronger than has been thought. They are certainly much stronger than with the Swedish Linnaeus. This on the proviso that if Darwin was paradoxically the more unpredictable character – on the one hand the self-appointed cultural attaché ministering to the Lunar Society from his Georgian English mansion in central Lichfield, on the other describing his plants and flowers in such lurid detail that his metaphors now appear an unprecedented verbal pornography – Blake was the figure with the most visionary artistic imagination of the century and the one most capable of first absorbing and then revolting against Newtonian science.[76]

Yet if Blake's psychic war against Newton energized his art and poetry, Darwin himself had no such need. His Virgil was the Swedish Linnaeus and not any of the local English physicists, providing a further reason for the disparity in their mental cosmologies. Their theologies, Swedish and English, indeed differed, as did the physical science on which these construed systems depended. Blake, moreover, harbored no "romantic" sensibility vis-à-vis love as did Darwin: Blake's central theme was a visionary mysticism based on personal experience and a paranoid response to it. If the dominant psychological response of aristocratic Byron (another contemporary) was the rage from which

---

[75] Raymond Lister, *Infernal Methods: A Study of William Blake's Art Techniques* (London: Bell, 1975); David Bindman, *Blake as an Artist* (Oxford: Phaidon, 1977); Morris Eaves, *William Blake's Theory of Art* (Princeton, NJ: Princeton University Press, 1982); Janet Warner, *Blake and the Language of Art* (Kingston: McGill-Queen's University Press, 1984).

[76] Donald Ault, *Visionary Physics: Blake's Response to Newton* (Chicago: University of Chicago Press, 1975); Nelson Hilton, *Literal Imagination: Blake's Vision of Words* (Berkeley: University of California Press, 1983).

Byron created the object he then exposed in his satires, Blake's was a crushing fear of authority that caused him to distrust everything: he saw himself as a type of anti-Christ of science, defying the knowledge of certainty on which all public institutions could claim to function. Rational knowledge and scientific system were his hated doppelgängers.

No other poet anywhere in Europe tapped so firmly as did Blake into the riddles of certainty and creativity, everywhere posing artistic solutions grounded in personal psychological response. For this reason he serves as crucible of the historical tensions in this realm and is worthy of extended discussion. His late poetic vision in *Jerusalem* is apocalyptic and thoroughly discontinuous from virtually all the received Enlightenment programs preceding it: the trumpet shall sound, the Earth will split open, the world will end; the meek are to inherit the kingdom of heaven, as in many radical sects from the days of the Civil War. Across the Channel, another, very different type of "civil war" was occurring that reinforced Blake's sense of all philosophical "Enlightenment" as a grotesque failure that merely fettered humanity further: "In chains of the mind locked up, / Like fetters of ice shrinking together / Disorganized, rent from Eternity" (*The (First) Book of Urizen*, 4.4.1, 37–39). This was a massive attack on virtually *all* the cults of progress and perfection I have discussed here.

Blake was a figure *sui generis* by virtue of his superfluity of talents in many fields (poetry, painting, mythology, mythography, graphic technology) and for having no precise parallel figure anywhere in Europe. All comparisons fail the test: the Scandinavian mystic Swedenborg, the Swiss visionary painter Füseli, the inspired French prophets stretching across the whole eighteenth century. Blake himself never held a coherent theory of the origins of the French Revolution but believed, as American Blake scholar David Erdman demonstrated long ago, that its political foundations were contingent on the Enlightenment science permitting *ancien régimes* to flourish. The best historians still debate these origins; few would uphold Blake's view. Yet there is something in it for a visionary imagination committed to the view that "I must Create a System, or be enslav'd by another Man's" (*Jerusalem*, pl. 10.1.20). Nor did Blake harbor any significant view of progress other than the apocalyptic one, dismissing most natural science as disguised evil.

Blake-the-phenomenon is unimaginable without Enlightenment science. Compare him to his French *manqué* approximations (given that there are no parallels, the task must be imaginary) and even, for further contrast, to the French Erasmus Darwins – Rousseau, Cuvier, and Lamarck – and the differences are so pronounced as to require no comment. These are not merely *ad hominem* disparities but also are national and socio-political ones, and they underline the matter of *ars combinatoria*. French science and French literature had never coexisted, as science and literature did, across the Channel owing to a rational (French) versus empirical (British) tradition. Who is the French John Locke or David Hume? The notion of "demanding the muse" of Newton, as

had Alexander Pope and his English wits, is unthinkable. It was the rational
Descartes, rather than an empirical Gallican Newton, whom they invoked.
This is also why the division of knowledge into the specialities (disciplines)
of the late eighteenth century arrived in Scotland and England before anywhere else – a British rather than a French phenomenon – and it also explains
why there is nothing even proximate in the eighteenth-century French literary
tradition to a Blake massively revolting against "the woof of Newton and
Locke." If countries and national traditions matter so far as the question
about the organization of the arts *and* sciences (foregrounded in this essay)
is concerned, and in the implications of this organization for the development of each branch of knowledge, then Blake must be seen in his genuine
biographical context: first as an Englishman, second as a Londoner, and only
third and fourth as a creative artist and visionary thinker.

Blake's radical rupture from the main traditions of the empirical and rational Enlightenments was also predicated on an another version of progress
and perfection than the ones we have been exploring, versions that could be
characterized as follows: rational, logical, social, civic, perhaps plodding but
nevertheless calculating benefits for all humanity, and – more than anything
else – sunny in their optimism that in the end human progress, perhaps even
a version of human perfection, could be attained. Blake routinely conjured
the opposite, as in his childlike but bleak *Songs of Innocence and Experience*.
Unlike many of his contemporaries and despite a healthy frame and long life
of seventy years, he never traveled, never left England, took the Grand Tour,
crossed the Alps, or saw other lands. His journeys were interior: the interior
voyage, the interior traveler, the apocalyptic imagination of desperate endings
mitigated only slightly by the promise of the trumpet and the redemption of
the *unholy* in the New Jerusalem.[77] Newtonian and post-Newtonian sciences
found in him their gloomiest commentator. He was the anti-Newtonian par
excellence who loathed Newton not merely for the literalist he was (as did
Lamb) but also for the Leviathan Newton had set in motion in his rampage
to scientize all human endeavor. Blake condemned Newton and Locke and
Voltaire for the harm they inflicted on mankind rather than praised any good
that came of their work. He thought that they represented crimes against humanity despite the fact that he continued to plunder their goods for his own
purposes.

Blake was not alone in his view of the English Newtonians, even if he was
the fiercest in his opposition. Certain contemporary Romantic poets and
painters, with whom Blake shared varied eclectic affinities, shared his recoil
(it was hardly a doubt). Progress, for them, was an illusion. Nothing had
changed, least of all the economic poverty and international war that constituted the best proof of the constancy of human nature, and certainly nothing

---

[77] For this tradition, see Stock, *The Holy and the Demonic*.

as the result of a beneficent science or applied technology.[78] Science applied to society was (to them) more corrupt than pure, forever prone to corrupting. And the developing scientific professions – in medicine and technology, on the seas, in universities – were as culpable as other increasingly bureaucratic institutions of government. The notion that science somehow enhanced art or liberated the human spirit was nonsense, they reasoned. It deadened it. It enslaved it. It crippled the creative urge, not quickened it. It froze it by systematizing it. The only possible perfection for humankind for Blake – if such a thing there were – was perfection of the soul, and science had not touched that. Science was silent about the soul, as it has continued to be.

New doubts about these primary reciprocities between the sciences and the arts arose as 1799 turned to 1800, and not merely for the forty-three-year-old Blake, whose apocalyptic *Jerusalem* would not be published for another four years (1804) in that centennial world in which (for example) Napoleon was now consul; heroic figures – such as Peruvian conqueror Francisco Pizarro and newly revived Thirty-Year's War icon, general Count Wallenstein – beckoned English writers (Sheridan and Coleridge respectively) to bring them, through translation, before British audiences. A similar rupture occurred through Wordsworth and Coleridge, who issued the second volume of their revolutionary "lyrical ballads" in 1800 containing a preface in the form of a literary manifesto saluting not the expert reader but the common man in the street. The doubt entailed the future in the new nineteenth century. It also embraced a new set of problematics between the sciences and the arts under the weight of growing specialization: in universities, in academies of learning, in the developing professions and their publications, and in society at large in the unstable political aftermath of the French Revolution.

Rarely were these shackles of the past articulated so violently (almost as if a curse) as they were by Blake, who yearned "To cast off Bacon, Locke and Newton from Albion's covering / To take off his filthy garments and clothe him with Imagination" (*Milton*, 2.5–6). What pronouncement could have been more unequivocal? Further east, beyond the Rhineland, the German Romantics – especially Schiller, Novalis, and Schlegel – and the French Mme de Staël were spinning their own myths about "the arts and the sciences," and such painters as François Boucher were drawing their visions in Paris of their union (as in his "The Arts and Sciences" sculpture), but never with Blake's steel edge of rebuke. Artistic doubts about science and the imagination were more routinely balanced than in Blake's version, tempered by the new worlds, especially in astronomy and space, science had opened for poets and painters. If virtually all these artists doubted that eighteenth-century progress, contingent on its new sense of nationhood, empire, luxury, and consumption,

[78] Despite new criticism and scholarship, Wylie Sypher's *Literature and Technology* (New York: Random House, 1968) has remained one of the shrewdest comments on these matters.

had paved the way for perfection of any sort, they still conceded the Whig view that things were getting better and that the common woman or man in the street was much better off than in 1700. Even Wordsworth and Coleridge were now writing poetry for them, a state of affairs unthinkable among the Scriblerians and Wits of the world of Addison. Hence the paradoxes flew. Society was still class-riven, east of Paris no less than west. The ordinary person knew little science, may never have heard of Newton or his fluxions. But even such a person's imagination had altered, perhaps through science's technological applications.

The realms of the interior mind had been opened up to the commoner, as Wordsworth and Beethoven (*both* deeply rooted in the eighteenth century, each thirty years old when the century changed) each acknowledged in his own idiosyncratic way despite profound personal skepticism about the possibility of human and artistic progress. A small glimpse of the difference is garnered by consulting the extremities of these centuries. As 1699 lapsed to 1700, English poet-playright John Dryden (1631–1700), who was himself to die that year, composed Janus, "The Secular Masque," to commemorate the new century, permeated with an optimism rarely found at the turn of other centuries:

> Chronos, Chronos, mend thy Pace:
> An hundred Times the rowling Sun
> Around the Radiant Belt has run
>   In his revolving Race.
> Behold, behold, the Goal in sight;
> Spread the Fans, and wing thy flight. (1–6)

Exactly one hundred years later, in 1800 – just as English poet William Cowper died and English historian Thomas Babington Macaulay was born – an anonymous editorial in the *Gentleman's Magazine* chastised the ignominious eighteenth century: "Into the opening [nineteenth] century have entered all these horrors [of the eighteenth century]. If the rising generation suffer a total debasement under their influence, adieu to Names and Characters eminent for virtuous and heroic achievements! Adieu to Honesty, Benevolence and every sound principle . . ."[79]

Such were some of the centennial anxieties of those who had lived through the terrifying 1790s on both sides of the Channel, with its French political upheavals, new American presidents, radical religious revivals throughout Europe, and predictions of doom and gloom. Nor was the French menace eradicated or put to sleep as the clock turned to 1800. Yet somehow, through all this welter of confusion and castigation, virtuous science, inspirer of poets and liberator of arts, had triumphed. Or so it seemed to some observers. Perhaps it was a liberator after all. Would it continue to be so? In 1800, or thereabout, one guess was as good as the next, and in the historical process of

---

[79] *Gentleman's Magazine*, 70 (1800), iv.

retrieval informing this essay one begins to peep into the hinterland of the mind of Lamb, with whom we started this exploration. The fogs lift when we see him number himself among the skeptics for whom Newton and Locke and their tribe had *precipitated*, rather than discouraged, these cannibalistic events. Literalist science, especially empirical Newtonian science, as we suggested at the start, was, for Lamb, nothing on which freedom, imagination, and creativity were built.

The charges in the early nineteenth century were different. They dwelled on the *kind of learning* science was and asked what kind of knowledge science is (Lamb deemed it to be dead knowledge). How, they asked, does it compare with other forms of knowledge, and what aspects of it will endure? The Lambites (if they can be lumped together) continued to hammer away about science's deadening literalism: it was hardly a liberator of imaginative freedom, they claimed. Science enslaves the imagination, they argued, and deifies reason and empirical experience (rather than the many other forms: metaphysical, transcendental, vadic) at the expense of the imagination. We seem to have reached an impasse: on the one hand those then touting science the liberator, on the other those decrying its deadening hand; and a third group, not yet as vocal as its colleagues, inquiring about its moral dimensions. Today, historians wishing to retrieve these strains of Enlightenment science and literature (however Enlightenment is construed), by performing a magical amalgamation of these positions may inevitably be sacrificing a key element of each. In the end, the attitude summoned in deciding counts for much, and it compels us in our own millennial moods to rethink the meanings of Hiroshima, Sputnik, and men on the moon. Had Pope and Lamb, for example, they would have disagreed on every facet. Herein lies one of the fierce winds of change that blew through their intellectually turbulent times.

34

SCIENCE, PHILOSOPHY, AND THE MIND

Paul Wood

During the eighteenth century, men and women of letters throughout the Atlantic world repeatedly celebrated the revolution they had witnessed in all the many branches of philosophy. Drawing on the rhetoric and historical vision of those who had championed the achievements of the "new science" of the seventeenth century, apologists for the Enlightenment claimed that humankind had finally been able to progress far beyond the narrow intellectual horizons of antiquity and the "dark ages" thanks to the new methods of inquiry forged by Sir Francis Bacon (1561–1626), René Descartes (1596–1650), John Locke (1632–1704), and Sir Isaac Newton (1642–1727). In this heroic reading of the genesis of modernity, Bacon was cast as the father of the experimental method, and Descartes played the tragic role of the flawed genius who used reason to liberate humankind from the shackles of scholasticism only to foist yet another false system of philosophy on the learned world. Locke was assigned the part of the humble reformer of metaphysics, who replaced meaningless verbal disputes with the patient empirical investigation of the mechanisms of mind and language and who carefully mapped the limits of human knowledge. But to the *siècle des lumières* it was Newton – apostrophized in Alexander Pope's (1688–1744) couplet, "Nature and Nature's Laws lay hid in Night./GOD said, *Let Newton be!* and all was Light." – who towered above the other founders of the Enlightenment. Not only had Newton divined the secrets of Nature by demonstrating that his theory of universal gravitation explained the motions of both celestial and terrestrial bodies, but he had also taught the salutary lesson that philosophers could discover the truth only by eschewing arbitrary hypotheses in order to focus their attention on what could be proved using the combined tools of geometry and experiment. The question of method was thus central to the *philosophes'* genealogy of the emergence of modern philosophy; and in their narratives of the growth of knowledge in their own day, they identified the spread of enlightenment with the growing adoption of the methodological principles established by the patriarchs of the Age of Reason.[1]

Subsequent accounts of the relations between what we now call "science" and "philosophy" in the Enlightenment have hitherto been framed largely by the *philosophes'* interpretation of the intellectual history of early modern Europe and, consequently, have focused primarily on methodological issues. For example, Victor Cousin (1792–1867), the philosophical *maître* of French Liberalism during the second quarter of the nineteenth century, praised the Scottish "school" of philosophy founded by Francis Hutcheson (1694–1746) for having promoted the true method of philosophizing by using empirical procedures in its inquiries. Across the Atlantic, James McCosh (1811–1894), the expatriate Scot and President of Princeton College, honored his Scottish forbears for applying the inductive method formulated by Bacon and Newton to the study of the human mind.[2] Writers in our own century have sometimes displayed sounder scholarship or added greater historical detail, but they have told essentially the same story regarding the formation of an empirical "science of man." In one of the most influential discussions of the *Zeitalter der aufklärung*, Ernst Cassirer maintained that the defining characteristic of Enlightenment philosophy was its adoption of the method of analysis championed by Newton; similar accounts are to be found in Preserved Smith as well as in more recent surveys by Lester G. Crocker, Peter Gay, and Norman Hampson, among others.[3] Newton's impact on Enlightenment philosophical thought is also the *leit-motif* of numerous specialist studies, from Elie Halévy's exploration of the roots of Jeremy Bentham's utilitarianism in "moral Newtonianism" to Henry Guerlac's examination of divergent interpretations of

[1] The classic examples of the Enlightenment's historical self-fashioning are François Marie Arouet de Voltaire, *Philosophical Letters*, trans. Ernest Dilworth (Indianapolis: Bobbs-Merrill, 1961); Anne-Robert-Jacques Turgot, "A Philosophical Review of the Successive Advances of the Human Mind," in *Turgot on Progress, Sociology, and Economics*, trans. Ronald L. Meek (Cambridge University Press, 1973), pp. 41–59; Jean Le Rond d'Alembert, *Preliminary Discourse to the Encyclopedia of Diderot*, trans. Richard N. Schwab (Chicago: University of Chicago Press, 1995); Marie Jean Antoinne Nicolas Caritat, Marquis de Condorcet, *Sketch for a Historical Picture of the Progress of the Human Mind*, trans. June Barraclough (London: Weidenfeld and Nicolson, 1955). Pope's couplet comes from his intended epitaph for Newton; see Alexander Pope, *Poetical Works*, ed. Herbert Davis (Oxford: Oxford University Press, 1966), p. 651.

[2] Victor Cousin, *Philosophie écossaise*, 3rd ed., revised and enlarged (Paris: Librairie nouvelle, 1857), pp. 33, 34, 237–41, 484; James McCosh, *The Scottish Philosophy, Biographical, Expository, Critical, from Hutcheson to Hamilton* (London: Macmillan, 1875), pp. 2–4. For a later variation on McCosh's theme, see Gladys Bryson, *Man and Society: The Scottish Inquiry of the Eighteenth Century* (Princeton, NJ: Princeton University Press, 1945). Compare their interpretation of the Scottish school with the assessment in Henry Thomas Buckle, *On Scotland and the Scotch Intellect*, ed. H. J. Hanham (Chicago: University of Chicago Press, 1970), pp. 235–44.

[3] Ernst Cassirer, *The Philosophy of the Enlightenment*, trans. Fritz C. A. Koelln and James P. Pettegrove (Princeton, NJ: Princeton University Press, 1951), chap. 1; Preserved Smith, *A History of Modern Culture* (New York: Collier Books, 1962), vol. 2: *The Enlightenment, 1687–1776*, pp. 117–21; Lester G. Crocker, *Nature and Culture: Ethical Thought in the French Enlightenment* (Baltimore, MD: Johns Hopkins Press, 1963); Peter Gay, *The Enlightenment: An Interpretation*, 2 vols. (New York: Knopf, 1966–69); Norman Hampson, *The Enlightenment* (Harmondsworth: Penguin, 1968).

Newton in eighteenth-century France.[4] As these representative examples suggest, accounts of the interaction of science and philosophy during the Enlightenment typically claim that philosophy was transformed from being a largely speculative to a predominantly empirical enterprise, inspired primarily by Newton's achievements in the physical sciences.

In what follows, I reassess the validity of this standard interpretation of the period and in so doing fashion a more balanced understanding of how the union of the natural and the human sciences gave birth to a science of the mind in the Enlightenment. I begin with a survey of the intellectual legacy of the seventeenth century, highlighting themes found in the works of Bacon, Descartes, Locke, and the natural law tradition, themes that were subsequently developed and varied by aspiring scientists of the mind. I then address the question of Newton's influence on the study of human nature during the *siècle des lumières* and suggest that Newton's impact has been exaggerated and that his writings were read in such radically different ways that it is difficult to identify a unified Newtonian tradition in the moral sciences. I next examine other sources of methodological inspiration available to enlightened savants and consider the appeal of mathematical models as well as the use of quantitative techniques in the analysis of mental and moral phenomena. I also point to the pervasive Enlightenment fascination with the notion of anatomizing the mind, and I conclude by outlining the emergence of a natural historical approach to the study of human nature that gained increasing favor as the eighteenth century progressed.[5]

SEVENTEENTH-CENTURY EXEMPLARS

Although Sir Francis Bacon was widely celebrated in the Enlightenment as the father of the experimental method, his contribution to the development of the study of the mind extended well beyond methodology, insofar as his map of learning also shaped definitions of the aims and scope of the human sciences throughout the eighteenth century. In *Of the Proficience and Advancement of Learning Divine and Human* (1605) and the revised Latin version, *De Dignitate et Augmentis Scientiarum* (1623), Bacon outlined his proposal for the formation of a new, practical adjunct to moral philosophy that he variously christened "the *Regiment* or *Culture of the Mind*" or, in an allusion to the writings of the Roman poet Virgil (70–19 B.C.), "the *Georgics of the Mind*." Liken-

---

[4] Elie Halévy, *The Growth of Philosophic Radicalism*, trans. Mary Morris (London: Faber and Faber, 1934); Henry Guerlac, *Essays and Papers in the History of Modern Science* (Baltimore, MD: Johns Hopkins University Press, 1977).

[5] Although the concept of human nature is treated in what follows as being relatively unproblematic, its meaning was by no means univocal or transparent in the eighteenth century; see Roger Smith, "The Language of Human Nature," in Christopher Fox, Roy Porter, and Robert Wokler (eds.), *Inventing Human Science: Eighteenth-Century Domains* (Berkeley: University of California Press, 1995), pp. 88–111.

ing the task of moralists to those of husbandmen or physicians, Bacon argued that just as farmers must know how to deal with different soils and climates or doctors with the various tempers and bodily constitutions of patients, so, too, must moralists base their ethical precepts on a thorough knowledge of human nature. Bacon therefore called for the initiation of new forms of inquiry focusing on the classification and description of distinct human character types; the analysis of how those characters are molded by our physical and social environments, the vagaries of fortune, and our bodily states; and the investigation of the origins of our actions in our passions and affections. Once such information was at hand, he was confident that we would then learn how to shape human conduct through education and the inculcation of virtuous habits and customs. For Bacon, the knowledge of human nature truly held out the promise of power. He believed that the findings of the culture of the mind could be applied beneficially in both the moral and the political realms, and his vision of our ability to manipulate individual behavior had a lasting impact on educational and social reformers during the Enlightenment.[6]

But how was an understanding of human nature to be gained? Bacon said little about the methods appropriate to the "Georgics of the Mind" in the *Advancement of Learning* beyond indicating that moralists had to develop an empirical science rooted in everyday experience, poetry, and the annals of history. Further guidance eventually came in his major tracts on methodological issues published together in 1620: the *Novum Organum* and the *Parasceve ad historiam naturalem et experimentalem*. Like many Renaissance thinkers, Bacon was fascinated by the idea of a universal method, and the inductive logic he sketched out in the *Novum Organum* was intended to guide inquiries in all branches of knowledge. Yet modern historians often ignore the fact that there was more to Bacon's method than the use of experiment or induction. Bacon himself emphasized that the human mind stood in need of "helps" to arrive at genuine knowledge, and his method was designed to combat the failings of our memory, senses, and reason. He therefore prescribed the compilation of comprehensive natural histories to complement experimentation and inductive reasoning, and he claimed that these histories constituted the true foundation of the sciences. Hence, the natural historical enterprise was an integral part of Bacon's projected reformation of learning, and, in the *Parasceve,* he extended its scope to encompass the many branches of the study of humankind. Bacon also pointed to an overlap between civil and natural history when reflecting on the literature generated by the voyages of exploration of his day, writing that these works typically combined data on physical

---

[6] *The Philosophical Works of Francis Bacon,* ed. John M. Robertson (London: George Routledge and Sons, 1905), pp. 141–8, 563, 571–8; the classical reference is to Virgil's poetic celebration of rural pursuits, the *Georgics.* Bacon's allusion to this poem served to underline his point that just as the soil requires careful cultivation for it to be fruitful, the mind, too, must be actively cultivated and trained for the pursuit of virtue.

environment and climate with information regarding "the habitations, regiments, and manners of the people." The science of human nature for Bacon was thus intrinsically natural historical in its approach and, as the "mixed histories" found in the travel literature showed, ethnography fused with natural history to produce accounts of the peoples and places encountered in the "new worlds" across the seas. Consequently, natural history (taken as both a body of knowledge and a methodological program) was at the heart of the Baconian legacy in the human sciences, and it was arguably of greater practical import than Bacon's notion of induction, insofar as his description of the inductive method in the *Novum Organum* was left radically incomplete.[7]

Likewise possessed by the dream of a universal method, René Descartes first gave the learned world a sample of his methodology in his *Discours de la méthode* (1637). Given that Descartes was later thought to have effected a revolution in the science of the mind through his use of the introspective method, it is significant that, unlike Thomas Hobbes (1588–1679), Descartes himself nowhere offered either an explicit rationale for, or any guidance on, the use of introspection. It must also be said that the "true method" of inquiry outlined in his writings found little favor in the Enlightenment.[8] Of far greater long-term importance was his absolute distinction between mind and matter and his mechanistic interpretation of the anatomy and physiology of the human body, which provided a stimulus for the rise of materialism in the eighteenth century. Dualism appealed to Descartes because it armed him with a defense of Christian orthodoxy, but it also posed the intractable problem of accounting for the interaction of such categorically different substances as mind and body. Notwithstanding the conceptual puzzles involved, he insisted that it is simply a "fact" that "being united to the body, [the soul] can act and be acted upon along with it" and effectively left it to posterity to make sense of the mind-body relationship.[9]

Recent scholarly interest in the physiological underpinnings of Descartes's account of our mental functions has brought into sharper focus one of the most influential features of his approach to the science of the mind – namely the integration of what we now identify as the separate fields of philosophy, psychology, and physiology into a seamless analytical whole. Descartes achieved this integration largely through his reformulation of the theory of animal spir-

---

[7] Bacon, *Works*, pp. 85, 251–2, 410–11, 437.

[8] Condorcet, *Sketch*, p. 132; Thomas Reid, *An Inquiry into the Human Mind, On the Principles of Common Sense*, ed. Derek R. Brookes (Edinburgh: Edinburgh University Press, 1997), pp. 205, 208–9; Thomas Hobbes, *Leviathan; or, The Matter, Forme, & Power of a Common-wealth Ecclesiastical and Civill* (Oxford: Clarendon Press, 1909), p. 9. For a useful introduction to Descartes and the scholarly literature devoted to him, see John Cottingham (ed.), *The Cambridge Companion to Descartes* (Cambridge University Press, 1992).

[9] Descartes to Princess Elizabeth of Bohemia, 21 May 1643, in *The Philosophical Writings of Descartes*, trans. John Cottingham, Robert Stoothoff, Dugald Murdoch, and Anthony Kenny, 3 vols. (Cambridge University Press, 1984–91), 3:217–18. The classic account of the Cartesian origins of eighteenth-century French materialism is Aram Vartanian, *Diderot and Descartes: A Study of Scientific Naturalism in the Enlightenment* (Princeton, NJ: Princeton University Press, 1953).

its, which he invoked to explain a wide range of physiological and mental phenomena, including voluntary and involuntary muscular motions, sensory perception, the imagination, dreaming, memory, individual temperaments, and the passions. In his system, animal spirits mediate between the physical and the mental, as can be seen in his discussion of visual perception in *L'homme*, which was written c. 1629–33 and published posthumously in 1664. Here he carefully distinguished between the "figures" impressed on our sensory organs or our brains in perception, and those "ideas" inscribed by the animal spirits on the surface of the pineal gland, "ideas" that the soul "consider[s] directly" when it perceives the external world.[10] This passage also shows that in Descartes's philosophical lexicon the term "idea" can refer either to the physical patterns of animal spirits or to the purely mental contents of thought. Insofar as he is recognized as the founding father of the "way of ideas," we should remember this duality of meaning when considering his intellectual legacy.[11] Descartes' writings blended physiological and philosophical considerations, and his style of analysis defined one of the major analytical traditions within the science of the mind, a tradition that can be traced through the works of Locke, Nicolas Malebranche (1638–1715), David Hartley (1705–1757), and the *Idéologue* Pierre-Jean-Georges Cabanis (1757–1808).

John Locke's map of the limits of human knowledge in his *Essay concerning Humane Understanding* (1690) was drawn according to the grid of his heterodox brand of Christianity. The basic tenet of his epistemology, that all our ideas are derived from either sensation or reflection, was founded on his view that God has "*fitted Men with faculties and means, to discover, receive, and retain Truths, accordingly as they are employ'd*" and consequently expects us to acquire knowledge through our own initiative. It followed for Locke that one of our basic moral obligations to the Creator is to examine critically our beliefs and to increase the stock of human knowledge. Hence the *Essay* was designed to help us achieve that end by teaching the proper use of reason. This was a lesson of paramount importance because Locke maintained that "*Reason* is natural *Revelation*, whereby the eternal Father of Light, and Fountain of all Knowledge communicates to Mankind that portion of Truth, which he has laid within the reach of their natural Faculties." Religious considerations also emerge in his refutation of the Cartesian thesis that God has implanted in each of us an ensemble of innate ideas, for Locke argued that one of the major faults of this thesis was that it could be used to suppress free inquiry in

---

[10] Descartes, *Philosophical Writings*, 1:106. On the physiological basis of Descartes's theory of the mind, see Cottingham, *Companion*, chaps. 11 and 12; Richard B. Carter, *Descartes' Medical Philosophy: The Organic Solution to the Mind-Body Problem* (Baltimore, MD: Johns Hopkins University Press, 1983); G. A. Lindeboom, *Descartes and Medicine* (Amsterdam: Rodopi, 1979).

[11] Edward S. Reed, "Descartes' Corporeal Ideas Hypothesis and the Origin of Scientific Psychology," *Review of Metaphysics*, 35 (1982), 731–52. The evolution of the theory of ideas is traced in John W. Yolton, *Perceptual Acquaintance from Descartes to Reid* (Minneapolis: University of Minnesota Press, 1984), and Yolton, *Perception and Reality: A History from Descartes to Kant* (Ithaca, NY: Cornell University Press, 1996).

matters of religion. Whereas Descartes apparently believed that the human mind could obtain an exhaustive understanding of the book of nature, Locke emphasized that the human condition was such that we could acquire certain knowledge about only a highly circumscribed range of subjects directly related to our practical concerns as God's creatures here on earth. To understand both the genesis and the reception of Locke's epistemology, therefore, we must see how his depiction of human nature registers his basic religious and moral preoccupations.[12]

Yet Locke's empiricism brought with it some uncomfortable consequences that threatened to undermine a number of his moral and religious assumptions. First, although he affirmed that morality could be turned into a demonstrative science, his discussion of "mixed modes" in the *Essay* indicated that our moral ideas are ultimately conditioned by language and hence are to some extent relative to the societies in which we live. His dream of creating a system of morals akin to Euclidean geometry was thus shattered by the realization that our social life shapes our experience and consequently our ideas. But although it served to destabilize his projected reconstruction of the science of morality, his recognition that knowledge and belief are at least partly conditioned by the society in which they are produced opened the possibility of fusing the science of the mind with history and ethnography, and this had a positive impact on the study of human nature in the eighteenth century.[13]

Second, Locke's thesis that experience is the source of all our ideas implied that human nature was malleable and that it could be improved (or corrupted) by education, habit, and custom. His works could be read as holding out the promise that human nature is perfectible, and they were later interpreted thus by Joseph Priestley (1733–1804) and Richard Price (1723–1791), among others. It is not clear, however, that Locke himself would have sanctioned such a reading. A number of passages in his writings suggest that he saw human nature as fallen, and his view of the limits of human knowledge precluded any overly optimistic hopes for human progress.[14] Ultimately, Locke's version of empiricism was something of a two-edged sword. It could be used to shore up a rational, irenic brand of Christianity and to attack Catholic and Protestant opponents of religious tolerance who wanted to limit free inquiry in

---

[12] John Locke, *An Essay concerning Humane Understanding*, ed. Peter H. Nidditch (Oxford: Clarendon Press, 1975), I.i.4–7, I.iv.22, I.iv.24, IV.xvii.24, IV.xix.4. The most detailed account of the development of Locke's thought to date is John Marshall, *John Locke: Resistance, Religion, and Responsibility* (Cambridge University Press, 1994). The reception of the *Essay* is reconstructed in John W. Yolton, *John Locke and the Way of Ideas* (Oxford: Clarendon Press, 1956). The moral and religious dimensions of Locke's epistemology are discussed in Nicholas Wolterstorff, *John Locke and the Ethics of Belief* (Cambridge University Press, 1996).

[13] Locke, *Essay*, II.xxii.5–7, 10, IV.iii.18–20; G. A. J. Rogers, "Locke, Anthropology, and Models of the Mind," *History of the Human Sciences*, 6 (1993), 73–87.

[14] David Spadafora, *The Idea of Progress in Eighteenth-Century Britain* (New Haven, CT: Yale University Press, 1990), especially chap. 6. Locke's view of our moral corruption is discussed in W. M. Spellman, *John Locke and the Problem of Depravity* (Oxford: Clarendon Press, 1988); see also Marshall, *Locke*, pp. 346–7.

religious matters. But it could also be deployed as a weapon against Christianity itself, for the polemics of John Toland (1670–1722), Anthony Collins (1676–1729), and Matthew Tindal (1657–1733) revealed that the "way of ideas" rendered fundamental aspects of Christian doctrine vulnerable to corrosive epistemological criticism. The eighteenth century bore witness to these divergent uses because Locke's theory of knowledge was appropriated both by defenders of the reasonableness of Christianity and by those whose relations with the Christian tradition were at best ambivalent, such as François-Marie Arouet de Voltaire (1694–1778), Etienne Bonnot de Condillac (1714–1780), and Claude-Adrien Helvétius (1715–1771).

Locke's approach to the science of the mind was likewise ambiguous. Harking back to Bacon and perhaps influenced by the procedures of Thomas Sydenham (1624–1689) and Robert Boyle (1627–1697), Locke cast himself in the role of the natural historian in the *Essay,* announcing at the beginning of Book I that he would employ an "Historical, plain Method" to trace out "the discerning Faculties of a Man" and claiming in Book II that he had "given a short, and, I think, true *History of the first beginnings of Humane Knowledge.*" Accordingly, he devoted himself to the description and classification of our ideas and the powers of the mind as well as to the temporal reconstruction of the genesis of our ideas and the sequential unfolding of our mental faculties.[15] He also signaled his intention to redefine the scope of the science of the mind when he stated that he proposed to "enquire into the Original, Certainty, and Extent of humane Knowledge; together, with the Grounds and Degrees of Belief, Opinion, and Assent." Implicitly rejecting Descartes's analytical style, Locke promised to avoid "the Physical Consideration of the Mind" and thus not to "examine, wherein [the mind's] Essence consists, or by what Motions of our Spirits, or Alterations of our Bodies, we come to have any Sensation by our Organs, or any *Ideas* in our Understandings; and whether those *Ideas* do in their Formation, any or all of them, depend on Matter or no," remarking that "[t]hese are Speculations, which, however curious and entertaining, I shall decline, as lying out of my Way, in the Design I am now upon."[16]

Yet Locke also indulged in precisely the kind of physiological theorizing that he professed to eschew. For instance, he affirmed that the brain is the

---

[15] Locke, *Essay,* I.i.2, II.i, II.xi.15. On Locke as a natural historian of the mind, see James G. Buickerood, "The Natural History of the Understanding: Locke and the Rise of Facultative Logic in the Eighteenth Century," *History and Philosophy of Logic,* 6 (1985), 157–90, and Neal Wood, *The Politics of Locke's Philosophy: A Social Study of* An Essay Concerning Human Understanding (Berkeley: University of California Press, 1983), chap. 4. Locke's natural historical approach to the mind should be compared with Boyle's codification of the "experimental life," as discussed in Steven Shapin and Simon Schaffer, *Leviathan and the Air Pump: Hobbes, Boyle, and the Experimental Life* (Princeton, NJ: Princeton University Press, 1985). On Locke's career as a physician and his connections with Sydenham, see Kenneth Dewhurst, *John Locke (1632–1704), Physician and Philosopher: A Medical Biography* (London: Wellcome Historical Medical Library, 1963).

[16] Locke, *Essay,* I.i.2.

seat of sensation and claimed that the motions of our animal spirits are the cause of our sensations. He further maintained that the association of ideas was explicable in terms of the behavior of the animal spirits, and he correlated various physical and mental states, notably with reference to memory.[17] There was thus a latent tension in the *Essay* between the kind of psychophysiological analysis exemplified in the works of Descartes and the natural historical approach that Locke was endeavoring to articulate, and this meant that his text conveyed a mixed methodological message to its eighteenth-century readers. Consequently, the *Essay* inspired thinkers such as Thomas Reid (1710–1796), who tried to disentangle the study of the mind from physiological and anatomical considerations, and materialists such as Julien Offray de La Mettrie (1709–1751) and Joseph Priestley, who followed up Locke's controversial suggestion that God could endow matter with the power of thought or used the way of ideas to legitimate materialism.[18]

Although the contributions of Bacon, Descartes, Locke, and (to a lesser extent) Hobbes have long been recognized, scholars have often ignored the ways in which the natural law tradition gave shape to the science of the mind in the Enlightenment. Those interested in the origins of scientism, for example, have not remarked on the fact that the founders of the modern natural law tradition – Hugo Grotius (1583–1645), Samuel Pufendorf (1632–1694), and Richard Cumberland (1631–1718) – all developed relatively sophisticated empirical methods to ascertain the laws governing human morality. Grotius combined what he called the a priori and a posteriori methods of proof; the former involved the deduction of natural laws from basic axioms concerning human nature, and the latter involved the discovery of the fundamentals of morals through the empirical and comparative study of human history and ethnography.[19] Pufendorf, too, adopted a comparative approach and was one of the first writers to suggest that moralists should employ methods similar to those used by natural philosophers, although he also acknowledged that there were important differences between "moral entities" such as human beings and "natural entities" such as material objects.[20] Methodological issues

---

[17] Locke, *Essay,* II.iii.1, II.viii.4, II.x.5, II.xxiii.6, II.xxix.3, III.vi.3. On Locke's use of the physiological ideas of the Oxford physician Thomas Willis, see John Wright, "Locke, Willis, and the Seventeenth-Century Epicurean Soul," in Margaret J. Osler (ed.), *Atoms, Pneuma, and Tranquillity* (Cambridge University Press, 1991), pp. 239–58.

[18] John W. Yolton, *Thinking Matter: Materialism in Eighteenth-Century Britain* (Oxford: Blackwell, 1983); Yolton, *Locke and French Materialism* (Oxford: Clarendon Press, 1991); Kathleen Wellman, *La Mettrie: Medicine, Philosophy, and Enlightenment* (Durham, NC: Duke University Press, 1992).

[19] Hugo Grotius, *De iure belli ac pacis libri tres,* trans. Francis W. Kelsey, 2 vols. in 4 (New York: Oceana Publications, 1964), 2:42; Joan-Paul Rubiés, "Hugo Grotius's Dissertation on the Origin of the American Peoples and the Use of Comparative Methods," *Journal of the History of Ideas,* 53 (1992), 221–44; Richard Tuck, *Philosophy and Government, 1572–1651* (Cambridge University Press, 1993), chap. 5.

[20] Samuel Pufendorf, *On the Natural State of Men,* trans. Michael Seidler (Lewiston: Edwin Mellen Press, 1990), p. 109; Pufendorf, *Of the Law of Nature and Nations,* trans. Basil Kennet, 3rd ed. (London: R. Sare, 1717), pp. 22–3.

were especially pressing for Cumberland because he aimed to refute Hobbes and insisted that moral philosophers must start from introspection, observation, and experiment (without specifying how experiments could be performed in the moral sciences) to derive sound moral precepts using methods patterned on the "analytic art" of algebra and the geometrical methods of analysis and synthesis.[21] Given that the works of these three natural law theorists and their followers became the staple of academic moral philosophy courses in the eighteenth century, there is little doubt that their methodological precepts and practice contributed to the empirical turn of the study of human nature.

Another key feature of the science of the mind derived from the natural law tradition was the stadial view of history initially outlined by Hugo Grotius. Prompted by his interest in the origins of private property, Grotius sketched the early evolution of human society in one of the seminal texts of the natural law canon, *De iure belli ac pacis* (1625). Using elements of the Christian historical narrative, he traced the move from a "primitive state," in which property was held in common and distributed according to need, to a more complex form of society in which private property was created on the basis of communal consent. To account for this shift, he appealed to the factors that subsequently became the theoretical trademark of philosophical historians in the Enlightenment: changing modes of subsistence, the division of labor, population growth, and related variations in manners and mores. Significantly, Grotius saw the gradual emergence of private property as being bound up with a profound transformation of human nature, for (echoing the Christian theme of the Fall) he contrasted the naive simplicity and ignorance of primitive subsistence societies with the refinement, moral depravity, and relative technological sophistication of agricultural and commercial ones. Retold by Pufendorf later in the century, this story ultimately provided the Enlightenment with what many observers believed was the key to explaining both the evolution of the mind and the transition from rudeness to refinement in the history of human society.[22]

NEWTONIAN LEGACIES

Now that we have identified some of the major intellectual trends of the seventeenth century that structured the Enlightenment science of the mind, we are in a better position to analyze the various elements of the methodological

---

[21] Richard Cumberland, *A Treatise of the Law of Nature*, trans. John Maxwell (London: J. Knapton et al., 1727), pp. 43, 54, 56, 184–8, 208, 278–80, 296.

[22] Grotius, *Laws*, pp. 186–90; Istvan Hont, "The Language of Sociability and Commerce: Samuel Pufendorf and the Theoretical Foundations of the 'Four-Stages Theory'," in Anthony Pagden (ed.), *The Languages of Political Theory in Early-Modern Europe* (Cambridge University Press, 1987), pp. 253–76.

revolution in the study of human nature that occurred at the turn of the eighteenth century. Although some of the historical claims made by the *philosophes* may have been exaggerated, there is no doubt that at least part of the inspiration for this revolution came from the writings of Sir Isaac Newton, and nowhere was this more apparent than in Scotland. From the 1690s onward, the Scottish universities led the way in the institutionalization of the Newtonian system, and those teaching moral philosophy north of the Tweed were quick to respond to this dramatic shift in the curriculum. In the classrooms of Marischal College Aberdeen during the 1720s, where Newton's protégé Colin Maclaurin (1698–1746) professed mathematics before moving to Edinburgh, the moralist George Turnbull (1698–1748) was perhaps the first Scottish academic to recommend a method for moral investigations patterned on those formulated by Bacon and Newton. Elaborating on his lectures and public pronouncements in *The Principles of Moral Philosophy* (1740), Turnbull proclaimed that "in order to bring moral philosophy . . . upon the same footing with natural philosophy . . . we must enquire into moral phenomena, in the same manner as we do into physical ones" and recalled that he had been "led long ago to apply myself to the study of the human mind in the same way as to that of the human body, or any other part of *Natural Philosophy*" by Newton's remark in Query 31 of the *Opticks* that "if natural Philosophy in all its Parts, by pursuing this method [of analysis and synthesis] shall at length be perfected the Bounds of Moral Philosophy will be also enlarged." Drawing on the *Opticks*, Turnbull duly stipulated that moral philosophers had to follow Newton's example in deploying the "double method of analysis and synthesis" to discover the system of laws governing the moral order. In terms that recalled Newton's rules of philosophizing and Roger Cotes's (1682–1716) preface to the second edition of the *Principia*, Turnbull also proscribed the use of hypotheses in the study of morality. Although other prominent academic moralists of the period, such as Francis Hutcheson, made similar assertions about the methodological unity of the two main branches of the philosophy curriculum, Turnbull stands out as being the earliest and most articulate Scottish spokesman for the adoption of the Newtonian method in the science of the mind. His brand of moral Newtonianism had a lasting impact on philosophical speculation in the Enlightenment through the works of his most distinguished pupil, Thomas Reid (1710–1796), who went so far as to claim that Newton's "*regulæ philosophandi* are maxims of common sense, and are practised every day in common life."[23]

---

[23] George Turnbull, *De Scientiæ Naturalis cum Philosophia Morali Conjunctione* (Aberdeen: James Nicol, 1723), p. 3; Turnbull, *The Principles of Moral Philosophy*, 2 vols. (London: John Noon, 1740), I:iii, 5–6, 9–10, 12–13, 19–20; Sir Isaac Newton, *Opticks; or, A Treatise on the Reflections, Refractions, Inflections and Colours of Light*, 4th ed. (London, 1730; New York: Dover, 1952), p. 405; William Leechman, "The Preface, Giving some Account of the Life, Writings, and Character of the Author," in Francis Hutcheson, *A System of Moral Philosophy*, 2 vols. (Glasgow: R. and A. Foulis and A. Millar, 1755), I:xxxvi; Reid, *Inquiry*, p. 12.

Alarmed by David Hume's (1711–1776) *Treatise of Human Nature* (1739–40), Reid sought to counter Humean skepticism by striking at what he maintained lay at its intellectual heart – namely, the "Cartesian system." According to Reid, philosophers from Descartes onward had been seduced into thinking that we directly perceive ideas rather than the external world, and consequently they had opened the conceptual door to the radical form of skepticism advanced by Hume. To avoid Hume's conclusions, Reid scrutinized this theory and its corollaries in *An Inquiry into the Human Mind, On the Principles of Common Sense* (1764), which is an outstanding example of the creative appropriation of Newton. Inspired by Newton's methodological pronouncements, Reid offered a battery of arguments, including a cleverly formulated *experimentum crucis,* designed to disprove the Cartesian system. Moreover, having imbibed Turnbull's antihypotheticalism as a young student, Reid targeted what he thought was the largely conjectural nature of the theory of ideas. Insisting that "no solid proof has ever been advanced of the existence of ideas," he contended that the theory was "a mere fiction and hypothesis, contrived to solve the phænomena of the human understanding" and repeatedly drove home the point that the "hypothesis of ideas or images of things in the mind, or in the sensorium" was especially dangerous because it was "the parent of those many paradoxes so shocking to common sense, and of that scepticism, which disgrace our philosophy of the mind, and have brought upon it the ridicule and contempt of sensible men."[24] Following his move to the University of Glasgow in 1764, Reid refined his critique of the hypothetical method in his lectures on the theory of ideas, and turned to Newton's first rule of philosophizing for further ammunition. As interpreted by Reid, Newton's "golden rule" that "we are to admit no more causes of natural things than such as are both true and sufficient to explain their appearances" proscribed the use of hypotheses in philosophy because it demanded of any causal explanation that the putative cause was sufficient to account for the effects *and* that its existence had been demonstrated. Consequently, for Reid the "ideal system" was to be banished from the science of the mind because the existence of ideas had never been proven. He likewise attacked the standard physiological theories of perception on the grounds that their proponents had failed to show that such entities as animal spirits actually existed.[25]

Reid's reading of Newton was highly influential in the latter part of the eighteenth century, but there were other versions of Newtonianism available in Britain that were at odds with Reid's austere antihypotheticalism and with his warnings against the "love of simplicity" and the misguided desire to build

[24] Reid, *Inquiry,* p. 28.
[25] Thomas Reid, *Essays on the Intellectual Powers of Man* (Edinburgh: J. Bell, 1785), pp. 46–52; *Sir Isaac Newton's Mathematical Principles of Natural Philosophy and His System of the World,* trans. Andrew Motte, ed. Florian Cajori, 2 vols. (Berkeley: University of California Press, 1934), 2:398; Paul Wood, "Reid on Hypotheses and the Ether: A Reassessment," in M. Dalgarno and E. Matthews (eds.), *The Philosophy of Thomas Reid* (Dordrecht: Kluwer, 1989), pp. 433–46.

philosophical systems. For example, Adam Smith (1723–1790), Reid's predecessor at Glasgow, identified the Newtonian method with the construction of simple, deductive systems in natural philosophy and saw the search for simplicity as a manifestation of the mind's attempt to impose order and coherence on experience.[26] The issue of simplicity also figured prominently in the major debate that erupted between Reid and Joseph Priestley over the merits of Reid's epistemology and Priestley's brand of materialism. Worried by the increasing reputation of Reid, James Beattie (1735–1803), and the Rev. James Oswald (1703–1793), Priestley took aim at the Scottish philosophy of common sense in the mid-1770s, and he singled out Reid's propensity to multiply the "*independent, arbitrary, instinctive principles*" of the mind as being particularly problematic. Invoking the notion that nature is simple, Priestley dismissed his opponent's account of our mental powers because it lacked "the recommendation of that agreeable *simplicity,* which is so apparent in other parts of the constitution of nature." Reid responded in kind when he came to assess Priestley's *Disquisitions Relating to Matter and Spirit* (1777). Reid's surviving manuscripts show that he was incensed by Priestley's attempt to justify materialism by appealing to Newton's second rule of philosophizing and that he accused Priestley of distorting both the wording and the sense of Newton's text.[27] What makes the differences between Smith, Priestley, and Reid over simplicity noteworthy is that they illustrate the fact that Newton's oeuvre generated a plurality of readings throughout the eighteenth century. No single "Newtonian method" was uniformly applied in the science of the mind during the period. Consequently, when thinking about the "moral Newtonianism" of the Enlightenment we must recognize that there were as many varieties of Newtonianism in the human sciences as in the natural.[28]

The historiographical problems posed by the concept of "moral Newtonianism" emerge with even greater clarity when we consider the *Weltanschauung* of German writers in the period. The godfather of *aufklärung,* Christian Wolff (1679–1754), owed more to scholasticism and to Gottfried Wilhelm

[26] Reid, *Inquiry,* pp. 210–12, 218; Adam Smith, *Lectures on Rhetoric and Belles Lettres,* ed. J. C. Bryce (Oxford: Clarendon Press, 1983), ii.132–4; Smith, *Essays on Philosophical Subjects,* ed. W. P. D. Wightman, J. C. Bryce, and I. S. Ross (Oxford: Clarendon Press, 1980), II.12, IV.19, and IV.67–76. Reid's assault on simplicity and system parallels the critique of the "spirit of system" found in the works of Voltaire, Condillac, and d'Alembert. See Condillac's seminal text of 1749, *Traité des systèmes,* in *Oeuvres philosophiques de Condillac,* ed. Georges Le Roy, 3 vols. (Paris: Presses universitaires de France, 1947–51), 1:324–67; Voltaire, *Philosophical Letters,* pp. 53, 64; d'Alembert, *Preliminary Discourse,* pp. 22–3, 94–5.

[27] Joseph Priestley, *An Examination of Dr. Reid's* Inquiry . . . *Dr. Beattie's* Essay . . . *and Dr. Oswald's* Appeal . . . (London: J. Johnson, 1774; New York: Garland Publishing, 1978), p. 6; Paul Wood (ed.), *Thomas Reid on the Animate Creation: Papers Relating to the Life Sciences* (Edinburgh: Edinburgh University Press, 1995), pp. 188–9.

[28] On the issue of the varieties of Newtonianism see, for example, Henry Guerlac, "Where the Statue Stood: Divergent Loyalties to Newton in the Eighteenth Century," in Guerlac, *Essays and Papers,* pp. 131–43, and Simon Schaffer, "Newtonianism," in R. C. Olby, G. N. Cantor, J. R. R. Christie, and M. J. S. Hodge (eds.), *Companion to the History of Modern Science* (London: Routledge, 1990), pp. 610–26.

Leibniz (1646–1716) than he did to Newton, and Wolff's vision of the myriad branches of human learning structured the academic pursuit of philosophy within Germany for much of the eighteenth century. Of particular significance was Wolff's distinction between the two separate yet interrelated fields of knowledge he christened "rational" and "empirical" psychology. According to Wolff, rational psychology encompassed the purely abstract and logical consideration of the nature of the soul, whereas its companion science was devoted to the introspective investigation of mental phenemona. The scope of the study of the mind was thus significantly different for Wolff than it was for enlightened savants elsewhere, because he fused empirical analysis with deductive a priori reasoning concerning "the things which are possible through the human soul." Moreover, even though he likened empirical psychology to experimental physics, Newtonian exemplars did not inform his understanding of the experimental method.[29] Consequently, the philosophical thrust of Wolff's works cannot be captured within the interpretative parameters of moral Newtonianism, and the distinctive cast of German versions of the science of the mind is further illustrated in the works of Immanuel Kant (1724–1804), the sage of Konigsberg.

Initially, Kant worked within the Wolffian framework of rational and empirical psychology, but he ultimately renounced both Wolff's methodolgical ideals and other established modes of inquiry to fashion his own unique system. As his lectures eventually published as *Anthropologie in pragmatischer Hinsicht* (1798) demonstrate, from the 1760s onward Kant commanded a broad range of factual information about the workings of our intellectual faculties and the natural history of the human species, but, like Hume, Kant raised serious doubts about the reliability of the introspective method and eventually denied that empirical psychology could be a genuine science like Newtonian mechanics. Kant's lectures also show that he had little time for those who followed Descartes in seeking correlations between mental and physiological states, for he remarked that such "theoretical speculation on the subject is a sheer waste of time." In the realm of rational psychology, once he was roused from his "dogmatic slumber" in the early 1770s, Kant's preoccupation with the delineation of the fundamental categories of human thought led him to reject the very possibility of a constructive form of rational psychology in his *Kritik der reinen Vernunft* (1781). Thus, even though Kant was second to none in his admiration for Newton's scientific achievements, the idiosyncratic style of his critical philosophy cannot be forced into a Procrustean Newtonian mold, and the highly individual stamp of his method underlines the limited analytical purchase of the notion of "moral

---

[29] Christian Wolff, *Preliminary Discourse on Philosophy in General*, trans. Richard J. Blackwell (Indianapolis: Bobbs-Merrill, 1963), pp. 34–5, 56–7; Robert J. Richards, "Christian Wolff's Prolegomena to Empirical and Rational Psychology: Translation and Commentary," *Proceedings of the American Philosophical Society*, 124 (1980), 227–39.

Newtonianism" on the development of the science of the mind in the German Enlightenment.[30]

QUANTIFICATION

In addition, there were other sources of methodological guidance to which moralists continued to turn. Given the considerable intellectual prestige mathematics had acquired through the seventeenth-century revolution in the physical sciences, it is hardly surprising to find instances of the adaptation of mathematical models to the study of morals and the mind. The dream initially conceived by Descartes, Hobbes, Benedict Spinoza (1632–1677), and Leibniz of constructing deductive systems of knowledge patterned on Euclid's *Elements* was kept very much alive in Germany by Christian Wolff; David Hartley serves as an example of those who cast their texts in a geometrical style, replete with propositions, corollaries, and scholia.[31] The methods of analysis and synthesis handed down by ancient geometers were likewise regarded as analogues of empirical inquiry. Newton's invocation of these procedures in the *Opticks* ensured that they would continue to figure in methodological discourse, and analysis took on special significance in the works of Condillac, Reid, and others.[32] There were also sporadic attempts to apply mathematical calculations to problems in allied fields, such as religion and natural theology, at the turn of the eighteenth century.[33] One of the earliest

[30] Immanuel Kant, *Anthropology from a Pragmatic Point of View,* trans. Mary J. Gregor (The Hague: M. Nijhoff, 1974), pp. 3, 13–15; Kant, *Critique of Pure Reason,* trans. Norman Kemp Smith (London: Macmillan, 1933), A 381–2/B 421–32, A 848–9/B 867–77; Kant, *Metaphysical Foundations of Natural Science,* trans. James Ellington (Indianapolis: Bobbs-Merrill, 1970), pp. 8–9; Kant, *Prolegomena to Any Future Metaphysics that will be able to Present Itself as a Science,* trans. Peter G. Lucas (Manchester: Manchester University Press, 1953), p. 9; David Hume, *A Treatise of Human Nature,* ed. L. A. Selby-Bigge, rev. P. H. Nidditch, 3rd ed. (Oxford: Clarendon Press, 1975), pp. 18–19; Lewis White Beck, *Early German Philosophy: Kant and His Predecessors* (Cambridge, MA: Belknap Press, 1969), chap. 17; Gary Hatfield, "Empirical, Rational, and Transcendental Psychology: Psychology as Science and as Philosophy," in Paul Guyer (ed.), *The Cambridge Companion to Kant* (Cambridge University Press, 1992), pp. 200–27.

[31] Wolff, *Preliminary Discourse,* pp. 76–8; Tore Frängsmyr, "The Mathematical Philosophy," in Tore Frängsmyr, J. L. Heilbron, and Robin E. Rider (eds.), *The Quantifying Spirit in the Eighteenth Century* (Berkeley: University of California Press, 1990), pp. 27–44; David Hartley, *Observations on Man, His Frame, His Duty, and His Expectations,* 2 vols. (London: S. Richardson, 1749; New York: AMS Press, 1971); David Hartley, *Various Conjectures on the Perception, Motion, and Generation of Ideas (1746),* trans. Robert E. A. Palmer (Los Angeles: William Andrews Clark Memorial Library, 1959). In his writings Hartley may have been specifically emulating the geometrical style of Newton's *Principia.*

[32] Newton, *Opticks,* pp. 404–5; Condillac, *La Logique,* in *Oeuvres,* 2:371–416; Reid, *Inquiry,* p. 15; see also Henry Guerlac, "Newton and the Method of Analysis," in Guerlac, *Essays and Papers,* pp. 193–216.

[33] Richard Nash, *John Craige's Mathematical Principles of Christian Theology* (Carbondale: Southern Illinois University Press, 1991); John Arbuthnot, "An Argument for Divine Providence, taken from the Constant Regularity Observ'd in the Births of both Sexes," *Philosophical Transactions of the Royal Society of London,* 27 (1710–12), 186–90; Abraham de Moivre, *The Doctrine of Chances; or, A Method of Calculating the Probability of Events in Play,* 2nd ed. (London: For the Author, 1738), p. v; Richard Price, "An Essay Towards Solving a Problem in the Doctrine of Chances. By the Late Mr. Bayes F.R.S. communicated by Mr. Price, in a Letter to John Canton, A.M. F.R.S.," *Philosophical Transactions of*

tentative steps toward the use of mathematical calculation in the realm of moral theory was taken in the 1690s at Halle by Christian Thomasius (1656–1728), who suggested that we can evaluate individual characters by measuring and comparing the degree to which they are marked by the four basic passions that control human behavior: sensuality, greed, ambition, and rational love. Another form of a moral calculus later appeared in Francis Hutcheson's *An Inquiry into the Original of our Ideas of Beauty and Virtue* (1725). Admitting that his initiative might "appear perhaps at first *extravagant* and *wild*," Hutcheson enumerated a set of "*Propositions, or Axioms*" intended to facilitate the computation of the virtue or evil of our actions. But the inclusion of a few simple equations outraged some of his readers, although we do not know the precise nature of their objections. The title page of the second edition no longer advertised the fact that the work contained "an Attempt to introduce a *Mathematical Calculation* in Subjects of *Morality*," and by the fourth edition the mathematical expressions were deleted altogether.[34]

Hutcheson's hasty retreat may have silenced most of his critics, but it did not satisfy the philosophical scruples of Thomas Reid, who publicly dismissed the whole exercise as "ring[ing] Changes upon Words" and "mak[ing] a Shew of mathematical Reasoning, without advancing one Step in real Knowledge."[35] Reid also claimed that, apart from calculating chances, "most Kinds of Probability" were "perhaps . . . not capable of Mensuration," and his surprisingly negative comment points to one of the areas of the science of the mind that in fact saw a significant degree of mathematization in the Enlightenment: the calibration of belief to evidence. With its complex roots in legal practice, the analysis of games of chance, political arithmetic, and the rise of "constructive scepticism" during the seventeenth century, the classical mathematical theory of probability was developed by Jean Le Rond d'Alembert (1717–1783), the Comte de Buffon (1707–1788), Condorcet, and Pierre-Simon de Laplace (1749–1827) to describe and regulate our assessments of empirical evidence, the testimony of witnesses, and our inductive inferences. Related qualitative treatments of probability also figure prominently in the works of philosophers

the Royal Society of London, 53 (1763), 370–418 (373–4); Price, *Four Dissertations*, 2nd ed. (London: A. Millar and T. Cadell, 1768), pp. 397–8n. Arbuthnot's argument was endorsed by William Derham, Bernard Nieuwentijt, and Willem 's Gravesande; see Eddie Shoesmith, "The Continental Controversy over Arbuthnot's Argument for Divine Providence," *Historia Mathematica*, 14 (1987), 133–46.

34 Richards, "Wolff's Prolegomena," p. 229n; Francis Hutcheson, *An Inquiry into the Original of our Ideas of Beauty and Virtue* (London: J. Darby, 1725; Hildesheim: G. Olms, 1971), pp. ix, 168–78, 265–70, 292–3; G. P. Brooks and S. K. Aalto, "The Rise and Fall of Moral Algebra: Francis Hutcheson and the Mathematization of Psychology," *Journal of the History of the Behavioural Sciences*, 17 (1981), 343–56. Brooks and Aalto point out (pp. 351–3) that Hutcheson's moral calculus was later taken up by the Scottish moralist Archibald Campbell (1691–1756) in his *An Enquiry into the Original of Moral Virtue* (Edinburgh: G. Hamilton, 1733), and by the anonymous author of *An Enquiry into the Origins of the Human Appetites Shewing how each Arises from Association* (Lincoln, UK: n.p., 1747).

35 Thomas Reid, "An Essay on Quantity; occasioned by Reading a Treatise, in which Simple and Compound Ratio's are applied to Virtue and Merit," *Philosophical Transactions of the Royal Society of London*, 45 (1748), 505–20 (513).

such as Joseph Butler (1692–1752), Hume, and Condillac, and in his *Observations* David Hartley combined quantitative and qualitative elements in his discussion of the grounds of assent. Thus, despite the widespread perception in the eighteenth century that moral issues were not amenable to formal mathematical analysis, probability theory was successfully combined with associationist psychology and a range of epistemological strategies designed to combat Pyrrhonian skepticism.[36]

Some attempts were also made to bring mathematical order to the manifold of experience by applying quantitative techniques to the investigation of the mechanisms of the mind. Christian Wolff's prediliction for mathematics manifested itself in his cultivation of "empirical psychology," for he maintained that mental phenomena can be understood quantitatively. Accordingly, he endeavored to establish simple measures that would facilitate the formulation of mathematical laws applicable to the faculty of memory. His follower Johann Gottlob Krüger (1715–1759) framed a set of equations that correlated the liveliness of our sensations with the tension of our nerves and the force of external objects acting upon our sensory organs, and Wolff's aim of mathematizing the study of the mind was more fully realized in Germany early in the nineteenth century by Johann Friedrich Herbart (1776–1841), who strove to turn psychology into a fully quantitative science.[37] Moreover, Krüger's work shows that the interaction of medicine, science, and philosophy (as we understand them) was at its most intense in the context of the study of the external senses and that the use of geometry was a standard feature of research on vision. In the seventeenth century, the subjects of light and vision were conjoined, and in the *siècle des lumières* writers on optics typically combined the physical and geometrical consideration of the nature and behavior of light with a treatment of the eye and various features of our visual experience, as can be seen in Robert Smith's (1689–1768) influential textbook, *A Compleat System of Opticks*. Similarly broad discussions are also found in the writings of George Berkeley (1685–1753) and Thomas Reid, whose *Inquiry* surveys such topics as the relationship between sensations and ideas, the parallel motion of the eyes, squinting, single and double vision, and the perception of distance. As well as dealing with the geometrical aspects of the phenomena he described, Reid outlined his remarkable non-Euclidean "geometry of visibles." The *Inquiry* thus exemplifies the manner in which geometrical considerations were routinely blended with philosophical, psycho-

---

[36] Reid, "Essay," p. 512; Hartley, *Observations*, 1:324–67; Lorraine Daston, *Classical Probability in the Enlightenment* (Princeton, NJ: Princeton University Press, 1988), passim.

[37] Gary Hatfield, "Remaking the Science of Mind: Psychology as Natural Science," in Fox et al. (eds.), *Inventing Human Science*, pp. 197–205, 213–14, 216; Geoffrey Cantor, *Optics after Newton: Theories of Light in Britain and Ireland, 1704–1840* (Manchester: Manchester University Press, 1983), pp. 20–1; David E. Leary, "The Historical Foundation of Herbart's Mathematization of Psychology," *Journal of the History of the Behavioural Sciences*, 16 (1980), 150–63.

logical, physiological, and anatomical concerns in eighteenth-century analyses of sensory perception.[38]

The question of mathematization in the science of the mind of the Enlightenment is by no means a simple one. Although geometry was an intrinsic part of the study of vision and the mathematical theory of probability was intertwined with epistemology and natural theology, quantification made only a limited contribution to the practical elucidation of most of our mental faculties, and the sporadic use of mathematics in morality was largely resisted. Moreover, a minority of scientists of the mind held to the Euclidean and deductive ideals championed by Descartes and others in the seventeenth century. Additionally, quantification was not a distinctively Newtonian methodological strategy. Rather, the origins of the move toward calculation and measurement were various, ranging from established analytical practices in fields such as optics to the mathematical dream of the major protagonists in the Scientific Revolution. Although Newton certainly provided some of the impetus, he was by no means the only nor even the most important inspiration behind attempts to mathematize the science of human nature.

ANATOMIZING THE MIND

The analogy between mind and body provided yet another methodological model for scientists of the mind – namely, anatomy. We have seen that Bacon likened the moralist to the physician, and he conjured up the practice of anatomists when he encouraged moral philosophers to engage in "a scientific and accurate dissection of minds and characters" to reveal "the secret dispositions of particular men."[39] The notion of the "anatomy of the mind" was then taken up in Britian during the first half of the eighteenth century by Anthony Ashley Cooper, third Earl of Shaftesbury (1671–1713) in his highly influential *Characteristicks* (1711), and Alexander Pope gave it added popular currency in his *An Essay on Man* (1733). In Scottish academic circles, George Turnbull also picked up on the idea and, like his English predecessors, he undoubtedly found it appealing because of the natural theological and moral meanings associated with anatomical dissections in the period.[40]

---

[38] Reid, *Inquiry*, pp. 77–202; Norman Daniels, *Thomas Reid's Inquiry: The Geometry of Visibles and the Case for Realism*, 2nd ed. (Stanford, CA: Stanford University Press, 1989); G. N. Cantor, "Berkeley, Reid, and the Mathematization of Mid-Eighteenth-Century Optics," *Journal of the History of Ideas*, 38 (1977), 429–48.

[39] Bacon, *Works*, p. 573; see also p. 574, where he states that poets and historians have "dissected" the affections.

[40] Anthony Ashley Cooper, Lord Shaftesbury, *Characteristicks of Men, Manners, Opinions, Times*, 4th ed., 3 vols. (London: J. Darby, 1727), 3:189; Pope, *Essay*, p. 239; Turnbull, *Principles*, 1:v. Anatomical illustrations figured prominently in natural theology texts of the period; see, for example, the first volume of Bernard Nieuwentijt, *The Religious Philosopher; or, The Right Use of Contemplating the*

The moralistic overtones of the concept were, however, stripped away by David Hume, who opposed his version of the anatomy of the mind to the highly influential philosophical and pedagogical style championed by Francis Hutcheson. After reading part of the manuscript of the *Treatise,* Hutcheson accused his younger contemporary of lacking "a certain Warmth in the Cause of Virtue." This provoked Hume to pen a letter to the Glasgow professor in which he distinguished the aspirations of the anatomist who wants "to discover [the mind's] most secret Springs & Principles" from those of the painter who tries "to describe the Grace & Beauty of its Actions." Still rankled by Hutcheson's rebuke, Hume elected to conclude the published version of the *Treatise* by repeating his belief that the "anatomist ought never to emulate the painter." He pointed out that, because of their close familiarity with the construction of human nature, anatomists of the mind are "admirably fitted to give advice" to practical moralists and stressed that "'tis even impracticable to excel" as a moralist "without the assistance" of the science of the mind. Implicitly rejecting Hutcheson's farrago of moral preaching and the study of our mental powers, Hume observed in closing that "the most abstract speculations concerning human nature, however cold and unentertaining, become subservient to *practical morality;* and may render this latter science more correct in its precepts, and more persuasive in its exhortations."[41] Thus, whereas Hutcheson gave precedence to the needs of moral inculcation, Hume privileged the science of human nature, claiming that it was "the only solid foundation for the other sciences," including that of morality.[42] Anatomizing the mind was therefore fundamental, and, despite some dissenting voices, Hume's view set the agenda for the teaching of moral philosophy in Scotland during the latter half of the eighteenth century. But even though the majority of Scottish academics agreed with his map of human knowledge, they differed from him in placing the anatomy of the mind within a providentialist and teleological framework.[43]

Works of the Creator, trans. J. Chamberlayne, 3 vols. (London: J. Senex, 1718–19). On the moral dimensions of anatomical dissections in Scotland, see Anita Guerrini, "Alexander Monro *primus* and the Moral Theatre of Anatomy," forthcoming.

[41] David Hume to Francis Hutcheson, 17 September 1739, in *The Letters of David Hume,* ed. J. Y. T. Greig, 2 vols. (Oxford: Clarendon Press, 1969), 1:32; Hume, *Treatise,* pp. 620–1. Compare Hume's later formulation in the *Enquiries concerning Human Understanding and concerning the Principles of Morals,* ed. L. A. Selby-Bigge, rev. P. H. Nidditch, 3rd ed. (Oxford: Clarendon Press, 1975), pp. 5–16. One of Hume's models may have been Bernard Mandeville, who implied that he was anatomizing human nature in the preface to *The Fable of the Bees; or, Private Vices, Publick Benefits,* ed. F. B. Kaye, 2 vols. (Oxford: Clarendon Press, 1924), 1:3.

[42] Hume, *Treatise,* p. xvi. Hume's view that all the sciences are founded on the study of human nature may have owed something to Hobbes; on the relations between Hobbes and Hume, see Paul Russell, "Hume's *Treatise* and Hobbes's *The Elements of Law,*" *Journal of the History of Ideas,* 46 (1985), 51–63.

[43] See, for example, Reid, *Inquiry,* pp. 11–12. The one major critic of Hume's view was James Beattie; see Paul Wood, "Science and the Pursuit of Virtue in the Aberdeen Enlightenment," in M. A. Stewart (ed.), *Studies in the Philosophy of the Scottish Enlightenment* (Oxford: Clarendon Press, 1990), pp. 127–49.

THE NATURAL HISTORY OF HUMAN NATURE

Although the introduction to the *Treatise* is usually read as a manifesto for moral Newtonianism, it is significant that Hume did not mention Newton in his introductory remarks and rarely appealed to Newton's methodological utterances to legitimize his own practice. Instead, he celebrated the fact that "some late philosophers in *England*", including "Mr. *Locke,* my Lord *Shaftsbury,* Dr. *Mandeville,* Mr. *Hutchinson,* Dr. *Butler,* &c.," had "put the science of man on a new footing" by applying "experimental philosophy to moral subjects." Thus, Hume's sense of how the "experimental philosophy" was defined does not seem reducible to Newtonian categories, and it is far more plausible to see his understanding of methodological issues as being shaped primarily by the canonical texts that he cited. Moreover, if we consider more carefully his list of moralists who had grounded their work in "experience and observation," it emerges that it was not some generic form of Newtonianism that Hume adopted as his methodological exemplar but rather the natural historical approach to the science of the mind initially articulated by Bacon and Locke.[44]

During the eighteenth century, among the first to champion this approach (which was seen as related to the anatomy of the mind) were Butler, who associated it with ascertaining the facts of human nature, and Voltaire, who paid tribute to Locke as both a historian and an anatomist of the mind.[45] Thanks to their efforts, as well as to the continued currency of the writings of Bacon and Locke and to the growing popularity of natural history itself, the model of the natural historian soon figured more prominently in the science of the mind, and it guided a wide range of authors, including George Turnbull, David Hartley, Julien Offray de La Mettrie, and Denis Diderot (1713–1784).[46]

---

[44] Hume, *Treatise,* pp. xvi–xvii. For a more detailed consideration of the argument of this paragraph, see Paul Wood, "Hume, Reid and the Science of the Mind," in M. A. Stewart and John P. Wright (eds.), *Hume and Hume's Connexions* (Edinburgh: Edinburgh University Press, 1994), pp. 119–39. I have also discussed the themes of the present section in "The Natural History of Man in the Scottish Enlightenment," *History of Science,* 28 (1990), 89–123; compare Paulette Carrive, "L'idée d'histoire naturelle de l'humanité' chez les philosophes écossais du XVIIIᵉ siècle," in O. Bloch, B. Balan, and P. Carrive (eds.), *Entre forme et histoire: La formation de la notion de développment à l'âge classique* (Paris: Meridiens Klincksieck, 1988), pp. 215–27.

[45] Joseph Butler, *Fifteen Sermons Preached at the Rolls Chapel and a Dissertation upon the Nature of Virtue,* ed. W. R. Matthews (London: Bell, 1964), pp. 34–5n; Voltaire, *Philosophical Letters,* pp. 53–4. Shaftsbury also made reference to the "natural History of Man," but he did not elaborate on the concept; *Characteristicks,* 2:186–7. D'Alembert later eulogized Locke in much the same way as Voltaire, and his notion of the "experimental physics of the soul" was closely related to the natural history approach; d'Alembert, *Preliminary Discourse,* p. 84.

[46] Turnbull, *Principles,* 1:ii; Martha Ellen Webb, "A New History of Hartley's *Observations on Man*," *Journal of the History of the Behavioural Sciences,* 24 (1988), 202–11; Ann Thomson, "From *L'Histoire Naturelle de L'Homme* to the Natural History of Mankind," *British Journal for Eighteenth-Century Studies,* 9 (1986), 73–80; Jacques-André Naigeon, *Mémoires historiques et philosophiques sur la vie et les oeuvres de Denis Diderot* (Paris, 1821), p. 291, quoted in Anthony Pagden, *European Encounters with the New World: From Renaissance to Romanticism* (New Haven, CT: Yale University Press, 1993), p. 156.

Whereas most of its practitioners were not overly self-conscious method-ologically, David Hume opted for the natural historical method as part of his solution to a major problem inherent in the use of introspection.

In the *Treatise,* Hume alerted his readers to the fact that moralists had yet to confront a practical issue that threatened to undermine the very project of constructing an empirical science of the mind – namely, that when we decide to observe the workings of our mental faculties introspectively, "this reflection and premeditation would so disturb the operation of my natural principles, as must render it impossible to form any just conclusion from the phænomenon." He proposed an alternative strategy, arguing in a Baconian vein that we should "glean up our experiments . . . from a cautious observation of human life, and take them as they appear in the common course of the world, by men's behaviour in company, in affairs, and in their pleasures." When he recast his thoughts in his *Philosophical Essays concerning Human Understanding* (1748), Hume reiterated the point that history and the experience "acquired by long life and a variety of business and company" enable us "to discover the constant and universal principles of human nature, by showing us men in all varieties of circumstances and situations, and furnishing us with materials from which we may form our observations and become acquainted with the regular springs of human action and behaviour." Furthermore, he recommended in the *Essays* that we should adopt the modest epistemological aims of the natural historian and map out a "mental geography, or delineation of the distinct parts and powers of the mind" in which the "different operations of the mind" would be identified and classified "under their proper heads."[47] Hence, Hume envisaged his "science of man" as resting on empirical materials derived from the annals of history as well as our collective experience of common life and as combining inductive inquiry with the descriptive and classificatory techniques of the natural historian.

Yet we must exercise care in defining the sense in which Hume was a natural historian of the mind, because his style of natural history differed from that of most of his Scottish contemporaries. Whereas Hume's *Natural History of Religion* (1757) suggests that he sometimes equated the "natural" with a rational or logical order in the manner of d'Alembert and Adam Smith, this normative usage of the term was largely absent from the works of other Scots. Thus, in Aberdeen a group of physicians and professors led by Alexander Gerard (1728–1795), John Gregory (1724–1773), Thomas Reid, and David Skene (1731–1770) drew on the writings of Bacon, Locke, Butler, and Turnbull in their histories of the mind. They defined "natural history" in terms of the empirical procedures employed by these philosophers as well as by natural historians such as Buffon and Carl von Linné, otherwise known as Linnaeus (1707–1778).[48]

---

[47] Hume, *Treatise,* p. xix, and *Enquiries,* pp. 13, 83–4.
[48] On Buffon and Linnaeus, see especially Phillip Sloan, "The Gaze of Natural History," in Fox et al. (eds.), *Inventing Human Science,* pp. 112–51.

Despite their stylistic differences, the Aberdonians shared Hume's sophisticated grasp of the methodological issues involved in writing natural histories, as can be seen in Reid's comments on the limitations of the natural historical and anatomical methods in the *Inquiry* or in the discourses on the natural history of the mind delivered by David Skene before the Aberdeen Philosophical Society. Perhaps most important of all, the researches of the Aberdeen men registered the shift toward charting the temporal development of human nature encouraged by the coalescence of natural and philosophical history, ethnography, and the science of the mind in the works of Charles-Louis de Secondat, Baron de Montesquieu (1689–1755) and especially Jean-Jacques Rousseau (1712–1778).[49]

Among the many responses to the theoretical challenges posed by Montesquieu's *L'esprit des lois*, Rousseau's *Discours sur l'origine et les fondemen[t]s de l'inégalité* (1755) stands out as the text that outlines the most comprehensive and controversial history of humankind. Like Montesquieu, Rousseau drew on the analytical tools of the natural law tradition and employed the framework of stadial history to construct his narrative of the "progress" of our species from savagery to civility. One of the most noteworthy features of this narrative is Rousseau's focus on the gradual unfolding of the powers of the mind and the emergence of new ideas, passions, and needs in the different stages of human development. According to Rousseau, in the earliest period of the "state of nature," humans lived a largely solitary existence, in which their needs were few and their mental life highly circumscribed. He maintained that in this state we possessed free will, the capacity for improvement (what he paradoxically called "perfectibility"), the fundamental desire for self-preservation, and the basic moral sentiment of pity or compassion, but he denied the claim made by Grotius and others that humankind is "naturally" sociable. He also thought that humans possessed strong, agile, and healthy bodies and that our faculties of sight, hearing, and smell were remarkably acute because survival depended on them. Remarkably, he further suggested that instincts were in no way fixed and that humans created a repertoire of instinctive behaviors through our imitation of animals.[50]

However, this idyllic state of simplicity, in which humankind was naturally good, was only transitory. Unlike earlier authors, whose depictions of the state of nature lacked a temporal dimension, Rousseau historicized the concept and underlined the fact that, even in this state, humans do not remain static. Although he echoed Pufendorf concerning the difficulties involved in accounting for the early history of humankind purely naturalistically, Rousseau favored the view that an increase in population prompted the invention of

---

[49] Reid, *Inquiry*, pp. 12–16; David Skene, Miscellaneous Papers, Aberdeen University Library, MS 37, fols. 168–79.

[50] Jean-Jacques Rousseau, *The First and Second Discourses*, ed. Roger D. Masters, trans. Roger D. Masters and Judith R. Masters (New York: St. Martin's Press, 1964), pp. 95, 105–7, 113–16, 193.

tools to satisfy our need for food, and he believed that this seemingly trivial technological advance initiated the process of domestication that corrupted our species. When humans began to hunt and fish, the first glimmerings of reflective thought appeared, and we began to form ideas of relations by comparing things around us. Moreover, humans were no longer solitary animals; they now associated in herds and established family units. Domestic life transformed human nature, bringing new needs, passions, and sentiments, as well as an intensification of emotions at odds with the previously tranquil life of the mind. Furthermore, the emergence of private property led to violence, and quarrels between individuals were exacerbated by the newly volatile character of the passions. But despite these elements of disorder in both our psychic and social existence, Rousseau affirmed that "this period of the development of human faculties, maintaining a golden mean between the violence of the primitive state and the petulant activity of our vanity, must have been the happiest and most durable epoch."[51]

Because of the continuing pressure of population growth and the onset of the division of labor, however, this last phase of the state of nature was short-lived. With the advent of agriculture, metal implements, and the related advance of the arts, humankind entered a new era wherein the trends observable in the savage state were taken to further extremes. The incessant proliferation of needs fueled the intensification and multiplication of our passions, the efflorescence of the faculties of the mind, and the advancement of learning. Human nature was, in effect, refashioned by the revolutionary effects of private property, luxury, and life in civil society. Once naturally moral and free, we were now vain as well as corrupted and enslaved by possessions and knowledge, and civilization enfeebled our bodies and rendered us prone to the ravages of disease. Where others saw the march of progress, Rousseau saw the fall of humankind. We had tasted the poisoned fruits of knowledge and were now suffering the fatal consequences.[52]

Rousseau's reconstruction of the path from rudeness to refinement was a brilliant realization of Locke's aim of writing a history of the mind and the origins of human knowledge, but his narrative also irrevocably changed the face of the science of human nature. Rousseau's fusion of the Lockean study of the temporal evolution of our mental powers with a stadial vision of the

---

[51] Rousseau, *Discourses*, pp. 141–51 (150–1). It must be said that Rousseau was inconsistent in his characterization of the initial state of humankind, because he wavered from stating that it was amoral to affirming that it was good (see pp. 128, 150, 193); his inconsistency reflects his ambivalence toward Hobbes's depiction of the state of nature. Pufendorf indicated that we could not explain the history of humankind solely in terms of natural causes; Pufendorf, *Natural State*, pp. 112–16.

[52] Rousseau, *Discourses*, pp. 108–11, 151–60, 193, 199. Rousseau's account of the transformation of human nature wrought by life in society echoes those of Bernard Mandeville and Denis Diderot; see Mandeville, *Fable*, 1:205–6, and Denis Diderot, "Supplément au Voyage de Bougainville," in *Political Writings*, ed. and trans. John Hope Mason and Robert Wokler (Cambridge University Press, 1992), pp. 71–3.

development of humankind rooted in ethnology and natural history posed probing moral and political questions about the meaning of the history of humankind, questions that scientists of the mind could not ignore. Moreover, the investigation of the natural history of our species raised disturbing questions about human nature. Most eighteenth-century savants maintained that human nature was uniform, but natural historical inquiry displayed a bewildering variety of physical characteristics and mental capacities, leading thinkers such as Henry Home, Lord Kames (1696–1782) to revive the idea that each of the varieties or races of humankind was created separately. This challenge to the view that our common nature derives from a shared descent from the original parents of the species was especially disconcerting because it threatened to undermine the cherished Enlightenment belief in a universal standard of either natural or revealed morality.[53] Anxieties about the existence of such standards and the uniformity of human nature were further exacerbated by the provocative philosophical history of humankind sketched out by Kant's ex-pupil Johann Gottfried Herder (1744–1803) in his *Ideen zur Philosophie der Geschichte der Menschheit* (1784–1791). Taking his cue from those, like Montesquieu, who correlated culture with the physical environment, Herder pushed this mode of analysis in a startlingly new direction, arguing that the divergent cultures nurtured by the differences in climate and terrain across the globe were all manifestations of God's providential design. Eschewing the search for moral absolutes (which had hitherto fueled the comparative study of human societies), Herder instead celebrated the plurality, specificity, and integrity of the range of cultures sustaining humankind. With a moral fervor equal to Rousseau's, he cautioned his fellow Europeans against imposing their norms on other peoples and contended that it "would be the most stupid vanity to imagine, that all the inhabitants of the World must be europeans, to live happily."[54] Writing the natural history of humankind therefore proved to be a highly subversive enterprise, because it paved the way for alternative conceptions of human nature by casting doubt on the Enlightenment shibboleths of cosmopolitanism, optimism, progress, and universalism.

---

[53] Henry Home, Lord Kames, *Sketches of the History of Man*, 2nd ed., 4 vols. (Edinburgh: W. Strahan, T. Cadell, and W. Creech, 1778), 1:72–7; Samuel Stanhope Smith, *An Essay on the Causes of the Variety of Complexion and Figure in the Human Species*, new ed. (Edinburgh: C. Elliot, 1788), pp. 164–5. On the uniformity of human nature, see, for example, Mandeville, *Fable*, 1:229, and Hume, *Enquiries*, pp. 83–4. A useful introduction to eighteenth-century debates over the issue of race is to be found in Emmanuel Chukwudi Eze (ed.), *Race and the Enlightenment: A Reader* (Cambridge, MA: Blackwell, 1997).

[54] John Godfrey Herder, *Outlines of a Philosophy of the History of Man*, trans. T. Churchill, 2 vols., 2nd ed. (London: T. Churchill, 1802), 1:393. Herder also invoked the four stages theory to explain the character of a culture, but he maintained that environmental factors conditioned the different stages and suggested that the stages were not necessarily distinct (1:33, 363). For a useful discussion of Herder's environmentalism see Clarence J. Glacken, *Traces on the Rhodian Shore: Nature and Culture in Western Thought from Ancient Times to the End of the Eighteenth Century* (Berkeley: University of California Press, 1967), pp. 537–43.

CONCLUSION

In Diderot's *Le neveu de Rameau* a discussion of pedagogical principles prompts the character "He" to ask, "where does method come from?"[55] We have seen that when it came to explaining the origins of the Age of Reason, enlightened men and women had a clear answer to this question and that their narrative continues to structure current accounts of the relations between natural and moral philosophy in the *siècle des lumières*. But the story of the genesis of the Enlightenment science of the mind has become highly simplified as scholars have increasingly focused on Newton's intellectual legacy. This single-minded obsession with Newton's influence and the fruitless search for a univocal Newtonian tradition in the natural and human sciences have done little justice to the complexities of the analysis of human nature in the eighteenth century or to the nuances of the *philosophes'* vision of their philosophical patrimony. As Voltaire's panegyric to Locke in his *Letters concerning the English Nation* illustrates, enlightened savants recognized a variety of methodological models, including anatomy and natural history, that supplemented or even supplanted methods derived from Newton's works. While acknowledging Newton's importance in the period, we must look to the canonical texts written by Bacon, Descartes, Grotius, Hobbes, and Locke to understand the competing definitions of the scope of that science and the various methods to be employed within it. As the *philosophes* themselves realized, the contours of their "science of man" were formed in the cataclysmic upheavals of the seventeenth century, and the analytical tools that they used to dissect human nature were forged in the crucible of the Scientific Revolution.

---

[55] Denis Diderot, *Rameau's Nephew and D'Alembert's Dream,* trans. Leonard Tancock (Harmondsworth: Penguin, 1966), p. 58. *Le neveu de Rameau* is one of the Enlightenment's most suggestive explorations of the effects of social life on human nature.

35

GLOBAL PILLAGE
Science, Commerce, and Empire

Larry Stewart

A map is a representation on paper – a picture – you understand picture? –
a paper picture – showing, representing the country – yes? – showing your
country in miniature – a scaled drawing on paper of – of – of –

Brian Friel, *Translations*

At the conclusion of the Peace of Paris in 1763, British blue-water policy bore
some strange fruit in exchanging the sugar island of Gaudeloupe for "quelques
arpents de neige" in the Canadian wilderness – leading to much consternation
and bitterness between the elder Pitt and the pliant Scotsman, Lord Bute.
This was surely the moment when an expansive British Empire was born and,
in response, a new wave of French adventures. Thus, we find the self-effacing
Louis de Bougainville soon to make his celebrated four-year circumnavigation (1766–9), a superb account of which was swiftly published – although
Bougainville lamented, "Ce n'est ni dans les forêts du Canada, ni sur le sein
des mers, que l'on se forme á l'art d'écrire." Nonetheless, unlike the fashionable experience of European naturalists and systematizers who constrained
"dans les ombres de leur cabinet . . . soumettent impérieusement la nature á
leurs imaginations," here was a self-described "voyageur & marin; c'est á dire,
un menteur, & un imbécille." Bougainville's brilliant tale is as much a romance
of rocky shoals, high seas, men overboard, and inevitable scurvy as much as
laying-to in sheltered Pacific coves and shallow bays, behind coral shoals and
the welcoming arms of Tahitians.[1]

THE PROGRESS OF TRADE AND LEARNING

The place of the scientist amid the grasping mercantile and imperialist
agendas of the European powers is the theme of this essay. It is a thread that

[1] Louis de Bougainville, *Voyage autour du monde, Par La Frégate du Roi, La Boudeuse, et La Flûte
L'Etoile; En 1766, 1767, 1768 & 1769* (Paris: Chez Saillant & Nyon, 1771), pp. 16–17.

stretched from the seventeenth-century notions of commerce triumphant to the imperious propaganda of eighteenth-century scientific potentates such as Joseph Banks. Its origins, however, lie not in much older Baconian propositions (however influential) but rather in the mercantilism of Jean-Baptiste Colbert and in the glorification of British merchants by Daniel Defoe. Trade brought power and distinction to entrepreneurs and naturalists as much as to would-be emperors. In the aftermath of the devastating South Sea Bubble of 1720, Defoe remained undaunted, convinced that England had important advantages over commercial and imperial rivals such as the French and the Dutch. To Defoe, the ascent to "prodigious heights, both in wealth and number" of the English gentry had followed precisely from the promotion of *"Trade and Learning."*[2] In an age when new lands were intensely hazardous to the health of Europeans, not to speak of the dangers imposed by voyages of discovery themselves, Defoe listed among English advantages (even over Continental rivals) a "climate [that] is the most agreeable climate in the world to live in."[3] Hyperbole aside, Defoe's proposition of the link between trade and learning would have been taken seriously by the many people he encountered in his haunts around London's Royal Exchange for, as he put it, "By trade we must be understood to include Navigation, and foreign discoveries, because they are generally speaking all promoted and carried on by trade, and even by tradesmen, as well as merchants."[4]

This relation between trade and scientific discoveries was the essence of the age of mercantilism and the expansion of global empires. But it was not simply a matter of entrepreneurs striking out into ill-charted waters. Long-standing European conflicts brought navies into play. As John Brewer has shown, the patterns of English military expansion inevitably followed the course of wars of succession and imperial squabbles. Here lay vast opportunities for the likes of Scottish professionals, politicians, and adventurers to secure their fortunes and their futures.[5] This was hardly a circumstance confined to the British. In this century of conflict, the Portuguese, the Spanish, the Dutch and the French felt the same imperative far beyond their continental circumstances and to which each of them would prove variously vulnerable. And it was widely believed that the promotion of a sophisticated natural philosophical or mathematical community was essential. Thus, the French were quick to assert the

---

[2] Daniel Defoe, *The Complete English Tradesman in Familiar Letters,* 2nd ed. (London, 1727; reprint, New York: Augustus M. Kelly Publishers, 1969), Letter XXII, "Of the Dignity of Trade in England More than Any Other Countries," p. 306. My italics. See also Ilse Vickers, *Defoe and the New Sciences* (Cambridge University Press, 1996), p. 96.

[3] On Defoe's view of the importance of science and reason, see Simon Schaffer, "Defoe's Natural Philosophy and the Worlds of Credit," in John Christie and Sally Shuttleworth (eds.), *Nature Transfigured: Science and Literature, 1700–1900* (Manchester: Manchester University Press, 1989), pp. 13–44, especially p. 26.

[4] Defoe, *Complete English Tradesman,* p. 307.

[5] John Brewer, *The Sinews of Power: War, Money and the English State, 1688–1783* (Cambridge, MA: Harvard University Press, 1988), pp. 40ff.; Linda Colley, *Britons: Forging the Nation 1707–1837* (New Haven, CT: Yale University Press, 1992), pp. 126–32.

navigational significance of variations in terrestrial magnetism, and in 1705 the Académie des Sciences set a competition for a method of determining that variation at sea. But it was an Englishman, Edmond Halley, who revealed the inadequacies of charts of variation in la Manche and who convinced the Royal Navy to give him a ship to command, thereby producing two impressive isogonic maps.[6]

Imperial rivalry magnified the urgency of navigational improvement. The disastrous sinking of most of Sir Cloudesley Shovell's Mediterranean fleet in 1707 impelled the British government in 1714 to offer a reward of £20,000 for the discovery of longitude at sea within half a degree.[7] Little, of course, was gained either by the many projectors demanding the reward or by the British government. For decades, even after the chronometer made its value evident, every imperial venture to the South Atlantic or to the Pacific was invariably fraught with fears of missing a destination or a replenishing anchorage even when charts were precise, something that was an almost certain result when charts were either carefully guarded secrets or were nothing more than a cultivated set of errors and rumor. To take only one legendary example, Robinson Crusoe's island (otherwise of the Scotsman Alexander Selkirk in 1704–09) at Juan Fernandez was a well-known and inevitably disputed refuge. Later, in 1741, during the war with Spain, Commodore George Anson's squadron, scattered by the seas, desperately searched for refuge and ultimately pressed British claims even while the ships' crews were ravaged by scurvy.[8]

If navigation was desperate, wars made matters even more urgent. Indeed, as imperial conflict emphasized the necessity of accurate sea charts they also revealed the vulnerability of colonial investment. Throughout the middle of the century, but especially during the Seven Years' War and its aftermath, trade pressed the imperial agenda and scientific investigation was along for the voyage. By December 1775, the search for a Northwest Passage was being justified by Crown and Royal Society by hopes for "many advantages both to commerce and science."[9] Within two years the "historiographer" William Robertson, D.D., principal of the University of Edinburgh, had made commerce part of

[6] Deborah Warner, "Terrestrial Magnetism: For the Glory of God and the Benefit of Mankind," in Albert Van Helden and Thomas L. Hankins (eds.), *Instruments, Osiris*, 9 (1993), 67–84, especially 74–5, 78.

[7] Larry Stewart, *The Rise of Public Science. Rhetoric, Technology, and Natural Philosophy in Newtonian Britain, 1660–1750* (Cambridge University Press, 1992), pp. 183–211; and A. J. Turner, "In the Wake of the Act, but Mainly Before," in William J. H. Andrewes (ed.), *The Quest for Longitude* (Cambridge, MA: Collection of Historical Scientific Instruments, 1996), pp. 115–32.

[8] Daniel A. Baugh, "Seapower and Science," in Derek Howse (ed.), *Background to Discovery: Pacific Exploration from Dampier to Cook* (Berkeley: University of California Press, 1990), p. 15; Boyle Sommerville, *Commodore George Anson's Voyage into the South Seas and Around the World* (London: Heinemann, 1934), pp. 63ff.; and Glyndwr Williams, "The Pacific: Exploration and Exploitation," in P. J. Williams (ed.), *Oxford History of the British Empire*, vol. 2, *The Eighteenth Century* (Oxford: Oxford University Press, 1998), pp. 554–75.

[9] J. C. Beaglehole, *The Life of Captain James Cook* (London: A. & C. Black, 1974), p. 484; Baugh, "Seapower and Science," p. 39.

the scientific vision that still links Columbus to Copernicus and Galileo. This was the modernist image of the benefits of the empirical, rather than theoretical, confrontation with the world.[10]

MERCHANTS AND IMPERIAL SCIENCE

A role for science was embedded in the imperialist doctrines of commercial and political advantage. Although new conquests made markets, the key transfer was that of wealth in a "cycle of accumulation" from the expanding periphery to the European center. Notably, this was not merely a matter of consumption but also of the existence, in places like the Jardin du Roi or Kew Gardens, of centers of naming, collecting, and display that solidified the link between natural knowledge and claims to power.[11] It is fundamental to a proper comprehension of the eighteenth-century adventures to eliminate artificial distinctions between trade and contemplation, assuming instead the mantle of *utilité* and *travail* that, from Colbert to Defoe and the Encyclo-pédistes, asserted advantage to the nation. The merchant was the agent of civilization. And, at least in the growing commercial empires of Britain and France, there was greater access to the vast archive of nature in which systematizers such as Linnaeus, Buffon, and their many international disciples happily rummaged.[12]

The implications, amid the rapidly capitalized food markets of the eighteenth century, were profound. Every new food and every new medicine demanded greater scientific comprehension, extended cultivation, and expansion of possible markets. In the wake of the Peace of Paris, Malachy Postlethwayt could see the bounty of Britain's newly secured colonies. Although he admired the French promotion of its chartered companies, the British might have done more. After 1763, the field was as open as it could be for colonies, companies, and scientists each to seek advantage.[13] Thus, in the Preface to his 1772 English

---

[10] Anthony Pagden, *European Encounters with the New World* (New Haven, CT: Yale University Press, 1993), pp. 99–100, 166; John Gascoigne, *Joseph Banks and the English Enlightenment: Useful Knowledge and Polite Culture* (Cambridge University Press, 1994), p. 264.

[11] See David Hume, "Of the Rise and Progress of the Arts and Sciences," in Thomas Hill Green and Thomas Hodge Grose (eds.), *The Philosophical Works*, vol. 3 (London, 1882; reprint, Darmstadt: Scientia Verlag Aalen, 1965), p. 185; David Philip Miller, "Joseph Banks, Empire, and 'Centres of Calculation' in Late Hanoverian London"; Michael Dettelbach, "Global Physics and Aesthetic Empire: Humboldt's Physical Portrait of the Tropics"; and Simon Schaffer, "Visions of Empire," in David Philip Miller and Peter Hans Reill (eds.), *Visions of Empire: Voyages, Botany, and Representation of Nature* (Cambridge University Press, 1996), pp. 22–3, 258, 336–7; Richard Drayton, "Knowledge and Empire," in P. J. Marshall, ed., *Oxford History of the British Empire*, vol. 7, *The Eighteenth Century*, pp. 231–52.

[12] Michael Nerlich, *The Ideology of Adventure: Studies in Modern Consciousness, 1100–1750*, vol. 2, trans. Ruth Crowley (Minneapolis: University of Minnesota Press, 1987), pp. 381–97; Anthony Pagden, *European Encounters*, pp. 170–2; and Gascoigne, *Joseph Banks and the English Enlightenment*, pp. 89–90.

[13] Malachy Postlethwayt, *Universal Dictionary of Trade and Commerce*, 4th ed. (London, 1774; reprint New York: Augustus M. Kelley, 1971), vol. 2, s.v. "Britain" (Remarks on island colonies); Bob Har-

translation of Bougainville's *Voyage,* the German naturalist Johann Reinhold Forster wished

> that our English East India Company, prompted by a noble zeal for the improvement of natural history, and every other useful branch of knowledge, might send a set of men properly acquainted with mathematics, natural history, physic, and other branches of literature, to their vast possessions in the Indies, and every other place where their navigations extend, and enable them to collect all kinds of useful and curious informations; to gather fossils, plants, seeds, and animals, peculiar to these regions; . . . to make observations on the climate and constitution of the various countries; the heat and moisture of the air, the salubrity and noxiousness of the place, the remedies usual in the diseases of hot countries, and various other subjects. A plan of this nature, once set on foot in a judicious manner, would not only do honour to the East India Company, but it must at the same time become a means of discovering many new and useful branches of trade and commerce.[14]

In the noises of South Sea ships slipping their moorings and in the racket of rat-tailed cranes on East India docks hauling bale upon bale of mysteriously scented goods, here was the edge of the world, loose upon the London quays, the only guard the sentry of the suspicious excise officer. In such places such as Le Havre, Marseilles, Lisbon, or Amsterdam worlds collided, futures were made and shattered, and voyages began and, for the lucky and the skillful, ended there. For thousands – many of them impressed in times of war against their will and others who hoped only to impress a Company director and gain a berth – here the whisper of fortunes began. So, in Britain, joint-stock companies such as the African Company employed learned surgeons to search for new crops and hints of gold upriver in the deadly slave coast, while, in France, countless schemes were promoted to find *Terres Australes* and the nonexistent Gonneville's Land. The intent of both projects, as in countless others, was the advance of trade.[15]

In far-flung European forts and factories, as at Macao or Vera Cruz, in Cartagena or Canton, commerce was surely a complex matter, especially when war threatened trade routes. Such was the case in Commodore Anson's arrival on

ris, "'American Idols': Empire, War and the Middling Ranks in Mid-Eighteenth-Century Britain," *Past and Present,* 150 (February 1996), 111–41, especially 127, 135, 138.

[14] Bougainville, *A Voyage Round the World. Performed by Order of His Most Christian Majesty, in the Years 1766, 1767, 1768, and 1769,* trans. John Reinhold Forster, F.A.S. (London, 1772; reprint Amsterdam: Israel, 1967), pp. viii–ix; see also Simon Schaffer, "The Earth's Fertility as a Social Fact in Early Modern Britain," in Mikuláš Teich, Roy Porter, and Bo Gustafson (eds.), *Nature and Society in Historical Context* (Cambridge University Press, 1997), pp. 124–47, especially p. 125.

[15] James Houstoun, *Memoirs of the Life and Travels of James Houstoun, M.D. (Formerly Physician and Surgeon-General to the Royal African Company's Settlements in Africa, and late Surgeon to the Royal Assiento Company's Factories in America),* (London, 1747), pp. 126–7; O. H. K. Spate, "De Lozier Bouvet and Mercantilist Expansion in the Pacific in 1740," in John Parker (ed.), *Merchants & Scholars: Essays in the History of Exploration and Trade* (Minneapolis: University of Minnesota Press, 1965), pp. 223–37, especially pp. 225–7.

the Chinese coast in search of provisions, having recently captured the significant prize of a Spanish galleon. Behind his guns Anson made his authority from King George quite plain, and the East India Company officers of various nations found it wise to abandon their obstructions to ocean provisions. But in such companies European scientists frequently found useful employment. The British East India Company, for example, on the recommendation of Joseph Banks would have in its service the botanist Johann Koenig of Schleswig-Holstein, whose brief was to locate "Drugs and Dying [sic] materials fit for the European market but above all [to put] the Company in possession of articles proper for the Chinese investment such as that nation at present receives from other people."[16] When Anson was First Lord of the Admiralty, British objectives were to secure a foothold from which it would be possible to challenge French claims in the Pacific.[17] This determined not only a naval policy but also one in which the ships of chartered companies and the voyages of discovery would be crucial. It is significant that when Banks later promoted commercial imperialism he did so by exploiting political and mercantile connection. After 1784, Banks's associate Henry Dundas was secretary of the Board of Control of the East India Company, and Banks was thereby able to exercise influence over Company policy on botanical matters. In his improving spirit, he attempted to marshal the scientific research of Humphry Davy on tanning in 1801.[18]

It would be misleading to regard the botanical or zoological projects of the various chartered companies as simply an extension of early modern national strategies. Because of the growing European import and reexport trade, as with that in coffee from Britain to the Continent, matters of biological discovery and classification and the biological transfers of the coffee plant from Africa to Brazil, or of tobacco to China and Japan, were actually far more complex phenomena than bilateral or even multilateral international competition would suggest. Early modern commercial empires were peopled by the detritus of domestic ambitions thwarted, of opportunities grasped in moments often of desperation or, at least, of hope of making a mark. Consequently, none of the chartered companies, large or small, was dominated by one language. Even after the Portuguese empire had shifted its focus to Brazil, Portuguese remained the dominant European language in Asia until the end of the eighteenth century. Germans successfully assumed much of the Dutch East India Company's activities, so that by the 1770s only about one-third of the Company's servants were Dutch. The Ostend, Swedish, and Prussian companies were apparently as multinational as were those of the English and Dutch. This diversity

---

[16] Sommerville, *Commodore Anson's Voyage*, pp. 249–50; Mackay, "Agents of Empire," p. 50; Richard Grove, "The Island and the History of Environmentalism: The Case of St. Vincent," in Teich, Porter, and Gustafson (eds.), *Nature and Society of Historical Context*, pp. 148–62, especially p. 156.
[17] Baugh, "Seapower and Science," pp. 32–3; see also Harris, "American Idols," p. 121.
[18] Gascoigne, *Joseph Banks and the English Enlightenment*, pp. 220–1.

is certainly pronounced among the naturalists. Whereas the Swedish East India Company cooperated in providing free passage to Linnaeus's students, undoubtedly seeing potential advantage in doing so, this scattering of botanists meant that his pupil Pehr Kalm would be found in North America in 1747, and, likewise, the remarkable Daniel Solander took advantage of Cook's *Endeavour* voyage in 1768 to cement his ties to Joseph Banks. The result was the sowing of the Linnean system throughout the commercial and imperial networks of the Europeans by pupils such as Pehr Lofling in South America, Jonas Dryander in Britain, Osbeck in China, and the German Johann Forster on Cook's second voyage.[19]

Alexander Pope had once suggested that in the apparent chaos of the world there was merely harmony not understood. Such a programmatic physicotheology, however, was only the beginning in an age that would be tested repeatedly by imperial and commercial conflict. While Pope interpreted the ways of God to man, Daniel Defoe just missed making the crucial commercial connection in his argument that "it is poverty fills armies, mans navies, and peoples Colonies." This was equally true of necessity and ambition, and people in that situation were not always fussy. Hence, there were numerous medical men who would find themselves on the Slave Coast at Wydah, at Sierra Leone looking for ways of making potash, or in the West Indies or Cartagena, where James Houstoun was surgeon to the Royal Asiento Company. Houstoun and many others depended, often desperately, on connection as their fortunes were buffeted by war and the collapse of contracts. Houstoun claimed to have lost £9,000 by the canceling of the Asiento.[20] Spain's efforts to protect the remnants of its trade routes from the vultures circling the carcass after the War of the Spanish Succession made things immensely difficult for botanical collectors. Such was the sorry tale of Robert Millar, an Edinburgh medical student, who left for Jamaica in 1734 on behalf of the trustees of the Georgia colony. Millar's brief was to locate promising botanical specimens, seeds, and plants that might ultimately be propagated in Georgia, and this required him to travel on South Sea Company ships to Porto Bello and Panama. From there he went to Jamaica and to Vera Cruz, but he was denied entrance to Mexico. He managed to hitch a ride to Havana and returned, bitterly frustrated, to

---

[19] Mary Louise Pratt, *Imperial Eyes: Travel Writing and Transculturation* (London: Routledge, 1992), pp. 25–8; Gascoigne, *Joseph Banks and the English Enlightenment*, passim; Holden Furber, *Rival Empires of Trade in the Orient, 1600–1800* (Minneapolis: University of Minnesota Press, 1976), pp. 299–305; Jan de Vries and Ad van der Woude, *The First Modern Economy: Success, Failure, and Perseverance of the Dutch Economy, 1500–1815* (Cambridge University Press, 1997), pp. 486–7; James Walvin, *Fruits of Empire: Exotic Produce and British Taste, 1660–1800* (New York: New York University Press, 1997), pp. 43–4; A. J. R. Russell-Wood, *A World on the Move: The Portuguese in Africa, Asia, and America, 1415–1808* (New York: St. Martin's Press, 1993), pp. 161ff.

[20] James Houstoun, *The Works of James Houstoun, M.D. Containing Memoirs of his Life and Travels to Asia, Africa, America and most Parts of Europe. From the Year 1690, to the present time . . .* (London, 1753), pp. 182, 216, 438–9, 449; Defoe, *Complete English Tradesman*, p. 317.

England in 1739. Despite all efforts no meaningful position could be obtained for him, and he died in 1742. His herbarium ultimately found its way into the hands of Joseph Banks.[21]

There were numerous such stories, especially as scientists as much as sailors were caught in the riptides of imperial conflict. Just as the Spanish were zealous in protecting their knowledge of their American territories, Dutch agents at Batavia long obstructed interlopers who might threaten the privileges of established trade in Austral-Asia. Certainly the wars of the middle century were a serious obstacle to scientific exploitation and settlement, but colonies did produce Fellows of the Royal Society. It would be a mistake simply to dismiss many of these entrepreneurs as "birds of passage"; indeed they were, but that is just the point. They took ship whenever and wherever they could and with whatever nation or company was willing to carry them. And it was precisely the rival claims to empire of the European powers that made this possible.

When Bougainville put in at Boero in September, 1768, he immediately encountered Dutch soldiers, who were nervous about his intentions; having been allowed provisions his next port in the Dutch territories was Batavia, where the chief factor was a Dutchman actually born there and married to a Creole. Although Bougainville was treated tolerably, he nonetheless became well aware of how desperate French sailors might become in their Pacific navigation from the Moluccas to Batavia, because French charts were woefully inaccurate and the Dutch kept theirs a closely guarded secret. This would hardly have happened were there not vast riches at stake in the spice trade centered on Batavia – where the wealthy, according to Bougainville, drank nothing but seltzer water imported at vast expense from Holland – in an area of the world that nonetheless could be highly treacherous for navigators.

The successful prosecution of a rich imperial trade demanded charts and instruments on which sailors could depend. Consequently, it is not very surprising that Bougainville should remark on an encounter at Batavia by M. Verron, Bougainville's astronomer:

> Je ne dois pas oublier un monument qu'un particulier y a élevé aux Muses. Le sieur Mohr, premier Curé de Batavia, homme riche à millions, mais plus estimable par ses connoissances & son goût pour les sciences, y a fait construire sans un jardin d'une de ses maisons, un observatoire qui honoreroit toute maison royale. Cet édifice, qui est à peine fini, lui a coûté, des sommes immenses. Il a tiré d'Europe les meilleurs instruments en tout genre, nécessaires aux observations les plus delicates, & il est en état de s'en servir. Cet Astronome, le plus riche sans contre dit des enfans d'Uranie, a été enchanté de voir M. Verron. Il a voulu qu'il passât les nuits dans son observatoire;

---

[21] Pratt, *Imperial Eyes*, p. 16; Anthony Pagden, "Heeding Heraclides: Empire and Its Discontents, 1619–1812," in Richard L. Kagan and Geoffrey Parker (eds.), *Spain, Europe and the Atlantic World* (Cambridge University Press, 1995), pp. 316–33, especially pp. 326ff.; Raymond Phineas Stearns, *Science in the British Colonies of America* (Urbana: University of Illinois Press, 1970), pp. 329–33.

malheureusement il n'y en a pas eu un seule qui ait été favorable a leurs desirs. M. Mohr a observé le dernier passage de Venus, & il a envoyé ses observations a l'Académie de Harlem; elles servoiront á déterminer avec la longitude de Batavia.[22]

At virtually this same moment, the Royal Society in London was able to overcome those political sycophants who lived in fear of Portuguese and Spanish reaction, even after the events of 1763 had really decided much of the issue, and petitioned for an expedition to the South Seas, ostensibly to observe the transit of Venus. Scientific and mercantile agendas merged increasingly, and Bougainville's assessment of the Indies was a bolt that shattered a great deal of complacency. It was his view that Dutch dominance of the South Seas made the Dutch East India Company "plus semblable á une puissante République, qu'à une société de Marchands," the establishment of which existed essentially in the "l'ignorance du reste de l'Europe sur l'état veritables de ces iles, & le nuage mystérieux qui enveloppe ce jardin des Hesperides." The time was ripe for a mortal stroke; simply in desiring the end of exclusivity, it would begin to crumble. As such views gained credence, after the 1760s it became fundamental that not only could scientific observers of various sorts determine the value of commodities, but also they themselves were a valuable commodity.[23]

THE BOTANIC EMPIRE

Breaking Nature's hidden botanic codes through the emerging nomenclatures of the naturalists; applying the systems of the zoologists and anatomists; transporting plants and animals; the need to make sense out of chaos – all these followed the movements of men. The scholarly ambitions of scientific linguists – themselves often well aware of the classification conundra of the systematists – and the contracts of the peripatetic draftsmen, such as the Forsters, who drew, described, named, and claimed, meant that Europeans would define what Europeans could ultimately trade.[24]

An expanding botanic empire promoted a vast and growing commerce in the eighteenth century. The strategy of rummaging through Nature's vast storehouse was hardly an attitude peculiar to the British. The Portuguese and the Dutch moved plant life quite as quickly as did anyone else. It is certainly the case that the British exploited the scientific connections that had been fostered by the Royal Society from the late seventeenth century. From London both the apothecary James Petiver, from the 1690s, and later the society physician

[22] Bougainville, *Voyage autour du monde,* pp. 354–5.
[23] Baugh, "Seapower and Science," pp. 36–7; Grove, "The Island and the History of Environmentalism," p. 156; Bougainville, *Voyage autour du monde,* pp. 367–8.
[24] Furber, *Rival Empires of Trade,* pp. 327–9.

Hans Sloane were able to cultivate the kinds of correspondents who could describe and send seemingly endless specimens from the limits of European trade routes. For example, Captain Thomas Walduck of Barbados had suggested that the Royal Society should develop correspondents in the West Indies. Petiver heard of the proposition from Walduck's nephew and immediately wrote to encourage collecting and reporting of "ye Natural productions of your Island in respect to its Animals, Vegetables, & Minerals as also to ye politicall & Trading part wch you have already so well begun."[25] It is an interesting problem of the extent to which scientific and imperial interests were then served by the settlers and mariners engaged in trade. This Walduck clearly understood when he suggested that he had learned most of what he knew of the uses of plants "from our Physicians (shall I call them) nurses, old women and Negroes, and for the future, I will take care by some Experiment or other not to be imposed upon."[26] If European scientists were dependent on the knowledge of traders, the merchants and medics studiously acquired the knowledge of the otherwise unworthy.

Of the relations with trade, the Royal Society was acutely aware. The trustees of the Georgia colony, many of whom were connected with the Royal Society at some point, established a garden from which to supply the plantations. Sloane's friend Dr. William Houstoun (not to be confused with the surgeon James Houstoun) was appointed its manager in 1732. Similarly, Sloane had a long association with the naval surgeon Henry Barham at Jamaica. Not surprisingly, they were especially interested in the medicinal qualities of the minerals and plants. Notably, after Barham retired to Chelsea in 1716 he was soon elected F.R.S. and, by 1718, apparently completed his "Hortus Americanus Medicianalis," which described the "known vertues and experienced Qualityes as I gained them from Spaniards, Indians, and Negroes." This approach was essential to the empire of science because it meant that scientific knowledge could be gathered without a vast diaspora of the highly educated or of numberless voyages of discovery, which obviously were constrained by expense and distance.[27]

In the diverse discoveries of lost botanists and wandering astronomers, the shifting scientific periphery sought solace in the center: in the museums, observatories, laboratories, cabinets, and botanical gardens of London and Paris, which feasted on the explorers' reports and specimens. To measure, describe, and compare specimens seen (and some only imagined), to list and to cultivate, laid the foundation of comprehension that Buffon promoted in 1749. In his *Histoire naturelle* it was more than Europeans who were revealed by comparisons with the exotic, with animals and with "savages" in a myriad of forms, but a scientific process was ultimately shown as dependent upon the

---

[25] Quoted in Stearns, *Science in the British Colonies*, p. 353.
[26] Ibid., pp. 344–55, especially p. 354.
[27] Ibid., pp. 385–8.

observational objectivity of the human, often distant, witnesses throughout empires remote in their reach. Hence, Bougainville in the Malouines (Falklands) found striated bivalves that, he asserted, were hitherto known only in their fossil form, which "peut servir de preuve a cette assertion que les coquilles fossiles trouvées à des niveaux beaucoup au-dessus de la mer, ne sont point des jeux de la nature & du hazard, mais qu'elles ont été la demure d'êtres vivans dans le tems que les terres étoient encouves couvertes par les eux." More than three years later, Bougainville obtained at the Cape of Good Hope drawings of a newly discovered genus, "lequel tient du taureau, du cheval & du cerf," and another of a quadruped seventeen feet high that Buffon told him was a giraffe, which "On n'en avoit pas revù depuis celui fut apporté à Rome du tems de César, & montré a l'amphithéatre." Along the routes of trade, in this case of the French and the Dutch at the Cape, Bougainville encountered evidence of an Africa that was then "la mere des montres," of a new biology and of the barely remembered.[28]

In the thicket of these promiscuous encounters with nature, the eighteenth century imposed order. Notwithstanding Linnaeus, Buffon, Bonnet, and many others, this would produce a structure not only of classification but also for exchange. Kew Gardens, therefore, would become "the great exchange house of the Empire," from which Banks could send plants for cultivation to numerous satellite gardens throughout the sphere of British influence. Centers such as Kew or, after 1793, the Jardin des Plantes were systems of dispersal and of gathering. And, faced with the vast "confused mingling of beings that seem to have been brought together by chance," botanists such as Michel Adanson could argue thirty years earlier that "this mixture is indeed so general and so multifarious that it appears to be one of nature's laws." Consequently, there needed to be what has been described as "European-based patterns of global unity and order." If these objects had merely been matters of curious complexity there would have been little such need; but when the pressure of utility confronted immense variety, some imposition of structure would prove urgent. And the more the army of collectors gathered in their nets, the more the complex overwhelmed, the more evident was the romanticizing of the utilitarian, promoted by Linnaeus and his disciples, that would make value out of the explosion of biological variety.[29]

Natural diversity implied the augmentation of nature. A good example of just how much governments thought seriously about such affairs occurred in 1785, when the British House of Commons considered the possibility of

---

[28] Pagden, *European Encounters with the New World*, p. 148; Bougainville, *Voyage autour du monde*, pp. 64, 383.

[29] Pratt, *Imperial Eyes*, pp. 30–1; Michel Foucault, *The Order of Things: An Archaeology of the Human Sciences* (London: Tavistock, 1970), p. 148, passim; Mackay, "Agents of Empire," Alan Frost, "The Antipodean Exchange: European Horticulture and Imperial Designs," and Lisbet Koerner, "Purposes of Linnaean Travel: A Preliminary Research Report," in Miller and Reill (eds.), *Visions of Empire*, pp. 38–9, 75, 127.

African settlement. Under intense secrecy, for fear of alerting the French, Spanish, and Portuguese, vessels were outfitted for an expedition to explore the suitability of colonizing the west coast of Africa. As botanist to the expedition, Banks ultimately recommended a Pole called Au, who, despite objections to his foreign origins, was ultimately accepted as an able and educated young man – and who ultimately changed his name to Hove. In early 1786 Hove cruised along the African coast, collecting the urgently sought-after specimens, which, upon his return, would languish in British customs. It was clear, however, within a matter of weeks of his arrival at Spithead in July that the African coast was unsuitable, even if convict labor were to be supplied to the East India Company. An imperial strategy would depend on the discoveries and advice of naturalists. This attitude was to result in the kind of policy exhibited most dramatically in the well-rehearsed efforts of Captain Bligh, scion of the customs service and agent of an entrepreneurial manager of prison hulks on the Thames, to transplant breadfruit from Tahiti to the West Indies. Even with Banks's patronage, these were not propitious beginnings to botanical expropriation.[30]

The intensity of the efforts of eighteenth-century classifiers implied the colonization and commodification of Nature. So, in the early eighteenth century, factors in Sierra Leone were instructed "to sett, sow and plant all things that may be found out in those parts that may be improvable for trade, as cotton, indigo, ginger, sugar canes, pepper, spice, gumm trees, druggs, etc. . . . to put a stock of cattle on the island of Torsus, clear the island of wood trees and make plantations thereon . . . and to carry on the indigo and potash works."[31] Cultivation, however, was only part of the brief. When the surgeon James Houstoun went out in the service of the African Company in 1722 he acknowledged that his own knowledge of the natural world was limited by his acquaintance with the materia medica. He understood that the Company had already appointed a botanist "who was to make a particular Collection of all Herbs, *Aromatick* Plants, Butterflies, Cockle-Shells, etc. which would have contributed to the Advantage of the learned World, and for ought I know, like wise to Advantage of the Company."[32] After virtually half a century of shells and cotton pods, of dye-stuffs and the deadly miasma of the littoral, the heart of darkness not yet breached, Bougainville faced the Falklands (or Malouines) with much the same view as his British predecessors. Although

---

[30] David Mackay, *In the Wake of Cook: Exploration, Science & Empire, 1780–1801* (New York: St. Martin's Press, 1985), pp. 18, 32–6; Greg Dening, *Mr Bligh's Bad Language: Passion, Power and Theatre on the Bounty* (Cambridge University Press, 1992), p. 12; Gascoigne, *Joseph Banks and the English Enlightenment*, pp. 203–4.

[31] Walter Rodney, *A History of the Upper Guinea Coast, 1545–1800* (Oxford: Clarendon Press, 1970), p. 170.

[32] James Houstoun, M.D., *Some New and Accurate Observations Geographical, Natural and Historical. Containing a true and impartial Account of the Situation, Product, and Natural History of the Coast of Guinea, So far As relates to the Improvement of that Trade, for the Advantage of Great Britain in general, and the Royal African Company in particular* (London, 1725), pp. 4–5.

the islands were claimed in 1765 by Commodore Byron, Bougainville at anchor read the barren "horizon terminé par des montagnes pelées; des terreins entre coupés par la mer, & dont elle sembloit se disputer l'empire" but, nonetheless, understood that his rival's "goût pour l'"Histoire naturelle" might more properly assess the benefits of settlement in such a remote part of the world. Despite first appearances, long before the bounty of Tahiti, which still awaited him, Bougainville noted many benefits: "une quantité innombrables d'amphibies des plus utiles, d'oiseaux & de poissons du meilleru gout; une matire combustible pour supléer au défaut du boit; des plantes reconnues spécifiques aux maladies des navigateurs." Each similar encounter, as of the beavers and the whales that abounded in North America, meant determinations of biological definition. Such was the deliberation of the faculty of theology in Paris, recounted by Pierre Francois-Xavier de Charlevoix in 1744, that the beaver's tail made it of the class of the mackerel and consequently fit to be eaten on fast days. Theologians often have had problems with biology. Their views were overtaken by the commercial concerns of this world.[33]

In the expanse of empires came dangers and delights; hazards and pleasures mixed like puddles on the shores of a Rio, Madagascar, or Macao. Each new environment brought apprehension and fears that often turned to substance, insinuating themselves unseen in the miasma of plague, smallpox, yaws, or a myriad of mortal ailments yet unnamed. The management of scurvy, endemic on any lengthy sea voyage, likewise required vigilance, and the constant search of new antiscorbutics preoccupied many a captain. Bougainville's gathering of antiscorbutic plants in the Falklands and in Tahiti and, when he could, of daily ordering a pint of lemonade prepared from a powder for each of his sailors speak of the seriousness of the problem, which he attributed, however, to the moistness sailors of necessity must face. Mistaken causes may have made the cure more elusive, but it was management more than remedy that pressed empires to expand.[34]

By the eighteenth century, new colonies meant a new regimen. From George Cheyne in 1724 to the democratic doctor Thomas Beddoes in the 1790s, the recommended response was temperance to ward off the excessive seductions of alcohol or sexuality, which, as many a voyage, colony, or remote paradise had often proved, were incubators of disease and death. Along the vast reach of the European trade routes, there were transported not only goods and specimens, settlers or convicts, but also illness. Hans Sloane's efforts on behalf of

[33] Bougainville, *Voyage autour du monde*, pp. 54–5; Gordon M. Sayre, *Les Sauvages Americains: Representations of Native Americans in French and English Colonial Literature* (Chapel Hill: University of North Carolina Press, 1997), pp. 221–3.

[34] Bougainville, *Voyage round the World*, p. 211; Christopher Lawrence, "Disciplining Disease: Scurvy, the Navy, and Imperial Expansion, 1750–1825," and Gascoigne, "The Ordering of Nature and the Ordering of Empire: A Commentary," in Miller and Reill (eds.), *Visions of Empire*, pp. 80–106, 112–13.

the African Company, in attempts by way of inoculation to secure the health of slaves on the long voyage amid a trade on the verge of collapse in "a Country So different from their Own, surrounded with the melancholy & repeated instances of Morality" – here was enough to focus European minds.[35] Disease often proved part of African cargoes, but likewise in such trade there were hints that many slaves had already been inoculated and "as they Show the *Marks* of it, So wee as well as they reap the *fruits* of it, in their Secure attendance upon our Sick." Where they had not had the inoculation, it wreaked havoc in the plantations in the West Indies, underlying the vulnerability of planters to diseases that made no distinction between slave and master, between investment and investor. There were many malignant fevers that cut through the planter classes, particularly those most recently arrived and presumably least resistant. In 1776 J.-B. Dazille wrote in *Observations sur les maladies des negres* that careful observation of the diseases of blacks "is to occupy oneself with that which is useful to the Colonists in particular, to the Commerce of the Nation in general, and to the prosperity of the State." In this respect, slaves were more than an investment; for the observant they were, in effect, a laboratory that clearly tied personal health to an animal economy and commercial strength.[36]

THE TRANSPORT OF NATURE

This was a century bracketed by Defoe's publication in 1719 of *Robinson Crusoe,* who had at least returned, and Fletcher Christian's mutineers, many of whom did not – or, if they did, were hanged for it. As Simon Schaffer reminds us, there was a curious and important message in that rational providence that let Crusoe discover, to his "Astonishment and Confusion" in a spot where he had shaken out a bag of chicken meal, the few green shoots of English barley "that was directed purely for my Sustenance, on that wild miserable Place." The message was Biblical in its certainty, for it was Crusoe's

---

[35] Public Record Office, London, T70/53/105. Africa Company to James Phipps, at Cape Coast Castle, 12 September, 1721; L. Stewart, "The Edge of Utility: Slaves and Smallpox in the Early Eighteenth Century," *Medical History,* 29 (January 1975), 54–70, especially 60–3.

[36] Royal Society, Inoculation Letters and Papers, "A Letter from Dr. Cotton Mather in Boston . . . to Dr. James Jurin," May 21, 1723, p. 45; Rev. Robert Robertson, *A Detection of the State and Situation of the Present Sugar Planters* (London, 1732), quoted in Richard B. Sheridan, "Africa and the Caribbean in the Atlantic Slave Trade," *American Historical Review,* 77 (February 1972), 15–35, especially 21; Philip Rose, M.D., *An Essay on the Small-Pox, Whether Natural, or Inoculated,* 2nd ed. (London, 1727), p. 85; Mons. de la Condamine, *A Succinct Abridgment of a Voyage Made within the Inland Parts of South-America; from the Coasts of the South-Sea, to the Coasts of Brazil and Guiana, down the River of Amazons* (London, 1747), p. 93. (Banks's copy, BL. 978.k.31.) On doctors in the West Indies, see Sheridan, "Mortality and the Medical Treatment of Slaves in the British West Indies," in Stanley L. Engerman and Eugene D. Genovese (eds.), *Race and Slavery in the Western Hemisphere: Quantitative Studies* (Princeton, NJ: Princeton University Press, 1975), pp. 285–310; Sean Quinlan, "Colonial Encounters: Colonial Bodies, Hygiene and Abolitionist Politics in Eighteenth-Century France," *History Workshop Journal,* 42 (Autumn 1996), 107–25, especially 108–11, 116.

solitary cultivation that bore him result. In the century between the marooned and the mutinous a great battle had been fought over the critical sentiment of improvement, which provided employment for naturalists and schemes for the enlightened to promote. It may not be coincidental that among the first casualties of Christian's mutiny were Bligh's breadfruit plants, which were flung into the sea. And as Diderot might challenge a Bougainville, at virtually the same moment there were heard the complaints of the poet John Langthorne in *The Country Justice* about those rapacious traders and monied men, encouraged by the Crown, "whose antic Taste, / Would lay the Realms of Sense and Nature waste." That may say more about the reasons for the lamentations of the relatives of Bligh's unhappy crew, shackled before the court and the yard, than many people then understood.[37]

Colonization descended on lonely islands like a cloudburst, sweeping away all those, as Defoe put it a very few years after *Crusoe,* who "by their own folly and treachery raising war against us, [had then] been destroy'd and cut off." Such was the fate of the defiant in those places, whose lot it was to be turned by trade into "a prodigy of Wealth and Opulence," where planters rose "to immense estates, riding in their coaches and six, especially at Jamaica with twenty or thirty negroes on foot running before them whenever they please to appear in publick." Colonists and cultivators followed in the wake of collectors, who expanded their cabinets and their systems of Nature and who, like the Linnean Anders Sparrman, gloried in discoveries for "medical and oeconomical purposes."[38]

Botanical transport certainly intended to achieve more than the prosperity of colonies. There were celebrated successful transfers by the end of the century, such as the discovery that European fruits and vegetables flourished spectacularly in the soil around Sydney. In the other direction, significant efforts in the European cultivation of exotic plants, even in hothouses, produced far from an unparalleled bounty and even the Linneans became a laughingstock. The encouragement of European planting was a program of import substitution, such as the many efforts, some of them fraudulent, to find an alternative to the vast import trade in tea, coffee, or olive oil. The list is endless. The essential criterion was utility, and in this the botanic gardens in Europe and in the colonies formed the crucial base of assessment, acclimatization, and dispersal necessary to transfer and cultivation. In 1777, J.-N. Thierry de Menonville, once a pupil of Jussieu, engaged in biological espionage in Mexico to obtain samples of cochineal, which at the time brought handsome prices at dyeworks in France, Holland, and England. Thierry de Menonville did not succeed, but he did become Botaniste du Roi in Saint Dominigue.

[37] Daniel Defoe, *Robinson Crusoe,* ed. Michael Shinagel (New York: W. W. Norton, 1975), pp. 62–63, 83, 91–2; Schaffer, "The Earth's Fertility," pp. 136–7; Raymond Williams, *The Country and the City* (London: Hogarth, 1993), pp. 62, 79–80.
[38] Defoe, *Complete English Tradesman,* p. 316; Pratt, *Imperial Eyes,* pp. 34, 55.

In this curiously scientific colony, botany was very much the rage. It served a major function in the worldwide network of gardens established by the French government to encourage the cultivation of spices and, of course, the breadfruit.[39]

It is the veneration of nature that can obscure our understanding of the tidal force of eighteenth-century trade, which overwhelmed in the new worlds the twin disasters of religion and conquest. Colonists might need to turn naturalist if survival and trade were the issue, setting aside – if only temporarily and superficially – the historical imperatives that had once driven European soldiers and missionaries into the dangerous interiors of the Americas. And the demands of settlement compounded demands for goods, such as timber or hemp, that were essential to the maritime empires. Settlement itself magnified markets for manufactures, thus closely tying European imports and re-exports to the circulation of goods from the colonies. No wonder, then, that by the latter half of the eighteenth century there were many, like Diderot, who perceived in the merchant the agent of civilization.[40]

This was a rather optimistic view, to say the least. Few merchants articulated it, but the fact was that by the middle of the eighteenth century many European shops stocked the vast variety of goods only empires could deliver. But it wasn't only the European shops. Even at the limits of trade, such as at Bance Island, fifteen miles up the Sierra Leone River, in 1773 merchants and botanists could feast on the products of the world. As David Hancock has shown, London merchants might invest the world over and especially, after 1763, in factories and plantations in India, the West Indies, and the American South. Leases could be taken on lands in Jamaica to expand the trade in rum and sugar, but all this depended on the knowledge and the skill to exploit "site, soil, and climate." By the same means, although much more closely directed by the French Crown, Haiti (Saint Dominigue) could become the world leader in the 1780s in sugar and coffee production along with the export of significant amounts of cotton and indigo. Imperial expansion was a matter of scientific management. Such a view explains the concordance of the anti-clericalism of Diderot and Joseph Banks, although their advocacy of Enlightenment diverged on most other points. Diderot put more hope in the merchant and reached more radical conclusions. The Lincolnshire gentleman Joseph Banks elevated the scientist and the scholar to governance by exploiting his commercial and political connections, thus confirming the maintenance of social order and of oligarchy.[41] Botany and trade would make his empire.

---

[39] Frost, "The Antipodean Exchange," and Koerner, "Purposes of Linnaean Travel," in Miller and Reill (eds.), *Visions of Empire*, pp. 59–63, 132–7; James E. McClellan III, "Science, Medicine and French Colonialism in Old-Regime Haiti," in Teresa Meade and Mark Walker (eds.), *Science, Medicine and Cultural Imperialism* (New York: St. Martin's Press, 1991), pp. 36–59, especially pp. 43–4.

[40] Postlethwayt, *Universal Dictionary of Trade and Commerce*, s.v. "Colonies"; Pagden, *European Encounters*, pp. 169–71.

[41] David Hancock, *Citizens of the World: London Merchants and the Integration of the British Atlantic*

INSTRUMENTS OF EMPIRE

Trade turned the exotic into the commonplace. But there were other devices, such as ships, chronometers and compasses, charts, and telescopes, whereby European empires and companies could manage the world. Throughout the long eighteenth century, instruments were emblematic of European power just as they were crucial to the expansive enterprise. In the waning of the seventeenth century, naturalists and astronomers were supercargo, hitching rides on frigates plying between Canton, Macao, and home. One such hitchhiker was the Reverend James Pound, who went to Madras in 1699 as chaplain to Fort St. George but was soon on the move again, with instructions from the Astronomer Royal Flamsteed, to check the latitude and longitude of ports and make observations of any astronomical phenomena. Armed, in 1700, with quadrants, cross-staff, and a 16-foot telescope, he made observations of Jupiter's satellites at Chusan (probably at the entrance to Hangzhou Bay) until forced away by the Chinese and unhelpful Jesuits. By 1702, while anxiously awaiting instruments from Flamsteed, Pound reported on a comet seen on his voyage to Batavia. His efforts to make precise navigational observations, especially of the southern stars, ultimately came to disaster. In 1705, his fort at Condore (likely in the Timor Sea) was set upon in the dead of night by its own soldiers, recruited in the Celebes, and only Pound and ten others managed to escape to Batavia, leaving behind all their clothes, books, and instruments.[42] Pound returned to patronage in England, but his case illustrates that even to astronomers charting new empires the risks were hardly small.

It is instructive that among Crusoe's first efforts at survival was his recovery of books of navigation and instruments such as dials, perspectives, and charts as much as pen and paper. Defoe's legend thus reinforces the value, even for the shipwrecked, of determining where one was marooned. The recruitment of those capable of mathematical calculation was fundamental to the certainty of navigation and trade. In 1714, in a letter from Whitehall, the Persia merchant Samuel Palmer heard complaints about the misfortune "that those yt travel in yr. parts are not well provided with Mathematical Knowledge."[43] Certainly, astronomical observations were limited to those who had access to instruments and knew how to make proper navigational calculations. We forget this was difficult enough in Europe, but such knowledge was crucial to those who would map the uncharted world. In the 1730s, observations were

Community, 1735–1785 (Cambridge University Press, 1995), pp. 1–2, 121, 144, 148; McClellan, "Science, Medicine and French Colonialism," pp. 36–7; Gascoigne, *Joseph Banks and the English Enlightenment*, p. 43; Gascoigne, "The Ordering of Nature," p. 110; and Gascoigne, *Science in the Service of Empire: Joseph Banks, the British State and the Uses of Science in the Age of Revolution* (Cambridge University Press, 1998).

[42] Cambridge University Library, RGO MSS. 1/37/91, December 18, 1700; 1/37/93, November 19, 1701; 1/37/100, June 9, 1702; 1/37/102, July 7, 1705. Pound to Flamsteed. My thanks to Dr. Ron Love for his assistance in identifying these sites.

[43] S.P.C.K., Society Letters, CS2/4/21. To Samuel Palmer, March 16, 1713/14.

made at Cartagena by John Gray of the Navy Office, who had obtained from the celebrated instrument-maker George Graham an isochronous clock. In the 1740s, the hazards of war interrupted the efforts of the Condamine Expedition to discover variations in a degree of the meridian between the Arctic and the Equator. In 1745, during the War of the Austrian Succession, the English were able to capture the Spaniard Don Antonio de Ulloa, who had in his possession papers from Condamine's effort, which were then turned over to the Royal Society. The Society ultimately elected Don Antonio to membership. Such information was of invaluable scientific importance; the French-Swedish expedition in Lapland was to be guided, and the dispute over the shape of the earth was to be determined, by the exactness of English instruments, especially the new Graham zenith sector. By the 1740s, it was increasingly obvious that precision instruments and their users were essential to the determination of scientific issues, but they were also crucial to the improvement of navigation. It is for this reason that by 1751 Gowin Knight, F.R.S., noted for his artificial magnets, was able to convince Admiral Anson and thus the Royal Navy to purchase his improved compasses.[44]

Each expedition, each settlement, each port of call demanded precise navigational determination. Without it, the measurement of Empire, literally, was not possible. Hence, at the request of the Royal Society, Neville Maskeleyne, F.R.S. and future Astronomer Royal, would make observations of the moons of Jupiter at Barbados and of the transit of Venus in 1761 at St. Helena. Bougainville's expedition used the latest Dollond achromatic telescope to observe the eclipse of the sun in 1768, and shortly thereafter the mathematicians Charles Mason and Jeremiah Dixon were employed in the determination of the highly contested boundary between Pennsylvania and Maryland. The Mason-Dixon line was achieved by use of the best possible British scientific instruments, which also afforded the opportunity to observe the transits of Venus and Mercury in 1769. What is striking about the employment of such apparatus throughout the entire reach of empire was the way in which practical requirements met the image of philosophical enlightenment. Instruments were, in effect, the devices by which scientific knowledge was not only achieved but also demonstrated. Thus, Thomas Jefferson would reflect his own republican image in the purchase of enlightened instruments.[45]

[44] Vickers, *Defoe and the New Sciences*, pp. 99–100; Stearns, *Science in the British Colonies*, pp. 389–92; Mary Terrall, "Representing the Earth's Shape: The Polemics Surrounding Maupertuis's Expedition to Lapland," *Isis*, 83 (June 1992), 218–37; Rob Iliffe, "'Aplatisseur du monde et de Cassini': Maupertuis, Precision Instrument, and the Shape of the Earth in the 1730s," *History of Science*, 31 (September 1993), 335–75; Richard Sorrenson, "The State's Demand for Accurate Astronomical and Navigational Instruments in Eighteenth-century Britain," in Ann Bermingham and John Brewer (eds.), *The Consumption of Culture, 1600–1800: Image, Object, Text* (London: Routledge, 1995), pp. 263–71; Deborah Warner, "Terrestrial Magnetism: For the Glory of God and the Benefit of Mankind," in Albert Van Helden and Thomas L. Hankins (eds.), *Instruments, Osiris*, 9 (1993), 67–84, especially 79–80; Patricia Fara, *Sympathetic Attractions: Magnetic Practices, Beliefs, and Symbolism in Eighteenth-Century England* (Princeton, NJ: Princeton University Press, 1996), pp. 79ff.

[45] Stearns, *Science in the British Colonies*, p. 362; Silvio A. Bedini, "The Transit in the Tower: English

CONCLUSION

The clash of navies, chartered companies, naturalists, and instrument-makers was the conflict of self-defined enlightened scientific cultures over the vast spoils of imperial pillage. But there was also a clash with members of alien cultures, who had difficulty regarding the often desperate European sailors as the harbingers of anything particularly edifying. Bougainville had found that Tahiti was a paradise of filchers, something that could be alarming if a pistol went missing, as it did, or potentially disastrous when a paper containing four days of exact longitudinal observations was stolen. But if that could be excused as a misunderstanding or as a failure of reciprocity that ultimately resulted in the death of one of the Tahitians, it could not also be said of the reaction of the Chinese in the 1790s, whose culture they felt could learn little from the British nuisance.[46]

The Macartney embassy to China in 1794 was an expedition that encapsulated many of the themes that had emerged throughout the century. Originating in the long-standing connection between Lord Bute and Banks, the embassy was under the command of Viscount George Macartney, Bute's son-in-law. It was designed to impress the Chinese emperor with highly sophisticated manufactures and thereby to create a great demand in the Chinese empire for British goods. On the advice of agents of the East India Company, who had the most to gain by the opening of China to trade beyond Canton, some of the best British scientific instruments were aboard; the list included a planetarium, celestial and terrestrial globes by Dudley Adams, telescopes by the Dollonds and Ramsden, and, significantly, a chronometer, which was essential for accurate navigation. These were showpieces, designed to impress, the best instruments of navigation and display in the world by those who intended to dominate it. The Chinese were studiously indifferent. Much of the apparatus was not even presented, finding its way back to the East India Company. Surveys of the globe meant little to those who feigned no interest in it. It is remarkable that the chronometer brought back to England was among those that might have been most useful to Chinese cartography. In the superiority of the Chinese emperor, the search for markets by the European barbarians had temporarily found its limit.[47]

If Tahitians had been amused, the mandarins were not at all impressed by the instumental and emblematic gadgets of the European empires. The fact that navigational devices were critical to imperial ventures made no difference.

Astronomical Instruments in Colonial America," *Annals of Science,* 54 (March 1997), 161–96; Bedini, *Thomas Jefferson: Statesman of Science* (New York: Macmillan, 1990), pp. 162ff., 419; see also Richard Sorrenson, "The Ship as a Scientific Instrument in the Eighteenth Century," in Henrika Kuklick and Robert E. Kohler (eds.), *Science in the Field, Osiris,* 11 (1996), 221–36.

[46] Bougainville, *Voyage autour du monde,* pp. 193–4, 197, 209; Pagden, *European Encounters,* p. 156.

[47] Gascoigne, *Joseph Banks and the European Enlightenment,* p. 82; Mackay, "Agents of Empire," p. 42; J. L. Cranmer-Byng and Trevor H. Levere, "A Case Study in Cultural Collision: Scientific Apparatus in the Macartney Embassy to China, 1793," *Annals of Science,* 38 (1981), 503–25.

But, on another level, these instruments were only curiosities to the uninitiated, whether of Canton or Cartagena. At the limits of the reach of empires,
the value of the natural or the manufactured was determined by the market.
It is fair to say that the European search for commodities, the control of and
access to new markets, the indentification of new medicines and useful plants,
the expansion of the state and the promotion of the public interest and glit-
tering, private wealth, all were a piece in the scientific pillage of the empires
of the Enlightenment.

36

TECHNOLOGICAL AND
INDUSTRIAL CHANGE

A Comparative Essay

Ian Inkster

Everyone now seems to agree that eighteenth-century industrialization was strongly associated with qualitative changes in the ways in which such formal productive inputs as fixed capital or skilled labor were combined, organized, and exploited by new agencies operating in novel physical sites. Although the analytical details for any one nation are hotly debated and the histories of different nation-states and regions are varied even within Europe itself, it is now increasingly conceded that the story of industrial modernization is at heart a story of institutions and technologies. Without informed reference to both institutional and technological features, it is no longer feasible to argue that the rise of new industries in the eighteenth century was a clear function of, say, new sources of investment funds or higher levels of demand, even when such conventional "factors" can be shown to have themselves arisen or altered or increased as a consequence of prior, prerequisite institutional and technological changes. This is not to say that anything goes. This chapter will consider the real problems of interpretation regarding the sources of technological change, the relations between scientific and technological changes and institutional innovations, and the interactions among national and even continental systems. For instance, however haphazard may have been the technological interaction between national systems, the fact that it insidiously, uncontrollably, and chaotically occurred means that a story of creativity in one place cannot in itself be the story of technological and industrial change throughout, say, Europe. How and why did novel machines or solutions move from one location to another? Are we content to define "location" only in terms of physical geography, or do we require knowledge of the social or perhaps even cultural siting of new technologies?

Second, we might also posit that there is now a consensus that Western Europe and the developing "Atlantic economy" did not stand in supreme technological isolation from the rest of the globe. Our period saw technological change, challenge, and response in many corners beyond that especially energetic one of Western Europe. This essay, therefore, makes something of the character of technological change in regions such as Russia, China, India, and Japan. Because we are befuddled over how to measure it (and, often, over how to recognize it), it is difficult to assign points for creativity to differing national settings, to rank the world in terms of attainments in invention or innovation. Given that the knowledge of machines and the machines themselves, alongside the skills of artificers, mechanics, and business agents, were transferred among nations and regions, then the historian studying the eighteenth century from the very late twentieth century tends to catch the problematical results of an only partially perceived complex process rather than the clear evidence of a simple one. In addition, even an elementary understanding of the exigencies, resource requirements, and conceptual difficulties involved in technological adoption, adaptation, refinement, and standardization serves to encourage either an abandonment of the term "creativity" or, perhaps more reasonably, an acknowledgment of the applicability of the term to many processes and mechanisms in many social and physical locations.

Thus, happily, our two levels of generalization or relative consensus provide some coherence to a point of departure. We will not encyclopedically itemize the technological breakthroughs of the century, judge with authority their relative "dependence upon the sciences," apportion slices of "creativity" between contending agents, or descend into sweeping cultural affirmations concerning Britain versus France or Europe versus Asia. Such tutorial-like topics may be of utility in their place, but here, for reasons briefly outlined, we focus on the social and institutional settings of technological change, the role of different types of agency (particularly, those of the state versus those of the contending interests or of the marketplace), and on the mechanisms whereby improved techniques were settled into locations far removed from their places of origin.

EUROPE: THE STRENGTH OF WEAK TIES

Legal impositions on the export of advanced machinery (such as that set by Britain in 1774) could not prohibit the transfer of technologies between nations or regions. Law could never forbid the purchase of books and manuals, visitations to installations, and inspections of layouts and machines in use, nor could it halt the two-way movements of key mechanisms and manufacturers. Although Samuel Slater, an employee of Arkwright and Strutt, was unable to remove models of textile technique from Britain, he and his more humble colleague, the Belper mechanic Sylvanus Brown, did convince the New World textile adventurers that their combined experience warranted financial backing. By 1790 a modernized water-powered mill (three 18-inch carding machines, draw heads with two rollers, roving cards and winders, thistle-spinning frames with seventy-two spindles) had effectively crossed the Atlantic. At a site in Pawtucket, Rhode Island, mechanical ingenuity and craft experience, rather than blueprints and formal knowledge, had established a successful embodiment of a stream of eighteenth-century British technological breakthroughs, including Kay's flying shuttle (1733), Hargreaves's spinning jenny (1764), and Arkwright's water frame (1769), as well as the subsequent improvements of Crompton and Cartwright. In a setting of enormous contrast, French rulers exerted a nominal hegemony over the Spanish economy throughout the early years of the century and actively discouraged French entrepreneurs and technicians from introducing new industries into Spain. Nevertheless, despite a sophisticated system of controls, there was little halt to the flow of French enterprise into the Spanish woolen, iron, glass, gunpowder, paper, and silk industries.[1]

Although most technology transfer between nations failed, it was the many points of success and development that helped to confirm an overall European technological advancement during the century.[2] Perhaps more so than in Asia or elsewhere, Europe was a large place of "chance meetings," international migrations and loose ties between centers of change, little of which was adversely affected by open warfare. British citizens brought iron foundries, blast furnaces, and textile machinery to France and Sweden. Textile centers in Bohemia, Moravia, and Lower Austria obtained their machinery from Britain and benefited by exploitation of markets left open during Napoleon's ascendancy, but Italy's new technology came from France, Estonia's techniques from German commercial groups, and Spanish woolen textile innovations from a collection of English, Irish, French, and Dutch artisans.

Such loose but significant ties among individuals and regions were complemented by a density of relationships within nations, particularly in the newer and growing urban settings of northwestern Europe and along the Atlantic coast. By 1800 the number of cities with populations of more than ten thousand reached 363, a reflection of an increase in the proportion of the urban to the total population in the north and west extending from the early seventeenth century. The public sector – military techniques of the great European arsenals and ports, and sites of bronze, copper, and iron metallurgy employing outworkers in small artisan workshops in surrounding areas – benefited particularly from the sturdy growth of the larger administrative and capital cities in the seventeenth and early eighteenth centuries. On the other

---

[1] B. Fay, "Learned Societies in Europe and America in the Eighteenth Century," *American Historical Review*, 37 (1932), 255–66; F. B. Tolles, *Meeting House and Counting House: The Quaker Merchants of Colonial Philadelphia, 1682–1763* (New York: W. W. Norton, 1963); J. W. Oliver, *History of American Technology* (New York: Norton, 1956); Nathan Rosenberg, "Technology," in G. A. Porter (ed.), *Encyclopaedia of American Economic History: Studies of the Principal Movements and Ideas* (New York: Scribner's, 1980), vol. I, pp. 294–308; H. Kamen, *The War of Succession in Spain, 1700–15* (London: Macmillan, 1969), chap. 6; J. V. Vivies, *An Economic History of Spain* (Princeton, NJ: Princeton University Press, 1969).

[2] Ian Inkster, "Mental Capital: Transfers of Knowledge and Technique in Eighteenth Century Europe," *Journal of European Economic History*, 19 (1990), 403–41; Inkster, *Science and Technology in History: An Approach to Industrial Development* (London: Macmillan, 1991), pp. 32–59.

hand, urban growth in the later eighteenth century was characterized by a disproportionate increase in the size of smaller urban areas; during the second half of the century, European cities of more than five thousand population increased in number by some 50 percent. The early stages of modernized metallurgy and water-powered textiles tended to be sited in such smaller urban areas and in nearby rural locations. During the century, quite conventional investments in infrastructural improvements (roads, canals, and coastal trafficking) in the Dutch Republic, in Britain, and in the Atlantic ports of France and Germany offered a real increase in the speed and frequency of passenger travel, together with significant reductions in the cost of freight carriage, even in the absence of any spectacular transport innovations.[3]

It was in such new centers that commerce and artisanship mixed more densely, new cultural functions and associations developed, and cultural societies, booksellers, newspapers, printing companies, and novel forms of intellectual and technological discourse (such as coffeehouses and public lecture courses) abounded. While such densities built assets and audiences for intellectual association and debate, they also provided an information system for technology, a competitive site for incremental technological emulation, and a social space for the construction and reconstruction of individual status and civic identity. Together, such features defined a public, institutionalized environment for technological transfer and diffusion. Technological breakthroughs were reported in scientific journals almost as frequently as in trade-oriented outlets. The first international publication of R. J. Eliot's invention for smelting iron from black magnetic sand, demonstrated at Killingworth, Connecticut, was in the *Philosophical Transactions* of the Royal Society of London in 1762. The chemical engineer John Roebuck gained vital scientific information in lecture courses at Edinburgh and Leyden. The knowledge gained by James Keir when translating P. J. Macquer's *Dictionary of Chemistry* inspired him to set up his alkali works at Dudley.[4] The thin links *between* such new sites, as well as the chaos of ideas, information, and investments, comprised the chance meetings of artifacts and skills that characterised such places, perturbed hitherto settled understandings, and held back the onset of diminishing returns not only to political debate and cultural ambitions but also to the techniques and organizations of material production. So it might be suggested that intraregional forces of diffusion and emulation and competition were juxtaposed with international mechanisms of technology transference to produce a mainly haphazard process of European technological advancement. Instabilities arose not only from rootlessness and discontent but also from the communications with and interventions of the foreign and

[3] H. Schmal (ed.), *Patterns of European Industrialisation since 1500* (London: Croom Helm, 1981); Jan de Vries, *Economy of Europe in an Age of Crisis* (Cambridge University Press, 1976).

[4] A. E. Musson and E. Robinson, *Science and Technology in the Industrial Revolution* (Manchester: Manchester University Press, 1969).

the novel, whether in the form of new products, new processes, or new ideas.[5] Such an approach might better explore the relocation of technological advancements in manufacturing, transport, and general civil engineering toward the North and away from the South (where the seventeenth-century innovations in navigation and commerce had centered, Amsterdam's banking emulating that of Venice) than would those perspectives that emphasize "technological and political accidents combined with favourable economic circumstances."[6] To Braudel's too casual formulation we may add a logic of site and agency informed by notions of social distance and urban association.

But if Europe was an unusual place of contending interests and new patterns of living, so, too, was it a collection of competing states. Any approach to technological change in eighteenth-century Europe must move from arguments about individual and group motivations and resources toward some acknowledgment of the location of all such elements within nations bounded by the mercantilist, interfering state. Whatever its exact intentions (and they were mostly confused), the state impinged on technological change through affecting the passage of information and artifacts, through offering inducements to or imposing limitations on transfer of techniques, artisans, and entrepreneurs, through its demand for military and strategic equipment and products and tools of expansion and empire, and through its impacts on migration of skills, internal colonization, and the settlement of foreign nationals.

The many futile attempts by European states to keep to themselves their craft and industrial production techniques (prohibitions on migration in Russia and Austria, Sardinia's harsh penalties on disclosure, the general constraints on machinery exports, and the imprisonment of spies) represented the repressive aspect of interventionist policies that were designed to encourage the development of strategic industries by whatever means. Although at times concerned with the flow of information, European states were more likely to be involved with inducements to entrepreneurs and skilled workers, as in the rewards and resources offered to John Milne (Arkwright frame), William Wilkinson (ironworks), John Holker (spinning jenny), and Michael Alcock (metalworks) by the mercantilist French state. John Holker gained official assistance in the form of expenses, salary, subsidies, and the granting of a royal "privilege" (patent of monopoly) when he established his factory at Rouen.

[5] S. Pollard, *Peaceful Conquest: The Industrialisation of Europe, 1760–1970* (Oxford: Oxford University Press, 1981); D. Gregory, "The Production of Regions in England since the Industrial Revolution," *Journal of Historical Geography,* 14 (1988), 14–32; B. Hoselitz, "Generative and Parasitic Cities," *Economic Development and Cultural Change,* 3 (1954–5), 278–94; T. Hagerstrand, "Quantitative Techniques for Analysis of the Spread of Information and Technology," in C. A. Anderson and M. J. Bowman (eds.), *Education and Economic Development* (Chicago: Aldine, 1963), pp. 244–80; D. S. Kaufer and K. M. Carley (eds.), *Communication at a Distance: The Influence of Print on Sociocultural Organisation and Change* (Hillsdale, NJ: Lawrence Erlbaum, 1993); M. Mulkay, *The Social Process of Innovation* (London: Macmillan, 1973).

[6] Fernand Braudel, *Civilization and Capitalism: 15–18th Century,* vol. 2: *The Wheels of Commerce* (London: Fontana, 1985), quotation at p. 570.

This site was devoted to the weaving of woolens and cotton material using British machinery and citizens. The establishments of Rouen then became sites of further transfers within France through the later movement of British workers, the training of French citizens who became foremen of works at other locations, and the building of machinery and setting up of works by former British employees of Rouen, such as Daniel Hall at Sens or James Morris at Rouen. Furthermore, French employees of Rouen copied the original model: Pierre Fouguier at Bernay, Thomas Leclerc at Bourges. Again, Holker offered advice to the French authorities on how to avoid restrictions on machine exporting and organized the recruitment of skilled workers from Scotland and from Irish regiments stationed in France (see the later discussion).[7] Generally, the royal privileges seemed to have been relatively effective instruments of transfer into France, especially in lines such as hardware, plated wares, and copper works, which produced goods in direct competition with Britain. The ministries of Turgot, Necker, and Calonne distributed awards, subsidies, and privileges to a variety of French entrepreneurs on the technical advice of Britons such as Holker and Milne.[8]

In more backward economies, war and aggression could result in sustained impacts on technology transfer. The success of Prussia in Silesia prior to 1786 stimulated a program of technological transfers under the paternalist management of Frederick the Great. This began with policies that encouraged internal development: the erection of state shipyards, the establishment of agencies for the marketing of Silesian iron ore, tariff protection and the granting of selective monopolies in salt and timber, and privileges in cutlery and munitions manufacture, sugar refineries, and metallurgical works. At a second stage, Prussian officials in Silesia organized technique transfers into the region from France, Belgium, Switzerland, and Britain. The visit of the government agent G. von Reden to Britain resulted in the introduction of steam engines, coke furnaces, and iron-puddling. It was Reden who first employed William Wilkinson as manager of an enterprise, as well as a Scot, John Baildon, to manage the newly introduced coke furnaces.

Among all such expensive efforts, few of which showed commercial profit, the impact of the state on the movements of individuals and large groups was probably of greatest importance to the location of technological innovation and the success of technology transfers. Special inducements to entrepreneurs and key agents may have created improved arsenals, and capital city handicrafts and metalworks, but in most of the arenas in which the private enter-

[7] W. O. Henderson, *Britain and Industrial Europe, 1750–1870* (Leicester: Leicester University Press, 1972).

[8] John Harris, "Spies Who Sparked the Industrial Revolution," *New Scientist,* 22 May 1986, 42–7; Harris, "Michael Alcock and the Transfer of Birmingham Technology to France before the Revolution," *Journal of European Economic History,* 15 (1986), 7–59; E. A. Allen, "Business Mentality and Technology Transfer in Eighteenth Century France: *The Calandre Anglaise* at Nimes, 1752–92," *History and Technology,* 8 (1990), 9–23.

prise system was thriving, the lasting impact of the state was almost certainly through the migration of human capital.

In a variety of ways, warfare and aggression displaced many of the peoples of Europe. During the years 1691 to 1745, perhaps forty thousand Irishmen served in French armies. Contrariwise, attempts to expand French control over Spain resulted in the settlement of some sixty-five thousand French citizens and mechanics in Spain, a principal conduit for the transfer of advanced technologies into textiles and metallurgy. Policies of Russian governments after 1762 included the establishment of "frontier" settlements of Germans, Moldavians, Belgians and Armenians. In 1764 regulations governing settlement of the Volga region concentrated on increased colonial immigration from Germany and elsewhere in order to improve agricultural technique. At the end of the century, during the Napoleonic Wars, the effectiveness of the Continental system, by which many products were blocked off, encouraged the movement of Swiss hand spinners and machine weavers into Alsace to satisfy the demands of Mulhouse for skilled workers. The French wars led to the takeover of areas of Germany during the 1790s, an initial effect of which was a loosening of guild control over industry, a reduction of internal customs duties, and an increase in the size of the market for local products, all of which encouraged innovation. Industrialists in areas such as Aachen were yet able to introduce British machinery during the first years of the nineteenth century. Similarly, the wars did not seem to hinder Casper Voght (Germany) or J. C. Fisher (Switzerland) in their investigations of British techniques in agriculture and metallurgy.[9]

The best-known and perhaps most pervasive example of the international technological impacts of state-instigated migration was that of the French Huguenots. The revocation of the Treaty of Nantes (1685) had resulted in the resettlement of some eighty thousand Huguenots in England, as many as seventy-five thousand in the Dutch United Republic, thirty thousand in the German States, twenty-five thousand in Geneva and Switzerland, and perhaps as many as ten thousand in Ireland.[10]

The movement of Huguenot skills was determined as much by the policies of European receptor states such as Switzerland or Prussia as it was by the decisions of Louis XIV of France. This is well evidenced in the policies of Elector Frederick William, who through the Potsdam decree (issued less than one

[9] O. Crisp, *Studies in the Russian Economy before 1914* (London: Macmillan, 1976); W. L. Blackwell, *The Beginnings of Russian Industrialisation, 1800–60* (Princeton, NJ: Princeton University Press, 1968); C. Trebilcock, *The Industrialisation of the Continental Powers* (London: Longman, 1981); D. J. Jeremy, "Damming the Flood: British Government Efforts to Check the Outflow of Technicians and Machinery, 1780–1830," *Business History Review*, 5 (1977), 64–91.

[10] J. G. Lorimer, *An Historical Sketch of the Protestant Churches of France* (Edinburgh: J. Johnstone, 1841); C. Weiss, *History of the French Protestant Refugees* (London: Hall, 1854); W. C. Scoville, *The Persecution of Huguenots and French Economic Development, 1680–1720* (Berkeley: University of California Press, 1960).

month following the revocation) ordered agents in Amsterdam and Hamburg to assist Huguenots traveling to Berlin or other Brandenburg (Prussian) cities. Money, passages, and passports were made available, but so too were occupations and guilds opened, materials freely provided, and grants offered for the establishment of new manufactures, all in addition to support through voluntary subscriptions. The great majority of Prussian Huguenots were craftspeople and industrialists, and through such human capital, Frederick William stimulated the establishment of fulling mills, presses and dyeing shops, silk mills and ribbon making, calico printing, and soap and oil manufactures and linen fabrication, as well as leather production, tapestry weaving, and plate glass production. Recent authorities have identified the Huguenots as fundamental to the introduction of knitting frames and to the more general foundation of mechanized woolen and textile manufacture in Prussia.

Even in nations of advanced technique, such as England and Holland, Huguenot influence was significant; in England, in a range of luxury handicrafts in London, in the silk industry, in fine linen, and manufacture of white paper; similarly, the silk, velvet, and linen manufactures of Holland were improved by French technique as well as by improved supplies of capital. Within the city of London alone, perhaps the single greatest urban center for density of scientific and technical communications across a range of interests and groups, Huguenot influence was increased through a working symbiosis of state and voluntary support. An initial government relief grant of £64,000 (sufficient to construct around twenty Arkwright water-frame mills in the late eighteenth century) was followed by metropolitan private subscriptions totaling some £200,000. Within London's urban density, Huguenots benefited both from channels of induction and from paths of acceptance. The former involved Charles II's proclamation of 1681, which offered England as a place of refuge, voluntary house-to-house collections in aid of settlement, the use of the funds of the civil list and parliamentary grants, the establishment of soup kitchens, and so on. The paths of acceptance toward assimilation were strewn with innovation, including the award of rights to insurance of property and assets, the lodging of patents, admittance to and success within the livery companies, activities within the burgeoning charity industry (from factory schools to poor relief), university matriculation, and entry into and leadership of distinguished urban intellectual associations, including those of the Rainbow Coffee House, the Society for the Encouragement of Learning, and the Society for the Promotion of Arts. The marks of success and acceptance were clear enough: intermarriage and tremendous social mobility from the crafts and technologies into brokerage, banking, warehousing, merchandising, and silk manufacturing.[11] Thus, in the longer run of the century, despite the smaller gap between

---

[11] W. C. Scoville, "Minority Migrations and the Diffusion of Technology," *Journal of Economic History*, 11 (1951) 63–84; S. B. Hamilton, "Continental Influence on British Civil Engineering to 1800," *Archives internationales d'histoire des sciences*, 11 (1958), 347–55; A. F. W. Papillon, *The Papillons of*

Huguenot textile and handicraft techniques and good practice in London and elsewhere, English industry may well have benefited much more from the Huguenot and other flows of human capital than was the case in areas in which technique was more backward. In the former case, recently earned advantages of urban density, public association, and social infrastructures meant that initial transfers might emerge as significantly diffused and assimilated techniques and skills. In the latter case, the prolonged and expensive efforts of enlightened state regimes may well have served to spread a European culture of improved technologies without creating a Europeanwide process of industrialization.

THE CASE OF BRITAIN

For the reasons already suggested, inventive activity seems to have been common in much of Europe. No nation-state monopolized significant breakthroughs in areas such as training methods (e.g., the first technical college was the Schemnitz Mining Academy in Hungary in 1733), medical operations (e.g., Claudius Aymand's successful appendicitis operation in 1736), industrial organization (e.g., the social unrest in Lyons during the early 1740s arising from attempts to implement the new labor rules suggested in the work of Jacques Vaucanson), technical publication (e.g., the French Academy's *Descriptions des arts et metiers,* seventy-six volumes between 1761 and 1789), instrumentation (e.g., the Swede Samuel Klingenstierna's method of constructing an optical instrument free of chromatic aberration, exhibited at the Russian Academy of Sciences in 1762), food processing (e.g., air tight sealing to preserve, suggested by the Italian Lazzaro Spallanzani in 1765), nor in agriculture generally (e.g., the American Eli Whitney's cotton gin of 1793). But it is also true that from around the 1730s, Britain became a center of technical innovation in such fields as manufacturing machine tools and devices, new materials, and energy production. Europeanwide social and institutional features may have generated a great wave of European inventive endeavors, but attributes, peculiar in their specific combination to Britain, encouraged a certain trajectory of manufacturing inventions there rather than elsewhere.

There was seemingly no lack of key technological innovations in eighteenth-century Britain. The list is well known to historians and to schoolchildren with careful teachers and good memories: Abraham Darby's coke smelting (1709), Newcomen's pumping engine (1712), Kay's "flying shuttle" (1733), Ward's sulphuric acid process (1736), Roebuck's lead chamber process (1746), Paul's carding machine (1748), Huntsman's steel making (1749), Bakewell's stockbreeding

London, Merchants 1623–1702 (London: Arnold, 1887); Tessa Murdoch (ed.), *The Quiet Conquest: The Huguenots, 1688 to 1988,* A Museum of London Exhibition in Association with the Huguenot Society of London, May–October 1985 (Catalogue, London: Museum of London, 1985).

(1760), Hargreaves's "jenny" (1764), Watt's steam engine (separate condenser, 1769), Arkwright's water frame (1769), Ramsden's lathe (1770), Wilkinson's boring device (1774), Watt's improved steam engine (1776), Crompton's "mule" (1779), Hornblower's compound engine (1780), Watt's "parallel motion" (1782), Tull's geared seed drill (1782), Cort's puddling process (1779), Watt's rotary motion (1781), Cartwright's power loom (1785), Murdoch's steam carriage (1786), Wilkinson's iron boat (1787), Macadam's and Telford's improved road-building technologies (1788–95), Cartwright's wool-combing machine (1790), Bramah's hydraulic press (1795), Maudslay's carriage lathe (1797), Tennant's bleaching (1799), Trevithick's high-pressure steam engine (1800), Maudslay's screw-cutting lathe (1800).

One or two points can be noted at the outset. First, it appears that several significant "breakthroughs" occurred long before any associated industrial development. Second, the dates just itemized are merely those of known (or hazarded), first recognizable invention, and they exclude the subsequent in-novation of originators or emulators. For example, Newcomen's pumping engine was first working in 1712 but not in use until the 1720s; Ramsden's lathe of 1770 made little real impact until the Maudslay improvements of the 1790s; Cort's puddling was first achieved around 1779 but was not put to use commercially until the second decade of the next century. Theories of "lag" are abundant. For instance, it can be noted that Crompton's power loom was first patented in 1786–8, but its adoption seems to have depended on distinct investment booms during 1823–5 and 1832–4, so that by 1833 there were some one hundred thousand power looms in operation.

This confused process of advancement can to an extent be explored in terms of a logic of technical challenge and response, which might have been more applicable to manufacturing and construction technologies than to others. Thus, Kay's loom was applied to cotton rather than woolen textiles because the workers in the former industry had less in the way of assets of skill and tradition to protect. The subsequent increase in the speed of cotton weaving induced the use of James Hargeaves's spinning jenny, which encouraged the use of Richard Arkwright's water frame as a machine that could produce a complementary yarn. In turn, the bulkiness and expense of the water frame stimulated the use of centralized power and factory organization. Factory pro-duction called up large energy needs at new sites, needs that were satisfied by a combination of various improvements in Newcomen and Savery engines and the post-1781 developments of the Watt engine, which by supplying direct rotary power created the first steam engine to power other machinery. So in 1785 the first steam-powered cotton mill opened in Nottinghamshire, one date for the beginning of the Industrial Revolution.[12]

---

[12] C. Singer et al. (eds.), *A History of Technology* (Oxford: Oxford University Press, 1988); A. P. Usher, *A History of Mechanical Inventions* (Cambridge, MA: Harvard University Press, 1954); F. Klemm, *A History of Western Technology*, trans. D. Singer (London: Allen & Unwin, 1959).

Note the key roles here being played by fuzzy terms such as "induced," "encouraged," and "stimulated." Demand for improved technique may arise at different times for very different reasons, and only seldom is it the result of a sudden great change in wants, a significant increase in national incomes, or a radical alteration in the distribution of consumers' incomes. In the eighteenth century an increase in the demand for a particular type of machine improvement was as likely to arise from a change in policy of the state apparatus (e.g., Peter the Great's decision to be less beholden to Sweden) or as a result of intersectoral shifts of demand by producers wishing to speed up or cheapen or perfect a production process. But at this level of analysis there is no convincing theoretical determination that any such increased demand will ever be met by changes of technique within that production system. Demand may be met by other measures (by introducing cheaper or different-quality labor or material resources from elsewhere), by transfers of superior techniques from elsewhere, by technological change in arenas beyond that of the increased demand (transport improvements may reduce enterprise costs), by imports of the final product from other regions. Or demand may not be met at all, resulting in inflation and a slowdown of the industrial process. What is truly interesting about eighteenth-century Britain that cannot be clearly explained in demand terms is the timing of key breakthrough technological change and its culmination in the superb precision machinery and metallurgical innovations at the end of the century, the breadth of both inventive and transfer activity across industries and sectors, mechanical and civil, and the depth of social involvement in the process of invention and improvement. When harnessed to production for whatever reason (including those of demand), such features reduced the cost and increased the speed of producing goods and thereby operated as determinants of British industrial advancement at the end of the century.

Eighteenth-century Britain witnessed a series of innovations of a nontechnological sort, to which the names of "inventors" cannot be readily attached. The list might include the increased momentum of parliamentary enclosure (1730); the fivefold increase in turnpike road mileage (1750–70); the completion of the Bridgewater canal in 1761 at an approximate cost of £250,000, at least one hundred times the cost of an Arkwright factory at the end of the century; the spread of stagecoaches during the 1780s; and the General Enclosure Act of 1801. Such "institutional innovations" may well have helped determine the conditions of both demand and supply, conditions that influenced the genesis and direction of technical invention, its application and diffusion, and the transfer of technique to Britain from other nations. There is good reason to argue, indeed, that institutional innovation, rather than the relative abundance of capital, was the major factor influencing technical change as well as ensuring that Britain became the greatest beneficiary of the chaos of technology transfers. Large amounts of fixed capital were rarely required for manufacturing industries prior to the 1830s; new techniques could be brought into production as part of replacement and conversion investment, and most fixed capital investment occurred outside the new manufactures – in canals, mines and roads – and by 1790 at least £1.5 million had been spent on the canal system alone. Capital was probably not a constraint on technical change and its implementation and therefore was not the major determinant of the character and trajectory of technological progress in eighteenth-century Britain. Contemporaries pointed to other, institutional features of eighteenth-century Britain, such as the strength of its artisanal and urban cultures and its relative openness and freedom from persecution and associated security of property and income. These political and sociological characteristics are increasingly acknowledged by modern economic historians of this period.[13]

As David Hume observed in one of his most succinct and brilliant essays, the commercial openness of Britain brought more than the simple mercantilist reward of revenue or gold, for it introduced into Britain a challenge and promoted response. The commercial city became a place of social experiment and competitive individualism. Entering the city of Birmingham in 1741, William Hutton proclaimed that individuals there possessed "a vivacity I have never beheld; I had been among dreamers, but now I saw men awake."[14] Wide-ranging institutional innovations, commercial openness, and a diverse urbanism might have been at the core of Britain's technological advantage, but less systematic forces cannot be ignored. Random shocks or stimulants may have been of great importance in invoking any or all of these seemingly "core" features of eighteenth-century Britain. Thus, in the case of London as an emporium of social change and experiment, we might note the stimulative impacts of such long-term forces as the impact of the French wars on London as a center of power, the extended effects of the conversion of erstwhile religious property to civic and commercial uses, and the consequent rise of a new aristocracy and the livery companies, to which we could add the better-known and fairly direct impact of the great fire of 1666 and the subsequent emergence of a restored metropolis. We could postulate that such varied local and historical events and processes created a shift in the role of the leading English city away from that which was essentially "orthogenetic" (i.e., the elaboration of what is already existent) to that which was "heterogenetic" (involving the generation of unorthodox and original modes of thought, new institutional forms, or new spatial arrangements).[15]

Several commentators accepted the advantages of Britain but advocated some program of state interventions to promote an acceleration of change in

---

[13] Joel Mokyr (ed.), *The British Industrial Revolution: An Economic Perspective* (Boulder, CO: Westview Press, 1993).

[14] William Hutton, *A History of Birmingham to the Year 1780* (Birmingham: White, 1781), quotation p. 44; David Hume, "On the Jealousy of Trade," *Essays Moral, Political and Literary* (Edinburgh 1741–2), in vol. 33 of *Works of David Hume* (London: Grant Richards, 1903).

[15] B. F. Hoselitz, "Generative and Parasitic Cities," *Economic Development and Cultural Change*, 3 (1954–5), 78–94.

agricultural and industrial techniques. Thus, in 1757 the neo-mercantilist Malachy Postlethwayt, a great advocate of agricultural improvement, suggested a six-point program for official policy relating to increases in efficiency. The British state should establish manufacturing sites overseas, which would eventually increase trade; promote the diffusion of existing better agrarian technologies to bring down the costs of industrial labor; regulate industrial skilling or training through apprenticeship and by reduction of the power of the guilds; induce the immigration of foreign skill in all cases when "workmen have been molested in their liberties, fortunes and religion"; improve rewards to all "inventions tending to abridge or ease the labour of men"; and generally promote "rivalship" within the nation while inhibiting the movement of skills and machines to other countries. So any foreigners allowed to lodge patents within Britain "ought to be obliged to bring from abroad, and maintain, a certain number of foreign workmen, and likewise take a certain number of national apprentices." It is noteworthy that Postlethwayt focused entirely on improvements in those institutions that were already undergoing significant changes throughout the eighteenth century.[16] Inward transfers of technique to Britain meant that by the end of the century there existed a vibrant, hybrid technological culture, the immediate backdrop to the technical applications associated with the first industrial revolution.

The earlier years of the century were characterized by an open, two-way flow of specific techniques between Europe and Britain, with the latter gaining disproportionately from the permanent settlement of French Huguenot, Dutch, and German migrants. Whereas technicians brought skills, foreign intellectuals and savants who were settled in England, such as Desaguliers (1683–1744), kept in constant touch with European academics and translated the new works of Continental engineers. Such connections not only carried specific knowledge but also were important in the spreading of foreign language sources as well as in the founding of institutions (e.g., the German mathematician John Müller at Woolwich Academy from 1741). Although such technology and related transfers were undoubtedly encouraged by the greater freedoms and opportunities of eighteenth-century England, they were also encouraged by deliberate practices of the British state. An enormous amount of time was spent by the British government on legislation concerning agricultural improvement, navigation, and the development of harbors, rivers, turnpike roads, and canals; of more than three thousand private and public Acts of the 1740s to 1760s, 20 percent were for turnpike schemes alone. The British state was also involved in the stimulation of colonial technology transfers. In 1749, for instance, Parliament passed an act exempting from duty all raw silk that was certified to be the product of Georgia or Carolina. A bounty was offered for the production of spun silk, and an Italian, G. Ortolengi, was hired to proceed

---

[16] M. Postlethwayt, *Britain's Commercial Interest Explained and Improved* (London: D. Browne and J. Whiston, 1757), quotation at vol. 2, p. 414.

to Georgia to instruct the colonists in the Italian methods of sericulture and silk throwing. In 1761 the London Society of Arts offered additional premiums for the production of good-quality American cocoons. Under Italian methods such stimuli led to a filature complex for reeling, doubling, cleaning, and twisting.[17]

Finally, British social institutions encouraged a long-term process of technological improvement, of incremental adjustments to key inventions. Nowhere else within eighteenth-century Europe was technological investigation and experiment so open to the talents and needs of such a large cross section of the population. This might be most concisely indicated in the patent data for the last decade of the century. More than six hundred patentees can be identified for the 1790s, representing much if not the great bulk of the incremental improvements surrounding early industrial revolution techniques. More than 50 percent of such patents were directly concerned with the improvement of the new machinofacture, from the generation and conversion of motive power through the industrial processing of raw materials to the invention of new machine tools and the development of new manufacturing chemical processes. Of the patentees, the majority were either skilled artisans or were new manufacturers and engineers, many of whose backgrounds would have included an apprenticeship to the trades.[18]

Thus, incremental technical improvements originated at relatively humble levels of British society, a phenomenon that might well lie at the explanatory hub of McCloskey's claim that "ordinary inventiveness was widespread in the British economy."[19] Of all the European nations of the eighteenth century, Britain was the one that exhibited the least in the way of barriers to continuous and accumulative technological change. Wars did not close off the economy, transfer mechanisms abounded, social distances were surely less than in any European contender, and the new techniques were still relatively simple and did not demand enormous amounts of fixed capital, a deep knowledge of abstract science or mathematics, or an extensive transport and communications infrastructure to support their diffusion and application.

EUROPEAN LIMIT: RUSSIA AND TECHNOLOGICAL PROGRESS

The contrast between Russian industry and society prior to the reign of Peter the Great (1672–1725) and the condition of the empire during and after

---

[17] L. P. Brockett, *The Silk Industry in America* (New York: Harper, 1876).

[18] Inkster, *Science and Technology*, pp. 80–6; Patrick O'Brien, Trevor Griffiths, and Philip Hunt, "Technological Change during the First Industrial Revolution: The Paradigm Case of Textiles, 1688–1851," in Robert Fox (ed.), *Technological Change* (Amsterdam: Harwood, 1996), pp. 155–76; W. H. G. Armytage, *A Social History of Engineering* (London: Faber and Faber, 1961).

[19] D. McCloskey, "The Industrial Revolution, 1780–1860: A Survey," in *The Economic History of Britain since 1700* (Cambridge University Press, 1981), vol. 1, *1700–1860*, ed. R. Floud and D. McCloskey, pp. 103–27, quotation at p. 117.

the reign of Catherine II (1762–96) was probably as great as could be found anywhere else in Europe. At the onset of the century, Russia represented the limit of Europe in a sense that goes far beyond location, topography and climate. The many pre-Petrine attempts at industrial and technological modernization, including those of Peter's father, Alexis, had failed to disturb traditional social institutions or influence the general course of the economy. Special favors granted to English, Dutch, and German settlements at Archangel and elsewhere cost far more than they returned, as did the sporadic importations of foreign officers and technicians into the Russian armed forces (associated with the construction of large-scale armaments production at Tula under Dutch expertise) or the foundation of a special German settlement in Moscow. Indeed, by the late seventeenth century, the foreigners of the latter settlement (the Sloboda), although possessed of workshops, mills, and ironworks, were increasingly alienated from Russian industry and were marked out as strangers. The severe curtailment of freedoms and liberties demonstrated in Muscovite society, extreme even for later seventeenth-century Europe, and the absence of any systemic diplomatic presence in Europe added to problems of distance and terrain in inhibiting significant technological modernization.[20]

Nevertheless, it has been cogently argued that early attempts by foreign entrepreneurs under the auspices of state inducements did have some longer-term impacts of relevance to any analysis of the greater success of the eighteenth century. English, Dutch, and Danish owners of Russian metal-producing concerns, especially the furnaces and ships of the Serpukhovo-Tula producing area, oversaw the installation of power-operated iron manufacturing, encouraged by the military demands of a Russian government concerned with its Baltic rivals: Poland and Sweden. The prime task of foreign entrepreneurship was to manage the combination of a core of skilled foreign specialists and a mass of relatively unskilled Russian workers. Although the failure of the Tula enterprises at the beginning of the century can be best explained in terms of the shortage of timber for fuel, it seems reasonable to judge that the early encouragement of foreign ownership and management of more advanced technologies yielded a "European" yardstick for future Russian industrialists and offered something of an object lesson to Russian rulers.

In the arena of modernized but enclavist industrialization, eighteenth-century Russia was to exhibit a motley agency of foreign entrepreneurship, the Russian state, and private initiative, about which few secure generalizations can be construed; the precise character of ownership and production depended on all of regional history, physical locale, and production sector as well as on the contrariness of state policy. As an example, Peter's reasonable but fairly sudden decision not to rely on foreign manufacturers to equip the Russian army and navy meant that the major metallurgical project of the century, the

[20] J. T. Fuhrmann, *The Origins of Capitalism in Russia: Industry and Progress in the Sixteenth and Seventeenth Centuries* (Chicago: Quadrangle Books, 1972); P. T. Lyashchenko, *History of Economy of the USSR*, 3 vols. (Moscow: GIPL, 1956), vol. 1.

development of the Urals, excluded *direct* foreign ownership and participation in favor of a collusive combination of a monopolistic Russian private
enterprise dynasty (the Demidovs) with state participation. The greater efficiency of the later eighteenth century tended to emerge from neither state
industry nor foreign enterprise, but instead from the association of private
Russian capital and hired foreign expertise. This was a more astute partnership, depending for much of its potency on the earlier interventions of the
state in the areas of military production, transport improvement, and mobi-
lization of serf, frontier, and free labor forces.[21]

On several levels, the reign of Catherine and beyond can be fairly compared with the history of large areas of Western Europe. The movement of
townsfolk had been freed up at the beginning of the century, internal customs
removed in the 1750s. Russia was no more strictly "mercantilist" than most
of Europe; there was a national market for staple products, guilds were no
stronger than elsewhere, and Catherine's liberalism compared well with the
European governance of the ancient regime. The Russian export sector was
buoyant, producing supplies of iron, textiles, timber and other naval stores to
the major nations of France and England. The empire was self-reliant in
basics, and there is a great deal of evidence to suggest that in Russia the emergent "bourgeois" class was represented in serf mobility, wealth, and innovativeness and that most industries were *decreasingly* dependent on direct aid.
Furthermore, the general European enlightenment had passed into Russia by
the end of the century. The empire published some 8,500 titles in the second
half of the century and boasted a savant class linked in a multitude of ways
to the intelligentsia of Western Europe and America.[22] Is it possible to distill from this seeming success story – a story that would have to acknowledge
the progressive impacts of massive population growth, territorial expansion,
and a much accelerated internal colonization process – an argument about
technological progress? Did European technological progress transfer effectively to the new Russia, and was this ancillary to a more general social and
cultural advancement or a partial cause of such advance?

Following a tradition set by Voltaire and the French physiocrats, the Russian technological watershed is conventionally demarcated in the young czar
Peter's Dutch trip and European missions of 1697 and in his "sensible concern"

[21] W. L. Blackwell, *The Beginnings of Russian Industrialisation* (Princeton, NJ: Princeton University Press, 1968); J. P. McKay, "Foreign Enterprise in Russian and Soviet Industry: A Long Term Perspective," *Business History Review,* 48 (1974), 336–56; Roger Bartlett, *Human Capital: The Settlement of Foreigners in Russia, 1762–1804* (Cambridge University Press, 1979); Alexander Gerschenkron, "Soviet Marxism and Absolutism," *Slavic Review,* 30 (1971), 853–69.
[22] A. Lentin, *Russia in the Eighteenth Century* (London: Heinemann, 1973); Jerome Blum, *Lord and Peasant in Russia* (Princeton, NJ: Princeton University Press, 1961), especially pp. 277–441; R. Portal, "Manufactures et classes sociales en Russie au XVIIIe Siècle," *Revue historique,* 201 (1949), 79–97; A. A. Zvorikine, "Inventions and Scientific Ideas in Russia: Eighteenth and Nineteenth Centuries," in G. S. Métraux and F. Crouzet (eds.), *The Nineteenth Century World* (New York: Mentor, 1963), pp. 254–79; P. Dukes, "Russia and the Eighteenth Century Revolution," *History,* 56 (1971), 371–86.

that in the conflict of the preceding year against the Turks his ships "had been built entirely by the hands of foreigners."[23] The result was a bundle of reforms designed to encourage the movement of foreign experts into Russia to aid the modernization of an independent army and navy, to build a symbolic and effective capital of St. Petersburg, and to use the technologies of the Atlantic to hold back the onslaughts of both the Baltic (the Northern Wars of 1700–21) and the Ottomans (1711–13). A result was the Manifesto of 1702, which invited foreigners to Russia and offered interest-free loans, subsidies, and monopolies (e.g., in silk production), tax exemption, and the legal formation of foreign companies of merchants and artisan guilds; there was also the later protection of the tariff of 1724. Of greater significance to the two hundred or so factories thus established were the state's forced labor interventions, including the assigning of peasants to factories and the purchase of serfs and even whole villages. By 1725, some fifty-four thousand peasants had been assigned to the metallurgical works alone. Although this system operated alongside similar encouragements to Russian entrepreneurship itself (cheap labor, introduction of government trading monopolies and restrictions on foreign trading, raising of the prices of the government's iron, copper, and other staple requirements, and transfers of some state-operated mines and enterprises to private hands), there is little evidence that the Petrine efforts yielded substantial overall technological or commercial results.

As Voltaire originally stressed, Peter's success lay rather in attuning "his people to the manners and customs of the nations which he had visited in his travels," in the establishment of training and organizational facilities for longer-term advancement (from mathematical schools to the *Manufakturkollegiya* and the *Berg* [mining]-*kollegiya* in 1719, both designed to finance new enterprises and regulate production levels and standards), and in the strengthening of the military arm of the state. Again, the ultimate impacts of public projects on the private sector will forever be uncertain. Thus, the great project of building St. Petersburg itself, coinciding with the failures at Olonets and in the Tula region, meant that the new Urals metallurgy needed to be linked to the new capital by water routes. A vast number of metal factories using migrant labor emerged along the many routes of the canal and river system during 1709–29, and modernized production here meant that early in the century Russia became the major producer of iron in Europe, approximately 60 percent of the total coming from state and private enterprises in the Urals.[24] Furthermore, Peter undoubtedly created nodes of technological modernity. The Petrovsky factory (state owned from 1703) became the largest in the Olonets iron region, with four blast furnaces, a water-powered cannon-

---

[23] M. de Voltaire, *The History of the Russian Empire Under Peter the Great*, 2 vols. (Aberdeen: J Boyle, 1777), vol. I, quotations at pp. III–12.

[24] B. H. Sumner, *Peter the Great and the Emergence of Russia* (New York: Collier Books, 1962); I. M. Matley, "Defence Manufactures of St Petersburg, 1703–1730," *Geographical Review*, 71 (1981), 411–26.

borer, an anchor workshop, wire-drawing machinery, and a weapons work-shop, in total employing more than one thousand workers and producing 60 percent of the region's pig iron by 1725. Moscow and, from 1710, St. Peters-burg were centers of nonmetal industry innovations. Manufacturers of sail, uniforms, rope, and linen for export all prospered under state encourage-ment. The tardiness of the entry of private capital into new ventures may not have been for technological or even institutional reasons but simply a result of the war with Sweden and its allies. The war created many alternative in-vestments for private wealth, a factor exacerbated by a lack of legislation to secure contracts and property rights in certain key industries. For instance, in iron manufacture every entrepreneur had to enter into a separate contractual agreement with the state, and only with the publication of the *Berg-kollegiya* privilege of 10 December 1719 was the property of iron manufacturers guar-anteed. From 1722, owners of ironworks were permitted to buy serfs. Such leg-islative changes meant that most of the water-driven machines and works associated with blast furnaces and metal rolling, or with crushing and grinding equipment, appeared in Russia in the post-Petrine years. After the death of Peter in 1725, Russian merchant and ex-serf entrepreneurs were increasingly joined by nobles in a variety of industrial activities, especially during the reign of Catherine II. Some evidence suggests that both the public and the private sectors experimented with the advanced industrial models first introduced by foreign technical experts in earlier years.[25]

Catherine's Manifesto of 22 July 1763, distributed throughout all European countries, resulted in a period of intensive skill migration; for example, be-tween 1763 and 1769 some thirty thousand people migrated from Germany alone, establishing more than one hundred *kolonii*. European migration was encouraged through a renewed round of privileges, such as the free exercise of religious faith, as well as the more material inducements of land tax exemptions for ten to thirty years, interest-free loans, exemption from military service, membership of guilds and citizenship, land awards, and so on. Catherine made special allowance for those foreigners "who undertake to build factories and plants," who could obtain loans for the establishment of "factories such as have not yet been built in Russia," gain permission to purchase serfs and peasants, and export produce without export duty.[26]

It is difficult to evaluate the effect of Catherine's system on technological progress in Russia. Foreign expertise appeared in a variety of forms. Foreign entrepreneurship served to transfer advanced techniques into cotton printing

[25] A. Kahan, "Entrepreneurship in the Early Development of Iron Manufacturing in Russia," *Economic Development and Cultural Change*, 10 (1961–2), 395–422; Kahan, "Continuing in Economic Activity and Policy during the Post-Petrine Period in Russia," *Journal of Economic History*, 25 (1965), 54–73; B. D. Wolfe, "Backwardness and Industrialisation in Russian History and Thought," *Slavic Review*, 26 (1967), 177–203.

[26] Karl Stump, *The Emigration from Germany to Russia in the Years 1763–1862* (Nebraska City, NE: Amer-ican Historical Society of Germans from Russia, 1978).

(Christian Lieman) and machine construction (Charles Baird). Associations of foreign and Russian capitalists were not unknown, as in the case of the silk manufacturing of Benjamin Müller. Foreign entrepreneurs, unlike the artisans, were placed under the authority of the Chancellery of Guardianship, whose official records show foreign factories established in ribbon, silk, lace, and cotton manufacturing, chemical dyeing, wax manufacture, velvet cloths, and so on, mostly on the understanding that the manufacturing would actively assist in the training of Russian apprentices and workers. However, few such ventures seemed well planned, and none was independently successful in the long term. Only at the turn of the century did individuals such as Charles Baird (the iron foundry at St Petersburg) begin to prosper and transfer modernized techniques more efficiently.[27] But, generally, transfer of techniques through the use of skilled artisans was probably of more importance than the short-term opportunistic adventures of European merchant entrepreneurs. For instance, German artisans dominated luxury handicrafts around St Petersburg, and the transfer of tacit knowledge and attitudes was then effected through local communities, guilds, and associations rather than model establishments. The wealth of foreign artisans best demonstrated the prosperity that might come with advanced methods and organizations.[28]

A brief focus on the most favored arena, but one that truly lay at the very edge of Europe, may yield some insight into the limitations inherent in the Russian case. In the Urals metallurgical project, iron ore, charcoal, and water power were in abundance, as was a supply of cheap, often fugitive peasantry, so that by 1800 some 80 percent of Russian iron came from the region. The informed exploitation of the region was initiated in the ore analyses of European scientists, sustained by the migration of European expertise to the Urals works and mines, and maintained there through adoption by both the Russian state and private enterprise. Natural resource expeditions prior to the 1740s included those of Daniel G. Messerschmidt (1685–1736), Vitus Bering (1681–1741), and George Steller (1709–1746), expeditions that together helped to define the metallurgical potentials of the Urals. Swedish exiles from the Northern War (1700–21) and copper-smelting and assaying specialists from Saxonia populated the region in the early part of the century, and in the 1720s and 1730s they were joined by groups of Saxon and Dutch specialists commissioned by the government. Contractual obligations in such state-owned works as Kamensky and Neviansky included that of teaching technical skills to Russian workers in exchange for high salaries, free fuel, and residence. Tests carried out at the Perm board of mining from 1731 seem to have identified a reasonably successful transfer process: foreigners were removed from mining

---

[27] Bartlett, *Human Capital*, pp. 164–79; S. J. Tomkireff, "The Empress Catherine and Matthew Boulton," *The Times Literary Supplement*, 22 December 1950.

[28] Bartlett, *Human Capital*, pp. 143–79; Olga Crisp, *Studies in the Russian Economy before 1914* (London: Macmillan, 1976); R. Bartlett, "Diderot and the Foreign Colonies of Catherine II," *Cahiers du monde russe et soviétique*, 23 (1982), 221–32.

activity in specific sites, and an inspection was made of all operations under solely Russian management. European experts operated in private mines also, as in the case of the German metallurgist and scientist G. E. Gellhert at Demidov's works in the 1750s; Gellhert acted as an employment agent for foreign specialists. Similar tasks were undertaken during the 1730s by the Dutchman Wilhelm de Gennin (1676–1750) and the Hungarian Simon Kachka. Under such occidental auspices, more than seventy enterprises were erected in the Urals in the first half of the century, almost equally producing copper and iron; in the following half century an additional one hundred plants were erected. Here the role of the State was essential in at least one respect: thirty thousand peasants were ascribed to the Urals metallurgical plants under Peter alone. Private enterprises, such as R. F. Nabatov's Irginsky Works, could also exploit state peasants or ascribed peasants from state works and could at times purchase landless peasants as well as employ recent settlers and fugitives, including the "Old-Believers" of Nizhny Novgorod and elsewhere. Settlers were trained in metallurgical techniques by skilled workers removed from more established concerns around Ekaterinburg. So Ekaterinburg and Yagoshikhinsky copper makers were brought in to train the motley groups of laborers organized at Irginsky.[29]

The technical history of the area is complex. Many of the early enterprises, such as the Pyskorsky copper-smelting works at the Ekaterinburgsky works (1723–30), had several independent production units (shops, storehouses, dams, waterwheels, etc.) the core of which shifted as particular ore sites became exhausted. Nizhne Tagilsky works produced its first cast iron from 1725, and by the second half of the century some thirty different production systems functioned at the site.

Among the variety there seem to have been several general technological characteristics of the region. Self-resourcing was common; each enterprise supplied and owned its own ore, timber, workshops, fuel supplies, allotments for workers, transportation, and shipping. Many enterprises organized iron forges and copper-smelting furnaces in one and the same shop, and several of them diversified into gold and semiprecious stone mining. Significantly, in the eighteenth century, Urals enterprises exhibited a fairly closed technological cycle. Ore extraction and fabrication into metal goods were undertaken within the same district, with the basic blast furnaces and finery works being divided between specialized enterprises. Finery works were situated downstream from blast furnaces, although the river transportation of pig iron was not a major constraint in the system because of the cheapness of labor and cartage by

---

[29] R. Portal, *L'Oural au XVIIIe siècle* (Paris: Institut d'études slaves, 1950); J. M. Crawford (ed.), *The Industries of Russia* (St. Petersburg: Published for the World's Columbian Exhibition, Chicago, 1893); W. L. Blackwell (ed.), *Russian Economic Development from Peter the Great to Stalin* (New York: Scribner's, 1974); M. C. Kaser, "Russian Entrepreneurship," in P. Mathias and M. M. Postan (eds.), *The Cambridge Economic History of Europe: VII, The Industrial Economies, Part 2* (Cambridge University Press, 1978), pp. 416–93.

assigned peasants and serfs. The organization of Urals techniques suggests strongly that optimization of the effective water power was always of greater concern than the efficient use of labor. Thus, a typical modernized Urals metallurgical works ran as a string falling away from the major dam works: blast furnace, copper-smelting shop, forge shops, followed by tin, iron rolling, anchor, hammer, and sheet-iron shops, perhaps followed by a wire mill. Essential nodes on such a string were the outcome of European experts, beginning with the low breast or undershot wheels of German, French, and Swedish technicians. Most dam and reservoir construction was under Russian management and tended to be sited on high banks to permit the building of short, wide, and high dams and a system of reserve upper works to regulate the main water levels. From the 1730s, waterwheels tended to be of Swedish design, with the local Russian carpenters working under the guidance of craftsmen trained in Sweden or under Swedish supervision. Early blast furnaces were English or Swedish, constructed either following European "proportions" or under European supervision; thus, the Kamensky blast furnace was erected by the English engineers Jarton and Parkhurst. Most blast furnace shops were equipped with one or two blast furnaces, and few iron-smelting units operated with more than one shop, each connected with wooden bridges along which ore, wood, and fluxes were transported. Generally, the dimensions of blast furnaces increased throughout the century, with average daily output from a furnace reaching around 400 poods (about 7 tons) by the end of the century. Initially, iron processing was dominated by the so-called German forge process, using a water-driven hammer, and thereby depended for its success on the strength and workmanship of the finers, who had to judge the blast, add charcoal, separate out the spongy semirefined iron *(zhuki)*, and test the halfbloom, mostly by eye and entirely by experience. The evidence seems to be that by the end of the century the basic German forging and hammering process had been significantly modified, with each works having its own version of the fundamental European technology, differing mostly in mode of operation rather than an altered manner of construction. In 1782 a British mechanician, Joseph Gill, erected a sheet-iron rolling shop at the Chermozsky works and later (1798) built a steam ore-lifting machine in the Gumeshevsky copper mines.[30]

Clearly, European metal working technologies were introduced periodically through such key agents as Gennin or Kachka (mentioned earlier), but in the private sector some transference appears to have also come about through the European visitations of members of the Demidov dynasty, especially during the 1750s, when the three brothers Demidov studied metallurgy at Freiberg,

[30] Russian Academy of Science, *Industrial Heritage of the Ural* (Ekaterinburg: Ural Branch Institute of History and Archaeology, 1993), pp. 1–37; Russian Academy of Sciences, *Conservation of the Industrial Heritage: World Experience and Russian Problems, Final Proceedings of International Intermediate Conference, September 1993* (Ekaterinburg: Institute of History and Archaeology, Ural Branch, 1994).

examined mines, saltworks, and iron manufacturers, and reported on bellows technology and smelting furnaces. In England, among other things, the brothers attended a three-week course on experimental philosophy delivered by James Bradley at Oxford, and followed this with fruitless attempts to penetrate the well-kept secrets of Birmingham and Sheffield steel production. The transfer line was clearly drawn; common steel and razor, saw, and hammer shops were demonstrated, but cast steel production remained a closed book.[31]

In the later eighteenth century, Urals metallurgical technique represented the effective geographical frontier of the European system but appears to have been to an extent successful, contributing to the greater part of Russian output and to metal exports to the more-sophisticated economies of the West, as well as an intermediate product to the building of St. Petersburg and other grandiose imperial projects. The main ingredient of success seems to have been the combination of abundant materials and labor with an array of transfer mechanisms, ranging from the innovations of Russian entrepreneurs to the migration of skilled artisans from all over Europe and from the other metallurgical areas within the empire. Thus, the diaspora of skills from the early Westernized technique at Tula was possibly a key element in the first years of the century.

BEYOND EUROPE, I: JAPAN

Eighteenth-century Japan both inherited and strengthened the policies of national stability set out by the early Tokugawa rulers from 1603, following the long years of civil war. The early Pax Tokugawa was forged from war and tempered in peace. The rights of the Emperor, which had been delegated during the civil war, were transferred as the rights of the new lay government. Will Adams, English pilot of a Dutch ship stranded in Japan, could write of the nation in 1611 that "there is not a land better governed in the world by the civil police." But this was prior to the reforms of 1634–43, which entailed the suppression of Christianity, the relegation of the Emperor to Kyoto, forced alternate residence *(sankin kotai)* or periodic hostage of the daimyo lords at the capital, Edo (Tokyo), and a series of enactments that enforced an internal passport system, monitored the marriage alliances of the daimyo, limited castle building, forced large, debilitating construction projects on individual lords, and finally, in 1643, prohibited the alienation of land as a check on han power and as an affirmation of faith in the moral economy arising from the small peasant proprietor. The reforms of the Kyoho era (1716–36) further restricted dissent by suppressing free speech and publishing. So the first clause

---

[31] A. S. Cherkasova, "The Urals and Europe in the Eighteenth Century," in S. V. Ustiantsev et al. (eds.), *Russia and West Europe: Interaction of Industrial Cultures, 1700–1950* (Ekaterinburg: Institute of History of Material Culture, 1996), pp. 22–7.

of the ordinance of 1722 dictated that "all books, whether they be Confucian, Buddhist, Shinto, medical, or poetic, and which include excessively mendacious or heterodox discourses, shall be strictly forbidden."[32]

Although hardly an open, innovative national system for technical change along European lines, eighteenth-century Japan possessed the advantages of a very large politically and demographically stable population (around thirty million), densely settled on alluvial island plains and seldom victim to marauding, frontier invaders, whether foreign or Japanese. This may well have been important to the establishment of a firm land tax base for the regime. At the same time, relative stability meant that the *sankin kotai* system at the level of the domain itself created a complex of large cities composed of lower lords, middling samurai, merchants, and temple populations, surrounded beyond the outer moats by an increasingly commercialised agriculture and large groups of lower merchants, artisans, and lower samurai. So within this somewhat involuted urbanism, the merchant class of the advanced regions progressed with the regime rather than against it: early official requests ensured that merchants set up business in the city to provide financial services for hostages and retainers and to import rice, silk, oil, and other necessities. In many cases, a fusion of merchant and samurai interests took place. The most centrally located merchants, the *goyo shonin,* could gain tax exemption, some autonomy, and something of social advantage (legally they remained at the foot of the social ladder) from the rising consumerism within the inner, upper circles of the city. Within the city, the financial pressure on samurai, dispossessed of warfare, also produced important social results. Many of those who could not shift to a bureaucratic function entered commerce or agriculture or converted their debts to the chonin commoners into marriage alliances or adoption contracts. The educated middling samurai also chose a role as educators of the administrative class, thereby strengthening a rise of the *jinzai* (men of talent) and the notion of an urban meritocracy *(jinzaishugi).* Thus, both despite and to an extent because of the control instruments of the centralized state, Togugawa society by the eighteenth century did contain novel sites for mobility, ambition, knowledge diffusion, and innovative behavior, set within a very large population, the great majority of whom lived a life of traditional activity and local imagination.[33]

The principal technical improvements of the Tokugawa years were found

[32] B. Hanley and Kozo Yamamura, *Economic and Demographic Change in Preindustrial Japan, 1600–1868* (Princeton, NJ: Princeton University Press, 1977); C. R. Boxer, *The Christian Century in Japan, 1549–1650* (Berkeley: University of California Press, 1951); Mitsutoshi Nakano, *Edo bunka hyobanki: Gazoku yuwa no sekai* (Directory of Edo Culture: A World of Harmony between the Refined and the Popular) (Tokyo: Chuo Koronsha, 1992); Bito Masahide, "Society and Social Thought in the Tokugawa Period," *Japan Foundation Newsletter,* 11 June 1981, 1–9.

[33] David Kornhauser, *Urban Japan* (London: Longman, 1976); E. H. Norman, *Origins of the Modern Japanese State* (Washington, DC: Institute of Pacific Relations, 1940); T. G. Tsukahira, *Feudal Control in Tokugawa Japan: The Sankin Kotai System* (Cambridge, MA: Havard University Press, 1970); C. P. Sheldon, *The Rise of the Merchant Class in Tokugawa Japan, 1600–1868* (New York: Russell, 1958).

not in the great *jokomachi* but rather in the rural areas, in the increased use of commercial fertilizers, irrigation, new threshing machinery *(semba-koki),* multiple cropping, and crop diversification. The wooden framed thresher appeared first around Osaka prior to 1700 and replaced the *koki-hashi* (large chopsticks) in the drawing of rice stalks. Any labor so saved at harvest appears to have been used to plant a winter crop immediately after the late harvest. It seems a general rule that most technical improvement in the eighteenth century represented diffusion of existing best methods, required little in the way of fixed capital, was based on small units of production, and was land-rather than labour-saving. There is no convincing evidence of any general shortage of labor in Japan.

Although much in the way of irrigation, drainage, and flood control technologies was constructed from human labor at the village level or under orders from the lords, multiple cropping, seed selection, fertilizer use, and the use of improved hoes, water pumps, and plows, as well as the introduction of draft animals, remained with the peasant family. Primarily, it was the family farm of one hectare or less that also provided the skilled labor for the by-employments of textiles production, tea and oil pressing, grain milling, pottery, and the manufacture of rice wine, paper, and ink. Fertilizer, from night soil and sewer mud to purchased oil cake and dried sardines, might involve the expenditure of as much as one-half of an eighteenth-century peasant's total annual cash expenditure. Fertilizers undoubtedly reduced the labor time previously devoted to tramping of leaves, grasses, or ash into the fields. But when combined with better irrigation, the productive use of fertilizers demanded new tasks of labor: seed treatment and selection, row planting, intensive weeding, the application of insecticides (oil), attention to frost protection, and so on. Considering these techniques together with the conversion of dry field to paddy and a move to higher-yielding rice varieties and double cropping, it seems evident that acreage, yields, and labor intensity all increased significantly in the absence of breakthroughs in mechanical technology.[34]

The intensification of wet rice paddying was complementary to the growth of cash crops (silk and cotton, rapeseed, indigo, tobacco, tea, and sugar) and to handicraft production in rural areas. In many regions, handicrafts were organized by merchants using both familial and wage labor. Merchants supplied spinning frames, reeling machinery, and looms to farmers and paid them for their work by the piece. Such a putting-out system dominated the specialized regions of handicraft production, such as the Kinai cotton region, the silk production of Fukushima, and the sugar of Satsuma. The imperatives of the factory were missing, and the very rare use of water power was mostly for

---

[34] T. C. Smith, *The Agrarian Origins of Modern Japan* (New York: Athenaeum, 1966); T. Furushima, *Kinsei Nihon nogyo no tenkai* (Tokyo: University of Tokyo Press, 1963); Francesca Bray, *The Rice Economies* (Oxford: Blackwell, 1986).

grinding and crushing of rice, wheat, and some minerals.[35] As in agriculture, technological improvement in the many rural industries resulted primarily from the internal diffusion of existing techniques.

Although apprenticeship and urban guilds were of some importance, during the eighteenth century the freedoms of village production were probably of greatest significance, where putting-out systems spread silk-reeling, cotton-weaving, and paper-making skills from region to region. This relatively uncomplicated level of interaction and inducement was sufficient to promote some spread of water-powered multi-spindled wheels in silk throwing, and around the same time it fostered the spread of ditch and dike techniques (replacing hand labor and tides) in salt manufacture on evaporation fields. Within fiefs, spread effects could be quite dramatic, as in Choshu, where there was a move from peasant household production of cotton for domestic use (using raw cotton imported from other regions) to the export to Odessa and elsewhere – all within a period of about fifty years. This increase in quantity and quality of supply was a result of the diffusion of cotton cultivation, commercial ginning, spinning, and weaving across the entire fief.[36] Periodically, the financial imperatives of the domain added to this more-pervasive technological system. Han authorities commonly required the villages to report the industrial productions of each family as a routine aspect of fiscal policy and as a measure of production potential if needed. For instance, financial pressures in the domain of Yonezawa in northern Japan led to the borrowing of money from Edo merchants to finance the invitation of silk producers from neighboring areas to advise on best practice, the provision of loans to farmers to establish mulberry bush growing, and the establishment of silk nurseries and indigo plantations.[37] It is difficult, however, to establish the overall relative importance to national production of han projects, urban groupings, peasant productions, and merchant-artisan combinations in the promotion of learning and the diffusion of knowledge. In the transfer of knowledge and technique into Japan from other places, it is almost certain that the court, the administration, and the cities figured most prominently.

The years of seclusion, or *Sakoku* (1639–1720), had allowed little in the way of coherent technological learning from other cultures, confining formal interactions with the West to a small group of *roju* or government inner

---

[35] Ryoshin Minami, "Water Wheels in the Pre-industrial Economy of Japan," *Hitotsubashi Journal of Economics,* 22 (1982), 1–5; Minami, *Power Revolution in the Industrialisation of Japan* (Tokyo: Kunokuniya, 1987); H. K. Takahashi, "Quelques Remarques Historique Sur la Repartitition Sociale de la Propriete Funcier Au Japan Depuis Le XVI Siece," *International Conference of Economic History,* 3rd section, Munich, 1965, 421–34; M. Araki, *Bakuhan taisei shakai no seirsu to kozo* (Tokyo: University of Tokyo Press, 1959).

[36] Thomas C. Smith, "Farm Family By-Employments in Preindustrial Japan," *Journal of Economic History,* 29 (1969), 687–715.

[37] Tessa Morris-Suzuki, *The Technological Transformation of Japan* (Cambridge University Press, 1994), pp. 28–9.

counselors. There is no evidence of any linkage between agriculture and rural industry and the spread of Western knowledge, and indeed the new treatises on agricultural techniques, especially the *Nogyo Zencho* and *Nogu Benri Ron* of the years after 1697, owed everything to developments in Chinese agriculture.[38] During the years of *Rangaku*, or Dutch learning (1720–1830), a more effective interaction emerged. In an attempt to solve administrative problems, the regime relaxed the exclusion rules and permitted some inflow of Western medicine, science (including Newtonian dynamics), and technologies. With an increased emphasis on *Jitsugaku* (the study of real things) a flow of discrete information took place not only through Dutch settlement but also through European publications imported directly from China. In 1796 the first Dutch-Japanese dictionary was published, and this was followed by a series of academy and school foundations in which Western learning was included. An important component of Rangaku that developed from that time was *Bussangaku*, or "knowledge of production," which emphasised practical rather than abstract themes in Western culture. The formal text was by no means the only artifact of Rangaku. From the 1720s Dutch ships brought into Japan almanacs, dictionaries, lexicons, foreign flora and fauna, globes, maps, paintings, telescopes, magnets, microscopes, sextants, scientific instruments, magnifying glasses, electrical equipment, thermometers and barometers, and clocks, as well as fire engines, lathes, air pumps, and cannon. Together with books introducing the work of Newton or Linnaeus and Western techniques of copper mining and mine drainage, many such symbols of the West were formally requested of the Dutch by not only the shogunal court but also by the College of Interpreters in Nagasaki, local officials, commissioners, and governors.[39]

The avenues of information dispersal were certainly widened after the early eighteenth century. At times, the libraries of Deshima surgeons and officials were ceded or auctioned to the College of Interpreters or to the physicians of the Edo court. The Japanese translation of the work on anatomy by J. A. Kulmus, produced in five volumes in 1773 as *Kaito Shinsko*, probably owed its origin to the availability of such literature. By the time of the residency of K. P. Thunberg (1743–1822) in the 1770s, some fifty interpreters were associated with the Deshima outpost, several of whom acted as go-betweens for Japanease scholars of astronomy and medicine. Without doubt, this thin thread of Rangaku spread to Edo and through the urban cultural system more generally. In 1788 Otsuki Gentaku (1759–1827) published his *Rangaku Kaitei*

[38] C. R. Boxer, *Portuguese Merchants and Missionaries in Feudal Japan* (London: Macmillan, 1986); Bray, *Rice Economies*, pp. 210–17; George Sansom, *A History of Japan*, vol. 3: *1615–1867* (London: Cresset Press, 1963).

[39] M. Muruyama, *Studies in the Intellectual History of Tokugawa Japan* (Tokyo: University of Tokyo Press, 1974); J. Maclean, "The Introduction of Books and Scientific Instruments into Japan, 1712–1854," *Japanese Studies in the History of Science*, 13 (1974), 9–68; G. K. Goodman, *Japan: The Dutch Experience* (London: Macmillan, 1986).

(The Ladder to Dutch Learning) and established an academy in Edo. By the late eighteenth century Rangaku academies and han schools ranged from science and the fine arts to medicine and military technology, and from the middle of the century *Ranpeki,* or "Dutch mania," embraced a number of scholars, whose educational institutions were to spawn many of the intellectuals of the later period of Meiji industrialization (1868–1912). Thereby, the "indiscriminate culture vultures" of the cities had available to them at least some avenues for the absorption of Western knowledge and technique.[40]

But until the early nineteenth century the spread of Western knowledge seems to have had little impact on either techniques of production or general intellectual and ethical discourse. Western chemistry and mathematics were all but absent from the eighteenth-century scene, translations of scientific texts omitted Western mathematical technicalities, and the small coteries of intellectuals, physicians, and officials who pursued Western ideas were not supported by a system of information dispersal or social recognition. It is debatable whether the notion of *kyuri* (the purpose of things) effectively captured the discernment of a few Japanese scholars that behind the new Western practices and results lay a wider and confused ethos of experimental investigation and intellectual challenge. Thus, Shizuki Tadao's *Rekisho Shinsho,* which was designed to introduce elements of the Newtonian system, wrestled with an absence of clear meanings for such terms as "particle" or "mechanics."[41]

BEYOND EUROPE, II: INDIA AND CHINA

From the point of view of knowledge of technique, the great civilizations of India and China were hardly inferior to the peripheral island system of Japan. In contrast to the effective isolationism of Japan, both these major systems were periodically interactive with European technologies, and recent research has emphasized the complexity of their agrarian and industrial sectors. In late Mughal India, differentiation of the peasantry was significant and evolved from an agriculture that yielded a great range of crops in both spring and autumn harvests. Over large areas, perhaps some forty crops might be produced. New crops such as tobacco or maize were spreading and in the case of sericulture the Hindu traditions that formally forbade the taking of life did not prevent the inevitable death of the silkworm. It seems that the diffusion of superior techniques – ranging from the bamboo seed drill to pin drum gearing and the parallel worm for the cotton gin to roller crushing in sugar mills – was not substantially hindered by features of the traditional value system. Seventeenth-

---

[40] Jeane-Pierre Lehmann, *The Roots of Modern Japan* (London: Macmillan, 1982), quotation at p. 110.

[41] S. Nakayama, *Academic and Scientific Traditions in China, Japan and the West* (Tokyo: University of Tokyo Press, 1984); S. Nakayama, D. Swain, and N. Yagi (eds.), *Science and Society in Modern Japan* (Tokyo: University of Tokyo Press, 1974).

century Indian craftsmen had established the working of alloys, soldering, lacquerwork, oil distillation, the use of saltpeter to cool water, riveting, and sophisticated haulage techniques to a level as advanced as in most places in Europe. Water wheels were at times used for corn grinding and even for powering trip-hammers in rice milling, although here the lack of cast iron metallurgy (and so of metal gearing) limited the potential of power transmission, which in turn confined normal development to the horizontal mill. Yet the textile sector was sophisticated enough. The earliest of the European merchant companies' trading stations ("factories") were established in the weaving villages of South India from the sixteenth century, and there is much evidence of skilled textile workers scattered among the port towns, interior centers and temple towns. Village fairs *(sandai)* also prospered, possibly stimulated by the Muslim influence on consumption standards among the upper groups and by state policies that encouraged foreign and domestic trade: as early as 1630 some four million yards of cloth were being exported to Portugal.[42] Inheritance of skills combined with cheap labor to keep techniques simple. The crank-handle spinning wheel and Indian weaving loom with foot treadles were well established, and production remained organized in a caste, merchant, and guild structure wherein merchant capital was tied strongly to landholding.

Larger-scale techniques can be illustrated in Indian shipbuilding. Again, the absence of a strong iron-making tradition meant that Indian shipping was rarely armed, with the exceptions using heavy and expensive bronze cannon, and navigation fell behind that of Europe. It relied principally on the astrolabe and mariners' cards (to plot the course) rather than on the telescope. But in construction techniques, Indian shipping was in many respects a match for that of Europe. Indeed, during the century there is evidence of European adoption of Indian methods of riveting planks, of wooden water tank construction, of lime compound protection of hulls, and of haulage.[43] Surveying the field, Habib has concluded that "there was no inbuilt resistance in the economic system to technological changes."[44]

[42] Surajit Sinha (ed.), *Science, Technology and Culture* (New Delhi: Research Council of Cultural Studies, 1970); Shri Dharampal, *Indian Science and Technology in the Eighteenth Century: Some Contemporary European Accounts* (New Delhi: Impex India, 1971); Stephen Hay (ed.), *Sources of Indian Tradition*, vol. 2: *Modern India and Pakistan* (New York: Columbia University Press, 1988); A. K. Bag, "Technology in India in the Eighteenth and Nineteenth Centuries," in G. Kuppuram and A. Kunnadamani (eds.), *History of Science and Technology in India* (New Delhi: Sundeep Prakashan, 1990), pp. 223–56.
[43] W. R. Moreland, "The Ships of the Arabian Sea about A.D. 1500," *Journal of the Royal Asiatic Society*, 172 (1939), 63–74 and 173–92; Frank Broeze, "Underdevelopment and Dependency: Maritime India during the Raj," *Modern Asian Studies*, 18 (1984), 432–67; Satpal Sangwan, *Science, Technology and Colonisation: An Indian Experience, 1757–1857* (New Delhi: Sage, 1991).
[44] Ifan Habib, "The Technology and Economy of India," *Indian Economic and Social History Review*, 17 (1980), 1–34, quotation at p. 32. See also Habib, "Potentialities of Capitalistic Development in the Economy of Mughal India," *Inquiry*, 3 (1971), 1–56; Habib, "Potentialities of Change in the Economy of Mughal India," *Socialist Digest*, 6 (September 1972), 123–36.

In a similar manner, it is impossible to dismiss eighteenth-century China as a bamboo economy. In agriculture, yields per hectare were high, and the empire was self-sufficient in most handicrafts and manufactures and possessed a low-cost system of water transport over large areas. The economy was highly monetarized and the prosperity of the merchant class arose from a truly massive interregional trade, equivalent in per capita terms to that found in Europe and in absolute terms far greater. Population growth was satisfied through the introduction of new cereal and starchy root crops such as sweet potatoes, which increased calorie production per acre.[45]

Within the vastness of China there was great variety. Thus the coastal strips of provinces such as Fukien and Hunan were bustling with commercial life, and there were large areas of advancement in spinning and weaving, glass-making, brewing, and metalworking around Peking and Shandong and in the expanding cities of Hankon (commerce) and Jingdezhen (porcelain) to the southeast. In contrast, massive inland areas were frequently poverty-stricken by famine and flood and reduced to survival techniques only. Such debilitating disparities were addressed but could not be removed by government taxation policy. It is difficult to generalize briefly about such a complex system. The enormous growth of population, from around 180 million to some 300 million in little more than fifty years, principally as a result of the annexation of Xinjiang (or the New Territories) in the second half of the century, was associated with a shrinkage in the size of landholdings – exacerbated by partible inheritance – and a large increase in the number of landless households. The exploitation of crops from the New World, such as sweet potatoes, maize and peanuts, boosted caloric intake but narrowed the focus of investment and state requirements to large irrigation projects, which indeed were not well maintained during the century. Representing almost one-third of the entire globe's labor force, Chinese workers could draw the plow, distribute water, husk grain by handmilling, lift, carry, and transport heavy burdens (including bureaucratic travelers), turn millstones in the manufacture of paper, and lift boats from one level to another using simple capstans. From such clear evidence of a static equilibrium most analysts conclude that China's early technological lead[46] had been eroded over a very lengthy period and, furthermore, that by the eighteenth century China's technological decline was not merely relative to the technological rise of Europe but also relative to the greatness of its own past: given a historicist view of technological progress derived from the experience of Western Europe, the eighteenth century was one of technological decline in China and, at best, stagnation in India.

---

[45] Jonathon D. Spence, *The Search for Modern China* (New York: W. W. Norton, 1990), especially chaps. 5 and 6; D. H. Perkins, *Agricultural Development in China, 1368–1968* (Chicago: Aldine, 1969); Colin A. Ronan and Joseph Needham, *The Shorter Science and Civilisation in China* (Cambridge University Press, 1986).

[46] Joseph Needham, *The Grand Titration* (Toronto: University of Toronto Press, 1969); Needham, *Science in Traditional China* (Cambridge, MA: Harvard University Press, 1981).

Explanations of Chinese and Indian history then tend to become danger-ously monochronal and frequently collapse the three themes of the character of technological change, the relations between technological change and in-dustrialization, and the seeming inability of the Asian systems to respond to the European challenge by transferring in the best techniques of Europe. If not the creation of new technology, why not its adoption? The great majority of recent historical accounts elaborate the classical liberal analyses of David Hume and Adam Smith, whose own decisive verdicts quite happily encompassed both cases.

Writing in the 1740s, Hume saw little difficulty in China concerning the stock of knowledge. China's problem was institutional:

> China is the one empire, speaking one law, and sympathising in the same man-ners. The authority of any teacher, such as Confucius, was propagated easily from one corner of the empire to the other. None had courage to resist the tor-rent of popular opinion: and posterity was not bold enough to dispute what had been universally received by their ancestors. This seems to be one natural reason why the sciences have made so slow a progress in that mighty empire.[47]

Compared with the authorities of China, with examinations and the man-darin class, European nationalist conflicts and republicanism created tensions and interruptions that were "rather favourable to the arts and sciences by breaking the progress of authority, and dethroning the tyrannical usurpers over human reason." As with his good friend, Adam Smith also recognized the importance of the ancient advances in both agriculture and industry within the two Asian civilizations, but he saw the large-scale improvements in inland transportation as detracting from foreign commerce and so reducing the like-lihood of both furthering the specialization or division of labor and gaining advantages through technological learning. Thus, the Chinese were settled on a high-level equilibrium because of natural endowments, but this had "been long stationary" due to the character of the country's "laws and institutions." The latter operated in five principal ways, which together ensured technolog-ical stasis or decline. The neglect of foreign commerce was a function of the control requirements of the state, and this was gained at the expense of the learning process associated with competitive trading: "By a more extensive navigation, the Chinese would naturally learn the art of using and construct-ing themselves all the different machines made use of in other countries, as well as the other improvements of art and industry which are practised in all the different parts of the world." Second, investment in improvement was greatly inhibited because of the low security afforded small merchant capital, which might be "pillaged and plundered at any time by the inferior man-

---

[47] Hume's general position is drawn from various parts of *Essays, Moral, Political and Literary* (Edin-burgh, 1741–2), the quotation from "On the Rise and Progress of the Arts and Sciences," in vol. 33 of the *Works of David Hume* (London: Grant Richards, 1903), p. 123.

darines"; the consequent high levels of interest were in this sense artifacts of inefficient institutions. In addition, the rice-growing base of the economy yielded as many as three crops each year (in contrast to European cereals), encouraging a large population. But because of the power structure there existed an unequal distribution of incomes, so the "grandees" were conspicuous consumers of precious stones and employers of large retinues; this altogether misallocated resources to food and navigation, to the detriment of investment in industry. Fourth, the sheer size of the home market, "not much inferior to the markets of all the different countries of Europe put together," led to a vicious circle of agrarian orientation in the absence of external commercial intercourse. Finally, in the two principal hydraulic economies of Asia, the overwhelming task of local elites and of the authorities at large was to maintain irrigation systems to protect land tax revenues, which were directly linked to the "annual produce of the land." So, in contrast to Europe, the "great interest of the sovereign" lay solely in land. This created the need for extreme centralism, corvee labor, and so on, and resulted in corruption and abuse. All this was, therefore, favored by "the mandarins and other tax gatherers."[48]

In the original liberal presentation, Asia's failure was neither cognitive nor cultural but rather political and institutional. This has remained the hallmark of later scholarly accounts. Geographers have elaborated on the effects of settlement on rich alluvial lands where there "was an abundance, and so long as the canals and dykes were maintained, there was less stimulus to the renewed practice of ingenuity," and they have sought to confirm the theory that distance and isolation ensured that China "elaborated its ideals in absolute segregation from alien thought" and that climate and river settlement conditioned the evolution of Hindu theology.[49] Global historians such as Braudel have concluded that centralism (*contra* Japan's seeming devolution prior to Tokugawa) and the prohibitive role of the mandarin meritocracy, which erected a systematic cultural barrier to technical change that could not be breached, represent a sufficient institutional argument. China may have originated movable characters in printing and used coal in cast iron production long before the Europeans, but the combination of such "social inertia" with cheap labor ensured that China's breakthroughs became the incremental advances of Europe rather than the causes of industrial advance in China itself.[50]

---

[48] Adam Smith, *The Wealth of Nations*, 1776, quotation from Everyman's Library, edited with introduction by Edwin R. A. Seligman (London: J. M. Dent and Sons, 1937), vol. 1, pp. 63–4, 85, vol. 2, pp. 173–4. 218–7, 319.

[49] Ray H. Whitbeck and Olive J. Thomas, *The Geographic Factor* (New York: Kennikat Press, 1932), quotations pp. 234–42, 287–9, 368–9; Karl Wittfogel, *Oriental Depotism: A Comparative Study of Total Power* (New Haven, CT: Yale University Press, 1957); Eric Jones, *Growth Recurring: Economic Change in World History* (Oxford: Clarendon Press, 1988).

[50] Fernand Braudel, *Civilisation and Capitalism: 15–18th Century,* vol. 1: *The Structure of Everyday Life* (London: Collins, 1981), pp. 338–77; vol. 2: *The Wheels of Commerce* (London: Fontana, 1985), pp. 117–20, 585–95; Norman Jacobs, *The Origin of Modern Capitalism and Eastern Asia* (Hong Kong: University of Hong Kong, 1958).

Specialist scholarship serves to confirm the broad liberal position. Thus, for Habib, the technology of Mughal India was often encouraged by the Mughal court and ruling class (artillery, horticulture) or by merchant groupings (ship-building) or by craftsmen (textiles and machinery, such as the belt drive), but an overall "technological inertia" centered on failures to adopt iron metallurgy and tools. Tentative explanations for this include demographic and institutional factors working in combination: a surplus of caste-skilled labor "would militate against labor-saving techniques," luxury demands of the small but immensely wealthy ruling class "might also deter such inventions as were designed to introduce greater uniformity in products," and merchant capital may indeed have been too richly rewarded in trade and land, leading to a paucity of resources in more risky or more demanding industrial adventures.[51] Similarly, Joseph Needham's immense and complex studies seem to conclude that China's bureaucratic feudalism (contra the aristocratic militarism of Europe) might well have encouraged scientific and other advances in earlier centuries but inhibited visible breaks with tradition of the sort commonly associated with the European scientific revolution and with the rise of industrial capitalism. In the end, "there was something in Chinese society which continually tended to restore it to its original character after all disturbances, whether these were caused by civil wars, foreign invasions, or inventions and discoveries." Again, the "institution of the mandarinate" meant taxation levels and sumptuary laws that retarded the growth of an independent merchant culture; the failure of the institutions of mercantile capitalism ensured that China would not use the technologies of eighteenth-century Europe.[52]

Any account of the failure of both India and China to follow the European technological advancement of the eighteenth century must clearly recognize the long tenability of alternative, non-iron-centered techniques of production, the strong imperatives of geography, and the character of the interactions between these two complex systems and the expanding, intruding Atlantic commercial and moral economy. Much of even the radical liberal approach seems to depend on the second of these three elements, although not in the conventional manner of an emphasis on the constraints of raw material scarcities. Rather than this, to claim some determination of size, location, and climate on institutional and power structures, and so on the distribution of incomes, seems a relevant departure point. Certainly, South India appears to offer as good an example of the hydraulic society as does China in the eighteenth century, with the enormous diversion of resources to anicuts, dams, and tanks under the instigation of chiefs, subregional governors, and village-organized landowning groups and temples. Again, it is worth emphasizing that

---

[51] Habib, "Technology and Economy," pp. 32–4.
[52] Joseph Needham, "Science, Technology, Progress and the Break-through: China as a Case Study in Human History," in Tord Ganelius (ed.), *Progress in Science and its Social Conditions* (Oxford: Pergamon Press and Nobel Foundation, 1986), pp. 5–22, quotation at p. 14.

in both India and China, much of this basic hydraulic infrastructure was in decline during this century.

The nature of resource commitment, together with the large continental, rather than maritime, locus of the two Asian economies, undoubtedly conditioned the character of their commercial and technological interactions with Europe. The eighteenth century was precisely the period when increased European resources were allocated to imperial and aggressive commercial expansions, when company rule over Asian outposts were replaced by direct state interventions, and when the maritime powers began to view national status in global terms.[53]

From the middle of the eighteenth century, European success globally required definite efforts other than the strictly military or technological: the sheer output of arms increased vastly, shipping costs fell, and new bases in Asia acted as strategic advanced outposts for amassing men and money. The tools of empire were as much those of transport, communication, and organization as they were of warfare per se.[54] Thus, the establishment of East India Company rule in India entailed a considerable loss of economic sovereignty, a decline in merchant capital investments, and a change in the rules – all of which could only put at higher risk the technological and manufacturing activities of Indian craftsmen. Between 1700 and 1825, imports of silk manufactures from India to Britain were prohibited by law, and heavy duties on cotton textiles were established in the 1790s, precisely when Samuel Crompton's invention of 1779 allowed machine spinning of fine yarn for muslins, which had previously been imported from the hand spinners of the subcontinent. Indian domestic industries (just the sort of by-employment, rural-based industries that were at the core of Meiji Japan's system of technology transfer and learning after 1868) were faced with a flood of colonial imports and were then burdened with excise and other taxation, the revenue from which was spent on administrative facilities which bore little relationship to the needs of the indigenous industrial sector.[55] These were not favorable conditions for the successful transfer of European technologies in either the eighteenth or in later centuries.

[53] Vincent Harlow, *The Founding of the Second British Empire, 1763–1793*, vol. 1: *Discovery and Revolution* (London: Macmillan, 1952); D. C. Mackay, "Directon and Purpose in British Imperial Policy 1783–1801," *Historical Journal*, 17 (1974), 481–96; T. Hopkins and I. Wallerstein, *Processes of the World System* (Beverly Hills, CA: Sage, 1979); G. Modelski and W. R. Thompson, *Sea Power in Global Politics, 1493–1933* (Seattle: University of Washington Press, 1988); Ursula Lamb (ed.), *The Globe Encircled and the World Revealed* (Aldershot: Variorum, 1996); Michael Mann, *The Sources of Social Power*, vol. 2: *The Rise of Classes and Nation States, 1760–1914* (Cambridge University Press, 1993); Paul Kennedy, *The Rise and Fall of the Great Powers* (New York: Vintage Books, 1989); Ian Inkster, "Global Ambitions: Science and Technology in International Historical Perspective 1450–1800," *Annals of Science*, 54 (1997), 498–509.

[54] G. Parker, *The Military Revolution: Military Innovation and the Rise of the West, 1500–1800* (Cambridge University Press, 1988); Daniel Headrick, *The Tools of Empire: Technology and European Imperialism in the Nineteenth Century* (New York: Oxford University Press, 1981).

[55] Deepak Kumar (ed.), *Science and Empire: Essays in Indian Context* (New Delhi: Anamika Prakashan, 1991); Roy Macleod and Deepak Kumar (eds.), *Technology and the Raj: Western Technology and Technical Transfers to India, 1750–1947* (New Delhi: Sage, 1995).

CONCLUSIONS

One of the most common problems faced by historians when essaying comparative studies is to avoid the pitfalls of unconscious hindsight and presentism. It might be tempting to contrast the dynamism of England with the backwardness of Russia, the novelties of Japan with the traditions of China or India. We know that from the late eighteenth century, British industrialization changed the world and that from the late nineteenth century Japan was the only large non-Western nation to undergo a sustained process of industrialization, and this knowledge colors our perceptions of the eighteenth century. We have tried to illustrate that the technological story is not as clear as hindsight would have us suppose.

Certainly, notions of a clear distinction or a linear relationship between science on the one hand and technique on the other are not given much support in eighteenth-century experience, whatever a retrospective review of the uniqueness of Western scientific rationalism might sometimes claim.[56] In the very long term of the modern era, rigid categories might allow such firm judgments. The eighteenth century does not. At times the analytical terrain has been muddied by the unjustified search for direct cognitive or inspirational links between new science and novel technique. Clearly, improvements in the steam engine or the fuel revolution were not closely associated with advances in crystallography or the discovery of the electrical current or the resolution of planetary inequalities. But all this may represent a potential reply to the wrong form of the query.[57] For example, forces of rising or changing demand may well have been of importance in stimulating an innovation industry that drew on existing scientific knowledge as integral to a shared milieu. In other words, demand or other stimulii may have determined the timing of events whose continuation and sustenance lay in the cheap supply of either basic science (e.g., knowledge concerning the properties of materials) or rational forms of inquiry and experimentation. The relations of science and technology may have been important and positive, and yet may not have been either unambiguously necessary or nearly sufficient to an overall interpretation of technological changes in Europe or elsewhere. In this paper we have pointed to the importance of social interactions within systems of mutual acquaintance and contact, which at certain sites enmeshed theory and practice and produced a wide domain of discourse. Recent work on the Watt steam engine emphasizes the great complexities in the microsocial network linking advances of technique to knowledge, information, instrumentation, and tacit skills.[58]

---

[56] Kurt Mendelsohn, *Science and Western Domination* (London: Thames and Hudson, 1976).

[57] For recent wrestles, see Margaret C. Jacob, *Scientific Culture and the Making of the Industrial West* (Oxford: Oxford University Press, 1997), and Robert McC. Adams, *Paths of Fire: An Anthropologist's Inquiry into Western Technology* (Princeton, NJ: Princeton University Press, 1996).

[58] Especially see the recent and forthcoming work of Richard L. Hills, as in his "James Watt, Mechanical Engineer," *History of Technology*, 18 (1997), 59–79. See also in the same volume, Ian Inkster, "Dis-

That we may not now go much further than this becomes evident when we take the larger comparative view. In many eighteenth-century places, the breakthrough into industrialization arose from technique and organizational transfers from more-advanced regions, or it was an outcome of a move toward best technical practice at more points of production. Rising expectations could be engendered and greater efficiencies captured through the spread of existing best techniques to more users of technology. In turn, in some places the spread of technique benefited from the existence or deliberate creation of social institutions or forms of association that encouraged varied discourse across social strata. Other things being equal, a site of social and geographical mobility, or competition, emulation and calculable risk, was simply more likely to accommodate and exploit new ideas or artifacts than was a site of tradition and hierarchy.

Our treatments of Japan, China, and India suggest a fairly common experience of handicraft techniques spreading more or less effectively from region to region under the auspices of various merchant and artisan combinations. Most agricultural improvements were biological rather than mechanical. In general, new technologies or ideas stemming from the Atlantic had little discernible impact on the output of economies or even of particular industries. In these three extremely large and important economic systems there is more of technological accommodation to increasingly difficult circumstances than there is of the creation, spread, or fuller use of novel technologies in such a manner that output per person or the range of outputs notably increased. Varying degrees of isolation encouraged an integration of merchants into the commanding mores and patterns of their societies. Traditional usury was, then, always a source of financial and social advantage, as well as a sink for profits originally generated in production. In Japan, *sankin kotai* could cost as much as 70 percent or even more of a han's cash outlays, and usury increasingly filled the gap between such forced, but habit-forming, conspicuous consumption and the taxpaying capacity of the peasantry. All three cultures shared large hydraulic expenses during the eighteenth century, periodic corvée labor projects, and land taxation and educational systems that emphasized tradition and classical texts. Thus, Hayashi Shihei's formulation of the samurai educational code instructed, "do not lose your dignity . . . do not fall in love with novelty" in a Chinese manner, but in this case the strictures applied to schools for commoners as well as the fief schools. With its comparatively numerous aristocracy, around 6 percent of the population, Japan does not emerge from such a comparison as starkly different from China or even India.

Rather than in strident terms of differences in kind across all levels, much might be gained from an emphasis on differences in kind at the mundane levels of demography and geography together with differences of degree

coveries and Industrial Revolutions: On the Varying Contribution of Technologies and Institutions from an International Historical Perspective," 39–58.

elsewhere. During the eighteenth century, the large population of Japan was relatively isolated and lay beyond the ambitions of most great powers; it was densely settled and obviously maritime, and a mild population growth had ceased by around 1720. This meant that communications were easier, risks lower, information more accessible, urbanism more developed, and taxes more collectible than in China or India. Even the monsoon was kinder.

Perhaps such obvious differences in kind permitted the differences in degree. Japan's polity was centralized, but more effectively so than in India, less nervously than in China, and at less cost than in both. The multiple authorities of emperor, shogun, and han emerged from pragmatism more than from ideology, and such multipolarity in turn permitted a degree of contradiction, of sustained social dislocations. Thus, it is generally recognized that under the Tokugawa regime the peaceful existence of most samurai did not accord with their militant ethos and stance. Whereas in China, classical learning not only dictated status but also represented belief, in Japan the educational system was deliberately managed to reduce or ameliorate the contradictions between traditional ideals and Tokugawa realities. It can be argued, then, that the Meiji Restoration of 1868 released pent-up energies and "provided the opportunity to overcome the long-standing internal contradictions of the Tokugawa regime."[59]

History is rarely even this clear, and to historians of technology the contrasting developments in the iron industry might appear to be of decisive importance. Whereas in China and India the eighteenth century witnessed little in the way of new developments, in parts of Japan the iron industry grew rapidly in the last decades of the century, and in several han the leading industry changed from rice culture or gold mining to iron mining and ironworking. Different techniques were used to produce cast iron, pig iron, and steel, and there is some evidence of a merging of traditional methods with Iberian metallurgical techniques. A wide-ranging production of threshing forks, scissors, and carpenters' tools might have been particularly significant during the incorporation of European methods of iron smelting in the nineteenth century.

At the level of mundane but vital differences in kind, Japan may well have shared more with advanced technological sites than did China or India. In large parts of both Japan and Europe population was densely settled in and around trading cities, iron technology allowed efficient power transmission, and distances between regional markets and advanced technological sites were relatively small; and in both systems there is evidence of multiple tensions between traditional values and new practices. Again, the expenses of the state in maintaining sovereignty were surely less in Japan and advanced Europe than in

---

[59] Reinhard Bendix, "A Case Study in Cultural and Educational Mobility: Japan and the Protestant Ethic," in Neil J. Smelser and Seymour Martin Lipset (eds.), *Social Structure and Mobility in Economic Development* (Chicago: Aldine, 1996), pp. 262–79, quotation at p. 273.

China and India, and the methods less draconian, permitting more secure market transactions. Such comparisons may bring us closer to radical liberal eighteenth-century interpretations of contrasting technological histories, as expressed by writers such as David Hume and Adam Smith, than to more recent emphases on contrasting cultures.

INDEX

Aalto, S. K., 815n34

Abbot, Sarah, 792

Abgar Tibir of Tokat, 657

Académie des Sciences (Paris): and chemistry, 379; establishment of, 88–9, 512; and image of virtuous scientist, 164–5; and mathematical astronomy, 352; as model of scientific academy, 87, 92, 93; and navigation, 827; and Newtonianism, 299–300; and observatories, 99; and physics, 355–7; printing and publication, 542; proceedings of, 93; and prosopography, 224; and science education in universities, 61, 83–4, 85, 98; and scientific expeditions, 332, 333, 362, 621; and scientific experiments, 104; and scientific journals, 95, 96; and state, 104, 107–8; and technology, 104, 114–16; and women, 187, 196

Academy of Mathematics (China), 690

Accademia del Cimento (Florence), 88, 512

Accademia dei Lincei (Rome), 88

Adams, Dudley, 529, 530f

Adams, George, 505, 529

Adams, John Couch, 347

Adams, Will, 866

Adanson, Michel, 409, 568, 835

Addison, Joseph, 8–9, 172

Adorno, Theodor T., 1n2, 436n3

advertising, and medicine, 547

Aepinus, Franz, 370–1, 373

aesthetics, and landscape art, 601

affinities and affinity tables, 384–9

Africa: trade and imperialism, 836; and voyages of discovery, 639

Age of Reason, 742, 824

Agnesi, Maria Gaetana, 103, 185

agriculture: and China, 873, 875, 879; and government patronage of science, 117; and India,

683–4, 879; and Japan, 868, 870, 879; natural knowledge and economic interests, 139–46. *See also* gardening; horticulture

Ahmed III, Sultan, 650, 657

Akenside, Mark, 767, 783–4

Alaska, and voyages of discovery, 634

'Alavī Khān, Mirza, 662, 663, 682

Albin, Eleazar, 574–5

Albinus, Bernard, 71, 578

alchemy: art and literature, 787; and chemistry, 378; and marginalized practices, 485, 499–503, 506–7; and natural knowledge, 136n15

Alcock, Michael, 849

Aldrovandi, Ulisse, 168

Alembert, Jean Le Rond d': and astronomy, 331, 335, 336, 338–42; and Baconianism, 282; and Cartesianism, 301, 302, 344; and classification of sciences, 254–6; and the *Encyclopédie,* 248n16–17, 250–1, 260–3, 264, 400, 667; and evaluation of scientific explanatory systems, 439n12; and Locke, 930n45; and materialism, 415; and mathematics, 4, 281, 305, 314; and mechanics, 361, 362; and probability theory, 815; and reform of education, 53–4; and skeptical critique of natural philosophy, 32

D'Alembert's Principle, 339, 362

Algarotti, Francesco, 137–8, 157, 188n9, 300, 546

algebra and algebraic analysis, 321, 323, 324, 325–6, 703–4

algorithms, 315, 334

Alkon, Paul, 895, 896

allegory: and gendered representations of science, 194f; and images of nature, 591–2

alphabetic arrangement, of encyclopedias, 249–50, 252, 255

Alzate, José Antonio de, 788, 789

amateurs, and scientific communities, 221, 222–4

American Academy of Arts and Sciences, 100

American colonies: and navigational expeditions, 842; and scientific academies, 100; and textile industry, 846–7, 857–8; and trade in scientific instruments, 534; and universities, 44, 53, 64, 100, 517–18, 523, 527, 534. *See also* Alaska; Franklin, Benjamin; Native Americans

American Philosophical Society (APS), 100, 755

Ampère, André-Marie, 246

analogical reasoning, and vitalism, 38

analysis: and human sciences, 439; mathematics and theorems of, 317–18; and specimens from voyages of discovery, 644

analytical philosophy, and mathematics, 318–20

anatomy: and Japan, 712–13; and medical education, 50, 51, 71n49, 85, 466; of mind, 817–18; and representations of body, 778; and scientific illustration, 576, 577–8; and sexual difference, 198, 200. *See also* medicine; physiology

al-Andalusi, Said, 669–60

Anderson, John, 79

Anderson, William, 633

Andrew, John, 192n18

Anglesea, Martyn, 606n57

Anglican Church, 134–5, 277, 505. *See also* Protestantism

animal magnetism, 491–5, 506–7

animism, and life sciences, 404–6

Annals of Agriculture, The (journal), 143

Anne, Queen of England, 132

Ansari, S. M. R., 677n30

Anson, George, 620, 624, 827, 829–30

al-Antākī, Dā'ūd, 662

anthropology: and human nature, 776; and professionalization in nineteenth century, 461–2; race and background of physical, 450–1; use of term, 437; and voyages of discovery, 644. *See also* human sciences

anti-Newtonianism, 503–4, 764–6, 796

apprenticeship, and medicine in Japan, 714. *See also* guilds

Arabian peninsula, and European contact, 650

Arabic script, and printing, 658–9

Arago, François, 350

Aramata, Hiroshi, 574

Arbuthnot, John, 447

Archimedes, 83

Arconville, Marie-Geneviève-Charlotte Thiroux de, 199f

Arduino, Giovanni, 425–6, 428, 432, 585n7

Aristotle: and classification of sciences, 244, 246, 254; and natural philosophy, 268, 269, 271, 285, 288; and physics, 46, 48, 67, 80, 83, 297, 354; and physiognomy, 496; and political theory, 447–8; and scientific method, 511

Arkwright, Richard, 854, 855

Armstrong, John, 767

Arnold, John, 636

art: and dualistic view of science, 617; *écoles de dessin* and decorative, 73–4; geological sections and diagrams, 609–17; history painting and cosmogonies, 587–92; impact of science on, 775, 781, 782–5; and natural history, 585–7, 592–609. *See also* illustration, scientific

Arteaga, Ignacio, 642

artisanal workshops, and women, 189–90

Arzani, Akbar, 681

Asada Goryu, 706, 707

associationist psychology, 457–61

astrology, 491–2, 497–9, 506–7

astronomy: and China, 689–90, 693; cosmology and nebular hypothesis, 348–50; and figure of the Earth, 332–3; and Halley's comet, 342–3; and India, 670, 672–3, 674–8; and Islamic world, 659–60; and Japan, 699, 700–1, 704, 706–10, 716, 717; and Laplacian synthesis, 351–3; and lunar orbit problem, 386–9; and nebular hypothesis, 348–50; and organization of science, 98–9; and perturbational problem, 334–7; and scientific instruments, 521, 523–5; and secular inequalities, 344–8; of solar system in 1700, 329–32; star positions and physical theory, 338–9; status of in eighteenth century, 328; and universities, 82; and women, 190. *See also* cosmography; cosmology; observatories; Venus

atheism, 759–60

attraction: and chemistry, 382–3, 384–6; and electricity, 369

Aubriet, Claude, 565

Australia, and voyages of discovery, 639, 640

Austria and Austro-Hungarian Empire: and government support for mining, 120; and Ottoman Empire, 651–2, 655; and prosopographic studies, 227–8, 230; and universities, 55, 230

authorship, and print culture, 555–60

autopsy, and pathology, 479, 481

Avicenna, 466, 468, 662, 663

Aymand, Claudius, 853

Ayurveda, and medicine in India, 681

al-'Azīz, Subhī-zāde 'Abd, 662

Aztecs, 718, 729

Baader, Franz Xaver von, 36

Bacon, Francis: and classification of sciences, 245, 246, 252, 253–4, 258, 259, 261f, 262, 264; and concepts of fire and heat, 367; and experimental method, 86, 512, 800, 802; and gendered concepts of nature, 193, 195; on human nature, 802–4; influence of on development of science, 35; and methodology, 281; and philosophy of science, 271; and religion, 741; and vitalism, 38

Baier, Johann Jacob, 419–20

Baildon, John, 850

Baily, Francis, 13

Baird, Charles, 863

Baker, Henry, 410, 517, 522–3, 532, 573, 579

Bakewell, Robert, 144, 145, 157, 853–4

Bandinus, Domenicus, 249

Banks, Joseph: and alchemy, 485; and botany, 152, 571; and class, 166; and Kew Gardens, 627; and Royal Society, 128, 282; and scientist as civic expert, 180, 181–2; trade and imperialism, 830, 836, 840; and voyages of discovery, 619–20, 623–4, 627, 629, 639–40

Banneker, Benjamin, 534

Barber, G., 538n3

Baret, Jeanne, 625

Barham, Henry, 834

Barker, Robert, 601

Barloeuf, Charles-Ariège, 60

Baroque polymaths, and Spanish America, 721, 722, 738

Barrett, Francis, 499, 502–3

Barrington, Daines, 631

Barrow, Isaac, 308–9

Barthes, Roland, 249

Barthez, Paul-Joseph, 34n35, 37, 72, 405–6, 475

Bartholinus, Caspar, 61

Bartin, François-Xavier, 492

Bartmann, Sarah. *See* Hottentot Venus

Bartram, John, 163, 181

Bartram, William, 566–7

basalt controversy, 426–8, 606–9

Bassi, Laura, 77, 103, 185, 186f, 187

Baudin, Nicolas, 619, 639

Bauer, Ferdinand & Franz, 571

Baugh, Daniel, 182

Bayezīd II, Sultan, 656

Bayle, Pierre, 195, 284, 487, 552

Bayly, William, 624, 629, 630–1

Beattie, James, 812

Beaumont, Juan Blas, 731n42

beauty, and landscape art, 587–8, 601

Beccaria, Caesar, 455, 460

Becher, Johann, 452, 456

Beddoes, Thomas, 175–6, 506, 837

Beethoven, Ludwig von, 780, 798

Behmenism, 504

Behn, Aphra, 773

Bell, Charles, 578

Bell, John, 494

Bella, Giovanni Antonio della, 519

belles-lettres and literary associations, 92

Benard, Robert, 260

Benedict XIV, Pope, 185

Benoit, Michel, 684, 690

Bentham, Jeremy, 64–5, 143, 459, 801

Bentley, Richard, 135, 291, 348, 742, 747

Berch, Andre, 457

Bergman, Torbern, 36, 118, 384, 585n7

Bering, Vitus, 863

Berkeley, George, 284, 322, 503, 816

Berlin Academy, 94, 109, 229, 362. *See also* *Societas regia scientiarum*

Bernier, François, 671, 682

Bernoulli, Daniel, 287, 305, 362

Bernoulli, Jakob, 305, 308, 309, 362

Bernoulli, Jean, 527

Bernoulli, Johann, 305, 308, 313, 315–16, 321, 334, 361

Berthaut, Henry Marie Auguste, 126n61

Berthollet, Claude-Louis, 85, 115–16, 393

Berthollet, Guyton, 393

Berthoud, Ferdinand, 636

Bertin, Henri L.-J.-B., 126n62

Besoigne, Jerome, 48

Betancourt, Agustín de, 123, 124n52

Beuer, Johann Ambrosius, 570

Bézout, Etienne, 167

Bhava Misra, 682

Bhuwah, Mian, 670, 681

Bichat, Marie-François-Xavier, 72, 416, 484

al-Bihari, Mushibullah, 686

Bildt, Jan van der, 532

Biographia Britannica (1747–66), 251

biology, use of term, 416. *See also* life sciences; physiology; reproduction

Biran, Maine de, 760

Birch, Thomas, 557

Bird, John, 524f, 528

Bitterli, Urs, 443n19

Black, Joseph, 36, 68, 97, 180, 367–8, 390, 475, 507, 525

Blaeu, Joan, 654

Blagden, Charles, 485

Blake, William, 7, 496, 502, 507, 559, 767, 778, 784, 794–5

Blakey, Nicholas, 590, 591f

Bligh, William, 640

Bliss, Nathaniel, 64–5

Bloch, Marcus Elieser, 574

Blondel, James, 150

Bloom, Harold, 784–5

Blumenbach, Johann Friedrich, 33–4, 37, 38, 40, 57, 207, 236, 236n55, 404, 406, 416, 432, 434, 451

Blunt, Thomas, 529

Boccone, Paolo, 419
Bodega y Quadra, Juan Francisco de la, 642
Bodin, Jean, 447
body: analogy between mind and, 817–18; science and representations of, 777–82. *See also* breasts; dualism
Boerhaave, Herman: and biography of Swammerdam, 579; and chemistry, 258, 378, 389, 525; and electricity, 369; and mechanistic preformation, 408; and medicine, 465–6, 468–70, 475–6, 477, 483, 666; and Newtonianism, 296, 401; and physics, 367; and physiology, 402; and print culture, 550; as professor of botany, 51; and textbooks, 531
Boisguilbert, Pierre de, 453
Boissier de Sauvages, François, 474
Bolingbroke, Viscount (Henry St. John), 743–4
Bolognese Academy of Sciences, 98
Bonet, Théophile, 479
Bonnet, Charles, 409, 413, 414, 423
Bonnier de la Masson, 516–17
Bonpland, Aimée, 737
Book of Hours, 563
Boot, Adrián, 724f
Bordeu, Théophile de, 405, 474–5
Borelli, Giovanni, 398
Boscovich, Roger, 61, 180, 278, 361
Bossut, Charles, 121
botany: and botanical gardens, 101–2, 108, 732; and medical education, 50, 51, 69, 467; and natural knowledge, 151–6; and scientist as civic expert, 181–2; and sexual reproduction in plants, 201–5; and Spanish America, 730–1; and taxonomic system of Linnaeus, 5–6, 78; trade and imperialism, 833–40; and voyages of discovery, 626–7, 644; and women, 196, 791–3. *See also* life sciences; reproduction
Bottoni, Federico, 731n42
Boucher, François, 797
Boudier, Claude, 675
Bougainville, Louis Antoine, comte de, 125, 620, 625, 825, 832, 835, 836–7, 842, 843
Bouguer, Pierre, 333, 364, 735
Boulliau, Ismael, 348
Boulton, Matthew, 124, 176, 180, 754
Bourguet, Louis, 421
Bourrit, Marc Théodore, 599, 601, 602f
Bowditch, Nathaniel, 352
Boyer, Carl, 311n9
Boyle, Robert: and chemistry, 381–2; and endowment of lectures, 134; and experimental method, 86, 358, 512; and gendering of science, 195; influence of on Locke, 807; and natural philosophy, 27, 291, 469; and philosophy of science, 270, 271, 272; and print cul-

ture, 556–7, 558; and religion, 742, 747, 751; and Royal Society of London, 172; and writing of science, 18n66
Bradley, James, 65, 328, 338–9, 635, 866
Brahe, Tycho, 98, 348, 694
Bramah, Joseph, 854
Brander, Georg Friedrich, 519, 520f, 532
Brandt, Georg, 118
Braudel, Fernand, 849, 875
breasts: and allegorical representations of nature, 591–2; and gender politics of science, 205, 206–7, 208f. *See also* body; gender
Breithaupt, Johann Christian, 532
Bret, Patrice, 116n29
Brewer, John, 790
Brewster, David, 521
Brissot, Jacques-Pierre, 174–5
Broc, Numa, 125
Brock, William H., 5n17
Brockliss, Laurence W. B., 2, 8, 60n25, 86n78, 96, 216, 222, 467n8
Brongniart, Alexandre, 432n39
Brooke, Henry, 506
Brooke, John H., 12, 435n48
Brooks, G. P., 815n34
Brougham, Henry, 375
Brown, John, 484
Brown, Sylvanus, 846
Brown, Theodore, 401
Bucareli y Ursúa, Antonio María, 642
Buchan, Alexander, 628
Buchan, William, 146, 148–9, 151
Buck, Peter, 457n39
Buffon, Comte de (Georges-Louis Le Clerc): and astronomy, 340; and class, 166; and classification of sciences, 246, 263; on concept of species, 451; and earth sciences, 421–3, 427, 599, 752; imperialism and scientific knowledge, 834–5; and materialism, 414; and natural history, 417; and natural philosophy, 29–30, 31; and probability theory, 815; representations of nature in works of, 588–91; and theory of reproduction, 411–12; and vitalism, 39, 40, 73
Bugge, Thomas, 529
Bulkeley, Richard, 606
Buonanni, Filippo, 419
Burdach, Karl Friedrich, 416
Burke, Edmund, 12, 175, 461, 547–8, 602–3
Burke, Peter, 247–8
Bürmann, Johannes, 566
Burnet, Thomas, 418–19, 544, 588, 589f, 745–6, 751
Burney, Charles, 185
Burney, Fanny, 791

Burtin, François-Xavier, 432
Bute, Earl of, 166
Butler, Joseph, 743, 819
Butterfield, Herbert, 5, 375, 381
Byrom, John, 504
Byron, Lord George Gordon, 762, 780
Byron, John, 624

Cabanis, Pierre-Jean-Georges, 37, 39, 439, 460, 483, 805
Calancha, Antonio, 728
calculus: and analytical functions, 322; and analytical philosophy, 318–20; and Berkeley's critique of Newton, 284; and graphical methods, 308–12; and Japan, 703, 704; Leibniz's notation for, 287, 351–2; and theorems of analysis, 317–18
Caldas, Francisco José de, 729
calendar, Japanese, 709
calorimetry, 372
Cambridge University, 49, 60, 65–6, 68, 85, 215, 216, 289, 290, 291, 333
Camden, Lord, 556
Campbell, Archibald, 815n34
Camper, Peter, 451
Canada, and voyages of discovery, 627, 640–1
canals, and industrialization, 855, 856, 857
Cannon, John, 498
Cannon, Susan, 15
Cantillon, Richard, 454
Caramuel, Juan, 721
Cardero, José, 643
Carl, Ernst Ludwig, 453–4
Carlisle, Richard, 755
Carlos III, King of Spain, 117, 641–2
Carlos IV, King of Spain, 643
Carmichael, Gershom, 441
Carnot, Lazare, 73, 122
Caroline, Princess of Wales, 292
Carozzi, Albert V. & Marguerite, 428n28
Carter, Elizabeth, 133
Carteret, Philip, 625
Cartesianism: and dualism, 27, 913; and life sciences, 402; and natural philosophy, 286; and Newtonianism, 287–8, 299, 300, 359–60. *See also* Descartes, René
cartography: and China, 690; and government support for science, 126–7; Islamic world and Western techniques of, 653–5; and voyages of discovery, 645. *See also* geography; navigation
Cartwright, Edmund, 854
case histories, and pathology, 480
Cassini, Gian Domenico, 332, 621, 651
Cassini, Jacques, 332, 337, 651
Cassini de Thury, César-François, 126

Cassirer, Ernst, 801
caste system, in India, 686
Castel, Louis Bertrand, 49
Castle, Terry, 772n28
catastrophism, 433–4
Catcott, Alexander, 277
Catesby, Mark, 565–6
Catherine the Great, Tsarina of Russia, 623, 632–3, 860, 862–3
Catholic Church: and natural religion, 743; and *philosophes*, 754–5; Reformation and German universities, 230; and relations between science and religion, 742, 754–5; and science education in universities, 53, 81; and scholarship in Spanish America, 724–7. *See also* Jesuits; religion
Catrou, François, 442
Cauchy, Augustin, 324
causality and causation: in nature and religion, 759; philosophy of science and critiques of, 276; and skepticism, 29, 30–2
Cavazza, Marta, 185n6
Cavendish, Henry, 166, 373, 390, 393
Cavendish, Margaret (Duchess of Newcastle), 187
Celsius, Anders, 82, 430–1
Ceranski, Beate, 187
Cessart, Louis-Alexandre, 122
Chambers, Ephraim, 248, 250, 252, 254–6, 256–60, 264
Channing, John, 665
Chappe d'Auteroche, Jean-Baptiste, 623, 624, 642
Chaptel, Jean-Antoine-Claude, 72, 83, 115
characterology, and images of scientists, 160
Charaka, 681n44
Charles II, King of England, 107, 538, 852
Charles III, King of Spain, 61, 730
Charles XII, King of Sweden, 110
Charlevoix, Pierre-François Xavier de, 443, 837
Charpentier, Johann Friedrich Wilhelm, 609, 611f, 612
Chateaubriand, Vicomte de, 760
Châtelet, Madame du (Gabrielle-Émilie le Tonnelier de Breteuil), 103, 170, 187, 188, 301
Chelsea Physic Garden, 627
chemical revolution, 35, 36–7, 392–6
Chemintz, J. H., 576
chemistry: affinities and composition, 384–8; and chemical revolution, 35, 36–7, 392–6; and classification of sciences, 256, 258, 259–60, 262–3; economic development and government support for, 118–19; and Enlightenment, 375, 377–80; and gases, 388–91; and German scientific community, 229–31; and history of

chemistry (*cont.*): science, 375–7; and matter,
381–4; and medical education, 50, 68, 70–2,
84, 467; and natural philosophy, 286; and
scientific illustration, 582–3; and scientific
instruments, 525; and scientist as civic expert,
180; and technical schools, 75; and textile
industry, 115–16
Chen Yuanyao, 693
Chéseaux, J.-P. L. de, 349
Cheselden, William, 578
Chesterfield, Earl of, 171n21, 172, 173
Cheyne, George, 147, 401, 402, 466, 837
chi'i, and medicine in Japan, 711–13
children: and Newtonianism, 138–9; and reading,
555. *See also* education; family; motherhood
China: and agriculture, 873, 875, 879; and
astronomy, 689–90, 693; and evidential
scholarship, 691–5; and Jesuits, 688–91;
and medicine, 695–7, 716; and navigational
instruments, 843; and technological develop-
ment, 871–7, 879–81; and voyages of discovery,
639
Chirikov, Aleksei Il'ich, 632
Christina, Queen of Sweden, 188
chronology, and geological history, 616
chronometers, 342, 635–6, 827, 843
Church of England. *See* Anglican Church
Churchill, Awnsham, 541, 556
Cígala, Francisco Ignacio, 734n52
civil service, and science as occupation in Japan,
699–700
Clairaut, Alexis-Claude, 121, 300, 305, 333, 336,
339–42, 362
Clark, George, 250
Clark, J. C. D., 10
Clark, William, 19, 181, 463n1
Clarke, Edward, 200
Clarke, John, 291
Clarke, Samuel, 135, 274, 276, 277, 291, 292, 357,
361, 742, 744–5, 748–9
class, socioeconomic: and education in India,
680; gardening and horticulture in England,
153, 154–5; and image of scientist as moral
philosopher, 166–7; and industrial development in Russia, 860; and natural knowledge,
158; science and technology in Japan, 699,
867; and scientific academies, 226; and scientific community, 234; and Spanish America,
720; and university students, 215–18; and
women in science, 188–9. *See also* caste
system; working class
classification, of sciences: and Baconian division,
253–4; and earth sciences, 426; and encyclopedias, 249–52, 254–66; and history of science,
241–5; practice of, 245–8

Clement V, Pope, 464n2
clergy, and natural history, 163–4. *See also*
Anglican Church; Catholic Church
Clerke, Charles, 631
Clerq, Peter de, 531
Clifford, George, 570
Cochin, Charles-Nicolas, 260
coffeehouses, and print culture, 553
Coke, Thomas William, 143
Colbert, Jean-Baptiste, 107–8, 122, 540
Cole, Christopher, 606, 607
Colegio de Minería (Mexico), 733
Coleridge, Samuel Taylor, 763, 765, 797
Collection Académique (journal), 95–6
collective biographies, 212
Collège de Navarre, 66
Collège Royal (France), 66, 97, 98, 108
Collège du Sorbonne-Plessis, 48
College of William and Mary, 44, 53, 100
collèges de plein exercice, 53, 62
Collins, Anthony, 276, 746, 775, 807
Collinson, Peter, 626
Colnett, James, 642
colonialism: and botanical transport, 839; gen-
der politics of science and racism, 207–10;
and scientist as civic expert, 181, 182; and
Spanish America, 718–19, 730–5. *See also*
imperialism; voyages of discovery
Columbia University, 100
comets, and astronomy, 342–3, 590
commerce. *See* mercantilism; trade
Commerson, Philibert, 625, 626
commodification, and botany, 151, 153, 155, 157–8
comparative analysis, and vitalism, 38
Complete Library of the Four Treasuries
(China), 693
Comte, Auguste, 246, 461
Comtian Positivism, 461
Condamine, Charles-Marie de La, 735
Condamine Expedition (1740s), 735, 842
Condillac, Étienne Bonnet de, 18, 282, 319, 394,
438, 439, 455, 459, 483, 807, 812n26
Condorcet, Marquis de, 6, 13, 32, 94, 165, 167,
440, 457–8, 460, 486, 551–2, 558, 815
Condorcet, Mme., 460
Confucianism, 688, 691, 692, 699, 710. *See also*
neo-Confucianism
Congreve, William, 685
Conservatoire des Arts et Métiers, 79
consumer culture: imagination and cultural
change, 790–9; and natural knowledge, 157,
158
contraceptives and abortifacients, and botany,
208–10
Cook, James, 125, 136, 443, 523, 620, 626–36

Cooper, Anthony Ashley. *See* Shaftesbury, Earl of

Cooter, Roger, 9, 17, 486

Copernican revolution and Copernicanism, 666, 709–10

copyrights, 539, 557–8

Cort, Henry, 854

cosmogonies, and landscape art, 587–92

cosmography, and China, 694. *See also* star catalogs and maps

cosmology: and astrology, 499; and Japan, 700; and Kant on metaphysics, 280

cosmopolitanism, and scientific community, 234

Costa, E. Mendes da, 491

Cosway, Richard, 502

Cotes, Roger, 86, 291, 335, 810

Coulomb, Charles-Augustin, 122, 167, 180, 371–2, 373

Cousin, Victor, 801

Cowles, Henry Chandler, 566

Cowper, William, 775

craft tradition, and women in science, 233

Crell, Lorenz, 57, 229, 234, 235

Creoles, and scholarship in Spanish America, 722–5, 727, 729, 735, 737–8

Crocker, Lester G., 1n2, 436n3, 801

Croll, Oswald, 661

Crompton, Samuel, 854, 877

Cudworth, Ralph, 275

Cuff, John, 527, 533

Cullen, William, 68, 142, 144, 145, 378, 379–80, 389–90, 479, 481, 754

Culpeper, Edmund, 526

Culpeper, Nicholas, 146

culture: and doctrines of optimism and doubt, 764–71; and forms of representation, 777–82; gendering of scientific, 196; and history of science, 237; and imaginations of consumers, 790–9; and impact of science on art, 782–5; and impact of science on literature, 771–7; and progress, 785–90; and scientific illustration, 562; and social roles of scientists, 159–83; travelers and change in Spanish America, 735–8. *See also* art; literature; music; popular culture; society

Cumberland, Richard, 808–9

Cunningham, Andrew, 14

Curie, Marie, 187

Curll, Edmund, 544

Curtis, William, 152

Cuthbertson, John, 529–31

Cuvier, Georges, 165n11, 200, 431, 432n39, 433, 619, 752

Cyclopaedia (Chambers, 1728), 248, 250, 252, 256–60, 264

Cyril Lucaris, Patriarch, 657

Dabney, John, 534

Dagoumer, Guillaume, 356

Dai Zhen, 693

Dalle Donne, Maria, 185n6

Dalrymple, Alexander, 627

Dalton, John, 166, 176, 395

Dampier, William, 619

Daniel, William, 782

Danishmand Khan, 671

Darby, Abraham, 853

Darnton, Robert, 262, 492, 493, 545, 549, 754n35, 789n60

Darwin, Charles, 760–1

Darwin, Erasmus, 153, 177, 603–4, 752, 767, 774, 792, 793, 794

Das, Govind, 681

Davy, Humphry, 78, 86, 396, 489, 760, 830

Dazille, J.-B., 838

Dedekind, Richard, 320

Defoe, Daniel, 490, 547, 831, 838, 839, 841

Degérando, Joseph-Marie, 619

Deijl, Jan & Harmanus van, 532

deism, 276, 620, 748

Delambre, Jean-Baptiste Joseph, 352

Delisle, Joseph-Nicolas, 332, 344, 622, 632

Deluc, Jean-André, 433, 434

Demainbray, Stephen, 77, 124–5, 513

Demidov, Sergei, 312n10, 319–20

democratization, of science in nineteenth century, 306

demography, and China, 695, 873, 879–80

de Moivre, Abraham, 305, 457

Denmark, and observatories, 98

Derby Philosophical Society, 177

Derham, William, 742

Desaguiliers, John Theophilus, 63–4, 135, 157, 252, 281, 290, 357, 513, 550, 749, 857

Descartes, René: and dualism, 751, 804; and earth sciences, 418, 589n17; and mechanical philosophy, 27, 46; and mechanical physiology, 398, 406–7; and medicine, 72; and natural philosophy, 269–70, 286, 469, 800, 804–5; and Newton's critique of vortex theory, 47; and physics, 359, 361, 364, 367; Queen Christina of Sweden as patron of, 188; and religion, 671n17, 741; and scientific revolution, 23. *See also* Cartesianism

Description des arts et métiers (Paris Academy of Sciences), 104

Deslon, Charles, 493

Desmarest, Nicolas, 423, 424, 425n20, 427, 428, 582, 606, 609

Destutt de Tracy, Antoine-Louis-Claude, 242, 245–6, 278, 460

determinism, and human sciences, 440

Deutsche Encyclopädie (1778), 263–4

Deutsches Museum (Munich, Germany), 519, 520f

Dezallier d'Argentville, Antoine-Joseph, 417–18, 431, 491, 575

Dezhnev, Semon, 632

Dharampal, A., 673, 683

dictionaries, of arts and sciences, 248, 251–2

Dictionnaire historique et critique (Bayle, 1697), 251

Diderot, Denis: and classification of sciences, 254–6, 264; and the *Encyclopédie*, 248n16–17, 250, 260–3, 400, 677; and human sciences, 439; Industrial Revolution and optimism, 775; and materialism, 414–15; natural history and study of the mind, 819; and Newtonianism, 282; and philosophical method, 824; and religion, 748, 752; and Rousseau on human nature, 819, 822n52; trade and imperialism, 840

differentiation, and mathematics, 312–15

diffraction, mathematical theory of, 365–6

diluvialism, and fossils, 419–21, 480. *See also* flood

al-Din Muhammad, Diya, 674

Dissenters, English, 176–7

Dixon, Jeremiah, 623, 624, 842

Dobbs, Arthur, 622

Dobbs, Betty Jo, 36

doctrine of signatures, and medicinal plants, 147

Dollond, John, 364, 372, 523

Dolland, Peter, 524

Dolomieu, Déodat, 418, 424, 429n30, 434

Dombey, Joseph, 642

Donaldson, Alexander, 545

Donati, Vitaliano, 418

Donovan, Edward, 575

Doody, Margaret A., 770n25

Dortous de Mairan, Jean-Jacques, 422n15, 429

doubt, and cultural change, 774–7

Douthwaite, Julia, 206n51

Drury, Dru, 575

Drury, Susanna, 609, 610f

Dryander, Jonas, 831

Dryden, John, 780, 785, 798

dualism: and mind-body relationship, 804; and natural religion, 751; and Newtonianism, 292, 294; and relationship between art and science, 617. *See also* Cartesianism

Duchesnau, François, 400n7

Duchet, Michelle, 443n20

Dufay, Charles-François, 368–9, 370, 514

Dumont, Etienne, 459

Dundas, Henry, 830

Dunnell, Jack, 527

Du Pont de Nemours, Pierre Samuel, 455

Dürer, Albrecht, 561, 562

Du Yu, 693

Earnshaw, Thomas, 636

Earth, and astronomy, 332–3, 343

earthquakes: and earth sciences, 422, 427; and landscape art, 602–3; and lightning rods, 755; and natural religion, 748

earth sciences: and Buffon's synthesis, 421–3; fossils and the flood, 419–21; fossils and time scales, 431–5; and new approaches in mid-eighteenth century, 423–6; roles of fire and water in, 426–31; and science curriculum in universities, 70, 72. *See also* geology

East India Company, 182, 658, 829, 830, 831, 833, 843, 877

ecofeminism, 193

écoles de dessin, 73–4

Écoles des Ponts et Chaussées, 121–3

economics: and botany, 102, 151, 153, 154, 155, 157–8; and government patronage of science, 111–28; and natural knowledge, 139–46, 178; and voyages of discovery, 620. *See also* agriculture; industrialization; labor; political economy; technology; trade

Eden, Frederick Morton, 143

education, scientific: and expansion in provision of instruction, 73–9; and India, 678–80; and reforms in eighteenth century, 53. *See also* lectures; textbooks; universities

Edwards, George, 566

Edwards, Jonathan, 753

Eguiara y Eguren, Juan José, 734

Egypt: and Islamic world, 650; and Napoleonic Expedition of 1798, 649, 664, 650; and printing press, 658

Ehrenvärd, Ulla, 612n70

Ehret, Georg Dionysius, 568, 569–71

Eisenstein, E. L., 538n2

electricity: and experimental physics, 368–71; Franklin's theory of, 373; and medicine, 147; Newtonianism and study of, 277–8, 280; and quantification, 372; and scientific instruments, 523

Eliot, R. J., 848

Elizabeth, Czarina of Russia, 110

Eller, J. T., 392

Ellicott, Andrew, 534

Ellis, John, 575, 626

Encyclopedia Britannica (1768–71), 248, 252, 263, 264–6, 285, 769

encyclopedias: and China, 696–7; and classification of knowledge, 248, 249–52; and maps of knowledge, 263–6. *See also Encyclopedia Britannica; Encyclopédie*

Encyclopédie (1751), 248, 250–1, 252, 260–3, 264, 486, 554, 667

Encyclopédie Methodique (1782), 265

engineering, 121–3, 654–5. *See also* military; technology

England: and animal magnetism, 493–4; and astrology, 497–9; and botanical gardens, 102; government and longitude, 341–2; government and technological innovations, 116; Huguenots and technology transfer, 851–3; imperialism and trade, 836; and mathematical physics, 86; and Newtonianism, 289–95; and private courses in sciences, 77–9; prosopography and universities, 227; and relationship between science and technology, 104–5; science and polite society, 171–3; and scientific academies, 91; and scientific instruments, 527–31; and state regulation of print, 537, 538–40; technology and industrialization, 853–8; theology and Newtonian philosophy, 276–7, 284; and transits of Venus, 344; and voyages of discovery, 619, 621–2, 623–5, 626–31, 633–8, 639–41. *See also* American colonies; Royal Society of London

Enlightenment: authorship and end of, 555–60; and chemistry, 375, 377–80; cultures of print at onset of, 536–40; historiography of, 1, 799; and human nature, 776–7; marginalized practices and rhetoric of, 482–92; and mathematics, 305; and medical education, 71; and modernization of sciences, 243; and prosopography, 235–7; and relationship between science and philosophy, 800–2, 817; and relationship between science and religion, 741–3, 758–61; and scientific revolution, 23–43

Entrecasteaux, Bruni de, 639

entymology, and scientific illustration, 574, 575

environmentalism, and human sciences, 444–50

epigenesis, and reproduction, 416

epistemology, and mechanical philosophy, 27

Erdman, David, 795

Erskine, Charles, 47

Escuela de Caminos y Canales, 123

ether: and chemistry, 389; and experimental physics, 366

ethics, and study of natural sciences in universities, 45

ethnography, and travel accounts, 443

ethnology, and voyages of discovery, 644

Euclid, 49, 83, 704

Eulenburg, Franz, 214, 217

Euler, Leonhard: and astronomy, 334–7, 338–42, 345–8; and mathematics, 305, 307–20, 321; and physics, 277–8, 283, 361, 364, 365–6, 373; and science for women, 188n9; and scientific

academies, 94, 108, 109; and theory of light, 288

Europe: and botanical gardens, 101; development of science compared to non-Western traditions, 17; differences between England and, 149–50; and diffusion of Newtonianism, 295–302; human sciences and non-European cultures, 442–4; impact of science on literature and art, 767–9; and Islamic world, 649–53, 665, 666–81; and Japan, 698, 700–3; and legal systems, 445; scientific institutions and expansion of, 100–1; and state regulation of print, 536–7, 540–1, 558–9; technology and industrialization, 846–53. *See also specific countries*

Eustachio, Bartolommeo, 572, 577

evolution, theories of, 5–6

examination system, and science as occupation in Japan, 699–700

expeditions, scientific: and China, 691; and Islamic world, 665, 649, 650, 664; and scientific academies, 93, 300, 332, 333, 362, 621, 623, 624; and Spanish America, 729, 730–1, 735, 737. *See also* navigation; voyages of discovery

experimental philosophy: and classification of sciences, 258; and natural philosophy, 303; and science education, 48, 51, 59, 83

experimental physics: and Boerhaave's research on fire, 418–19; and frictional electricity, 419–22; and Newton's research on light, 363–7; and public lectures, 358; and rational mechanics, 374; and science curriculum in universities, 64, 81

experimental science: and Bacon, 86, 512, 800, 802; emergence of in seventeenth century, 511–12. *See also* experimental philosophy; methodology; scientific method

extinction, and fossils, 432

Eyles, Victor A., 419n4

Eze, Emmanuel Chukwudi, 823n53

Fabris, Pietro, 604, 605f

Fagnano, C. G., 308

Fahrenheit, Daniel, 367

Falck, Chaim Schmul, 501–2

family, and privatization, 192, 235. *See also* children; motherhood

Fang Yizhi, 692

Fara, Patricia, 13

Faraday, Michael, 86

Farías, Manuel Ignacio, 727

Farris, Nancy, 718

Fattori, Marta, 253n29

Faujas de St.-Fondq, Barthélemy, 433

Fazl, Abul, 670–1, 687
Fechner's law, 349
Feder, Johann Georg Heinrich, 39
Felipe V, King of Spain, 116
feminism: and critiques of Enlightenment, 24; and human sciences, 461. *See also* ecofeminism
Ferber, Johann Jakob, 426, 428
Ferguson, Adam, 43, 438, 442, 444, 448
Ferguson, James, 78, 521
Ferrone, Vincenzo, 300
Feuillée, Louis, 737n57
Feyerabend, Paul, 281
Fielding, Henry, 542, 771, 772
Fielding, Sarah, 773
Figuerado, Emmanuel de, 675
Filipo, Jacopo, 562
Findlen, Paula, 10n31, 187, 192n20
fire: and chemistry, 387–8, 389; and earth sciences, 426–31; and physics, 420. *See also* heat
firearms, and military technology, 653
Fisher, J. C., 851
Fissell, Mary, 9, 17, 486
Flamsteed, John, 330, 331, 338, 343, 541, 556
Flemyng, Malcolm, 767
Flinders, Matthew, 619, 639
flood: and diluvialism, 419–21; and fossils, 433; and landscape art, 594, 599
Foley, Samuel, 606–7
Fontaine des Bertins, Alexis, 333
Fontenelle, Bernard de: and *Academie des Sciences,* 8; and astronomy, 173; and chemistry, 384–6; and forms of knowledge, 137, 138; and humanism, 7; and mechanistic preformation, 406, 408; and moral philosophy, 167; and Newtonianism, 299; and physics, 356, 361; and prosopography, 235–6; and scientific academies, 90n6; and scientific revolution, 1n3; and women in science, 187–8
force, and physics, 360–1. *See also* gravity; inertia; motion
Ford, Brian, 18
Fores, Michael, 2n3
Forster, Georg, 443, 629, 630, 829, 831
Forster, Johann, 629, 630
Fortin, Nicolas, 533
Fortis, Alberto, 425, 428, 434
fossils, and earth sciences, 419–21, 431–5
Foucault, Michel, 12, 24, 399, 400, 436n3–4, 478n35, 777
Fouguier, Pierre, 850
Fourcroy, Antoine François de, 393
Fox, Robert, 6, 8
Foxe, Luke, 621–2
France: and animal magnetism, 493; and astrology, 499; and astronomy, 352; and botanical gardens, 101–2; and emigration of Huguenots, 851–2; and government patronage of science, 121–3, 126; and medical education, 71–2; and Middle East, 649, 650, 658; mining and geology, 424; natural knowledge and court culture, 133; and Newtonianism, 299–300, 301–2; and novel, 770–1; and observatories, 99; and philosophical history, 449; and physics, 405; and political economy, 454–5; and private courses in sciences, 86; prosopography and universities of, 222–5; and reform of science education, 55, 56; and science curriculum in universities, 66–7, 96–7; and scientific academies, 91, 127–8, 512; and scientific expeditions to Spanish America, 735, 737; and scientific instruments, 533; scientists and polite society, 171; and state regulation of print, 536, 540, 541; and technical schools, 74–5, 121–3; and textile industry, 850; and transits of Venus, 344; and voyages of discovery, 619–20, 621, 622–3, 625–6, 638–9. *See also Académie des Sciences;* Europe; French Revolution
Francisco de Torres, Cayetano, 728
Francke, August Hermann, 56
Franklin, Benjamin, 100, 161, 165, 179, 181, 277, 278, 369, 373, 493, 523, 550, 754, 773
Fraser, Craig, 4, 319n23, 324n28
Fraunhofer, Joseph von, 532–3
Frederick II (the Great), King of Prussia, 108, 109, 259, 850
Frederick William I, King of Prussia, 120–1, 851–2
Freemasonry, 136n15, 283, 491, 493
Freind, John, 260n41, 382–3, 665
French Revolution: and philosophical history, 461; and print culture, 558; and reform of science education, 56, 84–5; and religion, 760; and social role of scientist, 174–5, 176; and women's rights, 207. *See also* France
Fresnel, Augustin, 365
Frijhoff, Willem, 52n12, 215, 217
Füßli, Melchior, 593, 595f, 596f, 597, 598f
Fuchs, Georg, 665
functions, and mathematics, 308–12, 335
Furneaux, Tobias, 625
Furtado, Francisco, 700
Fuseli, Henry, 496, 782
Fyfe, Andrew, 778n39

Gabled, Christophe, 66
Gagnier, John, 665
Galen and Galenism, 197–8, 468, 671
Galiani, Ferdinando, 440, 455–6
Galileo, Galilei, 23, 88, 191, 270–1, 272–3, 308
Galvani, Luigi, 475

García Fernández, Domingo, 118
gardening, and natural knowledge, 153, 154, 158.
    See also agriculture
Garrison, Fielding H., 2
Garth, Samuel, 767
Gascoigne, John, 14, 28, 97n17, 216, 220–1, 222, 226
gases, and chemistry, 388–91
Gassendi, Pierre, 27, 47, 458, 469
Gassendist atomism, 47
Gaub, Hieronymus David, 478–9, 481, 483
Gay, John, 791
Gay, Peter, 1n2, 801
Gellert, Christlieb Ehregott, 119, 864
Gemini, Thomas, 526
gender: and concepts of nature, 193, 195; and images of science and scientists, 159n1, 194f; literature and influence of science, 793–4; and medicine, 149; and science of woman, 197–201; and scientific culture, 222; and scientific knowledge, 201–7. *See also* breasts; sexual difference; women
General Enclosure Act (1801), 855
Gennin, Wilhelm de, 864, 865
Geoffrey, Etienne François, 384
Geoffroy, Claude-Joseph, 202
geography: France and mapping, 126; and industrialization in Japan, 879–80; and print cultures, 545; and scientist as civic expert, 181. *See also* cartography; longitude
Geological Society of London, 105
geology: landscape art and scientific illustration, 582, 584–617, 781; and natural religion, 751–2; and scientist as civic expert, 180; use of term, 5, 418–19, 586n10. *See also* basalt controversy; earthquakes; earth sciences; mining industry and mining schools; mountains; volcanoes
geometry, and analytical philosophy, 319–20
George III, King of England, 124, 144
Gerard, Alexander, 820
Gerard, Philip, 792
Germany: and chemistry, 379; government and manufacturing, 116; and higher education for women, 186; and Newtonianism, 282–3; and printing houses, 545; and prosopography, 258–62; and scientific academies, 94; and state regulation of print, 540; universities and science education, 56, 57–9, 84, 228–31; and women in science, 190. *See also* Europe
Geyer-Kordesch, Johanna, 480n43
Ghiyasuddin, Abul-Khair, 677–8
Giant's Causeway (Ireland), 429, 606–9
Gibbon, Edward, 749
Gill, Joseph, 865
Gillispie, Charles, 13–14, 122

Glacken, Clarence J., 823n54
Glick, Thomas F., 737–8
Gobelins textile factory, 114, 115–16
Godfrey, Thomas, 635
Godwin, William, 459, 773
Goethe, Johann Wolfgang von, 7, 40, 167, 180, 282, 363, 385–6n16, 585, 587, 601–2, 759, 762
Gohau, Gabriel, 601n38
Golinski, Jan, 5, 13
González, Pedro María, 643
González Laguna, Francisco, 729
Goodman, Dena, 232–3
Goto Gonzan, 711
Goudin, Antoine, 63
Gough, J. B., 36
Gould, Stephen Jay, 435
government: botanical research and pronatalist policies, 237–8; and patronage of science, 107–28; and scientist as civic expert, 179, 182–3. *See also* state; *specific countries*
Grabianka, Tadeusz, 502
Grabiner, Judith V., 324n28
Graham, George, 528, 817
graphical methods: and mathematics, 308–12; and scientific illustration, 562–3
Grattan-Guinness, Ivor, 312n10
Graunt, John, 456
Gravesande, Willem Jacob 's, 63, 65, 81, 97, 281, 289, 297–8, 299, 340, 357–8, 361, 513–14
gravity, and Newtonianism, 275, 366, 472
Gray, Henry, 578
Gray, John, 842
Gray, Stephen, 166, 368
Gray, Thomas, 775, 784
Gray's Anatomy (Henry Gray), 578
Great Historical, Geographical and Poetical Dictionary, The (Moreri, 1694), 251
Green, Charles, 624, 628
Greene, John C., 426n21
Greene, Robert, 277, 293
Greenwich Observatory. *See* Royal Observatory
Gregory, David, 65, 86, 289–90, 349, 401
Gregory, Dominic, 681
Gregory, James, 348
Gregory, John, 820
Grew, Nehemiah, 202, 408
Groethuysen, Bernard, 247
Grotius, Hugo, 808, 809
Gruzinski, Serge, 719n2
Gualtieri, Niccolo, 575
Guerlac, Henry, 288, 801–2
Guerrini, Anita, 401, 402
Guettard, Jean-Étienne, 126, 420n7, 427, 582, 604, 606, 612

guilds: and printing houses, 540; and women in sciences, 190, 192. *See also* apprenticeship

Guío, José, 643

Gundelscheimer, Andreas, 565

Gustav III, King of Sweden, 111

Gustav IV, King of Sweden, 111

Gu Yanwu, 692

Gwosdev, Mikhail, 632

Haas, Jacob Bernhard, 531

Habermas, Jürgen, 9, 482n46, 551n30

Habib, Irfan, 674, 872, 876

Hackaert, Jan, 596

Hacket, Jakob Philipp, 604

Hadley, John, 635

Haënke, Tadeo, 643

Hahn, Roger, 167n15, 223, 224

Haiti, and scientific society, 100–1

Hakki, Erzurumlu İbrahim, 654

Hales, Stephen, 163, 383–4, 387–8

Halévy, Elie, 801

Hall, A. R., 16n57

Hall, Daniel, 850

Hall, James, 166, 434n45

Hall, Rupert, 2, 16

Haller, Albrecht von, 57, 77, 180, 203, 402–3, 413–14, 464, 466, 472–4, 475, 483, 753

Halley, Edmond, 183n54, 331–2, 342, 343–4, 346, 349, 456–7, 556, 619, 827

Halley's comet, 342–3

Hamilton, William, 427–30, 603–4

Hampson, Norman, 243, 801

Hamy, E. T., 737n57

Hanbury, William, 155, 157

Hancock, David, 840

Hankins, Thomas, 16, 243, 397–8

Hannaway, Owen, 378

Hans, Nicholas, 213

Haq, Nurul, 682

Hargreaves, James, 854

Harilal, 678

Harriot, Thomas, 321

Harris, John, 248, 254–6, 769

Harris, Moses, 575

Harris, Steven, 218, 219–20

Harrison, John, 125, 341, 523, 635–6

Hartley, David, 173–4, 278, 438, 459, 555, 751, 784, 805, 814, 816, 819

Harvard University, 44, 53, 64, 100, 517–18, 523, 527, 534

Harvey, William, 452, 469, 661, 666

Hattí Efendi, Mustafa, 651–2

Hauch, Adam Wilhelm, 519

Hauksbee, Francis, 357, 368, 513, 522, 749

Hawaii, and voyages of discovery, 633

Hawkesworth, John, 628

Hayashi Shihei, 879

Hayat Khan, Muhammad, 680

Haydon, Robert, 763

Hayes, Walter, 526

Hays, Mary, 773

Haywood, Eliza, 773

heat: and chemistry, 390, 392; and earth sciences, 422–3, 430; Newtonianism and study of, 277–8. *See also* fire

Hegel, Georg, 196, 761

Heilbron, John, 219–20, 306–7

Heister, Lorenz, 71

Hellot, Jean, 114–15

Helvétius, Claude Adrien, 439, 441, 449, 460, 807

Henckel, Johann Friedrich, 423n16

Henderson, John, 695

Henry, John, 3n7, 4

herbals, and scientific illustration, 563

Herbart, Johann Friedrich, 816

Herder, Johann Gottfried, 40, 440, 441, 446, 462, 630, 823

Hermann, Jakob, 305, 334

Hernández, Francisco, 729

Herschel, William, 350, 691

Hertz, Deborah, 232

Hertz, Markus, 229

Hess, Volker, 478n35

Hevelius, Elisabetha and Johannes, 191f

Hiärne, Urban, 118

Hill, John, 431, 548, 773

Hills, Henry, Jr., 544

Hindenburg, Karl Friedrich, 57

Hippocrates, 35, 36, 468

Hire, Philippe de la, 675–6

history: and classification of sciences, 253; and Enlightenment, 799; and human sciences, 442–4; natural philosophy and hierarchy of knowledge, 31; and Newtonianism, 812–13; of philosophy of science, 303–4. *See also* history of science

history of science: and chemistry, 375–7; and China, 688; and classification of sciences, 241–5; and colonial Latin America, 719; and cultural history, 237; England as focus of studies on, 132–3; Enlightenment rhetoric and marginalized practices, 486–92; and geology, 586n8; and historiography, 1–20; and image of scientist, 162n6; and Islamic world, 649; and life sciences, 397–400; and prosopography, 213–14; and struggle between Cartesianism and Newtonianism, 359–60; and studies of Newtonianism, 134–9; and Whiggism, 264, 375–7, 388, 396, 488, 489. *See also* history

History of Science (journal), 242
Hobbes, Thomas, 246, 355, 437, 458
Hobsbawm, Eric, 1n2
Hodges, William, 629, 630
Hodgson, James, 357
Hoffman, Friedrich, 402, 465, 468, 469–70, 475, 477–8, 483
Hoftijzer, P. G., 539n4
Hogarth, William, 150–1
Holbach, Paul-Henri Thiry baron de, 414, 415, 429, 441, 747
Holker, John, 849–50
Holmes, Frederic L., 386
Holwarda, J. P., 348
Holwell, J. Z., 682
Homberg, Wilhelm, 386
Home, Rod, 5
Homo sapiens: introduction of term, 206; and physical anthropology, 450
Hong Liangji, 695
Hooke, Robert, 130, 131, 338, 419, 420, 512, 575
Hooker, Richard, 786
Hooykaas, R., 423n16
Horkheimer, Max, 1n2, 23–4, 25, 33, 41, 436n3
Hornblower, Jonathan, 854
Horne, George, 277, 505, 750
Hornsby, Thomas, 524f
Horrocks, Jeremiah, 329
horse, and anatomy, 576, 577f
horticulture, and natural knowledge, 153. *See also* agriculture
Hoskin, Michael, 348n37
Hottentot Venus, 200
Houston, William, 834
Houstoun, James, 831, 836
Hsu Kuang-ch'i, 706–7
Huang Yuanyu, 697
Hudson Bay Company, 621
Huerta, Alonso, 729
Hufbauer, Karl, 229–30
Hufeland, Christoph Wilhelm, 466
Huff, Toby, 695
Hugh of St. Victor, 248
Hughes, Griffith, 566
Hugo, Gustav, 446
Huguenots, emigration and technology transfer, 851–2
human nature: human sciences and theories of, 440–1, 447, 776–7, 818; and natural history, 819–23; and relationship between science and philosophy, 802–4, 809
human sciences: attempts to establish, 436–7; history and travel literature, 442–4; and notions of "human," 440–1; and notions of science, 437–40; psychology, utility, and

political science, 457–61; and quantification, 456–7; race and physical anthropology, 450–1; schools of thought and theories of social change, 444–50; state and political economy, 451–6; subject matter of in nineteenth century, 461–2. *See also* anthropology; sociology
Humboldt, Alexander von, 475, 644, 737
Humboldt, Wilhelm von, 39, 58, 74, 180, 181, 182–3
Hume, David: and anatomy of mind, 818; and Bacon, 282; and Cartesian system, 811; and China, 874–5, 881; on commerce, 856; and culture of optimism, 775; and human sciences, 436, 437, 444, 458–9; influence of Newton on, 12, 281–2; and natural history, 164, 418, 820–1; and religion, 742, 749, 756–8, 759; and skepticism, 29, 30–2, 317, 770
Hume, Robert, 779
Hungary, and scientific academies, 228
Hunt, Lynn, 549
Hunter, John, 37, 404, 406, 578
Hunter, Michael, 226
Hunter, William, 77, 151, 578, 678
Huntsman, Benjamin, 853
Hurter, Johann Heinrich, 531
Hutcheson, Francis, 448, 801, 810, 815, 818
Hutchinson, John, 277, 293–4, 489, 504, 765
Hutchinsonianism, 503–5, 506–7, 765
Hutton, James, 180, 263, 294, 396, 421, 430, 434, 751
Hutton, William, 856
Huxley, Thomas Henry, 760–1
Huygens, Christiaan, 308–9, 321, 332, 364–5
Hyder, Ali, 685
hypnosis. *See* animal magnetism

Iazzaro, Anton, 479
Ibarra Salazán, Andrés, 734–5
İbraham Paşa, 650–1
ichthyology, and scientific illustration, 574
ideology: and connection between science and technology, 104; cosmopolitanism and Republic of Letters, 234; and influence of science on culture, 11–12; and scientific academies and societies, 94; and sexual science, 226
Iliffe, Robert, 6, 9–10
illustration, scientific: acknowledged and unacknowledged reuse of, 572–4; and anatomy, 576–8; and early technical problems, 572; and geology, 582, 584–617; physics and chemistry, 582–3; and realism, 639, 641–6, 652–6; traditions of before eighteenth century, 563–4; and transition from woodcuts to metal engraving, 568–72; and zoology, 574–7. *See also* art
imagination, and cultural change, 780–1, 790–9

immigration, and technology transfer, 850, 851–2

Imperial Astonomical Bureau (China), 689–90

imperialism: and navigation, 827–8, 841–2; and scientist as civic expert, 181, 182; and technological development in China, India, and Japan, 877; and trade, 828–38; and voyages of discovery, 620, 624–6. *See also* colonialism

imponderable fluids, and study of gases, 388–91

Inca, 718, 728–9

Inchbald, Elizabeth, 773

index, and Harris's *Lexicon,* 255–6

India: and agriculture, 683–4, 879; and astronomy, 670, 672–3, 674–8; and education, 678–80; and Islamic world, 650; and medicine, 670, 680–3; philosophical and scientific traditions of, 669–71; and print culture, 658; and status of science in eighteenth century, 671–4, 685–7; technology and industrial development, 683–5, 871–7, 879–81

Indonesia. *See* Java

industrialization: and China, 871–7, 879–81; England and technological development, 853–8; Europe and transfer of technology, 846–53; and India, 879–81; institutional and technological features of, 845; and Japan, 866–71, 879–81; and Russia, 858–66; and scientific knowledge, 878. *See also* manufacturing; metallurgy; mining industry and mining schools; textile industry

Industrial Revolution: and ideology of science, 104; and literature, 775; scientific academies and technological innovations, 128; and scientist as civic expert, 179–80

inertia, and mechanical philosophy, 26

information revolution, 247

inoculations, and medicine, 192, 682, 838

insanity. *See* madness

Institute of Bologna, 97–8

institutions, scientific: and botanical gardens, 101–2; and consolidation of science, 7–8; and European expansion, 100–1; and industrialization, 858; and organizational revolution of seventeenth century, 87–90, 119–20; and Spanish America, 719–22; women and gender in science, 184–92. *See also* organization; scientific academies; universities

Instituto delle Scienze (Bologna), 185

instruments, scientific: and astronomy, 660; and chemistry, 394, 396; and imperialism, 841–2; and lecture demonstrations, 516–17, 521–2; and longitude, 634–6; methods, materials, and makers of, 525–31; and scientific research, 522–5; survival of eighteenth-century examples of, 517–18, 520f; and trade in Europe and

North America, 531–4. *See also* chronometers; navigation

Iraq, and Islamic world, 650

Ireland: and Giant's Causeway, 606–9; and printing houses, 544; and science fiction, 788

Irvine, William, 390

Isfahani, Mir Hussain, 686

Islam and Islamic world: and astronomy, 659–60; clocks and watches, 655–6; and Europe, 649-53, 665–8; and India, 650, 670; and medicine, 661–5; military technology and cartography, 653–5; and printing culture, 656–9

Italy: and Newtonianism, 300–1; and political economy, 455–6; Renaissance and science, 511–12; and science curriculum in universities, 64; and scientific academies, 512; women and universities, 185. *See also* Europe

Ivan III, Tsar of Russia, 631

Ives, Edward, 682n53

al-İzniki, Ömer Sinan, 662

Jackson, Marie, 792

Jacob, François, 399–400

Jacob, Margaret C., 4–5, 9n30, 11n34, 28, 134, 290–1

Jacobins, 176, 352, 540. *See also* French Revolution

Jacquin, Nicholas, 120

Jagannath Samrat, 678

Jago, Richard, 780–1

Jai Singh, Sawai, 675–8, 685, 686

James II, King of England, 290, 538

Japan: and agriculture, 868, 870, 879; and astronomy, 699, 700–1, 704, 706–10, 716, 717; banning of Westerners and Western scientific knowledge, 700–2; industrialization and technological development, 866–71, 878, 879–81; and mathematics, 703–6; and medicine, 710–16; and science as occupation, 699–700, 705, 708; and translations of Western literature, 702–3; and zoological illustrations, 574

Jardine, Nicholas, 418n1

Jardin des Plantes (Paris), 835

Jardin du Roi (Paris), 101–2, 108, 379

Jars, Gabriel, 124

Jaunpuri, Mulla Mahmud, 677

Java, and scientific society, 101

Jefferson, Thomas, 181, 761, 842

Jesuits: and China, 688–91, 692–3, 694; and cosmopolitanism, 234; and Japan, 698, 700–1, 710; and observatories, 99; and prosopographic studies, 218–20; and science education, 49, 54–5, 60, 74; and Spanish America, 720, 731. *See also* Catholic Church

Jews and Judaism: and scientific academies in

Germany, 229; and university students, 216–17; women and salons, 232
Jiang Yong, 694
Joblot, Louis, 410
Johns, Adrian, 18
Johnson, Samuel, 172n23, 245, 258, 786
Johnson, William, 750
Jones, Colin, 467n8
Jones, Peter, 225–6
Jones, William, 505, 678n34
José I, king of Portugal, 127
Journal des Sçavans, 90
journals, scientific: and agriculture, 143; and chemistry in Germany, 229; and classification of sciences, 247, 252; development of in seventeenth century, 102; and organizational revolution of seventeenth century, 90; and print culture, 522–3; and scientific societies, 95–6; and transfer of technology, 848
Juan, Jorge, 641
Julia, Dominique, 215, 217
Jurin, James, 192n18, 284
Jussieu, Antoine de, 421, 422n13
Justi, Johann Heinrich Gottlob von, 454

Kachka, Simon, 864, 865
Kaestner, Abraham-Gotthelf, 57
Kahler, Erich, 773n31
Kalm, Pehr, 831
Kamchatka, and voyages of discovery, 632
Kames, Lord (Henry Home), 141–2, 144, 449, 451, 823
Kangxi, emperor of China, 689–90
Kant, Immanuel: and astronomy, 349; and classification of sciences, 246, 263; and cosmopolitanism, 234; and definition of "schemata," 40; and Enlightenment, 1, 29; and mathematics, 322; and Newtonianism, 813–14; and philosophical history, 375–6; and philosophy of science, 279–80, 284; on reading, 554; religion and science, 164, 756, 758, 759; and teaching of physics, 57–8; and vitalism, 37n49; on women and science, 184
Karstens, W. J. C., 32n29
Kay, John, 853, 854
Keene, Henry, 524f
Keill, John, 63, 73, 290, 357, 382–3, 401, 513, 710
Keir, James, 17–18, 176, 511, 848
Keller, Evelyn Fox, 192
Kelley, Donald, 444
Kendall, Larcum, 125, 636
Kepler, Johannes, 23, 274, 320, 329, 348, 694
Kevalverma, 678
Kew Gardens. *See* Royal Gardens
Keynes, John Maynard, 507

Khairullah Khan, Mirza, 677
Kielmeyer, Carl Friedrich, 38
Kiernan, Colm, 10n32
King, James, 633, 639
King's College (Aberdeen), 53
Kino, Eusebio Francisco, 721
Kircher, Athanasius, 168, 603, 721–2, 788–9
Kirilov, Ivan, 632
Klein, Ursula, 386
Klonk, Charlotte, 18
Klose, Carl, 198n35
Knight, Gowin, 372, 842
Knight, James, 622
knowledge: ban on Western in Japan, 700–2; classification of scientific, 241, 243, 251, 257, 260, 261f, 263–6; criticisms of social forms, 168; and gender politics of science, 201–7, 210; natural philosophy and hierarchy of, 34; print cultures and commerce, 544; religion and Locke's theory of, 806–7; and scientific revolution, 242. *See also* natural knowledge
Koenig, Johann, 830
Koerner, Lisbet, 197n28
Kögler, Ignatius, 694
Korea, and civil service examination system, 699
Körner, Stephen, 58n22
Kovalevskaia, Sofia, 186
Koyré, Alexandre, 375
Krüger, Johann Gottlob, 816
Kuhn, Thomas S., 23n1, 244–5, 259, 327
Kulmus, J. A., 870
Kurushima Yoshihiro, 704

Labillardière, Jacques Julian de, 639
labor: political economy and division of, 237; science and sexual divisions of, 193; and technological development in Japan, 868
La Breton, Restif de, 789
Labrière Lapaute, Nicole-Reine Etable de, 342
Lacaille, Nicolas-Louis de, 338
L'Admiral, Jacob, 575
Laënnec, Guillaume-François, 79
La Fayette, Mme. de, 770
Lafitau, Joseph-François, 443
Lafuente, Antonio, 732
Lagrange, Joseph-Louis, 32, 94, 283, 302, 305, 307, 320–5, 337, 345–8, 362
La Hontan, Louis Armond de Lom d'Arce, Baron de, 443
Lalande, Joseph-Jérôme Lefrançais de, 342, 344
Lamarck, Jean-Baptiste, 416, 752
Lamb, Anthony, 534
Lamb, Charles, 7, 11, 763–4, 799
Lambert, Aylmer Bourke, 571

Lambert, Johann Heinrich, 305, 321, 346, 349n41, 364

La Mettrie, Julian Offray de, 202, 404, 439, 473, 753, 808, 819

landscape. *See* art

Langlois, Claude & Canivet, 533

Langthorne, John, 839

language: and physical anthropology, 450–1; and scientific journals, 95–6; and Spanish America, 728–9; and voyages of discovery, 628–9. *See also* Nahua; Quechua

Lanning, John Tate, 731–2n43

La Pérouse, Jean-François Galoup de, 638–41

La Peyrère, Isaac, 746

Laplace, Pierre-Simon de: and astronomy, 302, 333, 336–7, 345–8, 350, 351–3, 745; and Lavoisier, 393; and mathematics, 305; and mechanics, 362; and Newtonianism, 85, and physics, 66, 86, 279; and probability theory, 815; on religion and natural laws, 756; and skepticism, 32

Laquer, Thomas, 197

La Roche, Sophie V., 515

Latin America. *See* Spanish America

Laudan, Rachel, 429n29, 586n7

Lavater, Johann Caspar, 495–7, 507

Lavoisier, Antoine-Louis: and chemistry, 35, 375, 376, 390, 391, 392–3, 395, 396, 525; and earth sciences, 430, 612; and French Revolution, 13; and government support for science, 126, 180; and physics, 366, 367–8; and physiology, 475; scientific career of, 83, 167; and scientific discourse, 20; and scientific illustrations, 580

Law, John, 454

Law, William, 504

law and legal systems, and human sciences, 445, 808

Lawrence, Jenny, 792

Lawrence, Susan C., 465n3

Lazcano, Francisco X., 734n52

Lear, Edward, 566

Le Clerc, Georges-Louis. *See* Buffon, Comte de

Leclerc, Thomas, 850

lectures: and agriculture, 142, 144; and experimental science, 514; and physics, 356–7, 358; and popularization of science, 516; and public presentations of Newtonianism, 134–5; and scientific instruments, 516–17, 521–2

Ledran, Henri-François, 77

Lee, James, 152

Leeuwenhoek, Anton van, 97, 163, 172, 409, 410, 579

Lefanu, William, 2

legal localism, and human sciences, 444–50

Legaut, François, 561

Legendre, Adrien Marie, 305, 333

Le Gentil de la Galaisière, Guillaume-Hyacinthe-Jean-Baptiste, 623, 624

Lehmann, Johann Gottlob, 425–6, 585, 599, 613, 614f

Leibniz, Gottfried Wilhelm: and astronomy, 334, 351–2; and Cartesian duality, 27; and causation, 276; and classification of sciences, 243, 247; and earth sciences, 427; and government service, 167, 180; and human sciences, 437; influence on science of eighteenth century, 3; and life sciences, 408; and mathematics, 309, 316; and mechanical philosophy of nature, 26, 47; and natural philosophy, 286–7; and Newton, 273–4, 275, 292; and physics, 361; and physiology, 471; and relationship of religion and science, 164, 744–5, 748–9; and role of man of science, 161; and scientific academies, 89, 108; and Scientific Revolution, 23

Leibniz's rule, 360

Leith, James A., 54n15

Lemaire, Jacques & Pierre, 533

Lemery, Nicholas, 381, 422n15

Le Monnier, Pierre-Charles, 331, 335

Lennox, Charlotte, 773

Lenoir, Etienne, 533

Leopold, Prince of Tuscany, 512

LeRoy, Pierre, 636

Lewis, Meriwether, 181

Lexicon Technicum (Harris, 1704 and 1710), 248, 254–6

L'Héritier de Brutelle, Charles Louis, 571–2

L'Hôpital, Marquis de, 313–14

Lhwyd, Edward, 419

Libavius, Andreas, 378, 379

liberty, European concept of, 444

libraries: and China, 690, 693, 782; and classification of sciences, 248; and Japan, 709; and scientific academies and societies, 95n15

Lichtenberg, Georg, 57, 497

life sciences: and animism, vitalism, and mechanism, 404–6; and history of science, 397–400; and materialism, 414–16; and medical education, 70; and Newtonianism, 400–4, 411–13; and theories of reproduction, 406–14. *See also* botany; zoology

light: and chemistry, 392; Newtonianism and study of, 277–8, 280, 288, 294, 363–7

lightning rods, 370, 754, 755

Ligozzi, Giacomo, 562

Lind, James, 629

Lindroth, Sten, 110n8

Lindsay, George, 527

Linnaeus, Carl: and image of godly naturalist, 163–4; and religion, 743; and scientific expeditions, 626, 629, 665; and scientific illustration, 567–8, 570, 592; and sexual reproduction of plants, 202, 203–5; and taxonomy, 5, 69, 152, 197n28, 246, 417, 431, 450, 773; on wet-nursing, 206–7; on women and science, 184
Lister, Joseph Jackson, 525
Lister, Martin, 419
Literary and Philosophical Society, 91
literature: and impact of science on culture, 766–77, 784–5; and marginalized practices, 506–7. *See also* poetry; printing and print culture; travel and travel literature
lithographic printing, 583, 659
Liu Zhi, 696
Lizars, D., 580
Llano y Zapata, José Eusebio de, 729n29
Locke, John: and alchemy, 787; and classification of sciences, 243, 246; and education in natural sciences, 171n21; and human sciences, 438, 439, 458; and influence of science on art, 769–70; and Newton, 284; and physiology, 472; and print culture, 539, 555, 557; and relationship between religion and science, 741–2; theology and matter theory, 275–6
Loescher, M. G., 48
Lofling, Pehr, 831
logic, and scientific education, 45, 48. *See also* reason
London Mechanics' Institute, 79
longitude, and scientific expeditions, 341–2, 634–6, 827. *See also* navigation
Lonsdale, Kathleen, 187n8
Lorrain, Claude, 587, 601, 602, 604n51
Louis XIV, King of France, 28–9, 512, 748
Louis XV, King of France, 357
Loutherbourg, Philippe de, 136, 573
Luc, Jean André de, 523
lunar orbit, 339–42
Lunar Society of Birmingham, 91, 176
Lutheranism, and science education in German universities, 56. *See also* Protestantism
Luxborough, Lady, 792
Lyell, Charles, 760

Macadam, John, 854
Macartney, George, 843
Macartney expedition (China, 1797), 691, 843
Macaulay, Catherine, 461
Machiavelli, Niccoló, 448
Maclaurin, Colin, 86, 291, 305, 361, 810
McClellan, James E., III, 8, 111n12, 127, 221, 730n25

McCloskey, D., 858
McCosh, James, 801
Macquer, Pierre Joseph, 114, 115, 388
McRae, Robert, 319n23
Madan, Martin, 793
Madden, Samuel, 788
Madhava, 681
madness, and representations of body, 778–81
magnetism, 277–8, 280, 370–1, 521. *See also* animal magnetism
Mahadeva, Pandit, 682
Mahdī, Muhammad, 659
Maillet, Benoît de, 431n34, 589n17
Mainauduc, John de, 494, 507
Maistre, Joseph de, 760
Maitland, Charles, 192
Malaspina, Alejandro, 643–4, 730n36
Malebranche, Nicolas, 276, 299, 319, 365, 407–8, 805
Malpighi, Marcello, 408
Malthus, Thomas, 11
Mammalia, and gender in Linnaeus's taxonomy, 207
Manchester Literary and Philosophical Society (England), 177, 178, 227
Mandeville, Bernard, 784, 818n41, 822n52
Manifesto of 1702 (Russia), 861, 862
Manucci, Niocolao, 682
manufacturing, science and technological innovations in, 116–18. *See also* industrialization; Industrial Revolution; metallurgy; textile industry
mapping. *See* cartography
Maqul, and education in India, 678–80
marginalized practices: and alchemy, 499–503; and animal magnetism, 492–5; and astrology, 497–9; and Hutchinsonianism, 503–5; impact of on science and literature, 506–7; and physiognomy, 495–7; and rhetoric of Enlightenment, 486–92
Maria Theresa, Empress of Austria, 55, 227, 228
Marshall, John, 806n12
Marshall, P. J., 443n20
Marsigli, Luigi Ferdinando, 659
Martin, Benjamin, 168–9, 244n6, 252, 516, 517, 527, 534
Martín, Luis, 720
Martínez, Esteban José, 642
Martini, F. H. W., 576
Martyn, Thomas, 575–6
Maskelyne, Nevil, 341, 623, 635, 636, 842
Mason, Charles, 623, 624, 842
Mason, Stephen, 2
Mason-Dixon line (U.S.), 842

materialism: and human sciences, 502; and life sciences, 399, 414–16; and Newtonianism, 402; and secularization of science, 753

mathematical mechanism, and natural philosophy, 25

mathematics: and China, 693; development of, 4–5; and Japan, 703–6; and natural philosophy, 29, 269, 289, 296, 339; and physics, 63, 65–6, 85, 371–4; and practitioners of analytical, 307–27; and state-funded specialist schools, 74; and summary of activity, 305–7; and teaching in colleges and universities, 49–50. *See also* algebra and algebraic analysis; algorithms; calculus; geometry; quantification

matter: and activity in life sciences, 398–9; and chemistry, 381–4; definition of and mechanical philosophy of nature, 25–7; and experimental physics, 369–70; Newton's theory of, 293–4; philosophy of science and theories of, 275–80

Mattioli, Pietro-Andrea, 661

Maudslay, Henry, 854

Maupertuis, Pierre-Louis, 108, 300, 362, 402, 411, 752–3

Mauritius, and voyages of discovery, 626

Mavrocordato, Alexander, 661

Mawer, John, 762

Maya, 718

Mayer, Tobias, 337, 338, 339–42, 635

Mead, Richard, 401, 499

mechanical philosophy: and chemistry, 381, 382; and Islamic world, 666; and pathology, 477; and physics, 359; and physiology, 470–1, 475–6; and scientific revolution, 27–31; skepticism and critiques of, 31–5; and Spanish America, 734; universities and science education, 46–7; and vitalism, 35–48. *See also* mechanics

mechanical physiology, and life sciences, 397–8

mechanics: and analytical philosophy, 319–20; astronomy and Newtonian, 334; and chemistry, 387; and life sciences, 404–6; and mathematics, 305; and physics, 360–3; and scientific instruments, 521, 522. *See also* mechanical philosophy; mechanical physiology

Mechanics' Institutes, 79

mechanistic preformation, and life sciences, 462–8

medicine: and advertising, 547; and cameral science, 452; and China, 695–7, 716; imperialism and trade, 837–8; and India, 670, 680–3; and Islamic world, 661–5; and Japan, 699–700, 710–16, 870–1; and natural knowledge, 146–51; and natural philosophy, 463; and Newtonianism in Continental Europe, 296; and pathology, 476- 81; and physiology, 468–76; and science education, 50–1, 68,

69–72, 84; and scientist as civic expert, 179–80; scurvy and voyages of discovery, 636–8, 837–8; and Spanish America, 728, 738; status of in 1790s, 481–4; and universities, 465–8; and women as midwives, 190. *See also* anatomy; inoculations; physiology

Meek, Ronald, 448n29

Megharatna, 674

Mehmed Efendi, Yirmisekiz Çelebi, 651

Meidinger, Carl von, 574

Mei Wending, 693, 702

Mendelssohn, Moses, 39, 229

Mendes da Costa, E., 431

Menonville, J.-N. Thierry de, 839

Menzies, Archibald, 641

mercantilism: and gender politics, 237; and Spanish America, 733. *See also* trade

Mercator, Gerard, 526, 659

Mercator, Nicholas, 329, 331

Merchant, Carolyn, 26n5, 193, 195

Mercier, Louis-Sébastien, 789

Merian, Joanna, 564

Merian, Maria Sibylla, 190, 209, 564, 566

meritocracy, and Jesuit universities, 218, 234

Mersenne, Marin, 27

Merton, Robert, 213–14

Meslier, Jean, 742

Mesmer, Franz Anton, 169, 492–5, 507

mesmerism. *See* animal magnetism

Messerschmidt, Daniel G., 863

Messier, Charles, 343, 390

metal engraving, and scientific illustration, 568, 570–2, 583

metallurgy: in China and Japan, 880; and India, 684–5, 880; and Prussia, 850; and Russia, 861–2, 863–6. *See also* mining industry and mining schools

metaphysics: and classification of sciences, 253; and philosophy of science, 275–80, 283; and university study of natural sciences, 45

Metaxas, Nicodemus, 657

Meteorological Society of Mannheim, 93

methodology: and philosophy of science, 280–3; and universal method, 803–5. *See also* experimental science; scientific method

Metzger, Hélène, 381, 471n21

Mexico: and botanical gardens, 102; universities and science education, 78–9. *See also* Aztecs; Maya; Spanish America

Meyer, Felix, 596–7

Michelangelo, 590

Michell, John, 277, 278, 350

microscope and microscopy: and microscopic organisms, 410; and scientific instruments, 525, 526, 579–80, 581f

Middleton, Christopher, 622
Mikami Yoshio, 703
military: and engineering schools, 122–3; Islamic
world and Western technology, 653–5, 671;
and Jesuits in China, 690; and scientific
development in Spanish America, 732; and
scientist as civic expert, 179; and technology
in India, 671, 685. *See also* war
Mill, James, 459
Mill, John Stuart, 459
Millar, Andrew, 545
Millar, John, 443–4, 448–9
Millar, Robert, 831–2
Miller, David Philip, 177n44, 181n51
Miller, Philip, 570, 627
Millington, Thomas, 512
Milne, John, 849
Milton, John, 786
Minggantu, 694
mining industry and mining schools: and gov-
ernment patronage of science, 118–20, 424;
and India, 684; and Russia, 863–4; and sci-
ence education, 74, 75; and Spanish America,
100, 733. *See also* metallurgy
Mitchell, Timothy F., 586n11
modernity and modernization: and Enlighten-
ment, 243; and German universities, 231; and
medical sciences, 481; and printing in Islamic
world, 656; and scientific developments, 3
Molière, Jean-Baptiste, 189
Molière, Jean-Pierre de, 299
Molyneux, Thomas, 606–7
Monboddo, James Burnett, Lord, 130, 444,
450–1
Monceau, H. L. Duhamel de, 568
Monge, Gaspard, 122
Monnet, Antoine, 126, 612
Montagu, Lady Mary Wortley, 192, 666
Montchretien, Antoine de, 451
Montesquieu, Baron de (Charles-Louis Secondat),
438, 446–8, 462, 746, 821
morality: and human nature, 818; and human
sciences, 441; and literature, 771–4; and medi-
cine, 148–9; and philosophical empiricism,
806; and quantification, 815–16; and relation-
ship between religion and science, 748, 770;
and scientific community, 235
moral philosophy: and human sciences, 441,
444–50; and image of scientist, 164–7, 202;
and natural philosophy, 285
Morandi Manzolini, Anna, 185
Moravia, Sergio, 438n7
More, Hannah, 497
Moreau de Maupertuis, P. L., 333
Moreri, Louis, 251

Morgan, John, 527
Morgani, Giovanni Battista, 479, 480, 481, 483
Moro, Anton Lazzaro, 421, 427
Morrell, Jack, 214, 225
Morris, James, 850
Morveau, Louis Bernard Guyton de, 392
motherhood: and science education for women,
189; and sexual division of labor, 193. *See also*
children; women
motion: and mechanical natural philosophy, 26;
Newton and planetary, 311; and physics,
360–1. *See also* force; inertia
Motoki Ryoei, 709
mountains: and geology, 585–7; and landscape
art, 599
Mozart, Wolfgang, 780
Mulla Nizamuddin Sahalwi, 679
Müller, Benjamin, 863
Müller, John, 857
Murad III, Sultan, 656, 659
Muratori, Lodovico Antonio, 767
Murdock, William, 854
Musawi, Walih, 686
Museo di Storia della Scienza (Florence), 519
Museum of the History of Science (Oxford),
524f
music, and influence of science on art, 780
Musschenbroek, Jan van, 514, 531
Musschenbroek, Johan & Samuel van, 531
Musschenbroek, Pieter van, 64, 281, 289, 297–8,
358, 369, 371–2, 514
Mustafa III, Sultan, 652, 662
Müteferrika, İbrahim, 657–8, 660

Nabatov, R. F., 864
Nādir Shāh, 662–3
Nahua language, 728–9
Nairne, Edward, 529
Nakane Genkei, 701–2
Namier, Lewis, 159
Nantes, Treaty of (1685), 851
Naonobu, Ajima, 704
Napoleonic Expedition of 1798 (Egypt), 649,
650, 664
Narborough expedition (1669–71), 621
Narsimha, 674–5
nationalism: and human sciences, 445; and
India, 673–4; and scientific community, 234;
universities and science education, 85–6. *See*
also imperialism; patriotism
National Museums of Scotland, 525
Native Americans, 442, 443–4, 643. *See also*
Aztecs; Inca; Maya
natural disasters, and natural theology, 748. *See*
also earthquakes; volcanoes

natural history: and classification of sciences, 254, 256, 258, 262–3, 265–6; definition of, 417; development of, 324–8; gender politics and racism, 236; and human nature, 819–23; and materia medica, 716; and medical education, 69, 70–2; religion and image of scientist, 162–4; and Royal Society, 282; and Spanish America, 738; and universal method, 803–5; and voyages of discovery, 627; and women, 217. *See also* nature

natural knowledge: and agriculture, 139–46; and botany, 151–6; contemporary model of, 129–30; forms, sites, and social meanings of, 131–2, 156–8; and medicine, 146–51. *See also* knowledge

natural philosophy: approaches to in seventeenth century, 268–72; and Aristotle, 244; and astrology, 499; authors and authorship, 556; and classification of sciences, 254, 256, 258–9, 262–3, 266; definition of, 14; development of, 285–91; and Enlightenment, 25–8; and medical sciences, 463, 476–7; and Newton, 14, 27–8, 244, 272–3, 287; and philosophy, 344–5; and physics, 354–5; and practice of reading, 554; and religion, 744; and skepticism, 28–41; and vitalism, 41–3. *See also* nature; Newtonianism

natural religion, 743–4. *See also* theology

nature: allegory and representations of, 591–2; Galileo and mathematical concept of, 270–1; gender and concepts of, 191–2, 193, 195, 208f; and landscape art, 592–609; and natural philosophy, 285; Spanish America and emblematic readings of, 727–8; and trade, 838–40. *See also* natural history; natural philosophy

navigation: and imperialism, 827, 841–4; and Islamic world, 654. *See also* cartography; chronometers; longitude; shipbuilding

Nayansukha, 678

nebular hypothesis, 348–50

Née, Luis, 643

Needham, John Turberville, 410–11, 412–13, 415, 753

Needham, Joseph, 688, 876

Negri, Salomon, 665

Nelson, David, 640

neo-Confucianism, 692, 694, 710. *See also* Confucianism

neo-mechanical philosophy, 33

neoplatonism, and Spanish America, 722–9

neptunism, 424, 428, 430

Netherlands: and botanical gardens, 102; and Japan, 702–3, 716, 870; and Newtonianism, 298–9, 302; and physics, 357–8; and printing houses, 540–1, 545; and reform of science education, 55; and scientific instruments, 531; and state regulation of print, 540; trade and imperialism, 822–3; and voyages of discovery, 621. *See also* Europe

Newbery, John, 546

Newcomen, Thomas, 853, 854

Newton, Isaac: and alchemy, 499–500; and astronomy, 334–7, 342–3, 348–50, 351, 621; on Cartesianism, 287–8; and chemistry, 382; as cultural icon, 766; and graphical methods in calculus, 309; and human sciences, 499; influence of on culture, 10, 11–12, 762–3; influence of on science, 3–4; and mechanical physiology, 398, 472; and methodology, 281–3; and natural philosophy, 14, 27–8, 244, 287; and nature of matter, 28; and philosophy of Enlightenment, 800, 801, 809–14, 824; and philosophy of science, 267, 272–5, 283, 284; and physics, 359, 360–7; and print culture, 545, 555–6, 559–60; and religion, 741, 744–6, 748; and science education, 47, 60, 61, 62, 68, 80, 85, 86; and scientific revolution, 23; and social role of scientist, 161; and university posts, 49. *See also* Newtonianism

Newtonianism: and Cartesianism, 287–8, 359–60; and chemistry, 384; diffusion of in Europe, 295–302; history of science and studies of, 134–9; and Japan, 709–10; and natural philosophy in England, 289–95; and physiology, 400–4; and reproduction, 411–13; and Spanish America, 734, 738; and university culture, 97. *See also* anti-Newtonianism; natural philosophy; Newton, Isaac

New Zealand, and voyages of discovery, 628, 630

Nicolson, Marjorie Hope, 598n33

Niebuhr, Carsten, 650, 665

Niebuhr, Hermann, 231

Nieto-Galan, Agustí, 117n32

Nishikawa Masayasu, 706

Nixon, Robert, 547

Noble, David, 195

noble savage, image of, 442–3, 620

Nollet, Jean Antoine, 169, 170, 302, 358, 359, 369, 371, 514, 515f, 522, 754

nomenclature, and chemistry, 394

normal science, and dynamics of scientific revolution, 23n1

North Africa: and Islamic world, 650; and medicine, 663

Northern Wars of 1700–1721, 861, 863

Northwest Passage, and voyages of discovery, 618, 621–2, 629, 631–4, 643

Norton, John F., 783n45

Nouvelles de la République des Lettres (journal), 552

obscurantism, and scientific illustration, 563–4
Observatoire Royal (Paris), 99, 108, 352
observatories, astronomical: and India, 675; and Islamic world, 652; and organization of science, 98–9; and scientific instruments, 523–5; and universities, 82; and women, 190
obstetrics, and scientific illustration, 578
occult powers, and vitalism, 34, 38
oceans, and sea levels, 430–1. *See also* neptunism
Oersted, Hans, 280, 323
Ogiu Sorai, 707
Oken, Lorenz, 42–3
Oldenburg, Henry, 89
Olson, Richard, 11
optics: and experimental physics, 363–7; and scientific instruments, 522. *See also* microscope and microscopy; telescope
optimism, and cultural change, 764–77
organization, of science: and organizational revolution of seventeenth century, 87–90; and second organizational revolution in early nineteenth century, 105–6; in society, 103–5; and universities, 96–8. *See also* institutions
Orientalism, and Western imagination, 781
ornithology, and scientific illustration, 574–5
Orrio, Xavier, 734
Orta, Garcia da, 191–2, 682
Ortega, Casimiro Gómez, 731
Ortolengi, G., 857–8
Ortous de Mairan, Jean-Jacques de, 299, 411, 481n15
Osasunasco, Desiderio de, 734
Osorio Romero, Ignacio, 722n11–12
Oswald, James, 812
Otsuki Gentaku, 715, 870–1
Ottoman Empire: and contacts with Europe, 650–3; and Islamic civil and religious timekeeping, 655–6; and medicine, 662, 663–4; and military technology, 653–4; and print culture, 656–7; and Russia, 861. *See also* Islam and Islamic world
Ovington, John, 682
Owen, Robert, 459
Oxford University, 49, 60, 82, 215, 216, 289–90
oxygen, and chemistry, 391, 393–4, 580, 582f

Paauw, Jan, 532
Padel, Ruth, 782n44
Padmanabha, 674
Paley, William, 163, 760
Palitzsch, J. G., 343, 390
Pallas, Peter Simon, 428, 586, 633

Palmer, Samuel, 841
Pannini, Giovanni Paolo, 782
Paracelsus, 35, 378, 379, 452, 661
paradigm: and dynamics of scientific revolution, 23n1; and mathematics in Japan, 704
Parennin, Dominique, 695
Paris, Claude, 533
Paris Academy. *See Académie des Sciences*
Paris Observatory. *See Observatoire Royal*
Park, Mungo, 639
Parkinson, Sydney, 628
Pascal, Blaise, 161, 271
Passemant, Claude Siméon, 533
passions, and human sciences, 441, 523
patents, and industrialization, 858
pathognomy, 496, 497
pathology, and medical sciences, 476–81
patriotism, and scholarship in Spanish America, 722–5, 727, 737–8. *See also* nationalism
Paul, Lewis, 853
Paulze, Marie-Anne, 580
Pavón, José Antonio, 642
Payen, Jacques, 124n52
Peace of Paris (1763), 825
Peacock, George, 325–7
Pemberton, Henry, 291
Pennant, Thomas, 628, 633
Pereira, Andrea, 694
Peréz, Juan, 642
perfection, and concept of progress, 785–90
periodicals. *See* journals
Perrault, Claude, 107
Perronet, Jean-Rodolphe, 121, 122
Persia: and astronomy, 660; and Islamic world, 650; and medicine, 662–3; and military technology, 653–4, 655; and printing press, 658
Peru. *See* Inca; Spanish America
Peset, José Luis, 732
Peter the Great, Czar of Russia, 123, 632, 855, 859–62
Petiver, James, 833
Petty, William, 452, 456
Pfaff, Johann Friedrich, 57
Pfeffel, Johann Andreas, 593
Philip II, King of Spain, 729
Philip III, King of Spain, 720
philosophes: and Catholic Church, 754–5; and classification of sciences, 247; and method of modern philosophy, 800; and natural theology, 164
Philosophical Transactions (Royal Society of London), 90, 95, 289, 542, 549, 553
philosophy: and analogy between mind and body, 817–18; and classification of sciences, 253; and evidential scholarship in

philosophy (*cont.*): China, 691–5; human sciences and history of, 444–50, 456; and natural history of human nature, 819–23; and Newtonianism, 809–14; and quantification, 814–17; and reform of science education, 56; relationship of to natural philosophy, 43, 303–4; relationship between science and, 801–2; and science in seventeenth century, 802–9. *See also* analytical philosophy; experimental philosophy; mechanical philosophy; moral philosophy; natural philosophy; philosophy of science

philosophy of science: history of, 267–8; metaphysics, theology, and matter theory, 275–80; and methodology, 280–3; and natural philosophy in seventeenth century, 268–72; and Newton's legacy, 267

Phipps, Constantine, 628, 629

phlogiston, and chemistry, 387, 389, 391, 392, 396

phrenology, 497

physical anthropology, 450–1

physics: advances in eighteenth century, 5; and classification of sciences, 253, 262; and mathematics, 371–4; and mechanics, 360–3; and natural philosophy, 286, 295, 297, 354–5; Newtonianism and Cartesianism, 359–60; and private courses, 77–8; and public lectures, 356–7; and quantification, 422–6; and science education in universities, 45–9, 50, 60–8; 79–80, 83–4, 354–5, 357–8; and scientific academies, 355–7; and scientific illustration, 582–3; Wolff and definition of, 336. *See also* experimental physics

physiognomy, 495–7, 506–7

physiology: and classification of sciences, 265–6; and mechanism, vitalism, and materialism, 399; and medicine, 468–76; and Newtonianism, 400–4; and reading, 555; status of in 1790s, 481–4; and study of reproduction, 454, 455. *See also* anatomy; medicine

Piarists, and science education, 61

Picard, Jean, 332, 621

Picon, Antoine, 121n43

Pictet, Marc Auguste, 528

Pietism, and science education in German universities, 56

Pietsch, Theodore W., 573

Pine, Robert, 779f

Pineda y Ramírez, Antonio, 643

Pingré, Alexandre-Gui, 622–3, 624

piracy, and production of print, 540, 541–2, 556, 557–8

Piscopia, Elena Cornaro, 185

Pitcairne, Archibald, 401

Pizan, Christine de, 189–90

plagiarism, and scientific illustration, 573

planetary motion, 274, 336–7, 343

Plato, 206n49

Pledge, H. T., 2

Pliny, 35, 201

Plot, Robert, 417

Pluche, Noel Antoine, 163

Plumer, Charles, 566

Pocock, Richard, 607

poetry, and classification of sciences, 253. *See also* literature

Poggendorff, J. C., 213, 236

Poisson, Siméon-Denis, 373

Poivre, Pierre, 626

Poland, and modernization of science education, 54

Poleni, Giovanni, 64, 332

Polhem, Christopher, 119

Polinière, Pierre, 63, 168–9, 356–7, 359, 514

political economy, and human sciences, 449, 451–6, 462

political science, and human sciences, 457–61

Pombal, Marques de, 67, 127, 518–19

Pons, Francis, 676

Pontedera, Giulio, 202n43

Pope, Alexander, 7, 136, 548, 762–3, 766–7, 775, 781, 785, 800, 817

Pope, Joseph, 534

popular culture: natural knowledge and reform of, 153, 156; and reading in Japan, 700

population. *See* demography; immigration

porcelain manufacturing, 114–15

pornography, and printing industry, 549

Porset, Charles, 249–50

Porter, Roy, 166n14, 242n2

Portugal: and government patronage of science, 127; and science curriculum in universities, 61. *See also* Europe

Positivism, and human sciences, 461

Postlethwayt, Malachy, 828, 857

postmodernism, and critiques of Enlightenment, 24, 43

Potsdam decree (1685), 851–2

Potter, Elizabeth, 195

Pound, James, 841

Pourchot, Edmond, 660

Poussin, Nicolas, 587, 602, 756

Pratt, Mary Louise, 644

precision, in physics, 371, 394, 396. *See also* quantification

preexistence theories, and life sciences, 413–14

preformation, and life sciences, 406–9

Preiβler, Johann Daniel, 593, 595f

Press Act (1662), 538, 539

Preston, John, 606n57
Price, James, 485–6, 502
Price, Richard, 757, 806
Priestley, Joseph: and chemistry, 389, 390–1, 392, 394; and egalitarianism, 175; and foundations of modern science, 489; and Hartley on social reform, 459; and human nature, 806, 808; and image of godly naturalist, 164; and materialist philosophy, 278, 812; and natural knowledge, 143; and Newtonianism, 294, 812; and print culture, 555; science and future of society, 6; on science and religion, 742, 744, 749, 751, 754, 757, 760, 761; scientific education and class, 171n21; and scientific illustrations, 580, 582f; and vitalism, 36; and voyages of discovery, 629
Primus, Alexander Munro, 71
Prince, Thomas, 755
Princeton University, 52, 64, 100
printing and print culture: authorship and end of Enlightenment, 555–60; cultures of at onset of Enlightenment, 536–40; and Islamic world, 656–9; and Japan, 700, 702–3, 705–6, 709; and medicine in China, 696–7; property and piracy in production of, 540–50; reading and redefinition of reason, 550–5. *See also* illustration, scientific; libraries; literature; textbooks
private courses, and science education, 76–9
probability theory: and philosophy, 815–16; and religion, 756
professionalization, of science, 162, 192, 234–5
progress: cultural change and concepts of, 782n42, 785–90, 796; and Rousseau on human nature, 821, 822
Prony, Gaspard Riche de, 138
property, and production of print, 540–50, 557
prosopography: characteristics of studies, 212–13; definition of, 212; and Enlightenment, 235–7; and history of science, 213–14, 266–9; and Jesuits, 218–20; and science in Austro-German lands, 227–31; and science communities, 220–2, 233–5; and universities, 214–18, 222–7; and women, 232–3
Protestantism: and differences between England and Continental countries in eighteenth century, 132; nonconforming sects and science in England, 200–1; and universities in Germany, 260–1. *See also* Anglican Church; Dissenters; Lutheranism; religion
Proust, Joseph-Louis, 117
Prussia, and technological transfer, 850. *See also* Germany
psychology: and human sciences, 457–61; and philosophy, 804–5; and reading, 554–5. *See also* madness

Ptolemy, 269, 339, 348
public sphere: and culture of print, 538, 550–1; and medical sciences, 482; and Spanish America, 732, 733–4
Puerto Sarmiento, Francisco, 733
Pufendorf, Samuel, 444, 808, 822n51
Putter, J. S., 446
Pyenson, Lewis, 212, 213

Qaim Khan, 680
Qian Daxin, 693, 694
quantification: and human sciences, 456–7; and philosophy, 814–17; and physics, 422–6
Quechua language, 728–9
Quesnay, François, 454–5
Quli Shah, 681
Qur'ān, 659, 667

race and racism: and gender politics of science, 207–10; and human sciences, 450–1; and "sexual science," 200–1; and Spanish America, 720; and voyages of discovery, 644
Radcliffe, Anne, 773, 781
Radcliffe Observatory (Oxford University), 82, 524f
Rahman, A., 673, 677
Rainford, Charles, 502
Ramazzini, Bernardino, 776
Ramchandrapant, 685
Ramchandra Vajpeyin, 674
Ramsay, Allan, 775
Ramsden, Jesse, 528, 854
Ramus, Petrus, 253
Rappaport, Rhoda, 5, 422n14
Raspe, Rudolf Erich, 418, 606
rationality, and reevaluation of Enlightenment, 488. *See also* reason
rational mechanics, 374, 439
Ray, John, 163, 230, 398–9, 417, 565, 742, 752
Ray, P. C., 686
Rayner, William, 546
al-Rāzī, Muhammad ibn Zakarīyā (Albucasis), 665
reading, and print cultures, 550–5, 697, 700
realism and reality: and scientific illustration, 562, 564–8; social action and image, 159–60; and vitalism, 41
reason: and human sciences, 441, 523; print cultures and redefinition of, 550–5; and religion, 755–8. *See also* Age of Reason; analogical reasoning; rationality
Réaumur, René-Antoine de, 408, 421, 514
Redouté, Pierre-Joseph, 571–2
reductionism, and mechanical physiology, 455
Rees, Graham, 253n29

Index

Reeve, Clara, 773, 787

refraction, sine law of, 414–15

Reid, Thomas, 264, 281–2, 808, 810–12, 815, 816–17, 820

Reiger, Christian, 720n7

Reill, Johann Christian, 11, 483

religion: and Dissenters, 176–7; diversity of natural, 743–4; Enlightenment critiques of, 758–61; and human sciences, 441; and image of scientist as godly naturalist, 162–4; and limitations of reason, 755–8; and Locke's theory of knowledge, 806–7; providence and utility of science, 753–5; relationship between science and, 741–3, 744–9, 770; and secularization of science, 753–5. *See also* Anglican Church; atheism; Catholic Church; clergy; deism; Islam and Islamic world; Jesuits; Jews and Judaism; Lutheranism; Protestantism; theology

Renaissance: and experimental science, 511–12; and landscape art, 587; and legal humanism, 444–5; and scientific academies, 88

Renard, Louis, 573

Renaudot, Théophraste, 89

representation, culture and forms of, 777–82

reproduction: mechanism and preformation, 406–11; and Newtonianism, 411–13; physiology and studies of, 399–400; and preexistence theories, 413–14

repulsion, and electricity, 368, 370

research: and organizational revolution in nineteenth-century German universities, 106; and science teaching in universities, 83; and scientific instruments, 522–5

Revel, Jacques, 215, 217

Reynolds, Joshua, 587–8, 594, 601

Ricardo, David, 440

Ricci, Matteo, 700

Richards, Graham, 436–7n4

Richardson, Samuel, 771, 772

Richarz, Monika, 216–17

Richer, Jean, 332, 621

Riera, J., 731n42

Rinssen, Anthony, 531

Riolan, Jean, II, 51

Rittenhouse, Benjamin & David, 534

Ritter, Johann Wilhelm, 280

Rivière, Mercier de la, 440

roads, and industrialization, 855, 856, 857

Robertson, William, 827–8

Robinet, Jean Baptiste, 40

Robinson, Tancred, 621

Robison, John, 265, 281

Roche, Daniel, 90n6, 222–4, 235–6, 418n1

Rodríguez, Diego, 725–7, 728

Rodríguez de Mendoza, Toribio, 61

Roe, Shirley, 5, 13, 472n24, 475

Roebuck, John, 180, 848, 853

Roger, Jacques, 402, 408, 416

Rohault, Jacques, 63, 355, 357

Rolle, Michel, 167

Romanticism: and critique of Enlightenment, 237; and natural philosophy, 41–3; and notion of genius, 764

Romé de l'Isle, J.-B., 429

Rosa, Salvator, 587

Rosenberg, Charles, 148

Rosenhof, Roesel von, 575

Rouelle, Guillaume-François, 387, 425–6, 432

Rousseau, George S., 7, 242n2

Rousseau, Jean-Jacques, 196, 438, 443, 449–50, 620, 783, 821–3

Roussel, Pierre, 37, 198

Rouyer, Pierre-Charles, 664–5

Rowbottom, Margaret, 491n13

Rowden, Frances Arabella, 153

Royal Academy of History (Spain), 732

Royal Academy of Sciences (Sweden), 542

Royal Academy of Turin, 111–13

Royal Astronomical Society, 105

Royal Botanical Garden (Mexico), 102, 732

Royal Gardens (Kew), 101–2, 627, 835

Royal Observatory (Greenwich), 99, 124, 330

Royal Society of Edinburgh (RSE), 226–7

Royal Society of London: and class, 226; establishment of, 88–9; and experimental science, 512; and fossils, 420; and marginalized practices, 485, 490–1, 502; as model of scientific academy, 87, 92–3; and natural history, 282; and natural knowledge, 131; and physics, 355, 357, 406–7; printing and publication, 541–2, 549, 553; and relationship between science and technology, 104–5, 128; and Royal Observatory, 113; and scientific expeditions, 621, 623, 624; and scientific journals, 90, 95; and state, 104; trade and imperialism, 833, 834; and women, 187–8, 218

Rozier, François, 96

Rozier's Journal, 96

Ruan Yuan, 693, 694

Rubens, Peter Paul, 587

Rudwick, Martin J. S., 419n5, 586–7n11, 593n26, 612

Ruestow, Edward G., 60n25, 164n9

Ruini, Carlo, 562

Ruiz, Hipólito, 642

Rumpf, Georg Eberhardt, 568

Rush, Benjamin, 459, 484

Russell, Alexander, 663–4

Russell, Colin A., 118n36

Russia: government and scientific academies, 109–10; industrial and technological development, 851, 858–66; and transits of Venus, 344; and voyages of discovery, 631–3, 642–3
Russian Imperial Academy of Sciences, 123, 523. *See also* St. Petersburg Academy
Rutherford, Thomas, 66
Rutherford, William, 762, 766, 790
Ruysch, Frederick, 71n49

Sabean, David, 211
Sade, Marquis de, 770, 783
St. André, Nathanael, 149
Saint-Fond, Barthélemie Faujas de, 606
Saint-Germain, Comte de, 501
St. Petersburg Academy, 76, 94, 111, 109, 340, 432n39, 597
Saldaña, Juan José, 733
Sallūm, Ibn, 666
salons, women and scientific, 232–3, 235
Sánchez, Miguel, 724–5
Sandby, Paul, 612n70
Sandifort, Eduard, 578
Sandys, Edwin, 607, 608f
Şānizāde, 'Atā'ullāh, 667
Sano Antei, 713
Sarkar, B. K., 673
Sarma, S. R., 674n23
Sarton, George, 672n17
Saunderson, Nicholas, 65, 291
Saussure, Horace-Bénédict de, 418, 599–602
Saussure, Théodore, 600
Sauvages, François Boissier de, 405
Say, Jean-Baptiste, 460
Schaffer, Simon, 3, 4, 14, 136, 175, 807n15, 838
Scheele, Carl Wilhelm, 36, 118, 391
Schelling, Friedrich, 42, 279, 280, 283, 284
schemata, Kant's concept of, 40–1n58
Schemnitz Academy (Austria), 120
Scheuchzer, Johann Jacob, 593–7
Schiebinger, Londa, 6, 19, 232, 233, 235
Schiller, Friedrich, 40
Schott, Gaspar, 721
Schubert, Franz, 780
Schwab, Richard, 439n12
science. *See* astronomy; biology; botany; chemistry; classification, of sciences; earth sciences; education, scientific; geology; history of science; human sciences; institutions; journals; life sciences; medicine; philosophy of science; physics; scientific academies; Scientific Revolution; scientists; zoology
science fiction, 788, 789
Science Museum (London), 519

scientia: and classification of sciences, 245; gendered representations of, 194f
scientific academies: creation of in seventeenth century, 88; development of, 90–4; French provincial, 222–4; governments and patronage of science, 107–28; influence of on science, 8; and lecture courses, 75–6; and mathematics, 306; and medicine, 464; and physics, 355–7; scientific societies compared to, 221; technology and industrial development, 111–28; and women, 187–8. *See also* *Académie des Sciences;* Royal Society of London; scientific societies
scientific communities, and prosopographic studies, 220–2, 233–5
scientific method, in India, 675. *See also* experimental science; methodology
Scientific Revolution: and Enlightenment, 23–43; and explosion of knowledge, 242; and natural philosophy, 27–31, 286; and polite society, 171; and science education, 80–1; science in seventeenth century compared to, 1–3; and scientific sexism, 210. *See also* chemical revolution; Copernican revolution and Copernicanism
scientific societies: and botanical gardens, 102; in colonial America, 100; in early nineteenth century, 105; and ideology, 94, 235; influence on science in eighteenth century, 8–9; in provincial England, 177; and scientific academies, 92–3, 221; and scientific journals, 95–6. *See also* scientific academies
scientists: culture and social roles of, 159–83; and image of civic expert, 178–83; and image of godly naturalist, 162–4, 178; and image of moral philosopher, 164–7, 178; and image of polite philosopher of nature, 167–78; origins of term, 106
Scilla, Agostino, 419
Scopoli, J. A., 120
Scotland: agriculture and Scottish Enlightenment, 141–2; and chemistry, 379–80, 388–90; and Newtonianism, 291–2, 810–12; philosophical history and political economy, 456; and printing houses, 544; and prosopography, 225–7; religion and Scottish Enlightenment, 760–1; and social concern about scientific specialization, 173–4
scurvy, 477–8, 636–8, 837–8
Secord, Anne, 155
Secord, James, 138–9
secularization, of science, 749–53
Seki Takakazu, 703–4, 705
Selīm I, Sultan, 656
Selkirk, Alexander, 827

semiotics, and pathology, 478n35
Semler, Johann, 501
Senefelder, Alois, 583, 659
Sennert, Daniel, 468–9, 661
sensationalist psychology, 457–61
sensibility, and physiology, 473
Sepp, J. C., 575
Sessé, Martin de, 643, 729
Sève, Jacques de, 590–1, 592f
Seven Years' War, 827
sexual difference: and botany, 202–4; and "sexual science," 197–201, 793. *See also* gender
Shaftesbury, Earl of (Anthony Ashley Cooper), 173, 817
Shao Yong, 694
Shapin, Steven, 2n3, 3n9, 4, 13, 17, 18n66, 19, 35, 161n3, 183n55, 195, 212, 214, 225, 225–6, 227, 807n15
Sharif Khan, Hakim, 682
Shaw, Peter, 78, 379, 557
Shelley, Mary, 507, 794
Shelley, Percy Bysshe, 507
Sheng Bai'er, 694
Shiba Kokan, 709
Shibukawa Harumi, 701, 707
shipbuilding, in India, 872
al-Shīrāzī, Qutb al-Dīn, 661–2
Shizuki Tadao, 709–10, 871
Short, James, 524–5, 527–8
Siberia, and voyages of discovery, 632
Sibley, Ebenezer, 494, 503
Sibthorp, Humphrey, 571
Siegfried, Robert, 36
Şifai, Ömer, 662
Sigorgne, Pierre, 62
Sigüenza y Góngora, Carlos, 721, 722
Sinclair, John, 144n26
sine and cosine functions, 312, 335
Siraisi, Nancy G., 464n2, 466
Sirhindi, Ahmed, 671n10
Sisson, Jonathan, 528
Skene, David, 820
skepticism, and mechanical natural philosophy, 28–41
Slater, Samuel, 846
Sloane, Hans, 282, 548, 626, 834, 837
smallpox, and inoculations, 192
Smart, Christopher, 775, 783
Smellie, William, 203, 506–7
Smith, Adam, 165n12, 211, 237, 439, 440, 444, 448, 452, 456, 462, 812, 874, 881
Smith, Cyril Stanley, 429n29
Smith, Eugene, 703
Smith, James Edward, 571, 575
Smith, Robert, 816

Smith, Roger, 436–7n4
Smith, Samuel, 543
Smith, William, 166, 180, 582–3
Smollett, Tobias, 771–2, 781
Snell, Bruno, 782n44
Sobirov, G., 676
social sciences. *See* human sciences
Societas regia literaria et scientiarum Sueciae (Sweden), 110–11
Societas regia scientiarum (Berlin), 89, 108, 187
Société Typographique de Neuchâtel (Switzerland), 541
society: and Newtonianism in Augustan England, 139; and organization of science, 103–5; and stadial theory of social change, 444–50. *See also* class; culture; government; sociology
Society of the Incognito (Austria), 227
sociology, 461–2. *See also* human sciences
Solander, Daniel, 626, 629, 831
solar system and solar theory, 329–32, 343–4
Sonnenfels, Joseph von, 55, 454
Soroe Academy (Denmark), 519
Sorrenson, Richard, 177n44
Soulavie, Jean-Louis Giraud, 432, 613, 615–16
South America, and scientific academies, 100. *See also* Spanish America
Southey, Robert, 495
Sowerby, James, 152
space, Kant and concept of, 279
Spain: and engineering schools, 123; and science curriculum in universities, 61, 62; and scientific and technical support for manufacturing, 116–17; and voyages of discovery, 641–4, 730–1. *See also* Europe; Spanish America
Spalding Gentlemen's Society, 135, 156–7
Spallanzani, Lazaro, 68, 180, 413, 414
Spanish America: and botanical gardens, 117; colonialism and precontact social structures of, 718–19; and influences on scholarly traditions, 722–9; and institutions, 719–22; and trade, 730–5; travelers and cultural change in, 735–8; and universities, 100, 720, 731. *See also* Mexico
Sparrman, Anders, 626, 629, 839
specialization: and classification of sciences, 266; and scientific community, 234–5; and university students, 217, 219, 220–1, 226
species, and physical anthropology, 451. *See also* Homo sapiens
Spectator, The (magazine), 179
Spellman, W. M., 806n14
Spence, Jonathan, 689
Spencer, Herbert, 246
Spencer, Mary, 576

Spinoza, Benedictus, 292, 437, 458
Spöring, Herman, 626
Stafford, Barbara Maria, 586n11
Stahl, Georg, 27n15, 35–7, 38, 39, 381, 386–7, 402, 405, 471–2, 475, 483
Stahlian revolution, in chemistry, 36–7
Stapleton, Michael, 117
star catalogs and maps, 328, 338–9, 348–9, 352–3. *See also* cosmography
state: human sciences and political economy, 451–6; Kant on university and, 58; and natural philosophy, 290; and regulation of print, 536–7, 540–1, 558–9; and science education, 53, 74, 81; and scientific societies, 92. *See also* government
Stationers' Company, 538
statistics and statistical studies: and human sciences, 456–7; and prosography, 212–13. *See also* quantification
steel industry, in India, 684. *See also* metallurgy
Steele, Richard, 172
Stein, Howard, 330n5
Steitz, Adam, 532
Steller, Georg, 632, 863
Steno, Nicolaus, 418–19
Stephenson, George, 79
Stephenson, Marjory, 187n8
Sterne, Laurence, 772–3, 782
Steuart, James, 451–2
Stewart, Larry, 9–10, 135, 139
Stewart (David M.) Museum (Montreal, Canada), 517–18
Stirling, James, 305
Stone, Lawrence, 212, 213, 215, 216
Strachey, John, 613
Strange, John, 424–5, 428
Streete, Thomas, 329, 331
Stuart, John, 516
Stubbs, George, 576–7
Stukeley, William, 498
Sturdy, David, 224
sublime, and landscape art, 592–609
Sugita Genpaku, 703, 712, 713
Süleyman I, Sultan (Ottoman Empire), 655
Suri, Mahendra, 670
Suría, Tomás de, 643
Susskind, Charles, 491n13
Sutton, Geoffrey V., 103n30, 170
Svanberg, Jons, 333
Swammerdam, Jan, 161, 408, 579
Sweden: and book trade, 542; economic development and government patronage of science, 118–19; government and scientific academies, 110–11; government and statistics, 457; and research on sea levels, 430–1; Russia

and war with, 862; trade and imperialism, 831; universities and science education, 54, 68. *See also* Europe
Swedenborg, Emanuel, 502
Swift, Jonathan, 763–4, 772, 788, 789
Switzer, Stephen, 140–1, 144–5, 154
Sydenham, Thomas, 807
Symonds, John, 143
Sypher, Wylie, 797n78
Syria: and Islamic world, 650; and medicine, 663–4

Tafalla, Juan, 729
Tahiti, and voyages of discovery, 625, 843
Takahashi Yoshitoki, 708
Takebe Katahiro, 703
Talmon, J. L., 1n2
Taton, René, 16
Tavernier, Melchior, 659
Taylor, Brook, 305, 323
Taylor, Kenneth L., 418n2
technical schools, 74–5, 121–3
technology: and China, 871–7, 879–81; economics and government patronage of science, 111–28; and ideology of science, 104; and India, 671, 683–5, 879–81; and industrialization in England, 853–8; and industrialization in Europe, 846–53; Islamic world and Western, 653; and Japan, 866–71, 879–81; and print culture, 556; and Russia, 858–66; and scientific knowledge, 878. *See also* industrialization; Industrial Revolution
telescope, and scientific instruments, 523, 528, 676
Telford, Thomas, 79, 854
terra australis incognita, and voyages of discovery, 618, 619, 625, 626–31, 641
Terrall, Mary, 195
textbooks: and chemistry, 380, 394, 395; and classification of sciences, 252; and medicine, 465–6, 468; and physics, 358
textile industry: and government support for sciences, 112, 114, 115–16, 117; and India, 684, 872, 877; and industrialization, 846–7, 857–8; and Japan, 869
Teyler Museum (Netherlands), 519, 522, 525, 530
Thackray, Arnold, 27, 177–8, 212, 214, 225, 227
theater, and natural knowledge, 136–7
theology: concept of science and Islamic, 667; and encyclopedias, 249, 262; and life sciences, 398–9; and natural philosophy, 276–7, 290, 499; and Newtonianism, 276–7, 284, 292; and philosophy of science, 275–80, 284; and prosopographic studies of university students, 215, 216; and science education in universities, 56. *See also* religion

Theophrastus, Philipus, 35, 160n2, 201
Thomas, Keith, 152
Thomasius, Christian, 815
Thompson, E. P., 10, 12
Thomson, James, 7, 767, 780, 782
Thomson, Thomas, 265
Thornton, Robert, 202, 203, 572
Thrale, Hester, 490
Thuillier, Denis, 623
Thunberg, Carl Peter, 568, 870
time scales, and earth sciences, 431–5
Tindal, Matthew, 747, 807
Tipu Sultan, 685
Tissot, Samuel, 466, 774, 776
Toft, Mary, 146, 149–51, 157
Tokugawa Yoshimune, 701–2
Toland, John, 276, 292, 746–7, 807
Tolozan, Jean-François, 114
Tompion, Thomas, 528
topographical art, 594, 601, 607, 609
Tott, Baron de, 652–3
Toulmin, Stephen, 24, 33
Tournefort, Joseph Pitton de, 203, 419, 565
Trabulse, Eliás, 725–7
trade: and botany, 833–8; and imperialism, 828–38; and scientific discoveries, 825–8; in scientific instruments, 531–4; and Spanish America, 730–5; and transport of nature, 838–40. *See also* economics; mercantilism
transportation. *See* canals; roads
travel and travel literature: and cultural change in Spanish America, 735–8; and human sciences, 442–4; and imagination, 781; national interests and technological development, 125; and scientist as civic expert, 179
Trembley, Abraham, 409, 412, 580, 753
Treviranus, Gottfried Reinhold, 416
Trevithick, Richard, 854
Trew, Christophe Jacob, 570, 571
trigonometry, 704
Troughton, Edward & John, 528–9
Trudaine, Daniel, 114
Tucker, Josiah, 452–3
Tughlaq, Sultan Muhammad, 670
Tull, Jethro, 140–1, 144, 145, 164, 854
Turgot, Anne-Robert-Jacques, 114, 392, 448, 455
Turnbull, George, 810, 817, 819
Turner, Gerard, 9, 17–18

Ulloa, Antonio de, 641, 842
Umar, Muhammad, 670n5
Unánué, José Hipólito, 75
United States. *See* Alaska; American colonies
universities: and chemistry, 230–1, 379; and libraries, 95n15; and medicine, 464, 465–8; and organization of science, 88, 96–8; and physics, 354–5; and prosopographic studies, 214–18, 218–27, 230–1; and reform of science education, 52–9, 79–86; and science curriculum, 59–73; in Spanish America, 100, 720, 731; and status of science education in 1700, 44–51; and women, 184–8. *See also* education
University of Berlin, 58
University of Bologna, 98, 185, 187
University of Castile, 62
University of Coimbra, 61, 67, 127, 518–19
University of Edinburgh, 226, 289–90
University of Göttingen, 52, 56–7, 98, 231
University of Halle, 47, 56, 120–1, 471
University of Königsberg, 57–8
University of Oxford, *see* Oxford University
University of Mexico, 70, 113
University of Paris, 46, 51, 216
University of Pavia, 68, 82
University of San Marcos, 100
Unzer, Johanna Charlotte, 188n9
Uranus, discovery of, 350
urbanization, and industrialization, 847–8
utilitarianism, and human sciences, 457–61

Vaez de Torres, Luis, 641
Vaillant, Sébastien, 202
Valentijn, François, 573
Valla, Joseph, 62
Vallisneri, Antonio, 421, 430
Valsalva, Antonio Maria, 479
Vancouver, George, 640–1
van Helmont, Franciscus Mercurius, 35
van Helmont, Jean Baptist, 35
van Marum, Martinus, 516, 522, 529
Vanpaemel, Geert, 60n25
van Rymsdyk, Jan, 578
Varignon, Pierre, 309, 311, 334
Vartanian, Aram, 28
Vaucanson, Jacques, 116, 853
Venegas, Juan Manuel, 734
Venel, Gabriel François, 83, 387
Venus, transits of, 343–4, 622–4, 842
Verbiest, Ferdinand, 690
Veri, Petro, 455
Vernet, Claude-Joseph, 603
Verron, M., 832–3
Vesalius, Andreas, 577, 661
Vesuvius, Mt., 427, 603, 604, 605f
Vico, Giambattista, 440, 441, 445–6, 447, 462
Vicq d'Azyr, Félix, 71, 72
Vietnam, and civil service examination system, 699
Virgil, 802
vitalism: and life sciences, 404–6; and mechanical

philosophy, 33–41; and medical sciences, 73, 84, 475–6; and romantic *naturphilosophie,* 41–3

Vivarès, François, 607

Voght, Casper, 851

Voigtlander, Johann Christoph, 532

Volaire, Pierre-Jacques, 603

volcanoes: and earth sciences, 422–3, 426, 486–9; and landscape art, 603–6, 616

Volder, Burchardus de, 512–13

Volta, Alessandro, 64, 82, 372

Voltaire, François-Marie, 8, 135, 300, 487, 656, 742, 743, 748, 807, 819, 824, 861

voluntarism, and theology, 748

von Eggesberg, Anton Reprecht, 120

von Herbert, Thomas, 662

von Petrasch, Josef, 227, 228

von Reden, G., 850

vortexes, Cartesian theory of, 334–5, 367

voyages of discovery: background to scientific aspects of, 621–2; and Cook's voyages, 626–36; in final years of eighteenth century, 638–41; and governmental support for science, 125; and impact on science, 618–20, 644–5; and imperialism, 624–6; and scientific illustration, 564; and Spain, 641–4; and transits of Venus, 622–4. *See also* expeditions, scientific

Wada, Yasushi, 704

Wagner, Monika, 598–9n34

Wagner, Peter, 793n74

Wakefield, Priscilla, 153

Walch, Johann Ernst Immanuel, 432

Walduck, Thomas, 834

Wales, William, 624, 629, 630–1

Waliullah, Shah, 679

Walker, Adam, 505, 516

Walker, John, 69, 163

Wallerius, Johann Gottschalk, 68

Wallis, John, 512

Wallis, Samuel, 624–5

Walpole, Horace, 497, 765

Wandelaar, Jan, 570

Wang Fuzhi, 692

Wang Mingsheng, 693

war, and technology transfer, 850, 851. *See also* military

War of the Austrian Succession, 842

wars of independence, and Spanish America, 737–8

War of the Spanish Succession, 831

water, and earth sciences, 426–31. *See also* diluvialism; flood; neptunism; oceans

Watt, James, 104–5, 124, 176, 180, 367, 516, 754, 854

Watts, Isaac, 251–2

Watts, John, 546

Webber, John, 633

Wedgwood, Josiah, 104, 105, 176

Wegmann, Peter, 597n29

Weinman, J. W., 570

Wellington, Duke of, Arthur Wellesley, 685

Werner, Abraham Gottlob, 74, 119, 180, 423–4, 424, 430, 599, 607

Wesley, John, 146–8, 757, 765

Wheatland, David, 517

Whewell, William, 373, 583

Whiggism, and history of science, 264, 375–7, 388, 396, 488, 489

Whiston, William, 150, 259, 277, 291, 421, 513, 514, 522, 553, 746

White, Andrew Dickson, 754, 755

White, Gilbert, 130, 163, 417, 752

Whitehurst, John, 426, 428, 434, 603–4

Whitney, Eli, 853

Whytt, Robert, 403–4, 478, 479, 483

Wilkes, Benjamin, 575

Wilkinson, William, 849, 850, 854

Williams, Glyndwr, 443n20

Williams, John, 418

Willis, Thomas, 777, 786

Willughby, Francis, 565

Wilson, Benjamin, 523

Wilson, Curtis, 4

Wilson, James, 534

Winkelmann, Maria Margaretha, 187, 190

Winslow, Jacques-Bénigne, 71

Winthrop, John, 517, 523, 755

Wise, Henry, 154

Withering, William, 152

Wolff, Caspar Friedrich, 414

Wolff, Christian, 31n25, 47, 56, 120–1, 246, 259, 264, 286–7, 295, 812–13, 816

Wollaston, William, 743

Wollstonecraft, Mary, 195, 441, 461, 773

Wolterstorff, Nicholas, 806n12

women: and botany, 152–3, 791–3; and French places of scientific conversation, entertainment, and instruction, 170; and gender in science, 133, 184–210; and medical knowledge, 168; natural knowledge and Newtonianism, 136–7; and Newtonianism in France and Italy, 301; and prosopography, 232–3, 235; and reading, 554, 697; and scientific education, 77; society and organized science, 103. *See also* gender

Wood, Paul, 14, 281

woodcut printing, and scientific illustration, 569f, 583, 700

Woodhouse, Robert, 325–7

Woodward, John, 259, 417, 420, 424
Woolf, Harry, 93n10, 181n48, 523
Wordsworth, William, 775, 797, 798
working class, and science education, 79, 80.
  See also class
Worm, Ole, 573
Woulfe, Peter, 502
Wright, Joseph, 500, 603
Wright, William, 349
Wu Qian, 696–7
Wundt, Wilhelm, 461
Wyatt, James, 524f

Xu Dachan, 697

Yale University, 44, 100
Yamawaki Toyo, 711

Yar Khan, Jafar, 681
Yeo, Richard, 11, 15, 165n11, 486
Yolton, John W., 805n11
Yoshimasu Todo, 710
Young, Arthur, 121, 142–5, 156–7
Young, Thomas, 78, 86, 365
Youschkevitch, A. P., 312n10

Zedler, Johann, 244
Zhang Xuecheng, 695
Zhao Xuemin, 696
Zhu Xi, 692
Zimmerman, Johann Georg, 768
zoology: and fossils, 432; and gender politics
  of science, 205–7; and scientific illustration,
  574–7; and voyages of discovery, 644. *See also*
  life sciences; reproduction

NORMANDALE COMMUNITY COLLEGE
LIBRARY
9700 FRANCE AVENUE SOUTH
BLOOMINGTON, MN 55431-4399